KIRK-OTHMER

ENCYCLOPEDIA OF
CHEMICAL
TECHNOLOGY

FOURTH EDITION

VOLUME **24**

THIOGLYCOLIC ACID
TO
VINYL POLYMERS

EXECUTIVE EDITOR
Jacqueline I. Kroschwitz

EDITOR
Mary Howe-Grant

KIRK-OTHMER

ENCYCLOPEDIA OF CHEMICAL TECHNOLOGY

FOURTH EDITION

VOLUME **24**

THIOGLYCOLIC ACID
TO
VINYL POLYMERS

A Wiley-Interscience Publication
JOHN WILEY & SONS

New York • Chichester • Weinheim • Brisbane • Singapore • Toronto

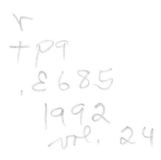

This text is printed on acid-free paper. ∞

Library of Congress Cataloging-in-Publication Data

Encyclopedia of chemical technology/executive editor, Jacqueline
   I. Kroschwitz; editor, Mary Howe-Grant.—4th ed.
     p. cm.
   At head of title: Kirk-Othmer.
   "A Wiley-Interscience publication."
   Contents: v. 24, Thioglycolic acid to vinyl polymers
   ISBN 0471-52693-2 (v. 24)
   1. Chemistry, Technical—Encyclopedias. I. Kirk, Raymond E.
   (Raymond Eller), 1890–1957. II. Othmer, Donald F. (Donald
   Frederick), 1904–1995. III. Kroschwitz, Jacqueline I., 1942– .
   IV. Howe-Grant, Mary, 1943– . V. Title: Kirk-Othmer encyclopedia
   of chemical technology.
   TP9.E685 1992   91-16789
   660'.03—dc20

Printed in the United States of America

10 9 8 7 6 5 4 3 2 1

CONTENTS

EDITORIAL STAFF
FOR VOLUME 24

Executive Editor: **Jacqueline I. Kroschwitz**
Editor: **Mary Howe-Grant**
Associate Managing Editor: **Lindy Humphreys**
Copy Editors: **Lawrence Altieri**
            Jonathan Lee
Assistant Managing Editor: **Brendan A. Vilardo**

CONTRIBUTORS
TO VOLUME 24

Satinder Ahuja, *Ahuja Consulting, Monsey, New York*, Trace and residue analysis
Bryan Ballantyne, *Union Carbide Corporation*, Toxicology
S. L. Bean, *General Chemical Corporation, Claymont, Delaware*, Thiosulfates
Roop S. Bhakuni, *The Goodyear Tire & Rubber Company, Akron, Ohio*, Tire cord
Karyn A. Booth, *Donelan, Cleary, Wood & Maser P.C., Washington, D.C.*, Transportation

Surendra K. Chawla, *The Goodyear Tire & Rubber Company, Akron, Ohio,* Tire cord

David L. Clark, *Glenn T. Seaborg Institute for Transactinium Science, Los Alamos National Laboratory, Los Alamos, New Mexico,* Thorium and thorium compounds; Uranium and uranium compounds

Antoine P. Cobb, *Donelan, Cleary, Wood & Maser P.C., Washington, D.C.,* Transportation

Cajetan F. Cordeiro, *Air Products and Chemicals, Inc., Allentown, Pennsylvania,* Vinyl acetate polymers (under Vinyl polymers)

Joseph A. Cowfer, *The Geon Company, Avon Lake, Ohio,* Vinyl chloride

Thomas W. Del Pesco, *E. I. du Pont de Nemours & Co., Inc., Deepwater, New Jersey,* Organic (under Titanium compounds)

Phillip T. DeLassus, *The Dow Chemical Company, Freeport, Texas,* Vinylidene chloride monomer and polymers

Terry A. Egerton, *Tioxide Group Services Limited, Billingham, U.K.,* Inorganic (under Titanium compounds)

Lawrence J. Esposito, *Rhône-Poulenc, Cranbury, New Jersey,* Vanillin

William George Fong, *Florida Department of Agriculture and Consumer Services, Tallahassee,* Toxicology

K. Formanek, *Rhône-Poulenc, Cranbury, New Jersey,* Vanillin

Richard D. Fortin (deceased), *Donelan, Cleary, Wood & Maser P.C., Washington, D.C.,* Transportation

Lance S. Fuller, *Synthetic Chemicals Limited, Wolverhampton, U.K.,* Thiophene and thiophene derivatives

Charles C. Gaver, Jr., *Consultant, Mt. Laural, New Jersey,* Tin and tin alloys

D. S. Gibbs, *The Dow Chemical Company, Midland, Michigan,* Vinylidene chloride monomer and polymers

Maximilian B. Gorensek, *The Geon Company, Avon Lake, Ohio,* Vinyl chloride

Stanley Hoffman, *Consultant, Danbury, Connecticut,* Transportation

Robert A. Howell, *Central Michigan University,* Vinylidene chloride monomer and polymers

Chia-Lung Hsieh, *Wyeth-Lederle Vaccine and Pediatrics, Pearl River, New York,* Vaccine technology

D. Webster Keogh, *Glenn T. Seaborg Institute for Transactinium Science, Los Alamos National Laboratory, Los Alamos, New Mexico,* Thorium and thorium compounds; Uranium and uranium compounds

G. Kientz, *Rhône-Poulenc, Cranbury, New Jersey,* Vanillin

D. K. Kim, *The Goodyear Tire & Rubber Company, Akron, Ohio,* Tire cord

Jerome W. Knapczyk, *Monsanto Company, Indian Orchard, Massachusetts,* Vinyl acetal polymers (under Vinyl polymers)

Ranga Komanduri, *Oklahoma State University, Stillwater,* Tool materials

Ralf Kuriyel, *Millipore Corporation, Bedford, Massachusetts,* Ultrafiltration

Y. Labat, *Elf Atochem, Artix, France,* Thioglycolic acid

Robert B. Login, *Sybron Chemicals Inc., Wellford, South Carolina,* Vinyl ether monomers and polymers; N-Vinylamide polymers (both under Vinyl polymers)

Robert P. Lukens, *American Society for Testing and Materials, Woodbury, New Jersey,* Units and conversion factors

F. Lennart Marten, *Air Products and Chemicals, Inc., Allentown, Pennsylvania,* Vinyl alcohol polymers (under Vinyl polymers)

F. Mauger, *Rhône-Poulenc, Cranbury, New Jersey,* Vanillin

V. Maureaux, *Rhône-Poulenc, Cranbury, New Jersey,* Vanillin

Norman Milleron, *SEN Vac Services, Berkeley, California,* Vacuum technology

Jeffrey O. Moreno, *Donelan, Cleary, Wood & Maser P.C., Washington, D.C.,* Transportation

Mary P. Neu, *Glenn T. Seaborg Institute for Transactinium Science, Los Alamos National Laboratory, Los Alamos, New Mexico,* Thorium and thorium compounds; Uranium and uranium compounds

B. E. Obi, *The Dow Chemical Company, Midland, Michigan,* Vinylidene chloride monomer and polymers

E. Dickson Ozokwelu, *Amoco Chemical Company, Naperville, Illinois,* Toluene

Thomas W. Penrice, *Consultant, Mt. Juliet, Tennessee,* Tungsten and tungsten alloys; Tungsten compounds

Donald E. Putzig, *E. I. du Pont de Nemours & Co., Inc., Deepwater, New Jersey,* Organic (under Titanium compounds)

Mary B. Ritchey, *Wyeth-Lederle Vaccine and Pediatrics, Pearl River, New York,* Vaccine technology

G. Robert, *Rhône-Poulenc, Cranbury, New Jersey,* Vanillin

John S. Roberts, *Phillips Petroleum Company/CH&A Corporation, Kingwood, Texas,* Thiols

Wolfgang Runde, *Glenn T. Seaborg Institute for Transactinium Science, Los Alamos National Laboratory, Los Alamos, New Mexico,* Thorium and thorium compounds; Uranium and uranium compounds

Stan R. Seagle, *Consultant, Warren, Ohio,* Titanium and titanium alloys

D. Shuttleworth, *The Goodyear Tire & Rubber Company, Akron, Ohio,* Tire cord

James W. Summers, *The Geon Company, Avon Lake, Ohio,* Vinyl chloride polymers (under Vinyl polymers)

Dean Thetford, *Zeneca Specialties, Manchester, U.K.,* Triphenylmethane and related dyes

F. Truchet, *Rhône-Poulenc, Cranbury, New Jersey,* Vanillin

Henri Ulrich, *Consultant, Guilford, Connecticut,* Urethane polymers

A. Dinsmoor Webb, *University of California, Davis,* Vinegar

R. A. Wessling, *The Dow Chemical Company, Midland, Michigan,* Vinylidene chloride monomer and polymers

Mike Woolery, *U.S. Vanadium, Hot Springs, Arkansas,* Vanadium compounds

NOTE ON CHEMICAL ABSTRACTS SERVICE REGISTRY NUMBERS AND NOMENCLATURE

Chemical Abstracts Service (CAS) Registry Numbers are unique numerical identifiers assigned to substances recorded in the CAS Registry System. They appear in brackets in the *Chemical Abstracts* (CA) substance and formula indexes following the names of compounds. A single compound may have synonyms in the chemical literature. A simple compound like phenethylamine can be named β-phenylethylamine or, as in *Chemical Abstracts*, benzeneethanamine. The usefulness of the *Encyclopedia* depends on accessibility through the most common correct name of a substance. Because of this diversity in nomenclature careful attention has been given to the problem in order to assist the reader as much as possible, especially in locating the systematic CA index name by means of the Registry Number. For this purpose, the reader may refer to the CAS Registry Handbook—Number Section which lists in numerical order the Registry Number with the *Chemical Abstracts* index name and the molecular formula; eg, **458-88-8**, Piperidine, 2-propyl-, (S)-, $C_8H_{17}N$; in the *Encyclopedia* this compound would be found under its common name, coniine [*458-88-8*]. Alternatively, this information can be retrieved electronically from CAS Online. In many cases molecular formulas have also been provided in the *Encyclopedia* text to facilitate electronic searching. The Registry Number is a valuable link for the reader in retrieving additional published information on substances and also as a point of access for on-line data bases.

In all cases, the CAS Registry Numbers have been given for title compounds in articles and for all compounds in the index. All specific substances indexed in *Chemical Abstracts* since 1965 are included in the CAS Registry System as are a large number of substances derived from a variety of reference works. The CAS Registry System identifies a substance on the basis of an unambiguous computer-language description of its molecular structure including stereochemical detail. The Registry Number is a machine-checkable number (like a Social Security number) assigned in sequential order to each substance as it enters the registry system. The value of the number lies in the fact that it is a concise and unique means of substance identification, which is independent of, and therefore bridges, many systems of chemical nomenclature. For polymers, one Registry Number may be used for the entire family; eg, polyoxyethylene (20) sorbitan monolaurate has the same number as all of its polyoxyethylene homologues.

Cross-references are inserted in the index for many common names and for some systematic names. Trademark names appear in the index. Names that are incorrect, misleading, or ambiguous are avoided. Formulas are given very frequently in the text to help in identifying compounds. The spelling and form used, even for industrial names, follow American chemical usage, but not always the usage of *Chemical Abstracts* (eg, *coniine* is used instead of *(S)-2-propylpiperidine*, *aniline* instead of *benzenamine*, and *acrylic acid* instead of *2-propenoic acid*).

There are variations in representation of rings in different disciplines. The dye industry does not designate aromaticity or double bonds in rings. All double bonds and aromaticity are shown in the *Encyclopedia* as a matter of course. For example, tetralin has an aromatic ring and a saturated ring and its structure

appears in the *Encyclopedia* with its common name, Registry Number enclosed in brackets, and parenthetical CA index name, ie, tetralin [*119-64-2*] (1,2,3,4-tetrahydronaphthalene). With names and structural formulas, and especially with CAS Registry Numbers, the aim is to help the reader have a concise means of substance identification.

CONVERSION FACTORS, ABBREVIATIONS, AND UNIT SYMBOLS

SI Units (Adopted 1960)

The International System of Units (abbreviated SI), is being implemented throughout the world. This measurement system is a modernized version of the MKSA (meter, kilogram, second, ampere) system, and its details are published and controlled by an international treaty organization (The International Bureau of Weights and Measures) (1).

SI units are divided into three classes:

BASE UNITS

length	meter[†] (m)
mass	kilogram (kg)
time	second (s)
electric current	ampere (A)
thermodynamic temperature[‡]	kelvin (K)
amount of substance	mole (mol)
luminous intensity	candela (cd)

SUPPLEMENTARY UNITS

plane angle	radian (rad)
solid angle	steradian (sr)

[†]The spellings "metre" and "litre" are preferred by ASTM; however, "-er" is used in the *Encyclopedia*.

[‡]Wide use is made of Celsius temperature (t) defined by

$$t = T - T_0$$

where T is the thermodynamic temperature, expressed in kelvin, and $T_0 = 273.15$ K by definition. A temperature interval may be expressed in degrees Celsius as well as in kelvin.

DERIVED UNITS AND OTHER ACCEPTABLE UNITS

These units are formed by combining base units, supplementary units, and other derived units (2–4). Those derived units having special names and symbols are marked with an asterisk in the list below.

Quantity	Unit	Symbol	Acceptable equivalent
*absorbed dose	gray	Gy	J/kg
acceleration	meter per second squared	m/s^2	
*activity (of a radionuclide)	becquerel	Bq	1/s
area	square kilometer	km^2	
	square hectometer	hm^2	ha (hectare)
	square meter	m^2	
concentration (of amount of substance)	mole per cubic meter	mol/m^3	
current density	ampere per square meter	$A//m^2$	
density, mass density	kilogram per cubic meter	kg/m^3	g/L; mg/cm^3
dipole moment (quantity)	coulomb meter	C·m	
*dose equivalent	sievert	Sv	J/kg
*electric capacitance	farad	F	C/V
*electric charge, quantity of electricity	coulomb	C	A·s
electric charge density	coulomb per cubic meter	C/m^3	
*electric conductance	siemens	S	A/V
electric field strength	volt per meter	V/m	
electric flux density	coulomb per square meter	C/m^2	
*electric potential, potential difference, electromotive force	volt	V	W/A
*electric resistance	ohm	Ω	V/A
*energy, work, quantity of heat	megajoule	MJ	
	kilojoule	kJ	
	joule	J	N·m
	electronvolt[†]	eV[†]	
	kilowatt-hour[†]	kW·h[†]	
energy density	joule per cubic meter	J/m^3	
*force	kilonewton	kN	
	newton	N	$kg·m/s^2$

[†]This non-SI unit is recognized by the CIPM as having to be retained because of practical importance or use in specialized fields (1).

Quantity	Unit	Symbol	Acceptable equivalent
*frequency	megahertz	MHz	
	hertz	Hz	1/s
heat capacity, entropy	joule per kelvin	J/K	
heat capacity (specific), specific entropy	joule per kilogram kelvin	$J/(kg \cdot K)$	
heat-transfer coefficient	watt per square meter kelvin	$W/(m^2 \cdot K)$	
*illuminance	lux	lx	lm/m^2
*inductance	henry	H	Wb/A
linear density	kilogram per meter	kg/m	
luminance	candela per square meter	cd/m^2	
*luminous flux	lumen	lm	cd·sr
magnetic field strength	ampere per meter	A/m	
*magnetic flux	weber	Wb	V·s
*magnetic flux density	tesla	T	Wb/m^2
molar energy	joule per mole	J/mol	
molar entropy, molar heat capacity	joule per mole kelvin	$J/(mol \cdot K)$	
moment of force, torque	newton meter	N·m	
momentum	kilogram meter per second	kg·m/s	
permeability	henry per meter	H/m	
permittivity	farad per meter	F/m	
*power, heat flow rate, radiant flux	kilowatt	kW	
	watt	W	J/s
power density, heat flux density, irradiance	watt per square meter	W/m^2	
*pressure, stress	megapascal	MPa	
	kilopascal	kPa	
	pascal	Pa	N/m^2
sound level	decibel	dB	
specific energy	joule per kilogram	J/kg	
specific volume	cubic meter per kilogram	m^3/kg	
surface tension	newton per meter	N/m	
thermal conductivity	watt per meter kelvin	$W/(m \cdot K)$	
velocity	meter per second	m/s	
	kilometer per hour	km/h	
viscosity, dynamic	pascal second	Pa·s	
	millipascal second	mPa·s	
viscosity, kinematic	square meter per second	m^2/s	
	square millimeter per second	mm^2/s	

Quantity	Unit	Symbol	Acceptable equivalent
volume	cubic meter	m^3	
	cubic diameter	dm^3	L (liter) (5)
	cubic centimeter	cm^3	mL
wave number	1 per meter	m^{-1}	
	1 per centimeter	cm^{-1}	

In addition, there are 16 prefixes used to indicate order of magnitude, as follows:

Multiplication factor	Prefix	Symbol	Note
10^{18}	exa	E	
10^{15}	peta	P	
10^{12}	tera	T	
10^{9}	giga	G	
10^{6}	mega	M	
10^{3}	kilo	k	
10^{2}	hecto	h[a]	[a] Although hecto, deka, deci, and centi
10	deka	da[a]	are SI prefixes, their use should be
10^{-1}	deci	d[a]	avoided except for SI unit-multiples
10^{-2}	centi	c[a]	for area and volume and nontech-
10^{-3}	milli	m	nical use of centimeter, as for body
10^{-6}	micro	μ	and clothing measurement.
10^{-9}	nano	n	
10^{-12}	pico	p	
10^{-15}	femto	f	
10^{-18}	atto	a	

For a complete description of SI and its use the reader is referred to ASTM E380 (4) and the article UNITS AND CONVERSION FACTORS which appears in Vol. 24.

A representative list of conversion factors from non-SI to SI units is presented herewith. Factors are given to four significant figures. Exact relationships are followed by a dagger. A more complete list is given in the latest editions of ASTM E380 (4) and ANSI Z210.1 (6).

Conversion Factors to SI Units

To convert from	To	Multiply by
acre	square meter (m^2)	4.047×10^3
angstrom	meter (m)	$1.0 \times 10^{-10\dagger}$
are	square meter (m^2)	$1.0 \times 10^{2\dagger}$

[†] Exact.

To convert from	To	Multiply by
astronomical unit	meter (m)	1.496×10^{11}
atmosphere, standard	pascal (Pa)	1.013×10^5
bar	pascal (Pa)	$1.0 \times 10^{5\dagger}$
barn	square meter (m²)	$1.0 \times 10^{-28\dagger}$
barrel (42 U.S. liquid gallons)	cubic meter (m³)	0.1590
Bohr magneton (μ_B)	J/T	9.274×10^{-24}
Btu (International Table)	joule (J)	1.055×10^3
Btu (mean)	joule (J)	1.056×10^3
Btu (thermochemical)	joule (J)	1.054×10^3
bushel	cubic meter (m³)	3.524×10^{-2}
calorie (International Table)	joule (J)	4.187
calorie (mean)	joule (J)	4.190
calorie (thermochemical)	joule (J)	$4.184^\dagger$
centipoise	pascal second (Pa·s)	$1.0 \times 10^{-3\dagger}$
centistokes	square millimeter per second (mm²/s)	$1.0^\dagger$
cfm (cubic foot per minute)	cubic meter per second (m³/s)	4.72×10^{-4}
cubic inch	cubic meter (m³)	1.639×10^{-5}
cubic foot	cubic meter (m³)	2.832×10^{-2}
cubic yard	cubic meter (m³)	0.7646
curie	becquerel (Bq)	$3.70 \times 10^{10\dagger}$
debye	coulomb meter (C·m)	3.336×10^{-30}
degree (angle)	radian (rad)	1.745×10^{-2}
denier (international)	kilogram per meter (kg/m)	1.111×10^{-7}
	tex‡	0.1111
dram (apothecaries')	kilogram (kg)	3.888×10^{-3}
dram (avoirdupois)	kilogram (kg)	1.772×10^{-3}
dram (U.S. fluid)	cubic meter (m³)	3.697×10^{-6}
dyne	newton (N)	$1.0 \times 10^{-5\dagger}$
dyne/cm	newton per meter (N/m)	$1.0 \times 10^{-3\dagger}$
electronvolt	joule (J)	1.602×10^{-19}
erg	joule (J)	$1.0 \times 10^{-7\dagger}$
fathom	meter (m)	1.829
fluid ounce (U.S.)	cubic meter (m³)	2.957×10^{-5}
foot	meter (m)	$0.3048^\dagger$
footcandle	lux (lx)	10.76
furlong	meter (m)	2.012×10^{-2}
gal	meter per second squared (m/s²)	$1.0 \times 10^{-2\dagger}$
gallon (U.S. dry)	cubic meter (m³)	4.405×10^{-3}
gallon (U.S. liquid)	cubic meter (m³)	3.785×10^{-3}
gallon per minute (gpm)	cubic meter per second (m³/s)	6.309×10^{-5}
	cubic meter per hour (m³/h)	0.2271

†Exact.
‡See footnote on p. xiii.

To convert from	To	Multiply by
gauss	tesla (T)	1.0×10^{-4}
gilbert	ampere (A)	0.7958
gill (U.S.)	cubic meter (m³)	1.183×10^{-4}
grade	radian	1.571×10^{-2}
grain	kilogram (kg)	6.480×10^{-5}
gram force per denier	newton per tex (N/tex)	8.826×10^{-2}
hectare	square meter (m²)	$1.0 \times 10^{4\dagger}$
horsepower (550 ft·lbf/s)	watt (W)	7.457×10^{2}
horsepower (boiler)	watt (W)	9.810×10^{3}
horsepower (electric)	watt (W)	$7.46 \times 10^{2\dagger}$
hundredweight (long)	kilogram (kg)	50.80
hundredweight (short)	kilogram (kg)	45.36
inch	meter (m)	$2.54 \times 10^{-2\dagger}$
inch of mercury (32°F)	pascal (Pa)	3.386×10^{3}
inch of water (39.2°F)	pascal (Pa)	2.491×10^{2}
kilogram-force	newton (N)	9.807
kilowatt hour	megajoule (MJ)	$3.6^{\dagger}$
kip	newton (N)	4.448×10^{3}
knot (international)	meter per second (m/S)	0.5144
lambert	candela per square meter (cd/m³)	3.183×10^{3}
league (British nautical)	meter (m)	5.559×10^{3}
league (statute)	meter (m)	4.828×10^{3}
light year	meter (m)	9.461×10^{15}
liter (for fluids only)	cubic meter (m³)	$1.0 \times 10^{-3\dagger}$
maxwell	weber (Wb)	$1.0 \times 10^{-8\dagger}$
micron	meter (m)	$1.0 \times 10^{-6\dagger}$
mil	meter (m)	$2.54 \times 10^{-5\dagger}$
mile (statute)	meter (m)	1.609×10^{3}
mile (U.S. nautical)	meter (m)	$1.852 \times 10^{3\dagger}$
mile per hour	meter per second (m/s)	0.4470
millibar	pascal (Pa)	1.0×10^{2}
millimeter of mercury (0°C)	pascal (Pa)	$1.333 \times 10^{2\dagger}$
minute (angular)	radian	2.909×10^{-4}
myriagram	kilogram (kg)	10
myriameter	kilometer (km)	10
oersted	ampere per meter (A/m)	79.58
ounce (avoirdupois)	kilogram (kg)	2.835×10^{-2}
ounce (troy)	kilogram (kg)	3.110×10^{-2}
ounce (U.S. fluid)	cubic meter (m³)	2.957×10^{-5}
ounce-force	newton (N)	0.2780
peck (U.S.)	cubic meter (m³)	8.810×10^{-3}
pennyweight	kilogram (kg)	1.555×10^{-3}
pint (U.S. dry)	cubic meter (m³)	5.506×10^{-4}
pint (U.S. liquid)	cubic meter (m³)	4.732×10^{-4}

†Exact.

To convert from	To	Multiply by
poise (absolute viscosity)	pascal second (Pa·s)	$0.10^{\dagger}$
pound (avoirdupois)	kilogram (kg)	0.4536
pound (troy)	kilogram (kg)	0.3732
poundal	newton (N)	0.1383
pound-force	newton (N)	4.448
pound force per square inch (psi)	pascal (Pa)	6.895×10^3
quart (U.S. dry)	cubic meter (m^3)	1.101×10^{-3}
quart (U.S. liquid)	cubic meter (m^3)	9.464×10^{-4}
quintal	kilogram (kg)	$1.0 \times 10^{2\dagger}$
rad	gray (Gy)	$1.0 \times 10^{-2\dagger}$
rod	meter (m)	5.029
roentgen	coulomb per kilogram (C/kg)	2.58×10^{-4}
second (angle)	radian (rad)	$4.848 \times 10^{-6\dagger}$
section	square meter (m^2)	2.590×10^6
slug	kilogram (kg)	14.59
spherical candle power	lumen (lm)	12.57
square inch	square meter (m^2)	6.452×10^{-4}
square foot	square meter (m^2)	9.290×10^{-2}
square mile	square meter (m^2)	2.590×10^6
square yard	square meter (m^2)	0.8361
stere	cubic meter (m^3)	$1.0^{\dagger}$
stokes (kinematic viscosity)	square meter per second (m^2/s)	$1.0 \times 10^{-4\dagger}$
tex	kilogram per meter (kg/m)	$1.0 \times 10^{-6\dagger}$
ton (long, 2240 pounds)	kilogram (kg)	1.016×10^3
ton (metric) (tonne)	kilogram (kg)	$1.0 \times 10^{3\dagger}$
ton (short, 2000 pounds)	kilogram (kg)	9.072×10^2
torr	pascal (Pa)	1.333×10^2
unit pole	weber (Wb)	1.257×10^{-7}
yard	meter (m)	$0.9144^{\dagger}$

†Exact.

Abbreviations and Unit Symbols

Following is a list of common abbreviations and unit symbols used in the *Encyclopedia*. In general they agree with those listed in *American National Standard Abbreviations for Use on Drawings and in Text* (*ANSI Y1.1*) (6) and *American National Standard Letter Symbols for Units in Science and Technology* (*ANSI Y10*) (6). Also included is a list of acronyms for a number of private and government organizations as well as common industrial solvents, polymers, and other chemicals.

Rules for Writing Unit Symbols (4):

1. Unit symbols are printed in upright letters (roman) regardless of the type style used in the surrounding text.
2. Unit symbols are unaltered in the plural.
3. Unit symbols are not followed by a period except when used at the end of a sentence.
4. Letter unit symbols are generally printed lower-case (for example, cd for candela) unless the unit name has been derived from a proper name, in which case the first letter of the symbol is capitalized (W, Pa). Prefixes and unit symbols retain their prescribed form regardless of the surrounding typography.
5. In the complete expression for a quantity, a space should be left between the numerical value and the unit symbol. For example, write 2.37 lm, *not* 2.37lm, and 35 mm, *not* 35mm. When the quantity is used in an adjectival sense, a hyphen is often used, for example, 35-mm film. *Exception:* No space is left between the numerical value and the symbols of degree, minute, and second of plane angle, degree Celsius, and the percent sign.
6. No space is used between the prefix and unit symbol (for example, kg).
7. Symbols, not abbreviations, should be used for units. For example, use "A," not "amp," for ampere.
8. When multiplying unit symbols, use a raised dot:

$$\text{N·m} \quad \text{for} \quad \text{newton meter}$$

In the case of W·h, the dot may be omitted, thus:

$$\text{Wh}$$

An exception to this practice is made for computer printouts, automatic typewriter work, etc, where the raised dot is not possible, and a dot on the line may be used.
9. When dividing unit symbols, use one of the following forms:

$$\text{m/s} \quad or \quad \text{m·s}^{-1} \quad or \quad \frac{\text{m}}{\text{s}}$$

In no case should more than one slash be used in the same expression unless parentheses are inserted to avoid ambiguity. For example, write:

$$\text{J/(mol·K)} \quad or \quad \text{J·mol}^{-1}\text{·K}^{-1} \quad or \quad \text{(J/mol)/K}$$

but *not*

$$\text{J/mol/K}$$

10. Do not mix symbols and unit names in the same expression. Write:

$$\text{joules per kilogram} \quad or \quad \text{J/kg} \quad or \quad \text{J·kg}^{-1}$$

but *not*

$$\text{joules/kilogram} \quad nor \quad \text{joules/kg} \quad nor \quad \text{joules·kg}^{-1}$$

ABBREVIATIONS AND UNITS

A	ampere	
A	anion (eg, HA)	
A	mass number	
a	atto (prefix for 10^{-18})	
AATCC	American Association of Textile Chemists and Colorists	
ABS	acrylonitrile–butadiene–styrene	
abs	absolute	
ac	alternating current, *n.*	
a-c	alternating current, *adj.*	
ac-	alicyclic	
acac	acetylacetonate	
ACGIH	American Conference of Governmental Industrial Hygienists	
ACS	American Chemical Society	
AGA	American Gas Association	
Ah	ampere hour	
AIChE	American Institute of Chemical Engineers	
AIME	American Institute of Mining, Metallurgical, and Petroleum Engineers	
AIP	American Institute of Physics	
AISI	American Iron and Steel Institute	
alc	alcohol(ic)	
Alk	alkyl	
alk	alkaline (not alkali)	
amt	amount	
amu	atomic mass unit	
ANSI	American National Standards Institute	
AO	atomic orbital	

AOAC	Association of Official Analytical Chemists	
AOCS	American Oil Chemists' Society	
APHA	American Public Health Association	
API	American Petroleum Institute	
aq	aqueous	
Ar	aryl	
ar-	aromatic	
as-	asymmetric(al)	
ASHRAE	American Society of Heating, Refrigerating, and Air Conditioning Engineers	
ASM	American Society for Metals	
ASME	American Society of Mechanical Engineers	
ASTM	American Society for Testing and Materials	
at no.	atomic number	
at wt	atomic weight	
av(g)	average	
AWS	American Welding Society	
b	bonding orbital	
bbl	barrel	
bcc	body-centered cubic	
BCT	body-centered tetragonal	
Bé	Baumé	
BET	Brunauer-Emmett-Teller (adsorption equation)	
bid	twice daily	
Boc	*t*-butyloxycarbonyl	
BOD	biochemical (biological) oxygen demand	
bp	boiling point	
Bq	becquerel	

C	coulomb	DIN	Deutsche Industrie Normen
°C	degree Celsius		
C-	denoting attachment to carbon	*dl*-; DL-	racemic
		DMA	dimethylacetamide
c	centi (prefix for 10^{-2})	DMF	dimethylformamide
c	critical	DMG	dimethyl glyoxime
ca	circa (approximately)	DMSO	dimethyl sulfoxide
cd	candela; current density; circular dichroism	DOD	Department of Defense
		DOE	Department of Energy
CFR	Code of Federal Regulations	DOT	Department of Transportation
cgs	centimeter-gram-second	DP	degree of polymerization
CI	Color Index	dp	dew point
cis-	isomer in which substituted groups are on same side of double bond between C atoms	DPH	diamond pyramid hardness
		dstl(d)	distill(ed)
		dta	differential thermal analysis
cl	carload		
cm	centimeter	(*E*)-	entgegen; opposed
cmil	circular mil	ϵ	dielectric constant (unitless number)
cmpd	compound		
CNS	central nervous system	*e*	electron
CoA	coenzyme A	ECU	electrochemical unit
COD	chemical oxygen demand	ed.	edited, edition, editor
coml	commercial(ly)	ED	effective dose
cp	chemically pure	EDTA	ethylenediaminetetra-acetic acid
cph	close-packed hexagonal		
CPSC	Consumer Product Safety Commission	emf	electromotive force
		emu	electromagnetic unit
cryst	crystalline	en	ethylene diamine
cub	cubic	eng	engineering
D	debye	EPA	Environmental Protection Agency
D-	denoting configurational relationship		
		epr	electron paramagnetic resonance
d	differential operator		
d	day; deci (prefix for 10^{-1})	eq.	equation
d	density	esca	electron spectroscopy for chemical analysis
d-	*dextro*-, dextrorotatory		
da	deka (prefix for 10^1)	esp	especially
dB	decibel	esr	electron-spin resonance
dc	direct current, *n*.	est(d)	estimate(d)
d-c	direct current, *adj*.	estn	estimation
dec	decompose	esu	electrostatic unit
detd	determined	exp	experiment, experimental
detn	determination	ext(d)	extract(ed)
Di	didymium, a mixture of all lanthanons	F	farad (capacitance)
		F	faraday (96,487 C)
dia	diameter	f	femto (prefix for 10^{-15})
dil	dilute		

FAO	Food and Agriculture Organization (United Nations)	hyd	hydrated, hydrous
		hyg	hygroscopic
		Hz	hertz
fcc	face-centered cubic	i (eg, Pr^i)	iso (eg, isopropyl)
FDA	Food and Drug Administration	i-	inactive (eg, i-methionine)
		IACS	International Annealed Copper Standard
FEA	Federal Energy Administration	ibp	initial boiling point
FHSA	Federal Hazardous Substances Act	IC	integrated circuit
		ICC	Interstate Commerce Commission
fob	free on board		
fp	freezing point	ICT	International Critical Table
FPC	Federal Power Commission	ID	inside diameter; infective dose
FRB	Federal Reserve Board		
frz	freezing	ip	intraperitoneal
G	giga (prefix for 10^9)	IPS	iron pipe size
G	gravitational constant = 6.67×10^{11} N·m^2/kg^2	ir	infrared
		IRLG	Interagency Regulatory Liaison Group
g	gram		
(g)	gas, only as in $H_2O(g)$	ISO	International Organization Standardization
g	gravitational acceleration		
gc	gas chromatography	ITS-90	International Temperature Scale (NIST)
gem-	geminal		
glc	gas–liquid chromatography	IU	International Unit
g-mol wt; gmw	gram-molecular weight	IUPAC	International Union of Pure and Applied Chemistry
GNP	gross national product	IV	iodine value
gpc	gel-permeation chromatography	iv	intravenous
		J	joule
GRAS	Generally Recognized as Safe	K	kelvin
		k	kilo (prefix for 10^3)
grd	ground	kg	kilogram
Gy	gray	L	denoting configurational relationship
H	henry		
h	hour; hecto (prefix for 10^2)	L	liter (for fluids only) (5)
ha	hectare	l-	$levo$-, levorotatory
HB	Brinell hardness number	(l)	liquid, only as in $NH_3(l)$
Hb	hemoglobin	LC$_{50}$	conc lethal to 50% of the animals tested
hcp	hexagonal close-packed		
hex	hexagonal	LCAO	linear combination of atomic orbitals
HK	Knoop hardness number		
hplc	high performance liquid chromatography	lc	liquid chromatography
		LCD	liquid crystal display
HRC	Rockwell hardness (C scale)	lcl	less than carload lots
		LD$_{50}$	dose lethal to 50% of the animals tested
HV	Vickers hardness number		

LED	light-emitting diode	N-	denoting attachment to
liq	liquid		nitrogen
lm	lumen	n (as n_D^{20})	index of refraction (for
ln	logarithm (natural)		20°C and sodium light)
LNG	liquefied natural gas	^{n} (as Bu^n),	
log	logarithm (common)	n-	normal (straight-chain
LOI	limiting oxygen index		structure)
LPG	liquefied petroleum gas	n	neutron
ltl	less than truckload lots	n	nano (prefix for 10^9)
lx	lux	na	not available
M	mega (prefix for 10^6);	NAS	National Academy of
	metal (as in MA)		Sciences
M	molar; actual mass	NASA	National Aeronautics and
$\overline{M}_w$	weight-average mol wt		Space Administration
$\overline{M}_n$	number-average mol wt	nat	natural
m	meter; milli (prefix for	ndt	nondestructive testing
	10^{-3})	neg	negative
m	molal	NF	*National Formulary*
m-	meta	NIH	National Institutes of
max	maximum		Health
MCA	Chemical Manufacturers'	NIOSH	National Institute of
	Association (was		Occupational Safety and
	Manufacturing Chemists		Health
	Association)	NIST	National Institute of
MEK	methyl ethyl ketone		Standards and
meq	milliequivalent		Technology (formerly
mfd	manufactured		National Bureau of
mfg	manufacturing		Standards)
mfr	manufacturer	nmr	nuclear magnetic
MIBC	methyl isobutyl carbinol		resonance
MIBK	methyl isobutyl ketone	NND	New and Nonofficial Drugs
MIC	minimum inhibiting		(AMA)
	concentration	no.	number
min	minute; minimum	NOI-(BN)	not otherwise indexed (by
mL	milliliter		name)
MLD	minimum lethal dose	NOS	not otherwise specified
MO	molecular orbital	nqr	nuclear quadruple
mo	month		resonance
mol	mole	NRC	Nuclear Regulatory
mol wt	molecular weight		Commission; National
mp	melting point		Research Council
MR	molar refraction	NRI	New Ring Index
ms	mass spectrometry	NSF	National Science
MSDS	material safety data sheet		Foundation
mxt	mixture	NTA	nitrilotriacetic acid
μ	micro (prefix for 10^{-6})	NTP	normal temperature and
N	newton (force)		pressure (25°C and 101.3
N	normal (concentration);		kPa or 1 atm)
	neutron number		

NTSB	National Transportation Safety Board	qv	quod vide (which see)
O-	denoting attachment to oxygen	R	univalent hydrocarbon radical
o-	ortho	(*R*)-	rectus (clockwise configuration)
OD	outside diameter	*r*	precision of data
OPEC	Organization of Petroleum Exporting Countries	rad	radian; radius
o-phen	*o*-phenanthridine	RCRA	Resource Conservation and Recovery Act
OSHA	Occupational Safety and Health Administration	rds	rate-determining step
		ref.	reference
owf	on weight of fiber	rf	radio frequency, *n.*
Ω	ohm	r-f	radio frequency, *adj.*
P	peta (prefix for 10^{15})	rh	relative humidity
p	pico (prefix for 10^{-12})	RI	Ring Index
p-	para	rms	root-mean square
p	proton	rpm	rotations per minute
p.	page	rps	revolutions per second
Pa	pascal (pressure)	RT	room temperature
PEL	personal exposure limit based on an 8-h exposure	RTECS	Registry of Toxic Effects of Chemical Substances
		^{s} (eg, Bu^s);	
pd	potential difference	*sec*-	secondary (eg, secondary butyl)
pH	negative logarithm of the effective hydrogen ion concentration	S	siemens
		(*S*)-	sinister (counterclockwise configuration)
phr	parts per hundred of resin (rubber)	*S*-	denoting attachment to sulfur
p-i-n	positive-intrinsic-negative		
pmr	proton magnetic resonance	*s*-	symmetric(al)
p-n	positive-negative	s	second
po	per os (oral)	(s)	solid, only as in $H_2O(s)$
POP	polyoxypropylene	SAE	Society of Automotive Engineers
pos	positive		
pp.	pages	SAN	styrene-acrylonitrile
ppb	parts per billion (10^9)	sat(d)	saturate(d)
ppm	parts per million (10^6)	satn	saturation
ppmv	parts per million by volume	SBS	styrene–butadiene–styrene
ppmwt	parts per million by weight	sc	subcutaneous
PPO	poly(phenyl oxide)	SCF	self-consistent field; standard cubic feet
ppt(d)	precipitate(d)		
pptn	precipitation	Sch	Schultz number
Pr (no.)	foreign prototype (number)	sem	scanning electron microscope(y)
pt	point; part		
PVC	poly(vinyl chloride)	SFs	Saybolt Furol seconds
pwd	powder	sl sol	slightly soluble
py	pyridine	sol	soluble

soln	solution	*trans-*	isomer in which substituted groups are on opposite sides of double bond between C atoms
soly	solubility		
sp	specific; species		
sp gr	specific gravity		
sr	steradian		
std	standard	TSCA	Toxic Substances Control Act
STP	standard temperature and pressure (0°C and 101.3 kPa)	TWA	time-weighted average
		Twad	Twaddell
sub	sublime(s)	UL	Underwriters' Laboratory
SUs	Saybolt Universal seconds	USDA	United States Department of Agriculture
syn	synthetic		
^{t} (eg, Bu^t), *t-, tert-*	tertiary (eg, tertiary butyl)	USP	*United States Pharmacopeia*
		uv	ultraviolet
T	tera (prefix for 10^{12}); tesla (magnetic flux density)	V	volt (emf)
		var	variable
t	metric ton (tonne)	*vic-*	vicinal
t	temperature	vol	volume (not volatile)
TAPPI	Technical Association of the Pulp and Paper Industry	vs	versus
		v sol	very soluble
		W	watt
TCC	Tagliabue closed cup	Wb	weber
tex	tex (linear density)	Wh	watt hour
T_g	glass-transition temperature	WHO	World Health Organization (United Nations)
tga	thermogravimetric analysis	wk	week
THF	tetrahydrofuran	yr	year
tlc	thin layer chromatography	(Z)-	zusammen; together; atomic number
TLV	threshold limit value		

Non-SI (Unacceptable and Obsolete) Units		Use
Å	angstrom	nm
at	atmosphere, technical	Pa
atm	atmosphere, standard	Pa
b	barn	cm^2
bar†	bar	Pa
bbl	barrel	m^3
bhp	brake horsepower	W
Btu	British thermal unit	J
bu	bushel	m^3; L
cal	calorie	J
cfm	cubic foot per minute	m^3/s
Ci	curie	Bq
cSt	centistokes	mm^2/s
c/s	cycle per second	Hz

†Do not use bar (10^5 Pa) or millibar (10^2 Pa) because they are not SI units, and are accepted internationally only for a limited time in special fields because of existing usage.

Non-SI (Unacceptable and Obsolete) Units		Use
cu	cubic	exponential form
D	debye	C·m
den	denier	tex
dr	dram	kg
dyn	dyne	N
dyn/cm	dyne per centimeter	mN/m
erg	erg	J
eu	entropy unit	J/K
°F	degree Fahrenheit	°C; K
fc	footcandle	lx
fl	footlambert	lx
fl oz	fluid ounce	m^3; L
ft	foot	m
ft·lbf	foot pound-force	J
gf den	gram-force per denier	N/tex
G	gauss	T
Gal	gal	m/s^2
gal	gallon	m^3; L
Gb	gilbert	A
gpm	gallon per minute	(m^3/s); (m^3/h)
gr	grain	kg
hp	horsepower	W
ihp	indicated horsepower	W
in.	inch	m
in. Hg	inch of mercury	Pa
in. H_2O	inch of water	Pa
in.-lbf	inch pound-force	J
kcal	kilo-calorie	J
kgf	kilogram-force	N
kilo	for kilogram	kg
L	lambert	lx
lb	pound	kg
lbf	pound-force	N
mho	mho	S
mi	mile	m
MM	million	M
mm Hg	millimeter of mercury	Pa
mμ	millimicron	nm
mph	miles per hour	km/h
μ	micron	μm
Oe	oersted	A/m
oz	ounce	kg
ozf	ounce-force	N
η	poise	Pa·s
P	poise	Pa·s
ph	phot	lx
psi	pounds-force per square inch	Pa
psia	pounds-force per square inch absolute	Pa
psig	pounds-force per square inch gage	Pa
qt	quart	m^3; L
°R	degree Rankine	K
rd	rad	Gy
sb	stilb	lx
SCF	standard cubic foot	m^3
sq	square	exponential form
thm	therm	J
yd	yard	m

BIBLIOGRAPHY

1. The International Bureau of Weights and Measures, BIPM (Parc de Saint-Cloud, France) is described in Appendix X2 of Ref. 4. This bureau operates under the exclusive supervision of the International Committee for Weights and Measures (CIPM).
2. *Metric Editorial Guide (ANMC-78-1)*, latest ed., American National Metric Council, 5410 Grosvenor Lane, Bethesda, Md. 20814, 1981.
3. *SI Units and Recommendations for the Use of Their Multiples and of Certain Other Units (ISO 1000-1981)*, American National Standards Institute, 1430 Broadway, New York, 10018, 1981.
4. Based on *ASTM E380-89a (Standard Practice for Use of the International System of Units (SI))*, American Society for Testing and Materials, 1916 Race Street, Philadelphia, Pa. 19103, 1989.
5. *Fed. Reg.*, Dec. 10, 1976 (41 FR 36414).
6. For ANSI address, see Ref. 3.

R. P. LUKENS
ASTM Committee E-43 on SI Practice

Continued

THIOGLYCOLIC ACID

Thioglycolic acid (2-mercaptoacetic acid [68-11-1]), HSCH$_2$COOH, molecular weight 92.11, is the first member of the mercaptocarboxylic acids series. It was prepared and identified in 1862 by Carius (1) and studied around 1900 (2). Thioglycolic acid was first developed commercially in the early 1940s in the field of cosmetology as an active material for cold wave permanents and depilatories (see HAIR PREPARATIONS). The advent of the PVC industry in the 1950s brought thioglycolic acid a significant application as a raw material in the manufacture of organotin stabilizers (see HEAT STABILIZERS; VINYL POLYMERS). These stabilizers improved thermal stability and prevented discoloration during polymer processing. As large-scale commercial chemical production grew, thioglycolic acid began to be used more and more as a raw material in the manufacture of fine and specialty chemicals in the pharmaceutical and agricultural sectors.

Properties

Pure thioglycolic acid is a water-white liquid that freezes at $-16°C$ and distills under reduced pressure. Reported constants are bp at 3.9 kPa (29 mm Hg), 123°C, bp at 1.33 kPa (10 mm Hg), 96°C (3); d$_4^{20}$, 1.325 g/cm^3; viscosity at 20°C, 6.55 mPa·s(=cP); refractive index n_D^{20}, 1.5030; heat of vaporization, 627.2 J/g (149.9 cal/g); heat of combustion, 1446 kJ/mol (345.6 kcal/mol); electrical conductivity at 20°C, 2×10^6 (Ω·cm)$^{-1}$; dielectric constant, 7.4×10^{-30} C·m (2.25 debye); dipole moment, 7.6×10^{-30} C·m (2.28 debye); flash point in closed cup, 132°C.

Both the carboxyl and the mercapto moieties of thioglycolic acid are acidic. Dissociation constants at 25°C are for pK_1, 3.6; pK_2, 10.5. Thioglycolic acid is miscible in water, ether, chloroform, dichloroethane and esters. It is weakly soluble in aliphatic hydrocarbons such as heptane, hexane. Solvents such as alcohols and ketones can also react with thioglycolic acid.

Reactions

Thioglycolic acid is altered by self-esterification. This alteration depends on temperature and concentration of aqueous solutions. Thioglycolic acid is almost stable at room temperature in 70 wt % aqueous solution. At room temperature, the loss in assay for pure product under correct storage conditions is about 1% of the initial value in a month. At a lower (5°C) temperature this figure is below 0.3% in a month; at higher temperatures the self-esterification increases. Under an air atmosphere oxidation to disulfide occurs. Many self-esterification products, called thioglycolides, have been detected by nmr. Examples are S-mercaptoacetylthioacetic acid [99-68-3], $HSCH_2COSCH_2COOH$, which represents more than 90% of thioglycolides, some polythioglycolides, $HS(CH_2COS)_xCH_2COOH$, cyclic thioglycolides such as 1,4-dithioglycolide [4835-42-6], and also solid ortho thioesters such as tetracarboxylmethylmercapto-1,4-dithiane [52959-43-0]. Formation of the last is promoted by acid species. The self-esterification products can be reversed in the presence of dilute acids or alkalies, and aged thioglycolic acid can be completely recovered by hydrolysis.

Thioglycolic acid undergoes reactions typical of carboxylic acids, forming salts, esters, amides, and reactions typical of mercaptans, and forming thioethers with olefins or halogenated compounds, disulfides by oxidation, and metal mercaptides by reaction with metal oxides or metal chlorides. Thioglycolic acid or its salts in aqueous solutions are rapidly oxidized by air or hydrogen peroxide to the disulfide, dithiodiglycolic acid [505-73-7], $HOOCCH_2SSCH_2COOH$. More vigorous oxidative conditions, eg, with dilute nitric acid, produce sulfoacetic acid [123-43-3], HO_3SCH_2COOH. Thioglycolic acid is a powerful reducing agent in neutral or alkaline solutions. One of the most widely used reactions of thioglycolic acid is the thiol disulfide interchange reaction, particularly with the disulfide bond of cystine [923-32-0] (**1**) in protein material, eg, wool and hair, to form the amino acid cysteine (**2**). The rate at which equilibrium is established is a function of pH; it is low at pH <6 and rapid at pH 8–10.

$$2\ HSCH_2COOH + HOOCCH(NH_2)CH_2SSCH_2CH(NH_2)COOH \rightleftharpoons$$
$$(\mathbf{1})$$

$$HOOCCH_2SSCH_2COOH + 2\ HOOCCH(NH_2)CH_2SH \qquad (1)$$
$$(\mathbf{2})$$

Long-chain thioacetic acids are obtained by reaction of primary alkenes with thioglycolic acid, by using uv lamps or radical initiators. This is the case for dodecylthioacetic acid [13753-71-4], $C_{12}H_{24}SCH_2COOH$, prepared from dodecene, or carboxymethylthiosuccinic acid [99-68-3], $HOOCCH_2SCH(CH_2-COOH)COOH$, prepared from maleic acid. Similar products have been described for the reaction of chloroacetic acid [79-11-8], $ClCH_2COOH$, and corresponding mercaptans.

Other products of commercial value, such as laurylthiopropionic acid [1462-52-8], $C_{12}H_{24}SCH_2CH_2COOH$, are produced starting from 3-mercaptopropionic acid [107-96-0], $HSCH_2CH_2COOH$, and unsaturated products. S-Alkylthiocarboxylic acids and their potassium salts have been described and evaluated as surfactants (qv). They provide excellent thermally stable behavior and good surface activity for their alkaline salts (4).

Because of its two active functions, thioglycolic acid is an ideal reagent for a variety of chemical reactions, including addition, elimination, and cyclization. The methyl [2365-48-2] and ethyl [623-51-8] esters of thioglycolic acid, $HSCH_2COOR$, have shown promise as raw materials in several fine chemicals. These block the carboxylic group, leaving the mercapto group free to react. In the presence of bases, the mercapto group becomes a strong nucleophile able to react with halogenated compounds by a substitution reaction. This property is applied to the manufacture of many valuable pharmaceutical and agrochemical intermediates, such as ethyl trifluoromethane thioglycolate [75-92-9], $F_3CSCH_2CO_2C_2H_5$. The methylene group of thioglycolic acid esters, having a particular position between a mercapto and an ester function, displays a significant ability to react. Under alkaline conditions, interesting routes are available to a variety of heterocyclic compounds, such as thiophene, thiazole and other N−S heterocyclic compounds. Methyl thioglycolate [2365-48-2], $HSCH_2COOCH_3$, can be used as the starting material to obtain methyl 3-amino-2-thiophenecarboxylate [22288-78-4] (**3**) by reaction with 2-chloroacrylonitrile [920-37-6]. Compound (**3**) is a key intermediate to drugs and agrochemicals (5).

$$HSCH_2COOCH_3 + CH_2{=}C \overset{\displaystyle Cl}{\underset{\displaystyle CN}{}} \longrightarrow \quad (2)$$

$$(\mathbf{3})$$

Several kinds of products can be obtained by reaction of thioglycolic acid and its esters with aldehydes to form mercaptals, $RCH(SCH_2COOH)_2$, or with ketones to form thiolketals, $RR'C(SCH_2COOH)_2$. Reaction with formaldehyde (qv) yields di-n-butylmethylene-bisthioglycolate [1433882-0] (MBT ester):

$$2\, HSCH_2COOC_4H_9 + CH_2O \longrightarrow C_4H_9OOCCH_2SCH_2SCH_2COOC_4H_9 + H_2O \quad (3)$$

Thioglycolic acid forms a multiplicity of stable complexes with metal ions. Depending on the particular metal and on experimental conditions, the thioglycolate complexes can be mononuclear or polynuclear and have the structure of carboxylate salts, metal mercaptides, or true chelates. Complexes studied include various metal and rare-earth metals, mercaptocarboxylic acids, and thiodiglycolic acid [123-93-3], $HOOCCH_2SCH_2COOH$. Potentiometric investigations and structural identifications have been carried out to determine properties and stability (6).

Manufacturing, Processing, and Storage

Thioglycolic acid is manufactured by the reaction of monochloracetic acid [79-11-8] or its salts with alkali hydrosulfides, eg, NaSII or NH_4SH, in aqueous medium, under controlled conditions of pressure, temperature, pH, concentration, to give a higher yield of thioglycolate salt and minimize the formation of such by-products as thiodiglycolic and dithiodiglycolic acids. The reaction mixture is acidified to liberate thioglycolic acid, which is extracted from the aqueous solution into an organic solvent, such as an ether, and then purified by vacuum distillation (7) (Fig. 1).

Commercial monochloroacetic acid contains many other organic acids, particularly dichloroacetic acid [79-43-6], $Cl_2CHCOOH$, which has to be completely converted into sulfur derivatives to avoid residual chlorine compounds which are harmful for cosmetic applications (8). Thioglycolic acid, which has to meet cosmetic specifications, must be free of metal impurities, and must be pure enough to avoid color and odor problems.

Many other routes to produce thioglycolic acid have been investigated (9). To try to minimize by-products, nucleophilic agents other than alkali sulfhydrates have been claimed, eg, thiosulfates, sodium disulfides, thiourea, xanthogenic acid derivatives, and sodium trithiocarbonates (10). These alternative methods, which require reduction of the disulfides or hydrolysis of carboxymethylthio derivatives, seem less competitive than those using alkali sulfhydrates.

All the processes starting from chloroacetic acid are characterized by the formation of 2 moles of salt per mole of thioglycolic acid produced:

$$ClCH_2COOH + 2\,NaSH \longrightarrow HSCH_2COONa + NaCl + H_2S \qquad (4)$$

$$HSCH_2COONa + HCl \longrightarrow HSCH_2COOH + NaCl \qquad (5)$$

Therefore, manufacture of thioglycolic acid is associated with the production of aqueous salt waste, so problems of waste disposal have to be resolved in each plant.

Where thioglycolic acid is manufactured mainly as an intermediate for conversion to higher alkyl esters, the alcohol of the ester can be used as solvent to extract thioglycolic acid. It is also convenient to manufacture the esters of thioglycolic acid directly and to use the ester of chloroacetic acid and alkali sulfhydrates (10) as raw material.

$$ROOCCH_2Cl + NaSH \longrightarrow ROOCCH_2SH + NaCl \qquad (6)$$

This is a waste-reducing process in comparison with the classical processes, which proceed by the thioglycolic acid esterification route.

Esters of thioglycolic acid are to a large extent manufactured by conventional esterification processes. Manufacture at a larger scale of 2-ethylhexyl thioglycolate [7659-86-1] and isooctyl thioglycolate [25103-09-7] by a continuous process (12) gives esters of higher consistency. Thioglycolic acid is marketed as pure product or at 80–85 wt % aqueous solution. The ammonium salts are available in aqueous solutions containing 50–60 wt % thioglycolic acid; the monoethanolamine salts are available as solutions containing 40–50 wt % thioglycolic acid. Glycerol monothioglycolate is supplied in anhydrous form. Calcium thioglycolate is supplied as a crystalline powder. Potassium thioglycolate

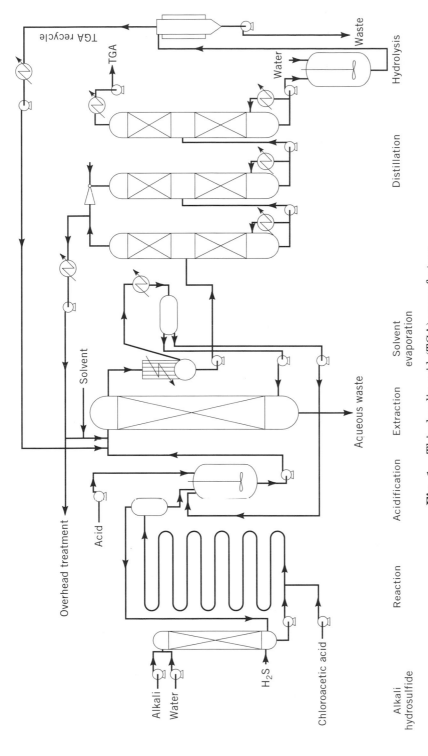

Fig. 1. Thioglycolic acid (TGA) manufacture.

[*34452-51-2*] and sodium thioglycolate [*367-51-1*] are also available as aqueous solutions.

Thioglycolic acid is stored in reinforced polyethylene or polypropylene tanks or containers. It is advisable to keep thioglycolic acid at low (<10°C) temperature to slow down self-esterification. The same care must be taken with drums or tank trucks. Drums are made of polyethylene or polyethylene-lined steel. For transport, thioglycolic acid is classified as a corrosive and toxic liquid. The handling of thioglycolic acid requires the usual precautions observed for strong acid and corrosive chemicals.

Among other mercaptocarboxylic acids, the mercaptopropionic acids have undergone a promising development. Thiolactic acid or 2-mercaptopropionic acid [*79-42-5*], $HSCH(CH_3)COOH$, is manufactured by using 2-chloropropionic acid [*598-78-7*]. 3-Mercaptopropionic acid [*107-96-0*], $HSCH_2CH_2COOH$, competes with thioglycolic acid in plastic additives or as modifiers in various polymers. Mercaptopropionic acid is produced by using acrylic monomers, eg, acrylic acid, acrylonitrile, methyl acrylate, as raw materials (13–19) (Table 1).

Table 1. Mercaptopropionic Acid Syntheses

Process/raw materials	Conversion method	References
$CH_2{=}CHCOOH + HCl \rightarrow ClCH_2CH_2COOH$		
$ClCH_2CH_2COOH + NaSH$	acidification	13
$CH_2{=}CHCOOH + H_2S$		14
$CH_2{=}CHCOOH + NaSH + CS_2$	acidification	15
$CH_2{=}CHCOOCH_3 + H_2S + S$	reduction hydrolysis	16
$CH_2{=}CHCN + NaSH + S$	reduction hydrolysis	17
$CH_2{=}CHCN + NaSH \rightarrow NCCH_2CH_2SCH_2CH_2CN$		
$NCCH_2CH_2SCH_2CH_2CN + NaSH + NaOH$	acidification	18
$HOOCCH_2CH_2SCH_2CH_2COOH + NaOH + Na_2S$	acidification	19

Economic Aspects

Since its development in cosmetics in the 1940s, thioglycolic acid has become a widespread thiochemical, used all over the world as the acid or in the form of its salts or esters. Because of the several derivatives of thioglycolic acid used, the market size for this chemical is often expressed in terms of thioglycolic acid equivalents. In 1994 the total world market was estimated at around 15,000 to 20,000 metric tons of thioglycolic acid equivalents. Some mercaptocarboxylic acid and esters available, are compiled in Table 2.

The worldwide growth of thioglycolic acid and its derivatives is primarily driven by the market growth of PVC, and more particularly by the demand for the tin stabilizers used in this thermoplastic. The demand for cold wave permanents and chemical depilatories is the second driving force for production of thioglycolic acid and its derivatives. The progress of thioglycolic acid within this sector is closely linked to local hair fashion, changes in cultural habits, and the development of average purchasing power. The world market for thioglycolic acid and its derivatives is thought to experience an average annual growth of 2%.

Table 2. Mercaptocarboxylic Esters Derivatives

Compound	CAS Registry Number	Structural formula	Bp,[a] °C
methyl thioglycolate	[2365-48-2]	$HSCH_2COOCH_3$	148
methyl 3-mercaptopropionate	[2935-90-2]	$HSCH_2CH_2COOCH_3$	166
ethyl thioglycolate	[623-51-8]	$HSCH_2COOC_2H_5$	155
n-butyl thioglycolate	[10047-28-6]	$HSCH_2COOC_4H_9$	$110_{5.3}$
n-butyl 3-mercaptopropionate	[16215-21-7]	$HSCH_2CH_2COOC_4H_9$	$101_{1.6}$
2-ethylhexyl thioglycolate	[7659-86-1]	$HSCH_2COOC_8H_{17}$	$106_{0.4}$
2-ethylhexyl 3-mercaptopropionate	[50448-95-8]		$85-87_{0.27}$
isooctyl thioglycolate	[25103-09-7]		$107_{0.53}$
isooctyl 3-mercaptopropionate	[30774-01-7]		$95_{0.30}$
ethylene glycol dimercaptoacetate	[123-81-9]	$(CH_2OOCCH_2SH)_2$	$137_{0.26}$
ethylene glycol dimercaptopropionate	[22504-50-3]	$(CH_2OOCCH_2CH_2SH)_2$	$101_{0.04}$
pentaerythritol-tetrakis-thioglycolate	[10193-99-4]	$C(CH_2OOCCH_2SH)_4$	[b]
pentaerythritol-tetrakis-3-mercaptopropionate	[7575-23-7]		[b]
di-n-butyl methylene bis thioglycolate	[14338-82-0]	$CH_2(SCH_2COOC_4H_9)_2$	[c]
thiodiglycolic acid	[123-93-3]	$S(CH_2COOH)_2$	128^{d}
dithiodiglycolic acid	[505-73-7]	$S_2(CH_2COOH)_2$	$100-102^{d}$
dithiodipropionic acid	[1119-62-6]	$S_2(CH_2CH_2COOH)_2$	$152-156^{d}$
thiodipropionic acid	[111-17-1]	$S(CH_2CH_2COOH)_2$	130^{d}
dimethyl thiodipropionate	[4131-74-2]	$S(CH_2CH_2COOCH_3)_2$	$130_{0.26}$
dilauryl thiodipropionate	[123-28-4]	$S(CH_2CH_2-COOC_{12}H_{25})_2$	40^{d}
distearyl thiodipropionate	[693-36-7]	$S(CH_2CH_2COOC_{18}H_{37})_2$	$63-68^{d}$
3-mercaptopropionic acid	[107-96-0]	$HSCH_2CH_2COOH$	15.5^{d} 111_2
2-mercaptopropionic acid	[79-42-5]	$HSCH(CH_3)COOH$	$85_{0.66}$
thiomalic acid	[70-49-5]	$HSCH(CH_2COOH)COOH$	150^{d}

[a]Pressure other than atmospheric given as subscript in kPa. To convert to mm Hg, multiply by 7.5.
[b]Material assays at 95%.
[c]Viscosity at 20°C is 14-18 mPa·s (=cP).
[d]Value given is melting point.

Manufacturing plants for thioglycolic acid and derivatives are found in Europe, the United States, and Asia. Producers in Europe are B. Bock (Germany), Elf Atochem (France), and Merck (Germany); in the United States Elf Atochem, Hampshire, and Witco; and in Japan, Daicel. Production capacity is expected to be sufficient to supply world demand for five years.

Specifications

Table 3 lists some properties and commercial specifications of thioglycolic and its salts.

Table 3. Specifications for Thioglycolic Acid and Some Derivatives

Property	Thioglycolic acid	Thioglycolic acid	Ammonium thioglycolate	Glyceryl monothioglycolate	Monoethanolamine thioglycolate	Calcium thioglycolate
CAS Registry Number	[68-11-1]	[68-11-1]	[5421-46-5]	[30618-84-9]	[126-97-6]	[814-71-1]
EINECS[a]	2006774	2006774	2265409	2502648	2048154	
assay as TGA, %	98–99	80.0–80.3	60.0–60.3	75–76	50.0–50.4	50
specific gravity, 20°C	1.32	1.26–1.27	1.205	1.285	1.25	
iron[b], ppm	0.4	0.3	0.4	2	0.4	
pH			7.0–7.2		6.9–7.3	11–12
refractive index, 20°C	1.5030			1.5020–1.5030		
appearance	water-white	water-white	clear	clear, colorless	clear	white crystalline powder

[a]EINECS = European Inventory of Existing Commercial Chemical Substances.
[b]Value is maximum.

Analytical and Test Methods

Thioglycolic acid can be identified by its ir spectrum or by gas chromatography. Most of the by-products and self-esterification products are also detected by liquid chromatography, eg, thiodiglycolic acid, dithiodiglycolic acid, linear dimers, and polymers. Iron content can be assayed by the red sensitive complex of 1,10-phenanthroline [66-71-7] and ferrous ion of a mineralized sample. Ferric ion turns an aqueous ammonia solution deep red-violet.

    Assay. Thioglycolic acid is determined by its quantitative reaction with iodine. Depending on the storage conditions and time, thioglycolic acid tends to form thioglycolides with formation of water. This process is reversible by hydrolysis at elevated temperature or by alkalies or acids. Thioglycolides can be completely reconverted to thioglycolic acid. Therefore, for the assay determination of the initial thioglycolic acid, it is necessary to carry out hydrolysis with ammonia prior to iodometric titration.

    Titration of thioglycolate esters is also realized by iodine in alcoholic solution. Titration of thioglycolic acid (acid number) in thioglycolate esters is effected by potentiometric titration with potassium hydroxide.

Health and Safety Factors

A safety assessment of ammonium and glyceryl thioglycolates and thioglycolic acid has been published (20). Hair products containing ammonium thioglycolate and glyceryl thioglycolate may be used safely, at infrequent intervals, at concentrations of ammonium thioglycolate and glyceryl thioglycolate up to 14.5% (as thioglycolic acid). Hairdressers should avoid skin contact and minimize consumer skin exposure.

    Metabolism. Absorption, distribution, metabolism, and excretion of thioglycolic acid have been reviewed (20). In summary, ^{35}S-thioglycolic acid was absorbed significantly after application to the skin of rabbits. After intravenous injection, the greatest counts of radioactivity were found in the kidneys, lungs, and spleen of monkey and in the small intestine and kidneys of rat. Most of the radioactivity was rapidly excreted in the urine in the form of inorganic sulfate and neutral sulfur.

    Acute Toxicity. Thioglycolic acid, its esters and 3-mercaptopropionic acid are considered moderately toxic on the basis of acute toxicity studies. These are presented in Table 4.

    Local Effects. Pure thioglycolic acid and 3-mercaptopropionic acid are stronger acids than acetic acid, and they must be handled with precautions appropriate for strong acids. Application of thioglycolic acid in a single patch test in rabbits resulted in necrosis in 5 min. This was accompanied by hyperemia and edema (23). 3-Mercaptopropionic acid induced corrosive dermatolysis after 24-h patch on rabbit skin (21). Thioglycolic acid esters induced at most a moderate irritation after 4-h contact on rabbit skin and are not classified as irritants under European regulations.

    Instillation of pure thioglycolic acid into the eyes of a rabbit resulted in severe pain, severe conjunctival inflammation, dense corneal opacity, and severe iritis. These effects had not improved at the end of 14 d after exposure. Washing

Table 4. Toxicity of Thioglycolic Acid and Derivatives[a]

Sex[b]	Route	Dosing sol'n, %	LD_{50}, mg/kg	Clinical signs	References
Thioglycolic acid					
M,F	oral	0.2–1	73	ptosis, decrease respiratory rate, prostaration	21
M	oral	2.5	114		22
M,F	inhalation/4 h (aerosol)		0.21[c]	irritation of the eyes and the lungs	21
[d]	dermal	10	848		23
Methyl thioglycolate					
F	oral	10	209		22
M,F	dermal	100	>2000[e]	no deaths, hypoactivity, piloerection, no cutaneous reaction	21
Ethyl thioglycolate					
F	oral	10	178		22
Isooctyl thioglycolate					
	oral		348–391		24
M,F	dermal	100	>2000[e]	no deaths, no clinical signs, no cutaneous reactions	21
2-Ethylhexyl thioglycolate					
M	oral	10	303		21
M,F	inhalation/6 h		[f]		21
M,F	dermal	100	>2000	20% mortality, hypoactivity, no cutaneous reactions	21
Glyceryl thioglycolate					
M,F	oral	68	>25 <200	sedation, coma, dyspnea, piloerection	21
Mercaptopropionic acid					
	oral		96–400		24
[g]	inhalation/1 h		[h]		21

[a]Tests carried out on rats, unless otherwise noted. [b]Of animals used. F = female; M = male. [c]Value is LC_{50} and units are mg/L. [d]Rabbits; no sex information given. [e]Value is LD_0. [f]No mortality at 0.51 mg/L. [g]Mice. [h]No death at saturated vapor.

immediately after exposure did not modify the response (23). These kinds of injuries correspond to those observed in one accident in which part of the contents of a bottle of concentrated thioglycolic acid were splashed into the eyes (25). Diluted (10% in water) thioglycolic acid is considered only as irritating according to the European criteria (26).

3-Mercaptopropionic acid induced immediate necrosis of the eye (21). Thioglycolic esters, eg, ethyl, isooctyl, 2-ethylhexyl, and glyceryl, are far less dangerous to rabbit eyes than free thioglycolic acid. They are not classified as irritants, whereas methyl thioglycolate is classified as an irritant according to European regulations (21).

Sensitization. The skin irritation and sensitization potentials of 9.0% thioglycolic acid were evaluated using the open epicutaneous test. Reactions were not observed during the challenge phase. Thioglycolic acid was an irritant, but not a sensitizer (20).

The potential of 3-mercaptopropionic acid and thioglycolic acid esters to induce delayed-contact hypersensitivity following intradermal injection and cutaneous application was evaluated in guinea pigs according to the maximization method of Magnusson and Kligman. Mercaptopropionic acid and methyl thioglycolate did not induce sensibilization. Glyceryl thioglycolate is a very weak sensitizer; isooctyl thioglycolate is a weak sensitizer; and 2-ethylhexyl thioglycolate is a moderate sensitizer (21).

Genotoxicity. Thioglycolic acid (27), sodium thioglycolate (28), 2-ethylhexyl thioglycolate (21), isooctyl thioglycolate (21), and glyceryl thioglycolate (21) were not mutagenic in the Ames test when tested with and without metabolic activation. 3-Mercaptopropionic acid showed mutagenic activity in the Ames test strain TA1535, both with and without metabolic activation (21). In the sex-linked recessive lethal mutation test, thioglycolic acid and sodium thioglycolate were not mutagenic (20,28). Thioglycolic acid (21) and sodium thioglycolate (28) also were not clastogenic when evaluated in the *in vitro* human lymphocytes chromosomal aberrations assay and the *in vivo* micronucleus test in bone marrow of mice, respectively.

Carcinogenicity. Very few data are available on the carcinogenic potential of thioglycolic acid and derivatives. There was no evidence of carcinogenicity in mice and rabbits that received dermal application of 1.0% sodium thioglycolate (in acetone) twice per week throughout the study. Mice were allowed to die spontaneously, whereas rabbits were killed during the 85th week of treatment (29).

Ecotoxicity. Thioglycolic acid is harmful to fish. The LC_{50}, 96 h for *Pimephales promelas* was found to be 30 mg/L. A biodegradation of 21% was achieved within 28 d in a closed bottle test based on the consumption of oxygen (21); thus, thioglycolic acid cannot be considered as readily biodegradable. But the acid was found to be inherently biodegradable in a MITI (Japanese) test based on the biological oxygen demand with a biodegradation rate of 70% after 28 days (30). It is not expected to bioaccumulate, owing to its low (log $P = 0.06$) n-octanol/water partition coefficient.

In the case of mercaptopropionic acid, biodegradation of 96% was achieved within 28 d in a closed bottle test based on the consumption of oxygen. The pass level of 60% was reached within 10 d of exceeding the 10% level; thus, mercaptopropionic acid can be considered as readily biodegradable. It is not expected to bioaccumulate, owing to the low (log $P = 0.09$) n-octanol/water partition coefficient.

2-Ethylhexyl thioglycolate is toxic to fish. The LC_{50}, 48 h to Leuciscus idus was found to be 9 mg/L (21). As a biodegradation of 22% was achieved within 28 d in a closed bottle test based on the biological oxygen demand (21), 2-ethylhexyl thioglycolate cannot be considered as readily biodegradable. From its n-octanol/water partition coefficient, log $P = 2.43$, low bioaccumulation potential is expected.

Uses

Thioglycolic Acid in Cosmetics.   In the 1940s the first patents were issued for the use of mercaptans in cold waving of human hair, and some time later thioglycolic acid-based formulations were put on the market. Thioglycolic acid rapidly became preferred, and as of the mid-1990s the thioglycolates continue to be the most commonly used active ingredient in permanent waving, in hair straightening lotions, and in depilatory creams.

The greater part of hair is made of an insoluble protein material, called keratin, composed of amino acids (qv) that form large condensed polymeric structures by the formation of amide links. The polypeptide chains are cross-linked by the disulfide bond of cystine. The purpose of the reducing step, in which thioglycolic acid is involved, is the cleavage of the disulfide bond of cystine (see eq. 1), so that hair becomes deformable. In a further step, the bonds are restored by a neutralizer, and curls are made permanent. Aqueous and dry neutralizers contain hydrogen peroxide, sodium perborate, and sodium and potassium bromate; as well as citric, tartaric, or phosphoric acids for pH adjustments; wetting agents, eg, fatty alcohols; protective agents; specifiers; and perfumes (31). Because the thioglycolates are not effective at low pH, the balance between buffer and ammonium hydroxide to get the right (from 7–9.5) pH is very important. The maximum thioglycolic acid concentration permitted under European regulations (32) is 8% for home formulations and 11% for professional use.

The majority of available depilatories are based on mercaptans like thioglycolic acid, which is used in the presence of alkaline reacting material. These preparations have less odor and are safer on the skin than other sulfides. Like waving lotions, the depilatory creams use various raw materials, thickening agents, fat compounds, chalk, alkaline salt of thioglycolic acid (2.5–5% as thioglycolic acid), chelating agent, perfume. The maximum concentration permitted under the European regulations is 5% and the pH must be in the range from 7 to 12.7 (32).

Many attempts have been made to reduce the ammoniacal and sulfurous odor of the standard thioglycolate formulations. As the cosmetics market is very sensitive to the presence of impurities, odor, and color, various treatments of purification have been claimed to improve the olfactory properties of thioglycolic acid and its salts, such as distillation (33), stabilization against the formation of H_2S using active ingredients (34), extraction with solvents (35), active carbon (36), and chelate resin treatments (37).

Polymerization Catalysts, Modifiers, and Chain-Transfer Agents. Properties of polymers obtained by free-radical polymerization depend on their molecular weight and molecular weight distribution, and substances capable of regulating these are called chain transfer agents. Owing to the thiol function, thioglycolic acid and its esters, or mercaptopropionic acids and their esters, can play such a role. Depending on the purpose for which a thioglycolate additive is used in polymerization, it is called a catalyst, a promoter, an accelerator, or a chain transfer agent.

Mercaptocarboxylic acids are also an easy means of introducing a polar function into the polymer for a further grafting, to influence the properties of the final polymer, and sometimes to obtain processing advantages (38). Thioglycolic acid, used as a chain transfer agent in various polymers, is useful for improving the dispersion properties of pigments and basic fillers in paints (39), or for increasing the affinity of polymer latexes with fillers in coatings for paper (40). Thioglycolic acid can also be used as a water-soluble chain transfer agent in aqueous solution polymerization, which mainly concerns acrylic and acrylamide polymerization, as retention agents (flocculants) in paper and water treatments, or as charge dispersants for paper.

Various polymers, such as polythiourethanes, polythioethers, and polythioacrylates, are used to produce resins which are transparent, colorless and have a high refractive index and good mechanical properties, useful for the production of optical lenses. Higher refractive indices are promoted by sulfur compounds and especially by esters of mercaptocarboxylic acids and polyols such as pentaerythritol (41) (see POLYMERS CONTAINING SULFUR).

The mercaptans are known to be efficient curing agents for epoxy resins for adhesives applications. Esters of thioglycolic acid or mercaptopropionic acid and polyols (42) are used as epoxy hardeners. They have several advantages such as low temperature curing properties, fast curing rate, lower toxicity than amines, and excellent color. They are also considered to be coupling agents, improving adhesion between metals and resins (43). 1,2,6-Hexanetriol trithioglycolate [*19759-80-9*] is known to exert a chelating action (44).

The stabilizing of halogen resins against the adverse effects of ionizing radiation has been obtained by using an ester of glycerol and thioglycolic acid (45).

Tin Stabilizers for Vinyl Chloride Polymers. The most important development of thioglycolic acid, and especially its isooctyl and 2-ethylhexyl esters, concerns its use as raw material for tin stabilizers, to prevent discoloration during thermal processing of PVC, and also to assure good compatibility and diffusion of the stabilizers through the resins (46). Commercial organotin stabilizers are typically tetravalent tin having one or two alkyl groups and two or three mercaptides or other basic ligands. The alkyl groups are usually methyl, butyl, or octyl. The choice of type and number depends largely on the use. Methyl and octyl derivatives are often used in food contact applications, but this varies by local custom and governmental regulations (FDA approval).

The alkyl group also produces subtle changes in the processing of the PVC, the use level and cost of the stabilizer, and in some cases even the final properties of the article, especially the heat distortion temperature or Vicat softening point. Overall, methyl derivatives are most widely used. Butyls are second and octyls a distant third.

Mercaptides are unchallenged as the ligand of choice for the other entities bonded to the tin, but carboxylates can also be used. Whereas a variety of mercaptans are used, the thioglycolic acid derivatives remain the largest single mercaptan. Dibutyltin bis(isooctyl thioglycolate) [*25168-24-5*] and butyltin tris-(isooctyl thioglycolate) [*25852-70-4*] are two common examples. These materials are produced by the reaction of the appropriate alkyltin chloride or oxide, and the mercaptan.

$$R'_p Sn(Cl)_{4-p} + (4-p)\ HSCH_2COOR \longrightarrow R'_p Sn(SCH_2COOR)_{4-p} + (4-p)\ HCl \tag{7}$$

$$R'_2 SnO + 2\ HSCH_2COOR \longrightarrow R'_2 Sn(SCH_2COOR)_2 + H_2O \tag{8}$$

Whereas other metal salts, especially lead stearates and sulfates, or mixtures of Groups 2 and 12 carboxylates (Ba–Cd, Ba–Zn, Ca–Zn) are also used to stabilize PVC, the tin mercaptides are some of the most efficient materials. This increased efficiency is largely owing to the mercaptans. The principal mechanism of stabilization of PVC, in which all types of stabilizers participate, is the adsorption of HCl, which is released by the PVC during degradation. This is important because the acid is a catalyst for the degradation, thus, without neutralization the process is autocatalytic.

However, because PVC is largely composed of alternating CH_2 and $CHCl$ groups, each elimination of HCl also produces an allylic chlorine (eq. 9). The allylic chlorine is less stable than the alkyl chlorine, ie, the production of allylic chlorine lowers the overall stability.

$$(CH_2 — CHCl — CH_2 — CHCl — CH_2)_n \longrightarrow (CH_2 — CH = CH — CHCl — CH_2)_n + HCl$$

$$(9)$$

Degradation then continues at adjacent sites to form conjugated olefins. This does not, however, eliminate the allylic chlorine (eq. 10). Rather the sequence leads to the zipper effect (46), in which long series of conjugated olefins are quickly produced. These series begin to adsorb light in the visible region when there are about 6 alternating double bonds and thus quickly degrade the appearance of the article.

$$(CH_2 — CH = CH — CHCl — CH_2 — CHCl)_n \longrightarrow$$
$$(CH_2 — CH = CH — CH = CH — CHCl)_n + HCl \quad (10)$$

Unlike other stabilizers, tin mercaptide, or the mercaptan that is formed after the HCl reacts with the mercaptide, can react with the allylic chlorine to produce a sulfide (47), thus eliminating the labile chlorine groups and stopping the unzipping.

$$(11)$$

The mercaptan or mercaptide can also add directly to the olefin to break up or shorten the conjugation.

Additionally, organotin mercaptides can act as antioxidants, as they can sequester free-radical degradation mechanisms (48). The one drawback of mercaptide-based tin stabilizers is the discoloration of the sulfur after exposure to uv-radiation. Special precautions or formulations need to be developed for outdoor applications.

The type of stabilizer used is greatly influenced by the specifics of the final products and the customs in the region of use. Lead- and tin-based stabilizers are mostly used in rigid opaque applications such as building materials (see BUILDING MATERIALS, PLASTIC). Lead-based stabilizers predominate in Europe because the PVC formulations are low in uv-blockers (titanium dioxide). However, the higher uv-blocker needed in the US allows the use of tin mercaptides, eliminating the hazards associated with the handling of the lead-based products and affording a larger processing latitude than that available with the lead-based materials.

Because of the stringent U.S. requirements on handling lead-based powders, these stabilizers are only found in flexible electrical wire insulation in the United States because no other stabilizers have been developed which have the low conductivity afforded by these materials.

However, most recently, mixed metal stabilizers are being introduced into this area. Whereas some mixed metal stabilizers are used in rigid applications, it is difficult for these newer materials to compete with lead and tin based stabilizers on a cost/efficiency basis. These materials find their largest applications in flexible compounds, especially those requiring good stability to uv exposure.

Miscellaneous. In the leather industry, dehairing is a specific step in the treatment of animal skins, where the reducing properties of thioglycolic salts for the disulfide links of keratins are exploited along side those of other reducing agents such as sodium sulfide and sodium hydrosulfide. Thioglycolic acid formulations have the advantage over simple sodium sulfide formulations regarding process safety issues, better quality of the final leather, reduction of waste, and recovery of hair (49).

Wool treatment by thioglycolic acid improves elongation, elasticity, and strength properties of fibers; prevents shrinkage; and makes dyeing easier. These improvements result from the interchange reactions between disulfides and the HS group of thioglycolic acid, observed on the fiber (50).

In wood industries, addition of thioglycolic acid to pulping liquor during alkaline cooking of pinewood led to an increase in the degree of delignification of wood. Thioglycolic acid has facilitated the cleavage of lignins, gives higher pulp yields, and leads to lower alkali consumption (51).

Thioglycolic acid is recommended as a cocatalyst with strong mineral acid in the manufacture of bisphenol A by the condensation of phenol and acetone. The effect of the mercapto group (mercaptocarboxylic acid) is attributed to the formation of a more stable carbanion intermediate of the ketone that can alkylate the phenol ring faster. The total amount of the by-products is considerably reduced (52).

Thioglycolic acid is described in descaling compositions for iron oxide removal. It can also be used as ammonium or ethanolamine salts to remove rust without attacking the metal substrate (53) (see METAL SURFACE TREATMENTS). Thioglycolic acid can modify the fluidity of concretes and cements in a new class of superplasticizers, where the polymeric melamine structures are grafted with a sulfur atom instead of nitrogen in the case of the well-known sulfanilic acid [121-57-3], 4-$(H_2N)C_6H_4SO_3H$, (54). This newer way of linking using an S atom modifies the structure of the resin and is responsible to a great extent for high fluidity values, a water-reducing content in concrete, and an improvement in mechanical properties (see CEMENT).

In the oil field industry, in drilling activities the sequestering properties of thioglycoic acid for iron are reflected in its use as an iron-controlling agent, for acidizing in well stimulation, and also as corrosion inhibitors in high density completion fluids (55). The oxazoline derivative of thioglycolic acid has been claimed, in water-based drilling muds, as a lubricant additive (56). In the catalytic cracking of hydrocarbons, in petroleum refining activities, mercaptides of thioglycolic acid are effectively added as a heavy metal passivator to counteract the adverse effects of metal (Ni, V, Fe) contaminants on catalysts. This is the case for antimony tris(2-ethylhexyl thioglycolate) [26888-44-4] or the isooctyl analogue [27288-44-4]. Other antimony, stannous, or phosphorous derivatives in various combinations have also been used (57).

In the lightening of petroleum hydrocarbon oil, esters of mercaptocarboxylic acids can modify radical behavior during the distillation step (58). Thioesters of dialkanol and trialkanolamine have been found to be effective multifunctional antiwear additives for lubricants and fuels (59). Alkanolamine salts of dithiodipropionic acid [1119-62-6] are available as water-soluble extreme pressure additives in lubricants (60).

Sulfur compounds are traditionally used as rubber and plastic additives. Esters of thioglycolic acid and various glycols, eg, ethylene glycol, propylene glycol, pentaerythritol, and particularly the butyl ester of methylene bisthioglycolic acid, $(C_4H_9OOCCH_2S)_2CH_2$, are used as a polar plasticizer and softener for synthetic rubber, especially for nitrile rubber and chloroprene. Dodecylthioacetic acid, $C_{12}H_{24}SCH_2COOH$ is a viscosity modifier (61). Esters of thiodipropionic acid [111-17-1], $(HOOCCH_2CH_2)_2S$, and long-chain alcohols, in the $C_{12}-C_{18}$ range, are largely used as thioester antioxidant additives in polyolefins and styrene–butadiene latex resins.

The pentaerythritol ester of dodecylthiopropionic acid [1462-52-8], $C_{12}H_{25}SCH_2CH_2COOH$, is marketed for the same purpose.

Thioglycolic acid and esters are used in the manufacture of sulfur dyes (qv), thioindigo pigments, and as additives for dyeing baths (62). Examples are 2,5-dichlorophenylthioglycolic acid [6274-27-7] (63) for thioindigo pigment and 2-naphtylthioglycolic acid as dyestuff intermediate (64).

Pharmaceuticals and Agrochemicals.   Thioglycolic acid and its esters are useful as a raw material to obtain biologically active molecules. In cephalosporine syntheses, (4-pyridyl)thioacetic acid [10351-19-8] (65) and trifluoromethane (ethyl) thioglycolate [75-92-9] (66) are used as intermediates. Methyl-3-amino-2-thiophene carboxylate can be used as intermediate for herbicidal sulfonylureas (67) and various thiophenic structures (68).

Various other cyclic compounds can be built using thioglycoic acid, eg, thiazolidinone, thiazole, isothiazole, and thiazine-type structures, leading to intermediates for the agricultural and pharmaceutical industries (69). Fungicidal organotin mercaptocarboxylates have also been claimed (70).

BIBLIOGRAPHY

"Thioglycolic Acid" in *ECT* 1st ed., Vol. 14, pp. 78–83, by S. D. Gershon and R. M. Rieger, Lever Brothers Co.; in *ECT* 2nd ed., Vol. 20, pp. 198–204, by S. D. Gershon, Lever Brothers Co., and R. M. Rieger, Warner Lambert Co.; in *ECT* 3rd ed., Vol. 22, pp. 933–946, by O. S. Kauder, Argus Chemical Corp.

1. L. Carius, *Ann.* **124**, 43 (1862).
2. P. Klason and T. Carlson, *Ber.* **39**, 732–738 (1906).
3. E. Bilmann, *Justus Liebigs Ann. Chem.*, 339–357 (1905).
4. K. Kamio and co-workers, *J. Am. Oil Chem. Soc.* **72**(7), 805–809 (1995).
5. P. R. Huddleston and J. M. Barker, *Synthet. Communica.* **9**, 731 (1979).
6. H. Matsui and H. Ohtaki, *Polyhedron* **2**(7), 631–633 (1983); J.-D. Joshi and P. K. Bhattacharya, *J. Indian Chem. Soc.* **57**(3), 336–337 (1980); U.S. Pat. 4,208,398 (June 17, 1980), T. F. Bolles and D. O. Kubiatowicz (to Hoffmann-LaRoche); A. Napoli, *Gaz. Chim. Ital.* **102**, 273–280 (1972).
7. U.S. Pat. 3,927,085 (Dec. 10, 1975), H. G. Zenkel and co-workers (to Akzo); Jpn. Pat. 55,145,663 (Nov. 14, 1980), K. Tamashima and S. Nagasaki (to Denki Kagaku Kogyo); Jpn. Pat. 56,097,264 (Aug. 5, 1981), K. Tamashima and S. Nagasaki (to Denki Kagaku Kogyo); U.S. Pat. 5,023,371 (June 11, 1991), M. E. Tsui and M. B. Sherwin (to W. R. Grace).
8. P. Darles, Y. Labat, and Y. Vallee, *Phosphor. Sulfur Silicon*, **46**(99), 43–45 (1989); Brit. Pat. 2,164,939 (Sept. 28, 1984), M. E. Tsui and M. B. Shervin (to W. R. Grace).
9. G. T. Walker, *Seifen Ole Fette Wachse* **13**, 402–404 (1962); **14**, 431 (1962).
10. U.S. Pat. 3,860,641 (Jan. 14, 1975), M. Bergfeld and H. G. Zengel (to Akzo); S. Afr. Pat. 7,404,097 (May 21, 1975), M. F. Werneke (to American Cyanamid).
11. R. M. Acheson, J. A. Barltrop, M. Hichens, and R. E. Hichens, *J. Chem. Soc.*, 650–660 (1961); Jpn. Pat. 6,310,755 (Jan. 18, 1988), H. Itsuda, M. Kawanura, K. Kato, and S. Kimura (to Seitetsu Chemicals); Jpn. Pat. 2,304,061 (May 17, 1989) (to Nippon Chemical Industries); Fr. Pat. 2,723,737 (Aug. 19, 1994), Y. Labat and J.-P. Muller (to Elf Atochem).
12. Eur. Pat. 421831 (June 16, 1993), Y. Labat, D. Litvine, and J.-P. Muller (to Elf Aquitaine).
13. U.S. Pat. 3,927,085 (Dec. 10, 1975), H. G. Zenkel and co-workers (to Akzo).
14. Eur. Pat. 208333 (July 11, 1985), R. J. Scott (to Phillips Petroleum).
15. U.S. Pat. 4,490,307 (Dec. 25, 1984), H. Kienk (to Degussa).
16. Eur. Pat. 485139 (May 13, 1992), D. R. Chisholm and G. A. Seubert (to Witco).
17. Jpn. Pat. 2,121,962 (Apr. 9, 1990) (to Daicel Chemical Industries).
18. U.S. Pat. 5,391,820 (Feb. 21, 1995), R. P. Woodburry and D. Wood (to Hampshire Chemicals).
19. Jpn. Pat. 4,009,363 (Jan. 14, 1992), T. Tetsuzo and T. Tomiaka (to Tomioka).
20. *J. Am. Coll. Toxicol.* **10**(1), 135–192 (1991).
21. Technical data, Elf Atochem, 1976, 1988, 1989, 1992, 1994, 1995.
22. P. Von Schmidt, G. Fox, K. Hollenbach, and R. Rothe, *Zeitsch. f. Gesamte Hygiene Grenzgebiete* **20**, 575–578, 1974.
23. *Documentation of the Threshold Limit Values and Biological Exposure Indices*, 5th ed., American Conference of Governmental Industrial Hygienists (ACGIH), Cincinnati, Ohio, 1986.
24. J. G. A. Luijten and O. R. Klimmer, *A Toxicological Evaluation of the Organotin Compounds*, Tin Research Institute, 1973.
25. W. M. Grant, *Toxicology of the Eye*, 3rd ed., C. C. Thomas, Springfield, Ill., 1986.
26. G. A. Jacobs, *J. Am. Coll. Toxicol.* **11**(6), 740 (1992).
27. E. Zeiger and co-workers, *Environ. Mutagen.* **9**(Suppl. 9), 1–109 (1987).
28. E. Gocke, M.-T. King, K. Eckhardt, and D. Wild, *Mutat. Res.* **90**, 91–109 (1981).
29. F. G. Stenbäck, J. C. Rowland, and L. A. Russell, *Food Cosmet. Toxicol.* **15**(6), 601–606 (1977).
30. Citi, Chemicals Inspection and Testing Institute, eds., *Biodegradation and Bioaccumulation Data of Existing Chemicals Based on the CSCL (Chemical Substance Control Law)*, Japan, Oct. 1992.

31. Eur. Pat. 302265 (Feb. 8, 1989), P. Hartmann, Y. Greiche, and J. Kohler (to Wella AG).
32. *EEC Regulations for Cold Wave Hair Straightening Lotions and Depilatories*, Commission Directive 88/233 CEE adapting to Technical Progress Council Directive 76/768/CEE, Mar. 2, 1988.
33. Jpn. Pat. 02,072,155 (Mar. 3, 1990), S. Kimura and Y. Harano (to Daicel Chemical Industries).
34. DE 3,243,959 (Aug. 18, 1983), I. Sandler and co-workers (to Merck).
35. Jpn. Pat. 59,027,866 (Feb. 14, 1984), J. Nakayama (to Lion Corp.).
36. Jpn. Pat. 55,064,569 (May 16, 1980), K. Tamashima, C. Fujii, and S. Nagasaki (to Denki Kagaku Kogyo).
37. Jpn. Pat. 55,051,055 (Apr. 15, 1980), K. Tamashima and S. Nagasaki (to Denki Kagaku Kogyo).
38. U.S. Pat. 4,758,626 (July 19, 1988), T. Ishira and N. Boutnio (to General Electric).
39. Jpn. Pat. 58,164,656 (Sept. 29, 1983) (to Toa Gosei Chemical Industries); Jpn. Pat. 05,005,076 (Jan. 14, 1993), Y. Yasutaro, H. Kato, T. Miki, and S. Kimata (to Nippon Paint).
40. Jpn. Pat. 05,140,207 (May 8, 1993) (to Sumitomo Dow KK).
41. Jpn. Pat. 02,036,216 (Feb. 6, 1990) (to Mitsui Toatsu Chemicals); U.S. Pat. 5,059,673 (Oct. 22, 1991), Y. Kanamura and co-workers (to Mitsui Toatsu); Eur. Pat. 490778 (June 17, 1992), S. Lavault and G. Velleret (to Rhone-Poulenc); U.S. Pat. 5,236,907 (Aug. 17, 1993), S. Ohkawa and S. Saito (to Asahi Denka Kygyo); U.S. Pat. 5,424,472 (Oct. 6, 1993), M. Bader, P. Hartmann, and G. Schwinn (to Rohm GmbH); U.S. Pat. 5,422,422 (Oct. 12, 1993), M. Bader and V. Kerscher (to Rohm GmbH).
42. U.S. Pat. 3,352,810 (1976), A. J. Duke and G. Mclay (to Ciba); U.S. Pat. 4,861,863 (Aug. 29, 1989), R. A. Gonzalez and co-workers (to Henkel).
43. U.S. Pat. 4,812,363 (Mar. 14, 1989), J. P. Bell and R. G. Schmidt (to Bell).
44. Can. Pat. 1,006,074 (Jan. 3, 1977), R. L. Leroy (to Noranda Mines); *Corrosion* **34**(4), 113–119 (1988).
45. U.S. Pat. 4,412,897 (Nov. 1, 1983), S. Kornbaum and J. Y. Chenard (to Elf Atochem Chimie).
46. H. Andreas, *Plastics Additives Handbook*; 3rd ed., Hanser Publishers, New York, 1990, pp. 271–325.
47. W. H. Starnes, I. M. Plitz, *Makromol.*, **9**(4) 633–640 (1976); A. H. Frye, R. W. Horst, and M. A. Paliobagis, *J. Polym. Sci. Part A-2*, 1765 (1964).
48. J. S. Brooks and co-workers, *Poly. Deg. Stab.* **4**, 359–363 (1982).
49. Y. Nakamura and co-workers, *Seni Kogyo Shisetsu Hokoku* **7**, 89–101 (1969).
50. C. Kraniecki and A. Sochacka, *Przegl. Skorzany* **46**, (4) 85–88 (1991); U.S. Pat. 4,484,924 (Nov. 27, 1984), E. Pfleiderer, T. Taeger, and R. Monsheimer (to Rohm GmbH).
51. U.S. Pat. 3,490,991 (Jan. 20, 1970), W. E. Fisher and A. S. Hider (to Owens-Illinois).
52. U.S. Pat. 4,822,923 (1989), K. Simon (to Shell Oil Co.); U.S. Pat. 4,931,594 (June 5, 1990), J. Knebel, V. Kerscher, and W. Ude (to Rohm GmbH); Jpn. Pat. 01,157,926 (June 21, 1989), H. Sakemoto and K. Shigemaksu (to Idemitsu Kosan); U.S. Pat. 5,414,152 (Apr. 9, 1995), M. J. Cipuelo (to General Electric).
53. Jpn. Pat. 05,014,028 (1993), J. Ieiri, Y. Nagayama, and T. Tsukada (to Mitsubishi); Jpn. Pat. 60,218,488 (Nov. 1, 1985), T. Suzuki and Y. Hirozawa (to Rinrei); Jpn. Pat. 54,048,643 (Apr. 17, 1979), S. Watamabe.
54. Eur. Pat. 557211 (Feb. 18, 1992), A. Sers and co-workers (to Chryso).
55. Fr. Pat. 2,677,074 (Dec. 4, 1992), P. Dejeux, J.-P. Feraud, and H. Perthuis (to Dowell-Schlumberger); U.S. Pat. 4,784,778 (Nov. 15, 1988), C. Shin (to Great Lakes Chemicals).
56. U.S. Pat. 4,491,524 (Jan. 1, 1985), A. Gutierrez and co-workers (to Exxon).

57. U.S. Pat. 4,830,731 (Apr. 16, 1989), F. Kotaroh, W. Toshio, and N. Hitomi (to Sakai Chemical); U.S. Pat. 4,562,167 (Dec. 31, 1985), B. J. Berthus and D. L. McKay (to Phillips Petroleum); U.S. Pat. 4,324,648 (Apr. 13, 1982), J. S. Roberts and co-workers (to Phillips Petroleum).

58. U.S. Pat. 4,931,170 (June 5, 1990), T. Sasaki (to Seibu Oil).

59. U.S. Pat. 5,198,131 (Mar. 30, 1993), A. G. Horodysky and co-workers (to Mobil Oil); U.S. Pat. 5,405,545 (Apr. 11, 1995), A. G. Horodysky and co-workers (to Mobil Oil).

60. U.S. Pat. 4,880,552 (Nov. 14, 1989), P. Guesnet and co-workers (to Elf Aquitaine).

61. U.S. Pat. 5,130,363 (July 4, 1992), U. Eholzer and co-workers (to Bayer); U.S. Pat. 4,395,349 (July 26, 1983), K. Kinoshita (to Osaka Yuki Kagaku).

62. U.S. Pat. 3,014,654 (Apr. 18, 1979) (to Asahi Chem).

63. U.S. Pat. 5,210,291 (Apr. 11, 1993), H. Goda and co-workers (to Sumitomo); U.S. Pat. 4,461,911 (July 24, 1984), D. I. Schvetze (to Bayer).

64. Eur. Pat. 832273 (May 10, 1985), W. Bayer (to Cassella).

65. U.S. Pat. 3,644,377 (1971), C. Sapino and P. D. Sleezer (to Bristol Myers-Squibb).

66. U.S. Pat. 4,491,547 (Jan. 1, 1985), T. Tsuji and co-workers (to Shionogi Seyaku).

67. Eur. Pat. 30142 (Nov. 28, 1980), G. Levitt (to E. I. du Pont de Nemours & Co., Inc.).

68. Fr. Pat. 2,689,129 (Oct. 1, 1993), S. Rault and co-workers (to Elf Atochem).

69. U.S. Pat. 4,479,792 (Apr. 12, 1983), E. H. Blaine (to Merck); U.S. Pat. 4,346,094 (Aug. 24, 1982), J. R. Beck and co-workers (to Eli Lilly).

70. Eur. Pat. 91148 (Oct. 12, 1983), P. Tenhaken and S. B. Welle (to Shell International Research); U.S. Pat. 3,846,459 (Nov. 5, 1974), C. Stapfer (to Cincinnatti Milacron).

Y. Labat
Elf Atochem

THIOLS

The chemistry of organic sulfur compounds is very rich and organosulfur compounds are incorporated into many molecules. Thiols, or mercaptans as they were originally called, are essential as feedstocks in the manufacture of many types of rubber (qv) and plastics (qv). They are utilized as intermediates in agricultural chemicals, pharmaceuticals (qv), in flavors and fragrances, and as animal feed supplements. Many reviews have been undertaken on the chemistry of the thiols, regarding both their preparation and their reactions (1–7).

Nomenclature

Thiols are still commonly named as mercaptans, although the proper nomenclature is that established by the International Union of Pure and Applied Chemists (IUPAC). A listing of the IUPAC name, common name, and structure, is shown in Table 1.

Table 1. Nomenclature of Thiols

IUPAC name	Common name	Structure
methanethiol	methyl mercaptan	CH_3SH
ethanethiol	ethyl mercaptan	CH_3CH_2SH
2-propanethiol	isopropyl mercaptan	$CH_3CH(SH)CH_3$
2-butanethiol	*sec*-butyl mercaptan	$CH_3CH(SH)CH_2CH_3$
1-butanethiol	*n*-butyl mercaptan	$CH_3CH_2CH_2CH_2SH$
2-methyl-2-propanethiol	*t*-butyl mercaptan	$(CH_3)_3CSH$
2,4,4-trimethyl-2-pentanethiol	*tert*-octyl mercaptan	$(CH_3)_2C(SH)CH_2C(CH_3)_3$
cyclohexanethiol	cyclohexyl mercaptan	$C_6H_{11}SH$
benzenethiol	thiophenol	C_6H_5SH
α-toluenethiol	benzyl mercaptan	$C_6H_5CH_2SH$

Occurrence

Cysteine [*52-90-4*] is a thiol-bearing amino acid which is readily isolated from the hydrolysis of protein. There are only small amounts of cysteine and its disulfide, cystine, in living tissue (7). Glutathione [*70-18-8*] contains a mercaptomethyl group, $HSCH_2-$, and is a commonly found tripeptide in plants and animals. Coenzyme A [*85-61-0*] is another naturally occurring thiol that plays a central role in the synthesis and degradation of fatty acids.

Methanethiol has been found in sewer gases (8,9) and is thought to be produced by the bacterial degradation of methionine. Methanethiol, 1-butanethiol, and 2-methyl-1-propanethiol are produced by the bacterial degradation of blue-green algae. Some yeasts produce ethanethiol, 2-propanethiol is produced by actively growing *Microcystis flosaquae* (10). Methanethiol, ethanethiol, 1-propanethiol, 2-propanethiol, 2-methyl-1-propanethiol, 2-methyl-2-propanethiol, and 1-butanethiol are found in the environment near domestic animal pens (11).

Skunks excrete 1-butanethiol and 2-methyl-1-butanethiol [*1878-18-8*] as a natural defense mechanism (12). Methanethiol is found in cheese, milk, coffee, and oysters (13–16). It is also found in the kurrin fruit, which is endemic to Southeast Asia.

Physical Properties

The physical characteristic of thiols that most distinguishes them is their odor. Thiols have been used since the late 1800s as malodorants in combustible gases. In the 1930s the U.S. Bureau of Mines published a series of reports regarding the intensity of odors of warning agents that put this practice on a scientific basis (17). There have been innumerable studies since that time regarding the odor intensity of thiols, primarily about their role in the odorization of natural gas and propane gas. All of the definitive studies on thiol threshold odor levels have been brought together and normalized to give a standardized odor scale (18). Even the heavier mercaptans have a significant odor, particularly when heated.

Thiols range over the gamut of physical properties. Most of the important thiols are liquids, however methanethiol is a gas, and 1-hexadecanethiol and 1-octadecanethiol are waxy solids. Tables 2 and 3 list a variety of physical properties for the more important thiols.

Table 2. Properties of Thiols

Compound	CAS Registry Number	Molecular weight	Melting point, K	Boiling point, K
methanethiol	[74-93-1]	48.11	150.18	279.11
ethanethiol	[75-08-1]	62.14	125.26	308.15
2-mercaptoethanol	[60-24-2]	78.14	200.00	430.90
1,2-ethanedithiol	[540-63-6]	94.20	231.95	419.20
mercaptoacetic acid	[68-11-1]	92.12	256.65	493.00
1-propanethiol	[107-03-9]	76.16	159.95	340.87
2-propanethiol	[75-33-2]	76.16	142.61	325.71
3-mercaptopropionic acid	[107-96-0]	106.15	290.65	501.00
1-butanethiol	[109-79-5]	90.19	157.46	371.61
2-butanethiol	[513-53-1]	90.19	133.02	358.13
2-methyl-1-propanethiol	[513-44-0]	90.19	128.31	361.64
2-methyl-2-propanethiol	[75-66-1]	90.19	274.26	337.37
2,2'-oxybisethanethiol	[2150-02-9]	138.26	193.15	490.15
1-pentanethiol	[110-66-7]	104.22	197.45	399.79
cyclohexanethiol	[1569-69-3]	116.23	189.64	431.95
1-hexanethiol	[111-31-9]	118.24	192.62	425.81
benzenethiol	[108-98-5]	110.18	258.26	442.29
1-heptanethiol	[1639-09-4]	132.27	229.92	450.09
α-toluenethiol	[100-53-8]	124.21	243.95	472.03
1-octanethiol	[111-88-6]	146.30	223.95	472.19
2,4,4-trimethyl-2-pentanethiol	[141-59-3]	146.30	199.00	428.65
1-nonanethiol	[1455-21-6]	160.32	253.05	492.95
1-decanethiol	[143-10-2]	174.35	247.56	512.35
1-undecanethiol	[5332-52-5]	188.38	270.15	530.55
1-dodecanethiol	[112-55-0]	202.40	265.15	547.75
tert-dodecanethiol[a]	[25103-58-6]	202.40		515.65
1-hexadecanethiol	[2917-25-2]	258	291–293	$396-401_{0.07}$[b]
1-octadecanethiol	[2885-00-9]	286	301	$461_{0.1}$[b]

[a] tert-Dodecanethiol is a mixture of isomers.
[b] Subscripted value represents pressure in kPa at which boiling point was taken. To convert kPa to mm Hg, multiply by 7.5.

The heat capacities and entropies of organic compounds, including many thiols, have been compiled (19,20). The thermochemistry of thiols and other organosulfur compounds has been extensively reviewed (21).

Preparation of Thiols

Thiols can be prepared by a variety of methods. The most-utilized of these synthetic methods for tertiary and secondary thiols is acid-catalyzed synthesis; for normal and secondary thiols, the most-utilized methods are free-radical-initiated, alcohol substitution, or halide substitution; for mercaptoalcohols, the most-utilized method is oxirane addition; and for mercaptoacids and mercaptonitriles, the most-utilized methods are Michael-type additions.

Acid-Catalyzed Synthesis. The acid-catalyzed reaction of alkenes with hydrogen sulfide to prepare thiols can be accomplished using a strong acid (sulfuric or phosphoric acid) catalyst. Thiols can also be prepared continuously

Table 3. Physical Properties of Thiols

Compound	Flash point, K	Lower flammability limit, %	Heat of formation,[a] kJ/mol[b]	Gibbs heat of formation,[a] kJ/mol
methanethiol	217	3.9	−22.9	−9.80
ethanethiol	225	2.8	−46.3	−4.81
2-mercaptoethanol	340.15	2.3	−197	−134
1,2-ethanedithiol	317.15	1.9	−9.70	26.7
mercaptoacetic acid	398	3.1	−394	−344
1-propanethiol	253.15	1.8	−67.5	2.58
2-propanethiol	238.15	1.8	−75.9	2.18
3-mercaptopropionic acid	366.15	2.2	−406	−344
1-butanethiol	274.82	1.4	−87.8	11.4
2-butanethiol	250.15	1.4	−96.6	5.12
2-methyl-1-propanethiol	264.15	1.4	−96.9	5.98
2-methyl-2-propanethiol	<253	1.4	−109	1.01
2,2′-oxybis(ethanethiol)	371.15	1.3	−173,000	−59,900
1-pentanethiol	291	1.2	−110	18.0
cyclohexanethiol	316.15	1.1	−96.0	36.7
1-hexanethiol	293.15	1	−129	27.6
benzenethiol	346.15	1.2	112	148
1-heptanethiol	319.15	0.9	−150	36.2
α-toluenethiol	343.15	1.1	93.3	163
1-octanethiol	341.15	0.8	−170	44.6
2,4,4-trimethyl-2-pentanethiol	304	0.8	−202	31.1
1-nonanethiol	351.15	0.7	−191	52.8
1-decanethiol	371.15	0.6	−211	61.7
1-undecanethiol	382	0.6	−233	69.0
1-dodecanethiol	360.15	0.5	−253	77.2
tert-dodecanethiol	363.15	0.6	−273	68.2

[a]To gaseous product.
[b]To convert kJ to kcal, divide by 4.814.

over a variety of solid acid catalysts, such as zeolites, sulfonic acid-containing resin catalysts, or aluminas (22). The continuous process is utilized commercially to manufacture the more important thiols (23,24). The acid-catalyzed reaction is commonly classed as a Markownikoff addition. Examples of two important industrial processes are 2-methyl-2-propanethiol and 2-propanethiol, given in equations 1 and 2, respectively.

$$(CH_3)_2C{=}CH_2 + H_2S \longrightarrow (CH_3)_3CSH \tag{1}$$

$$CH_3CH{=}CH_2 + H_2S \longrightarrow (CH_3)_2CH(SH) \tag{2}$$

These reactions tend to give few by-products. The main by-product in each case is the sulfide, RSR, which amounts to less than 3–5% of the thiols produced. Some of the sulfides produced have applications, although they tend to be much smaller-volume requirements than the amounts produced. The sulfides can be incinerated for disposal, assuming that the incineration facility can handle high sulfur feedstocks.

Free-Radical-Initiated Synthesis. Free-radical-initiated reactions of hydrogen sulfide to alkenes are commonly utilized to prepare primary thiols. These reactions, where uv light is used to initiate the formation of hydrosulfuryl (HS) radicals, are utilized to prepare thousands of metric tons of thiols per year. The same reaction can be performed using a radical initiator, but is not as readily controlled as the uv-initiated reaction. These types of reactions are considered to be anti-Markownikoff addition reactions.

This synthesis method can be utilized by any alkene or alkyne, but steric hindrance on internal double bonds can cause these reactions to be quite slow. Conjugated dienes and aromatic alkenes are not suited for the ultraviolet light-initiated process. The use of other free-radical initiators is required in free-radical-initiated reactions involving these species.

Examples of thiols prepared using this type of technology are 2-methyl-1-propanethiol, 1-butanethiol, and 2-butanethiol, in equations 3–5, respectively.

$$(CH_3)_2C = CH_2 + H_2S \longrightarrow (CH_3)CHCH_2SH \qquad (3)$$

$$CH_3CH_2CH = CH_2 + H_2S \longrightarrow CH_3CH_2CH_2CH_2SH \qquad (4)$$

$$CH_3CH = CHCH_3 + H_2S \longrightarrow CH_3CH(SH)CH_2CH_3 \qquad (5)$$

The main by-products of this synthesis type are sulfides and the isomer resulting from the Markownikoff addition to the alkene. For example, in the synthesis of 1-butanethiol (eq. 4), 5-thiononane, $C_4H_9SC_4H_9$, and 2-butanethiol are produced as by-products. The 2-butanethiol has uses as a herbicide intermediate and a gas odorant blend component and is further processed. The 5-thiononane is incinerated or reprocessed for fuel value. Sulfides account for up to 10% of the thiols produced. Another 2–5% is the Markownikoff addition product.

Alcohol Substitution. In the early period of normal thiol production, the normal alcohols were utilized as feedstocks. The use of a strong acid catalyst results in the formation of a significant amount of secondary thiol, along with other isomers resulting from skeletal isomerization of the starting material. This process has largely been replaced by uv-initiation because of the higher relative cost of alcohol vs alkene feedstock.

Methanethiol (eq. 6) and cyclohexanethiol (eq. 7) are the only commercially important thiols prepared using alcohol substitution. In most cases, when the alcohol is utilized, less control over the substitution patterns is obtained. Only one isomer is obtainable in the case of methanol and cyclohexanol.

$$CH_3OH + H_2S \longrightarrow CH_3SH + H_2O \qquad (6)$$

$$C_6H_{11}OH + H_2S \longrightarrow C_6H_{11}SH + H_2O \qquad (7)$$

The main by-product of this type of reaction is the sulfide. For the synthesis of methanethiol (eq. 6), the main by-product is 2-thiapropane, CH_3SCH_3. This material has a variety of uses and is further processed.

Oxirane Ring Opening. Mercaptoalcohols are prepared by the reaction of hydrogen sulfide and a suitable oxirane. This reaction is readily extendable to many oxiranes, but in general, there are only three oxiranes that are readily available: oxirane (ethylene oxide), 1-methyloxirane (propylene oxide), and α-chloromethyloxirane (epichlorhydrin). Epichlorhydrin is too corrosive and the resulting thiol is too unstable for general use. The oxiranes made from long-chain normal alkenes ($C_{10}-C_{18}$) also work in this type of reaction. There are, however, no known applications utilizing the mercaptoalcohols prepared from the higher alkyl oxiranes of this type.

The oxirane ring-opening reaction requires the presence of a basic catalyst. An acidic catalyst also works, but the polymerization of the oxirane limits its usefulness. In the case of 2-mercaptoethanol (eq. 8), the product has been found to be autocatalytic, ie, the product is a catalyst for the reaction.

These reactions are extremely exothermic and there is also the potential of initiating oxirane self-polymerization, which is an extremely exothermic reaction. In general, these reactions are run by adding oxirane to a large excess of thiol for heat control purposes and to minimize the multiple addition of oxirane to the resulting mercaptoalcohol. 2-Mercaptoethanol (eq. 8) is the only product prepared in large scale using this technology, although 2-mercapto-1-propanol can also be prepared in this manner (eq. 9):

$$H_2S + \overline{CH_2-CH_2}O \longrightarrow HSCH_2CH_2OH \qquad (8)$$

$$H_2S + CH_3\overline{CH-CH_2}O \longrightarrow HOCH_2CH(SH)CH_3 \qquad (9)$$

The by-products of these reactions are sulfides. The sulfide formed in the synthesis of 2-mercaptoethanol, 3-thia-1,5-pentanediol (thiodiglycol), has a variety of uses ranging from lubricant additive intermediates to textile finishing.

Michael-Type Additions.   Michael additions are generally used to prepare methyl 3-mercaptopropionate (eq. 10) and mercaptopropionitrile (eq. 11) by the reaction of methyl acrylate or acrylonitrile and hydrogen sulfide using a basic catalyst. This reaction proceeds as shown:

$$CH_2 \!=\! CHCOOCH_3 + H_2S \longrightarrow HSCH_2CH_2COOCH_3 \qquad (10)$$

$$CH_2 \!=\! CHCN + H_2S \longrightarrow HSCH_2CH_2CN \qquad (11)$$

The catalyst for this type of reaction is generally an amine. These reactions are quite exothermic, and the selection of the amine plays an important role in the rate of the reaction as well as in the amount of heat generated. The processes utilized to make these types of products are either batch or continuous (25–27).

The main by-products of this type of process are sulfides and disulfides. The disulfides are formed by the inclusion of an oxidizing agent (generally oxygen) that may be present in the reaction mixture or upon purification. Some of the sulfides formed in this fashion are useful as intermediates for the production of antioxidants. Other mercaptopropionates can be made in similar fashion, if the alkyl acrylate is available.

Halide Displacement. Halide displacement is a method used to prepare thiols that are not readily available by normal means. It requires a two-phase, water–organic system, that can be quite corrosive. Normally, this type of reaction, a classic S_N2 type, is undertaken in Hastelloy or glass-lined reactors. The syntheses of the two most important thiols prepared using this technology, α-toluenethiol and 1,2-ethanedithiol, are shown in equations 12 and 13, respectively.

$$C_6H_5CH_2Cl + NaHS \longrightarrow C_6H_5CH_2SH + NaCl \qquad (12)$$

$$ClCH_2CH_2Cl + 2\,NaHS \longrightarrow HSCH_2CH_2SH + 2\,NaCl \qquad (13)$$

The main by-products of this type of reaction are sulfides. By exercising careful control of stoichiometry, these can be minimized quite readily. By-products are disposed of by incineration or by reprocessing for fuel value.

Esterification and Transesterification to Produce Alkyl Mercaptopropionates. The other methods used to produce many of the alkyl mercaptopropionates are either transesterification of methyl 3-mercaptopropionate or esterification of 3-mercaptopropionic acid. The process to obtain 3-mercaptopropionic acid is shown in equation 14. A transesterification is shown in equation 15 and an esterification in equation 16, both to the n-butyl 3-mercaptopropionate.

$$HSCH_2CH_2CN + H_2O \longrightarrow HSCH_2CH_2COOH + NH_3 \qquad (14)$$

$$HSCH_2CH_2COOCH_3 + C_4H_9OH \longrightarrow HSCH_2CH_2COOC_4H_9 + CH_3OH \qquad (15)$$

$$HSCH_2CH_2COOH + C_4H_9OH \longrightarrow HSCH_2CH_2COOC_4H_9 + H_2O \qquad (16)$$

These reactions, catalyzed using sulfuric acid or a titanate ester, tend to be free of by-products.

Arenethiols. Most of the chemistry described for alkanethiols is not applicable to benzenethiols and other arenethiols. The synthesis of benzenethiol is best carried out by reduction of the benzene sulfonyl chloride (eq. 17). Other methods for synthesizing arenethiols are discussed at greater length in the literature (28).

$$C_6H_5SO_2Cl + 3\,H_2 \longrightarrow C_6H_5SH + 2\,H_2O + HCl \qquad (17)$$

Reactions of Thiols

Oxidation. Disulfides are prepared commercially by two types of reactions. The first is an oxidation reaction utilizing the thiol and a suitable oxidant as in equation 18 for 2,2,5,5-tetramethyl-3,4-dithiahexane. The most common oxidants are chlorine, oxygen (29), elemental sulfur, or hydrogen peroxide. Carbon tetrachloride (30) has also been used. This type of reaction is extremely exothermic. Some thiols, notably tertiary thiols and long-chain thiols, are resistant to oxidation, primarily because of steric hindrance or poor solubility of the oxidant in the thiol. This type of process is used in the preparation of symmetric disulfides, RSSR. The second type of reaction is the reaction of a sulfenyl halide with a thiol (eq. 19). This process is used to prepare unsymmetric disulfides, RSSR′ such as 4,4-dimethyl-2,3-dithiahexane. Other methods may be found in the literature (28).

$$2 \, (CH_3)_3CSH + 1/2 \, O_2 \longrightarrow (CH_3)_3CSSC(CH_3)_3 + H_2O \qquad (18)$$

$$(CH_3)_3CSH + Cl_2 \longrightarrow (CH_3)_3CSCl + HCl$$

$$(CH_3)_3CSCl + CH_3SH \longrightarrow (CH_3)_3CSSCH_3 + HCl \qquad (19)$$

When oxygen is used as the oxidant, a basic catalyst is required for the lighter thiols (31) and a transition metal co-catalyst may be required for the heavier thiols (32). Oxidation using sulfur as the oxidant requires a basic catalyst.

Gas-phase oxidation of thiols has been discussed in some depth (33). This review mainly emphasizes atmospheric processes, but a section on nitrogen oxides and thiols appears to be broadly applicable. The atmospheric oxidation chemistry of thiols is quite different from that of alcohols.

Thiols can be utilized to prepare polysulfides such as 2,2,6,6-tetramethyl-3,4,5-trithiaheptane. These are prepared (eq. 20) using sulfur as the only oxidant (34,35).

$$2 \, (CH_3)_3CSH + 1/4 \, S_8 \longrightarrow (CH_3)_3CSSSC(CH_3)_3 + H_2S \qquad (20)$$

Alkyl sulfonic acids are prepared by the oxidation of thiols (36,37). This reaction is not quite as simple as would initially appear, because the reaction does not readily go to completion. The use of strong oxidants can result in the complete oxidation of the thiol to carbon dioxide, water, and sulfur dioxide.

Formation of Sulfides.   Thiols react readily with alkenes under the same types of conditions used to manufacture thiols. In this way, dialkyl sulfides and mixed alkyl sulfides can be produced. Sulfides are a principal by-product of thiol production. Mixed sulfides can be formed by the reaction of the thiol using a suitable starting material, as shown in equations 21, 22, and 23. Vinyl sulfides can be produced by the reaction of alkynes with thiols (38).

$$RSH + R'CH{=}CH_2 \longrightarrow RSCH_2CH_2R' \qquad (21)$$

$$RSH + R'Cl + NaOH \longrightarrow RSR' \qquad (22)$$

$$RSH + CH_2CH_2O \longrightarrow RSCH_2CH_2OH \qquad (23)$$

Mixed sulfides are prepared in the flavor industry by the reaction of thiols, zinc oxide, and a bromoalkane (39). Some of these mixed sulfides are constituents of allium, asafetida, coffee, and meat flavors. A representative reaction is represented in equation 24.

$$RSH + ZnO + R'Br \longrightarrow RSR' + ZnBr_2 \qquad (24)$$

Reactions with Aldehydes and Ketones. Thiols react with carbonyl compounds to form dithioacetals and thioacetals. The dithioacetals have been utilized as protecting groups for carbonyl compounds. They generally require stringent conditions to remove the blocking groups. The thioacetals are not overly stable. At least two thioacetals, 2,2,2-trichloro-1-thioethylethanol (40) and 1-dodecanethiomethanol (41), are stable.

This type of chemistry also functions for hydroxyketones and aldehydes. The process using 1,2-ethanedithiol or 2-mercaptoethanol results in cyclic structures (eq. 25). The 1,3-ditholenes (X = S) and 1,3-thioxalanes (X = O) resulting from these reactions have been shown to be of interest commercially.

$$RCH(OH)COR' + HXCH_2CH_2SH \longrightarrow RC{=}C(R')SCH_2CH_2X + 2 H_2O \qquad (25)$$

A wide variety of products can be obtained by thioalkylation (42). The reactants are usually an aldehyde, a thiol, and either a phenol, a sulfone, an amine, or a heterocyclic compound. Phenols primarily react with formaldehyde in a process known as thiomethylation (eq. 26). Other types of reactions are depicted in equations 27 and 28.

$$C_6H_5OH + CH_2O + RSH \longrightarrow RSCH_2C_6H_4OH \qquad (26)$$

$$C_6H_5SO_2CH_2C_6H_5 + RSH + CH_2O \longrightarrow C_6H_5SO_2CH(SR)C_6H_5 \qquad (27)$$

$$CH_3NH_2HCl + C_6H_5CH_2SH + CH_2O \longrightarrow CH_3NHCH_2SCH_2C_6H_5HCl \qquad (28)$$

Decomposition of Thiols. Thiols decompose by two principal paths (43–45). These are the carbon–sulfur bond homolysis and the unimolecular decomposition to alkene and hydrogen sulfide. For methanethiol, the only available route is homolysis, as in reaction 29. For ethanethiol, the favored route is formation of ethylene and hydrogen sulfide via the unimolecular process, as in reaction 30.

$$CH_3SH \longrightarrow CH_3{\cdot} + HS{\cdot} \qquad (29)$$

$$C_2H_5SH \longrightarrow C_2H_4 + H_2S \qquad (30)$$

The preferred route depends upon the availability of a hydrogen atom in the beta-position to the thiol group. In other words, α-toluenethiol (in toluene) decomposes to give 1,2-diphenylethane and hydrogen sulfide, via the homolytic route, whereas 2-methyl-2-propanethiol decomposes to give 2-methyl-1-propene and hydrogen sulfide.

Photolysis of Thiols. Thiols undergo photolytic reactions (46–48). These reactions proceed as shown in equations 31 and 32. A secondary mechanistic pathway is also observed. This is shown in equations 33 and 34.

$$CH_3SH + h\nu \longrightarrow CH_3S{\cdot} + H{\cdot} \qquad (31)$$

$$CH_3SH + H{\cdot} \longrightarrow CH_3S{\cdot} + H_2 \qquad (32)$$

$$CH_3SH + h\nu \longrightarrow CH_3 + HS{\cdot} \qquad (33)$$

$$CH_3SH + CH_3{\cdot} \longrightarrow CH_4 + CH_3S{\cdot} \qquad (34)$$

A review of the role of thiols as electron donors in photoinduced electron-transfer reactions has been compiled (49).

Thiol–Disulfide Interchange Reactions.   The interchange between thiols and disulfides has been reviewed (50). This reaction is base-catalyzed. It involves the nucleophilic attack of a thiolate ion on a disulfide. This is shown in equations 35, 36, and 37.

$$RSH \rightleftharpoons RS^- + H^+ \tag{35}$$

$$RS^- + R'SSR' \rightleftharpoons RSSR' + R'S^- \tag{36}$$

$$H^+ + R'S^- \rightleftharpoons R'SH \tag{37}$$

The effect of pH and the pK_a of the thiol has been discussed. This reaction is not of great synthetic interest, primarily because it yields a mixture of products, but it is of commercial consequence. It is also applicable in polysulfide synthesis, where the presence of small amounts of thiols can cause significant problems for the stability of the polysulfide (51). A similar reaction between thiols and sulfides has also been described (52). In this instance, the process is heterogenous and acid-catalyzed.

Metal Ion-Promoted Reactions of Thiols.   Metal ion-promoted reactions of thiols have been reviewed (53). The bulk of the coverage concerns metal ion promoted aspects of sulfur chemistry. The main topics of interest are the formation of sulfenamides, sulfides, and disulfides using metal-mediated reactions.

Applications of Alkanethiols

The principal impetus behind the synthesis of thiols came from the need to produce synthetic rubber in the early 1940s. These rubbers, styrene–butadiene rubbers (SBRs), were produced by many companies at that time. Originally, 1-dodecanethiol was utilized, but the most important thiol became *tert*-dodecanethiol, which was made from propylene tetramer, using an acid-catalyzed process (54,55).

In rubber production, the thiol acts as a chain transfer agent, in which it functions as a hydrogen atom donor to one rubber chain, effectively finishing chain growth for that polymer chain. The sulfur-based radical then either terminates with another radical species or initiates another chain. The thiol is used up in this process. The length of the rubber polymer chain is a function of the thiol concentration. The higher the concentration, the shorter the rubber chain; and the softer the rubber. An array of thiols have subsequently been utilized in the production of many different polymers. Some of these applications are as follow:

Thiol	Polymer
1-dodecanethiol	polybutadiene–acrylonitrile (PBAN)
	neoprene
	nitrile
	acrylic resins

Thiol	Polymer
tert-dodecanethiol	SBR
	SBR latex
	polyacrylates
	polymethacrylates
	nitrile rubber
	ABS
	SB latex
	expandable polystyrene
2-mercaptoethanol	PVC
	polyacrylates
	polymethacrylates
alkyl mercaptopropionates	polyacrylates
	polymethacrylates

In general, rubber manufacturers balance thiol reactivity and odor. The structure of the thiol plays a significant role in its ability to be transported within the polymer matrix, particularly in emulsion polymerizations, ie, mixed water–monomer emulsion. The odor of light thiols is generally too strong for most rubber manufacturers, as it is generally hard to remove residual odors from polymers.

Another area in which sulfur compounds have long found use is in the area of agricultural chemicals. Many of these materials had been produced by the manufacturer of the agricultural chemicals, but difficulties in containing odor and the use of hydrogen sulfide in heavily populated areas again pushed toward specialization by several companies. A list of agricultural chemicals, and the thiol that is used or has been used in production, follows:

Thiol	Agricultural chemical
ethanethiol	butylate
	cycloate
	demeton
	disulfoton
	oxydemeton-methyl
	oxydeprofos
	dipropetryn
	EPTC
	molinate
	phenothiol
	phorate
	sethoxydim

Thiol	Agricultural chemical
1-propanethiol	ethoprop
	pebulate
	profenophos
	prothiophos
	sulprofox
	vernolate
1-butanethiol	DEF
	merphos
	terbufos
1-octanethiol	MGK R-874
	pyridate
2-mercaptoethanol	Vitavax
1,2-ethanedithiol	Harvade
α-toluenethiol	IBP
	tiocarbazil
	Londax
methyl 3-mercaptopropionate	Kathon

Over the years, a diverse group of products made use of the rather unique chemistry of organosulfur chemicals. These include the following:

Thiol	Application
methanethiol	DL-methionine
ethanethiol	propane odorant
2-methyl-2-propanethiol	natural gas odorant components
1-octanethiol	water repellants
cyclohexanethiol	prevulcanization inhibitor
2-propanethiol	flocculent
	natural gas odorant component
tert-nonanethiol	polysulfides for oil additives
tert-dodecanethiol	polysulfides for oil additives
	surfactants
2-mercaptoethanol	PVC stabilizers
	solvent for acrylic fibers

Organosulfur chemicals serve a diverse group of industries. In these applications ways have been found to control the odor of the thiols, thus utilizing these compounds effectively in a wide range of products.

Toxicology and Handling

The toxicology and procedures for handling thiols used in gas odorants have been reviewed (56), as have thiols as a whole (57). A partial listing of toxicological results is shown in Table 4.

Table 4. Toxicological Studies of Thiols[a]

Species	Route[b]	Concentration, mg/kg	LD$_{50}$, mg/kg[c]
		Methanethiol	
mice	respiratory (4)	1200–2200	1664[d]
		Ethanethiol	
rat	respiratory (4)	2600–5125	4870[d] (1)
			4420[d] (15) d
rat	respiratory (0.25)	27,000–38,000	
rat	oral	210–3360	1034 (15) d
rat	intraperitoneal	105–1680	450 mg/kg
mice	respiratory (4)	2600–4812	2770[d]
		Propanethiol	
rat	respiratory (4)	3050–11260	7300[d]
rat	oral	1327–3344	2360
rat	intraperitoneal	209–1672	1028
		Butanethiol	
mice	respiratory (4)	3050–11260	4950[d] (24)
			4010[d] (15) d
rat	oral	1039–3344	2475 (1)
			4020 (15) d
rat	intraperitoneal	209–672	2475 (1)
			1500 (15) d
		t-Butanethiol	
rat	respiratory (4)		97[d,e]
rat	oral		8.4[f]
rabbit	dermal		20.8[f]
		Dodecanethiol	
mice	intragastric		4225

[a]Refs. 56, 58.
[b]Value in parentheses is exposure time in hours.
[c]Value in parentheses is time in days.
[d]Value is LC$_{50}$.
[e]Value is in mg/L.
[f]Value is in g/kg.

A listing of known TLV values and odor thresholds for a variety of thiols is given in Table 5.

Thiol spills are handled in the same manner that all chemical spills are handled, with the added requirement that the odor be eliminated as rapidly as possible. In general, the leak should be stopped, the spill should be contained, and then the odor should be reduced. The odor can be reduced by spraying the spill area with sodium hypochlorite (3% solution), calcium hypochlorite solution (3%), or hydrogen peroxide (3–10% solution). The use of higher concentrations of oxidant gives strongly exothermic reactions, which increase the amount of thiol in the vapor, as well as pose a safety hazard. The application of an adsorbent prior to addition of the oxidant can be quite helpful and add to the ease of cleanup.

Thiols interact readily with many rubber-containing materials. For this reason, care should be taken in the selection of gasket and hose materials. Teflon,

Table 5. Odor Threshold Levels and Threshold Limit Values (TLV)

Thiol	Odor threshold,[a] ppb	TLV,[b] ppm
methanethiol	1.05	0.5
ethanethiol	1.07	0.5
1,2-ethanedithiol	19.50	
1-propanethiol	1.26	
2-propanethiol	0.35	
2-propene-1-thiol	0.40	
1-butanethiol	1.41	0.5
2-butanethiol	0.18	0.5
2-butene-1-thiol	0.13	
2-methyl-1-propanethiol	1.12	0.5
2-methyl-2-propanethiol	0.33	0.5
1-pentanethiol	0.12	0.5
2-methyl-1-butanethiol	0.26	
2-methyl-2-butanethiol	0.72	
benzenethiol	0.31	0.1
4-methylbenzenethiol	1.70	
α-toluenethiol	1.58	
1-dodecanethiol	0.25	
2-methyl-1-undecanethiol	190.55	

[a]Ref. 18.
[b]Refs. 59 and 60.

Kel-F, Viton, or other suitable fluoroelastomers function as gasket materials. Viton is suitable for hoses. Carbon steel is useful for many thiols, although some thiols become very discolored when carbon steel is utilized. In these cases, the use of stainless steel is very desirable. Isolation from air and water also minimizes color formation. 2-Mercaptoethanol and 1,2-ethanedithiol should be stored in stainless steel (61).

Thiols are shipped in every conceivable container size. Drums and cans can be of carbon steel for most thiols, provided color is not a determining factor. Truck, rail, and isocontainer shipments should be set up to utilize a vapor return line from the tank to the shipping container. This substantially minimizes the amount of odor that escapes. Phillips Petroleum Company and Atochem North America can supply further information regarding the handling and properties of many thiols.

BIBLIOGRAPHY

"Mercaptans" in *ECT* 1st ed., Vol. 8, pp. 858–868, by E. E. Reid, Johns Hopkins University; "Thiols" in *ECT* 2nd ed., Vol. 20, pp. 205–218, by S. D. Turk, Phillips Petroleum Co.; in *ECT* 3rd ed., Vol. 22, pp. 946–964, by J. Norell and R. P. Louthan, Phillips Petroleum Co.

1. E. E. Reid, *Organic Chemistry of Bivalent Sulfur*, Chemical Publishing Co., Inc., New York, 1958–1962.
2. E. Block, *Reactions of Organic Sulfur Compounds*, Academic Press, New York, 1978.
3. S. Oae, ed., *Organic Chemistry of Sulfur*, Plenum Press, New York, 1977.

4. S. Patai, ed., *The Chemistry of the Thiol Group*, John Wiley & Sons, Ltd., London, 1974.

5. S. Patai, ed., *The Chemistry of Functional Groups, Supplement E*, John Wiley & Sons, Inc., New York, 1980.

6. S. Patai and Z. Rappoport, eds., *The Chemistry of Sulphur-Containing Functional Groups, Supplement S*, John Wiley & Sons, Inc., New York, 1993.

7. P. C. Jocelyn, *Biochemistry of the SH Groups*, Academic Press, New York, 1978, p. 8.

8. M. Rinken, M. Aydin, S. Sievers, and W. A. König, *Fresenius' Z. Anal. Chem.* **318**, 27 (1984).

9. M. Phol, E. Bock, M. Rinken, M. Aydin, and W. A. König, *Z. Naturforsch. Sect. C, Biosciences* **39C**, 240 (1984).

10. D. Jenkins, L. L. Medsker, and J. F. Thomas, *Environ. Sci. Tech.* **1**, 731 (1967).

11. T. Jankowski, J. Gawlik, and S. Zimnal, *VDI Ber. (Ver. Dtsch. Ing.)* **226**, 123 (1975).

12. E. E. Beach and A. White, *J. Biol. Chem.* **127**, 87 (1939).

13. L. M. Libbey and E. A. Day, *J. Dairy Sci.* **46**, 859 (1963).

14. D. B. Emmons, J. A. Elliot, and D. C. Becket, *J. Dairy Sci.* **49**, 1325 (1966).

15. S. Segall and B. Proctor, *Food Technol.* **13**, 679 (1959).

16. A. P. Ronald and W. A. Thomson, *J. Fish. Res. Board Can.* **21**, 1481 (1964).

17. S. H. Katz and E. J. Talbert, *Intensity of Odors and Irritating Effects of Warning Agents for Inflammable and Poisonous Gases*, Technical Paper 480, U.S. Department of Commerce, U.S. Government Printing Office, 1930.

18. M. Devos, F. Patte, J. Rouault, P. Laffort, and L. J. Van Gemert, eds., *Standardized Human Olfactory Thresholds*, IRL Press, New York, 1990.

19. E. S. Domalski and E. D. Hearing, *J. Phys. Chem. Ref. Data* **19**, 881 (1990).

20. E. S. Domalski, W. H. Evans, and E. D. Hearing, *J. Phys. Chem. Ref. Data* **13**(Suppl. 1), 1 (1984).

21. J. F. Liebman, K. S. K. Crawford, and S. W. Slayden, in Ref. 6, p. 197.

22. E. Tanaka, *Kagaku to Kogyo* **43**, 355 (1990).

23. W. A. Schlze, J. P. Lyon, and G. H. Short, *Ind. Eng. Chem.* **40**, 2308 (1948).

24. C. F. Fryling in G. S. Whitby and co-workers, eds., *Synthetic Rubber*, John Wiley & Sons, Inc., New York, 1954, Chapt. 8.

25. U.S. Pat. 5,008,432 (Apr. 16, 1991), J. S. Roberts (to Phillips Petroleum Company).

26. U.S. Pat. 4,433,134 (Feb. 1, 1984), R. P. Louthan (to Phillips Petroleum Company).

27. J. S. Roberts, A. Pauwels, and P. A. Aegerter, *Proceedings of the Chemical Specialties USA 1993 Symposium, Oct. 1993*.

28. L. Field, *Sulfur Rep.* **13**, 197 (1993).

29. J. Xan, E. A. Wilson, L. D. Roberts, and N. H. Horton, *J. Amer. Chem. Soc.* **63**, 1130 (1941).

30. E. Wenschuk, M. Heydenreich, R. Runge, and S. Fischer, *Sulfur Lett.* **8**, 251 (1988).

31. U.S. Pat. 4,721,813 (Jan. 26, 1988), H. W. Mark and J. S. Roberts (to Phillips Petroleum Company).

32. J. D. Hopton, C. J. Swann, and D. L. Trimm, *Adv. Chem. Ser.*, 216 (1986).

33. J. Heicklen, *Rev. Chem. Intermed.* **6**, 175 (1985).

34. B. D. Vineyard, *J. Org. Chem.* **32**, 3833 (1967).

35. U.S. Pat. 2,983,747 (May 9, 1961), P. F. Warner (to Phillips Petroleum Company).

36. P. Allen and J. W. Brook, *J. Org. Chem.* **27**, 1019 (1962).

37. W. A. Savige and J. A. Maclaren in J. E. Hesse, ed., *J. Chem. Eng. Data* **12**, 266 (1967).

38. N. A. Nedolya and B. A. Trofimov, *Sulfur Reports* **15**, 237 (1994).

39. B. Ravindranath, *Perfumer Flavorist* **10**, 39 (1985).

40. R. L. Frank, S. S. Drake, P. V. Smith, Jr., and C. Stevens, *J. Polymer Sci.* **3**, 50 (1948).

41. U.S. Pat. 4,939,302 (July 3, 1990), J. S. Roberts (to Phillips Petroleum Company).

42. D. J. R. Massy, *Synthesis*, 589 (1987).

43. T. Shono in Ref. 5, p. 327.
44. G. Martin in Ref. 6, p. 395.
45. I. Safarik and O. P. Strausz, *Rev. Chem. Intermed.* **6**, 143 (1985).
46. R. L. Failes, J. S. Shapiro, and V. R. Stimson, in Ref. 5, p. 450.
47. C. von Sonntag, in C. Chatgilialoglu and K.-D. Asmus, eds., *Sulfur-Centered Reactive Intermediates in Chemistry and Biology*, Plenum Press, New York, 1990, p. 359.
48. C. von Sonntag and H.-P. Schuchmann, in Ref. 5, p. 923.
49. N. J. Pienta, in M. A. Fox and M. Chanon, eds., *Photoinduced Electron Transfer*, Elsevier, Amsterdam, the Netherlands, 1988, p. 421.
50. R. Singh and G. M. Whitesides, in Ref. 6, p. 633.
51. Technical data, Phillips Petroleum Company.
52. U.S. Pat. 4,537,994 (Aug. 27, 1985), J. S. Roberts (to Phillips Petroleum Company).
53. D. P. N. Satchell and R. S. Satchell, in Ref. 6, p. 599.
54. W. A. Schulze, J. P. Lyon, and G. H. Short, *Ind. Eng. Chem.* **40**, 2308 (1948).
55. C. F. Fryling, in G. S. Whitby and co-workers, eds., *Synthetic Rubber*, John Wiley & Sons, Inc., New York, 1954, Chapt. 8.
56. J. S. Roberts and D. W. Kelly, in G. G. Wilson and A. A. Attari, eds., *Odorization III*, Institute of Gas Technology, Chicago, Ill., 1993, p. 29.
57. P. Jacques, *Cah. Med. Trav.* **20**, 169 (1983).
58. NIOSH, *Criteria for a recommended standard: Occupational Exposure to n-Alkane Monothiols*, U.S. Dept. of Health, Education and Welfare, Public Health Service, Center for Disease Control, National Institute for Occupational Safety and Health, Sept. 1978.
59. ACGIH, *Documentation of the Threshold Limit Values for Substances in Workroom Air*, American Conferences of Governmental Industrial Hygienists, 1977.
60. ACGIH, *TLVs, Threshold Limit Values for Chemical Substances in Workroom Air*, American Conferences of Governmental Industrial Hygienists, 1982.
61. Technical literature, Phillips Petroleum Co., Bartlesville, Okla., 1995.

JOHN S. ROBERTS
Phillips Petroleum Company
CH&A Corporation

THIOPHENE AND THIOPHENE DERIVATIVES

Thiophene [*110-02-1*] and a number of its derivatives are significant in fine chemical industries as intermediates to many products for pharmaceutical, agrochemical, dyestuffs, and electronic applications. This article concentrates on the industrial, commercial, and economic aspects of the production and applications of thiophene and thiophene derivatives and details the main synthetic schemes to the parent ring system and simple alkyl and aryl derivatives. Functionalization of the ring and the synthesis of some functional derivatives that result, not from the parent ring system, but by direct ring cyclization reactions are also considered. Many good reviews on the chemistry of thiophene and thiophene derivatives are available (1–7).

Nomenclature, Numbering, and Structure

The basic nomenclature of the thiophene ring system and its derivatives is indicated by the following: the sulfur atom is number 1, positions 2 and 5 are equivalent in the parent ring, as are the 3 and 4 positions.

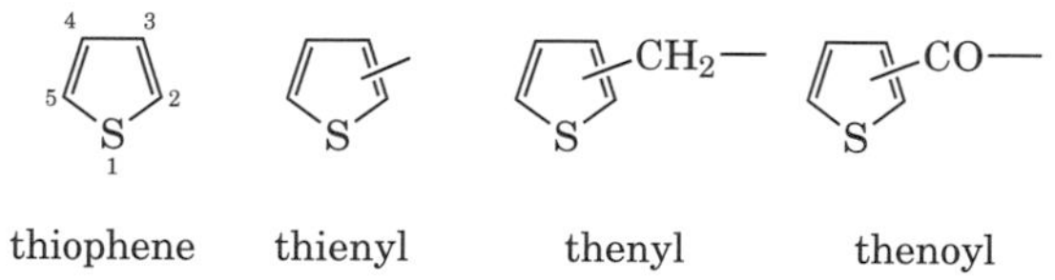

thiophene   thienyl   thenyl   thenoyl

The structure of the parent ring has been studied by numerous techniques. The precise data from microwave studies (8) on the dimensions and bond angles is given in Fig. 1.

In valence bond terms the mesomers indicated by (1–7) reflect the ground-state position of thiophene. Mesomer (1) is the principal contributor to the ring structure; (2) and (3) are significant; (4–7) contribute in a minor way to the structure.

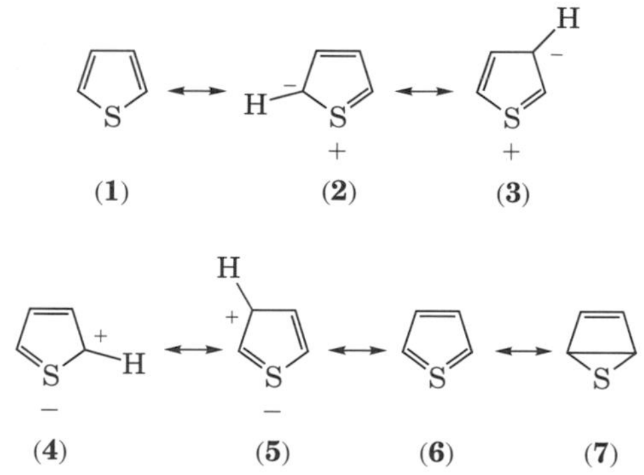

(1)   (2)   (3)

(4)   (5)   (6)   (7)

With its sextet of π electrons, thiophene possesses the typical aromatic character of benzene and other similarly related heterocycles. Decreasing

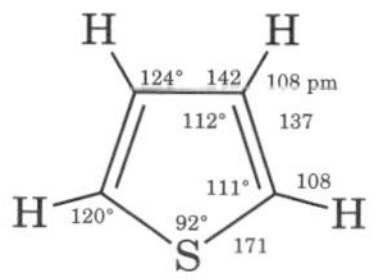

Fig. 1. Bond lengths and bond angles in thiophene.

orders of aromaticity have been suggested to reflect the strength of this aromatic character: benzene > thiophene > pyrrole > furan (9); and benzene > thiophene > selenophene > tellurophene > furan (10).

Physical Properties and Spectroscopy

Table 1 indicates the significant physical properties of thiophene and 2- and 3-methylthiophene; Table 2, the toxicological and ecotoxicological properties.

Table 1. Physical Properties of Thiophene and Methylthiophenes[a]

Property	Thiophene	2-Methylthiophene	3-Methylthiophene
CAS Registry Number	[110-02-1]	[554-14-3]	[616-44-4]
freezing point, °C	−38.3	−63.4	−68.9
boiling point at 101.3 kPa[b], °C	84.16	113	115
flash point, °C	−7	16	15.5
density, d_4^0, kg/m^3	1087.3	1025.0	1025.0
density, d_4^{25}, kg/m^3	1057.3	1014.0	1016.2
refractive index, n_D^{25}	1.52572	1.5174	1.5172
viscosity at 25°C, mPa·s(=cP)	0.621	0.669	0.642
surface tension at 20°C, mN/m(=dyn/cm)	31.34	30.95	32.37
vapor pressure, kPa[b], at			
0°C	2.86		1.33 at 11°C
20°C	8.36	2.66 at 21°C	3.99 at 31°C
50.1°C	31.16	10.66 at 49.1°C	10.66 at 51.5°C
84.16°C	101.3	101.3 at 112.6°C	101.3 at 115.4°C
95.9°C	143.3	133.3 at 122.4°C	133.3 at 125.4°C
119.8°C	270.1	199.9 at 138°C	199.9 at 141°C
critical constants			
temperature, °C	306.2	335.8	339.8
pressure, MPa[c]	5.70	4.91	4.91
volume, mL/mol	219	275	276
density, kg/m^3	385		
heat of vaporization at bp, kJ/mol[d]	31.47	33.90	34.25
heat of formation at 298.16 K, kJ/mol[d]			
liquid	81.67	45.44	43.89
gaseous	115.44	84.35	83.43
heat of combustion at 101.3 kPa[b] and 25°C, kJ/mol[d]	−2435.2	−3471.3	−3469.0
dielectric constant at 20°C	2.74		

[a]Ref. 11.
[b]To convert kPa to mm Hg, multiply by 7.5.
[c]To convert MPa to psi, multiply by 145.
[d]To convert kJ/mol to kcal/mol, divide by 4.184.

Table 2. Toxicological and Ecotoxicological Properties of Thiophene and Methylthiophenes

Property	Thiophene	2-Methylthiophene	3-Methylthiophene
Log P_{ow}	1.81[a]		
LD$_{50}$, rat (oral), mg/kg	1400	3200 (mouse)	2300[b]
LD$_0$, rat (dermal), mg/kg	>2000		
LD$_{50}$, rabbit (dermal), mg/kg	830		>2000[b]
LC$_0$, rat (inhalation), 1 h mg/L	>27		
LC$_{50}$, mouse (inhalation), 2 h, mg/m^3	9500	>2000 (rat, 4 h)	>2000 (rat, 4 h)
skin irritancy	irritant		slight irritant[b]
eye irritancy	irritant		minimal irritant[b]
bacterial mutagenicity, Ames	negative[c]	negative	negative[b]
EC$_{50}$, *Daphnia magna*, mg/kg	13[d]		
LC$_{50}$, fish (*Oryzias latipes*), 48 h, mg/kg	15.6[e]		
biotic degradation	minimal[d]		
biological accumulation	no		
U.N. No.	2414	1993	1993
hazard symbol	F, T[f]	F	F

[a]Ref. 12.
[b]Ref. 13.
[c]Ref. 14.
[d]Ref. 15.
[e]Ref. 16.
[f]Toxic, only if >0.1% benzene is present.

Uv Spectroscopy. The commonly used solution uv spectrum of thiophene consists of a broad band, 220–250 nm (log ϵ = 3.9), caused by an overlap of transition states and two low intensity bands at 313 and 318 nm. Alkyl derivatives and similarly related +M substituents, at the 2-position, give a single band; similar 3-substituents give a double peak. Negative, ie, −I, −M, substituents at the 2-position give two absorptions, whereas 3-substituents of this type give only one band of lower intensity. A useful bathochromic shift is noted for thiophene derivatives compared to benzene analogues, thus making certain thiophene derivatives of significant commercial interest as dyes and optical brightening agents.

Ir Spectroscopy. Significant absorptions can be identified as characteristic of particular substitutions within families of thiophene derivatives. The most widely studied in this connection are probably the halothiophenes, where absorption bands have been characterized. This is useful for qualitative analysis, but has also been used quantitatively in association with the standard spectrum of materials of known purity.

Nuclear Magnetic Resonance Spectroscopy. Nmr is a most valuable technique for structure determination in thiophene chemistry, especially because spectral interpretation is much easier in the thiophene series compared to benzene derivatives. Chemical shifts in proton nmr are well documented;

for thiophene ($CDCl_3$), $\delta = H_2$ 7.12, H_3 7.34, H_4 7.34, and H_5 7.12 ppm. Coupling constants occur in well-defined ranges: $J_{2-3} = 4.9-5.8$; $J_{3-4} = 3.45-4.35$; $J_{2-4} = 1.25-1.7$; and $J_{2-5} = 3.2-3.65$ Hz. The technique can be used quantitatively by comparison with standard spectra of materials of known purity. ^{13}C-nmr spectroscopy of thiophene and thiophene derivatives is also a valuable technique that shows well-defined patterns of spectra. ^{13}C chemical shifts for thiophene, from tetramethylsilane (TMS), are C_2 127.6, C_3 125.9, C_4 125.9, and C_5 127.6 ppm.

Reactions of Thiophene and Alkylthiophenes

Electrophilic substitution of thiophene occurs largely at the 2-position and the reactivity of the ring is greater than that of benzene. 3-Substituted derivatives are generally prepared by indirect means or through ring cyclization reactions.

Alkylation. Thiophenes can be alkylated in the 2-position using alkyl halides, alcohols, and olefins. Choice of catalyst is important; the weaker Friedel-Crafts catalysts, eg, $ZnCl_2$ and $SnCl_4$, are preferred. It is often preferable to use the more readily accomplished acylation reactions of thiophene to give the required alkyl derivatives on reduction. Alternatively, metalation or Grignard reactions, on halothiophenes or halomethylthiophenes, can be utilized.

Acylation. To achieve acylation of thiophenes, acid anhydrides with phosphoric acid, iodine, or other catalysts have been widely used. Acid chlorides with $AlCl_3$, $SnCl_4$, $ZnCl_2$, and BF_3 also give 2-thienylketones. All reactions give between 0.5 and 2.0% of the 3-isomer. There has been much striving to find catalyst systems that minimize the 3-isomer content attempting to meet to customer specifications. The standard procedure for formylation is via the Vilsmeier-Haack reaction, using phosphorus oxychloride/N,N-dimethylformamide ($POCl_3$/DMF) or N-methylformanilide.

Halogenation. Many different halogenating reagents have been used to accomplish halogenation of the thiophene ring. Excess of reagent gives di-, tri-, and even tetrahalogenation of thiophene itself. The bromothiophenes are often the preferred target. Bromine in acetic acid or chloroform has been the traditional route, but utilizes just one of the atoms of bromine. Addition products are often observed, particularly when proceeding through to di- and tribromothiophenes.

N-Bromosuccinimide and N,N'-dibromo-5,5-dimethylhydantoin have also been used successfully, which makes possible recycling of succinimide or the hydantoin and utilizes all the bromine atoms. A mixture of sodium bromide–sodium bromate in aqueous acid has also been used commercially.

Introduction of a 3-bromosubstituent onto thiophene is accomplished by initial tribromination, followed by reduction of the α-bromines by treatment with zinc/acetic acid, thereby utilizing only one of three bromines introduced. The so-called halogen dance sequence of reactions, whereby bromothiophenes are treated with base, causing proton abstraction and rearrangement of bromine to the produce the most-stable anion, has also been used to introduce a bromine atom at position 3. The formation of 3-bromothiophene [872-31-1] from this sequence of reactions (17) is an efficient use of bromine. Vapor-phase techniques have also been proposed to achieve this halogen migration (18), but with less specificity. Table 3 summarizes properties of some brominated thiophenes.

Table 3. Physical, Toxicological, and Ecotoxicological Properties of 2-Bromothiophene, 3-Bromothiophene, and 2-Bromo-3-Methylthiophene

Property	2-Bromo-thiophene	3-Bromo-thiophene	2-Bromo-3-methylthiophene
CAS Registry Number	[1003-09-4]	[872-31-1]	[14282-76-9]
description	liquid	liquid	liquid
flash point, °C	60	56	
boiling point at 101.3 kPa, °C	150	158	181–182
density at 20°C, kg/m^3	1684	1740	1571 (25°C)
refractive index, n_D^{25}	1.5860	1.5910	1.5680
LD$_{50}$, rat (oral), mg/kg	200–250[a]	66–160[a]	1923[a]
LD$_{50}$, rabbit (dermal), mg/kg	134[a]	173–694[a]	>2000[b]
LC$_{50}$, 4 h, mg/L	1.04[a]		
skin irritancy	irritant[a]	slight irritant[a]	moderate irritant[b]
eye irritancy	irritant[a]	slight irritant[a]	moderate irritant[b]
sensitization	no[a]		
Ames test	negative[a]	negative[a]	negative[c]
EC$_{50}$, *Daphnia magna*, 48 h, mg/L	5.0[d]		0.79[d]
hazard symbol	T, N	T	Xn, N
TSCA status	listed	listed	not listed
U.N. No.	2810	2810	3082

[a]Ref. 19.
[b]Ref. 20.
[c]Ref. 21.
[d]Ref. 22.

Nitration.   It is difficult to control nitration of thiophene, which yields 2-nitrothiophene [609-40-9]. The strongly electrophilic nitronium ion leads to significant yields (12–15%) of 3-isomer. A preferred procedure is the slow addition of thiophene to an anhydrous mixture of nitric acid, acetic acid, and acetic anhydride.

Metalation.   Direct reaction of thiophene and butyllithium in diethyl ether gives 2-thienyllithium [2786-07-4], a valuable intermediate and source of many further derivatives. Some examples are given in Table 4. Thienyllithiums can also be prepared by halogen exchange from bromothiophenes and butyllithium, but this requires low temperature to avoid any possible proton–lithium exchange, or rearrangement by the halogen dance mechanism. Thienyl Grignard

Table 4. Reactions of 2-Thienyllithium

Reagent	Product	CAS Registry Number of product
carbon dioxide	2-thiophenecarboxylic acid	[527-72-0]
sulfur	thiophene-2-thiol	[7774-74-5]
N,N-dimethylformamide	2-thiophenecarboxaldehyde	[98-03-3]

reagents are also prepared from the bromothiophenes and can be used in reaction schemes to give many other derivatives.

Oxidation.   Strong oxidizing agents can rupture the thiophene ring structure, eg, nitric acid gives maleic acid. In the vapor phase, oxidation can lead to loss of aromaticity, producing thiomaleic anhydride [6007-87-0]. Oxidation of alkylthiophenes can lead to carboxylic acids. A useful reagent for this is neutral sodium dichromate under autogenous pressure at 270°C (23). Oxidation at the sulfur atom with peroxides and peracids leads to the sulfone. This material readily undergoes Diels Alder reaction and a dimer has been isolated.

Reduction and Hydrodesulfurization.   Reduction of thiophene to 2,3- and 2,5-dihydrothiophene and ultimately tetrahydrothiophene can be achieved by treatment with sodium metal–alcohol or ammonia. Hydrogen with Pd, Co, Mo, and Rh catalysts also reduces thiophene to tetrahydrothiophene [110-01-0], a malodorous material used as a gas odorant.

Rigorous hydrogenating conditions, particularly with Raney Nickel, remove the sulfur atom of thiophenes. With vapor-phase catalysis, hydrodesulfurization is the technique used to remove sulfur materials from crude oil. Chemically hydrodesulfurization can be a valuable route to alkanes otherwise difficult to access.

Side-Chain Derivatization.   Reaction of thiophene with aqueous formaldehyde solution in concentrated hydrochloric acid gives 2-chloromethylthiophene [765-50-4]. This relatively unstable, lachrymatory material has been used as a commercial source of further derivatives such as 2-thiopheneacetonitrile [20893-30-5] and 2-thiopheneacetic acid [1918-77-0] (24). Similar derivatives can be obtained by peroxide, or light-catalyzed (25) halogenation of methylthiophenes, eg, N-bromosuccinimide/benzoylperoxide on 2-, and 3-methylthiophenes gives the corresponding bromomethylthiophenes.

Nucleophilic Reactions.   Useful nucleophilic substitutions of halothiophenes are readily achieved in copper-mediated reactions. Of particular note is the ready conversion of 3-bromoderivatives to the corresponding 3-chloroderivatives with copper(I)chloride in hot N,N-dimethylformamide (26). High yields of alkoxythiophenes are obtained from bromo- and iodothiophenes on reaction with sodium alkoxide in the appropriate alcohol, and catalyzed by copper(II) oxide, a trace of potassium iodide, and in more recent years a phase-transfer catalyst (27).

Manufacture of Thiophenes

Preparative Methods.   The thiophene ring system has been synthesized through numerous reaction types. A systematic review of these has been made (28) and is based on the number of components utilized in the construction of the ring. Some 19 combinations are possible, utilizing five method types. Not all combinations have been reported and some would be of only minor benefit.

Manufacture of thiophene on the commercial scale involves reactions of the two component method type wherein a 4-carbon chain molecule reacts with a source of sulfur over a catalyst which also effects cyclization and aromatization. A range of suitable feedstocks has included butane, n-butanol, n-butyraldehyde, crotonaldehyde, and furan; the source of sulfur has included sulfur itself, hydrogen sulfide, and carbon disulfide (29–32).

Process Description.   Reactors used in the vapor-phase synthesis of thiophene and alkylthiophenes are all multitubular, fixed-bed catalytic reactors operating at atmospheric pressure, or up to 10^3 kPa and with hot-air circulation on the shell, or salt bath heating, maintaining reaction temperatures in the range of 400–500°C. The feedstocks, in the appropriate molar ratio, are vaporized and passed through the catalyst bed. Condensation gives the crude product mixture; noncondensable vapors are vented to the incinerator.

Reactions tend to deposit carbon on the catalyst, which ultimately leads to loss of activity. At this stage reactions are stopped, the catalyst purged and regenerated by a controlled oxygenating vapor. All the vapors from the regeneration process are also passed to the incinerator. A catalyst can undergo many regenerations before replacement becomes necessary.

Distillation System.   The crude condensate consists of the desired product, some low boiling constituents, and a smaller quantity of high boiling tar. Distillation separates the low boiling components, which are invariably incinerated, followed by the product fraction. Tar accumulates in the still kettles, from which it is periodically removed, again to incineration. Stills work at atmospheric pressure and are vented to the incinerator.

Releases and Abatements.   Organic chemical reactions under conditions as severe as 400–500°C are inherently less than completely selective. Although the synthesis of thiophene under such conditions is considered the best practical commercial environmental option, it nevertheless produces waste gases in quantities roughly equal to those of saleable products. It is generally not considered practical to treat the various off-gases individually. The scale and limited revenue of the process could not support the operation of such an ancillary plant. The incinerator handles all vents. All the components of the streams are combusted under controlled conditions to give a single discharge, which is readily controlled and monitored. The overall environmental impact is minimal.

Quality.   The plant operated by Synthetic Chemicals Ltd. operates within a quality management system, which complies with British Standard (BS) 5750 and ISO 9001, Part 1.

Dependent on the C-4 feedstock, the process outlined in Figure 2 gives a product containing low levels of benzene. The Elf Atochem process using the furan–H$_2$S route to give benzene-free material, but the process has the disadvantage of coproducing small amounts of mercaptans. Raw materials are not a problem for either process; market demand is the limiting factor in production capacity.

Manufacturing Processes for Thiophene Derivatives

Halothiophenes.   The bromothiophenes, commercially the most important of the halothiophenes, are readily made and can be further derivatized. Manufacture of 2-bromothiophene involves the reaction of thiophene with a solution of sodium bromide/sodium bromate in acid solution. Such a reaction is controlled by the rate of addition of the acid. The two-phase system is stirred throughout the reaction; the heavy product layer is separated and washed thoroughly with

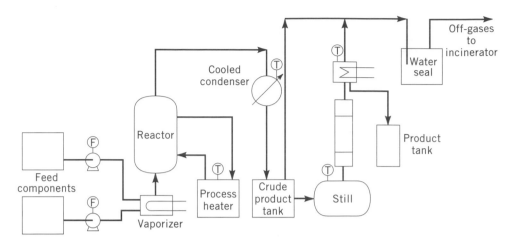

Fig. 2. Schematics of a vapor-phase process that generates thiophene and alkyl thiophenes, where the controlled parameters F = flow and T = temperature.

water and alkali before distillation (Fig. 3). The alkali treatment is particularly important and serves not just to remove residual acidity but, more importantly, to remove chemically any addition compounds that may have formed. The wash-water must be maintained alkaline during this procedure. With the introduction of more than one bromine atom, this alkali wash becomes more critical as there is a greater tendency for addition by-products to form in such reactions. Distillation of material containing residual addition compounds is hazardous, because traces of acid become self-catalytic, causing decomposition of the still contents

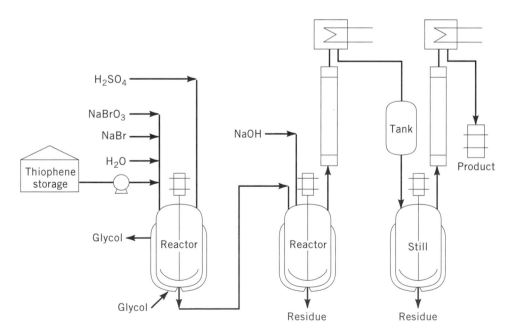

Fig. 3. Schematics of a liquid two-phase reactor system for 2-bromothiophene.

and much acid gas evolution. Bromination of alkylthiophenes follows a similar pattern.

The route to 3-bromothiophene utilizes a variation of the halogen dance technology (17). Preferably, 2,5-dibromothiophene [3141-27-3] is added to a solution of sodamide in thiophene containing the catalyst tris(2-(2-methoxyethoxy)ethyl)amine (TDA-1) (33) at temperatures marginally below reflux. On completion, quenching exothermically liberates ammonia gas; the organic phase is separated, washed, and distilled, and forerunning thiophene is recycled. Material of 97–98% purity is isolated.

Acylthiophenes. Manufacturing methods introducing the carboxaldehyde group into the 2- or 5-positions of thiophene and alkylthiophenes utilize the Vilsmeier-Haack reaction. To synthesize 2-thiophenecarboxaldehyde (Table 5), a controlled addition of phosphorus oxychloride to thiophene in N,N-dimethylformamide is carried out, causing the temperature to rise. Completion of the reaction is followed by an aqueous quench, neutralization, and solvent extraction to isolate the product.

3-Thiophenecarboxaldehyde [498-62-4] has been commercially available (35) via carbonylation of 2,5-dimethoxy-2,5-dihydrofuran, followed by treatment with hydrogen sulfide, which introduces the sulfur atom with loss of methanol, inducing aromaticity and producing 3-thiophenecarboxaldehyde directly.

Manufacture of 2-acetylthiophenes involves direct reaction of thiophene or alkylthiophene with acetic anhydride or acetyl chloride. Preferred systems use acetic anhydride and have involved iodine or orthophosphoric acid as catalysts. The former catalyst leads to simpler workup, but has the disadvantage of leading to a higher level of 3-isomer in the product. Processes claiming very low levels of

Table 5. Physical, Toxicological, and Ecotoxicological Properties of 2-Thiophenealdehyde and 2-Acetylthiophene

Property	2-Thiophenecarboxaldehyde	2-Acetylthiophene
CAS Registry Number	[98-03-3]	[88-15-3]
description	liquid	liquid
melting point, °C		10–11
boiling point, °C	198	214
flash point, °C	77	96
density at 20°C, kg/m^3	1215 (21°C)	1171
refractive index, n_D^{25}	1.5920 (20°C)	1.5650
LD$_{50}$, rat (oral), mg/kg	1100	50 (mouse)
LD$_{50}$, rabbit (dermal), mg/kg	>2000	320 (rat)
LC$_{50}$, rat, 1 h, mg/m^3		1460[a]
skin irritancy	slight irritant	nonirritant
eye irritancy	slight irritant	slight irritant
sensitization	yes	
Ames test	negative	negative[a]
hazard symbol	none	T
TSCA status	listed	listed
U.N. No.	none	2810

[a]Ref. 34.

3-isomer operate with catalysts that are proprietary, though levels of less than 0.5% are not easily attained.

The need for low levels of 3-isomer in 2-thiophenecarboxylic acid [527-72-0], which is produced by oxidation of 2-acetylthiophene [88-15-3] and used in drug applications, has been the driving force to find improved acylation catalysts. The most widely used oxidant is sodium hypochlorite, which produces a quantity of chloroform as by-product, a consequence that detracts from its simplicity. Separation of the phases and acidification of the aqueous phase precipitate the product which is filtered off. Alternative oxidants have included sodium nitrite in acid solution, which has some advantages, but, like the hypochlorite method, also involves very dilute solutions and low throughput volumes.

The long-standing manufacturing route to 2-thiopheneacetic acid [1918-77-0] has also involved 2-acetylthiophene. Oxidation with potassium permanganate under controlled conditions leads to 2-thiopheneglyoxylic acid [4075-59-6], which may be isolated as ammonium salt. The salt is then carried through a reduction stage involving the Wolff-Kishner reaction in aqueous solution and utilizing hydrazine hydrate. Workup via acidification gives the unpleasant smelling 2-thiopheneacetic acid.

Methyl 3-aminothiophene-2-carboxylate. Synthesis of this amino ester [22288-78-4] has been variously described in the literature (see Table 6); it is a key intermediate to both pharmaceutical and agrochemical products. The main synthetic schemes use thioglycollate esters as starting materials.

Table 6. Physical, Toxicological, and Ecotoxicological Properties of Methyl 3-Aminothiophene-2-Carboxylate and Methyl 3-Amino-4-Methylthiophene-2-Carboxylate

Property	Methyl 3-aminothiophene-2-carboxylate	Methyl 3-amino-4-methylthiophene-2-carboxylate
CAS Registry Number	[22288-78-4]	[85006-31-1]
description	detached crystals	detached crystals
melting point, °C	65	84
boiling point, °C	100–102 at 0.1 mm Hg	
flash point, °C	>100	>100
bulk density, kg/L	0.4	0.4
solubility in water	0.2% (18°C)	
LD_{50}, rat (oral), mg/kg	406[a]	2300[a]
LD_{50}, rabbit (dermal), mg/kg	>2000[a]	
skin irritancy	nonirritant[a]	nonirritant[a]
eye irritancy	slight mechanical irritant[a]	slight mechanical irritant[a]
sensitization	no[a]	
bacterial mutagenicity, Ames	negative[a]	negative[a]
LC_{50}, *Daphnia magna*, 48 h, mg/L	18[b]	
LC_{50}, fish (rainbow trout), 96 h, mg/L	43[b]	
biotic degradation	nonbiodegradable[b]	
transport classification	nonhazardous	nonhazardous

[a]Ref. 36.
[b]Ref. 37.

One reaction (38) employs 2,3-dichloropropionitrile (23DCPN) as the second component in early work, though this material was never commercialized. 2-Chloroacrylonitrile was available commercially for a time in the 1980s and was used to give this same product (39). This route was later upscaled by Synthetic Chemicals Ltd. following pharmaceutical and later agrochemical demand.

$$
\underset{CH_2=\overset{\displaystyle \overset{Cl}{|}}{C}-X}{} + HSCH_2CO_2CH_3 \xrightarrow{CH_3ONa} \quad
$$

When X = CN, Y = NH$_2$; X = CO$_2$CH$_3$, Y = OH.

Chemical processing is carried out in the liquid phase, in glass-lined, stirred-batch reactors fitted with a heating/cooling jacket. The product is filtered off and dried. Handling the dry product requires grounded equipment to prevent static charges from building up, a particular hazard with this crystalline dust.

Economic Aspects

Sales of thiophene in the 1990s amount to hundreds of metric tons per year. Supplies are available worldwide from Synthetic Chemicals Ltd. (SCL) in the United Kingdom and Elf-Atochem SA in France. There is currently no U.S. producer of thiophene or the principal thiophene derivatives. At these levels of demand, material is shipped in 200-liter drums and in bulk quantities. Market price is dependent on the level of off-take. 3-Methylthiophene is also available from SCL, but demand is low and even lower in the case of 2-methylthiophene; lower production and lower market demand have led to higher prices for these derivatives.

2-Bromothiophene is produced in Europe by Solvay and SCL, at up to 50 metric tons per year, with a 98%-pure specification and prices commensurate with production levels. 3-Bromothiophene is still a specialty product as of the mid-1990s, produced in multipurpose plant by SCL in hundreds of kilos per year, but at this level of market demand and also on account of the complexity of the synthesis, it commands a relatively high price.

The principal source of 2-thiophenecarboxaldehyde is Great Lakes Fine Chemicals Ltd. in the United Kingdom, whereas 2-acetylthiophene is produced by a number of manufacturers. Some of the 2-acetylthiophene producers continue derivatization to 2-thiophenecarboxylic acid and 2-thiopheneacetic acid.

The intermediates of type (**8**), wherein G is an electron withdrawing group, which are used in the dyestuffs industry, are usually produced by the user companies themselves and used directly. The type (**9**) amino ester is another product from the SCL range of thiophene derivatives, produced in metric ton quantities for specific outlets.

(8)   (9)

Transport

Thiophene itself is classified as a highly flammable liquid for transportation, designated by U.N. No. 2414, Class 3, Packaging Group II, Hazard Symbol F. Classification for use is additionally exacerbated by a low volume of benzene found in commercial material. This is a component of manufacture from a cracking side reaction over the catalyst system. Although benzene levels are minimized in optimization of the process and in workup, typical product contains from 0.1 to <0.3% benzene. In Europe this demands the use of a risk phrase, R45: May Cause Cancer, being added to the label and included in the material safety data sheet. The subsidiary hazard symbol T, for toxic, is then also required. A benzene-free grade is available from Elf Atochem. The methylthiophenes are also highly flammable liquids; U.N. No. 1993, Flammable liquid, not otherwise specified (NOS) Class 3, Packaging Group II, Hazard Symbol F is again required. All of these products are regulated by the U.S. Department of Transport.

Bromothiophenes, if not stored and treated correctly, may decompose, liberate HBr gas, and lead to pressurizing of containers. Prior treatment with alkali, avoiding metal contaminants, and keeping a cool temperature can avert any problems. Acylthiophenes are transported under U.N. No. 2810, Toxic liquids, organic, N.O.S., Class 6.1, Packaging Group III.

Toxicity

Thiophene is harmful by inhalation, in contact with skin, or if swallowed; it is also a skin-irritant. Studies (40) indicate acute oral toxicity to rat in the range $1000 < LD_{50} < 3000$ mg/kg, and a dermal toxicity $LD_0 > 2000$ in rats. Thiophene may act as a central nervous system depressant; some evidence for liver damage in rats also exists. The methylthiophenes are irritating to skin, eyes, and the respiratory system. Thiophene and the methylthiophenes give a negative response in the Ames test for bacterial mutagenicity. Reports (41,42) concerning neurotoxicity indicate that thiophene induces cerebellar degeneration, mediated via a disturbance of cerebellar blood vessels and hepatic injury. Thiophene is harmful in the aqueous environment and is nonbiodegradable (see Table 2).

Thiophene and 3-methylthiophene are listed on the TSCA chemical substances inventory. Thiophene is regulated as a hazardous material under OSHA and also regulated under the Clean Air Act, Section 110, 40 CFR 60.489, but there are no exposure limits or controls set for 3-methylthiophene. Both materials are regulated under sections 311/312 of the Superfund Amendments and Reauthorization Act, 1986 (SARA), as materials with an acute health and fire hazard, and under the Resource Conservation and Recovery Act, as ignitable hazardous wastes (D001).

Bromothiophenes are toxic materials by all routes. Inhalation toxicity of 2-bromothiophene is significant. Ecotoxicity is also noted for these materials, particularly for 2-bromo-3-methylthiophene. 2-Thiophenecarboxaldehyde and the 3-methyl derivative can cause minor irritation to the skin and eyes of rabbits. The former is a sensitizer to guinea pig skin, the latter is not. 2-Acetylthiophene is toxic in all modes of contact. Severe exposure causes serious inflammation of the lung, damage to many organs, and depression of the central nervous system.

Uses of Thiophene and Derivatives

Pharmaceuticals. Thiophene and its derivatives find applications in the pharmaceutical area over a wide range of drug types (43), which can be divided into four main groups.

Nonsteroidal Antiinflammatory Rheumatoid and Osteoarthritis Drugs. Of the many drugs in this area on the market, a number contain the thiophene moiety, and new ones are coming through trials, prior to entering the market. Tenoxicam [59804-37-4] (Hoffmann La Roche) (44), a cyclooxygenase/lipoxygenase inhibitor developed as an antiinflammatory, is also used to treat arthritis. Although it has been launched in over 70 countries, it has not been introduced into the United States. A near relative, Lornoxicam [70374-39-9] (Haflund/Nycomed) (45), is in trials as a nonsteroidal antiinflammatory (NSAI) analgesic and for the treatment of post-operative pain. Lornoxicam is said to be 10 times more effective than tenoxicam and is to be marketed to control severe pain (see ANALGESICS, ANTIPYRETICS, AND ANTIINFLAMMATORY AGENTS).

An established product, Surgam [33005-95-7] (Roussel UCLAF) (46), is still in production in the 1990s. One of the newer drugs coming into this area is Tenidap [100599-27-7] (Pfizer) (47), derived from 2-thiophenecarboxylic acid. The promising trial results of this drug indicate NSAI activity and disease-modifying effects, which minimizes joint damage by inhibiting the destruction mechanism. Its use would therefore constitute a novel arthritis treatment, however, continuing trials have indicated side effects and the drug has been withdrawn (Scrip No. 2169, 1996).

Hypertension and Heart Drugs. Heart disease and failure has a number of causes: hypertension, coronary artery disease, congestive heart failure, and arrhythmia of the heart. There are drugs that assist and control one or more of these conditions: angiotensin-converting enzyme (ACE) inhibitors, calcium antagonists, diuretics, and beta-blockers, respectively (see CARDIOVASCULAR AGENTS). Eprosartan [133040-01-4] (SB) (48) in phase-three clinical trials is one of a new class of drugs that can be used to control both congestive heart failure and hypertension. Ticlopidine [55142-85-3] (Sanofi/Syntex) (49) and Clopidogrel [90055-48-4] (Sanofi/BMS) (50) are both antithrombotic drugs that prevent heart attacks, strokes, and peripheral arterial disease by acting as platelet aggregation inhibitors. Ticlopidine was the first such drug on the U.S. market following FDA approval in 1990. Clopidogrel is a similar product for the same purpose, with reported increased potency and fewer side effects. It is expected to have a bright future.

Antibiotics. Cephaloridine and Cephalothin (Glaxo) were early thiophene-containing, cephalosporin antibiotics. They have largely been replaced by later products, of which Cefoxitin [35607-66-0] (Merck) (51) is the principal thiophene-containing example (see ANTIBIOTICS).

Ticarcillin [34787-01-4] (SB) (52) is a significant penicillin antibiotic that incorporates the thiophene ring system. A number of routes to the required intermediate, 3-thiophenemalonic acid [21080-92-2], have been used over the years. Those from thiophene-based starting materials have involved 3-methylthiophene and 3-bromothiophene.

Other Pharmaceuticals. Other pharmaceutical products incorporating the thiophene ring include the antiasthmatic drug Ketotifen [34580-13-7] (Sandoz) (53), which is particularly marketed in Japan. The antifungal drug Tioconazole [65899-73-2] (Pfizer) (54) is based on 2-chloro-3-methylthiophene [14345-97-2]. The antiglaucoma drug Dorzolamide [120279-96-1] is made from a range of fused thiophene derivatives developed by Merck (55). A group of thiophene-containing drugs from the Japanese pharmaceutical industry includes Tipepidine [5169-78-8] (Tanabe) (56), an antitussive; Tiquizium Bromide [71731-58-3] (Hokuriku) (57), an antispasmodic; and Timepidium Bromide [35035-05-3] (Tanabe) (58), an anticholinergic.

Veterinary Products. Principal users of thiophene are the anthelmintics Pyrantel [15686-83-6] and Morantel [20574-50-9] (Pfizer) (59), based on 2-thiophenecarboxaldehyde and 3-methyl-2-thiophenecarboxaldehyde [5834-16-2], respectively. Tioconazole, one of a range of fungicidal products incorporating thiophene, has also found veterinary applications.

Agrochemical Products. The principal thiophene derivative in herbicidal protection, one of a range of sulfonylurea herbicides, is Harmony [79277-27-3] (Du Pont) (60), based on the intermediate methyl 3-aminothiophene-2-carboxylate (9). The product is characterized by a rapid biodegradability in the soil. Many other thiophene derivatives have been shown to have agrochemical activity, but few of these have been developed to the commercial level.

Dyestuffs. The use of thiophene-based dyestuffs has been largely the result of the access of 2-amino-3-substituted thiophenes via new cyclization chemistry techniques (61). Intermediates of type (8) are available from development of this work. Such intermediates act as the azo-component and, when coupled with pyrazolones, aminopyrazoles, phenols, 2,6-dihydropyridines, etc, have produced numerous monoazo disperse dyes. These dyes impart yellow–green, red–green, or violet–green colorations to synthetic fibers, with excellent fastness to light as well as to wet- and dry-heat treatments (62–64).

Conjugated Polythiophenes. Because of their potential electrical conductivity, conjugated polythiophenes, along with other conjugated polymeric systems, have been extensively studied since the pioneering works in the early 1980s (65,66) (see ELECTRICALLY CONDUCTIVE POLYMERS). Poly-3-alkylthiophenes, in particular, have attracted much attention. The use of 3-n-alkyl (C_{6-12}) side chains improves significantly the solubility and processibility of these polymers (67–73). Branched-chain-substituted thiophenes are difficult to polymerize and are generally less conductive. It is not surprising that many variations of the 3-substitution have been studied. In the laboratory this poses no significant problem, as synthetic routes to 3-substituted thiophenes are readily available (74). However, these routes tend to involve multistep syntheses, which add considerably to the cost of such monomers when scaleup is considered.

Applications of polythiophenes being considered utilize either the electrical properties of the doped conducting state with either anionic or cationic species, the electronic properties of the neutral material, or the electrochemical reversibility of the transition between the doped and undoped state of these materials (71).

The development of polythiophenes since the early 1980s has been extensive. Processible conducting polymers are available and monomer derivatization has extended the range of electronic and electrochemical properties associated with such materials. Problem areas include the need for improved conductivity by monomer manipulation, involving more extensive research using structure–activity relationships, and improved synthetic methods for monomers and polymers alike, which are needed to bring the attractive properties of polythiophenes to fruition on the commercial scale.

Another group of conjugated thiophene molecules for future applications are those being developed as nonlinear optical (NLO) devices (75). Replacement of benzene rings with thiophene has an enormous effect on the molecular nonlinearity of such molecules. These NLO molecules are able to switch, route, and modulate light. Technology using such materials should become available by the turn of the twenty-first century.

BIBLIOGRAPHY

"Thiophene" in *ECT* 1st ed., Vol. 14, pp. 95–102, by D. E. Badertscher and H. E. Rasmussen, Socony Mobil Oil Co., Inc.; in *ECT* 2nd ed., Vol. 20, pp. 219–226, by O. Meth-Cohn, University of Salford; "Thiophene and Thiophene Derivatives" in *ECT* 3rd ed., Vol. 22, pp. 965–973, by B. Buchholz, Pennwalt Corp.

1. S. Gronowitz, ed., *Thiophene and Thiophene Derivatives*, Vols. 1–4, Wiley-Interscience, New York, 1985.
2. H. D. Hartough, *Thiophene and Thiophene Derivatives*, Interscience Publishers, New York, 1952.
3. S. Gronowitz, *Adv. Heterocyclic Chem.* **1**, 1 (1963).
4. S. Gronowitz, *Org. Chem. Sulphur Selenium Tellurium*, London, **3**, 400 (1975); S. Gronowitz, *Org. Chem. Sulphur Selenium Tellurium*, **4**, 244 (1977).
5. O. Meth-Cohn, *Comp. Org. Chem.* **4**, 789 (1979).
6. R. M. Kellog, *Comp. Heterocyclic Chem.* **4**, 713 (1984); S. J. Rajappa, *Comp. Heterocyclic Chem.* **4**, 741 (1984); E. Campaigne, *Comp. Heterocyclic Chem.* **4**, 863 (1984).
7. J. B. Press and R. K. Russell, *Prog. Heterocyclic Chem.* **2**, 50 (1990); **3**, 70 (1991); **4**, 62 (1992); **5**, 82 (1993); **6**, 88 (1994); **7**, 82 (1995).
8. B. Bak, D. Christensen, and L. H. Nygaard, *J. Mol. Spectroscopy*, **7**, 58 (1961).
9. M. J. Cook, A. R. Katritzky, and P. Linda, *Adv. Heterocyclic Chem.* **17**, 255 (1974).
10. F. Fringuelli and co-workers, *J. Chem. Soc. Perkin Trans. II*, (4), 332 (1974).
11. Data obtained on-line from *Beilstein Handbook of Organic Chemistry*, Beilstein Informationssysteme, Frankfurt, Germany.
12. K. Verschueren, *Handbook of Environmental Data on Organic Chemicals*, 2nd ed., Van Nostrand Reinhold, Co., Inc., New York, 1983, p. 1097.
13. Technical data, Synthetic Chemicals, Ltd., Huntingdon Research Centre Reports, July 1987.
14. Zeiger and co-workers, *Environ. Mutagen.* **9**(Suppl. 9), 1 (1987).
15. Technical data, Synthetic Chemicals Ltd., Binnie Environmental Report ENV161, 1994.

16. Technical data, Elf Atochem, Safety Data Sheet, November 4, 1994.
17. Eur. Pat. 299,586 (July 14, 1987), P. R. Grosvenor and L. S. Fuller (to Inspec Group plc).
18. Austral. Pat. 42,881 (Nov. 28, 1985), K. Eichler and E. I. Leupold (to Hoechst AG).
19. Technical data, Synthetic Chemicals Ltd, Huntingdon Research Centre Reports, 1979, 1986, 1988.
20. Technical data, Synthetic Chemicals Ltd, Safepharm Laboratories Ltd. Reports, 1994.
21. Technical data, Shell Japan Ltd, Hita Research Laboratories Report, T-3554, 1993.
22. Technical data, Synthetic Chemicals Ltd., Binnie Environmental Ltd. Reports, 1994.
23. L. Friedman, D. L. Fishel, and H. Schecter, *J. Org. Chem.* **30**, 1453 (1965).
24. B. F. Crowe and F. F. Nord, *J. Org. Chem.* **15**, 81 (1950).
25. Brit. Pat. 1,483,349 (Aug. 17, 1974), J. A. Clark and O. Meth-Cohn (to Synthetic Chemicals Ltd.).
26. S. Conde and co-workers, *Synthesis*, (6), 412 (1976).
27. S. Gronowitz, *Arkiv Kemi*, **12**, 239 (1958).
28. O. Meth-Cohn, *Org. Chem. Sulphur Selenium Tellurium*, **4**, 828 (1979).
29. U.S. Pat. 3,197,483 (July 27, 1965), B. Buchholz, T. E. Deger, and R. H. Goshorn (to Pennwalt Corp.).
30. U.S. Pat. 3,822,289 (July 2, 1974), Brit. Pat. 1,345,203, (Jan. 30, 1974), N. R. Clark and W. E. Webster (to Synthetic Chemicals Ltd.).
31. U.S. Pat. 3,939,179 (Feb. 17, 1976), T. R. Bell and P. G. Smith (to Pennwalt Corp.).
32. Brit. Pat. 1,585,647 (Mar. 11, 1981), J. Barrault, L. Lucien, and M. Guisnet (to Societe Nationale, Elf Aquataine).
33. Eur. Pat. 43,303 (July 1, 1980), G. Soula (to Rhone Poulenc Specialties Chimiques); G. Soula, *J. Org. Chem.* **50**, 3717, 1985.
34. Technical data, Synthetic Chemicals Ltd., Huntingdon Research Centre Reports, 1976.
35. Brit. Pat. 1,523,650 (Sept. 6, 1978), H. Koenig and U. Ohnsorge (to BASF AG).
36. Technical data, Synthetic Chemicals Ltd., Huntingdon Research Centre Reports, 1987.
37. Technical data, Synthetic Chemicals Ltd., Binnie Environmental Ltd., Reports, 1994.
38. Ger. Pats. 1,055,007 (Aug. 29, 1957) and 1,083,830, (Aug. 2, 1958); Brit. Pat. 837,086 (Aug. 29, 1958), H. Feisselmann (to Hoechst AG).
39. P. R. Huddleston and J. M. Barker, *Synthetic Comm.* **9**(8), 731 (1979).
40. Elf-Atochem, Springborn Labs. Report No., 3255.14, and 3255.15, 1994.
41. F. Mori and co-workers, *J. Toxicol. Pathol.* **5**, 21 (1992).
42. F. Mori and co-workers, *J. Toxicol. Pathol.* **6**, 213 (1993).
43. L. S. Fuller, J. W. Pratt, and F. S. Yates, *Manufactur. Chem. Aerosol News*, **49**(5), 67 (1978).
44. Ger. Pat. 2,537,070 (Mar. 18, 1976), O. Hromatka and co-workers (to Hofmann La Roche AG).
45. Eur. Pat. 313,935 (May 3, 1989), D. Binder, F. Rovenszky, and H. P. Ferber (to Hafslund Nycomed Pharma AG).
46. Ger. Pat. 2,055,264 (May 19, 1971), F. Clemence and O. Le Martret (to Roussel UCLAF).
47. U.S. Pat. 4,556,672 (Dec. 3, 1985), S. B. Kadin (to Pfizer Inc.).
48. Eur. Pat. 403,159 (Dec. 19, 1990), J. A. Finkelstein, R. M. Keenan, and J. Weinstock (to SmithKline Beecham Corp.).
49. U.S. Pat. 4,127,580 (Nov. 28, 1978), E. Braye (to Parcor/Sanofi).
50. Eur. Pat. 99,802 (June 13, 1982), D. Aubert, C. Ferrand, and J.-P. Maffrand (to Parcor/Sanofi).
51. Brit. Pat. 1,348,984 (Mar. 27, 1974), B. G. Christensen and co-workers (to Merck and Co. Inc).
52. Brit. Pat. 1,004,670 (Apr. 23, 1963), E. G. Brain and J. H. Naylor (to Beecham Group Ltd.).

53. Ger. Pat. 2,111,071 (Sept. 23, 1971), J. P. Bourquin, G. Schwarb, and E. Waldvogel (to Sandoz Ltd.).
54. U.S. Pat. 4,062,966 (Dec. 13, 1977), G. E. Gymer (to Pfizer Corp.).
55. Eur. Pat. 296,879 (Dec. 28, 1988), J. J. Baldwin, G. S. Ponticello, and M. E. Christy (to Merck and Co. Inc.).
56. Brit. Pat. 924,544 (Apr. 24, 1963), Y. Yamamoto (to Tanabe Seiyaku Co. Ltd.).
57. Belg. Pat. 866,988 (May 17, 1977), H. Kato and co-workers (to Hokuriku Pharmaceutical Co. Ltd.).
58. Brit. Pat. 1,358,446 (July 3, 1974), T. Kanno, S. Saito, and H. Tamaki (to Tanabe Seiyaku Co. Ltd.).
59. Brit. Pat. 1,120,587 (July 17, 1968), W. C. Austin, L. H. Conoverand, and J. W. McFarland (to Pfizer Ltd.).
60. Eur. Pat. 41,404 (Dec. 9, 1981), G. Levitt (to E. I. Du Pont de Nemours & Co., Inc.); *Proceedings of Meeting of Weed Society of America*, Seattle, Washington, 1980.
61. K. Gewald, *Chem. Ber.* **98**, 3571 (1965); *Org. Chem. Sulphur Selenium Tellurium*, **3**, 401 (1975).
62. Brit. Pats. 1,394,365, 1,394,367, and 1,394,368 (May 14, 1975), D. B. Baird and co-workers (to ICI Industries Ltd.).
63. Brit. Pat. 1,434,654 (May 5, 1976), W. Groebke and A. Jotterand (to Sandoz AG).
64. Brit. Pat. 1,461,738 (Jan. 19, 1977), D. Von der Brueck and G. Wolfrum (to Bayer AG).
65. A. F. Diaz, *Chem. Scr.* **17**, 142 (1981).
66. G. Tourillon and F. Garnier, *J. Electroanal. Chem.* **135**, 173 (1982).
67. M. Sato, S. Tanaka, and K. Kaeriyama, *J. Chem. Soc. Chem. Commun.* (11), 873 (1986).
68. Fr. Pat. 2,596,566 (Oct. 2, 1987), F. Garnier and co-workers (to Solvay & Cie).
69. S. Hotta and co-workers, *Macromolecules*, **20**, 212 (1987).
70. G. Tourillon, in T. A. Skotheim, ed., *Handbook of Conducting Polymers*, Vol. 1, Marcel Dekker, Inc,. New York, 1986, p. 294.
71. J. Roncali, *Chem. Rev.* **92**, 711 (1992).
72. M. Schott and M. Nechtschein, in J.-P. Farges, ed., *Organic Conductors*, Marcel Dekker, Inc., New York, 1994, p. 495.
73. M. Schott, in Ref. 72, p. 539.
74. K. Schulz, K. Fahmi, and M. Lemaire, *Acros Organics Acta*, **1**, 10 (1995).
75. K. J. Drost, A. K.-Y. Jen, and V. P. Rao, *Chemtech.* **25**(9), 16 (1995).

LANCE S. FULLER
Synthetic Chemicals Limited

THIOSULFATES

The thiosulfate ion, $S_2O_3^{2-}$, is a structural analogue of the sulfate ion where one oxygen atom is replaced by one sulfur atom. The two sulfur atoms of thiosulfate thus are not equivalent. Indeed, the unique chemistry of the thiosulfate ion is dominated by the sulfide-like sulfur atom which is responsible for both the reducing properties and complexing abilities. The ability of thiosulfates to dissolve silver halides through complex formation is the basis for their commercial application in photography (qv).

Physical Properties

Structure.   The thiosulfate sulfur atoms have been shown to be nonequivalent by radioactive sulfur exchange studies (1). When a sulfite is treated with radioactive sulfur and the resulting thiosulfate decomposed to sulfur and sulfite by acids, the radioactivity appears in the sulfur:

$$^{35}S + SO_3^{2-} \longrightarrow {}^{35}SSO_3^{2-}$$
$$^{35}SSO_3^{2-} + H^+ \longrightarrow {}^{35}S + HSO_3^-$$

When the silver salt of the radioactive thiosulfate [83682-20-6] is decomposed, the radioactivity appears in the sulfide:

$$Ag_2^{35}SSO_3 + H_2O \longrightarrow Ag_2^{35}S + H_2SO_4$$

If both sulfur atoms were equivalent, the radioactivity would be equally divided between the two products.

Calculations from binding-force measurements (2) indicate the limiting structural forms of the thiosulfate ion:

$$\begin{array}{cc} (1) & (2) \end{array}$$

Structure (**1**) explains the formation of sulfur and sulfite in the presence of acid; structure (**2**) is consistent with the formation of sulfide and sulfate in the presence of heavy metals. The bonding in thiosulfate complexes and the chemistry of thiosulfates are normally explained on the basis of (**2**) (see also SULFUR COMPOUNDS).

The thiosulfate ion has tetrahedral C_{3v} symmetry and the six fundamental modes are both infrared and Raman active. The calculated frequencies (3) are in good agreement with experimental values (4).

Calculated N, cm^{-1}	Experimental N, cm^{-1}
995	1002
669	672
435	451
1123	1125
541	541
335	339

The low value of the S–S force constant compared to that of S–O is consistent with the ease of cleavage of the S–S bond. Spectral data indicate that the structure of the thiosulfate ion in solid thiosulfates is the same as that of the ion in solution.

X-ray crystallographic analysis of the sodium thiosulfate pentahydrate [10102-17-7] crystal indicates a tetrahedral structure for the thiosulfate ion. The S–S bond distance is 197 pm; the S–O bond distance is 148 pm (5). Neutron diffraction of a barium thiosulfate monohydrate [7787-40-8] crystal confirms the tetrahedral structure and bond distances for the thiosulfate ion (6).

Thermodynamic Properties. The heat of formation of the thiosulfate ion, -5.75 kJ/g (-1.37 kcal/g), was determined by studying the equilibrium of the following reaction:

$$2\,\text{Ag} + \text{S}_2\text{O}_3^{2-} \longrightarrow \text{Ag}_2\text{S} + \text{SO}_3^{2-}$$

and by direct calorimetric experiments (7). The standard free energy of formation is -4.58 kJ/g (-1.09 kcal/g). The partial molal entropy is 62.8 ± 25.1 J/K (15 ± 6 cal/K).

Electrochemical Properties. The oxidation potential for the reaction

$$2\,\text{S}_2\text{O}_3^{2-} \rightleftharpoons \text{S}_4\text{O}_6^{2-} + 2\,e^-$$

ranges from 0.2 to 0.4 V in neutral solution, depending on the method of measurement (8). The electrolytic oxidation of thiosulfate solutions yields tetrathionate, $\text{S}_4\text{O}_6^{2-}$, as the principal product; HSO_3^- is the by-product in acid solutions and $\text{S}_3\text{O}_6^{2-}$ is in alkaline solutions. The electrolytic oxidation is catalyzed by trace amounts of iodide (9).

Electrolytic reduction with a mercury or platinum electrode produces equimolar amounts of sulfide and sulfite:

$$\text{S}_2\text{O}_3^{2-} + 2\,e \longrightarrow \text{S}^2 + \text{SO}_3^{2-}$$

Chemical Properties

Thiosulfuric Acid. Thiosulfuric acid [14921-76-7] is relatively unstable and thus cannot be recovered from aqueous solutions. In laboratory preparation, a lead thiosulfate [26265-65-6] solution is treated with H_2S to precipitate PbS, or a concentrated solution of sodium thiosulfate [7772-98-7] is treated with HCl and cooled to $-10°\text{C}$ to crystallize NaCl. Aqueous solutions of thiosulfuric acid spontaneously decompose to yield sulfur, SO_2, and polythionic acids, $\text{H}_2\text{S}_n\text{O}_6$. Thiosulfuric acid is a strong acid comparable to sulfuric acid. Dissociation constants, $K_1 = 0.25$, $K_2 = 0.018$, have been determined from pH measurements using a glass electrode (10).

Pure thiosulfuric acid has been prepared in liquid CO_2 at $-50°\text{C}$ (11) or in diethyl ether at $-78°\text{C}$ (12). It decomposes at $-30°\text{C}$ to $\text{H}_2\text{S}_3\text{O}_6$ [27621-39-2] and H_2S, and rapidly at higher temperatures to H_2O, SO_2, and sulfur (13).

Thiosulfates.   The ammonium, alkali metal, and alkaline-earth thiosulfates are soluble in water. Neutral or slightly alkaline solutions containing excess base or the corresponding sulfite are more stable than acid solutions. Thiosulfate solutions of other metal ions can be prepared, but their stability depends on the presence of excess thiosulfate, the formation of complexes, and the prevention of insoluble sulfide precipitates.

Acidification of thiosulfate with strong acid invariably leads to decomposition with the formation of colloidal sulfur and sulfur dioxide. The mechanism of this reaction is complex and depends on the thiosulfate concentration and the pH (14). The following reaction explains the formation of the main products:

$$H^+ + S_2O_3^{2-} \longrightarrow [HS_2O_3^-] \longrightarrow HSO_3^- + S^0 \longrightarrow SO_2 + S^0 + H_2O$$

By-products are also formed:

$$3\ S_2O_3^{2-} + 3\ H^+ \longrightarrow H_2S + S_4O_6^{2-} + HSO_3^-$$
$$S_2O_3^{2-} + H_2O \longrightarrow SO_4^{2-} + H_2S$$
$$5\ S_2O_3^{2-} + 6\ H^+ \longrightarrow 2\ S_5O_6^{2-} + 3\ H_2O$$

In dilute aqueous solution, the following equilibrium is established (15):

$$S_2O_3^{2-} + H^+ \rightleftharpoons HSO_3^- + S \qquad K = 0.013 \text{ at } 11°C$$

This equilibrium explains the stabilization of thiosulfate solutions using sulfite or bisulfite as one of the components of acid photographic fixing baths.

The existence of anhydrothiosulfuric acid [83682-21-7] has been proposed to explain the apparent stability of thiosulfate in concentrated hydrochloric acid solution (16):

$$2\ H_2S_2O_3 \rightleftharpoons H_2S_4O_5 + H_2O$$

Reactions.   Catalytic amounts of arsenic, antimony, or tin salts promote the formation of pentathionate (16):

$$S_2O_3^{2-} + H_2AsO_3^- \rightleftharpoons HSO_3^- + HAsSO_3^{2-}$$
$$S_2O_3^{2-} + 4\ HSO_3^- + 2\ H^+ \rightleftharpoons 2\ S_3O_6^{2-} + 3\ H_2O$$
$$S_3O_6^{2-} + HAsSO_3^{2-} + H^+ \rightleftharpoons S_4O_6^{2-} + H_2AsO_3^-$$
$$S_4O_6^{2-} + HAsSO_3^{2-} + H^+ \rightleftharpoons S_5O_6^{2-} + H_2AsO_3^-$$

Mild oxidizing agents such as hydrogen peroxide in acid solutions produce tetrathionates and trithionates (17):

$$2\ S_2O_3^{2-} + H_2O_2 \rightleftharpoons S_4O_6^{2-} + 2\ OH^-$$
$$3\ S_2O_3^{2-} + 4\ H_2O_2 \rightleftharpoons 2\ S_3O_6^{2-} + 2\ OH^- + 3\ H_2O$$

The presence of Fe^{2+} promotes oxidation to the sulfate:

$$S_2O_3^{2-} + 4\,H_2O_2 \underset{}{\overset{Fe^{2+}}{\rightleftharpoons}} 2\,SO_4^{2-} + 2\,H^+ + 3\,H_2O$$

The reaction with iodine in neutral or slightly acid solution is the basis of a volumetric analytical procedure.

$$2\,S_2O_3^{2-} + I_2 \longrightarrow S_4O_6^{2-} + 2\,I^-$$

Stronger oxidizing agents such as chlorine, bromine, permanganate, chromate, or alkaline hydrogen peroxide oxidize thiosulfate quantitatively to sulfate:

$$S_2O_3^{2-} + 2\,OH^- + 4\,H_2O_2 \rightleftharpoons 2\,SO_4^{2-} + 5\,H_2O$$

Hypochlorite, hypobromite, and hypoiodite are also strong enough to oxidize thiosulfate to sulfate:

$$S_2O_3^{2-} + 4\,BrO^- + 2\,OH^- \longrightarrow 2\,SO_4^{2-} + 4\,Br^- + H_2O$$

Thiosulfates are reduced to sulfides by metallic copper, zinc, or aluminum:

$$S_2O_3^{2-} + 2\,Cu \longrightarrow Cu_2S + SO_3^{2-}$$

The thiosulfate reaction with cyanide to give thiocyanate is the basis for the use of thiosulfate as an antidote in cyanide poisoning:

$$S_2O_3^{2-} + CN^- \longrightarrow SO_3^{2-} + SCN^-$$

Thiosulfates form complex ions with a number of metal ions by the coordination of more than one thiosulfate ion. The stability constants for the lead thiosulfate complexes $[Pb(S_2O_3)_2]^{2-}$ and $[Pb(S_2O_3)_3]^{4-}$ have been determined (18). Mercury forms a thiosulfate complex [83682-22-3], $K_6[Hg(S_2O_3)_4]$, by reaction of mercuric oxide and potassium thiosulfate [10294-66-3] (19). The stability of the silver complex thiosulfates $[Ag(S_2O_3)_2]^{3-}$ and $[Ag(S_2O_3)_3]^{5-}$ is the basis for the use of thiosulfates to dissolve the residual silver chloride in photographic gelatin coatings. The structure of some thiosulfate complexes has been determined (20).

The addition of thiosulfate to aqueous solutions of silver, lead, and copper precipitates the corresponding thiosulfates, which, on heating, decompose to the sulfides. In this manner, thiosulfate can be used as a reagent for most metals having insoluble sulfides. Details of the reactions of other metals with thiosulfate are available (21).

Corrosion. Copper-base alloys are seriously corroded by sodium thiosulfate (22) and ammonium thiosulfate [7783-18-8] (23). Corrosion rates exceed

10 kg/(m²·yr) at 100°C. High silicon cast iron has reasonable corrosion resistance to thiosulfates, with a corrosion rate <4.4 kg/(m²·yr) at 100°C. The preferred material of construction for pumps, piping, reactors, and storage tanks is austenitic stainless steels such as 304, 316, or Alloy 20. The corrosion rate for stainless steels is <440 g/(m²·yr) at 100°C (see also CORROSION AND CORROSION CONTROL).

Preparation

Thiosulfates are normally prepared by the reaction of sulfur and sulfite in neutral or alkaline solution:

$$S^0 + SO_3^{2-} \longrightarrow S_2O_3^{2-}$$

Polysulfides react similarly:

$$S_x^{2-} + SO_3^{2-} \longrightarrow S_2O_3^{2-} + S_{x-1}^{2-}$$

Sulfides react with sulfur dioxide, sulfite, or bisulfite:

$$2\,S^{2-} + 2\,SO_2 + 2\,HSO_3^- \longrightarrow 3\,S_2O_3^{2-} + H_2O$$
$$2\,S^{2-} + 3\,SO_2 + SO_3^{2-} \longrightarrow 3\,S_2O_3^{2-}$$
$$2\,HS^- + 4\,HSO_3^- \longrightarrow 3\,S_2O_3^{2-} + 3\,H_2O$$

These three methods are employed commercially. In addition, decomposition of polythionates in alkaline solution or their reaction with sulfide or sulfite gives thiosulfates:

$$4\,S_4O_6^{2-} + 6\,OH^- \longrightarrow 5\,S_2O_3^{2-} + 2\,S_3O_6^{2-} + 3\,H_2O$$
$$S_3O_6^{2-} + S^{2-} \longrightarrow 2\,S_2O_3^{2-}$$
$$S_5O_6^{2-} + SO_3^{2-} \longrightarrow S_2O_3^{2-} + S_4O_6^{2-}$$

The high temperature hydrolysis of sulfur in alkaline solutions also produces thiosulfates:

$$S_8 + 6\,OH^- \longrightarrow 2\,S_3^{2-} + S_2O_3^{2-} + 3\,H_2O$$

Sodium Thiosulfate

Sodium thiosulfate, either the anhydrous salt, $Na_2S_2O_3$, or the crystalline pentahydrate, is commonly referred to as hypo or crystal hypo. When a concentrated sodium thiosulfate solution (50–60 wt %) is cooled to <48°C, the pentahydrate, containing 63.7% $Na_2S_2O_3$, crystallizes in monoclinic transparent prisms as shown in the equilibrium phase diagram (Fig. 1). The monohydrate [55755-19-6] and the heptahydrate [36989-91-0] are also known.

Although 16 different crystalline modifications have been identified (24,25), the α-pentahydrate is the stable form below 48°C. Solutions of sodium thiosulfate in the absence of seed crystals can be easily supercooled below their normal crystallization temperatures. The dotted line extension of the dihydrate phase in Figure 1 is an indication that, if supercooling takes place below this line, solutions normally giving the pentahydrate may form the dihydrate [36989-90-9] instead.

Selected physical properties of sodium thiosulfate pentahydrate are shown in Table 1. The crystals are relatively stable, efflorescing in warm, dry air and deliquescing slightly in moist air. They melt in their water of hydration at 48°C and can be completely dehydrated in a vacuum oven at this temperature, or at atmospheric pressure at 105°C. Anhydrous sodium thiosulfate can also be

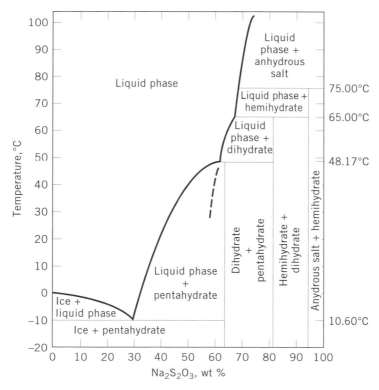

Fig. 1. Equilibrium phase diagram, $Na_2S_2O_3$ and H_2O, where the dashed line is an extension of dihydrate crystal phase (24).

Table 1. Physical Properties of Sodium Thiosulfate Pentahydrate

Property	Value	Reference[a]
refractive index, n_D^{20}	1.4886	27
density, d_4^{25}, g/cm^3	1.750	28
heat of solution in water at 25°C, J/g[b]	-187	29
heat of formation, kJ/g[b]	-10.48	30
heat of fusion, J/g[b]	200	31
specific heat J/(g·K)[b]		
solid	1.84	28
molten salt	2.38	28
dissociation pressure, kPa[c]		
20°C	0.796	
25°C	1.154	
30°C	1.679	
35°C	2.395	
vapor pressure of saturated solutions, kPa[c]		28
33°C	1.33	
57°C	5.60	
90°C	31.06	
120°C	100.4	
density of aqueous solutions, d_{20}^{20}, g/cm^3 at wt % $Na_2S_2O_3$		32
1.00	1.0083	
10.00	1.0847	
20.00	1.1760	
30.00	1.2762	
40.00	1.3851	

[a]Ref. 26 gives a comprehensive summary of the properties of sodium thiosulfate and its aqueous solutions.
[b]To convert J to cal, divide by 4.184.
[c]To convert kPa to mm Hg, multiply by 7.5.

crystallized directly from a 72% solution above 75°C. It decomposes at 233°C:

$$4 \, Na_2S_2O_3 \longrightarrow 3 \, Na_2SO_4 + Na_2S_5$$

Further heating to 440–500°C gives sodium sulfide and sulfur dioxide (33).

Aqueous sodium thiosulfate solutions are neutral. Under neutral or slightly acidic conditions, decomposition produces sulfite and sulfur. In the presence of air, alkaline solutions decompose to sulfate and sulfide. Dilute solutions can be stabilized by small amounts of sodium sulfite, sodium carbonate, or caustic, and by storage at low temperatures away from air and light. Oxidation is inhibited by HgI_2 (10 ppm), amyl alcohol (1%), chloroform (0.1%), borax (0.05%), or sodium benzoate (0.1%).

Manufacture. Sodium thiosulfate has been produced commercially by the air oxidation of sulfides, hydrosulfides, and polysulfides.

Sodium thiosulfate is a by-product of the manufacture of Sulfur Black and other sulfur dyes (qv), where organic nitro compounds are treated with a solution of sodium polysulfide to give thiosulfate. The dyes are insoluble and are recovered by filtration. The filtrate is treated with activated carbon and filtered to obtain a sodium thiosulfate solution. After concentration and crystallization, the final product assays ca 96% $Na_2S_2O_3 \cdot 5H_2O$ (34) (see DYES AND DYE INTERMEDIATES).

Other commercial processes are based on the reaction of sodium sulfide or hydrogen sulfide with sulfur dioxide, and caustic or soda ash:

$$2\,Na_2S + Na_2CO_3 + 4\,SO_2 \longrightarrow 3\,Na_2S_2O_3 + CO_2$$
$$2\,Na_2S + 3\,SO_2 \longrightarrow 2\,Na_2S_2O_3 + S^0$$

Excess sulfur is filtered before evaporation and crystallization. In one modification, excess sulfur is preground in the sodium sulfide solution and an equivalent amount of sodium sulfite added (35).

Another procedure utilizes a slurry of sodium sulfite, produced by the reaction of soda ash with sulfur dioxide, which is digested with excess sulfur until all of the sulfite is used up:

$$Na_2SO_3 + S^0 \longrightarrow Na_2S_2O_3$$

Cationic surface-active agents promote wetting of the sulfur and thereby increase the reaction rate (36). The quality of the product is improved by using photographic-grade sodium sulfite or bisulfite. Excess sulfur is filtered before evaporation (qv) and crystallization (qv). Evaporation is energy-intensive; thus it is important to produce the thiosulfate solution at the highest possible concentration. The purity of the product is typically >99%; sulfite and sulfate are the main impurities.

Economic Aspects. As of 1995, two manufacturers produced sodium thiosulfate in the United States with an estimated total capacity of 30,000 metric tons. Despite declining volume, prices have increased, reflecting increased raw material, energy, and labor costs. Production outside of the United States is at least 25% of the U.S. production, mainly in Germany.

Specifications. Sodium thiosulfate pentahydrate and the anhydrous salt are available in various grades, as shown in Tables 2 and 3.

Analytical and Test Methods. An aqueous solution of sodium thiosulfate forms a white precipitate with hydrochloric acid and evolves sulfur dioxide gas which is detected by its characteristic odor. The white precipitate turns yellow, indicating the presence of sulfur. The addition of ferric chloride to sodium thiosulfate solutions produces a dark violet color which quickly disappears.

Sodium thiosulfate is determined by titration with standard iodine solution (37). Sulfate and sulfite are determined together by comparison of the turbidity produced when barium chloride is added after the iodine oxidation with the turbidity produced by a known quantity of sulfate in the same volume of solution. The absence of sulfide is indicated when the addition of alkaline lead acetate produces no color within one minute.

Health and Safety Factors. The LD_{50} of anhydrous sodium thiosulfate for mice is 7.5 ± 0.752 g/kg (40). Because of low toxicity, it can be safely used in veterinary medicine. Sodium thiosulfate pentahydrate is affirmed as a GRAS

Table 2. Specifications for Sodium Thiosulfate Pentahydrate

Specifications	ACS reagent grade[a]	USP grade[b]
assay, %[c]	99.5	99.0
water, %		32–37
insoluble matter, %[d]	0.005	
sulfate and sulfite, as SO_4, %[d]	0.1	
sulfide, as S, ppm[d]	1	
pH of 5% solution at 25°C	6.0–8.4	
arsenic, ppm[d]		3
heavy metals, ppm[d]		20[e]
nitrogen compounds, as N, %[d]	0.002	

[a]Ref. 37.
[b]Ref. 38.
[c]Value is minimum.
[d]Value is maximum.
[e]Also calcium content must be so as to pass the test.

Table 3. American National Standard Specifications for Photographic-Grade Sodium Thiosulfate[a]

Specifications[b]	Crystalline pentahydrate	Anhydrous
assay, %		
	99.0[c]	97.0[c]
	101.0	
insoluble matter[d], %	0.2	0.4
alkalinity, as NaOH, %	0.02	0.06
acidity, as H_2SO_4, %	0.01	0.01
sulfide, as S, ppm	4	6
heavy metals, as Pb, ppm	10	20
iron, as Fe, ppm	30	50
appearance of solution	to pass test	to pass test

[a]Ref. 39.
[b]Values are maximum unless otherwise noted.
[c]Value is minimum.
[d]Including calcium, magnesium, and ammonium hydroxides, ppt.

indirect and direct human food ingredient under the Federal Food, Drug, and Cosmetic Act (41) (see FOOD ADDITIVES).

It has been reported that humans can consume up to 12 g sodium thiosulfate orally per day with no effect other than catharsis (42).

Uses. The principal use for sodium thiosulfate continues to be as fixative in photography (qv) to dissolve undeveloped silver halide from negatives or prints. In applications where rapid processing is required, such as the processing of x-ray film, sodium thiosulfate has been largely replaced by ammonium thiosulfate.

Sodium thiosulfate is still used in chrome leather tanning as a reducing agent in two-bath processes to reduce dichromate (hexavalent chromium) to chrome alum (trivalent chromium) (see LEATHER).

In paper (qv) and textiles (qv) manufacture, sodium thiosulfate removes residual bleach before dyeing. It inhibits fermentation in dyeing baths, and is a source of sulfur dioxide in the bleaching of wool. It is also used as a dechlorinating agent for waste streams. Sodium thiosulfate reduces the residual free chlorine in chlorinated potable water to protect the taste of a beverage (43). Both for the dechlorination of potable water and in the various treatments of textiles sodium thiosulfate is a favored dechlorinating agent. The other widely used reducing agents such as sodium metabisulfite and sodium sulfite have pHs of 4.3–4.6 and 9.3–9.5, respectively, whereas the ideal pH for sodium thiosulfate is 7.0–8.0. Using sodium thiosulfate with an inadvertent overdose would not result in the pH being too high or too low.

Minor and potential new uses include flue-gas desulfurization (44,45), silver-cleaning formulations (46), thermal-energy storage (47), cyanide antidote (48), cement additive (49), aluminum-etching solutions (50), removal of nitrogen dioxide from flue gas (51), concrete-set accelerator (52), stabilizer for acrylamide polymers (53), extreme pressure additives for lubricants (54), multiple-use heating pads (55), in soap and shampoo compositions (56), and as a flame retardant in polycarbonate compositions (57). Moreover, precious metals can be recovered from difficult ores using thiosulfates (58). Use of thiosulfates avoids the environmentally hazardous cyanides.

Ammonium Thiosulfate

Ammonium thiosulfate, $(NH_4)_2S_2O_3$, commonly referred to as ammo hypo, has displaced sodium thiosulfate in photography. It is normally sold in the United States only as the aqueous solution. In addition, a crystal slurry and anhydrous crystal are available in Europe.

The anhydrous monoclinic crystalline form has a density of 1.679 g/cm^3 (59); no hydrates are known. Solubility in water is given in Table 4. Ammonium thiosulfate solutions decompose slowly below 50°C and more rapidly at higher

Table 4. Solubility of Ammonium Thiosulfate

Temperature, °C	$(NH_4)_2S_2O_3$ Solubility, wt %
−30	53.5
−20	56.5
−10	58.5
0	60.5
10	62.5
20	64.0
30	65.5
40	67.2
50	68.5
60	69.6
70	70.9
80	72.2

temperatures. The anhydrous salt decomposes above 100°C to sulfite and sulfur (60):

$$(NH_4)_2S_2O_3 \longrightarrow (NH_4)_2SO_3 + S^0$$

At 230–245°C, the ammonium sulfite decomposes.

Manufacture. Ammonium thiosulfate has been produced by the reaction of ammonium sulfite with sulfur, sulfides, or polysulfides:

$$(NH_4)_2SO_3 + (NH_4)_2S_8 \longrightarrow (NH_4)_2S_2O_3 + (NH_4)_2S_7$$

$$(NH_4)_2SO_3 + (NH_4)_2S_7 \longrightarrow (NH_4)_2S_2O_3 + (NH_4)_2S_6$$

This reaction series continues until the last polysulfide is ammonium sulfide and the process is completed by reaction with sulfur dioxide:

$$3\ SO_2 + (NH_4)_2SO_3 + 2\ (NH_4)_2S \longrightarrow 3\ (NH_4)_2S_2O_3$$

Ammonium bisulfite can be used in place of the sulfur dioxide. The solution is treated with activated carbon and filtered to remove traces of sulfur. Excess ammonia is added and the solution evaporated if the anhydrous crystalline form is desired. The crystals are dried at low temperature in the presence of ammonia to prevent decomposition (61–63).

Other commercial processes are based on the direct reaction of ammonium sulfite and sulfur (64,65).

$$(NH_4)_2SO_3 + S \longrightarrow (NH_4)_2S_2O_3$$

Both batch and continuous processes employ excess sulfur and operate at 85–110°C. Trace amounts of polysulfides produce a yellow color which indicates that all the ammonium sulfite has been consumed. Ammonium bisulfite is added to convert the last polysulfide to thiosulfate and the excess ammonia to ammonium sulfite. Concentrations of at least 70% $(NH_4)_2S_2O_3$ are obtained without evaporation. Excess sulfur is removed by filtration and color is improved with activated carbon treatment or sodium silicate (66). Upon cooling the aqueous concentrated solution, ammonium thiosulfate crystallizes.

Agricultural grades of ammonium thiosulfate are prepared by similar processes and contain some excess sulfur. The sulfur can be removed by washing with carbon disulfide. A typical sulfur-free product contains 87% $(NH_4)_2S_2O_3$, 3.4% $(NH_4)_2SO_3$, and 9.6% $(NH_4)_2SO_4$ (67).

Ammonium thiosulfate, stable as a solution, is produced in the form of a 56–60% solution from ammonia and solid sulfur or an H_2S-rich gas stream or both solid sulfur and H_2S gas streams (68). As a result of availability, only development of solutions for processing x-ray and color film and prints has been encouraged. The evolution of automatic processors to develop and print color reinforced the trend toward use of solutions. Most x-ray laboratories and automatic film and print processors require almost immediate results.

Specifications. The specifications for photographic-grade ammonium thiosulfate are shown in Table 5. There are no corresponding specifications for the agricultural-grade material.

Analytical and Test Methods. Analysis and test methods are similar to those for sodium thiosulfate. Sulfite is determined by an indirect method based on the titration of the acid liberated when both the sulfite and thiosulfate are oxidized with iodine solution (69).

Health and Safety Factors (Toxicology). The toxicological properties of ammonium thiosulfate are generally considered to be the same as those of sodium thiosulfate and thiosulfates in general (42).

Uses. The use distribution of ammonium thiosulfate in 1995 was estimated to be photography, 48%; agricultural applications, 50%; and others, including dechlorination, 2%.

The principal use of photochemical-grade ammonium thiosulfate continues to be in photography, where is dissolves undeveloped silver halides from negatives and prints. It reacts considerably faster than sodium thiosulfate, and the fixing solutions can be used about twice as long as sodium thiosulfate solutions; the washing period to remove residual thiosulfate is shorter.

Agricultural uses for ammonium thiosulfate take advantage of both the sulfur and ammonium content by blending with other nitrogen fertilizers such as urea (71). Some foliar-spray fertilizers contain ammonium thiosulfate together with other metal micronutrients (72,73). Ammonium thiosulfate or mixtures with ammonium nitrate can also be used as desiccants and defoliants in crop-bearing plants such as cotton (qv), soybean, alfalfa, rice, and peppers (74,75).

Minor and potential new uses for ammonium thiosulfate include fluc-gas desulfurization (76,77), removal of nitrogen oxides and sulfur dioxide from flue gases (78,79), converting sulfur in hydrocarbons to a water-soluble form (80), and converting cellulose to hydrocarbons (81,82) (see SULFUR REMOVAL AND RECOVERY).

Table 5. American National Standard Specifications for Photographic-Grade Ammonium Thiosulfate Solution

Specifications[a]	Aqueous solution[b]	Crystalline solid[c]
assay as $(NH_4)_2S_2O_3$, %	56.0–60.0	97.0[d]
insoluble matter[e], %	0.2	0.4
alkalinity, as NH_4OH, %	0.3–1.5	0.4
sulfide, as S, ppm	20	10
sulfite, as SO_3, %	0.70	1.4
heavy metals, as Pb, ppm	10	20
iron, as Fe, ppm	2.5	50
residue after ignition, %	0.1	0.2
specific gravity, at 15/15°C	1.310–1.335	

[a]Values are maximum unless range is given.
[b]Ref. 69.
[c]Ref. 70.
[d]Value is minimum.
[e]Including calcium and magnesium, precipitated by ammonium hydroxide.

Other Thiosulfates

Many other metal thiosulfates, eg, magnesium thiosulfate [10124-53-5] and its hexahydrate [13446-30-5], have been prepared on a laboratory scale, but with the exception of the calcium, barium [35112-53-9], and lead compounds, these are of little commercial or technical interest. Although thallous [13453-46-8], silver, lead, and barium thiosulfates are only slightly soluble, other metal thiosulfates are usually soluble in water. The lead and silver salts are anhydrous; the others usually form more than one hydrate. Aqueous solutions are stable at low temperatures and in the absence of air. The chemical properties are those of thiosulfates and the respective cation.

Thiosulfates are generally prepared by treating aqueous solutions of either calcium or barium thiosulfate with the corresponding carbonate or sulfate of the desired metal. The insoluble calcium or barium sulfates or carbonates are filtered and the thiosulfate recovered from the filtrate by vacuum evaporation.

Other method thiosulfates have been prepared by reaction of suspensions of the metal sulfide with sulfur dioxide. However, these thiosulfates are usually contaminated with polythionates (83).

Some metal thiosulfates are inherently unstable because of the reducing properties of the thiosulfate ion. Ions such as Fe^{3+} and Cu^{2+} tend to be reduced to lower oxidation states, whereas mercury or silver, which form sulfides of low solubility, tend to decompose to the sulfides. The stability of other metal thiosulfates improves in the presence of excess thiosulfate by virtue of complex thiosulfate formation.

The most common form of calcium thiosulfate is the hexahydrate [10035-02-6], $CaS_2O_3 \cdot 6H_2O$, which has triclinic crystals and a density of 1.872 g/cm^3 at 16°C (84). Heating, however, does not give the anhydrous salt because of decomposition at 80°C. At lower temperatures, dehydration stops at the monohydrate [15091-91-5]. The solubility of calcium thiosulfates in water is as follows:

temperature, °C	0	10	20	30	40
CaS_2O_3, wt %	25.8	29.4	33.0	36.6	40.3

Aqueous solutions decompose on heating as low as 60°C with formation of sulfur.

Calcium thiosulfate has been prepared from calcium sulfite and sulfur at 30–40°C, or from boiling lime and sulfur in the presence of sulfur dioxide until a colorless solution is obtained. Alternatively, a concentrated solution of sodium thiosulfate is treated with calcium chloride; the crystalline sodium chloride is removed at low temperature. Concentrated solutions of calcium thiosulfate are prepared from ammonium thiosulfate and lime; the liberated ammonium ion is recycled to the ammonium thiosulfate process (85).

Calcium thiosulfate is not produced commercially in the United States. Uses include fungicide formulations (86), a noncorrosive concrete-set accelerator (87), and a catalyst for polyolefin manufacture (88).

Complexes and Organic Thiosulfates

Gold thiosulfate complexes of the form $Na_3[Au(S_2O_3)_2]\cdot2H_2O$ [*19153-98-1*] are prepared by addition of gold trichloride to concentrated sodium thiosulfate solution (89). The gold is completely reduced and some thiosulfate is oxidized to tetrathionate. This complex has been used in the treatment of rheumatoid arthritis.

Other complex thiosulfates have been prepared to study crystal properties, eg, cadmium ammonium thiosulfates (90), $NaAgS_2O_3\cdot H_2O$ [*37954-66-8*] (91), $K_2Mg(S_2O_3)_2\cdot6H_2O$ [*64153-76-0*] (92), and $(NH_4)_9[Ag(S_2O_3)_4]Cl_2$ [*12040-89-0*] (93).

Organic thiosulfate salts are usually prepared by the reaction of alkyl chlorides with sodium thiosulfate:

$$RCl + Na_2S_2O_3 \longrightarrow RSSO_3Na + NaCl$$

Sodium ethyl thiosulfate [*26264-37-9*] is also known as Bunte's salt after the name of its discoverer. Bunte salts may be thought of as esters of thiosulfuric acid (94–96). In essentially all of their chemical reactions, the cleavage is between the divalent and hexavalent sulfur atom. For example, acid hydrolysis produces a thiol and the acid sulfate:

$$RCH_2SSO_3Na + H_2O \longrightarrow RCH_2SH + NaHSO_4$$

Bunte salts have bacterial, insecticidal, and fungicidal properties, and are also used as chelating agents (qv) or surfactants (qv) (97,98). Bunte salts have been tested for preirradiation protection for mammals exposed to lethal radiation doses (99,100) (see RADIOPROTECTIVE AGENTS).

BIBLIOGRAPHY

"Thiosulfuric Acid and Thiosulfates" in *ECT* 1st ed., Vol. 14, pp. 102–115, by R. V. Townend, Allied Chemical Corp.; "Thiosulfates" in *ECT* 2nd ed., Vol. 20, pp. 227–247, by C. A. Warnser, Allied Chemical Corp.; in *ECT* 3rd ed., Vol. 22, pp. 974–989, by J. W. Swaine, Jr., Allied Corp.

1. T. Moeller, *Inorganic Chemistry*, John Wiley & Sons, Inc., New York, 1959, p. 543.
2. H. Siebert, *Z. Anorg. Allg. Chem.* **275**, 225 (1954).
3. U. Agarwala, C. E. Rees, and H. G. Thode, *Can. J. Chem.* **43**, 2802 (1965).
4. Y. Y. Kharitonov, N. A. Knyazeva, and L. V. Goeva, *Opt. Spektrosk.* **24**, 639 (1968).
5. P. G. Taylor and C. A. Beevers, *Acta Crist.* **5**, 341 (1952).
6. L. Manojlovic-Muir, *Acta Crystallogr. Sect. B*, 135 (1975).
7. H. C. Mel, Z. Z. Hugus, Jr., and W. M. Latimer, *J. Am. Chem. Soc.* **78**, 1822 (1956).
8. *Gmelins Handbuch der Anorganischen Chemie*, 8th ed., Schwefel, Part B, No. 2, Verlag Chemie, G.m.b.H., Weinheim/Bergstrasse, 1960, p. 868.
9. I. P. Chernobaev and M. A. Loshkarev, *Ukr. Khim. Zh.* **33**, 253 (1967).
10. F. M. Page, *J. Chem. Soc.*, 1719 (1953).
11. J. Piccard and E. Thomas, *Helv. Chim. Acta* **6**, 1032 (1923).
12. M. Schmidt, *Z. Anorg. Allg. Chem.* **289**, 141 (1957).

13. M. Schmidt and G. Talsky, *Angew. Chem.* **70**, 312 (1958).
14. Ref. 8, pp. 875–894.
15. F. Foerster and R. Vogel, *Z. Anorg. Allg. Chem.* **155**, 161 (1926).
16. H. Bassett and R. G. Durrant, *J. Chem. Soc.*, 1401 (1927).
17. F. Yokosuka, T. Kurai, A. Okuwaki, and T. Okabe, *Nippon Kagaku Kaishi*, 1901 (1975).
18. G. M. Vol'dman, *Zh. Fiz. Khim.* **44**, 2066 (1970).
19. P. Ray and J. Das-Gupta, *J. Indian Chem. Soc.* **5**, 483 (1928).
20. A. N. Freedman and B. P. Straughan, *Spectrochim. Acta, Part A* **27**, 1455 (1971).
21. Ref. 8, pp. 919–947.
22. *Metals Section, Corrosion Data Survey*, 5th ed., National Association of Corrosion Engineers, Houston, Texas, 1974, pp. 172–173.
23. Ref. 22, pp. 18–19.
24. S. W. Young and W. E. Burke, *J. Am. Chem. Soc.* **26**, 1413 (1904); technical data, Allied Chemical Corp., Syracuse, N.Y., 1978.
25. S. W. Young and W. E. Burke, *J. Am. Chem. Soc.* **28**, 315 (1906).
26. *Gmelins Handbuch der Anorganischen Chemie*, 8th ed., Verlag Chemie, GmbH, Weinheim/Bergstrasse, 1966, pp. 1162–1174.
27. *Gmelins Handbuch der Anorganischen Chemie*, 8th ed., Verlag Chemie, GmbH, Weinheim/Bergstrasse, Berlin, 1965, p. 614.
28. Ref. 26, p. 1166.
29. Ref. 26, p. 1168.
30. E. W. Washburn, ed., *International Critical Tables*, Vol. 5, McGraw-Hill Book Co., New York, 1929, p. 201.
31. Ref. 30, p. 131.
32. R. C. Weast, ed., *Handbook of Chemistry and Physics*, 63rd ed., CRC Press, Boca Raton, Fla., 1982–1983, p. D-268.
33. T. Golgotiu and V. Rotaru, *Bul. Inst. Politeh. Iasi*, 37 (1972).
34. F. A. Lowenheim and M. K. Moran, *Faith, Keyes, and Clark's Industrial Chemicals*, 4th ed., John Wiley & Sons, Inc., New York, 1975, pp. 769–773.
35. USSR Pat. 779,297 (Nov. 15, 1980), T. G. Akhmetov, A. A. Murav'ev, E. V. Polyakov, and I. N. Smirnov.
36. U. S. Pat. 2,763,531 (Sept. 18, 1956), G. I. P. Levenson (to Eastman Kodak Co.).
37. *American Chemical Society Specifications—Reagent Chemicals*, 6th ed., American Chemical Society, Washington, D.C., 1981, p. 551.
38. *United States Pharmacopeia XX* (*USP XX-NF XV*), United States Pharmacopeial Convention, Inc., Rockville, Md., 1980, p. 738.
39. *ANSI PH 4.250-1980*, American National Standards Institute, New York, Nov. 16, 1979.
40. G. Plume, *Latv. Lopkopibas Vet. Zinat. Petnieciska Inst. Raksti* **24**, 73 (1970).
41. *Fed. Regist.* **43**, 22937 (May 30, 1978).
42. N. I. Sax, *Dangerous Properties of Industrial Materials*, 5th ed., Van Nostrand Reinhold Company, New York, 1979, pp. 372, 991, 1030.
43. U.S. Pat. 5,192,571 (Mar. 9, 1993), Ehud Levy.
44. Ger. Offen. 2,700,549 (July 13, 1978), H. Hoelter, H. Gresch, and H. Igelbuescher.
45. W. I. Nissen and R. S. Madenburg, *Inf. Circ. U.S. Bur. Mines*, IC 8806, 1979, p. 16.
46. Can. Pat. 926,554 (May 22, 1973), S. Young and K. Mueller (to Pioneer Marketing Associates).
47. U.S. Pat. 4,280,553 (July 28, 1981), S. L. Bean, J. W. Swaine, Jr., and P. R. Crawford (to Allied Chemical Corp.).
48. U.S. Pat. 4,292,311 (Sept. 30, 1981), S. J. Sarnoff.
49. USSR Pat. 814,930 (Mar. 23, 1981), N. Buchvarov, I. Botev, and N. Teneva.

50. Australian Pat. 461,061 (June 27, 1974), R. Lowe.

51. Jpn. Kokai 76 03,387 (Jan. 12, 1976), T. Yamada, H. Yamamoto, A. Kita, and T. Iwata (to Osaka Soda Co.).

52. Fr. Demande 2,222,327 (Oct. 18, 1974), A. Binick (to Société Française Lanco).

53. U.S. Pat. 3,753,939 (Aug. 21, 1973), H. S. Von Euler-Chelpin (to Kemanord AG).

54. Fr. Pat. 1,555,358 (Jan. 24, 1969), L. M. Niebylski (to Ethyl Corporation).

55. Ger. Offen. 2,917,192 (Nov. 6, 1980), G. Arrhenius (to Kay Laboratories).

56. U.S. Pat. 4,295,985 (Oct. 20, 1981), H. G. Petrow and M. L. Weissman.

57. U.S. Pat. 4,028,297 (June 7, 1977), J. L. Webb (to General Electric Co.).

58. U.S. Pat. 4,369,061 (Jan. 18, 1983), Bernard J. Kerley, Jr.

59. Y. Elerman, A. Aydin Uraz, N. Armagan, and Y. Aka, *J. Appl. Crystallogr.* **11**(6), 709 (1978).

60. L. Erdey, S. Gal, and G. Liptay, *Talanta* **11**, 913 (1964).

61. U.S. Pat. 2,219,258 (Oct. 22, 1940), W. H. Hill (to American Cyanamid).

62. U.S. Pat. 2,412,607 (Dec. 17, 1946), H. V. Farr and J. R. Ruhoff (to Mallinckrodt Chemical Co.).

63. U.S. Pat. 2,586,459 (Feb. 19, 1952), H. V. Farr and J. R. Ruhoff (to Mallinckrodt Chemical Co.).

64. Ger. Offen. 2,635,649 (Feb. 9, 1978), K. H. Henke and G. Weiner (to Hoechst AG); Brit. Pat, 1,175,069 (Dec. 23, 1969), (to Farbwerke Hoechst AG).

65. U.S. Pat. 3,473,891 (Oct. 21, 1969), E. Mack (to Th. Goldschmidt AG).

66. U.S. Pat. 3,890,428 (June 17, 1975), M. D. Jayawant (to E. I. du Pont de Nemours & Co., Inc.).

67. U.S. Pat. 3,524,724 (Aug. 18, 1970), R. L. Every and P. F. Cox (to Continental Oil Co.).

68. U.S. Pat. 4,478,807 (Oct. 23, 1984), Clifford J. Ott.

69. *ANSI PH 4.252-1980*, American National Standards Institute, New York, Aug. 15, 1980.

70. *ANSI PH 4.253-1960*, American National Standards Institute, New York, Dec. 21, 1960.

71. Brazil Pedido PI 79 00,040 (Aug. 14, 1979), J. E. Sansing and T. M. Parham (to Allied Chemical Corp.).

72. U.S. Pat. 4,191,550 (Mar. 4, 1980), E. F. Hawkins and T. M. Parham (to Allied Chemical Corp.).

73. U.S. Pat. 4,210,437 (July 1, 1980), L. E. Ott and R. J. Windgassen (to Standard Oil Co., Indiana).

74. U.S. Pat. 3,730,703 (May 1, 1973), R. H. Daehnert (to Allied Chemical Corp.).

75. U.S. Pat. 3,689,246 (Sept. 5, 1972), D. C. Young (to Union Oil Co. of California).

76. A. Zey, S. White, and D. Johnson, *Chem. Eng. Prog.* **76**(10), 76 (1980).

77. U.S. Pat. 4,163,776 (Aug. 7, 1979), M. D. Kulik and E. Gorin (to Conoco, Inc.).

78. Fr. Demande 2,299,070 (Aug. 27, 1976), (to Exxon Research and Engineering Co.).

79. Can. Pat. 1,022,728 (Dec. 20, 1977), A. B. Welty, Jr. (to Exxon Research and Engineering Co.).

80. U.S. Pat. 4,201,662 (May 6, 1980), R. L. Horton (to Phillips Petroleum Co.).

81. U.S. Pat. 3,864,096 (Feb. 4, 1975), P. Urban (to Universal Oil Products Co.).

82. U.S. Pat. 3,864,097 (Feb. 4, 1975), P. Urban (to Universal Oil Products Co.).

83. W. E. Henderson and H. B. Weiser, *J. Am. Chem. Soc.* **35**, 239 (1913).

84. *Gmelins Handbuch der Anorganischen Chemie*, 8th ed., Part B, No. 3, Verlag Chemie, GmbH, Weinheim/Bergstrasse, 1961, pp. 785–793.

85. U.S. Pat. 4,105,754 (Aug. 8, 1978), J. W. Swaine, Jr., W. W. Low, and S. L. Bean (to Allied Chemical Corp.).

86. Jpn. Kokai 77 38,018 (Mar. 24, 1977), Y. Miyahara (to Sankei Chemicals Co.).

87. Jpn. Pat. 73 06,039 (Feb. 22, 1973), H. Tanaka and S. Nakagawa.
88. Ger. Offen. 2,364,170 (July 25, 1974), M. Galliverti, M. B. Ghirga, and B. Calcagno.
89. M. Windholtz, ed., *The Merck Index*, 9th ed., Merck and Co., Rahway, N.J., 1976, pp. 586–587.
90. Z. Gabelica, *Bull. Cl. Sci. Acad. R. Belg.* **59**, 1164 (1973).
91. L. Cavalca, A. Mangia, C. Palmieri, and G. Pelizzi, *Inorg. Chim. Acta* **4**, 299 (1970).
92. Z. Gabelica, *Rev. Chim. Miner.* **14**, 269 (1977).
93. F. Bigoli, A. Tiripicchio, and M. Tiripicchio Camellini, *Acta Crystallogr. Sect. B* **28**, 2079 (1972).
94. B. Milligan and J. M. Swan, *Rev. Pure Appl. Chem.* **12**, 72 (1962).
95. H. Distler, *Angew. Chem. Int. Ed. Engl.* **6**, 554 (1967).
96. D. L. Klayman and R. Shine, *Quarterly Reports on Sulfur Chemistry*, Vol. 3, No. 3, Intra-Science Research Foundation, Santa Monica, Calif., 1968, pp. 191–309.
97. U.S. Pat. 3,364,247 (Jan. 16, 1968), M. H. Gollis (to Monsanto Co.).
98. R. D. Westland, E. R. Karger, B. Green, and J. R. Dice, *J. Med. Chem.* **11**, 84 (1967).
99. J. Tulecki and co-workers, *Ann. Pharm. (Poznan)* **13**, 139 (1978).
100. R. D. Westland and co-workers, *J. Med. Chem.* **11**, 1190 (1968).

General References

References 8, 26, 27, and 84 are also general references.
Gmelins Handbuch der Anorganischen Chemie, 8th ed., Natrium, Verlag Chemie, GmbH, Weinheim/Bergstrasse, 1964, pp. 247–252.

S. L. Bean
General Chemical Corporation

THORIUM AND THORIUM COMPOUNDS

Thorium [7440-29-1], a naturally occurring radioactive element, atomic number 90, atomic mass 232.0381, is the second element of the actinide (5*f*) series (see ACTINIDES AND TRANSACTINIDES; RADIOISOTOPES). Discovered in 1828 in a Norwegian mineral, thorium was first isolated in its oxide form. For the light actinide elements in the first half of the 5*f* series, there is a small energy difference between $5f^n7s^2$ and $5f^{n-1}6d7s^2$ electronic configurations. Atomic spectra of neutral thorium atoms in the gas phase indicate that the Th 6*d* levels are lower in energy than the 5*f* levels, so that neutral thorium atoms have a diamagnetic electronic ground state of $[Rn]6d^27s^2$. For actinide elements following Th, the 5*f* shell appears to be lower in energy than the 6*d* shell. Thorium is, therefore, rather unique, generally having an oxidation state of +4, and thus no 5*f* valence electrons in any of its compounds.

The chemistry of thorium, dominated by the +4 oxidation state, is similar to that of the lanthanides (qv) and the Group 4 (IVB) elements, Ti, Zr, and Hf.

Compounds containing thorium in lower oxidation states are known, but exceedingly rare in contrast to its lanthanide homologue, cerium. Reduction of Th(IV) to lower oxidation states is very difficult, especially in solution. Standard potentials of Th(IV) in aqueous solutions have been estimated based on spectroscopic measurements, or determined indirectly from thermodynamic data. The electrochemistry of actinide elements has been reviewed (1), and the data span a wide range. The values most often quoted are those calculated using a modified ionic model to obtain the Gibbs free energy of formation for the Th(IV) aquo ion (2). The predicted standard electrode potentials from this approach are as follow:

$$Th^{4+} \mid Th^{3+} \qquad -3.0 \text{ V}$$
$$Th^{3+} \mid Th^{2+} \qquad -2.8 \text{ V}$$
$$Th^{2+} \mid Th^{0} \qquad -0.77 \text{ V}$$
$$Th^{3+} \mid Th^{0} \qquad -1.44 \text{ V}$$
$$Th^{4+} \mid Th^{0} \qquad -1.82 \text{ V}$$

A full discussion of thorium electrochemistry is available (3). Thorium is generally more acidic than the lanthanides but less acidic than other light actinides, such as U, Np, and Pu, as expected from the larger Th^{4+} ionic radius (108 pm).

Twenty-five isotopes of thorium have been observed having masses ranging from 212 to 236. Radioactive half-lives range from 0.1 μs for ^{218}Th to 1.405 $\times$ 10^{10} yr for ^{232}Th. The latter is the predominant isotope in nature. The light thorium isotopes having masses of 212 to 232 decay by emission of α-particles; the heavy thorium isotopes, ie, 231 and 233–236, decay by β-emission. Thorium is a member of all three naturally occurring decay series of the long-lived isotopes, ^{232}Th, ^{235}U, and ^{238}U, as well as the synthetic Np series. ^{232}Th is the progenitor of the $4n$ decay series (Fig. 1), which includes ^{228}Th as a transient. ^{234}Th and ^{230}Th ($4n + 2$), ^{231}Th and ^{227}Th ($4n + 3$), and ^{229}Th ($4n + 1$) are daughters in other decay series. The most commonly used isotopes after ^{232}Th include ^{228}Th(1.91 yr, α), ^{230}Th(7.54 $\times$ 10^4 yr, α), and ^{234}Th (24.5 d, β).

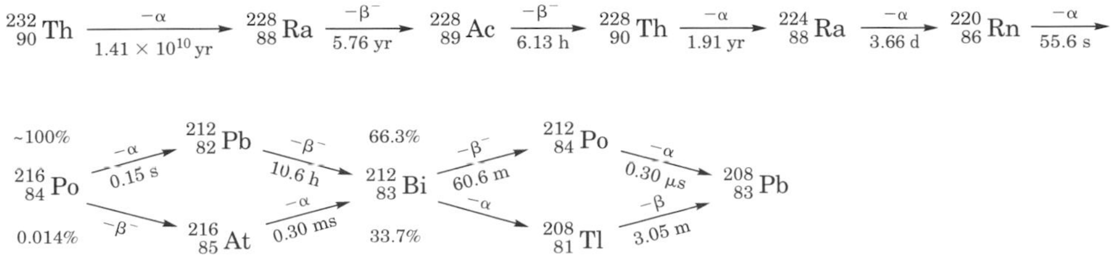

Fig. 1. The $4n$ decay series.

Occurrence

Thorium has a wide distribution in nature and is present as a tetravalent oxide in a large number of minerals in minor or trace amounts. Thorium is significantly more common in nature than uranium, having an average content in the earth's crust of approximately 10 ppm. By comparison, Pb is approximately 16 ppm. Thorium has a seawater concentration of $<0.5 \times 10^{-3}$ g/m^3. Thorium refined from ores free of uranium would be almost monoisotopic ^{232}Th, ie, the ^{228}Th from its own decay chain would be one part in 10^{10}. The presence of uranium in the ore introduces infinitesimal amounts of short-lived ^{231}Th and ^{227}Th from the decay of ^{235}U, and ^{234}Th from ^{238}U decay. The isotopes ^{232}Th and ^{228}Th occur in thorianite, TiO$_2$, and thorite, ThSiO$_4$; ^{234}Th and ^{230}Th are present in naturally occurring uranium; ^{231}Th and ^{227}Th occur in uranium minerals as members of the ^{235}U decay chain. The remaining isotopes are formed upon neutron bombardment of those isotopes discussed, or by charged particle bombardment of various targets.

Thorium isotope concentrations and ratios, as well as parent and daughter isotope concentrations, are used to date and study the formation and metamorphosis of rocks and sediments. For example, ^{230}Th/^{234}U has been used to date coral reef terraces (4). ^{238}U/^{230}Th disequilibria and ^{230}Th/^{232}Th ratios are used to determine the source and history of basalts (5). In addition to dating and petrogenetic studies, thorium isotopes are used as magma source tracers (6).

There are only a few minerals where thorium occurs as a significant constituent. The commercially important ore is the golden-brown, lanthanide phosphate, monazite [1306-41-8], LnPO$_4$, where Ln = Ce, La, or Nd, in which thorium is generally present in a 1–15% elemental composition (7,8). Monazite is widely distributed around the world. Some deposits are quite large. Beach sands from Australia and India contain monazite from which concentrates of lanthanides, titanium, zirconium, and thorium are produced (7). The Travancore deposits in India are the most famous, and have been perhaps one of the most significant sources of commercial thorium. Additional information on the occurrence of thorium in minerals can be found in the literature (7). A review of the mineralogy of thorium is also available (9).

Recovery from Ores

There are a number of minerals in which thorium is found. Thus a number of basic process flow sheets exist for the recovery of thorium from ores (10). The extraction of monazite from sands is accomplished via the digestion of sand using hot base, which converts the oxide to the hydroxide form. The hydroxide is then dissolved in hydrochloric acid and the pH adjusted to between 5 and 6, affording the separation of thorium from the less acidic lanthanides. Thorium hydroxide is dissolved in nitric acid and extracted using methyl isobutyl ketone or tributyl phosphate in kerosene to yield Th(NO$_3$)$_4$, which can then be removed from the organic solvent (11). An exhaustive compilation of flow sheets covering caustic soda, sulfuric acid, ammonium fluoride, chlorination, and other processes can be found in the literature (10).

Uses

Thorium is mainly used in the production of commercial lantern mantles, refractory materials (see REFRACTORIES), electronic components, alloys utilized for components of jet engines, and as a catalyst in the chemical industry. The isotope ^{232}Th is used in nuclear reactor fuels. The oxide finds application in electrodes for arc welding, in the manufacturing of ceramics, and as a minor component in a catalyst for the production of liquid fuel (12). Over a ton of thorium is produced annually, approximately half of which is devoted to the production of gas mantles. By 1891, the thoria gas mantle had been perfected to improve the low luminosity of the coal-gas flames then used for lighting. Fabric of the required shape was soaked in an aqueous metal nitrate solution and the fiber burned off to convert the nitrates into oxides. A mixture of 99% ThO_2 and 1% CeO_2 was used and has not since been bettered. The CeO_2 catalyzes the combustion of the gas. Apparently, because of the poor thermal conductivity of the ThO_2, particles of CeO_2 become hotter, making the flame brighter than would otherwise be possible. The commercial success of the gas mantle was immense and produced a worldwide search for thorium. In the process of mining thorium, the lanthanides, found to be more plentiful than had been previously thought, were recovered in large quantities (13).

Thorium is also used industrially in the catalytic production of hydrocarbon mixtures for use as liquid motor fuel (14). Thorium is combined in the catalyst at a concentration level of 0.1–25 wt % based on the catalyst weight. For direct conversion of synthetic gas to liquid fuels, thorium acts as a promoter in an Al_2O_3-supported catalyst to increase olefin and liquid hydrocarbon production (15). A Ni–ThO_2/Al_2O_3 catalyst was developed for the oxidative cracking of hydrocarbons by steam (16). Loading of the alumina carrier with 16.2% thorium resulted in an increased resistance of the catalyst to inactivation by coking. Thorium supported on dehydroxylated γ-alumina is an outstanding heterogeneous catalyst for arene hydrogenation that rivals the most active platinum metal catalysts in activity (17,18).

In the area of superconductivity, tetravalent thorium is used to replace trivalent lanthanides in n-type doped superconductors, $R_{2-x}Th_xCuO_{4-\delta}$, where R = Pr, Nd, or Sm, producing a higher T_c superconductor. Thorium also forms alloys with a wide variety of metals. In particular, thorium is used in magnesium alloys to extend the temperature range over which structural properties are exhibited that are useful for the aircraft industry. More detailed discussions on thorium alloys are available (8,19).

By far the most important thorium compound is ThO_2 owing to its high chemical and thermal stability. Moreover, this oxide has a high melting point of nearly 3000°C, the highest for any metal oxide. The inherent radioactivity of ^{232}Th (the most important isotope) and the formation of radioactive daughter products are important limiting factors in the uses of thorium. The radiological protection and necessary permission required by most countries to handle large quantities of thorium result in a steadily decreasing usage of thorium on an industrial scale. ^{232}Th has a high neutron reaction yield, producing the fissile uranium isotope, ^{233}U. Thorium is therefore an excellent breeding material, especially for high temperature reactors. The use of ThO_2 and mixed oxide

(Th,U)O_2 fuels in a nuclear reactor is well understood (12), and a breeder cycle involving ^{233}U and ^{232}Th as its fissile and fertile components may be summarized as follows:

$$^{232}_{90}\text{Th} \xrightarrow{+n'_0} {}^{233}_{90}\text{Th} \xrightarrow[t_{1/2}\ =\ 23.3\ \text{min}]{-\beta^-} {}^{233}_{91}\text{Pa} \xrightarrow[t_{1/2}\ =\ 27.4\ \text{d}]{-\beta^-} {}^{233}_{92}\text{U}$$

This reaction offers the advantage of a superior neutron yield of ^{233}U in a thermal reactor system. The ability to breed fissile ^{233}U from naturally occurring ^{232}Th allows the world's thorium reserves to be added to its uranium reserves as a potential source of fission power. However, the ^{232}Th/^{233}U cycle is unlikely to be developed in the 1990s owing both to the more advanced state of the ^{238}U/^{239}Pu cycle and to the availability of uranium. Thorium is also used in the production of the α-emitting radiotherapeutic agent, ^{213}Bi, via the production of ^{229}Th and subsequent decay through ^{225}Ac (20).

Thorium Metal

Properties.   Pure thorium metal is a dense, bright silvery metal having a very high melting point. The metal exists in two allotropic modifications. Thorium is a reactive, soft, and ductile metal which tarnishes slowly on exposure to air (12). Having poor mechanical properties, the metal has no direct structural applications. A survey of the physical properties of thorium is summarized in Table 1. Thorium metal is diamagnetic at room temperature, but becomes superconducting below 1.3–1.4 K.

Thorium metal alloys readily with a large number of metals, including Fe, Co, Ni, Cu, Au, Ag, B, Pt, Mo, W, Ta, Zn, Bi, Pb, Hg, Na, Be, Mg, Si, Se, and Al. Like many electropositive metals, finely divided thorium metal is pyrophoric in air, and burns to give the oxide. Massive metal, chips, and turnings are stable under ambient conditions, although the surface darkens with time as the oxide forms. Hydrogen, nitrogen, halogens, and sulfur all undergo energetic reactions with thorium at a variety of temperatures.

Most mineral acids react vigorously with thorium metal. Aqueous HCl attacks thorium metal, but dissolution is not complete. From 12 to 25% of the metal typically remains undissolved. A small amount of fluoride or fluorosilicate is often used to assist in complete dissolution. Nitric acid passivates the surface of thorium metal, but small amounts of fluoride or fluorosilicate assists in complete dissolution. Dilute HF, HNO_3, or H_2SO_4, or concentrated $HClO_4$ and H_3PO_4, slowly dissolve thorium metal, accompanied by constant hydrogen gas evolution. Thorium metal does not dissolve in alkaline hydroxide solutions.

Preparation.   Pure thorium metal is very difficult to prepare owing to high reactivity with H_2, O_2, N_2, and C at the high temperatures necessary for production. Thorium metal can be produced by a variety of reduction techniques, all of which have unique difficulties. The molten salt reduction of thorium chloride [10026-08-1], $ThCl_4$, with Mg or NaHg, or of thorium oxide [1314-20-1], ThO_2, with Ca, and the thermal decomposition of thorium iodide [7790-49-0], ThI_4, are examples (8,12). By far, the most advantageous and common method of

Table 1. Physical Properties of Thorium Metals[a]

crystal structure	
fcc to 1360°C, a_0, pm	508.42
density, g/cm^3	11.724
atoms per unit cell, Z	4
bcc, 1360–1750°C, a_0, pm	411
density, g/cm^3	11.10
atoms per unit cell, Z	2
melting point, °C	1750
boiling point, °C	~3800
enthalpy of vaporization, 25°C, kJ/mol[b]	598
enthalpy of fusion, kJ/mol[b]	14
vapor pressure, 1757–1956 K	$\log(p/\text{atm}) = -28{,}780\,(T/\text{K})^{-1} + 5.991$
thermal conductivity, 25°C, W/(cm·K)	0.6
work function, eV	3.49
Hall coefficient, 24°C, cm^3/C	-11.2×10^{-5}
elastic constants	
Young's modulus, kPa[c]	7.2×10^7
shear modulus, kPa[c]	2.8×10^7
Poisson's ratio	0.265
compressibility, cm^2/dyn	17.3×10^{-13}

[a]Refs. 21 and 22.
[b]To convert kJ to kcal, divide by 4.184.
[c]To convert kPa to psi, multiply by 0.145.

preparation is the molten salt reduction of thorium fluoride [*13709-56-6*], ThF$_4$, in a blend of Ca and ZnCl$_2$ in a dolomite-lined reactor at 660°C (8). The highly exothermic reaction produces a massive alloy of Th and Zn that settles to the bottom of the slag (CaF$_2$–CaCl$_2$–ZnF$_2$). The crude alloy contains 4.1–7.0% Zn. Yields are up to 95%. Distillation of the Zn finally leads to the production of a high purity thorium sponge.

Thorium Compounds

Oxo Ion Salts. Salts of oxo ions, eg, nitrate, sulfate, perchlorate, hydroxide, iodate, phosphate, and oxalate, are readily obtained from aqueous solution. Thorium nitrate is readily formed by dissolution of thorium hydroxide in nitric acid from which, depending on the pH of solution, crystalline Th(NO$_3$)$_4$·5H$_2$O [*33088-17-4*] or Th(NO$_3$)$_4$·4H$_2$O [*33088-16-3*] can be obtained (23). Thorium nitrate is very soluble in water and in a host of oxygen-containing organic solvents, including alcohols, ethers, esters, and ketones. Hydrated thorium sulfate, Th(SO$_4$)$_2$·nH$_2$O, where $n = 9$, 8, 6, or 4, is easily crystallized from thorium and sulfuric acid and can be readily dehydrated by heating to 350–400°C. The crystal structure of the octahydrate has been determined to be a bicapped square antiprism (24). Thorium perchlorate [*16045-17-3*] forms upon dissolution of thorium hydroxide in perchloric acid and crystallizes as Th(ClO$_4$)$_4$·4H$_2$O.

Coordination Complexes. The coordination and organometallic chemistry of thorium is dominated by the extremely stable tetravalent ion. Except in a few cases where large and sterically demanding ligands are used, lower thorium oxidation states are generally unstable. An example is the isolation of a molecular Th(III) complex [107040-62-0], Th[η-C$_5$H$_3$(Si(CH$_3$)$_3$)$_2$]$_3$ (25). Reports (26) on the synthesis of soluble Th(II) complexes, such as ThI$_2$(NCCH$_3$)$_2$ [85613-74-7], have become suspect (27).

The chemistry of Th(IV) has expanded greatly since the mid-1980s (14,28,29). Being a hard metal ion, Th(IV) has the greatest affinity for hard donors such as N, O, and light halides such as F$^-$ and Cl$^-$. Coordination complexes that are common for the d-block elements have been studied for thorium. These complexes exhibit coordination numbers ranging from 4 to 11.

Nitrogen Donors. Tetravalent thorium is a relatively strong Lewis acid, and, as such, forms compounds with a wide variety of ligands containing nitrogen donors, ranging from neutral mono-, bi-, and polydentate ligands to anionic ligands such as amides and thiocyanates. The most common coordination number is 8; however, some complexes have been observed to have coordination numbers as low as 4.

Adducts of ThCl$_4$ with ammonia or primary, secondary, and tertiary amines having coordination numbers ranging from 6 to 8 have been characterized. Examples include the six-coordinate ThCl$_4$(N(C$_2$H$_5$)$_3$)$_2$ (**1**), where L = N(C$_2$H$_5$)$_3$, seven-coordinate ThCl$_4$(NH$_3$)$_3$ (**2**) (30), and eight-coordinate ThBr$_4$(C$_2$H$_5$NH$_2$)$_4$ and ThCl$_4$(C$_6$H$_5$(CH$_3$)NH)$_4$. Cationic metal hydrates coordinated with primary, secondary, and tertiary amines have also been isolated with acetylacetonate, nitrate, or oxalate as counterions. Another common class of ligands for thorium halides are nitriles, RCN (R = alkyl, aryl). These complexes are predominately eight-coordinate species, eg, ThCl$_4$(CH$_3$CN)$_4$ [17499-62-6] (**3**), where L = CH$_3$CN. However, six-coordinate isocyanide complexes are also known.

          (**1**)   (**2**)   (**3**)

Halides, nitrates, and perchlorates of Th(IV) also bind *N*-heterocyclic ligands, ie, pyridine (py), substituted pyridines, quinoline, and isoquinoline. In the case of these heterocyclic ligands, the typical coordination numbers are 6 and 8. An interesting case is ThI$_4$(py)$_6$. Instead of the 10-coordinate complex, as indicated by the formula unit, the compound is actually an eight-coordinate salt, [ThI$_2$(py)$_6$]I$_2$. In some other salts of the general formula [ThL$_x$](ClO$_4$)$_4$, where x = 6, L = 2-H$_2$N-2,6-(CH$_3$)$_2$-pyridine, or x = 8, L = py, the thorium coordination number shows an obvious dependence on the steric bulk of the ligands.

A broad spectrum of multidentate ligand–Th(IV) complexes have been isolated and characterized. The simplest bidentate ligands utilized are diaminoalkanes, $H_2N(CH_2)_nNH_2$, where $n = 2$–4, eg, $ThBr_4(H_2N(CH_2)_xNH_2)_y \cdot nH_2O$, where $y = 2$, $x = 3$ or 4, or $y = 4$, $x = 3$. Diaminoarenes, diaminobenzene, benzidine, o-tolidine, and o-dianisidine have also been found to coordinate to Th(IV) halides and nitrates. The chlorides have been isolated with all of the aforementioned ligands in the form of $ThCl_4L_2$, and the nitrates have been identified as complex salts, eg, $[Th(NO_3)_2(1,2\text{-diaminobenzene})_2](NO_3)_2$.

Classic N-heterocyclic ligands, eg, bipyridyl (bipy), terpyridyl, imidazole, pyrazine, phenanthroline, piperazine (including alkyl- and aryl-substituted derivatives), and polypyrazol-1-yl-borates (bis, tris, and tetra), have all been found to coordinate Th(IV) chlorides, perchlorates, and nitrates. The tripodal hydrotris(pyrazolyl)borates, $HBPz_3$, have been used to stabilize organometallic complexes (31). Bis-porphyrin Th(IV) "sandwich" complexes have been synthesized. These are formally eight-coordinate. Because thorium is electrochemically inactive, these metal compounds serve as structural models for photosynthetic reaction centers where the particular emphasis is on the porphyrin-based electrochemistry (32).

Thorium compounds of anionic nitrogen-donating species such as $[Th(NR_2)_4]_x$, where R = alkyl or silyl, are well-known. The nuclearity is highly dependent on the steric requirements of R. Amides are extremely reactive, readily undergoing protonation to form amines or insertion reactions with CO_2, COS, CS_2, and CSe_2 to form carbamates. Tetravalent thorium thiocyanates have been isolated as hydrated species, eg, $Th(NCS)_4(H_2O)_4$ [17837-16-0] or as complex salts, eg, $M_4[Th(NCS)_8] \cdot xH_2O$, where M = NH_4, Rb, or Cs.

A tripodal ligand that has more recently gained attention is $N(CH_2CH_2NSi(CH_3)_2)_3$. Thorium complexes with this ligand are dramatically more stable than those of $N(Si(CH_3)_3)_2$ owing to the combination of the chelate effect and the necessarily facial configuration of the ligand. The dimeric chloro species, $[Th(N(CH_2CH_2NSi(CH_3)_2)_3)Cl]_2$, has been prepared. This complex can then serve as a convenient starting material for monomeric cyclopentadienyl or solvated tetrahydroborate complexes and dimeric alkoxide or tetrahydroborate complexes (33). Another amido ligand that has the possibility of a tripodal arrangement is $N(CH_2CH_2P(i\text{-}C_3H_7)_2)_2$. This ligand forms $\{ThCl_2[N(CH_2CH_2P(i\text{-}C_3H_7)_2)_2]_2\}$, where only one of the phosphines from each ligand is coordinated to the thorium (34).

Phosphorus Donors. Phosphine coordination complexes of thorium are rare because the hard Th(IV) cation favors harder ligand donor types. The only stable thorium–phosphine coordination complexes isolated as of the mid-1990s contain the chelating ligand, 1,2-(bis-dimethylphosphino)ethane (DMPE). $ThCl_4(DMPE)$ and $ThI_4(DMPE)$ have been synthesized directly from the tetrahalides at 80°C and low temperature, respectively. Attempts to employ other chelating phosphines, eg, 1,2-(bis-diphenylphosphino)ethane (DPPE), have failed.

The phosphido complex, $Th(PPP)_4$ [145329-04-0], where PPP = $P(CH_2CH_2P(CH_3)_2)_2$, has been prepared and fully characterized (35) and represents the first actinide complex containing exclusively metal–phosphorus bonds. The x-ray structural analysis indicated 3-3-electron donor phosphides and 1-1-electron phosphide, suggesting that the complex is formally 22-electron. Similar to the amido system, this phosphido compound is also reactive toward insertion reactions, especially with CO, which undergoes a double insertion (35,36).

Oxygen Donors. A variety of *O*-donors have been used to complex thorium. The majority of the complexes have coordination numbers from 6 to 12, depending mostly on the steric bulk of the ancillary ligands. Owing to the prevalence of *O*-donating ligands in natural systems, ie, aquo, hydroxo, carbonate, phosphates, carboxylates, and catecholates, an understanding of the complexation of thorium and other radioactive nuclides is crucial for environmental and bioinorganic chemistry, as well as for waste processing and storage.

Oxides. Owing to the importance as nuclear fuel material, actinide oxides have been intensively investigated. These are very complicated compounds because of the formation of nonstoichiometric or polymorphic materials. Actinide oxides are very heat-resistant and ThO_2 is the highest (3390°C) melting of any metal oxide. Thorium dioxide has the cubic fluorite structure where $a_0 = 558.63(6)$ pm and $d = 10.001$ g/cm^3. It can be readily obtained by ignition of thorium hydroxide, oxalate, carbonate, peroxide, nitrate, and other oxyacid salts. Several binary compounds have been synthesized by fusing the respective thorium and alkali or alkaline-earth oxides to form double oxides of the form M_2ThO_3, where M = Na [12058-67-9], K, Rb, or Cs, and $BaThO_3$ [12230-90-9]. Ternary thorium oxides with lanthanides have been reported, eg, cerium, $(ThCe)O_2$ or $(ThCe)O_{2-x}$ ($x < 0.25$); niobium, $Th_{0.25}NbO_3$; tantalum, $ThTa_2O_7$ or $Th_2Ta_2O_9$; molybdenum, $Th(MoO_4)_2$ or $ThMo_2O_8$; germanium, $ThGeO_4$; titanium; and vanadium. Superconducting properties have been observed in the δ-compound $Nd_{2-x}Th_xCuO_4$ at $x = 0.16$. Two thorium tantalum oxides, $Th_4Ta_{18}O_{53}$ and $Th_2Ta_6O_{19}$, both representatives of the Jahnberg structural family, have been synthesized (37,38). The formula unit $(Th_2O_3)_2(Ta_3O_8)_6$ corresponds to an octahedral layer arrangement of the Th−O coordination polyhedra in $Th_4Ta_{18}O_{53}$. Thorium is eight-coordinate, having a trans-bicapped octahedral geometry. The quaternary compound, $K_3NaTh_2O_6$, has also been synthesized (39).

Hydroxides. Thorium(IV) is generally less resistant to hydrolysis than similarly sized lanthanides, and more resistant to hydrolysis than tetravalent ions of other early actinides, eg, U, Np, and Pu. Many of the thorium(IV) hydrolysis studies indicate stepwise hydrolysis to yield monomeric products of formula $Th(OH)_n^{(4-n)+}$, where n is integral between 1 and 4, in addition to a number of polymeric species (40–43). More recent potentiometric titration studies indicate that only two of the monomeric species, $Th(OH)^{3+}$ and thorium hydroxide [13825-36-0], $Th(OH)_4$, are important in dilute ($\leq 10^{-3}$ M Th) solutions (43). However, in a ThO_2 [1314-20-1] solubility study, the best fit to the experimental data required inclusion of the species. $Th(OH)_3^+$ (44). In more concentrated ($\geq 10^{-3}$ M) solutions, polynuclear species have been shown to exist. For example, a more recent model includes the dimers $Th_2(OH)_2^{6+}$ and $Th_2(OH)_4^{4+}$, the tetramers $Th_4(OH)_8^{8+}$ and $Th_4(OH)_{12}^{4+}$, and two hexamers, $Th_6(OH)_{14}^{10+}$ and $Th_6(OH)_{15}^{9+}$ (43).

Carbonates. There has been a great deal of interest in carbonate complexes of thorium owing to their environmental relevance (45). Solution studies for thorium have been reported (44,46–48). For example, the solubility of microcrystalline ThO_2, examined as a function of pH and CO_2 partial pressure (44), gave results consistent with the presence of the mixed hydroxocarbonatothorium complex, $Th(OH)_3(CO_3)^-$ [*154789-49-8*], and pentacarbonatotho-rium(IV) [*12364-90-8*], $Th(CO_3)_5^{6-}$ (**4**). Solids of formula $ThO(CO_3)$ [*49741-19-7*] and $Th(OH)_2(CO_3)\cdot 2H_2O$ [*12538-65-7*] have been reported, but these materi-als are not well characterized. The pentacarbonato salts of thorium(IV) and uranium(IV) are the most well studied of the tetravalent actinide carbonate solids. Salts of the formula $M_6Th(CO_3)_5\cdot nH_2O$, where M = Na, K, or Tl, or M_6 = $[Co(NH_3)_6]_2$, have all been reported (49–51). These hydrated salts con-tain bidentate carbonate ligands and no water molecules are bound directly to the central metal atom. The only single-crystal x-ray diffraction studies avail-able are those for salts of (**4**) (52–54) and the mineral tuliokite [*128706-42-3*], $Na_6BaTh(CO_3)_6\cdot 6H_2O$, which contains the unusual $Th(CO_3)_6^{8-}$ anion (**5**) (55).

(**4**)   (**5**)

Phosphates. Thorium phosphates are of considerable interest because of their potential as radioactive waste forms (56,57) and as xerogel thin films for light waveguides (58,59). Binary and ternary thorium phosphates have been synthesized having varying ratios of nonthorium metal, thorium, and phosphate. Binary compounds having $ThO_2:P_2O_5$ ratios of 1:2, 1:1, 3:2, and 3:1 have been reported (60,61). The 3:2 thorium(IV) phosphate [*15578-50-4*], $Th_3(PO_4)_4$, has been identified in two allotropic modifications having a tran-sition temperature around 1250°C, and its solubility in low ionic strength waters has been determined (62). More recently, two distinct thorium types in $Th_4(PO_4)_4(P_2O_7)$ [*171845-49-1*] have been reported: one is eight-coordinate with oxygen from five phosphate and one diphosphate group around the tho-rium atom (63). Ternary compounds of the general formula $M(I)Th_2(PO_4)_3$ and $M(II)Th(PO_4)_2$, where M(I) = alkali metal, Tl, Ag, or Cu (64,65), and M(II) = Ca, Sr, Cd, or Pb (61,66), have been studied. In the solid-state struc-ture of $NaTh_2(PO_4)_3$, each thorium atom is eight-coordinate, and the local co-ordination environment can be described as $[Th(\eta^2\text{-}PO_4)_2(\eta^1\text{-}PO_4)_4]$ (**6**). For $KTh_2(PO_4)_3$ [*15653-56-2*] (**7**), each Th(IV) ion is nine-coordinate, forming a lo-cal coordination environment described as $[Th(\eta^2\text{-}PO_4)_2(\eta^1\text{-}PO_4)_5]$ and having both bridging and bidentate phosphate groups. In $Na_2Th(PO_4)_3$ [*56467-86-8*], two different thorium atoms are identified as having 8 and 10 neighboring oxy-gen atoms (67).

Limited data on Th(IV) phosphate complexation in aqueous solution are available (68,69). Owing to the low solubility of thorium orthophosphate at neutral pH, complexation studies have been performed in acidic phosphate media and the results suggest the formation of solution complexes of the general formula $Th(H_3PO_4)_m(H_2PO_4)_n^{(4-n)+}$, where $m/n = 1/0$; $0/1$ and $0/2$; or $1/1$. At neutral pH, $H_2PO_4^-$ and HPO_4^{2-} are the main complexing ligands and the complexes $ThO(HPO_4)_3(H_2PO_4)^{5-}$ and $ThO(HPO_4)_3(H_2PO_4)_2^{5-}$ (pH 6–7) (70), as well as $Th(HPO_4)_m^{(4-2n)+}$ ($m = 1$–3; pH 8–9), have been shown to form (71). Solubility measurements of microcrystalline ThO_2 in phosphate solutions indicate that the phosphate may have an insignificant effect on the solubility of thorium in most natural waters (69).

Oxygen-Containing Organics. Neutral and anionic oxygen-containing organic molecules form complexes with thorium. Recent work has focused on alkoxides (72), aryloxides, and carboxylates; however, complexes with alcohols, ethers, esters, ketones, aldehydes, ketoenolates, and carbamates are also well known.

Carboxylates, Oxalates, and Catecholates. Complexes of Th(IV) with mono-, di-, tri-, and polycarboxylates have been extensively studied. Monocarboxylates, $RCOO^-$, have been complexed with Th(IV), eg, $Th(RCOO)_4$, where R = H, CH_3, CCl_3, or 2-$CH_3C_6H_4$; and $M_xTh(HCO_2)_{4+x}$, where $x = 1$, M = K–Cs, or $x = 2$, M = Rb, Cs, or NH_4. The formate complex, R = H, has been isolated both as the anhydrous species and as a trihydrate complex. The structural analysis of the latter revealed a bicapped trigonal prism geometry having eight different bridging formate groups.

The simplest dicarboxylate ligand is oxalate, $C_2O_4^{2-}$. Thorium oxalate complexes have been used to produce high density fuel pellets, which improve nuclear fuel processes (73). The stability of oxalate complexes and the relevance to waste disposal have also been studied (74). Many thorium oxalate complexes are known, ranging from the simple $Th(C_2O_4)_2 \cdot xH_2O$ to complex salts such as $Th(C_2O_4)_n^{2n-4}$, where $n = 4$, 5, or 6 and where the counterions can be alkali cations; complex transition-metal cations, eg, $Co(NH_3)_6^{3+}$ or $Cr(NH_3)_6^{3+}$; and the ammonium ion. Other dicarboxylate thorium complexes are of the form $Th[R(CO_2)_2]_2 \cdot xH_2O$, where R may be methylene units, CHCH, or aryls. The dicarboxylate complex, $Th(PDC)_2 \cdot xH_2O$, where $x = 0$ or 4 and PDC = NC_5H_3–$2,6$-$(CO_2)_2$, also has a

10-coordinate bicapped square antiprism geometry about the thorium atom. The pyridinyl N-atoms occupy the faces, and the O-atoms of the carboxylates occupy the corners.

The most well known of the tri- and polycarboxylates is the tetracarboxylic acid ethylenediaminetetraacetic acid (EDTA), $(HOOCCH_2)_2NCH_2CH_2N(CH_2CO-OH)_2$. Examples of thorium$-$EDTA complexes are $Th(EDTA)\cdot xH_2O$, $[Th(EDTA)-(OH)]^-$, and $[Th(EDTA)F_3]^{3-}$. The structure of $(CN_3H_6)_3[Th(EDTA)F_3]$ is based on a nine-coordinate distorted capped tetragonal antiprism geometry, having three F-atoms, four O-atoms (one from each of the carboxylate groups), and two N-atoms (one in the capping position). Two other polycarboxylate ligands, which complex thorium, are 1,4,7,10-tetraazacyclodocedcane-N,N',N'',N'''-tetraacetic acid (DOTA) and 1,4,7,10,13,16-hexaazacyclooctadecane-$N,N',N'',N''',N'''',N'''''$-hexaacetic acid (HEHA) (75). For both $Th(DOTA)$ and $Th(HEHA)^{2-}$, nmr studies have led to the assignment of an icosahedral geometry.

Catecholate-type ligands complexing with thorium include catechol (1,2-dihydroxybenzene), resorcinol (1,3-dihydroxybenzene), hydroquinone (1,4-dihydroxybenzene), orcinol (2,5-dihydroxytoluene), and phloroglucinol (1,3,5-trihydroxybenzene). For ligands other than catechol, the only complexes that have been isolated are $ThCl_2L$. Studies of the catecholates have led to the isolation of $ThCl_2(C_6H_4O_2)$, $[Th(C_6H_4O_2)_n]^{2n-4}\cdot xH_2O$ ($n = 3$, 4), and $[Th_3(C_6H_4O_2)_7]^{2-}$. A structural determination of $Na_4[Th(C_6H_4O_2)_4]\cdot21H_2O$ revealed a trigonal-faced dodecahedral geometry about the thorium atom where the water molecules produce a hydrogen-bonding network. The use of polycatecholates as sequestering agents for the actinides has also been studied (76).

Alkoxides and Aryloxides. Studies of alkoxide and aryloxide ligands have been extended to thorium, focusing on determining which ligand systems yield crystalline compounds and provide useful starting materials. Oligomerization in thorium alkoxide complexes, as well as many of the solution properties, is highly dependent on the steric requirements of the alkoxide ligands. In the case of the sterically demanding ligand O-2,6-t-$(C_4H_9)_2C_2H_3$, monomeric $Th(O$-2,6-$t(C_4H_9)_2C_6H_3)_4$ (**8**) can be readily isolated (77). As the steric bulk of the alkoxide ligand decreases, dimers such as $Th_2(OCH(-i-C_3H_7)_2)_8$ [140684-41-9] (**9**) (78) and $Th_2(O-t-C_4H_9)_8L$ (**10**) (79), trimers such as $Th_3O(O-t-C_4H_9)_{10}$ [147361-65-7] (**11**) (79), and tetramers such as $Th_4(O-i-C_3H_7)_{16}(py)_2$ [157440-79-4], etc, are observed (80).

(8)   (9)   (10)   (11)

Halides. *Fluorides.* Thorium(IV) fluoride [*13709-59-6*] is widely used as a source of elemental thorium through electrochemical reduction. Anhydrous ThF_4 is insoluble in water and has been isolated a number of ways. One of the first methods involved the dehydration of $ThF_4 \cdot nH_2O$ at high temperatures in atmospheres of either HF or $CO_2/HF/CCl_4$. More recent techniques have employed the reactions of fluorochlorohydrocarbons with ThO_2 [*1314-20-1*], F_2 with thorium carbides and tetrahalides, and HF with thorium hydroxides, oxides, carbonates, and tetrahalides. The isolated hydrates indicate a stability of ThF_4 against hydrolysis at room temperature. This characteristic is not shared by other tetrahalides. Pyrohydrolysis does occur at higher (>700°C) temperatures, producing first the oxofluoro species, $ThOF_2$ [*13597-30-3*], and finally ThO_2. The oxohalo species, $ThOF_2$, has been produced by direct interaction of ThF_4 and O_2. This reactivity has been exploited to remove O_2 from thorium metal.

Complex salts of thorium fluorides have been generated by interaction of ThF_4 with fluoride salts of alkali or other univalent cations under molten salt conditions. The general forms of these complexes are $[ThF_5]^-$ [*15891-02-8*], $[ThF_6]^{2-}$ [*17300-48-0*], and $[ThF_7]^{3-}$ [*56141-64-1*], where typical countercations are Li^+, Na^+, K^+, Cs^+, NH_4^+, and $N_2H_5^+$. Additional information on thorium fluorides can be found in the literature (81).

Chlorides. Anhydrous $ThCl_4$ [*10026-08-1*] has usually been prepared by direct interaction of thorium metal, hydride, or carbide with chlorine. An alternative to this approach is the reaction of anhydrous HCl with the metal or the hydride at elevated temperatures (700–900°C). One of the difficulties of these processes is the production of $ThOCl_2$ [*13637-74-6*] from either residual H_2O or O_2 in the reactants, but the tetrachloride can be purified by sublimation.

Complex ions of Th(IV) have been studied and include $M_2[ThCl_6]$ [*21493-66-3*], where M = Li–Cs, $(CH_3)_4N$, or $(C_2H_5)_4N$. Under more extreme conditions, eg, molten KCl or vapor phase, $ThCl_5^-$ [*51340-85-3*], $ThCl_7^{3-}$ [*51340-84-2*], $ThCl_8^{4-}$ [*53565-25-6*], and $ThCl_9^{5-}$ are known to be important. Additional information on thorium chlorides can be found in the literature (81).

Bromides and Iodides. Anhydrous $ThBr_4$ [*13453-49-1*] and ThI_4 [*7790-49-0*] have been prepared in a similar fashion as the chloride, ie, interaction of thorium metal or the hydride with the elemental halide of choice, or at high (700–900°C) temperatures with HX, where X is Br or I. Both the tetrabromide and the tetraiodide are light-sensitive and more readily hydrolyzed or oxidized than the tetrachloride to $ThOX_2$, where X = Br [*13596-00-4*] or I [*13841-21-9*]. The tetrabromide and tetraiodide are polymeric and thus problematic synthetic starting materials. Organic solvent-soluble tetrahalides $ThX_4(THF)_4$, where X = Br [*140361-04-2*] or I [*140361-05-3*], have been obtained through the interaction of thorium metal turnings with elemental halides in THF at 0°C (27). These complexes exhibit distorted dodecahedral coordination geometry. Dissolution of these complexes in the presence of Lewis bases such as CH_3CN, py, and 1,2-dimethoxyethane (DME) results in substitution of the four THF molecules, producing ThX_4L_y (X = Br, $y = 4$, L = py [*79086-83-2*], CH_3CN [*17499-64-8*]; $y = 2$, L = DME [*140361-07-5*]; X = I, $y = 4$, L = py [*140361-06-4*]) (27).

Comparable with the chloride system, complex ions of the form $M_2[ThX_6]$ (X = Br [*44490-06-4*], M = $(CH_3)_4N$, $(C_2H_5)_4N$; X = I [*44490-18-8*], M = $(C_2H_5)_4$-N, $(CH_3)_3C_6H_5N$) are known where the metal center is octahedral. Additional information on thorium bromides and iodides can be found in the literature (81).

Organometallic Complexes. The organometallic chemistry of thorium (14,28,29) has been widely studied owing to potential utility in homogeneous and heterogeneous catalysis. Activities range from the hydrogenation and polymerization of olefins to the selective activation of alkanes (8). Although there are no examples of thorium carbonyl complexes, hydrocarbyls, allyls, arenes, cyclooctatetraenyl, and a host of cyclopentadienyl-based ligand complexes have been reported.

Cyclopentadienyl and Substituted Cyclopentadienyl Complexes. Thorium complexes containing cyclopentadienyl rings (Cp), $C_5H_5^-$, and its modified analogues, Cp*, $C_5(CH_3)_5^-$; Cp†, $(CH_3)_3SiC_5H_4^-$; Cp‡, $((CH_3)_3Si)_2C_5H_3^-$; and Cp', $CH_3C_5H_4^-$, are among the most common organothorium complexes known. Electron-deficient compounds of the type (η-ring)ThX$_3$, where X is halogen, are not stable and have only been isolated as Lewis base adducts, such as CpThX$_3$L$_n$ (**12**) ($n = 2$, X = Cl or Br, and L = THF, DME, amides, organonitriles; $n = 3$, X = Cl, and L = CH_3CN [91947-79-4]) or Cp*ThX$_3$L$_2$ (X = Cl, Br; L = $(C_6H_5)_3PO$, THF, and organonitriles). Single-ring complexes can be stabilized through the use of sterically demanding ancillary ligands, as in the case of Cp*ThBr$_{3-x}$(OAr)$_x$ ($x = 1,2$), where Ar is an aryl system, and serve as starting materials for the synthesis of thorium alkyl complexes (82). Three-legged piano stool complexes of the form (η-ring)ThR$_3$ can also be stabilized by ligands capable of π-donation to the metal center such as amides, allyl, or benzyl ligands to form complexes of the type CpTh(N(C$_2$H$_5$)$_2$)$_3$ [108187-12-8], Cp*Th(C$_3$H$_5$)$_3$, or Cp*Th(C$_7$H$_7$)$_3$ [82511-73-7]. The benzyl analogue exists in a rapid $\eta^1-\eta^3$ equilibrium in solution, indicating the preference for higher coordination environments.

Whereas Cp$_2$MX$_2$ compounds are ubiquitous in early transition-metal organometallic chemistry, the thorium analogues are rather unstable. Cp$_2$ThX$_2$ compounds have been stabilized against ligand redistribution by adding ancillary ligands such as the chelating phosphine 1,2-bis-(dimethylphosphino)ethane (DMPE) to form Cp$_2$ThX$_2$(DMPE) (**13**) for X = Cl [108089-88-9], CH$_3$ [108089-89-0], or CH$_2$C$_6$H$_5$ [108678-62-3]. Stabilization against redistribution has also been achieved through the use of Cp*, sterically demanding X groups, or ligands capable of both σ- and π-donation to the actinide metal center, thereby saturating the metal coordination sphere. For example, treatment of Cp$_2$Th(N(C$_2$H$_5$)$_2$)$_2$ with bulky acidic ligands gives a variety of stable Cp$_2$ThX$_2$ complexes for X = OC(t-C$_4$H$_9$)$_3$, O-2,6-(CH$_3$)$_2$C$_6$H$_3$, etc. The reaction of Cp$_2^*$ThCl$_2$ [67506-88-1] with dialkyl phosphido lithium reagents yields stable Cp$_2^*$Th(PR$_2$)$_2$ complexes for R = C$_6$H$_5$ [93943-04-5], cyclo-C$_6$H$_{11}$ [98720-32-2], and C$_2$H$_5$ [98720-31-1]. X-ray diffraction analysis suggested that there is no Th$-$P multiple bond character. Complexes of the form, Cp$_2^*$ThR$_2$, where R= Cl, alkoxide, amide, phosphide, acyl, alkyl, thiolate, vinyl, phosphorylide, butadiene, etc, have been synthesized and fully characterized. The syntheses of Cp$_2^*$Th(C$_6$H$_5$)$_2$ [79301-39-6] and Cp*$_2$ThCl(aryl) have been improved, and the first examples of mixed alkyl–aryl complexes, Cp$_2^*$Th(CH$_3$)(aryl), where the aryl is tolyl [156956-31-9] or xylyl [156956-32-0], have been isolated (83). These compounds exhibit a host of interesting properties and reactivities, including C$-$H activation and migratory insertion reactions. A hydride complex, [Cp$_2^*$ThH$_2$]$_2$ [79735-38-9], has also been isolated, which engages in alkene insertion and migration. Double insertion of CO to produce the first metalloxy ketene complex has been observed for a mixed chloro–silyl complex, $Cp_2^*ThCl(Si-t-C_4H_9(C_6H_5)_2)$ [163851-90-9] (84).

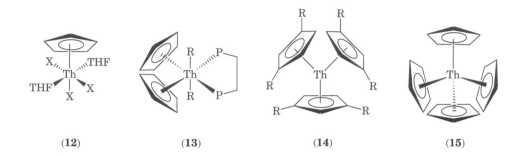

(12)   (13)   (14)   (15)

Similar stability and reactivity have also been observed for bridged-Cp^* systems. The catalytically active $(CH_3)_2Si(C_5(CH_3)_4)_2ThR_2$, where R = Cl [89597-06-8], alkyl, $CH_2C_6H_5$ [89597-10-4], aryls, or H [89597-11-5]. Similar to Group 4 transition-metal Zeigler-Natta catalysts, stable cationic Th(IV) species, eg, $[Cp_2^*ThCH_3]^+$ [108834-69-1], have been isolated with a host of noncoordinating/nonreactive anions. Metallacycle formation has also been shown to stabilize bis-Cp complexes, eg, $Cp_2'Th(CH_2Si(CH_3)_2CH_2)$ (85). These complexes are structurally similar to the Group 4 and 6 transition metallacycle complexes, but show a dramatically reduced reactivity.

In contrast to the instability of simple mono- and bis-Cp complexes, simple systems of the form, $(\eta$-ring$)_3$ThX (ring = Cp, X = F [61288-97-9], Cl [1284-82-8], alkyl, allyl; ring = indenyl or alkyl-substituted indenyl, X = Cl [11133-05-4], Br [57034-55-6], I [66775-24-4]) are well known. More recently, $[(RC_5H_4)_3Th]B(C_6H_5)_4$, where R = $Si(CH_3)_3$ [168024-92-8] or $t-C_4H_9$ [168024-95-1], have been prepared. These compounds react with alkyl lithium reagents to produce stable $(\eta$-ring$)_3$ThH species (86). A tris-indenyl Th(IV) monohydride species has also been prepared, utilizing the sterically bulky trimethylsilylindenyl ligand.

The synthesis of the first stable Th(III) organometallic complex, $Cp_3^{\ddagger}Th$ [107040-62-0] (14) was achieved via reduction of $Cp_2^{\ddagger}ThCl_2$ [87654-17-9] (25). This air-sensitive complex is epr-active, having a $6d^1$ ground-state configuration as opposed to a $5f^1$ ground-state (87).

There is little known chemistry of tetrakis-Cp thorium complexes. Pseudotetrahedral molecules, $(\eta$-ring$)_4$Th (15), where ring = Cp [1298-75-5] or Ind [11133-17-6], have a measurable dipole, resulting from deviation of T_d symmetry (88). Difference of coordination environments, eg, η^3 in the indenyl system and η^5 in the Cp system, appears to indicate great steric crowding about the thorium center, which probably limits the reactivity and synthetic derivatization of these complexes.

Cyclooctatetraenyl Compounds. Sandwich-type complexes of cycloocta-
tetraene (COT), $C_8H_8^{2-}$, are well known. The chemistry of thorium–COT complexes is similar to that of its Cp analogues in steric number and electronic
configurations. Thorocene [12702-09-9], COT_2Th, (**16**), the simplest of the COT
derivatives, has been prepared by the interaction of $ThCl_4$ [10026-08-1] and two
equivalents of $K_2C_8H_8$. Thorocene derivatives with alkyl-, silyl-, and aryl-
substituted COT ligands have also been described. These compounds are thermally stable, air-sensitive, and appear to have substantial ionic character.

Mono-COT compounds that have been studied include $(COT)ThX_2(THF)_2$
(**17**), where X = Cl [73652-04-7] or BH_4 [73643-98-8], $(COT)Th[N(Si(CH_3)_3)_2]_2$
[117097-70-8], and $Cp^*(COT)ThCl(THF)$ [119390-83-9]. The chloro complexes
have been used to generate mixed-ring Cp^*/COT complexes, eg, $Cp^*(COT)ThX$,
where X = $N(Si(CH_3)_3)_2$ [119390-81-7] or $CH(Si(CH_3)_3)_2$ [119390-80-6], as well as
organo/COT complexes (89). The unsaturated complex, $Cp^*(COT)Th(CH_2Si-$
$(CH_3)_3)$ (**18**), is an example of an organo derivative stabilized by an agostic
interaction with one of the methyl groups of the trimethylsilylmethyl ligand.

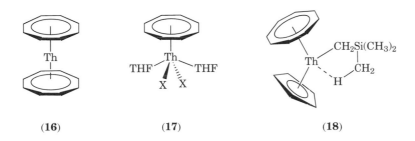

(**16**)   (**17**)   (**18**)

Allyl Complexes.   Allyl complexes of thorium have been known since the
1960s and are usually stabilized by cyclopentadienyl ligands. Allyl complexes
can be accessed via the interaction of a thorium halide and an allyl grignard.
This synthetic method was utilized to obtain a rare example of a naked allyl
complex, $Th(\eta^3\text{-}C_3H_5)_4$ [144564-74-9], which decomposes at 0°C. This complex,
when supported on dehydroxylated γ-alumina, is an outstanding heterogeneous
catalyst for arene hydrogenation and rivals the most active platinum metal
catalysts in activity (17,18).

Hydrocarbyl Complexes.   Stable homoleptic and heteroleptic thorium hydrocarbyl complexes have been synthesized. Two common homoleptic species are [Li-
$TMEDA]_3[Th(CH_3)_7]$ [92366-18-2] (**19**), where TMEDA = tetramethyl ethylenediamine, and $Th(CH_2C_6H_5)_4$ [54008-63-8] (**20**). The benzylic complex has been
synthesized at low temperatures and may owe its stability to possible multihapto
coordination, $\eta^1-\eta^3$, of the benzyl ligand. The methyl complex is stable even up
to room temperature. Six of the methyl groups are hydrogen-bonded to the Li
atom to stabilize this highly charged species. This compound is very reactive
with H_2 and CO. However, there is no concrete structural data for the final
products of such reactions.

(19)   (20)

Comparable to the homoleptic complexes, the heteroleptic systems are dominated by methyl and benzyl groups. Examples of these materials are $ThR_3R'(DMPE)$ (R = $CH_2C_6H_5$, R' = CH_3; R = R' = CH_3, $CH_2C_6H_5$) and $ThX_2(pyr)_2$ (X = Cl, 2,6-di-*t*-butylphenoxide, pyr = 2-(6-methylpyridyl)). Interestingly, these compounds coordinate soft chelating phosphine ligands, a rarity for the hard Th(IV) atom.

Another class of heteroleptic complexes contains π-donors and is of the form $ThR[N(Si(CH_3)_3)_2]_3$, where R = CH_3 [69517-43-7], H [70605-07-1], or BH_4 [69532-06-5]. The methyl compound has exhibited insertion reactivity, including aldehydes, ketones, nitriles, and isocyanides (29). Stable metallacycle compounds are also known, eg, $[((CH_3)_3Si)_2N]_2Th(CH_2Si((CH_3)_2)NSi(CH_3)_3)$, and can be obtained by pyrolysis of triamido–hydrido–thorium complexes (90).

Bimetallic Complexes. There are two types of bimetallic organometallic thorium complexes: those with, and those without, metal–metal interactions. Examples of species containing metal–metal bonds are complexes with Fe or Ru carbonyl fragments. $Cp^*ThX(CpRu(CO)_2)$, where X = Cl or I, and $Cp'_3Th(CpM-(CO)_2)$, where M = Fe or Ru, have both been prepared by interaction of Cp^*_2-ThX_2 or Cp'_3ThCl [62156-90-5], respectively, with the anionic metal carbonyl fragment. These complexes contain very polar metal–metal bonds that can be cleaved by alcohols.

The chemistry of nonmetal–metal-bonded species is more extensive, especially with the use of bridging ligands such as phosphido, polyoxo, and σ-bonded ferrocenyl groups. Phosphides have been found to bridge thorium fragments with nickel and platinum. The two examples are $Cp^*_2Th(\mu$-$P(C_6H_5)_2)Ni(CO)_2$ and $Cp^*_2Th(\mu$-$P(C_6H_5)_2)Pt(P(CH_3)_3)$. The polyoxoanions are represented by [*N*-*n*-$(C_4H_9)_4$][$Cp_3Th(MW_5O_{19})_2$], where M = Nb or Ta. Compounds of this type are also known for uranium. A single thorium σ-bonded ferrocenyl complex, $Th(ACAC)_3(FcN)$ [143687-88-1], where ACAC is acetyl acetone and FcN is (dimethylaminomethyl)ferrocenyl, has been isolated. This complex is eight-coordinate, having six oxygens from the ACAC ligands and two nitrogens from the ferrocenyl group.

Economic Aspects

Total reserves of thorium at commercial price in 1995 was estimated to be >2×10^6 metric tons of ThO_2 (11). Thorium is a potential fuel for nuclear power reactors. It has a 3–4 times higher natural abundance than U and the separation of the product ^{233}U from ^{232}Th is both technically easier and less expensive than the enrichment of ^{235}U in ^{238}U. However, side-reaction products, such as ^{232}U, and the intense α- and γ-active decay products lead to a high buildup of radioactive products during the nuclear reaction of ^{232}Th.

Health and Safety Factors

Thorium is potentially hazardous. Finely divided thorium metal and hydrides can be explosive or inflammatory hazards with respect to oxygen and halogens. Finely divided ThO_2 and other inorganic salts also present an inhalation and irritation hazard. The use of standard precautions, skin covering, and a conventional dust respirator should be sufficient for handling thorium materials.

The long half-life of ^{232}Th makes it a minimal radiation hazard. External radiological hazards are generally not of concern because γ-rays from thorium daughter products have low abundance and many are in the x-ray range, ie, <85 keV, where they are absorbed in the source material. There are some residuals in the 240 and 727 keV energy range that can lead to radiation effects from ton quantities of materials. Such sources merit safety precautions in terms of distance and shielding.

When inhaled, ingested, or adsorbed through the skin, thorium isotopes are potentially harmful because of ionizing radiation and chemical toxicity. If Th(IV) is injected as the soluble nitrate form, it can cause hemolysis. Thorium is generally insoluble in the presence of organic complexing agents and mainly retained in the bones, lungs, lymph nodes, and parenchymatous tissues. In animal experiments, not more than $10^{-4}\%$ of thorium was absorbed from the gastrointestinal tract (91). Prolonged retention of thorium, particularly in a fixed location, allows the possibility of radiation damage by the energetic α-particles of some of the short-lived daughter members of the decay chain, of which there are five for every thorium alpha.

Thorotrast (colloidal ThO_2) was once used as a radiopaque agent in medicine (see RADIOPAQUES). Its injection in a dose of 2.0–15.0 g caused rises in body temperature, nausea, and injury to tissues at the injection site, followed by anemia, leukopenia, and impairment of the reticuloendothelial system. After intravenous administration, thorotrast particles are taken up by reticuloendothelial cells of the liver and spleen. Thorotrast is virtually not eliminated from the body (91). Between 1947 and 1961, 33 cases of cancer of the liver, larynx, and bronchi and sarcoma of the kidneys, developing from 6 to 24 years after thorotrast administering, have been described in the literature (92).

Thorium compounds are legally classified as source materials for nuclear energy and thus are regulated by various government agencies, eg, the Nuclear Regulatory Commission. The relevant regulations cover licensing and safety aspects.

BIBLIOGRAPHY

"Thorium and Thorium Compounds" in *ECT* 1st ed., Vol. 14, pp. 116–124, by H. E. Kremers, Lindsay Chemical Co.; in *ECT* 2nd ed., Vol. 20, pp. 248–259, by W. L. Silvernail and

J. B. McCoy, American Potash & Chemical Corp.; in *ECT* 3rd ed., Vol. 22, pp. 989–1002, by L. I. Katzin, Consultant.

1. L. Martinot, in A. J. Bard, ed., *Encyclopedia of Electrochemistry of the Elements*, Vol. III, Marcel Dekker, Inc., New York, 1978, pp. 153.
2. S. G. Bratsch and J. J. Lagowski, *J. Phys. Chem.* **90**, 307 (1986).
3. S. Ahrland and co-workers, eds., *Gmelin Handbook of Inorganic Chemistry, Thorium, Suppl. Vol. D1, Properties of Thorium Ions in Solutions*, 8th ed., Springer-Verlag, Berlin, 1988.
4. G. Faure, *Principles of Isotope Geology*, 2nd ed., John Wiley & Sons, Inc., New York, 1986, pp. 370–373.
5. C. Gariépy and B. Dupré, in L. Heaman and J. N. Ludden, eds., *Proceedings of the Mineralogical Association of Canada Short Course on Radiogenic Isotope Systems to Problems in Geology*, Toronto, 1991, pp. 191–214.
6. J. B. Gill, R. W. Williams, and D. M. Pyle, in Ref. 5, pp. 310–335.
7. R. Ditz and co-workers, eds., *Gmelin Handbook of Inorganic Chemistry, Thorium, Suppl. Vol. A1a, Natural Occurrence, Minerals (Excluding Silicates)*, Springer-Verlag, Berlin, 1990.
8. L. I. Katzin, in J. J. Katz, G. T. Seaborg, and L. R. Morss, eds., *The Chemistry of the Actinide Elements*, Chapman and Hall, London, 1986.
9. C. Frondel, in *Proceedings of International Conference on Peaceful Uses of Atomic Energy*, Vol. 6, Geneva, 1955, p. 568.
10. H. W. Kirby and co-workers, eds., *Gmelin Handbook of Inorganic Chemistry, Thorium, Suppl. Vol. A2, History, Isotopes, Recovery of Thorium*, 8th ed., Springer-Verlag, Berlin, 1986.
11. G. Choppin, J. Rydberg, and J. O. Liljenzin, *Radiochemistry and Nuclear Chemistry*, 2nd ed., Butterworth-Heinemann Ltd., Oxford, 1995; J. B. Hedrick in *U.S. Bureau of Mines, Mineral Commodity Summaries*, Washington, D.C., Jan. 1995.
12. J. R. Allen and co-workers, eds., *Gmelin Handbook of Inorganic Chemistry, Thorium, Suppl. Vol. A3, Technology, Uses, Irradiated Fuel, Reprocessing*, 8th ed., Springer-Verlag, Berlin, 1988.
13. N. N. Greenwood and A. Earnshaw, *Chemistry of the Elements*, Pergamon Press, Oxford, 1988.
14. G. Folcher, *J. Less-Comm. Met.* **122**, 139 (1986).
15. H. W. Pennline and co-workers, *Div. Fuel Chem.* **28**, 164 (1983).
16. S. Licka, J. Macak, and J. Malecha, *Technol. Paliv.* **D40**, 113 (1979).
17. M. S. Eisen and T. J. Marks, *J. Am. Chem. Soc.* **114**, 10358 (1992).
18. M. S. Eisen and T. J. Marks, *Organometallics*, **11**, 3939 (1992).
19. H. U. Borgstedt and H. Wedemeyer, eds., *Gmelin Handbook of Inorganic and Organometallic Chemistry, Thorium, Suppl. Vol. B2, Alloys of Thorium with Metals of Main Groups I to IV*, Springer-Verlag, Berlin, 1992.
20. Eur. Pat. Appl. EP 443479 Al 910828, J. N. C. Van Geel, J. Fuger, and L. Kock.
21. J. F. Smith and co-workers, *Thorium: Preparation and Properties*, Iowa State University Press, Ames, Iowa, 1975.
22. F. L. Oetting, M. H. Rand, and R. J. Ackermann, *The Chemical Thermodynamics of Actinide Elements and Compounds, Part 1, The Actinide Elements*, IAEA, Vienna, 1976.
23. J. R. Ferraro, L. I. Katzin, and G. Gibson, *J. Am. Chem. Soc.* **76**, 909 (1954).
24. J. Habash and A. J. Smith, *Acta Crystallogr.* **29C**, 413 (1983).
25. P. C. Blake and co-workers, *J. Chem. Soc., Chem. Commun.* **15**, 1148 (1986).
26. N. Kumar and D. Tuck, *Inorg. Chem.* **22**, 1951 (1983).
27. D. L. Clark and co-workers, *Inorg. Chem.* **31**, 1628 (1992).
28. D. L. Clark and A. P. Sattelberger, in R. B. King, ed., *Encyclopedia of Inorganic Chemistry*, Vol. 1, Wiley-Interscience, New York, 1994, p. 24.

29. M. Ephritikhine, *New J. Chem.* **16**, 451 (1992).
30. M. G. B. Drew and G. R. Willey, *J. Chem. Soc., Dalton Trans.* **4**, 727 (1984).
31. A. Domingos, J. Marcalo, and A. Pires De Matos, *Polyhedron* **11**, 909 (1992).
32. G. S. Girolami and co-workers, *J. Coord. Chem.* **32**, 173 (1994).
33. P. Scott and P. B. Hitchcock, *J. Chem. Soc., Dalton Trans.* **4**, 603 (1995).
34. S. J. Coles and co-workers, *J. Chem. Soc., Dalton Trans.* **20**, 3401 (1995).
35. P. G. Edwards and co-workers, *J. Chem. Soc., Chem. Commun.* **19**, 1469 (1992).
36. P. G. Edwards, J. S. Parry, and P. W. Read, *Organometallics*, **14**, 3649 (1995).
37. J. Busch, R. Hofmann, and R. Z. Gruehn, *Anorg. Allg. Chem.* **622**, 67 (1996).
38. J. Busch and R. Z. Gruehn, *Anorg. Allg. Chem.* **622**, 640 (1996).
39. P. Kroeschell and R. Z. Hoppe, *Anorg. Allg. Chem.* **509**, 127 (1984).
40. C. F. Baes and R. E. Mesmer, *The Hydrolysis of Cations*, John Wiley & Sons, Inc., New York, 1976.
41. Y. P. Davydov and I. G. Toropov, *Dokl. Akad. Nauk Belarusi.* **36**, 229 (1992).
42. I. Engkvist and Y. Albinsson, *Radiochim. Acta*, **58/59**, 109 (1992).
43. I. Grenthe and B. Lagerman, *Acta Chem. Scand.* **45**, 231 (1991).
44. E. Oesthols, J. Bruno, and I. Grenthe, *Geochim. Cosmochim. Acta*, **58**, 613 (1994).
45. D. L. Clark, D. E. Hobart, and M. P. Neu, *Chem. Rev.* **95**, 25 (1995).
46. J. Bruno and co-workers, *Inorg. Chim. Acta*, **140**, 299 (1987).
47. A. Joao, S. Bigot, and F. Fromage, *Bull. Soc. Chim. France*, **1**, 42 (1987).
48. A. Joao and co-workers, *Radiochim. Acta*, **68**, 177 (1995).
49. I. I. Chernyaev, V. A. Golovnya, and A. K. Molodkin, *Russ. J. Inorg. Chem.* **3**, 100 (1958).
50. J. Dervin, J. Faucherre, and P. Herpin, *Bull. Soc. Chim. France*, **7**, 2634 (1973).
51. J. Dervin and J. Faucherre, *Bull. Soc. Chim. France*, **3**, 2930 (1973).
52. P. S. Voliotis and E. A. Rimsky, *Acta Crystallogr.* **B31**, 2615 (1975).
53. S. Voliotis and co-workers, *Rev. Chim. Minérale*, **14**, 441 (1977).
54. P. S. Voliotis, *Acta Crystallogr.* **B35**, 2899 (1979).
55. N. A. Yamnova, D. Y. Pushcharovskii, and A. V. Voloshin, *Soc. Phys. Dokl.* **35**, 12 (1990).
56. J. A. Fortner and J. K. Bates, *Mater. Res. Soc. Symp. Proc.* **412**, 205 (1996).
57. M. Genet and co-workers, *Mater. Res. Soc. Symp. Proc.* **353**, 799 (1995).
58. L. Lou and co-workers, *J. Non-Cryst. Solids*, **171**(2), 115 (1994).
59. L. Lou and co-workers, *Proc. SPIE-Int. Soc. Opt. Eng.* **1758**, 83 (1992).
60. C. Merigou and co-workers, *New J. Chem.* **19**, 275 (1995).
61. C. Merigou and co-workers, *New J. Chem.* **19**, 1037 (1995).
62. N. Baglan and co-workers, *New J. Chem.* **18**, 809 (1994).
63. P. Benard and co-workers, *Chem. Mater.* **8**, 181 (1996).
64. M. Louer, R. Brochu, and D. Louer, *Acta Cryst.* **B51**, 908 (1995).
65. B. Matkovic, B. Prodic, and M. Sljukic, *Croat. Chem. Acta*, **40**, 147 (1968).
66. M. Quarton, M. Zouiri, and W. Freundlich, *C. R. Acad. Sci. Ser. 2* **299**, 785 (1984).
67. N. Galesic and co-workers, *Croat. Chem. Acta*, **57**, 597 (1984).
68. J. Fuger and co-workers, *The Actinide Aqueous Inorganic Complexes*, Part 12, International Atomic Energy Agency, Vienna, 1992.
69. E. Oesthols, *Radiochim. Acta*, **68**, 185 (1995).
70. B. Fourest and co-workers, *J. Alloys Comp.* **213/214**, 219 (1994).
71. I. Engkvist and Y. Albinsson, *Fourth International Conference on the Chemistry and Migration Behavior of Actinides and Fission Products in the Geosphere*, R. Oldenburg Verlag, Charleston, S.C., 1993; p. 139.
72. A. P. Sattelberger and W. G. Van der Sluys, *Chem. Rev.* **90**, 1027 (1990).
73. T. Shiratori and K. Fukuda, *J. Nucl. Mater.* **202**, 98 (1993).
74. H. N. Erten, A. K. Mohammed, and G. R. Choppin, *Radiochim. Acta*, **66**, 123 (1994).

75. V. Jacques and J. F. Desreux, *J. Alloys Comp.* **213**, 286 (1994).
76. D. W. Whisenhunt and co-workers, *Inorg. Chem.*, **35**, 4128 (1996).
77. J. M. Berg and co-workers, *J. Am. Chem. Soc.* **114**, 10811 (1992).
78. D. M. Barnhart and co-workers, *Inorg. Chem.* **34**, 5416 (1995).
79. D. L. Clark and J. G. Watkin, *Inorg. Chem.* **32**, 1766 (1993).
80. D. M. Barnhart and co-workers, *Inorg. Chem.* **33**, 3939 (1994).
81. N. P. Freestone, H. Geckeis, and J. H. Holloway, eds., *Gmelin Handbook of Inorganic and Organometallic Chemistry, Thorium, Suppl. Vol. C4, Compounds with F, Cl, Br, I*, 8th ed., Spring-Verlag, Berlin, 1993.
82. R. J. Butcher and co-workers, *Organometallics*, **15**, 1488 (1996).
83. A. F. England, C. J. Burns, and S. L. Buchwald, *Organometallics*, **13**, 3491 (1994).
84. N. S. Radu and co-workers, *J. Am. Chem. Soc.* **117**, 3621 (1995).
85. E. Ciliberto and co-workers, *Organometallics*, **11**, 1727 (1992).
86. M. Weydert and co-workers, *Organometallics*, **14**, 3942 (1995).
87. W. K. Kot and co-workers, *J. Am. Chem. Soc.* **110**, 986 (1988).
88. R. Maier and co-workers, *J. Alloys Comp.* **190**, 269 (1993).
89. T. M. Gilbert, R. R. Ryan, and A. P. Sattelberger, *Organometallics*, **8**, 857 (1989).
90. S. J. Simpson, H. W. Turner, and R. A. Andersen, *Inorg. Chem.* **20**, 2991 (1981).
91. V. A. Filov, B. A. Ivin, and A. L. Bandman, eds., *Harmful Chemical Substances*, Vol. 1, *Elements in Group I-IV of the Periodic Table and Their Inorganic Compounds*, Ellis Horwood, New York, 1993, pp. 344–350.
92. N. Y. Tarasenko, *Encyclopedia of Occupational Health and Safety*, Vol. 2, International Labor Office, Geneva, 1983, p. 2173.

General References

Refs. 3, 7, 8, 11, 12, 13, 19, 21, 28, 29, 40, 45, 72, and 81 are general references.
K. W. Bagnal, in G. Wilkinson, R. D. Gillard, and J. A. McCleverty, eds., *Comprehensive Coordination Chemistry: the Synthesis, Reactions, Properties, and Applications of Coordination Compounds*, 1st ed., Pergamon Press, New York, 1987, pp. 1120–1130.
F. T. Edelman, in E. W. Abel, F. G. A. Stone, and G. Wilkinson, eds., *Comprehensive Organometallic Chemistry II: A Review of the Literature 1982–1994*, 1st ed.; Pergamon Press, New York, 1995, pp. 12–192.

DAVID L. CLARK
D. WEBSTER KEOGH
MARY P. NEU
WOLFGANG RUNDE
Glenn T. Seaborg Institute for Transactinium Science
Los Alamos National Laboratory

THROMBOLYTIC AGENTS.   See BLOOD COAGULANTS AND ANTICOAGULANTS.

THULIUM.   See LANTHANIDES.

THYROID AND ANTITHYROID PREPARATIONS

The main role of the human thyroid gland is production of thyroid hormones (iodinated amino acids), essential for adequate growth, development, and energy metabolism (1–6). Thyroid underfunction is an occurrence that can be treated successfully with thyroid preparations. In addition, the thyroid secretes calcitonin (also known as thyrocalcitonin), a polypeptide that lowers excessively high calcium blood levels. Thyroid hyperfunction, another important clinical entity, can be corrected by treatment with a variety of substances known as antithyroid drugs.

Related substances include thyroid-stimulating hormone [9002-71-5] (TSH) or thyrotropin, secreted by the pituitary gland; thyrotropin-releasing hormone [24305-27-9] or thyroliberin (TRH), a hypothalamic tripeptide; D-thyroxine [51-49-0] (dextrothyroxine), the synthetic unnatural enantiomer of one of the thyroid hormones which has blood-cholesterol lowering activity; various radioactive iodine-containing preparations used to destroy excessive thyroid tissue or to measure thyroid function; and long-acting thyroid stimulator [9034-48-4] (LATS), an immunoglobulin.

Thyroid Function and Malfunction

Human life without thyroid hormones is possible but of minimal quality. In the fetus, thyroid hormones affect growth and differentiation; in the mature human, they regulate metabolism. The two principal thyroid hormones, L-thyroxine [51-48-9] (L-thyroxine, 3,5,3′,5′-tetraiodo-L-thyronine, T_4) (1) and L-triiodothyronine [6893-02-3] (3,5,3′-triiodo-L-thyronine, T_3) (2) are produced by the thyroid gland and secreted into the blood stream. The minute amounts secreted are regulated by a complex system (Fig. 1) that originates in the central nervous system (CNS)

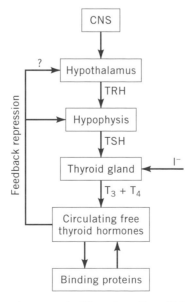

Fig. 1. Mechanisms controlling free thyroid-hormone levels.

and is amplified by both the hypothalamus and the anterior pituitary. These amounts can, however, be diminished by feedback loops in which circulating levels of free T_3 and T_4 repress production of the pituitary TSH, and perhaps of this hormone itself by inhibiting release in the hypothalamus of its liberating hormone, TRH. In addition, amounts of hormones reaching the cells to preserve an optimal (euthyroid) condition are regulated by two plasma proteins, ie, thyroid hormone-binding globulin [9010-34-8] (TBG) and thyroid hormone-binding prealbumin [632-79-1] (TBPA). Only a small fraction (<0.3%) of the total hormones in circulation is free. Finally, tissue deiodinases convert T_4 (possibly a prohormone) into the fivefold more active T_3.

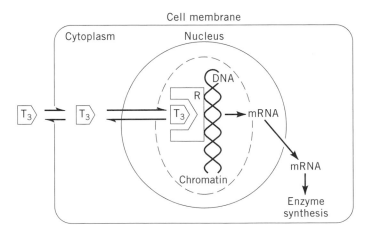

(**1**) R = I
(**2**) R = H

Thyroid hormones affect growth and development by stimulating protein synthesis. It is thought that a specific receptor protein that strongly binds the hormones is present in cell nuclei (7) (Fig. 2). This protein is closely associated with nuclear deoxyribonucleic acid (DNA), a complex involved in DNA transcription. The binding of the hormones is a specific signal to a DNA template that, when activated, stimulates the synthesis and release of a specific messenger ribonucleic acid (mRNA). The latter stimulates the synthesis of astructural and functional proteins, eg, enzymes and other hormones, which then bring about growth and development.

Fig. 2. Early events in thyroid-hormone action. Interaction of T_3 with cell nuclear receptors (6).

In mature animals, the main action of the thyroid hormones is their calorigenic effect which is caused by an increase in the basal metabolic rate (BMR). Although many theories have been advanced (8), the mechanism of action of this effect at the molecular level is not understood. Given the importance of the thyroid hormones in bringing about and then maintaining a normal metabolic state, it is not surprising that malfunctions of the thyroid gland have grave consequences (2,9).

Thyroid underfunction results in a series of hypothyroid states clinically known as cretinism if present in a fetus or an infant, and myxedema in an adult. If the hypothyroidism is owing to insufficient iodine intake, it is known as simple goiter, a state characterized by an enlarged but functionally underactive thyroid gland. Goiter can be avoided by adding iodine to the diet in a convenient form, eg, iodate. In the United States, iodized table salt contains 100 μg of iodate per gram of NaCl (2) (see SODIUM COMPOUNDS, SODIUM HALIDES–SODIUM CHLORIDE). Even so, endemic goiter is still an important health problem in many areas of the world, especially in those where underdevelopment coincides with remoteness from oceans.

Myxedema and goiter are the main conditions for which thyroid preparations are indicated. The treatment of cretinism is difficult because it is recognized only at or after birth. Even if this disease could be diagnosed *in utero*, thyroid hormones do not readily cross the placental barrier. In addition, the fetus, as does a premature infant, rapidly deactivates the thyroid hormones. The halogen-free analogue DIMIT [26384-44-7] (3), which is resistant to fetal deiodinases, may prove useful for fetal hypothyroidism (cretinism).

Thyroid hyperfunction occurs as diffuse toxic goiter, also known as Graves' disease, seen mainly in young adults and premenopausal women. This disease is considered to be caused by an immune disorder. It is characterized by protruding eyeballs (exophthalmos). Another form of thyroid hyperfunction is thyrotoxicosis, ie, a collection of symptoms caused by excessive production of thyroid hormones, including hyperthermia, rapid heart rate, increased appetite and loss of weight, insomnia, anxiety, etc. Toxic nodular goiter (Plummer's disease) is less common. Severe cases are treated by partial surgical removal of the thyroid gland or its partial destruction with radioactive iodine. Milder cases are controlled with antithyroid drugs.

Thyromimetic Compounds

Thyroidal Amino Acids.   Toward the end of the nineteenth century, it was discovered that the consumption of fresh sheep thyroid glands was beneficial in hypothyroidism. In an attempt to isolate the active principle, an extract was prepared (10) and commercialized (11). In the course of this work, it was discovered that thyroid glands were rich in iodine. In 1914, a biologically active pure compound was isolated from thyroid extracts and was called thyroxin on the mistaken assumption that it had an oxyindole structure. Some years later, the correct structure (1) was established by degradation and synthesis. About 25 years later two groups simultaneously identified another biologically active compound that is recognized as the main thyroid hormone. It is the 5'-desiodo analogue of thyroxine, T_3 (2). Two more iodinated thyronines have been found in the thyroid (Fig. 3). They are 3,3′,5′-triiodothyronine [5817-39-0] (reverse-T_3) (**4**) and 3,3′-diiodothyronine [4604-41-5] (T_2) (**5**). These compounds are hormonally inactive and are secreted by the thyroid or arise by partial deiodination of T_3 and T_4. Their physiological significance is not clear. Some properties of these compounds are listed in Table 1.

For many years it was believed that iodine, or some other halogen, had to be present to endow these compounds with thyromimetic activity. This was shown to be incorrect when a halogen-free analogue, DIMIT (**3**), was found to have 20% of the potency of T_4 in a variety of *in vivo* tests (12).

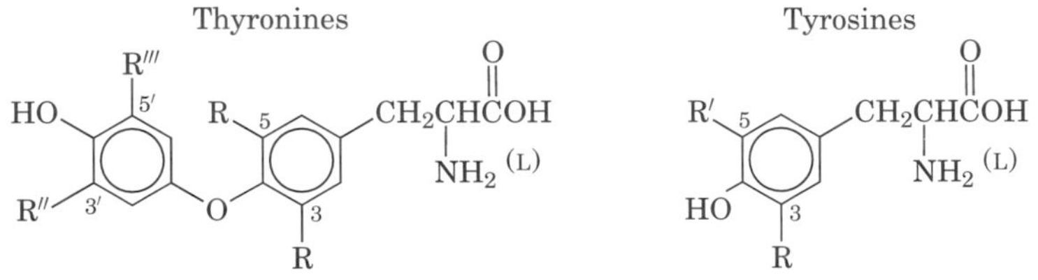

Compound	R	R′	R″	R‴	Compound	R	R′
(**1**) T_4	I	I	I	I	(**6**) TYR	H	H
(**2**) T_3	I	I	I	H	(**7**) MIT	I	H
(**3**) DIMIT	CH_3	CH_3	i-C_3H_7	H	(**8**) DIT	I	I
(**4**) r-T_3	I	H	I	H			
(**5**) 3,3′-T_2	I	H	I	I			

Fig. 3. Structures of the thyroidal iodinated amino acids and the halogen-free analogue DIMIT (**3**). Compound (**4**) is reverse-T_3.

Table 1. Thyroidal Iodinated Amino Acids[a]

Name	CAS Registry Number	Compound	Mol wt	I, %	pK_a (OH)	$[\alpha]_D$
3-iodo-L-tyrosine	[70-78-0]	(**7**)	307.1	41.3	8.70	-4.4^b
3,5-diiodo-L-tyrosine	[66-02-4]	(**8**)	433.0	58.6	6.48, 6.36	2.75^b
3,5-diiodo-L-thyronine	[1041-01-6]	(**9**)	525.1	48.3	9.29	26.0^c
3,3′-diiodo-L-thyronine	[4604-41-5]	(**5**)	525.1	48.3		18.8^c
3,5,3′-triiodo-L-thyronine	[6893-02-3]^d	(**2**)	650.9	58.5	8.45	21.5^c
3,3′,5′-triiodo-L-thyronine	[5817-39-0]	(**4**)	650.9	58.5	6.5^e	16.7^c
3,5,3′,5′-tetraiodo-L-thyronine	[51-48-9]^f	(**1**)	776.8	65.3	6.73, 6.45	17.5^c

[a]Data mainly from Refs. 6 and 10. These compounds decompose; their melting points are indistinct.
[b]In 4.8% HCl.
[c]In 1 N HCl–C_2H_5OH.
[d]Anhydrous Na salt [55-06-7].
[e]Value is estimated.
[f]Anhydrous Na salt [55-03-8]; Na salt, 5 H_2O [25416-65-3].

Structure–Activity Relationships.   In spite of the considerable synthetic and bioassay effort involved in establishing the thyromimetic potency of thyroid-hormone analogues, more than 100 compounds have been studied (Table 2). The main structural requirements for thyromimetic activity can be summarized as follows (6,12–16).

(*1*) Two aromatic rings insulated electronically from each other by connecting oxygen, sulfur, or carbon bridges, forming a central lipophilic core in which the two rings are angled 120°.

(*2*) Substitution at the 3 and 5 positions with alkyl groups or with halogens large enough to force the diphenyl ether nucleus to adopt a minimum energy conformation in which the two rings are approximately in mutually perpendicular planes.

(*3*) At position 1, an acidic side chain two or three carbons long should be present. The natural L-alanyl side chain reduces receptor binding but enhances *in vivo* activity by increasing access to the receptor and by retarding metabolism and excretion. The enantiomeric D-analogues retain considerable activity in contrast to other bioactive substances (17).

(*4*) The presence of a small substituent capable of forming hydrogen bonds in the 4′-position. Isosteric groups such as NH_2 reduce activity, whereas any

Table 2. Thyromimetic Compounds: Relative Binding Affinities (BA)[a] and Antigoiter Potencies (AG)[b]

R	R′	CAS Registry Number	BA	AG
I	H	[6893-02-3]	1	1
Br	H	[58437-19-7]	0.16	0.24
Cl	H	[4299-63-2]	0.04	0.05
F	H	[348-94-7]	0.02	ca 0.01
i-C_3H_7	H	[51-23-0]	0.89	1.42
s-Bu	H	[3415-06-3]	0.78	0.80
n-C_3H_7	H	[72468-99-6]	0.24	0.40
t-Bu	H	[857-98-7]	0.08	0.22
CH_3	H	[2378-96-3]	0.03	0.14
I	I	[51-48-9]	0.14	0.18
Br	Br	[2500-09-6]	0.05	0.02
Cl	Cl	[4299-64-3]	0.04	0.04
i-C_3H_7	I	[3458-12-6]	0.12	
i-C_3H_7	Br	[75628-30-7]	0.22	
i-C_3H_7	Cl	[75628-29-4]	0.53	
i-C_3H_7	i-C_3H_7	[30804-63-8]	0.01	

[a]To solubilized rat hepatic nuclear protein receptor
[b]Affinities and potencies are relative to L-T$_3$ taken as 1 (6).

other group that cannot be converted metabolically to a 4'-OH group results in inactive compounds.

(5) The minimal activity residing in the core structure so far described is greatly enhanced by one lipophilic substituent ortho to the 4'-OH group. High activity imparted by iodine or alkyl groups of similar size, eg, isopropyl. Inspection of structure (1) shows that for T_3 two atropisomers (preferred conformers owing to restricted rotation about single bonds) are possible. In the one shown, the 3'-I is distal to the other ring whereas a second one is obtained by rotating the phenolic ring 180° about the C-1'–O bond, in which the lone iodine is proximal to the other ring. Both atropisomers have been detected in a variety of compounds by x-ray crystallography (18). In their interaction with biomacromolecules, however, affinity is increased when the 3'-substituent is distal. A second lipophilic group at the other position ortho to the 4'-OH (ie, at C-5') always reduces activity because of steric hindrance at the binding site. In a variety of *in vivo* test systems, one synthetic analogue (3,5-diiodo-3'-isopropyl-L-thyronine, see Table 2) has been shown to be more potent than T_3.

Quantitative structure–activity relationships have been established using the Hansch multiparameter approach (14). For rat antigoiter activities (AG), the following (eq. 1) was found, where, as in statistical regression equations, n = number of compounds, r = regression coefficient, and s = standard deviation

$$\log \text{AG} = 1.354\pi35 + 1.344\pi3' - 1.324\,[(\text{size-}3') > \text{I}]$$
$$- 0.359\pi5' - 0.658\sigma3'5' - 0.890\,(\text{OCH}_3\text{-}4') - 2.836$$
$$n = 36 \qquad r = 0.938 \qquad s = 0.304 \tag{1}$$

of the dependent variable. In equation 1, AG is the relative antigoiter potency, π and σ are lipophilic and electronic substituent parameters, $(\text{OCH}_3\text{-}4')$ is a dummy parameter indicating the presence (1) or absence (0) of this substituent, and (size-3' > I) is a computed estimate of the size of this substituent beyond that of an iodine atom. This equation does not include a π^2-term (optimal lipophilicity), and therefore the substituent pattern is unimportant in the overall partitioning behavior. Analogous equations have been derived for interactions of thyroid hormone analogues with other systems (TBG, TBPA, or nuclear receptors) (14). Slight but significant differences in the regression constants point to differences in the shape of the recognition site. Nevertheless, good correlations exist between them and *in vivo* activities (see Table 2), and they can be used with confidence to predict intact animal activity (19).

Biosynthesis, Distribution, and Metabolism. Although iodine is a trace element in the environment (0.006% in ocean water) and the diet, the thyroid gland avidly extracts it from the blood as iodide ion via an active transport system. In the thyroid cells it is converted by a peroxidase to a form capable of iodinating tyrosyl residues present in a large glycoprotein called thyroglobulin [9010-34-8]. It is believed that the resulting mono- and diiodinated residues react with each other in the protein matrix (possibly by a free-radical coupling mechanism) to form all the possible di-, tri-, and tetraiodothyronines (20–22). Thyroglobulin is a glycoprotein having several subunits. Its molecular weight is 660,000

(19S), and it contains ca 10% carbohydrate (corresponding to 300 residues) and ca 5500 amino acid residues of which only two to five are thyroxine. Its amino acid composition and further details of its structure and physical properties are available (23).

The iodinated thyroglobulin is stored as a colloid in thyroidal follicular cells, and T_3 and T_4 are liberated from it by proteolysis as required. It is estimated that ca 90 μg of T_4 and 6 μg of T_3 are secreted daily by the thyroid gland, giving mean plasma concentrations of 80 and 2 μg/L, respectively, of which only 0.03 and 0.3% are in the free form, ie, not protein bound (21). The half-life of T_4 in the body is long (6–7 d) (2); that of T_3 is somewhat shorter (2 d).

The biosynthesis and release of the hormones can be interfered with in various ways, which is the basis of action of certain antithyroid preparations.

Only the small amounts of T_4 and T_3 that are free in the circulation can be metabolized. The main route is deiodination of T_4 to T_3 and r-T_3, and from these to other inactive thyronines (21). Most of the liberated iodide is reabsorbed in the kidney. Another route is the formation of glucuronide and sulfate conjugates at the 4'-OH in the liver. These are then secreted in the bile and excreted in the feces as free phenols after hydrolysis in the lower gut.

Synthesis. In the syntheses of T_4 and its congeners, formation of the sterically hindered diaryl ether core is difficult, as is the introduction of the alanyl side chain (or the preservation of its L (S) absolute configuration) and iodination to the desired degree (T_3 or T_4).

The most widely employed route is the so-called Glaxo method (24) (Fig. 4). The starting material is tyrosine, which is readily available in the L-form and accessible in the D-form. Nitration and protection of the side chain give the key intermediate, N-acetyl-3,5-dinitrotyrosine, ethyl ester [29358-99-4]. The activating effect of the two ortho nitro groups allows the phenolic OH to be displaced readily by pyridine. The resulting quaternary pyridinium adduct, in turn, displays a high reactivity toward nucleophilic displacement by phenoxides, which is the key factor in the successful formation of the ether link in spite of the formidable steric hindrance presented by the nitro groups. In the original procedure the pyridinium tosylate was isolated and purified (24). A modification, in which methanesulfonyl chloride is used, obviates this isolation and results in a faster reaction with higher yields (25). The dinitro ether is then subjected to a Sandmeyer procedure in which the nitro groups are converted to iodines, followed by removal of the blocking groups to give 3,5-diiodothyronine. Finally, one or two additional iodines are introduced to give T_3 and T_4, respectively. In the original procedure, L-T_4 was obtained in an overall yield of 26% based on L-tyrosine (20). By starting with D-tyrosine, D-T_4, which is used clinically to reduce high cholesterol blood levels, can be obtained (26).

This versatile synthetic route has been used extensively with a great variety of phenols and thiophenols to establish structure–activity relationships for thyromimetic activity. Other routes can be summarized as follows (13).

(1) In the first synthesis of T_4, the diphenyl ether was formed from p-methoxyphenol and 3,4,5-triiodonitrobenzene. The nitro group was replaced by a nitril which was then built up into the alanyl side chain by a series of steps (10).

(2) In a biomimetic synthesis, two DIT (**8**) molecules have been coupled to give T_4 (27).

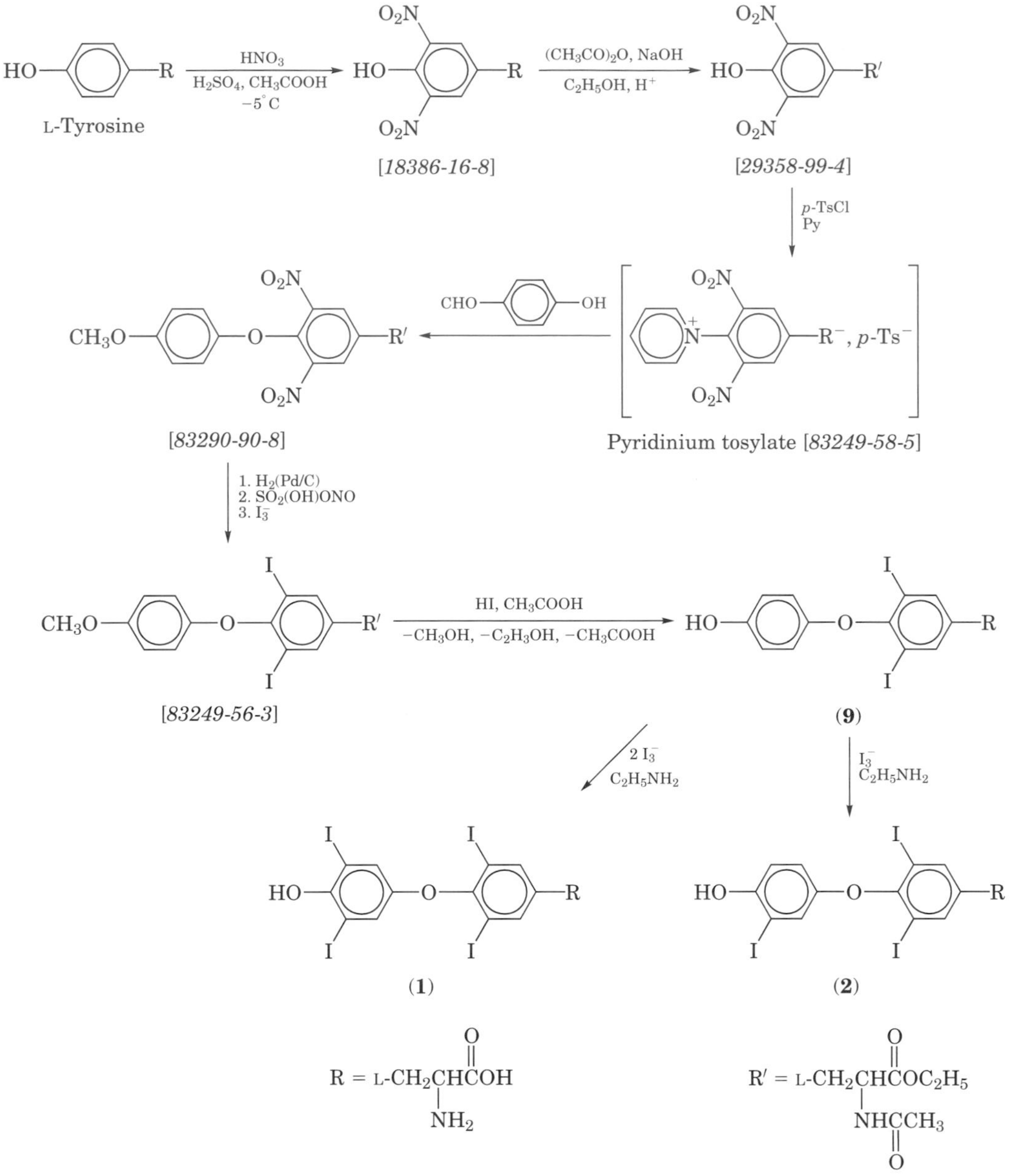

Fig. 4. Glaxo synthesis of T_3 and T_4 where p-TsCl = p-toluenesulfonyl chloride and Py = pyridine (24).

(3) In a procedure known as the iodonium condensation, substituted di-aryliodonium compounds react with tyrosines. Thyronines with substituents other than NO_2 or I at the 3 and 5 positions are obtained.

(4) Many other routes have been used to prepare analogues with structures that differ more from T_4, eg, a methylene, carbonyl, or no bridge in place of the ether linkage (13). The halogen-free analogue DIMIT was synthesized by an

ingenious route involving the replacement of the iodines at the 3- and 5-positions with cyano groups and their reduction to methyl groups (28).

Chemical Assay.   In view of the similarity of their chemical and physical properties (see Table 1) (29), the main problem in the chemical analysis of the thyroid hormones is their separation. A USP procedure gives the details of a paper chromatographic separation in which T_3 is examined for contamination by T_4 and 3,5-diiodothyronine (30). Other systems are also employed (29).

When the purity of the preparation has been ascertained, both T_3 and T_4 are assayed on the basis of their iodine content after combustion in an oxygen flask (29,30).

Body fluids are analyzed for T_3 and T_4 by a variety of radioimmunoassay procedures (31) (see IMMUNOASSAYS). The important clinical parameter for estimating thyroid function, the protein-bound iodine (PBI), is measured as described in treatises of clinical chemistry. High performance liquid chromatographic (hplc) methods have replaced tlc (32,33).

Bioassay.   Although the chemical assays described above have replaced bioassays for the determination of T_3 and T_4, several *in vivo* and *in vitro* bioassays are used to determine the potency of thyroglobulin preparations and to establish the thyromimetic or antithyroid potency of new compounds.

In Vivo Tests.   The rat antigoiter assay is the most common test for thyromimetic activity. Rats are fed an antithyroid compound, eg, propylthiouracil, for 10 days. At the end of this period, they have developed a goiter of such size that the thyroid weighs ca six times that of control rats. A group of rats is injected daily with standard doses of T_4 (2.5 $\mu g/100$ g body wt) or T_3 (0.5 $\mu g/100$ g body wt) which are sufficient to prevent goiter formation. Other groups of rats are treated with the appropriate amounts of the thyromimetic compound. Comparison of the equiactive dose with that of T_3 or T_4 establishes its relative potency. Putative antagonists are administered concurrently with T_3 or T_4 and their activity is assessed from the weight of the goiter formed. Strictly speaking, this assay is based on the relative efficacy of the analogues in their interaction with a thyroid-hormone receptor found in anterior pituitary cells which modulates the secretion of TSH (see Fig. 1).

The mouse anoxia or oxygen-consumption test is based on the stimulation of the basal metabolic rate by thyromimetic compounds. Mice or other small animals are placed in airtight containers of known volume and their survival time is determined (34).

The amphibian metamorphosis test is based on the ability of thyroid hormones to induce precocious transformation of a tadpole into a frog or of the axolotl into a salamander. It is rarely used because of solubility problems and the difficulty of applying the results to humans.

In Vitro Biological Tests.   The inherent complexities, vagaries, and high cost of whole animal assays spurred the development of a series of *in vitro* binding assays to various macromolecules that avidly bind thyroid hormones. In general, and allowing for differences in metabolism, excellent agreement with *in vivo* assays was found, and studies of thyroid-hormone structure–activity relationships (SAR) have been greatly simplified.

Using any of the carrier proteins available in highly purified form, eg, TBG or TBPA, a convenient and accurate quantitative determination of T_3 and T_4 is possible by displacement of radioiodinated T_3 or T_4. This procedure enables their quick determination at low concentrations even in the presence of countless other substances that occur in body fluids (31). In a similar fashion, intact cell nuclei or solubilized proteins from rat liver cell nuclei, which display high affinities for thyroid hormones, especially T_3, have been used to establish relative binding affinities of many thyromimetic compounds (7).

Antithyroid Substances

In principle, antithyroid effects (35–39) can be produced by destroying excess thyroid gland tissue surgically or by treatment with radioiodine; blocking synthesis of thyroid hormones with goitrogens such as certain thionamides; inhibition of thyroid-hormone release with lithium; inhibition of the peripheral deiodination of T_4 to the more active T_3 with thiouracils; increasing excretion of thyroid hormone (n-butyl-3,5-diiodo-4-hydroxybenzoate [51-38-7] as a result of displacing T_3 and T_4 from serum proteins; and competitive antagonism at the receptor level (r-T_3).

Intrathyroidal Inhibitors. *Iodide and Other Inorganic Anions.* When large doses of iodide ion are administered, a transient inhibition of synthesis and release of the thyroid hormones is brought about by the so-called Wolff-Chaikoff effect.

The selective uptake of iodide ion by the thyroid gland is the basis of radioiodine treatment in hyperthyroidism, mainly with ^{131}I, although various other radioactive isotopes are also used (40,41). With a half-life of eight days, the decay of this isotope produces high energy β-particles which cause selective destruction within a 2 mm sphere of their origin. The γ-rays also emitted are not absorbed by the thyroid tissue and are employed for external scanning.

Certain inorganic monovalent anions, similar in size to I, are also taken up by the thyroid gland and competitively inhibit active iodide transport with the following decreasing potencies:

$$TcO_4^- \gg ClO_4^- > ReO_4^- > BF_4^- > I^-$$

Clinical use of perchlorate salts (Na or K) is limited because of side effects.

Thiocyanate ion, SCN^-, inhibits formation of thyroid hormones by inhibiting the iodination of tyrosine residues in thyroglobulin by thyroid peroxidase. This ion is also responsible for the goitrogenic effect of cassava (manioc, tapioca). Cyanide, CN^-, is liberated by hydrolysis from the cyanogenic glucoside linamarin it contains, which in turn is biodetoxified to SCN.

Thionamides. A large group of compounds incorporating thionamide,

$$\underset{\displaystyle -CN}{\overset{\displaystyle S}{\overset{\displaystyle \|}{}}}$$

or thiourea,

$$\begin{array}{c} \text{S} \\ \parallel \\ \diagdown\text{NCN}\diagup \\ \diagup \qquad \diagdown \end{array}$$

moieties are potent antithyroid agents. These inhibit the peroxidases which catalyze the iodination of tyrosine residues in thyroglobulin and their coupling. Although several hundred such compounds are known (42), only four (Fig. 5, Table 3) are used clinically and only two are accepted by the USP XX.

The imidazoles methimazole [60-56-0] (MMI) (12) and Carbimazole [22232-54-8] (13) act by inhibiting intrathyroidal hormone synthesis, whereas the thiouracils (10) [51-52-5] and (11) [56-04-2] also inhibit the peripheral deiodination of T_4 to T_3. Thus, the latter are preferred in the treatment of thyroid storm (thyrotoxic crisis) where a quick drop in circulating T_3 is desired (2,9). In general, the imidazoles are 10 times as active as the thiouracils.

The synthesis of these compounds is shown in Figure 5. Extensive compilations of the chemical, and chromatographic and spectral properties of compounds (10) and (12) are given in References 43 and 44, respectively.

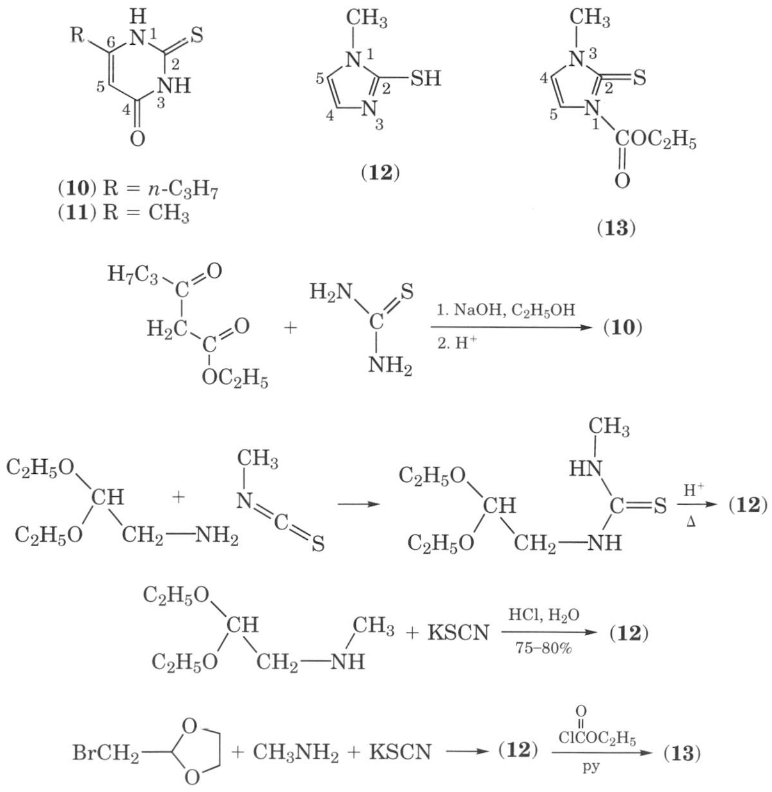

Fig. 5. Structures and syntheses of the clinically employed thionamides, where py = pyridine.

Table 3. Antithyroidal Thionamides

Name	CAS Registry Number	Structure	Composition	Mol wt	S, %	Mp, °C
6-propyl-2-thiouracil (propylthiouracil, PTU)	[51-52-5]	(10)	$C_7H_{10}N_2OS$	170.23	18.84	219–221
6-methyl-2-thiouracil	[56-04-2]	(11)	$C_5H_6N_2OS$	142.18	22.55	325 dec
1-methyl-2-mercaptoimidazole (methimazole, MMI)	[60-56-0]	(12)	$C_4H_6N_2S$	114.16	28.09	146–148
3-methyl-1-carbethoxy-2-thio-imidazoline (carbimazole)	[22232-54-8]	(13)	$C_7H_{10}N_2O_2S$	186.23	17.22	122–125

Although several metabolites of propylthiouracil have been found (36,44), it is mainly excreted in urine as the glucuronide. Its relatively short plasma half-life requires that it be administered four times daily.

Extensive studies have been carried out on the metabolic fate of compounds (**12**) and (**13**) (36). After initial accumulation in the thyroid gland, the unchanged drugs and various metabolites appear in the urine. The carbethoxy group in carbimazole, which was introduced to mask the bitter taste of methimazole, is metabolically removed, and therefore carbimazole can be considered a prodrug of methimazole.

Recommended daily maintenance doses are 50–200 mg for PTU, and 5–20 mg (three times per day) for the imidazoles. The incidence of side effects is low.

It has been known for a long time that some foodstuffs, eg, turnips and rutabaga, are goitrogenic because of the presence of progoitrin. This substance is hydrolyzed to goitrin, or (S)-5-vinyl-2-oxazolidinethione [500-12-9] (**14**), which is goitrogenic when iodine intake is low.

(**14**)

Aromatic Amines and Phenols.   The discovery that sulfaguanidine [57-67-0] was goitrogenic to rats was serendipitous. Many related compounds were then examined, and the aniline moiety was usually present (2,6). Such compounds, as well as resorcinol-like phenols, may act as goitrogens by inhibiting thyroid peroxidases. These are not used clinically.

Lithium.   In the lithium carbonate treatment of certain psychotic states, a low incidence (3.6%) of hypothyroidism and goiter production have been observed as side effects (6,36) (see PSYCHOPHARMACOLOGICAL AGENTS). It has been proposed that the mechanism of this action is the inhibition of adenyl cyclase. Lithium salts have not found general acceptance in the treatment of hyperthyroidism (see LITHIUM AND LITHIUM COMPOUNDS).

Peripheral Antagonists.   The relatively long duration of action of the thyroid hormones makes it desirable to have compounds capable of blocking them competitively at their site of action. This is desirable in the treatment of thyroid storm where the reduction of circulating hormone levels brought about by the inhibition of their synthesis is too slow.

A large number of thyroid hormone analogues have been tested for this effect (6). Among others, r-T$_3$ (**3**) and 3,3'-T$_2$ (**5**) and their propionic acid side-chain analogues decrease oxygen consumption at molar ratios of 50–200:1 of T$_4$. Nevertheless, no potent or clinically useful peripheral antagonists have been found.

The level of circulating hormones is lowered indirectly by *n*-butyl 3,5-diiodo-4-hydroxybenzoate which displaces them from their carriers (TBG and TBPA) and thus accelerates their metabolism and excretion (6).

Calcitonin

Several years ago, it was discovered that the thyroid gland was also the source of a hypocalcemic hormone having effects in general opposition to those of the parathyroid hormone. This hormone is produced in mammals by the parafollicular C-cells and in other vertebrates by the ultimobrachial bodies (45). Originally called thyrocalcitonin, it is now referred to as calcitonin (CT).

Calcitonins from several species have been characterized and synthesized. They are all single-chain 32-residue polypeptides (ca 3600 mol wt), although a disulfide link between the first and seventh cysteine residues results in a cyclic structure that is indispensable for activity (Fig. 6).

Calcitonin is secreted when abnormally high calcium levels occur in plasma. Although plasma concentrations are normally minute (<100 pg/mL), they increase two- to threefold after calcium infusion. Calcitonin has a short plasma half-life (ca 10 min). Certain thyroid tumors are the result of CT concentrations 50–500 times normal. The mechanism of action is a direct inhibition of bone resorption. Calcitonin is used clinically in various diseases in which hypercalcemia is present, eg, Paget's disease (46).

(H) {
H$_2$N-Cys-Gly-Asn-Leu-Ser-Thr-Cys-Met-Leu-Gly-Thr-Tyr-Thr-Gln-Asp-Phe-
5   10   15

Asn-Lys-Phe-His-Thr-Phe-Pro-Gln-Thr-Ala-Ile-Gly-Val-Gly-Ala-Pro-CONH$_2$
20   25   30   32
}

(P) {
H$_2$N-Cys-Ser-Asn-Leu-Ser-Thr-Cys-Val-Leu-Ser-Ala-Tyr-Trp-Arg-Asn-Leu-
5   10   15

Asn-Asn-Phe-His-Arg-Phe-Ser-Gly-Met-Gly-Phe-Gly-Pro-Glu-Thr-Pro-CONH$_2$
20   25   30   32
}

Fig. 6.   Amino acid sequence of human (H) and porcine (P) calcitonins.

Commercial Preparations

Sodium Levothyroxine. As one of the active principles of the thyroid gland, sodium levothyroxine [55-03-8] (levothyroxine sodium) can be obtained either from the thyroid glands of domesticated animals (10) or synthetically. It should contain 61.6–65.5% iodine, corresponding to 100 ±3% of the pure salt calculated on an anhydrous basis. Its chiral purity must also be ascertained because partial racemization may occur during synthesis and because dl-T$_4$ is available commercially. Sodium levothyroxine melts with decomposition at ca 235°C. It is prepared as pentahydrate [6106-07-6] from L-thyroxine and sodium carbonate (47).

Sodium L-thyroxine is a light yellow or buff-colored odorless, tasteless, hygroscopic powder that is stable when dry and protected from light. It is slightly soluble in water (1 g/700 mL) and ethanol (1 g/300 mL) and insoluble in most organic solvents. It is soluble in aqueous alkaline solutions (48). The sodium salt is reported to be better absorbed than the free acid although its bioavailability is still low (50%). Its plasma half-life is five days. An extensive compilation of its chemical, spectroscopic, and chromatographic characteristics is given in Reference 29.

Sodium Liothyronine. Sodium liothyronine [55-06-1] is the sodium salt of L-3,5,3′-triiodothyronine. It is made by the controlled iodination of L-3,5-diiodothyronine. It may be contaminated by starting material or L-T$_4$. The USP assay (49) describes a chromatographic separation specifying 3,5-T$_2$ 2% max and T$_4$ 5% max. Iodine content is specified at 95–101%. Chiral purity must also be ascertained. Detailed information on its chromatographic behavior is available (29).

Thyroglobulin. Thyroglobulin is obtained by fractionating hog thyroid glands until a preparation is obtained containing not less than 0.7% of organically bound iodine (50). It is a cream- to tan-colored powder with a characteristic odor and taste. It is stable in air but sensitive to light and is insoluble in water, alcohol, and other organic solvents (51). It is standardized by chemical and biological assay to contain a T$_4$:T$_3$ ratio of 2.5:1.

Thyroid. *Glandulae Thyroideae siccatae* is the cleaned, dried, and powdered thyroid gland previously deprived of connective tissue and fat from domesticated animals used for food by humans (52). It contains 0.20 ±0.03% iodine in organically bound form and is free of inorganic iodine. Batches of high or low iodine content should be adjusted to the specified concentration by blending or with a suitable diluent. It is dispensed in tablets of various strengths (15–300 mg).

Calcitonin. Calcitonin is available commercially from pork and salmon extracts (Calcimar, Armour) as well as by synthesis. Preparations are bioassayed on the basis of their calcium-lowering activity in comparison to the potency of pure pork calcitonin of which ca 4 μg is equivalent to 1 MRC unit (Medical Research Council, U.K.). For clinical use, vials containing 400 units in 4 mL are available. The recommended daily dosage is 100 units to be administered subcutaneously or intramuscularly because its plasma half-life is short (4–12 min).

Antithyroid Drugs. *Propylthiouracil.* This compound is a white, powdery, crystalline substance of starch-like appearance with a bitter taste. It is slightly soluble in water, chloroform, and ethyl ether, sparingly soluble in ethanol, and soluble in aqueous alkaline solutions (53). An extensive compilation of its chemical, spectral, and chromatographic properties is available (43). It is assayed titrimetrically with NaOH (53).

Methimazole. This compound is a white to pale buff crystalline powder with a faint characteristic odor. It is soluble in water, ethanol, and chloroform (1 g/5 mL) and only slightly soluble in other organic solvents. A detailed chemical, analytical, spectral, and chromatographic description is available (44). It is assayed titrimetrically with NaOH (54).

BIBLIOGRAPHY

"Thyroid and Antithyroid Preparations (Antithyroid Substances)" in *ECT* 1st ed., Vol. 14, pp. 132–135, by R. G. Jones, Eli Lilly and Co.; in *ECT* 2nd ed., Vol. 20, pp. 260–272, by R. G. Jones, Eli Lilly and Co., and J. B. Lesh, Armour Pharmaceutical Co.; in *ECT* 3rd ed., Vol. 23, pp. 1–17, by P. A. Lehmann, Centro de Investigación y de Estudios Avanzados del IPN.

1. S. C. Werner and S. H. Ingbar, eds., *The Thyroid*, 4th ed., Harper & Row Publishers Inc., New York, 1978.
2. R. C. Haynes, Jr. and F. Murad, in A. Goodman Gilman and co-workers, eds., *The Pharmacological Basis of Therapeutics*, 6th ed., Macmillan Publishing Co., New York, 1980, pp. 1397–1419.
3. F. Neuman and B. Schenk, in W. Forth and co-workers, eds., *Allgemeine und Spezielle Pharmakologie und Toxikologie*, 2nd ed., Wissenschaftsverlag, Mannheim, Germany, 1977, pp. 349–359.
4. M. A. Greer and D. H. Solomon, *Thyroid*, Vol. III, Section 7 of *Handbook of Physiology*, American Physiological Society, Washington, D.C., 1974.
5. J. Robbins and L. E. Braverman, eds., *Thyroid Research: 7th International Thyroid Conference*, Excerpta Medica, Amsterdam, the Netherlands, 1976.
6. E. C. Jorgensen, in M. E. Wolff, ed., *Burger's Medicinal Chemistry*, Pt. III, John Wiley & Sons, Inc., New York, 1981, pp. 103–145.
7. J. D. Baxter and co-workers, *Recent Prog. Horm. Res.* **35**, 97 (1979).
8. P. A. Lehmann F., *J. Med. Chem.* **15**, 404 (1972).
9. H. F. Conn, ed., *Current Therapy 1978*, W. B. Saunders Co., Philadelphia, Pa., 1978, pp. 479–501.
10. R. Pitt-Rivers and J. R. Tata, *The Thyroid Hormones*, Pergamon Press, New York, 1959.
11. C. L. Lautenschläger, *50 Jahre Arzneimittelforschung*, Thieme, Stuttgart, Germany, 1955, pp. 229–238.
12. E. C. Jorgensen, W. J. Murray, and P. Block, Jr., *J. Med. Chem.* **17**, 434 (1974); the tetramethyl analogue has been shown to have 1–5% activity of T_4 by J. A. Pittman and co-workers, *J. Med. Chem.* **16**, 306 (1973).
13. E. C. Jorgensen, in C. H. Li, ed., *Hormonal Proteins and Peptides*, Vol. VI, Academic Press, Inc., New York, 1978, pp. 57–105, 107–204.
14. S. W. Dietrich and co-workers, *J. Med. Chem.* **20**, 863 (1977).
15. E. C. Jorgensen, *Pharmacol. Ther. B* **2**, 661 (1976).
16. M. Bolger and E. C. Jorgensen, *J. Biol. Chem.* **255**, 10271 (1980).
17. P. A. Lehmann F., *Trends Pharmacol. Sci.* **3**, 103 (1982).
18. V. Cody, *Recent Prog. Horm. Res.* **34**, 437 (1978).
19. P. L. Ballard and co-workers, *J. Clin. Invest.* **65**, 1407 (1980).
20. H. L. Schwartz and J. H. Oppenheimer, *Pharmacol. Ther. B* **3**, 349 (1978).

21. J. J. DiStefano III and D. A. Fisher, *Pharmacol. Ther. B* **2**, 539 (1976).

22. J. Robbins and co-workers, *Recent Prog. Horm. Res.* **34**, 477 (1978).

23. S. Lissitzky, *Pharmacol. Ther. B* **2**, 219 (1976).

24. J. R. Chalmers and co-workers, *J. Chem. Soc.*, 3424 (1949); U.S. Pat. 2,823,164 (Feb. 11, 1958), R. Pitt-Rivers and J. Gross (to National Research Development Corp.); Brit. Pat. 671,070 (Apr. 30, 1952), G. T. Dickson (to Glaxo Laboratories, Ltd.).

25. R. I. Meltzer and co-workers, *J. Org. Chem.* **22**, 1577 (1957); **26**, 1977 (1961).

26. Ref. 2, p. 844.

27. U.S. Pat. 2,889,363 (June 2, 1959), L. G. Ginger and P. Z. Anthony (to Baxter Laboratories, Inc.).

28. P. J. Block, Jr., and D. H. Coy, *J. Chem. Soc. Perkin Trans. 1*, 633 (1972).

29. A. Post and R. J. Wagner, in K. Florey, ed., *Analytical Profiles on Drug Substances*, Vol. 5, Academic Press, Inc., New York, 1976, pp. 225–281.

30. *The United States Pharmacopeia XX* (*USP XX–NF XV*), The United States Pharmacopeial Convention, Inc., Rockville, Md., 1980, p. 412.

31. B. M. Jaffe and H. R. Behrman, eds., *Methods of Hormone Radioimmunoassay*, 2nd ed., Academic Press, Inc., New York, 1979.

32. B. Hepler and co-workers, *Anal. Chim. Acta* **113**, 269 (1980).

33. B. v. D. Walt and H. J. Cahmann, *Proc. Natl. Acad. Sci. USA* **79**, 1492 (1982).

34. A. Osol and co-eds., *Remington's Pharmaceutical Sciences*, 16th ed., Mack Publishing Co., 1980, p. 130.

35. P. Langer and M. A. Greer, *Antithyroid Substances and Naturally Occurring Goitrogens*, S. Karger, Basel, Switzerland, 1977.

36. B. Marchant, J. F. H. Lees, and W. D. Alexander, *Pharmacol. Ther. B* **3**, 305 (1978).

37. D. H. Solomon, in Ref. 1, p. 814.

38. T. Yamada and co-workers, in Ref. 4, pp. 345–357.

39. W. L. Green, in Ref. 1, p. 77.

40. Ref. 30, pp. 407–411.

41. Ref. 34, pp. 480–481.

42. G. W. Anderson in *Medicinal Chemistry*, Vol. 1, John Wiley & Sons, Inc., New York, 1951, pp. 1–150.

43. H. Y. Aboul-Enein, in Ref. 29, Vol. 6, 1977, pp. 457–486.

44. H. Y. Aboul-Enein and A. A. Al-Badr, in Ref. 29, Vol. 8, 1979, pp. 351–370.

45. H. Rasmussen and M. Pechet, in H. Rasmussen, ed., *International Encyclopedia of Pharmacology and Therapeutics*, Vol. 1, Sect. 51, Pergamon Press, Oxford, U.K., 1970, pp. 237–260.

46. Ref. 2, pp. 1536–1538.

47. Price list, Sigma Chemical Co., St. Louis, Mo., Feb. 1982.

48. Ref. 30, p. 446.

49. Ref. 30, p. 452.

50. Ref. 10, p. 189.

51. Ref. 30, p. 799.

52. Ref. 30, p. 800.

53. Ref. 30, p. 686.

54. Ref. 30, p. 505.

TIN AND TIN ALLOYS

Tin [*7440-31-5*] is one of the world's most ancient metals. When and where it was discovered is uncertain, but evidence points to tin being used in 3200–3500 BC. Ancient bronze weapons and tools found in Ur contained 10–15 wt % tin. In 79 AD, Pliny described an alloy of tin and lead now commonly called solder (see SOLDERS AND BRAZING ALLOYS). The Romans used tinned copper vessels, but tinned iron vessels did not appear until the fourteenth century in Bohemia. Tinned sheet for metal containers and tole (painted) ware made its appearance in England and Saxony about the middle of the seventeenth century. Although tinplate was not manufactured in the United States until the early nineteenth century, production increased rapidly and soon outstripped that in all other countries (1).

In most cases, tin is used on or in a manufactured material in small amounts, much out of proportion to the purpose it serves. Nevertheless, tin in some form has been associated with the economic and cultural growth of civilization. Food preservation and canning developed rapidly with the invention of tin-coated steel; transportation and high speed machinery became a reality with the invention of tin-base bearing metals; the casting of type metal was an important advance in printing technology; bronze alloys became weapons, tools, and architectural objects; tin alloys are used in organ pipes and bells; and telecommunications and electronic equipment depend upon the tin–lead soldered joint. In modern technology, new uses of tin include the plating of protective coatings, nuclear energy, plastics and other polymers, agriculture, biochemistry, electronic packaging, and glassmaking.

Of the nine different tin-bearing minerals found in the earth's crust, only cassiterite [*1317-45-9*], SnO_2, is of importance. Over 80% of the world's tin ore occurs in low grade alluvial or eluvial placer deposits where the tin content of the ore can be as low as 0.015%. Complex tin sulfide minerals such as stannite [*12019-29-3*], $Cu_2S \cdot FeS \cdot SnS_2$; teallite [*12294-02-9*], $PbSnS_2$; cylinderite [*59858-98-9*], $PbSn_4FeSb_2S_{14}$; and canfieldite [*12250-27-0*], Ag_8SnS_6, are found in the lode deposits of Bolivia and Cornwall associated with cassiterite and granitic rock. In the lode mines, the ores often contain 0.8–1 wt % of tin metal. No workable tin deposits have been found in the United States.

Tin-mining methods depend on the character of the deposit. Primary deposits are embedded in underground granitic rock and recovery methods are complex. The more important secondary deposits are in the form of an alluvial mud in the stream beds and placers and the recovery is simpler than lode mining. Cassiterite is recovered from alluvial deposits by dredging, hydraulicking where a head of water permits it, jets and gravel pumps on level ground, or open-pit mining.

Gravel-pump mining is widely used in southeast Asia and probably accounts for 40% of the world's tin production. Powerful jets of water are directed onto the mine face to break down the tin-bearing soil, which is allowed to collect in a sump. A gravel pump in the sump elevates the watery mud to a wooden trough termed a palong in the trade. The palong has a gentle slope and as the ore flows down the slope, the tin oxide particles, which are 2.5 times heavier than sand, are trapped behind wooden slats or riffles. Periodically, the preliminary concentrates are collected and transferred to the dressing shed for final concentration. Hydraulicking and open-pit mining methods also involve gravity separation with water in palongs (see MINERALS RECOVERY AND PROCESSING).

Dredging is mining with a floating dredge on an artificial pond in a placer. Chain buckets or suction cutters, digging at depths of 46 m, transfer the tin-bearing mud to revolving screens, hydrocyclons, jigs, shaking tables, classifiers, and similar equipment. The sand and dirt are removed in these preliminary roughing steps. The mineral is further beneficiated in dressing sheds on shore with modern techniques such as flotation (qv), heavy-media separation, and electrostatic, magnetic, and spiral separators for the removal of associated minerals (see SEPARATION, MAGNETIC). Final concentrates ready for direct smelting contain 70–77 wt % tin, which is almost pure cassiterite.

Underground-lode deposits in Bolivia are located at very high altitudes, 4000–5000 m above sea level, whereas the lode deposits in Cornwall are ca 430 m below sea level. Access to the lodes is by shaft sinking or by adits, ie, passages driven into the side of a mountain, depending on the terrain. The ore is broken from the working face by blasting and drilling. Further crushing and grinding above-ground is necessary to produce the finely divided ore capable of being concentrated by the various gravity-concentration methods and mechanical separations commonly used for alluvial deposits.

Tin concentrates from the lode deposits are 40–60 wt % tin and must be further upgraded before smelting. Roasting the ore removes sulfur and arsenic; the sulfides of iron, copper, bismuth, and zinc are converted to oxides and lead sulfide is oxidized to sulfate. When the concentrates contain considerable quantities of sulfides, the impurities are sometimes removed by a chloridizing roast, followed by leaching. Advances in froth-flotation methods offer another alternative for the removal of unwanted sulfide minerals. When the concentrates are roasted with 1–5 wt % salt, NaCl, in an oxidizing atmosphere, sodium sulfate and the chlorides of the metals are formed without attack on the tin oxide. Many chlorides are volatile; bismuth, lead, arsenic, antimony, and silver may be partially removed in the form of fume; leaching with water removes the remaining chlorides which are readily soluble (see TIN COMPOUNDS).

Physical Properties

Physical, mechanical, and thermal constants of tin are shown in Table 1.

Although the pure metal has a silvery-white color, in the cast condition it may have a yellowish tinge caused by a thin film of protective oxide on the surface. When highly polished, it has high light reflectivity. It retains its brightness well during exposure, both outdoors and indoors.

The melting point (232°C) is low compared with those of the common structural metals, whereas the boiling point (2625°C) exceeds that of most metals except tungsten and the platinum group. Loss by volatilization during melting and alloying with other metals is insignificant. Tin is a soft, pliable metal easily adaptable to cold working by rolling, extrusion, and spinning. It readily forms alloys with other metals, imparting hardness and strength. Only small quantities of some metals can be dissolved in pure liquid tin near its melting point. Intermetallic compounds are freely formed, particularly with metals of high melting

Table 1. Physical Properties of Tin[a]

Property	Value
mp, °C	231.9
bp, °C	2625
sp gr	
α-form (gray tin)	5.77
β-form (white tin)	7.29
liquid at mp	6.97
transformation temp $\beta \rightleftharpoons \alpha$, °C	13.2
vapor pressure, Pa[b]	
at 1000 K	986×10^{-6}
1300 K	1.1
1500 K	22.6
2000 K	4.08×10^3
2550 K	91×10^3
surface tension, at mp, mN/m (=dyn/cm)	544
viscosity, at mp, mPa·s (=cP)	1.85
specific heat, at 20°C, J/(kg·K)[c]	222
latent heat of fusion, kJ/(g·atom)[c]	7.08
thermal conductivity, at 20°C, W/(m·K)	65
coefficient of linear expansion, $\times 10^{-6}$	
at 0°C	19.9
100°C	23.8
shrinkage on solidification, %	2.8
resistivity of white tin, $\mu\Omega$·cm	
at 0°C	11.0
100°C	15.5
200°C	20.0
mp (solid)	22.0
mp (liquid)	45.0
volume conductivity, % IACS	15
Brinell hardness, 10 kg, 5 mm, 180 s	15
at 20°C	3.9
220°C	0.7
tensile strength, as cast, MPa[d]	
at 15°C	14.5
200°C	4.5
−40°C	20.0
−120°C	87.5
latent heat of vaporization, kJ/mol[c]	296.4

[a]Ref. 2.
[b]To convert Pa to mm Hg, multiply by 0.0075.
[c]To convert J to cal, divide by 4.184.
[d]To convert MPa to psi, multiply by 145.

point, and some of these compounds are of metallurgical importance. Copper, nickel, silver, and gold are appreciably soluble in liquid tin. A small amount of tin oxide dispersed in tin has a hardening effect.

Molten tin wets and adheres readily to clean iron, steel, copper, and copper-base alloys, and the coating is bright. It provides protection against oxidation of the coated metal and aids in subsequent fabrication because it is ductile and solderable. Tin coatings can be applied to most metals by electrodeposition (see ELECTROPLATING).

Tin exists in two allotropic forms: white tin (β) and gray tin (α). White tin, the form which is most familiar, crystallizes in the body-centered tetragonal system. Gray tin has a diamond cubic structure and may be formed when very high purity tin is exposed to temperatures well below zero. The allotropic transformation is retarded if the tin contains small amounts of bismuth, antimony, or lead. The spontaneous appearance of gray tin is a rare occurrence because the initiation of transformation requires, in some cases, years of exposure at $-40°C$. Inoculation with α-tin particles accelerates the transformation.

Chemical Properties

Tin, at wt 118.69, falls between germanium and lead in Group IV A of the periodic table. It has ten naturally occurring isotopes, which, in order of abundance, are 120, 118, 116, 119, 117, 124, 122, 112, 114, and 115.

Tin is amphoteric and reacts with strong acids and strong bases, but is relatively resistant to nearly neutral solutions. Distilled water has no effect on tin. Oxygen greatly accelerates corrosion in aqueous solutions. In the absence of oxygen, the high over-potential of tin (0.75 V) causes a film of hydrogen to be retained on the surface which retards acid attack. The metal is normally covered with a thin protective oxide film which thickens with increasing temperature.

A reversal of potential of the tin–iron couple occurs when tin-coated steel (tin-plate) is in contact with acid solutions in the absence of air. The tin coating acts as an anode; it is the tin that is slowly attacked and not the steel. This unique property is the keystone of the canning industry because dissolved iron affects the flavor and appearance of the product. Thus, the presence of tin protects the appearance and flavor of the product.

Tin does not react directly with nitrogen, hydrogen, carbon dioxide, or gaseous ammonia. Sulfur dioxide, when moist, attacks tin. Chlorine, bromine, and iodine readily react with tin; with fluorine, the action is slow at room temperature. The halogen acids attack tin, particularly when hot and concentrated. Hot sulfuric acid dissolves tin, especially in the presence of oxidizers. Although cold nitric acid attacks tin only slowly, hot concentrated nitric acid converts it to an insoluble hydrated stannic oxide. Sulfurous, chlorosulfuric, and pyrosulfuric acids react rapidly with tin. Phosphoric acid dissolves tin less readily than the other mineral acids. Organic acids such as lactic, citric, tartaric, and oxalic attack tin slowly in the presence of air or oxidizing substances.

Dilute solutions of ammonium hydroxide and sodium carbonate have little effect on tin, but strong alkaline solutions of sodium or potassium hydroxide, cold and dilute, dissolve tin to form stannates.

Neutral aqueous salt solutions react slowly with tin when oxygen is present but oxidizing salt solutions, such as potassium peroxysulfate, ferric chloride and sulfate, and aluminum and stannic chlorides dissolve tin. Nonaqueous organic solvents, lubricating oils, and gasoline have little effect.

Processing

Smelting.   Although the metallurgy of tin is comparatively simple, several complicating factors must be dealt with in tin smelting: The temperature necessary for the reduction of tin dioxide with carbon is high enough to reduce the oxides of other metals which may be present. Thus, reduced iron forms troublesome, high melting compounds with tin, the so-called hard head of the tin smelter. Tin at smelting temperatures is more fluid than mercury at room temperature. It escapes into the most minute openings and soaks into porous refractories. Furthermore, tin reacts with either acid or basic linings, and the slags produced contain appreciable quantities of tin and silica and must be retreated.

Because of the high tin content of the slag, a primary smelting is used to effect a first separation, followed by a second stage to process the slag and hardhead from the first smelting plus refinery drosses.

In primary smelting, carbon (in the form of coal or fuel oil) is the reducing agent. During heat-up, carbon monoxide is formed by reaction with carbon dioxide of the furnace atmosphere. The carbon monoxide reacts with the solid cassiterite particles to produce tin and carbon dioxide:

$$2\ CO + SnO_2 \longrightarrow Sn + 2\ CO_2 \tag{1}$$

As the temperature rises, silica (which is present in nearly all concentrates) also reacts with cassiterite under reducing conditions to give stannous silicate:

$$SnO_2 + CO + SiO_2 \longrightarrow SnSiO_3 + CO_2 \tag{2}$$

Iron, also present in all concentrates, reacts with silica to form ferrous silicate:

$$Fe_2O_3 + CO + 2\ SiO_2 \longrightarrow 2\ FeSiO_3 + CO_2 \tag{3}$$

The silicates formed in reactions 2 and 3 fuse with the added fluxes to form a liquid slag at which point carbon monoxide loses its effectiveness as a reducing agent. Unreacted carbon from the fuel then becomes the predominant reductant in reducing both stannous silicate to tin and ferrous silicate to iron. The metallic iron, in turn, reduces tin from stannous silicate:

$$SnSiO_3 + Fe \rightleftharpoons FeSiO_3 + Sn \tag{4}$$

This is the equilibrium established at the end of each smelting cycle. By this time, a considerable proportion of the tin produced in the heat-up stages has been drained from the furnace and only the metal remaining in the furnace at the final tapping time comes into equilibrium with the slag (3).

Primary smelting can be carried out in a reverberatory, rotary, or electric furnace. The choice depends more on economic circumstances than on technical considerations (3). Thus, in the Far East, reverberatory furnaces fired with anthracite coal as the reductant were and still are widely used. Indonesia and Singapore use slow-speed rotary furnaces. Both Malaysia and Thailand have added new electric-furnace smelting capacity in order to improve smelting efficiencies. Reverberatory and rotary furnaces are also used in Indonesia. On the other hand, the smelters in Central Africa, including those in Zaire and Rwanda as well as those in South Africa, which are far away from coal sources, use electric furnaces because of the availability of electric power.

The development of satisfactory processes for the fuming of tin slags has been one of the greatest contributions to tin smelting in recent years. This work, stimulated by the need for better metal recoveries, relies on the formation and volatilization of tin as SnO_2 in a type of blast furnace (4). The process requires the addition of pyrites (FeS_2) to the tin-rich slag where it reacts to produce $FeSiO_3$ and SnS. The tin sulfide vapor oxidizes to SnO_2 and is carried out in the furnace exhaust gases where it is collected and recycled.

Fuming is also an alternative to roasting in the processing of low grade concentrates (5–25 wt % tin). This procedure yields a tin oxide dust, free of iron, which is again fed back to a conventional smelting furnace.

Other variations of the fuming process under development by the Commonwealth Scientific and Industrial Research Organization in Clayton, Australia, promise even greater efficiencies of metal recovery (5).

Refining.   The crude tin obtained from slags and by smelting ore concentrates is refined by further heat treatment or sometimes electrolytic processes.

The conventional heat-treatment refining includes liquidation or sweating and boiling, or tossing.

In liquidation, tin is heated on the sloping hearth of a small reverberatory furnace to just above its melting point. The tin runs into a so-called poling kettle, and metals that melt sufficiently higher than tin remain in the dross. Most of the iron is removed in this manner. Lead and bismuth remain, but arsenic, antimony, and copper are partly removed as dross.

In the final refining step, the molten tin is agitated in the poling kettles with steam, compressed air, or poles of green wood which produce steam. This process is referred to as boiling. The remaining traces of impurities form a scum which is removed and recirculated through the smelting cycle. The pure tin is cast in iron molds in the form of 45-kg ingots. Purity is guaranteed to exceed 99.8%.

Iron, copper, arsenic, and antimony can be readily removed by the above pyrometallurgical processes or variations of these (3). However, for the removal of large quantities of lead or bismuth, either separately or together, conventional electrolysis or a newly developed vacuum-refining process is used. The latter is now in use in Australia, Bolivia, Mexico, and the CIS (5).

Electrolytic refining is more efficient in regard to both the purity of the product and the ratio of tin to impurities in by-products. However, a large stock of the crude-tin anodes is tied up in the cells, requiring a high capital investment for equipment. An electrolytic plant working with an acid stannous salt at ca 108 A/m^2 requires ca 25 metric tons of working anodes for every ton of refined tin produced per day. Because of these high costs, fire refining should be used as much as possible. The by-products containing high lead, bismuth, and other metal impurities can then be treated in a modest electrolytic plant (6).

An electrorefining plant may operate with either an acid or an alkaline bath. The acid bath contains stannous sulfate, cresolsulfonic or phenolsulfonic acids (to retard the oxidation of the stannous tin in the solution), and free sulfuric acid with β-naphthol and glue as addition agents to prevent tree-like deposits on the cathode which may short-circuit the cells. The concentration of these addition agents must be carefully controlled. The acid electrolyte operates at room temperature with a current density of ca $86-108$ A/m^2, cell voltage of 0.3 V, and an efficiency of 85%. Anodes (95 wt % tin) have a life of 21 d, whereas the cathode sheets have a life of 7 d. Anode slimes may be a problem if the lead content of the anodes is high; the anodes are removed at frequent intervals and scrubbed with revolving brushes to remove the slime (7).

The alkaline bath contains potassium or sodium stannate and free alkali and operates without addition agents. The solution must, however, be heated to 82°C. Stannous tin must be absent because it passivates the anodes. The tin dissolves as stannate, SnO_3^{2-}, but only if the anodes are initially coated with a film of yellow-green hydrated oxide, $SnO_2 \cdot 2H_2O$. This so-called filming is accomplished by passing a high current through the anode after insertion in the cell; the current density is reduced to normal once the film is formed. Slow insertion of the anode with the current turned on gives the same result. The advantages of the alkaline electrolyte are ease of operation and the capability of using a lower grade of anode. The disadvantages are that the solution must be heated and that the current-carrying species is Sn^{4+}, giving an electrochemical equivalent half that of the acid electrolyte. At equilibrium, the lead plumbite is almost completely precipitated as hydroxide in the slime, but some antimony remains in solution. The spent anode is returned to the fire-refining process for recovery of lead, antimony, etc. The plated cathode sheets, weighing ca 90 kg, are melted in a holding pot and cast into ingots. Part of the metal from the holding pots is used to form the starter cathode sheets weighing ca 7.2 kg each.

Secondary Tin. In 1990, >7700 metric tons of tin were recovered in the United States from scrap (14). Sources include bronze rejects and used parts, solder in the form of dross or sweepings, dross from tinning pots, sludges from tinning lines, babbitt from discarded bearings, type-metal scrap, and clean tinplate clippings from container manufacturers. High purity tin is recovered by detinning clean tinplate (see RECYCLING).

Alloy scrap containing tin is handled by secondary smelters as part of their production of primary metals and alloys; lead refineries accept solder, tin drosses, babbitt, and type metal. This type of scrap is remelted, impurities such as iron, copper, antimony, and zinc are removed, and the scrap is returned to the market as binary or ternary alloy. The dross obtained by cleaning up the scrap metal is returned to the primary refining process.

Economic Aspects

Tin has long been regarded as a strategic metal because of its importance in canning, electrical, and transportation applications. Accordingly, it is stockpiled by the General Services Administration (GSA) at various locations in the country. On December 31, 1979, the U.S. Government stocks of pig tin totaled 203,691 metric tons, which included 170,670 metric tons above the goal of 33,021 t (9). On May 2, 1980, the Federal Emergency Agency set the new National Defense Stockpile Goal for tin at 42,700 metric tons. On January 2, 1980, the Strategic and Critical Materials Transaction Authorization Act became effective. This authorizes the President to dispose of materials determined to be excessive to the current needs of the stockpile. This act provides for the sale of up to 35,600 metric tons of tin, including a contribution of up to 5100 metric tons of tin to the International Tin Council (ITC) buffer stock (see below). The GSA set up a schedule to offer about 500 metric tons of Grade A tin, for domestic sales and consumption only, every other Tuesday beginning July 1, 1980. On December 14, 1981, the restrictions on exporting the GSA tin sold were lifted; sales increased immediately. Thus, from July 1, 1980, through December 11, 1981, the total GSA sales were 3170 metric tons. An additional 1815 metric tons were sold soon thereafter, mostly to traders (10).

Since the mid 1950s, the price of tin has been subject to an international agreement between producing and consuming nations. Under the International Tin Agreement (ITA), the ITC seeks to deal effectively with situations where a shortage or surplus of tin arises and prevent excessive price fluctuations (11). It attempts to ensure that a fair price is paid for tin on the world market and tries to stimulate export earnings from tin, especially for developing producing countries. To these ends, the ITC establishes floor and ceiling prices for tin. Prices within this range are subject to buffer-stock management. The floor price is maintained by ITC buffer stock purchases, or by applying export controls on member producing nations. The ceiling price is maintained through buffer-stock sales. The United States was a member of the Fifth ITA which expired on June 30, 1982. On October 17, 1981, the ITC price range was the equivalent of $11.92/kg for the floor price and $15.51/kg for the ceiling price (12).

World trading in tin occurs mostly at Penang, London, and New York. As shown in Table 2, most of the world's tin is produced in Southeast Asia, and generally, the marketing at Penang establishes the world tin price. The Penang price is determined daily at two Malaysian smelters by comparing bids from dealers and consumers with the available metal supply. The London Metal Exchange (LME) and New York market offer cash and forward metal prices. The LME price is based on the tin smelted in the U.K., whereas the New York price is an average of dealer-quoted prices.

Table 2. World Tin Production, Metric Tons[a]

Year	World	Malaysia	Thailand	Indonesia	Bolivia	Brazil	Zaire	UK
1925[b]	165,600	57,746	8,335	33,831	37,763		982	2,700
1950	164,800	58,694	10,530	32,617	31,714	183	11,947	904
1960	138,700	52,813	12,275	22,958	20,543	1,581	9,350	1,218
1965	154,400	64,692	19,353	14,935	23,407	1,220	6,311	1,334
1970	185,800	73,794	21,779	19,092	30,100	3,610	6,458	1,722
1975	177,700	64,364	16,406	25,346	28,324	5,000	4,562	3,330
1978	196,900	62,650	30,186	27,410	30,881	6,320	3,450	2,802
1979	200,700	62,995	33,962	29,440	27,781	6,645	3,300	2,374
1980	199,300	61,404	33,685	32,527	27,271	6,756	3,159	3,028
1985[c]	198,000	36,900	16,600	21,800	16,100	26,400	3,100	5,200
1990[c]	210,700	28,500	14,600	31,700	17,300	39,100	1,600	3,400

[a]Tin-in-concentrates (tin content) as compiled by the International Tin Council (13).
[b]From Ref. 1.

The United States is by far the largest consumer of tin, followed by Japan and Germany. A more detailed breakdown of U.S. tin uses in 1980 and 1990 is given in Table 3. The United States uses ca 6300 t/yr in the form of tin compounds (qv).

Tinplate provides an outlet for over one-third of the primary tin used in the United States. In 1980, ca 3.7×10^6 t of tinplate were produced in U.S. steel mills. Total world production in 1980 was 13.6×10^6 t. In the United States in 1980, ca 56×10^6 base boxes was used to make food containers (one base box comprises an area of 202,000 cm^2 or ca 31 in.^{2}).

Table 3. Distribution of U.S. Tin Consumption, Metric Tons

Product	1980[a]		1990[b]	
	Primary	Secondary	Primary	Secondary
tinplate	16,346		11,750	
solder	11,653	3,965	11,567	4,011
babbitt	1,537	843	552	211
bronze and brass	2,147	5,331	1,160	2,003
collapsible tubes and foil	526			
tinning	2,531	46	1,707	
bar tin and anodes	486		603	
tin powder	1,098		563	
alloys (misc)		134		
white metal	914		1,045	
chemicals			6,275	
other	7,104	1,701	1,394	1,522
Total	*44,342*	*12,020*	*36,616*	*7,747*
Grand Total	*56,362*		*44,363*	

[a]Ref. 15.
[b]Ref. 14.

Specifications and Analytical Methods

The ASTM Classification of Pig Tin B 339 lists three grades as shown in Tables 4 and 5 (16).

Table 4. Classification of Pig Tin[a]

ASTM designation	Tin, %[b,c]	General applications
ultrapure	99.95	analytical standards and research, pharmaceuticals, fine chemicals
A standard	99.85	collapsible tubes, unalloyed (block) tin products, electrotinning, tin-alloyed cast iron, high grade solders
grade A tinplate	99.85	food containers, foil

[a]Ref. 16.
[b]A more complete description of these grades is given in the full ASTM Standard B339-93.
[c]Percent shown is minimum.

Table 5. Chemical Composition and Impurity Contents

Element[a]	Composition, wt %		
	Grade A	Grade A for the manufacture of tinplate	Ultra pure grade
tin	99.85[b]	99.85[b]	99.95[b]
antimony	0.04	0.04	0.005
arsenic	0.05	0.05	0.005
bismuth	0.030	0.030	0.015
cadmium	0.001	0.001	0.001
copper	0.04	0.04	0.005
iron	0.010	0.010	0.010
lead	0.05	0.020	0.001
nickel + cobalt	0.01	0.01	0.010
sulfur	0.01	0.01	0.010
zinc	0.005	0.005	0.005
silver	0.01	0.01	0.010
other impurities[c]		0.010	0.010

[a]Value is maximum unless otherwise noted.
[b]Value is minimum.
[c]Maximum per impurity not listed above.

In the field, cassiterite ore is usually recognized by its high density (7.04 g/cm^3), low solubility in acid and alkaline solutions, and extreme hardness. Tin in solution is detected by the white precipitate formed with mercuric chloride. Stannous tin in solution gives a red precipitate with toluene-3,4-dithiol.

The tin content of ores, concentrates, ingot metal, and other products is determined by fire assay, fusion method, and volumetric wet analysis.

In a fire-assay method used at the smelters, a weighed quantity of concentrate is mixed with sodium cyanide in a clay or porcelain crucible and heated in a muffle furnace at red heat for 20–25 min. The tin oxide is reduced to metal, which is cleaned and weighed. Preliminary digestion of the concentrate with hydrochloric and nitric acids to remove impurities normally precedes the sodium cyanide fusion.

Tin ores and concentrates can be brought into solution by fusing at red heat in a nickel crucible with sodium carbonate and sodium peroxide, leaching in water, acidifying with hydrochloric acid, and digesting with nickel sheet. The solution is cooled in carbon dioxide, and titrated with a standard potassium iodate–iodide solution using starch as an indicator.

The determination of tin in metals containing over 75 wt % tin (eg, ingot tin) requires a special procedure (17). A 5-g sample is dissolved in hydrochloric acid, reduced with nickel, and cooled in CO_2. A calculated weight of pure potassium iodate (dried at 100°C) and an excess of potassium iodide (1:3) are dissolved in water and added to the reduced solution to oxidize 96–98 wt % of the stannous chloride present. The reaction is completed by titration with 0.1 N KIO$_3$–KI solution to a blue color using starch as the indicator.

Several ASTM methods are available for the determination of tin in tin-containing alloys such as solder, babbitt, and bronze (18).

The purity of commercial tin is under strict control at the smelters. Photometric, chemical, atomic absorption, fluorimetric, and spectrographic methods are available for the determination of impurities (17).

Health and Safety Factors

Tests have shown that considerable quantities of tin can be consumed without any effect on the human system. Small amounts of tin are present in most liquid canned products; the permitted limit of tin content in foods is 300 mg/kg in the United States and 250 ppm in the UK, which far exceed the amount in canned products of good quality (19) (see also TIN COMPOUNDS).

Uses

Tin is used in various industrial applications as cast and wrought forms obtained by rolling, drawing, extrusion, atomizing, and casting; tinplate, ie, low carbon steel sheet or strip rolled to 0.15–0.25 mm thick and thinly coated with pure tin; tin coatings and tin alloy coatings applied to fabricated articles (as opposed to sheet or strip) of steel, cast iron, copper, copper-base alloys, and aluminum; tin alloys; and tin compounds.

Cast and Wrought Forms. Thousands of tons of tin ingots are cast into anodes for plating processes. Tin foil is used for electrical condensers, bottle-cap liners, gun charges, and wrappings for food. Tin wire is used for fuses and safety plugs. Extruded tin pipe and tin-lined brass pipe are the first choice for conveying distilled water and carbonated beverages. Sheet tin is used to line storage tanks for distilled water. Tin powder is used in powder metallurgy, the largest use going to tin powder mixtures with copper to form bronze parts. It is also used for coating paper and for solder pastes. In the float-glass process, adopted by all leading plate-glass manufacturers, the molten glass is allowed to float and solidify on the surface of a pool of molten tin which provides an ideally flat surface. The endless glass ribbon has a surface so smooth that costly grinding and polishing are unnecessary.

Tinplate. The development of tinplate was associated with the need for a reliable packaging material for preserving foods. It comprises in one inexpensive material the strength and formability of steel and the corrosion resistance, solderability, absence of toxicity, and good appearance of tin. The tin coating is applied by electroplating in a continuous process or by passing cut sheet through a bath of molten tin. In the United States, tinplate is now made mainly by the electrolytic process with less than 1% of production from hot-tinning machines.

The electrolytic process is flexible and capable of applying tin coatings from 250 nm to 2.5 μm on each face. A thick coating can be applied to one side of the sheet and a thinner coating to the other (differential tinplate) providing a cost savings to the can manufacturer if less protection is needed on the outside of the can. The thinner coatings usually require a baked enamel coating over the tin, except when packaging dry foods and nonfood products.

In addition to the packaging of foods and beverages in regular containers, a large quantity of tinplate is used in the form of aerosol containers for cosmetics, paint, insecticides, polishes, and other products. Decorative trays, lithographed boxes, and containers of unusual shape are additional outlets for tinplate.

A tinplate container has an energy advantage over some of the competitive container materials, as shown in Table 6. By virtue of its magnetic properties, tinplated material is more readily separated by other forms of industrial or domestic waste.

Tin Coatings. The coating may be applied by hot-dipping the fabricated article in liquid tin, by electroplating using either acid or alkaline electrolytes, and by immersion tinning (see below). The hot-tinned coating is bright. Electrodeposited coatings are normally dull as plated, but may be flow-brightened by heating momentarily to the melting point by induction or in hot air or oil. Proprietary bright tin-plating processes have been developed and are in commercial use. These incorporate long-chain organic molecules as brightening and leveling agents in a stannous sulfate electrolyte. Brighteners for other electrolytes have not been developed.

The coating thickness may range from 0.0025 to 0.05 mm, depending on the type of protection required. Pure tin coatings are used on food-processing equipment, milk cans, kitchen implements, electronic and electrical components, fasteners, steel and copper wire, pins, automotive bearings, and pistons.

For articles that require only a very thin film of tin, seldom exceeding 0.8 μm immersion tin coatings are applied. The process is based on chemical displacement by immersion in a solution of tin salts. Recently, a new autocatalytic tin-deposition process was developed at the research level. It promises to be useful to coat any base material including plastics, in addition to providing coatings of any thickness desired (21).

Tin Alloys. *Coatings.* Tin-alloy coatings provide harder, brighter, and more corrosion-resistant coatings than tin alone. Tin–copper electrodeposited coatings (12 wt % tin) have the appearance of 24-carat gold and provide a bronze finish for furniture hardware, trophies, and ornaments (see COATINGS). They also provide a stop-off coating (resist) for nitriding.

Tin–lead coatings (10–60 wt % tin) can be applied by hot-dipping or electrode position to steel and copper fabricated articles and sheet. A special product is terne plate used for roofing and flashings, automobile fuel tanks and fittings, air filters, mufflers, and general uses such as covers, lids, drawers, cabinets, con-

Table 6. Energy Consumption in Container Manufacturing[a]

Container type	Energy consumed in producing raw material, GJ/t[b]	Number of containers per ton	Energy consumed per container, kJ[b]
tinplate can	49	16,500	3,010
aluminum can	395	44,500	8,660
bimetallic can	77	18,400	4,210
glass bottle			
returnable	54.8	2,000	27,540
nonreturnable	54.8	4,000	13,770

[a]Ref. 20.
[b]To convert J to Btu, divide by 1054.

soles for instruments, and for radio and television equipment. Terne plate is low carbon steel, coated by a hot-dip process with an alloy of tin and lead, commonly about 7–25 wt % tin, remainder lead. Electroplating is another possibility.

Because of the ease with which they can be soldered, electroplated tin–lead coatings of near eutectic composition (62 wt % tin) are extensively used in the electronics industry for coating printed circuit boards and electrical connectors, lead wires, capacitor and condenser cases, and chassis.

Tin–nickel electrodeposited coatings (65 wt % tin) provide a bright decorative finish with a corrosion resistance that exceeds that of nickel and copper–nickel–chromium coatings. A new and expanding use for tin–nickel is for printed circuit boards. Such coatings deposited on the copper-clad boards are etch-resistant and provide protection for the conducting path. The well-established decorative and functional uses are for watch parts, drawing instruments, scientific apparatus, refrigeration equipment, musical instruments, and handbag frames.

Tin–zinc coatings (75 wt % tin) have application as a solderable coating for radio, television, and electronic components. They also provide galvanic protection for steel in contact with aluminum.

A ternary tin–copper–lead coating, 2 Cu–8 Sn–90 Pb, is a standard overplate for steel-backed copper–lead automotive bearings.

Tin–cadmium coatings are particularly resistant to marine atmospheres and have applications in the aviation industry.

Solder.　Tin and lead combine easily to form a group of alloys known generally as soft solders. The joining of metals with tin-containing solders can be attributed to several properties. Their low melting point allows simple equipment to be used for melting and joining, the alloys are unsurpassed in wetting and adhering to clean metal surfaces and flowing into small spaces, and they are relatively cheap. The tin–lead solders have no serious competitors in the field of low temperature joining (see SOLDERS AND BRAZING ALLOYS).

Tin is the important constituent in solders because it is the element that wets the base metal, such as copper and steel, by alloying with it. Solders are used mainly in auto radiators, air conditioners, heat exchangers, plumbing and sheet-metal joining, container seaming, electrical connections in radio and television, generating equipment, telephone wiring, electronic equipment and computers, and aerospace equipment (see Table 7). Lead-free solder alloys, where the tin is alloyed individually with antimony, silver, gold, zinc, or indium, are available for special joining applications where properties such as high strength, absence of toxicity, and special corrosion resistance are required.

Low melting or fusible alloys (mp, 20–176°C) may be loosely described as solders and are employed for sealing and joining materials which may be damaged in ordinary soldering practice. They have applications in automatic safety devices, foundry patterns, electroforming, tube bending, tempering baths, molds for plastics, and denture models. Fusible alloys, mostly eutectic alloys, are usually two-, three-, four-, or even five-component mixtures of bismuth, tin, lead, cadmium, indium, and gallium.

Bronze.　Copper–tin alloys, with or without other modifying elements, are classed under the general name of bronzes. They can be wrought, sand-cast, or continuously cast into shapes. Binary tin–copper alloys are difficult to

Table 7. Solders and Their Uses

Alloy content, wt %	Melting range, °C	Typical tensile strength of cast solder, MPa[a]	Uses
General engineering			
60 Sn, 40 Pb	183–188	53	electronics and instruments
50 Sn, 50 Pb	183–212	45	sheet-metal work and light engineering
40 Sn, 60 Pb	183–234	43	general engineering and capillary fittings
30 Sn, 70 Pb	183–255	43	plumbers' solder, cable joining, automobile radiators
20 Sn, 80 Pb	183–276	42	automobile radiators
40 Sn, 57.8 Pb, 2.2 Sb	185–227	51	similar to 40 Sn, 60 Pb solder
Special purpose			
2 Sn, 98 Pb	315–322	28	tinplate can side seams
10 Sn, 90 Pb	267–301	37	cryogenics
5 Sn, 93.5 Pb, 1.5 Ag	296–301	39	
62 Sn, 36 Pb, 2 Ag	178	43	
95 Sn, 5 Sb	236–243	40	creep resistance
95 Sn, 5 Ag	221–225	59	
98 Sn, 2 Ag	221–235	26	food and beverage containers
100 Sn	232	14	
52 Sn, 30 Pb, 18 Cd	145	43	low melting solder
80 Sn, 20 Zn	200–265	70	soldering aluminum

[a]To convert MPa to psi, multiply by 145.

cast because they are prone to gassing, which can be alleviated by additions of phosphorus and zinc. The phosphor bronzes (5–10 wt % tin) are preferred because they have superior elastic properties, excellent resistance to alternating stress and corrosion fatigue and to corrosive attack by the atmosphere and water, and superior bearing properties. So-called gun metals containing 1–6 wt % zinc and 5–10 wt % tin are gas-free, pressure-tight alloys used for valves and fittings for water and steam lines. These alloy types may be further modified with lead to improve machinability and, in the case of the phosphor bronzes, to obtain a more conformable bearing alloy. The 85 Cu–5 Sn–5 Zn–5 Pb alloy is a popular composition. Bronzes are especially applicable to marine and railway engineering pumps, valves and pipe fittings, bearings and bushings, gears and springs, and ship propellers. Included in special bronze alloys is bell metal, known for its excellent tonal quality, containing 20 wt % tin, and statuary bronze (see COPPER ALLOYS).

Bearing Metals. Metals used for casting or lining bearing shells are classed as white bearing alloys, but are known commercially as babbitt (see BEARING MATERIALS). The term white metal was used by Isaac Babbitt in 1839 in his description of tin-base bearing metals supported by a stronger shell. Although white metal is a general term for many white-colored alloys of relatively low melting point, the white-metal product mentioned in Table 4 is made from pewter, britannia, or jeweler's metal (alloys containing ca 90–95 wt % tin, 1–8 wt % antimony and 0.5–3 wt % copper).

The term babbitt includes high tin alloys (substantially lead-free) containing >80 wt % tin, and high lead alloys containing ≥70 wt % lead and ≤12 wt % tin. Both have the characteristic structure of hard compounds in a soft matrix, and although they contain the same or similar types of compounds, they differ in composition and properties of the matrix.

The common high tin babbitts are all based upon the tin–antimony–copper system. Compositions and properties of the more widely used tin-base bearing alloys are given in Table 8. Antimony up to 8.0% strengthens the bearing alloy matrix by dissolving in the tin. Above 8.0%, hard particles of tin–antimony compound (SnSb) are formed which tend to float. Additions of copper secure a uniform distribution of these hard particles in a soft but rigid bearing matrix.

The lead-base babbitts are based upon the lead–antimony–tin system, and, like the tin-base, have a structure of hard crystals in a relatively soft matrix. The lead-base alloys are, however, more prone to segregation, have a lower thermal conductivity than the tin-base babbitts, and are employed generally as an inexpensive substitute for the tin-base alloys. Properly lined, however, they function satisfactorily as bearings under moderate conditions of load and speed.

Both types of babbitt are easily cast and can be bonded rigidly to cast iron, steel, and bronze backings. They perform satisfactorily when lubricated against a soft steel shaft, and occasional corrosion problems with lead babbitt can be corrected by increasing the tin content or shifting to high tin babbitt.

Babbitt alloys are suitable for hundreds of types of installations involving the movement of machinery, eg, the main, crankshaft, connecting rod big end, camshaft, and journal bearings associated with marine propulsion, railroad

Table 8. Composition and Properties of ASTM B 23 Bearing Alloys[a]

Alloy no.	Nominal composition, wt %	Compressive strength, MPa[b] Yield[c]	Ultimate[d]	Brinell hardness[e]
1	91 Sn, 4.5 Sb, 4.5 Cu	30	89	17.0
2	89 Sn, 7.5 Sb, 3.5 Cu	42	103	24.5
3	84 Sn, 8 Sb, 8 Cu	46	121	27.0
4	75 Sn, 12 Sb, 10 Pb, 3 Cu	38	111	24.5
5	65 Sn, 15 Sb, 18 Pb, 2 Cu	35	104	22.5

[a]Ref. 22.
[b]To convert MPa to psi, multiply by 145.
[c]Determined at a total deformation of 0.125% reduction in gauge length.
[d]Determined from unit load required to produce a 25% deformation of specimen length.
[e]Average of three values using a 10 mm ball and applying a 500-kg load for 30 s.

and automotive transportation, compressors, motors, generators, blowers, fans, rolling-mill equipment, etc.

The field of bearing metals also includes bronze and aluminum alloys. The aluminum–tin alloys have fatigue properties comparable with white-metal alloys at ordinary temperatures but at the temperatures encountered in automobile engines (up to 250°C) they are much stronger. An aluminum–tin alloy containing 6.5 wt % Sn, 1 wt % Cu, and 1 wt % Ni is used in applications such as bushings and solid bearings in aircraft landing gear assemblies subject to shock loads of 48 MPa (ca 7000 psi), as tracks for vertical boring mills, and as floating bearings in gas-turbine engines, diesel locomotive and tractor engines, and cold-rolling mills. A modification of this alloy contains also 2.5 wt % silicon. Increasing the tin content of 20 wt % gives a good compromise between fatigue strength and softness, and such steel-backed aluminum–tin alloys are used for connecting rod and main bearings for passenger cars, automatic transmission bushings, camshaft bearings, and thrust washers.

Pewter.   Modern pewter may have a composition of 90–95 wt % tin, 1–8 wt % antimony, and 0.5–3 wt % copper. Lead should be avoided by contemporary craftsman because it causes the metal surface to blacken with age. Pewter metal can be compressed, bent, spun, and formed into any shape, as well as being easily cast. A wide variety of consumer articles are available from domestic and foreign manufacturers. Reproductions of pewter objects from colonial times, some cast from the original molds, are popular. The annual U.S. production of pewter exceeds 1100 t.

Type Metals.   The printing trade still requires some amounts of lead-based alloys containing 10–25 wt % antimony and 3–13 wt % tin. By varying the tin and antimony content, type suitable for each printing process can be obtained. Linotype machines demand a fluid and mobile metal with a short freezing range. Casting stereotype plates requires cool metal and a hot box with progressive solidification from the bottom up. Monotype, with the higher percentages of tin and antimony, provides a type with a fine face and superior wear resistance. Foundry type is extra hard-wearing and has extra long life, which reduces the need for recasting for duplicating jobs.

Alloyed Iron.   Tin-alloyed flake and nodular cast irons are widely used throughout the world. Estimated 1980 consumption was ca 1,200 t. As little as 0.1% tin when added to flake and spheroidal graphite cast irons in the pouring ladle gives the iron a structure that is completely pearlitic. Tin-inoculated iron has a uniformity of hardness, improved machinability, wear resistance, and better retention of shape on heating. Where pearlitic and heat-resistant cast irons are required, such as for engine blocks, transmissions, and automotive parts, tin additions may provide a suitable material.

Special Alloys.   Alloys of tin with the rarer metals, such as niobium, titanium, and zirconium, have been developed. The single-phase alloy Nb_3Sn [*12035-04-0*] has the highest transition temperature of any known superconductor (18 K) and appears to keep its superconductivity in magnetic fields up to at least 17 T (170 kG) (see SUPERCONDUCTING MATERIALS (SUPPLEMENT)). Niobium–tin ribbon, therefore, is of practical importance for the construction of high field superconducting solenoid magnets.

Tin is an important addition to titanium. As a nominal addition (2–4% Sn), tin is a solid-state strengthener, retards interstitial diffusion, and promotes plasticity and free-scaling. Some of the more widely used commercial alloys include 92.5 Ti–5 Al–2.5 Sn, 86 Ti–6 Al–6 V–2 Sn, and 86 Ti–6 Al–2 Sn–4 Zn–2 Mo.

Because of its low neutron absorption, zirconium is an attractive structural material and fuel cladding for nuclear power reactors, but it has low strength and highly variable corrosion behavior. However, Zircalloy-2, with a nominal composition of 1.5 wt % tin, 0.12 wt % iron, 0.05 wt % nickel, 0.10 wt % chromium, and the remainder zirconium, can be used in all nuclear power reactors that employ pressurized water as coolant and moderator (see NUCLEAR REACTORS).

Dental amalgams, mainly silver–tin–mercury alloys, have been used as fillings for many years (see DENTAL MATERIALS). The most common alloy contains 12 wt % tin.

Other Uses.   The production of finished shapes from iron powder by compacting and sintering utilizes about 100,000 t of iron powder annually; copper powder (2–10 wt %) is normally added as a sintering aid. Addition of 2% tin powder or equal amounts of tin and copper powder considerably lowers the sintering temperature and time of sintering at a cost saving. The tin addition also improves dimensional control. Iron powder plus 10 wt % powdered lead–tin metal is pressed and sintered to make pistons for use in automotive hydraulic brake cylinders.

The electronics and aerospace industries have for a number of years used gold-plated printed circuit boards and component leads where highest reliability is desired. Problems in the use of gold coatings have plagued the industry and the trend is toward the substitution of tin–lead or tin coatings for the gold coatings. Tin–nickel coatings with a thin flash of tin or gold are also used as a substitute for heavy gold coatings (see ELECTRICAL CONNECTORS).

BIBLIOGRAPHY

"Tin and Tin Alloys" in *ECT* 2nd ed., Vol. 20, pp. 273–293, by R. M. MacIntosh, Tin Research Institute, Inc.; in *ECT* 3rd ed., Vol. 23, pp. 18–35, by D. J. Maykuth, Tin Research Institute, Inc.

1. C. L. Mantell, *Tin: Its Mining, Production, Technology, and Applications*, 2nd ed., Reinhold Publishing Corp., New York, 1949.

2. C. J. Faulkner, *The Properties of Tin*, Publication 218, International Tin Research Institute, London, 1965, 55 pp.

3. S. C. Pearce in J. Cigan, T. S. Mackey, and T. O'Keefe, eds., *Proceedings of a World Symposium on Metallurgy and Environmental Control at 109th AIME Annual Meeting*, The Metallurgical Society of AIME, Warrendale, Pa., 1980, pp. 754–770.

4. S. M. Kolodin, *Vetoritznoe Olova (Secondary Tin)*, Moscow, 1964, p. 207.

5. T. S. Mackey, *J. Met.* **34**, 72 (Apr. 1982).

6. P. A. Wright, *Extractive Metallurgy of Tin*, American Elsevier Publishing Co., New York, 1966.

7. C. L. Mantell, *J. Met.* **15**, 152 (1963).

8. U.S. Bureau of Mines, *Metal Statistics 1981*, American Metal Market Fairchild Publications, New York, 1981, p. 219.
9. J. F. Carlin, Jr., *Tin, Bulletin 671*, a chapter from *Mineral Facts and Problems*, 1980 ed., Superintendent of Documents, Washington, D.C., 1980.
10. *Am. Met. Mark.* **89**(248), 2 (Dec. 24, 1981).
11. P. M. Dinsdale, *A Guide to Tin*, Publication No. 540, International Tin Research Institute, London.
12. J. F. Carlin, Jr., *Tin in October 1981*, *Tin Industry Monthly*, Mineral Industry Surveys, U.S. Bureau of Mines, Washington, D.C., Dec. 31, 1981.
13. *Monthly Statistical Bulletin*, International Tin Council, London, U.K.
14. *Metal Statistics, 1992*, 84th ed., American Metal Market, New York, 1992.
15. U.S. Bureau of Mines, *Metal Statistics 1982*, American Metal Market Fairchild Publications, New York, 1982, p. 211.
16. *Nonferrous Metals*; *Electrodeposition Coatings*; *Metal Powders*; *Surgical Implants, Part 7, 1972 Annual Book of ASTM Standards*, American Society for Testing Materials, Philadelphia, Pa., 1972.
17. J. W. Price and R. Smith in W. Fresenius, ed., *Handbook of Analytical Chemistry*, Part III, Vol. 4a, Springer-Verlag, Berlin, Heidelberg, New York, 1978.
18. *Nonferrous Metals*; *Book of ASTM Standards*, Sect. 2, Vol. 2.04, American Society for Testing Materials, Philadelphia, Pa., 1985.
19. G. W. Monier-Williams, *Trace Elements in Food*, John Wiley & Sons, Inc., New York, 1950, p. 138.
20. *Environmental Data Handbook*, American Can Company, Technical Information Center, Barrington, Ill.
21. M. E. Warwick and B. J. Shirley, *Trans. Inst. Met. Finish.* **58**, 9 (1980).
22. *Metals Handbook*, *Vol. 1*, *Properties and Selection of Metals*, 8th ed., American Society for Metals, Metals Park, Ohio, 1961, pp. 843–864.

CHARLES C. GAVER, JR.
Consultant

TIN COMPOUNDS

Tin is between germanium and lead in Group 14(IVA) of the Periodic Table. As a bronze component, tin was used as early as 3500 BC. The pure metal was not used until 600 BC (see TIN AND TIN ALLOYS). The history of tin compounds dates to the Copts of Egypt, who reportedly used basic tin citrate [59178-29-9] in dye preparation. In 1605 stannic chloride was prepared. The alchemists' symbol for tin is ♃. Tin occurs in the earth's crust to the extent of 40 grams per metric ton and is present in the form of nine different minerals from two types of deposits. The most commercially significant ore is cassiterite [1317-45-9] (tinstone), SnO_2. The more economically valuable deposits of cassiterite are heavily concentrated in bands and layers of varying thickness, in Malaysia, Thailand, Indonesia, and the People's Republic of China. Complex sulfidic ores, combinations of the sulfides of base metals and pyrites, are in lode deposits. These, economically significant only in Bolivia, are stannite [12019-29-3], $SnS_2 \cdot Cu_2S \cdot FeS$; herzenbergite [14752-27-3], SnS; teallite [12294-02-9], SnS·PbS; franckeite [12294-04-1], $2SnS_2 \cdot Sb_2S_3 \cdot 5PbS$; cylindrite [12294-05-0, 59858-98-9], $Sn_6Pb_6Sb_2S_{11}$; plumbostannite, $2SnS_2 \cdot 2PbS \cdot 2(Fe,Zn)S \cdot Sb_2S_3$; and canfieldite [12250-27-0], $4Ag_2S \cdot SnS_2$. The important tin-producing countries are Malaysia, Bolivia, Indonesia, Nigeria, Thailand, Zaire, and the People's Republic of China. Smaller quantities are produced in the United Kingdom, Burma, Japan, Canada, Portugal, Spain, and Australia. Tin is also normally present in natural waters, in soil, in marine organisms and animals, in the milk of lactating animals, in meteorites, in the tissues of animals, and in minor amounts in human organs.

Tin, having valence of +2 and +4, forms stannous (tin(II)) compounds and stannic (tin(IV)) compounds. Tin compounds include inorganic tin(II) and tin(IV) compounds; complex stannites, $MSnX_3$, and stannates, M_2SnX_6, and coordination complexes, organic tin salts where the tin is not bonded through carbon, and organotin compounds, which contain one-to-four carbon atoms bonded directly to tin.

Of the large volume of tin compounds reported in the literature, possibly only ca 100 are commercially important. The most commercially significant inorganic compounds include stannic chloride, stannic oxide, potassium stannate, sodium stannate, stannous chloride, stannous fluoride, stannous fluoroborate, stannous oxide, stannous pyrophosphate, stannous sulfate, stannous 2-ethylhexanoate, and stannous oxalate. Also important are organotins of the dimethyltin, dibutyltin, tributyltin, dioctyltin, triphenyltin, and tricyclohexyltin families.

Inorganic Tin Compounds

Because of its amphoteric nature, tin reacts with strong acids and strong bases but remains relatively resistant to neutral solutions. A thin oxide film forms on tin exposed to oxygen or dry air at ordinary temperatures; heat accelerates this oxide formation. Tin is not attacked by gaseous ammonia even when heated. Chlorine, bromine, and iodine react with tin at normal temperatures, and fluorine reacts at 100°C, forming the appropriate stannic halides. Tin is easily attacked by hydrogen iodide and hydrogen bromide, but less readily by hydrogen chloride; it is weakly attacked by gaseous hydrogen fluoride, and it slowly dissolves in aqueous hydrochloric acid. Hot concentrated sulfuric acid reacts with tin, forming stannous sulfate, whereas dilute sulfuric acid reacts only slowly with tin at room temperature. Reaction of tin with dilute nitric acid yields soluble tin nitrates; in concentrated nitric acid, tin is oxidized to insoluble hydrated stannic oxide. No reaction occurs upon direct union of tin with hydrogen, nitrogen, or carbon dioxide.

If tin and sulfur are heated, a vigorous reaction takes place with the formation of tin sulfides. At 100–400°C, hydrogen sulfide reacts with tin, forming stannous sulfide; however, at ordinary temperatures no reaction occurs. Stannous sulfide also forms from the reaction of tin with an aqueous solution of sulfur dioxide. Molten tin reacts with phosphorus, forming a phosphide. Aqueous solutions of the hydroxides and carbonates of sodium and potassium, especially when warm, attack tin. Stannates are produced by the action of strong sodium hydroxide and potassium hydroxide solutions on tin. Oxidizing agents, eg, sodium or potassium nitrate or nitrite, are used to prevent the formation of stannites and to promote the reactions.

Stannic and stannous chloride are best prepared by the reaction of chlorine with tin metal. Stannous salts are generally prepared by double decomposition reactions of stannous chloride, stannous oxide, or stannous hydroxide with the appropriate reagents. Metallic stannates are prepared either by direct double decomposition or by fusion of stannic oxide with the desired metal hydroxide or carbonate. Approximately 80% of inorganic tin chemicals consumption is accounted for by tin chlorides and tin oxides.

Halides.   The tin halides of the greatest commercial importance are stannous chloride, stannic chloride, and stannous fluoride. Tin halides of less commercial importance are stannic bromide [7789-67-5], stannic iodide [7790-47-8], stannous bromide [10031-24-0], and stannous iodide [10294-70-9] (1).

Stannous Chloride.   Stannous chloride is available in two forms: anhydrous stannous chloride, $SnCl_2$, and stannous chloride dihydrate [10025-69-1], $SnCl_2 \cdot 2H_2O$, also called tin crystals or tin salts. These forms are sometimes used interchangeably; however, where stability, concentration, and adaptability are important, anhydrous stannous chloride is preferred. Even after long storage, changes in the stannous tin content of anhydrous stannous chloride are extremely low. Physical properties of the tin chlorides are listed in Table 1.

Anhydrous stannous chloride, a water-soluble white solid, is the most economical source of stannous tin and is especially important in redox and plating reactions. Preparation of the anhydrous salt may be by direct reaction of chlorine and molten tin, heating tin in hydrogen chloride gas, or reducing stannic chloride solution with tin metal, followed by dehydration. It is soluble in a number of organic solvents (g/100 g solvent at 23°C): acetone 42.7, ethyl alcohol 54.4, methyl isobutyl carbinol 10.45, isopropyl alcohol 9.61, methyl ethyl ketone 9.43; isoamyl acetate 3.76, diethyl ether 0.49, and mineral spirits 0.03; it is insoluble in petroleum naphtha and xylene (2).

Solutions of anhydrous stannous chloride are strongly reducing and thus are widely used as reducing agents. Dilute aqueous solutions tend to hydrolyze and oxidize in air, but addition of dilute hydrochloric acid prevents this hydrolysis; concentrated solutions resist both hydrolysis and oxidation. Neutralization of tin(II) chloride solutions with caustic causes the precipitation of stannous oxide or its metastable hydrate. Excess addition of caustic causes the formation of stannites. Numerous complex salts of stannous chloride, known as chlorostannites, have been reported (3). They are generally prepared by the evaporation of a solution containing the complexing salts.

Table 1. Physical Properties of Tin Chlorides

Property	$SnCl_2$	$SnCl_2 \cdot 2H_2O$	$SnCl_4$	$SnCl_4 \cdot 5H_2O$
mol wt	189.60	225.63	260.50	350.58
mp, °C	246.8	37.7	−33	ca 56 dec
bp, °C	623		114	
density, at 25°C, g/cm^3	3.95	2.63	2.23^a	2.04

^{a}At 20°C.

Anhydrous stannous chloride is used extensively in the plating industry, eg, in the high speed electrotinning of continuous strip steel by the halogen process involving an aqueous solution of stannous chloride and alkali–metal fluorides, in a variety of formulations in immersion tinning processes, and in tin alloy plating (4,5) (see ELECTROPLATING). The strongly reducing nature of this chloride has contributed to its commercial development and success. Established applications of this property include its use as an analytical reducing reagent, a reducing agent in inorganic and organic chemicals manufacture and in the photoleaching of dyes, and as a sensitizing agent for nonconductive surfaces before silver coating or other metallization processes. Originally, surface sensitization with stannous chloride solutions was used only on glass prior to silvering (6). It has come to be used to sensitize plastics prior to their electroless coating with metals, eg, nickel or copper (7–9) (see ELECTROLESS PLATING).

Stannous chloride is also used as a food additive, for which use it has FDA GRAS approval (10). Other approvals include its use as a preservative for canned soda water, a color-retention agent in canned asparagus, and a component in food-packaging materials (10–14). It catalyzes a variety of organic reactions, eg, condensation, curing of resins and rubbers, esterification, halogenation, hydrogenation, oxidation, polymerization, hydrocarbon conversion, etc. Minor applications include the use of stannous chloride as an additive to drilling muds, an antisludge agent for oils, in tin coating of sensitized paper, and to improve the dyeing fastness of synthetic fibers, eg, polyamides (see PETROLEUM, DRILLING FLUIDS; POLYAMIDES, FIBERS).

Stannous Chloride Dihydrate. A white crystalline solid, stannous chloride dihydrate is prepared either by treatment of granulated tin with hydrochloric acid followed by evaporation and crystallization or by reduction of a stannic chloride solution with a cathode or tin metal followed by crystallization. It is soluble in methanol, ethyl acetate, glacial acetic acid, sodium hydroxide solution, and dilute or concentrated hydrochloric acid. It is soluble in less than its own weight of water, but with much water it forms an insoluble basic salt.

Stannic Chloride. Stannic chloride is available commercially as anhydrous stannic chloride, $SnCl_4$ (tin(IV) chloride); stannic chloride pentahydrate, $SnCl_4 \cdot 5H_2O$; and in proprietary solutions for special applications. Anhydrous stannic chloride, a colorless fuming liquid, fumes only in moist air, with the subsequent hydrolysis producing finely divided hydrated tin oxide or basic chloride. It is soluble in water, carbon tetrachloride, benzene, toluene, kerosene, gasoline, methanol, and many other organic solvents. With water, it forms a number of hydrates, of which the most important is the pentahydrate. Although stannic chloride is an almost perfect electrical insulator, traces of water make it a weak conductor.

Stannic chloride is made by the direct chlorination of tin at 110–115°C. Any stannous chloride formed in the process is separated from the stannic chloride by volatilization and subsequently chlorinated to stannic chloride. The latter is inert to steel in the absence of moisture and is shipped in plain steel drums of special design. Because prolonged contact with the skin causes burns, goggles and protective clothing should be used in the handling of stannic chloride. Stannic chloride, like stannous chloride, also forms many complexes (3).

The main uses of stannic chloride are as a raw material for the manufacture of other tin compounds, especially organotins, and in the surface treatment of glass (qv) and other nonconductive materials, whereby deposition of stannic oxide from stannic chloride solutions onto the nonconductive substrate gives it strength, abrasion-resistance, and conductivity. Very thin stannic oxide films (less than 100 nm thick) are thus used to strengthen glassware for returnable and nonreturnable foodstuff bottles and jars and for restaurant and catering glasses, which are subject to rigorous use (16). Glass treated in this way can also be made considerably lighter, which is advantageous to packing, shipping, and handling. The process involves passing freshly formed glassware through an oven maintained in an atmosphere containing stannic chloride vapor. The chloride breaks down, leaving at the glass temperature a stannic oxide deposit on the glass surface (17,18).

Where electrical conductivity and optical transparency are required, thicker (greater than 1 μm) films of stannic oxide are necessary. Such treated glasses are used in low intensity lighting panels and display signs, fluorescent lights, electron-beam control in cathode-ray tubes, and deicing windshields in aircraft. The deposition of stannic oxide films on glass surfaces is accomplished by the decomposition of stannic chloride vapor at 500–600°C and depositing it onto the glass surface or by spraying it from aqueous or mixed-organic solutions.

Stannic chloride is also used widely as a catalyst in Friedel-Crafts acylation, alkylation and cyclization reactions, esterifications, halogenations, and curing and other polymerization reactions. Minor uses are as a stabilizer for colors in soap (19), as a mordant in the dyeing of silks, in the manufacture of blueprint and other sensitized paper, and as an antistatic agent for synthetic fibers (see DYES, APPLICATION AND EVALUATION; ANTISTATIC AGENTS).

Stannic Chloride Pentahydrate.   Stannic chloride pentahydrate [*10026-06-9*] is a white, crystalline, deliquescent solid that is soluble in water or methanol and stable at 19–56°C. It is used in place of the anhydrous chloride where anhydrous conditions are not mandatory. It is easier to handle than the fuming anhydrous liquid form. The pentahydrate is prepared by dissolving stannic chloride in hot water, thereby forming the pentahydrate at a temperature above the melting point and crystallizing by cooling. The cake is broken into small lumps for packaging.

A stannic chloride pentahydrate–ammonium bifluoride formulation for fireproofing wool is commercially available and used in New Zealand and Australia (20) (see FLAME RETARDANTS FOR TEXTILES).

Stannous Fluoride.   Stannous fluoride [*7783-47-3*] (tin(II) fluoride), mol wt 156.7, mp 219.5°C, occurs as opaque, white, lustrous crystals, which are soluble in potassium hydroxide, fluorides, and water (31 g/100 g H_2O at 0°C, 78.5 g/100 g H_2O at 106°C) and practically insoluble in methanol, ether, and chloroform. Dilute aqueous solutions hydrolyze unless stabilized with excess acid. The specific gravity of a saturated aqueous solution at 25°C is 1.51. Commercially, stannous fluoride is produced by the reaction of stannous oxide and aqueous hydrofluoric acid or by dissolving tin in anhydrous or aqueous hydrofluoric acid.

The principal commercial use of stannous fluoride is in toothpaste formulations and other dental preparations, eg, topical solutions, mouthwash, chewing gum, etc, for preventing demineralization of teeth (21–23) (see DENTIFRICES).

Oxides. *Stannous Oxide.* Stannous oxide, SnO ((tin(II) oxide), mol wt 134.70, sp gr 6.5) is a stable, blue-black, crystalline product that decomposes at above 385°C. It is insoluble in water or methanol, but is readily soluble in acids and concentrated alkalies. It is generally prepared from the precipitation of a stannous oxide hydrate from a solution of stannous chloride with alkali. Treatment at controlled pH in water near the boiling point converts the hydrate to the oxide. Stannous oxide reacts readily with organic acids and mineral acids, which accounts and for its primary use as an intermediate in the manufacture of other tin compounds. Minor uses of stannous oxide are in the preparation of gold–tin and copper–tin ruby glass.

Stannous Oxide Hydrate. Stannous oxide hydrate [*12026-24-3*], $SnO \cdot H_2O$ (sometimes erroneously called stannous hydroxide or stannous acid), mol wt 152.7, is obtained as a white amorphous crystalline product on treatment of stannous chloride solutions with alkali. It dissolves in alkali solutions, forming stannites. The stannite solutions, which decompose readily to alkali-metal stannates and tin, have been used industrially for immersion tinning.

Stannic Oxide. Stannic oxide tin(IV) oxide, white crystals, mol wt 150.69, mp >1600°C, sp gr 6.9, is insoluble in water, methanol, or acids but slowly dissolves in hot, concentrated alkali solutions. In nature, it occurs as the mineral cassiterite. It is prepared industrially by blowing hot air over molten tin, by atomizing tin with high pressure steam and burning the finely divided metal, or by calcination of the hydrated oxide. Other methods of preparation include treating stannic chloride at high temperature with steam, treatment of granular tin at room temperature with nitric acid, or neutralization of stannic chloride with a base.

In the ceramics (qv) and glass industries, stannic oxide is used for the production of opaque glasses; as an opacifier for glazes and, to a lesser extent, enamels for metals (eg, cast iron) as used in bathtubs, sinks, tile, and other sanitary ware; as a base for certain ceramic colors, eg, chrome–tin pink, vanadium–tin yellow, and antimony–tin blue; and as a component of ceramic capacitor dielectrics. More than 15 large glass-melting furnaces in the world use stannic oxide electrodes in the electromelting of lead glass (24).

Other important uses of stannic oxide are as a putty powder for polishing marble, granite, glass, and plastic lenses and as a catalyst. The most widely used heterogeneous tin catalysts are those based on binary oxide systems with stannic oxide for use in organic oxidation reactions. The tin–antimony oxide system is particularly selective in the oxidation and ammoxidation of propylene to acrolein, acrylic acid, and acrylonitrile. Research has been conducted for many years on the catalytic properties of stannic oxide and its effectiveness in catalyzing the oxidation of carbon monoxide at below 150°C has been described (25).

Transparent electroconductive coatings of stannic oxide are deposited on nonconductive substrates for electrical and strengthening applications. However, the agents used to deposit the oxide film are actually stannic chloride. More recently, some organotin compounds have been employed.

Hydrated Stannic Oxide. Hydrated stannic oxide of variable water content is obtained by the hydrolysis of stannates. Acidification of a sodium stannate solution precipitates the hydrate as a flocculent white mass. The colloidal solution, which is obtained by washing the mass free of water-soluble ions and peptization with potassium hydroxide, is stable below 50°C and forms the basis for the patented Tin Sol process for replenishing tin in stannate tin-plating baths. A similar type of solution (Stannasol A and B) is prepared by the direct electrolysis of concentrated potassium stannate solutions (26).

Metal Stannates. *Soluble Stannates.* Many metal stannates of formula $M_n Sn(OH)_6$ are known. The two main commercial products are the soluble sodium and potassium salts, which are usually obtained by recovery from the alkaline detinning process. They are also produced by the fusion of stannic oxide with sodium hydroxide or potassium carbonate, respectively, followed by leaching and by direct electrolysis of tin metal in the respective caustic solutions in cells using cation-exchange membranes (27). Another route is the recovery from plating sludges.

Potassium stannate, $K_2Sn(OH)_6$ (mol wt 298.93), and sodium stannate [12058-66-1], $Na_2Sn(OH)_6$, mol wt 266.71, are colorless crystals and are soluble in water. The solubility of potassium stannate in water is 110.5 g/100 mL water at 15°C and that of sodium stannate is 61.5 g/100 mL water at 15°C. The solubility of sodium stannate decreases with increasing temperature, whereas the solubility of potassium stannate increases with increasing temperature. The solubility of either sodium or potassium stannate decreases as the concentration of the respective free caustic increases. Hydrolysis of stannates yields hydrated stannic oxides and is the basis of the Tin Sol solution, which is used to replenish tin in stannate tin-plating baths (28,29).

Although sodium stannate formed the basis for the first successful alkaline tin-electroplating bath, both stannates are used in these baths, with potassium stannate being favored in the United States. Potassium stannate is used for an alkaline tin-electroplating bath yielding higher cathode efficiencies and higher conductivities than any other alkaline bath. The stannates are also used in immersion tinning, particularly the immersion plating of aluminum pistons and other parts for the automotive industry, and in the electroplating of alloy coatings, especially tin–zinc and tin–copper alloys from mixed stannate–cyanide baths. Reviews of the use of stannates in tin plating are given in References 4, 30, and 31.

Other. Insoluble alkaline-earth metal and heavy metal stannates are prepared by the metathetic reaction of a soluble salt of the metal with a soluble alkali–metal stannate. They are used as additives to ceramic dielectric bodies (32). The use of bismuth stannate [12777-45-6], $Bi_2(SnO_3)_3 \cdot 5H_2O$, with barium titanate produces a ceramic capacitor body of uniform dielectric constant over a substantial temperature range (33). Ceramic and dielectric properties of individual stannates are given in Reference 34. Other typical commercially available stannates are barium stannate [12009-18-6], $BaSnO_3$; calcium stannate [12013-46-6], $CaSnO_3$; magnesium stannate [12032-29-0], $MgSnO_3$; and strontium stannate [12143-34-9], $SrSnO_3$.

Certain anhydrous stannates are effective as smoke suppressants in glass-reinforced polyester, especially $Na_2Sn(OH)_6$ [12058-66-1] and $ZnSnO_3$ [12036-37-2]. This use has not yet been commercialized (35).

Salts. *Stannous Sulfate.* Stannous sulfate (tin(II) sulfate), mol wt 214.75, $SnSO_4$, is a white crystalline powder which decomposes above 360°C. Because of internal redox reactions and a residue of acid moisture, the commercial product tends to discolor and degrade at ca 60°C. It is soluble in concentrated sulfuric acid and in water (330 g/L at 25°C). The solubility in sulfuric acid solutions decreases as the concentration of free sulfuric acid increases. Stannous sulfate can be prepared from the reaction of excess sulfuric acid (specific gravity 1.53) and granulated tin for several days at 100°C until the reaction has ceased. Stannous sulfate is extracted with water and the aqueous solution evaporates *in vacuo*. Methanol is used to remove excess acid. It is also prepared by reaction of stannous oxide and sulfuric acid and by the direct electrolysis of high grade tin metal in sulfuric acid solutions of moderate strength in cells with anion-exchange membranes (36).

The main use for stannous sulfate is in tin plating. The sulfate bath is widely used commercially for general plating, especially barrel plating. Significant tin-plating processes involving stannous sulfate baths include flow melting, ie, momentarily melting the coating to attain a bright finish on tin or tin–lead alloy deposits, which is used primarily in the production of printed circuit boards; bright-acid tin plating, which is used in finishing electrical contacts, radio chassis, domestic articles, and kitchen utensils; electrotinning steel strip by the vertical acid process, and liquor finishing, ie, immersion plating of steel wire with tin or copper–tin alloy prior to drawing. All tin-plating and tin-alloy-plating processes are reviewed in Reference 4.

Stannous Fluoroborate. Stannous fluoroborate, $Sn(BF_4)_2$, is available only in solution, as the solid form has not been isolated. It is prepared by dissolving stannous oxide in fluoroboric acid or by direct electrolysis of tin metal in fluoroboric acid in cells with anion-exchange membranes (36). The commercially available 47-wt % solution is widely used in tin and tin–lead alloy plating, especially in the deposition of tin–lead solder alloys for the electronics industry (37,38). Stannous fluoroborate solutions are important in plating because of their good throwing and covering power and high solubility, which promote high rates of deposition. They are used in the tin plating of copper wire, backing of electrotypes, and barrel tin plating of components for subsequent soldering (4).

Stannous Pyrophosphate. Stannous pyrophosphate [*15578-26-4*], $Sn_2P_2O_7$, mol wt 411.32, sp gr 4.009 at 16°C, is an amorphous white powder, decomposing at above 400°C. It is insoluble in water and soluble in concentrated mineral acids and sodium pyrophosphate. It is prepared from stannous chloride and sodium pyrophosphate, and used as a caries preventative in toothpaste and as a diagnostic aid in radioactive bone scanning and red-blood-cell labeling (23,39–41).

Other Inorganic Tin Compounds. Other inorganic tin compounds, which have been used industrially in the past but are not used as of this writing (ca 1997) or which have limited use, include stannic sulfide [*1315-01-1*] as a pigment and bronzing agent, stannous sulfide [*1314-95-0*] as a pigment, stannic vanadate [*66188-22-5*] as an oxidation catalyst and in ceramic pigments, stannic molybdate [*34782-17-7*] as a source of gamma rays in Mössbauer spectroscopy of tin compounds, stannic arsenate [*35568-59-3*] as an anthelmintic, and tin naphthenate. Stannic phosphate [*15142-98-0*] gels are effective ion exchangers.

Toxicology. Inorganic tin and its compounds are generally of a low order of toxicity, largely because of the poor absorption and rapid excretion from the tissues of the metal (42–49). The acidity and alkalinity of their solutions make assessment of their parenteral toxicity difficult. The oral LD_{50} values for selected inorganic tin compounds are listed in Table 2. It is estimated that the average U.S. daily intake of tin, which is mostly from processed foods, is 4 mg (see FOOD PROCESSING).

Tin is normally present in animals, including humans. In the human body, it is present in small amounts in nearly all organs (53). Tin is eliminated almost completely by the alimentary tract and is scantily absorbed by the alimentary tissues. The output and intake of tin balances during adult life (54). In human subjects fed packaged C (canned) rations for 22 successive days, all tin ingested was accounted for in fecal excretion (55). Tin may be an essential trace element for the growth of the mammalian organism (see MINERAL NUTRIENTS). Many inorganic tin compounds have been approved for human contact or use by the FDA (56).

The inorganic tin compound that has received the most study from a toxicological viewpoint is stannic oxide. Autopsies performed on workers in the tin mining and refining industry, who inhaled tin oxide dust for as long as 20 yr, disclosed no pulmonary fibrosis (57). Inhalation for long periods produces a benign, symptomless pneumoconiosis with no toxic systemic effects (58).

Stannous chloride, an FDA-approved direct food additive with GRAS status, has also been extensively studied (59–62). In three FDA-sponsored studies, it was determined that stannous chloride is nonmutagenic in rats; when administered orally up to 50 mg/kg to pregnant mice for ten consecutive days, stannous chloride has no discernible effect on nidation or on maternal or fetal survival; and, when administered orally at 41.5 mg/kg to pregnant rabbits for 13 consecutive days, it produced no discernible effect on nidation or on maternal or fetal survival (63–65).

Other studies of the toxicity of stannous fluoride, sodium pentafluorostannite, sodium pentachlorostannite, sodium chlorostannate, stannous sulfide [1314-95-0], stannous and stannic oxides, stannous pyrophosphate [15578-26-4], stannous tartrate [815-85-0], and other inorganic tin compounds are reviewed in References 63–72. The OSHA TLV standard for inorganic tin compounds is two milligrams of inorganic tin compounds as tin per cubic meter of air averaged over an eight-hour work shift (47).

Table 2. Acute Oral Toxicity of Selected Inorganic Tin Compounds

Compound	CAS Registry Number	LD_{50}, mg/kg	Test animal	Reference
stannous chloride	[7772-99-8]	700	rat	50
		1,200	mouse	50
stannous ethylene glycoxide	[68921-71-1]	>10,000	rat	51
stannous 2-ethylhexanoate	[301-10-0]	5,870	rat	51
stannous fluoride	[7783-47-3]	128.4	mouse	52
		188.2	rat	52
stannous oxalate	[814-94-8]	3,400	rat	51
stannous oxide	[21651-19-4]	>10,000	rat	51
sodium pentafluorostannite	[22578-17-2]	595	male mouse	52
		221	male rat	52
		227	female rat	52

Organotin Compounds

In an organotin compound, there is at least one tin–carbon bond. The oxidation state of tin in most organotin compounds is +4, although organotin compounds having bulky groups bonded to divalent tin have been reported (73). Five classes of organotin compounds are known: R_4Sn (tetraorganotins), R_3SnX (triorganotins), R_2SnX_2 (diorganotins), $RSnX_3$ (monoorganotins), and R_6Sn_2 (hexaorganoditins) (see ORGANOMETALLICS). Of commercial importance are those organotins where R is methyl, butyl, octyl, cyclohexyl, phenyl, or β,β-dimethylphenethyl (neophyl). The noncarbon-bonded anionic group is commonly halide, oxide, hydroxide, carboxylate or mercaptide.

It was not until the 1940s that the commercial potential of organotins was realized. Organotins first were used as stabilizers for poly(vinyl chloride), which is normally processed just below its decomposition temperature (see HEAT STABILIZERS). The high biocidal activity of the triorganotins is one of the most applied areas of their usefulness. In addition, organotins are widely used as catalysts and curing agents and in the treatment of glass. A number of organotin subjects, including structural organotin chemistry and industrial applications, are discussed in References 74–78.

Properties. As a member of Group 14(IVA) of the Periodic Table, tin has four valence electrons available for bonding. In its usual tetravalent state, tin assumes a typical sp^3 hydrization and the configuration of its covalent bonds is tetrahedral. In the tin atom, d orbitals are available and are utilized in the formation of pentacoordinate and hexacoordinate complexes by Lewis bases with organotin halides. These complexes are frequently trigonal bipyramidal or octahedral. Tin forms predominately covalent bonds to other elements, but these bonds exhibit a high degree of ionic character, with tin usually acting as the electropositive member.

Although the mean dissociation energy of tin–carbon bonds is less than that normally associated with carbon–carbon bonds ($\overline{D}_{Sn-C} = 188-230$ kJ/mol (45–55 kcal/mol), $\overline{D}_{C-C} = 335-380$ kJ/mol (80–90 kcal/mol) (75), the difference is not great enough to render the tin–carbon bond very reactive. The bond is stable to water and atmospheric oxygen at normal temperatures. The tin–carbon bond is also quite stable to heat and many organotins can be distilled under reduced pressure with little decomposition. Strong acids, halogens, and other electrophilic reagents readily cleave the tin–carbon bond, although other reactions that are common with other organometallics, eg, Grignard and organolithium reagents, do not occur with organotins. For example, the tin–carbon bond does not add to the carbonyl group, nor does it react with alcohols.

The ionic nature of organotins leads to dissimilar chemical properties. For example, triorganotin hydroxides behave not as alcohols, but more like inorganic bases, although strong bases remove the proton in certain triorganotin hydroxides because tin is amphoteric. The bis(triorganotin) oxides, $(R_3Sn)_2O$, are strong bases and react with inorganic and organic acids forming normal saltlike but nonconducting and water-insoluble compounds. They do not in the least resemble organic ethers, though they can occasionally form peroxides. Tin doubly bonded to oxygen, which is analogous to an organic ketone, does not exist and diorganotin oxides, R_2SnO, are polymers, ie, $-\!(Sn(R_2)O\!)_{\overline{n}}$, and usually are highly cross-linked via intermolecular tin–oxygen bonds. Unlike the halocarbons, organotin halides are reactive compounds and, because of their ionic character, readily enter into metathetical substitution reactions resembling the inorganic metal halides. Tin–hydrogen bonds are unlike carbon–hydrogen bonds and, although essentially covalent, their partial ionicity makes them true hydrides with hydrogen as the formal electronegative partner. Organotin hydrides are strong reducing agents and are similar to lithium aluminum hydride. Many are organic-soluble and easily distilled and are used increasingly in organic syntheses. Unlike carbon, tin shows much less tendency to catenate, ie, form chains of Sn atoms bonded to each other. Although tin–tin-bonded compounds are known, the tin–tin bond is easily cleaved by oxygen, halogens, and acids.

Tetraorganotins. *Physical Properties.* Physical properties of typical tetraorganotin compounds are shown in Table 3. All tetraorganotin compounds are insoluble in water but are soluble in many organic solvents.

Chemical Properties. The most important reactions which tetraorganotins undergo are heterolytic, ie, electrophilic and nucleophilic, cleavage and Kocheshkov redistribution (81–84). The tin–carbon bond in tetraorganotins is easily cleaved by halogens, hydrogen halides, and mineral acids:

$$R_4Sn + Br_2 \longrightarrow R_3SnBr + RBr$$

$$R_3SnBr + Br_2 \longrightarrow R_2SnBr_2 + RBr$$

$$R_4Sn + HCl \longrightarrow R_3SnCl + RH$$

$$R_3SnCl + HCl \longrightarrow R_2SnCl_2 + RH$$

With tetraaryltin compounds, the reaction can proceed further to the aryltin trihalides:

$$(C_6H_5)_2SnBr_2 + Br_2 \longrightarrow C_6H_5SnBr_3 + C_6H_5Br$$

        [4713-59-1]   [7727-17-5]

Table 3. Physical Properties of Typical Tetraorganotin Compounds[a]

Compound	CAS Registry Number	Mp, °C	Bp, °C	n_D^{20}	d^{20}, g/cm^3
$(CH_3)_4Sn$	[594-27-4]	−54	78	1.4415	1.2905[b]
$(C_4H_9)_4Sn$	[1461-25-2]	−97	$127_{1.3\ kPa}$[c,d]	1.4727	1.0541
$(C_8H_{17})_4Sn$	[3590-84-9]			1.4677[d]	0.9609[d]
$(C_6H_5)_4Sn$	[595-90-4]	228			1.521
$(C_6H_{11})_4Sn$	[1449-55-4]	261	160–163		
$(CH_2{=}CH)_4Sn$	[1112-56-7]		$70_{0.59\ kPa}$[c]	1.4914[b]	1.257
$(CH_3)_2(C_4H_9)_2Sn$	[1528-00-3]		$73{-}75_{0.5\ kPa}$[c]	1.4640[b]	1.124[b]
$(C_2H_5)_3(C_4H_9)Sn$	[17582-53-5]			1.4736	1.1457

[a]Ref. 79, except where noted.
[b]At 25°C.
[c]To convert kPa to mm Hg, multiply by 7.5.
[d]Ref. 80.

$$(C_6H_5)_2SnCl_2 + HCl \longrightarrow C_6H_5SnCl_3 + C_6H_6$$
$$[1135\text{-}99\text{-}5] \qquad\qquad\qquad [1124\text{-}19\text{-}2]$$

In practice, these cleavage reactions are difficult to control, and usually mixtures of products form, even with stoichiometric quantities of reagents. Selectivity improves at lower temperatures, higher dilutions, and in the presence of polar solvents, eg, pyridine. This method is not used to prepare the lower alkylated−arylated organotins outside the laboratory.

The most widely utilized reaction of tetraorganotins is the Kocheshkov redistribution reaction, by which the tri-, di-, and in some cases the monoorganotin halides can be readily prepared:

$$R_4Sn + SnCl_4 \longrightarrow 2\,R_2SnCl_2$$

$$R_2SnCl_2 + H_4Sn \longrightarrow 2\,R_3SnCl$$

$$3\,R_4Sn + SnCl_4 \longrightarrow 4\,R_3SnCl$$

$$R_4Sn + 3\,SnCl_4 \longrightarrow 4\,RSnCl_3$$

These reactions proceed rapidly and in good yield with primary alkyl and phenyl organotin compounds at ca 200°C. The reactions proceed at lower temperatures if anhydrous aluminum chloride is used as a catalyst.

If the reaction temperature is controlled through the use of a low boiling solvent or other means, it is possible to isolate equimolar quantities of monoalkyltin trichloride and trialkyltin chloride using a 1:1 ratio of tetraorganotin and tin tetrachloride:

$$R_4Sn + SnCl_4 \xrightarrow{<100°C} R_3SnCl + RSnCl_3$$

When R is a lower alkyl, the organotin trichloride can be easily separated from the reaction mixture by extraction with dilute aqueous hydrochloric acid, in which it is soluble. This reaction also works well with unsymmetrical tetraorganotins and has been practiced commercially (85).

With tetraaryltins, the redistribution reaction can be made to proceed to the monoorganotin stage with the proper stoichiometry of reactants:

$$Ar_4Sn + 3\,SnCl_4 \longrightarrow 4\,ArSnCl_3$$

Preparation. The tetraorganotins, although of little commercial utility by themselves, are important compounds since they are the starting materials for many of the industrially important mono-, di-, and triorganotins. Among the most widely used preparations of tetraalkyl- and tetraaryltin compounds is the reaction of stannic chloride with tetrahydrofuran-based Grignard reagents or organoaluminum compounds:

$$4\,RMgX + SnCl_4 \xrightarrow[\text{or } R_2O]{\text{THF}} R_4Sn + 4\,MgXCl$$

$$4\,R_3Al + 3\,SnCl_4 \xrightarrow[\text{or } R_2O]{R_3N} 3\,R_4Sn + 4\,AlCl_3$$

Excess alkylating reagent is required if the tetraorganotin is desired as the exclusive product. In commercial practice, the stoichiometry is kept at or below 4:1, since the crude product is usually redistributed to lower organotin chlorides in a subsequent step and an ether is used as the solvent (86). The use of diethyl ether in the Grignard reaction has been generally replaced with tetrahydrofuran.

Organolithium and organosodium reagents can also be used to prepare tetraorganotins:

$$4 \text{ RLi} + \text{SnCl}_4 \longrightarrow \text{R}_4\text{Sn} + 4 \text{ LiCl}$$

$$4 \text{ RNa} + \text{SnCl}_4 \longrightarrow \text{R}_4\text{Sn} + 4 \text{ NaCl}$$

The Wurtz reaction, which relies on *in situ* formation of an active organosodium species, is also useful for preparing tetraorganotin compounds and is practiced commercially. Yields are usually only fair and a variety of by-products, including ditins, also form:

$$\text{SnCl}_4 + 8 \text{ Na} + 4 \text{ RCl} \longrightarrow \text{R}_4\text{Sn} + 8 \text{ NaCl}$$

A variant of the Wurtz reaction is the preparation of tetrabutyltin from activated magnesium chips, butyl chloride, and stannic chloride in a hydrocarbon mixture. Only a small amount of tetrahydrofuran is required for the reaction to proceed in high yield (86).

The use limitations of an active metal organometallic, eg, Grignard or organolithium reagents, allow preparation of only tetraorganotins, which have no functional groups reactive to the organometallic reagent on the molecule. The preparation of tetraorganotins with functional groups, eg, hydroxyl, amino, nitrile, etc, bonded to the organic group requires special measures, eg, blocking the functional group with an inert function then deblocking, usually mildly, after the formation of the tin–carbon bonds. The nitrile derivative, tetrakis(cyanoethyl)tin [15961-16-7], is prepared in good yield via a unique electrochemical reaction of tin metal with acrylonitrile (87). Unsymmetrical tetraorganotins can be prepared from the mono-, di-, or triorganotin halides and the appropriate organometallic reagent of magnesium, lithium, sodium, or aluminum:

$$\text{RSnCl}_3 + 3 \text{ R}'\text{MgX} \longrightarrow \text{RR}'_3\text{Sn} + 3 \text{ MgXCl}$$

$$\text{R}_2\text{SnCl}_2 + 2 \text{ R}'\text{MgX} \longrightarrow \text{R}_2\text{R}'_2\text{Sn} + 2 \text{ MgXCl}$$

$$\text{R}_3\text{SnCl} + \text{R}'\text{MgX} \longrightarrow \text{R}_3\text{R}'\text{Sn} + \text{MgXCl}$$

$$\text{R}_2\text{SnCl}_2 + 2 \text{ R}'\text{Li} \longrightarrow \text{R}_2\text{R}'_2\text{Sn} + 2 \text{ LiCl}$$

Unsymmetrical functional tetraorganotins are generally prepared by tin hydride addition (hydrostannation) to functional unsaturated organic compounds (88) (see HYDROBORATION). The realization that organotin hydrides readily add to aliphatic carbon–carbon double and triple bonds forming tin–carbon bonds led to a synthetic method which does not rely on reactive organometallic reagents for tin–carbon bond formation and, thus, allows the synthesis of organofunctional tetraorganotins containing a wide variety of functional groups. Typical compounds which undergo such a reaction include tributyltin hydride and triphenyltin hydride, which can be prepared by the reaction of the chlorides with lithium aluminum hydride or sodium borohydride (89,90). Representative organic substrates include acrylonitrile, acrylate and methacrylate esters, allyl alcohol, vinyl ethers, styrene, and other olefins:

$$R_3SnH + \; >\!\!C\!\!=\!\!CHX \; \longrightarrow \; R_3Sn\overset{|}{\underset{|}{C}}CH_2X$$

Compounds with active halogens, eg, allyl chloride, also undergo reduction. Diorganotin dihydrides, monoorganotin trihydrides, and even stannane [2406-52-2], SnH_4, undergo analogous reactions, but the stability of the organotin hydrides decreases with increasing number of hydride groups, so these hydrostannation reactions generally proceed in poorer yield with more by-products.

Other methods for preparing tetraorganotin compounds include the use of diorganozinc compounds, halomethylzinc halides, electrolysis of organoaluminum reagents with a tin anode, and the electrolysis of diethyl sulfate with a zinc cathode and a tin anode (91–94). The latter method probably involves the *in situ* generation of an organozinc intermediate.

The reaction of an organotin–lithium, organotin–sodium, or organotin–magnesium reagent is occasionally useful for the preparation of tetraorganotins in the laboratory (78). These reagents or organostannylanionoids are air- and moisture-sensitive and can be prepared from most triorganotin halides and some tetraorganotins:

$$(CH_3CH_2CH_2)_3SnCl \; \xrightarrow[\text{or Li in THF}]{\text{Na in liquid NH}_3} \; (CH_3CH_2CH_2)_3SnNa \; \text{or} \quad (Li)$$
$$[84474\text{-}09\text{-}9] \qquad\qquad [84474\text{-}08\text{-}8]$$

$$(CH_3CH_2CH_2)_3SnNa(Li) + RX \; \longrightarrow \; (CH_3CH_2CH_2)_3SnR + Na(Li)X$$

$$(CH_3)_4Sn \; \xrightarrow{\text{Na in liquid NH}_3} \; (CH_3)_3SnNa + NaNH_2 + CH_4$$
$$[16643\text{-}09\text{-}7]$$

Primary and secondary alkyl halides react well, but *tert*-alkyl halides are preferentially dehydrohalogenated by the tin reagents.

Uses. The main use for tetraorganotin compounds is as (usually captive) intermediates for the tri-, di-, and monoorganotins. Although there have been reports in the patent literature of the use of tetraorganotins as components of Ziegler-Natta-type catalysts for the polymerization of olefins, there is no evidence that such catalysts are used commercially.

Triorganotins. Triorganotins and diorganotins constitute by far the most important classes of organotins.

Physical Properties. Physical properties of some typical triorganotin halides are listed in Table 4 and those of commercially important triorganotin

Table 4. Physical Properties of Typical Triorganotin Halides[a]

Compound	CAS Registry Number	Mp, °C	Bp, °C	n_D^{20}	d^{20}, g/cm^3
$(CH_3)_3SnCl$	[1066-45-1]	37.5	154–156		
$(CH_3)_3SnBr$	[1066-44-0]	26–27	163–165		
$(C_4H_9)_3SnCl$	[1461-22-9]		$152–156_{1.9\ kPa}{}^b$	1.4930	1.2105
$(C_4H_9)_3SnF$	[1983-10-4]	218–219 (dec)			1.27^c
$(C_6H_5)_3SnCl$	[639-58-7]	106			
$(C_6H_5)_3SnF$	[379-52-2]	357 (dec)			
$(C_6H_{11})_3SnCl$	[3091-32-5]	129–130			

[a]Ref. 110.
[b]To convert kPa to mm Hg, multiply by 7.5.
[c]At 25°C.

compounds are listed in Table 5. The triorganotin halides are insoluble in water, except for $(CH_3)_3SnCl$ which is completely water soluble, but are soluble in most organic solvents. The fluorides are insoluble in most organic solvents because of their highly associated structure resulting from strong SnF–Sn interactions.

Reactions. The utility of triorganotin chlorides and their application as starting materials for most other triorganotin compounds results from the ease of nucleophile displacement, as indicated in Figure 1. The commercially important triorganotin compounds are most frequently the oxides or hydroxides, the fluorides, and the carboxylates.

The basic hydrolysis of trialkyltin halides and other salts forms bis(oxide)s since, except for trimethyltin, hydroxides are unstable towards dehydration at room temperature. With tin aryl, aralkyl, and cycloalkyltin compounds, the hydroxides can be isolated. Although quite stable, they exist in mobile equilibrium with the bisoxide and water and are easily dehydrated. Trimethyltin hydroxide is exceptionally stable towards dehydration.

Table 5. Physical Properties of Commercially Important Triorganotin Compounds

Compound	CAS Registry Number	Mp, °C	Bp, °C	n_D^{20}	d^{20}, g/cm^3
$[(C_4H_9)_3Sn]_2O$	[56-35-9]	<-45	$210–214_{1.3\ kPa}{}^a$	1.488	1.17
$(C_4H_9)_3SnF$	[1983-10-4]	218–219 dec			1.27^b
$(C_4H_9)_3SnOCOC_6H_5$	[4342-36-3]		$166–168_{0.13\ kPa}{}^a$	1.5157	1.1926
$(C_4H_9)_3SnOCOCH_3$	[56-36-0]	80–85			1.27
$(C_6H_5)_3SnOH$	[76-87-9]	118–120 dec			1.552^b
$(C_6H_5)_3SnF$	[379-52-2]	357 dec			1.53
$(C_6H_5)_3SnOCOCH_3$	[900-95-8]	119–120			
$(C_6H_{11})_3SnOH$	[13121-70-5]	c			
$(C_6H_{11})_3SnN_3C_2H_2{}^d$	[41083-11-8]	218.8			
$(Neoph_3Sn)_2O^e$	[13356-08-6]	$138–139^f$			

[a]To convert kPa to mm Hg, multiply by 7.5.
[b]At 25°C.
[c]No true melting point; converts to bis-oxide at >120°C.
[d]$N_3C_2H_2$ = 1,2,4-triazole.
[e]Neoph = neophyl = β,β-dimethylphenethyl.
[f]Technical material.

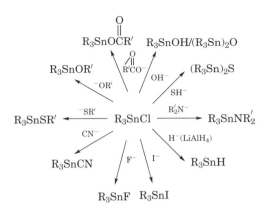

Fig. 1. Reactions of triorganotin chlorides (95).

Triorganotin oxides and hydroxides are moderately strong bases and react readily with a wide variety of acidic compounds:

$$R_3SnOH + HX \longrightarrow R_3SnX + H_2O$$

$$(R_3Sn)_2O + 2\ HX \longrightarrow 2\ R_3SnX + H_2O$$

This reaction is useful in the preparation of anionic derivatives from the chlorides when the nucleophilic displacement route is unsatisfactory. Even weak acids, eg, phenols, mercaptans, and cyclic nitrogen compounds, can be made to undergo reaction with triorganotin hydroxides or bisoxides if the water of reaction is removed azeotropically as it forms.

Triorganotin compounds of strong acids are generally quite stable to hydrolysis under neutral conditions. Under basic conditions, the hydroxide or bisoxide forms. Strong acids, halogens, and other electrophiles can cause cleavage of tin–carbon bonds with the formation of diorganotins. The triorganotin oxides of lower alkyl groups (C_1–C_4) are sufficiently basic to react with carbon dioxide in air, resulting in the precipitation of triorganotin carbonates.

Preparation.   Triorganotin chlorides of the general formula R_3SnX are the basic starting materials for other triorganotins. They are generally prepared by Kocheshkov redistribution from the crude tetraorganotin:

$$4\ R_4Sn + SnCl_4 \longrightarrow 4\ R_3SnCl$$

The stoichiometric reaction of Grignard or alkylaluminum reagents with stannic chloride to give the trialkyltin chloride usually gives a mixture of products. Only in a very few cases is it possible to alkylate tin tetrachloride directly to the triorganotin chloride in good yield with few by-products using a Grignard reagent. In such cases, the formation of the triorganotin is favored because of steric hindrance (96):

$$3\ C_6H_{11}MgCl + SnCl_4 \longrightarrow (C_6H_{11})_3SnCl + 3\ MgCl_2$$

Acid, hydrogen halide, or halogen cleavage of tetraorganotins is not used except on a laboratory scale because they are wasteful of tin–carbon bonds and uneconomical on a commercial scale.

Tribenzyltin chloride [3151-41-5] is a unique example of a triorganotin chloride that can be prepared directly from the organic halide and tin metal:

$$3 \; C_6H_5CH_2Cl + 2 \; Sn \xrightarrow[\text{reflux}]{H_2O} (C_6H_5CH_2)_3SnCl + SnCl_2$$

This reaction only proceeds in water. In a solventless system, only organic condensation products of benzyl chloride form, including dibenzyl. In toluene, dibenzyltin dichloride [3002-01-5] is the principal reaction product (97).

The production of triphenyltin hydroxide [76-87-9] and triphenyltin acetate [900-95-8] start with triphenyltin chloride, which is prepared by the Kocheshkov redistribution reaction from tetraphenyltin and tin tetrachloride. The hydroxide is prepared from the chloride by hydrolysis with aqueous sodium hydroxide. The acetate can be made directly from the chloride using sodium acetate or from the hydroxide by neutralization with a stoichiometric quantity of acetic acid.

For the preparation of tricyclohexyltin chloride, the Kocheshkov redistribution reaction is not suitable, since tetracyclohexyltin decomposes in the presence of stannic chloride at the normal redistribution temperatures. Two alternative routes are practiced for the manufacture of tricyclohexyltin chloride. The closely controlled reaction of cyclohexylmagnesium chloride and stannic chloride in a three-to-one molar ratio can be made to give the desired product in a good yield (96). Another method involves two steps for the preparation of tricyclohexyltin chloride (98). In the first step, butyltin trichloride [1118-46-3] reacts with three moles of cyclohexylmagnesium chloride forming butyltricyclohexyltin [7067-44-9]. This tetraorganotin then reacts with stannic chloride under mild conditions in an inert solvent, cleaving a butyl group and yielding tricyclohexyltin chloride and butyltin trichloride. The latter is recovered and recycled. The reactions are shown below:

$$C_4H_9SnCl_3 + 3 \; C_6H_{11}MgCl \longrightarrow (C_6H_{11})_3SnC_4H_9$$

$$(C_6H_{11})_3SnC_4H_9 + SnCl_4 \longrightarrow (C_6H_{11})_3SnCl + C_4H_9SnCl_3$$

Tricyclohexyltin chloride is converted to the hydroxide with sodium hydroxide. The triazole can be prepared from the chloride with sodium or potassium hydroxide and 1,2,4-triazole.

Bis(trineophyltin) oxide [60268-17-4] is prepared from the chloride in the normal manner. The chloride can either be prepared directly from the reaction of three moles of neophylmagnesium chloride and stannic chloride or by the butyl transfer reaction between butyltrineophyltin and stannic chloride. The hydroxide derivative initially formed on hydrolysis of the chloride is readily dehydrated to the bis(oxide) at ca 100°C.

Uses. Triorganotin compounds are widely used as industrial biocides, agricultural chemicals, wood preservatives, and marine antifoulants. Although the *in vitro* fungicidal biological activity of the triorganotins was recognized in the mid-1950s, commercial development was not seriously undertaken until the early 1960s (99,100). The triorganotins that are most useful as biological control agents, in general, are the tributyltins, triphenyltins, and tricyclohexyltins.

The lower trialkyltins from trimethyl to tri-*n*-pentyl show high biological activity. The trimethyltins are highly insecticidal and the tripropyl-, tributyl-, and tripentyltin compounds have a high degree of fungicide and bactericide activity. Dialkyltin compounds are less active than the analogous trialkyltins. The maximum activity towards bacteria and fungi is exhibited by the tripropyl and tributyltin compounds, with the tributyltins providing the optimum balance between fungicidal and bactericidal activity and mammalian toxicity. Tributyltin compounds, especially the oxide and benzoate, are used as antimicrobials and slimicides for cooling-water treatment and as hard-surface disinfectants. These and similar compounds have been used as laundry sanitizers and mildewcides to prevent mildew formation in the dried film of water-based emulsion paints. In most microbiocide applications, the tributyltin compound is used in conjunction with another biocide, usually a quaternary ammonium compound, to complement the activity of the organotin which is most effective against gram-positive bacteria.

Although the lower trialkyltins show high fungicide activity, they are unlikely candidates for agricultural fungicide use because of their high phytotoxicity to the host plant. Various attempts have been made to moderate the phytotoxicity of the lower trialkyltins by changing the anion portion of the molecule. These have not been successful because the nature of the anionic group has little influence on the spectrum of biological activity, provided that the anion is not biologically active and it confers a sufficient minimal solubility on the compound.

In the early 1960s, the first organotin-based agricultural fungicide, triphenyltin acetate, was introduced in Europe commercially by Farbwerke Hoechst AG as Brestan. Brestan, which is a protectant foliar fungicide, was recommended for the control of *phytophthora* (late blight) on potatoes and *cercospora* on sugar beets at application rates of a few ounces per acre (101). Shortly thereafter, triphenyltin hydroxide was introduced as Du-Ter by Philips-Duphar, NV, with about the same activity and spectrum of disease control as Brestan. Du-Ter is registered with the EPA in the United States as a fungicide for potatoes, sugar beets, pecans, and peanuts. Both compounds also exhibit a strong antifeedant effect on some insects and are fly sterilants at sublethal concentrations. Triphenyltin hydroxide formulations are also supplied by Griffin Corp. in the United States.

Tricyclohexyltin hydroxide was introduced into the U.S. market by the Dow Chemical Company as Plictran. Plictran was originally recommended for the control of phytophagous (plant-feeding) mites on apples and pears. It is also registered in the United States and in many European and Asian countries for this use as well as for mite control on citrus, stone fruits, and hops. This product has since been joined in the market by similar-acting, competitive products marketed by Shell and Bayer. Other triorganotin compounds with significant agricultural uses are tricyclohexyltin hydroxide (Plictran, Dow Chemical), and hexaneophyldistannoxane (Vendex, Shell U.S.A.; Torque, Shell International Chemical).

Bis(tributyltin) oxide [56-35-9] is widely used in Europe for the preservation of timber, millwork, and wood joinery, eg, window sashes and door frames. It is applied from organic solution by dipping or vacuum impregnation. It imparts resistance to attack by fungi and insects but is not suitable for underground use. An advantage of bis(tributyltin) oxide is that it does not interfere with subsequent painting or decorative staining and does not change the natural color of the wood. Tributyltin phosphate, $((C_4H_9)_3Sn)_3PO_4$, has also been suggested as a wood preservative.

Most surfaces in prolonged contact with seawater and freshwater are susceptible to the attachment of marine growths, eg, algae and barnacles.

The most common method for preventing marine fouling has been to paint the underwater structure of the vessel with an antifouling paint containing a toxicant. For many years, the antifouling agent of choice was cuprous oxide, but there has been a strong trend towards the use of triorganotin compounds, both alone and in combination with cuprous oxide (102). Preferred compounds for use in this application are tributyltin fluoride, triphenyltin hydroxide, and triphenyltin fluoride because they are highly active against a wide range of fouling species. Bis(tributyltin) oxide, tributyltin acetate, and other tributyltin carboxylates have also been successfully used as antifoulants. Triorganotin compounds offer many advantages over cuprous oxide. Because they are colorless, they can be used in the preparation of paints of a variety of colors. Unlike cuprous oxide, they do not contribute to galvanic corrosion on steel or aluminum hulls. The triorganotins are rapidly degraded into lower alkylated species and then to nontoxic inorganic tin once released from the coating. Inorganic copper, on the other hand, is toxic in all its forms (see COATINGS, MARINE).

There has been much interest in eroding antifouling paints that are based on tributyltin acrylate [13331-52-7] or methacrylate [2155-70-6] copolymers with various organic acrylate esters as the combined toxicant and paint binder resin (102). Such paints erode in moving seawater because the triorganotin portion slowly hydrolyzes from the acrylic backbone in normally basic seawater, releasing the active species tributyltin chloride and bis(tributyltin) oxide. The depleted surface layer of the paint film, containing hydrophilic-free carboxylic acid groups, becomes water-swollen and is easily eroded by moving seawater. A fresh surface of triorganotin acrylate polymer is thereby exposed and the process repeats. Coatings based on organotin polymers can be formulated to release the toxicant at a rate which is linear with time. Such coatings are claimed to reduce fuel costs over and above the savings resulting from a clean hull by providing a surface which becomes smoother with time. M&T Chemicals, Inc., is a worldwide supplier of a variety of tributyltin methacrylate copolymers with different hydrolysis and erosion rates (bioMeT 300 series antifoulant polymers). Paints based on organotin copolymers are offered by the principal marine paint companies, including Hempel's Marine Paints (Nautic Modules), International Paint Company, Ltd. (Intersmooth SPC), Nippon Oil & Fats Company, Ltd., and Jotun Marine Coatings (Takata LLL) (see COATINGS, MARINE).

The advantages claimed for organotin polymer-based antifouling paints include constant toxicant delivery vs time, erosion rate and toxicant delivery are controllable, no depleted paint residue to remove and dispose, 100% utilization of toxicant, polishing at high erosion rates, surface is self-cleaning, and function is continuously reactivated.

Triorganotin compounds have also been used experimentally in controlled-release formulations to control the infective snail vector in the debilitating tropical disease schistosomiasis (bilharzia) and to control mosquitoes in stagnant ponds (103). As yet, the large-scale use of such methods has little support in the host third world countries where these problems are most severe. Tributyltin chloride has been used to confer rodent-repellent properties on wire and cable coatings (104).

Diorganotins. *Physical Properties.* Physical properties of some typical diorganotin compounds are shown in Table 6. The diorganotin chlorides, bromides, and iodides are soluble in many organic solvents and, except for dimethyltin dichloride, are insoluble in water.

Commercial grades of diorganotin carboxylates frequently have wider melting ranges because of the use of less pure grades of carboxylic acids in their manufacture which, for many applications, permits more facile handling of the liquids.

Reactions. Although there are few industrial applications for the diorganotin halides, these compounds are the basic intermediates for the preparation of all the commercially important diorganotin derivatives. They are prepared by nucleophilic displacement similar to that used for triorganotin derivatives (see

Table 6. Physical Properties of Diorganotin Compounds[a]

Compound	CAS Registry Number	Mp, °C	Bp, °C	n_D^{20}	d^{20}, g/cm³
$(CH_3)_2SnCl_2$	[753-73-1]	107–108	185–190		
$(C_4H_9)_2SnCl_2$	[683-18-1]	41–42	$140–143_{1.3 \text{ kPa}}$[b]		
$(C_4H_9)_2SnBr_2$	[996-08-7]	21–22	$90–92_{0.04 \text{ kPa}}$[b]	1.5400	1.3913[c]
$(C_4H_9)_2SnI_2$	[2865-19-2]		$145_{0.8 \text{ kPa}}$[b]	1.6042	1.996[c]
$(C_6H_5)_2SnCl_2$	[1135-99-5]	42–44	$180–185_{0.7 \text{ kPa}}$[b]		
$(CH_3OC(O)CH_2CH_2)_2SnCl_2$	[10175-01-6]	132			
$(CH_3)_2Sn(SC_4H_9)_2$	[1000-40-4]		$81_{0.013 \text{ kPa}}$[b]	1.5400	1.280
$(C_4H_9)_2Sn(OCH_3)_2$	[1067-55-6]		$126–128_{7 \text{ Pa}}$[b]	1.4880	
$[(C_4H_9)_2SnS]_3$	[15220-82-3]	63–69			
$(C_4H_9)_2Sn(OCCH_3)_2$ (O‖)	[1067-33-0]	8.5–10	$142–145_{1.3 \text{ kPa}}$[b]	1.4706	
$(C_4H_9)_2Sn(C_{11}H_{23}CO)_2$ (O‖)	[77-58-7]	22–24		1.4683	1.05
$(C_4H_9CHCO)_2Sn(C_4H_9)_2$ (O‖, C₂H₅)	[2781-10-4]	54–60	$215–220_{0.3 \text{ kPa}}$[b]	1.4653	1.070[c]
$[(C_2H_5O)_2PS]_2Sn(C_6H_5)_2$ (S‖)	[74097-03-3]	149.5			

[a]Refs. 95 and 105.
[b]To convert kPa to mm Hg, multiply by 7.5.
[c]At 25°C.

Fig. 1). Basic hydrolysis of the diorganotin halides gives the diorganotin oxides in high yield. Except in rare cases, dihydroxide derivatives are unknown. As with the triorganotins, diorganotin oxides are sometimes used as intermediates from which other derivatives can be obtained by neutralization with strong or weak acids:

$$R_2SnCl_2 \xrightarrow{OH^-} R_2SnO$$

$$R_2SnO + 2\,HY \longrightarrow R_2SnY_2 + H_2O$$

Diorganotin dihalides are moderately strong Lewis acids and form stable complexes with ammonia and amines. The commercially important diorganotin compounds are most frequently the oxides, carboxylates, and mercaptocarboxylic acid esters. The oxides are amorphous or polycrystalline, highly polymeric, infusible, and insoluble solids. They are moderately strong bases and react readily with a wide variety of strongly and weakly acidic compounds. Their insolubility in all nonreactive solvents makes the choice of proper reaction conditions for such a neutralization reaction an important consideration for optimum yields.

Diorganotin esters of strong acids are relatively stable to hydrolysis under neutral conditions, but generally, diorganotin compounds are more reactive chemically than the triorganotins. Diorganotin esters of weak acids are somewhat susceptible to hydrolysis, even under neutral conditions, but this reactivity is moderated somewhat by their hydrophobicity.

On partial hydrolysis, diorganotin halides and carboxylates may form basic salts having a rather complicated structure:

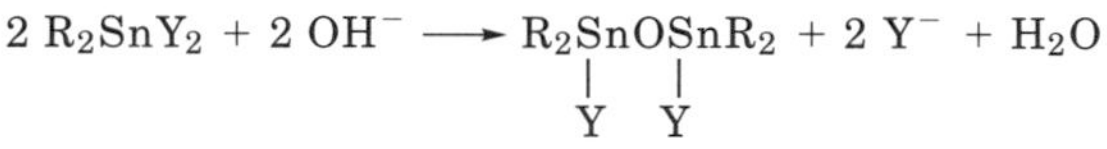

$$2\,R_2SnY_2 + 2\,OH^- \longrightarrow \underset{\underset{Y}{|}\ \underset{Y}{|}}{R_2SnOSnR_2} + 2\,Y^- + H_2O$$

Diorganotin sulfides can be prepared from the chlorides or oxides by the exchange of a reactive substituent for sulfur:

$$R_2SnCl_2 + S^{2-} \longrightarrow R_2SnS + 2\,Cl^-$$

$$R_2SnO + CS_2 \longrightarrow R_2SnS + COS$$

The sulfides are associated like the oxides, but to a lesser degree. They are crystalline, sharp-melting, soluble in many organic solvents, and resistant to hydrolysis. Most are cyclic trimers (106).

Some diorganotin compounds, eg, the alkoxides, add to heterounsaturated systems, eg, isocyanates. This reaction is believed to occur in stages (107).

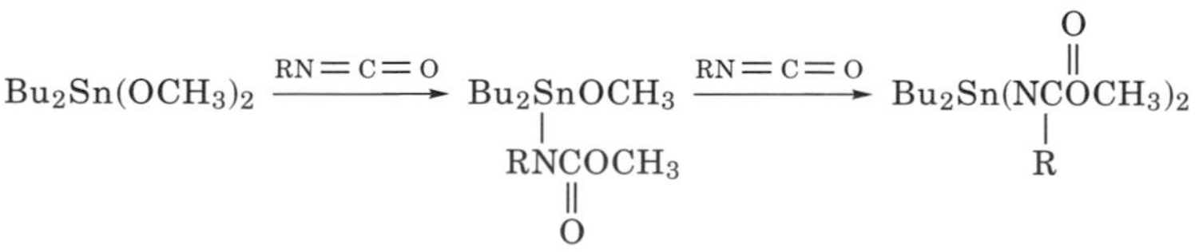

$$Bu_2Sn(OCH_3)_2 \xrightarrow{RN=C=O} \underset{\underset{O}{\overset{\|}{\underset{RNCOCH_3}{|}}}}{Bu_2SnOCH_3} \xrightarrow{RN=C=O} \underset{\underset{R}{|}}{Bu_2Sn(N\overset{\overset{O}{\|}}{C}OCH_3)_2}$$

Preparation.   Diorganotin dichlorides are the usual precursors for all other diorganotin compounds; three primary methods of manufacture are practiced. Dibutyltin dichloride is manufactured by Kocheshkov redistribution from crude tetrabutyltin and stannic chloride and usually is catalyzed with a few tenths of a percent aluminum trichloride:

$$(C_4H_9)_4Sn + SnCl_4 \longrightarrow 2\ (C_4H_9)_2SnCl_2$$

Yields are almost quantitative and product purity is good with formation of only minute amounts of mono- and tributyltin by-products.

Many organic halides, especially alkyl bromides and iodides, react directly with tin metal at elevated temperatures (>150°C). Methyl chloride reacts with molten tin metal, giving good yields of dimethyltin dichloride, which is an important intermediate in the manufacture of dimethyltin-based PVC stabilizers. The presence of catalytic metallic impurities, eg, copper and zinc, is necessary to achieve optimum yields (108):

$$2\ CH_3Cl + Sn \xrightarrow[Cu]{235°C} (CH_3)_2SnCl_2$$

The reaction of higher alkyl chlorides with tin metal at 235°C is not practical because of the thermal decomposition which occurs before the products can be removed from the reaction zone. The reaction temperature necessary for the formation of dimethyltin dichloride can be lowered considerably by the use of certain catalysts. Quaternary ammonium and phosphonium iodides allow the reaction to proceed in good yield at 150–160°C (109). An improvement in the process involves the use of amine–stannic chloride complexes or mixtures of stannic chloride and a quaternary ammonium or phosphonium compound (110). Use of these catalysts is claimed to yield dimethyltin dichloride containing less than 0.1 wt % trimethyltin chloride. Catalyzed direct reactions under pressure are used commercially to manufacture dimethyltin dichloride.

The direct reaction of tin metal with higher haloalkanes is less satisfactory even when catalysts are used, except with alkyl iodides. The reaction of butyl iodide with tin metal is used commercially in Japan to prepare dibutyltin diiodide, from which dibutyltin oxide is obtained on hydrolysis with base:

$$2\ C_4H_9I + Sn \longrightarrow (C_4H_9)_2SnI_2 \xrightarrow{OH^-} (C_4H_9)_2SnO + 2\ I^-$$

The economics of this process depend on near-quantitative recovery and recycle of the iodine to prepare butyl iodide.

Tin metal also reacts directly with a number of activated organic halides, including allyl bromide, benzyl chloride, chloromethyl methyl ether, and β-halocarboxylic esters and nitriles giving fair-to-good yields of diorganotin dihalides (97,111–114).

The facile reaction of metallic tin in the presence of hydrogen chloride with acrylic esters to give high yields of bis(β-alkoxycarbonylethyl)tin dichlorides is reported in References 115 and 116. This reaction proceeds at atmospheric pressure and room temperature and has been practiced commercially. Halogenostannanes have been postulated as intermediates (105).

Uses. *Poly(Vinyl Chloride) Stabilizers.* The largest single industrial application for organotin compounds is in the stabilization of PVC. Of the estimated 30,000-t world production of organotins, it is believed that 20,000 t or two thirds of production, is accounted for by PVC stabilization (7). The estimated 1981 U.S. consumption of organotins as PVC stabilizers was 10,650 t, representing 27% of the market. Organotins are added to PVC to prevent its degradation by heat (180–200°C) during processing and by long-term exposure to sunlight (117–139).

Dialkyltin compounds are the best general-purpose stabilizers for PVC, especially if colorlessness and transparency are required. Commercial organotin stabilizers include the carboxylates, especially the maleates, laurates, and substituted maleates; the mercaptide, the mercaptoacid, and mercaptoalcohol ester derivatives; and the estertins, 2-carboalkoxyethyltin derivatives. The common industrial organotin stabilizers are listed in Table 7. U.S. producers of organotin stabilizers and the trade names of their products are: Argus (Witco), Mark; Cardinal, Cardinal Clear; Thiokol, Carstab; Ferro, Thermchek and Polychek; Interstab (Akzo), Interstab and Stanclere; M&T Chemicals, Thermolite; Tenneco, Nuostabe; and Synthetic Products (Dart), Synpron.

Table 7. Typical Commercially Significant Organotin PVC Stabilizers

Compound	CAS Registry Number	Structure
dibutyltin bis(isooctyl mercaptoacetate)	[25168-24-5]	$(C_4H_9)_2Sn(SCH_2CO_2C_5H_{17}\text{-}i)_2$
dioctyltin bis(isooctyl mercaptoacetate)	[26401-97-8]	$(C_8H_{17})_2Sn(SCH_2CO_2C_8H_{17}\text{-}i)_2$
dimethyltin bis(isooctyl mercaptoacetate)	[26636-01-1]	$(CH_3)_2Sn(SCH_2CO_2C_8H_{17}\text{-}i)_2$
bis(2-carbobutoxyethyltin) bis(isooctyl mercaptoacetate)	[63397-60-4]	$(C_4H_9OCOCH_2CH_2)_2Sn(SCH_2CO_2\text{-}C_8H_{17}\text{-}i)_2$
dibutyltin sulfide	[4253-22-9]	$(C_4H_9)_2SnS$
dibutyltin bis(lauryl mercaptide)	[1185-81-5]	$(C_4H_9)_2Sn(SC_{12}H_{25})_2$
dibutyltin β-mercaptopropionate	[27380-35-4]	$+\!\!-(C_4H_9)_2SnSCH_2CH_2COO\!-\!\!\frac{}{}\!\!_n$ $(n = 1\text{--}3)$
dibutyltin bis(mercaptoethyl-decanoate) (also other esters)	[28570-24-3]	$(C_4H_9)_2Sn(SCH_2CH_2OC(O)C_{11}H_{25})_2$
butylthiostannoic acid anhydride	[15666-29-2]	$\overset{\displaystyle S}{\overset{\displaystyle \|}{(C_4H_9Sn)_2S}}$
butyltin tris(isooctyl-mercaptoacetate)	[25852-70-4]	$C_4H_9Sn(SCH_2CO_2C_8H_{17}\text{-}i)_3$
dibutyltin dilaurate	[77-58-7]	$(C_4H_9)_2Sn(OOCC_{11}H_{23})_2$
dibutyltin maleate (dioctyltin derivative)	[32076-99-6] [16091-18-2]	$+\!\!-(C_4H_9)_2SnOOCCH\!=\!CHCOO\!-\!\!\frac{}{}\!\!_n$ $(n = 1\text{--}3)$
dibutyltin bis(monoisooctyl-maleate) (also other alkyl maleate esters)	[25168-21-2]	$(C_4H_9)_2Sn(OOCCH\!=\!CHCOO\text{-}C_8H_{17}\text{-}i)_2$

Sulfur-containing organotins impart excellent heat stability to PVC, but nonsulfur-containing organotins are used when resistance to light and weathering are required. The two main markets for organotin stabilizers are in the packaging and building industries. In the packaging industry, certain organotin stabilizers are used in PVC food packaging and drink containers. In the United States and Germany, dioctyltin maleate [16091-18-2], dioctyltin bis(isooctylmercaptoacetate), and butylthiostannoic acid [26410-42-4] are approved for use in PVC food packaging; in Germany, dimethyltin bis(isooctylmercaptoacetate), 2-carbobutoxyethyltin tris(isooctylmercaptoacetate) [63438-80-2], and bis(2-carbobutoxyethyltin) bis(isooctylmercaptoacetate) are also approved. These uses reflect the low toxicity of these organotin stabilizers.

In the building industry, rigid PVC is stabilized with diorganotin carboxylates, especially dibutyltin maleate, for use in floorings and light fixture glazing and with diorganotin mercaptides and mercaptoacid esters for use in sidings, profiles, roofing, fencing, window frames, and piping. The dibutyltin, dimethyltin, and estertin sulfur-containing derivatives are used for these nonfood applications as well as in PVC potable-water piping.

Polyurethane Foam Catalysts. Early production of polyurethane foams involved a two-step reaction in which a polyether glycol reacted with toluene diisocyanate forming a urethane prepolymer having reactive isocyanate end groups. Water was then added to condense the neighboring isocyanate groups to urethane linkages. In the process, carbon dioxide formed, which acted on the gelling polymer to produce a rigid or elastomeric foam. Inorganic tin compounds and diorganotin compounds, eg, dibutyltin diacetate [1067-33-0], dilaurate [77-58-7], and di(2-ethylhexanoate) [2781-10-4], catalyze the glycol–isocyanate reaction as well as the urethane condensation step and enable the preparation of foams in one step in a semicontinuous process (140,141). In the United States, dibutyltin compounds are used mostly in the catalysis of rigid foams and the laurate has been the catalyst of choice (142) (see URETHANE POLYMERS).

Diorganotin compounds have been used increasingly as catalysts for high resiliency foam in automotive seating. In high resiliency foam, diorganotin mercaptocarboxylates and mercaptides as catalysts improve some key physical properties (143). Some diorganotins, eg, the mercaptocarboxylates and mercaptides, are stable enough to be used in the preparation of masterbatches containing premixed polyol, water surfactant, amine, and organotin catalyst which are stored for up to six months (144).

Esterification Catalysts. Dibutyltin compounds as well as monobutyltins are used increasingly as esterification (qv) catalysts for the manufacture of organic esters used in plasticizers (qv), lubricants, and heat-transfer fluids (see LUBRICATION AND LUBRICANTS; HEAT-EXCHANGE TECHNOLOGY). Although esterification reactions catalyzed by organotins require higher temperatures (200–230°C) than those involving strong acid catalysts, eg, p-toluenesulfonic acid, side reactions are minimized and the products need no extensive refining to remove acidic ionic catalyst residues. Additionally, equipment corrosion is eliminated and the products have better color and odor properties because fewer by-products form (145). Usual catalyst levels are 0.05–0.3 wt % based on the total reactants charged. Dibutyltin compounds are also useful in catalyzing the transesterification and polycondensation of dimethyl terephthalate to poly(ethylene terephthalate) for packaging applications and in the manufacture of polyester-based alkyd resins (146). Both solid and liquid and insoluble and soluble organotin-based esterification catalysts are marketed by M&T Chemicals, Inc., as Fascat (see POLYESTERS).

Other. Dibutyltin dilaurate [77-58-7] has been successfully used for many years as a coccidiostat in the treatment of intestinal worm infections in chickens and turkeys (see ANTIPARASITIC AGENTS).

In Japan and Europe, dimethyltin dichloride that has been purified to remove all traces of trimethyltin chloride is used to provide a thin coating of stannic oxide on glass upon thermal decomposition at 500–600°C. Thin deposits of stannic oxide improve the abrasion resistance and bursting strength of glass bottles. Dimethyltin dichloride is manufactured and marketed as Glahard by Chugoku Toryo Company, Ltd., Shiga, Japan. Electroconductive films can be formed with thick coatings of tin oxide that are deposited in this manner.

Dibutyltin and dioctyltin diacetate, dilaurate, and di-(2-ethylhexanoate) are used as catalysts for the curing of room-temperature-vulcanized (RTV) silicone elastomers to produce flexible silicone rubbers used as sealing compounds, insulators, and in a wide variety of other applications. Diorganotin carboxylates also catalyze the curing of thermosetting silicone resins, which are widely used in paper-release coatings.

In addition, diorganotin compounds are used as transesterification catalysts for the curing of cathodic, electrocoated paints (147). The biological activity and toxicity of diorganotins are much less than of analogous triorganotins with the same carbon-bonded organic groups.

Monoorganotins. *Physical Properties.* Properties of some monoorganotin trihalides are listed in Table 8. The monoorganotin trihalides are hygroscopic, low melting solids or liquids which are to varying extents hydrolyzed in water or moist air, liberating the hydrogen halides. They are soluble in most organic solvents and in water that contains enough acid to retard hydrolysis.

Chemical Properties. The monoorganotin trihalides are strong Lewis acids and form complexes with ammonia, amines, and many other oxygenated organic compounds, eg, ethers. In many ways, they resemble acid chlorides. As with the diorganotin dichlorides, the halogens on the molecule are easily replaced by a wide variety of nucleophilic reagents, making these trihalides useful intermediates for other monoorganotins. Typical compounds, which are easily formed by displacement reactions, include tris(alkoxides), tris(carboxylates), tris(mercaptides), and tris(mercaptocarboxylate esters). These compounds are generally more easily hydrolyzed than the analogous diorganotins.

Table 8. Physical Properties of Typical Organotin Trihalides

Compound	CAS Registry Number	Mp, °C	Bp, °C	n_D^{20}
CH_3SnCl_3	[993-16-8]	45–46		
CH_3SnBr_3	[993-15-7]	53	211	
$C_4H_9SnCl_3$	[1118-46-3]		$102–103_{1.6\ \mathrm{kPa}}{}^a$	1.5233
$C_6H_5SnCl_3$	[1124-19-2]		$142–143_{3.3\ \mathrm{kPa}}{}^a$	1.5871

^{a}To convert kPa to mm Hg, multiply by 7.5.

The oxide monobutyltin oxide [*51590-67-1*], is a sesquioxide, $C_4H_9SnO_{1.5}$, from which it is difficult to remove the last traces of water. It is an infusible, insoluble, amorphous white powder that forms when butyltin trichloride is hydrolyzed with base. The partially dehydrated material, butylstannoic acid [*2273-43-0*], is slightly acidic and forms alkali metal salts. These salts, ie, alkali metal alkylstannonates, form when excess alkali is used to hydrolyze the organotin trichloride:

$$RSnCl_3 + 4\,NaOH \longrightarrow RSnO_2Na + 3\,NaCl + 2\,H_2O$$

Partially hydrolyzed products of the form $RSn(OH)_2Cl$ are believed to be mixtures in most cases.

When organotin trihalides are treated with alkali metal sulfide, the sesquisulfides form (148):

$$2\,RSnCl_3 + 3\,Na_2S \longrightarrow 2\,RSnS_{1.5} + 6\,NaCl$$

At least one, the monobutyl compound, is a tetramer in benzene (148).

Preparation and Manufacture. Monoorganotin halides are the basic raw materials for all other triorganotin compounds and are generally prepared by Kocheshkov redistribution from the tetraorganotin, eg, tetrabutyltin or the higher organotin halides:

$$R_4Sn + 3\,SnCl_4 \longrightarrow 4\,RSnCl_3$$

$$R_2SnCl_2 + SnCl_4 \longrightarrow 2\,RSnCl_3$$

The oxidative addition of aliphatic organic halides to stannous chloride has long been of interest for the preparation of monoorganotin trihalides:

$$SnCl_2 + RCl \longrightarrow RSnCl_3$$

This reaction gives fair-to-good yields of monoorganotin tribromides and trichlorides when quaternary ammonium or phosphonium catalysts are used (149). Better yields are obtained with organic bromides and stannous bromide than with the chlorides. This reaction is also catalyzed by trialkylantimony compounds at 100–160°C, bromides are more reactive than chlorides in this preparation (150,151). α,ω-Dihaloalkanes also react in good yield giving ω-haloalkyltin trihalides when catalyzed by organoantimony compounds (152).

A significant advance in the synthesis of monoorganotin trihalides was the preparation of β-substituted ethyltin trihalides in good yield from the reaction of stannous chloride, hydrogen halides, and α,β-unsaturated carbonyl compounds, eg, acrylic esters, in common solvents at room temperature and atmospheric pressure (153,154). The reaction is believed to proceed through a solvated trichlorostannane intermediate (155):

$$SnCl_2 + HCl \xrightarrow{(C_2H_5)_2O} HSnCl_3 \cdot 2(C_2H_5)_2O \xrightarrow{\quad} Cl_3SnCH_2CH_2COR$$

This reaction can be extended to unsaturated nitriles, eg, acrylonitrile, which can give trihalostannyl-functional carboxylic acids, esters, and amides by the proper choice of solvents and reaction conditions (156).

Uses.   Poly(Vinyl Chloride) Stabilizers.   Although generally less effective as PVC stabilizers than dialkyltin derivatives, monoalkyltin compounds added to the dialkyltin compounds in amounts of 5–20 wt % exert a synergistic effect on stabilizer effectiveness, preventing early yellowing. They supposedly function by reacting more quickly and at lower processing temperatures than the dialkyltin species, thus preventing the early onset of yellowing; conversely, diorganotins are more effective in retarding the long-term degradation of the polymer. Butylthiostannoic acid anhydride is used as a sole stabilizer for certain grades of PVC in Germany, but elsewhere is rarely used alone (157). It is approved in Germany and in the United States for food packaging (158). In Germany, the following monoorganotins alone or in mixtures are also approved for this use: butylthiostannoic acid anhydride with either dioctyltin compounds or 2-carbobutoxyethyltin compounds, 2-carbobutoxyethyltin tris(isooctylmercaptoacetate) alone or mixed with its dicounterpart, monomethyltin tris(isooctylmercaptoacetate) [56225-49-1] plus its dicounterpart in a 24:76 wt % ratio, and monooctyltin tris[alkyl (C_{10}–C_{16}, isooctyl) mercaptoacid esters] with their dicounterparts.

Treatment of Glass.   The use of monobutyltin trichloride in the hot-end coating of glass to improve the abrasion resistance and bursting strength of glass bottles has been patented, and the deposition process variables and product advantages have been described (159–161). Highly efficient utilization of tin is one of the main benefits.

Compounds with Tin–Tin Bonds.   The most important class of catenated tin compounds is the hexaorganoditins. The ditin compounds are usually prepared by reductive coupling of a triorganotin halide with sodium in liquid ammonia:

$$2\ R_3SnCl + 2\ Na \xrightarrow{\text{NH}_3(l)} R_3SnSnR_3 + 2\ NaCl$$

This reaction proceeds in stages via an organostannylsodium compound:

$$R_3SnCl + 2\ Na \longrightarrow R_3SnNa + NaCl$$

$$R_3SnNa + R_3SnCl \longrightarrow R_3SnSnR_3 + NaCl$$

Lithium metal in tetrahydrofuran can also be used as the coupling reagent, and unsymmetrical ditins can be prepared when the reaction is conducted in stages (162,163).

Hexaorganoditins with short-chain aliphatic groups are colorless liquids, distillable under vacuum, soluble in organic solvents other than the lower alcohols, and insoluble in water. They are generally unstable in air, undergoing ready oxidation to a mixture of organotin compounds. Hexaary{}ditins are usually crystalline solids and are much more stable towards oxidation.

The ditins as of yet are insignificant commercially, although there has been interest in hexamethylditin [661-69-8] (Pennwalt TD-5032) as an insecticide (164,165).

Salts. Organic tin salts are tin compounds containing an organic radical in which the tin is bonded with an element other than carbon. The most common of these are the tin carboxylates, especially the stannous carboxylates. The latter are manufactured by reaction of stannous oxide or chloride with the appropriate acid. The most commercially significant of the stannous carboxylates is stannous 2-ethylhexanoate [301-10-0]. It is estimated that in 1979, worldwide annual consumption of tin-based catalysts for polyurethanes was ca 2500 t with stannous 2-ethylhexanoate accounting for 95% of this usage (140). The second most important industrial organic tin salt is stannous oxalate [814-94-8]. Other commercially available organic tin salts that are of minor commercial importance are listed in Table 9.

Stannous 2-Ethylhexanoate. Stannous 2-ethylhexanoate, $Sn(C_8H_{15}O_2)_2$ (sometimes referred to as stannous octanoate, mol wt 405.1, sp gr 1.26), is a clear, very light yellow, and somewhat viscous liquid that is soluble in most organic solvents and in silicone oils (166). It is prepared by the reaction of stannous chloride or oxide with 2-ethylhexanoic acid.

The primary use for stannous 2-ethylhexanoate is as a catalyst with certain amines for the manufacture of one-shot polyether urethane foams (167). Resulting foams exhibit good dry-heat stability over a wide range of catalyst concentrations. Food-grade stannous 2-ethylhexanoate is approved by the FDA for use in polymers and resins used in food packaging (168). Other industrial applications include its use as a catalyst in silicones, including room-temperature-vulcanizing (RTV) silicone rubbers and silicone–oil emulsions; in epoxy formulations; and in various urethane coatings and sealants (qv) (169,170). Proprietary catalyst formulations based on stannous 2-ethylhexanoate are also available.

Stannous Oxalate. Stannous oxalate, $Sn(C_2O_4)$ (mol wt 206.71, dec 280°C, sp gr 3.56 at 18°C), is a white crystalline powder, is soluble in hot concentrated hydrochloric acid and mixtures of oxalic acid and ammonium oxalate, and is insoluble in water, toluene, ethyl acetate, dioctyl phthalate, THF, isomeric heptanes, and acetone (171). It is prepared by precipitation from a solution of stannous chloride and oxalic acid and is stable indefinitely.

Stannous oxalate is used as an esterification and transesterification catalyst for the preparation of alkyds, esters, and polyesters (172,173). In esteri-

Table 9. Physical Properties of Organic Tin Salts of Minor Commercial Importance

Salt	CAS Registry Number	Mp, °C	Density, g/cm³	Use
stannous acetate	[638-39-1]	182.5–183	2.31	promotes dye uptake by fabrics
stannous ethylene glycoxide	[68921-71-1]	dec >300	2.87	esterification catalyst
stannous formate	[2879-85-8]	dec >100		catalyst for hydrogenation of liquid fuels
stannous gluconate	[35984-19-1]		1.35	silicone catalyst
stannous oleate	[1912-84-1]		1.06	silicone catalyst
stannous stearate	[6994-59-8]	90; dec 340	1.05	catalyst
stannous tartrate	[815-85-0]	dec 280	2.6	dyeing and printing of textiles

fication reactions, it limits the undesirable side reactions responsible for the degradation of esters at preparation temperatures. The U.S. Bureau of Mines conducted research on the use of stannous oxalate as a catalyst in the hydrogenation of coal (174) (see COAL).

Toxicology.   The toxicological properties of organotin compounds are reviewed in References 43, 74–77, and 175. The toxicity of organotin compounds is a reflection of their biological activity. Thus, the most toxic to mammals, including man, are the lower trialkyltin compounds, ie, trimethyltin and triethyltin. As with the fungicidal activity, the toxicity seems little affected by the nature of the anionic group bonded to the trialkyltin moiety. There is some evidence that triorganotin compounds that are five-coordinate and intramolecularly chelated are less toxic than similar unchelated four-coordinate compounds (176,177). As a general rule, the toxicity of the trialkyltins decreases with increasing chain length of the alkyl group.

The acute oral mammalian toxicities of typical triorganotin compounds, including some which are not used commercially, are listed in Table 10. In some

Table 10. Acute Oral Toxicities of Triorganotin Compounds

Compound	CAS Registry Number	LD_{50}, mg/kg	Test animal	Reference
$(CH_3)_3SnOCCH_3$ (O ‖)	[1118-14-5]	9	rat	178
$(C_2H_5)_3SnOCCH_3$ (O ‖)	[1907-13-7]	4	rat	178
$(C_3H_7)_3SnOCCH_3$ (O ‖)	[3267-78-5]	118	rat	178
$(C_4H_9)_3SnOCCH_3$ (O ‖)		133	rat	175
		380	rat	
$[(C_4H_9)_3Sn]_2O$	[56-35-9]	ca 200	rat	175
$(C_4H_9)_3SnF$		200	rat	175
$(C_6H_{13})_3SnOCCH_3$ (O ‖)	[2897-46-3]	1,000	rat	178
$(C_8H_{17})_3SnOCCH_3$ (O ‖)	[919-28-8]	>1,000	rat	178
$(C_6H_5)_3SnOCCH_3$ (O ‖)		136	rat	175
		491	rat	175
$(C_6H_5)_3SnOH$		108	rat	179
		209	mouse	179
$(C_6H_{11})_3SnOH$		540	rat	179
		780	guinea pig	179
$(Neoph_3Sn)_2O^a$		1,450	mouse	179
		>1,500	dog	179
		2,630	rat	179

^{a}Neoph = neophyl = β,β-dimethylphenethyl.

cases, two or more substantially different values are reported in the literature for the same test animal. In these cases, both high and low values are tabulated. The toxicity of triorganotins is strongly dependent on the nature of the organic groups bonded to tin. The toxicity varies from the highly toxic lower alkyl trimethyl and triethyltins, which are not used in any commercial applications, to the substantially less toxic trineophyl and trioctyl derivatives. The widely used tributyl-, triphenyl-, and tricyclohexyltin derivatives are intermediate in mammalian oral toxicity. The highest trialkyl and triaryltins are less toxic when given orally than when given parentally because of their poor absorption from the gastrointestinal tract (180). Uncoupling of oxidative phosphorylation in cellular mitochondria has been suggested as one of the mechanisms of lower trialkyltin toxicity (181).

Most triorganotins that have been studied and all commercial ones are eye and skin irritants. Animal studies have shown that, particularly with tributyl and triphenyltin compounds, untreated eye contact can result in permanent corneal damage. If allowed to remain in contact with the skin for prolonged periods, these compounds can produce severe irritation and, in some cases, severe chemical burns (182). Thus, eye and skin protection must be worn when handling triorganotin compounds. Sometimes the irritant effect is delayed and may not be apparent for several hours. In the event of acute local dermal contact episodes with tributyltin compounds, pruritis, minor edema, and follicular pustules in hirsute areas occur. Systemic effects are observed in percutaneous tests of tributyltin iodide, bromide, chloride and bis-oxide in tests with rabbits, so it could be assumed that these compounds are absorbed through the skin (183). Bis(tributyltin) oxide produces a typical lower trialkyltin response dermally, characterized by redness, swelling, and skin discoloration in test animals. Its effects on the eyes are serious with damage to the cornea (183).

Among the most widely studied triorganotin compounds are triphenyltin hydroxide, triphenyltin acetate, and tricyclohexyltin hydroxide because of their use as agricultural chemicals. Triphenyltin hydroxide is a severe eye irritant in rabbits but is nonirritating to dry rabbit skin (184). In contrast, triphenyltin chloride on rabbits produces erythema and edema with tissue damage. The injuries are worsened by washing with organic solvent (183). In feeding tests on rats and mice, triphenyltin hydroxide shows no evidence of carcinogenicity (185).

Diorganotin compounds as a class are substantially less toxic than the analogous triorganotins. Some compounds of this class are used as additives in plastics intended to be in contact with food or potable water or used as PVC stabilizers. The acute oral toxicities of common commercial diorganotin compounds are given in Table 11. The dialkyltin chlorides and oxides generally show decreasing oral toxicity with increasing length of the alkyl chain. The toxicity of the lower dialkyltins is believed to be related to their ability to combine with enzymes containing two thiol groups in a suitable stereochemical conformation and thereby inhibiting the oxidation of α-ketoacids in the cell (188). 2,3-Dimercapto-1-propanol, $HSCH_2CH(SH)CH_2OH$, has been reported as an effective antidote for lower dialkyltin poisoning (68). The lower dialkyltin halides are somewhat less irritating to the skin than the analogous triorganotins, but skin contact should be avoided. Other studies of the toxicities of specific diorganotin compounds are reported in References 178, 189, and 190.

Table 11. Acute Oral Toxicities of Diorganotin Compounds

Compound	CAS Registry Number	LD_{50} (rat), mg/kg	Reference
$(CH_3)_2SnCl_2$		74	175
$(CH_3)_2Sn(SCH_2\overset{\displaystyle O}{\overset{\|}{C}}OC_8H_{17}\text{-}i)_2$	[26636-01-1]	1380	175
$(C_4H_9)_2SnCl_2$		126	175
$(C_4H_9)_2SnO$	[818-08-6]	600–800	186
$(C_4H_9)_2Sn(O\overset{\displaystyle O}{\overset{\|}{C}}C_{11}H_{23})_2$		175	175
$(C_4H_9)_2Sn(SCH_2\overset{\displaystyle O}{\overset{\|}{C}}OC_8H_{17}\text{-}i)_2$	[25168-24-5]	500	175
$(C_8H_{17})_2SnCl_2$	[3542-36-7]	5500	175
$(C_8H_{17})_2SnO$	[870-08-6]	2500	175
$(C_8H_{17})_2Sn(O\overset{\displaystyle O}{\overset{\|}{C}}C_{11}H_{23})_2$	[3648-18-8]	6450	175
$(C_8H_{17})_2Sn(SCH_2\overset{\displaystyle O}{\overset{\|}{C}}OC_8H_{17}\text{-}i)_2$	[26401-97-8]	2000	175
$(RO\overset{\displaystyle O}{\overset{\|}{C}}CH_2CH_2)2\ SnCl_2{}^a$		2350	187
$(RO\overset{\displaystyle O}{\overset{\|}{C}}CH_2CH_2)_2Sn(SCH_2\overset{\displaystyle O}{\overset{\|}{C}}OC_8H_{17}\text{-}i)_2{}^a$		1430	187

^{a}R is undefined; probably C_2H_5.

Monoorganotin compounds present no special toxicological problems. In general, they show the familiar trend of decreasing toxicity with increasing alkyl chain length, but of a lower order of toxicity than the diorganotins. As with most organotin compound classes, there are conflicting toxicity data and exceptions to general rules. Monobutyltin sulfide [15666-29-2] (butylthiostannoic anhydride, BTSA, poly[(1,3-dibutyldistannthiondiylidene)-1,3-dithiol] is allowed as a stabilizer in semirigid or rigid PVC used in food packaging. Typical LD_{50} values for monoorganotin compounds are

$$CH_3Sn(SCH_2\overset{\displaystyle O}{\overset{\|}{C}}OC_8H_{17}\text{-}i)_3$$

920 mg/kg (rats) (187);

$$C_4H_9Sn(SCH_2\overset{\displaystyle O}{\overset{\|}{C}}OC_8H_{17}\text{-}i)_3$$

1063 mg/kg (rats) (187). $(C_4H_9SnS_{1.5})_n$, >20,000 mg/kg (rats) (186);

$$C_8H_{17}Sn(SCH_2\overset{\displaystyle O}{\overset{\|}{C}}OC_8H_{17}\text{-}i)_3$$

3400 mg/kg (rats) (187). The lower monoorganotin trihalides can present special problems, however, because of their facile reaction with moisture, resulting in the liberation of hydrochloric acid.

The toxicity of the tetraorganotins has been little studied. Available literature indicates that tetrabutyltin and the higher tetraalkyltins are substantially less toxic than triorganotins to mammals if taken orally (175). The high toxicity reported for tetraethyltin (LD_{50} = 9–16 mg/kg) appears to be caused by its rapid conversion in the liver to a triethyltin species.

The inhalation toxicities (50% fatality in rats) of dimethyltin dichloride, monomethyltin trichloride, and dibutyltin dichloride are 1070, 600, and 73 mg/(L·h) (191).

The current OSHA TLV standard for exposure to all organotin compounds is 0.1 mg of organotin compounds (as tin)/m^3 air averaged over an 8-h work shift (192). NIOSH has recommended a permissible exposure limit of 0.1 mg/m^3 of tin averaged over a work shift of up to 10 h/d, 40 h/wk; Reference 193 should be consulted for more detailed information. Additional information on the health effects of organotin compounds is given in Reference 48.

Economic Aspects

Commercially available tin compounds having annual production or gross sales of >2.3 metric tons or $5,000.00 are listed in References 194 and 195. Principal U.S. producers of inorganic tin compounds include M&T Chemicals, Inc., Vulcan Materials Company, and Allied Corporation. M&T Chemicals, Inc., is the largest U.S. producer of organotin compounds, followed by Carstab Corporation, Witco Chemical Corporation, and Cardinal Chemical Company; minor producers are Interstab, Synthetic Products Company, Tenneco Chemicals Company, and Ferro Chemical Company

Tin compounds are used in a wide variety of industries as catalysts and stabilizers for many materials, including polymers (see HEAT STABILIZERS); and as biocidal agents, eg, bactericides, insecticides, wood preservatives, acaricides, and antifouling paints; ceramic opacifiers; textile additives; in metal finishing operations; as food additives; and in electroconductive coatings (see CATALYSIS; COATINGS, MARINE; DENTIFRICES; INDUSTRIAL ANTIMICROBIAL AGENTS; INSECT CONTROL TECHNOLOGY).

In 1975, the estimated world annual production of tin chemicals represented the consumption of 12,000–14,000 t of tin metal or 5% of total tin consumption (196). In 1978, ca 20,000 t/yr was consumed worldwide, with equal amounts represented by inorganic tin and organotin compounds (197). It is established that worldwide production of organotins rose from ca 50 t in 1950 to a possible 30,000–35,000 t in 1980.

Consumptions of primary tin for chemical applications by the five reporting countries from 1975 to 1980 is shown in Table 12. The use of primary tin in chemical applications increased sharply in the United States and Germany, but declined in the United Kingdom and Italy. In 1979, ca 11% of the total U.S. demand for tin was accounted for by tin chemicals. In 1977, the tin consumption in tin chemicals production was based on ca 80% primary tin usage and 20% secondary tin usage. Since 1979, these statistics have been withheld by the U.S.

Table 12. Consumption of Primary Tin for Chemical Application, t[a]

Year	France	Germany	Italy[b]	U.K.[c]	United States Primary	United States Secondary
1975	345	492	650	1210	2735	1263
1976	330	998	700	1463	4718	903
1977	420	1137	650	1468	4655	1072
1978	540	1414	600	1374	4557	
1979	590	1894	500	1305	4797	
1980	700[d]	2050[d]	450[d]	1282	4800[d]	

[a]Ref. 198.
[b]Includes secondary tin.
[c]Includes tin powder.
[d]Estimates from Ref. 199.

Bureau of Mines (BOM). The BOM estimates that the use of tin in tin chemicals in the year 2000 will be 11,000 t. The approximate use of tin chemicals in the Western world has been estimated as follows for inorganic tins: ca 6000 t including ca 3000 t stannic oxide; 1000 t stannic chloride; 500 t stannous sulfate and sodium and potassium stannates; 200 t stannous fluoride, fluoroborate, and pyrophosphate; and 1000 t stannous chloride, oxide, and octanoate (200). The approximate use of organotins is ca 30,000 t, including ca 21,000 t for mono- and diorganotins (representing 20,000 t for PVC stabilizers and 1000 t for homogeneous catalysts) and 9000 t for triorganotins (representing 1000 t for agricultural fungicides and 8000 t in other uses). More detailed approximate world consumption data of inorganic tin compounds are reported in Reference 201. Other sources estimate the consumption of stannous octoate as a catalyst in flexible urethane foams at 1000 t/yr, stannic chloride as a perfume stabilizer in soaps at 250 t/yr, and stannic oxide as an opacifier for ceramic glazes and vitreous enamels at 2500 t/yr (198).

The most rapidly increasing use for tin is in chemicals, particularly for organotin compounds. In Germany, the amount of tin consumed in organotin production rose from 691 t in 1973 to 1760 t in 1979 (198). In 1979, organotins accounted for ca 93% of tin consumption in chemicals in Germany. In 1980, the estimated world production of organotin compounds was 30,000–35,000 t, with 75% used in the manufacture of PVC stabilizers. The market for PVC stabilizers is large and, depending on the country, organotins represent 10–25% of the market. At the beginning of 1980, the organotin share of the market for PVC stabilizers was ca 20% in the United States, 15% in Japan, 25% in Germany, and 12% in the rest of Western Europe (198). The estimated 1981 U.S. consumption of organotins as PVC stabilizers was 10,650 t or 27% of the market (202). The estimated 1981 U.S. consumption of tin compounds as urethane catalysts was 808 t or 34% of the market (203). The 1978 U.S. market for stannous fluoride in dental preparations was 186 t valued at 3.72×10^6 (204). An important earlier report with market data on organotins was issued in early 1976 (205).

The price of tin chemicals depends to a large extent on the fluctuating price of tin. The 1982 prices and the CAS Registry Numbers for selected inorganic tin compounds and organotins are listed in Table 13.

Table 13. Prices of Selected Inorganic Tin and Organotin Compounds, 1982[a]

Compound	CAS Registry Number	Price, $/kg
	Inorganic[b]	
potassium stannate	[12142-33-5]	9.66
stannic chloride, anhydrous	[7646-78-8]	9.66
stannic oxide	[18282-10-5]	26.40
stannous chloride, anhydrous	[7772-99-8]	14.10
stannous fluoroborate	[13814-97-6]	5.39
stannous oxide	[21651-19-4]	20.30
stannous sulfate	[7488-55-3]	16.13
	Organic[c]	
bis(tributyltin) oxide	[56-35-9]	17.51
dibutyltin dichloride	[683-18-1]	11.55
tributyltin acetate	[56-36-0]	19.93
tributyltin chloride	[1461-22-9]	17.12
tributyltin fluoride	[1983-10-4]	18.44

[a]Based on largest quantity price.
[b]Ref. 206.
[c]Ref. 207.

BIBLIOGRAPHY

"Tin Compounds" in *ECT* 1st ed., Vol. 14, pp. 157–165, by H. Richter, Metal & Thermit Corp.; in *ECT* 2nd ed., Vol. 20, pp. 304–327, by C. Kenneth Banks, M&T Chemicals, Inc.; in *ECT* 3rd ed., Vol. 23, pp. 42–77, by M. N. Gitlitz and M. K. Moran, M&T Chemicals, Inc.

1. *Toxic Substances Control Act Chemical Substances Inventory*, Vol. 3, U.S. Environmental Protection Agency, Washington, D.C., May 1979, Cumulative Suppl., July 1980.
2. *Stannochlor Compound*, Technical Bulletin 161, M&T Chemicals Inc., Rahway, N.J., 1976.
3. *Gmelin Handbuch der Organischen Chemie*, Band 46, Parts C5 and C6, Springer-Verlag, New York, 1977, and 1978.
4. F. A. Lowenheim, ed., *Modern Electroplating*, 3rd ed., John Wiley & Sons, Inc., New York, 1974, pp. 377–417.
5. U.S. Pat. 2,407,579 (Sept. 10, 1946), E. W. Schweikher (to E. I. du Pont de Nemours & Co., Inc.).
6. S. Wein, *Glass Ind.*, 367 (1954).
7. C. J. Evans, *Tin Its Uses* **98**, 7 (1973).
8. J. I. Duffy, ed., *Electroless and Other Nonelectrolytic Plating Techniques*, Noyes Data Corp., Park Ridge, N.J., 1980.
9. F. A. Domino, *Plating on Plastics—Recent Developments*, Noyes Data Corp., Park Ridge, N.J., 1979, p. 108.
10. CFR Title 21, 182.3845, pp. 617, 631 (1977); *Fed. Reg.* **33**, 5619 (Apr. 11, 1968).
11. *Fed. Reg.* **34**, 12087 (July 18, 1969).
12. *Fed. Reg.* **33**, 3375 (Feb. 27, 1968); **35**, 15372 (Oct. 27, 1970).
13. *Fed. Reg.* **35**, 8552 (June 3, 1970).
14. U.S. Pat. 2,785,076 (March 12, 1957), G. Felton (to Hawaiian Pineapple Co.).
15. U.S. Pat. 3,414,429 (Dec. 3, 1968), H. G. Bruss, W. J. Schlientz, and B. E. Wiens (to Owens-Illinois Co.); U.S. Pat. 3,554,787 (Jan. 12, 1971), C. E. Plymale (to Owens-Illinois Co.).

16. T. Williamson, *Tin Its Uses* **129**, 14 (1981).

17. U.S. Pat. 3,623,854 (Nov. 20, 1971), C. A. Frank (to Owens-Illinois Co., Inc.).

18. U.S. Pat. 3,952,118 (April 20, 1976), M. A. Novice (to Dart Industries, Inc.).

19. U.S. Pat. 2,162,255 (June 13, 1939), R. F. Heald (to Colgate-Palmolive-Peet Co.).

20. P. A. Cusack, P. J. Smith, J. S. Brooks, and R. Smith, *J. Text Inst.* **7**, 308 (1979).

21. J. C. Muhler, W. H. Nebergall, and H. G. Day, *J. Dent. Res.* **33**, 33 (1954).

22. U.S. Pat. 2,876,166 (Mar. 3, 1959), W. H. Nebergall (to Indiana University Foundation).

23. U.S. Pat. 2,946,725 (July 26, 1960), P. E. Norris and H. C. Schweizer (to Procter & Gamble Co.).

24. W. B. Hampshire and C. J. Evans, *Tin Its Uses* **118**, 3 (1978).

25. M. J. Fuller and M. E. Warwick, *J. Catal.* **42**, 418 (1976).

26. U.S. Pat. 3,723,273 (Mar. 27, 1973), H. P. Wilson (to Vulcan Materials Co.).

27. U.S. Pat. 4,066,518 (Jan. 3, 1978), R. E. Horn (to Pitt Metals and Chemicals, Inc.).

28. U.S. Pat. 3,346,468 (Oct. 10, 1967), J. C. Jongkind (to M&T Chemicals, Inc.).

29. U.S. Pat. 3,462,373 (Aug. 19, 1969), J. C. Jongkind (to M&T Chemicals, Inc.).

30. S. Karpel, *Tin Its Uses* **124**, 3 (1980).

31. J. C. Jongkind, *Metal Finishing Guidebook & Directory*, Metals and Plastics Pub. Inc., Hackensack, N.J., 1982, p. 346.

32. W. W. Coffeen, *J. Am. Ceram. Soc.* **36**, 207, 215 (1953); **37**, 480 (1954).

33. U.S. Pat. 2,658,833 (Nov. 10, 1953), W. W. Coffeen and H. W. Richter (to M&T Chemicals, Inc.).

34. *Electronic Ceramie Stannates*, Technical Data Sheet CER-322, M&T Chemicals, Inc., Rahway, N.J., 1969.

35. P. A. Cusack, P. J. Smith, and L. T. Arthur, *J. Fire Retardant Chem.* **7**, 9 (1980).

36. U.S. Pat. 3,795,595 (Mar. 5, 1974), H. P. Wilson (to Vulcan Materials Co.).

37. H. Silman, *Prod. Finish. (London)* **34**(1), 15 (1981).

38. N. J. Spiliotis, *Metal Finishing Guidebook & Directory*, Metals and Plastics Pub. Inc., Hackensack, N.J., 1982, pp. 333, 338.

39. U.S. Pat. 4,075,314 (Feb. 21, 1978), R. G. Wolfangel and H. A. Anderson (to Mallinckrodt, Inc.).

40. A. M. Zimmer, *Am. J. Hosp. Pharm.* **34**(3), 264 (1977).

41. D. G. Pavel, A. M. Zimmer, and V. N. Patterson, *J. Nucl. Med.* **18**, 305 (1977).

42. R. A. Hiles, *Toxicol. Appl. Pharmacol.* **27**, 366 (1974).

43. J. M. Barnes and H. B. Stoner, *Pharmacol. Rev.* **11**, 211 (1959).

44. *Toxicants Occurring Naturally in Foods*, 2nd ed., Natural Academy of Sciences, Washington, D.C., 1973, pp. 63–64.

45. E. Browning, in *Toxicity of Industrial Metals*, 2nd ed., Butterworths, London, 1969, pp. 323–330.

46. H. Cheftel, *Tin in Food*, Joint FAO/WHO Food Standards Program, 4th Meeting of the Codex Committee on Food Additives, PEPT, 1967, Joint FAO/WHO Food Standards Branch (Codex Alimentarius), FAO, Rome.

47. *Occupational Health Guidelines for Inorganic Tin Compounds (as Tin)*, U.S. Department of Labor, Occupational Safety and Health Administration, Washington, D.C., Sept. 1978.

48. *Environmental Health Criteria, Vol. 15: Tin and Organotin Compounds: A Preliminary Review*, WHO, Geneva, 1980.

49. H. E. Stokinger in G. D. Clayton and F. F. Clayton, eds., *Patty's Industrial Hygiene and Toxicology*, 3rd rev. ed., Vol. 2A, Wiley-Interscience, New York, 1981, p. 1945.

50. H. O. Calvery, *Food Res.* **7**, 313 (1942).

51. Unpublished data, M&T Chemicals, Inc., Rahway, N.J., 1976, 1978.

52. J. K. Lim, G. J. Renaldo, and P. Chapman, *Caries Res.* **12**(3), 177 (1978).

53. R. A. Kehoe and co-workers, *J. Nutr.* **19**, 597 (1940); **20**, 85 (1940).

54. H. A. Schroeder, J. J. Balassa, and I. H. Tipton, *J. Chronic. Dis.* **17**, 483 (1964).

55. D. H. Calloway and J. J. McMullen, *Am. J. Clin. Nutr.* **18**, 1 (1966).

56. CFR Title 21, 172.180, 175.105, 175.300, 176.130, 176.170, 177.120, 177.1210, 177.2600, 178.3910, 181.29, 182.3845, 1981.

57. C. C. Dundon and J. P. Hughes, *Am. J. Roentgenol.* **63**, 797 (1950).

58. G. E. Spencer and W. C. Wycoff, *Arch. Ind. Hyg. Occup. Med.* **10**, 295 (1954).

59. H. A. Schroeder and co-workers, *J. Nutr.* **96**, 37 (1968).

60. A. P. DeGroot, *Voeding* **37**(2), 87 (1976).

61. M. Kanisawa and H. A. Schroeder, *Cancer Res.* **29**, 892 (1969).

62. O. J. Stone and C. J. Willis, *Toxicol. Appl. Pharmacol.* **13**, 332 (1968).

63. *Mutagenic Evaluation of Compound FDA 71-33, Stannous Chloride*, PB 245,461, prepared by Litton Bionetics Inc. for the FDA, National Technical Information Service, Springfield, Va., Dec. 31, 1974.

64. *Teratologic Evaluation of FDA 71-33, Stannous Chloride in Mice*, PB 221,780, prepared by Food & Drug Research Labs. for FDA, National Technical Information Service, Springfield, Va., Oct. 1972.

65. *Teratologic Evaluation of Compound FDA 71-33, Stannous Chloride in Rabbits*, PB 267,192, prepared by Food & Drug Research Labs. for FDA, National Technical Information Service, Springfield, Va., Sept. 17, 1974.

66. J. K. J. Lim, G. K. Hensen, and O. H. King, Jr., *J. Dent. Res.* **54**, 615 (1975).

67. R. C. Theuer, A. W. Mahoney, and H. P. Sarett, *J. Nutr.* **101**, 525 (1971).

68. H. G. Stoner, J. M. Barnes, and J. I. Duff, *Br. J. Pharm. Chemother.* **10**(1), 16 (1955).

69. F. J. C. Roe, E. Boyland, and K. Millican, *Food Cosmet. Toxicol.* **3**, 277 (1965).

70. M. Walters and F. J. C. Roe, *Food Cosmet. Toxicol.* **3**, 271 (1965).

71. A. P. DeGroot, V. J. Feron, and H. P. Til, *Food Cosmet. Toxicol.* **11**, 19 (1973).

72. D. M. Smith and co-workers, *A Preliminary Toxicological Study of Silastic 386 Catalyst*, Report LA-7367-MS, Los Alamos Scientific Laboratory, Los Alamos, N.M., June 1978.

73. M. P. Bigwood, P. J. Corvan, and J. J. Zuckerman, *J. Am. Chem. Soc.* **103**, 7643 (1981).

74. A. Sawyer, ed., *Organotin Compounds*, Marcel Dekker, New York, 1971.

75. R. C. Poller, *The Chemistry of Organotin Compounds*, Academic Press, New York, 1970.

76. W. Neumann, *The Organic Chemistry of Tin*, John Wiley & Sons, Inc., New York, 1970.

77. A. G. Davies and P. J. Smith, *Adv. Inorg. Chem. Radiochem.* **23**, 1 (1980).

78. H. G. Kuivila in J. J. Zuckerman, ed., *Organotin Compounds—New Chemistry and Applications, Advances in Chemistry Series No. 157*, American Chemical Society, Washington, D.C., 1976, p. 41.

79. Ref. 75, pp. 19–20.

80. Ref. 74, p. 666.

81. C. Eaborn, *J. Organomet. Chem.* **100**, 43 (1975).

82. O. A. Reutov, *J. Organomet. Chem.* **100**, 219 (1975).

83. K. A. Kocheshkov, *Ber.* **62**, 996 (1929).

84. K. A. Kocheshkov and co-workers, *Ber.* **67**, 717, 1348 (1934).

85. M. H. Gitlitz, Ref. 78, p. 169.

86. U.S. Pat. 2,675,398 (Apr. 13, 1954), H. E. Ramsden and H. Davidson (to Metal & Thermit Corp.).

87. A. P. Tomilov and I. N. Brago in A. N. Frumkin and A. B. Ershler, eds., *Progress in Electrochemistry of Organic Compounds*, Plenum Press, London, 1971.

88. G. J. M. van der Kerk, J. G. A. Luijten, and J. Noltes, *J. Appl. Chem.* **356** (1957); *Angew. Chem.* **70**, 298 (1958).

89. A. F. Finholt, A. C. Bond, Jr., K. E. Wilzbach, and H. J. Slesinger, *J. Am. Chem. Soc.* **69**, 2692 (1947).
90. E. R. Birnbaum and P. H. Javora, *Inorg. Synth.* **12**, 45 (1970).
91. R. F. Chambers and P. C. Scherer, *J. Am. Chem. Soc.* **48**, 1054 (1926).
92. D. Seyferth and S. B. Andrews, *J. Organomet. Chem.* **18**, P21 (1969).
93. U.S. Pat. 3,028,320 (Apr. 3, 1962), P. Kobetz and R. C. Pinkerton (to Ethyl Corp.).
94. G. Mengoli and S. Daolio, *J. Chem. Soc. Chem. Commun.*, 96 (1976).
95. Ref. 75, p. 59.
96. U.S. Pat. 3,355,468 (Nov. 28, 1967), J. L. Hirshman and J. G. Natoli (to M&T Chemicals, Inc.).
97. K. Sisido, Y. Takeda, and Z. Kinugawa, *J. Am. Chem. Soc.* **83**, 538 (1961).
98. U.S. Pat. 3,607,891 (Sept. 21, 1971), B. G. Kushlefsky, W. J. Considine, G. H. Reifenberg, and J. L. Hirshman (to M&T Chemicals, Inc.).
99. G. J. M. van der Kerk and J. G. A. Luijten, *J. Appl. Chem.* **4**, 314 (1954).
100. *Ibid.*, **6**, 56 (1956).
101. K. Hartel, *Agric. Vet. Chem.* **3**, 19 (1962).
102. M. H. Gitlitz, *J. Coat. Technol.* **53**(678), 46 (1981).
103. N. F. Cardarelli, *Controlled Release Pesticide Formulations*, CRC Press Inc., Boca Raton, Fla., 1976.
104. C. J. Anthony, Jr., and J. R. Tigner, *Wire Wire Prod.* **43**(2), 72 (1968).
105. J. Burley, P. Hope, R. E. Hutton, and C. J. Groenenboom, *J. Organomet. Chem.* **170**, 21 (1979).
106. H. Schumann and M. Schmidt, *Angew. Chem.* **77**, 1049 (1965).
107. A. G. Davies and P. G. Harrison, *J. Chem. Soc. C*, 298, 1313 (1967).
108. U.S. Pat. 2,679,506 (May 25, 1954), E. R. Rochow (to M&T Chemicals, Inc.).
109. U.S. Pat. 3,519,665 (July 7, 1970), K. R. Molt and I. Hechenbleikner (to Carlisle Chemical Works, Inc.).
110. U.S. Pat. 3,901,824 (Aug. 26, 1975), V. Knezevic, M. W. Pollock, K.-L. Liauu, and G. Spiegelman (to Witco Chemical Corp.).
111. V. V. Pozdaev and V. E. Gel'fan, *J. Gen. Chem. USSR* **43**(5), 1201 (1973).
112. V. I. Shiryaev, E. I. M. Stepina, and V. F. Mironov, *Zh. Prikl. Khim.* (*Leningrad*), **45**(9), 2124 (1972).
113. S. Matsuda, S. Kikkawa, and I. Omae, *J. Organomet. Chem.* **18**, 95 (1969).
114. U.S. Pat. 3,440,255 (Apr. 22, 1969), S. Matsuda and S. Kikkawa.
115. R. E. Hutton and V. Oakes, Ref. 78, p. 123.
116. U.S. Pat. 4,130,573 (Dec. 19, 1978), R. E. Hutton and J. Burley (to Akzo, N.V.).
117. B. B. Cooray and G. Scott, *Dev. Polym. Stab.* **2**, 53 (1980).
118. G. Ayrey and R. C. Poller, *Dev. Polym. Stab.* **2**, 1 (1980).
119. H. O. Wirth and H. Andreas, *Pure Appl. Chem.* **49**, 627 (1977).
120. L. I. Nass, *Encyclopedia of PVC*, Vol. 1, Marcel Dekker, New York, 1976, pp. 313–384.
121. P. Klimsch, *Plaste Kautsch.* **24**, 380 (1977).
122. U.S. Pat. 4,028,337 (June 7, 1977), W. H. Starnes, Jr. (to Bell Telephone Laboratories, Inc.).
123. K. Figge and W. Findeiss, *Angew. Makromol. Chem.* **47**, 141 (1975).
124. W. H. Starnes, Jr., *Dev. Polym. Degrad.* **3**, 135 (1981).
125. D. Braun, *Dev. Polym. Degrad.* **3**, 101 (1981).
126. U.S. Pats. 2,219,463 (Oct. 29, 1940), 2,267,777 (Dec. 30, 1941), 2,267,779 (Dec. 20, 1941), and 2,307,092 (Jan. 5, 1943), V. Yngve (to Carbide & Carbon Chemicals Corp.).
127. U.S. Pat. 2,307,157 (Jan. 5, 1943), W. M. Quattlebaum, Jr., and C. A. Noffsinger (to Carbide and Carbon Chemicals Corp.).
128. U.S. Pat. 2,731,484 (Jan. 17, 1956), C. E. Best (to Firestone Tire & Rubber Co.).

129. U.S. Pat. 2,648,650 (Aug. 11, 1953), E. L. Weinberg and E. W. Johnson (to Metal & Thermit Corp.).
130. J. G. A. Luijten and S. Pezarro, *Br. Plast.* **30**, 183 (1957).
131. *Thermolite 813 PVC Stabilizer*, Technical Information Bulletin 273, M&T Chemicals, Inc., Rahway, N.J., Feb. 1979.
132. *Thermolite 831 PVC Stabilizer*, Technical Information Bulletin 316, M&T Chemicals, Inc., Rahway, N.J., Feb. 1979.
133. *Fed. Regist.* **33**, 16334 (Nov. 7, 1968); **37**, 5019 (March 9, 1972); **39**, 28899 (Aug. 12, 1974); **40**, 2798 (Jan. 16, 1975).
134. U.S. Pat. 3,424,717 (Jan. 28, 1969), J. B. Gottlieb and W. E. Mayo (to M&T Chemicals, Inc.).
135. U.S. Pat. 3,769,263 (Oct. 30, 1973), W. E. Mayo and J. B. Gottlieb (to M&T Chemicals, Inc.).
136. U.S. Pat. 3,810,868 (May 14, 1974), L. B. Weisfeld and R. C. Witman (to Cincinnati Milacron Chemicals, Inc.).
137. J. W. Burley, *Tin Its Uses* **111**, 10 (1977).
138. D. Lanigan and E. L. Weinberg, ref. 93, p. 134.
139. U.S. Pat. 4,080,363 (Mar. 21, 1978), R. E. Hutton and J. W. Burley (to Akzo, N.V.).
140. S. Karpel, *Tin Its Uses* **125**, 1 (1980).
141. F. G. Willeboordse, F. E. Critchfield, and R. L. Mecker, *J. Cell. Plast.* **1**, 3 (1965).
142. L. R. Brecker, *Plast. Eng.*, 39 (Mar. 1977).
143. R. V. Russo, *J. Cell. Plast.* **12**, 203 (1976).
144. J. Kenney, Sr., *Plast. Eng.*, 32 (May 1978).
145. *Fascat Esterification Catalysts from M&T*, Bulletin Quick Facts No. 45, M&T Chemicals, Inc., Rahway, N.J., 1978.
146. *M&T Fascat 4201 Organotin Esterification Catalyst*, Technical Data Sheet No. 345, M&T Chemicals, Inc., Rahway, N.J., 1978.
147. U.S. Pat. 4,170,579 (Oct. 9, 1979), J. F. Bosso and M. Wismer (to PPG Industries, Inc.).
148. M. Komura and R. Okawara, *Inorg. Nucl. Chem. Lett.* **2**, 93 (1966).
149. Brit. Pat. 1,146,435 (Oct. 7, 1966), P. A. Hoye (to Albright & Wilson, Ltd.).
150. U.S. Pat. 3,824,264 (July 16, 1974), E. J. Bulten (to Cosan Chemical Corp.).
151. E. J. Bulten, *J. Organomet. Chem.* **97**, 167 (1975).
152. E. J. Bulten, H. F. M. Gruter, and H. F. Martens, *J. Organomet. Chem.* **117**, 329 (1976).
153. J. W. Burley, R. E. Hutton, and V. Oakes, *J. Chem. Soc. Chem. Commun.*, 803 (1976).
154. U.S. Pat. 4,105,684 (Aug. 8, 1978), R. E. Hutton, V. Oakes, and J. Burley (to Akzo, NV).
155. R. E. Hutton, J. W. Burley, and V. Oakes, *J. Organomet. Chem.* **156**, 369 (1978).
156. U.S. Pat. 4,195,029 (Mar. 25, 1980), E. Otto, W. Wehner, and H. O. Wirth (to CIBA-GEIGY Corp.).
157. U.S. Pat. 3,021,302 (Feb. 13, 1962), H. H. Frey and C. Dorfelt (to Farbwerke Hoechst A.G.).
158. *Fed. Regist.* **35**, 13124 (Aug. 18, 1970); **37**, 4077 (Feb. 26, 1972).
159. U.S. Pat. 4,130,673 (Dec. 19, 1978), W. Larkin (to M&T Chemicals, Inc.).
160. U.S. Pat. 4,144,362 (Mar. 13, 1979), W. Larkin (to M&T Chemicals, Inc.).
161. G. H. Lindner, *Verres Refract.* **35**, 262 (1981).
162. C. Tamborski, F. E. Ford, and E. J. Soloski, *J. Org. Chem.* **28**, 181 (1963).
163. Ref. 162, p. 237.
164. K. Harrendorf and R. E. Klutts, *J. Econ. Entomol.* **60**, 1471 (1967).
165. K. R. S. Ascher and J. Moscowitz, *Int. Pest Control*, 17 (Jan.–Feb. 1969).
166. *Stannous Octanoate (Stannous 2-Ethylhexanoate)*, Technical Data Sheet 176, M&T Chemicals, Inc., Rahway, N.J., Dec. 1981.

167. S. L. Axelrood, C. W. Hamilton, and K. C. Frisch, *Ind. Eng. Chem.* **53**, 889 (1961).

168. *Fed. Regist.* **39**, 43390 (Dec. 13, 1974).

169. C. J. Evans, *Tin Its Uses* **89**, 5 (1971).

170. U.S. Pat. 3,578,616 (May 11, 1971), L. D. Harry and G. A. Sweeney (to the Dow Chemical Co.).

171. *Stannous Oxalate*, Technical Data Sheet 227, M&T Chemicals, Inc., Rahway, N.J., 1975.

172. U.S. Pat. 3,153,010 (Oct. 13, 1964), L. L. Jenkins and N. R. Congiundi (to Monsanto Co.).

173. U.S. Pat. 3,194,791 (July 13, 1965), E. W. Wilson and J. E. Hutchins (to Eastman Kodak Co.).

174. Hydrogenation of Coal in the Batch Autoclave, U.S. Bureau of Mines Bulletin 622, U.S. Bureau of Mines, Washington, D.C., 1965.

175. P. J. Smith, *Toxicological Data on Organotin Compounds*, ITRI Publication No. 538, International Tin Research Institute, Perivale, UK, 1977.

176. Ger. Pat. 63,490 (June 10, 1968), A. Tzschach, E. Reiss, P. Held, and W. Bollmann.

177. W. N. Aldridge, J. E. Casida, R. H. Fish, E. C. Kimmel, and B. W. Street, *Biochem. Pharmacol.* **26**, 1997 (1977).

178. J. M. Barnes and H. B. Stoner, *Br. J. Ind. Med.* **15**, 15 (1958).

179. C. R. Worthington, ed., *The Pesticide Manual*, 6th ed., British Crop Protection Council, Croydon, U.K., 1979, pp. 142, 259, 266, 267.

180. R. D. Kimbrough, *Environ. Health Perspectives* **14**, 51 (1976).

181. H. G. Verschuuren, E. J. Ruitenberg, F. Peetoom, P. W. Helleman, and G. J. Van Esch, *Toxicol. Appl. Pharmacol.* **16**, 400 (1970).

182. Ref. 49, p. 1964.

183. Ref. 49, p. 1962.

184. Ref. 49, p. 1959.

185. *Fed. Regist.* **43**, 49575 (1978).

186. Ref. 76, p. 233.

187. D. Lanigan and E. W. Weinberg, Ref. 78, p. 135.

188. W. N. Aldridge, Ref. 78, p. 189.

189. J. M. Barnes and L. Magos, *Organometal. Chem. Rev.* **3**, 137 (1968).

190. Ref. 49, p. 1956.

191. L. B. Weisfeld, *Kunststoffe* **65**, 298 (1975).

192. *Occupational Health Guidelines for Organic Tin Compounds (as Tin)*, U.S. Department of Labor, Occupational Safety and Health Administration, Washington, D.C., 1978.

193. *Occupational Exposure to Organotin Compounds*, DHEW (NIOSH) Publication No. 77-115, U.S. Department of Health, Education, and Welfare, National Institute for Occupational Safety and Health, Washington, D.C., Nov. 1976.

194. *SRI International, 1981 Directory of Chemical Producers—United States of America*, SRI International, Palo Alto, Calif., 1981, pp. 459, 471, 493, 798, 855, 892, 897, 960–962.

195. *SRI International, 1981 Directory of Chemical Producers—Western Europe*, 4th ed., Vol. 2, SRI International, Palo Alto, Calif., 1981, pp. 1555, 1567, 1636, 1690, 1760–1763.

196. M. J. Fuller, *Industrial Uses of Inorganic Tin Chemicals*, ITRI Publication 499, International Tin Research Institute, Middlesex, UK, 1975.

197. *Annual Report 1978*, International Tin Research Institute, Middlesex, UK, p. 20.

198. *The Economics of Tin*, 3rd ed., Roskill Information Services, Ltd., London, July 1981, pp. 110–117, 129–130, 349–354.

199. *Annual Review of the World Tin Industry 1981–1982*, Hargreaves & Williamson, London, 1981, pp. 11–12, 15.

200. *The World Tin Industry—Supply and Demand*, Australian Mineral Economics Pty., Ltd., Sydney, Australia, 1980, pp. 176–180, 202–206.

201. P. A. Cusack and P. J. Smith in R. Thomson, ed., *Specialty Inorganic Chemicals*, Royal Society of Chemistry, London, 1981, pp. 285–310.

202. *Mod. Plast.*, 70 (Sept. 1981).

203. *Mod. Plast.*, 82 (Sept. 1981).

204. *Cosmetic and Toiletry Raw Materials*, Charles H. Kline & Co., Inc., Fairfield, N.J., 1979, pp. 275–278.

205. T. W. Lapp, *The Manufacture and Use of Selected Alkyltin Compounds—Final Report*, EPA 5601-6-76-011, U.S. Environmental Protection Agency, Washington, D.C., 1976, 123 pp.

206. *Chem. Mark. Rep.*, 39 (Jan. 11, 1982).

207. Price lists, M&T Chemicals, Inc., Rahway, N.J.; Bis(tributyltin) oxide, Jan. 4, 1982; Dibutyltin dichloride (solid), Mar. 1, 1981; Tributyltin acetate, Jan. 4, 1982; Tributyltin chloride, Jan. 4, 1982; Tributyltin fluoride, Jan. 4, 1982.

General References

J. D. Donaldson, *A Review of the Chemistry of Tin(II) Compounds*, ITRI Publication 348, 1964.

Gmelin Handbook of Inorganic Chemistry, System No. 46, Pts. A–D (1971–1978); Pts. 1–8 (1975–1981), Springer-Verlag, New York.

C. L. Mantell, *Tin—Its Mining, Production, Technology and Applications*, 2nd ed., Hafner Publishing Co., New York, 1970, pp. 407–449.

J. W. Mellor, *A Comprehensive Treatise on Inorganic and Theoretical Chemistry*, Vol. 7, Longmans, Green & Co., New York, 1937.

Tin Times, International Tin Research Institute, Middlesex, England, Nos. 1–6, 1981.

A. F. Trotman-Dickenson, ed., *Comprehensive Inorganic Chemistry*, Vol. 2, Pergamon Press, Oxford, 1973, pp. 43–104.

J. W. Price and R. Smith in W. Fresenius, ed., *Tin, Handbook of Analytical Chemistry*, Vol. 4a, Pt. 3, Springer-Verlag, New York, 1978.

TIRE CORD

Since the introduction of the pneumatic tire (1–3) the reinforcing strength or tensile member has been some form of textile or steel fiber, usually in the form of a cord. Early tires used fabrics in the form of square woven fabrics; however, this was replaced by a unidirectional arrangement of tire cords utilizing woven fabric or creel calendering. The development of textile reinforcements for tire cords has been driven by the tire need for better mechanical properties and therefore better tire performance (4,5). Tire cord development has been paralleled by the emergence of other markets for new fibers (qv) in the apparel and industrial textile sectors. Thus, tire cords have progressed from natural fibers (cotton), through man-made (rayon) to the present totally synthetic suite of reinforcement candidates (nylons and polyesters). Furthermore, high performance fibers have found use in specialty areas (aramids and carbon fiber) and as new fibers are developed, the potential for tire use will be examined. Although each of these fibers is referred to herein as a single type, there has been and continues to be significant product development associated with each of the man-made or synthetic fibers, leading to stages of product development. In common with the rest of technology, the time between generations continues to diminish as a result of market pressure and research and development by the fiber producers and tire manufacturers. The evolution of the various tire cords in terms of annual consumption in North America is illustrated in Figure 1.

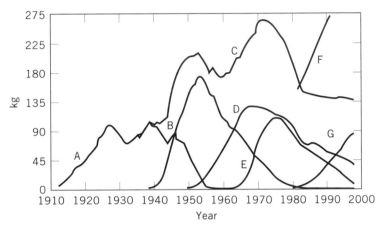

Fig. 1.   U.S. reinforcement consumption for tires, where A shows cotton; B, rayon; C, total all-fibers; D, nylon; E, conventional poly(ethylene terephthalate) (PET); F, steel cord; and G, high modulus low shrink (HMLS) PET (6,7).

Cotton

Cotton (qv) was the first material to be used in mass production of tires. Its relative abundance and well-established handling make the fiber naturally amenable to industrial usage. Also, the chemical structure of cotton allows for good adhesion to rubber thus facilitating a useful composite structure. However, as tire usage and performance demands rose, specifically in the areas of strength and fatigue resistance, the need for better reinforcement arose. This need was met by a number of organic and inorganic tire cords (6,8,9).

Rayon

Introduced commercially in 1938 and developed through generations of product to the present Super III rayon (6), rayon is a man-made regenerated cellulose fiber derived from wood pulp (qv). The process, which must be rigorously controlled to be environmentally acceptable, involves extraction of alkali cellulose with carbon disulfide, maturation, and spinning into fibers. The fiber, which also has wide use in textile markets, has outstanding dimensional stability but only moderate strength-to-weight ratio. Rayon fibers do not possess the intrinsic ability to bond with rubber and thus require development of suitable adhesive systems in the form of resorcinol–formaldehyde–latex (RFL) dip systems (*vide infra*). Although in declining use due to the increased cost associated with environmental aspects of the fiber production, rayon continues to be useful in specialty high performance tires.

Nylon

Nylon, an aliphatic polyamide, was introduced as a commercial tire cord in 1947 and grew in usage to ~5.4 billion kg/yr (~2 billion lb/yr) in the 1990s (10,11). Nylon-reinforced tires use nylon-6 polymer (polycaprolactam) fibers as well as nylon-6,6 (poly(hexamethylenediamine adipamide)) fibers. Nylon tire cords are characterized by extremely good fatigue resistance in compression and good adhesion to most rubber compounds with simple RFL adhesives.

Polyester

Polyester is the newest of the tire-reinforcement fibers to achieve volume usage of ~2.7 billion kg (~1 billion lbs) in the form of poly(ethylene terephthalate) (PET) fiber introduced in 1962 (12). As such, it has developed through several product generations beginning with a so-called standard modulus product up to the generation of high modulus low shrink (HMLS) or dimensionally stable polyester (DSP) forms of the 1990s (13). With respect to nylon, polyester fibers display lower thermal shrinkage and higher modulus. More complex adhesive systems tend to be required for polyesters to bond to rubber. Developers of polyester have explored other monomer types such as the liquid crystal polyesters (14) and the naphthalate-based copolymer with ethylene (15).

Aramid

Introduced in 1972, the wholly aromatic polyamide, poly(*para*-phenylene teraphthalamide), termed aramid, was the subject of extensive evaluation as a tire cord in all types of tires (8,14). As of the late 1990s, however, only specialized applications have emerged for aramid tire cord that draw on their high strength-to-weight ratio to produce tires with lower weight (16).

Glass

Introduced successfully for tires in 1967, glass fibers had properties that made them very attractive for use in tires (5,8). The brittleness of glass fibers, however, imposed some limitations on the final tire cord properties because of the requirement that each fiber be individually coated with a rubbery adhesive to avoid interfilament damage during fabrication and use. This additional treatment step is introduced at the fiber manufacturing stage. For several years fiber glass was used extensively in bias-belted and radial tires, but was ultimately replaced by steel belts in radial tires.

New Fibers

Whenever a new high strength fiber is developed, its potential for tire cord use is always explored because of the commercial attraction of large volumes available in the tire market. Few materials have emerged to displace the current two major fibers, nylon and polyester (14). Nonetheless, many examples of fibers offering attractive properties for tire cords have been reported in the literature, eg, polyethylene ketone (17), poly(paraphenylene benzobisoxazole) (18), acrylics (19), and high strength poly(vinyl alcohol) (20) (see VINYL POLYMERS).

Fiber Development for Tire Cord Use

The fundamental requirements for a useful tire cord fiber are high strength and modulus coupled with good dimensional stability (ie, resistance to deformation under temperature and load), and durability (fatigue and chemical stability) at favorable economics (21–23). The search for new fibers is driven by a wide array of potential nontire applications such as protective textiles (eg, against fire), ballistic protection, apparel, ropes, netting, geotextiles (qv), boat sails, and composite reinforcement. For load-bearing applications, the goal is to devise or utilize molecular structures that take full advantage of the inherently high strength of the C–C bond (24). One approach includes the area of molecular architecture, ie, polymer chemistry, to control features such as molecular weight, chain alignment, and chain–chain interactions. The aspects of processing (25) must also be considered when forming a fiber with suitable properties that will take full advantage of the polymer's potential and maintain viable economics. Although a wide array of reinforcement options appear to be available based on technical feasibility, the underlying features of cost vs value and the ease of integration into a manufacturer's process will dictate what is used in mass-produced tires. Thus, the market for organic tire cord remains divided between nylon and polyester fibers and will likely remain so for the medium-term future.

Fiber Properties

A tire reinforcement's use is dependent on several physical properties (26). Some of the most important are tabulated in Table 1. These properties effectively screen candidates for use in tires. The secondary features define a fiber's potential for tire use.

A key feature implicitly included in the use of organic fibers for tire reinforcement is the ability to retain strength and modulus characteristics at elevated temperatures (80–120°C for most tires) sufficient to sustain service demands. Thus, a tire which may appear to be overdesigned at room temperature is, in many cases, reflecting the changes in properties experienced at operating temperatures.

Fatigue Resistance.   Although tensile strength and modulus governs the amount of material required to reinforce a particular tire design, a critical service parameter is fatigue resistance. The extent of service fatigue (compressive

Table 1. Mechanical Properties for Reinforcing Fibers[a]

Name	Strength, cN/Tex[b]	Modulus, cN/Tex[b]	Thermal shrinkage	Density	T_g, °C	Mp, °C	Compression fatigue resistance
rayon(III)	52	1,200	0.2	1.53		210	good
nylon-6	84	300	7.5	1.13	20	220	excellent
nylon-6,6	86.5	400	7.4	1.14	50	254	excellent
polyester							
standard modulus	90	1,000	13.0	1.38			good
high modulus	70	1,000	3.9	1.38	75	260	good
aramid	200	4,900	0.1	1.44		454[c]	poor
carbon fiber	300–370	33,500– 15,250	0.0	1.83		3700[c]	poor
high tensile steel	43.4	2,680	0.0	7.85		1600	good
fiber glass (E-glass)	100	2,200	0.0	2.55		1180	poor

[a]Strength and modulus normalized to linear density (Tex = (g·wt)/km), an appropriate basis for materials of similar density; however, this breaks down for dense materials (eg, steel) because the basis for tire use is better described by properties per unit cross section (MPa) rather than weight.
[b]To convert cN/Tex to MPa (N·mm^{-2}), multiply by density (g·cm^{-3}) by 10.
[c]Decomposition temperature.

stress) is governed by tire design and service conditions (5,6,27–29). Thus laboratory fatigue testing performance has a critical impact on materials selection depending on tire design and expected service. For example, the design of a bias tire imposes more fatigue stress on a tire cord than a typical radial tire. Fatigue testing data are highly dependent on the nature of the testing geometry, and many different tests exist (see Table 1).

Dimensional Stability. Dimensional stability refers to how a fiber changes length under the influence of load or heat. Conventionally described in terms of fiber shrinkage (ASTM D885-64) at a defined temperature, the term has also come to mean time dependent length change or creep. In general, more highly oriented and therefore higher modulus fibers tend to exhibit lower shrinkage and less creep. Creep is an important factor in the control of tire dimensions during service and in certain aspects of tire appearance (30).

Toughness. Toughness is generally used to describe a fiber's ability to absorb energy before failure, defined as the area under the load vs elongation curve (energy to break). This parameter is important in controlling how well a tire cord resists impact damage, such as curb or pothole strikes. Although intrinsic strength controls quasi-static strength, for example, tire burst strength (as affected by the total load exerted on a tire cord) energy is a more significant parameter for impact damage.

Heat Generation. During tire operation, ie, rolling, reinforcing cords experience cyclic deformations (31). Because textile reinforcements are not perfectly elastic, this deformation results in energy loss that is mostly in the form of heat buildup. This process is known as hysteresis and such losses contribute to rolling resistance. Excessive heat buildup can accelerate tire cord thermochemical degradation, thus shortening tire lifetime.

Thermal and Chemical Stability. In addition to load-bearing properties, tire reinforcement must be able to resist degradation by chemicals in cured rubber and heat generation. The most critical degradant depends on the material in use. Most thermoplastic reinforcements are either modified directly or stabilized with additives to offset some, mostly thermal, degradation (32,33).

Processing

The basis for reinforcement of a pneumatic tire requires placing the strength or tensile member in a preferred direction, depending on the location and cord function in the tire. An overview of the tire production process, including essential elements of transforming a continuous yarn into a useful embodiment for tire reinforcement, is shown in Figure 2.

Linear Density. Linear density is defined as the weight per unit length (usually defined as denier), weight in grams of 9000 m or Tex, gram weight of 1000 m of yarn or cord (26). Tex is often used as dTex (Tex $\times$ 10) and is related to denier because dTex = denier $\times$ 1.111.

Twist. Twist (23) is measured in numbers of turns per unit original length, turns per inch, cm, or meter.

Construction. Reinforcing cords are normally constructed by combining single bundles of twisted yarns (plies) in an operation known as cabling (23). The final cord is defined by the linear densities of the individual plies and the twist applied during the ply and cabling operations. For example, a common polyester construction is 1100/1/3 315 $\times$ 315. That is, an 1100 dTex yarn is twisted 315 turns per meter (tpm), three of which are then cabled together by twisting in the opposite direction 315 tpm.

Twisting. The function of twisting (23) is to combine individual fiber filaments to improve the tensile properties by the cooperative interactions between the filaments, and to form into a bundle that is more easily manipulated in manufacturing processes. More importantly, the function of twisting fibers to form a ply (single) and then a cable or cord is to impart required properties of fatigue resistance and abrasion resistance. Cords are formed by twisting the fiber

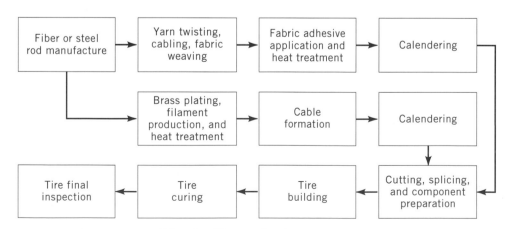

Fig. 2. Tire production process.

bundle into a ply, usually in the Z direction (anticlockwise spiral), then combining two or more plies by twisting together in the S direction (clockwise spiral). By convention and for ease of manufacture, the numerical values of twist in ply and cable are usually matched. The process of ply twisting and cabling may be accomplished in two discrete steps or, with modern direct cabling equipment, in a single operation. Proper handling of the sometimes fragile filaments is critical at the twisting stage, before fibers are transformed into the more robust cord structure. This is especially true of high modulus fibers such as aramids. In general, as the degree of twist (sometimes characterized by the twist factor index) is increased, the tensile properties of strength and modulus decrease. In contrast, the fatigue resistance rises as a cord is twisted. The amount of property gain or loss is governed by the fiber type, linear density, construction, and twist. Therefore, selection of the appropriate twist is a compromised decision guided by the needs of the final application. Some typical cord constructions are provided in Table 2.

Processing Oils. Individual fibers used for reinforcement purposes are quite small, on the order 10–20 micrometers in diameter, and are sensitive to damage resulting from abrasion by contact with machine parts during manufacture. Oils are applied during this process and serve to lubricate the filament bundle and act as wetting agents to assist the spreading and adhesion of coatings applied at later stages of processing (34). As such, processing oils must be amenable to a wide range of uses. Their formulation is highly proprietary based in general on the use of modified fatty acid esters. Other types of finish can also act as adhesive activators for later processing.

Weaving. After cord forming, a loose fabric is made by weaving the cords (as the warp) at a density in the range 6 to 12 cords/cm, depending on the construction using weft or pick cords to hold the fabric together. Pick cords are generally cotton, rayon, or polyester single-ply yarns used at densities of 50–75 threads per meter. Later, the pick cords may be broken to allow the fabric to expand laterally when formed into a tire, thus requiring that pick cords be relatively weak. Another method used, which avoids the need to break pick cords, is to use high elongation-fill pick cords with low modulus for ease of stretching. In addition, pick cords should be compatible with all later stages of processing, such as adhesive dipping and tire curing.

Adhesive Application and Heat Treating. This step represents perhaps the most critical phase in preparing a tire cord where a suitably strong adhesive is applied to bond the fiber to the rubber, thus forming a useful composite (35,36). In addition, the cords are heat treated to impart desired physical (tensile) characteristics depending on the final tire performance requirements. Adhesive formulations for tire cord use have been developed since rayon was introduced as a tire cord. The basis of application for almost all types is the so-called resorcinol formaldehyde latex (RFL) system (35–38). Such adhesives are usually

Table 2. Constructions for Tire Cord Use

Name	Construction	Typical application
polyester (1100 dTex 350 tpm twist)	1100/3 350 × 350	radial tire carcass
nylon (1400 dTex 390 tpm twist)	1400/2 390 × 390	bias tire carcass
nylon (940 dTex 470 tpm twist)	940/2 470 × 470	radial tire overlay

waterborne dispersions and solutions of ingredients designed to form a network or matrix to bond to the reinforcement and the rubber, and serves to transmit stress from the rubber to the reinforcement. In general, the role of the resin component (RF) is to develop adhesion to the fiber, and the latex (L) adheres to the rubber. The mechanisms of RFL-to-rubber adhesion are quite complex and have been the subject of many years of development and manipulation (37). After dipping and fabric/adhesive curing, the coated fabrics become completely dry and develop adhesion to the rubber during the tire cure cycle. Although basic RFL formulations are similar, many variations have been made to suit particular cord and rubber requirements. For instance, polyester can be usefully treated with a single RFL formulation or by two discrete formulas (pre-dip and post-dip). An alternative for polyester is the application of an adhesive-activating material, usually an epoxy, by the fiber producer which enables the tire cord producer to apply a single, more simple dip system. Furthermore, highly inert materials such as aramid require an additional pre-dip non-RFL adhesive or similar modification techniques before an RFL-type adhesive can be applied to bond with rubber.

A typical formulation for nylon tire cord adhesives are given in Table 3. Modifications to the formulations can, for example, include the use of different resin-forming agents, various latexes and mixtures thereof, and different adhesives applied in more than one step. Some processes have been developed to incorporate a fiber surface modification as part of the adhesive application using surface treatments (39,40). Special adhesive systems have been developed for fiber glass, using agents to bond to the glass surface and a softer, more rubbery matrix to separate and protect the individual filaments in a bundle (36,41).

Because organic tire cords are mostly thermoplastic polymers (except rayon and aramids) they can be modified by further heat stretch during dipping steps (42,43). This is in addition to the principal fiber alignment operation that takes place during the fiber spinning steps. The process conditions used are highly dependent on fiber properties and therefore are the subject of detailed study and optimization by individual tire cord producers. In addition to heat setting, process conditions must be suitable for (1) applying the correct amount of adhesive and (2) exposure to the correct temperature for completion of adhesive curing reactions.

Creel Calendering and Nonconventional Tire Manufacturing. An alternative to the fabric weaving and dipping approach is to assemble the final in-rubber treatment directly from the individual cords. Such processing is known as creel

Table 3. RFL Dip System for Nylon[a,b]

Product	Value
water	407.7
sodium hydroxide, 10% (aq) solution	8.0
penacolite resin R2170, 75% (aq) solution	26.7
formalin, 37% (aq) solution	20.3
vinylpyridine latex, 40%	250.0

[a]Ref. 5.
[b]This system results in a 18% solids by weight dispersion.

calendering and is commonly used in coating steel cord with rubber for tire use. An additional complication for organic tire cords over steel involves adhesive and cord processing that must be included in the calendering process. Although such processing is colloquially referred to as single end, production facilities process many cords or ends (ca 50) simultaneously. Such processing eliminates any requirement for weaving fabric and the need for and subsequent inclusion of pick cords in tires.

The basic principles and approaches to manufacturing pneumatic tires have been in place for many years, and because of the scale of modern tire production, radical change is slow. However, developments of new tire production processes continue (44,45) and as new methods take hold, it is likely that changes in tire cord handling and preparation will be required.

Steel Tire Cord

Steel tire cord provides a unique combination of strength, ductility, dimensional stability, resistance to fatigue, rubber adhesion, and consumer value that led to a dramatic increase in steel cord consumption for the last decade (46,47).

Materials and Process. The steel chosen for tire cord is a eutectoid carbon steel containing 0.7% carbon, 0.5% manganese, 0.2% silicon, and a very low amount of sulfur and phosphorus (9,48). The steel rod is cleaned with acid, rinsed, drawn through tungsten carbide dies to reduce its diameter from 5.5 to ~3.0 mm, heat treated (patented) to increase ductility for further drawing to ~1 mm, then patented again.

The patented wire is again cleaned with acid, rinsed, and brass plated just before the second drawing. The brass acts as a drawing lubricant and as well as an adhesive to rubber. The brass composition is typically 60–70% copper with zinc as the remainder. The patented, brass-plated wire is drawn into filaments of 0.15–0.38 mm diameter.

Steel Cord Construction. Filaments or wires are twisted into strands and then combined into cords. The lay length is the axial distance required to make a 360° revolution of any component in a strand or cord and is expressed in millimeters. Direction of lay is the helical disposition of components of a strand or cord. Strands or cords have an S (left-hand) lay if, when held vertically, the spirals around the central axis of the strands conform in direction of slope to the central portion of the letter "S". Strands have a Z (right-hand) lay if the spirals conform in direction of slope to the central portion of the letter "Z" (49). Lay lengths of the strands or cords, ranging from 2.5 to 25.0 mm, are chosen to hold the filaments together and yet obtain the smallest degree of fretting and the highest fatigue life during tire service. A spiral wrap of a single filament is applied to some large-diameter tire cords in order to increase the cord's compression resistance and bending rigidity. However, this wrap decreases elasticity and increases fretting of cord during tire service.

The way the filament, strand, and spiral wrap are assembled, and the filament diameters used (usually in millimeters), determine the cord construction. The tire industry has adopted an ASTM wire nomenclature. The description of the construction follows the sequence of manufacturing of the cord, ie, starting with the innermost strand or filament and moving outward. A plus (+) sign separates each layer. If the filament diameters are the same in two or more components, the diameter is omitted for all but the last component. If a strand is a single filament, the numerical designation "1" is omitted. The full description of the construction is given by the following formula:

$$(N \times F) \times D + (N \times F) \times D + F \times D \qquad (1)$$

where N = number of strands, F = number of filaments, and D = nominal diameter of filaments, in mm (7,49). The cords can be classified into (1) regular, (2) Lang's lay, (3) open, (4) compact, and (5) high elongation. The most common cord is the regular cord in which the direction of lay in strands is opposite to direction of lay in closing the cord. Lang's lay cord is formed such that the direction of lay in the strands is the same as the direction of lay in closing the cord. Open cords are loosely twisted and movable relative to each other to enable rubber penetration into the cord (49,50). Open cord can be also obtained by using preformed filaments. Compact cord is bunched so that the filaments have mainly linear contact with each other; rubber penetration can be promoted by using filaments of different diameters (49,51). High elongation cord is a Lang's lay cord in which the strands are loosely associated and movable relative to each other, to allow the cord substantial stretchability under load (49). Examples of the various cords used in tires are shown in Figure 3.

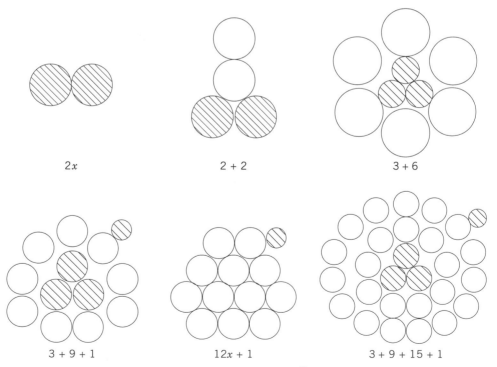

Fig. 3. Cross sections of cords used in tires where ⬳ represent inner- and outermost strands (first and/or last number in description).

Adhesion. A thin layer of brass coating on steel cord facilitates adhesion between the metal and rubber compound. The mechanism of rubber–brass adhesion has been the topic of much speculation and fundamental research. Advanced techniques such as auger electron spectroscopy (aes), electron spectroscopy for chemical analysis (esca), and scanning transmission electron microscopy (stem) have contributed to a better understanding of adhesion formation and degradation (52,53). The interfacial reaction products, which consist of Cu_xS where x ranges from 1.8 to 2.0, depend on copper and ZnO content of the brass surface. There is a minimum critical thickness of Cu_xS for maximum adhesion as well as a maximum thickness above which adhesion begins to drop. In general, adhesion degradation is prevented by a lower copper content or thinner brass layer. Additional elements in copper–zinc alloy improve adhesion, especially aged adhesion. Additional elements under study include cobalt, nickel, and iron (46,52–54).

Stress in a Cord. Twisting two or more filaments together to form a cord reduces the overall modulus and strength of the construction. The degree of the reduction depends on the individual filaments (55). If the lay length is significantly higher than the filament diameter, the stress is $\sigma_a = 4P/n\pi d^2$, under axial loading, and $\sigma_b = Ed/D$ under bending load, where σ_a and σ_b are axial and bending stresses, respectively; P is axial load; n is the number of filaments in the cord; E is the modulus; d is the filament diameter; and D is the diameter of bending. The bending load equation shows that the smaller the filament, the lower the bending stress of the cord, resulting in increased bending fatigue life (48).

Cord Mechanics

Cord Elasticity. Tire cords, consisting of several yarns twisted together, are one-dimensional structural members. The cord properties depend on yarn filament properties and their geometrical organization (56). The filament itself possesses an internal structure at the microscopic level. The way in which cord properties are related to filament properties, geometries, and other variables is the subject of textile mechanics (21,56). In application to tires, interest lies in cord properties obtained experimentally without relating to the microstructure in formulation of cord composite properties. Cords, as the reinforcing member of cord–rubber composite, have certain strength, modulus, and dimensional, thermal, and chemical stability properties. Cords may be characterized by linear relationship between stress and strain tensors as in the following:

$$\sigma_i = Q_{ij}\epsilon_j \quad \text{or} \quad \epsilon_i = S_{ij}\sigma_j i, j = 1, 2 \ldots 6 \tag{2}$$

where σ, ϵ, Q, and S are stress component, strain component, stiffness matrix, and compliant matrix, respectively. In general, cords are considered transversely isotropic materials, ie, in one plane, mechanical properties are equal in all directions. This application leads to five independent constraints that characterize cord, namely (1) extensional Young's modulus, (2) extensional Poisson's ratio, (3) transverse Young's modulus, (4) transverse Poisson's ratio, and (5) shear modulus. However, most applications of cord–rubber composite require only extensional Young's modulus, extensional Poisson's ratio, and shear modulus. The influence of transverse Young's modulus and transverse Poisson's ratio are negligible.

Cord materials such as nylon, polyester, and steel wire conventionally used in tires are twisted and therefore exhibit a nonlinear stress–strain relationship. The cord is twisted to provide reduced bending stiffness and achieve high fatigue performance for cord–rubber composite structure. The detrimental effect of cord twist is reduced tensile strength. Analytical studies on the deformation of twisted cords and steel wire cables are available (22,56–59). The tensile modulus E_c of the twisted cord having diameter D and pitch p is expressed as follows (60):

$$E_c = \frac{E_f}{1 + \tan^2 \theta} = \frac{E_f}{1 + \frac{\pi^2 D^2}{p^2}} \tag{3}$$

where θ is the helix angle of the outermost filament and E_f is the filament modulus. The Young's modulus of twisted cord E_c decreases with increasing twist, whereas the Poisson's ratio v_c increases with increasing twist, according to equation 4 (60).

$$v_c = \frac{p^2}{\pi^2 D^2} \tag{4}$$

The cord fatigue phenomenon and effect of twist has been discussed (61). Values of Young's modulus of some belt and carcass cords used in passenger tires are given in Table 4 (62). Poisson's ratio v_c of tire cords is often in excess of 0.5 due to twist; the higher the twist, the larger Poisson's ratio of cords.

Cord Viscoelasticity.   Cords used in tires exhibit time-dependent viscoelastic property. Viscoelasticity means the loss of energy during stress–strain cycle. This loss of energy is dissipated in the form of heat which affects cord material properties and therefore tire performance. Viscoelastic behavior of tire cord contributes to the total power loss of the tire and therefore its rolling resistance. When a tire cord is subjected to sinusoidal strain of small amplitude, the resulting stress–strain curve is elliptical, exhibiting linearly viscoelastic characteristics, and the material properties are represented by real and imaginary moduli E' and E'', and $\tan \delta$ the ratio E''/E'. The properties are dependent on temperature and frequency but independent of strain amplitude. However, all

Table 4. Values of Young's Modulus for Tire Cord[a]

Cord construction	Young's modulus, E_c, GPa[b]
belt ply	
5 × 1 × 0.025 mm steel	110
1670/2 Kevlar	25
1840/3 rayon	11
body ply	
1110/2 polyester	4.0
940/2 nylon	3.4

[a]Ref. 62.
[b]To convert GPa to psi, multiply by 145,000.

tire cords, including steel, exhibit nonlinearity, especially when the strain amplitude is relatively large, eg, tire cord operating under normal tire conditions (63). The stress–strain loop is no longer elliptical. Material properties in the nonlinear region are no longer represented by real and imaginary moduli, and the viscoelastic properties in this nonlinear region are characterized by effective dynamic modulus and mechanical loss. The most important aspect of viscoelasticity is that organic cord properties change as cords are processed for adhesive application through the dip unit. Here organic cord is subjected to tension under high temperature conditions; the organic cords in the new cured tires may have considerably different properties from that of cord pulled from used tires (64,65). A typical hysteresis loop indicating nonlinear viscoelastic behavior of a tire cord is shown in Figure 4. Demonstrated approaches of designing tire cords for minimum mechanical loss are available (66).

Cord–Rubber Composite. The pneumatic tire is a commonly used fiber-reinforced rubber composite product that is subjected to severe punishment during the course of its life. The main characteristic of fiber-reinforced rubber composites is a low stiffness property ratio of rubber matrix to that of reinforcing cords. The geometrical and design complexities of tires, along with the heterogeneous, anisotropic material properties and complicated load application and distribution, necessitate better understanding of these composite properties. Such information is valuable for tire engineering application, analysis, and performance, as well as continuously increasing requirements for smooth ride, handling, and durability.

The characteristic features of a cord–rubber composite have produced the netting theory (67–70), the cord–inextensible theory (71–80), the classical lamination theory, and the three-dimensional theory (67,81–83). From structural considerations, the fundamental element of cord–rubber composite is unidirectionally reinforced cord–rubber lamina as shown in Figure 5. From the principles of micromechanics and orthotropic elasticity laws, engineering constants of tire T cord composites in terms of constitutive material properties have been expressed (72–79,84). The most commonly used Halpin-Tsai equations (75,76)

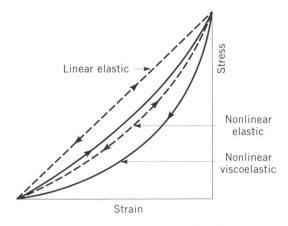

Fig. 4. Nonlinear elastic and viscoelastic response (66).

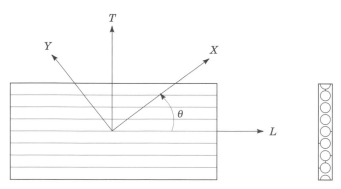

Fig. 5. Calendered unidirectionally reinforced single-ply cord–rubber lamina where θ is the helix angle, L is lamina, and T is tire.

for cord–rubber single-ply lamina L are expressed in equation 5:

$$E_L = E_c V_c + E_t V_t$$

$$E_\tau = E_r \frac{(1 + 2V_c)}{(1 - V_c)}$$

$$G_{LT} = \frac{G_r[G_c + G_r + (G_c - G_r)V_c]}{[G_c + G_r - (G_c - G_r)V_c]} \tag{5}$$

$$V_L = V_c V_c + V_r(1 - V_c)$$

$$V_T = V_L \left[\frac{E_T}{E_L} \right]$$

where V_c is the volume fraction of cord $(\pi r^2/t)$epi (r is the radius of cord, t is the thickness of ply, and epi is cords per inch of ply); V_r is the volume fraction of rubber; v_c and v_r are Poisson's ratio of cord and rubber, respectively; and G is shear modulus. It is important to realize that the Halpin-Tsai equations are valid only for small deformations. Attempts to predict effective elastic properties of orthotropic cord–rubber composites from the principle of virtual work have been made (85–87).

The direction of applied load in the cord rubber composite is not, in general, coincident with the principal axis of the system. The response of such a system is complicated by the development of shear strains due to off-axis loadings and therefore display complex behavior in terms of coupling between shear strain and normal stress, and normal strain and shear stress, resulting in generally orthotropic ply behaving as anisotropic lamina. The effect of cord angle on elastic constants for nylon cord–rubber lamina are shown in Figures 6 and 7 (88).

The tire is a system of laminated composites composed of numbers of orthotropic laminae. The elastic engineering properties of each ply vary due to material constituents, cord end count, cord angle, and rubber thickness. The stress–strain relations or engineering properties of laminated composites are complicated by the fact that the tensile load applications cause twisting, bending, and stretching. The derivative of in-plane and out-of-plane displacements define

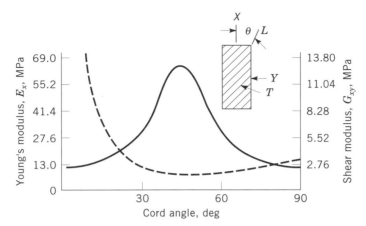

Fig. 6. Variation of Young's modulus (– – –) and shear modulus (—), with cord angle θ for one-ply nylon–rubber system (88).

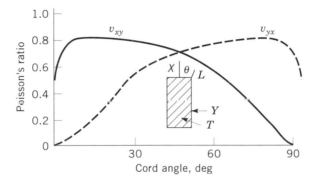

Fig. 7. Variation of Poisson's ratio, v, with cord angle θ for one-ply nylon–rubber system (88).

stretching and bending strains, respectively. The mathematical formulations of stress–strain relations for laminated composites with application to tires have been discussed (83,89), with some experimental verification.

Inter-ply shear is prominently featured in cord–rubber composite laminates, and may relate to delamination-induced failures. Studies utilizing experimental, analytical, and finite element tools, with specific application to tires, are significant in compliant cord–rubber composites (90–95).

Cord Impact on Tire Performance

The major load-bearing member of cord–rubber composites is the cord, which provides strength and many other critical properties essential for tire performance. Cords in plies form the structural backbone of the tire. The rubber plays the important but secondary role of transmitting load to the cords via shearing stresses at the cord–rubber interface. Other expected performance characteristics of the tire are due to design and manufacturing processes. Table 5 (96)

Table 5. Tire Performance Characteristics[a]

Tire performance and processability	Knowledge of relationship				Properties
	None	Poor	Fair	Good	
burst strength				*	tensile, uniformity
bruise and cut resistance			*		high speed, hot tensile modulus, uniformity, toughness
endurance (separation resistance)			*		fatigue resistance, adhesion, adhesion degradation resistance in tire environment, dynamic properties (hysteresis), uniformity
for radial high speed			*		modulus, density
power loss		*			dynamic properties, density
tire size and shape			*		dimensional stability, modulus, creep
tire uniformity and flatspotting			*		dimensional stability (T_g), modulus, creep
tread wear			*		modulus
tire ride and handling		*			modulus, density
spring rate		*			modulus
noise		*			modulus, density
groove cracking resistance			*		creep
processability			*		dimensional stability (T_g), hot creep and shrinkage, moisture regain, adhesion, environmental degradation resistance, uniformity, stiffness, toughness, moisture swelling and shrinkage, compaction, tensile

[a]Ref. 96.

identifies several tire performance characteristics and how they are dependent on tire cord properties.

Burst Strength. The burst strength of the tire is derived from the cord strength and is related by equation 6 (97,98):

$$\text{burst strength} = N t_u K = \frac{\Pi P_b (r_c^2 - r_{\max}^2)}{\sin \alpha} \qquad (6)$$

where N is total number of cords in a tire, t_u is average ultimate tensile strength of cord, P_b is burst pressure, r_c is radius from the center of rotation to the crown of tire, $r_{\max}$ is radius from the center of rotation to the maximum section width of a tire, α is the crown angle between cord path and circumferential plane through the crown of a tire, and K is the efficiency factor which depends on the distribution of ultimate cord strength and is always less than one.

Tires are designed with a very high factor of safety, about six for radial medium truck tires and ten or higher for radial passenger tires. The reason for such a higher factor of safety is that the cord tensile strength is measured at 23°C and 55% relative humidity (ASTM D885-64). However, tires can reach short-term operating temperatures up to 150°C in some applications. At these temperatures some organic tire cord materials can degrade resulting in loss of strength ≤50% of room temperature value. Adding 10–20% degradation in cord strength during the life cycle of the tire, drops the effective tire strength to 150–200% of designed strength, which may be sufficient for burst but may fail to meet several important performance criterias. Therefore, selecting cords only on burst strength requirements may not be the most desirable approach for overall tire design.

Bruise Resistance.   Bruise resistance of a rolling tire describes its ability to resist impact failure. Bruise resistance is tested by measuring the energy required to break a passenger tire under inflation at room temperature when a 19-mm diameter plunger is pushed through the crown at 51 mm/min crosshead speed (DOT test 49CFR571-109). The area under the load-deflection curve is measured as bruise energy. Bruise resistance should be measured at operating temperature conditions of the tire for an accurate prediction of cord contribution because its material properties exhibit different performance characteristics at elevated temperatures (97,99,100).

Tire Endurance.   The interlaminar shear stress and deformation in the composite laminate of tire can produce delamination-induced failures, especially at the belt edges. The tire composite can be designed to minimize interlaminar shear stress through proper selection of cord properties, cord orientation, cord end count, and rubber properties. Another separation mechanism can involve cord–rubber adhesion. This adhesion depends on chemical and mechanical interaction between adhesive-treated cord–rubber surface. Another failure phenomenon is fatigue which is mostly observed on sidewall near the shoulder and bead regions due to high temperature and stresses (101–103). To some extent fatigue performance can be controlled through proper selection of cord material and twist.

High speed performance tires may generate enough heat to cause tread separation. At higher speeds, a tire can go into resonance (standing wave) which distorts the tire and results in shoulder growth and excessive heat buildup. The onset of resonance for a passenger radial tire generally begins at speeds of about 120 km/h. This onset speed can be raised through tire design changes such as shorter or stiffer sidewall or through lower belt weight by selecting lighter cord materials.

Cornering and Ride.   The stiffness or modulus properties of cord–rubber composites used in tires strongly influence the ultimate performance of tires (104). The stiffness of the belt package primarily determines the cornering and ride characteristics of a radial passenger tire. The belt package in contact with the road is a fairly complex composite consisting of tire components including the innerliner, carcass, belts, and tread. The in-plane flexural rigidity of the belt package is the most important parameter controlling cornering, whereas out-of-plane flexural rigidity controls ride. Flexural rigidity, in turn, depends on the stiffness of the belt package. From the principles of composite mechanics, the circumferential modulus or stiffness per unit area in the hoop direction of a cord–rubber laminate with identical plies at equal and opposite cord angles is expressed as in equation 7 (62,104):

$$E_x = E_c V_c \cos^4\theta + G_r(1 - V_c) - \frac{(E_c V_c \sin^2\theta \cos^2\theta + 2G_r(1 - V_c)^2)}{(E_c V_c \sin^4\theta + 4G_r(1 - V_c))} \quad (7)$$

where E_x is the circumferential belt modulus, θ is cord angle with respect to circumference, V_c is cord volume fraction, G_r is rubber shear modulus, and E_c is cord modulus in tension. Through the analysis of equation 6, it has been shown that, in order of importance, belt modulus depends on rubber modulus, cord angle, cord volume fraction, and cord modulus (62). However, equation 6 does not take into account the effect of additional stiffening due to internal pressure, double curvature of belt, and most importantly the presence of body ply or plies in tires, which can significantly affect overall stiffness. Again, cord properties play an important role in providing necessary stiffness for tire performances.

Treadwear.   Treadwear is a complex physical–chemical process driven by the frictional energy developed at the interface between tread pattern elements and the pavement. The rate of wear is influenced by the loss modulus property of the tread compound, the microstructure (or abrasiveness) of the pavement, environmental conditions, and vehicle operation (105,106). The wear of passenger tires occurs mainly during cornering maneuvers (104), where the tread center line distortion in the footprint impacts slippage of the tread rubber relative to road surface mainly in the region at the rear of the footprint. Belted radial tires experience less slippage than nonbelted tires in cornering. Even at straight driving, ie, at 0° slip angle, the footprint pressure distribution is much more uniform for radial belted than nonbelted constructions. This difference in performance has been explained by testing the tire footprint as a laminated anisotropic beam subjected to cornering force at the plane of footprint (83,107). The Gough stiffness, S, which has been shown to correlate well with actual wear rate of tires (108), can be expressed as in equation 8:

$$S = \frac{E_x G_{xy}}{C_1 E_x + C_2 G_{xy}} \quad (8)$$

where E_x and G_{xy} are circumferential and shear moduli of the cord–rubber laminate in the tread region (which depend on cord properties and tread package design), and C_1, C_2 are constants. However, it should be observed that the loss modulus of tread compound has been demonstrated as a single property to characterize compound wear characteristics. The loss modulus is dependent on the glass-transition temperature, T_g, the carbon black level and morphology, process oil content, and polymer microstructure properties. The proper use of these components opens up the possibility of developing compounds that can accurately characterize treadwear performance under selected environmental conditions.

Flatspotting.   It is generally assumed that the tire cord is primarily responsible for flatspotting (109–114). The mechanism of flatspotting in tires is based on the viscoelastic behavior of tire cords. Tire cords such as nylon and polyester tend to shrink when heated above their glass-transition temperature,

T_g, as in the case of a running tire. The cord strain in the footprint is much smaller than in other parts of the tire. When the tire stops rotating, the cord elements in the footprint cool to ambient temperature because they are under much less strain, and therefore shrink more than the cords in the remainder of tire which are in considerable tension due to inflation pressure. When the tire starts to rotate again, persistence of this difference causes flatspot, which remains until the tire is reheated to a temperature at which the flatspot was introduced.

Power Loss on Tire Rolling Resistance. About 95% of rolling resistance or power loss is due to the viscoelastic loss of the tire. However, the relative contribution of cord and rubber to tire rolling resistance has been the subject of numerous studies (113–119). It is estimated that 20 to 40% of a tire's power loss is due to the behavior of cords in tires. Tires having carcass constructed with aramid cords have a lower rolling resistance than those constructed with steel cords (116). Similar findings are quoted for tires constructed with aramid and steel belts. Also, tire design, tire weight, cord deformation in tire, as well as dynamic properties of cords affect tire rolling resistance. There have been significant changes in tire construction and materials which have markedly reduced power loss, eg, change from bias to radial tire construction, use of higher inflation pressure, and improved rubber compounds, specifically tread compounds with better dynamic properties (120). However, the challenge is to understand the exact contribution of cord to tire power loss in radial tires and to develop cord material with properties that significantly reduce power loss.

Test Methods

Tire cords are characterized for their physical, adhesion, and fatigue properties for use in tires. These characterizations are conducted under normal and varying test conditions to predict their performance during tire operation. Various test methods used to characterize tire cords are described.

Standard Test Methods for Steel Tire Cord. ASTM standard D2969-92 includes test methods for steel cords that are specifically designed for use in the reinforcement of pneumatic tires. It describes test methods determining steel cord construction, break strength, elongation at break, modulus, flare, linear density, straightness, residual torsion, brass coating composition, and mass of steel cords.

Standard Test Method for Adhesion Between Steel Tire Cords and Rubber. Steel cords are vulcanized into a block of rubber and the force necessary to pull the cords linearly out of the rubber is measured as adhesive force. ASTM method D2229-93a can be used for evaluating rubber compound performance with respect to adhesion to steel cord. The property measured by this test method indicates whether the adhesion of the steel cord to the rubber is greater than the cohesion of the rubber, ie, complete rubber coverage of the steel cord or less than the cohesion of rubber (lack of rubber coverage).

Steel Cord Impact Test. A transverse impact test method for steel cords has been designed to determine the resistance to cutting (puncture resistance) when used as a tire belt reinforcement (121). The test is a modified charpy test, and the sample consists of a 3-mm diameter rubber cylinder reinforced in the center with a steel cord. The impact force and total amount of energy absorbed is measured. This test has demonstrated that high elongation cord possesses greater capacity to absorb energy compared with regular cord. This is one reason why high elongation cord constructions are used as a protection layer in truck and off-the-road tires.

Rotating Beam Fatigue Test for Steel Cords.   The purpose of this test method is to evaluate steel cord for pure bending fatigue (121). The test sample consists of a 3-mm diameter rubber embedded with steel cord. Different bending stress levels are applied and the time to failure is recorded. The test stops at 1.44 million cycles. The fatigue limit is calculated from $S - N$ (stress–number of cycles) curve.

Rotoflex Test for Steel Cords.   The purpose of this test is to determine the bending fatigue limit as a function of pretension in the carcass cords of truck tires (121). The test sample consists of rubber strip 5 mm thick, 12 mm wide, and 450 mm long, reinforced with steel cord in the longitudinal direction. The dead weight determines the pretension in the cord, whereas the mold diameter determines the applied alternating bending stress in the filaments of the cord. Repeating the procedure for different pretensions leads to the determination of the Smith Goodman diagram, which reveals differences in cord construction from bending fatigue considerations.

Standard Test Methods for Tire Yarns, Cords, and Woven Fabrics.   ASTM standard D885M-94 includes test methods for characterizing tire cord twist, break strength, elongation at break, modulus, tenacity, work-to-break, toughness, stiffness, growth, and dip pickup for industrial filament yarns made from organic base fibers, cords twisted from such yarns, and fabrics woven from these cords that are produced specifically for use in the manufacture of pneumatic tires. These test methods apply to nylon, polyester, rayon, and aramid yarns, tire cords, and woven fabrics.

Standard Test Method for Thermal Shrinkage of Yarn and Cord.   ASTM test method D4974-89 is used for measuring thermal shrinkage of yarn and cords with linear density ranging from $20-700 \times 10^{-6}$ kg/m (20–700 tex) using the Testrite thermal shrinkage oven. A relaxed, conditioned specimen of yarn or cord is subjected to a tension of 4.4 ±0.88 mN/tex to dry heat at a temperature of 177°C for two minutes. The percent shrinkage is read from a scale on the instrument.

Static Adhesion Tests for Organic-Based Yarns, Cords, and Fabrics. The most commonly used static adhesion tests are the H-test (ASTM D4776-88), U-test (ASTM D4777-88), T-test (122), I-test (123), and strip–peel test (ASTM D4393-85). These tests derive their name from the shape of the test specimen. In the H, U, and T tests, adhesion is reported by the force required to pull an embedded cord through and out of the rubber. The force of adhesion is affected by embedded length of cord, rate of loading, and temperature. In the I-test, both ends of an I-shaped test specimen are pulled. The force–deflection curve is recorded and adhesion is measured as the second peak force. The strip–peel adhesion test is used to determine peel adhesion force of reinforcing fabrics bonded to rubber compounds. This method is applicable to either woven or parallel cord textile structures.

Dynamic Adhesion for Organic-Based Yarns, Cords, and Fabrics. Adhesion gradually deteriorates with repeated deformation. Dynamic adhesion evaluation is characterized by number of cycles of deformation to reach limiting value. Various dynamic adhesion test methods have been developed based on the type of deformation. The two most commonly used methods are the Goodrich Disk Fatigue test (124) and the Dynamic Flex Strip Adhesion (ASTM D430-59). In the Goodrich Disk Fatigue test, samples can be subjected to both compression and extensional deformations. In the Dynamic Flex Strip Adhesion test, a two-ply strip test piece is subjected to compressive fatigue by flexing it over a spindle. Adhesion is estimated by either comparison of strip adhesion before and after flexing or number of cycles for ply separations.

Fatigue Tests for Organic-Based Tire Cords. Tire cords experience large number of stress–strain cycles which lead to their degradation. There are a number of laboratory tests used to estimate fatigue life of tire cords (eg, ASTM D885-64). Some of the most commonly used test methods include the following. (*1*) In the Firestone Compression Flex test (ASTM D885-64) (125) the test specimen consists of two plies of cords in parallel plane, one ply of test cords and the other of steel cords. Compressive fatigue is produced by flexing the test piece over a spindle for a certain length of time. The test cords are then extracted and tested for retained strength. (*2*) In the Goodrich Disk Fatigue test (ASTM D885-64) (125,126) the test specimen consists of a rectangular rubber block reinforced with cords parallel to a long axis. The test piece is firmly secured into the periphery of two canopied disks. Cords are subjected to simple longitudinal extension and compression during testing. Strength loss after a certain number of revolutions is used as a measure of fatigue life. (*3*) In the Mallory Tube Fatigue test (ASTM D885-64) (126,127) the test sample consists of a hollow rubber cylinder in which cords are parallel to each other and parallel to the axis of the test cylinder and have certain ends per inch. The cords in the tube are subjected to alternate compression and tension. The number of revolutions until failure is the measure of fatigue resistance. (*4*) In the Dunlop Fatigue test a test specimen is an endless belt made up of five plies of cords in rubber at certain ends per inch. The second and fifth plies are comprised of cords that are to be tested. The test consists of running belt for a known time, then extracting cords from two test plies and measuring their retained strength.

Tire Cord Status

As a load-carrying member, the type of tire cord used for a specific tire depends on the use requirements. For example, the requirements for original vehicle fitment are more extensive than the replacement market. In addition, these requirements also vary with the type of vehicle, eg, passenger, light truck, or medium truck, vehicle design, and service expectations. One important consideration in selecting tire cord remains value to customer. Therefore, the total cost of tire from materials to conversion to customer delivery may play a significant role in the final preference for tire cord. It is therefore not surprising that volume usage of tire cord varies in different regions around the world. To some extent the preference for a tire cord is dictated by tire designs, ie, bias vs radial. Also, within the same design, the belt cord could be different than carcass cord due to different functionality in these components. The belt material for radial tires requires high stiffness/modulus vs carcass material requirements of flexibility and fatigue resistance for comfort and durability.

The automotive industry is becoming more global in nature due to various trade pacts and the free market economy push. An apparent change from globalization is tire design conversion from bias to radial. From a historical perspective, the transition from bias to radial led to relative change in the volume of tire cord material used in that market. For bias applications, nylon tire cord remains dominant, particularly in heavy tires with high load-carrying capacity. The use of nylon-6 vs nylon-6,6 follows the basic criteria: value to customer in that market, with the exeptions of the airplane and large earth-mover tires. In this tires, the wider temperature performance range for nylon-6,6, due to its higher melting temperature, may offer an advantage in certain applications. Polyester has also been used in bias tires (passenger and light trucks). For radial tires the choice of tire cord for belt application varies among glass, rayon, and steel. Fiber glass, once a dominant belt material in North America, has not made much headway in Europe and Japan, due to value glass provided in these regions vs steel, which remains the dominant belt material. Whereas polyester quickly became the choice for carcass material in passenger tires in North America and Japan, rayon and nylon continued to be used in Europe and Asia. Value to the customer remains a significant factor in choice of tire cord in specific markets at specific times. In this respect, the total cost structure of tire cord material as influenced by nontire market requirements and also plays an important role in determining the intrinsic value to the customer.

Requirements for tire cord material will to some extent be driven by new vehicle trends. For example, the clean air emphasis in North America places lightweight vehicles and materials at a premium. For tire cord the fuel economy or rolling resistance provided by the cord–rubber composite may shift the pattern of usage. A common requirement for all types of tire cord surfaces is a high strength-to-weight ratio.

For steel cord, to meet the higher strength-to-weight ratio, continuous improvements have been made to increase steel cord strength, including steel composition, low levels of impurities, controlled nonmetallic inclusion, minimized segregation during metallurgical solidification processes, and minimum surface defects and decarbonization (128), in conjunction with an increase in the total amount of drawing.

A new approach for high strength organic fibers includes polymer compositions capable of forming in the liquid crystalline state (thermotropic and lyotropic). High modulus, high strength fibers from aromatic polyamides and aromatic copolyesters have been manufactured utilizing conventional dry-jet wet spinning or melt spinning technology (129).

BIBLIOGRAPHY

"Tire Cords" in *ECT* 2nd ed., Vol. 20, pp. 328–346, by L. Skolnik, The B. F. Goodrich Co.; in *ECT* 3rd ed., Vol. 23, pp. 78–97, by L. Skolnik, BF Goodrich.

1. E. S. Thompkins, *Prog. Rubb. Plast. Technol.* **6**(3), 203 (1990).
2. *Tyres & Access* (2), 21 (1992).
3. F. J. Kovac, *Tire Technology*, The Goodyear Tire & Rubber Co., Akron, Ohio, 1978.
4. R. S. Bhakuni, S. K. Mowdood, W. H. Waddell, I. S. Rai, and D. L. Knight, in J. I. Kroschwitz, ed., *Encyclopedia of Polymer Science and Engineering*, Vol. 16, John Wiley & Sons, Inc., New York, 1989, p. 834.
5. S. K. Clark, ed., *Mechanics of Pneumatic Tires*, DOT HS No. 805,952, U.S. Dept. of Transportation, Washington, D.C., 1981.
6. G. Corallo, "Historical Review of Light-Duty Tyre Carcass Reinforcement and the State of Current and Next Generation Technology," *Conference Preprints ACS, Rubber Division*, Philadelphia, Pa., May 1995, p. 13.
7. R. M. Shemenski, *Wire J. Int.*, 70 (Sept. 1994).
8. C. S. Slaybaugh, *Tyre Cord, Rubb. Plast. News* **18**(2), 86 (Aug. 1988).
9. A. G. Causa, D. K. Kim, and R. S. Bhakuni, "Advances In Metallic and Polymeric Fibre Reinforcement For Tyres," International Rubber Conference, IRC 86, 1986, p. 236.
10. *Plast. Technol.* **18**, 58 (Oct. 1994).
11. R. S. Williams and T. Daniels, *Polyamides, Rapra Rev.* **3**(3), 33/1 (1990).
12. D. Brunnschweiler and J. Hearle, *Textile Horizons*, **12**(6), 22 (June 1992).
13. G. S. Rogowski, *Ind. Rubb. J.* (6), 60 (May 1994).
14. J. Mack, *Mat. Edge* (16), 15 (Mar. 1990).
15. P. B. Rim, "Properties and Applications of PEN Fibres," *Conference Preprints ACS, Rubber Division*, Paper 27, Philadelphia, Pa., May 1995, p. 14.
16. E. Dommershuijzen, "Super Single Truck Tyre Carcasses: A Challenge For P-Aramid Fibers," *Conference Proceedings ACS, Rubber Division*, Orlando, Fla., Paper 53, Oct. 1993, p. 7.012.
17. U.S. Pats. 5115003-A and 5122565-A, J. H. Coker, E. R. George, J. M. Machado, and L. H. Slaugh (to Shell Research Ltd.).
18. Jpn. Pat. 7026029 (1995) (to Yokahama Rubber Co. Ltd.).
19. Jpn. Pat. 05272005-A and 931019 (9346), (to Mitsubishi Rayon Co., Ltd.).
20. Jpn. Pat. 06212513-A and 940802 (9435), (to Toray Ind. Inc.).
21. "High Performance Polymers: Their Origin and Development," *Proceedings of the ACS Symposium, New York, Apr. 15–18, 1986*, Elsevier Science Publishing Co. Inc., New York, 1986.
22. J. W. S. Hearle, P. Grosberg, and S. Backer, *Structural Mechanics of Fibers Yarns & Fabrics*, Wiley Interscience, New York, 1969.
23. J. W. S. Hearle and W. E. Morton, *Physical Properties of Textile Fibres*, Heinemann Ltd., London, 1975.
24. W. B. Adams, R. K. Eby, and D. E. McLemore, eds., *Mater. Res. Soc. Symp. Proc.*, 134 (1989).
25. A. Ziabicki and H. Kawai, eds., *High Speed Fiber Spinning—Science and Engineering Aspects*, John Wiley & Sons, Inc., New York, 1985.
26. *Dictionary of Fiber & Textile Technology*, Hoechst Celanese Corp., Charlotte, N.C., IZ 503.
27. M. Borowczak and A. G. Causa, "Fatigue Behaviour of Cord-Reinforced Rubber Composites," *Conference Proceedings ACS, Rubber Division*, Louisville, Ky., May 1989.
28. B. L. Lee and D. S. Liu, *J. Composite Mat.* **28**(13), 1261 (1994).
29. K. Yabuki, *Sen'i Gakkaiishi* **42**(11), 460 (1986).
30. P. B. Rim, C. J. Nelson, and D. S. Liu, *Rubber World*, 209 (Oct. 1, 1993).
31. D. C. Prevorsek, S. Murthy, and Y. D. Kown, *Rubber Chem. Technol.* **60**(4), 659 (1987).

32. *Nippon Gomu Kyokaishi*, **64**(4), 260 (1992).

33. H. Hisaki and S. Suzuki, *Rubber World*, **201**(2), 19 and 22 (1989).

34. M. Mayer, *Plastic Extrusion Technology*, Hanser Publisher, Munich, Germany, 1988, p. 561.

35. R. Iyengar, *Rubber World*, **197**(2), 24 (Nov. 1987).

36. T. S. Solomon, *Rubber Chem. Technol.* **58**(3), 561 (1985).

37. N. K. Porter, *J. Coated Fabrics* **23**, 34 (July 1993).

38. D. C. Prevorsek and R. K. Sharma, *Rubber Chem. Technol.* **54**(1), 72 (1981).

39. W. H. Waddell, L. R. Evans, J. G. Gillick, and D. Shuttleworth, *Rubber Chem. Technol.* **65**(3), 687, July–Aug. 1992.

40. M. A. Doherty, B. Rijpkema, and W. Weening, *Conference Preprints ACS, Rubber Division*, Paper 17, Philadelphia, Pa., 12, May 1995.

41. R. F. Seibert, *Conference Proceedings ACS, Rubber Division*, Paper 52, Orlando, Fla., Oct. 1993.

42. C. A. Litzler, *Rubber Age* **105**(3), 37 (Mar. 1973).

43. J. Wagenmakers, H. Stuut, and J. Noordman, *Conference Proceedings AFICEP, International Rubber Conference Communications*, No. 223-5.012, Paris, June 1990.

44. B. Davis, *Rubb. Plast. News* **2**, 3(17), 8 (June 15, 1992).

45. *Rubb. Plast. News* **19**(18), 21 (Mar. 1990).

46. W. D. Havens and C. L. Laun, *Wire J. Int.*, 36 (June 1970).

47. D. Chambaere, L. Bourgois, and W. Meersseman, *Akron Rubber Group Technical Symposia, Fall Meeting*, Akron, Ohio, 1988, p. 21.

48. A. Prakash, D. K. Kim, and R. M. Shemenski, *International Conference on Fatigue, Corrosion Cracking, Fracture Mechanical and Failure Analysis, ASM Proceedings*, Dec. 1985.

49. *Bekaert Steel Cord Catalogue*, NV Bekaert SA, Zwevegem/Belgium, 1987.

50. D. Chambaere, *Eur. Rubb. J.*, 41 (Sept. 1995).

51. U.S. Pat. 4,829,760 (May 16, 1989), P. Dampre (to NV Bekaert SA).

52. G. Hamers, *Rubber World*, **182**(6), 26 (1980).

53. W. J. Van Ooij, *Rubber Chem. Technol.* **57**, 421 (1984).

54. U.S. Pat. 4,446,198 (May 1984), R. M. Shemenski, D. K. Kim, and T. W. Starinshak (to The Goodyear Tire & Rubber Co.).

55. G. A. Costello and J. W. Phillips, *J. Eng. Mech. Div. (ASLE)*, **102**, 171 (Feb. 1976).

56. F. Tabaddor, *Finite Element Analysis of Tire & Rubber Products, Continuing Education*, The University of Akron, Akron, Ohio, 1984, Chapt. VI.

57. S. Ohwada, *Rep. Inst. Ind. Sci. Univ. Tokyo*, **4**(6), 238 (1955).

58. K. Kabe, M. Koishi, and T. Akasaka, "Stress Analysis for Twisted Cord and rubber of FRR," presented at *6th Japan–U.S. Conference on Composite Materials*, Orlando, Fla., June 1992.

59. S. Backer, in F. A. McClintock and A. S. Argon, eds., *Mechanical Behavior of Materials*, Addison-Wesley, Reading, Mass., 1966, p. 56.

60. J. O. Wood and G. B. Redmond, *J. Textile Inst.* **56**, T191 (1965).

61. T. Takeyama and J. Matsui, *Rubber Chem. Technol.* **42**, 159 (1969).

62. J. D. Walter, *Rubber Chem. Technol.* **51**, 524 (1978).

63. P. R. Willett and co-workers, *J. Appl. Polym. Sci.* **19**, 2005 (1975).

64. P. R. Willett, *Rubber Chem. Technol.* **46**, 425 (1973).

65. Y. D. Kwon, R. K. Sharma, and D. C. Prevorsek, in R. A. Fleming and D. I. Livingston, eds., *Tire Reinforcement and Tire Performance*, ASTM STP 694, American Society for Testing and Materials, Philadelphia, Pa., 1979, p. 239.

66. D. C. Prevorsek, C. W. Beringer, and Y. D. Kwon, *Polymers for Fibers and Elastomers*, Vol. 24, American Chemical Society, Washington, D.C., 1984, p. 371.

67. T. Akasaka, "Keynote Address," *Composite Materials Conference*, Orlando, Fla., 1989.

68. R. B. Day and S. D. Gehman, *Rubber Chem. Technol.* **36**(1), 11 (1963).

69. T. Akasaka, K. Kabe, and H. Togawa, *J. Jpn. Soc. Comp. Matls.* **10**(1), 24 (1984).

70. T. Akasaka and S. Yamazaki, *J. Soc. Rubb. Indust. Jpn.* **59**(7), 24 (1986).

71. T. Akasaka, S. Yamazaki, and K. Asano, *Trans. Jpn. Soc. Comp. Matls.* **10**(1), 24 (1984).

72. J. E. Adkins and R. S. Rivlin, *Philos. Trans. A.* **248**, 20 (1955).

73. J. D. Walter and H. P. Patel, *Rubber Chem. Technol.* **52**, 710 (1979).

74. S. K. Clark, *Rubber Chem. Technol.* **56**, 372 (1982).

75. J. C. Halpin and S. W. Tsai, *AFML-TR*, 67 (June 1969).

76. J. E. Ashton, J. C. Halpin, and P. H. Petit, *Primer on Composite Materials*, Stamford, Conn., 1969.

77. V. E. Gough, *Rubber Chem. Technol.* **41**, 988 (1968).

78. G. Tangorra, *Proc. Int. Rubber Conf. (Moscow)* (1971).

79. T. Akasaka and M. Hirano, *Fukuggo Zairyo (Composite Materials)* **1**, 70 (1972).

80. R. A. Ridha, *Rubber Chem. Technol.* **53**(4), 849 (1980).

81. C. T. Sun and L. Sijian, *J. Comp Matls.* **22**(7), 629 (1988).

82. T. Akasaka and K. Kabe, *J. Jpn. Soc. Comp. Matls.* **3**(2), 76 (1977).

83. J. D. Walter, in S. K. Clark, ed., *Mechanics of Pneumatic Tires*, DOT HS 805 952, U.S. Dept. of Transportation, National Highway Traffic Safety Administration, Washington, D.C., 1981, p. 123.

84. S. K. Chawla, in H. Brody, ed., *Synthetic Fibre Materials*, Longman Scientific & Technical, Longman Group U.K. Ltd., London, 1994, p. 203.

85. O. Po'sfalvi, *Tire Sci. Tech.* **4**, 219 (1976).

86. O. Po'sfalvi and P. Szor, *Periodica Polytechnica* **1**, 189 (1973).

87. O. Po'sfalvi and P. Szor, *Muszaki Tudomauy* **48**, 401 (1974).

88. H. P. Patol, J. L. Turner, and J. D. Walter, *Rubber Chem. Technol.* **49**, 1095 (1976).

89. T. Akasaka, *Plenary Lecture, Eighth Annual Tire Society Meeting*, Akron, Ohio, 1989.

90. J. D. Walter, *Rubber Chem. Technol.* **51**, 524 (1978).

91. J. L. Turner and J. L. Ford, *Rubber Chem. Technol.* **55**(4), 1078 (1982).

92. A. Y. C. Lou and J. D. Walter, *Rubber Chem. Technol.* **52**, 792 (1979).

93. D. O. Stalmaker, R. M. Kennedy, and J. L. Ford, *Exp. Mech.* **20**, 87 (1980).

94. M. Hirano and T. Akasaka, *Fukugo Zairyo (Composite Materials)* **2**(3), 6 (1973).

95. J. DeEskinazi and R. J. Cembrole, *Rubber Chem. Technol.* **57**, 168 (1983).

96. C. Z. Draves and L. Skolnik, in Ref. 65, p. 3.

97. B. D. Coleman, *J. Mech. Phys. Solids* **7**, 60 (1958).

98. E. W. Lothrop, *Appl. Polym. Symp.* **1**, 111 (1965).

99. C. Z. Draves, T. P. Kuebler, Jr., and S. F. Vukan, *ASTM Matl. Res. Stand.* **10**(6), 26 (1970).

100. R. L. Guslister, *Tr. Nauchuo-Issled. Inst. Shimoi Promsti.* **3**, 154 (1957); J. R. Mosely, *Res. Assoc. Brit. Rubb. Manu.* (Transl. 817).

101. S. Eccher, *Rubber Chem. Technol.* **40**, 1014 (1967).

102. G. Butterwirth, *Chem. Eng. News*, 45 (1967).

103. A. G. Causa, in Ref. 65, p. 200.

104. J. D. Walter, *Technical Symposium*, Akron Rubber Group, Akron, Ohio, 1988, p. 15.

105. A. G. Veith, *Polym. Test.* **7**, 177 (1987).

106. A. Schallamach and K. Grosch, in Ref. 83, p. 365.

107. V. E. Gough and G. R. Shearer, *Inst. Mech. Engr., Pro. Auto. Div.*, 171 (1955).

108. V. E. Gough, *Rubber Chem. Technol.* **41**, 988 (1968).

109. B. K. Daniels, *SAE Paper*, 667A (1973).

110. W. H. Howard and M. L. Williams, *Rubber Chem. Technol.* **40**, 1139 (1967).

111. G. W. Rye and J. E. Martin, *Rubber World* **149**(1), 75 (1963).

112. P. V. Papero, R. C. Wincklhofer, and H. J. Oswald, *Rubber Chem. Technol.* **38**, 999 (1965).

113. W. E. Claxton, M. J. Forster, J. J. Robertson, and G. R. Thurman, *Textile Res. J.* **36**, 903 (1966).
114. J. M. Collins, W. L. Jackson, and P. S. Oubridge, *Trans. Inst. Rubber Ind.* **40**, 239 (1964).
115. P. R. Willett, *Rubber Chem. Technol.* **46**, 425 (1973).
116. J. J. Vorachek, R. J. Dill, and R. J. Montag, *SAE Conference Proceedings*, No. P-74, 1977, p. 169.
117. P. D. Shepherd, *SAE Paper*, 770333 (1977).
118. D. J. Shuring, in Ref. 117, p. 31.
119. D. C. Prevorsek, Y. D. Kwon, and R. L. Sharma, in Ref. 117, p. 75.
120. R. L. Sharma, Y. D. Kwon, and D. C. Prevorsek, in Ref. 65, p. 263.
121. L. Bourgois, in Ref. 65, p. 19.
122. G. S. Fielding-Russell, D. I. Livingston, and D. W. Nicholson, *Rubber Chem. Technol.* **53**, 950 (1980).
123. G. M. Doyle, *Trans. IRI* **36**, 177 (1960).
124. D. Kenyon, *Trans. IRI* **40**, 67 (1964).
125. M. W. Wilson, *Textile Res. J.* **21**, 47 (1954).
126. D. Kenyon, *Proc. IRI* **11**, 67 (1964).
127. U.S. Pat. 241,524 (1946), G. D. Mallory.
128. K. Benedens and co-workers, *Conference Proceedings of the Wire Association International Inc., 63rd Annual Convention*, Mar. 1993, p. 55.
129. A. G. Causa, *Tire Technol. Int.*, 28 (1993).

ROOP S. BHAKUNI
SURENDRA K. CHAWLA
D. K. KIM
D. SHUTTLEWORTH
The Goodyear Tire & Rubber Company

TITANIUM AND TITANIUM ALLOYS

Titanium [*7440-32-6*], a metal element of Group IVB, has a melting point of 1675°C and an atomic weight of 47.90. Titanium metal has become known as a space-age metal because of its high strength-to-weight ratio and inertness to many corrosive environments. Its principal use, however, is as TiO_2 as paint filler (see PAINT; PIGMENTS). The whiteness and high refractive index of TiO_2 are unequaled for whitening paints, paper, rubber, plastics, and other materials. A small amount of mineral-grade TiO_2 is used in fluxes and ceramics.

Titanium is the ninth most abundant element in the earth's crust, at approximately 0.62%, and the fourth most abundant structural element. Its elemental abundance is about five times less than iron and 100 times greater than copper, yet for structural applications titanium's annual use is ca 200 times less than copper and 2000 times less than iron. Metal production began in 1948; its principal use was in military aircraft. Gradually the applications spread to commercial aircraft, the chemical industry, and, more recently, consumer goods.

Titanium mineral occurs in nature as ilmenite, $FeTiO_3$; rutile, tetragonal TiO_2; anatase, tetragonal TiO_2; brookite, rhombic TiO_2; perovskite, $CaTiO_3$; sphene, $CaTiSiO_5$; and geikielite, $MgTiO_3$. Ilmenite is by far the most common, although rutile has been an important source of raw material. Although some deposits of anatase and perovskite are rich enough to be of commercial interest, the abundance and availability of high grade deposits of ilmenite and rutile have postponed the development of these minerals.

The principal titanium mineral, ilmenite, is found in either alluvial sands or hard-rock deposits. After concentrating, the titanium ore color is black. This is the black sand often found concentrated in bands along sandy beaches. The density of this concentrate is ca $4-5$ g/cm^3. The concentrate is processed to either pigment-grade TiO_2 or metal. The titanium metal is won from the ore in a physical form called sponge, a name derived from its appearance. The sponge is consolidated to an ingot and further processed to mill products in a manner similar to steel. In metallic form, titanium has a dull silver luster and an appearance similar to stainless steel.

Titanium was first identified as a constituent of the earth's crust in the late 1700s. In 1790, an English clergyman and mineralogist discovered a black magnetic sand (ilmenite) which he called menaccanite after his local parish. In 1795, a German chemist found that a Hungarian mineral, rutile, was the oxide of a new element which he called titan, after the mythical Titans of ancient Greece. In the early 1900s, a sulfate purification process was developed to obtain commercially high purity TiO_2 for the pigment industry, and titanium pigment became available both in the United States and Europe. During this period, titanium was also used as an alloying element in irons and steels. In 1910, 99.5%-pure titanium metal was produced at General Electric from the reaction of titanium tetrachloride with sodium in an evacuated steel container (Hunter process). Because the metal did not have the desired high melting point necessary for incandescent-lamp filaments, further work was discouraged. However, this reaction formed the basis of three commercial sodium reduction plants that were in operation between the 1950s and 1992. In the 1920s, high purity ductile titanium was prepared using an iodide dissociation method combined with Hunter's sodium reduction process.

The 1990s reduction process was based on work started in the early 1930s. A magnesium vacuum reduction process was developed for reduction of titanium tetrachloride to metal. Based on this process, the U.S. Bureau of Mines (BOM) initiated a program in 1940 to develop commercial production. Some years later, the BOM publicized its work on titanium and made samples available to the industrial community. By 1948, the BOM produced batch sizes of 104 kg. In the same year, Du Pont announced commercial availability of titanium, thus beginning the modern titanium metals industry (1).

By the mid-1950s, this new metals industry had become well established, including six producers, two other companies having tentative production plans, and more than 25 institutions engaged in research projects. Titanium, termed the wonder metal, was billed as the successor to aluminum and stainless steels.

When in the late 1950s, the U.S. Department of Defense (DOD) (titanium's most staunch supporter) shifted emphasis from aircraft to missiles, the demand for titanium sharply declined. None of the original titanium sponge plants are still in use in the 1990s. The Titanium Metals Corporation of America's plant in Henderson, Nevada, utilizes the original plant only in high demand markets.

Optimism followed by disappointment has characterized the titanium metals industry. In the late 1960s, the future again appeared bright. Supersonic transports and desalination plants were intended to use large amounts of titanium (see WATER, SUPPLY AND DESALINATION). Oregon Metallurgical Corporation, a titanium melter, decided at that time to become a fully integrated producer, ie, from raw material to mill products. However, the supersonic transports and the desalination industry did not grow as expected. Nevertheless, in the late 1970s and early 1980s, the titanium metal demand again exceeded capacity and companies in both the United States and Japan expanded capacities. This growth was stimulated by greater acceptance of titanium in the chemical process industry, by requirements of the power industry for seawater cooling, and by commercial and military aircraft demands. However, as a result of the economic recession of 1981–1983, the demand dropped well below capacity and the industry was again faced with hard times. A large demand for wide-body commercial aircraft led the industry to a new high in production in the late 1980s. In the early 1990s, the industry suffered another retrenchment resulting from not only a worldwide recession but also the collapse of the former Soviet Union, which impacted military budgets and opened the Soviet's large titanium production capability to Western markets. In 1995 the titanium metal industry again rebounded on account of increased demand for commercial aircraft and new consumer markets such as golf club heads (2).

Titanium ore bodies are uniformly distributed throughout the continents of the world (Table 1). They occur either as hard-rock deposits, magnetic in origin, or as secondary placer deposits. Titanium processing from placer deposits is shown in Figure 1. The titanium oxide contained in known deposits of ilmenite is close to 1 billion metric tons, whereas only about 50 million metric tons of titanium oxide in rutile are known to exist. The largest known reserves of titanium are in Canada and China. However, a significant amount of these reserves may be marginally economical.

Table 2 shows the TiO_2 mineral production in both 1990 and 1994; Table 3, the analyses of selected titanium mineral concentrates. Titanium ore concentrated from ilmenite ranges from 45% TiO_2 in hard-rock and some placer deposits to 64% in some Florida placer deposits. A value of 55% is typical for ilmenite; for rutile, ca 95%. Some ilmenites, particularly those of lower TiO_2 content, are smelted in Norway, Canada, and South Africa to produce pig iron and a 71% TiO_2 slag (Sorel slag). This slag is called titaniferous slag.

Future demand for titanium raw materials will continue to depend on the requirements for TiO_2 pigments. The pigment industry requires 94% of the world's mineral production, whereas the metal industry only requires 4% (5). The future supply may be influenced by several factors. Traditionally, Australia, Canada, and Norway have been the largest exporters of titanium minerals. More recently, however, newly established exporters such as Sierra Leone and South Africa are making an impact. Because the mines in the CIS, China, Brazil,

Table 1. Ilmenite and Rutile Resources in the World's Major Deposits[a]

Country	Ilmenite[b], 10^6 t		Rutile[c], 10^6 t		% of world Ti resources
	Economically exploitable	Other[d]	Economically exploitable	Other[d]	
Australasia					
Australia	27	3	6.7	0.6	1.5
New Zealand		31			1.1
North America					
Canada	75	541	1.8		22.1
Mexico					0.0
United States	19	77	1.5	5.5	3.9
South America					
Brazil		24			0.9
Europe					
Finland	5				0.2
Italy			6.0	12.0	1.1
Norway	128				4.6
Romania		8		1.7	0.4
CIS	211	118			11.7
Africa					
Mozambique		30		2.0	1.2
South Africa	85	50	5.1	2.4	5.3
Sierra Leone			2.5	23.0	1.6
Tanzania	13				0.5
Madagascar					0.0
Asia					
Bangladesh					0.0
China	216	910	0.2		40.1
India	58	33	4.2	1.8	3.6
Malaysia		10			0.4
Sri Lanka	2		0.3		0.1
Total	*839*	*1,835*	*28.3*	*49.0*	*100.0*

[a]Ref. 3.
[b]Ilmenite includes equivalent titanomagnetite, leucoxene, and perovskite.
[c]Rutile does not include anatase from Brazil.
[d]Marginally economical, economically exploitable, and subeconomic.

and India are government-owned, the viability of these mines in a free market is uncertain. In particular, Brazil could emerge as a significant producer of titanium-bearing minerals through its vast resources of anatase (90% TiO_2).

Manufacture

Ore-Concentrate Refining. The TiO_2 content of ore concentrates determines further processing steps. High grade ore such as rutile, synthetic rutile, or slag from Richard's Bay is refined to pigment-grade TiO_2 via chlorination (Du Pont chlorinates 60% TiO_2 concentrate). Lower grade ore is processed via

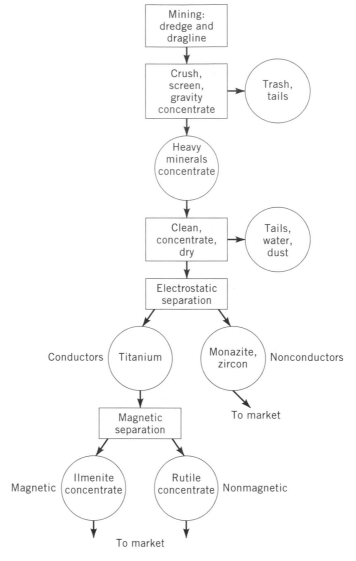

Fig. 1. Typical processes and products of a titanium beach-sand mining and beneficiating operation (4).

the sulfate route. The chlorination process, commercialized by Du Pont in the early 1960s, which produces a better quality pigment, requires less processing energy than the sulfate process (1800 kWh/t compared to 2500 kWh/t), and has less waste discharge (7–9). The sulfate process produces approximately six tons of waste per ton of TiO_2, whereas only one ton of waste is produced through the chloride process (10). However, high grade ore is required for the latter process, ie, TiO_2 content >70%, with <1% MgO and 0.2% CaO, because ores that have high MgO and CaO can clog the chlorinator. Environmental problems have forced the industry either to shut down sulfate plants or to install expensive pollution-control equipment. Because of the shortage of high grade

Table 2. World Titanium Mineral Production[a, b], 10^3 t

Country	Rutile 1990	Rutile 1994	Titaniferous slag[c] 1990	Titaniferous slag[c] 1994	Ilmenite[d] 1990	Ilmenite[d] 1994
Australia	245	223			1619	1805
Sierra Leone	144	137			55	47
South Africa[e]	64	78	672	744		
Ukraine	9	3			430	150
China					150	155
Malaysia					530	116
India and Sri Lanka	16	16			346	360
Norway					814	700
Canada[f]			1050	764		
Brazil[g]	2	2			114	91
Total	*480*	*459*	*1722*	*1508*	*4058*	*3424*

[a]Ref. 5.

[b]Includes lecoxene in ilmenite production; data for United States are withheld to avoid disclosing proprietary information.

[c]Slag is also produced in Norway but is not included under slag to avoid duplication. Beginning in 1990, about 25% of Norway's ilmenite production was used to produce a slag containing 75% TiO_2.

[d]Ilmenite is also produced in Canada and in South Africa, but this output is not included here because an estimated 90% of it is duplicated output reported under slag.

[e]Slag contains 85% TiO_2.

[f]Refined Sorel slag contained 80% TiO_2 in 1990.

[g]Excludes production of unbeneficiated anatase ore.

Table 3. Analyses of Selected Titanium Mineral Concentrates, Wt %[a]

Constituent	Ilmenite Placer Florida	Ilmenite Placer Western Australia	Ilmenite Placer Sri Lanka	Ilmenite Hard rock New York	Ilmenite Sorel slag Canada	Rutile Eastern Australia
TiO_2	64.10	55.30	53.45	44.40	71.00	96.40
ZrO_2		0.10	0.16	0.01		0.30
FeO	4.70	26.70	20.45	36.70	13.00	
Fe_2O_3	25.60	15.40	22.18	4.40		0.25
P_2O_5	0.21	0.04	0.21	0.07	0.33	0.02
SiO_2	0.30	0.20	0.52	3.20		0.56
Cr_2O_3	0.10	0.03	0.09	trace	0.20	0.15
Al_2O_3	1.50	0.38	0.58	0.19	5.70	0.17
V_2O_5	0.13	0.08		0.24	0.59	0.61
MnO	1.35	1.64	0.93	0.35	0.22	trace
CaO	0.13	0.17		1.00	1.00	0.05
MgO	0.35	0.29	1.46	0.80	5.00	0.04

[a]Refs. 4 and 6.

TiO_2 reserves, the pigment industry must adapt the ore to the chloride process. As of this writing (1997), the trend is toward ore beneficiation. Ore containing 50–60% TiO_2 content is beneficiated by partial reduction, then leached with sulfuric or hydrochloric acid to yield a concentrate containing >90% TiO_2, the so-called synthetic rutile (7,11).

Sulfate Process.   In the sulfate process (Fig. 2), ilmenite ore is treated with sulfuric acid at 150–180°C:

$$5\ H_2O + FeTiO_3 + 2\ H_2SO_4 \longrightarrow FeSO_4 \cdot 7H_2O + TiOSO_4$$

The undissolved solids are removed and the liquid is evaporated under vacuum and cooled. The precipitated $FeSO_4 \cdot 7H_2O$ is filtered and the filtrate concentrated to ca 230 g/L. Heating to 90°C hydrolyzes titanyl sulfate to insoluble titanyl hydroxide.

$$TiOSO_4 + 2\ H_2O \longrightarrow TiO(OH)_2 \downarrow + H_2SO_4$$

To ensure the rutile crystal form, seed crystals are added, otherwise anatase is obtained. The precipitate is thoroughly washed using water and sulfuric acid to remove all traces of discoloring elements, eg, iron, chromium, vanadium, and manganese. The $TiO(OH)_2$ is finally calcined at 1000°C to TiO_2 (8).

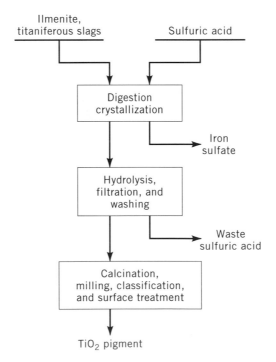

Fig. 2.   The sulfate process (7).

Chloride Process.   In the chloride process (Fig. 3), a high grade titanium oxide ore is chlorinated in a fluidized-bed reactor in the presence of coke at 925–1010°C:

$$TiO_2 + 2\ C + 2\ Cl_2 \longrightarrow 2\ CO + TiCl_4 \qquad \Delta G_{1300°C} = -125\ kJ\ (30\ kcal)$$

The volatile chlorides are collected and the unreacted solids and nonvolatile chlorides are discarded. Titanium tetrachloride is separated from the other chlorides by double distillation (12). Vanadium oxychloride, $VOCl_3$, which has a boiling point close to $TiCl_4$, is separated by complexing with mineral oil, reducing with H_2S to $VOCl_2$, or complexing with copper. The $TiCl_4$ is finally oxidized at 985°C to TiO_2 and the chlorine gas is recycled (8,11) (see also TITANIUM COMPOUNDS; PIGMENTS, INORGANIC).

Tetrachloride-Reduction Process.   Titanium tetrachloride for metal production must be of very high purity. The required purity of technical-grade $TiCl_4$ for pigment production is compared with that for metal production in Table 4. Titanium tetrachloride for metal production is prepared by the same process

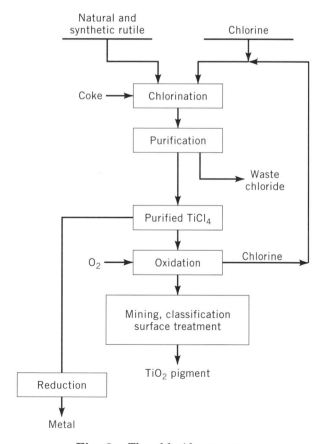

Fig. 3.   The chloride process.

Table 4. Chemical Composition of Titanium Tetrachloride Grades[a], Wt %

Impurity	Technical[b]	Purified[c]
$VOCl_3$	0.33	0.0034
$AlCl_3$	0.02	0.05
$SiCl_4$	0.4	0.006
Si_2OCl_6	0.04	0.003
$FeCl_3$	0.012	0.0029
CCl_3COCl	0.005	0.0002
CS_2	0.01	0.00002
COS	0.009	0.00002
$Si_3O_2Cl_8$	0.007	0.002
$COCl_2$	0.5	0.00002
other[d]	0.175	0.001

[a]Refs. 4 and 11.
[b]Pigment grade.
[c]Sponge grade.
[d]Includes oxychlorides, CO_2, Cl_2, CCl_4, and C_6Cl_6.

as described above, except that a greater effort is made to remove impurities, especially oxygen- and carbon-containing compounds.

Magnesium-Reduction (Kroll) Process.   In the 1990s, nearly all sponge is produced by the magnesium reduction process (Fig. 4).

$$TiCl_4(g) + 2\ Mg(l) \longrightarrow Ti(s) + 2\ MgCl_2(l) \qquad \Delta G_{900°C} = -301\ kJ\ (-72\ kcal)$$

$TiCl_4(g)$ is metered into a carbon-steel or 304 stainless-steel reaction vessel that contains liquid magnesium. An excess of 25% magnesium over the stoichio-

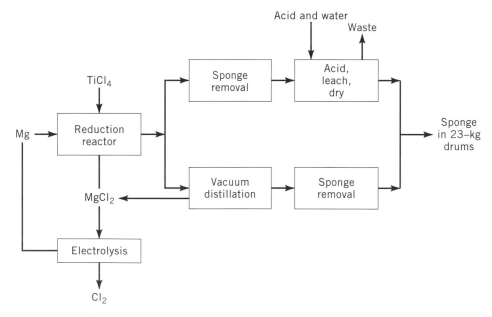

Fig. 4.   Flow diagram for titanium sponge production.

metric amount ensures that the lower chlorides of titanium ($TiCl_2$ and $TiCl_3$) are reduced to metal. The highly exothermic reaction ($\Delta H_{900°C} = -420$ kJ/mol (-100 kcal/mol)) is controlled by the feed rate of $TiCl_4$ at ca 900°C. The reaction atmosphere is helium or argon. Molten magnesium chloride is tapped from the reactor bottom and recycled using conventional magnesium-reduction methods, including L. G. Farben, Alcan, and USSR VAMI cells. The production is in batches up to 10 metric tons of titanium. The product, the so-called sponge (Fig. 5), is further processed to remove the unreacted titanium chlorides, magnesium, and residual magnesium chlorides. These impurities, which can be as much as 30 wt %, are removed by either acid leaching in dilute nitric and hydrochloric acids at low energy requirement of ca 0.3 kWh/kg of sponge but effluent production of 8 L/kg of sponge; vacuum distillation at 960–1020°C for as much as 60 h; or the argon sweep at 1000°C used by the Oregon Metallurgical Plant. After purification, the sponge is crushed, screened, dried, and placed in air-tight, 23-kg drums to await consolidation. The energy required to convert $TiCl_4$ to sponge, which is ready for further processing by the leaching routes, is ca 37 kWh/kg of sponge (9), of which ca 97% is required for magnesium production. The Japanese have claimed an energy consumption of approaching 15 kWh/kg of sponge using vacuum disillation instead of acid leaching for purification (13). In 1992, Timet installed in Henderson, Nevada, a new facility adjacent to their original acid leach facility, which utilized Japanese vacuum distillation technology from Toho Titanium.

Sodium-Reduction Process. The sodium-reduction process was employed in Japan, United States, and England for several years as an alternative to magnesium reduction. The last large production plant was closed in the early 1990s. Although the process was more costly than magnesium reduction, the product contained less metallic impurities, ie, Fe, Cr, and Ni. This product is

Fig. 5. Vacuum-distilled titanium sponge produced by magnesium reduction at Teledyne Wah Chang (Albany, Oregon).

desirable for a growing titanium market in the electronics industry. As a result, a small plant having a yearly capacity of 340 metric tons was opened in 1996 by Johnson-Matthey in Salt Lake City.

Comparison of purity of sponge produced by magnesium reduction and acid leach, magnesium reduction and vacuum distillation, and sodium reduction is given in Table 5. Hardness, indicating the degree of purity, is affected both by the interstitial impurities, ie, oxygen, nitrogen, and carbon, and by the noninterstitial impurity, ie, iron. Hardness numbers range from 80 to 150 HB units; typical commercial sponge is characterized by 110–120 HB units. Some developmental processes, eg, electrolysis reduction, produce sponge having 60–90 HB units. Iron impurities in Kroll sponge are difficult to control because of diffusion into the sponge from the reactor wall. In the sodium-reduction process, the sponge is protected from the wall by sodium chloride. The other impurities originate from tetrachloride, residual gases in the reactor, helium or argon impurities, and magnesium or sodium residues.

Other Reduction Processes. Other methods of producing titanium have been studied in the hope of finding another reduction route (16), collecting the metal as an ingot instead of as a sponge, and designing a continuous process. The most successful and widely studied noncommercialized process is the electrolytic reduction of $TiCl_4$. Primary electrical energy reduces $TiCl_4$ to titanium metal at the cathode and chlorine gas at the anode. This process was first conceived and developed at the U.S. Bureau of Mines in the mid-1950s. Since that time Timet, Dow-Howmet Titanium Company (D-H Titanium), and RMI Titanium Company have all built large-scale pilot plants to study the commercial feasibility. In all processes, $TiCl_4$ is fed into a molten-salt electrolyte, which for Timet and RMI is

Table 5. Comparison of ASTM Specifications for Titanium Sponge[a], Wt % on a Dry Basis

Property	ASTM B299 69			Electrolytic
	MD 120 type A[b]	ML 120 type B[c]	SL 120 type C[d]	
nitrogen, max	0.015	0.015	0.010	0.003
carbon, max	0.020	0.025	0.020	0.011
sodium, total max			0.190	
magnesium, max	0.08	0.50		
chlorine, max	0.12	0.20	0.20	0.035
iron, max	0.12	0.10	0.05	0.02
silicon, max	0.04	0.04	0.04	
hydrogen, max	0.005	0.03	0.05	0.005
oxygen, max	0.10	0.10	0.10	0.065
all other impurities	0.05	0.05	0.05	
titanium balance, nominal	99.3	99.1	99.3	
Brinell hardness (HB), max	120	120	120	60–90

[a]Refs. 14 and 15.
[b]Type A magnesium reduced and finished by vacuum distillation.
[c]Type B magnesium reduced and finished by acid leaching on inert gas sweep distillation.
[d]Type C sodium reduced and finished by acid leaching.

NaCl operated at 830–900°C and for Dow-Howmet KCl–LiCl eutectic operated at 520°C (17). The feed, which is insoluble in the electrolyte, is immediately reduced to soluble $TiCl_2$ at a feed electrode. The anode is isolated from the bulk of the electrolyte by a diaphragm, which minimizes reaction of titanium ions by using chlorine to liberate $TiCl_4$. The uniqueness of the cells lies in the design of the diaphragm (18). The main problem is suppression of reactions of titanium ions using chlorine. Each cell has a capacity of ca 1–2 t per run, which requires ca 20 kAh. The RMI cell was designed for 140 t/yr (18). The titanium electrodeposit is transferred from the salt bath into an argon chamber, removed from the cathode (after cooling by Timet and RMI and hot by D-H Titanium), and finally acid-leached. The energy requirement is considerably lower than that of other reduction processes at ca 18 kWh/kg of leached sponge equivalent (17). All pilot facilities have been closed, however, as of this writing.

Other methods include hydrogen reduction of $TiCl_4$ to $TiCl_3$ and $TiCl_2$; reduction above the melting point of titanium metal with sodium, which presents a container problem; plasma reduction, in which titanium is collected as a powder, and ionized and vaporized titanium combine with chlorine gas to reform $TiCl_2$ on cool-down; and aluminum reduction, which reduces $TiCl_4$ to lower chlorides (19,20).

Methods that do not utilize $TiCl_4$ include reduction of TiO_2 with, for instance, Al, Ca, or C. The problems are the purity of the TiO_2, the amount of reductant remaining in the metal, and the interstitial elements remaining in the metal. Ductile metal has not been produced by direct TiO_2 reduction (21–23) (see ELECTROCHEMICAL PROCESSING).

Sponge Consolidation. The next step is the consolidation of the sponge into ingot. The crushed sponge is blended with alloying elements or other sponge. Consumable electrodes are produced by welding 45–90-kg sponge compactions (electrode compacts) in an inert atmosphere, and then double-vacuum-arc-remelted (VAR). A portion of the elemental sponge compacts are often replaced with bulk scrap. The ingots are ca 71–91-cm dia and long enough to weigh 4.5 to 9.0 t. The double melt, included in aerospace specifications, is required for thorough mixing of alloying elements, scrap, and titanium sponge, and for improving yields because vaporization of volatiles during the first melt leaves a rough, porous surface. The double melt removes residual volatiles such as Mg, $MgCl_2$, Cl_2, and H_2. Triple melts are specified for critical applications such as rotating components on gas turbine engines. The third melt allows more time to dissolve high melting point inclusions that infrequently occur. This is often referred to as rotating quality titanium.

A two-station VAR furnace for double melting has an annual production capacity of ca 1400–3000 t, depending on product mix, ie, alloy and number of remelts. The energy requirement is ca 1.1 kWh/kg per single melt. Plasma-hearth melting and electron-beam-hearth melting have been employed more recently for both consolidation and final melting. Table 6 shows a comparison of these processes and split-crucible induction melting; the latter process has, as of this writing, not yet been scaled up to large ingots (15). The hearth processes are well suited for utilizing scrap in various shapes and forms and for avoiding the costly electrode fabrication inherent in consumable vacuum arc melting. In addition, these processes can produce cast metal into shapes such as slabs. For

Table 6. Titanium Melting Processes[a]

Component	Consumable vacuum-arc melt	Plasma melt[b]	Plasma cold-hearth melt	Electron-beam cold-hearth melt	Induction melt[c]
atmosphere	vacuum	Ar, He	Ar, He	vacuum	Ar, He, vacuum
Operational					
rate, kg/h	1200	200		400	300
power, kWh/kg	2.0	2.2		3.0	1.4
cost, \$/kg	<1.00	<2.00	<2.00	<2.00	
degassing of H	yes	no	no	yes	no/yes
liquid residence time control	no	no	yes	yes	yes
Material input					
scrap input					
large	yes	no	yes	yes	
small	no	yes	yes	yes	yes
sponge input					
high chlorides	yes	yes	yes	no	yes
Output quality					
ingot structure uniformity	no	no	yes	yes	no
high density elimination	no	no	yes	yes	no
nitride elimination	no	no	yes	yes	no
Al vaporization	no	no	no	yes	no

[a]Ref. 15.
[b]No hearth.
[c]Small-diameter ingots only.

many industrial applications, a single hearth melt is acceptable. The hearth process can be designed to trap high density inclusions such as carbide tool bits and nitrogen-rich TiN (Type I inclusions) in the hearth skull.

Alloys

Titanium alloy systems have been extensively studied. A single company evaluated over 3000 compositions in eight years (Rem-Cru sponsored work at Battelle Memorial Institute). Alloy development has been aimed at elevated-temperature aerospace applications, strength for structural applications, biocompatibility, and corrosion resistance. The original effort has been in aerospace applications to replace nickel- and cobalt-base alloys in the 250–600°C range. The useful strength and corrosion-resistance temperature limit is ca 550°C.

In pure titanium, the crystal structure is close-packed hexagonal (α) up to 882°C and body-centered cubic (β) to the melting point. The addition of alloying elements alters the $\alpha-\beta$ transformation temperature. Elements that raise the transformation temperature are called α-stabilizers; those that depress the transformation temperature, β-stabilizers; the latter are divided into β-isomorphous and β-eutectoid types. The β-isomorphous elements have limited α-solubility and increasing additions of these elements progressively depresses the transformation temperature. The β-eutectoid elements have restricted β-solubility and form intermetallic compounds by eutectoid decomposition of the β-phase. The binary phase diagram illustrating these three types of alloy systems is shown in Figure 6.

The important α-stabilizing alloying elements include aluminum, tin, zirconium, and the interstitial alloying elements, ie, elements that do not occupy lattice positions, oxygen, nitrogen, and carbon. Small quantities of interstitial alloying elements, generally considered to be impurities, are always present, and have a great effect on strength. In sufficient amounts they can embrittle the titanium at room temperature (24). Oxygen is often used as an alloying element, ranging from as low as 500 ppm to as high as 3000 ppm, whereas carbon and nitrogen are maintained at their residual level. Oxygen additions increase strength and serve to identify several commercially pure grades. This strengthening effect diminishes at elevated temperatures. For cryogenic service, low oxygen content (<1300 ppm) is specified because high concentrations of interstitial impurities increase sensitivity to cracking, cold brittleness, and fracture toughness. Alloys having low interstitial content are identified as extra-low interstitial (ELI) after the alloy name. Nitrogen has the greatest effect and commercial alloys specify its limit to be less than 0.05 wt %. It may also be present concentrated as high melting point nitride inclusions (TiN), referred to as Type 1 defects, which are detrimental to critical aerospace structural applications. Carbon does not affect strength at concentration above 0.25 wt % because carbides (TiC) are formed.

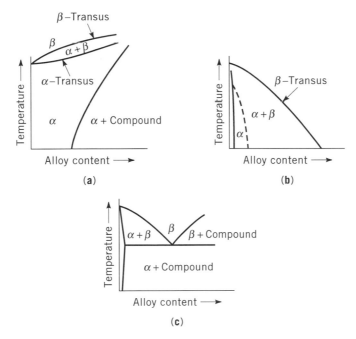

Fig. 6.   Effect of alloying elements on the phase diagram of titanium: (**a**) α-stabilized system, (**b**) β-isomorphous system, and (**c**) β-eutectoid system.

Carbon content is usually specified at 0.08 wt % max (25). The relative effect of these elements on specific room temperature strength is expressed in terms of an oxygen equivalency, O_e (26):

$$\%O_e = \%O + 2(\%N) + 0.67(\%C)$$

The most important α-stabilizing alloying element is aluminum, which is inexpensive and has an atomic weight less than that of titanium. Hence, aluminum additions slightly lower the density. The mechanical strength of titanium can be increased considerably by aluminum additions. Even though the solubility range of aluminum extends to 27 wt %, above 7.5 wt % the alloy becomes too difficult to fabricate and is embrittled. The embrittlement is caused by a coherently ordered phase based on Ti_3Al [12635-69-7]. Other α-stabilizing elements also cause phase ordering. An empirical relationship (27) that describes the compositional ranges where ordering does not occur is

$$[\text{wt } \% \text{ Al}] + [\text{wt } \% \text{ Sn}]/3 + [\text{wt } \% \text{ ZR}]/6 + 10[\text{wt } \% \text{ O}] = /<9$$

The important β-stabilizing alloying elements are the bcc elements vanadium, molybdenum, tantalum, and niobium of the β-isomorphous type and manganese, iron, chromium, cobalt, nickel, copper, and silicon of the β-eutectoid type. The β-eutectoid elements, arranged in order of increasing tendency to form compounds, are shown in Table 7. The elements copper, silicon, nickel, and cobalt are termed active eutectoid formers because of a rapid decomposition of β to α and a compound. The other elements in Table 7 are sluggish in their eutectoid reactions and thus it is possible to avoid compound formation by careful control of heat treatment and composition. The relative β-stabilizing effects of these elements can be expressed in the form of a molybdenum equivalency, Mo_e (29):

$$\% \text{ Mo}_e = [\% \text{ Mo}] + [\% \text{ Nb}]/3.6 + [\% \text{ Ta}]/5 + [\% \text{ W}]/2.5 + [\% \text{ V}]/1.5$$

$$+ 1.25[\% \text{ Cr}] + 1.25[\% \text{ Ni}] + 1.7[\% \text{ Mn}] + 1.7[\% \text{ Co}] + 2.5[\% \text{ Fe}]$$

Alloys of the β-type respond to heat treatment, are characterized by higher density than pure titanium, and are more easily fabricated. The purpose of

Table 7. β-Eutectoid Elements in Order of Increasing Tendency to Form Compounds[a]

Element	Eutectoid composition, wt %	Eutectoid temperature, °C	Composition for β-retention on quenching, wt %
manganese	20	550	6.5
iron	15	600	4.0
chromium	15	675	8.0
cobalt	9	685	7.0
nickel	7	770	8.0
copper	7	790	13.0
silicon	0.9	860	

[a]Ref. 28.

alloying to promote the β-phase is either to form an all-β-phase alloy having commercially useful qualities, to form alloys that have duplex α- and β-structure to enhance heat-treatment response, ie, changing the α and β volume ratio, or to use β-eutectoid elements for intermetallic hardening. The most important commercial β-alloying element is vanadium.

Physical Properties

The physical properties of titanium are given in Table 8. The most important physical property of titanium from a commercial viewpoint is the ratio of its strength (ultimate strength >690 MPa (100,000 psi)) at a density of 4.507 g/cm^3. Titanium alloys have a higher yield strength to density rating, between -200 and 540°C, than either aluminum alloys or steel. Titanium alloys can be made to have strength equivalent to high strength steel, yet having density ca 60% that of iron alloys. At ambient temperatures, titanium's strength-to-weight ratio

Table 8. Physical Properties of Titanium

Property	Value
melting point, °C	1668 ± 5
boiling point, °C	3260
density, g/cm^3	
α-phase at 20°C	4.507
β-phase at 885°C	4.35
allotropic transformation, °C	882.5
latent heat of fusion, kJ/kg[a]	440
latent heat of transition, kJ/kg[a]	91.8
latent heat of vaporization, MJ/kg[a]	9.83
entropy, at 25°C, J/mol[a]	30.3
thermal expansion coefficient, at 20°C, per °C	8.41×10^{-6}
thermal conductivity, at 25°C, W/(m·K)	21.9
emissivity	9.43
electrical resistivity, at 20°C, nΩ·m	420
magnetic susceptibility, mks	180×10^{-6}
modulus of elasticity, GPa[b]	
tension	ca 101
compression	103
shear	44
Poisson's ratio	~0.41
lattice constants, nm	
α, 25°C	$a_0 = 0.29503$
	$c_0 = 0.46531$
β, 900°C	$a_0 = 0.332$
vapor pressure, kPa[c]	$\log P_{kPa} = 5.7904 - 24644/T$
	$- 0.000227\, T$
specific heat, J/(kg·K)[d]	$C_p = 669.0 - 0.037188\, T$
	$- 1.080 \times 10^7/T^2$

[a]To convert J to cal, divide by 4.184.
[b]To convert GPa to psi, multiply by 145,000.
[c]To convert $\log P_{kPa}$ to $\log P_{atm}$, add 2.0056 to the constant.
[d]$T > 298$ K.

is equal to that of magnesium, 1.5 times greater than that of aluminum, two times greater than that of stainless steel, and three times greater than that of nickel. Alloys of titanium have much higher strength-to-weight ratios than alloys of nickel, aluminum, or magnesium, and stainless steel. Because of its high melting point, titanium can be alloyed to maintain strength well above the useful limits of magnesium and aluminum alloys. This property gives titanium a unique position in applications between 150–550°C when the strength-to-weight ratio is the sole criterion.

Solid pure titanium exists in two allotropic crystalline forms. The α-phase, stable below 882.5°C, is a close-packed hexagonal structure, whereas the β-phase, a bcc crystalline structure, is stable between 882.5°C and the melting point of 1668°C. The high temperature β-phase can be found at room temperature when β-stabilizing elements are present as impurities or additions. The α- and β-phases can be distinguished by examining an unetched polished mount with polarized light: α is optically active and changes from light to dark as the microscope stage is rotated. The microstructure of titanium is difficult to interpret without knowledge of the alloy content, working temperature, and thermal treatment (28,30).

The heat-transfer qualities of titanium are characterized by the coefficient of thermal conductivity. Even though the coefficient is low, heat transfer in service approaches that of admiralty brass (thermal conductivity seven times greater) because titanium's greater strength permits thinner-walled equipment, relative absence of corrosion scale, erosion–corrosion resistance that allows higher operating velocities, and the inherently passive film.

Corrosion Resistance.   Titanium is immune to corrosion in all naturally occurring environments. It does not corrode in air, even if polluted or moist with ocean spray. It does not corrode in soil and even the deep salt-mine-type environments where nuclear waste might be buried. It does not corrode in any naturally occurring water and most industrial wastewater streams. For these reasons, titanium has been termed the metal for the earth, and 20–30% of consumption is used in corrosion-resistance applications (see CORROSION AND CORROSION INHIBITORS).

Even though titanium is an active metal, it resists decomposition because of a tenacious protective oxide film.

$$Ti^{2+} + 2\,e^- \longrightarrow Ti \qquad\qquad E^0 = -1.63\ V$$

$$TiO_2 + 4\,H^+ + 4\,e \longrightarrow Ti + 2\,H_2O \qquad E^0 = -0.86\ V$$

The titanium oxide film consists of rutile or anatase (31) and is typically 250-A thick. It is insoluble, repairable, and nonporous in many chemical media and provides excellent corrosion resistance. The oxide is fully stable in aqueous environments over a range of pH, from highly oxidizing to mildly reducing. However, when this oxide film is broken, the corrosion rate is very rapid. Usually the presence of a small amount of water is sufficient to repair the damaged oxide film. In a seawater solution, this film is maintained in the passive region from ca 0.2 to 10 V versus the saturated calomel electrode (32,33).

Titanium is resistant to corrosion attack in oxidizing, neutral, and inhibited reducing conditions. Examples of oxidizing environments are nitric acid, oxidizing chloride ($FeCl_3$ and $CuCl_2$) solutions, and wet chlorine gas. Neutral conditions include all neutral waters (fresh, salt, and brackish), neutral salt solutions, and natural soil environments. Examples of inhibited reducing conditions are in hydrochloric or sulfuric acids with oxidizing inhibitors and in organic acids inhibited with small amounts of water. Corrosion resistances to a variety of media are given in Table 9. Titanium's resistance to aqueous chloride solutions and chlorine accounts for most of its use in corrosion-resistance applications.

Titanium corrodes very rapidly in acid fluoride environments. It is attacked in boiling HCl or H_2SO_4 at acid concentrations of >1% or in ca 10 wt % acid concentration at room temperature. Titanium is also attacked by hot caustic solutions, phosphoric acid solutions (concentrations >25 wt %), boiling $AlCl_3$ (concentrations >10 wt %), dry chlorine gas, anhydrous ammonia above 150°C, and dry hydrogen–dihydrogen sulfide above 150°C.

Titanium is susceptible to pitting and crevice corrosion in aqueous chloride environments. The area of susceptibility for several alloys is shown in Figure 7 as a function of temperature and pH. The susceptibility depends on pH. The susceptibility temperature increases parabolically from 65°C as pH is increased from zero. After the incorporation of noble-metal additions such as in ASTM Grades 7 or 12, crevice corrosion attack is not observed above pH 2 until ca 270°C. Noble alloying elements shift the equilibrium potential into the passive region where a protective film is formed and maintained.

Titanium does not stress-crack in environments that cause stress-cracking in other metal alloys, eg, boiling 42% $MgCl_2$, NaOH, sulfides, etc. Some of the aluminum-rich titanium alloys are susceptible to hot-salt stress-cracking. However, this is a laboratory observation and has not been confirmed in service. Titanium stress-cracks in methanol containing acid chlorides or sulfates, red fuming nitric acid, nitrogen tetroxide, and trichloroethylene.

Titanium is susceptible to failure by hydrogen embrittlement. Hydrogen attack initiates at sites of surface iron contamination or when titanium is galvanically coupled with iron (37,38). In hydrogen-containing environments, titanium may absorb hydrogen above 10°C or in areas of high stress. The rate of absorption depends on the alloy and temperature. The alloys containing β-phase show less embrittlement. If the surface oxide is removed by vacuum-annealing or abrasion, pure dry hydrogen reacts at lower temperatures. Small amounts of oxygen or water vapor repair the oxide film and prevent this occurrence. Titanium resists oxidation in air up to 650°C. Noticeable scale forms at temperatures above 650°C. Because oxygen is soluble in both the α- and β-phases, oxygen can gradually diffuse into the metal surface and cause embrittlement. Surface contaminants accelerate oxidation. In the presence of oxygen, the metal does not react significantly with nitrogen. Spontaneous ignition occurs in gas mixtures containing more than 40% oxygen under impact loading or abrasion. Ignition occurs in dry halogen gases.

Titanium resists erosion–corrosion by fast-moving sand-laden water. In a high velocity, sand-laden seawater test (8.2 m/s) for a 60-d period, titanium performed more than 100 times better than 18 Cr–8 Ni stainless steel, Monel, or 70 Cu–30 Ni. Resistance to cavitation, ie, corrosion on surfaces

Table 9. Corrosion Data for ASTM Grade-2 Titanium[a]

Media	Concentration, wt %	Temperature, °C	Corrosion rate, mm/yr
acetaldehyde	100	149	0.0
acetic acid	5–99.7	124	0.0
adipic acid	67	232	0.0
aluminum chloride, aerated	10	100	0.002
	10	150	0.03
	20	149	16
	25	20	0.001
	25	100	6.6
	40	121	109
ammonia + 28% urea + 20.5% H_2O + 19% CO_2 + 0.3% inerts + air	32.2	182	0.08
ammonium carbamate	50	100	0.0
ammonium perchlorate aerated	20	88	0.0
aniline hydrochloride	20	100	0.0
aqua regia	3:1	RT	0.0
	3:1	79	0.9
barium chloride, aerated	5–20	100	<0.003
bromine–water solution		RT	0.0
calcium chloride		RT	0.0
	5	100	0.005
	10	100	0.007
	20	100	0.02
	55	104	0.0005
	60	149	<0.003
	62	154	0.05–0.4
	73	177	2.1
calcium hypochlorite	6	100	0.001
chlorine gas, wet	>0.7 H_2O	RT	0.0
	>1.5 H_2O	200	0.0
chlorine gas, dry	<0.5 H_2O	RT	may react
chlorine dioxide in steam	5	99	0.0
chloracetic acid	100	189	<0.1
chromic acid	50	24	0.01
citric acid	25	100	0.0009
copper sulfate + 2% H_2SO_4	saturated	RT	0.02
cupric chloride, aerated	1–20	100	<0.01
cyclohexane (plus traces of formic acid)		150	0.003
ethylene dichloride	100	boiling	0.005–0.1
ferric chloride	10–30	100	<0.1
formic acid, nonaerated	10	100	2.4
hydrochloric acid, aerated	5	35	0.04
	20	35	4.4
HCl, chlorine saturated	5	190	<0.03
HCl + 10% HNO_3	5	38	0.0
HCl + 1% CrO_3	5	93	0.03
hydrofluoric acid	1–48	RT	rapid
hydrogen peroxide	3	RT	<0.1

Table 9. (*Continued*)

Media	Concentration, wt %	Temperature, °C	Corrosion rate, mm/yr
hydrogen sulfide, steam and 0.077% mercaptans	7.65	93–110	0.0
hypochlorous acid + Cl_2O and Cl_2	17	38	0.00003
lactic acid	10	boiling	<0.1
manganous chloride, aerated	5–20	100	0.0
magnesium chloride	5–40	boiling	0.0
mercuric chloride, aerated	1	100	0.0003
	5	100	0.01
	10	100	0.001
	55	102	0.0
mercury	100	RT	0.0
nickel chloride, aerated	5–20	100	0.0004
nitric acid	17	boiling	0.08–0.1
	70	boiling	0.05–0.9
nitric acid, red fuming	<about 2% H_2O	RT	ignition-sensitive
	>about 2% H_2O	RT	nonignition-sensitive
oxalic acid	1	37	0.3
oxygen, pure			ignition-sensitive
phenol	saturated	21	0.1
phosphoric acid	10–30	RT	0.02–0.05
	10	boiling	10
potassium chloride	saturated	60	<0.0002
potassium dichromate			0.0
potassium hydroxide	50	27	0.01
	50	boiling	2.7
seawater, 10-year test			0.0
sodium chlorate	saturated	boiling	0.0
sodium chloride	saturated	boiling	0.0
sodium chloride, titanium in contact with Teflon	23	boiling	crevice attack
sodium dichromate	saturated	RT	0.0
sodium hypochlorite + 12–15% sodium chloride + 1% sodium hydroxide + 1–2% sodium carbonate	1.5–4	66–93	0.03
stannic chloride	5	100	0.003
	24	boiling	0.04
sulfuric acid	1	boiling	2.5
sulfuric acid + 0.25% $CuSO_4$	5	93	0.0
terephthalic acid	77	218	0.0
urea–ammonia reaction mass		elevated temperature and pressure	no attack
zinc chloride	20	104	0.0
	50	150	0.0
	75	200	0.5
	80	200	203

[a]Refs. 34 and 35.

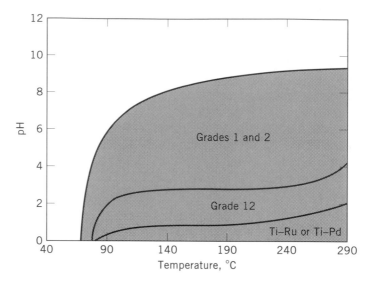

Fig. 7. Temperature–pH limits for crevice corrosion of titanium alloys in naturally aerated sodium chloride-rich brines. The shaded areas indicate regions where alloys are susceptible to attack (36).

exposed to high velocity liquids, is better than by most other structural metals (34,35).

In galvanic coupling, titanium is usually the cathode metal and consequently not attacked. The galvanic potential in flowing seawater in relation to other metals is shown in Table 10. Because titanium is a cathode metal, hydrogen absorption may be of concern, as it occurs with titanium complexed to iron (38).

Casting.   Consolidated titanium is cast either by precision castings or investment casting. The metal is melted using a consumable titanium electrode in a protected atmosphere, usually in a water-cooled copper crucible. In precision casting, rammed graphite molds are used; in investment casting, ceramic molds. Hot isostatic pressing (HIP) of castings promotes property optimization and porosity closure. Casting companies in the United States include Titech International, Inc., Tiline Corporation, Oregon Metallurgical Corporation, Howmet Turbine Components Corporation, Precision Cast Parts Corporation, Wyman Gordon, and Selmet. Because of the popularity of titanium golf club heads, several new casters are serving this market, such as Ruger, Coastcast, Cast Alloys, and Commercial Ti Castings.

Powder.   In the 1940s, powder metallurgy (qv) was investigated before the advent of VAR consumable electrode melting. In the early 1960s, Du Pont had an extensive program studying powder consolidation. However, the company gave up the program in 1964, partly because of technical difficulties associated with sodium and chlorine residues in sponge powder which cause porosity in weldments. The powder fines used were a product of the sodium-reduction process and were termed elemental powder. Up to one million pounds per year were consumed until all the large sodium-reduction plants were closed. The applications include powder metallurgy parts, pyrotechnic, metal spray coatings, and alloy element for Al-, Ni-, and Fe-base alloys. The magnesium reduction process

Table 10. Galvanic Series in Flowing Seawater, 4 m/s at 24°C[a]

Metal	Potential, V[b]
T304 stainless steel, passive	0.08
Monel alloy	0.08
Hastelloy alloy C	0.08
unalloyed titanium	0.10
silver	0.13
T410 stainless steel, passive	0.15
nickel	0.20
T430 stainless steel, passive	0.22
70–30 copper–nickel	0.25
90–10 copper–nickel	0.28
admiralty brass	0.29
G bronze	0.31
aluminum brass	0.32
copper	0.36
naval brass	0.40
T410 stainless steel, active	0.52
T304 stainless steel, active	0.53
T430 stainless steel, active	0.57
carbon steel	0.61
cast iron	0.61
aluminum	0.79
zinc	1.03

[a]Ref. 37.
[b]Steady-state potential, negative to saturated calomel half-cell.

does not yield any significant amount of high quality powders. As a result of the closing of the large volume sodium-reduction plants, the availability of low cost powders has diminished.

Other, but high cost, powder-making processes include the hydride–dehydride process, which gives blocky powder particles that are often cracked; the electron-beam, rotating-disk method (39); the Crucible Research Center's colt titanium process, which uses hydrogen charged into the metal to disintegrate the latter when melted; the Battelle Columbus Laboratory's pendent drop process; and the rotating-electrode process from Nuclear Metals, Inc.

Contamination limits the amount of handling and atmospheric exposure the powder can be subjected to before consolidation. The consolidation method most widely used is cold compaction and sintering.

A principal motive for developing consolidation techniques is to reduce cost by improving yields. For aircraft parts, the so-called buy-to-fly ratio is ca 6:1. Product yields based on sponge are shown in Figure 8. The potential improvements in yield using powder are obvious. However, reliable, low cost powder-production techniques have not yet been fully developed as of this writing.

Metal Working. The ingots are further processed by the conventional methods of forging, hot-rolling, cold-rolling, extrusion, etc. The mill product forms include billet, bar, plate, sheet, strip, foil, extrusion, wire, pipe, and welded tubing. Mill practices differ somewhat from those of other metal products. Minimum heating time at the lowest practical temperatures generally attains the

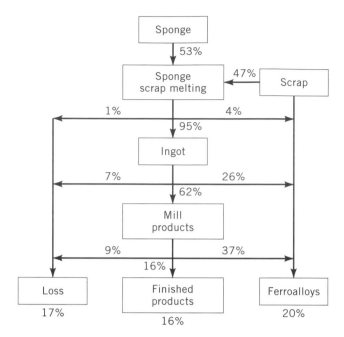

Fig. 8.   Titanium metal utilization (6).

best mechanical properties, minimizes contamination and oxidation, and avoids excess grain growth. Hydrogen absorption must be minimized because it forms hydrides when the solubility is exceeded (depending on alloy, ie, β-content), thus rendering the metal brittle. An oxidizing atmosphere minimizes absorption. The furnace must be kept clean to avoid contact with other metal oxides that can be reduced by titanium, thereby absorbing oxygen and damaging the metal surface. Primary working generally takes place above the phase-transition temperature (β-transus); secondary working generally below. Typical hot-working temperatures for common titanium alloys are given in Table 11.

The surface is conditioned by lathe turning, grit blasting, belt grinding, centerless grinding, and caustic or acid pickling. Although other metals are similarly treated, in the case titanium, lathe speeds are usually slower, grinding fines is a potential fire hazard, and acid pickling is in 2–4 wt % hydrofluoric acid, 15–30 wt % nitric acid, and the remainder water. The amount of HF and the temperature determine the pickling rate. Hydrofluoric acid etches titanium by reacting with the oxide surface to form titanium fluoride complex ions. Nitric acid is added to minimize hydrogen pickup (see METAL SURFACE TREATMENTS).

Sheet, thin plate, welded tubing, and small-diameter bar of commercially pure titanium are manufactured into parts by conventional cold-working techniques. The formability of titanium, when worked at room temperature, is like that of cold-rolled stainless steel. At 65°C the formability compares with stainless steel annealed at room temperature. Cold-working may be difficult for some titanium alloys and heat may be required, especially for severe forming operations. Generally, titanium and its alloys are worked between 200 and 300°C.

Table 11. Typical Titanium Alloy Processing Temperatures[a]

Alloy	β-Transus, °C	Ingot forging temperature, °C	Rolling temperature, °C		
			Bar	Plate	Sheet
CP Ti					
Grades 1–4	887–949	950–980	760–815	760–790	705–760
α- and near-α-alloys					
Ti–5Al–2.5Sn	1038	1120–1180	1010–1065	980–1040	980–1010
Ti–6Al–2Sn– 4Zr–2Mo	993	1090–1150	955–1010	955–980	925–980
Ti–8Al–1Mo–1V	1038	1120–1170	1010–1040	980–1040	980–1040
α–β-alloys					
Ti–6Al–4V	995	1090–1150	955–1010	925–980	900–925
Ti–6Al–6V–2Sn	946	1040–1090	900–955	870–925	870–900
β-alloys					
Ti–3Al–8V– 6Cr–4Mo–4Zr	790	1090–1200	955–1065	980–1040	730–900

[a]Ref. 30 and 40.

Lubricants reduce friction and galling. Slow forming speeds at controlled rates improve workability and are recommended for more difficult operations.

Fabrication. Fabrication of titanium into useful parts, such as tanks, heat exchangers, and pressure vessels, is comparable to the fabrication of austenitic steel in method, degree of difficulty, and cost. Commercial-grade titanium can be bent 105° without cracking around a radius of 2–2.5 times the sheet thickness. The bend radius for alloys is as high as five times the sheet thickness. A loss of 15–25 degrees in the included bend angles is normal because of springback at room temperature related to the low elastic modulus. Heat is required to form most titanium alloy parts. Super plastic forming (SPF) is used to form complex shapes in α–β-type alloys such as Ti–6Al–4V. The forming is conducted at ca 900°C where the alloy becomes super plastic, ie, elongates without necking. The process is sometimes combined with diffusion bonding (SPF/DB) to form complex structures (41).

Welding (qv) of titanium requires a protected atmosphere of inert gas. Furthermore, parts and filler wire are cleaned with acetone (trichloroethylene is not recommended). The pieces to be welded are clamped, not tacked, unless tacks are shielded with inert gas. A test sample should be welded. Coated electrodes are excluded and higher purity metal (lower oxygen content) is preferred as filler. Titanium cannot be fusion-welded to other metals because of formation of brittle intermetallic phases in the weld zone.

In some applications, a titanium cladding, especially Detaclad, is desirable for cost reduction. In this process, ca 2 mm of ASTM Grade-1 titanium sheet is explosively bonded to steel plate (see METALLIC COATINGS, EXPLOSIVELY CLAD METALS). The cladding is cost effective for wall thicknesses of >1.5 cm. Another common cladding method is loose lining. Experimental methods include roll cladding and resistance welding (42).

Table 12. Properties of Common Aerospace Titanium Alloys[a]

Nominal composition, wt %	Common name	CAS Registry Number	β-Content, % Mo equivalency	ASTM B265 Grade	Condition
		α-Alloys[c]			
99.5 Ti	Ti		0.4	1	annealed
99.2 Ti	Ti		0.4	2	annealed
99.1 Ti	Ti		0.4	3	annealed
99.0 Ti	Ti		0.4	4	annealed
Ti–5Al–2.5Sn	Ti-5-2.5	[11109-19-6]	0.4	6	annealed
		$\alpha - \beta$-Alloys			
Ti–5.8Al–4Sn–3.5Zr– 0.7Nb–0.5Mo–0.35Si	IMI 834		0.7		annealed
Ti–3Al–2.5V	Ti-3-2.5	[11109-23-2]	2.0	9	annealed
Ti–8Al–1Mo–1V	Ti-811	[39303-55-4]	2.0		annealed
Ti–6Al–2Sn–4Zr– 2Mo–0.1Si	Ti-6242S		2.4		annealed
Ti–6Al–4V	Ti-6-4	[12743-70-3]	3.1	5	annealed
Ti–6Al–4V			3.1		aged
Ti–4Al–4Mo–2Sn–0.5Si	Ti-550		4.4		aged
Ti–6Al–6V–2Sn	Ti-662	[12606-77-8]	4.5		annealed
Ti–6Al–6V–2Sn			4.5		aged
Ti–6Al–2Sn–2Zr–2Mo– 2Cr–0.1Si	Ti-6-2222S		4.9		aged
Ti–6Al–2Sn–4Zr–6Mo	Ti-6246		6.4		aged
Ti–4.5Al–3V–2Mo–2Fe	SP-700		7.0		annealed
Ti–5Al–2Sn–2Zr–4Mo–4Cr	Ti-17		7.8		aged
		β-Alloys			
Ti–10V–2Fe–3Al	Ti-10-2-3		11.7		aged
Ti–15V–3Cr–3Al–3Sn	Ti-15-3		13.8		annealed
Ti–15V–3Cr–3Al–3Sn			13.8		aged
Ti–3Al–8V–6Cr–4Mo–4Zr	Beta C		18.4	19	annealed
Ti–3Al–8V–6Cr–4Mo–4Zr			18.4		aged
Ti–3Al–2.8Nb–15Mo–0.2Si	Beta 21S		20.8		aged

[a]Refs. 30 and 43.
[b]Room temperature.
[c]Commercially pure.

Aerospace Alloys.   The alloys of titanium for aerospace use can be divided into three categories: an all-α structure, a mixed $\alpha-\beta$ structure, and an all-β structure. The $\alpha-\beta$-structure alloys are further divided into near-α alloys (<2% β-stabilizers). Most of the ca 100 commercially available alloys, including ca 30 in the United States, 40 in the CIS, and 10 in Europe and Japan, are of the $\alpha-\beta$- structure type (4). Some of these, produced in the United States, are given in Table 12, along with some wrought properties. The most important commercial alloy is Ti–6Al–4V, an $\alpha-\beta$-alloy having a good combination of strength and

Table 12. (*Continued*)

β-Transus, °C	Density, g/cc^3	Young's modulus, E Tensile, 10^{10} N/m^2	Compressive, 10^{10} N/m^2	Shear modulus, G, 10^{10} N/m^2	Ultimate tensile strength, 10^8 N/m^2	0.2% yield strength, 10^8 N/m^2	Elongation, %
			Average physical properties[b]			Average tensile properties[b]	
			α-Alloys[c]				
887	4.5	10.5	11.0	4.5	2.7	2.1	40.0
912	4.5	10.5	11.0	4.5	4.1	3.3	30.0
921	4.5	10.5	11.0	4.5	4.8	4.1	25.0
949	4.5	10.5	11.0	4.5	6.2	5.5	20.0
1038	4.5	10.9		4.8	8.5	8.0	15.0
			α – β-Alloys				
1045	4.5	11.2			10.5	9.3	12.0
935	4.5	10.0	10.3		6.5	6.2	22.0
1038	4.4	12.1	12.4	4.6	10.0	9.3	12.0
993	4.5	10.8	11.1	4.6	10.0	9.3	15.0
995	4.4	11.0	11.1	4.2	9.6	9.0	17.0
995	4.4	11.4	11.4	4.2	11.7	11.0	12.0
975	4.6				11.3	10.2	12.0
946	4.5	11.0		4.5	10.7	10.0	14.0
946	4.5	11.7	12.1	4.5	12.8	12.1	10.0
970	4.6	11.4	12.4		11.2	10.1	14.0
938	4.7	11.4	12.4		12.1	11.4	8.0
900	4.5	11.0			10.1	9.6	14.0
890	4.7	11.2			11.4	10.7	8.0
			β-Alloys				
800	4.6	11.0	11.2		13.0	12.0	7.0
760	4.8	8.3			7.9	7.7	21.0
760	4.8	10.2			12.2	11.2	10.0
790	4.8	8.6			8.7	8.0	14.0
790	4.8	11.5	10.3	4.0	12.4	11.7	7.0
800	4.9	10.0			12.4	11.0	10.0

ductility. This alloy represents over 65% of all the titanium alloys produced (44). It can be age-hardened in thin sections and has both moderate ductility and an excellent record of successful applications. It is mostly used for compressor blades, fan blades, and rotating disks in aircraft gas-turbine engines, but also used for rocket motor cases, structural airframe forgings, steam-turbine blades, and cryogenic parts for which ELI grades are usually specified.

Other commercially important α – β-alloys used in aircraft applications include Ti–3Al–2.5V, Ti–6Al–6V–2Sn, Ti–6Al–2Sn–4Zr–6Mo, Ti–5Al–2Sn–2Zr–4Mo–4Cr (Ti-17), Ti–6Al–2Sn–2Zr–2Mo–2Cr–0.15Si, and Ti–10V–2Fe–3Al (see Table 12). As a group, these alloys have good combination of strength and ductility. Weldability becomes more difficult as β-constituents increase.

Application temperatures are lower than those of the α- or near-α-alloys. The alloy Ti–3Al–2.5V, called one-half Ti–6Al–4V, is easier to fabricate than Ti–6Al–4V and is used primarily as seamless aircraft hydraulic tubing. The alloy Ti–10V–2Fe–3Al is easier to forge at lower temperatures than Ti–6Al–4V because it contains more β-alloying constituents and has good fracture toughness. This alloy can be hardened to high strengths, ie, 1.24–1.38 GPa $((1.8–2) \times 10^5$ psi), and is used as forgings, primarily in landing gears for airframe structures to replace steel below about 300°C (45). The alloy Ti–6Al–2Sn–2Zr–2Mo–2Cr–0.15Si is used in more recent military aircraft airframe components because of its excellent combination of strength and toughness (46). The β-21S, which has excellent corrosion resistance to high temperature hydraulic fluid, is used in secluded areas of the structure where hydraulic fluid may leak (45). The alloy Ti–6Al–6V–2Sn is used for some aircraft forgings because it has a higher strength than Ti–6Al–4V. However, the alloy is not being designed into new airframe components because of the preference for newer alloys such as Ti–6Al–2Sn–2Zr–2Mo–2Cr–0.15Si and Ti–10V–2Fe–3Al.

Besides Ti–6Al–4V, significant quantities of the $\alpha – \beta$-alloys Ti–6Al–2Sn–4Zr–6Mo and Ti–17 are used in the compressor section of gas turbine engines. These alloys have higher strength capability than Ti–6Al–4V.

The only α-alloy of commercial importance is Ti–5Al–2.5Sn. This alloy is weldable, has good elevated temperature stability and good oxidation resistance to about 600°C, and is used for forgings and sheet-metal parts, such as aircraft engine compressor cases because of its weldability. The ELI version of this alloy is used in the cryogenic area of rocket engines. However, the alloy is difficult to produce and has not been designed into more recent aerospace systems.

The commercially important near-α-alloys used in the United States are Ti–8Al–1Mo–1V and Ti–6Al–2Sn–4Zr–2Mo–0.1Si. Both of these alloys exhibit good creep resistance and the excellent weldability and high temperature strength of α-alloys; temperature limit is ca 470°C. The Ti–8Al–1Mo–1V alloy has not been designed into more recent applications because of the lack of producibility, and also because of the excellent properties of the newer Ti–6Al–2Sn–4Zr–2Mo–0.1Si. Ti–8Al–1Mo–1V is used for compressor blades on account of its high elastic modulus and creep resistance; however, the alloy may suffer from ordering embrittlement. Ti–6Al–2Sn–4Zr–2Mo–0.1Si, also used for blades and disks in aircraft engines, has a service temperature limit of 470°C, which is ca 70°C higher than that of Ti–8Al–1Mo–1V (47). In Europe, a series of alloys developed by IMI in conjunction with Rolls Royce are used in place of the Ti–6Al–2Sn–4Zr–2Mo–0.1Si. These alloys slightly extend the temperature capability of titanium by another 60°C.

β-Titanium alloys represent only 1% of the commercial titanium production. Even though the alloys have high strength, up to 1.5 GPa (217,500 psi), the properties of hot-worked product can be variable on account of differences in grain size. These alloys are metallurgically unstable above 290°C. They are fabricable because of the dominant β-phase. Good welding properties are difficult to achieve but can be accomplished by careful selection of heat-treatment parameters. This alloy type is used successfully in the cold-drawn or cold-rolled condition, where uniform grain size is achieved, and finds application in spring manufacture, eg, Ti–3Al–6V–6Cr–4Mo–4Zr (Beta C) (48). Also, the β-21S (Ti–3.0Al–2.8Nb–15Mo–0.2Si) is used in the plug and nozzle as well as the aft cowl of the Boeing 777 airplane because of its excellent corrosion resistance to areas that may be subject to hot hydraulic oil exposure (44). There is one commercially available alloy of the β-eutectoid type, ie, Ti–2.5 Cu [37270-40-9], that uses a true precipitation-hardening mechanism to increase strength. The precipitate is Ti_2Cu [12054-13-6]. This alloy is only slightly heat-treatable and is produced and used primarily in Europe, as an alternative to commercially pure titanium, in engine castings and flanges (49).

Nonaerospace Alloys. The nonaerospace alloys are used primarily in industrial applications where titanium's corrosion resistance is important. In order to optimize corrosion resistance, many new alloys and modification of existing alloys have been developed (42,50). The four basic grades of pure titanium, ie, ASTM Grades 1 through 4, differ primarily in oxygen and iron content (Table 13) and represent about 24% of the titanium production (44). ASTM Grade 1 has the highest purity and the lowest strength. Modifications of these alloys are available, using Pd or Ru additions to enhance corrosion resistance in hot aqueous halide and reducing acid services. The α-alloys in this group are distinguished by excellent weldability, formability, and corrosion resistance. The strength, however, is not maintained at elevated temperatures. The primary use of alloys in this group is in industrial processing equipment, eg, tanks, heat exchangers, pumps, and electrodes, though there is also some use in airframes and aircraft engines. The ASTM Grade 1 is used where higher purity is desired, eg, as weld wire for Grade 2 fabrication and as sheet for explosive bonding to steel. Grade 1 is manufactured from high purity sponge. ASTM Grade 2 is the most commonly used grade of commercially pure titanium. ASTM Grades 3 and 4 are higher strength versions of Grade 2. The basic grades of pure titanium (Grades 1–4) have poor elevated temperature strength. Grade 12, Ti–0.3Mo–0.8Ni, has a room temperature strength similar to Grade 3 but better elevated temperature strength.

The $\alpha - \beta$- and β-alloys are used where higher strengths are required, such as in shafts, oil and gas wells, and medical implants. Again, Pd and Ru variations of the basic alloys are available where improved corrosion resistance is needed. Several of the listed β-alloys were developed for implants. These alloys were designed to be free of aluminum and vanadium, which have created some concern related to potential toxicity when used in implants (50).

Other Alloys. Other alloying ranges include the aluminides, TiAl [12003-96-2] and Ti_3Al; the superconducting alloys, Ti–Nb type; the shape-memory alloys, Ni–Ti type; and the hydrogen storage alloys, Fe–Ti (see SUPERCONDUCTING MATERIALS; SHAPE-MEMORY ALLOYS). The aluminides TiAl and Ti_3Al have excellent high temperature strengths, comparable to those of nickel- and cobalt-base alloys, but having less than half the density. The Ti_3Al-type alloys exhibit ultimate strengths of 1 GPa (145,000 psi), 800 MPa (116,000 psi) yield, 4–5% elongation, and 7% reduction in area. The TiAl type alloys have lower ductility and toughness but maintain their strengths to 800–900°C. The modulus of elasticity is high, at 125–165 GPa ((18–24) $\times 10^6$ psi), and oxidation resistance is good (51). The aluminides are intended for both static and rotating parts in the turbine section of newer gas-turbine aircraft engines.

Table 13. Commercial Industrial Titanium Alloys

Alloy base	Rare-earth addition, %	ASTM Grade	Condition	Density, g/cc^3	Typical mechanical properties, RT			
					Ultimate tensile strength, MPa	Yield strength, MPa	Elongation, %	Elastic modulus, GPa
α-alloys								
99.5Ti		1	annealed	4.5	270	210	40	105
	0.15Pd	11	annealed	4.5	270	210	40	105
	0.05Pd	17	annealed	4.5	270	210	40	105
	0.10Ru	27	annealed	4.5	270	210	40	105
99.2Ti		2	annealed	4.5	410	330	30	105
	0.15Pd	7	annealed	4.5	410	330	30	105
	0.05Pd	16	annealed	4.5	410	330	30	105
	0.10Ru	26	annealed	4.5	410	330	30	105
99.1Ti		3	annealed	4.5	480	410	25	105
99.0Ti		4	annealed	4.5	620	550	20	105
Ti–0.3Mo–0.8Ni		12	annealed	4.5	550	410	22	106
α − β-alloys								
Ti–3Al–2.5V		9	annealed	4.5	690	550	22	105

Alloy	Addition	Ref.	Condition	Density			Elong.	
Ti–6Al–4V	0.05Pd	18	annealed	4.5	690	550	22	105
	0.10Ru	28	annealed	4.5	690	550	22	105
		5	annealed	4.4	960	900	17	110
	0.05Pd	24	annealed	4.4	960	900	17	110
	0.10Ru	29	annealed	4.4	960	900	17	110
Ti–6Al–4V ELI		23	annealed	4.4	900	815	20	110
β-alloys								
Ti–3Al–8V–6Cr–4Zr–4Mo		19	aged	4.8	1210	1140	7	110
Ti–15Mo–2.7Nb–3Al–0.2Si	0.05Pd	19	annealed	4.8	862	827	18	
		20	aged	4.8	1210	1140	7	110
		21	aged	4.9	1210	1140	10	100
Ti–12Mo–6Zr–2Fe[a]			annealed	5.0[b]	1080	1030	20	80
Ti–13Nb–13Zr[a]			aged	5.0[b]	1005	872	13	82
Ti–15Mo–2.7Nb–0.2Si[a]			annealed	5.0[b]	989	966	17	83
Ti–16Nb–9.5Hf[a]			aged	5.6[b]	851	736	10	81
Ti–15Mo[a]			annealed	4.9[b]	874	544	21	78

[a]Medical alloys.
[b]Calculated.

Titanium alloyed with niobium exhibits superconductivity, and a lack of electrical resistance below 10 K. Composition ranges from 25 to 50 wt % Ti. These alloys are β-phase alloys having superconducting transitional temperatures at ca 10 K. Their use is of interest for power generation, propulsion devices, fusion research, and electronic devices (52).

Titanium alloyed with nickel exhibits a memory effect, ie, the metal form switches from one specific shape to another in response to temperature changes. The group of Ti–Ni alloys (nitinol) was developed by the U.S. Navy in the early 1960s and its first use was for couplings on hydraulic tubing in F-14 fighter jets. The compositions are typically Ti with 55 wt % Ni. The transition temperature ranges from -10 to $>100°C$ and is controlled by additional alloying elements. These alloys are of interest for thermostats, recapture of waste heat, pipe joining, etc. The nitinols have not been extensively used because of high price and fabrication difficulties (53).

Titanium alloyed with iron is a candidate for solid-hydride energy storage material for automotive fuel. The hydride, $FeTiH_2$, absorbs and releases hydrogen at low temperatures. This hydride stores 0.9 kWh/kg. To provide the energy equivalent to a tank of gasoline would thus require about 800-kg $FeTiH_2$ (54).

Specifications, Standards, and Quality Control

The alloys of titanium utilized in industrial applications have compositional specifications tabulated by ASTM. The ASTM specification number is given in Table 14 for the commercially important alloys. Military specifications are found under MIL-T-9046 and MIL-T-9047, and aerospace material specifications for bar, sheet, tubing, and wire under specification numbers 4900–4980. Every large aircraft company has its own set of alloy specifications.

The alloy name in the United States can include a company name or trademark in conjunction with the composition for alloyed titanium or the strength, ie, ultimate tensile strength for Timet and yield strength for other U.S. producers, for unalloyed titanium. The common alloys and specifications are shown in Table 14.

Because titanium alloys are used in a variety of applications, several different material and quality standards are specified. Among these are ASTM, ASME, Aerospace Materials Specification (AMS), U.S. military, and a number of proprietary sources. The correct chemistry is basic to obtaining mechanical and other properties required for a given application. Minor elements controlled by specification include carbon, iron, hydrogen, nitrogen, and oxygen. In addition, control of thermomechanical processing and subsequent heat treatment is vital to obtaining desired properties. For extremely critical applications, such as rotating parts in aircraft gas turbines, raw materials, melting parameters, chemistry, thermomechanical processing, heat treatment, testing, and finishing operations must all be carefully and closely controlled at each step to ensure that required characteristics are present in the products supplied.

Health and Safety Factors

Titanium and its corrosion products are nontoxic. A safety problem does exist with titanium powders, grindings, turnings, and some corrosion products that

Table 14. Common Name and Typical Specifications for Titanium Alloys

Nominal composition, wt %	Common name	UNS	AMS	ASTM	Military
		α-Alloys			
99.5Ti	Ti	R50250	4591E	1	9046J
99.2Ti	Ti	R50400	4902E	2	9046J
99.1Ti	Ti	R50550	4900J	3	9046J
99.0Ti	Ti	R50700	4901L	4	9046J
Ti–5Al–2.5Sn	Ti-5-2.5	R54520	4926H	6	9047J
		α–β-Alloys			
Ti–5.8Al–4Sn–3.5Zr–0.7Nb– 0.5Mo–0.35Si	IMI 834				
Ti–3Al–2.5V	Ti-3-2.5	R56320	4943D	9	9047G
Ti–8Al–1Mo–1V	Ti-811	R54810	4972C		9047G
Ti–6Al–2Sn–4Zr–2Mo–0.1Si	Ti-6242S	R54620	4976C		9047G
Ti–6Al–4V	Ti-6-4	R56400	4911F	5	9046J
Ti–4Al–4Mo–2Sn–0.5Si	Ti-550				
Ti–6Al–26V–2Sn	Ti-662	R56620	4918F		9046J
Ti–6Al–2Sn–2Zr–2Mo– 2Cr–0.1Si	Ti-6-2222S				
Ti–6Al–2Sn–4Zr–6Mo	Ti-6246	R56260	4981B		9047G
Ti–4.5Al–3V–2Mo–2Fe	SP-700		4899		
Ti–5Al–2Sn–2Zr–4Mo–4Cr	Ti-17	R58650	4955		
		β-Alloys			
Ti–10V–2Fe–3Al	Ti-10-2-3		4984		
Ti–15V–3Cr–3Al–3Sn	Ti-15-3		4914		
Ti–3Al–8V–6Cr–4Mo–4Zr	Beta C	R58640	4957	19	9046J
Ti–3Al–2.8Nb–15Mo–0.2Si	Beta 21S	R58210			

are pyrophoric. Powders can ignite at about 250°C and should be handled in small quantities at room temperature in electrically grounded, nonsparking equipment made from materials, such as monel, aluminum, and stainless steel. Grindings and turnings should be stored in a closed container and not left on the floor. Smoking must be prohibited in areas where titanium is ground or turned. If a fire occurs, it must be extinguished with a Class-D extinguisher, specifically used against metal fires. Dry common salt can also be used to smother a fire. Water or other liquids must not be used as they could react with the titanium and release hydrogen. The larger the surface area, the more pyrophoric the titanium fines. When titanium equipment is being worked on, all flammable and corrosive products must be removed, and the area well ventilated. A pyrophoric corrosion product has been observed in environments of dry Cl_2 gas and in dry red-fuming nitric acids.

Uses

Titanium is primarily used in the form of high purity titanium oxide as a pigment in surface coatings. Other uses reflect its special properties, which include high refractory index that inputs good hiding power; high reflectivity that inputs great brightness and brilliant whiteness; chemical inertness that contributes to excellent color retention; and thermal stability over a wide range of temperatures. Although the principal application of high purity (pigment-grade) TiO_2 is in paint pigments, other important uses are in plastics for color in floor-covering products and to help protect plastic products and foodstuffs contained in plastic bags from uv radiation deterioration; in paper as a filler and whitener; and in rubber (Table 15). Future applications areas could include TiO_2 single-crystal electrodes for water decomposition for the production of hydrogen fuel, flue-gas denitrification catalysts, and high purity TiO_2 for the manufacture of barium titanate thermistors.

Titanium metal was first used as a material for aerospace. In the late 1970s, new applications developed that utilize the metal's excellent corrosion resistance. The Japanese market development goals pursued nonaerospace uses, whereas the former Soviet Union and the U.S. markets were mainly aerospace and military. Western European markets have been equally divided between aerospace and nonaerospace.

In the United States, the high strength–weight ratio of titanium accounts for ca 70% of its uses. Before 1980, the high strength–weight ratio was the basis of over 90% of applications, such as engines, where the advantage of light weight is translated to higher flying, faster, and more fuel efficient planes. Aerospace applications have shaped and controlled the titanium metal industry.

The use of titanium in aircraft is divided about equally between engines and airframes. For engine components, titanium is limited because of temperature constraints at the compressor area (540°C) where it is used as blades, casings, and disks. In the frame, it is used in bulkheads, firewall, flap tracks, landing-gear parts, wiring pivot structures, fasteners, hydraulic tubing, and hot-area skins. In the U.S. Air Force F-15 and F-22 fighter jets, titanium accounts for over 35% of the structural weight. Each new generation of commercial airframes includes more titanium; the Boeing 777 contains nearly 10% by weight. Titanium has seen increased usage following the introduction of graphite–epoxy composite aircraft structures. These structures galvanically increase the corrosion of aluminum but are compatible with titanium. Furthermore, the thermal expansion coefficients of these composites are similar to those of titanium alloys.

The other outstanding property of titanium metal is its corrosion resistance. The largest single application is related to heat-exchanger pipes and tubing (ca 800 mm or 22 gauge welded) for the power industry, as well as marine and desalination applications, where titanium provides protection against corrosion by seawater, brackish water, and other estuary waters containing high concentrations of chlorides and industrial wastes (see HEAT-EXCHANGE TECHNOLOGY).

Table 15. Distribution of TiO_2 in U.S. Market[a], %

Industry	1978	1995
paint	52	47
plastics	12	18
papers	22	24
other	14	11

[a]Refs. 55 and 56.

Titanium metal is especially utilized in environments of wet chlorine gas and bleaching solutions, ie, in the chlor–alkali industry and the pulp and paper industries, where titanium is used as anodes for chlorine production, chlorine–caustic scrubbers, pulp washers, and Cl_2, ClO_2, and $HClO_4$ storage and piping equipment (see ALKALI AND CHLORINE PRODUCTS; PAPER; PULP).

In the chemical industry, titanium is used in heat-exchanger tubing for salt production, in the production of ethylene glycol, ethylene oxide, propylene oxide, and terephthalic acid, and in industrial wastewater treatment. Titanium is used in environments of aqueous chloride salts, eg, $ZnCl_2$, NH_4Cl, $CaCl_2$, and $MgCl_2$; chlorine gas; chlorinated hydrocarbons; and nitric acid.

In metal recovery, titanium is used for ore-leaching solutions and as racks for metal plating. The leaching solutions contain HCl or H_2SO_4, with enough ferric or cupric ions to inhibit the corrosion of titanium. In metal-plating applications, titanium is anodically protected against H_2SO_4 and chrome-plating solution corrosion.

In oil and gas refinery applications, titanium is used as protection in environments of H_2S, SO_2, CO_2, NH_3, caustic solutions, steam, and cooling water. It is used in heat-exchanger condensers for the fractional condensation of crude hydrocarbons, NH_3, propane, and desulfurization products using seawater or brackish water for cooling.

In energy extraction, titanium alloys are being used in deep-water hydrocarbon and geothermal wells for risers. Corrosion resistance, high strength, low modulus (flexible), and low density can result in risers one-fourth the weight and three times the flexibility of steel.

In consumer applications, titanium is used in golf club heads, jewelry, eyeglass frames, and watches. The Japanese have promoted the use of titanium in roofing and monuments. Other application areas include nuclear-waste storage canisters, pacemaker castings, medical implants, high performance automotive applications, and ordnance armor.

Economic Aspects

Titanium raw-material utilization can be broken down as illustrated in Figure 9. About 4% of the titanium mined is used as metal, 94% is used as pigment-grade TiO_2, and 2% as ore-grade rutile for fluxes and ceramics. In 1995, the estimated U.S. TiO_2 pigment production was valued at $2.6 billion and was produced by five companies at 11 plants in nine states. About 47% was used in paint, 18% in plastics, 24% in paper, and 18% in other miscellaneous applications (56).

In 1995, titanium sponge was produced in two plants in the United States. The value of the domestic and imported sponge was about $180 million, assuming a selling price of $4 per pound. Ingot was produced by the two sponge producers and nine other firms in seven states. About 30 companies produce mill products, forgings, and castings. The mill products manufactured in the United States had a value of about $400 million, assuming an average selling price of $9 per pound. Approximately 65% was consumed in aerospace applications; the remainder was used in chemical process industry, power generation, marine, ordinance, and consumer applications.

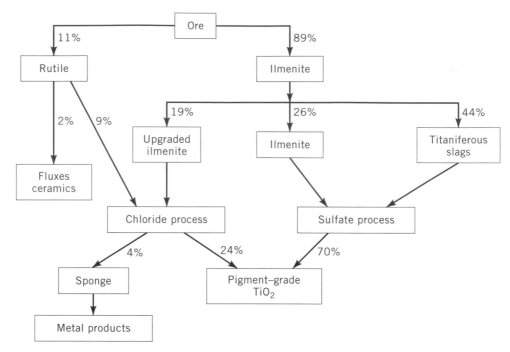

Fig. 9. U.S. titanium raw material utilization (3,6).

The principal world producers of pigment-grade TiO_2 are the United States, Western Europe, and Japan (Table 16). The growth rate from 1960 to 1973 was ca 8% annually. Consumption decreased sharply after the 1973 oil crisis, from 2×10^6 to 1.5×10^6 t. The demand has since recovered and growth rates of 5–8% were reported in the early 1990s. Global growth is expected to be between 3 and 4% annually, with Latin America and the Asia–Pacific areas experiencing slightly higher growth rates (5).

The principal sponge producers are United States, Japan, and the CIS countries of Russia and Kazakstan (see Table 16). U.S. metal demand has been greater than sponge production, which has been supplemented by imports, primarily from Japan, Russia, Kazakstan, and, to a lesser extent, China. Imports supply ca 25–40% of the U.S. demand. The United States does not supply its own demands because of the closing of capacity in the weak markets of the early 1990s and the reluctance to add capacity for peak demand periods when capital requirement is large ($22/kg of annual capacity). The price history of titanium sponge is given in Table 17.

The U.S. titanium market distribution is shown in Table 18. Before 1970, more than 90% of the titanium produced was used for aerospace, which fell to ca 70–80% by 1982. Military use has continually decreased from nearly 100% in the early 1950s to 20% in the 1990s. In contrast to the United States, aerospace uses in Western Europe and Japan account for only 40–50% of the demand (58). The CIS's consumption of titanium metal prior to the breakup was about

Table 16. World Sponge and Pigment Capacities[a]

Country	Sponge, 10^3 t	Pigment, 10^3 t
United States	29.5	1330
Australia		150
Belgium		80
Canada		74
China	7	40
Finland		80
France		230
Germany		350
Italy		80
Japan	25.8	320
Kazakstan	35	1
Russia	35	20
Spain		65
Ukraine		120
United Kingdom		275
other countries		585
Total	*132.3*	*3800*

[a]Ref. 56.

Table 17. Price History of Titanium Sponge

Year	$/kg
1948–1955	11.02
1964–1975	2.91
1981	16.86
1982	12.20
1991–1994	8.80
1996	11.00

Table 18. U.S. Titanium Market Distribution[a]

Market	1955	1961	1966	1975	1979	1990	1995
aerospace, %							
military	94	72	74	54	45	27	20
civilian	3	20	19	20	35	50	50
industrial, %	3	6	7	26	20	23	30
Total mill products, t	*910*	*2,940*	*6,580*	*7,030*	*9,750*	*23,600*	*19,840*

[a]Refs. 2 and 57.

one-half of the world consumption. In the 1980s, considerable amounts were used for submarine construction. Since the breakup of the former Soviet Union, the internal consumption of titanium in the CIS is believed to be a modest fraction of its former capacity, thus leaving a large capacity available for export. The world production facilities for titanium metal and extraction are given in Table 19.

Table 19. Principal Titanium Metal Producers

Region	Company	Sponge, 10^3 t	Ingot Vacuum arc, 10^3 t	Ingot Cold hearth, 10^3 t
United States	RMI Titanium Co.	0	16.4	0
	Timet	10	12.7	0
	Oremet	5.5	7.3	0
	Wyman-Gordon	0	1.8	1.4[a]
	Wah Chang	0	0.9	0
	Howmet	0	2.3	0
	Allvac	0	2.7	1.4[a]
	Lawrence Aviation	0	1.4	0
	Axel Johnson	0	0	2.3[b]
	Titanium Hearth Technologies	0	0	9.1[b]
	Alta Titanium (Johnson-Matthey)	0.3	0	0.1[b]
Total		*15.8*	*45.5*	*14.3*
Japan	Suimitomo Sitix Corp.	15	5	0
	Toho Titanium	11	6.4	0
	Daido	0	0.5	0
	Kobe	0	7.3	0
	Sumitomo	0	3.7	0
Total		*26*	*22.9*	*0*
Europe	IMI (England)	0	5.5	0
	Deutsche Titan (Germany)	0	1.4	0
	Cezus (France)	0	1.4	0
Total		*0*	*8.3*	*0*
CIS	Ust Kamen (Kazakhstan)	36.4	0	0
	Avisma (Russia)	25.9	0	0
	VSMPO (Russia)	0	100	0
	Vils (Russia)	0	5	0
Total		*62.3*	*105*	*0*
China	Metals Ministry	2.7	3.6	0
Total world		*106.8*	*185.3*	*14.3*

[a]Plasma.
[b]Electron beam.

BIBLIOGRAPHY

"Titanium and Titanium Alloys" in *ECT* 1st ed., Vol. 14, pp. 190–213, by C. H. Winter, Jr., and E. A. Gee, E. I. du Pont de Nemours & Co., Inc.; in *ECT* 1st ed., 2nd Suppl., pp. 866–873, by H. R. Ogden, Battelle Memorial Institute; in *ECT* 2nd ed., Vol. 20, pp. 347–379, by H. B. Bomberger, Reactive Metals, Inc.; in *ECT* 3rd ed., Vol. 23, pp. 98–130, by D. Knittel, Cabot Corp.

1. S. C. Williams, *Report on Titanium*, J. W. Edwards, Inc., Ann Arbor, Mich., 1965.
2. S. R. Seagle, *Mater. Sci. Eng. A*, **213** 1 (Aug. 1996).
3. R. Towner, J. Gray, and L. Porter, *International Strategic Minerals Inventory Summary Report–Titanium*, U.S. Geological Survey Circular 930-G, 1989.
4. R. A. Wood, *The Titanium Industry in the Mid-1970s*, Battelle Report MCIC-75-26, Battelle Memorial Institute, Columbus, Ohio, 1975.

5. *Titanium Annual Review-1994*, Mineral Industry Surveys, U.S. Department of the Interior, Bureau of Mines, Washington, D.C., 1995.

6. L. E. Lynd, in *Mineral Commodity Profiles MCP-18*, U.S. Department of the Interior, Bureau of Mines, Washington, D.C., 1978.

7. N. Ohta, *Chem. Eco. Eng. Rev.* **13**, 22 (1981).

8. G. E. Haddeland and S. Morikawa, *Titanium Dioxide Pigment*, Process Economics Program Report No. 117, Stanford Research Institute International, Menlo Park, Calif., 1978.

9. *Energy Use Patterns in Metallurgical and Nonmetallic Mineral Processing*, Report No. PB-246 357, Battelle Columbus Laboratories, U.S. Department of Commerce, Washington, D.C., 1975.

10. J. A. Slatnick and co-workers, *Availability of Titanium in Market Economy Countries*, IC 9413, Bureau of Mines Information Circular, Washington, D.C., 1994.

11. W. W. Minkler and E. F. Baroch, *The Production of Titanium, Zirconium and Hafnium*, 1981.

12. R. C. Weast, ed., *CRC Handbook of Chemistry*, 62nd ed., CRC Press, Inc., Boca Raton, Fla., 1982.

13. T. Ikeshima, *Proceedings of the 5th International Conference on Titanium Science and Technology*, Munich, Germany, 1984, p. 3.

14. ASTM Standard Specification for Titanium Sponge, ANSI–ASTM B265-74. American Society for Testing and Materials, Philadelphia, Pa., 1981.

15. S. R. Seagle and R. L. Fisher, in P. Lacombe, R. Tricot, and G. Beranger, eds., *Critical Review: Raw Materials*, Société Française de Metallurgie, Cedex France, 1988, p. 565.

16. J. L. Henry and co-workers, *Bureau of Mines Development of Titanium Production Technology*, Bulletin 690, Washington, D.C., 1984.

17. G. Cobel, J. Fisher, and L. K. Snyder, in R. I. Jaffee and H. Kimura, eds., *Titanium '80 Science and Technology*, The Metallurgical Society of AIME, Warrendale, Pa., 1980, p. 1969.

18. M. V. Ginatta and co-workers, in Ref. 15, p. 753.

19. Brit. Pat. 1,355,433 (June 5, 1974), P. D. Johnson, J. Lawton, and I. M. Parker (to the Electricity Council).

20. Fr. Demande 2,002,771 (Oct. 31, 1969), A. G. Halomet.

21. U.S. Pat. 3,429,691 (Feb. 25, 1969), W. J. McLauglin (to Aerojet General Corp.).

22. U.S. Pat. 3,794,482 (Feb. 26, 1974), R. N. Anderson and N. A. D. Parlee (to Parlee Anderson Corp.).

23. G. V. Samsonov and V. S. Sinelinikova, in A. T. Dogvinenko, ed., *Mealloterm. Protsessy Khim. Met. Masters. Konf., Nauka. Sib Otd., Novosibirsk*, U.S.S.R., 1971, p. 32.

24. A. D. McQuillan and M. K. McQuillan, in H. M. Finniston, ed., *Metallurgy of the Rarer Metals*, Academic Press, Inc., New York, 1956, p. 335.

25. ASTM Standard Specification for Titanium and Titanium Alloy Strip, Sheet, and Plate, ANSI–ASTM B265-95, American Society for Testing and Materials, Philadelphia, Pa., 1995.

26. R. I. Jaffee, F. C. Holden, and H. R. Ogden, *J. Metals*, **6**, (Nov. 1954).

27. H. Rosenberg, in R. I. Jaffee and N. E. Promisel, eds., *Titanium Alloying in Theory and Practice*, Pergamon Press, Oxford, U.K., 1970, p. 851.

28. *Facts About the Metallography of Titanium*, RMI Company, Nile, Ohio, 1975.

29. M. Molchanova, *Phase Diagrams of Titanium Alloys*, Israel Program for Scientific Translations, Jerusalem, Israel, 1965.

30. R. Boyer, G. Welsch, and E. Collings, eds., *Materials Property Handbook: Titanium Alloys*, ASM International, Materials Park, Ohio, 1994.

31. V. V. Andreeva, *Corrosion*, **20**, 35t (1964).

32. T. R. Beck, in R. W. Staehle and co-workers, eds., *Localized Corrosion*, Vol. NACE-3, National Association of Corrosion Engineers, Houston, Tex., 1974, p. 644.

33. E. E. Millaway, *Mater. Prot.* **4**, 16 (1965).

34. *Titanium for industrial Brine and Sea Water Service*, Titanium Metal Corporation of America, Denver, Colo., 1968.

35. L. C. Covington, R. W. Schutz, and I. A. Fronson, *Chem. Eng. Prog.* **74**, 67 (1978).

36. R. W. Schutz, *Plat. Metals Rev.* **40**(2), 54–61 (1996).

37. L. C. Covington and R. W. Schutz, in E. W. Kleefisch, ed., *Industrial Applications of Titanium and Zirconium*, STP 728, American Society for Testing and Materials, Philadelphia, Pa., 1981, p. 163.

38. G. Venkataraman and A. D. Goolsby, *Proceedings of Corrosion Conference '96*, No. 554, NACE, Houston, Tex., 1996.

39. R. Ruthardt, H. Stephan, and W. Dietrich, in Ref. 17, p. 2289.

40. M. J. Donachie, *Titanium, A Technical Guide*, ASM International, Metals Park, Ohio, 1988.

41. J. W. Brooks, P. J. Bridges, and D. Stephen, in F. H. Froes and I. L. Caplan, eds., *Titanium '92 Science and Technology*, TMS, Warrendale, Pa., 1992, pp. 1 and 319.

42. R. W. Schutz, *Proceedings of Corrosion Conference '95*, No. 244, NACE, Houston, Tex., 1995.

43. *Structural Alloys Handbook*, 1982 Suppl. Vol., Battelle Memorial Institute, Columbus, Ohio, 1982.

44. P. J. Bania, *J. Metals*, 16 (July 1994).

45. R. R. Boyer, *J. Metals*, 20 (July 1994).

46. H. R. Phelps and C. Harberg, *Proceedings of Titanium Development Association Conference*, San Diego, Calif., 1994.

47. P. A. Blenkinsop, in Ref. 13, p. 2323.

48. S. R. Seagle, C. F. Pepka, and R. Bajoraitis, *Proceedings of the 1986 International Conference on Titanium Products and Applications*, 1986, p. 523.

49. R. M. Duncan, P. A. Blenkinsop, and R. E. Goosey, in G. W. Meethan, ed., *The Development of Gas Turbine Materials*, John Wiley & Sons, Inc., New York, 1981, p. 63.

50. K. Wang, *Mater. Sci. Eng. A*, **213**, 134 (Aug. 1996).

51. M. Yamaguhchi, in Ref. 41, p. 959.

52. E. M. Savitskiy, M. I. Bychkova, and V. V. Baron, in R. I. Jaffee and H. Kimura, eds., *Titanium '80 Science and Technology*, Metallurgical Society of AIME, Warrendale, Pa., 1980, p. 735.

53. C. M. Wayman, *J. Met.* **32**, 129 (1980).

54. R. I. Jaffee, in Ref. 52, p. 53.

55. H. Schmidt and P. Eggert, *Unterschungen uder Angebot und Nachfrage Mineral ischer Roh Stoffe Titan*, Deutsches Institut für Wirtschaftsforschary, Berlin, Germany, 1980.

56. *Mineral Commodity Summaries*, Department of the Interior Geological Survey and Bureau of Mines, Washington, D.C., 1996.

57. W. W. Minkler, in *Assessment of Selected Materials Issues*, National Materials Advisory Board, National Academy of Sciences, Washington, D.C., 1981.

58. S. R. Seagle and J. R. Wood, in F. Froes and T. Khan, eds., *Synthesis, Processing, and Modeling of Advanced Materials*, Trans Tech Publications, Aedermannsdorf, Switzerland, 1993, p. 91.

STAN R. SEAGLE
Consultant

TITANIUM COMPOUNDS

INORGANIC

Titanium, Ti, atomic number 22, relative atomic mass 47.90, is the ninth most common element (ca 0.6% by weight) and is widely distributed in the earth's crust. It is found particularly in the ores rutile, TiO_2, and ilmenite, $FeTiO_3$.

Titanium is the first member of the d-block transition elements. Its electron configuration is [Ar] $4s^2 3d^2$, and successive ionization potentials are 6.83, 13.57, 27.47, and 43.24 eV. Its technologically important chemistry is predominantly that of oxidation states (II), (III), and (IV), in order of increasing importance. Ti(II) and Ti(III) compounds are readily oxidized to the tetravalent state by air, water, and other oxidizing agents. The energy for the removal of four electrons is so high that the Ti^{4+} ion does not itself have a real existence, and compounds of Ti(IV) are generally significantly covalent. Titanium, capable of utilizing the $3d$ shell, forms a large group of addition compounds in which the Ti is coordinated by donor atoms such as oxygen, sulfur, or chlorine.

The titanium compounds of greatest technological importance are titanium dioxide, the predominant white pigment, which sold ca 4×10^6 metric ton worldwide in 1996 (see PIGMENTS; INORGANIC); titanium esters, eg, titanium isopropoxide, and derived compounds, which are used for applications such as structuring agents in paint (qv); titanium metal, which has excellent corrosion resistance and a high strength–weight ratio (see TITANIUM AND TITANIUM ALLOYS); and barium titanate, an important electroceramic (see ADVANCED CERAMICS; CERAMICS AS ELECTRICAL MATERIALS). The general properties and chemistry of titanium are reviewed elsewhere (1–14).

Thermochemical Data

Data relating to changes of state of selected titanium compounds are listed in Table 1. Heats of formation, free energy of formation, and entropy of a number of titanium compounds at 298 and 1300 K are collected in Table 2.

Titanium–Hydrogen System

Titanium metal readily absorbs hydrogen. Absorption rates above 400°C are normally high, but at lower temperatures the rates depend critically on the cleanliness of the surface. There is, for example, significant inhibition by surface oxides. The hydrogen dissociates prior to absorption. Because the absorption is a reversible process, there is thus an equilibrium pressure of hydrogen at each temperature and composition. The limiting stoichiometry of the system is normally accepted as TiH_2, although higher hydrides have been reported under special conditions. For example, TiH_4 has been reported in the products formed when mixtures of titanium tetrachloride and hydrogen were irradiated with uv radiation at 254 nm.

Table 1. Thermal Data for Changes of State of Titanium Compounds

Compound	CAS Registry Number	Change[a]	Temperature, K	ΔH, kJ/mol[b]
TiCl$_4$	[7550-45-0]	mp	249.05	9.966
		bp	408	35.77
TiCl$_3$	[7705-07-9]	sublimation	1103.3	166.15
TiCl$_2$	[10049-06-6]	sublimation	1580.4	248.5
TiI$_4$	[7720-83-4]	mp	428	19.83 ± 0.63
		bp	651.8	56.48 ± 2.09
TiF$_4$	[7783-63-3]	sublimation	558.3	97.78 ± 0.42
TiBr$_4$	[7789-68-6]	mp	311.4	12.89
		bp	503.5	45.19
TiO$_2$	[13463-67-7]	anatase to rutile		ca −12.6
TiN	[25583-20-4]	mp	3220 ± 50	66.9
TiB$_2$	[12405-63-5]	mp	3193	100.4

[a]Melting and sublimation temperatures are generally based on those given in Ref. 15.
[b]To convert kJ/mol to kcal/mol, divide by 4.1814.

Table 2. Thermochemical Data for the Formation of Titanium Compounds[a]

Compound[b]	ΔH_f, kJ/mol[c]		ΔG_f, kJ/mol[c]		S, J/(mol·K)[c]	
	298.15 K	1300 K	298.15 K	1300 K	298.15 K	1300 K
TiO(α)	−542.7	−540.3[d]	−513.3	−427.9[d]	34.8	110.9[d]
anatase	−938.7	−934.1	−883.3	−703.4	49.9	152.2
rutile	−944.7	−940.7	−889.4	−709.2	50.3	151.7
TiB$_2$	−279.5	−287.5	−275.3	−257.5	28.5	125.3
TiC	−184.1	−188.3	−180.4	−168.8	24.2	92.25
TiN	−337.6	−337.6	−308.9	−214.7	30.2	100.8
TiCl$_2$	−515.5	−504.2	−465.8	−310.2	87.4	204.3
TiCl$_3$	−721.7	−706.5	−654.5	−445.4	139.8	290.9
TiCl$_4$[e]	−804.2	−771.3[f]	−728.1	−573.7[f]	221.9	400.8[f]
TiCl$_4$[g]	−763.2	−765.8	−726.8	−606.3	354.9	507.5
TiBr$_4$	−618	−654.8	−590.7	−347	243.6	440.6
TiBr$_3$	−550.2	−546.5	−525.6	−340.8	176.4	369
TiF$_4$	−1649.3	−1611	−1559.2	−1289.9	134	338.7
TiI$_4$	−375.7	−453.3	−370.7	−164.6	246.2	486.6

[a]ΔH_f and ΔG_f refer to the formation of the named substances in the specified states at 298.15 and 1300 K from their elements in the standard states of those elements at these temperatures. Refs. 9 and 15.
[b]In the crystalline state unless otherwise specified.
[c]To convert J to cal, divide by 4.1814.
[d]Values at 1200 K. Above this temperature the α-form of TiO transforms to β.
[e]Liquid.
[f]Values at 1000 K.
[g]Gas.

A phase diagram, with notation for the titanium−hydrogen system, is available, as is a review of kinetic data and physical properties (16). The phases may be conveniently described by reference to pure metallic titanium, which at room temperature and pressure has a hexagonal close-packed (hcp) structure

(α-titanium). At high temperatures, this converts to a body-centered cubic (bcc) structure (β-titanium). At high pressures, the hcp α-phase converts to a hexagonally distorted bcc structure, ω. In the titanium–hydrogen system, the α-phase, ie, the solid solution of H in hcp titanium, exists up to ca 0.12 atom % at 25°C and up to 7.9 atom % ($TiH_{0.09}$) between 300 and 600°C. Its formation has negligible effect on the titanium's lattice parameters, but whether the hydrogen atoms (radius 41 pm) occupy tetragonal (34 pm) or octahedral (62 pm) interstices within the lattice remains unclear. The region of stability of the β-phase H solution in bcc titanium extends from 0 atom % H at 882°C to ca 50 atom %, ie, a stoichiometry corresponding to TiH, at ca 300°C. In this case, hydrogen solubility expands the β-titanium lattice and the hydrogens have been shown to occupy tetrahedral sites. An fcc δ-phase exists as a mixture with the α or β solid solutions over a wide range of hydrogen concentrations, but only as a single phase region between 57 and 64 atom % hydrogen ($TiH_{1.5-1.94}$) at room temperature. Above 64 atom %, and at relatively low temperatures, variously reported as 20–41°C, transformation from this fcc phase to a tetragonally distorted fcc structure, the ϵ-phase, takes place in near stoichiometric TiH_2. In the ϵ-phase, the H atoms are symmetrically located in tetrahedral interstices. In addition, a metastable γ-phase with a stoichiometry TiH has been identified and is formed when the α-phase containing 1–3 atom % H is cooled. A phase transition from α to the hexagonally distorted bcc ω structure occurs under high pressures of hydrogen, eg, 4 GPa (40 kbar) for $TiH_{0.33}$. By quenching this phase to 90 K, a metastable superconducting phase ($T_c = 4.3$ K) is formed.

Titanium hydrides are grey powders. Their density decreases with increasing hydrogen content from 4410 kg/m^3 for pure titanium metal to ca 3800 kg/m^3 for TiH_2. The applications of titanium hydrides are related to their ability to store hydrogen reversibly (qv). Their possible use as a hydrogen storage medium in hydrogen-fueled vehicles has received particular attention (see HYDROGEN ENERGY). Titanium hydrides may also be used to purify cylinder hydrogen or as high purity hydrogen sources for, eg, plasma guns. Suppliers of titanium hydride include Atomergic Chemmetals Corp., Micron Metals, Morton International, and Noah Technologies in the United States, and Hladik-Skalick VOS in Europe.

Titanium Borides

Five phases of titanium boride have been reported. TiB_2 [12405-65-35], Ti_2B [12305-68-9], TiB [12007-08-8], Ti_2B_5 [12447-59-5], and TiB_{12} [51311-04-7]. The most important of these is the diboride, TiB_2, which has a hexagonal structure and lattice parameters of $a = 302.8$ pm and $c = 322.8$ pm. Titanium diboride is a gray crystalline solid. It is not attacked by cold concentrated hydrochloric or sulfuric acids, but dissolves slowly at boiling temperatures. It dissolves more readily in nitric acid/hydrogen peroxide or nitric acid/sulfuric acid mixtures. It also decomposes upon fusion with alkali hydroxides, carbonates, or bisulfates.

Research-grade material may be prepared by reaction of pelleted mixtures of titanium dioxide and boron at 1700°C in a vacuum furnace. Under these conditions, the oxygen is eliminated as a volatile boron oxide (17). Technical grade (purity >98%) material may be made by the carbothermal reduction of titanium dioxide in the presence of boron or boron carbide. The endothermic reaction is carried out by heating briquettes made from a mixture of the reactants in electric furnaces at 2000°C (11,18,19).

$$TiO_2 + B_2O_3 + 5\,C \longrightarrow TiB_2 + 5\,CO$$

Titanium diboride, typically 96–98% pure, may also be made by the electrolysis of mineral rutile dissolved in mixed electrolytes, $TiO_2/Na_2CO_3/Na_3AlF_6/NaCl/Na_2B_4O$, at 1050°C (20). Very fine titanium diboride may be made by a gas-phase plasma process in which titanium tetrachloride and boron trichloride are reacted in a hydrogen gas heated by a d-c plasma (21).

$$TiCl_4 + 2\,BCl_3 + 5\,H_2 \longrightarrow TiB_2 + 10\,HCl$$

This reaction, operated at pilot plant scale, has not as of this writing (ca 1997) been commercialized. The same reaction may be used for chemical vapor deposition of titanium boride.

The applications of titanium diboride depend on such properties as its hardness, electrical conductivity, and ability to be wetted by metals. The actual values of these properties attained in sintered bodies depend on the grain structure and degree of densification and, ultimately, on both the fabrication method and the quality of the starting powder (22). Some typical values are given in Table 3. In monolithic form, sintered titanium diboride can be used as a strong but lightweight armor. It is an important constituent of cermets, eg, composites based on TiB_2–TiC–Fe mixtures show excellent performance as cutting tools for aluminum alloys. Titanium diboride also shows excellent wettability and stability in liquid metals, eg, liquid aluminum, and many tons of the powder are used in the production of vacuum metallizing boats, eg, TiB_2, AlN, and BN composites. These are resistance-heated to about 1400°C and used for the continuous evaporation of aluminum to produce thin metal films (see THIN FILMS). The use of titanium diboride as an inert solid cathode, which would replace the molten aluminum cathode in aluminum reduction refining cells, has been proposed.

Suppliers of titanium diboride include Micron Metals, Atomergic Chemmetals, Cerac, and Noah Technologies, in the United States, and Elektroschmelzwerk Kempten, Herman C. Starck, and RTZ Chemicals in Europe.

Titanium Carbide

Titanium carbide [12070-08-05], TiC, has the fcc, NaCl structure in which the carbon atoms can be regarded as occupying the octahedral interstices in a slightly expanded cubic close-packed arrangement of titaniums. The system is stable from $TiC_{1.0}$ to $TiC_{0.47}$ (23,24), ie, ca 50% of the carbon sites can be vacant. As the carbon content is reduced, the lattice parameter ($a = 432.8$ pm (25)) and density decrease. From $TiC_{0.47}$ to $TiC_{0.08}$, TiC and Ti phases coexist; below $TiC_{0.08}$, only the α-Ti phase is present. Titanium carbide also forms solid solutions with TiO and TiN. Sintered titanium carbide is light gray when fractured but can be polished to a silver gray. The maximum melting point in the Ti–C system is 3067°C for $TiC_{0.8}$, that of $TiC_{1.0}$ is slightly lower. The boiling point of titanium carbide is 4800°C. Representative values of physical properties are given in Table 3.

Table 3. Physical Properties of Titanium Borides, Carbides, and Nitrides[a]

Compound	Structure	Lattice parameter, pm	Density, kg/m³	Melting point, °C	Electrical resistivity, 25°C, $\Omega \cdot m \times 10^{-8}$		Hardness, Mohs' scale	Microhardness, GPa
TiB$_2$	hexagonal		4520	2980[b]	10–30	28.4[b]	9	2600[c]
	a	302.8						
	c	322.8						
TiC	fcc (NaCl)	432.8	4910	3000	50–200	180–250[b]	9–10	3200[d]
								–31.4
TiN	fcc (NaCl)	423.5	5213	2950	20–50	21.7[b]		1770[c]
								15–20

[a]Values are indicative only. Precise values depend on sintering conditions and degree of densification.
[b]Figures based on Ref. 22.
[c]Knoop hardness, 1-N load, Kgf/mm.
[d]Vickers hardness.

Titanium carbide is resistant to aqueous alkali except in the presence of oxidizing agents. It is resistant to acids except nitric acid, aqua regia, and mixtures of nitric acid with sulfuric or hydrofluoric acid. In oxygen at 450°C, a nonprotecting anatase coating forms. The reaction

$$3 \; CO_2 + TiC \longrightarrow 4 \; CO + TiO_2$$

occurs readily at 1200°C, and oxidation by CO also occurs. It is stable to hydrogen to 2400°C, but reacts with nitrogen at 1000°C.

Annual world production of titanium carbide is thousands of metric tons. It is manufactured mainly in-house by cutting-tool manufacturers by the reduction of titanium dioxide with carbon:

$$TiO_2 + 3 \; C \longrightarrow TiC + 2 \; CO$$

An intimate mixture of titanium dioxide is heated to ca 2000°C in an electric-arc furnace or graphite tube in a hydrogen atmosphere. An important mixed carbide constituent of cemented carbides is made by similarly treating a mixture of tungsten metal, TiO_2, and Ta_2O_5. The product is a carbide powder having a grain size of 1–5 μm, depending on the reaction conditions and starting materials (26,27). Suppliers of titanium carbide powder include Cerac and Adams Carbide in the United States, London and Scandinavian Metallurgical, Treibacher Chemische Werke, and Herman C. Starck in Europe, and Nippon Soda in Japan (see CARBIDES, CEMENTED CARBIDES).

A number of high temperature processes for the production of titanium carbide from ores have been reported (28,29). The aim is to manufacture a titanium carbide that can subsequently be chlorinated to yield titanium tetrachloride. In one process, a titanium-bearing ore is mixed with an alkali-metal chloride and carbonaceous material and heated to 2000°C to yield, ultimately, a highly pure TiC (28). Production of titanium carbide from ores, eg, ilmenite [12168-52-4], $FeTiO_3$, and perovskite [12194-71-7], $CaTiO_3$, has been described (30). A mixture of perovskite and carbon was heated in an arc furnace at ca 2100°C, ground, and then leached with water to decompose the calcium carbide to acetylene. The TiC was then separated from the aqueous slurry by elutriation. Approximately 72% of the titanium was recovered as the purified product. In the case of ilmenite, it was necessary to reduce the ilmenite carbothermally in the presence of lime at ca 1260°C. Molten iron was separated and the remaining $CaTiO_3$ was then processed as perovskite.

Titanium carbide may also be made by the reaction at high temperature of titanium with carbon; titanium tetrachloride with organic compounds such as methane, chloroform, or poly(vinyl chloride); titanium disulfide [12039-13-3] with carbon; organotitanates with carbon precursor polymers (31); and titanium tetrachloride with hydrogen and carbon monoxide. Much of this work is directed toward the production of ultrafine (<1 μm) powders. The reaction of titanium tetrachloride with a hydrocarbon–hydrogen mixture at ca 1000°C is used for the chemical vapor deposition (CVD) of thin carbide films used in wear-resistant coatings.

The primary commercial applications of titanium carbide are in wear-resistant components and cutting tools. Titanium carbide is added to alumina-based cutting tools to give improved thermal conductivity and hence improved resistance to thermal shock, as well as mechanical performance, such as strength, hardness, and toughness, compared with pure alumina ceramics. In cemented carbides, 20–50% of a carbide, typically a mixed TiC–WC–TaC phase, is bonded by a metal, usually a cobalt, a nickel, or an iron alloy. TiC-based cemented carbides bonded with nickel–molybdenum are also used. To reduce tool wear, many cemented carbide tool tips used for steel cutting are coated with a 5–10 μm TiC-containing layer produced by chemical vapor deposition. There is significant research interest in the production of titanium carbide fibers for composite materials. Manufacturers of TiC-based hard material include GTE, Greenleaf, and Kennametal in the United States, Sandvik in Europe, and Kyocera, Mitsubishi Metal, and Nippon Tungsten in Japan.

Titanium–Nitrogen Compounds

Nitrogen dissolves in metallic titanium up to a nitrogen content of <20 atom % (TiN$_{0.23}$). Above 30 atom % (TiN$_{0.42}$), a cubic titanium nitride phase is stable.

Titanium Nitride. Titanium nitride [25583-20-4] has the cubic NaCl structure, but the structure is stable over a wide range of either anion or cation deficiency (TiN$_{0.42}$, $a = 421$ pm; TiN$_{1.0}$, $a = 423.5$ pm). The nitride is a better conductor of electricity than titanium metal. It becomes superconductive at 1.2–1.6 K. Titanium nitride has a density of 5213 kg/m^3 and melts at 2950°C. Thermodynamic data are given in Table 2.

The powder, usually described as chocolate-brown, is actually bluish black when sufficiently fine (32). The ease of sintering depends on particle size. Normal powders require hot pressing at 1800°C or higher, and full density is difficult to achieve. Ultrafine powders (<0.1 μm) can be pressureless-sintered at temperatures as low as 1400°C without sintering aids. Sintered titanium nitride is a bronze color and can be polished to a golden lustre. TiN is thermodynamically unstable with respect to the oxide and, under normal conditions, titanium nitride powder is covered by a layer of chemisorbed oxygen. When it is heated in air, oxygen, nitric oxide, or carbon dioxide, it rapidly oxidizes. The susceptibility to oxidation depends on the degree of subdivision. It is resistant to attack by acids except boiling aqua regia, but is decomposed by alkalies with the evolution of ammonia.

Direct synthesis from nitrogen and finely divided titanium metal can be achieved at temperatures of >ca 1200°C (4). Typically, titanium sponge or powder is heated in an ammonia- or nitrogen-filled furnace and the product is subsequently milled and classified.

$$2\ \text{Ti} + \text{N}_2\ (\text{or NH}_3) \longrightarrow 2\ \text{TiN}\ (+1.5\ \text{H}_2)$$

A stoichiometric product can be obtained by repeated grinding and reaction. Alternatively, carbothermal reduction of titanium dioxide can be used (33). The reaction is carried out in an inert atmosphere at ca 1600°C.

$$2\ \text{TiO}_2 + 4\ \text{C} + \text{N}_2 \longrightarrow 2\ \text{TiN} + 4\ \text{CO}$$

The resulting titanium nitride forms a sintered mass, which must be subsequently milled to form a powder having a wide size distribution. The powders produced by these routes are typically $0.5-10$ μm, with a wide size distribution. Very fine powders ($0.005-0.5$ μm) have been prepared at pilot-plant scale by the reaction of $TiCl_4$ with a large excess of gaseous ammonia (32).

$$2\ TiCl_4 + 4\ NH_3 \longrightarrow 2\ TiN + 8\ HCl + N_2 + 2\ H_2$$

The reactants are fed into the tail flame of a d-c nitrogen plasma. The reaction occurs rapidly at temperatures around 1500°C and the HCl reacts with excess ammonia to form ammonium chloride. Similar reactions have been carried out using furnaces, lasers, and r-f plasmas (34) as the source of heat. Other routes using titanium tetrachloride starting material include

$$2\ TiCl_4 + N_2 + 4\ Fe \longrightarrow 2\ TiN + 4\ FeCl_2$$

$$2\ TiCl_4 + N_2 + 4\ H_2 \longrightarrow 2\ TiN + 8\ HCl$$

$$2\ TiCl_4 + N_2 + 8\ Na\ (or\ 4\ Mg) \longrightarrow 2\ TiN + 8\ NaCl\ (or\ 4\ MgCl_2)$$

A liquid-phase reaction in which $TiCl_4$ is reacted with liquid ammonia at -35°C to form an adduct that is subsequently calcined at 1000°C has also been proposed (35). Preparation of titanium nitride and titanium carbonitride by the pyrolysis of titanium-containing polymer precursors has also been reported (36).

TiN is used in cutting tools and wear parts, often in conjunction with TiC. Thus, the thermal conductivity, strength, toughness, and hardness of alumina-based cutting tools are improved by adding titanium carbide and titanium nitride at ca 30%. Because of its excellent electrical conductivity, TiN can be used as an additive to confer electrical conductivity on silicon nitride and sialon ceramic components. As a result, these can be shaped by spark erosion instead of by costly diamond grinding. Surface treatment of metals, often by chemical vapor deposition, as well as tungsten carbide cutting inserts, is widely employed to form a wear-resistant, low friction, titanium nitride surface layer.

Titanium Nitrate.   Titanium nitrate [*13860-02-1*], $Ti(NO_3)_4$, is a white powder, mp 58°C, in which the titanium achieves a coordination number of eight by being surrounded by four bidentate nitrate groups. It is stable in a sealed tube at room temperature, is less sensitive to moisture than titanium tetrachloride, and does not fume in air but reacts vigorously with water, liberating nitrogen oxides. It may be prepared by prolonged reaction of N_2O_5 on titanium tetrachloride at -60 to -20°C. This produces an intermediate compound that decomposes on warming in the presence of excess nitrogen dioxide to form NO_2Cl and titanium nitrate.

A more recent patent describes the production of titanyl nitrate by electrolysis of titanium tetrachloride or titanyl chloride (37). Other titanium nitrogen compounds that have been described include titanous amide [*15190-25-9*], $Ti(NH_2)_3$, titanic amide [*15792-80-0*], $Ti(NH)_2$, and various products in which amines have reacted with titanium tetrachloride (38).

Titanium Oxides

The titanium oxygen-phase diagram has been discussed (4). Metallic α-titanium can dissolve oxygen up to a composition of $TiO_{0.42}$, retaining the hexagonal structure but showing an increase in the lattice parameters. One consequence of the absorption of oxygen is that the transition temperature from α-titanium to the high temperature β-phase increases steeply from ca 900°C at 0 atom % to ca 1750°C at ca 15 atom %. As the oxygen content is increased above $TiO_{0.42}$, titanium oxides of increasing oxygen content are formed. The electronic structure of titanium oxides has been described (39,40).

Lower Oxides of Titanium. The properties of lower oxides of titanium are summarized in Table 4.

Titanium Monoxide. Titanium monoxide [12137-20-1], TiO, has a rock-salt structure but can exist with both oxygen and titanium vacancies. For stoichiometric TiO, the lattice parameter is 417 pm, but varies from ca 418 pm at 46 atom % to 416_2 pm at 54 atom % oxygen. Apparently, stoichiometric TiO has ca 15% of the Ti and O sites vacant. At high temperatures (>900°C), these vacancies are randomly distributed; at low temperatures, they become ordered. Titanium monoxide may be made by heating a stoichiometric mixture of titanium metal and titanium dioxide powders at 1600°C

$$Ti + TiO_2 \longrightarrow 2\ TiO$$

Alternative methods of production include reduction of TiO_2 with magnesium, which yields TiO only. When titanium monoxide is heated in air at 150–200°C, titanium sesquioxide, Ti_2O_3, forms, and at 250–350°C, it changes to Ti_3O_5.

Titanium Sesquioxide. Ti_2O_3 has the corundum structure. At room temperature it behaves as a semiconductor having a small (0.2 eV) band gap. At higher temperatures, however, it becomes metallic. This is associated with

Table 4. Properties of the Lower Oxides of Titanium

Property	TiO	Ti_2O_3	Ti_3O_5
color	golden yellow	violet	blue-black
density, kg/m³	4888	4486	421(0)
melting point, °C	1737	2127	
structure	fcc	hexagonal	monoclinic
lattice parameters, pm			
a	417	515.5	975.2
b		515.5	380.2
c		1316.2	944.2
solubility			
HF[a]	dissolves rapidly	dissolves	
HCl[b]	slow attack	no action	
H_2SO_4[b]	slow attack	slow attack	
HNO_3[b]	surface attack	no action	
NaOH[b]	slow attack	no action	

[a]Hot 40 wt %.
[b]Hot concentrated.

marked change in the mean Ti–Ti distance. As with TiO, titanium sesquioxide, Ti_2O_3, may be made by heating a stoichiometric mixture of titanium metal and titanium dioxide powders at 1600°C under vacuum in an aluminum or molybdenum capsule.

$$Ti + 3\ TiO_2 \longrightarrow 2\ Ti_2O_3$$

Trititanium Pentoxide. Trititanium pentoxide, Ti_3O_5, may be made by the reduction of titanium dioxide by hydrogen at 1300°C. The low temperature form is monoclinic. Above 177°C, Ti_3O_5 has a distorted pseudobrookite structure, which can be stabilized at lower temperatures by small amounts of iron.

Crystallographic Shear Structures. As with all oxides, the rutile form of titanium dioxide can lose oxygen from the lattice, eg, when heated in vacuum at >1000°C. The white oxide becomes first blue-grey and then a very dark blue or blue-black. At all but the very lowest oxygen deficiencies, such as for $x > 0.0005$ in TiO_{2-x}, the resulting point defects (oxygen vacancies) cluster along specific planes in the rutile lattice in such a way that a plane of oxygen atoms is eliminated. The resulting planar defects, which have replaced the point defects, are called crystallographic shear (CS) planes. Low degrees of reduction are accommodated on the {132} CS planes and greater degrees of reduction on the {121} CS planes. The transition occurs at compositions between $TiO_{1.93}$ and $TiO_{1.90}$. At higher degrees of reduction, these CS planes become ordered to give the oxide series Ti_nO_{2n-1}, sometimes called Magnelli phases, where n is from 4 to 9 or 10. These phases have been extensively studied in both pure rutile and rutile doped with transition metal ions (39,41–43). At temperatures above 150 K, Ti_4O_7 acts as a metallic conductor, and reduction of rutile to oxides such as Ti_4O_7 gives a conducting ceramic known as Ebonex, which has been proposed as an electrode material in a wide variety of applications (44).

Hydrated Titanium Oxides. Hydroxides of Ti(II) (black) and Ti(III) (brown) are precipitated when an alkali metal hydroxide is added to a solution of the corresponding salt. These precipitates, though difficult to purify (45), are powerful reducing agents and readily oxidize in air to form a hydrated titanium dioxide.

Hydrolysis of solutions of Ti(IV) salts leads to precipitation of a hydrated titanium dioxide. The composition and properties of this product depend critically on the precipitation conditions, including the reactant concentration, temperature, pH, and choice of the salt (46–49). At room temperature, a voluminous and gelatinous precipitate forms. This has been referred to as orthotitanic acid [20338-08-3] and has been represented by the nominal formula $TiO_2 \cdot 2H_2O$ ($Ti(OH)_4$). The gelatinous precipitate either redissolves or peptizes to a colloidal suspension in dilute hydrochloric or nitric acids. If the suspension is boiled, or if precipitation is from hot solutions, a less-hydrated oxide forms. This has been referred to as metatitanic acid [12026-28-7], nominal formula $TiO_2 \cdot H_2O$ ($TiO(OH)_2$). The latter precipitate is more difficult to dissolve in acid and is only soluble in concentrated sulfuric acid or hydrofluoric acid.

Precipitation of a hydrated titanium oxide by mixing aqueous solutions of titanium chloride with alkali forms the precipitation seeds, which are used to initiate precipitation in the Mecklenburg (50) variant of the sulfate process for the production of pigmentary titanium dioxide. Hydrolysis of aqueous solutions of titanium chloride is also used for the preparation of high purity (>99.999%) titanium dioxide for electroceramic applications (see CERAMICS). In addition, hydrated titanium dioxide is used as a pure starting material for the manufacture of other titanium compounds.

The properties of hydrated titanium dioxide as an ion-exchange (qv) medium have been widely studied (51–55). Separations include those of alkali and alkaline-earth metals, zinc, copper, cobalt, cesium, strontium, and barium. The use of hydrated titanium dioxide to separate uranium from seawater and also for the treatment of radioactive wastes from nuclear-reactor installations has been proposed (56).

Titanium Dioxide. *Physical and Chemical Properties.* Titanium dioxide [13463-67-7] occurs in nature in three crystalline forms: anatase [1317-70-0], brookite [12188-41-9], and rutile [1317-80-2]. These crystals are essentially pure titanium dioxide but contain small amounts of impurities, such as iron, chromium, or vanadium, which darken them. Rutile is the thermodynamically stable form at all temperatures and is one of the two most important ores of titanium. Large deposits of anatase-bearing ore occur in Brazil.

Anatase and rutile are produced commercially, whereas brookite has been produced by heating amorphous titanium dioxide, which is prepared from an alkyl titanate or sodium titanate [12034-34-3] with sodium or potassium hydroxide in an autoclave at 200–600°C for several days. Only rutile has been synthesized from melts in the form of large single crystals. More recently (57), a new polymorph of titanium dioxide, $TiO_2(B)$, has been demonstrated, which is formed by hydrolysis of $K_2Ti_4O_9$ to form $H_2Ti_4O_9 \cdot H_2O$, followed by subsequent calcination/dehydration at 500°C. The relatively open structure may be considered to be formed by the removal of K_2O from $K_2Ti_4O_9$. In addition, a high pressure polymorph, $TiO_2(ii)$, has been made, which possesses the orthorhombic α-PbO_2 structure and has a density about 2.5% greater than that of rutile. A metastable product has been formed from rutile crystals shocked to pressures greater than 330 kbar, or by heating rutile at pressures from 40 to 120 kbar at temperatures of 400–1500°C (58,59). The phase has also been prepared when Ti_3O_5 is dissolved in sulfuric acid at elevated temperatures (60). The existence of a fluorite or distorted fluorite-type structure at very high pressures, ie, >20 GPa (200 kbar), has also been postulated (61), and structural parameters of rutile and anatase have been reviewed (62). Crystallographic information on the different forms of titanium dioxide is summarized in Table 5.

It is accepted that, at normal pressures, rutile is the thermodynamically stable form of titanium dioxide at all temperatures. Calorimetric studies have demonstrated that rutile is more stable than anatase and that brookite and $TiO_2(ii)$ have intermediate stabilities, although the relative stabilities of brookite and $TiO_2(ii)$ have not yet been defined. The transformation of anatase to rutile is exothermic, eg, 12.6 KJ/mol (9), although lower figures have also been reported (63). The rate of transformation is critically dependent on the detailed environment and may be either promoted or retarded by the presence of other substances. For example, phosphorus inhibits the transformation of anatase to rutile (64).

Table 5. Different Forms of Titanium Dioxide

Property	TiO$_2$ (B)	Anatase	Brookite	Rutile	α-PbO$_2$ form
crystal structure	monoclinic	tetragonal	orthorhombic	tetragonal	orthorhombic
number of TiO$_2$/unit cell	8	4	8	2	4
space group		141/amd	Pbca	P42/mnm	Pbcm
lattice parameters, pm					
a	1216.4	378.45	917.4	459.37	455(0)
b	373.5		544.9		547(0)
c	651.3	951.43	513.8	295.87	490(0)
volume per TiO$_2$, nm^3	0.03527	0.03407	0.03211	0.03122	0.03049
theoretical density, kg/m^3		3895	4133	4250	4350
Ti–O distance, pm		2 at 191, 4 at 195	184–203	4 at 194.4, 2 at 198.8	4 at 191, 2 at 205
hardness, Mohs' scale		5.5–6	5.5–6	7–7.5	

The commercially important anatase and rutile both have tetragonal structures; consequently, the values of physical properties such as refractive index and electrical conductivity depend on whether these are being measured parallel or perpendicular to the principal, ie, c, axis. However, in most applications, this distinction is lost because of random orientation of a large number of small crystals. It is thus the mean value that is significant. Representative physical properties are collected in Table 6.

Both anatase and rutile are broad band gap semiconductors in which a filled valence band, derived from the O $2p$ orbitals, is separated from an empty conduction band, derived from the Ti $3d$ orbitals, by a band gap of ca 3 eV. Consequently the electrical conductivity depends critically on the presence of impurities and defects such as oxygen vacancies (7). For very pure thin films, prepared by vacuum evaporation of titanium metal and then oxidation, conductivities of 10^{-13} S/cm have been reported. For both single-crystal and ceramic samples, the electrical conductivity depends on both the state of reduction of the TiO$_2$ and on dopant levels. At 300 K, a maximum conductivity of 1 S/cm has been reported at an oxygen deficiency of 10^{19} cm^{-3}. Reduction can be brought about by heating under vacuum and is accompanied by a color change through light to dark blue, which is reversed by heating in air. Partial reduction may also be achieved by heating in very low partial pressures of oxygen. Five valent dopants, such as niobium, are able to act as electron donors and increase the conductivity of both single-crystal rutile and rutile ceramic at dopant levels of ca 1%. Factors of between 20 and 1000 have been reported, depending on the measurement temperature and sample type. Dopants such as chromium and aluminum reduce the conductivity by a factor of ca 0.5 (5).

Table 6. Physical Properties of Anatase and Rutile[a]

Property	Anatase	Rutile
refractive index, 550 nm	2.54[b]	2.75[b]
dielectric constant		
static	48[b]	114[b]
parallel to c axis		170
perpendicular to c axis		86
high frequency		7.37[b]
parallel to c axis		8.43
perpendicular to c axis		6.84
band gap, eV	3.25	3.05
melting point, °C	converts to rutile	1830–1850
electrical conductivity, S/cm		
parallel to c axis		
30°C		10^{-13}
227°C		10^{-6}
perpendicular to c axis		
30°C		10^{-10}
227°C		10^{-7}
breakdown voltage, mV/m		15.2/17.8
hardness, Mohs' scale	5.5–6	7–7.5

[a]Refs. 5, 7, 65, and 66.
[b]Weighted mean values.

The reactivity of titanium dioxide toward acid is dependent on the temperature to which it has been heated. Freshly precipitated titanium dioxide is soluble in concentrated hydrochloric acid. However, titanium dioxide that has been heated to 900°C is almost insoluble in acids except hot concentrated sulfuric, in which the solubility may be further increased by the addition of ammonium sulfate to raise the boiling point of the acid, and hydrofluoric acid. Similarly, titanium dioxide that has been calcined at 900°C is almost insoluble in aqueous alkalies but dissolves in molten sodium or potassium hydroxide, carbonates, or borates.

Preparation. Normally, the first stage in the preparation of pure titanium dioxide is repeated distillation of titanium tetrachloride. A number of different routes may then be followed, the choice depending on the use for which the titanium dioxide is required.

Hydrolysis in aqueous solution precipitates hydrated titanium dioxide which, after washing and drying, can be calcined at 800°C to remove water and residual Cl. This method has been the basis of producing titanium dioxide of 99.999% purity. If retaining a high specific surface area is important, it may be convenient to reduce the residual Cl content by Sohxlet extraction rather than calcination. For many electroceramic uses, the millability of the product is as important as the absolute purity. A number of process variants designed to modify particulate morphology have been described (67). Titanium dioxide with silica, magnesium, and iron contents as low as 10^{-5} wt % can be achieved by converting the initial hydrous titanium dioxide precipitated from $TiCl_4$ into the double oxalate [*10580-02-06*] $(NH_4)_2[TiO(C_2O_4)_2]$, recrystallizing this from methanol, and then forming the oxide by calcination (68). Alternatively, the titanium tetrachloride may be converted into an alkoxide, eg, the often-used titanium isopropoxide [*546-89-9*] (isopropyl titanate), which is then hydrolyzed, washed, and dried. The hydrolysis products have a well-defined spherical morphology but may retain considerable amounts of organic residues (69).

A high purity titanium dioxide of poorly defined crystal form (ca 80% anatase, 20% rutile) is made commercially by flame hydrolysis of titanium tetrachloride. This product is used extensively for academic photocatalytic studies (70). The gas-phase oxidation of titanium tetrachloride, the basis of the chloride process for the production of titanium dioxide pigments, can be used for the production of high purity titanium dioxide, but, as with flame hydrolysis, the product is of poorly defined crystalline form unless special dopants are added to the principal reactants (71).

Nonpigment Uses. Sales of titanium oxide for various nonpigmentary applications are of the order of 100,000 t/yr. The traditional ceramic applications of titanium dioxide are in vitreous enamels, thread guides for the fibers industry, and electroceramics. In enamels used on steel and cast iron, titanium dioxide makes up part of the raw-material batch and dissolves during frit smelting. As the enamel cools and fuses to its substrate, the titanium dioxide recrystallizes and confers both improved opacity and acid resistance. To avoid color effects, elements such as tungsten and niobium must be carefully controlled. For ease of handling, it is convenient to use a coarse grade of TiO_2. Thread guides for the fiber industry are made by firing anatase or rutile at ca 1300°C to produce a dense ceramic that can be polished to allow easy drawing of the fiber over the guides during spinning. Friction between the threads and the guides generates static electricity. To dissipate this, the electrical conductivity of the fired pieces is increased by a second firing under reducing conditions. In a separate application area, reduction of rutile to lower oxides such as Ti_4O_7 gives a conducting ceramic known as Ebonex, which has potential as an electrode material (44). The preparation, stability, and properties of titania-based ceramic membranes constitute an area of active research (72).

Titanium dioxide may be used in varistors, electrical circuit elements for which $I = KV^a$, where $a > 1$. TiO_2 varistors find particular use for low voltage applications in arc prevention in small motors. Titanium dioxide may also be used as a ceramic sensor for oxygen in automotive exhaust systems. Feedback from the detector controls the air–fuel ratio to minimize pollution. The basis of the device is the sensitivity of the resistance of titania to the oxygen partial pressure, because the removal of oxygen from the titania increases the number of electrons in the conduction band. The rapid response time, about 40 ms, over the full range of oxygen pressures suggests that the measured changes are dominated by surface effects (see SENSORS) (73).

The main electroceramic applications of titanium dioxide derive from its high dielectric constant (see Table 6). Rutile itself can be used as a dielectric in multilayer capacitors, but it is much more common to use TiO_2 for the manufacture of alkaline-earth titanates, eg, by the cocalcination of barium carbonate and anatase. The electrical properties of these dielectrics are extremely sensitive to the presence of small (<20 ppm) quantities of impurities, and high performance titanates require consistently pure (eg, >99.9%) TiO_2. Typical products are made by the hydrolysis of high purity titanium tetrachloride.

Titanium dioxide is also used as an inorganic sunblock (74,75). For this application, the particle size is controlled to minimize the scattering of visible light, hence the whiteness of the product, and to maximize the scattering and absorption of ultraviolet radiation. This requires particles ca 50 nm in size, called ultrafine TiO_2. A particular advantage of titanium dioxide as a sunblock is that it is a broad-spectrum uv attenuator and therefore allows the formulation of skin-care products that protect not only from the uv-B radiation, which causes sunburn and erythema, but also from uv-A, which causes skin wrinkling and aging. Ultrafine titanium dioxide may also be used as a transparent uv barrier in applications such as food packaging, where the resistance to migration of the inorganic absorber is particularly desirable. Ultrafine titanium dioxide may be made by both aqueous- and gas-phase routes.

Titanium dioxide finds limited, specialist application as a catalyst or catalyst support. Vanadia/titania has been used as a selective oxidation catalyst, eg, for the oxidation of o-xylene to phthalic anhydride (76,77). Tungsten oxide on TiO_2 is a preferred catalyst for the selective catalytic reduction of nitrogen oxides, NO_x, in the gaseous emissions from power stations (78). Activity for the removal of hydrogen sulfide and carbon disulfide has been claimed (79). Anatase has received much research interest in the photocatalytic degradation of pollutants, eg, to remove trace amounts of pesticide residues from water (80,81) (see PHOTOCHEMICAL TECHNOLOGY, PHOTOCATALYSIS). Academic research on photo-assisted electrochemical splitting of water to form hydrogen, a nonpolluting fuel, and oxygen (82,83) has waned. However, in 1991, this general area was given a new and different emphasis by the report of a titanium dioxide-based photo-voltaic device capable of quantum efficiencies of ca 10% (84).

Mineral rutile is used as an ingredient of welding-rod coatings, for which impurities such as iron are acceptable. Annual U.S. sales are several thousand metric tons.

Pigmentary Titanium Dioxide. The most important commercial use of titanium dioxide is as a white pigment in a wide range of products, including paint (qv), plastics, paper (qv), and inks (qv) (8,85,86). Because of its opacifying properties, TiO_2 is used not just in whites but to opacify colored systems also. Titanium dioxide is the predominant white pigment both because of its high refractive index and because technology has been developed to allow the right size range and the necessary chemical purity.

In order to appear white, a pigment must have minimal optical absorption at visible wavelengths. For titanium dioxide, this in turn requires high chemical purity; in particular, transition-metal impurities must be eliminated as far as possible. Because titanium dioxide pigments do not significantly absorb visible light, their opacity derives from their ability to scatter light. Light scattering by particles depends on their refractive index. A measure of the effect of refractive index on opacity can be achieved by using a simplified form of an equation introduced by Fresnel, which gives the reflectivity, R, of a coating by

$$R = \frac{(n_p - n_m)^2}{(n_p + n_m)^2}$$

where n_p is the refractive index of the pigment and n_m the refractive index of the medium in which the pigment is dispersed. This is 1 for air and typically ca 1.5 for an alkyd paint resin. Hence, the higher the value of the pigment refractive index, the higher its potential opacity. Both anatase and rutile are used as white pigments, and scatter light more efficiently than other oxides (Table 7). Because of its higher refractive index, rutile has a higher opacity than anatase and is more widely used. Anatase is, however, used for many specialized applications, eg, in paper and in fibers, where the lower abrasivity of anatase causes less wear on thread guides and other spinning equipment. In order to realize the maximum scattering power, the pigment particles must be of the correct size, ie, light-scattering is a function of d/λ, where d is the diameter of the scattering particle and λ the wavelength in a particular medium of the light. Maximum scattering of visible (400–700 nm) light occurs for pigment particles of ca 300 nm and the scattering efficiency decreases rapidly as the particle size varies from the optimum. This decrease is by about 60% for particles twice the optimum size. Therefore careful control of particle size is necessary. Two further desirable properties are that titanium dioxide is nontoxic and that its density (ca 4000 kg/m^3) is sufficiently low as not to cause excessively rapid settling of the pigment particles in paint. Although titanium dioxide does not absorb visible light, it is a strong absorber of uv radiation. In paints opacified with titanium dioxide, the pigment protects the organic film-forming molecules from photochemical degradation by solar uv. However, absorption of uv by TiO_2 can lead to the generation of hydroxyl radicals, which cause the pigment to act as a photocatalyst. The net effect of the pigments on the uv stability, ie, the durability, of a paint is a balance between the protective and photocatalytic effects. In order to ensure that the actual effect is beneficial, the photocatalytic effect of the pigment is minimized by coating the pigment with hydrous oxides, eg, silica and alumina, during manufacture. This coating treatment is also applied to facilitate good pigment dispersion in the medium of application. Good dispersion ensures that the pigment scatters light efficiently and does not contribute to any surface roughness that would otherwise reduce the gloss of, eg, the final paint.

Table 7. Refractive Indexes and Reflectivities of White Pigments

Pigment	Refractive index, n_p	Reflectivity, R, in air ($n_m = 1$)		Reflectivity, R, in alkyd resin ($n_m = 1.5$)	
		$R \times 100$	relative to rutile = 100	$R \times 100$	relative to rutile = 100
rutile	2.72	21.4	100.0	8.36	100.0
anatase	2.55	19.1	89.2	6.72	80.4
antimony oxide	2.20	14.1	65.8	3.58	42.8
zinc oxide	2.01	11.3	52.7	2.11	25.3
lithopone	1.84	8.7	40.9	1.04	12.4
kaolin	1.56	4.8	22.4	0.04	0.5

Two pigment production routes are in commercial use. In the sulfate process, the ore is dissolved in sulfuric acid, the solution is hydrolyzed to precipitate a microcrystalline titanium dioxide, which in turn is grown by a process of calcination at temperatures of ca 900–1000°C. In the chloride process, titanium tetrachloride, formed by chlorinating the ore, is purified by distillation and is then oxidized at ca 1400–1600°C to form crystals of the required size. In both cases, the raw products are finished by coating with a layer of hydrous oxides, typically a mixture of silica, alumina, etc.

A hydrochloric acid process for the manufacture of anatase has been proposed but has not been developed. Other routes include fluoride, bromide, nitrate, sulfide, and chloroacetate processes (1). None of these, however, has been used successfully on a commercial scale.

Mineral Feedstocks. The titanium-bearing ores ilmenite [12168-52-4], natural rutile, and leucoxene [1358-95-8] used in the production of titanium dioxide pigment occur as mineral sands and massive hard rock in many parts of the world (87). Increasingly, mining companies process these natural ores to extract iron and other minerals to produce slags, or synthetic rutile having higher TiO_2 contents than the original ore. World production of the various types of titanium ore is summarized in Table 8. About 97% of these TiO_2 concentrates were used for pigment production. The choice of ore depends on the production process used. Ilmenites can be attacked by sulfuric acid, the first step in the sulfate process. Ilmenite is ideally ferrous titanate, $FeTiO_3$, having an ideal composition of 52.7% TiO_2 and 47.3% FeO, but usually includes some ferric iron. It can occur either in massive form, for instance, in Norway and Finland, or as a constituent of ancient beach sands, which are not necessarily on the present coast line, for example, in Florida, southern India, and Australia. Ilmenite in sands is often much altered by oxidation and leaching. The ratio of ferric to ferrous iron increases and the total amount of iron falls. Consequently, the amount of titanium dioxide increases. On further weathering, the form of titanium dioxide also changes, passing through leucoxene [1358-95-8] and pseudorutile [1310-39-0] stages to rutile. Typical analyses of TiO_2 process feedstocks are collated in Table 9. Ilmenites from rock deposits typically contain 45% TiO_2, but beach sand ilmenite deposits, because of oxidation/hydrolysis and expulsion of Fe_2O_3, can contain more than 55% TiO_2. These are widely used in the sulfate process, although extreme weathering (>60% TiO_2) can reduce their reactivity in sulfuric acid, and hence their usefulness. Ores containing high levels of transition-metal impurities are not used because of the deleterious effects of these impurities on color. Strict limits are imposed on thorium and uranium levels for radiological reasons. Neither leucoxene, pseudorutile, nor rutile are attacked by sulfuric acid under the conditions normally used for pigment manufacture.

Although ilmenites and leucoxenes can be used in the chloride process, ores having higher TiO_2 contents, eg, mineral rutile, which is not readily attacked by sulfuric acid, are preferred in order to minimize loss of chlorine in iron chloride by-product.

Both processes also use up-graded ilmenite (slags). About 30% of the world's titanium feedstocks are supplied by titanium slag producers in Canada, South Africa, and Norway. Slags are formed by the high temperature reduction of

Table 8. Production of Titanium Minerals, t $\times$ 10³

Country	1980	1985	1990	1995	Total 1995 production, %
		Rutile			
Australia	294	261	245	246	4.50
CIS	10	10	10	10	0.18
India and Sri Lanka	16	15	17	17	0.31
Sierra Leone	28	145	145	10	0.18
South Africa	48	75	75	110	2.01
United States	25	25	25	25	0.46
Total	*421*	*531*	*517*	*418*	*7.64*
		Titaniferous slag			
Canada	876	845	860	710	12.98
Norway	0	0	40	162	2.96
South Africa	344	700	658	820	14.99
Total	*1220*	*1545*	*1558*	*1692*	*30.94*
		Synthetic rutile			
Australia	40	54	358	515	9.42
India	20	20	60	62	1.13
Japan	46	46	46	12	0.22
United States	80	90	100	125	2.29
Total	*186*	*210*	*564*	*714*	*13.06*
		Ilmenite			
Australia	1203	989	728	768	14.04
Brazil	15	76	145	145	2.65
China	140	140	140	184	3.36
CIS	400	400	300	280	5.12
Finland	159	53	0	0	0.00
India and Sri Lanka	216	223	230	213	3.89
Malaysia	189	300	250	186	3.40
Norway	830	642	678	514	9.40
Sierra Leone	0	0	62	5	0.09
Thailand	5	5	10	10	0.18
United States	340	340	340	340	6.22
Total	*3497*	*3168*	*2883*	*2645*	*48.36*
Total world	*5324*	*5454*	*5522*	*5469*	*100.00*

ilmenites in electric furnaces. Much of the iron oxide content is reduced to metallic iron and separated as a saleable by-product. Magnesium and other impurities may also be incorporated in the following equations.

$$2\ FeTiO_3 + C^0 \longrightarrow FeTi_2O_5 + Fe^0 + CO$$

$$3\ FeTi_2O_5 + 5\ C^0 \longrightarrow 2\ Ti_3O_5 + 3\ Fe^0 + 5\ CO$$

The resultant slag, a complex mixture of titanates, may contain 70–85% TiO_2. The slag route is particularly useful when ilmenite is closely associated with haematite, from which it cannot economically be separated mechanically.

Table 9. Analyses of Titanium Dioxide Process Feedstocks

Component, wt %	Ilmenite			Slag		Rutiles		Synthetic rutiles	
	Massive	Beach sand							
	Norway	West Australia	United States, Florida	Canada	South Africa	West Australia	South Africa	West Australia	West Australia
TiO_2	44.3	54.77	64.7	77.5	85.8	95.5	94.5	93	92.3
FeO	34.8	19.27		10.9	10.5				
Fe_2O_3	11.5	21.2	30				0.8		
Fe metal				0.5	0.1			0.85	
total iron				9	8.3			3	2.48
Cr_2O_3	0.075	0.04	0.17	0.17	0.17	0.15	0.14	0.07	0.12
V_2O_5	0.2	0.14	0.15	0.56	0.44	0.55	0.33	0.3	0.26
SiO_2	2.6	0.78	0.45	3	2.1	0.8	2	0.9	1.14
Al_2O_3	0.6	0.55	0.8	3.5	1.3	0.3	0.6	1.2	1.04
MgO	4.3		0.25	5.3	1	0.03	0.03	0.35	0.46
CaO	0.25	0.01	0.1	0.6	0.15	0.01	0.04	0.01	0.04
MnO	0.3	1.57	1.6	0.25	1.7	0.01	0.01	1	0.87
ZrO_2	0.02	0.08	0.13			0.95	1.1	0.5	0.14
Nb_2O_5	0.01	0.15		0.01		0.35	0.4	0.2	0.25
P_2O_5	0.018	0.04	0.14			0.02	0.04	0.05	0.02
Th^a	<10	100		1.2	16		40		
U^a	<10	10		0.7	2		65		

^{a}Values are ppm.

243

Because the iron content of the slag is low, its use reduces the quantity of iron sulfate in the liquid effluent of sulfate process plants. Slag used as a feedstock for $TiCl_4$ production must be low in magnesium and calcium. A variety of other ilmenite beneficiation or synthetic rutile processes have been pursued, primarily to provide alternative chloride process feedstocks. Low grade ilmenite is purified mainly by leaching at high temperatures with either sulfuric or hydrochloric acid to remove iron. Typically, synthetic rutile contains about 94% TiO_2.

Manufacture. There are two routes for the manufacture of raw pigmentary titanium dioxide. Both processes provide high quality products having good color (chemical purity) and opacity (mean particle size and narrow size distribution). Almost all pigments are then finished by coating with inorganic oxides and phosphates in order to control such properties as dispersion, dispersion stability, gloss, and durability.

The Sulfate Process. A flow diagram for the sulfate process is shown in Figure 1. The strongly exothermic digestion of the dried, milled feedstock in 85–95% sulfuric acid converts metal oxides into soluble sulfates, primarily titanium and iron.

$$FeTiO_3 + 2\ H_2SO_4 \longrightarrow TiOSO_4 + FeSO_4 + 2\ H_2O$$

$$Fe_2TiO_5 + 4\ H_2SO_4 \longrightarrow TiOSO_4 + Fe_2(SO_4)_3 + 4\ H_2O$$

More than stoichiometric amounts of acid are used and approximately 95% of the titanium is extracted. The excess acid stabilizes the resultant titanium solutions and influences the rate of subsequent hydrolysis. Unreacted ore residues are removed by settling and filtration after any ferric iron in solution has been reduced to ferrous by treatment with metallic iron, a step that is not necessary for slag-derived liquor. A portion of the iron in solution is then removed as $FeSO_4 \cdot 7\ H_2O$ (copperas) by crystallization and filtration. Subsequent concentration by evaporation yields a clarified liquor of carefully controlled composition (Ti, Fe, acid, and water). Controlled heating and hydrolysis then initiates precipitation of finely divided titanium dioxide (pulp). In one variant (Mecklenburg) of the process, this precipitation is seeded with externally prepared nuclei. In another (Blumenfeld) variant, seeds are generated *in situ* but further, rutile nuclei are added prior to calcination. In both the Mecklenburg and the Blumenfeld processes, the initial precipitate is microcrystalline anatase but the nuclei control both crystal growth during calcination and the ultimate crystal form (anatase or rutile) of the product. The precipitation stage is followed by leaching, filtration, and washing to remove residual metal impurities, eg, to reduce Fe to below 30 ppm. Inorganic salts are added at this point to modify crystallization, crystal growth, and durability. The additioned pulp is then calcined in large rotary, oil- or gas-fired kilns having discharge temperature of ca 900°C. Residence times may be 18–24 hours, during which water and adsorbed acid/sulfates are removed, while crystal growth and, in some cases, crystal transformation to rutile occur. After calcination, the cooled product is milled and may be either coated to improve its application properties or (decreasingly) packed directly. There are some signs that the comparatively slow crystal growth in the sulfate process is able to generate narrower particle size distributions than the chloride process.

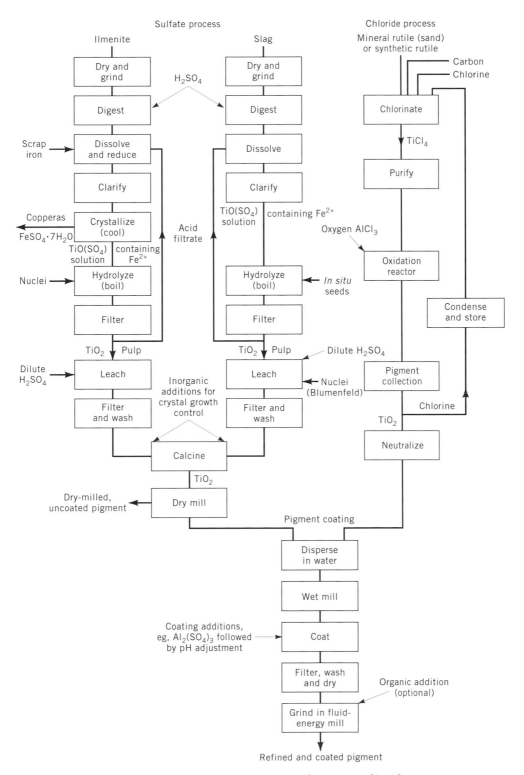

Fig. 1. Flow diagram for the manufacture of titanium dioxide pigments.

245

Since 1970, there has been virtually no change in the global sulfate-route pigment production capacity (Table 10). One reason for this is that wastes from the TiO_2 industry have received special and unfavorable attention from environmental and governmental agencies, especially in the United States and Western Europe. The main manufacturers of TiO_2 by the sulfate route have, in the late 1980s and early 1990s, made significant and successful efforts to introduce a variety of processes to meet the resulting environmental requirements. These have included concentration and recycle of waste acids, roasting of metal sulfates to recover the SO_3 values, and neutralization of waste acid with chalk or lime. Corresponding co-products have included production and sale of copperas for water treatment and production and sale of white gypsum for, eg, wall boards. As a result of these developments, there is no longer an inherent difference between the environmental acceptability of the chloride and the sulfate processes.

Chloride Process. A flow diagram for the chloride process is shown in Figure 1. The first stage in the process, carbothermal chlorination of the ore to produce titanium tetrachloride, is carried out in a fluid-bed chlorinator at ca 950°C. If mineral rutile is used as the feedstock, the dominant reaction is chlorination of titanium dioxide.

$$2\ TiO_2 + 4\ Cl_2 + 3\ C^0 \longrightarrow 2\ TiCl_4 + 2\ CO + CO_2$$

The exothermic oxidation reaction is carried out in the gas phase at temperatures of 1200°C or higher. Relevant thermodynamic data are given in Table 11.

$$TiCl_4\ (g) + O_2\ (g) \longrightarrow TiO_2\ (s) + 2\ Cl_2\ (g)$$

Table 10. Annual World Capacity for Titanium Dioxide Pigment Production, t $\times$ 10^3

Region	1971		1981		1996	
	Sulfate	Chloride	Sulfate	Chloride	Sulfate	Chloride
Central and South America	34		34	35	50	100
Canada and United States	497	227	306	705	113	1392
European Union	737	103	795	121	812	405
Finland and Eastern Europe	179		264		317	
China	12		15		83	
Saudi Arabia						56
Japan and the Pacific Rim	210		258	24	371	155
Australia	43		68			159
South Africa	22		28		43	
Total	*1734*	*330*	*1768*	*885*	*1789*	*2267*

Table 11. Energy Changes in the Exothermic Oxidation Reaction^a

Temperature, °C	ΔH, kJ/mol^b	ΔG, kJ/mol^b
827	−174.7	−116.0
1027	−174.7	−105.4
1327	−170.9	−90.0

^{a}TiCl$_4$ (g) + O$_2$ (g) → TiO$_2$ (rutile, solid) + 2 Cl$_2$ (g).
^{b}To convert kJ to kcal, divide by 4.184.

Aluminum chloride is added at this stage to ensure that the product is rutile; in its absence a mixture of anatase and rutile is obtained. Unlike the slow calcination stage in the sulfate process, the nucleation, growth, and rutilization steps are essentially complete in ca 10 ms. Careful design of the oxidation burner is critical for preventing excessive growth of product on the reactor walls and for achieving satisfactory control of crystal mean size. The hot gas stream is cooled and the product is filtered and slurried. Chlorine is recycled to the initial chlorination stage. As with the sulfate process, the raw pigment is further treated.

Since 1970, the global production capacity by the chloride route has increased approximately eightfold, largely because of the lower capital costs associated with this process. A 1996 trade press report estimated the cost of every additional ton of capacity as $2000–$2500 (88). The process is continuous and the product normally has a better color than that from the sulfate process. The disadvantages of the chloride process include the need for high grade feedstocks and the potential hazards associated with large amounts of chlorine and titanium tetrachloride. The chloride process is not able to produce anatase pigment on a large scale and in certain applications the greater abrasivity of chloride pigments can be a disadvantage. Chlorination of low Ti ores leads to iron chloride wastes that must be disposed of. In the United States, deep wells are often used but this may not always be allowed.

Pigment Finishing. Rutile pigments produced by the chloride and sulfate routes are basically similar and require coating for the same reasons, ie, to optimize dispersibility, dispersion stability, opacity, gloss, and durability. The coating techniques are essentially common to sulfate and chloride-route base pigments. Because the treatments are tailored to the requirements of the final application, the details are specific to the needs of the different market sectors, eg, paints, plastics, and paper. Generally, rutile pigments have between 1 and 15% inorganic coating, the higher coating levels typically used for applications such as mat emulsion paints. Anatase pigments generally have lower coating levels of 1 to 5%. The first finishing stage is to disperse the base pigment in water, generally with phosphate, silicate, and/or organic dispersants. The resulting suspensions may then be milled and/or classified to remove oversize particles. The dispersed particles are then coated by selective precipitation of (usually) small quantities of colorless hydrous oxides, eg, P$_2$O$_5$, SiO$_2$, Al$_2$O$_3$, TiO$_2$, and ZrO$_2$ deposited via controlled changes in pH and temperature. Essentially, the processes are of the type

$$\text{aluminum salt + alkali} \longrightarrow \text{hydrated aluminum oxide}$$

or

$$silicate + acid \longrightarrow hydrated\ silica$$

but the detailed chemistry to ensure that deposition occurs on the pigment surface and not as a bulk precipitate is more complex (89) and not completely understood. Specific examples of coating types are available (90,91). In this way, the pigment surface properties are modified to give the required application properties (92,93). After the coating stage, the pigment that has, perforce, flocculated during the coating process, is filtered, washed, and dried. It is then milled once more, usually in a steam-fluid-energy mill (micronizer) to break up the flocculates. At the same time, the pigment is additioned with an organic surface treatment, eg, polyol or alkanolamine on pigments designed for use in paints and printing inks or siloxane on pigments for plastics. Finally, the pigment is filtered and packed.

Significant effort has been directed toward encapsulating titanium dioxide using an organic polymer surface layer instead of an inorganic coating. Although there are numerous patents and research papers (94) on this topic, no product has been commercialized as of this writing (1997). Manufacturers produce many grades of titanium dioxide pigments and their literature should be consulted for recommendations on the grade to be used for any particular purpose.

Colored Pigments.   Colored pigments may be derived from titanium dioxide by substituting some of the titanium in the rutile lattice by small amounts of transition-metal ions. The most important are a yellow pigment formed by introducing nickel ions, and a buff pigment formed by introducing chromium. Of lesser importance are cobalt titanate green and iron titanate brown, both of which have the spinel structure (95,96).

For rutile-derived pigments, it is important to maintain charge balance. This is accomplished by adding antimony or niobium, both of which can be thought of as forming $5+$ ions to compensate for the charge deficit that would otherwise result from the replacement of Ti^{4+} by either Ni^{2+} or Cr^{3+}. Thus, a typical composition for nickel titanate yellow is $NiO \cdot Sb_2O_5 \cdot 20TiO_2$; for chrome titanate buff, $Cr_2O_3 \cdot Sb_2O_5 \cdot 31TiO_2$. The pigments are made by mixing, either wet or dry, stoichiometric quantities of the oxides, or oxide precursors, calcining at ca 1000°C and then milling. Both pigments have excellent hiding power but poor tinting strength compared to organic alternatives. The principal advantage of these pigments is the excellent light-fastness, which makes them particularly suited for use in vinyl sidings. In this application, they have the additional advantage that deterioration of the plastic by weathering does not cause a change in color. In 1984, world sales of nickel titanate yellows and chrome titanate buffs were ca 1500 and 2000 t/yr, respectively. A number of pearlescent pigments based on mica coated with titanium dioxide have also been manufactured (97).

Production and Shipment.   A list of the manufacturers of titanium dioxide pigments and their trade names is given in Table 12, together with production estimates for the six largest manufacturers: Du Pont, Tioxide, Millenium, Kronos, Kemira, and Ishihara. Collectively, these six accounted for approximately two-thirds of the 1996 world pigment production, ca 4×10^6 t. The distribution of titanium dioxide pigments among industries in the United States is shown in Table 13.

Table 12. Manufacturers of Titanium Dioxide Pigment

Manufacturer	Trade name[a]	Country
Bayer AG	Bayertitan	Germany, Belgium, Brazil
China Metal & Chemical Co.	Tai Pai	Taiwan
Cinkarna Celje		Slovenia
The National Titanium Dioxide Co.	Cristal	Saudi Arabia
E. I. du Pont de Nemours & Co., Inc.	Ti-Pure (730)	United States, Taiwan, Mexico, South Korea
Fuji Titan Kogyo KK	Fujititan	Japan
Furukawa Kogyo KK	Furukawa	Japan
Guandong New Technology		China
Hankook Titanium Ind. Co.	Hankook	South Korea
Ishihara Sangyo Kaisha	Tipaque (220)	Japan, Singapore, Taiwan
Kemira Pigments Oy	Finntitan (300) Finntitan (was Unitane) Finntitan	Finland, United States, the Netherlands
Kerala Minerals and Metals		India
Kerr-McGee Chemical Corp.	Tronox	United States
Kronos	Kronos (400)	Germany, Canada, Norway, Germany, Belgium
Louisiana Pigment Co. (Kronos & Tioxide Joint)		United States
Millenium Chemicals Inc. (formerly SCM)	Tiona (430)	United Kingdom, Australia, United States
Precheza AS	Pretiox	Czech Republic
Rhone-Poulenc Chemicals	Tita-France	France
Sachtleben Chemie GmbH	Hombitan	Germany
Sakai Kagaku Kogyo KK	Titone	Japan
Soyuzkraska		CIS
Tayca Corp.	Teika	Japan
Tioxide Group Ltd. (ICI)	Tioxide (600)	United Kingdom, France, Italy, Spain, Malaysia, South Africa, Malaysia
Titan Kogyo KK	SunTiox	Japan
Tohkem	Dia White	Japan
Zaklady Chemiczne Police	Tytanpol	Poland

[a]Values in parentheses represent the approximate 1996 production capacity in thousands of metric tons.

Titanium dioxide is traditionally supplied in paper bags containing 25 kg (50 lb) of pigment. Evolutionary developments in packaging have included palletization, the use of multiple bags, and the addition of shrinkwrapping to prevent excessive moisture pickup during transportation. For specialized applications, such as papermaking or plastic masterbatch manufacture, pigment is supplied in bags that can be added directly to the processing machinery without

Table 13. Distribution of U.S. Titanium Dioxide Pigment Shipments and Production[a]

Industry distribution, wt %	1980	1988	1992
paints, varnishes, and lacquers	44.1	48.1	44.7
paper	24.3	24.2	26.2
plastics[b]	10.6	17	17.5
printing inks			1.7
rubber		1.7	1.8
coated fabrics and textiles		0.2	0.3
ceramics			0.4
other, including exports	21	7.1	7.4
Total production[c], $\times 10^3$	*605*	*116,431*	*1,261,812*

[a]Ref. 87.
[b]Excluding floor coverings, vinyl-coated fabrics, and textiles.
[c]Based on gross weight of pigment.

preliminary slitting and emptying. In the United States, titanium dioxide pigment, particularly pigment for paper and latex paints, is often delivered as an aqueous slurry by rail. By contrast, in Europe and particularly in the United Kingdom, the supply of dry pigment by road tanker has become the favored route for the delivery of large quantities of titanium dioxide pigment. At an intermediate (1-t) level, semibulk containers (big bags) constructed from, eg, woven polypropylene, have become increasingly popular in Europe, Japan, and North America. In the period from 1988 to 1992, typical prices of rutile varied from $2.05 to 2.31/kg; that of anatase, $2.09 to 2.27/kg (87).

Peroxidic Compounds. When hydrogen peroxide is added to a solution of titanium(IV) compounds, an intense, stable, yellow solution is obtained, which forms the basis of a sensitive method for determining small amounts of titanium. The color probably results from the peroxo complex $[Ti(O_2)(OH)(H_2O)_x]^+$, and crystalline salts such as $K_2[Ti(O_2)(SO_4)_2] \cdot nH_2O$ can be isolated from alkaline solutions. The peroxo ligand is bidentate; the two oxygen atoms are equidistant from the titanium (98).

The action of hydrogen peroxide on freshly precipitated hydrated Ti(IV) oxide or the hydrolysis of a peroxide compound such as $K_2[Ti(O_2)(SO_4)_2]$ yields, after drying, a yellow solid, stable below 0°C, of composition $TiO_3 \cdot 2H_2O$. There is one peroxo group per titanium, but the precise structure is not known. The yellow solid loses oxygen and water when heated and liberates chlorine from hydrochloric acid. When freshly prepared, it is stable in acid or alkali, giving peroxy salts.

Inorganic Titanates. Titanium forms a series of mixed oxide compounds with other metals. Only in one of these, Ba_2TiO_4, are there discrete $[TiO_4]^{2-}$ ions (99). Compounds of the general formula $M_2^+TiO_3$ or $M^{2+}TiO_3$ are known as metatitanates; those having $M_4^+TiO_4$ and $M_2^{2+}TiO_4$ are called orthotitanates. Metatitanates of the type $M^{2+}TiO_3$ crystallize with the ilmenite ($Fe^{2+}TiO_3$) structure if M and Ti are of similar size. In the ilmenite structure, the oxygens are octahedrally close-packed. One-third of the octahedral interstices are occupied by M^{2+} ions and a second third by Ti^{4+}. If M is bigger than Ti, the perovskite structure is usually adopted (100).

Alkali Metal Titanates. Alkali metatitanates may be prepared by fusion of titanium oxide with the appropriate alkali metal carbonate or hydroxide. Representative alkali metal titanates are listed in Table 14. The alkali metal titanates tend to be more reactive and less stable than the other titanates, eg, they dissolve relatively easily in dilute acids.

High temperature hydrogen reduction of sodium metatitanate, Na_2TiO_3, produces nonstoichiometric titanium bronzes, Na_xTiO_2, where x is ca 0.2. These are chemically inert, blue-black metallic in appearance, and electrically conducting. The compounds have the hollandite structure, in which the TiO_6 octahedra are linked via their edges to form square cross-section tunnels. A variable population of cations can occupy these tunnels to give the materials significant composition ranges. Hollandites have received considerable attention as possible host matrices for the storage of radioactive metal ions. The concept involves trapping radioactive ions in barium/aluminium/titanium hollandite tunnels and then encapsulating the matrix in an inert shell (43). Work on this material, Synroc, is being developed mainly in Australia (101).

The crystal structure and properties of potassium titanates $K_2O \cdot nTiO_2$ depend on the value of n. Those having high ($n = 2$ or 3) potassium content have a layer structure and show both an intercalation ability and a catalytic activity (102,103). Those having a low potassium content have a tunnel structure and exhibit high chemical stability and low thermal conductivity (104). The polytitanates, $K_2Ti_4O_9$ and $K_2Ti_6O_{13}$, are of considerable interest because they can be manufactured in fibrous form. The fibers have average diameters of ca 1 μm and may be several millimeters long (105,106), each fiber being made up of several fibrils that are discrete crystals. They are chemically stable and melt at 1370°C. Production methods include hydrothermal synthesis in which titanium dioxide reacts with aqueous potassium hydroxide at high (ca 20 MPa (200 atm) and 600–700°C) pressure and temperature. Preparation from alkali halide melts and, by slow cooling after calcination, has also been described (106,107). The product is produced in the form of lumps, which may be broken down under high shear to give a water-dispersed pulp, which in turn may be treated in the same way as paper pulp to give papers, felts, and mats. Potassium titanate has a high

Table 14. Alkali Metal and Alkaline-Earth Titanates

Compound	CAS Registry Number	Formula	Density, kg/m^3	Mp, °C
lithium metatitanate	[12031-82-2]	Li_2TiO_3	3418	
lithium dititanate	[12600-48-5]	$Li_2Ti_2O_5$	350(0)	
lithium orthotitanate	[12768-28-4]	Li_4TiO_4		
sodium metatitanate	[12034-34-3]	Na_2TiO_3	319(0)	1030
sodium dititanate	[12164-19-1]	$Na_2Ti_2O_5$		985
sodium trititanate	[12034-36-5]	$Na_2Ti_3O_7$		1128
sodium pentatitanate	[12034-52-5]	$4Na_2O \cdot 5TiO_2$		
sodium trititanate	[12503-05-8]	$2Na_2O \cdot 3TiO_2$		
potassium metatitanate	[12030-97-6]	K_2TiO_3		806
potassium dititanate	[12056-46-1]	$K_2Ti_2O_5$		980
potassium polytitanate	[12056-49-4]	$K_2Ti_4O_9$		
potassium polytitanate	[12056-51-8]	$K_2Ti_6O_{13}$		

refractive index and a low thermal conductivity. Moreover, its size is in the right range to scatter infrared radiation. Thus it has potential use as an insulating and ir-reflective material. Other potential applications of potassium titanate include its use as a filtration medium, a reinforcement material for organic polymers, and an asbestos replacement in friction brakes. Between 1965 and 1972, pigmentary potassium titanate was manufactured in the United States (108).

Alkaline-Earth Titanates. Some physical properties of representative alkaline-earth titanates are listed in Table 15. The most important applications of these titanates are in the manufacture of electronic components (109). The most important member of the class is barium titanate, $BaTiO_3$, which owes its significance to its exceptionally high dielectric constant and its piezoelectric and ferroelectric properties. Further, because barium titanate easily forms solid solutions with strontium titanate, lead titanate, zirconium oxide, and tin oxide, the electrical properties can be modified within wide limits. Barium titanate may be made by, eg, cocalcination of barium carbonate and titanium dioxide at ca 1200°C. With the exception of Ba_2TiO_4, barium orthotitanate, titanates do not contain discrete TiO_4^{4-} ions but are mixed oxides. Ba_2TiO_4 has the β-K_2SO_4 structure in which distorted tetrahedral TiO_4^{4-} ions occur.

Barium titanate [12047-27-7] has five crystalline modifications. Of these, the tetragonal form is the most important. The structure is based on corner-linked oxygen octahedra, within which are located the Ti^{4+} ions. These can be moved from their central positions either spontaneously or in an applied electric field. Each TiO_6^{8-} octahedron may then be regarded as an electric dipole. If dipoles within a local region, ie, a domain, are oriented parallel to one another and the orientation of all the dipoles within a domain can be changed by the

Table 15. Properties of Alkaline-Earth Titanates

Name	CAS Registry Number	Formula	Crystal structure	Mean refractive index	Density, kg/m^3	Mp, °C
magnesium metatitanate[a]	[1312-99-8]	MgTiO$_3$	hexagonal	2.19	3840	1565
magnesium orthotitanate	[12032-52-9]	Mg$_2$TiO$_4$	cubic		3530	1840
magnesium dititanate	[12032-35-8]	Mg$_2$TiO$_5$	orthorhombic			1645
calcium titanate[b]	[12049-50-2]	CaTiO$_3$		2.34	4020	1980
barium titanate	[12047-27-7]	BaTiO$_3$	cubic	2.42		1625
barium titanate		BaTiO$_3$	tetragonal	2.3	6000	1625
barium titanate		BaTiO$_3$	hexagonal	2.2		1625
strontium titanate	[12060-52-9]	SrTiO$_3$	cubic	2.41	5120	2080

[a]The mineral geikielite.
[b]The mineral perovskite.

application of an electric field, the material is said to be ferroelectric. At ca 130°C, the Curie temperature, the barium titanate structure changes to cubic. The dipoles now behave independently, and the material is paraelectric (see FERROELECTRICS).

The ease with which the titanium ions may be moved from their central positions in the TiO_6^{8-} octahedra results in a very high dielectric constant, the basis of the use of barium titanate in ceramic capacitors. The dielectric constant is ca 1200 at room temperature but can increase tenfold at the Curie temperature. By forming solid solutions of barium titanate with, eg, $SrTiO_3$, the Curie temperature can be lowered to room temperature, and in this way the high dielectric characteristic of the Curie maximum can be exploited. Alternatively, by forming solid solutions with $CaTiO_3$ or rare-earth titanates, capacitors having a lower dielectric constant but greater temperature stability can be made. When ferroelectric materials are polarized, significant changes in dimension can result, eg, ca 1% in $BaTiO_3$. This phenomenon is the basis of piezoelectric actuators. The parent material for such devices is normally a solid solution of lead titanate and lead zirconate.

$BaTiO_3$ is also used for positive temperature coefficient (PTC) resistors, which exploit a large increase of resistivity at the Curie temperature of ferroelectrics (110). The resistance is the result of barrier layers at the grain boundaries in these sintered ceramics, and is not found in single-crystal material. The $BaTiO_3$ is made semiconducting by doping, eg, by substituting some of the Ti^{4+} ions by pentavalent ions, such as Nb^{5+}, and the temperature at which the resistance change occurs is controlled by forming solid solutions, eg, with $SrTiO_3$, to lower the Curie temperature. Applications include self-regulating heaters, power controls, and motor starters.

Other alkaline-earth titanates may be synthesized by heating together stoichiometric amounts of the oxide or by decomposing double salts such as strontium titanium oxalate. The primary use of magnesium, calcium, and strontium titanates is as additives to modify the properties of electroceramic components. Gems cut from strontium titanate boules have good color and brilliance, as well as a refractive index close to that of diamond. Magnesium titanate has also been used as a gemstone (see GEMSTONES). All three titanates are broad-band-gap (ca 3–4 eV) semiconductors. Strontium titanate absorbs uv wavelengths shorter than ca 395 nm, but magnesium titanate has a band gap of 3.7 eV. Consequently, the absorption of magnesium titanate is sufficiently far into the ultraviolet that the compound can be used as a pigment in uv-cured systems.

Other Titanates. Aluminum Titanate. Aluminum titanate, Al_2TiO_5, a white solid, density 3700 kg/m^3, mp ca 1860°C, has an orthorhombic, pseudobrookite, structure. Al_2TiO_5 may be made by calcination of the stoichiometric amount of oxides at temperatures above 1280°C (111). Below this temperature the compound is thermodynamically unstable and tends to decompose into alumina and titanium dioxide (112). The uses of aluminum titanate are as a low thermal expansion ceramic which has a good thermal shock resistance. The thermal expansion coefficients are markedly anisotropic. The coefficient in the a direction is negative. The consequent stresses produced during the cooling of sintered polycrystalline bodies cause extensive microcracking and the application of aluminum titanate relies on the control of the consequent microstructure (113).

The material finds applications in thermally insulating exhaust port liners in engine cylinder heads and in burner nozzles and thermocouple sleeves.

Iron Titanates. Ferrous metatitanate [12168-52-4], $FeTiO_3$, mp ca 1470°C, density 472(0), an opaque black solid having a metallic luster, occurs in nature as the mineral ilmenite. This ore is used extensively as a feedstock for the manufacture of titanium dioxide pigments. Artificial ilmenite may be made by heating a mixture of ferrous oxide and titanium oxide for several hours at 1200°C or by reducing a titanium dioxide/ferric oxide mixture at 450°C.

Ferrous orthotitanate [12160-20-2], Fe_2TiO_4, is orthorhombic and opaque. It has been prepared by heating a mixture of ferrous oxide and titanium dioxide. Ferrous dititanate [12160-10-0], $FeTi_2O_5$, is orthorhombic and has been prepared by reducing ilmenite with carbon at 1000°C. The metallic ion formed in the reaction is removed, leaving a composition that is essentially the dititanate. Ferric titanate [1310-39-0] (pseudobrookite), Fe_2TiO_5, is orthorhombic and occurs to a limited state in nature. It has been prepared by heating a mixture of ferric oxide and titanium dioxide in a sealed quartz tube at 1000°C.

Lead Titanate. Lead titanate [12060-00-3], $PbTiO_3$, is a yellow solid having a density of 73(00). It can be made by heating the calculated amounts of the two oxides together at 400°C. It may also be made by sol-gel routes, which may proceed via an intermediate pyrochlore phase, $Pb_2Ti_2O_6$ (114) (see SOL-GEL TECHNOLOGY). Its former use as a pigment has been superseded (115). Like barium titanate, lead titanate is a ferroelectric and the static dielectric constant shows a strong maximum, at ca 500°C, associated with the phase change from ferroelectric to paraelectric behavior. Lead titanate zirconate [12626-81-2] (PZT) is widely used as a piezoelectric ceramic. A typical composition is $PbZr_{0.6}Ti_{0.4}O_3$. In a typical production process, the oxides, together with minor amounts of modifying additives, are wet-mixed, dried, prereacted at ca 925°C, and then fabricated by pressing and firing at ca 1300°C (116). PZT is also used in pyroelectric detectors (109).

Other Titanates. Nickel titanate [12035-39-1], $NiTiO_3$, is a canary-yellow solid having a density of 73(00). When a mixture of antimony oxide, nickel carbonate, and titanium dioxide is heated at 980°C, nickel antimony titanate [8007-18-9] forms, which is used as a yellow pigment (95).

Zinc orthotitanate [12036-69-0], Zn_2TiO_4, a white solid having a density of 512(0) and a spinel structure, is obtained by heating the calculated amounts of the two oxides at 1000°C. Zinc orthotitanate forms a series of solid solutions with titanium dioxide, extending to the composition $Zn_2TiO_4 \cdot 1.5TiO_2$. These solid solutions begin to dissociate at 775°C with the formation of the rutile form of titanium dioxide. The properties of the rare-earth titanates and their electronic structure are available (117–119).

Titanium Halides

The most important halides and oxyhalides are shown in Table 16. General introductions to the chemistry of the titanium halides are available (10,120). Thermodynamic data are given in Tables 1 and 2.

Titanium Fluorides. *Titanium Difluoride.* Unlike other titanium dihalides, titanium difluoride [13814-20-5] is known only from mass spectra of gases.

Table 16. Titanium Halides

Titanium oxidation state	Fluoride	Chloride	Bromide	Iodide
Ti(II)	TiF_2	$TiCl_2$	$TiBr_2$	TiI_2
Ti(III)	TiF_3, TiOF	$TiCl_3$, TiOCl	$TiBr_3$	TiI_3
Ti(IV)	TiF_4	$TiCl_4$, $TiOCl_2$	$TiBr_4$, $TiOBr_2$	TiI_4, $TiOI_2$

Titanium Trifluoride. The trifluoride (121) is a blue crystalline solid, density 2980 kg/m^3, in which the titanium atoms are six-coordinate at the center of a slightly distorted octahedron, where the mean Ti–F distance is 197 pm. Titanium trifluoride [13470-08-1] is stable in air at room temperature but decomposes to titanium dioxide when heated to 100°C. It is insoluble in water, dilute acid, and alkalies but decomposes in hot concentrated acids. The compound sublimes under vacuum at ca 900°C but disproportionates to titanium and titanium tetrafluoride [7783-63-3] at higher temperatures.

Titanium trifluoride may be prepared in 90% yield by the reaction of gaseous hydrogen fluoride, in practice in a 1:4 ratio of hydrogen:HF, with either titanium metal or titanium hydride at 900°C.

$$2\ Ti^0 + 6\ HF \longrightarrow 2\ TiF_3 + 3\ H_2$$

The temperature is chosen to maximize reaction rate and to avoid the competitive formation of the tetrafluoride. Alternative preparation methods include the reaction between $TiCl_3$ and gaseous HF,

$$TiCl_3 + 3\ HF \longrightarrow TiF_3 + 3\ HCl$$

as well as the hydrogen reduction of ammonium hexafluorotitanate.

Titanium Tetrafluoride. Titanium tetrafluoride [7783-63-3] is a white hygroscopic solid, density 2798 kg/m^3, that sublimes at 284°C. The properties suggest that it is a fluorine-bridged polymer in which the titanium is six-coordinate. The preferred method of preparation is by direct fluorination of titanium sponge at 200°C in a flow system. At this temperature, the product is sufficiently volatile that it does not protect the unreacted sponge and the reaction proceeds to completion. The reaction of titanium tetrachloride with cooled, anhydrous, liquid hydrogen fluoride may be used if pure hydrogen fluoride is available.

$$TiCl_4 + 4\ HF \longrightarrow TiF_4 + 4\ HCl$$

A patent (122) for the production of pigment-grade titanium dioxide describes preparation of titanium tetrafluoride by the reaction of SiF_4 and ilmenite.

$$SiF_4 + TiO_2(\text{ilmenite}) \longrightarrow TiF_4 + SiO_2$$

The impure product is dissolved in methanol. Subsequent reaction with ammonium fluoride forms ammonium hexafluorotitanate. This solid, obtained in high purity, reacts with water vapor in a stream of air at 250°C to give anatase.

$$(NH_4)_2TiF_6 + 2\,H_2O \longrightarrow TiO_2 + 2\,NH_4F + 4\,HF$$

Other preparative methods include direct synthesis from the elements, reaction between gaseous hydrogen fluoride and titanium tetrachloride, and decomposition of barium hexafluorotitanate [31252-69-6], $BaTiF_6$, or ammonium, $(NH_4)_2TiF_6$.

Fluorotitanates. Anhydrous potassium fluorotitanate [23969-67-7], K_2TiF_6, may be prepared by dissolving titanium dioxide in dilute hydrofluoric acid to form a clear solution of H_2TiF_6. The exothermic reaction is carried out in agitated, carbon-tiled, rubber-lined, mild steel vessels. The acid is neutralized with potassium hydroxide solution, through which brilliant white crystals of potassium fluorotitanate are formed (11). Alternatively, titanium dioxide may be fused with potassium fluoride and the melt extracted with water. The anhydrous salt forms sparingly soluble transparent crystals (density 3022 kg/m^3, mp 780°C), which decompose to titanium dioxide and hydrogen fluoride when heated in air at 500°C. The principal use of potassium fluorotitanate is as a grain-refining agent for aluminum and its alloys, ie, to promote the production of a small grain size as the molten metal cools and solidifies. There are minor uses in the preparation of dental fillings and in abrasive grinding wheels.

Titanium Chlorides.   *Titanium Dichloride.*   Titanium dichloride [10049-06-6] is a black crystalline solid (mp >1035 at 10°C, bp >1500 at 40°C, density 31(40) kg/m^3). Initial reports that the titanium atoms occupy alternate layers of octahedral interstices between hexagonally close-packed chlorines (analogous to titanium disulfide) have been disputed (120). $TiCl_2$ reacts vigorously with water to form a solution of titanium trichloride and liberate hydrogen. The dichloride is difficult to obtain pure because it slowly disproportionates.

$$2\,TiCl_2 \longrightarrow Ti^0 + TiCl_4$$

$TiCl_2$ may be prepared either by the disproportionation of $TiCl_3$ at 475°C in vacuum,

$$2\,TiCl_3 \longrightarrow TiCl_2 + TiCl_4$$

or by passing titanium tetrachloride vapor over titanium metal at a temperature of ca 1100°C, slightly above the titanium dichloride melting point.

$$TiCl_4 + Ti^0 \longrightarrow 2\,TiCl_2$$

Alternatively, the $TiCl_4$ may be reduced using hydrogen, sodium, or magnesium. It follows that $TiCl_2$ is the first stage in the Kroll process for the production of titanium metal from titanium tetrachloride. A process for recovery of scrap titanium involving the reaction of scrap metal with titanium tetrachloride at

>800°C to form titanium dichloride, collected in a molten salt system, and followed by reaction of the dichloride with magnesium to produce pure titanium metal, has been patented (122,123).

Titanium Trichloride.   Titanium trichloride [7705-07-9] exists in four different solid polymorphs that have been much studied because of the importance of $TiCl_3$ as a catalyst for the stereospecific polymerization of olefins (120,124). The α-, γ-, and δ-forms are all violet and have close-packed layers of chlorines. The titaniums occupy the octahedral interstices between the layers. The three forms differ in the arrangement of the titaniums among the available octahedral sites. In α-$TiCl_3$, the chlorine sheets are hexagonally close-packed; in γ-$TiCl_3$, they are cubic close-packed. The brown β-form does not have a layer structure but, instead, consists of linear strands of titaniums, where each titanium is coordinated by three chlorines that act as a bridge to the next Ti. The structural parameters are as follows:

Material	Lattice constants, pm		Ti–Ti distance, pm
	a	c	
α-$TiCl_3$	356	587	354
β-$TiCl_3$	627	582	291
γ-$TiCl_3$	614	1740	354

The δ-form is produced by grinding either the α or the γ, and the structure is believed to be disordered and intermediate between α and γ.

Titanium trichloride is almost always prepared by the reduction of $TiCl_4$, most commonly by hydrogen. Other reducing agents include titanium, aluminum, and zinc. Reduction begins at temperatures of ca 500°C and under these conditions α-$TiCl_3$ is formed. The product is cooled quickly to below 450°C to avoid disproportionation to the di- and tetrachlorides. β-$TiCl_3$ is prepared by the reduction of titanium tetrachloride with aluminum alkyls at low (80°C) temperatures; whereas γ-$TiCl_3$ is formed if titanium tetrachloride reacts with aluminum alkyls at 150–200°C. At ca 250°C, the β-form converts to α. δ-$TiCl_3$ is made by prolonged grinding of the α- or γ-forms.

The primary use of $TiCl_3$ is as a catalyst for the polymerization of hydrocarbons (125–129). In particular, the Ziegler-Natta catalysts used to produce stereoregular polymers of several olefins and dienes, eg, polypropylene, are based on α-$TiCl_3$ and $Al(C_2H_5)_3$. The mechanism of this reaction has been described (130). Suppliers of titanium trichloride include Akzo America and Phillips Petroleum in the United States, and Mitsubishi in Japan.

Titanium Trichloride Hexahydrate.   Titanium trichloride hexahydrate [19114-57-9] can be prepared by dissolving anhydrous titanium trichloride in water or by reducing a solution of titanium tetrachloride. Evaporation and crystallization of the solution yield violet crystals of the hexahydrate. The hydrated salt has had some commercial application as a stripping or bleaching agent in the dyeing industry, particularly where chlorine must be avoided.

Titanium Tetrachloride.   Titanium tetrachloride [7550-45-0] is a clear, colorless liquid, normally made by the chlorination of titanium dioxide at ca 1000°C in the presence of a reducing agent.

$$TiO_2 + 2\ Cl_2 + 2\ C^0 \longrightarrow TiCl_4 + 2\ CO$$

It is the most widely studied of the titanium halides both because of its commercial importance and because of its abilities to participate in substitution reactions and form addition compounds with a wide range of donor molecules. Its main uses are in the manufacture of titanium dioxide pigments and titanium metal, as a starting material in the manufacture of a wide range of commercially important titanium organic compounds, which are mostly alkoxides rather than true organometallic compounds, and as a starting material in the production of Ziegler-Natta catalysts. In 1996, world production was ca 6×10^6 t.

Properties. Physical properties of titanium tetrachloride are given in Table 17. In the vapor phase, the titanium tetrachloride molecule is tetrahedral and has a Ti–Cl bond length of 218 pm. The regular tetrahedral coordination is retained in the solid, although each of the chlorines is crystallographically different in the monoclinic lattice (131).

Titanium tetrachloride is completely miscible with chlorine. The dissolution obeys Henry's law, ie, the mole fraction of chlorine in a solution of titanium tetrachloride is proportional to the chlorine partial pressure in the vapor phase. The heat of solution is 16.7 kJ/mol (3.99 kcal/mol). The apparent maximum solubilities of chlorine at 15.45 kPa (116 mm Hg) total pressure follow.

Temperature, °C	Cl_2, mole fraction	Cl_2, in 100-g $TiCl_4$
−10	0.38	23.4
0	0.28	14.3
20	0.155	6.85

The system shows a eutectic that solidifies at −108°C at 87.5 atom % Cl.

Titanium tetrachloride's affinity for water is so high that it acts as a desiccating agent. It is readily hydrolyzed by water and fumes strongly when in contact with moist air. The dense white fumes, consisting of finely divided oxychlorides, are the basis of the use of titanium tetrachloride as white smoke. Hence $TiCl_4$ is used by the military for smoke screen purposes (see CHEMICALS IN WAR). At room temperature and below, liquid titanium tetrachloride dissolves exothermically in water to form clear mixtures that are acid because of hydrolysis. A series of oxychlorides, and ultimately titanium dioxide, are formed, the precise product depending on conditions of temperature, concentration, and pH. Thus, Raman spectra suggest that complexes of the type $TiO_2Cl_4^{4-}$ are formed in hydrochloric acid, whereas the hexachlorotitanate anion is reported to be formed in fuming hydrochloric acid (132).

Titanium tetrachloride is also miscible with other common liquids, including organic solvents such as hydrocarbons, carbon tetrachloride, and chlorinated hydrocarbons. With those containing hydroxyl, carboxyl, or diketone (in the enol form) groups, reaction occurs. Substitution products are formed with the elimination of hydrogen chloride. Thus, with alcohols, alkoxides, also called

Table 17. Physical Properties of Titanium Tetrachloride

Property	Value and description
color	none
density, 20°C, g/cm^3	1.7
freezing point, °C	−24.1
heat of fusion, kJ/mol^a	9.966
boiling point, °C	136.5
vapor pressure, kPa^b	
20°C	1.33
50°C	5.52
100°C	35.47
heat of vaporization, kJ/mol^a	
25°C	41.087
135.8°C	35.98
specific heat, 20°C, J/(k·mol)^a	145.21
critical temperature, °C	370
heat of liquid formation, 25°C, kJ/mol^a	−804.2 ± 4.2
viscosity, mPa·s(=cP)	0.079
refractive index, n	1.6985
magnetic susceptibility	-0.287×10^{-6}
dielectric constant, 20°C	2.79

^{a}To convert kJ to kcal, divide by 4.184.
^{b}To convert kPa to psi, multiply by 0.145.

esters of titanic acid, are formed (see ALKOXIDES, METAL).

$$TiCl_4 + ROH \longrightarrow TiCl_3OR + HCl \longrightarrow TiCl_2(OR)_2 + HCl \longrightarrow$$

$$TiCl(OR)_3 + HCl \longrightarrow Ti(OR)_4 + HCl$$

For the reaction to go to completion, it is necessary to remove the liberated hydrogen chloride, eg, by reaction with ammonia.

Addition compounds form with those organics that contain a donor atom, eg, ketonic oxygen, nitrogen, and sulfur. Thus, adducts form with amides, amines, and N-heterocycles, as well as acid chlorides and ethers. Addition compounds also form with a number of inorganic compounds, eg, POCl$_3$ (6,120). In many cases, the addition compounds are dimeric, eg, with ethyl acetate, in titanium tetrachloride-rich systems. By using ammonia, a series of amidodichlorides, Ti(NH$_2$)$_x$Cl$_{4-x}$, is formed (133).

Titanium tetrachloride is readily reduced. By using sodium, calcium, or magnesium, reduction to the metal occurs. By using hydrogen, on the other hand, complete reduction to the metal only occurs at very high temperatures and the resultant metal sponge takes up hydrogen on cooling. However, at 700°C, titanium trichloride is formed and can be collected as a crystalline solid in a receiver held above the boiling point of the tetrachloride.

Manufacture. Titanium chloride is manufactured by the chlorination of titanium compounds (1,134−138). The feedstocks usually used are mineral or synthetic rutile, beneficiated ilmenite, and leucoxenes. Because these are all oxygen-containing, it is necessary to add carbon as well as coke from either coal or fuel oil during chlorination to act as a reducing agent. The reaction is normally carried out as a continuous process in a fluid-bed reactor (139). The bed consists of a mixture of the feedstock and coke. These are fluidized by a stream of chlorine introduced at the base (see FLUIDIZATION). The amount of heat generated in the TiO_2 chlorination process depends on the relative proportions of CO_2 or CO that are formed (eqs. 1 and 2), and the mechanism that determines this ratio is not well understood.

$$TiO_2 + 2\ Cl_2 + C^0 \longrightarrow TiCl_4 + CO_2 \qquad \Delta H = -217.6\ \text{kJ/mol at 1100 K}$$
$$= -221.7\ \text{kJ/mol at 1300 K} \qquad (1)$$

$$TiO_2 + 2\ Cl_2 + 2\ C^0 \longrightarrow TiCl_4 + 2\ CO \qquad \Delta H = -50.1\ \text{kJ/mol at 1100 K}$$
$$= -54.2\ \text{kJ/mol at 1300 K} \qquad (2)$$

The second reaction (eq. 2) is relatively more favored as the temperature increases ($\Delta G_A = -280$ kJ/mol, $\Delta G_B = -302$ kJ/mol at 1100 K; $\Delta G_A = -290.8$ kJ/mol, $\Delta G_B = -305.7$ kJ/mol at 1300 K). It follows that the chlorinator temperature depends on the proportion of CO_2 and CO generated; it can be adjusted either by coke oxidation, with oxygen in the Cl_2 gas, or by cooling with cold $TiCl_4$.

Under typical chlorination conditions, most elements are chlorinated. Therefore, for every metric ton of titanium tetrachloride produced, lower grade feedstocks require more chlorine. Minor impurities such as alkaline-earths, where the chlorides are relatively involatile, may either inhibit bed-fluidization or cause blockages in the equipment and require particular consideration regarding feedstock specification.

If ores having a lower TiO_2 content, eg, high grade slags, leucoxene, or ilmenites, are used, the majority of metal (mainly iron) oxides are chlorinated and exit with the titanium tetrachloride. The reaction products are condensed and impurities are removed by a sequential process involving a solids separator and a liquid scrubbing system. The crude titanium tetrachloride is further purified by distillation. Vanadium oxychloride that boils at a similar temperature to titanium tetrachloride may be reduced at this stage and converted into a nonvolatile sludge. Many patents relate to the choice of suitable reductants. The final distilled product is extremely pure, a key feature in the subsequent production of pigment having good color.

Alternatives to the fluidized-bed method process include the chlorination of titanium slags in chloride melts, chlorination with hydrogen chloride, and flash chlorination. The last is claimed to be particularly advantageous for minerals having a high impurity content (133–135,140). The option of chlorinating titanium carbide has also been considered (30).

Producers and Economic Aspects. The main producers of titanium tetrachloride throughout the world are producers of titanium dioxide pigment by the chloride route. These include Bayer, Du Pont, Ishihara Sangyo Kaisha,

Kerr McGee, Kronos Titan, Millenium Chemicals, and the Tioxide Group. Other suppliers, not necessarily large-scale producers, include Akzo Nobel Chemicals, Aldrich Chemical, Atomergic Chemetals, Cerac, Janssen Chimica, Noah Technologies, Titanium Metals, and Toth Aluminium.

Hexachlorotitanates. H_2TiCl_6 may be made by dissolving anhydrous hydrogen chlorine in titanium tetrachloride (C13). Ammonium hexachlorotitanate [21439-26-9], potassium hexachlorotitanate [16918-46-0], rubidium hexachlorotitanate [16902-24-2], and cesium hexachlorotitanate [16918-47-1] are light-green to yellow crystalline solids. They may be prepared either by direct interaction of the alkali-metal chloride with titanium tetrachloride or by reaction in fuming HCl. Both the acids and its salts are more susceptible to hydrolysis than the corresponding fluorotitanates. They are also thermally unstable and decompose to the alkali metal chloride and titanium tetrachloride.

Oxychlorides. Hydrolysis of $TiCl_4$ yields a number of products, the composition of which depends on the hydrolysis conditions. In the $TiCl_3-HCl-H_2O$ system, species ranging from $Ti(H_2O)_6^{3+}$ through $TiCl(H_2O)_5^{2+}$ to $TiCl_5(H_2O)^{2-}$, as the acid concentration increases, has been reported (141).

Titanium oxide dichloride [13780-39-8], $TiOCl_2$, is a yellow hygroscopic solid that may be prepared by bubbling ozone or chlorine monoxide through titanium tetrachloride. It is insoluble in nonpolar solvents but forms a large number of adducts with oxygen donors, eg, ether. It decomposes to titanium tetrachloride and titanium dioxide at temperatures of ca 180°C (136).

A titanium monoxychloride, TiOCl, may be prepared by the action of TiO_2, Fe_2O_3, or oxygen on $TiCl_3$ (137). For example, at 650°C,

$$2\ TiCl_3 + TiO_2 \longrightarrow 2\ TiOCl + TiCl_4$$

Titanous oxychloride forms yellow tablets, is inert in mineral acids and water and also stable in air. When heated in air, it gives titanium tetrachloride and titanium dioxide.

Titanium Bromides. *Titanium Dibromide.* Titanium dibromide [13873-04-5], a black crystalline solid, density 4310 kg/m^3, mp 1025°C, has a cadmium iodide-type structure and is readily oxidized to trivalent titanium by water. Spontaneously flammable in air (142), it can be prepared by direct synthesis from the elements, by reaction of the tetrabromide with titanium, or by thermal decomposition of titanium tribromide. This last reaction must be carried out either at or below 400°C, because at higher temperatures the dibromide itself disproportionates.

$$2\ TiBr_3 \longrightarrow TiBr_2 + TiBr_4$$

Titanium Tribromide. Titanium tribromide [13135-31-4] crystallizes in two different habits: hexagonal plates or blue-black needles. It can be prepared by the reaction of $TiBr_4$ with either titanium or hydrogen.

$$3\ TiBr_4 + Ti^0 \longrightarrow 4\ TiBr_3$$

or

$$2 \, TiBr_4 + H_2 \longrightarrow 2 \, TiBr_3 + 2 \, HBr$$

A hexahydrate is also known and may be prepared by the electrolytic reduction of the tetrabromide in hydrobromic acid solution.

Titanium Tetrabromide. Titanium tetrabromide [7789-68-6] is an amber-yellow, easily hydrolyzed, crystalline solid, having a density of 3250 kg/m^3. The crystal structure depends on temperature. At 20°C, TiBr$_4$ is monoclinic (143) and has $a = 1017$ pm, $b = 709$ pm, $c = 1041$ pm, and $\beta = 101.97°$ (143). It melts at 39°C and may be purified by vacuum sublimation. The liquid boils at 233°C to give a monomeric vapor in which the Ti–Br distance is 231 pm. Titanium tetrabromide is soluble in dry chloroform, carbon tetrachloride, ether, and alcohol. Like titanium tetrachloride, TiBr$_4$ forms a range of adducts with molecules such as ammonia, amines, nitrogen heterocycles, esters, and ethers.

Analogous with titanium tetrachloride, the tetrabromide may be made by the carbothermal bromination of titanium dioxide at ca 700°C,

$$TiO_2 + 2 \, Br_2 + C^0 \longrightarrow TiBr_4 + CO_2$$

and also by direct bromination of titanium at 300–600°C in a flow system. Halogen exchange among titanium tetrachloride, hydrogen bromide, and boron tribromide has also been used.

Titanium Iodides. *Titanium Diiodide.* Titanium diiodide is a black solid ($\rho = 499(0)$ kg/m^3) that has the cadmium iodide structure. Titaniums occupy octahedral sites in hexagonally close-packed iodine layers, where $a = 411$ pm and $c = 682$ pm (144). Magnetic studies indicate extensive Ti–Ti bonding. TiI$_2$ reacts rapidly with water to form a solution of titanous iodide, TiI$_3$.

Titanium diiodide may be prepared by direct combination of the elements, the reaction mixture being heated to 440°C to remove the tri- and tetraiodides (145). It can also be made by either reaction of solid potassium iodide with titanium tetrachloride or reduction of TiI$_4$ with silver or mercury.

Titanium Triiodide. Titanium triiodide is a violet crystalline solid having a hexagonal unit cell (146). The crystals oxidize rapidly in air but are stable under vacuum up to 300°C; above that temperature, disproportionation to the diiodide and tetraiodide begins (147).

Titanium triiodide can be made by direct combination of the elements or by reducing the tetraiodide with aluminum at 280°C in a sealed tube. TiI$_3$ reacts with nitrogen, oxygen, and sulfur donor ligands to give the corresponding adducts (148).

Titanium Tetraiodide. Titanium tetraiodide [7720-83-4] forms reddish-brown crystals, cubic at room temperature, having reported lattice parameter of either 1200 (149) or 1221 (150) pm. TiI$_4$ melts at 150°C, boils at 377°C, and has a density of 440(0) kg/m^3. It forms adducts with a number of donor molecules and undergoes substitution reactions (151). It also hydrolyzes in water and is readily soluble in nonpolar organic solvents.

Titanium tetraiodide can be prepared by direct combination of the elements at 150–200°C; it can be made by reaction of gaseous hydrogen iodide with a solution of titanium tetrachloride in a suitable solvent; and it can be purified by vacuum sublimation at 200°C. In the van Arkel method for the preparation of pure titanium metal, the sublimed tetraiodide is decomposed on a tungsten or titanium filament held at ca 1300°C (152). There are frequent literature references to its use as a catalyst, eg, for the production of ethylene glycol from acetylene (153).

Titanium Silicon Compounds

Titanium Silicides. The titanium−silicon system includes Ti_3Si, Ti_5Si_3, TiSi, and $TiSi_2$ (154). Physical properties are summarized in Table 18. Direct synthesis by heating the elements *in vacuo* or in a protective atmosphere is possible. In the latter case, it is convenient to use titanium hydride instead of titanium metal. Other preparative methods include high temperature electrolysis of molten salt baths containing titanium dioxide and alkalifluorosilicate (155); reaction of $TiCl_4$, $SiCl_4$, and H_2 at ca 1150°C, using appropriate reactant quantities for both TiSi and $TiSi_2$ (156); and, for Ti_5Si_3, reaction between titanium dioxide and calcium silicide at ca 1200°C, followed by dissolution of excess lime and calcium silicate in acetic acid.

Titanium disilicide [12039-83-7] is a silvery-gray, crystalline material that oxidizes slowly in air when heated to 700−800°C. It is resistant both to mineral acids (except hydrofluoric) and to aqueous solutions of alkalies, but reacts with fused borax, sodium hydroxide, and potassium hydroxide. It reacts explosively with chlorine at high temperatures.

Titanium silicides are used in the preparation of abrasion- and heat-resistant refractories. Compositions based on mixtures of Ti_5Si_3, TiC, and diamond have been claimed to make wear-resistant cutting-tool tips (157). Titanium silicide can be used as an electric−resistant material, in electrically conducting ceramics (158), and in pressure-sensitive elastic resistors, the electric resistance of which varies with pressure (159).

Table 18. Structure and Physical Constants of Titanium Silicides[a]

Property	$TiSi_2$	TiSi	Ti_5Si_3	Ti_3Si
CAS Registry Number	[12039-83-7]	[12039-70-2]	[12067-57-1]	
Ti, atomic %	33	50	62.5	75
structure	orthorhombic	rhombic	hexagonal	
lattice parameters, nm[b]				
a	0.8253	0.654(4)	0.7448	1.039
b	0.4783	0.363(8)		
c	0.854	0.499(7)	0.5114	0.517
density, kg/m^3[b]	439(0)	434(0)	431(0)	
mp, °C	ca 1540	1760	2120	
hardness, 100-g load, kg/mm	850	1050	1000	
resistivity, $\mu\Omega\cdot cm$	123			

[a]Ref. 153.
[b]Numbers in parentheses are fourth digit of reduced accuracy.

Titanium Silicates. A number of titanium silicate minerals are known (160); examples are listed in Table 19. In most cases, it is convenient to classify these on the basis of the connectivity of the SiO_4 building blocks, eg, isolated tetrahedra, chains, and rings, that are typical of silicates in general. In some cases, the SiO_4 units may be replaced, even if only to a limited extent by TiO_4. For example, up to 6% of the SiO_4 in the garnet schorlomite can be replaced by TiO_4. In general, replacement of SiO_4 by TiO_4 building blocks increases the refractive indices of these minerals. Ti has also replaced Si in the framework of various zeolites. In addition, the catalytic activity of both titanium-substituted ZSM-5 (TS-1) and ZSM-11 (TS-2) has received attention (161), eg, the selective oxidation of phenol, with hydrogen peroxide, to hydroquinone and catechol over TS-1 has been operated at the 10,000 t/yr scale in Italy (162).

Table 19. Silicate Minerals

Mineral	Chemical formula	Crystal form, lattice parameters, pm	Structural unit
natisite	Na_2TiSiO_5	tetragonal; $a = 650$, $c = 507$	isolated SiO_4 tetrahedra
titanite (or sphene)	$CaTiSiO_5$	monoclinic; $a = 706.6$, $b = 870.5$, $c = 656.1$	isolated SiO_4 tetrahedra
benitoite	$BaTiSi_3O_9$	hexagonal	hexagonal rings of three SiO_4 tetrahedra
schorlomite	$Ca_3(Fe,Ti)_2(Si,Ti)_3O_{12}$	cubic; $a = 1212.8$	garnet structure, isolated SiO_4 tetrahedra
lorenzenite	$Na_2TiSi_2O_9$	orthorhombic; $a = 1449$, $b = 870$, $c = 523$	SiO chains
davanite	$K_2TiSi_6O_{15}$	triclinic; $a = 714$, $b = 753$, $c = 693$	silicate layers

Titanium Phosphorus Compounds

Titanium Phosphides. The titanium phosphides (154) include Ti_3P [12037-66-0], Ti_5P_3, and TiP (163). Titanium monophosphide [12037-65-9], TiP, can be prepared by heating phosphine with titanium tetrachloride or titanium sponge. Alternatively, titanium metal may be heated with phosphorus in a sealed tube. The gray metallic TiP is slightly phosphorus-deficient ($TiP_{0.95}$), has a density of 408(0) kg/m^3, and displays considerable mechanical hardness (700 kg/mm^2). It is oxidized on heating in air but is stable when heated to 1100°C in either vacuum or a protective atmosphere; it is resistant to concentrated acid (except aqua regia) and alkalies; and it is reported to act as a catalyst in polycondensation reactions.

Titanium Phosphates. Titanium(III) phosphate [24704-65-2] (titanous phosphate) is a purple solid, soluble in dilute acid, giving relatively stable solutions. It can be prepared by adding a soluble phosphate to titanous chloride or sulfate solution and raising the pH until precipitation occurs.

Titanium(IV) phosphate gel [17017-60-6] may be prepared by adding an alkali phosphate to titanium(IV) sulfate or chloride solution, followed by filtering, leaching, and drying the derived gel. The product is insoluble in dilute sulfuric acid. Titanium phosphate prepared in this way has been used in the dyeing and leather (qv) tanning industries. Polymeric titanium phosphate compositions for metal activation prior to phosphating have also been claimed (164). A process for the recovery of titanium phosphate pigment from wastes from the TiO_2−sulfate process has been proposed (165,166).

The α-form of titanium(IV) bis(hydrogen phosphate) dihydrate has a layer structure similar to that of the analogous zirconium compound (167,168). It is prepared by refluxing the gels in phosphoric acid for long (400 h or more) periods. The degree of crystallinity achieved depends on the acid concentration and treatment times. The derived compounds can act as ion-exchange materials. The catalytic activity of these compounds for alcohol dehydration, oxidative dehydrogenation, oxidation, and polymerization has been reviewed (169). Under special conditions, a second, γ, form may be prepared (168). The interlayer distance in the γ-form is significantly larger than that in the α, which makes γ-forms of greater potential use as ion exchangers. Processes for the removal of K^+ from seawater, $^{42}K^+$ and $^{37}Cs^+$ from strong mineral acids, and NH_4^+ and NH_3 from waste solutions have been described (170−172).

Titanium pyrophosphate [13470-09-2], TiP_2O_7, a possible uv reflecting pigment, is a white powder that crystallizes in the cubic system and has a theoretical density of 3106 kg/m^3. It is insoluble in water and can be prepared by heating a stoichiometric mixture of hydrous titania and phosphoric acid at 900°C.

Titanium Sulfur Compounds

Titanium Sulfides. The titanium sulfur system has been summarized (4). Titanium subsulfide [1203-08-6], Ti_2S, forms as a gray solid of density 4600 kg/m^3 when titanium monosulfide [12039-07-5], TiS, is heated at 1000°C with titanium in a sealed tube. It can also be formed by heating a mixture of the two elements at 800−1000°C. The sulfide, although soluble in concentrated hydrochloric and sulfuric acids, is insoluble in alkalies.

The structures of titanium monosulfide and titanium disulfide [12039-13-3], TiS_2, are both made up of layers of hexagonally stacked sulfur atoms (43). In TiS, all the octahedral sites between each layer are filled by titanium; in the disulfide, only the sites in alternate layers are filled. Partial occupation of the remaining octahedral sites in titanium disulfide then results in a series of nonstoichiometric phases, where the nominal formula varies from Ti_8S_9 through Ti_4S_5 to titanium sesquisulfide [12039-16-6], Ti_2S_3, and in which the ordering of the additional titaniums depends on the preparation conditions. Because the outer electrons are delocalized to an appreciable extent, the compounds have pseudometallic properties. TiS is a dark-brown solid (density 4050 kg/m^3), Ti_2S_3 is a black crystalline solid (density 3520 kg/m^3), and TiS_2 is a bronze-colored solid (density 3220 kg/m^3). Each can be prepared by direct combination of the elements. Other methods of preparation are also similar:

$$TiS_2 + H_2 \longrightarrow TiS + H_2S$$

$$2\,TiS_2 + H_2 \longrightarrow Ti_2S_3 + H_2S$$

Titanium disulfide can also be made by pyrolysis of titanium trisulfide at 550°C. A continuous process based on the reaction between titanium tetrachloride vapor and dry, oxygen-free hydrogen sulfide has been developed at pilot scale (173). The preheated reactants are fed into a tubular reactor at approximately 500°C. The product particles comprise orthogonally intersecting hexagonal plates or plate segments and have a relatively high surface area (>4 m^2/g), quite different from the flat platelets produced from the reaction between titanium metal and sulfur vapor. The powder, reported to be stable to storage for long periods under dry air or nitrogen at or below 20°C, is not attacked by hydrochloric acid but is soluble in both hot and cold sulfuric acid and also, unlike the other sulfides, in hot sodium hydroxide solution.

Titanium trisulfide [12423-80-2], TiS$_3$, a black crystalline solid having a monoclinic structure and a theoretical density of 3230 kg/m^3, can be prepared by reaction between titanium tetrachloride vapor and H$_2$S at 480–540°C. The reaction product is then mixed with sulfur and heated to 600°C in a sealed tube to remove residual chlorine. Sublimation may be used to separate the trisulfide (390°C) from the disulfide (500°C). Titanium trisulfide, insoluble in hydrochloric acid but soluble in both hot and cold sulfuric acid, reacts with concentrated nitric acid to form titanium dioxide.

The principal use of titanium sulfides is as a cathode material in high efficiency batteries (11). In these applications, the titanium disulfide acts as a host material for various alkali or alkaline-earth elements.

$$x\,Li^+ + x\,e^- + TiS_2 \longrightarrow Li_x TiS_2$$

The titanium sulfide is able to act as a lithium reservoir. On intercalation with lithium, the titanium lattice expands from ca 570 to 620 pm as the intercalation proceeds to completion on formation of TiLiS$_2$. Small button cells have been developed, incorporating lithium perchlorate in propylene carbonate electrolyte, for use in watches and pocket calculators (see BATTERIES).

Titanium disulfide has been proposed as a solid lubricant. The coefficient of friction between steel surfaces is 0.3, compared to only 0.2 for molybdenum disulfide. However, because it does not adhere strongly to metal surfaces, TiS$_2$ is generally less effective than molybdenum sulfide.

Titanium Sulfates. Solutions of titanous sulfate [10343-61-0] are readily made by reduction of titanium(IV) sulfate in sulfuric acid solution by electrolytic or chemical means, eg, by reduction with zinc, zinc amalgam, or chromium(II) chloride. The reaction is the basis of the most used titrimetric procedure for the determination of titanium. Titanous sulfate solutions are violet and, unless protected, can slowly oxidize in contact with the atmosphere. If all the titanium has been reduced to the trivalent form and the solution is then evaporated, crystals of an acid sulfate 3 Ti$_2$(SO$_4$)$_3$·H$_2$SO$_4$·25H$_2$O [10343-61-0] are produced. This purple salt, stable in air at normal temperatures, dissolves in water to give a stable violet solution. When heated in air, it decomposes to TiO$_2$, water, sulfuric acid, and sulfur dioxide.

If a solution of the acid sulfate in dilute sulfuric acid is evaporated at 200°C, green crystals of the anhydrous neutral salt form. The anhydrous salt is insoluble in water, alcohol, and concentrated sulfuric acid, but dissolves in dilute sulfuric or hydrochloric acid to give a violet solution. The salt is isomorphous with chromic sulfate and is used commercially as a reducing agent.

Titanium(IV) sulfate can be prepared by the reaction of titanium tetrachloride with sulfur trioxide dissolved in sulfuryl chloride.

$$TiCl_4 + 6\ SO_3 \longrightarrow Ti(SO_4)_2 + 2\ S_2O_5Cl_2$$

It is readily hydrolyzed by moisture and, when heated, changes first to titanyl sulfate [13825-74-6], $TiOSO_4$, and then to TiO_2. Because of the high charge/radius ratio, normal salts of Ti^{4+} cannot be prepared from aqueous solutions. Instead, these hydrolyze to form a titanyl species normally written as $[TiO]^{2+}$. Ion-exchange studies on aqueous solutions of Ti(IV) in perchloric acid are consistent with the existence of monomeric, doubly charged cationic species, although it is not clear whether these are $[TiO]^{2+}$ or $[Ti(OH)_2]^{2+}$ (174,175). Titanyl sulfate is the best known of the derived salts because of its importance as an intermediate in the manufacture of titanium dioxide pigments by the sulfate process. The white, needle-like powder of the hydrate [58428-64-1], $TiOSO_4 \cdot H_2O$, is formed when solutions of TiO_2 in sulfuric acid are evaporated. However, there is no evidence for the existence of the $[TiO]^{2+}$ ion in the solid. Rather, it is believed to contain $-Ti-O-Ti-O-$ chains, where each Ti is octahedrally coordinated by the two bridging oxygens (Ti−O distance ca 180 pm), a water molecule, and one oxygen from each of the three sulfates. In addition to its role as an intermediate in the production of titanium dioxide, titanyl sulfate is used for treatment of metals and in the dyeing industry for the preparation of titanous sulfate. Titanyl sulfate may also be used as a tanning agent for the production of leather (176). In this application, it is often complexed, ie, masked, with, eg, sodium gluconate to prevent unwanted precipitation of hydrous titania (177). Various double sulfates have also been described. For example, $(NH_4)_2[TiO(SO_4)_2]$, which can be prepared from sulfate liquors of titanium by precipitation with ammonium sulfate and sulfuric acid, is also used as a tanning agent (178).

Analytical Methods

The analytical chemistry of titanium has been reviewed (179–181). Titanium ores can be dissolved by fusion with potassium pyrosulfate, followed by dissolution of the cooled melt in dilute sulfuric acid. For some ores, even if all of the titanium is dissolved, a small amount of residue may still remain. If a full analysis is required, the residue may be treated by moistening with sulfuric and hydrofluoric acids and evaporating, to remove silica, and then fused in a sodium carbonate–borate mixture. Alternatively, fusion in sodium carbonate–borate mixture can be used for ores and a boiling mixture of concentrated sulfuric acid and ammonium sulfate for titanium dioxide pigments. For trace-element determinations, the preferred method is dissolution in a mixture of hydrofluoric and hydrochloric acids.

Titrimetric methods based on dissolution of the sample, reduction of the Ti(IV) to Ti(III), and reoxidation to Ti(IV) using a standard oxidizing agent are widely employed for the determination of significant levels of titanium. Reduction may be by means of either a solution of chromium(II) chloride, Nakozono, Jones reductors, or aluminum metal. In all cases, the reduced solutions, which must be protected from reoxidation by atmospheric oxidation, can be titrated using standard ferric iron solutions. Chromium chloride reduction followed by a potentiometric titration is preferred because it is not subject to interference by chromium or vanadium. For reductions using Nakozono, Jones reductors, or aluminum metal, the end points can be determined either potentiometrically or by using a potassium thiocyanate indicator but chromium and vanadium interfere. Gravimetric procedures for the determination of titanium are not widely employed, although precipitation with cupferron from an ammonium acetate-buffered solution containing ethylenediamine tetraacetic acid (EDTA) can be used.

Wavelength dispersive x-ray fluorescence spectrometric (xrf) methods using the titanium K_α line at 0.2570 nm may be employed for the determination of significant levels of titanium only by careful matrix-matching. However, xrf methods can also be used for semiquantitative determination of titanium in a variety of products, eg, plastics. Xrf is also widely used for the determination of minor components, such as those present in the surface coating, in titanium dioxide pigments.

Minor levels of titanium are conveniently measured by spectrophotometry, eg, by the 410-nm absorbance of the yellow-orange peroxide complex that develops when hydrogen peroxide is added to acidic solutions of titanium.

For trace levels, instrumental methods of analysis are normally used. Flame atomic absorption spectrometry using a reducing nitrous oxide–acetylene flame typically gives a sensitivity of 2 mg/L for 1% absorption at 364.3 nm. Much greater sensitivity (ca 1 μg/L) can be obtained using electrothermal atomic absorption spectrometry using Zeeman background correction, but many interfering elements can lead to the formation of stable titanium complexes in the furnace. Inductively coupled plasma–atomic emission spectrometry atomizes the titanium more effectively. For the 334.941-nm line, detection limits of 0.5 μg/L have been reported. Finally, inductively coupled plasma–mass spectrometry can be used to monitor the titanium isotopes of mass 46, 47, 48, 49, and 50, achieving detection limits in the range of 0.001–0.01 μg/L for the principal (~75% abundance) mass 48.

Health and Safety Aspects

The following discussion on health and safety aspects of titanium compounds is concerned only with the behavior of the titanium present in inorganic compounds and not with the effects of the compounds themselves. For example, titanium tetrachloride must be treated with care because of the effects of the hydrochloric acid and heat produced when it reacts with water, not because of the possible toxicity of titanium. Apart from very few exceptions, the inorganic compounds of titanium are generally regarded as having low toxicity. Because of the ubiquitous nature of the element and its compounds, average concentrations of titanium in blood have been determined at 130–160 $\mu g/L$ (182–184), with a typical value of 10 $\mu g/L$ in urine (185).

Titanium metal is frequently used as a surgical implant, and several titanium compounds have been used in medical, food, food-contact, and cosmetic products. In the human body, the highest concentrations of titanium are found in the lung (182,186,187), resulting from the slow clearance of insoluble inhaled particles that accumulate with time. Such particles are typically oxides or complex silicates. Exposures during occupational activities have been shown to increase the lung burden. For example, miners have reported values of 119 $\mu g/g$ dry weight titanium in lung, compared with 19 $\mu g/g$ dry weight for nonoccupationally exposed people (188).

Much work has been carried out on the toxicology of titanium, mainly as a result of the widespread use of titanium dioxide pigments. Animal experiments have demonstrated that titanium dioxide is not carcinogenic by the oral route (189), whereas inhalation experiments have demonstrated that an excess of tumors is formed on high exposure in rats, but not in other species, such as mice (190,191). It is considered that the tumors in rats are an exaggerated response to lung-overload effects, associated with any insoluble, low toxicity dust, which may be rat-specific. This is corroborated by the fact that an epidemiological study of pigment production workers did not demonstrate any excess of tumors (192).

Titanium dioxide, because of its insolubility and low toxicity, has been allocated occupational exposure limits by the ACGIH of 10-mg/m^3 total dust, or 3-mg/m^3 respirable dust averaged over an eight-hour exposure period (193). Guidance on the precautions to be taken while handling titanium dioxide pigments includes the avoidance of the generation of inhalable dust, preferably by engineering controls or, when necessary, by respiratory protection (194). Because titanium dioxide has the ability to dry and defat the skin by adsorption, prolonged skin contact should be avoided.

BIBLIOGRAPHY

"Titanium Compounds (Inorganic)" in *ECT* 1st ed., Vol. 14, pp. 213–237, by L. R. Blair, H. H. Beacham, and W. K. Nelson, National Lead Co.; in *ECT* 2nd ed., Vol. 20, pp. 380–424, by G. H. J. Neville, British Titan Products Co., Ltd.; in *ECT* 3rd ed., Vol. 23, pp. 131–176, by J. Whitehead, Tioxide Group PLC.

1. J. Barksdale, *Titanium, Its Occurrence, Chemistry and Technology*, 2nd ed., Ronald Press Co., New York, 1966.
2. R. Field and P. L. Crowe, *The Organic Chemistry of Titanium*, Butterworth & Co., (Publishers) Ltd., London, U.K., 1965.
3. *Gmelins Handbuch der Anorganischen Chemie*, 8th ed., Springer-Verlag, Berlin, Germany, 1979.
4. P. Pascal, *Nouveau Traite de Chimie Minerale*, Tome IX, Masson et Cie, Paris, France 1963.
5. F. A. Grant, *Rev. Mod. Phys.* **31**, 646–674 (1959).
6. R. J. H. Clark, *The Chemistry of Titanium and Vanadium*, Elsevier, Amsterdam, The Netherlands, 1968.
7. J. B. Goodenough and A. Hamnett, in O. Madelung, ed., *Landolt-Bornstein Semiconductors*, Group 3, Vol. 17, Springer-Verlag, Berlin, Germany, pp. 133–166, 1984.

8. T. A. Egerton and A. Tetlow, in R. Thompson, ed., *Industrial Inorganic Chemicals: Production and Uses*, The Royal Society of Chemistry, Cambridge, U.K., 1995, Chapt. 13.

9. JANAF Thermochemical Tables, 3rd ed., *J. Phys. Chem. Ref. Data*, **14**(Suppl. 1) (1985).

10. R. J. H. Clark, in A. F. Trotman Dickenson, ed., *Comprehensive Inorganic Chemistry*, Pergamon, London, U.K., 1973, Chap. 32.

11. G. F. Eveson, in R. Thompson, ed., *Speciality Inorganic Chemicals*, The Royal Society of Chemistry, London, U.K., 1980.

12. P. Ehrlich, in G. Brauer, ed., *Handbook of Preparative Inorganic Chemistry*, 2nd ed., Vol. 2, Academic Press, Inc., New York, 1965.

13. G. P. Luchinskii, *Chemistry of Titanium*, Khimiya, Moscow, Russia, 1971.

14. Y. G. Goreschenko, *Chemistry of Titanium*, Nankova Dumka, Kiev, Ukraine, 1970.

15. O. Knacke, O. Kubaschewski, and K. Hesselmann, *Thermochemical Properties of Inorganic Substances*, Springer-Verlag, Berlin, Germany, 1991.

16. I. Lewkowicz, *Diffus. Defect Data*, **B**(49–50), 239–279 (1996).

17. P. Peshov and G. Bliznakov, *J. Less Common Metals*, **14**, 23 (1968).

18. J. E. Hove and W. C. Riley, *Modern Ceramics, Some Principles and Concepts*, John Wiley & Sons, Inc., New York, 1965.

19. J. J. Kim and C. H. McMurthy, *Ceram. Eng. Sci. Prog.* 306–325 (1985).

20. J. M. Gomes, K. Uchida, and M. M. Wong, U.S. Bureau of Mines, R. I. 8053, Washington, D.C., 1975.

21. U.S. Pat. 4,022,872, J. F. Edd (to Alcoa).

22. C. C. Wang and co-workers, *J. Mat. Sci.* **30**, 1627 (1995).

23. E. K. Storms, *The Refractory Carbides*, Academic Press, Inc., New York, 1967, pp. 1–17.

24. E. G. Kendall, in J. E. Hove and W. C. Riley, eds., *Intermetallic Materials in Ceramics for Advanced Technologies*, John Wiley & Sons, Inc., New York, 1965, p. 143.

25. J. Hofmann, *Pokroky Praskove Metal*, **3**, 5–18 (1985).

26. D. H. Jack, in R. Brook, ed., *Refractory Carbides in Concise Encyclopedia of Advanced Ceramic Materials*, Pergamon, Oxford, U.K., 1991, pp. 391–393.

27. L.-M. Berger, *J. Hard. Mat.* **3**, 3–15 (1992).

28. U.S. Pat. 3,786,133 (Jan. 15, 1974), S.-T. Chiu (to Quebec Iron and Titanium Corp.).

29. U.S. Pat. 4,521,385 (Mar. 2, 1982) (to Ontario Research Foundation).

30. G. W. Elger, *Preparation and Chlorination of Titanium Carbide from Domestic Titaniferous Ores*, Report of Investigation 8497, U.S. Department of Interior, Bureau of Mines, Washington, D.C., 1980.

31. U.K. Pat. Appl. 2,172,276A (Mar. 4, 1986), M. A. Janey (to U.S. Department of Energy).

32. S. R. Blackburn, T. A. Egerton, and A. G. Jones, *Brit. Ceramics Proc.* **47**, 87 (1991).

33. A. I. Karasev and co-workers, *Vopr. Khim. Khim. Tekhnol.* **31**, 153 (1973).

34. G. J. Vogt and L. R. Newkirk, *Proc. Electrochem. Soc.* **86**, 164 (1986).

35. U.S. Pat. 4,196,178 (1980) T. Iwai, T. Kawahito, and T. Yamada (to Ube).

36. D. Seyferth and G. Mignani, *J. Mat. Sci. Lett.* **7**, 487 (1988).

37. Eur. Pat. 0,674,025A (Mar. 26, 1994) (to Merck Patent GmbH).

38. G. W. A. Fowles, *Prog. Inorg. Chem.* **6**, 1–36 (1964).

39. P. A. Cox, *Transition Metal Oxides*, Oxford University Press, New York, 1995.

40. E. P. Meagher and G. A. Lager, *Canad. Miner.* **17**, 77 (1979).

41. C. R. A. Catlow and R. James, in M. W. Roberts and J. M. Thomas, eds., *Chemical Physics of Solids and Their Surfaces*, Vol. 8, Royal Society of Chemistry, London, U.K., 1981, p. 108.

42. R. J. D. Tilley, in M. W. Roberts and J. M. Thomas, eds., *Chemical Physics of Solids and Their Surfaces*, Vol. 8, Royal Society of Chemistry Specialist Publications, London, U.K., 1980, p. 121.

43. R. J. D. Tilley, *Defect Crystal Chemistry and Its Applications*, Blackie, Glasgow, Scotland, 1987.

44. Belg. Pat. 890,744 (to I.M.I. Marston Ltd.).

45. Gutbier and co-workers, *Z. Anorg. Allgem. Chem.* **162**, 87 (1927).

46. C. de Rohden, *Chimie Industrie*, **75**, 287 (1956).

47. E. Santecesaria and co-workers, *J. Coll. Inter. Sci.* **111**, 44 (1986).

48. E. Matijevic, M. Budnik, and L. Meites, *J. Coll. Inter. Sci.* **61**, 302 (1977).

49. H. Becker, E. Klein, and H. Redman, *Farbe Lack*, **70**, 779 (1964).

50. U.S. Pat. 1,758,528 (1930), Mecklenburg.

51. Y. Inoue and M. Tsuji, *J. Nucl. Sci. Technol.* **13**, 85 (1976).

52. Y. Inoue and M. Tsuji, *Bull. Chem. Soc. Jpn.* **49**, 111 (1976).

53. Y. Inoue and M. Tsuji, *Bull. Chem. Soc. Jpn.* **51**, 479, 794 (1978).

54. G. R. Doshi and V. N. Sastry, *Indian J. Chem.* **15A**, 904 (1977).

55. U.S. Pat. 94,004,271 (Aug. 21, 1992) (to Engelhard Corp.).

56. A. M. Andrianov and co-workers, *J. Appl. Chem. USSR*, **51**, 1789 (1978).

57. R. Marchand, L. Brohan, and M. Tournaux, *Mat. Res. Bull.* **15**, 1129 (1980).

58. R. G. McQueen, J. C. Jamieson, and S. P. Marsh, *Science*, **155**, 1404 (1967).

59. A. Navrotsky, J. C. Jamieson, and O. J. Kleppa, *Science*, **158**, 388 (1967).

60. I. E. Grey and co-workers, *Mater. Res. Bull.* **23**, 743 (1988).

61. K. Kusaba and co-workers, *Phys. Chem. Minerals*, **154**, 238 (1988).

62. C. J. Howard, T. M. Sabine, and F. Dickson, *Acta. Cryst. B*, **47**, 462 (1991).

63. T. Mitsuhashi and O. J. Kleppa, *J. Am. Ceram. Soc.* **62**, 356 (1979).

64. J. Criado and C. Real, *J. Chem. Soc. Faraday Trans. I*, **79**, 2765 (1983).

65. J. R. De Vore and A. H. J. Pfund, *J. Opt. Soc. Am.* **37**, 826 (1947).

66. D. C. Cronmeyer, *Phys. Rev.* **87**, 876 (1952).

67. Eur. Pat. 0,611,039A (Feb. 10, 1993), D. G. Meina (to Tioxide Specialties).

68. W. Piekarczyk, *Int. Symp. Reinst. Wiss. Tech. Tagungsber.* **1**, 213 (1966).

69. J. H. Jean and T. A. Ring, in R. W. Davidge, ed., *Novel Ceramic Fabrication Processes and Applications*, Institute of Ceramics, Stoke-on-Trent, U.K., 1986, pp. 11–33.

70. R. I. Bickley and co-workers, *J. Solid State Chem.* **92**, 178 (1991).

71. S. Vemury and S. Pratsinis, *J. Am. Ceram. Soc.* **78**, 2984 (1995).

72. K.-N. P. Kumar, K. Keizer, and A. J. Burgraff, *J. Mater. Chem.* **3**, 1141 (1993).

73. E. M. Logothetis, L. H. Van Vlack, ed., *Resistive-Type Sensors in Automotive Sensors*, American Ceramic Society, Westerville, Ohio, 1980, pp. 281–301.

74. V. P. S. Judin, *Chem. Brit.* **29**, 503 (1993).

75. Jpn. Pat. 07,257,923 (Mar. 22nd, 1994) (to Ishihara Sangyo Kaisha Ltd.).

76. Eur. Pat. 0,539,878B (Oct. 25th, 1991) (to Nippon Shokubai Co. Ltd.).

77. M. S. Wainwright and N. R. Foster, *Catal. Rev. Sci. Eng.* **19**, 211 (1979).

78. Jpn. Pat. 04,310,240A (Apr. 8th, 1991) (to Sakai Chemical Industry Co. Ltd.).

79. U.S. Pat. 4,735,788 (Dec. 12th, 1984) (to Societe Nationale Elf Aquitaine).

80. M. R. Hoffman and co-workers, *Chem. Rev.* **95**, 69 (1995).

81. M. Schiavello, ed., *Photocatalysis and Environment, Trends and Applications*, Vol. 237C, D. Riedel, Dordrecht, Boston, 1984.

82. A. Fujishima and K. Honda, *Nature*, **38**, 238 (1971).

83. M. Schiavello, ed., *Photoelectrochemistry, Photocatalysis and Photoreactors*, Vol. 146C, D. Riedel, Dordrecht, Boston, 1984.

84. M. Gratzel and B. O'Regan, *Nature*, **353**, 737 (1991).

85. R. R. Blakey and J. E. Hall, in P. A. Lewis, ed., *Pigment Handbook*, 2nd ed., Vol. 1, John Wiley & Sons, Inc., New York, 1988.

86. D. H. Solomon and D. G. Hawthorne, *Chemistry and Pigments*, John Wiley & Sons, Inc., New York, 1983, Chapt. 2.
87. J. Gambogi, in *Minerals Year Book*, Vol. 1, U.S. Department of Interior, Bureau of Mines, Washington, D.C., 1992.
88. *Eur. Coatings J.* (July/Aug. 1996).
89. P. B. Howard and G. D. Parfitt, *Croatica Chemica Acta*, **50**, 15 (1977).
90. U.S. Pat. 3,437,502 (Apr. 8, 1969), A. J. Werner (to E. I. du Pont de Nemours & Co., Inc.).
91. U.S. Pat. 405,223 (Nov. 2, 1977), P. B. Howard (to Tioxide).
92. M. J. B. Franklin and co-workers, *J. Paint. Tech.* **42**, 551, 740 (1971).
93. M. Cremer, *Polym. Paint Colour J.* **173**, 86 (1983).
94. A. M. van Herk and co-workers, *Proc. Organic Coatings Sci. Tech.* **19**, 219 (1993).
95. J. R. Hackman, in P. E. Lewis, ed., *Pigment Handbook*, John Wiley & Sons, Inc., New York, 1985, pp. 375, 383, 403.
96. F. Hund, *Angew. Chem. Ind. Ed.* **1**, 41 (1962).
97. U.S. Pat. 5,456,749 (July 2, 1992) (to Merck Patent GmbH).
98. E. Wendling and J. de Lavillandre, *Bull. Soc. Chim. France*, 2142 (1967).
99. J. A. Bland, *Acta Cryst.* **14**, 875–881 (1961).
100. F. S. Galasso, *Structure, Properties and Preparation of Perovskite-Type Compounds*, Pergamon, Oxford, U.K., 1969.
101. K. D. Reeve, *Mater. Sci. Forum*, **34–36**, 567 (1988).
102. Y. Fujiki, Y. Komastsu, and T. Sasaki, *Ceram. Jpn.* **19**, 126 (1984).
103. H. Hukunaka and J. Pan, *Advanced Structural Materials*, Elsevier, Amsterdam, the Netherlands, 1991, p. 45.
104. S. Andersson and A. D. Wadsley, *Acta. Cryst.* **15**, 194 (1962).
105. T. Shimizu, *Kagaku Kogyo*, **31**, 752 (1980).
106. J.-K. Lee, K.-H. Lee, and H. Kim, *J. Mat. Sci.* **31**, 5493 (1996).
107. E. K. Ovechkin and co-workers, *Inorg. Mater.* **7**, 1000 (1968).
108. W. W. Riches, in T. C. Patton, ed., *Pigment Handbook*, Vol. 1, Wiley-Interscience, New York, 1973, p. 51.
109. D. Kolar, in R. J. Brook, ed., *Concise Encyclopedia of Advanced Ceramic Materials*, Pergamon, Oxford, 1991, p. 484.
110. B. Huybrechts, K. Ishizaki, and M. Takata, *J. Mat. Sci.* **30**, 2463 (1995).
111. B. Freudenberg and A. Mocellin, *J. Amer. Ceram. Soc.* **70**, 33 (1987).
112. R. A. Slepetys and P. A. Vaughan, *J. Phys. Chem.* **73**, 2157 (1969).
113. B. Freudenberg, in Ref. 109, p. 20.
114. D. Bersani and co-workers, *J. Mat. Sci.* **13**, 3153 (1996).
115. F. H. W. Wachholtz, *Chim. Peint.* **16**, 141 (1953).
116. S. C. Abrahams and K. Nassau, in Ref. 109, p. 351.
117. L. Shcherbakova and co-workers, *Usp. Khim.* **48**, 423 (1979).
118. V. A. Reznichenko and G. A. Menyailova, *Synthetic Titanates*, Nauka, Russia, 1977, p. 136.
119. P. Ganguly, D. Parkash, and C. N. R. Rao, *Phys. Status Solidi*, **36**, 669 (1976).
120. R. Colton and J. H. Canterford, *Halides of the First Row Transition Metals*, Wiley-Interscience, New York, 1969.
121. P. Ehrlich and G. Pietzka, *Z. Anorg. Allgem. Chem.* **275**, 121–140 (1954).
122. U.S. Pat. 5,225,178 (1993), T. A. O'Donnell, D. G. Wood, and T. K. Pong (to University of Melbourne).
123. Jpn. Pat. 76,28,599 (Aug. 20, 1976), K. Egi.
124. G. Natta, P. Corrodini, and G. Allegra, *J. Polym. Sci.* **51**, 399 (1961).
125. Ger. Pat. 2,110,380 (Oct. 7, 1971), J. P. Hermans and P. Henrique (to Solvay et Cie).
126. U.S. Pat. 3,681,256 (Feb. 16, 1972), H. W. Blunt (to Hercules Inc.).

127. Jpn. Pat. 74,20,476 (May 24, 1974), S. Okudaira (to Toho Titanium Co.).

128. Ger. Pat. 2,600,593 (July 15, 1976), N. Kuroda, T. Shiraishi, and A. Itoh (to Nippon Oil Co.).

129. U.S. Pat. 4,124,532 (Nov. 7, 1978), U. Giannini and co-workers (to Mondedison Spa).

130. J. M. Thomas and W. J. Thomas, *Heterogeneous Catalysis*, Academic Press, Inc., London, U.K., 1967.

131. P. Brand and H. Sackmann, *Z. Anorg. Allgem. Chem.* **321**, 262 (1963).

132. J. E. D. Davies and D. A. Long, *J. Chem. Soc.* **A**, 2560 (1968); D. M. Adams and D. C. Newton, *J. Chem. Soc.* **A**, 2262 (1968).

133. G. W. A. Fowles and F. H. Pollard, *J. Chem. Soc.* 258 (1955).

134. Ger. Pat. 1,043,290 (Nov. 13, 1958), R. H. Walsh (to Columbia Southern Chem. Corp.).

135. Nor. Pat. 92,999 (Dec. 8, 1958), A. G. Oppegaard, A. Helge, and H. Barth (to Titan Co. A/s).

136. U.S. Pat. 2,962,353 (Nov. 29, 1960), J. N. Haimsohn (to Stauffer Chem. Co.).

137. K. Dehnicke, *Angew. Chem. Intern. Ed. Engl.* **2**, 325 (1963).

138. H. Schafer, E. Weise, and F. Wartenpfuhl, *Z. anorg., allgem. Chem.* **295**, 268 (1985).

139. P. L. Vijay, C. Subramanian, and C. S. Rao, *Trans. Indian. Inst. Met.* **29**, 355 (1976); E. C. Perkins and co-workers, *Fluidised-Bed Chlorination of Ores and Slags*, U.S. Bureau of Mines Report Investigation 6317, Department of Interior, Washington, D.C., 1963.

140. A. Z. Bezukladnikov, *J. Appl. Chem. USSR*, **40**, 25 (1967).

141. N. M. Karpinskaya and S. N. Andreev, *Russ. J. Inorg. Chem.* **13**, 25 (1968).

142. P. Ehrlich, W. Gutsche, and H. J. Seifert, *Z. anorg. allgem. Chem.* **312**, 80 (1961).

143. P. Brand and J. Schmidt, *Z. Anorg. Allgem. Chem.* **348**, 257 (1966).

144. W. Klemm and L. Grimun, *Z. Anorg. Allgem. Chem.* **249**, 198 (1942).

145. L. Hock and W. Knauf, *Z. Anorg. Allgem. Chem.* **228**, 204 (1936).

146. H. G. Schnering, *Naturwissenschaften*, **53**, 359 (1966).

147. K. Funaki, K. Uchimura, and H. Matsunaga, *Kogyo Kagaku Zasshi*, **64**, 129 (1961).

148. G. W. A. Fowles, T. E. Lester, and B. J. Russ, *J. Chem. Soc.* **A**, 805 (1968).

149. O. Hassel and H. Kringstad, *Z. Phys. Chem. (Leipzig)*, **B15**, 274 (1932).

150. R. F. Rolsten and H. H. Siler, *J. Am. Chem. Soc.* **79**, 5891 (1957).

151. M. J. Frazer and B. Rimmer, *J. Chem. Soc.* **A**, 69 (1968).

152. I. E. Campbell and co-workers, *Trans. Electrochem. Soc.* **93**, 271 (1948).

153. Jpn. Pat. 78,21,107 (Feb. 27, 1978), T. Okano, N. Wada, and Y. Kobayashi (to Mitsubishi Chemical Industries Ltd.).

154. B. Aronsson, T. Lundstrom, and S. Rundqvist, *Borides, Silicides and Phosphides*, Methuen, London, U.K., 1965.

155. J. Beaudouin, *C.R. Acad. Sci. Paris, Ser. C*, **263**, 993 (1966); Brit. Pat. 901,402 (July 18, 1962) (to E. I. du Pont de Nemours & Co., Inc.).

156. I. V. Petrusevich, L. A. Nisel'sen, and A. I. Belyaev, *Izv. Aakd. Nauk. SSSR, Metally*, **6**, 52 (1965).

157. Jpn. Pat. 80,62,850 (May 12, 1980) (to Mitsubishi Metal Corp.).

158. Ger. Pat. 2,261,523 (June 28, 1973), H. Carbonnel and L. Hamon (to Groupement Atomique Alsacienne Atlantique).

159. U.S. Pat. 4,028,276 (June 7, 1977), J. C. Harden and S. V. R. Mastrangelo (to E. I. du Pont de Nemours & Co., Inc.).

160. A. M. Clark, *Hays Mineral Index*, 3rd ed., Chapman and Hall, London, U.K., 1993.

161. B. Notari, *Stud. Surf. Sci. Catal.* **37**, 413 (1988).

162. B. Notari, in D. D. Eley, W. O. Haag, and B. Gates, eds., *Advances in Catalysis*, Vol. 41, Academic Press, Inc., New York, 1996, p. 253.

163. M. Knausenberger, G. Brauer, and K. A. Gingerich, *J. Less Common Metals*, **8**, 136 (1965); K. A. Gingerich, *Nature*, **200**, 877 (1963).

164. Eur. Pat. Appl. 0,339,452A (Apr. 28, 1988), (to Henkel Kgaa).

165. Rus. Pat. 815,012 (Mar. 23, 1981), N. Z. Yaramenko and I. P. Dobrovolskii.

166. V. P. Titov and co-workers, *Izv. Vyssh. Uchebn. Zaved. Khim. Khim. Tekhnol.* **23**, 64 (1980).

167. A. Clearfield and G. D. Smith, *Inorg. Chem.* **8**, 431 (1969).

168. A. Clearfield, ed., *Inorganic Ion Exchange Materials*, CRC Press, Inc., Boca Raton, Fla., 1982.

169. A. Clearfield and D. Thakur, *Appl. Catal.* **26**, 1 (1986).

170. Jpn. Pat. 80,51,442 (Apr. 15, 1980), E. Kobayashi and T. Kanamaya (to Agency of Industrial Science and Technology).

171. A. Ludmany, G. Torok, and L. G. Nagy, *Radiochem. Radioanal. Lett.* **45**, 387 (1980).

172. G. Alberti and co-workers, *J. Inorg. Nucl. Chem.* **42**, 1637 (1980).

173. Brit. Pat. Appli. 53,781/77 (to Laporte Industries Ltd.).

174. J. D. Ellis and A. G. Sykes, *J. C. S. Dalton*, 537 (1973).

175. C. Baes and J. E. Mesmer, *The Hydrolysis of Cations*, Wiley-Interscience, New York, 1975.

176. U. Manivel, S. Bangaruswamy, and J. B. Rao, *Leather Science (Madras)*, **27**, 257 (1980).

177. U.S. Pat. 4,731,089 (Mar. 15, 1988), A. D. Covington (to Tioxide Group PLC).

178. Brit. Pat. 2,062,596A (May 2, 1980), D. L. Motov and co-workers.

179. E. R. Scheffer, in I. M. Kolthoff and P. J. Elving, eds., *Treatise on Analytical Chemistry*, Vol. 5, Pt. II, Interscience Publishers, New York, 1961, pp. 1–60.

180. W. T. Elwell and J. Whitehead, in C. L. and D. L. Wilson, eds., *Comprehensive Analytial Chemistry*, Vol. 1c, Elsevier, Amsterdam, the Netherlands, 1979, pp. 627–636.

181. J. D. Norris, in A. Townshend, ed., *Encyclopaedia of Analytical Science*, Academic Press, Inc., New York, 1995, pp. 5236–5240.

182. I. H. Tipton and M. J. Cook, *Health Phy.* **9**, 103 (1963).

183. L. C. Maillard and J. E. Horie, *Bull. Acad. Med. Paris*, **115**, 631 (1936).

184. N. P. Timakan and co-workers, *Vap. Teor. Khim. Tomsk. Med. Inst.*, 114 (1967).

185. H. M. Perry, Jr., and E. F. Perry, *J. Clin. Invest.* **38**, 1452 (1959).

186. J. Ferin, *Proceedings of an International Symposium, 1970*, 3rd ed., Unwin Brothers, Ltd., Old Woking, Surrey, U.K., 1971, pp. 283–292.

187. E. I. Hamilton, M. J. Minski, and J. J. Cleary, *Sci. Total Environ.* **1**, 341 (1972).

188. J. V. Crabb and co-workers, *Am. Ind. Hyg. Assoc.* **28**, 8 (1967); **29**, 106 (1968).

189. *Bioassay of Titanium Dioxide for Possible Carcinogenicity*, National Cancer Institute Technical Report No. 97, U.S. Department of Health, Education and Welfare, Washington, D.C., 1979.

190. K. P. Lee and co-workers, *Tox. App. Pharmacol.* **79**, 179–192 (1985).

191. U. Heinrich and co-workers, *Inhalation Toxicology*, **7**, 533–556 (1995).

192. J. L. Chen and W. E. Fayerweather, *J. Occup. Med.* **30**, 937–942 (1988).

193. "Threshold Limit Values for Chemical Substances and Physical Agents and Biological Exposure Indexes," *Proceedings of American Conference of Government Hygienists*, Cincinnati, Ohio, 1996.

194. Guidance Note EH40/96, Occupational Exposure Limits, Health and Safety Executive, Her Majesty's Stationery Office, London, 1996.

Terry A. Egerton
Tioxide Group Services Limited

ORGANIC

Organic titanium compounds are materials having a covalent bond between titanium and another atom that is also bonded to a carbon-based group. Titanium tetrachloride [7550-45-0], $TiCl_4$, the basic raw material from which organic titanate compounds are made (see TITANIUM COMPOUNDS, INORGANIC), is readily converted to tetraisopropyl titanate, TYZOR TPT [546-68-9], by the Nelles process. This ester can be converted by alkoxy exchange (transesterification) to a wide variety of tetraalkyl titanates, sold commercially worldwide. The tetraalkyl titanates react with other ligands and chelating agents (qv), such as glycols (qv), β-diketones and ketoesters, α-hydroxycarboxylic acids, and alkanolamines (qv), to give complexes having properties significantly different from the starting materials. These complexes are also important items of commerce.

True organometallic compounds having a titanium–carbon bond are prepared from $TiCl_4$ by reaction with main-group organometallics such as organomagnesium, sodium, or lithium reagents. Most simple C–Ti bonds are very unstable. The predominant exceptions are bis(cyclopentadienyl)titanium dichloride [1271-19-8] and its analogues. Many organometallics and complexes of trivalent titanium are stable at room temperature, but most are attacked by oxygen and moisture. Organic derivatives of divalent titanium are much less common. Titanium(II) and titanium(0) are potent reducing agents.

Titanium alkoxides (titanate esters) are superb catalysts for esterification, transesterification, and cross-linking of ester-containing resins and epoxides (see ALKOXIDES, METALS; CATALYSIS). Water-soluble titanium chelates are widely used in cross-linking (gelling) of dilute polysaccharide solutions, which are used in oilfield fracturing fluid applications (see PETROLEUM). Titanium's great affinity for oxygen atoms is reflected in its bonding to oxide surfaces, such as glass (qv) and plastic, to yield a scratch-resistant oxide coating (see COATINGS). Titanates are also used to cross-link silicone resins and as curing agents for wire coatings (see SILICONES). Titanates disperse pigments (qv) and often help bond resins to fillers and reinforcement agents such as fiber glass.

Alkoxides

The standard manufacturing method for tetraalkyl titanates, such as TYZOR TPT, or tetra-n-butyl titanate, TYZOR TBT [5593-70-4], involves the addition of $TiCl_4$ to an alcohol. In a series of reversible displacement reactions, the alkoxy substitution products and hydrochloric acid form as follows:

$$TiCl_4 + ROH \rightleftharpoons ROTiCl_3 + HCl$$

$$ROTiCl_3 + ROH \rightleftharpoons (RO)_2TiCl_2 + HCl$$

$$(RO)_2TiCl_2 + ROH \rightleftharpoons (RO)_3TiCl + HCl$$

$$(RO)_3TiCl + ROH \rightleftharpoons (RO)_4Ti + HCl$$

The reaction can be driven to the tetraalkoxide stage by addition of an amine or ammonia to scavenge the liberated hydrochloric acid. The amine or ammonium hydrochloride that forms can be filtered from the reaction mass and the tetraalkyl titanate purified by distillation. If the reaction is run in the starting alcohol as solvent, the chloride salts formed are in a finely divided state and difficult to filter. When the reaction is run in the presence of an inert hydrocarbon solvent such as heptane or toluene, a much more readily filterable salt is obtained. The solution of crude tetraalkyl titanate can be distilled to remove solvent and give a pure product (1,2).

TYZOR TPT and the tetraethyl titanate, TYZOR ET [3087-36-3], have also been prepared by direct electrochemical synthesis. The reaction involves anode dissolution of titanium in the presence of the appropriate alcohol and a conductive admixture (3).

High purity tetraalkyl titanates can be obtained by a process whereby the liquid organic titanate esters are subjected to partial hydrolysis to form a solid that can be separated from the reaction mass. The remaining liquid is distilled to give a high purity product (4). For example, tetraisopropyl titanate purified in this manner contains:

Element, ppm	Starting titanate	Purified titanate
Fe	880	2
Si	370	1
Na	480	3
V	110	2
U	1	<0.001

The partially alkoxylated chlorotitanates, $(RO)_{4-n}TiCl_n$, can be prepared in high purity by reaction of $TiCl_4$ with an organosilane ester, $Si(OR)_4$ (see SILICON COMPOUNDS). The degree of esterification of the titanium can be controlled by the amount of silane ester used. When n is 3 or 4, the addition of the appropriate alcohol and an amine receptor is required (5).

Higher alkoxides, such as tetra(2-ethylhexyl) titanate, TYZOR TOT [1070-10-6], can be prepared by alcohol interchange (transesterification) in a solvent, such as benzene or cyclohexane, to form a volatile azeotrope with the displaced alcohol, or by a solvent-free process involving vacuum removal of the more volatile displaced alcohol. The affinity of an alcohol for titanium decreases in the order: primary > secondary > tertiary, and branched > unbranched. Exchange processes are more convenient than direct synthesis of tetraalkoxide from $TiCl_4$, an alcohol, and a base because a metal chloride need not be handled. However, in general, traces of impurities of mixed tetraalkyl titanates can result.

$$(i\text{-}C_3H_7O)_4Ti + 4\ ROH \longrightarrow (RO)_4Ti + 4\ i\text{-}C_3H_7OH$$

Phenols react readily with tetraalkoxides to give highly colored (yellow to orange) titanium tetraphenoxides (6). TYZOR KTM [83897-99-8] is the bis-cresyl titanate derived from TYZOR TBT. The tetracresyl titanate [28503-70-0] is also available commercially.

Mixed esters can be prepared from $Cl_nTi(OR)_{4-n}$ and a second alcohol in the presence of a base or by mixing a tetraalkoxide and the second alcohol in the desired proportions and flash-evaporating the mixed alcohols. Mixing of two pure tetraalkoxides of titanium also leads, via a rapid ester interchange reaction, to a mixture of all possible combinations of tetraalkyl titanates.

A rarely used preparation involves the reaction of tetraamido titanates, such as tetradimethylamino titanate [*3275-24-9*], with an alcohol. This reaction goes to completion because of the greater affinity of titanium for oxygen over nitrogen.

$$((CH_3)_2N)_4Ti + 4\ ROH \longrightarrow (RO)_4Ti + 4\ NH(CH_3)_2$$

Fluoroalkyl-Substituted Titanates. Tetrahexafluoroisopropyl titanate [*21416-30-8*] can be prepared by the reaction of $TiCl_4$ and hexafluoroisopropyl alcohol [*920-66-1*], in a process similar to that used for TYZOR TPT (7). Alternatively, it can be prepared by the reaction of sodium hexafluoroisopropoxide and $TiCl_4$ in excess hexafluoroisopropyl alcohol (8). The fluoroalkyl material is much more volatile than its hydrocarbon counterpart, TYZOR TPT, and is used to deposit titanium on surfaces by chemical vapor-phase deposition (CVD).

β-Chloroalkoxy Titanates. The reaction of $TiCl_4$ with epoxides, such as ethylene or propylene oxide (qv), gives β-chloroalkyl titanates (8,9). One example is $Ti(OCH_2CH_2Cl)_4$ [*19600-95-5*]. The β-chloroalkoxy titanates can be used to bind refractory powders and in admixture with diethanolamine to impart thixotropy to emulsion paints (10).

$$4\ RCH\!\!\overset{O}{\overbrace{}}\!\!CH_2 + TiCl_4 \longrightarrow (RCHClCH_2O)_4Ti$$

Titanium Complexes of Unsaturated Alcohols. Tetraallyl titanate can be prepared by reaction of TYZOR TPT with allyl alcohol, followed by removal of the by-product isopropyl alcohol. Ebullioscopic molecular weight determinations support its being the dimeric product, octaalloxydititanium. A vinyloxy titanate derivative can be formed by reaction of TYZOR TPT with vinyl alcohol formed by enolization of acetaldehyde (11):

$$4\ CH_3CH\!\!=\!\!O \longrightarrow 4\ CH_2\!\!=\!\!CHOH + Ti(OR)_4 \longrightarrow Ti(OCH\!\!=\!\!CH_2)_4 + 4\ ROH$$

Properties and Reactions. *Associations.* Organic titanates tend to associate (12). Although a titanium (IV) atom strives to achieve a coordination number of 6 by sharing electron pairs from nearby ester molecules, this tendency may be opposed by stearic crowding. X-ray diffraction (xrd) experiments performed on tetramethyl and tetraethyl titanate single crystals show that, in the solid state, these are tetramers (13). In benzene solution, however, cryoscopic measurements suggest that tetraethyl and tetra-*n*-butyl titanate are trimeric; tetraisopropyl titanate is monomeric in nature (14). Titanium nmr experiments support a monomeric structure for tetraisopropyl and tetra-*t*-butyl titanate [*3087-39-6*] (15). These conclusions are supported by more recent x-ray absorption studies (xanes-exafs), which indicate only one kind of Ti–O bond distance (1.8 nm) for tetra-*t*-amyl and tetraisopropyl titanate, whereas there are bond distances of 18.0 nm and 20.5 nm for tetraethyl and tetra-*n*-butyl titanate. These latter have been attributed to terminal and bridging groups on trimeric oligomers, in which each titanium atom is five coordinate (Fig. 1).

Increased molecular association increases viscosity. Tetra-*t*-butyl titanate and tetraisopropyl titanate are mobile liquids at room temperature; tetra-*n*-butyl titanate and tetra-*n*-propyl titanate, TYZOR NPT [*3087-37-4*], are thick and syrupy. The boiling points of these materials also reflect association (Table 1).

Hydrolysis and Condensation. The rate of hydrolysis of the tetraalkyl titanates is governed by the nature of the alkoxy groups. The lower titanium alkoxides, with the exception of tetramethyl titanate [*992-92-7*], are rapidly hydrolyzed by moist air or water, giving a series of condensed titanoxanes, $(Ti-O-Ti-O-)_x$ (17). As the chain length of the alkyl group increases, the rate of hydrolysis decreases. Titanium methoxides, aryloxides, and C-10 and higher alkyl titanates are hydrolyzed much more slowly.

In a limited amount of water, the first product of hydrolysis is the monohydroxy ester, which cannot be isolated because it immediately reacts with another alkyl titanate to give a μ-oxo dimer:

$$(RO)_3Ti-OR + H_2O \longrightarrow (RO)_3Ti-OH$$

$$(RO)_3Ti-OH + (RO)_4Ti \longrightarrow (RO)_3Ti-O-Ti(OR)_3 + ROH$$

Further addition of water leads to the stepwise formation of a whole range of condensation polymers. Initially, condensation most likely occurs at the less-hindered terminal ends forming linear oligomers. For example, TYZOR BTP [*9022-96-2*] is believed to be a linear C-8 oligomer. It is a viscous liquid soluble in hydrocarbon and alcohol solvents (18–20). Further condensation most likely leads to branching and the products become insoluble. Ultimately, hydrous titanium oxides are formed, although complete hydrolysis to TiO_2 is difficult to achieve without heating the reaction mixture.

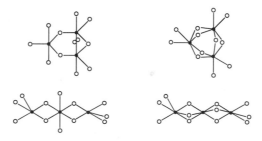

Fig. 1. Proposed structures for Ti $(OC_2H_5)_4$, where • and ∘ represent titanium and the ethoxy group, respectively.

Table 1. Titanium(IV) Tetraalkoxides and Tetraaryloxides

Titanate	CAS Registry Number	Formula	Mp, °C	Bp^a, °C	Other properties
			Alkoxides		
methoxide	[992-92-7]	Ti(OCH$_3$)$_4$	210	170$_{1.33}$ sublimes	dipole moment 5.37 × 10^{-30} C·m^b
ethoxide	[3087-36-3]	Ti(OC$_2$H$_5$)$_4$		103$_{13}$	n_D^{35} 1.5051, d_4^{35} 1.107 g/cm^3, η_{25} 44.45 mPa·s(=cP)
allyloxide	[5128-21-2]	Ti(OCH$_2$CH=CH$_2$)$_4$		141–142$_{133}$	n_D^{35} 1.5381
n-propoxide	[3087-37-4]	Ti(OCH$_2$CH$_2$CH$_3$)$_4$		124$_{133}$	d_4^{35} 0.9970 g/cm^3, n_D^{35} 1.4803, η_{25} 161.35 mPa·s(=cP)
isopropoxide	[546-68-9]	Ti(OCH(CH$_3$)$_2$)$_4$	18.5	49$_{133}$	n_D^{35} 1.4568, η_{25} 4.5 mPa·s(=cP), d_4^{20} 0.9711 g/cm^3
n-butoxide	[5593-70-4]	Ti(OC$_4$H$_9$)$_4$	ca –50	142$_{133}$	d_4^{35} 0.9927 g/cm^3, n_D^{35} 1.4863, η_{25} 67 mPa·s(=cP), dipole moment 3.84 × 10^{-30} C·m^b
isobutoxide	[7425-80-1]	Ti(OCH$_2$CH(CH$_3$)$_2$)$_4$		141$_{133}$	n_D^{54} 1.4749, d_4^{50} 0.9601 g/cm^3, η_{25} 97.40 mPa·s(=cP)
sec-butoxide	[3374-12-7]	Ti(OCH$_3$CHC$_2$H$_5$)$_4$	ca –25	81$_{133}$	n_D^{35} 1.4550, d_4^{35} 0.9196 g/cm^3
tert-butoxide	[3087-39-6]	Ti(OC(CH$_3$)$_3$)$_4$	107	62–63$_{133}$	n_D^{20} 1.4436, d_4^{20} 0.8893 g/cm^3
n-pentoxide	[10585-24-7]	Ti(OC$_5$H$_{11}$)$_4$		158$_{133}$	η_{25} 79.24 mPa·s(=cP), d_4^{25} 0.9735 g/cm^3, n_D^{35} 1.4813
cyclopentyloxide	[1517-19-7]	Ti(OC$_5$H$_9$)$_4$	45	200–201$_{667}$	
n-hexyloxide	[7360-52-3]	Ti(OC$_6$H$_{13}$)$_4$		176$_{133}$	η_{25} 64.90 mPa·s(=cP), d_4^{25} 0.9499 g/cm^3, n_D^{20} 1.4830, dipole moment 5.54 × 10^{-30} C·m^b
cyclohexyloxide	[6426-39-7]	Ti(OC$_6$H$_{11}$)$_4$		190.5–192$_{133}$	d_4^{25} 1.0589 g/cm^3, n_D^{35} 1.5155
benzyloxide	[103-50-4]	Ti(OCH$_2$C$_6$H$_5$)$_4$		250$_{267}$ dec	
n-octyloxide	[3061-42-5]	Ti(OC$_8$H$_{17}$)$_4$		214$_{133}$	n_D^{20} 1.4810, d_4^{20} 0.9339 g/cm^3, dipole moment 5.68 × 10^{-30} cm^b
2-ethylhexyl-oxide	[1070-10-6]	Ti(OCH$_2$C$_2$H$_5$CHC$_4$H$_9$)$_4$	< –25	248–249$_{1467}$	n_D^{35} 1.4750
nonyloxide	[6167-42-6]	Ti(OC$_9$H$_{19}$)$_4$		264–265$_{200}$	n_D^{20} 1.4785, d_0^{20} 0.9241 g/cm^3, dipole moment 5.60 × 10^{-30} C·m^b
n-decyloxide		Ti(OC$_{10}$H$_{21}$)$_4$	68	265$_{27}$	d_4^{25} 0.87 g/cm^3, η_{25} 76.0 mPa·s(=cP)
isooctyloxide		Ti(OC$_8$H$_{17}$)$_4$		235$_{13}$	d_4^{25} 0.940

Table 1. (Continued)

Titanate	CAS Registry Number	Formula	Mp, °C	Bp^a, °C	Other properties
			Aryloxides		
isobornyloxide	[84215-64-5]	Ti(OC$_{10}$H$_{17}$)$_4$			
benzhydryloxide	[84215-65-6]	Ti(OCH(C$_6$H$_5$)$_2$)$_4$	57–59		d_4^{20} 0.87 g/cm^3
oleyloxide	[26291-85-0]	Ti(OC$_{18}$H$_{35}$)$_4$		288$_{13}$	orange-red solid
phenoxide	[2892-89-9]	Ti(OC$_6$H$_5$)$_4$	153–154	267$_{40}$	red solid
o-chlorophen- oxide	[22922-75-4]	Ti(OC$_6$H$_4$Cl)$_4$	145.5–147		
p-chlorophen- oxide	[13438-75-0]	Ti(OC$_6$H$_4$Cl)$_4$	84–86		
o-nitrophen- oxide	[55535-59-6]	Ti(OC$_6$H$_4$NO$_2$)$_4$	154–158		shiny black solid
p-nitrophen- oxide	[22922-76-5]	Ti(OC$_6$H$_4$NO$_2$)$_4$	97		grayish black solid
o-methylphen- oxide	[22922-73-2]	Ti(OC$_6$H$_4$CH$_3$)$_4$	48–51		
m-methylphen- oxide	[22949-88-0]	Ti(OC$_6$H$_4$CH$_3$)$_4$		323–325$_{40}$	red crystalline solid
1-naphthyloxide	[84215-66-7]	Ti(OC$_{10}$H$_7$)$_4$		^{c}	
2-naphthyloxide	[36452-22-9]	Ti(OC$_{10}$H$_7$)$_4$	60–64	^{c}	
resorcinyloxide	[34075-40-6]	Ti(O$_2$C$_6$H$_4$)$_2$			black solid
stearyloxide	[84215-73-4]	Ti(OC$_{18}$H$_{37}$)$_4$			
2,4,6-trinitro- phenoxide	[84215-67-8]	Ti(OC$_6$H$_2$(NO$_2$)$_3$)$_4$			dark-brown solid

^{a}Subscripted values are pressure in Pa. To convert Pa to mm Hg, divide by 133.
^{b}To convert C·m to D, divide by 3.336 × 10^{-30}.
^{c}Does not distill.

The less branched, more associated tetraalkyl titanates are more slowly hydrolyzed because titanium is more fully coordinated. The hydrolysis of tetraethyl titanate has also been considered in terms of its trimeric form:

$$2 \ Ti_3(OC_2H_5)_{12} + 4 \ H_2O \longrightarrow Ti_3(OC_2H_5)_8(O)_4Ti_3(OC_2H_5)_{11}$$

Several excellent reviews discussing the hydrolysis of tetraalkyl titanates are available (19,21–23).

Higher aliphatic alcohol and phenolic group-containing polytitanates may be prepared by transesterification of TYZOR BTP (24).

$$(2n + 2) \ ROH + C_4H_9 \!\!\left[\!\! \begin{array}{c} OC_4H_9 \\ | \\ OTi \\ | \\ OC_4H_9 \end{array} \!\!\right]_{\!\!n} \!\!\!O \longrightarrow RO \!\!\left[\!\! \begin{array}{c} OR \\ | \\ OTi \\ | \\ OR \end{array} \!\!\right]_{\!\!n} \!\!\!OR + (2n + 2) \ C_4H_9OH$$

Titanoxanes can also be prepared by reaction of a tetraalkyl titanate and carboxylic acids (25). If the ratio of carboxylic acid to tetraalkyl titanate is 1:1, a simple polymeric titanate ester is formed. If two or more moles of acid are used per mole of tetraalkyl titanate, the resulting polymeric titanate ester contains ester carboxylate groups.

$$z \ Ti(OR)_4 + z \ R'COOH \longrightarrow \left[\!\! \begin{array}{c} OR \\ | \\ O\!-\!Ti \\ | \\ OR \end{array} \!\!\right]_{\!\!z} \!\!\! + z \ R'COOR + z \ ROH$$

$$z \ Ti(OR)_4 + 2z \ R'COOH \longrightarrow \left[\!\! \begin{array}{c} OR \\ | \\ O\!-\!Ti \\ | \\ OOCR' \end{array} \!\!\right]_{\!\!z} \!\!\! + z \ R'COOR + 2z \ ROH$$

$$x \ Ti(OR)_4 + 3x \ R'COOH \longrightarrow R \!\!\left[\!\! \begin{array}{c} OOCR' \\ | \\ O\!-\!Ti \\ | \\ OOCR' \end{array} \!\!\right]_{\!\!n} \!\!\! OH + x \ R'COOR + (3x - 1) \ ROH$$

Similar ester carboxylate group containing polymeric titanate esters are obtained by the reaction of titanoxanes with carboxylic acids (26), by reaction of a tetraalkyl titanate with a carboxylic acid and 1–2 moles of water (27), or by reacting a polymeric metal acylate with a higher boiling carboxylic acid and removing the lower boiling carboxylic acid by distillation (28).

Titanoxanes can also be prepared by pyrolysis of tetraalkyl titanates at 200–250°C. Higher temperatures, however, can lead to thermal decomposition.

$$2(n\text{-}C_4H_9O)_4Ti \longrightarrow (n\text{-}C_4H_9O)_3TiOTi(OnC_4H_9)_3 + n\text{-}C_4H_9OH + C_4H_8$$

Oxidation of Ti(III) alkoxides also gives pure dimers (29):

$$4\,(RO)_3Ti + O_2 \longrightarrow 2\,(RO)_3TiOTi(OR)_3$$

Alternatively, titanoxanes can be prepared by reaction of tetraalkyl titanates with carboxylic acid anhydrides (30):

$$n\,Ti(OR)_4 + (n-1)\,(R'\overset{\overset{\displaystyle O}{\|}}{C})_2O \longrightarrow R\!\!\left[\!OTi(OR)_2\!\right]_n\!\!OR + 2(n-1)\,R'\overset{\overset{\displaystyle O}{\|}}{C}OR$$

If more than one acid anhydride group per mole of tetraalkyl titanate is used, a polytitanyl acylate is formed:

$$n\,Ti(OR)_4 + (2n-1)\left(R'\overset{\overset{\displaystyle O}{\|}}{C}\right)_2O \longrightarrow R\!\!\left[\!O\!\!-\!\!\underset{\underset{\displaystyle OR}{|}}{\overset{\overset{\displaystyle OOCR'}{|}}{Ti}}\!\right]_n\!\!OR + (3n-2)\,R'COOR$$

The polymeric acyl titanate esters are viscous liquids or waxes that are soluble in hydrocarbon solvents and can be used as TiO_2-dispersing agents, water-repellent agents for textile fabrics, and rust inhibitors for steel.

Organic-solvent-soluble, higher molecular weight polytitanoxanes, having a proposed rudder-shaped structure, can be prepared by careful addition of an alcohol solution of 1.0–1.7 moles of water per mole of tetraalkyl titanate, followed by distillation of the low boiling alcohol components. Polytitanoxanes having molecular weights up to 20,000 have been prepared by this method (31).

Substitution of some of the alkoxy groups on the polytitanoxanes with glycols, β-diketones or β-ketoesters, fatty acids, diester phosphates or pyrophosphates, and sulfonic acids gives a group of products that are very effective surface-treating agents for carbon black, graphite, or fibers (32).

Reactions with Alcohols. The tendency of titanium(IV) to reach coordination number six accounts for the rapid exchange of alkoxy groups with alcohols. Departure of an alkoxy group with the proton is the first step in the ultimate exchange of all four alkoxyls. The four-coordinated monomer is expected to react

more rapidly than the dimer. The process is very fast, as evidenced by nmr studies at room temperature, which show no difference in chemical shift between free ROH and combined ROTi. At sufficiently lower temperatures, both kinds of alkoxyl can be seen (33).

For preparative purposes, the equilibrium must be shifted either by using an excess of the exchanging alcohol or by distilling the more volatile lower alcohol. TYZOR TPT is the preferred starting ester because of the low boiling point of isopropyl alcohol. Monohydric alcohols of varying complexity, diols and polyols, phenols, alkanolamines, and the enolic forms of β-diketones and β-ketoesters all react readily with TYZOR TPT. Often all four alkoxy groups can be interchanged.

Reactions with Esters. Ester interchange catalyzed by titanates is an important industrial reaction (34).

$$RCOOR' + R''OH \rightleftharpoons RCOOR'' + R'OH$$

One commercial example is the formation of basic methacrylates from dialkylaminoethanols and methyl methacrylate:

$$R_2NCH_2CH_2OH + CH_3OCOCCH_3{=}CH_2 \rightleftharpoons R_2NCH_2CH_2OCOCCH_3{=}CH_2 + CH_3OH$$

The same reaction can be used to convert one alkoxide to another by distillation of a lower boiling ester:

$$Ti(OCH(CH_3)_2)_4 + 4\ CH_3COOC_4H_9 \rightleftharpoons Ti(OC_4H_9)_4 + 4\ CH_3COOCH(CH_3)_2$$

Titanium-catalyzed ester interchange can be used to prepare polyesters from diester and diols as well as from diacids and diols at considerably higher temperatures. Polymer chains bearing pendant ester and hydroxy functions can be cross-linked with titanates.

Reaction with Lactones. Hydroxycarboxylic acid ester complexes of titanium are formed by reaction of a tetraalkyl titanate with a lactone, such as β-propiolactone, γ-butyrolactone, or valerolactone (35). For example,

$$4\ \lceil\quad\text{O} + \text{Ti(OR)}_4 \rightleftharpoons \text{Ti(O(CH}_2)_3\text{COOR)}_4$$

Reactions with Acids. Organic acids form acylates when heated with tetraalkyl titanates. Best results are obtained using only one or two moles of acid, as attempts to force the reaction with three or four moles of acid can yield polymers.

$$n\ \text{RCOOH} + \text{Ti(OR}')_4 \longrightarrow (\text{RCOO})_n\text{Ti(OR}')_{4-n} + n\ \text{R}'\text{OH}$$

$$n\ (\text{RCO})_2\text{O} + \text{Ti(OR}')_4 \longrightarrow (\text{RCOO})_n\text{Ti(OR}')_{4-n} + n\ \text{RCOOR}'$$

With acid anhydrides, the exothermic reaction yields similar products, subject to the limitation on n. For higher anhydride ratios, condensed acylates form. If $n = 3$, the result is $(\text{R}'\text{OTi(OOCR)}_2)_2\text{O}$; if $n = 4$, then $((\text{RCOO})_3\text{Ti})_2\text{O}$ forms. Phthalic anhydride does not give a cyclic product:

$$\text{Ti(OR)}_4 + \text{[phthalic anhydride]} \longrightarrow \text{[product with OTi(OR)}_3 \text{ and COR]}$$

Reactions with α-Amino Acids. On heating two moles of an α-amino acid, such as alanine, in the presence of a tetraalkyl titanate and an alcohol, reaction that gives a 2,5-piperazinedione and an oxytitanate occurs (36).

$$2\ \text{NH}_2\text{CHRCOOH} \longrightarrow \text{[2,5-piperazinedione]}$$

Reactions with Salicylaldehydes. Tetraalkyl titanates react in benzene with salicylaldehyde in a 1:1 or 1:2 molar ratio to give salicylaldehydotrialkoxy and dialkoxy products, which when heated at reflux seem to undergo a Meewein-Ponndorf reaction to give an aldehyde derived from the alcohol group on the titanate and a reduced titanate complex (37):

$$\text{Ti(OR)}_4 + n\ \text{C}_6\text{H}_4\text{OHCHO} \longrightarrow (\text{RO})_{4-n}\text{Ti(C}_6\text{H}_4\text{OCHO})_n + n\ \text{ROH} \longrightarrow$$

$$n\ \text{R}'\text{CHO} + (\text{RO})_{4-2n}\text{Ti(C}_6\text{H}_4\text{OCH}_2\text{O})_n$$

Reactions with Isocyanates. TYZOR TPT catalyzes the trimerization of isocyanates and polyisocyanates to isocyanurates and polyisocyanurates (38).

Titanium alkoxides of the type Cl_3TiOR initiate the living polymerization of isocyanates. Polyisocyanates possessing controlled molecular weights and narrow polydispersities can be synthesized using these catalysts (39):

$$R—N{=}C{=}O + Cl_3TiOR \longrightarrow {+}C{=}ONR{-)}_n$$

Thermolysis. Lower tetraalkyl titanates are reasonably stable and can be distilled quickly at atmospheric pressure. Protracted heating forms condensation polymers plus, usually, alcohol and alkene. Longer or more branched chains are less stable. Thus, tetra-*n*-pentyl titanate [10585-24-7] can be distilled at 314°C and 101.3 kPa (1 atm), whereas tetra-*n*-hexyl titanate [7360-52-3] must be distilled at below 18.7 kPa (140 mm Hg) and tetra-*n*-hexadecyl titanate [34729-16-3] cannot be distilled even under high vacuum. Unsaturated groups, as in the allyloxide, also decrease thermal stability.

Thermolysis is used in the coating of glass and other surfaces with a film of titanium dioxide. When a lower alkoxide, eg, TYZOR TPT, vaporizes in a stream of dry air and is blown onto hot glass bottles above ca 500°C, a thin, transparent protective coating of TiO_2 is deposited.

Production and Economic Aspects. The 1995 world production of organic titanates is estimated to be 8000–9000 metric tons, some of which is for captive use. Principal producers in the United States are Du Pont, Kenrich Petrochemicals, and Akzo Nobel; in the United Kingdom, Tioxide U.K.; in Japan, Nippon Soda, Matsumoto Trading, and Mitsubishi Gas Chemicals; and in India, Synthochem.

The products offered commercially in the United States include a variety of tetraalkly titanates, chelates thereof, and simple and complex acylates. Prices of some of these are listed in Table 2.

The alkoxides are manufactured by a modification of the classical Nelles process, in which $TiCl_4$ in an inert solvent, eg, heptane, is treated with a monohydric lower alcohol (40). If the hydrogen chloride by-product is expelled by heating or by sweeping with dry nitrogen, the product is the dialkoxydichlorotitanate (41,42). Expulsion of hydrogen chloride is facilitated by adding the required alcohol gradually with continuous sweep (43). An acid acceptor is required to obtain the tetraalkoxide; ammonia is preferred because of its low cost. In a variation,

Table 2. Titanium Alkoxides and Chelates

Products	CAS Registry Number
Alkoxides	
tetraisopropyl titanate	[546-68-9]
tetra-*n*-butyl titanate	[5593-70-4]
tetra-2-ethylhexyl titanate	[1070-10-6]
Chelates	
bis-triethanolamine titanate	[36673-16-2]
diisopropoxy bis(acetylacetone)titanate	[17927-72-9]
bis-ethylacetoacetate titanate	[27858-32-8]
bis-lactic acid titanate	[65104-06-5]

$TiCl_4$ in an inert solvent is converted to the ammine $TiCl_4 \cdot 8NH_3$. This ammine reacts with almost any alcohol, yielding the tetraalkoxide. This operation can be performed in a continuous process (44,45). Higher alkoxides are readily prepared from TYZOR TPT or TYZOR TBT by alcohol interchange (transesterification), with removal of the by-product isopropyl alcohol or n-butanol by distillation.

Chelated titanates are made simply by mixing the chelating agent with TYZOR TPT or another alkoxide. The liberated alcohol is usually left in the product to maintain the products fluidity. It may, however, be removed by distillation if desirable. Organic titanates are normally shipped in 208-L drums, totes, cylinders, or tank trucks. Most titanates are moisture-sensitive and must be handled with care, preferably under dry nitrogen.

Alkoxy Halides

Titanium alkoxy halides have the formula $Ti(OR)_nX_{4-n}$, where R may be alkyl, alkenyl, or aryl, and X is F, Cl, or Br, but not I.

Properties. Alkoxytitanium fluorides and chlorides are colorless or pale-yellow solids or viscous liquids that darken on standing, especially in the light. Bromides are yellow crystalline solids. Aryloxytitanium halides are orange-to-red solids. The alkoxy halides are hygroscopic. Most dissolve in water without immediate decomposition, but hydrolyze slowly to hydrogen halide, alcohol, alkyl halide, and hydrous titanium dioxide. Using less than one mole of water, poly(alkoxytitanium) compounds can be prepared if the liberated hydrogen halide is neutralized by ammonia. Primary alkoxy halides are thermally stable, although they disproportionate on heating. Secondary and tertiary alkoxy halides decompose gradually on standing (more rapidly on heating) yielding alkyl halides, polymers, and titanium oxychloride. Physical properties of some alkoxytitanium halides are shown in Table 3.

Synthesis. Titanium alkoxy halides are intermediates in the preparation of alkoxides from a titanium tetrahalide (except the fluoride) and an alcohol or phenol. If $TiCl_4$ is heated with excess primary alcohol, only two chlorine atoms can be replaced and the product is dialkoxydichlorotitanium alcoholate, $(RO)_2TiCl_2 \cdot ROH$. The yields are poor, and some alcohols, such as allyl, benzyl, and t-butyl alcohols, are converted to chlorides (46). Using excess $TiCl_4$ at 0°C, the trichloride $ROTiCl_3$ is obtained nearly quantitatively, even from sec- and tert-alcohols (47,48).

For phenols, the number of chlorines replaced depends in part on the acidity of the phenol. Under mild conditions, phenol or m-nitrophenol displaces only one chlorine from $TiCl_4$, whereas p-chlorophenol, o-nitrophenol, p-nitrophenol, picric acid, or 2-naphthol gives $(ArO)_2TiCl_2$. If the mixtures are heated sufficiently, eg, in refluxing phenol, all four chlorines can be displaced (49–52).

Tetraalkoxides can be cleaved by hydrogen chloride or bromide in an inert solvent. Dialkoxytitanium dichloride is obtained as an alcoholate (53):

$$Ti(OR)_4 + 2\,HX \longrightarrow Ti(OR)_2X_2 \cdot ROH + ROH$$

Table 3. Titanium(IV) Alkoxyhalides and Aryloxyhalides

Compound	CAS Registry Number	Formula	Mp, °C	Bp^a, °C	Other properties
ethoxytrifluoride	[1524-67-0]	$TiOC_2H_5F_3$	220	$185-186$	
ethoxytrichloride	[3112-67-2]	$TiOC_2H_5Cl_3$	30–81 dec		
ethoxytribromide	[2489-72-8]	$TiOC_2H_5Br_3$	indefinite		
diethoxydifluoride	[650-27-1]	$Ti(OC_2H_5)_2F_2$	115	142_{240}	
diethoxydichloride	[3582-00-1]	$Ti(OC_2H_5)_2Cl_2$	40–50	$95-105_{67}$	
diethoxydibromide	[3981-88-2]	$Ti(OC_2H_5)_2Br_2$	47–50	$162-163_{27}$	red solid
triethoxyfluoride	[1868-77-5]	$Ti(OC_2H_5)_3F$	75–78		
triethoxychloride	[3712-48-9]	$Ti(OC_2H_5)_3Cl$		176_{240}	dipole moment, 9.57×10^{-30} C·m[b]
propoxytrichloride	[7569-98-4]	$TiOC_3H_7Cl_3$	65–66	$83-85_{147}$	
isopropoxytrichloride	[3981-83-7]	$TiOC_3H_7\text{-}i\text{-}Cl_3$	78–79	65_{13}	
dipropoxydichloride	[1790-23-4]	$Ti(OC_3H_7)_2Cl_2$	53–57	159_{240}	yellow solid
diisopropoxydibromide	[37943-35-4]	$Ti(OC_3H_7\text{-}i)_2Br_2$		$100-102_{27}$	
triisopropoxyfluoride	[757-61-9]	$Ti(OC_3H_7\text{-}i)_3F$	83–85	$140-150_{80}$	
tripropoxychloride	[24287-11-4]	$Ti(OC_3H_7)_3Cl$		168_{160}	d_4^{25} 1.1348 g/cm^3
butoxytrichloride	[3112-68-3]	$TiOC_4H_9Cl_3$	67.5–70	$124-127_{540\text{-}567}$	
isobutoxytrichloride	[17754-61-9]	$TiOC_4H_9\text{-}i\text{-}Cl_3$	81–83	$92-94_{120}$	
dibutoxydichloride	[1790-25-6]	$Ti(OC_4H_9)_2Cl_2$	52–53	$146-147_{133}$	
diisobutoxydichloride	[14180-16-6]	$Ti(OC_4H_9\text{-}i)_2Cl_2$		184_{213}	
tributoxyfluoride	[84215-68-9]	$Ti(OC_4H_9)_3F$	45–48	$175-180_{27}$	
tributoxychloride	[4200-76-4]	$Ti(OC_4H_9)_3Cl$		$154-155_{27}$	
triisobutoxychloride	[52027-15-3]	$Ti(OC_4H_9\text{-}i)Cl$		125_{27}	
phenoxytrichloride	[4403-68-3]	$TiOC_6H_5Cl_3$			
diphenoxydichloride	[2234-06-2]	$Ti(OC_6H_5)_2Cl_2$	116	265_{53}	n_D^{20} 1.5169, d_4^{20} 1.0985 g/cm^3
triphenoxychloride	[4401-43-8]	$Ti(OC_6H_5)_3Cl$		480_{27}	n_D^{20} 1.5158, d_4^{20} 1.1043 g/cm^3; dark-red crystalline solid; dipole moment, 9.91×10^{-30} C·m[b]
tri-o-xylenoxychloride	[84501-82-6]	$Ti((CH_3)_2C_6H_5O)_3Cl$			reddish brown solid

^{a}Subscripted values are pressure in Pa. To convert Pa to mm Hg, divide by 133.

^{b}To convert C·m to D, divide by 3.336×10^{-30}.

The same products may be made from primary alkoxides by the violent reaction with elementary chlorine or bromine. A radical mechanism has been proposed to account for the oxidation of some of the alkoxy groups (54):

$$\text{Ti(OCH}_2\text{R)}_4 + \text{X}_2 \xrightarrow{\text{[O]}} \text{Ti(OCH}_2\text{R)}_2\text{X}_2\cdot\text{RCH}_2\text{OH} + 1/2 \text{ RCOOCH}_2\text{R}$$

More useful than the preceding methods is cleavage of alkoxides by acetyl chloride or bromide. One, two, three, or four alkoxyls can be replaced by chloride or bromide. Benzoyl chloride gives poor yields, however. The tri- and tetrachlorides, which are stronger Lewis acids than mono- and dichlorides, coordinate with the alkyl acetate formed and yield distillable complexes (46,55,56).

$$\text{Ti(OR)}_4 + n\text{ CH}_3\text{COX} \longrightarrow \text{Ti(OR)}_{4-n}\text{X}_n + n\text{ CH}_3\text{COOR}$$

A very useful method is the proportionation of alkoxides with a stoichiometric quantity of titanium tetrachloride or bromide, preferably in an inert hydrocarbon solvent (55,57), as follows, where n is 1, 2, or 3:

$$n\text{ Ti(OR)}_4 + (4-n)\text{ TiCl}_4 \longrightarrow 4\text{ Ti(OR)}_n\text{Cl}_{4-n}$$

Alkoxy fluorides are prepared using acetyl fluoride. Alternatively, antimony trifluoride can be used to replace one alkoxyl by fluorine (58):

$$3\text{ Ti(OR)}_4 + \text{SbF}_3 \longrightarrow 3\text{ Ti(OR)}_3\text{F} + \text{Sb(OR)}_3$$

Titanium tetrafluoride reacts with Ti(OR)$_4$ in a manner similar to TiCl$_4$, especially if pyridine is present.

A principal use for the alkoxy halides is their reaction with organolithium or -magnesium compounds, R'M, in a Wurtz-type reaction to form compounds having carbon–titanium bonds. For this purpose, it is inconvenient that proportionation is reversible, and that the products Cl$_2$Ti(OR)$_2$ and ClTi(OR)$_3$ disproportionate on vacuum distillation. This occurs because the alkoxyhalides tend to satisfy their coordination needs through bridge structures, in which the halogen and alkoxyl readily interchange:

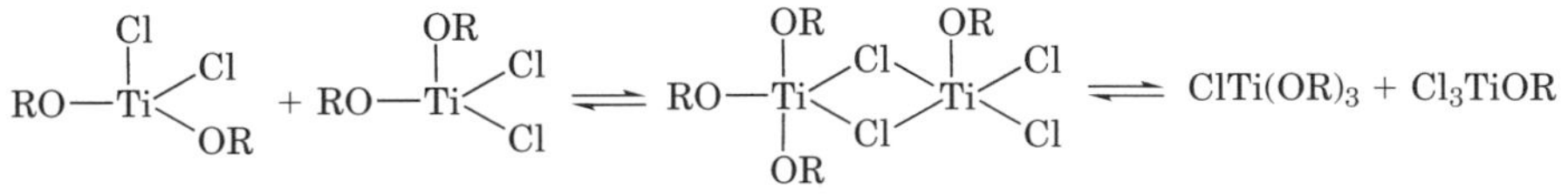

This can be prevented by supplying external electron pairs to satisfy the coordination needs of titanium. The strong base piperidine serves well, and a wide variety of compounds, Ti(OR)$_n$Cl$_{4-n}\cdot$C$_5$H$_{10}$NH, can be prepared (59). Alcohols and pyridine are also effective.

In a given $(RO)_n TiCl_{4-n}$, the alkoxy group can be exchanged by a higher alcohol if the resulting lower alcohol is removed by distillation (60). In the intermediate, HOR departs much more easily than HCl.

Chelates

Titanium chelates are formed from tetraalkyl titanates or halides and bi- or polydentate ligands. One of the functional groups is usually alcoholic or enolic hydroxyl, which interchanges with an alkoxy group, RO, on titanium to liberate ROH. If the second function is hydroxyl or carboxyl, it may react similarly. Diols and polyols, α-hydroxycarboxylic acids and oxalic acid are all examples of this type. β-Keto esters, β-diketones, and alkanolamines are also excellent chelating ligands for titanium.

Glycol Titanates.   Primary diols (HOGOH), such as ethylene glycol and 1,3-propanediol, react by alkoxide interchange at both ends, yielding insoluble, white solids that are polymeric in nature (18,61–63):

$$(RO)_4 Ti + HOGOH \longrightarrow (RO)_3 TiOGOTi(OR)_3 \longrightarrow (RO)_3 TiOGOTi(OR)_2 OGOTi(OR)_3$$
$$\longrightarrow (RO)_3 TiOGOROTi(OGOTi(OR)_3)_2 \longrightarrow etc$$

The 1:1 molar addition products of a primary diol and a tetraalkyl titanate, $Ti(OGO)(OR)_2$ may react with water to give either $Ti(OGO)(OH)_2$ or condensed products $(Ti(OGO)O)_n$, which can be used as esterification catalysts (64).

Where the glycol contains one or two secondary or tertiary hydroxyls, the products are more soluble and some are even monomeric cyclic chelates (65,66). Three compounds are obtained from 2-methylpentane-2,4-diol, depending on the mole ratio (67–70). Structure (**3**) represents an isolable but labile alcoholate of (**2**)

In a given $(RO)_n TiCl_{4-n}$

(**1**)   (**2**)   (**3**)

(69). The solvating glycol molecule can be driven off by heating. An alternative structure (**1**) has both hydroxyls of one HO–G–OH molecule involved in covalent bonds instead of one hydroxyl from each of two glycol molecules.

Solvent-soluble polymeric products of structures (**1–3**) can be obtained upon reaction of tetraalkyl titanate, 2-methyl-n-pentane-2,4-diol, and water in a 2:4:1 molar ratio (71). The tetraprimary glycol titanate complexes have been used as catalysts for the production of polyisocyanurates and polyoxazolidones (72).

Silanediols, eg, $(C_6H_5)_2 Si(OH)_2$ and $HOSi(C_6H_5)_2 OSi(C_6H_5)_2 OH$, yield four- and-six-membered rings with titanium alkoxides. Pinacols and 1,2-diols form chelates rather than polymers. The more branched the diol molecule, the more likely are its titanium derivatives to be soluble and even monomeric.

Reaction of 2,4-diorgano-1,3-diols, such as 2-ethylhexane-1,3-diol, with TYZOR TPT in a 2:1 molar ratio gives the solvent soluble titanate complex, TYZOR OGT [5575-43-9] (**4**) (73). If the reaction is conducted in an inert solvent, such as hexane, and the resultant slurry is treated with an excess of water, an oligomeric hydrolysis product, also solvent-soluble, is obtained (74).

(**4**)

Products similar to (**4**) are obtained if one starts with $TiCl_4$ instead of a tetraalkyl titanate (75). These products are useful as adhesives (qv), as textile-treating agents to impart water repellency, and as coatings and sizes for treating paper.

The reaction products of TYZOR TPT with 2–4 moles of 1,3-diols having two to three alkyl substituents, such as 2,2,4-trimethyl-1,3-pentanediol, gives complexes that could be used as cross-linking agents for hydroxy group containing powdered lacquer resins (76).

Highly cross-linked polyol polytitanates can be prepared by reaction of a tetraalkyl titanate with a polyol, such as pentaerythritol, followed by removal of the by-product alcohol (77). The isolated solids are high activity catalysts suitable for use in the preparation of plasticizers by esterification and/or transesterification reactions. The insoluble nature of these complexes facilitates their

removal from the plasticizer resins.

α-Hydroxycarboxylic Acid Complexes.   Water-soluble titanium lactate complexes can be prepared by reactions of an aqueous solution of a titanium salt, such as $TiCl_4$, titanyl sulfate, or titanyl nitrate, with calcium, strontium, or barium lactate. The insoluble metal sulfate is filtered off and the filtrate neutralized using an alkaline metal hydroxide or carbonate, ammonium hydroxide, amine, or alkanolamine (78,79). Similar solutions of titanium lactate, malate, tartrate, and citrate can be produced by hydrolyzation of titanium salts, such as $TiCl_4$, in strongly (>pH 10) alkaline water; isolation of the titanium hydrate produced; and reaction with an aqueous solution of the desired α-hydroxycarboxylic acid (80).

Solid, water-soluble α-hydroxycarboxylic acid and oxalic acid titanium complexes can be formed by reaction of the acid and a tetraalkyl titanate in an inert solvent, such as acetone or heptane. The precipitated complex is filtered, rinsed with solvent, and dried to give an amorphous white solid, which is water- and alcohol–water-soluble (81,82).

TYZOR LA [65104-06-5] (**5**), an aqueous solution of the ammonium salt of the titanium bis-lactate complex, is prepared from two equivalents of lactic acid to one of TYZOR TPT. The by-product isopropyl alcohol is removed by distillation and the resultant solution is neutralized with ammonium hydroxide.

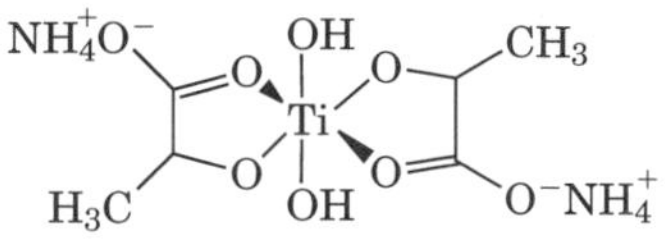

(**5**)

The structure of these products is uncertain and probably depends on pH and concentrations in solution. The hydroxyl or carboxyl or both are bonded to the titanium. It is likely that most, if not all, of these products are oligomeric in nature, containing Ti$-$O$-$Ti titanoxane bonds (81). Their aqueous solutions are stable at acidic or neutral pH. However, at pH ranges above 9.0, the solutions readily hydrolyze to form insoluble hydrated oxides of titanium. The alkaline stability of these complexes can be improved by the addition of a polyol such as glycerol or sorbitol (83). These solutions are useful in the textile, leather (qv), and cosmetics (qv) industries (see TEXTILES).

Water-soluble, alkaline-stable ammonium or metal titanium malates and citrates can be formed by adding a tetraalkyl titanate to an aqueous solution of the ammonium or metal titanium malate or citrate (84). A typical formula is $M_x TiO(citrate)_x$, where M is NH_4, Na, K, Ca, or Ba.

The addition of an α-hydroxycarboxylic acid to a tetraethylene, propylene, diethylene, or hexylene glycol titanate gives water-soluble complexes suitable for gelling aqueous solutions of hydroxyl polymers, such as poly(vinyl alcohol) (PVA), or cellulose (qv) derivatives. These are useful as binding agents for glass fibers, clays (qv), and paper coatings (85).

Oxalic acid behaves as an α-hydroxy acid, yielding crystalline ammonium or potassium salts from either aqueous titanium(IV) solutions or tetraalkyl titanates (86). These are written as:

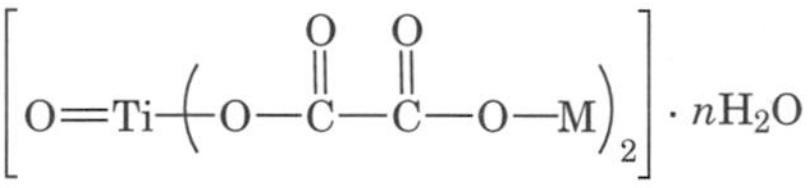

Dicarboxylic acids, eg, succinic or adipic, do not dissolve titanic acid. A phthalate has been prepared by adding acidic titanium sulfate solution to sodium phthalate solution.

β-Diketone Chelates. β-Diketones, reacting as enols, readily form chelates with titanium alkoxides, liberating in the process one mole of an alcohol. TYZOR AA [17927-72-9] (**6**) is the product mixture from TYZOR TPT and two moles of acetylacetone (acac) reacting in the enol form. The isopropyl alcohol is left in the product (87). The dotted bonds of structure (**6**) indicate electron

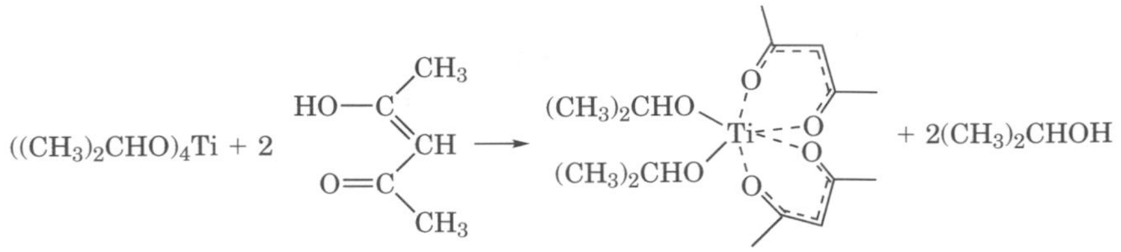

$$((CH_3)_2CHO)_4Ti + 2 \quad \begin{matrix} HO-C \overset{CH_3}{\diagup} \\ \| \quad CH \\ O=C \diagdown CH_3 \end{matrix} \longrightarrow \quad \begin{matrix} (CH_3)_2CHO \\ (CH_3)_2CHO \end{matrix} Ti \begin{matrix} O \\ O \\ O \end{matrix} \quad + 2(CH_3)_2CHOH$$

(6)

delocalization. The chelate may be isolated by careful vacuum stripping of the residual isopropyl alcohol, but if the distillation is pushed, an oligomeric titanoxane (**7**) forms.

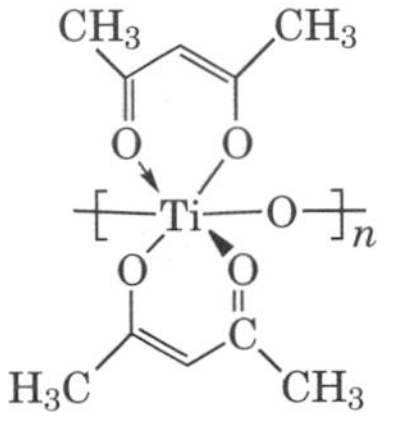

$$\begin{matrix} CH_3 \quad\quad CH_3 \\ O \quad\quad O \\ \overset{|}{+} Ti-O \overset{|}{+}_n \\ O \quad\quad O \\ H_3C \diagup \quad\quad \diagdown CH_3 \end{matrix}$$

(7)

The six coordinated titanium(IV) compounds, $Ti(acac)_2(X)_2$, where X is methoxy, ethoxy, isopropoxy, n-butoxy, or chloro, all adopt the cis-configuration. This is believed to result from the ligand-to-metal π-electron donation (88,89).

The orange-red titanium acetylacetone chelates are soluble in common solvents. These compounds are coordinately saturated (coordination number equals 6) and thus much more resistant to hydrolysis than the parent alkoxides (coordination number 4). The alkoxy groups are the moieties removed by hydrolysis. The initial product of hydrolysis is believed to be the bis-hydroxy bis-acetylacetone titanate, $(HO)_2Ti(acac)_2$, which oligomerizes to a polytitanaoxane, $HO(Ti(acac)_2\text{-}O)_nH$. Hydrolysis is slowest at ca pH 4.5 (90,91).

The addition of TYZOR AA to larger quantities of water results in precipitation of the polytitanaoxane as a yellow solid, which can be isolated and dried (92). The precipitation of this solid can be prevented by diluting the TYZOR AA with an equal volume of 10% aqueous acetic acid and then 13 times the amount of a 1:1 mixture of isopropyl alcohol and water (93).

Alternatively, a water-stable bis-acetylacetone titanate can be formed by replacing at least one of the alkoxy groups with an alkyloxyalkyleneoxy or alkyloxypolyalkyleneoxy group (94).

The TYZOR AA, which is a 75% isopropyl alcohol solution, is unstable in cold storage. The titanate complex precipitates from solution and causes handling difficulties. The addition of small amounts (0.05–0.15 mol water/mol titanate) of water gives a solution, TYZOR AA75, that is stable in cold-temperature storage (95).

The solvent-free mono-n-butyl, monoisopropyl bis-acetylacetone titanate, TYZOR AA95 [*9728-09-9*], is a liquid at room temperature. By carefully controlling the mixture of alcohols used as solvent, a product, TYZOR GBA can be obtained, which is cold-storage-stable even in the presence of nucleating agents (96).

β-Ketoester Chelates. β-Ketoesters react in a fashion similar to the β-diketones. TYZOR DC [27858-32-8] is the light-yellow liquid from TYZOR TPT and two moles of ethyl acetoacetate (eaa) after removal of the isopropyl alcohol. TYZOR BEAT, the bis-ethylacetoacetate [20753-28-0] derived from the tetra-n-butyl titanate, and TYZOR IBAY [83877-91-2], the isobutoxy analogue, perform similarly to TYZOR DC. Both, however, have better cold-storage stability.

$$((CH_3)_2CHO)_4Ti + 2 \quad \begin{array}{c} O=C \overset{OC_2H_5}{\underset{CH}{\diagdown}} \\ \overset{\|}{C} \\ HO \diagup \overset{\diagdown}{CH_3} \end{array} \longrightarrow ((CH_3)_2CHO)_2Ti(eaa)_2 + 2(CH_3)_2CHOH$$

Beginning with a mixture of tetraalkyl titanates, the resultant bis-ethyl acetoacetate contains a mixture of all possible combinations of alkoxy groups, resulting in a reduction in viscosity and improved cold-storage stability of the product (97).

Partial hydrolysis of TYZOR DC or the monoethylacetoacetate ester chelate, followed by removal of the isopropyl alcohol by-product, gives a dimeric μ-oxo chelate (**8**), which also has improved cold-temperature-storage stability (98).

(**8**)   (**9**)

Reaction of TYZOR DC and 1,3-propanediol gives titanium 1,3-propylenedioxide bis(ethyl acetoacetate) [36497-11-7], which can be used as a non-corrosive curing catalyst for room-temperature-vulcanizing silicone rubber compositions (99). Similar structures could be made, starting with titanium bis-acetylacetonates, such as that shown in structure (**9**).

Titanium Phosphorous Containing Chelates. The reaction of a mixture of mono(alkyl)diacid orthophosphate, di(alkyl)monoacid orthophosphate, and $TiCl_4$ in a high boiling hydrocarbon solvent such as heptane, with nitrogen-assisted evolution of liberated HCl, gives a mixture of titanium tetra(mixed alkylphosphate)esters, $(HO)(RO)O=PO)_nTi(OP=O(OR)_2)_{4-n}$ in heptane solution (100). A similar mixture can be prepared by the addition of two moles of P_2O_5 to one mole of $TiCl_4$ in the presence of six moles of alcohol:

$$2\ P_2O_5 + TiCl_4 + 6\ ROH \longrightarrow ((RO)_2P=OO)_2Ti(OP=O(OH)(OR))_2 + 4\ HCl$$

These mixed phosphate ester titanium complexes or their amine salts are useful as fuel additives to help maintain cleanliness of carburetors and inhibit surface corrosion. Chloride-free mixed alcohol phosphate esters can be obtained if a tetraalkyl titanate is used (101).

Addition of one mole of P,P'-diphenylmethylenediphosphinic acid to tetraisopropyl titanate gives a chelated product, the solutions of which can be used as a primer coat for metals to enhance the adhesion of topcoats, eg, alkyds, polyalkyl acylates, and other polymeric surface coating products, and improve the corrosion resistance of the metal to salt water (102).

$$Ti(OR')_4 + \begin{array}{c} R \\ HO-P \end{array}^{\!\!\!O}_{\!\!\!CH_2} \begin{array}{c} \\ \\ HO-P \end{array}_{R}^{\!\!\!O} \longrightarrow 2\,R'OH + (R'O)_2Ti$$

Alkoxy titanium acylate derivatives coordinated with a phosphite diester (phosphonate diester) can be prepared by reaction of a tetraalkyl titanate and an equal molar amount of a carboxylic acid, such as methacrylic acid or isostearic acid, and a phosphite or phosphonate diester, such as dibutyl hydrogen phosphite (103). These materials reduce the viscosity of a composite system, improve

$$(RO)_n Ti \text{---} (OOCR')_{4-n}$$
$$\uparrow$$
$$(HOP(OR'')_2)_m$$

the dispersion of the filler in the composite system, and improve the mechanical properties of the cured composite system by modification of the surface of the filler.

Alkanolamine Chelates.   Alkanolamine chelates, which are prepared by reaction of tetraalkyl titanates with one or more alkanolamines, are used primarily in cross-linking water-soluble polymers (qv) (see ALKANOLAMINES) (104). The products are used in thixotropic paint emulsion paints, in hydraulic fracturing and drilling of oil and gas wells, and in many other fields. The structure of

$$((CH_3)_2CHO)_4Ti + 2\,N(CH_2CH_2OH)_3 \longrightarrow ((CH_3)_2CHO)_2Ti \left[\begin{array}{c} O \\ \\ N \\ | \\ (CH_2CH_2OH)_2 \end{array} \right]_2 + 2(CH_3)_2CHOH$$

the product indicates that titanium has received electron pairs from both nitrogens to complete the coordination shell. No role is, however, ascribed to the four free hydroxyls. If the liberated isopropyl alcohol is left in place, the product is a mobile liquid; if the alcohol is removed, the product is a viscous, sticky, and doubtless oligomeric oil.

The 1-alkoxytitatranes can be synthesized by the reaction of equimolar amounts of tetraalkyl titanates and triethanolamine (105). X-ray crystallographic analysis of the solid isolated from the reaction of one mole of triethanolamine and one mole of TYZOR TPT confirms the structure as a centrosymmetric dimer having a Ti:isopropoxy:nitrilotriethoxy ratio of 1:1:1. The titanium atoms have achieved a coordination number of six via a rather unsymmetrical titanium–oxygen bridge (106).

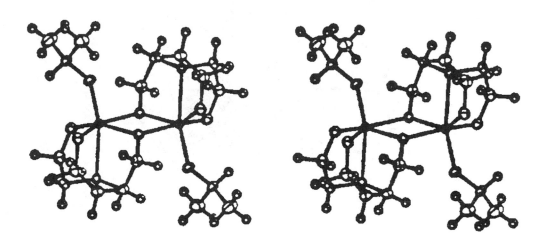

Mixtures of the monotriethanolamine titanate and polyols, such as fructose and sorbitol, imparted thixotropic properties to polymer-containing cement (107). A similar reaction of triethanolamine and tetraalkyl titanates previously treated with stearic acid gives the corresponding 1-acyloxytitatrancs (108). Aqueous dispersions of these materials can be used to treat fabrics to impart a high degree of water repellency.

The diisopropoxy-bis-(triethanolamine)-titanate, TYZOR TE [*36673-16-2*], is an excellent cross-linker for aqueous solutions of hydroxyl-containing polymers. The reaction product of TYZOR TPT with a mixture of trialkanolamines and dialkanolamines or monoalkanolamines can be used to cure polyester-based powder coatings (109). Other ligands of this type include triisopropanolamine [*122-20-3*],

$$N(CH_2CHCH_3)_3,$$
$$\overset{|}{OH}$$

and the general class

$$\overset{R'}{\overset{|}{HOCH_2CH_2N}}-R-\overset{R'}{\overset{|}{N}}-R',$$

where the R and R′ groups are alkyl, hydroxyalkyl, and aminoalkyl (110). The 1:1 molar addition product of TYZOR TPT and tetrakis[hydroxyisopropyl]-

ethylenediamine, Quadrol [*102-60-3*] (**10**), has been isolated and shown by x-ray crystallography to consist of [(1,1′,1″,1‴-(ethylenedinitrilo)tetra-2-propoxy)titanium(IV)] dimers centered on crystallographic inversion centers. The titanium atom is seven-coordinate, in which one oxygen from each triethanolamine chelate serves to bridge the titanium atoms (111).

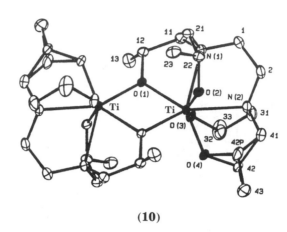

(10)

The alkoxy titanate compounds formed by reaction of one mole of tetraalkyl titanate with one mole of a dialkanolamine are excellent esterification catalysts for the manufacture of phthalate-based plasticizers (112). If a 1:1 molar mixture of alkanolamine and water is used in place of the alkanolamine, oligomeric titanate complexes are formed, which have high catalyst activity and can be used as thixotropic additives to paints and other aqueous coating formulations (113).

The mono- and dialkanolamine titanates are water-soluble and slowly hydrolyze at pH 9.0. Lowering the pH increases the rate of hydrolysis, which is shown by the development of turbidity. Turbidity also occurs above pH 11. The tetrahydroxyalkylethylenediamine titanate complexes form much more stable water solutions and can be used as dispersing agents for aqueous TiO_2 slurries (114).

Addition of secondary chelating agents, eg, polyols such as sorbitol or mannitol and the strongly chelating α-hydroxycarboxylic acids such as citric or oxalic, prevents development of turbidity outside the pH range of 9–11 (115–117).

These polyol-stabilized alkanolamine titanate solutions are used in a method to improve the wet strength of paper (115,118). The addition of 1–2 moles of a glycol ether, such as methoxyethoxyethanol to TYZOR TE, gives either monoalkoxy-, monoalkoxyalkylenoxy-, bis-triethanolamine titanates, or bis-alkoxyalkylenoxy- or bis-triethanolamine titanates, which also form stable aqueous solutions (119).

Similarly, the stability of aqueous solutions of monoacetylacetone or monotriethanolamine titanate complexes can be improved by the addition of

glycol ethers (120). These solutions are useful in applications in which the catalytic, cross-linking, or film-forming actions of titanates are desired to take place
in aqueous systems.

The reactions of simpler alkanolamines with tetraalkyl titanates are not
completely understood. Ethanolamine reacts with the lower tetraalkyl titanates
to give insoluble white solids. N,N-Dialkylethanolamines, $R_2NCH_2CH_2OH$, react with TYZOR TPT and, depending on the mole ratio, yield all members of the
family $((CH_3)_2CHO)_{4-n}Ti(OCH_2CH_2NR_2)_n$, where $R = CH_3$ or C_2H_5 (121,122).
When $R = CH_3$, all products are distillable (121); when $R = C_2H_5$, the products disproportionate on attempted distillation (122). All of these products are
monomeric (ebullioscopy in benzene), which suggests that electron donation from
nitrogen completes the coordination sphere of Ti. The compounds derived from
$CH_3NHCH_2CH_2OH$ are dimeric crystalline solids.

Products from aminoalcohols and TYZOR TPT were obtained by azeotroping
the isopropyl alcohol with benzene (121,122). From trimethylethylenediamine,
dimethylethanolamine, and dimethylisopropanolamine with TYZOR TPT, the
orange (**11**), the yellow (**12**), and the pale-green (**13**) were obtained, respectively.
The lithium salt of the ligand, derived from C_4H_9Li, combined with $(RO)_3TiCl$
in hexane has also been used (123).

(**11**)   (**12**)   (**13**)

Compounds (**11**), (**12**), and (**13**) are all five-coordinated according to spectroscopic data (85,124).

Acylates.   Titanium acylates are prepared either from $TiCl_4$ or tetraalkyl
titanates. Because it is difficult to obtain titanium tetraacylates, most compounds reported are either chloro- or alkoxyacylates. Under most conditions,
$TiCl_4$ and acetic acid give dichlorotitanium diacetate [*4644-35-3*]. The best
method involves passing preheated (136–170°C) $TiCl_4$ and acetic acid simultaneously into a heated chamber. The product separates as an HCl-free white
powder (125):

$$TiCl_4 + 2\ HOOCCH_3 \longrightarrow Cl_2Ti(OOCCH_3)_2 + 2\ HCl$$

Alternatively, $TiCl_4$ reacts with a cold mixture of acetic acid and acetic anhydride. If the mixture is heated, condensation occurs with elimination of acetyl

chloride to yield hexaacetoxydititanoxane [4861-18-1] (126,127).

$$ClTi(OOCCH_3)_3 + Cl_2Ti(OOCCH_3)_2 \longrightarrow O(TiCl(OOCCH_3)_2)_2 + CH_3COCl \xrightarrow{2\,(CH_3CO)_2O}$$

$$O(Ti(OOCCH_3)_3)_2 + 2\,CH_3COCl$$

Trichlorotitanium monoacylates form, during thermal decomposition of TiCl$_4$, ester complexes (128):

$$TiCl_4 \cdot RCOOR' \longrightarrow TiCl_3OOCR + R'Cl$$

Tetraacylates are prepared from titanium tetrabromide and excess carboxylic acid in an inert solvent. After solvent removal, the residue is heated to remove hydrogen bromide.

$$TiBr_4 + 4\,RCOOH \longrightarrow Ti(OOCR)_4 + 4\,HBr$$

Tetraacylates have been prepared in this way from stearic, benzoic, cinnamic, and other acids, as well as from diacids such as succinic and adipic acids (**14**) (129). In some cases, TiCl$_4$ may also be used.

(**14**)

The usual products from alkoxides and acids are dialkoxytitanium diacylates (130,131). The third acyl group, but not the fourth, can often be introduced by azeotroping the lower alcohol with benzene (132). Using acetic anhydride, the same hexaacetoxydititanoxane is prepared from the chloride forms.

An acylate group is potentially a bidentate ligand. It may bond once or twice to one titanium, or bridge two titanium atoms as shown.

Dimeric (ebullioscopy in benzene)

$$(CH_3)_2CHOTi(O\overset{O}{\overset{\|}{C}}R)_3$$

has eight-coordinate Ti. If structure (**15**) is correct, isopropoxy is a better bridging ligand than acylate (132). This saturation of titanium coordination may explain the difficulty in preparing titanium tetraacylates by mild methods. Dimeric dichloro- and dibromotitanium diacylates exhibit three different carbonyl frequencies in the ir region, ie, at 1650, 1540, and 1400/cm, associated with three different types of bonding (133).

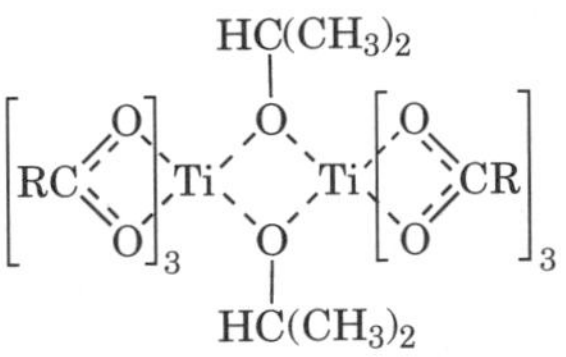

$$\left[RC \overset{O}{\underset{O}{<}} Ti \underset{O}{\overset{O}{<}} Ti \overset{O}{\underset{O}{>}} CR \right]$$

(15)

Titanium(IV) Complexes with Other Ligands

The d^0-titanium(IV) atom is hard, ie, not very polarizable, and can be expected to form its most stable complexes with hard ligands, eg, fluoride, chloride, oxygen, and nitrogen. Soft or relatively polarizable ligands containing second- and third-row elements or multiple bonds should give less stable complexes. The stability depends on the coordination number of titanium, on whether the ligand is mono- or polydentate, and on the mechanism of the reaction used to measure stability.

A partial list of ligands that bond to titanium(IV) includes sulfinates, $-OSOR$; sulfonates, $-OSO_2R$; peroxide, O_2^{2-}; superoxide, O_2; nitro groups, $-NO_2$; nitrites, $-ONO$; nitrates, $-ONO_2$; carbonate, CO_3^{2-}; phosphate, PO_4^{3-}; amide, $-NR_2$; acylamido, $-NRCOX$; Schiff base, $-N=CR_2$; N in a heterocyclic ring; azo, $-N=NR$; azido, $-N_3$; isocyanato, $-N=C=O$; isothiocyanato, $-N=C=S$; isoselenocyanato, $-N=C=Se$; alkanethio, $-SR$; arenethiol, $-SAr$; xanthato, $-SS=COR$; dialkyldithiocarbamato, $-SS=CNR_2$; tetrahydridoborato or borohydride, BH_4; trialkylsilyl, $-SiR_3$; and Ge and Sn analogues. Phosphine, arsine, and stibine ligands, as well as CO and hydride, associate only with lower valent titanium. Carbon ligands, eg, alkyl, aryl, dienyl, and cyclopentadienyl, give organometallic compounds. Early literature on many of these complexes has been reviewed (134).

Peroxide Titanate Complexes.   Titanates may influence reactions of organic peroxides (see PEROXIDES, ORGANIC). For example, t-butyl hydroperoxide epoxidizes olefins:

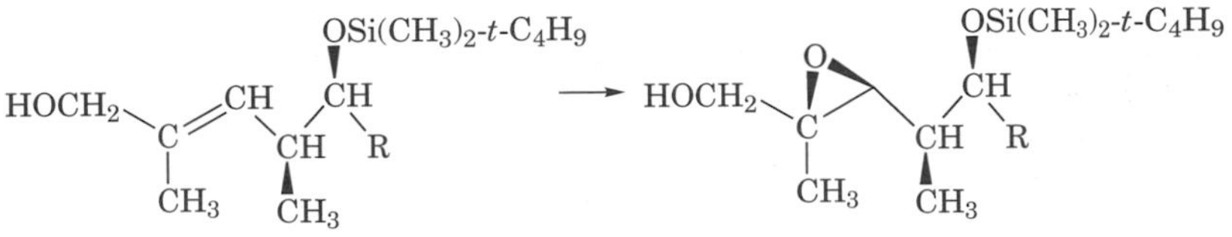

The ratio of *syn*-epoxide (shown above) to *anti*-epoxide is 10–25:1 with TYZOR TPT catalysis, whereas vanadylacetylacetonate is less selective and m-chloroperoxybenzoic acid gives the reverse 1:25 ratio. It is supposed that TYZOR TPT esterifies the free hydroxyl, then coordinates with the peroxide to favor *syn*-epoxidation (135). This procedure is related to that for enantioselective epoxidation of other allylic alcohols in 9–95% enantiomeric excess (135).

Titanates trigger peroxide-initiated curing of unsaturated polyesters to give products of superior color, compared to conventional cobalt-initiated curing (see INITIATORS) (136,137). Titanium coordinated to a porphyrin bonds hydrogen peroxides as $Ti{<}^{O}_{O}$ (138).

Hydrogen peroxide (qv) produces an intense yellow color with Ti(IV) in aqueous solution and has long been used as a qualitative test for Ti. Solid inorganic peroxides have been reported. The reaction of TYZOR TPT with a dilute solution of hydrogen peroxide and phosphoric acid generates a soluble titanate hydrolysis complex, which can be used to coat sheets of wood, paper, metal, glass, or plastics to form an hydrophilic surface (139). These hydrophilic sheets can be used in the formation of lithographic printing plates or color proofing elements.

Fluorocarbon Group Containing Titanium Complexes. Fluorocarbon groups containing carboxylic acids and alcohols, such as perfluorooctanoic acid or 1*H*,1*H*,5*H*-octafluoropentanol, react with tetraalkyl titanates to give complexes that are useful either in the treatment of fabrics to render them water-repellent (140,141), or as gasoline additives to minimize deposits and improve performance (142). Reaction of TYZOR TPT with a mixture of a fatty acid, such as isovaleric acid, and a fluorinated carboxylic acid, such as octafluoro-1*H*-pentanoic acid, produces a complex that is useful as a surface-treating agent for fillers used in polymeric composite applications (143).

Reaction of TYZOR TPT with polyperfluoroalkylene ethers containing a carbonyl group produces a complex that is an excellent surface-treating agent, imparting improved surface wettability and anticorrosion properties to metal surfaces (144). These complexes can be used by themselves, or as additives to perfluoropolyethers as vacuum pump oils, lubricant oils, or mold release agents.

The compound $C_8F_{17}CH_2CH_2O \cdot Ti(O\text{-}i\text{-}C_3H_7)_3$ is useful in car polishes to impart water repellency (145).

Chiral Titanium Complexes. Chiral titanium complexes are useful for the enantioselective addition of nucleophiles to carbonyl groups:

$$
\begin{array}{ccc}
\underset{R'}{\overset{O}{\underset{}{\parallel}}}\!\!\!\underset{}{C}\!\!\!\underset{R''}{} + TiL_4 & \longrightarrow & \underset{R'}{\overset{O\text{--}TiL_3}{\underset{R''}{C\text{---}L}}} & \longrightarrow & \underset{R'}{\overset{L\ OTiL_3}{C}}\underset{R''}{}
\end{array}
$$

The advantages of titanium complexes over other metallic complexes is high selectivity, which can be readily adjusted by proper selection of ligands. Moreover, they are relative inert to redox processes. The most common synthesis of chiral titanium complexes involves displacement of chloride or alkoxide groups on titanium with a chiral ligand, L^*:

$$TiL_4 + L^* \longrightarrow TiL_3L^* + L$$

The chemistry of complexes having achiral ligands is based solely on the geometrical arrangement on titanium. Optically active alcohols are the most favored monodentate ligands. Cyclopentadienyl is also well suited for chiral modification of titanium complexes.

One of the most famous chiral titanium complexes is the Sharpless catalyst (**16**), based on a diisopropyl tartarate complex. Nmr studies suggest that the complex is dimeric in nature (146). An excellent summary of chiral titanium complexes is available (147).

(**16**)

Other Complexes. The reaction of TYZOR TPT with two equivalents of a semicarbazone produces complex (**17**), the structure of which has been assigned the trans-configuration (148):

(**17**)   (**18**)

Azines form an interesting class of nitrogen donor ligands, which can react in a 1:1 or 2:1 ratio with a tetraalkyl titanate to form complexes of stereochemical interest (**18**) (149).

Composite Oxyalkoxides. Composite oxyalkoxides can be prepared by reaction of tetraalkyl titanates and alkaline-earth metal hydroxides. These oxyalkoxides and their derivatives can be hydrolyzed and thermally decomposed to give alkaline-earth metal titanates such as barium titanate (150).

$$M(OH)_2 + n\ Ti(OR)_4 \longrightarrow (MTiO_2(OR)_2)_n + 2n\ ROH$$

Barium titanate thin films can be deposited on various substances by treating with an aqueous solution containing barium salts and an alkanolamine-modified titanate such as TYZOR TE (151). In a similar fashion, reaction of a tetraalkyl titanate with an alkali metal hydroxide, such as potassium hydroxide, gives oxyalkoxide derivatives $(KTi_xO(OR)_y)_n$, which can be further processed to give alkali metal titanate powders, films, and fibers (152–155). The fibers can be used as adsorbents for radioactive metals such as cesium, strontium, and uranium (156).

Addition of lithium or sodium alkoxide to TYZOR TPT gives a double alkoxide derivative, $MTi_2(OR)_9$, the structure of which has been proposed (157) as follows, where M = Na or Li.

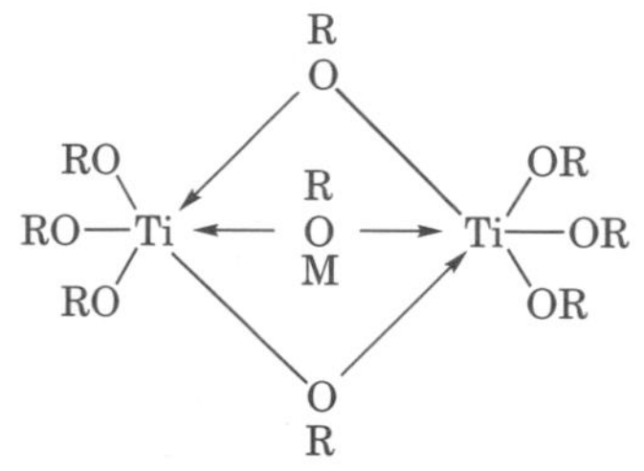

$$
\begin{array}{c}
\text{R} \\
\text{O} \\
\end{array}
$$

Hydrous metal oxide powders, such as sodium titanate, $NaTi_2O_5H$, can be prepared by treating TYZOR TPT with sodium hydroxide in methanol solvent to form a soluble intermediate, which is hydrolyzed in acetone–water to form an ion-exchange material useful in treating radioactive waste (158). Exchange of the sodium ion with an active metal such as Ni, Pt, or Pd gives heterogeneous catalysts useful in olefin polymerization, coal liquification, and hydrotreating.

Titanium–Vanadium Mixed Metal Alkoxides. Titanium–vanadium mixed metal alkoxides, $VO(OTi(OR)_3)_2$, are prepared by reaction of titanates, eg, TYZOR TBT, with vanadium acetate in a high boiling hydrocarbon solvent. The by-product butyl acetate is distilled off to yield a product useful as a catalyst for polymerizing olefins, dienes, styrenics, vinyl chloride, acrylate esters, and epoxides (159,160).

Titanium Amides. The reaction of lithium amides, $LiNR_2$, with $TiCl_4$ gives tetrakisdialkylaminotitanates, $(R_2N)_4Ti$, which can react with alcohols or other ligands to produce tetraalkyl titanates and chelated derivatives (161,162). The chlorotris(dialkylamino) titanates, $TiCl(NR_2)_3$, are useful initiators for the polymerization of acrylic monomers to living polymers (163).

Schiff Bases. The nitrogen of a Schiff base unit in a polydentate ligand coordinates readily. One example (**19**) involving a tridentate ONS ligand is salicylaldehyde dithiocarbazate, which is dimeric through double isopropoxy bridges (67). This compound is readily hydrolyzed. The related TiL_2, where L is salicylaldehyde dithiocarbazate, is monomeric and stable in hot water because the coordination number of Ti is six. The related ONO ligands (**20**) derived from enol ketones or aldehydes, eg, acetylacetone or salicylaldehyde, and hydroxyamines react similarly with TYZOR TPT. Other products are analogous, pentacoordinate,

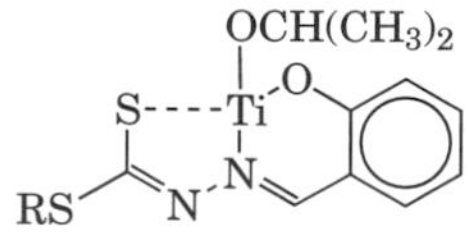

(19) 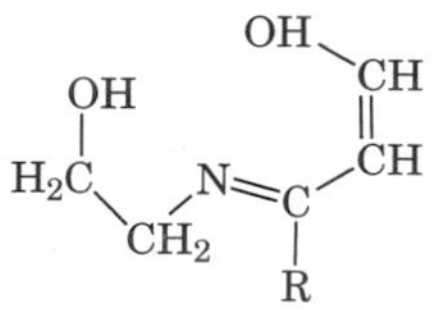 **(20)** 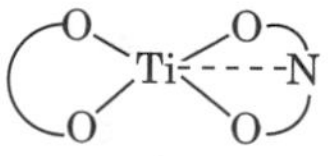 **(21)**

dimeric, 1:1 water-sensitive complexes and hexacoordinate, monomeric, 2:1 water-stable complexes. Both 1:1 complexes react with pyridine to become six-coordinate, and react with 2-methylpentane-2,4-diol by displacing isopropoxy to form pentacoordinate chelates (**21**), which also add pyridine (164). Several similar ligands and their five- and six-coordinate complexes with Ti have been reported (165–167). Complexes with 8-hydroxyquinoline are well known (168,169).

Polymetallocarbosilanes. Polymetallocarbosilanes having a number-average molecular weight of 700–100,000 can be prepared by reaction of poly-carbosilane, $-\!(\!Si(R)_2\!-\!CH_2\!)_{\overline{x}}$, where R is H, or lower alkyl, with a tetraalkyl titanate, to give a mono-, di-, tri-, or tetrafunctional polymer containing at least one Si–O–Ti bond. By firing polytitanocarbosilanes in a vacuum, an inert gas, or a nonoxidizing atmosphere, they can be converted to a molded article consisting mainly of SiC and TiC and having a higher mechanical strength and better oxidation resistance at higher temperatures than SiC itself (see CARBIDES; TOOL MATERIALS) (170).

Polytitanosiloxane (PTS) polymers containing Si–O–Ti linkages have also been synthesized through hydrolysis–polycondensation or hydrolysis–polycondensation–pyrolysis reactions involving clear precursor sol solutions consisting of monomeric silanes, TYZOR TET, methanol, water, and hydrochloric acid (Fig. 2). These PTS polymers could be used to form excellent corrosion protection coatings on aluminum substrates (171).

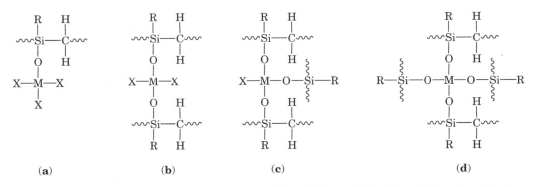

Fig. 2. Polytitanosiloxane polymers, where M is Ti/Zr and X is Cl. (**a**) Monofunctional, (**b**) difunctional, (**c**) trifunctional, and (**d**) tetrafunctional polymers.

Titanium Silicates

The synthesis of titanium silicate catalysts, such as TS-1, by controlled hydrolysis of a mixture of tetraethylorthosilicate (TEOS) and TYZOR TBT was first reported in the mid-1980s (172,173). Since that time, several alternative methods of synthesis have been developed. The addition of sodium hydroxide during formation of the TS-1 precursor gel gives a titanium silicate having fewer acidic sites, which modifies its catalytic activity (174). The use of chelating agents, such as acetylacetone, during the synthesis avoids formation of insoluble TiO_2 during gel/reaction mixture preparation before the formation of titanium silicate species (109,110,112–116,174). Prehydrolysis of the silicon precursor promotes homogeneous component mixing and minimizes formation of segregated silica and titania components (175). Titanium silicates are excellent catalysts for the selective oxidation of alkanes to alkenes (176) and for the hydroxylation of benzene and phenols (177,178). These have also found use as an ester-exchange catalyst in the formation of dimethyl carbonate from ethylene carbonate (179).

Lower Valent Titanates

Titanium(III) alkoxides can be produced by photoreduction of the tetraalkyl titanates in the presence of a base, such as pyridine (180), and by reduction of tetraalkyl titanates by organosilicon compounds containing Si–H groups (181).

TYZOR ET is reduced by sodium and ethanol to a dark-blue compound (182). Use of potassium as the reducing agent in the alcohol permits the isolation and identification of $Ti(OC_3H_7)_3$ [22922-82-3] and $Ti(OC_4H_9)_3$ [5058-41-3] (183,184). The products precipitate as solid alcoholates, $(RO)_3Ti\cdot2ROH$, which can be dried to solvent-free material. Air oxidation in a radical reaction yields solvent-soluble materials formulated as $(C_3H_7O)_2Ti=O$ [20644-85-3], mp 100–113°C, and $(C_4H_9O)_2Ti=O$ [30860-71-0], mp 112–115°C, plus C_3H_7OH or C_4H_9OH and the corresponding aldehydes. These materials may be oligomers. With potassium in ether, triethoxytitanium [19726-75-1] can be prepared as a pink-lilac compound, which, on exposure to oxygen, yields $(C_2H_5O)_3TiOTi(OC_2H_5)_3$ [84215-71-4] (9). The ethoxy groups in the titanoxane exchange with C_3H_7OH without breaking the Ti–O–Ti bond. Table 4 is a list of a few of the organotitanium compounds of valency lower than four.

A family of Ti(III) derivatives roughly parallels those of Ti(IV). Titanium(III) chelates are known, eg, titanium trisacetylacetonate [14284-96-9] prepared in benzene from titanium trichloride, acetylacetone, and ammonia (185). This deep-blue compound is soluble in benzene but insoluble in water.

$$TiCl_4 + 3\ Hacac + 3\ NH_3 \longrightarrow Ti(acac)_3 + 3\ NH_4Cl$$

The compound is oxidized by air to orange-red crystals, which are possibly $O=Ti(acac)_2$ or an oligomer. If, however, the mixture is refluxed in the absence of ammonia, a red dimer is formed (mp 214°C), to which the doubly bridged structure,

$$(acac)_2Ti\overset{Cl}{\underset{Cl}{<\ \ \ >}}Ti(acac)_2$$

[61436-17-7], is assigned (186). Titanium(III), β-diketonates can also be prepared by reduction of the Ti(IV) chelates (187).

α-Hydroxy acids are represented by oxalic acid. A titanous oxalate is prepared in water from $TiCl_3$ and oxalic acid, precipitating upon the addition of ethyl alcohol as a yellow solid (188). It forms double salts with metal oxalates, $MTi(C_2O_4)_2\cdot2H_2O$.

Table 4. Organotitanium Compounds of Lower Valence

Compound	CAS Registry Number	Type[a]	Appearance	Mp, °C	Other properties
$Ti(CH_3)_3$	[32835-60-2]	Ti(III) trialkyl	not isolated; green in THF solution		solutions give positive Gilman test; decompose above −20°C
$(C_5H_5)_2TiCl$	[60955-54-6]	Ti(III) Cp₂ halide	green crystals	279–281 sublimes in vacuo at 150	insoluble in hydrocarbons; very sensitive to oxygen; blue solution in acetonitrile
$C_5H_5TiCl_2$	[31781-62-1]	Ti(III) Cp halide	violet		
$(CH_2=CH)_2TiCl$		Ti(III) (divinyl) halide	infusible powder		
$(C_9H_7)_2TiCl$	[12113-02-9]	Ti(III) (indenyl) halide	yellowish red		
$(C_5H_5)_2TiCH_2CH=CH_2$	[12110-59-7]	Ti(III) Cp₂ allyl	purple-blue		monomeric; extremely air-sensitive
$(C_5H_5)_2TiOOCCH_3$	[12248-00-9] [56260-60-7]	Ti(III) Cp₂ carboxylate	blue	110	air-sensitive
$(C_5H_5)_2TiOOCC_9H_{19}$	[12248-77-0]		blue	ca −5	air-sensitive solution very sensitive to air
$(C_5H_5)_2TiBH_4$	[12772-20-2]	Ti(III) Cp₂ hydroborate	black-violet needles		very sensitive to air; aqueous alkali → Ti₂O₃ (hydrate)
$(C_5H_5)_2TiBF_4$	[83562-93-0]	Ti(III) Cp₂ fluoroborate	light blue		
$(C_5H_5)_3Ti$	[39333-58-9]	Ti(III) Cp₃	green	sublimes at 125	extremely air-sensitive; gives (C₅H₅)₂Ti(CO)₂ with CO under pressure
$Ti(C_6H_5)_2$	[14724-88-0]	Ti(II) diphenyl	black solid		pyrophoric; gives phenylmercury chloride with HgCl₂
$C_5H_5TiC_6H_5$	[12109-06-7]	Ti(II) Cp₂ C₆H₅	black solid		sensitive to air and moisture; thermally stable to 170°C
$(C_5H_5)_2Ti$	[1271-29-0]	Ti(II) Cp₂	dark green	200	pyrophoric; catalyst for polymerization of olefins and acetylenes

[a] Cp = cyclopentadienyl.

A group of violet titanium(III) acylates has been prepared from $TiCl_3$ and alkali carboxylates. All of the acylates are strong reducing agents similar to $TiCl_3$ (189). Studies of Ti(III) compounds include the reaction (190)

$$TiCl_3 + (RO)_3\,P{=}O \longrightarrow TiCl_2OPO(OR)_2 + RCl \longrightarrow Ti(OPO(OR)_2)_3$$

where R is either C_2H_5 [56170-51-5], C_3H_7 [56170-52-6], $CH_2CH{=}CH_2$ [56170-53-7], or C_4H_9 [56170-54-8].

The tris compounds are highly bridged three-dimensional polymers. Photoreduction of aqueous Ti(IV)-containing alcohols or glycols, but not of ethylene glycol, yields Ti(III) and the aldehyde or ketone corresponding to the alcohol (191,192). A possible mechanism is

$$\text{>Ti}\begin{smallmatrix}O\text{-}i\text{-}C_3H_7\\O\text{-}i\text{-}C_3H_7\end{smallmatrix} \longrightarrow \text{>Ti}\begin{smallmatrix}O\text{-}i\text{-}C_3H_7\\ {}^\bullet O\text{-}i\text{-}C_3H_7\end{smallmatrix}$$

$${}^\bullet O\text{-}i\text{-}C_3H_7 + {}^\bullet O\text{-}i\text{-}C_3H_7 \longrightarrow CH_3\overset{\overset{O}{\|}}{C}CH_3 + i\text{-}C_3H_7OH$$

A broad selection of Ti(III) compounds coordinated to α-hydroxy acids, diboric acids, and 8-hydroxyquinoline has been prepared by the reaction

$$TiCl_3 + n\ HL \xrightarrow[\text{DMF}]{(C_2H_5)_3N} TiCl_{3-n}L_n$$

The esr has been measured at 77 and 298 K. The compounds are dimers, but the Ti–Ti distances vary with the ligand (193,194).

Exposure of $TiCl_4$ in ethyl, propyl, and butyl alcohols for four weeks results in the precipitation of green octahedral Ti(III) complexes. Similar products form on irradiating $Ti(OR)_4$ in ROH containing an equivalent of a lithium halide (192).

Among the applications of lower valent titanium, the McMurry reaction, which involves the reductive coupling of carbonyl compounds to produce alkenes, is the most well known. An excellent review of lower valent titanium reactions is available (195). Titanium(II)-based technology is less well known. A titanium(II)-based complex has been used to mediate a sterio- and regio-specific reduction of isolated conjugated triple bonds to the corresponding polyenes (196).

Organometallics

Titanium(IV) Organometallics. In classical organometallic chemistry, Grignard reagents (qv) or organolithium compounds react with halides of less active metals to form new carbon–metal bonds. This type of reaction with titanium halides invariably failed, until it was realized that many simple titanium alkyls are extraordinarily unstable thermally as well as to moisture and air. This thermal instability is derived from the presence of unfilled, low lying $3d$ orbitals. In titanium metal, the electron configuration is $3s^23p^63d^24s^2$; in simple tetralkyltitaniums, it is $3s^23p^63d^64s^2$, and may also include hybrid $3d{-}4s$

orbitals. Supplying two extra electron pairs by coordination with, for example, pyridine, diamines, or strong donor ethers, gives stable molecules having a configuration of $3s^2 3p^6 3d^{10} 4s^2$ (85,197,198). Another source of instability is the availability of facile decomposition mechanisms, eg, β-elimination:

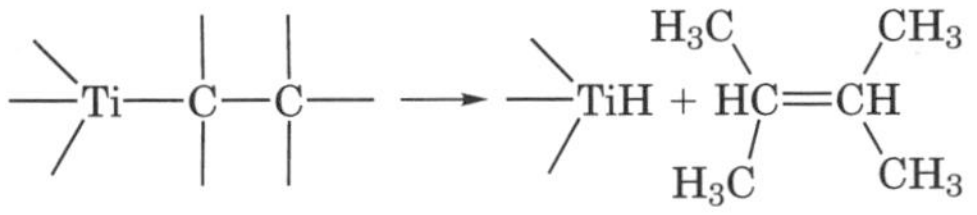

This can be circumvented by choosing alkyl groups with no β-H, eg, methyl, neopentyl, trimethylsilylmethyl, phenyl and other aryl groups, and benzyl. The linear transition state for β-elimination can also be made sterically impossible. The most successful technique for stabilization combines both principles. The pentahaptocyclopentadienyl ring anion η^5-$C_5H_5^-$ (Cp) has six π-electrons available to share with titanium. Biscyclopentadienyltitanium dichloride [1271-19-8] (titanocene dichloride), Cp_2TiCl_2, melts at 289°C, can be sublimed at 190°C at 267 Pa (2 mm Hg), and can be recovered almost quantitatively from its solution in boiling dilute hydrochloric acid (pH <1). The Cp ligand and its substitution products, abbreviated as Cp′, can also stabilize the otherwise labile Ti–R compounds. Thus, $Cp_2Ti(C_6H_5)_2$ [1273-09-2] is unchanged for several days at room temperature.

Thermal stability is enhanced in chelates; thus dimethyl-2-methylpentane-2,4-dioltitanium [23916-35-0] (**22**) is much more stable than $(CH_3)_3Ti$-$(OCH(CH_3)_2)_2$ (68). The structure of the former has been shown by x-ray diffraction to be dimeric and five-coordinate through oxygen bridges. The more highly substituted the six-membered ring, the more thermally stable the compound.

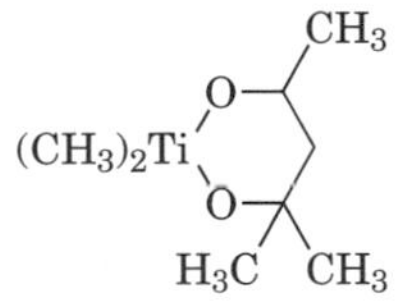

(**22**)

Covalent Noncyclopentadienyl Compounds. The general synthesis of covalent non-Cp compounds, $R_n TiX_{4-n}$, where R = alkyl or aryl and X = halogen, alkoxyl, or amido, involves a lithium, sodium, or magnesium organometallic with a titanium–halogen compound in an inert atmosphere. Solvents are usually either ethers, eg, $(C_2H_5)_2O$, THF, or glyme, or hydrocarbons, eg, hexane or benzene. In addition, a low temperature is required to compensate for the low thermal stability. Schlenk-tube techniques are commonly used. Grignard reagents and alkylaluminum compounds are reducing agents; organoalkalies generally give less reduction. The halides may be TiX_4 or a readily prepared alkoxytitanium halide $X_n Ti(OR)_{4-n}$. Because the R–Ti bond is generally broken by protic reagents (HA), an alkoxyl cannot be introduced by alcoholysis after the R–Ti bond forms.

An unusual reaction leading to a Ti–C bond is unrelated to those just discussed. Diphenylketene adds $Ti(OR)_4$ (199) as follows:

$$Ti(OR)_4 + n\ (C_6H_5)_2C{=}C{=}O \longrightarrow (RO)_{4-n}Ti(C(C_6H_5)_2COOR)_n$$

for $n = 1$, R = C_2H_5, [78319-02-5]; $n = 1$, R = $CH(CH_3)_2$, [78319-03-6]; $n = 2$, R = C_2H_5, [78319-05-8]; $n = 2$, R = $CH(CH_3)_2$, [78319-06-9]; and $n = 1$, R = C_6H_5, [78319-04-7]. Other heterocumulenes, eg, $C_6H_5N{=}C{=}O$, react similarly to yield carbamic esters (121,122).

The reaction of ketene itself with tetraalkyl titanates followed by a ketone $R^1R^2C{=}O$ gives β-hydroxy-esters, $R^1R^2COHCH_2CO_2R$. Polyinsertion of ketene and aldehyde into the Ti–O bond leads to di-, tri-, and tetraesters, eg, $HOCR^1R^2CH_2CO_2CR^1R^2CH_2CO_2R$ (200).

There are numerous alkyltitaniums, and many of their reactions resemble those of alkyllithiums and alkylmagnesium halides. They are protolyzed by water and alcohols, $R{-}Ti(R')_3 + HA \rightarrow RH + A{-}Ti(R')_3$; they insert oxygen, $R{-}TiR' + O_2 \rightarrow ROTiR'$; and they add to a carbonyl group:

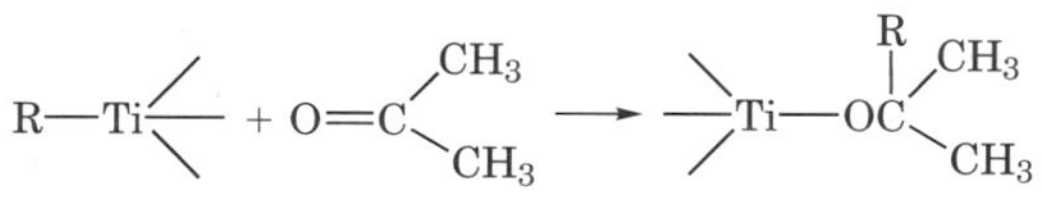

One reaction, disproportionation, has made it difficult to prepare di- and triaryltitanium alkoxides. It is easy to prepare phenyltitanium trialkoxide, $C_6H_5Ti(OR)_3$. However, the reaction

$$2\ C_6H_5Li + Cl_2TiOCH(CH_3)_2 \longrightarrow (C_6H_5)_2TiOCH(CH_3)_2$$
$$[762\text{-}99\text{-}2] \qquad\qquad\qquad\qquad [84215\text{-}72\text{-}5]$$

is followed by disproportionation, doubtless promoted by the intermolecular bridging of alkoxyl groups, which creates a molecule differing only slightly from a reasonable transition state for phenyl migration. Thermodynamics takes over and the products are $C_6H_5Ti(OR)_3$ and $(C_6H_5)_4Ti$.

The unstable CH_3TiCl_3 [12747-38-8] from $(CH_3)_2Zn + TiCl_4$ forms stable complexes with such donors as $(CH_3)_2NCH_2CH_2N(CH_3)_2$, THF, and sparteine, which methylate carbonyl groups stereoselectively. They give 80% of the isomer shown and 20% of the diastereomer; this is considerably more selective than the more active CH_3MgBr (201). Such complexes or $CH_3Ti(OC_3H_{7-i})_3$ methylate tertiary halides or ethers (202) as follows:

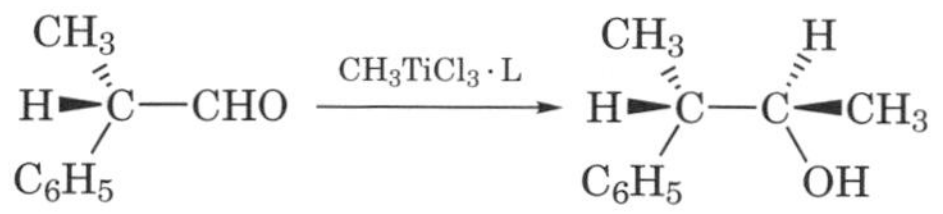

in which R_3CCl is methylated to R_3CCH_3; R_2CCl_2 to $R_2C(CH_3)_2$. Such reactions can even be performed with $(CH_3)_2Zn$ and catalytic amounts of $TiCl_4$ at $-78°C$. Gem dialkylation of a ketone has been achieved in a one-pot reaction (203,204).

Grignard reagents and lithium alkyls add to ester groups, but the CH_3Ti reagents do not. This selectivity has synthetic value (205). Titanium alkyls and aryls discriminate between aldehydes and ketones (206). Titanium alkyls minimize side reactions, eg, elimination, rearrangement, and enolization, which often occur with aluminum alkyls and other active organometallics (204). The solvent is important. Mesityl lithium (MesLi) and $TiCl_4$ in THF yield $LiTiMes_4 \cdot 4THF$ [63916-94-1], but Mes_3Zn and $TiCl_4$ in hydrocarbon produces $MesTiCl_3$ [77801-18-4], which reacts with THF to yield $TiCl_3$ (207).

Other $RTiX_3$ (X = Hal or OR') compounds are selective at $-20°C$. A chiral R'OH, eg, $(S)(-)C_2H_5CH_3CHCH_2OH$, permits asymmetric synthesis. A 98:2 or 99:1 selectivity for aldehyde over ketone has been reported (208–210). The activation energy difference is ca 42 kJ/mol (10 kcal/mol) for RTi, and only 4 kJ/mol (1 kcal/mol) for RLi or RMgX. The kinetic reactivity of $RTi(OR')_3$ decreases in the order of n $C_3H_7O > (CH_3)_3CHO > CH_3CH_2CHCH_3O > CH_3CH_2CH_3CHCH_2O$

for R'O; and $C_6H_5 > CH_3 > C_4H_9 > $ ⟨ ⟩— for R.

Titanium alkyls are prepared simply either as

$$TiCl_4 + 3\ TPT \longrightarrow ClTi(OCH(CH_3)_2)_3$$

where the product, [20717-86-6], is distillable, or as

$$CH_3COCl + TPT \longrightarrow ClTi(OCH(CH_3)_2)_3 + CH_3COOCH(CH_3)_2$$

followed by

$$RLi + ClTi(OCH(CH_3)_2)_3 \longrightarrow RTi(OCH(CH_3)_2)_3 + LiCl$$

Titanium alkyls, known as tamed Grignard reagents, do not add to esters, nitriles, epoxides, or nitroalkanes at low temperatures. Rather, they add exclusively in a 1,2 fashion to unsaturated aldehydes (208–210).

Tetraneopentyltitanium [36945-13-8], Np_4Ti, forms from the reaction of $TiCl_4$ and neopentyllithium in hexane at $-80°C$ in modest yield only because of extensive reduction of Ti(IV). Tetranorbornyltitanium [36333-76-3] can be prepared similarly. When exposed to oxygen, $(NpO)_4Ti$ forms. If it is boiled in benzene, it decomposes to neopentane. When dissolved in monomers, eg, α-olefins or dienes, styrene, or methyl methacrylate, it initiates a slow polymerization (211,212). Results from copolymerization studies indicate a radical mechanism (212). Ultraviolet light increases the rate of dissociation to radicals. The titanocycle [79953-32-5], Cl_2Ti⟨▷⟩, is planar and may be an intermediate in some forms of olefin metathesis (213).

Tetraneopentyltitanium and $((CH_3)_3SiCH_2)_4Ti$ [33948-28-6] react with nitric oxide to yield compounds having R groups bonded in two different ways, as shown by nmr (214). The reactions of NO with R_3TiCl and R_2TiCl_2 yield similar products that have Cl instead of the $-ONRNO$ substituent. However, Cp_2TiR,

$$R_4Ti + \text{excess NO} \longrightarrow$$

where R is C_6H_5 or $CH_2C_6H_5$ and Ti is paramagnetic Ti(III), reacts with NO, resulting in the loss of one Cp and one R group, to form the trinuclear, compound [76722-03-7] (**23**).

$$\text{(23)}$$

Carbometalation, an important reaction of RTi(IV) compounds in which RTi adds to a C=C or C≡C multiple bond and results in a net R–H addition, is involved in Ziegler-Natta polymerization as follows:

$$\text{RTi} + \text{C}{\equiv}\text{C} \longrightarrow \text{RC}{=}\text{CTi} \xrightarrow{H_2O} \text{RC}{=}\text{CH}$$

Solutions of RC triple-bond C–Ti(O-i-C_3H_7)$_3$ can be prepared by treating acetylenic compounds, such as phenylacetylene, with butyl lithium and then Cl–Ti(O-i-C_3H_7)$_3$. These materials can react with aldehydes and epoxides to give the expected addition products (215).

In the following cases, only those reactions in which there is no chain growth, or at most dimerization, are considered (see OLEFIN POLYMERS). Alkyltitanium halides can be prepared from alkylaluminum derivatives. The ring structure imparts regiospecificity to the ensuing carbometalation (216):

$$C_2H_5C{\equiv}CCH_2CH_2OH + (CH_3)_3Al + TiCl_4 \xrightarrow[-78°C]{CH_2Cl_2} \xrightarrow{quench} \overset{\overset{\displaystyle CH_3}{\displaystyle |}}{C_2H_5C}{=}CHCH_2CH_2OH$$

The metallocycle [*67719-69-1*] (**24**) undergoes an apparent β-elimination to a carbene-like reagent, which adds regiospecifically to terminal acetylenes (217).

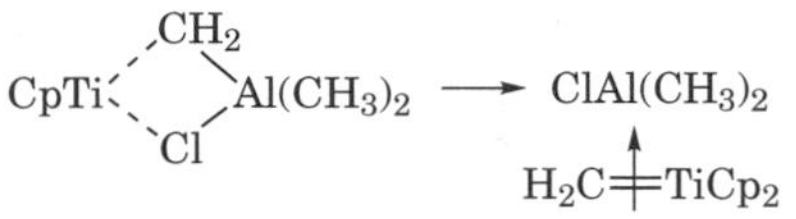

$$CpTi\underset{Cl}{\overset{CH_2}{<}}Al(CH_3)_2 \longrightarrow ClAl(CH_3)_2$$

$$H_2C{=}TiCp_2$$

(**24**)

$$RC{\equiv}CH + (CH_3)_3Al + Cp_2TiCl_2 \longrightarrow RC{=}C\underset{CH_3}{\overset{Al(CH_3)_2}{<}}TiCp_2Cl$$

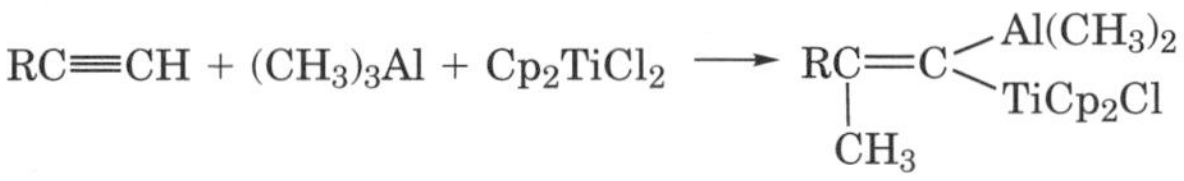

(**25**)

The intermediate (**25**) reacts with ketones ($R_2'CO$) to form cumulenes $RC(CH_3){=}C{=}CR_2$. The indenyl derivative reacts similarly.

Olefin isomerization is often catalyzed by titanium. An example is the conversion of vinylnorbornene to the comonomer ethylidenenorbornene (141). The catalyst is a mixture of a sodium suspension, $AlCl_3$, and $(RO)_4Ti$ or Cp_2TiCl_2. Although isomerization is slow, the yield is high. The active reagent is doubtless a Ti(III) compound.

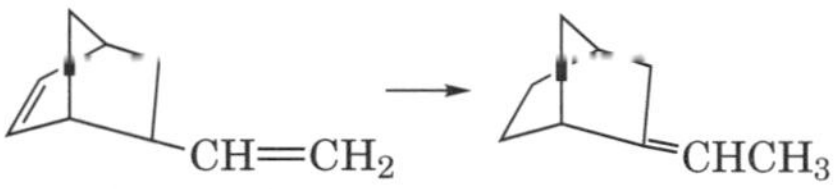

Complexes of titanium, such as 2,6-$(RNCH_2)_2NC_5H_3TiCl_2$, prepared by reaction of $TiCl_4$ with $2,6((CH_3)_3Si)RNCH_2)_2NC_5H_3$, can react with various Grignard reagents to prepare conformationally rigid diamide mono- and dialkyl titanate complexes (218,219).

Cyclopentadienyltitanium Compounds. Properties. The structure of Cp_2Ti-Cl_2 has been shown by x-ray diffraction to be a distorted tetrahedron: Ti–Cl = 236.4 Pm, Ti–centroid = 205.8 Pm, and C–C = 133.9–141.9 Pm (220,221). The compound has also been studied by electron diffraction (222). Changes in the structure are imposed by bridging the Cp rings with $-(CH_2)_n-$ or other groups (223). In $Cp_2Ti(C_6H_5)_2$, the C–C bonds in the Cp rings have different lengths, as do those in the phenyl rings (224). Chemical ionization mass spectrometry of Cp_2TiCl_2 gives mainly M+(−Cl) (225,226); ^{13}C-nmr results are available (227). The rings of Cp_2TiCl_2 spin rapidly, so the ^{1}H-nmr spectrum produces one peak. Bridging the rings prevents spinning and splitting can be detected (228,229). Electron spectroscopy for Cp_2TiX_2, where X is halogen, and other CpTi(IV) compounds has been reported (230–232). Infrared and Raman spectra are discussed elsewhere (233–236) and the photoelectron spectrum has been determined (237). Molecular orbital calculations have been reported for many Cp–Ti compounds (237–245). Vapor pressure equations for Cp_2TiCl_2 and TYZOR TBT are available, as are other thermodynamic quantities (246,247). Selected titanium(IV) cyclopentadienyls and biscyclopentadienyls are given in Tables 5 and 6, respectively.

Table 5. Organotitanium(IV) Compounds Cyclopentadienyls

Compound	CAS Registry Number	X in $C_5H_5TiX_3$	Appearance	Mp[a], °C	Bp[a], °C	Other properties
$C_5H_5TiCl_3$	[1270-98-0]	halide	orange	140–142		on hydrolysis gives (C_5H_5Ti-$ClO)_n$
$C_5H_5TiBr_3$	[12240-42-5]		orange	163–165		
$C_5H_5TiI_3$	[12240-43-6]		deep red	185–190		
$C_5H_5TiBrCl_2$	[70568-74-0]		orange	165–170		
$C_5H_5Ti(OCH_3)_3$		alkoxy		50–52	88_{200}	readily hydrolyzes; darkens on storage
$C_5H_5Ti(OC_2H_5)_3$	[1282-41-3]				$106–107_{400}$	very sensitive to moisture; monomeric in benzene
$C_5H_5Ti(OC_3H_7)_3$	[12242-59-0]				$106–107_{67}$	very sensitive to moisture; monomeric in benzene
$C_5H_5Ti(OC_4H_9)_3$	[84215-74-7]				$124.5–125.5_{67-133}$	n_D^{20} 1.5224
$C_5H_5Ti(OC(CH_3)_3)_3$	[12148-33-3]				102_{133}	n_D^{20} 1.5065
$C_5H_5Ti(OC_6H_{13})_3$	[22290-79-8]				$177–181_{133}$	n_D^{20} 1.5082
$C_5H_5TiOCH_3Cl_2$	[12192-52-8]	alkoxyhalide	yellow crystals	93–96		
$C_5H_5TiOC_2H_5Cl_2$	[1282-32-2]			49		
$C_5H_5TiOC_3H_7Cl_2$	[70046-23-0]		yellow-green		$159–161_{267}$ dec	
$C_5H_5Ti(OCH_3)_2Cl$	[84215-75-8]					
$C_5H_5Ti(OC_2H_5)_2Cl$	[1282-38-8]				$109–111_{133}$	n_D^{20} 1.5818
$C_5H_5Ti(OC_3H_7)_2Cl$	[84215-76-9]		yellow-green		$132–145_{133}$ dec	
$C_5H_5Ti(OC_4H_9)_2Cl$	[84215-77-0]				$145–150_{267-400}$	
$C_5H_5Ti(OC_4H_9)_2Br$	[84215-78-1]				$36–45_{107}$	
$C_5H_5Ti(OC_6H_5)Cl_2$	[12288-59-4]	aryloxyhalide				very sensitive to moisture

Compound	CAS No.	Type	Color	mp, °C	bp, °C (Pa)	Properties
$C_5H_5Ti(OC_2H_5)_2$-$OOCCH_3$	[84215-79-2]	alkoxyacetate			$106-108_{267}$	on heating $C_5H_5Ti(OC_2H_5)_3$ is formed
$C_5H_5Ti_5(OOCCH_3)_3$	[1282-42-4]	acetate		115–117		hydrolytically and thermally unstable
$C_5H_5Ti(SCH_3)_3$	[84215-80-5]	mercapto				
$C_5H_5Ti(SCH_3)_2Cl$	[84215-81-6]	mercaptohalide				
$C_5H_5Ti(OSi(CH_3)_3)_3$	[57665-25-5]	siloxy			$138-139_{133}$	n_D^{20} 1.45824; d_4^{20} 0.9436 g/cm^3
$CH_3C_5H_4TiCl_3$	[1282-31-1]	halide (substituted cyclopentadiene)		98–99		crystals liquefy in air
$C_2H_5C_5H_4TiCl_3$	[1282-33-3]		orange solid			
$(CH_3)_5C_4TiCl_3$	[12129-06-5]		red solid	136_{133} sub		
$CH_3C_5H_4TiOC_2$-H_5Cl_2	[1282-39-9]	alkoxyhalide(substituted cyclopentadiene)	green	225–227		viscous mass
$CH_3C_5H_4Ti(OC_2$-$H_5)_2Cl$	[1282-36-6]	alkoxy (substituted cyclopentadiene)			$143-145_{267}$	easily hydrolyzed in air; n_D^{20} 1.5401, d^{20} 1.0780 g/cm^3
$CH_3C_5H_4Ti(OC_2$-$H_5)_3$	[1282-47-9]				$80-81_{133}$	easily hydrolyzed; darkens in storage even at −500C
$C_2H_5C_5H_4Ti(OC_2$-$H_5)_3$	[1292-46-8]				$101-102_{267}$	n_D^{20} 1.5359; d^{20} 1.0717 g/cm^3

[a]Subscripted values are pressure in Pa. To convert Pa to mm Hg, divide by 133.

Table 6. Organotitanium(IV) Compounds Biscyclopentadienyls

Compound, $(C_5H_5)_2TiX_2$	CAS Registry Number	X	Appearance	mp °C	Other properties
$(C_5H_5)_2TiCl_2$	[1271-19-8]	halide	vivid-red crystals	289	slightly soluble in water; forms salts with acids
$(C_5H_5)_2TiBr_2$	[1293-73-8]		vivid-red crystals	314	diamagnetic, soluble in nonpolar solvents
$(C_5H_5)_2TiI_2$	[12152-92-0]		purple crystals	319	
$(C_5H_5)_2TiF_2$	[309-89-7]		yellow		
$(C_5H_5)_2Ti(OC_4H_9)_2$	[12303-65-0]	alkoxy			
$(C_5H_5)_2Ti(OC_6H_5)_2$	[12246-19-4]	aryloxy	yellow	142	thermally stable; hydrolyzed only when heating with concentrated NaOH
$(C_5H_5)_2Ti(o\text{-}ClC_6H_4O)_2$	[12309-06-7]		orange-yellow	145–147	thermally stable; hydrolyzed only when heating with concentrated NaOH
$(C_5H_5)_2Ti(o\text{-}CH_3C_6H_4O)_2$	[12309-37-4]		yellow	162	thermally stable; hydrolyzed only when heating with concentrated NaOH
$(C_5H_5)_2Ti(o\text{-}NO_2C_6H_4O)_2$	[12309-11-4]		red	122–124	thermally stable; hydrolyzed only when heating with concentrated NaOH
$(C_5H_5)_2Ti(p\text{-}ClC_6H_4O)_2$	[12309-07-8]		yellow	125–127	thermally stable; hydrolyzed only when heating with concentrated NaOH

Compound	CAS No.	Type	Color	mp, °C	Properties
(C$_5$H$_5$)$_2$TiOC$_2$H$_5$Cl	[12129-76-9]	alkoxyhalide		91–92	stable to air and water
(C$_5$H$_5$)$_2$TiOC$_3$H$_7$Cl	[12715-66-1]			57–58	
(C$_5$H$_5$)$_2$TiOC$_6$H$_5$Cl	[62652-01-1]	aryloxyhalide		71–73	
(C$_5$H$_5$)$_2$Ti(SH)$_2$	[12170-34-2]	mercapto	deep red	150–160 dec	
(C$_5$H$_5$)$_2$Ti(SCH$_3$)$_2$	[12089-78-0]	alkylmercapto		193–197	
(C$_5$H$_5$)$_2$Ti(SC$_2$H$_5$)$_2$	[1291-79-8]			107–110	
(C$_5$H$_5$)$_2$Ti(SC$_3$H$_7$)$_2$	[1292-07-5]			88–93	
(C$_5$H$_5$)$_2$Ti(SC$_6$H$_5$)$_2$	[1292-72-4]	arylmercapto		199–201	
(C$_5$H$_5$)$_2$Ti(SCH$_2$C$_6$H$_5$)$_2$	[1292-61-1]			172–174	
(C$_5$H$_5$)$_2$Ti(SCH$_2$CH$_2$C$_6$H$_5$)$_2$	[1292-47-3]			92–94	
(C$_5$H$_5$)$_2$Ti(OOCCH$_3$)$_2$	[1282-51-5]	acyl	orange	126–128	readily hydrolyzed and thermally unstable
(C$_5$H$_5$)$_2$Ti(OOCC$_3$H$_7$)$_2$	[12290-20-9]		red-orange	114–116	readily hydrolyzed and thermally unstable
(C$_5$H$_5$)$_2$Ti(OOCCH$_2$Cl)$_2$	[1282-44-6]		red-orange	98–99	readily hydrolyzed and thermally unstable
(C$_5$H$_5$)$_2$TiOOCCCl$_3$	[12212-37-2]		red-orange	192–194	
(C$_5$H$_5$)$_2$Ti(OOCCF$_3$)$_2$	[1282-45-7]		orange	178–180	soluble in benzene, ethyl acetate, ethyl alcohol; moderately soluble in chloroform; thermally stable
(C$_5$H$_5$)$_2$Ti(OOCC$_6$H$_5$)$_2$	[51178-00-8]	benzoyl	yellow	188	reacts with aqueous alkali to give benzoic acid
(C$_5$H$_5$)$_2$Ti(OSi(CH$_3$)$_3$)$_2$	[12319-01-6]	siloxy			n_D^{20} 1.4582, d_4^{20} 0.9436 g/cm^3
(C$_5$H$_5$)$_2$Ti(OSi(C$_6$H$_5$)$_3$)$_2$	[12321-33-4]				stable in air but not in acidic or basic media

Table 6. (Continued)

Compound, $(C_5H_5)_2TiX_2$	CAS Registry Number	X	Appearance	mp °C	Other properties
$(C_5H_5)_2TiOSi(CH_3)_3Cl$	[12319-01-6]	siloxyhalide	orange	137.5	stable in air but not in acidic or basic media
$(C_5H_5)_2TiOSi(C_6H_5)_3Cl$	[12320-99-9]			210–212	
$(C_5H_5)_2Ti(CO)_2$	[12129-51-0]	carbonyl	dark reddish-brown crystals	dec ca 90	extremely air-sensitive
$(C_5H_5)_2Ti(CH_3)_2$	[1271-66-5]	alkyl	yellow-orange	dec ca 100	
$(C_5H_5)_2TiCH_3Cl$	[1278-83-7]	alkylhalide	orange-red	168–170 dec	
$(C_5H_5)_2TiC_3H_7Cl$	[12715-66-1]		orange	160	
$(C_5H_5)_2Ti(C_6H_5)_2$	[1278-09-2]	aryl	orange-yellow	146–148 dec	
$(C_5H_5)_2Ti(m\text{-}CH_3C_6H_4)_2$	[12156-57-9]		red	137–139	
$(C_5H_5)_2Ti(p\text{-}CH_3C_6H_4)_2$	[12156-58-0]		yellow-orange	133–134	
$(C_5H_5)_2Ti\text{-}p\text{-}((CH_3)_2NC_6H_4)_2$	[12156-86-4]		maroon	137–139 dec	
$(C_5H_5)_2Ti(p\text{-}FC_6H_4)_2$	[12155-98-5]		orange	120 dec	
$(C_5H_5)_2Ti(p\text{-}ClC_6H_4)_2$	[12155-97-4]		orange	130 dec	
$(C_5H_5)_2Ti(p\text{-}BrC_6H_4)_2$	[12155-95-2]		orange	130 dec	
$(C_5H_5)_2Ti(m\text{-}CF_3C_6H_4)_2$	[12156-38-6]		orange-yellow	145–146	
$(C_5H_5)_2Ti(p\text{-}CF_3C_6H_4)_2$	[12156-39-7]			142–143	
$(C_5H_5)_2Ti(C{\equiv}CC_6H_5)_2$	[12303-93-4]		orange-brown crystals	141	
$(C_5H_5)_2Ti(C_6F_5)_2$	[12155-89-4]	fluorophenyl	orange needles	228–230	thermally stable in vacuo at 110°C
$(C_5H_5)_2TiClC_6F_5$	[50648-18-5]		pale-orange needles	201–203	
$(C_5H_5)_2TiOC_2H_5C_6F_5$	[84215-82-7]		yellow solid	117	stable in dry air; soluble in organic solvents
$(C_5H_5)_2TiOHC_6F_5$	[84215-83-8]		yellow solid	183–185 dec	soluble in organic solvents
$(C_5H_5)_2TiC_6F_5F$	[84501-83-7]		yellow solid	240 dec	

Titanium tetrachloride is a Lewis acid having many useful synthetic properties (248). Replacing Cl by OR weakens the acid. Acid–base complexes having dimethyl sulfoxide (DMSO), dimethylformamide (DMF), dimethylacetamide (DMAc), which are all O-bonded, as well as pyridine and ethylenediamine, have been reported (249). $CpTiCl_3$ forms stable, sublimable complexes having ditertiary amines or arsines, eg, $(CH_3)_2ECH_2CH_2E(CH_3)_2$, where E is N or As. The less-basic analogues, where E is O, S, or P, do not form stable complexes. The monodentates pyridine, $(CH_3)_2S$, $(CH_3)_3As$, and $(C_6H_5)_3P$ also do not form complexes. The bromide and iodide behave similarly (250). Biscyclopentadienyltitanium dichloride is virtually devoid of Lewis acidity. The properties of ring-substituted Cp derivatives differ from those of the parent Cp derivatives in a manner predictable from the electronic properties of the substituent.

Synthesis. The discovery of stable cyclopentadienyltitanium compounds, in particular Cp_2TiCl_2, stimulated the synthesis and study of a host of related compounds. These include both $CpTiX_3$ and Cp_2TiX_2.

The basic laboratory synthesis involves a salt of cyclopentadiene, eg, lithium, sodium, potassium, or magnesium, and a titanium(IV) halide, usually in an ether solvent (251–253). However, this is probably too expensive for industrial use. Different cyclopentadienyl groups can be introduced in two separate steps (254). One patent shows that merely heating CpH and $TiCl_4$ in dioxane at 60–80°C in the presence of diethylamine yields Cp_2TiCl_2 (255). Titanium alkoxides react with CpMgBr to yield $CpTiBr(OR)_2$ or Cp_2TiBr_2, depending on the mole ratio (256). Another patent describes the reaction (257)

$$CpH + Cl_2Ti(OR)_2 + (C_2H_5)_3N \text{ (or piperidine)} \longrightarrow CpTi(OR)_2Cl$$

The product forms in 80% yield and is distillable. However, no additional amine is necessary for the following reaction (162), where n is 1 or 2:

$$n\ CpH + Ti(NR_2)_4 \longrightarrow Cp_nTi(NR_2)_{4-n}$$

For laboratory use, cyclopentadienylthallium [*34822-90-7*] reacts cleanly with TiX compounds, where X is halogens (258–264). The cost and toxicity of thallium compounds are drawbacks to large-scale use. However, Cp_2Pb may, in certain cases, be useful (see LEAD COMPOUNDS; THALLIUM AND THALLIUM COMPOUNDS) (265).

Monocyclopentadienyl compounds can be prepared by the above techniques with appropriate control of the stoichiometric proportion of reagents or by use of reagents such as $ClTi(OR)_3$ (266). Biscyclopentadienyltitanium dichloride can be carefully chlorinolyzed with Cl_2 or with SO_2Cl_2 in refluxing $SOCl_2$ (254,267). A wide variety of ring-substituted cyclopentadienes have been converted to Cp′–Ti compounds. Methyl and other alkyl and aryl groups and $(CH_3)_3Si-$ are most common: Pentamethylcyclopentadiene is a popular, although expensive, ligand (268,269). Bridged bis(cyclopentadiene)s, eg, $Cp(CH_2)_nCp$ and $CpSi(CH_3)_2Cp$, have received much attention (228,270). The properties of derivatives have been compared with various length bridges (271,272). Indene and, to a very limited extent, fluorene function as ligands.

Titanium-containing polyethers have been prepared by the reaction of dicyclopentadienyltitanium dichloride with aromatic and aliphatic diols via an interfacial and/or aqueous solution polycondensation technique (273).

Cyclopentadienyltitanium Compounds with Other Carbon Titanium Links. Cyclopentadienyltitanium trichloride and, particularly, Cp_2TiCl_2 react with RLi or with RAl compounds to form one or more R–Ti bonds. As noted, the Cp groups stabilize the Ti–R bond considerably against thermal decomposition, although the sensitivity to air and moisture remains. Depending on the temperature, mole ratio, and structure of R, reduction of Ti(IV) may be a serious side reaction, which often has preparative value for $Cp_n Ti(III)$ compounds (268,274,275).

Methyl and aryl groups are most commonly used. Higher alkyltitaniums tend to decompose by β-elimination except when R is $CH_2Si(CH_3)_3$ or $CH_2C(CH_3)_3$ (276). Ring compounds or titanocycles, made from Cp_2TiCl_2 and 1,n-dilithioalkanes, $Li(CH_2)_n Li$, at $-78°C$, are quite thermolabile, though more stable than $Cp_2Ti(C_4H_9)_2$ [52124-69-3; 71297-31-9]. In the compound where n is 4, β-elimination is suppressed because the Ti–C–C–H dihedral angle is far from 0°. The compound inserts carbon monoxide to yield a titanoketone, which expels Ti at 25°C. These reactions do not occur when n is 5. Other titanocene

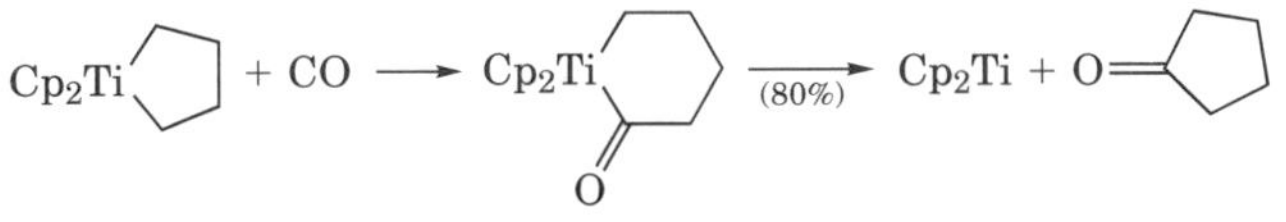

$$[52124\text{-}67\text{-}1] \qquad\qquad [52124\text{-}68\text{-}2]$$

synthons may undergo similar reactions with CO (277–279). Both vinylic and ethynylic groups can be attached to the Cp_2Ti framework. These tend to be stable thermally and to air and moisture (280).

A rather air-stable titanocycle [76933-94-3; 76933-97-6] has been prepared in meso form only (281):

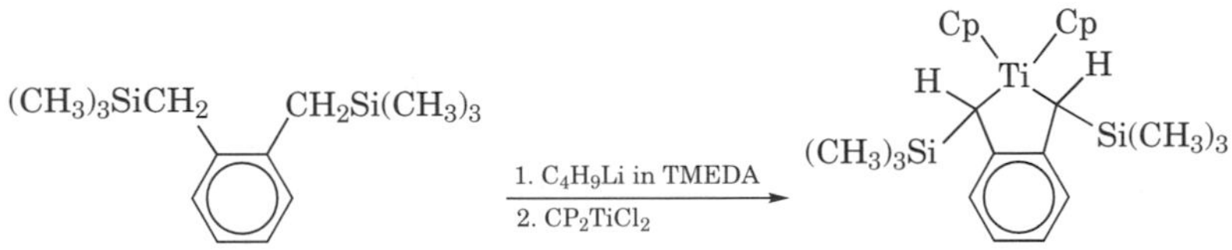

Related examples of silicon-containing titanocycles have also been prepared (282). These compounds are air-stable and not decomposed by methanol. When CpCp′TiClOAr reacts with Grignard reagents, the aryloxy group is replaced by R, not by Cl (283). Compounds of the formula $(Cp_2TiR)_2O$ prepared from $(CP_2TiCl)_2O$ + RLi, where R is CH_3, C_2H_5, C_6H_5, p-tolyl, $C_6H_5C\equiv C$, or $CH_2=CH$, are thermally stable and quite stable to air (283,284).

Alkyl and aryl groups are cleaved by iodine, but Cp groups are not affected (285).

Carbometalation of olefins and acetylenes is a useful reaction. For example, in

$$Cp_2TiCl_2 + R_nAlCl_{3-n} \longrightarrow Cp_2TiRCl \cdot AlR_{n-1}Cl_{4-n} \longrightarrow Cp_2TiR^+ + AlR_{n-1}Cl_{5-n}$$

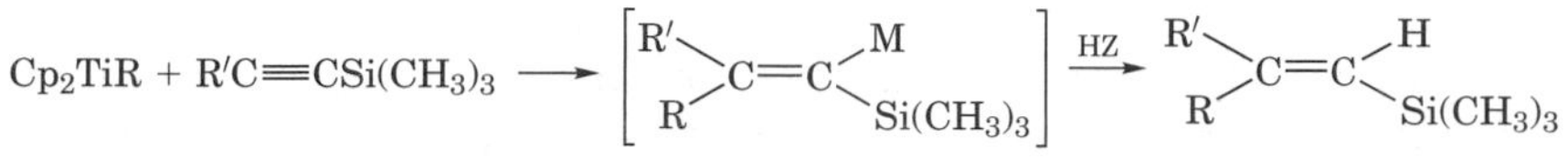

$$Cp_2TiR + R'C{\equiv}CSi(CH_3)_3 \longrightarrow \left[\begin{array}{c} R' \\ R \end{array}{>}C{=}C{<}\begin{array}{c} M \\ Si(CH_3)_3 \end{array} \right] \xrightarrow{HZ} \begin{array}{c} R' \\ R \end{array}{>}C{=}C{<}\begin{array}{c} H \\ Si(CH_3)_3 \end{array}$$

up to 90% trans-addition is obtained, where M is a complex of Ti and Al. The hydrogen in the product is acquired from the solvent or from elsewhere in the reaction mixture, because if the mixture is quenched with $(C_2H_5)_3N$ followed by D_2O, no deuterium occurs in the olefin. If R' in the acetylene is phenyl or 1-cyclohexenyl, equal amounts of cis- and trans-products are obtained. If the unsilylated parent $R'C{\equiv}CH$ is exposed to this catalyst system, it is cyclotrimerized to the $1,3,5-C_6H_3R'_3$. However, this system does not polymerize propylene (286,287).

Other authors disagree with the results of this carbometalation sequence. They find that the initial product results strictly from cis-carbometalation; but with a trace of base, H is abstracted from the medium homolytically with loss of stereochemistry (288). They show that M in the above structure must be Ti, not Al. If the reaction mixture is quenched with D_2O containing NaOD, 95 mol % D is incorporated in the olefin.

Alkynols are ethylated by $(C_2H_5)_2AlCl$ catalyzed by Cp_2TiCl_2 (289). $(CH_3Cp)_2TiCl_2$ [1282-40-2] is sometimes preferred because it is more soluble in nonpolar solvents. Ten-to-fifty percent of the titanium compound is required, because many alkynols rapidly deactivate the titanium. In one example, when a prereacted mixture of $(C_2H_5)_2AlCl$ and $HOCH_2CH_2C{\equiv}CH$ is treated with Cp_2TiCl_2, a 1:1 mixture of (**26**) and (**27**) forms in only 55% yield. Reducing the amount of Cp_2TiCl_2 to 10 mol % raises the yield to 80–90%. The corresponding

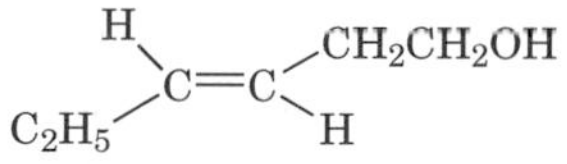

(**26**) 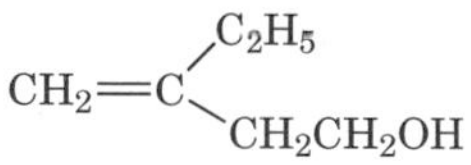   (**27**) 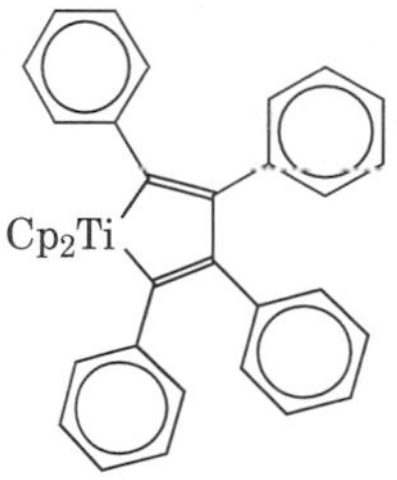   (**28**)

olefin is also carbometalated by ethylaluminum–titanium combinations. There are marked differences among the titanate esters of the unsaturated alcohol $HOCH_2CH_2CH{=}CH_2$ prepared with $TiCl_4$, $Cl_2Ti(acac)_2$, or Cp_2TiCl_2 (290). Olefin dimerization involves carbometalation. High selectivity (98%) for the ethylene-to-1-butane dimerization is achieved by adding 10 mol % of $Ti(OC_4H_9)_4$ and 0.5 mol % of Cp_2TiCl_2 to the $(C_2H_5)_3Al$ catalyst (291).

Photolysis has been intensively studied. For example, $Cp_2Ti(C_6H_5)_2$ yields a green polymer $(Cp_2TiH)_x$ [11136-22-4]. At low temperature in benzene or THF, a dark-green transient Cp_2Ti–solvent species forms and quickly dimerizes.

The phenyl groups appear as benzene and biphenyl. In the presence of CO or $C_6H_5C{\equiv}CC_6H_5$, $Cp_2Ti(CO)_2$ [12159-51-0] or (**28**) [1317-21-1] forms, respectively (292,293).

Photolysis of Cp_2TiAr_2 in benzene solution yields titanocene and a variety of aryl products derived both intra- and intermolecularly (293–297). Dimethyltitanocene photolyzed in hydrocarbons yields methane, but the hydrogen is derived from the other methyl group and from the cyclopentadienyl rings, as demonstrated by deuteration. Photolysis in the presence of diphenylacetylene yields the dimeric titanocycle (**28**) and a titanomethylation product [65090-11-1].

$$Cp_2Ti(CH_3)_2 + C_6H_5C{\equiv}CC_6H_5 \longrightarrow (\mathbf{28}) + \begin{array}{c} Cp_2Ti{\Big\langle}\genfrac{}{}{0pt}{}{CH_3}{} \\ C{=}C \\ C_6H_5 \end{array}\genfrac{}{}{0pt}{}{CH_3}{C_6H_5}$$

The fluorinated titanocycle related to (**28**) is not obtained from $C_6H_5C{\equiv}CC_6F_5$. Photolysis of Cp_2TiX_2 always gives first scission of a Cp–Ti bond. In a chlorinated solvent, the place vacated by Cp is assumed by Cl. In the absence of some donor, the radical dimerizes (298–299).

When $Cp_2Ti(CH_3)_2$ is photolyzed in toluene, two moles of CH_4 is produced; but in monomers, only one mole (1.08 mol in styrene, 0.90 mol in methyl methacrylate) of CH_4 is liberated as the monomers polymerize (300). $Cp_2Ti(CD_3)_2$ [65554-67-8] photolyzed in $C_6D_5CD_3$ gives CD_3H but not CD_4, thus ruling out free CD_3. In methyl methacrylate, $Cp_3Ti(^{14}CH_3)_2$ yields a polymer containing 0.8–1.1 ^{14}C per polymer chain, but no Ti. If tritiated $(Cp–T)_2Ti(CH_3)_2$ is used in methyl methacrylate, the polymer contains only traces of tritium. Clearly, the hydrogen for the methane comes from the Cp groups.

Pryolysis of solid $Cp_2Ti(CD_3)_2$ yields CD_3H but not CD_4. Pyrolysis of $(C_5D_5)_2Ti(CH_3)_2$ yields CH_3D. These results show that the radical attacks the Cp rings (301,302). Pyrolysis of $Cp_2Ti(C_6H_5)_2$ proceeds via a benzyne intermediate, as shown by trapping experiments involving cycloadditions (293,303–306).

The detailed mechanism of pyrolysis of Cp_2TiR_2 compounds has been studied (307–313). A useful titanocycle is formed from Cp_2TiCl_2 and trimethylaluminum; triethylaluminum gives a different product (314). The titanocycle adds to terminal olefins in the presence of 4-dimethylaminopyridine; the adduct expels olefin above 0°C to yield a bistitanocyclobutane (315). The titanocycle can also behave like a Wittig reagent, reacting with aldehydes and ketones to give olefins (314,316).

Displacement Reactions. Cyclopentadienyltitanium halides undergo displacements with a wide variety of nucleophiles. Hydroxylic reagents cleave Ti–R bonds (317,318):

$$Cp_2TiCH_3Cl + H_2O \longrightarrow \overset{.}{C}H_4 + (Cp_2TiCl)_2O$$

Amides are formed with amines, often with strong base assistance (319,320). Occasionally Cp groups are lost (321–332):

$$Cp_2TiCl_2 + NaNRAr \longrightarrow Cp_2TiClNRAr \longrightarrow Cp_2Ti(NRAr)_2$$
$$Cp_2TiCl_2 + LiN(CH_3)_2 \text{ (excess)} \longrightarrow CpTi(NCH_3)_3$$

In titanium acylates, the carboxylate ligands are unidentate, not bidentate, as shown by ir studies (333,334). The ligands are generally prepared from the halide and silver acylate (335). The benzoate is available also from a curious oxidative addition with benzoyl peroxide (335–338):

$$Cp_2TiCl_2 + AgO\overset{\overset{\displaystyle O}{\|}}{C}R \longrightarrow Cp_2Ti(O\overset{\overset{\displaystyle O}{\|}}{C}R)_2$$

$$Cp_2Ti(C_6H_5)_2 + (C_6H_5\overset{\overset{\displaystyle O}{\|}}{C}O)_2 + 2\,(CH_3)_2CHOH \longrightarrow Cp_2Ti(O\overset{\overset{\displaystyle O}{\|}}{C}C_6H_5)_2 + 2\,C_6H_6 + 2\,(CH_3)_2CO$$

The acylates undergo facile hydrolysis or alcoholysis with loss of one or both Cp groups (335).

A silyl group can be attached to titanium by using $Al(Si(CH_3)_3)_3$ or $KSi(C_6H_5)_3$; $LiSi(CH_3)_3$ causes reduction (339,340). Germyl and stannyl groups can also be attached to titanium (341,342). Organometallic ligands can be attached to titanium by displacement reactions involving Cp_2TiCl_2. Both chlorines are displaceable (343–346). Unusual titanocycles are formed from sulfur and phosphorus reagents (241,242,262,347–352).

Reaction of cyclopentadienyltitanium halides and oxygen-bonding reagents is confusing. On the one hand, alcohols cleave one Cp group readily, and the second more slowly, from Cp_2TiCl_2. Sodium alkoxides or aryloxides in aprotic solvents give less cleavage (353–364). Silanols and ambient nitrite and nitrate do not cleave Cp groups (365,366). Several studies have attempted to explain the often facile cleavage by O-bonding reagents, which contrasts with normal displacement by N- or S-bonding reagents (350,367–388).

On the other hand, Cp_2TiCl_2 and CpTiCl dissolve in boiling dilute hydrochloric acid, yielding aquo cations that retain the Cp groups, eg, $(Cp_2TiOH \cdot H_2O)^+Cl^-$ [11216-84-6] and $(Cp_2Ti_3 \cdot H_2O)^{3+} \cdot 3Cl^-$ [1270-90-0]. These can be isolated as salts (261,389–406).

Insertion into the CpTi–R Bond. Sulfur dioxide yields sulfones and ultimately sulfinates. The latter are available also from RSO_2Na, where R is CH_3, C_2H_5, C_4H_9, or C_6H_5, and X is F or Cl.

$$XCp_2TiR + SO_2 \longrightarrow XCp_2Ti\overset{\overset{\displaystyle O}{\|}}{\underset{\underset{\displaystyle O}{\|}}{S}}R \longrightarrow XCp_2TiO\overset{\overset{\displaystyle O}{\|}}{S}R$$

Such titanium sulfinates are reported to increase crop yields (407–409). Isocyanides insert to yield imines as follows (410,411):

Organic isocyanates and isothiocyanates as well as nitric oxide insert similarly (412). Carbon monoxide inserts to yield very stable acyltitaniums (412,413).

   Miscellaneous Reactions of Cp₂Ti Derivatives.   Coupling of fluxional pentadienide ion with allyl bromide is regiospecifically catalyzed by Cp_2TiCl_2 (414,415). In contrast, cuprous chloride gives the linear triene:

$$(C_5H_7)_2Mg + BrCH_2CH{=}CH_2 \xrightarrow{Cp_2TiCl_2}$$

$$
\begin{array}{c}
\text{CH}_2 \\
\parallel \\
\text{CH} \\
\mid \\
\text{CH} \quad \text{CH} \\
\text{CH}_3 \quad \text{CH} \quad \text{CH}_2 \\
\mid \\
\text{CH}_2 \quad \text{CH}_2 \\
\text{CH}
\end{array}
$$

Reactions of titanium alkyls with aldehydes and ketones are generally more stereospecific and selective than the corresponding Grignard reactions (416).

   Transition-metal-catalyzed polymerizations of β-propiolactone and $CH_2{=}CHOCH_2CH_2Cl$ are markedly accelerated by Cp_2TiCl_2 (417,418). Formation of dicyclopentadienylmagnesium is promoted by Cp_2TiCl_2, but not by $TiCl_4$ or $CpTiCl_3$ (419). In olefin metathesis, the potent WCl_6 and $WOCl_4$ catalysts are tamed sufficiently by $CpTiCl_2$ to permit metathesis with unsaturated esters (420).

Health and Safety Factors

Commercial titanates should be handled according to good industrial practice. The tetraalkyl titanates have a low acute oral toxicity, LD_{50}, of 7,500–11,000 mg/kg in rats. Because of their rapid hydrolysis, these titanates can cause severe eye damage. They cause mild-to-moderate irritation of guinea pig skin, but are not sensitizers. The titanium chelates possess the added toxicity of the chelating agent. For example, the LD_{50} (rat, oral) of acetylacetonate is 5000 mg/kg. The chelates containing isopropyl alcohol have flash points of 12–27°C. The toxicology of titanium compounds has been reviewed (421).

Uses

Organic titanates perform three important functions for a variety of industrial applications. These are: (*1*) catalysis, especially polyesterification and olefin polymerization; (*2*) polymer cross-linking to enhance performance properties; and (*3*) Surface modification for adhesion, lubricity, or pigment dispersion.

   Glass-Surface Coating.   A thin (<100 nm) film of $(TiO_2)_x$ is virtually transparent. On glass, these films are bonded by Ti–O–Si bridges. After application of a lubricant, they impart considerable scratch resistance to glass and consequently greatly reduce its fragility (422,423). The lower alkoxides, particularly TYZOR TPT, are preferred for glass treatment. They can be applied undiluted, ie, in a hot process, or in a solvent, which may be hot or cold. Chelates, usually with TYZOR TE, may be used in water solution and applied hot or cold. Mixtures of TYZOR TPT and chelates, eg, TYZOR TE and TYZOR AA, are said to give more uniform coatings (40).

The $(TiO_2)_x$ films are also applied to glass or vitreous enamel for decorative purposes. Thin films enhance brilliance; thicker films impart a silver-gray luster. Milk glass can be produced by mixing the titanate with a low melting enamel, which sinters when the coating is baked (424).

When the $(TiO_2)_x$ film is ca 150-nm thick, ie, one-quarter wavelength of average visible light, it is antireflective toward visible light, yet reflective toward heat-producing infrared radiation. Precisely coated window glass is used in hot countries to reduce solar heating of houses (425). Thicker coatings pass infrared light for solar cells and are valuable as antireflective coatings for lightwave guides, photodiodes, or semiconductors (426–428). In some of these applications, mixtures of titanate with a silicate ester are valuable. These protective coatings have been used in liquid-crystal devices and electronic products and sometimes as mixtures with silicates or zirconates (429–432) (see LIQUID CRYSTALLINE MATERIALS). The coatings adhere to polytetrafluoroethylene (PTFE) (430). In an unusual application, titanoxane–siloxane mixtures, which may contain a borate, are applied to glass or glass particles that are then fired in an oxidizing atmosphere to yield a crystalline glass (pyroceram) of superior thermal properties (433) (see GLASS-CERAMICS).

The bonding properties of $(TiO_2)_x$ have been used for size-reinforcing of glass fibers so that they adhere to asphalt or to a PTFE–polysulfide mixture to impart enhanced flex endurance (434–436). Poly(vinyl alcohol) (PVA) solutions mixed with sucrose can be cross-linked with the lactic acid chelate and used generally for glass-fiber sizing (437).

Nonemulsion Paints. Heat-resistant paint (up to 500–600°C) can only yield films containing little or no organic residues because most C–C or C–H bonds are pyrolyzed below those temperatures (see PAINT). Pyrolysis of oligomeric titanates, obtained by controlled hydrolysis of TYZOR TBT, furnishes adherent films of nearly inorganic $(TiO_2)_x$. These oligomers suspend pigments, particularly aluminum, alumina, and silica. Paints were formulated from these treated pigments, oligomers, ethylcellulose (to prevent pigment settling), and mineral-spirit solvent. Whereas encouraging results were reported for these paints on rocket launchers, smokestacks, motor exhaust systems, and fire doors (438,439), scant patent literature on such applications implies that the needs for heat-resistant paints are met by other formulations (see PAINT).

Titanates are valuable in other paint applications. Corrosion-resistant coatings have been described for tinplate, steel, and aluminum (440–444). Incorporation of phosphoric acid or polyphosphates enhances the corrosion resistance. Because titanates promote hardening of epoxy resins, they are often used in epoxy-based paint (445). Silicones (polysiloxanes) are often cured by titanates. Pigments, eg, TiO_2, SiO_2, Al_2O_3, and ZrO_2, are frequently pretreated with titanates before incorporation into paints (441,446). In these applications, the $Ti(OR)_4$ compounds are often mixed with $Si(OR)_4$, $Al(OR)_3$, $Zr(OR)_4$, and other metal alkoxides (12).

Titanates react with ester groups in paint vehicles, eg, linseed oil, tuna oil, and alkyds, and with hydroxy groups, eg, in caster oil and some alkyds, to prevent wrinkling of paint films (104,447).

Adhesives. Tetrafunctional titanates react with hydroxyl, ester, amide, imide, and other functions, and with oxide groups on metals and nonmetal oxides. Titanates bond such materials together (110). Titanates bond to polyethylene and fluorinated polymers (448). Packaging films, such as Mylar polyester or aluminized films, are coated with a titanate, then warmed in moist air to hydrolyze and oligomerize the titanate and evaporate volatile organics (449). The resulting film is not tacky and can be rolled and stored. Subsequently, polyethylene is extruded onto the surface and bonded by calendering. Printing inks and decorations can also be bonded.

Titanates have been instrumental in the bonding of fluorinated resins to packaging films, poly(hydantoin)–polyester to polyester wire enamel, polysulfide sealant to polyurethane (a phosphated titanate is recommended), polyethylene to cellophane using a titanated polyethylenimine, and silicone rubber sealant to metal or plastic support using polysilane (Si–H) plus polysiloxane (Si–OR) and titanate as the adhesive ingredients (450–454). Polyester film coated first with a titanium alkoxide, then with a poly(vinyl alcohol)–polyethylenimine blend, becomes impermeable to gases (455).

Water Repellents. Titanate–wax compositions have been used for the reproofing of textiles that have been dry-cleaned. Typically, a slowly hydrolyzable titanate ester from octylene glycol or 3,5,5-trimethyl-1-hexanol is dissolved in a dry-cleaning solvent with wax and applied to the garment before the final drying stage. Hydrolysis and bonding to the cellulosic textile occur during steam pressing. In a variation of this process, a silicone is applied with the titanate; the former furnishes repellency, the latter bonds it for durability. Leather waterproofed with silicones possesses improved properties when the resin is bonded with TYZOR TBT (456) (see WATERPROOFING AND WATER/OIL REPELLENCY).

Catalysts. Titanates accelerate many organic reactions and frequently provide significant advantages in product purity and yield over conventional catalysts (see CATALYSIS). Their polyfunctionality permits assembling oxygen-containing reactants at one location in a geometrical, usually octahedral, arrangement, which permits facile shuffling of groups to yield products.

Olefin Polymerization. Titanates having a carbon–titanium bond are extensively involved in Ziegler-Natta and metallocene polymerization of olefins (see METALLOCENES; see SUPPLEMENT).

Esterification. Esterification of an acid and an alcohol can be catalyzed by small quantities of tetraalkyl titanates (see ESTERIFICATION). Although the water that forms can hydrolyze and inactivate the titanate, titanoxane oligomers are cleaved by carboxylic acids to bicoordinated monomeric acylates. A simplified mechanism is illustrated:

$$((CH_3)_2CHO)_4Ti + R'OH \longrightarrow ((CH_3)_2CHO)_4Ti\text{- - -}\underset{\underset{H}{|}}{O}R' \longrightarrow (CH_3)_2CHOH + R'OTi(OHC(CH_3)_2)_3$$

$$\xrightarrow{RCOOH} ((CH_3)_2CHO)_2Ti\diamondsuit CR + (CH_3)_2CHOH \longrightarrow ((CH_3)_2CHO)_2Ti{=\!=}O + RCOR'$$

$$\xrightarrow[\substack{R'OH \\ 200°C}]{RCOOH} ((CH_3)_2CHO)_2Ti\diamondsuit CR + H_2O$$

In plasticizer manufacture, eg, of phthalates or sebacates, using sulfuric or p-toluenesulfonic acid catalysts, the temperature (140–150°C) required for rapid reaction and high conversion may dehydrate or oxidize the alcohol and may yield a dark or foul-smelling product. Neutral titanates do not cause such side reactions. Although a temperature of 200°C is required, esterifications can easily be forced to over 99% conversion without the formation of odors or discoloration. Preparations include long-chain esters from neopentyl glycol, trimethylolpropane, and pentaerythritol for synthetic lubricants, and triglycerides from mixed long-chain acids for suppositories (457–459).

In polyester manufacture from dibasic acids and diols, color is particularly important for fiber and film products. Moreover, destruction of diol by strong acids upsets the stoichiometric balance required for high molecular weight and forms by-products such as diethylene glycol or toxic dioxane from ethylene glycol or tetrahydrofuran from tetramethylene glycol. Titanate catalysts, eg, $MgTi(OR)_6$ and $(RO)_4Ti$, are devoid of these problems (460–463). Using hydroxyl-terminated prepolymers, chain extension by diisocyanates is also accelerated by titanates (460).

Ester Interchange. Ester interchange is exemplified by the following reactions, all of which are strongly promoted by titanates:

$$
\underset{\text{RCOR}'}{\overset{\text{O}}{\parallel}} + \text{R}''\text{OH} \rightleftharpoons \underset{\text{RCOR}''}{\overset{\text{O}}{\parallel}} + \text{R}'\text{OH}
$$

$$
\underset{\text{RCOR}'}{\overset{\text{O}}{\parallel}} + \underset{\text{R}''\text{COH}}{\overset{\text{O}}{\parallel}} \rightleftharpoons \underset{\text{R}''\text{COR}'}{\overset{\text{O}}{\parallel}} + \underset{\text{RCOH}}{\overset{\text{O}}{\parallel}}
$$

$$
\underset{\text{RCOR}'}{\overset{\text{O}}{\parallel}} + \underset{\text{R}'''\text{COR}'''}{\overset{\text{O}}{\parallel}} \rightleftharpoons \underset{\text{RCOR}'''}{\overset{\text{O}}{\parallel}} + \underset{\text{R}''\text{COR}'}{\overset{\text{O}}{\parallel}}
$$

The first is the most common synthetically, whereas the third is important in cross-linking polyesters. Classical catalysts, ie, sulfuric acid and NaOR, react with other functional groups to lower yield and product purity (464). In transesterification, the methyl ester of sebacic or other acid is heated with an equivalent amount of a long-chain alcohol and titanate (0.1–2.0 mol % of TYZOR TPT) at atmospheric or reduced pressure to distill methanol and drive the reaction to completion. A solvent, eg, benzene or cyclohexane, which forms a low boiling azeotrope with methyl alcohol, may be added. Esters prepared from methyl esters include diaryl carbonates from phenols, long-chain carbamates (urethanes), and diethylaminoethyl methacrylate (465–467). In the first example, the catalyst is the solid prepared by calcining the precipitate from cohydrolysis of $TiCl_4$ and $SiCl_4$. The third example is typical of the preparation of a host of methacrylate or acrylate esters from low cost methyl methacrylate (MMA) or ethyl acrylate and other alcohols under nonpolymerizing conditions. The product methacrylate is often pure enough for subsequent polymerization. To remove the titanate catalyst, if required, two moles of water should be added to each mole of titanate and the $Ti(OCH(CH_3)_2O$ [66593-86-0] should be removed by filtration through diatomaceous earth.

Polyesterification.   High molecular weight linear polyester resins, such as poly(ethylene terephthalate) (PET), poly(propylene terephthalate) (PPT), and poly(butylene terephthalate) (PBT), can be produced by either transesterification of dimethyl terephthalate (DMT) with an excess of the corresponding diol or by direct esterification of terephthalic acid (TPA). Tetraalkyl titanates, such as TYZOR TPT or –TYZOR TBT, have been found to be excellent catalysts for either of these reactions. However, in the case of PET, the residual titanate catalyst reacts with trace quantities of aldehydic impurities produced in the polymerization process to generate a yellow discoloration of the polymer (468,469). In the case of PPT and PBT, where the color of polymer is not as critical, organic titanates are the catalyst of choice because of their greater reactivity than antimony or tin (470). Numerous processing variations have been described in the literature to minimize formation of tetrahydrofuran in the PBT process (471–472).

The organic titanate catalysts are typically used at the 50–300-ppm Ti concentration range based on DMT or TPA. Higher concentrations, which give a higher polymerization rate, also lead to greater yellow discoloration in the final product. Phosphoric acid or other strong chelating ligands have been added to the polymerization system in an effort to control color formation. However, these ligands invariably slow down the rate of catalysis so that their use must be carefully regulated (473–474).

The addition of an alkanolamine, such as diethanolamine, to TYZOR TBT, as well as the use of a less moisture-sensitive alkanolamine titanate complex such as TYZOR TE, has been reported to prolong catalyst life and minimize haze formation in the polymer (475–476). Several excellent papers are available that discuss the kinetics and mechanism of titanate-catalyzed esterification and polycondensation reactions (477–484).

Alkyd Resins.   Polyesters bearing pendant ester groups react with such groups on adjacent chains, resulting in alkyd resin formation. These reactions are catalyzed by titanates. Acrylic polymers are cross-linked with glycols or to hydroxyl-containing alkyds (485,486) (see ALKYD RESINS).

Polycarbonate Resins.   Polycarbonate molding resins can be prepared by the titanate-catalyzed transesterification of diphenylcarbonate with a dihydroxy compound, such as bisphenol A. Polycarbonates can be blended with other polyesters, such as polybutylene terephthalate (PBT), to give molding resins having improved toughness and ductility. Under the high temperatures required for blending, ester–carbonate interchange reactions can occur, which result in reduced strength and increased mold cycle times. The use of complexes of tetraalkyl titanates with phosphorus compounds, such as phenylphosphinic acid or dialkyl- or diarylphosphites, minimizes this interchange reaction (487).

Methyl Esters.   The addition product of two moles of TYZOR TPT and one mole of ethylene glycol, GLY–TI, can be used as a transesterification catalyst for the preparation of methyl esters. The low solubility of tetramethyl titanate has prevented the use of them as a catalyst for methyl ester preparation (488).

$$2\,Ti(OC_2H_5)_4 + HO(CH_2)_2OH \longrightarrow (C_2H_5O)_3TiO(CH_2)_2OTi(OC_2H_5)_3 + 2\,C_2H_5OH$$

Unsaturated Polyester Resins.   Unsaturated polyester resins are widely used as fiber-reinforced plastics, coating materials, tire cords, films, and casting or molding resins. Organic titanates such as TYZOR TPT, TYZOR TBT, or TYZOR TOT can be used to catalyze the preparation of the resins, which involves the polyesterification of a mixture of α- and β-unsaturated polybasic acids, such as maleic or fumaric acid, and alicyclic polybasic acids, such as adipic or isophthalic acid with polyhydroxyalcohols (489).

Epoxy Resins. Titanates react with free hydroxy groups in epoxy resins or with the epoxy group itself:

$$\text{RCH}\!-\!\text{CH}_2 \quad \text{Ti(OR')}_4 \longrightarrow \text{RCH}\!-\!\text{CH}_2 \longrightarrow \underset{\underset{\text{OTi(OR')}_3}{|}}{\text{RCHCH}_2\text{OR'}}$$

with Ti(OR')$_4$ bridging under the epoxy O.

Sufficient titanate leads to a fully hardened polymer. Using only enough titanate to react with free hydroxyls, the resin may subsequently be cured at lower cost with conventional cross-linking agents. The titanated epoxy resin has a low power factor, which is important in electrical applications, eg, potting components and insulation (see EMBEDDING). Titanates improve adhesion of metals to epoxies.

Epoxy cross-linking is catalyzed by TYZOR TPT and TYZOR TBT, alone or with piperidine, and by TYZOR TE. The solid condensation product from 3 TPT:4 TEA (triethanolamine) has also been applied to epoxy curing (490). Titanate curing is accelerated by selected phenolic ethers and esters at 150°C; the mixtures have a long pot life at 50°C (491) (see EPOXY RESINS).

Copper powder suspended in titanate-cured epoxy resin shields electronic apparatus against electromagnetic interference (492). Magnetic particles, ie, iron, iron oxides, and magnetic alloys, suspended in resin containing a titanate yield a superior recording tape (493,494) (see INFORMATION STORAGE MATERIALS). Titanates are very effective in dispersing and suspending particles. Such compositions can be formed into larger magnets (495). Other titanate-curable resins useful for these purposes include silicones, polyesters, phenolics, polyurethanes, polyamides, and acrylics (495). The combination $(C_4H_9O)_4Ti$ and $(C_4H_9O)_3B$ provides a fast-curing system suitable for metal coating (475).

Thixotropic Paints. Water-based latex emulsion paints may be made thixotropic or nondrip by the addition of alkanolamine-based titanium chelates. Thixotropic paints are very viscous, yet thin out enough when applied to a surface. They remain thin long enough to allow leveling to occur. Such easy applicability is the result of shear forces.

It is believed that the thixotropic structure in these systems is produced by hydrogen bonding between the hydroxyl groups on the cellulose ether colloids present in the paint and the hydroxyl end groups formed by partial hydrolysis of the chelated titanates. This association is reversible and may take place repeatedly. However, if the alkoxy groups on the titanate undergo chemical reaction with the hydroxyl groups present in the colloid polymer, an irreversible covalent bond is formed, which in the case of poly(vinyl alcohol) (PVA) polymers can lead to the formation of rubber-like gels. With cellulose ethers this chemical reaction is somewhat slow and the physical H-bonded structure predominates, leading to the observed thixotropy. As a rule, the more reactive chelates, such as diethanolamine titanates, give rapid buildup in gel strength. Trialkanolamine titanates, such as TYZOR TE, and chelates based on use of an alkanolamine in combination with other chelating ligands and/or glycols, are less reactive, giving paints that have excellent thixotropic structure (496). The stability of the titanium chelates is dependent on the ratio of the total number of moles of glycol, alkanolamine, and other ligands to the number of moles of titanium metal. The lower this ratio, the stronger the gelling effect on the aqueous emulsion paints (497,498).

Incorporation of an α-hydroxycarboxylic acid in the formulation slows the rate of metal gelation of the paint, making it possible to fill more containers during the packing operation before the paint becomes too viscous to flow, and improves the ability to blend the chelated titanate into the paint without localized areas of gelation (499,500). These paints are also less likely to lose potential gel strength resulting from excessive shearing during the manufacturing process.

Metal complexes prepared by reacting less than one mole of an alkanolamine with an excess of a polyhydric alcohol, such as polyethylene glycol 200–400 or glycerol, reportedly impart a greater degree of thixotropy to systems containing protective organic colloids (501).

Printing Inks.   Organic titanates are useful for reducing the drying time of flexographic and gravure printing inks. In the case of uv-curable printing inks, tetraalkyl titanates such as TYZOR TPT or TYZOR TBT are believed to catalyze the polymerization of vinyl monomers, such as styrene, with unsaturated polyesters comprising the ink vehicle to form a hard surface in a matter of seconds (502) (see INKS).

For ink vehicles based on hydroxyl group containing binders such as nitrocellulose and cellulose acetate, the tetraalkyl titanates cross-link the binder prematurely, limiting the storage stability of the printing ink. Chelated organic titanates such as TYZOR AA and TYZOR TE are preferred for use in these cases because they only initiate cross-linking when the ink is heated to temperatures above 80°C (503).

The chelated organic titanates also function as adhesion promoters of the ink binder to printed substrates such as plastic films, paper, and aluminum foil (504). The acetylacetone complexes of titanium are the preferred products for promoting adhesion of printing inks to polypropylene films.

The intense reddish-brown color of the acetylacetone titanium complexes impart a yellow discoloration to white inks. This discoloration is accentuated when the inks are used to print substrates that contain phenol-based antioxidants. The phenolic compounds react with the organic titanate to form a highly colored titanium phenolate. Replacement of 0.25 to 0.75 moles of acetylacetone with a malonic acid dialkyl ester, such as diethyl malonate, gives a titanium complex that maintains the performance advantages of the acetyl acetone titanium complexes, but which is only slightly yellow in color (505). These complexes still form highly colored titanium phenolates.

Mixtures of a titanium complex of saturated diols, such as TYZOR OGT, and a titanium acylate, such as bis-n-butyl-bis-caproic acid titanate, do not have a yellowing or discoloring effect on white inks used to print polyolefin surfaces (506). The complexes formed by the reaction of one or two moles of diethyl citrate with TYZOR TPT have an insignificant color on their own and do not generate color with phenol-based antioxidants (507). The complexes formed by the addition of a mixture of mono- and dialkyl phosphate esters to TYZOR TBT are also low color-generating, adhesion-promoting additives for use in printing polyolefin films (508).

The addition of one mole of a diol, such as ethylene glycol, 1,2-propanediol, or 1,4-butanediol, to bis-acetylacetone titanate complexes gives a complex that is stable on dilution with water and that can be used in aqueous printing inks (509). An excellent review of the use of organic titanates in printing inks is available (510).

Other Inks. The alkanolamine titanates, such as TYZOR TE, when mixed with a coloring agent used to print fibrous materials such as cotton, wool, or silk, promote adhesion of the dye molecule to the fiber, thus minimizing bleeding of the printed design (511).

The titanium triethanolamine chelates, such as TYZOR TE, are excellent adhesion promoters for use in water-based laminating inks typically used to bond two similar or dissimilar plastic films together (512). The use of α-hydroxycarboxylic acid titanates, such as TYZOR LA, in nonaqueous or aqueous ink-jet printing ink minimizes premature destruction of bubble-jet printer nozzles by forming a protective coating on the printing head (513). To improve the de-inking capability of sheet- and web-fed lithographic printing ink systems, tetralinoleic alcohol titanate drying oils may be incorporated in the alkyd-resin-based inks (514).

Polysiloxane Resin Coatings. Various organopolysiloxane waterproofing compositions have been proposed, which use organic titanates as catalysts, curing agents, and adhesion-promoting materials. Solutions of TYZOR TPT or TYZOR TBT and a polysiloxane resin in hydrocarbon solvents have been used to impart water repellency to leather goods (515) and cotton, wool, and synthetic fabrics (516). Water-compatible, chelated organic titanates such as TYZOR TE, TYZOR AA, or TYZOR LA can be used in aqueous formulations to impart water repellency to fibrous organic materials (517). These aqueous emulsions can also be used to size paper for improving its water repellency and antiadhesion properties (529). An increased level of organic titanate can be used to ensure the rapid curing (≤ 5 s) for fast-paced paper or fiber curing operations (519).

Room-Temperature Vulcanizable Silicone Rubber. Organic titanates are incorporated into room-temperature vulcanizable (RTV) silicone rubber formulations to provide a one-component system that is stable in the absence of moisture, but that also cures spontaneously at room temperature on exposure to moisture (520,521). Chelated titanates such as TYZOR AA or TYZOR DC can be used to prevent the composition from curing prematurely and to increase the adhesion of the resulting elastomer (522,523). To avoid thickening during initial mixing of the ingredients and to provide a more fluid system, glycol complexes of TYZOR AA or TYZOR DC can be used (524).

Coupling Agents for Polymer Composites. Organic titanate esters and their chelates or functionalized derivatives are finding increasing utility as coupling agents for reinforcing fillers in polymeric resin composites. The main function of a coupling agent is to serve as a molecular bridge at the interface of two dissimilar surfaces, thereby promoting adhesion of the fillers to the polymer matrix. Fillers treated appropriately with organic titanates should wet out better and disperse more readily in the polymer matrix, thereby reducing viscosity and improving flow characteristics of the mix. Reduced water sensitivity at the filler polymer interface aids in maintenance of a composite's flexural and tensile strength.

Titanate coupling agents may be applied to an inorganic filler as a pretreatment, or added to the polymer resin, where they migrate to the interface of the two substances during mixing and compounding. They may be applied in a liquid or vapor state, or as a solution in a hydrocarbon or alcohol-based solvent. Amounts required vary between 0.1–5% based on the weight of filler.

The mechanism of titanate coupling is dependent on the type of titanate, substrate, solvent/plasticizer, and binder used. It falls into one or more of the following categories: alcoholysis (solvolysis), surface chelation; coordination exchange, coordination to salt formation, polymer ligand exchange catalysis, and organic ligand interaction (525).

Recommended for use as coupling agents are monoalkoxytriacyl titanates, such as isopropoxytriisostearyl titanate, Ken-React KR TTS or TYZOR ISTT [61417-49-0], which are useful in mineral-filled polyolefin resins (526); chelated titanates, such as di(dioctyl)pyrophosphatooxoethylene titanate, Ken-React KR 138S, or di(dioctyl)pyrophosphatoethylene titanate, Ken-React KR 238S, which maintain high activity even when applied to fillers containing free water (527); organophosphite-coordinated titanates, such as tetraiisopropyl di(dioctyl)phosphito titanate, Ken-React KR 41B, which are useful in PVC resins because they exhibit minimal interaction with the polyester-based plasticizers, and which also exhibit significant viscosity reduction with epoxy resin formulations without accelerating cure rates (528); quaternized titanate coupling agents, such as Ken-React KR 238T, the triethylamine adduct of KR 238S, which are water soluble and useful in controlling the viscosity, flow, and conductivity of many filled resin systems (529); neoalkoxytitanates, such as neopentyl(diallyl)oxy, tri(dioctyl)pyrophosphato titanate, LICA 38, which have improved thermal and/or solvolytic stability compared to their isopropoxy group containing counterparts and can therefore be used in polymeric systems requiring higher extrusion temperatures (530); cycloheteroatom type titanates, such as dicylo(dioctyl)pyrophosphato titanate, Ken-React KR OPP2, which have been developed for ultrahigh thermal and specialty applications; and organic titanates, such as TYZOR TBT, which may be used in combination with silane coupling agents containing ethylenically unsaturated groups, such as γ-methacryloxypropyl, trimethoxysilane to reduce the viscosity of an acrylic ester-based resin formulation containing finely dispersed inorganic fillers (531).

Other Uses.   *Cross-Linking of Polyols.*   Polyols such as natural polysaccharides, eg, cellulose, starch, guar gum and their derivatives, and polyvinyl alcohol and its derivatives can be cross-linked by organic titanates.

Gelled Explosives.   Aqueous solutions of water-dispersible, nonionic, natural hydroxylated polymers, such as galactomannans and their derivatives, can be cross-linked with organic titanates, such as TYZOR TE, TYZOR AA, and TYZOR LA to give water-bearing gels, which can be used to form gelled explosives (532–534).

Incorporation of a conductive salt into similar cross-linked gels can be used in preparing a devise to measure and transmit electrical signals (535).

Cellulose. Cellulose or starch xanthate cross-linked by titanates can adsorb uranium from seawater (536). Carboxymethylcellulose cross-linked with TYZOR ISTT is the bonding agent for clay, talc, wax, and pigments to make colored pencil leads of unusual strength (537).

Paper Sizing. Various materials are added to paper to improve its wet strength and ink acceptance, to make possible clay coating, etc (see PAPERMAKING ADDITIVES). Titanic acid precipitation of paper fibers can be controlled (121). Aqueous TYZOR TE, when neutralized, rapidly deposits titanic acid. Addition of a monosaccharide or derivative delays precipitation for several hours or months, depending on the stereochemistry of the sugar. Glucose, lacking cis-hydroxyl pairs, is less effective than mannose (2,3-cis-diol) and also lactose. Yet fructose is better than mannose, which indicates that the hemiketal structure, ie, the furanose ring, furnishes a cis-diol structure. Alditols are better than the corresponding aldose, and mannitol surpasses sorbitol. Handsheets prepared from semibleached kraft pulp treated with sugar-stabilized TYZOR TE solutions exhibit a considerably higher wet strength than those from untitanated pulp.

Poly(vinyl alcohol). Poly(vinyl alcohol) (PVA) is used extensively as a paper size, alone or in combination with dyes or pigments; it is rendered insoluble by cross-linking with titanates. A rayon-based paper has been made resistant to boiling water (538,539). A size of PVA and hydroxyethyl starch cross-linked with titanium citrate or lactate prevents liming in the sizing press (540). A mixed chelate (diol and hydroxyacid) has been recommended (541). A mixture of PVA and a silicone with TYZOR TE provides a release treatment for papers, though silicones alone or with an epoxyamine also provide release coatings or waterproofed paper (466,541–545). The patent literature includes many other examples of titanate-cross-linked PVA. A smokeable sausage casing has been prepared from PVA, hydrolyzed ethylene–vinyl acetate copolymer, and glycerol on paper (546). Poly(vinyl alcohol) and poly(oxyethylene)sorbitan laurate suspend kaolin in cosmetic mud packs (547). Cross-linked PVA is foamed and then dried to give porous material for an unspecified medicinal use (548). Cotton textiles sized with cross-linked PVA are wrinkle-resistant (549). Clay or humus soils are stabilized with PVA–titanate products (550,551). Poly(vinyl alcohol) titanates are used as capsule walls for microencapsulated dyes for copy paper (552). A PVA solution reinforced with lignin has been recommended as a temporary plug for leaks in equipment containing $TiCl_4$ (553).

Oilfield Hydraulic Fracturing Fluids. Hydraulic fracturing can be used to stimulate production of oil and gas from subterranean formations. In a typical fracturing operation, a thickened fluid is pumped down the well-bore under sufficient pressure to produce a fracture in the rock formation. The fluid is required to maintain sufficient viscosity to carry a propping agent, typically sand, into the fracture, where it is deposited once the fluid breaks down and loses viscosity. A typical fluid consists of an aqueous liquid containing a polysaccharide gelling agent; a borate, an organic titanate, or a zirconate cross-linking agent; pH buffers; thermal stabilizers (sodium thiosulfate); and, if necessary, an enzyme or peroxygen gel breaker (554). Water-compatible chelated titanates, such as TYZOR TE, TYZOR LA, and TYZOR AA, are excellent cross-linking agents. However, they also develop viscosity at too great a rate to allow pumping into the formation before they are shear-degraded. A delay in cross-linking rate can be achieved by the addition of a retarding agent, such as a polyol or α-hydroxycarboxylic acid, to the aqueous polymer solution prior to addition of the cross-linker (555–559). The reaction of a combination of an α-hydroxycarboxylic acid and a polyol sugar with $TiCl_4$, followed by neutralization with a base, such as sodium hydroxide, gives a water-soluble titanium complex, which shows a delayed rate of cross-linking (560,561).

Tetraalkyl titanates react with organic borates, $B(OR')_3$, to give complexes of the general formula $Ti(OR)_4 : B(OR')_3$, which are useful as catalysts and cross-linking agents (562). Mixtures of chelated organic titanates such as TYZOR TE and TYZOR LA with alkali metal borates, such as borax, or boric acid can be used to produce shear-stable fracturing fluids (563).

A more complete discussion of the rheology and molecular structure of HP Guar gels used in fracturing fluids has been published (564–566).

Oilfield Drilling Fluids.   A fluid-loss reduction agent for oil-based drilling fluids has been prepared by reacting an organic titanate, such as TYZOR TPT, TYZOR TBT, TYZOR AA, or TYZOR TE, with a fatty acid such as oleic or stearic acid and a metal oxide (567). The reaction product of an organic titanate, such as TYZOR TOT, TYZOR AA, or TYZOR TE, with an anionic emulsifying agent, such as calcium dodecylbenzene sulfonate, can be used to stabilize oil-based drilling fluids against inorganic salt contamination (568).

Enhanced Oil Recovery.   A hydrocarbon solution of TYZOR TPT, TYZOR TBT, or TYZOR TOT can be pumped into the porous zones of an oil-bearing formation; upon contact with water, an amorphous, gelatinous TiO_2 plug is formed, which allows water to be diverted to less porous zones (569).

The use of a strong chelating ligand, such as bis-hydroxyethyl glycine, gives a titanate complex, which has an extremely slow cross-linking rate. They can therefore be used at high temperatures or high pH and still effect cross-linking at acceptable rates (570).

Sol–Gel Technology.   The sol–gel process involves conversion of a metal alkoxide or mixture of metal alkoxides, dissolved in an organic solvent (generally the parent alcohol) into a hydroxooxyalkoxide sol, followed by gelation and sintering to give the desired ceramic material.

The steps involved in formation of the metal oxoalkoxide are hydrolysis, dehydration, and dealkoxylation:

hydrolysis $\qquad\qquad\qquad$ $M(OR)_n + HOH \longrightarrow M(OR)_{n-1} + ROH$

dehydration $\qquad\quad$ $(RO)_{n-1}MOH + HOM(OR)_{n-1} \longrightarrow (RO)_{n-1}MOM(OR)_{n-1}$

dealkoxylation $\quad$ $(RO)_{n-1}MOH + ROM(OR)_{n-1} \longrightarrow (RO)_{n-1}MOM(OR)_{n-1}$

Besides direct hydrolysis, heterometallic oxoalkoxides may be produced by ester elimination from a mixture of a metal alkoxide and the acetate of another metal. In addition to their use in the preparation of ceramic materials, bimetallic oxoalkoxides having the general formula $(RO)_{n-1}MOM'OM(OR)_{n-1}$, where M is Ti or Al, M' is a bivalent metal (such as Mn, Co, Ni, and Zn), n is 3 or 4, and R is Pr or Bu, are being evaluated as catalysts for polymerization of heterocyclic monomers, such as lactones, oxiranes, and epoxides. An excellent review of metal oxoalkoxides has been published (571).

Spherical, Fine-Particle Titanium Dioxide. Spherical, fine-particle titanium dioxide that has no agglomeration and of mono-dispersion can be manufactured by carrying out a gas-phase reaction between a tetraalkyl titanate vapor and methanol vapor in a carrier gas to form an initial fine particle, which can then be hydrolyzed with water or steam (572).

Titanium Dioxide Hollow Fibers. Hollow fibers of titanium dioxide can be manufactured by preparing a solution of a tetraalkyl titanate, an acid such as HCl, and an alcohol such as isopropyl alcohol, followed by spinning and drying the resultant fiber (573).

Analytical Chemistry

The use of organic titanates in analytical chemistry has been reviewed (389,574–594).

BIBLIOGRAPHY

"Titanium Compounds" in *ECT* 1st ed., Vol. 14, pp. 213–241, by L. R. Blair, H. H. Beacham, and W. K. Nelson, National Lead Co.; "Titanium Compounds (Organic)" in *ECT* 2nd ed., Vol. 20, pp. 424–503, by R. H. Stanley, Titanium Intermediates, Ltd.; in *ECT* 3rd ed., Vol. 23, pp. 176–245, by C. S. Rondestvedt, Jr., E. I. du Pont de Nemours & Co., Inc.

1. U.S. Pat. 3,119,852 (Jan. 28, 1961), R. T. Gilsdorf (to E. I. du Pont de Nemours & Co., Inc.).
2. U.S. Pat. 4,789,752 (Dec. 5, 1988), H. J. Kotzch, H. G. Srenby, and H. J. Vehlensieck (to Huels Troisdorf AG).
3. V. A. Shreider and co-workers, *Inorg. Chim. Acta*, **53**(2), 73–76 (1981).
4. U.S. Pat. 4,650,895 (Mar. 17, 1987), H. Kadokura, H. Umezaki, and Y. Higuchi (to Sumitomo Chemical Co.).
5. U.S. Pat. 4,824,979 (Apr. 25, 1989), H. J. Kotzch, H. G. Srenby, and H. J. Vehlensieck (to Huels Troisdorf AG).
6. B. D. Jain and R. Kumar, *Proc. Indian Acad. Sci. Sect.* **A60**, 265 (1964); *Indian J. Chem.* **1**, 317 (1963); V. Patrovsky, *Coll. Czech. Chem. Commun.* **27**, 1824 (1962).
7. P. N. Kapoor, R. N. Kapoor, and R. C. Mehrotra, *Chem. Ind.* **39**, 1314 (1968).
8. K. S. Mazdujasni, B. J. Schaper, and L. M. Burr, *Inorg. Chem.* **10**(5), 889–892 (1971).
9. U.S. Pat. 2,709,174 (May 24, 1955), J. B. Rust and L. Spialter (to Montclair Research and Ellis-Foster Co.); U.S. Pat. 2,883,238 (Apr. 21, 1959), A. Pechucas (to Columbia Southern Chemical Co.).
10. H. G. Emblem and S. E. Maskery, *J. Appl. Chem.* **20**, 183–187 (1970).
11. P. N. Kapoor, S. K. Mehrotra, and R. C. Mehrotra, *Inorg. Chim. Acta*, **12**(3), 273–276 (1975).
12. D. C. Bradley, R. C. Mehrotra, and D. P. Gaur, *Metal Alkoxides*, Academic Press, Inc., New York, 1978.
13. J. A. Ibers, *Nature (London)*, **197**, 686 (1963).
14. D. C. Bradley and C. E. Holloway, *J. Chem. Soc. A2*, 1316 (1968).
15. N. Hao and co-workers, *J. Magn. Reson.* **50**, 50 (1982).
16. F. Babonneau and co-workers, *Inorg. Chem.* **27**, 3166–3172 (1988).
17. U.S. Pat. 2,689,858 (Sept. 21, 1954), T. Boyd (to Monsanto Chemical Co.).
18. I. Kraitzer, F. K. McTaggart, and G. Winter, *J. Counc. Sci. Ind. Res.* **21**, 328 (1948).
19. T. Boyd, *J. Polym. Sci.* **7**, 591 (1951).

20. T. Ishino and S. Minami, *Tech. Rep. Osaka Univ.* **3**, 357 (1953).
21. D. C. Bradley, R. Gaze, and W. Wardlaw, *J. Chem. Soc.* **1**, 468–478 (1957).
22. D. C. Bradley, R. Gaze, and W. Wardlaw, *J. Chem. Soc.* **1**, 721–728 (1955).
23. D. C. Bradley, R. Gaze, and W. Wardlaw, *J. Chem. Soc.* **4**, 3973–3982 (1955).
24. U.S. Pat. 2,614,112 (Oct. 14, 1952), T. Boyd (to Monsanto Chemical Co.).
25. U.S. Pat. 2,621,193 (Dec. 9, 1952), C. Langkammerer (to E. I. du Pont de Nemours & Co., Inc.).
26. U.S. Pat. 2,666,772 (Jan. 19, 1954), T. Boyd (to Monsanto Chemical Co.).
27. U.S. Pat. 2,621,195 (Dec. 9, 1952), J. H. Haslam (to E. I. du Pont de Nemours & Co., Inc.).
28. U.S. Pat. 2,708,203 (May 10, 1955), J. Haslam (to E. I. du Pont de Nemours & Co., Inc.).
29. A. N. Nesmeyanov, O. V. Nogina, and R. K. Freidlina, *Bull. Acad. Sci. USSR Div. Chem. Sci.*, 355 (1956).
30. U.S. Pat. 2,980,719 (Apr. 18, 1961), J. Haslam (to E. I. du Pont de Nemours & Co., Inc.).
31. Jpn. Pat. 1 129,032 (May 22, 1989), A. Mori and co-workers (to Nippon Soda Co. Ltd.).
32. Jpn. Pat. 1 129,031 (May 22, 1989), A. Mori and co-workers (to Nippon Soda Co. Ltd.).
33. Ref. 12, p. 27ff.
34. D. Seebach and co-workers, *Synthesis*, 138 (1982).
35. U.S. Pat. 3,006,941 (Oct. 31, 1961), A. Murdrak and L. Stevick (to Harshaw Chemical Co.).
36. N. Yoshino and T. Yoshino, *Bull. Chem. Soc. Japan*, **46**(9), 2899–2903 (1973).
37. R. C. Mehrotra, V. D. Gupta, and P. C. Bharara, *Indian J. Chem.* **11**, 814–816 (1973).
38. U.S. Pat. 4,568,703 (Feb. 4, 1986), K. Ashida (to B.P. Chemicals).
39. B. M. Novak, S. M. Hoff, and Y. He, *Polym. Prep. Am. Chem. Soc. (Div. Poly. Chem.)*, **34**(11), 258–259 (1993).
40. Brit. Pat. 1,510,587 (May 10, 1978), (to St. Gobain Industries).
41. Ger. Pat. 934,352 (Oct. 20, 1955), D. F. Herman (to Titangesellschaft).
42. U.S. Pats. 3,091,625 (May 28, 1963) and 3,119,852 (Jan. 28, 1964), R. T. Gilsdorf (to E. I. du Pont de Nemours & Co., Inc.).
43. Brit. Pat. 997,892 (July 14, 1965), L. J. Lawrence (to British Titan).
44. U.S. Pat. 2,684,972 (July 27, 1954), J. H. Haslam (to E. I. du Pont de Nemours & Co., Inc.).
45. Brit. Pat. 787,180 (Dec. 4, 1957), J. H. Haslam (to E. I. du Pont de Nemours & Co., Inc.).
46. J. S. Jennings, W. Wardlaw, and W. J. Way, *J. Chem. Soc.* **1**, 637 (1936).
47. A. N. Nesmeyanov and co-workers, *Izv. Akad. Nauk SSSR Otd. Khim. Nauk*, 1037 (1952).
48. G. A. Razuvaev and co-workers, *Dokl. Akad. Nauk SSSR*, **122**, 618 (1958).
49. S. Prasad and J. B. Tripathi, *J. Indian Chem. Soc.* **35**, 177 (1958).
50. G. P. Luchinskii, *Zh. Obshch. Khim.* **7**, 2044 (1937).
51. O. C. Dermer and W. C. Fernelius, *Z. Anorg. Allg. Chem.* **221**, 83 (1934).
52. H. Funk and E. Rogler, *Z. Anorg. Allg. Chem.* **252**, 323 (1944).
53. R. C. Mehrotra, *J. Indian Chem. Soc.* **30**, 731 (1953).
54. A. N. Nesmeyanov, O. V. Nogina, and R. Kh. Freidlina, *Izv. Akad. Nauk SSSR Otd. Khim. Nauk.* 518 (1951).
55. D. C. Bradley, D. C. Hancock, and W. Wardlaw, *J. Chem. Soc.* **3**, 2773 (1952).
56. I. D. Verma and R. C. Mehrotra, *J. Less Common Met.* **1**, 263 (1959).
57. A. N. Nesmeyanov, E. M. Brainina, and R. K. Freidlina, *Dokl. Akad. Nauk. SSSR*, **94**, 249 (1954).

58. A. B. Bruker, R. I. Frenkel, and L. Z. Soborovskii, *Zh. Obshch. Khim.* **28**, 2413 (1958).
59. A. N. Nesmeyanov, E. M. Brainina, and R. K. Freidlina, *Izv. Akad. Nauk. SSSR Otd. Khim. Nauk.* 987 (1954).
60. O. V. Nogina, R. K. Freidlina, and A. N. Nesmeyanov, *Izv. Akad. Nauk. SSSR Otd. Khim. Nauk.* 74 (1952).
61. D. M. Puri and R. C. Mehrotra, *Indian J. Chem.* **5**, 448 (1967); R. C. Mehrotra and R. K. Mehrotra, *J. Indian Chem. Soc.* **39**, 635 (1962).
62. R. E. Reeves and L. W. Mazzeno, *J. Am. Chem. Soc.* **76**, 2533 (1954).
63. U.S. Pat. 2,920,089 (Jan. 5, 1980), C. M. Samoui (to Kendall Co.).
64. Brit. Pat. 1,586,671 (Mar. 25, 1981), M. S. Howarth and B. W. H. Terry (to Imperial Chemical Industries).
65. N. Baggett, D. S. P. Poolton, and W. B. Jennings, *J. Chem. Soc. Dalton Trans.* **6**, 1128 (1979); *J. Chem. Soc. Chem. Commun.* **7**, 239 (1975).
66. R. C. Fay and A. F. Lindmark, *J. Am. Chem. Soc.* **97**, 5928 (1975); P. Finocchiaro, *J. Am. Chem. Soc.* **97**, 4443 (1975).
67. R. V. Singh, R. V. Sharma, and J. P. Tandon, *Synth. React. Inorg. Met. Org. Chem.* **11**, 139 (1981).
68. H. Sugahara and Y. Shuto, *J. Organomet. Chem.* **24**, 709 (1970).
69. A. Yamamoto and S. Kambara, *J. Am. Chem. Soc.* **81**, 2663 (1959).
70. A. Yamamoto and S. Kambara, *J. Am. Chem. Soc.* **79**, 4344 (1957).
71. Brit. Pat. 975,452 (Nov. 18, 1964), R. Feld (to Laporte Titanium Co.).
72. U.S. Pat. 4,568,703 (Feb. 4, 1986), K. Ashida (to B.P. Chemicals Ltd.).
73. U.S. Pat. 2,628,171 (Feb. 10, 1953), L. Q. Green (to E. I. du Pont de Nemours & Co., Inc.).
74. U.S. Pat. 2,643,262 (June 23, 1953), C. O. Bostwick (to E. I. du Pont de Nemours & Co., Inc.).
75. U.S. Pat. 3,091,625 (May 28, 1963), R. T. Gilsdorf (to E. I. du Pont de Nemours & Co., Inc.).
76. U.S. Pat. 4,148,989 (Apr. 10, 1979), G. Tews, H. Wilff, and G. Shade (to Dynamit Nobel AG).
77. U.S. Pat. 4,705,714 (Oct. 11, 1987), S. Matsumoto (to Research and Development Corp. of Japan).
78. U.S. Pat. 2,926,183 (Feb. 23, 1960), C. A. Russell (to National Lead Co.).
79. Brit. Pat. 811,425 (Apr. 8, 1959) and Brit. Pat. 811,426 (Apr. 8, 1959), (to National Lead Co.).
80. Jpn. Pat. 81 46,835 (Apr. 28, 1981), I. Kishima and co-workers (to Matsumoto Pharmaceutical Co.).
81. Brit. Pat. 757,190 (Sept. 12, 1956), (to E. I. du Pont de Nemours & Co., Inc.).
82. U.S. Pat. 5,260,466 (Nov. 9, 1993), G. McGibbon (to Tioxide Specialties Ltd.).
83. U.S. Pat. 2,898,356 (Aug. 4, 1959), C. A. Russell (to National Lead Co.).
84. Brit. Pat. 1,191,480 (May 13, 1970), R. H. Stanley, D. W. Brock, and L. J. Lawrence (to British Titan Products Co. Ltd.).
85. R. J. H. Clark and A. J. McAlees, *J. Chem. Soc.* **A2**, 2026 (1970).
86. U.S. Pat. 3,699,137 (Oct. 17, 1972), E. Termin and O. Bleh (to Dynamit-Nobel A.G.); V. Zatka and O. Hoffman, *Analyst (London)*, **95**, 200 (1970).
87. U.S. Pat. 3,682,688 (Aug. 8, 1972), C. T. Hughes and D. A. Paulsen (to E. I. du Pont de Nemours & Co., Inc.).
88. D. C. Bradley and C. E. Holloway, *J. Chem. Soc.* **A1**, 282 (1969).
89. N. Serpone and R. C. Fay, *Inorg. Chem.* **6**(10), 1835–1843 (1967).
90. S. Minami, H. Takano, and T. Ishino, *J. Chem. Soc. Jpn.* **60**, 1406 (1957).
91. A. Yamamoto and S. Kambara, *J. Am. Chem. Soc.* **79**, 4344–4348 (1957).
92. Eur. Pat. 122,312 (Oct. 24, 1984), G. J. Rummo (to Kay-Fries Inc.).

93. U.S. Pat. 3,017,282 (Jan. 16, 1962), H. C. Brill (to E. I. du Pont de Nemours & Co., Inc.).

94. U.S. Pat. 4,647,680 (Mar. 3, 1987), D. Barfurth and H. Nestler (to Dynamit Nobel AG).

95. U.S. Pat. 4,313,851 (Feb. 2, 1982), D. Barfurth and H. Nestler (to Dynamit Nobel AG).

96. U.S. Pat. 4,551,544 (Nov. 5, 1985), G. B. Robbins (to E. I. du Pont de Nemours & Co., Inc.).

97. U.S. Pat. 4,478,755 (Oct. 23, 1984), G. B. Robbins (to E. I. du Pont de Nemours & Co., Inc.).

98. U.S. Pat. 4,924,016 (May 8, 1990), D. Barfurth and H. Nestler (to Huels Troisdorf AG).

99. Ger. Offen. 2,200,347 (Oct. 25, 1973), S. D. Smith and S. B. Hamilton, Jr. (to General Electric Co.).

100. U.S. Pat. 3,338,935 (Aug. 26, 1967), P. M. Kerschner and F. G. Hess (to Cities Service Oil Co.).

101. U.S. Pat. 3,422,126 (Jan. 14, 1969), C. R. Bauer (to E. I. du Pont de Nemours & Co., Inc.).

102. U.S. Pat. 3,415,762 (Dec. 10, 1968), B. P. Block and G. H. Dahl (to Pennwalt Chemicals).

103. U.S. Pat. 4,999,442 (Mar. 12, 1991), M. Sato and co-workers (to Nippon Soda Co. Ltd.).

104. Brit. Pat. 755,728 (Aug. 27, 1956), (to National Lead Co.).

105. M. G. Voronkov and V. P. Baryshok, *J. Organomet. Chem.* **239**, 228 (1982).

106. R. L. Harlow, *Acta Chrystal. Sect. C. Cryst. Struct. Commun.* **C39**(10), 1344–1346 (1983).

107. U.S. Pat. 4,618,435 (Oct. 21, 1986), D. E. Putzig (to E. I. du Pont de Nemours & Co., Inc.).

108. U.S. Pat. 2,845,445 (July 29, 1958), C. A. Russell (to National Lead Co.).

109. Brit. Pat. 2,290,795 (Oct. 1, 1996), J. Ridland (to Tioxide Specialties Ltd.).

110. U.S. Pat. 2,935,522 (May 3, 1960), C. M. Samour (to Kendall Co.); Brit. Pat. 786,388 (Nov. 20, 1957), (to National Lead Co.).

111. Technical data, R. L. Harlow, E. I. du Pont de Nemours & Co., Inc., Wilmington, Del.

112. U.S. Pat. 4,526,725 (July 2, 1985), to D. L. Deardorff.

113. U.S. Pat. 4,788,172 (Nov. 29, 1988), to D. L. Deardorff.

114. U.S. Pat. 2,524,115 (Feb. 18, 1958), H. H. Beacham and D. F. Herman (to National Lead Co.).

115. P. Lagally and H. Lagally, *TAPPI*, **39**, 747 (1956).

116. U.S. Pat. 2,894,966 (July 14, 1959), C. A. Russell (to National Lead Co.).

117. U.S. Pat. 2,950,174 (Aug. 23, 1960), to P. Lagally.

118. U.S. Pat. 3,028,297 (Apr. 3, 1962), P. Lagally (to Linden Laboratories Inc.).

119. U.S. Pat. 4,621,148 (Nov. 4, 1986), D. Barfurth and H. Nestler (to Dynamit Nobel AG).

120. U.S. Pat. 4,609,746 (Sept. 2, 1986), D. Barfurth and H. Nestler (to Dynamit Nobel AG).

121. P. C. Bharara, V. D. Gupta, and R. C. Mehrotra, *Z. Anorg. Allg. Chem.* **403**, 337 (1974).

122. P. C. Bharara, V. D. Gupta, and R. C. Mehrotra, *J. Indian Chem. Soc.* **51**, 859 (1974).

123. E. C. Alyea and P. H. Merrell, *Inorg. Nucl. Chem. Lett.* **9**, 69 (1973).

124. D. A. Baldwin and G. H. Leigh, *J. Chem. Soc.* **A2**, 1432 (1968).

125. U.S. Pat. 2,670,363 (Feb. 23, 1954), J. P. Wadington (to National Lead Co.).

126. K. C. Pande and R. C. Mehrotra, *J. Prakt. Chem.* **5**, 101 (1957).

127. K. C. Pande and R. C. Mehrotra, *Chem. Ind.* 114 (1957).

128. Y. A. Lysenko and O. A. Osipov, *Zh. Obshch. Khim.* **28**, 1724 (1958).

129. S. Prasad and R. C. Srivastava, *J. Indian Chem. Soc.* **39**, 9 (1962).

130. Brit. Pat. 787,180 (Dec. 4, 1957), J. H. Haslam (to E. I. du Pont de Nemours & Co., Inc.).

131. I. D. Varma and R. C. Mehrotra, *J. Prakt. Chem.* **8**, 235 (1959).

132. A. N. Solanki, K. R. Nahar, and A. M. Bhandari, *Synth. React. Inorg. Met. Org. Chem.* **8**, 335 (1978).

133. J. Amaudrut, *Bull. Soc. Chim. Fr.* **7–8**(1), 624 (1977); J. Amaudrut, B. Viard, and R. Mercier, *J. Chem. Res. Synop.* **4**, 138 (1979).

134. R. Feld and P. L. Cowe, *The Organic Chemistry of Titanium*, Butterworths, London, 1965.

135. M. Isobe and co-workers, *Tetrahedron Lett.* **23**, 221 (1982); B. E. Rossiter, T. Katsuki, and K. B. Sharpless, *J. Am. Chem. Soc.* **102**, 5976 (1980); **103**, 464 (1981).

136. Rus. Pat. 771,056 (Oct. 15, 1980), V. K. Skubin and co-workers.

137. Brit. Pat. 2,051,093 (Jan. 14, 1981) and Ger. Offen. 3,017,887 (May 9, 1979), H. Kamio, Y. Ogina, and K. Nakamura (to Nippon Mining Co.).

138. J. M. Latour, B. Galland, and J. C. Marchon, *J. Chem. Soc. Chem. Commun.* **13**, 570 (1979).

139. U.S. Pat. 3,694,251 (Sept. 26, 1972), J. F. Houle and G. R. VanNorman (to Eastman Kodak Co.).

140. Brit. Pat. 2,085,491 (Oct. 16, 1980), P. S. Collishaw and D. Bird (to Caligen Foam Ltd.).

141. U.S. Pat. 2,628,170 (Feb. 10, 1953), L. Q. Green (to E. I. du Pont de Nemours & Co., Inc.).

142. U.S. Pat. 3,478,088 (Nov. 11, 1969), A. Revukas (to Cities Service Oil Co.).

143. Jpn. Pat. 62 89,690 (Apr. 24, 1987), M. Sato, N. Kobayashi, and M. Aizawa (to Nippon Soda Co. Ltd.).

144. U.S. Pat. 4,978,389 (Dec. 18, 1990), M. Sato, A. Mori, and M. Aizawa (to Nippon Soda Co. Ltd.).

145. Jpn. Pat. 5 163,482 (June 29, 1993), A. Mori and H. Kobayashi (to Nippon Soda Co. Ltd.).

146. M. G. Finn and K. B. Sharpless, *Assy. Synthesis* **5**, 247–308 (1985).

147. R. O. Duthaler and A. Hafner, *Chem. Rev.* **92**, 807–832 (1992).

148. R. K. Sharma, R. V. Singh, and J. P. Tandon, *J. Pract. Chem.* **322**(3), 508–516 (1980).

149. R. K. Sharma and J. P. Tandon, *J. Pract. Chem.* **322**(1), 161–168 (1980).

150. U.S. Pat. 4,529,552 (July 16, 1985), I. Kato and co-workers (to Nippon Soda Co. Ltd.).

151. U.S. Pat. 5,328,718 (Sept. 27, 1996), Y. Abe, Y. Hamaji, and Y. Saakabe (to Myrata Manufacturing Co.).

152. Jpn. Pat. 58 77,889 (May 11, 1983), I. Kato and A. T. Sakamura (to Nippon Soda Co. Ltd.).

153. Jpn. Pat. 58 81,618 (May 17, 1983), S. Trukamura and I. Kato (to Nippon Soda Co. Ltd.).

154. Jpn. Pat. 58 217,429 (Dec. 17, 1983), S. Trukamura and I. Kato (to Nippon Soda Co. Ltd.).

155. Jpn. Pat. 58 172,395 (Nov. 11, 1983), I. Kato and S. Trukamura (to Nippon Soda Co. Ltd.).

156. Jpn. Pat. 58 156,347 (Sept. 17, 1983), S. Trukamura and I. Kato (to Nippon Soda Co. Ltd.).

157. R. C. Mehrotra and A. Mehrotra, *Inorg. Chem. Acta Rev.* **5**, 127–136 (1971).

158. U.S. Pat. Appl. 515,844 (Sept. 14, 1984), to R. G. Dosch, H. P. Stepens, and F. V. Stohl.

159. U.S. Pat. 4,324,736 (Apr. 13, 1982), W. Josten and H. J. Vehlensieck (to Dynamit Nobel AG).

160. U.S. Pat. 4,387,199 (June 7, 1983), W. Josten and H. J. Vehlensieck (to Dynamit Nobel AG).
161. D. C. Bradley and I. M. Thomas, *J. Chem. Soc.* **3**, 3854 (1960).
162. G. Chandra and M. F. Lappert, *Inorg. Nucl. Chem. Lett.* **1**(2), 83 (1965); *J. Chem. Soc.* **A2**, 1940 (1968).
163. U.S. Pat. 4,728,706 (Mar. 1, 1988), W. Farnham (to E. I. du Pont de Nemours & Co., Inc.).
164. R. K. Sharma, R. V. Singh, and J. P. Tandon, *Synth. React. Inorg. Met. Org. Chem.* **9**, 519 (1979).
165. E. C. Alyea, A. Malek, and P. H. Merrell, *Trans. Met. Chem.* **4**, 172 (1979).
166. E. C. Alyea and A. Malek, *Inorg. Nucl. Chem. Lett.* **13**, 587 (1977).
167. E. C. Alyea, A. Malek, and P. H. Merrell, *J. Coord. Chem.* **4**, 55 (1974).
168. I. R. Unny, S. Gopinathan, and C. Gopinathan, *Indian J. Chem.* **19A**, 598 (1980).
169. J. F. Harrod and K. R. Taylor, *Inorg. Chem.* **14**, 1541 (1975).
170. U.S. Pat. 4,359,559 (Nov. 16, 1992), S. Yajima and co-workers (to UBE Industries Ltd.).
171. T. Sugama, J. R. Fair, and A. P. Reed, *J. Coating Tech.* **65**(826), 27–36 (1993).
172. U.S. Pat. 4,401,501 (Oct. 18, 1983), M. Taramasso, G. Perego, and B. Notari (SNAM Progetti SpA).
173. U.S. Pat. 4,547,557 (Oct. 15, 1985), M. McDaniel (to Phillips Petroleum Co.).
174. J. Weitkamp and co-workers, eds., *Zeolite and Related Microprous Materials, State of the Art 1994, Studies in Surface Science and Catalysis*, Vol. 84, Elsevier Science Publishing Co., Inc., New York, 1994, pp. 653–659.
175. J. B. Miller, S. T. Johnston, and E. I. Ko, *J. Catal.* **150**, 310–320 (1994).
176. C. B. Khouw and co-workers, *J. Catal.* **149**, 195–205 (1995).
177. S. Vetter and co-workers, *Chem. Eng. Tech.* **17**, 348–353 (1994).
178. T. Tatsumi, K. Yuasa, and H. Tominaga, *J. Chem. Soc. Chem. Commun.* **19**, 1446–1447 (1992).
179. T. Tatsumi, Y. Watanabe, and K. A. Koyano, *J. Chem. Soc. Chem. Commun.* **19**, 2281–2282 (1996).
180. Jpn. Kokai Tokkyo Koho 80 139,392 (Oct. 31, 1980), I. Kijama.
181. E. Albizatti and co-workers, *Inorg. Chem. Acta*, **120**(2), 197–203 (1986).
182. D. W. MacCorquodale and H. Adkins, *J. Am. Chem. Soc.* **50**, 1938 (1928).
183. A. N. Nesmeyanov, O. V. Nogina, and R. K. Freidlina, *Dokl. Akad. Nauk. SSSR*, **95**, 813 (1954).
184. A. N. Nesmeyanov and co-workers, *Bull. Acad. Sci. USSR Div. Chem. Sci.* 1117 (1960).
185. B. N. Chakravarti, *Naturwissenschaften*, **45**, 286 (1958).
186. A. Pflugmacher and co-workers, *Naturwissenschaften*, **45**, 490 (1958).
187. Ger. Pat. 1,091,105 (Oct. 20, 1960), A. Gumboldt and W. Herwig (to Hoechst).
188. A. Stahler and H. Wirthwein, *Chem. Ber.* **38**, 2619 (1905).
189. A. Monnier, *Ann. Chim. Anal.* **20**, 1 (1915).
190. C. M. Mikulski and co-workers, *J. Inorg. Nucl. Chem.* **41**, 1671 (1979).
191. F. E. McFarlane and G. W. Tindall, *Inorg. Nucl. Chem. Lett.* **9**, 907 (1973).
192. M. R. Hunt and G. Winter, *Inorg. Nucl. Chem. Lett.* **6**, 529 (1970).
193. T. D. Smith, T. Lund, and J. R. Pilbrow, *J. Chem. Soc.* **A2**, 2786 (1971).
194. A. N. Glebov, P. A. Vasil'ev, and Y. I. Sal'nikov, *Russ. J. Inorg. Chem.* **26**, 140 (1981).
195. A. Furstner and B. Bogdanovic, *Angew. Chem. Int. Ed. Eng.* **35**, 2442–2469 (1996).
196. N. L. Hungerford and W. Kitching, *J. Chem. Soc. Chem. Commun.* **14**, 1697–1698 (1996).
197. R. J. H. Clark and A. J. McAlees, *Inorg. Chem.* **11**, 342 (1972).
198. R. J. H. Clark and A. J. McAlees, *J. Chem. Soc. Dalton Trans.* **5**, 640 (1972).
199. C. Blandy and D. Gervais, *Inorg. Chim. Acta*, **47**, 197 (1981).

200. R. Hofer, D. Evard, and A. Jacot-Guillarmod, *Helv. Chim. Acta*, **68**(4), 969–974 (1985).
201. M. T. Reetz and J. Westermann, *Synth. Commun.* **11**, 647 (1981).
202. M. T. Reetz, R. Steinbach, and B. Wenderoth, *Synth. Commun.* **11**, 261 (1981).
203. M. T. Reetz, J. Westermann, and R. Steinbach, *Angew. Chem. Int. Ed. Engl.* **19**, 900 (1980).
204. M. T. Reetz, J. Westermann, and R. Steinbach, *J. Chem. Soc. Chem. Commun.* **5**, 237 (1981).
205. M. T. Reetz, J. Westermann, and R. Steinbach, *Angew. Chem. Int. Ed. Engl.* **19**, 901 (1980).
206. M. T. Reetz and co-workers, *Angew. Chem. Int. Ed. Engl.* **19**, 1011 (1980).
207. W. Seidel and E. Riesenberg, *Z. Chem.* **20**, 450 (1980).
208. B. Weidmann and co-workers, *Helv. Chim. Acta*, **64**, 357 (1981).
209. A. G. Olivero, B. Weidmann, and D. Seebach, *Helv. Chim. Acta*, **64**, 2485 (1981).
210. B. Weidmann and D. Seebach, *Helv. Chim. Acta*, **63**, 2451 (1980).
211. P. J. Davidson, M. F. Lappert, and R. Pearce, *J. Organomet. Chem.* **57**, 269 (1973).
212. J. C. W. Chien, J. Wu, and M. D. Rausch, *J. Am. Chem. Soc.* **103**, 1180 (1981).
213. A. K. Rappe and W. A. Goddard, *J. Am. Chem. Soc.* **104**, 297 (1982).
214. A. R. Middleton and G. Wilkinson, *J. Chem. Soc. Dalton Trans.* **10**, 1888 (1980).
215. N. Krause and D. Seebach, *Chem. Ber.* **120**(11), 1845–1851 (1987).
216. M. D. Schiavelli, J. J. Plunkett, and D. W. Thompson, *J. Org. Chem.* **46**, 807 (1981).
217. T. Yoshida and E. Negishi, *J. Am. Chem. Soc.* **103**, 1276 (1981).
218. U.S. Pat. 3,694,517 (Sept. 26, 1972), W. Schneider (to B. F. Goodrich); Jpn. Kokai Tokkyo Koho 75 88,059 (July 15, 1975) (to Montedison); Ital. Pat. 932,191 (Nov. 15, 1972), (to Montedison).
219. F. Guerin, D. H. McConville, and N. C. Payne, *Organometallics*, **15**, 5085–5089 (1996).
220. A. Clearfield and co-workers, *Can. J. Chem.* **53**, 1622 (1975).
221. V. V. Tkachev and L. O. Atomyan, *Zh. Strukt. Khim.* **13**, 287 (1972).
222. I. A. Ronova and N. V. Alekseev, *Zh. Strukt. Khim.* **18**, 212 (1977).
223. E. F. Epstein and I. Bernal, *Inorg. Chim. Acta*, **7**, 211 (1973).
224. V. Kocman and co-workers, *J. Chem. Soc. Chem. Commun.* **21**, 1340 (1971).
225. D. F. Hunt, J. W. Russell, and R. L. Torian, *J. Organomet. Chem.* **43**, 175 (1972).
226. A. N. Nesmeyanov and co-workers, *J. Organomet. Chem.* **61**, 225 (1973).
227. A. N. Nesmeyanov and co-workers, *Zh. Strukt. Khim.* **13**, 1033 (1972); **16**, 759 (1975).
228. M. Hillman and A. J. Weiss, *J. Organomet. Chem.* **42**, 123 (1972).
229. A. N. Nesmeyanov and co-workers, *Dokl. Chem.* **163**, 704 (1965).
230. C. Cauletti and co-workers, *J. Electron Spectrosc. Relat. Phenom.* **18**, 61 (1980).
231. A. A. MacDowell and co-workers, *J. Chem. Soc. Chem. Commun.* **9**, 427 (1979).
232. O. S. Roshchupkina and co-workers, *Soviet J. Coord. Chem.* **1**, 1052 (1975).
233. E. Samuel, R. Ferner, and M. Bigorgne, *Inorg. Chem.* **12**, 881 (1973).
234. G. Balducci and co-workers, *J. Mol. Struct.* **64**, 163 (1980).
235. M. Spoliti and co-workers, *J. Mol. Struct.* **65**, 105 (1980).
236. N. N. Vyshinskii and co-workers, *Tr. Khim. Khim. Tekhnol.* 64 (1973); 119, 123 (1974).
237. G. Condorelli and co-workers, *J. Organomet. Chem.* **87**, 311 (1975).
238. D. W. Clack and K. D. Warren, *Theor. Chim. Acta*, **46**, 313 (1977); *Inorg. Chem. Acta*, **24**, 35 (1977); **30**, 251 (1978); *J. Organomet. Chem.* **162**, 83 (1978).
239. V. E. L'vovskii, *Soviet J. Coord. Chem.* **4**, 1266 (1978).
240. V. E. L'vovskii and G. B. Erusalimskii, *Soviet J. Coord. Chem.* **2**, 934, 1221 (1976).
241. E. G. Muller, S. F. Watkins, and L. F. Dahl, *J. Organomet. Chem.* **111**, 73 (1976).
242. E. G. Muller, J. L. Petersen, and L. F. Dahl, *J. Organomet. Chem.* **111**, 91 (1976).
243. J. L. Petersen and co-workers, *J. Am. Chem. Soc.* **97**, 6433 (1975).

244. A. M. McPherson, G. D. Stucky, and co-workers, *J. Am. Chem. Soc.* **101**, 3425 (1979).
245. V. E. Lvovsky, E. A. Fushman, and F. S. Dyachkovsky, *J. Mol. Catal.* **10**, 43 (1981).
246. T. D. Grabik and co-workers, *Russ. J. Phys. Chem.* **52**, 894 (1978).
247. V. I. Tel'noi and I. B. Rabinovich, *Conf. Int. Thermodyn. Chim.* **1**, 98 (1975).
248. T. Mukaiyama, *Angew. Chem. Int. Ed. Engl.* **16**, 817 (1977); J. Lange, J. M. Kanabus-Kaminska, and A. Kral, *Synth. Commun.* **10**, 473 (1980); M. T. Reetz, *Angew. Chem. Int. Ed. Engl.* **21**, 96 (1982).
249. R. C. Paul and co-workers, *J. Less Common Met.* **17**, 437 (1969).
250. A. M. Cardoso, R. J. H. Clark, and S. Moorhouse, *J. Chem. Soc. Dalton Trans.* **7**, 1156 (1980).
251. G. Wilkinson and J. M. Birmingham, *J. Am. Chem. Soc.* **76**, 4281 (1954).
252. L. Summers and R. H. Uloth, *J. Am. Chem. Soc.* **76**, 2278 (1954); **77**, 3604 (1955).
253. R. B. King, *Organometallic Syntheses*, Vol. 1, Academic Press, Inc., New York, 1965, pp. 75–78.
254. R. D. Gorsich, *J. Am. Chem. Soc.* **82**, 4211 (1960).
255. Rus. Pat. 825,534 (May 7, 1981), Y. A. Sorokin and co-workers (to Scientific Research Institute of Chemistry, Gorko).
256. Brit. Pat. 793,354 (Apr. 16, 1958), (to National Lead Co.).
257. Brit. Pat. 798,001 (July 9, 1958), (to National Lead Co.).
258. J. A. Marsella, K. G. Moloy, and K. G. Caulton, *J. Organomet. Chem.* **201**, 389 (1980).
259. L. E. Manzer, *J. Organomet. Chem.* **110**, 291 (1976).
260. F. H. Kohler and D. Cozek, *Z. Naturforsch. Tell B*, **33**, 1274 (1978).
261. D. Nath, R. K. Sharma, and A. N. Bhat, *Inorg. Chim. Acta*, **20**, 109 (1976).
262. K. Chandra and co-workers, *J. Indian Chem. Soc.* **58**, 10 (1981).
263. R. K. Tuli and co-workers, *Trans. Met. Chem.* **5**, 49 (1980).
264. K. Chandra and co-workers, *Inorg. Chim. Acta*, **37**, 125 (1979).
265. A. K. Holliday and co-workers, *J. Organomet. Chem.* **57**, C45 (1973).
266. A. N. Nesmeyanov and co-workers, *Bull. Acad. Sci. USSR Div. Chem. Sci.* 777 (1967).
267. K. Chandra and co-workers, *Chem. Industry (London)*, 288 (1980).
268. M. F. Lappert and co-workers, *J. Chem. Soc. Dalton Trans.* 805 (1981).
269. J. E. Bercaw, *J. Am. Chem. Soc.* **96**, 5087 (1974).
270. T. J. Katz, N. Acton, and G. Martin, *J. Am. Chem. Soc.* **95**, 2934 (1973).
271. J. A. Smith and co-workers, *J. Organomet. Chem.* **173**, 175 (1979).
272. J. A. Smith and H. H. Brintzinger, *J. Organomet. Chem.* **218**, 159 (1981).
273. C. E. Carraher, Jr. and S. T. Bajah, *Brit. Polym. J.* **7**(3), 155–159 (1975).
274. J. Jeffery and co-workers, *J. Chem. Soc. Dalton Trans.* **7**, 1593 (1981).
275. G. K. Barker and M. F. Lappert, *J. Organomet. Chem.* **76**, C45 (1974).
276. B. Wozniak, J. D. Ruddick, and G. Wilkinson, *J. Chem. Soc.* **A3**, 3116 (1971).
277. J. X. McDermott, M. E. Wilson, and G. M. Whitesides, *J. Am. Chem. Soc.* **98**, 6529 (1976).
278. J. X. McDermott and G. M. Whitesides, *J. Am. Chem. Soc.* **96**, 947 (1974).
279. N. J. Foulger and B. J. Wakefield, *J. Organomet. Chem.* **69**, 161 (1974).
280. R. Jimenez and co-workers, *J. Organomet Chem.* **174**, 281 (1979).
281. M. F. Lappert and C. L. Raston, *J. Chem. Soc. Chem. Commun.* **24**, 1284 (1980).
282. H. Sakurai and H. Umino, *J. Organomet. Chem.* **142**, C49 (1977); Jpn. Pat. 816 437 (Feb. 10, 1981), (to Mitsubishi Chemical Industries KK).
283. S. A. Ciddings, *Inorg. Chem.* **3**, 684 (1964).
284. H. Surer, S. Claude, and A. Jacot-Guillarmod, *Helv. Chim. Acta*, **61**, 2956 (1978).
285. T. P. Bryukhanova and co-workers, *Tr. Khim. Khim. Tekhnol.* 101 (1974).
286. J. J. Eisch and R. J. Manfre, *Fundam. Res. Homog. Catal.* **3**, 397 (1978).
287. J. J. Eisch, R. J. Manfre, and D. A. Komar, *J. Organomet. Chem.* **159**, C13 (1978).
288. B. B. Snider and M. Karras, *J. Organomet. Chem.* **179**, C37 (1979).
289. D. C. Brown and co-workers, *J. Org. Chem.* **44**, 3457 (1979).

290. H. E. Tweedy and co-workers, *J. Mol. Catal.* **3**, 239 (1977–1978).

291. U.S. Pat. 3,969,429 (July 13, 1976), G. P. Belov and co-workers.

292. M. Peng and C. H. Brubaker, *Inorg. Chim. Acta*, **26**, 231 (1978).

293. M. D. Rausch, W. H. Boon, and E. A. Mintz, *J. Organomet. Chem.* **160**, 81 (1978).

294. M. D. Rausch, W. H. Boon, and H. G. Alt, *J. Organomet. Chem.* **141**, 229 (1977).

295. W. H. Boon and M. D. Rausch, *J. Chem. Soc. Chem. Commun.* **11**, 397 (1977).

296. J. L. Atwood and co-workers, *J. Am. Chem. Soc.* **98**, 1454 (1976).

297. H. Alt and M. D. Rausch, *J. Am. Chem. Soc.* **96**, 5936 (1974).

298. H. G. Alt and M. D. Rausch, *Z. Naturforsch. Teil B*, **30**, 813 (1975).

299. R. W. Harrigan, G. S. Hammond, and H. B. Gray, *J. Organomet. Chem.* **81**, 79 (1974).

300. C. H. Bamford, R. J. Puddephatt, and D. M. Slater, *J. Organomet. Chem.* **159**, C31 (1978).

301. H. G. Alt and co-workers, *J. Organomet. Chem.* **107**, 257 (1976).

302. G. A. Razuvaev, V. P. Mar'in, and Yu. A. Andrianov, *J. Organomet. Chem.* **174**, 67 (1979); G. A. Razuvaev and co-workers, *J. Organomet. Chem.* **164**, 41 (1979).

303. J. Mattia and co-workers, *Inorg. Chem.* **17**, 3257 (1978).

304. E. G. Berkovich and co-workers, *Chem. Ber.* **113**, 70 (1980).

305. V. B. Shur and co-workers, *J. Organomet. Chem.* **78**, 127 (1974).

306. M. K. Grigoryan and co-workers, *Bull. Acad. Sci. USSR Div. Chem. Sci.* 1024 (1978).

307. M. D. Rausch and H. B. Gordon, *J. Organomet. Chem.* **74**, 85 (1974).

308. G. J. Erskine and co-workers, *J. Organomet. Chem.* **170**, 51 (1979).

309. G. J. Erskine, D. A. Wilson, and J. D. McCowan, *J. Organomet. Chem.* **114**, 119 (1976).

310. J. A. Waters, V. V. Vickroy, and G. A. Mortimer, *J. Organomet. Chem.* **33**, 41 (1971).

311. C. P. Boekel, J. H. Teuben, and H. J. de Liefde Meijer, *J. Organomet. Chem.* **102**, 161 (1975).

312. *Ibid.*, p. 317.

313. C. P. Boekel, J. H. Teuben, and H. J. de Liefde Meijer, *J. Organomet. Chem.* **81**, 371 (1974).

314. F. W. Hartner and J. Schwartz, *J. Am. Chem. Soc.* **103**, 4974 (1981).

315. K. C. Ott and R. H. Grubbs, *J. Am. Chem. Soc.* **103**, 5922 (1981); *Chem. Eng. News*, 34 (Apr. 19, 1982).

316. T. Yoshida and E. I. Negishi, *J. Am. Chem. Soc.* **103**, 1276 (1981).

317. Y. Le Page and co-workers, *J. Organomet. Chem.* **193**, 201 (1980).

318. A. Glivicky and J. D. McCowan, *Can. J. Chem.* **51**, 2609 (1973).

319. L. J. Baye, *Synth. React. Inorg. Met. Org. Chem.* **5**, 95 (1975).

320. L. J. Baye, *Synth. React. Inorg. Met. Org. Chem.* **2**, 47 (1972).

321. C. R. Bennett and D. C. Bradley, *J. Chem. Soc. Chem. Commun.* **1**, 29 (1974).

322. R. V. Bynum and co-workers, *Inorg. Chem.* **19**, 2368 (1980).

323. P. L. Maxfield and E. Lima, *Proceedings of the 16th International Conference on Coordination Chemistry*, Dublin, Ireland, 1974, p. R58.

324. A. Jensen and co-workers, *Acta Chem. Scand.* **26**, 2898 (1972).

325. J. L. Burmeister and co-workers, *Inorg. Chem.* **9**, 58 (1970).

326. S. J. Anderson, D. S. Brown, and K. J. Finney, *J. Chem. Soc. Dalton Trans.* **1**, 152 (1979).

327. S. J. Anderson, D. S. Brown, and A. H. Norbury, *J. Chem. Soc. Chem. Commun.* **23**, 996 (1974).

328. J. Besangon and D. Camboli, *Compt. Rend.* **288C**, 121 (1979).

329. A. Chiesi-Villa, A. G. Manfredotti, and C. Guastini, *Acta Cryst.* **B32**, 909 (1976).

330. K. Issleib, H. Kohler, and G. Wille, *Z. Chem.* **13**, 347 (1973).

331. D. Camboli, J. Besancon, and B. Trimaille, *Compt. Rend.* **290C**, 365 (1980).

332. J. Besangon and co-workers, *Compt. Rend.* **287C**, 573 (1978).

333. L. Saunders and L. Spirer, *Polymer*, **6**, 635 (1965).
334. R. B. King and R. N. Kapoor, *J. Organomet. Chem.* **15**, 457 (1968).
335. G. A. Razuvaev, V. N. Latyaeva, and L. I. Vyshinskaya, *Dokl. Chem.* **138**, 592 (1961).
336. T. S. Kuntsevich and co-workers, *Kristallografiya*, **21**, 80 (1976).
337. V. P. Nistratov and co-workers, *Tr. Khim. Khim. Tekhnol.* 54 (1975).
338. J. C. G. Calado and co-workers, *J. Chem. Soc. Dalton Trans.* 1174 (1981).
339. L. Rosch and co-workers, *J. Organomet. Chem.* **197**, 51 (1980).
340. E. Hengge and W. Zimmermann, *Angew. Chem. Intl. Ed.* **7**, 142 (1968).
341. G. A. Razuvaev and co-workers, *Bull. Acad. Sci. USSR Div. Chem. Sci.* 2310 (1978).
342. G. A. Razuvaev and co-workers, *J. Organomet. Chem.* **87**, 93 (1975).
343. B. Stutte and co-workers, *Chem. Ber.* **111**, 1603 (1978).
344. G. Schmid, V. Batzel, and B. Stutte, *J. Organomet. Chem.* **113**, 67 (1976).
345. H. G. Raubenheimer and E. O. Fischer, *J. Organomet. Chem.* **91**, C23 (1975).
346. E. O. Fischer and S. Fontana, *J. Organomet. Chem.* **40**, 159 (1972).
347. H. Kopf and R. Voigtlander, *Chem. Ber.* **114**, 2731 (1981).
348. K. Issleib, F. Krech, and E. Lapp, *Synth. React. Inorg. Met. Org. Chem.* **7**, 253 (1977).
349. K. Issleib, G. Wille, and F. Krech, *Angew. Chem. Intl. Ed.* **11**, 527 (1972).
350. J. M. McCall and A. Shaver, *J. Organomet. Chem.* **193**, C37 (1980).
351. E. Samuel, *Bull, Soc. Chim. France*, 3548 (1966).
352. J. L. Petersen and L. F. Dahl, *J. Am. Chem. Soc.* **96**, 2248 (1974).
353. J. C. LeBlanc, C. Moise, and T. Bounthakna, *Compt. Rend.* **278C**, 973 (1974).
354. M. B. Bert and D. Gervais, *J. Organomet. Chem.* **165**, 209 (1979).
355. G. V. Drozdov, A. L. Klebanskii, and V. A. Bartashev, *J. Gen. Chem. USSR*, **33**, 2362 (1963).
356. M. A. Chaudhari and F. G. A. Stone, *J. Chem. Soc.* **A1**, 838 (1966); M. A. Chaudhari, P. M. Treichel, and F. G. Stone, *J. Organomet. Chem.* **2**, 206 (1964).
357. K. Andra, *J. Organomet. Chem.* **11**, 567 (1968).
358. J. C. LeBlanc, C. Moise, and J. Tirouflet, *Nouv. J. Chim.* **1**, 211 (1977).
359. T. Marey and co-workers, *Compt. Rend.* **284C**, 967 (1977).
360. J. Besangon, S. Top, and J. Tirouflet, *Compt. Rend.* **281C**, 135 (1975).
361. J. Besangon, F. Hug, and M. Colette, *J. Organomet. Chem.* **96**, 63 (1975).
362. R. Sharan, G. Gupta, and R. N. Kapoor, *Indian J. Chem.* **20A**, 94 (1981).
363. R. Sharan, G. Gupta, and R. N. Kapoor, *J. Less Common Met.* **60**, 171 (1978).
364. A. R. Dias, M. S. Salema, and J. A. M. Simoes, *J. Organomet. Chem.* **222**, 69 (1981).
365. H. Suzuki and T. Takiguchi, *Bull. Chem. Soc. Jpn.* **48**, 2460 (1975).
366. R. S. Arora, M. B. Bhalla, and R. K. Multani, *Indian J. Chem.* **16A**, 169 (1978).
367. R. S. P. Coutts and co-workers, *Austr. J. Chem.* **19**, 1377 (1966).
368. M. Sato and T. Yoshida, *J. Organomet. Chem.* **67**, 395 (1974).
369. H. Kopf and M. Schmid, *Angew. Chem. Intl. Ed.* **4**, 953 (1965).
370. A. Kutoglu, *Acta Cryst.* **B29**, 2891 (1973).
371. A. Kutoglu, *Z. Anorg. Allgem. Chem.* **390**, 195 (1972).
372. D. Sen and U. N. Kantak, *Indian J. Chem.* **13**, 72 (1975).
373. S. K. Sengupta, *Indian J. Chem.* **20A**, 515 (1981).
374. R. S. Arora and co-workers, *J. Chinese Chem. Soc. (Taipei)*, **27**, 65 (1980).
375. W. L. Steffen, H. K. Chun, and R. C. Fay, *Inorg. Chem.* **17**, 3498 (1978).
376. K. Chandra and co-workers, *J. Inorg. Nucl. Chem.* **43**, 663 (1981).
377. K. Chandra and co-workers, *J. Inorg. Nucl. Chem.* **43**, 29 (1981).
378. K. Chandra and co-workers, *Trans. Met. Chem.* **5**, 209 (1980).
379. N. K. Kaushik, B. Bhushan, and G. R. Chhatwal, *Z. Naturforsch.* **34b**, 949 (1979).
380. B. Bhushan and co-workers, *J. Inorg. Nucl. Chem.* **41**, 159 (1979).
381. N. K. Kaushik, B. Bhushan, and G. R. Chbatwal, *Trans. Met. Chem.* **3**, 215 (1978).
382. N. K. Kaushik, B. Bhushan, and G. R. Chhatwal, *Synth. React. Inorg. Met. Org. Chem.* **8**, 467 (1978).

383. P. C. Bharara, *J. Organomet. Chem.* **121**, 199 (1976).

384. O. N. Suvorova, V. V. Sharutin, and G. A. Domrachev, *3rd Mater. Vses Semin.* **1977**, 132 (1978).

385. G. Fachinetti and co-workers, *J. Am. Chem. Soc.* **100**, 1921 (1978).

386. G. G. Dvoryantseva and co-workers, *Dokl. Chem.* **161**, 303 (1965).

387. A. N. Nesmeyanov, O. V. Nogina, and V. A. Dubovitskii, *Bull. Acad. Sci. USSR Div. Chem. Sci.* 1395 (1962).

388. A. N. Nesmeyanov, O. V. Nogina, and A. M. Berlin, *Bull. Acad. Sci. USSR Div. Chem. Sci.* 743 (1961).

389. G. Wilkinson and J. M. Birmingham, *J. Am. Chem. Soc.* **76**, 4281 (1954).

390. K. Doppert and R. Sanchez, *J. Organomet. Chem.* **210**, C9 (1980).

391. K. Doppert, *Makromol. Chem. Rapid Commun.* **1**, 519 (1980).

392. K. Doppert, *J. Organomet. Chem.* **178**, C3 (1979).

393. K. Chandra, R. K. Tuli, N. K. Bhatia, and B. S. Garg, *J. Indian Chem. Soc.* **58**, 122 (1981).

394. D. Pacheco and co-workers, *Inorg. Chim. Acta*, **18**, L24 (1976).

395. H. Klein and U. Thewalt, *Z. Anorg. Allgem. Chem.* **476**, 62 (1981).

396. P. M. Druce and co-workers, *J. Chem. Soc.* **A3**, 2106 (1969).

397. V. K. Jain and co-workers, *J. Indian Chem. Soc.* **57**, 6 (1980).

398. D. Nath and A. N. Bhat, *Indian J. Chem.* **14**, 281 (1976).

399. C. E. Carraher and J. D. Piersma, *Makromol. Chem.* **152**, 49 (1972).

400. C. E. Carraher and co-workers, *J. Macromol. Sci. Chem.* **A16**, 195 (1981).

401. C. E. Carraher and J. L. Lee, *J. Macromol. Sci. Chem.* **A9**, 191 (1975).

402. C. E. Carraher and J. L. Lee, *Am. Chem. Soc. Div. Org. Coat. Plast. Chem. Pap.* **34**, 478 (1974).

403. C. E. Carraher and R. Frary, *Br. Polym. J.* **0**, 255 (1974).

404. C. E. Carraher and R. A. Frary, *Makromol. Chem.* **175**, 2307 (1974).

405. C. E. Carraher and J. D. Piersma, *J. Macromol. Sci. Chem.* **A7**, 913 (1973).

406. C. E. Carraher and L. S. Wang, *Angew. Makromol. Chem.* **25**, 121 (1972).

407. P. C. Wailes, H. Weigold, and A. P. Bell, *J. Organomet. Chem.* **33**, 181 (1971).

408. U.S. Pat. 3,782,917 (Jan. 1, 1974), J. J. Mrowca (to E. I. du Pont de Nemours & Co., Inc.).

409. U.S. Pat. 3,728,365 (Apr. 17, 1973), J. J. Mrowca (to E. I. du Pont de Nemours & Co., Inc.).

410. G. Fachinetti and C. Floriani, *J. Organomet. Chem.* **71**, C5 (1974).

411. G. Fachinetti and C. Floriani, *J. Chem. Soc. Chem. Commun.* **11**, 654 (1972).

412. R. J. H. Clark, J. A. Stockwell, and J. D. Wilkins, *J. Chem. Soc. Dalton Trans.* **2**, 120 (1976).

413. K. W. Chiu and co-workers, *J. Chem. Soc. Dalton Trans.* **10**, 2088 (1981).

414. H. Yasuda and co-workers, *Bull. Chem. Soc. Jpn.* **53**, 1089 (1980).

415. S. Akutagawa and S. Otsuka, *J. Am. Chem. Soc.* **97**, 6870 (1975).

416. F. Sato and co-workers, *J. Chem. Soc., Chem. Commun.* **21**, 1140 (1981).

417. K. Kaeriyama, *Makromol. Chem.* **175**, 2285 (1974).

418. K. Kaeriyama, *Makromol. Chem.* **153**, 229 (1972).

419. T. Saito, *J. Chem. Soc. Chem. Commun.* **22**, 1422 (1971).

420. J. Tsuji and S. Hashiguchi, *Tetrahedron Lett.* **21**, 2955 (1980).

421. D. F. Williams, *Syst. Aspects Biocompat.* **1**, 169 (1981).

422. U.S. Pat. 4,272,588 (June 9, 1981), B. E. Yoldas, A. M. Filippi, and R. W. Buckman (to Westinghouse Electric Corp.).

423. Brit. Pat. 2,067,540 (July 30, 1981), J. H. Novak and G. L. Smay (to American Glass Research).

424. Jpn. Kokai Tokkyo Koho 81 88,843 (July 18, 1981), (to Matusushita Electric Works).

425. U.S. Pat. 3,094,436 (June 18, 1963), H. Schroder (to Jenaer Glaswerk Schott and Gen.).

426. Jpn. Kokai Tokkyo Koho 81 60,068 (May 23, 1981), (to Tokyo Shibaura Electric, Ltd.).

427. Jpn. Kokai Tokkyo Koho 81 114,904 (Sept. 9, 1981), (to Nippon Telegraph and Telephone).

428. Jpn. Kokai Tokkyo Koho 81 37,173 (Aug. 29, 1981), (to Sharp KK).

429. Jpn. Kokai Tokkyo Koho 81 116,015 (Sept. 11, 1981), (to Suwa Seikosha KK).

430. Jpn. Kokai Tokkyo Koho 81 116,016 (Sept. 11, 1981), (to Suwa Seikosha KK).

431. Jpn. Kokai Tokkyo Koho 81 63,846 (May 30, 1981), (to Suwa Seikosha KK).

432. Jpn. Kokai Tokkyo Koho 81 94,651 (July 31, 1981), (to Tokyo Denshi Kagaku).

433. U.S. Pat. 4,279,654 (July 21, 1981), S. Yajima and co-workers (to Research Institute for Special Inorganic Materials); S. Yajima and co-workers, *J. Mater. Sci.* **16**, 1349 (1981).

434. U.S. Pat. 4,246,314 (Jan. 20, 1981), A. Marzocchi, M. G. Roberts, and C. E. Bolen (to Owens Corning Fiberglas Corp.).

435. U.S. Pat. 4,269,756 (May 26, 1981), T. Y. Su (to Union Carbide).

436. Ger. Offen. 3,026,987 (Feb. 12, 1981), R. G. Adams and S. J. Milletari (to J. P. Stevens and Co.).

437. Rus. Pat. 235,907 (Jan. 24, 1969), L. V. Golosova and K. S. Zatsepin.

438. G. Winter, *J. Oil Color Chem. Assoc.* **36**(402), 689 (1953); **34**(367), 30 (1951); A. Hancock and R. Sidlow, *J. Oil Color Chem. Assoc.* **35**(379), 28 (1952).

439. G. Pagliara, *Pitture Vernici*, **40**, 279 (1964).

440. Jpn. Kokai Tokkyo Koho 80,11,147 (Mar. 22, 1980), (to Kansai Paint Co.).

441. U.S. Pat. 4,224,213 (June 9, 1978), S. D. Johnson (to Cook Paint & Varnish Co.).

442. Jpn. Kokai Tokkyo Koho 80,152,759 (Nov. 28, 1980), (to Dainippon Toryo Co.).

443. U.S. Pat. 3,524,799 (Aug. 18, 1970), K. H. Dale (to Reynolds Metals Co.).

444. Jpn. Kokai Tokkyo Koho 80,141,573 (Nov. 5, 1980), (to Showa Keikinzoku KK).

445. Rus. Pat. 744,014 (June 2, 1980), (to Enam Chem. EQP Res.).

446. Jpn. Kokai Tokkyo Koho 81,104,973 (Aug. 21, 1981), (to Toray Industries).

447. Brit. Pat. 786,388 (Nov. 20, 1957), (to National Lead Co.).

448. U.S. Pat. 2,888,367 (May 26, 1959), W. L. Greyson (to Hitemp Wires, Inc.).

449. U.S. Pat. 3,862,099 (Jan. 21, 1975), N. S. Marans (to W. R. Grace).

450. Jpn. Kokai Tokkyo Koho 77,10,322 (Jan. 26, 1977), T. Yoshimura and co-workers (to Kaikin Kogyo Co.).

451. Jpn. Kokai Tokkyo Koho 81 30,472 (Mar. 27, 1981), (to Furukawa Electric Co.).

452. A. M. Usmani and co-workers, *Rubber Chem. Technol.* **54**, 1081 (1981).

453. Jpn. Kokai Tokkyo Koho 77 132,082 (Nov. 5, 1977), I. Sugiyama, H. Doi, and Y. Takaoka (to Matsumoto Seiyaku Kogyo Co.).

454. Belg. Pat. 887,145 (July 20, 1981), (to Toray Silicone Co.).

455. Jpn. Pat. 71 22,878 (June 30, 1971), I. Honda (to Kuraray Co.).

456. N. V. Vakrameeva and co-workers, *Kozh. Obuvn. Promst.* **23**, 37, 43 (1981).

457. Brit. Pat. 1,374,263 (Nov. 20, 1974), T. Keating (to Imperial Chemical Industries).

458. U.S. Pat. 4,234,497 (Nov. 18, 1980), M. Honig (to Standard Lubricants).

459. Ger. Offen. 2,004,098 (Aug. 12, 1971), R. Tuma and R. Lebender (to Dynamit Nobel AG).

460. A. G. Okuneva and co-workers, *Plast. Massy*, 10 (1981).

461. U.S. Pat. 4,260,735 (Apr. 7, 1981), J. A. Bander, S. D. Lazarus, and I. C. Twilley (to Allied Corp.).

462. Jpn. Kokai Tokkyo Koho 80 125,120 (Sept. 26, 1980), (to Toray Industries, Inc.).

463. Ger. Offen. 2,751,385 (May 24, 1978), S. P. Elliot (to E. I. du Pont de Nemours & Co., Inc.).

464. *Organic Titanium Compounds as Acrylic Ester Alcoholysis Catalysts*, Titanium Intermediates, Ltd., London, 1967.

465. Jpn. Kokai Tokkyo Koho 79,125,617 (Sept. 29, 1979), T. Onoda, K. Tano, and Y. Hara (to Mitsubishi Chemical Industries).

466. Ger. Offen. 2,922,343 (Dec. 4, 1980), B. Luthingshauser and C. Lindzus (to Dynamit Nobel AG).

467. Jpn. Kokai Tokkyo Koho 74,95,918 (Sept. 11, 1974), K. Kimura and H. Ito (to Toa Gosei Chemical Industry Co.).

468. L. H. Buxbaum, *Angew. Chem.* **80**, 225 (1968).

469. H. Zimmermann and E. Leibnitz, *Faserforsh. Text. Tech.* **16**, 282 (1965).

470. G. Ratner and co-workers, *Akad. Polym.* **34**, 48–51 (1983).

471. U.S. Pat. 5,015,759 (May 14, 1991), D. J. Lowe (to E. I. du Pont de Nemours & Co., Inc.).

472. Eur. Pat. 46,670 (Mar. 3, 1982), H. K. Hall and A. B. Padias (to Celanese Corp.).

473. Eur. Pat. 634,435 (Jan. 18, 1995), W. H. F. Borman and T. G. Shannon (to General Electric Corp.).

474. Jpn. Pat. 5,117,379 (May 14, 1993), (to Kuraray Co.).

475. Jpn. Pat. 62,141,022 (June 24, 1987), (to Toray Industries).

476. Jpn. Pat. 62,225,523 (Oct. 3, 1987), (to Toray Industries).

477. F. Pilati and co-workers, *Polymer*, **24**(11), 1479–1483 (1983).

478. F. Pilati and co-workers, *Polymer*, **26**(11), 1745–1748 (1985).

479. B. Fortunato and co-workers, *Polym. Commun.* **27**(1), 29–31 (1986).

480. B. Fortunato and co-workers, *Polym. Commun.* **30**(2), 55–77 (1989).

481. H. L. Traub and co-workers, *Angew. Makromol. Chem.* **230**, 179–187 (1995).

482. J. Hsu and K. Y. Choi, *J. Appl. Poly. Sci.* **32**(1), 3117–3122 (1986).

483. Y. Yurramendi, M. J. Barandiaran, and J. M. Asua, *J. Macromol. Sci. Chem.* **A24**(11), 1357–1367 (1987).

484. P. Fritzsche, G. Rafler, and K. Tauer, *Akta. Polym. Commun.* **40**(30), 143–160 (1989).

485. Eur. Pat. 32,587 (July 29, 1981), M. J. Keogh (to Union Carbide Corp.).

486. Jpn. Kokai Tokkyo Koho 79 77,635 (June 21, 1979), M. Yoshiaki and co-workers (to Kansai Paint Co.).

487. U.S. Pat. 5,453,479 (Sept. 26, 1995), W. F. H. Borman and T. G. Shannon (to General Electric Co.).

488. P. Schnurrenberger, M. Zuger, and D. Seebach, *Helv. Chem. Acta*, **65**(4), 110, 1197–2001 (1982).

489. U.S. Pat. 5,371,172 (Dec. 6, 1994), E. Takayama, I. Niikura, and T. Hokari (to Showa Highpolymer Co.); U.S. Pat. 4,348,498 (Sept. 7, 1982), H. Kamio, Y. Ogino, and K. Nakamura (to Nippon Mining Co.).

490. Brit. Pat. 994,717 (June 10, 1965), (to Dr. Beck and Co.).

491. U.S. Pats. 4,297,447–4,297,449 (Oct. 27, 1981), C. J. Stark (to General Electric Co.).

492. Ger. Offen. 3,028,114 (Feb. 12, 1981), S. R. Stoelzer and R. E. Wiley (to Acheson Industries, Inc.).

493. Jpn. Kokai Tokkyo Koho 81,88,471 (July 17, 1981), (to Mitsui Toatsu Chem.).

494. Ger. Offen. 3,038,646 (Apr. 23, 1981), K. Kawasumi, H. Watanabe, and J. Seto (to Sony Corp.).

495. Jpn. Kokai Tokkyo Koho 81,75,544 (June 22, 1981), (to Suwa Seikosha KK).

496. Brit. Pat. 922,456 (Apr. 3, 1963), G. E. Westwood (to Berger, Jensen, and Nicholson, Ltd.).

497. U.S. Pat. 3,679,721 (July 25, 1972), D. W. Brook and R. Ward (to British Titan Products Co. Ltd.).

498. U.S. Pat. 3,694,475 (Sept. 26, 1972), D. W. Brook and P. D. Kay (to British Titan Ltd.).

499. U.S. Pat. 4,874,806 (Aug. 17, 1989), P. D. Kay and K. McDonald (to Tioxide Group PLC).

500. U.S. Pat. 5,076,847 (Dec. 31, 1991), P. D. Kay and K. McDonald (to Tioxide Group PLC).
501. U.S. Pat. 4,159,209 (June 26, 1979), P. Womersley (to Manchem., Ltd.).
502. U.S. Pat. 3,013,895 (Dec. 19, 1961), M. S. Agruss (to Miehle-Goss-Dexter, Inc.).
503. U.S. Pat. 3,682,688 (Aug. 8, 1972), C. T. Hughes and D. O. Paulsen (to E. I. du Pont de Nemours & Co., Inc.).
504. Jpn. Kokai Tokkyo Koho 47,29,571 (Aug. 3, 1972), (to Dainippon Ink and Chemicals, Inc.).
505. U.S. Pat. 4,617,408 (Oct. 14, 1986), H. Nestler and D. Barfurth (to Dynamit Nobel AG).
506. U.S. Pat. 4,909,846 (Mar. 20, 1990), D. Barfurth and H. Nestler (to Dynamit Nobel AG).
507. U.S. Pat. 4,931,094 (June 5, 1990), D. Barfurth, C. Lindzus, and H. Nestler (to Dynamit Nobel AG).
508. U.S. Pats. 4,659,848 (Apr. 21, 1987) and 4,705,568 (Nov. 10, 1987), P. D. Kay and M. C. Girot (to Tioxide Group PLC).
509. U.S. Pat. 5,286,774 (Feb. 15, 1994), G. McGibbon and J. E. Robinson (to Tioxide Specialties Ltd.).
510. C. Parducci, *Pitture Vernici*, **56**(4), 19–27 (1980); technical bulletin, *Titanium and Zirconium Chelates for the Printing Ink Industry*, E. I. du Pont de Nemours & Co., Inc., Wilmington, Del., 1997.
511. Brit. Pat. 759,570 (Oct. 17, 1956), (to National Lead).
512. U.S. Pat. 4,483,712 (Nov. 20, 1984), P. N. Murphy (to Crown Zellerbach Corp.).
513. Jpn. Koksai Tokkyo Koho 58,134,164 (Aug. 10, 1983), (to Dai Nippon Ink and Chemical Co.).
514. World Pat. 80,2292 (Oct. 30, 1980), M. L. Barstall and B. D. Podd (to National Research and Development).
515. U.S. Pat. 2,970,126 (Jan. 31, 1961), E. D. Brown Jr. (to General Electric Co.).
516. U.S. Pat. 2,732,320 (Jan. 24, 1956), C. J. Guillissen (to Union Chimique Belge).
517. U.S. Pat. 2,911,324 (Nov. 3, 1959), (to Bradford Dyers Association Ltd.).
518. Brit. Pat. 1,406,336 (Sept. 17, 1975), M. Camp and J. Dumoulin (to Rhone-Poulenc).
519. Eur. Pat. 164,470 (Dec. 18, 1985), (to Dow Corning).
520. U.S. Pat. 3,151,099 (Sept. 29, 1964), P. L. Brown and J. L. Hyde (to Rhone-Poulenc).
521. U.S. Pat. 3,161,614 (Dec. 15, 1964), L. F. Ceyzeriat and G. L. Pagni (to Dow Corning).
522. U.S. Pat. 3,378,520 (Apr. 16, 1968), H. Sattleger and co-workers (to Farbenfabriken Bayer).
523. U.S. Pat. 3,334,067 (Aug. 1, 1967), D. Weyenberg (to Dow Corning).
524. U.S. Pat. 3,856,839 (Dec. 24, 1974), S. D. Smith and S. Hamilton, Jr. (to General Electric Co.).
525. S. J. Monte, *Ken-React Reference Manual, Titanate, Zirconate and Aluminate Coupling Agents*, Kenrich Petrochemicals Inc., Bayonne, N.J., 1993.
526. U.S. Pat. 4,098,758 (July 4, 1978), S. J. Monte and P. F. Bruins (to Kenrich Petrochemicals Inc.).
527. U.S. Pat. 4,087,402 (May 2, 1978), S. J. Monte and G. Sugarman (to Kenrich Petrochemicals Inc.).
528. U.S. Pat. 4,261,913 (Apr. 14, 1981), S. J. Monte and G. Sugarman (to Kenrich Petrochemicals Inc.).
529. U.S. Pat. 4,277,415 (July 7, 1981), G. Sugarman and S. J. Monte (to Kenrich Petrochemicals Inc.).
530. U.S. Pat. 4,600,789 (July 15, 1986), G. Sugarman and S. J. Monte (to Kenrich Petrochemicals Inc.).
531. U.S. Pat. 5,079,286 (Jan. 7, 1992), H. Hanisch and co-workers (to Huels Troisdorf AG).

532. U.S. Pat. 3,301,723 (Jan. 31, 1967), J. D. Crisp (to E. I. du Pont de Nemours & Co., Inc.).

533. U.S. Pat. 3,355,366 (Nov. 28, 1967), W. M. Lyerley (to E. I. du Pont de Nemours & Co., Inc.).

534. U.S. Pat. 3,840,520 (Oct. 8, 1974), R. Nordgren and C. I. Carl (to General Mills Chemicals).

535. U.S. Pat. 4,692,273 (Sept. 8, 1987), K. Lawerence (to Hewlett-Packard).

536. T. Sakaguchi, A. Nakajima, and T. Horikoshi, *Nippon Kagaku Kaishi*, 788 (1979).

537. Jpn. Kokai Tokkyo Koho 81 109,266 (Aug. 29, 1981), (to Mitsubishi Pencil Co.).

538. Jpn. Pats. 71 12,085 (Mar. 27, 1971), K. Hamahiro, Y. Yoshioka, and H. Sakurai (to Kuraray Co.); 71 12,089 (Mar. 27, 1971), Y. Yoshioka, S. Kurokawa, and K. Hashita (to Kuraray Co.); 71 405 (Jan. 7, 1971), T. Ashikaga and U. Maeda (to Kuraray Co.).

539. Jpn. Pats. 71 38,410 (Nov. 12, 1971), T. Asikaga and S. Higashimori (to Kuraray Co.); 71 38,411 (Nov. 12, 1971), T. Ashikaga and U. Maeda (to Kuraray Co.).

540. U.S. Pats. 3,941,728–3,941,730 (Mar. 2, 1976), J. C. Solenberger (to E. I. du Pont de Nemours & Co., Inc.).

541. U.S. Pat. 4,113,757 (Sept. 12, 1978), P. D. Kay (to Tioxide Group).

542. Rus. Pat. 380,774 (May 15, 1973), A. F. Tishchenko and co-workers.

543. Ger. Offen. 2,326,828 (Nov. 29, 1973), M. Camp and J. Dumoulin (to Rhone-Poulenc S.A.).

544. Czech. Pat. 181,452 (Jan. 15, 1980), E. Wurstova.

545. U.S. Pat. 4,288,496 (Sept. 8, 1981), R. E. Reusser and B. E. Jones (to Phillips Petroleum Co.).

546. Jpn. Kokai Tokkyo Koho 78 24,408 (Mar. 7, 1978), T. Noguchi and co-workers (to Kureha Chemical Industry Co.).

547. Jpn. Kokai Tokkyo Koho 80 167,211 (to Matsumoto Seiyaku Kogyo Co.).

548. Rus. Pat. 443,859 (Sept. 25, 1974), T. P. Osipova and A. A. Kas'yanova.

549. Jpn. Kokai Tokkyo Koho 74 36,998 (Apr. 5, 1974), M. Komeyama, S. Murakami, and H. Ohnishi (to Kuraray Co.).

550. S. D. Voronkevich, L. K. Zgadzai, and M. T. Kuleev, *Plast. Massy*, 64 (1973).

551. A. A. Panasevich, S. P. Nichiporenko, and G. M. Nikitina, *Tr. Mezhvuz. Konf. Primen. Plastmass Stroit.*, **3**, 174 (1970).

552. Ger. Offen. 2,310,820 (Sept. 13, 1973), A. E. Vassiliades, D. N. Vincent, and M. P. Powell (to Champion Paper Co.).

553. Research Disclosure 20402, E. I. du Pont de Nemours & Co., Inc., Wilmington, Del., 1981.

554. U.S. Pat. 3,888,312 (June 10, 1975), R. L. Tiner and co-workers (to Halliburton Co.).

555. U.S. Pat. 4,462,917 (July 31, 1984), M. J. Conway (to Halliburton Services Co.).

556. U.S. Pat. 4,464,270 (Aug. 7, 1984), K. H. Hollenbeak and C. J. Githens (to Halliburton Services Co.).

557. U.S. Pat. 4,470,915 (Sept. 11, 1984), M. J. Conway (to Halliburton Services Co.).

558. U.S. Pat. 4,502,967 (Mar. 5, 1985), M. J. Conway (to Halliburton Services Co.).

559. U.S. Pat. 4,657,080 (Apr. 14, 1987), R. M. Hodge (to Dowell Schlumberger).

560. U.S. Pat. 4,609,479 (Sept. 2, 1986), K. C. Schmeltz (to E. I. du Pont de Nemours & Co., Inc.).

561. Eur. Pat. 195,531 (Sept. 24, 1986), D. E. Putzig and K. C. Schmeltz (to E. I. du Pont de Nemours & Co., Inc.).

562. U.S. Pat. 3,860,622 (Jan. 1, 1978), R. C. Wade (to Ventron Corp.).

563. U.S. Pat. 4,514,309 (Apr. 30, 1985), S. K. Wadhwa (to Hughes Tool Co.).

564. R. K. Prud'homme, *Polym. Mater. Sci. Eng.* **55**, 798 (1986).

565. J. Kramer and R. K. Prud'homme, *Polym. Mater. Sci. Eng.* **57**, 376–379 (1988).

566. D. I. Collias and R. J. Prud'homme, *J. Rheol.* **38**(2), 217–230 (1994).

567. U.S. Pat. 3,784,579 (Jan. 8, 1974), R. E. McGlothlin and co-workers (to Dresser Industries).
568. U.S. Pat. 3,878,111 (Apr. 15, 1975), R. E. McGlothin and T. E. Cox (to Dresser Industries).
569. U.S. Pat. 3,141,503 (July 21, 1964), N. Stein (to Secony Mobil Oil Co.).
570. U.S. Pat. 4,996,336 (Feb. 26, 1991), D. E. Putzig (to E. I. du Pont de Nemours & Co., Inc.).
571. R. C. Mehrotra and A. Singh, *Chem. Rev.* **96**(1), 1–13, 1996.
572. U.S. Pat. 5,200,167 (Apr. 6, 1993), U. Maeda, T. Imagawa, and M. Noriz (to Nippon Soda Co. Ltd.).
573. Jpn. Pat. 3 76,819 (Apr. 2, 1991), S. Miyama and co-workers (to Nippon Soda Co. Ltd.).
574. T.-C. Chai and C.-M. Wei, *Fen Hsi Hua Hsueh*, **7**, 327 (1979).
575. V. D. Bakalov, V. V. Dunina, and V. M. Potapov, *Zh. Anal. Khim.* **34**, 2138 (1979).
576. G. D. Brykina and T. A. Belyavskaya, *Vestn. Mosk. Univ. Khim.* **13**, 608 (1972).
577. C. G. Macarovici and E. Motiu, *Stud. Univ. Babes-Bolyai Ser. Chem.* **16**, 39 (1971).
578. R. M. Dranitskaya, A. I. Gavril'chenko, and L. A. Okhitina, *Zh. Anal. Khim.* **25**, 1740 (1970).
579. Y. I. Tur'yan and co-workers, *Zh. Neorg. Khim.* **23**, 2061 (1978).
580. Y. K. Tselinskii and V. K. Gadzhun, *Soviet J. Coord. Chem.* **4**, 1028 (1978).
581. E. I. Stepanovskii, G. M. Fofanov, and G. A. Kitaev, *Zh. Neorg. Khim.* **24**, 941 (1979).
582. F. G. Banica and L. Carlea, *Rev. Chim. (Bucharest)*, **30**, 640 (1979).
583. A. Y. Nazarenko and I. V. Pyatnitskii, *Zh. Neorg. Khim.* **23**, 2655 (1978).
584. A. A. Popel, A. N. Glebov, and Y. I. Saltnikov, *Zh. Neorg. Khim.* **24**, 2409 (1979).
585. I. V. Pyatnitskii and R. S. Kharchenko, *Ukr. Khim. Zh.* **33**, 734 (1967).
586. V. M. Savostina, F. I. Lobanov, and V. M. Peshkova, *Zh. Neorg. Khim.* **12**, 2162 (1967).
587. V. K. Zolotokhin, O. M. Gnatishin, and E. I. Senchishin, *Visn. L'uiv. Derzh. Univ. Ser. Khim.* **17**, 40 (1975).
588. I. V. Pyatnitskii and A. Y. Nazarenko, *Zh. Anal. Khim.* **32**, 853 (1977); *Zh. Neorg. Khim.* **22**, 1816 (1977).
589. Y. K. Tselinskii, L. Y. Kvyatokovskaya, and V. K. Gadzhun, *Zh. Fiz. Khim.* **50**, 3002 (1976).
590. R. S. Ramakrishna and D. T. A. Seneratyapa, *J. Inorg. Nucl. Chem.* **39**, 333 (1977).
591. Ger. Offen. 1,811,502 (June 26, 1969), R. H. Stanley, D. W. Brook, and L. L. Lawrence (to British Titan Products Co.).
592. C. G. Macarovici and L. Czegledi, *Reu. Roum. Chim.* **14**, 57 (1969); *Stud. Univ. Babes-Bolyai Ser. Chem.* **22**, 25 (1977).
593. F. I. Lobanov and co-workers, *Zh. Neorg. Khim.* **14**, 1077 (1969); *Vestn. Mosk. Uniu. Khim.* **24**, 121 (1969).
594. Jpn. Pat. 67 26,628 (Dec. 16, 1967), I. Sugiyama, K. Takahashi, and N. Takahashi (to Matsumoto Pharmaceutical Industry Co.).

General References

J. H. Clark, *The Chemistry of Titanium and Vanadium*, Elsevier, Amsterdam, the Netherlands, 1968.
R. J. H. Clark, in J. C. Bailar and co-workers, eds., *Comprehensive Inorganic Chemistry*, Vol. 3, Pergamon Press, London, 1973, pp. 355–417.
P. C. Wailes, R. S. P. Coutts, and H. Weigold, *Organometallic Chemistry of Titanium, Zirconium, and Hafnium*, Academic Press, Inc., New York, 1974.
R. J. H. Clark, D. C. Bradley, and P. Thornton, *Chemistry of Titanium, Zirconium, and Hafnium*, Pergamon Press, New York, 1975.

R. J. H. Clark, S. Moorhouse, and J. A. Stockwell, *J. Organometal. Chem. Library*, **3**, 223 (1977); literature covered through 1975.

E. Muller, ed., *Houben-Weyl's Methods in Organic Chemistry*, 4th ed., Vol. 13, Part 7, Georg Thieme, Stuttgart, Germany, 1975.

Gmelin, *Handbuch der Anorganischen Chemie*, 8th ed., Springer Verlag, Berlin, 1977.

Annual Surveys

K. S. Mazdiyasni, in K. Niedenzu and H. Zimmer, eds., *Annual Reports in Inorganic and General Synthesis*, Academic Press, Inc., New York, 1972, pp. 73–81; 1973, pp. 137–147.

J. J. Alexander, in K. Niedenzu and H. Zimmer, eds., *Annual Reports in Inorganic and General Synthesis*, Academic Press, Inc., New York, 1974, pp. 130–141; 1975, pp. 138–151; 1976, pp. 151–166.

F. Calderazzo, *Organomet. Chem. Rev.* **B4**, 12 (1968); **5**, 547 (1969); **6**, 1001 (1970); **9**, 137 (1972); *J. Organomet. Chem.* **53**, 179 (1973); **89**, 193 (1975); surveys cover the years 1967–1972.

P. C. Wailes, *J. Organomet. Chem.* **79**, 201 (1974); **103**, 475 (1975); **126**, 361 (1977); surveys cover the years 1973–1975.

J. A. Labinger, *J. Organomet. Chem.* **138**, 185 (1977); **167**, 19 (1979); **180**, 187 (1979); **196**, 37 (1980); **227**, 341 (1981); surveys cover the years 1976–1980.

R. C. Fay, *Coord. Chem. Rev.* **37**, 9 (1981); covers the literature on titanium for 1979; review for later years may appear subsequently.

Specialist Periodical Reports of the Chemical Society (London), Organometallic Compounds, contain numerous references on titanium compounds which can be located through the volume indexes.

Titanium Alkoxides, Amides, and Polyalkoxides

D. C. Bradley, in F. G. A. Stone and W. A. Graham, eds., *Inorganic Polymers*, Academic Press, Inc., New York, 1962, pp. 410–446.

D. C. Bradley, *Prep. Inorg. React.* **2**, 169 (1965).

D. C. Bradley, *Coord. Chem. Rev.* **2**, 299 (1967).

D. C. Bradley, *Inorg. Macromol. Rev.* **1**, 141 (1970).

Ziegler-Natta Polymerization

G. Henrici-Olivé and S. Olivé, *Chemtech*, 746 (1981); a readable overview of this complex topic.

J. Boor, *Ziegler-Natta Catalysts and Polymerization*, Academic Press, Inc., New York, 1979.

C. E. Schildknecht and I. Skeist, *Polymerization Processes*, John Wiley & Sons, Inc., New York, 1977.

J. C. W. Chien, *Coordination Polymerization*, Academic Press, Inc., New York, 1975.

H. Sinn and W. Kaminsky, *Adv. Organomet. Chem.* **18**, 99 (1980).

Lower Valent Titanium Compounds

R. S. P. Coutts and P. C. Wailes, *Adv. Organometal. Chem.* **9**, 136 (1970).

DONALD E. PUTZIG
THOMAS W. DEL PESCO
E. I. du Pont de Nemours & Company, Inc.

TOBIAS ACID.   See NAPHTHALENE DERIVATIVES.

TOCOPHEROLS.   See VITAMINS, VITAMIN E.

TOILET PREPARATIONS.   See COSMETICS.

TOLU BALSAM.   See PERFUMES.

TOLUENE

Toluene [108-88-3], C_7H_8, is a colorless, mobile liquid with a distinctive aromatic odor somewhat milder than that of benzene. The name toluene derives from a natural resin, balsam of Tolu, named for a small town in Colombia, South America. Toluene was discovered among the degradation products obtained by heating this resin.

Prior to World War I, the main source of toluene was coke ovens. At that time, trinitrotoluene (TNT) was the preferred high explosive and large quantities of toluene were required for its manufacture (see EXPLOSIVES AND PROPELLANTS). To augment the supply, toluene was obtained for the first time from petroleum sources by subjecting narrow-cut naphthas containing relatively small amounts of toluene to thermal cracking. The toluene concentrate so produced was then purified and used for the manufacture of TNT. Production from petroleum was discontinued shortly after World War I. Petroleum again became the source for toluene with the advent of catalytic reforming and the need for large quantities of toluene for use in aviation fuel during World War II. Since then, manufacture of toluene from petroleum sources has continued to increase, and manufacture from coke ovens and coal-tar products has continued to decrease.

Toluene is generally produced along with benzene, xylenes, and C_9-aromatics by the catalytic reforming of C_6–C_9 naphthas. The resulting crude reformate is extracted, most frequently with sulfolane or tetraethylene glycol and a cosolvent, to yield a mixture of benzene, toluene, xylenes, and C_9-aromatics, which are then separated by fractionation. There have been, ca 1997, recent technological developments to produce benzene, toluene, and xylenes from pyrolysis of light hydrocarbons C_2–C_5, LPG, and naphthas (see XYLENES AND ETHYLBENZENE). About 85–90% of the toluene produced annually in the United States is not isolated, but is blended directly into the gasoline pool as a component of reformate and of pyrolysis gasoline. Capacity exists to isolate ca 6.1×10^6 metric tons (1.9×10^9 gal) per year, of which about 75–80% is used for chemicals and solvents or exported. The remainder is blended into gasoline to increase octane number. However, with the Clean Air Act in place, methyl *tert*-butyl ether (MTBE) and other oxygenates, as of ca 1997, are beginning to take over the supply of octane and oxygen requirements in Reformulated Gasoline. The use of toluene in gasoline blending is therefore expected to decline slightly, except in special cases for vapor-pressure fine-tuning (see BTX PROCESSING; GASOLINE AND OTHER MOTOR FUELS).

Physical Properties

The physical properties of toluene have been well studied experimentally. Several physical properties are presented in Table 1 (1). Thermodynamic and transport properties can also be obtained, from other sources (2–7). The vapor pressure of toluene can be calculated as follows (8), where P is in kPa and T is in K.

$$\ln P = 14.01 - \frac{3103}{T - 53.36} \qquad 310K \leq T \leq 385K \tag{1}$$

The saturated liquid density can be calculated as follows (7), where ρ is in g/L and T is in K.

$$\rho = 12.415 - 0.009548\, T - \frac{65.155}{606.9 - T} \qquad 179K \leq T \leq 400K \tag{2}$$

There is a considerable amount of experimental data for properties of mixtures wherein toluene is a principal constituent. Compilations and bibliographies exist for vapor–liquid equilibrium measurements (9,10), liquid–liquid equilibrium measurements (11), and azeotropic data (12,13).

Chemical Properties

Toluene, an alkylbenzene, has the chemistry typical of each example of this type of compound. However, the typical aromatic ring or alkene reactions are affected by the presence of the other group as a substituent. Except for hydrogenation (qv) and oxidation, the most important reactions involve either electrophilic substitution in the aromatic ring or free-radical substitution on the methyl group. Addition reactions to the double bonds of the ring and disproportionation of two toluene molecules to yield one molecule of benzene and one molecule of xylene also occur.

The aromatic ring has high electron density. As a result of this electron density, toluene behaves as a base, not only in aromatic ring substitution reactions but also in the formation of charge-transfer (π) complexes and in the formation of complexes with super acids. In this regard, toluene is intermediate in reactivity between benzene and the xylenes, as illustrated in Table 2.

Table 1. Physical Properties of Toluene

Property	Value
molecular weight	92.14
melting point, K	178.15
normal boiling point, K	383.75
critical temperature, K	591.80
critical pressure, MPa[a]	4.108
critical volume, L/(g·mol)	0.316
critical compressibility factor	0.264
acentric factor	0.262
flash point, K	278
autoignition temperature, K	809
Gas properties, 298.15 K	
H_f, kJ/mol[b]	50.17
G_f, kJ/mol[b]	122.2
C_p, J/(mol·K)[b]	104.7
H_{vap}, kJ/mol[b]	38.26
H_{comb}, kJ/mol[b]	−3734.
viscosity, mPa·s(=cP)	0.00698
flammability limits, in air[c], vol %	
lower limit at 1 atm	1.2
upper limit at 1 atm	7.1
Liquid properties, 298.15 K	
density, L/mol	9.38
C_p, J/(mol·K)[b]	156.5
viscosity, mPa·s(=cP)	0.548
thermal conductivity, W/(m·K)	0.133
surface tension, mN·m(=dyn/cm)	27.9
Liquid properties, 178.15 K	
density, L/mol	10.49
C_p, J/(mol·K)[b]	135.1
viscosity, mPa·s(=cP)	1.47
thermal conductivity, W/(m·K)	0.162
surface tension, mN·m(=dyn/cm)	42.8
Solid properties	
density at 93.15 K, L/mol	11.18
C_p at 178.15 K, J/(mol·K)[b]	90.0
heat of fusion at 178.15 K, kJ/mol[b]	6.62

[a]To convert MPa to psi, multiply by 145.
[b]To convert J to cal, divide by 4.184.
[c]At 101.3 kPa (1 atm).

In the formation of π-complexes with electrophiles such as silver ion, hydrogen chloride, and tetracyanoethylene, toluene differs from either benzene or the xylenes by a factor of less than two in relative basicity. This difference is small because the complex is formed almost entirely with the π electrons of the aromatic ring; the inductive effect of the methyl group provides only minor enhancement. In contrast, with HF or BF_3 which form a sigma-type complex,

Table 2. Relative Basicity and Reactivity Toward Electrophiles

| Electrophile | Benzene | Toluene | Xylene | | |
			Ortho	*Meta*	*Para*
Ag^{+a}	0.90	1.00	1.08	1.13	0.98
HCl^{b}	0.66	1.00	1.23	1.37	1.09
TCE^{c}	0.54	1.00	1.89	1.62	2.05
HF-BF$_3$^{d}		1.00	200.00	2000.00	100.00
NO$_2^{+e}$	0.045	1.00			
Cl$_2$^{f}	0.003	1.00	13.1	1250.00	6.3

[a] Solubility in aqueous Ag$^+$ (14).
[b] K for Ar + HCl $\rightleftarrows$ Ar·HCl in *n*-heptane at $-78°C$ (15).
[c] K for association with tetracyanoethylene (TCE) in CH$_2$Cl$_2$ (16).
[d] Basicity by competitive protonation (17,18).
[e] CH$_3$COONO$_2$ in (CH$_3$C)$_2$OO at 24°C (19).
[f] Cl$_2$ in CH$_3$COOH at 24°C (20).

or in the case of reaction as with nitronium ion or chlorine where formation of the sigma bonds and complexes plays a significant role, the methyl group participates through hyperconjugation and the relative reactivity of toluene is enhanced by several orders of magnitude compared to that of benzene. Reactivity of xylenes is enhanced again by several orders of magnitude over that of toluene. Thus, when only the π electrons are involved, toluene behaves much like benzene and the xylenes.

When sigma bonds are involved, toluene is a much stronger base than benzene and a much weaker base than the xylenes. The reasons for this difference are readily shown by contrasting the complexes of toluene with hydrogen chloride in the absence and presence of aluminum chloride. In the absence of aluminum chloride, hydrogen chloride is loosely attached to the π-cloud of electrons above and below the plane of the ring. With aluminum chloride present, the electrophilicity is greatly enhanced and a sigma bond is formed with a specific electron pair; resonance structures involving the methyl group contribute to the stabilization.

For attack at either of the two ortho positions or the para position, three such structures can be written.

Hydrogenation Reactions. Hydrogen over a nickel, platinum, or palladium catalyst can partially or totally saturate the aromatic ring. Thermal hyrogenolysis of toluene yields benzene, methane, and biphenyl.

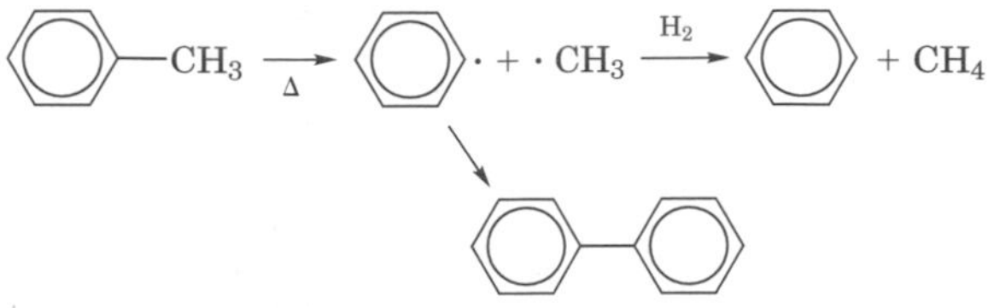

Oxidation Reactions. Although benzene and methane are quite unreactive toward the usual oxidizing agents, the benzene ring renders the methyl group susceptible to oxidation. With oxygen in the liquid phase and particularly in the presence of catalysts, eg, bromine-promoted cobalt and manganese, very good yields of benzoic acid are obtained.

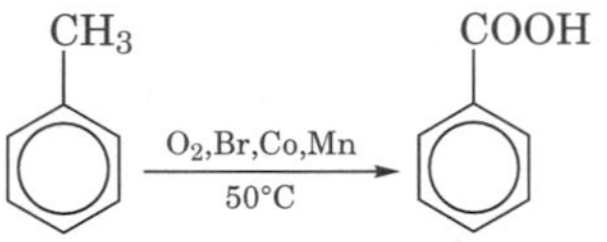

Partial oxidation of toluene yields stilbene:

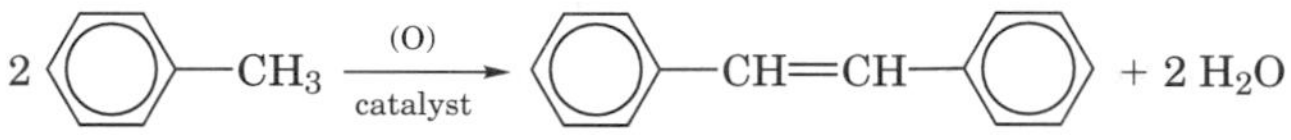

Substitution Reactions on the Methyl Group. The reactions that give substitution on the methyl group are generally high temperature and free-radical reactions. Thus, chlorination at ca 100°C, or in the presence of ultraviolet light and other free-radical initiators, successively gives benzyl chloride, benzal chloride, and benzotrichloride.

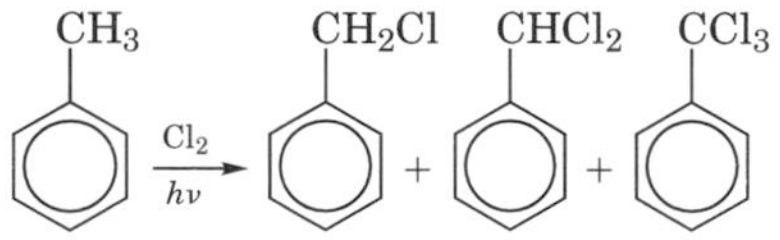

This oxidation reaction which yields benzoic acid is another example of this type of reaction.

In the presence of alkali metals such as potassium and sodium, toluene is alkylated with ethylene on the methyl group to yield, successively, normal propylbenzene, 3-phenylpentane, and 3-ethyl-3-phenylpentane (21).

$$\underset{\text{(toluene)}}{C_6H_5CH_3} \xrightarrow[195^\circ C]{C_2H_4, Na} C_6H_5CH_2CH_2CH_3 \longrightarrow C_6H_5CH(CH_2CH_3)_2 \longrightarrow C_6H_5C(CH_2CH_3)_3$$

These reactions occur on the benzylic hydrogens because these hydrogens are much more reactive. Competition experiments show, for example, that at 40°C a benzylic hydrogen of toluene is 3.3 times as reactive toward bromine atoms as the tertiary hydrogen of an alkane and nearly 100 million times as reactive as a hydrogen of methane.

In the presence of a potassium catalyst dispersed on calcium oxide, toluene reacts with 1,3-butadiene to yield 5-phenyl-2-pentane (22).

$$C_6H_5CH_3 \xrightarrow[100^\circ C]{C_4H_6, K, CaO} C_6H_5CH_2CH_2CH-CHCH_3$$

When lithium is used as a catalyst in conjunction with a chelating compound such as tetramethylethylenediamine (TMEDA), telomers are generally obtained from toluene and ethylene (23), where $n = 0-10$.

$$C_6H_5CH_3 \xrightarrow[110^\circ C]{C_2H_4, Li, TMEDA} C_6H_5CH_2-(C_2H_4)_n-CH_2CH_3$$

The intermediates in these base-catalyzed reactions are believed to be of the nature of a benzyl cation because the reaction product from toluene and propylene is isobutylbenzene, not n-butylbenzene, and the reaction rate is slower than with ethylene (24).

Substitution Reactions on the Aromatic Ring. To predict the location of electrophilic aromatic ring substitutions, the electrophilic reactions can be modeled as proceeding through an intermediate step in which a negative and positive charge are separated on the ring. The most stable intermediates are those in which the positive charge is localized on the carbon containing the methyl group (tertiary carbon). The resonance structures indicate that substitution will occur

$$\left[\underset{+}{\overset{CH_3}{\bigcirc}} \longleftrightarrow \underset{+}{\overset{CH_3}{\bigcirc}} \longleftrightarrow \underset{-}{\overset{CH_3}{\bigcirc}} \longleftrightarrow \overset{CH_3}{\bigcirc} \right]$$

at the ortho and para positions but not the meta position because this position cannot be resonance-stabilized by the carbonium–methyl hyperconjugate structures. The presence of the methyl group is therefore ortho- and para-directing. There is also a steric effect at the ortho position, as shown by the data in Table 3. These data clearly demonstrate that bulky groups cannot enter easily into the position adjacent to the methyl group and therefore attack selectively at the para position.

Substitution of the ring hydrogen atoms by electrophilic attack occurs with all of the same reagents that react with benzene. Some of the common groups with which toluene can be substituted directly are

$$-Cl, \quad -Br, \quad -\overset{O}{\overset{\|}{C}}CH_3, \quad -SO_3H, \quad -NO_2, \quad -C_nH_{2n+1}, \text{ and } -CH_2Cl$$

Typical electrophilic reactions are summarized in Tables 3 and 4. The reactivity ratios in Table 4 show that under the same conditions, toluene reacts more rapidly than benzene and that those reactions that exhibit the highest selectivity to the ortho and para positions also show the most greatly enhanced reactivity relative to benzene. In addition to these reactions, nitration can be performed with HNO_3 in H_2SO_4, sulfonation can be performed with H_2SO_4 and SO_3, alkylation can be performed with RX (X = Cl or Br) with $AlCl_3$, and halogenation can be performed with X_2 (X = Cl or Br) with FeX_3.

The halogenation reaction conditions can be chosen to direct attack to the methyl group (high temperature or light to form free-radicals) or the aromatic ring (dark, cold conditions with FeX_3 present to form electrophilic conditions).

Toluene itself does not undergo substitution by nucleophilic attack of anions but requires substitution by strongly electronegative groups, such as nitro groups, before the ring becomes sufficiently electrophilic to react with anions.

Table 3. Isomer Distributions in the Monoalkylations of Toluene[a], %

Entering group	Ortho	Meta	Para
methyl	53.8	17.3	28.8
ethyl	45	30	25
isopropyl	37.5	29.8	32.7
t-butyl	0	7	93

[a]Ref. 25.

Table 4. Isomer Distribution and Reactivity Ratio for Selected Reactions[a]

Reaction	Conditions	Isomer distribution			Reactivity ratio
		Ortho	Meta	Para	
chlorination	Cl_2 in $HOCCH_3$ at 24°C	58	<1	42	353
chloromethylation	CH_2O in $HOCCH_3$ at 60°C with HCl and $ZnCl_2$	34.7	1.3	64.0	112
nitration	90% $HOCCH_3$ at 45°C	56.5	3.5	40.0	24.5
mercuration	$Hg(OCCH_3)_2$ in $HOCCH_3$ with $HClO_4$ at 25°C	21.0	9.5	69.5	7.9
sulfonylation	CH_3SO_2Cl with $AlCl_3$ at 100°C	49	15	36	
isopropylation	C_3H_6 at 40°C with $AlCl_3$	37.0	28.5	33.9	2.1

[a]Ref. 25.

Addition Reactions to the Aromatic Ring. Additions to the double bonds in the aromatic ring of toluene result from both free-radical and catalytic reactions. Chlorination using free-radical initiators at temperatures <0°C saturates the ring. However, this reaction is not entirely selective, for in addition to saturating the ring to yield hexachlorohexane derivatives, the reaction also effects substitution on the methyl group (26). Hydrogenation with typical hydrogenation catalysts readily yields methylcyclohexane. However, rates for hydrogenation of toluene are only 60–70% of that for benzene (27). The commercial technology used for hydrogenating benzene to cyclohexane (28) can be applied directly to the manufacture of methylcyclohexane. Both of these ring-saturating reactions probably proceed stepwise, but since the initial reaction must overcome the high resonance energy of the aromatic ring, saturation of the second and the third double bond is much more rapid, with the result that partially saturated intermediates are not normally detected (29).

Manufacture and Processing

The principal source of toluene is catalytic reforming of refinery streams. This source accounts for ca 79% of the total toluene produced. An additional 16% is separated from pyrolysis gasoline produced in steam crackers during the manufacture of ethylene (qv) and propylene (qv). Other sources are an additional 1% recovered as a by-product of styrene manufacture and 4% entering the market via separation from coal tars. The reactions taking place in catalytic reforming to yield aromatics are dehydrogenation or aromatization of cyclohexanes, dehydroisomerization of substituted cyclopentanes, and the cyclodehydrogenation of paraffins. The formation of toluene by these reactions is shown.

$$
\begin{array}{c}
\text{CH}_3 \qquad\qquad \text{CH}_3 \\
\text{H}_2 + \bigcirc \rightleftharpoons \bigcirc + 3\,\text{H}_2 \\
\end{array}
$$

$$\text{CH}_3\text{CH}_2\text{CH}_2\text{CH}_2\text{CH}_2\text{CH}_2\text{CH}_3$$

$$
\text{H}_2 + \underset{\text{CH}_3}{\square}{}^{\text{CH}_3}
$$

Of the main reactions, aromatization takes place most readily and proceeds ca 7 times as fast as the dehydroisomerization reaction and ca 20 times as fast as the dehydrocyclization. Hence, feeds richest in cycloparaffins are most easily reformed. Hydrocracking to yield paraffins having a lower boiling point than feedstock proceeds at about the same rate as dehydrocyclization.

In order to obtain pure aromatics, crude reformate is extracted to separate the aromatics from unreacted paraffins and cycloparaffins. The aromatics are, in turn, separated by simple fractional distillation to yield high purity benzene, toluene, xylenes, and C_9 aromatics.

Catalytic reforming, which was introduced primarily to increase octane values for both aviation and automotive fuels, has since become the main source of benzene and xylenes as well as of toluene. Before 1940, both fixed-bed and fluidized-bed units, typically using a 10–15% Mo–Al_2O_3 catalyst or similar catalysts promoted with 0.5–2% cobalt, predominated. Improved operation was obtained in 1940 by the introduction of a 0.3–0.6% Pt–Al_2O_3 catalyst. Since ca 1970, further improvement has been obtained by promoting the Pt–Al_2O_3 catalyst with up to 1% chloride, by using bimetallic catalysts containing 0.3–0.6% of both platinum and rhenium to retard deactivation, and by using molecular sieves as part of the catalyst base to gain activity. Continuous catalytic reforming was introduced ca 1971.

Because catalytic reforming is an endothermic reaction, most reforming units comprise about three reactors with reheat furnaces in between to minimize kinetic and thermodynamic limitations caused by decreasing temperature. There are three basic types of operations, ie, semiregenerative, cyclic, and continuous. In the semiregenerative operation, feedstocks and operating conditions are controlled so that the unit can be maintained on-stream from 6 mo–2 yr before shutdown and catalyst regeneration. In cyclic operation, a swing reactor is employed so that one reactor can be regenerated while the other three are in operation. Regeneration, which may be as frequent as every 24 h, permits continuous operation at high severity. Since ca 1970, continuous units have been used commercially. In this type of operation, the catalyst is continuously withdrawn, regenerated, and fed back to the system. Flow sheets for representatives of each of the three types of processes, ie, Rheniforming (30), Ultraforming (31), and Platforming (32), are shown in Figures 1, 2, and 3, respectively.

The predominant feeds for reforming are straight-run naphthas from crude stills. Naphthas from catalyst crackers and naphthas from code stills are also

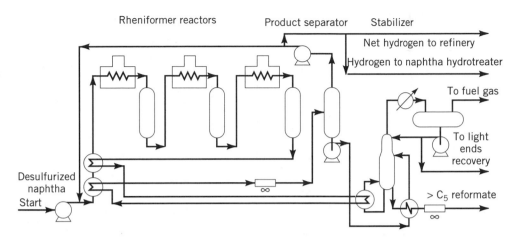

Fig. 1. Chevron Research Co. Rheniforming process (30). Courtesy of Gulf Publishing Co.

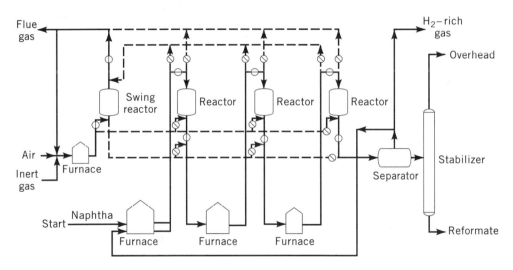

Fig. 2. Standard Oil (In) Co. Ultraforming process (31). Courtesy of Gulf Publishing Co.

used. Typical compositions are summarized in Table 5. Typical operating conditions for catalytic reforming are 1.135–3.548 MPa (150–500 psi), 455–549°C, 0.356–1.069 m³ H₂/L (2000–6000 ft³/bbl) of liquid feed, and a space velocity (wt feed per wt catalyst) of 1–5 h. Operation of reformers at low pressure, high temperature, and low hydrogen recycle rates favors the kinetics and the thermodynamics for aromatics production and reduces operating costs. However, all three of these factors, which tend to increase coking, increase the deactivation rate of the catalyst; therefore, operating conditions are a compromise. More

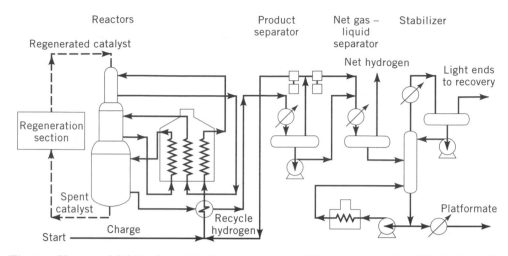

Fig. 3. Universal Oil Products Platforming process (32). Courtesy of Gulf Publishing Co.

Table 5. Composition of Typical 93–204°C Reformer Feeds, Vol %

Source	Paraffins	Cycloparaffins	Aromatics
crude still	40–55	40–30	10–20
catalytic cracker	30–40	15–25	40–50
coking still	50–55	30–35	10–15

detailed treatment of the catalysis and chemistry of catalytic reforming is available (33–35). Typical reformate compositions are shown in Table 6.

   Toluene, Benzene, and BTX Recovery. The composition of aromatics centers on the C_7- and C_8-fraction, depending somewhat on the boiling range of the feedstock used. Most catalytic reformate is used directly in gasoline. That part which is converted to benzene, toluene, and xylenes for commercial sale is separated from the unreacted paraffins and cycloparaffins or naphthenes by

Table 6. Composition of Typical Reformate, Vol %

Component	Value
paraffins	20–30
cycloparaffins	2–3
aromatics	67–77
C_6	2–3
C_7	15–20
C_8	20–28
C_9	15–25
C_{10}	1–10

liquid–liquid extraction or by extractive distillation. It is impossible to separate commercial purity aromatic products from reformates by distillation only because of the presence of azeotropes, although complicated further by the closeness in boiling points of the aromatics, *cyclo*-paraffin, and unreacted C_6-, C_7-, and C_8-paraffins.

Most of the technologies practiced for the recovery of toluene, benzene, and BTX are based on choice of solvent to dissolve the aromatics or nonaromatics in the case of liquid–liquid extraction, or to enhance the relative volatility of the nonaromatics in the case of extractive distillation. UOP and Dow Chemical in the 1950s developed the Udex process, which used glycol-based solvents, ie, ethylene glycol (EG), diethylene glycol (DEG), triethylene glycol (TEG), tetraethylene glycol (TTEG), dipropylene glycol, and diglycoamine, in a combined liquid–liquid extraction and extractive distillation to extract aromatics from wide boiling range reformates. Next, the Shell-developed sulfolane process also marketed by UOP (Fig. 4) increased the aromatic separation efficiency by using the solvent tetrahydrothiophene dioxide (sulfolane) in a combined liquid–liquid extraction and extractive distillation to dissolve the smaller fraction of nonaromatics in the feed mixture.

In the 1960s, the German engineering company Krupp Koppers used *N*-formylmorpholine (NFM) to develop two processes. In the first, morphylex is used to recover all BTX aromatics from a feedstock low in aromatics content. Morphylane, on the other hand, is an extractive distillation process used for the recovery of single aromatics, eg, toluene, benzene, from appropriate feedstocks. Octenar is a modified morphylane extractive distillation process used for recovering aromatics from catalytic reformates. Also in the 1960s, Union Carbide developed the Tetra process, using TTEG solvent (Fig. 5). In 1986, Union Carbide introduced the Carom process. Carom process inherited design improvements

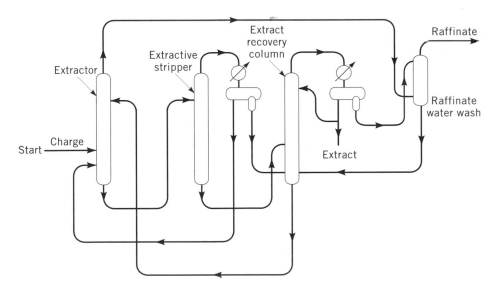

Fig. 4. Shell-UOP's Sulfolane extraction process (35). Courtesy of Gulf Publishing Co.

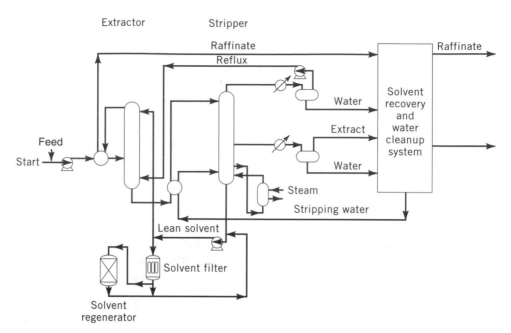

Fig. 5. Union Carbide Corp. Tetra extraction process (36). Courtesy of Gulf Publishing Co.

over Udex and Tetra processes and uses TTEG combined with a proprietary cosolvent that enhances the capacity of the solvent system.

So, Sulfolane and Carom, ca 1997, are two current rival processes. Sulfolane has a slight advantage over Carom in energy consumption, while Carom has 6–8% less capital for the same capacity Sulfolane unit. In 1995, Exxon (37) commercialized the most recent technology for aromatics recovery when it used copolymer hollow-fiber membrane in concentration-driven processes, pervaporation and perstraction, for aromatic–paraffin separation. Once the nonaromatic paraffins and cycloparaffins are removed, fractionation to separate the C_6 to C_9 aromatics is relatively simple.

Proper choice of feedstocks and use of relatively severe operating conditions in the reformers produce streams high enough in toluene to be directly usable for hydrodemethylation to benzene without the need for extraction.

Toluene is recovered from pyrolysis gasoline, usually by mixing the pyrolysis gasoline with reformate and processing the mixture in a typical aromatics extraction unit. Yields of pyrolysis gasoline and the toluene content depend on the feedstock to the steam-cracking unit, as shown in Table 7. Pyrolysis gasoline is hydrotreated to eliminate dienes and styrene before processing to recover aromatics.

Emerging Technologies for the Production of BTX from Light Hydrocarbons. Recent (ca 1997) technological developments have centered on high temperature pyrolysis of light hydrocarbons C_2 to C_5, LPG, and naphtha to form aromatics in higher yields. Conversions were traditionally low because they were accompanied by a high degree of degradation to carbon and hydrogen. Recent

Table 7. Toluene Content of Pyrolysis Gasoline, C$_5$ to 200°C

Feedstock	Wt % to pyrolysis gasoline	Wt % toluene in pyrolysis gasoline
C$_2$–C$_4$ paraffin	5–10	7–15
naphthas	15–21	11–22
gas oils	17–20	13–19

improvements include modification of the thermal cracking process to produce higher yields of liquid products rich in aromatics and the extension of the catalytic hydroforming process to promote oligomerization and dehydrocyclization of the lower olefins. The common core of these developments is the use of shape-selective zeolite catalysts to promote the various reactions. One example is the commercialization of the Alpha process by Asahi Chemical Industry Company in Tokyo, an affiliate of Sanyo Petrochemical Company. The Alpha process uses modified ZSM-5 type zeolite catalyst to convert C$_3$–C$_8$ olefins at 490°C to aromatics at 510°C and 5 kg/cm^2 pressure (see MOLECULAR SIEVES). Selectivity for toluene and xylenes peaks at 550°C but continues with increasing temperature for benzene. The Cyclar process (Fig. 6) developed jointly by BP and UOP uses a spherical, proprietary zeolite catalyst with a nonnoble metallic promoter to convert C$_3$ or C$_4$ paraffins to aromatics. The drawback to the process economics is the production of fuel gas, a low value by-product. BP operated a 1000-bpd demonstration unit in 1989–1991 in its refinery at Grangemouth, Scotland. UOP has agreement with Saudi Basic Industries Corporation (SABIC) to use Cyclar process in an aromatics plant at Yanbu, Saudi Arabia, expected on-stream in mid-1998. Mitsubishi Oil and Chiyoda's Z-forming process (Fig. 7), which has been

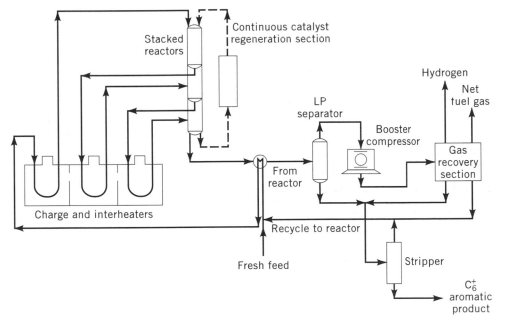

Fig. 6. UOP-BP Cyclar process for LPG aromatization (38). Courtesy of Chem Systems Inc.

proven in a demonstration unit, shut down in December 1991 at Mitsubishi Oil's Kawasaki refinery, uses a metallosilicate zeolite catalyst to promote dehydrogenation of paraffins, followed by oligomerization and dehydrocyclization reactions. Feedstock consists of light naphtha or LPG. The BTX component of the product is mostly toluene.

Mobil Oil has a process (Fig. 8) that uses ZSM-5 zeolite catalyst with palladium and zinc promoters to oligomerize C_2 or C_3 to cyclohexane, which in turn is dehydrogenated to toluene, benzene, and xylenes. Similarly to Mobil's process, the KTI's Pyroform process (Fig. 9) uses a shape-selective zeolite catalyst to convert C_2 and C_3 paraffins to aromatics. The unique feature of this process is the design of proprietary reactor furnace and the operating temperature and pressure profiles. IF and Salute are developing the Aroformer process (Fig. 10) to use C_3-C_5, LPG, and light naphtha feedstocks. Chevron's Aromax process (Fig. 11) is similar to conventional catalytic reforming process, except that its feedstock has high paraffinicity and it has extra sulfur-removal facilities to avoid deactivating its L-type zeolite catalyst, which is very sensitive to sulfur.

Table 8 summarizes the Chem Systems' analysis of the cost of production of BTX from these feeds, resulting in a recommendation of the best-suited technology for each feedstock.

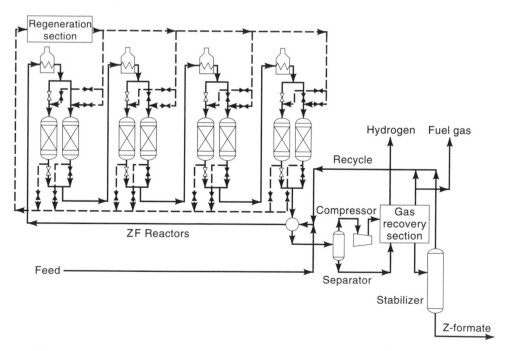

Fig. 7. Z-Forming process flow diagram (38). Courtesy of Chem Systems Inc.

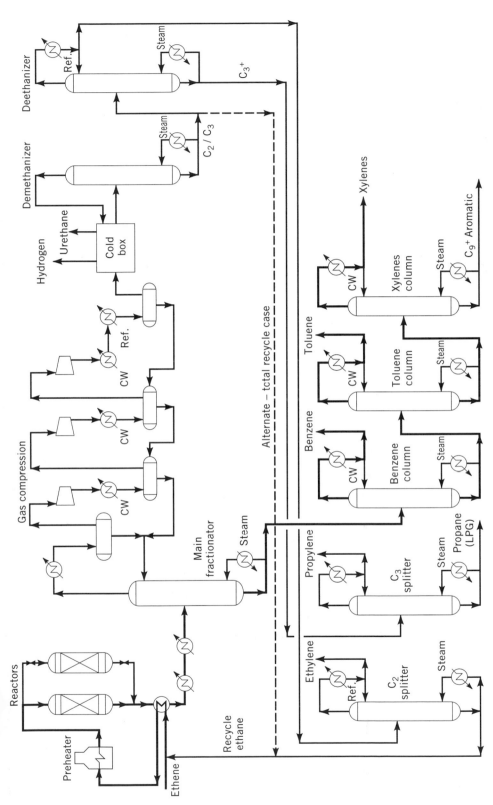

Fig. 8. Mobil's process for aromatics from ethane (38). Ref. = refrigeration; CW = cooling water; Stm. = steam. Courtesy of Chem Systems Inc.

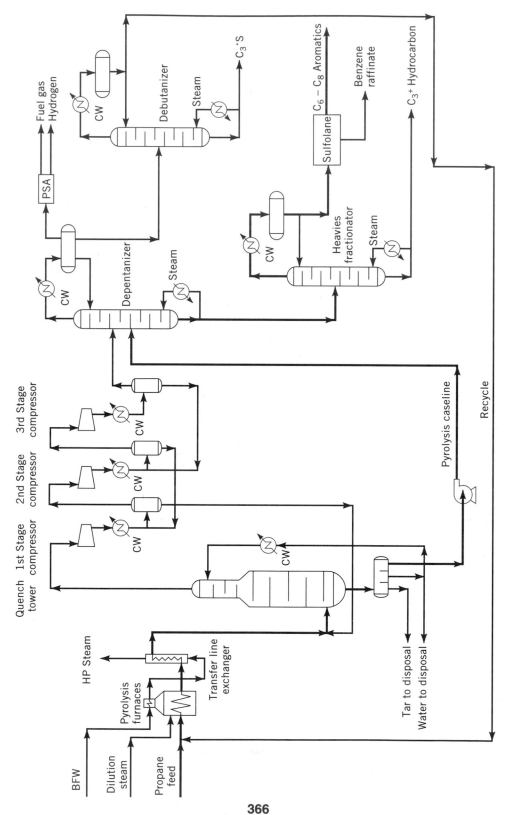

Fig. 9. Pyroform process flow diagram (38). Ref. = refrigeration, CW = cooling water, and BFW = boiling feed water. Courtesy of Chem Systems Inc.

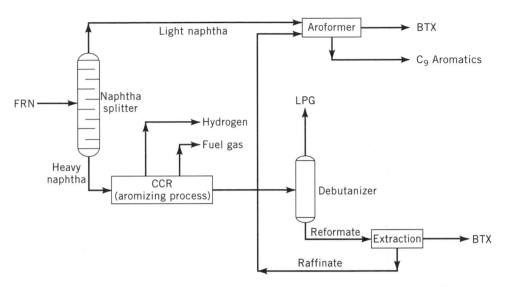

Fig. 10. Aromizing-Aroformer process overall flow plan (38). Courtesy of Chem Systems Inc.

Economic Aspects

Table 9 is a summary of world toluene supply and demand for 1996. North America, Asia, and Western Europe dominated the world's toluene business in 1996. The three regions together accounted for over 85% of world production, imports, exports, and actual consumption, respectively. North America led in production and consumption, while Asia led in imports and exports. Table 10 presents the world toluene supply and demand. The worldwide demand for toluene increased by 7% from 1993 to 1994 and from 1994 to 1995, consecutively, because of higher hydrodealkylation (HDA) and disproportionation (TDP) operations, plus strong demand for all other derivatives. Over 70% of toluene is derived from a single source, catalytic reformate.

Production of toluene in North America went up by 10% in 1993–1994 and 12% in 1994–1995 because of a large (16%) increase in demand for xylenes and all derivatives during the same period (Table 11). Main imports were from Europe, South Korea, and Taiwan, while exports went to South America and Western Europe.

A similar trend was observed in the United States, with toluene supply and demand in 1994 and 1995 up from preceding years 12 and 13%, respectively (Table 12). Historically, most of the toluene produced is used for gasoline blending. Only ca 10–15% is extracted as chemical-grade toluene. Usually about half of the extracted toluene is used for chemical manufacture, while the rest is returned to the gasoline pool. Chemical (nitration) grade toluene consumed in 1994–1995 went to either hydrodealkylation (HDA) operations to produce benzene or to disproportionation and Mobil Selective Toluene Disproportionation (MSTDP) units to produce benzene and xylenes, or to toluene diisocyanate (TDI) and other derivatives. The major demand pull came from para-xylene production

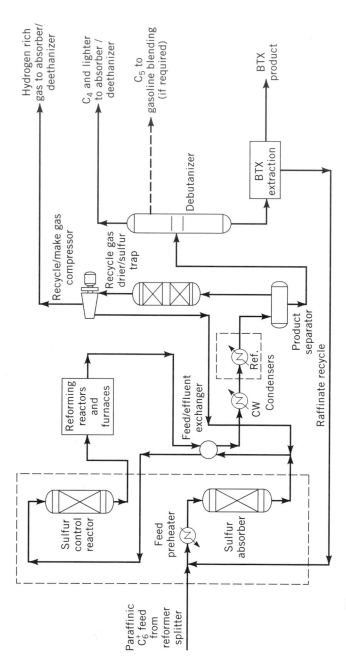

Fig. 11. Chevron's Aromax process for paraffins to BTX (38). Courtesy of Chem Systems Inc.

Table 8. Best-Suited Developing Technology for Various Low Hydrocarbon Feedstocks

Feedstock	Best-suited technology
C_2 paraffin (ethane)	Mobil technology
C_3 paraffin (propane)	KTI's technology (Pyroform process)
C_4 paraffin (butane)	Mitsubishi's Z-Forming process
LPG	BP/UOP's Cyclar technology
light naphtha	Chevron's Aromax

Table 9. World Toluene Supply and Demand Summary[a] for 1996, 10^3 t

Region	Capacity	Production	Imports	Exports	Actual Consumption
Africa	76	58	15	2	71
Asia	4,538	3,504	537	867	3,174
Eastern Europe	1,806	905	59	237	724
Middle East	229	91	84	50	125
North America	6,685	4,927	181	446	4,662
Oceania	28	10	16		26
South and Central America	1,093	664	38	31	651
Western Europe	3,077	2,471	529	650	2,405
World total	*17,532*	*12,609*	*1,458*	*2,282*	*11,837*

[a]Ref. 71. Courtesy of SRI International.

in the MSTDP units, especially in the second quarter of 1995. Despite the high benzene prices, HDA production levels remained steady, while a plentiful supply of pygas made up the supply of benzene. This is the case because the high percentage of benzene in pygas makes it easier to extract benzene from pygas, rather than use other processes. Unlike chemical-grade toluene, there was no noticeable increase in toluene demand for gasoline blending in 1995. The Clean Air Act led to the introduction of Reformulated Gasoline (RFG) in the fall of 1994. Consumption of RFG rose sharply in January 1995, fell slightly overall during the first quarter of 1995, and stayed flat through the remainder of 1995 and all through 1996. Consequently, with MTBE providing most of the octane and oxygen needs for RFG, the historical dependence of gasoline on toluene seems to have declined slightly. Table 13 summarizes U.S. annual toluene capacities by companies.

Toluene demand in 1996 increased because of the new Amoco and Mobil (Chalmette) disproportionation plants as well as other capacity changes at Coastal (Eagle Point), Phillips (Sweeney), Gulf Chemicals (Arochem plant, Puerto Rico), Koch, and Texaco (Huntsman, Port Arthur). Dewitt (71) forecasts continued increase for this application at the rate of about 14% between 1995 and the year 2000. These will have a significant effect on toluene price and

Table 10. World Toluene Supply and Demand[a], 10³ t

	1991	1992	1993	1994	1995	1996	1997
			Supply				
capacity							
reformate	11,536	11,591	11,254	11,613	11,841	12,466	12,704
coal	72	72	72	72	72	72	72
coke-oven light oil	241	241	265	265	255	255	255
pygas	3,117	3,290	3,444	3,299	3,298	3,332	3,621
unspecified raw material	1,258	1,258	1,255	1,276	1,276	1,276	1,276
styrene plant	76	76	76	109	118	118	118
xylenes isom. plant	13	13	13	13	13	13	13
capacity total	*16,313*	*16,541*	*16,379*	*16,647*	*16,873*	*17,532*	*18,059*
operating rate, %	65	64	64	69	73	72	72
production							
reformate	7,507	7,456	7,490	7,984	8,629	8,994	9,275
coal	30	24	24	24	25	25	24
coke-oven light oil	273	261	221	216	229	226	226
pygas	2,466	2,607	2,497	2,923	3,095	3,092	3,186
unspecified raw material	92	73	67	72	75	73	75
styrene plant	166	167	167	201	200	199	199
xylenes isom. plant	9						
production total	*10,543*	*10,589*	*10,465*	*11,419*	*12,252*	*12,609*	*12,984*
import total	*1,660*	*1,521*	*1,573*	*1,551*	*1,499*	*1,458*	*1,251*
supply total	*12,203*	*12,110*	*12,039*	*12,970*	*13,752*	*14,067*	*14,235*
			Demand				
consumption							
benzene	4,878	4,710	4,342	5,047	5,473	5,829	5,897
benzoic acid	68	69	76	92	89	89	91
caprolactam	15	15	15	15	18	18	18
phenol/acetone	212	277	287	307	323	325	335
solvent	1,799	1,773	1,718	1,771	1,834	1,805	1,806
TDI	653	684	699	761	785	824	831
gasoline	954	962	950	941	937	777	777
other	880	896	1,168	1,345	1,446	1,463	1,447
unspecified	986	894	728	627	662	708	764
consumption total	*10,444*	*10,280*	*9,983*	*10,905*	*11,566*	*11,837*	*11,966*
export total	*1,717*	*1,820*	*2,036*	*1,999*	*2,243*	*2,282*	*2,269*
demand total	*12,162*	*12,100*	*12,019*	*12,904*	*13,810*	*14,119*	*14,235*
inventory change/ other	41	10	19	66	−58	−52	
trade balance	57	299	463	448	744	824	1018

[a]Ref. 72. Courtesy of SRI Consulting.

Table 11. Toluene Supply and Demand, North America[a], 10^3 t

	1991	1992	1993	1994	1995	1996	1997
			Supply				
capacity							
reformate	5,952	6,024	5,945	6,163	6,284	6,347	6,495
coke-oven light oil	26	26	26	26	26	26	26
pygas	309	315	299	299	299	299	299
xylenes isom. plant	13	13	13	13	13	13	13
capacity total	*6,300*	*6,378*	*6,283*	*6,501*	*6,622*	*6,685*	*6,833*
operating rate, %	65	65	62	66	72	74	73
production							
reformate	3,766	3,813	3,557	3,792	4,260	4,421	4,509
coke-oven light oil	23	23	22	21	21	20	20
pygas	185	178	174	309	312	322	322
styrene plant	130	130	130	164	164	164	164
xylenes isom. plant	9						
production total	*4,113*	*4,144*	*3,883*	*4,286*	*4,757*	*4,927*	*5,015*
import total	*309*	*266*	*243*	*249*	*261*	*181*	*141*
supply total	*4,422*	*4,410*	*4,126*	*4,535*	*5,018*	*5,108*	*5,156*
			Demand				
consumption							
benzene	2,119	2,129	1,905	2,396	2,763	2,961	3,090
benzoic acid	37	40	45	48	51	51	51
phenol/acetone	73	49	36	39	39	39	39
solvent	581	561	468	478	517	477	457
TDI	250	267	266	294	290	323	333
gasoline	822	822	822	822	822	658	658
other	157	154	155	151	149	153	147
consumption total	*4,039*	*4,022*	*3,697*	*4,228*	*4,631*	*4,662*	*4,775*
export total	*430*	*388*	*429*	*307*	*432*	*446*	*381*
demand total	*4,469*	*4,410*	*4,126*	*4,535*	*5,063*	*5,108*	*5,156*
inventory change/ other	−47				−45		
trade balance	121	122	186	58	171	265	240

[a]Ref. 72. Courtesy of SRI Consulting.

Table 12. Toluene Supply and Demand, United States[a], 10^3 t

	1991	1992	1993	1994	1995	1996	1997
				Supply			
capacity							
reformate	4,960	5,032	4,953	5,171	5,292	5,355	5,503
coke-oven light oil	26	26	26	26	26	26	26
pygas	309	315	299	299	299	299	299
xylenes isom. plant	13	13	13	13	13	13	13
capacity total	*5,308*	*5,386*	*5,291*	*5,509*	*5,630*	*5,693*	*5,841*
operating rate, %	63	63	62	67	74	75	73
production							
reformate	3,017	3,041	2,929	3,175	3,671	3,748	3,781
coke-oven light oil	23	23	22	21	21	20	20
pygas	177	178	174	309	312	322	322
styrene plant plant	130	130	130	164	164	164	164
production total	*3,347*	*3,372*	*3,255*	*3,669*	*4,168*	*4,254*	*4,287*
import total	*247*	*250*	*233*	*224*	*230*	*66*	*66*
supply total	*3,594*	*3,622*	*3,488*	*3,893*	*4,398*	*4,320*	*4,353*
				Demand			
consumption							
benzene	1,737	1,772	1,585	2,016	2,390	2,433	2,466
HDA	1,473	1,443	1,190	1,361	1,315	1,184	986
TDP	264	329	395	655	1,075	1,249	1,480
benzoic acid	36	39	43	46	49	49	49
phenol/acetone	33	36	36	39	39	39	39
solvent	427	427	421	411	411	411	411
TDI	237	253	252	280	276	309	319
gasoline	822	822	822	822	822	658	658
other	105	109	112	115	115	115	115
consumption total	*3,397*	*3,458*	*3,271*	*3,729*	*4,102*	*4,014*	*4,057*
export total	*197*	*164*	*217*	*164*	*296*	*306*	*296*
demand total	*3,594*	*3,622*	*3,488*	*3,893*	*4,398*	*4,320*	*4,353*
inventory change/ other							
trade balance	−50	−86	−16	−60	66	240	230
product price (U.S. cents/kg)	26	28	23	25	24	23	

[a]Ref. 72. Courtesy of SRI Consulting.

Table 13. Toluene Supply and Demand—United States[a], 10³ t

Company	Year-end capacity
Mobil Chemical	378
Oxy Petrochemicals	82
Phillips Chemicals Company	394
Shell Chemical	100
Sun Refining and Marketing	447
Texaco Refining and Marketing	36
Dow Chemical	46
Exxon Chemical	704
Fina Oil and Chemical	171
Hess Oil Virgin Islands	295
Koch Refining	523
Lyondell Petrochemical	254
Marathon Petroleum	66
Amoco Oil	806
Ashland Petroleum	112
Basis Petroleum	46
BP Chemicals America	566
Chevron Chemical	131
Citgo: Corpus Christi	299
Coastal Eagle Point	175
Uno-ven: Lemont	62
Total Country Toluene Capacity	5693

[a]Ref. 72. Courtesy of SRI Consulting.

availability in the later 1990s. On the other hand, toluene demand for gasoline blending is expected to decline by about 283 million liters by 1997–1998.

In summary, beginning in 1993, toluene consumption for TDP has been increasing because of p-xylene demand for DMT/PTA, and in turn for polyester fiber production. This trend is expected to continue in the future and whenever the p-xylene market is stronger than the benzene market, TDP units will operate, and HDA units will be shut down or converted to TDP units.

Tables 14 and 15 show historical U.S. prices for nitration- and commercial-grade toluene, respectively, from 1976 to 1995. The minimum price for the toluene used in chemicals is set by its value in unleaded gasoline, which is the principal use. The ceiling price is set by the relative values of benzene and toluene. When the value of benzene is such that the differential between benzene and toluene exceeds the cost of converting toluene to benzene, then the price of toluene is set by its value for the conversion to benzene. A differential of $91.00/t (ca $0.30/gal) is generally needed to make conversion of toluene to benzene economically attractive.

Specifications, Standards, and Quality Control

Toluene is marketed mostly as nitration and industrial grades. The generally accepted quality standards for the grades are given by ASTM D841 and D362, respectively, which are summarized in Tables 16 and 17, with the appropriate ASTM test method specified for determining the specification properties (40).

Table 14. United States Toluene, Nitration-Grade DCI Historical Prices[a]

Year	Spot, $/Metric Ton		Contract, $/Metric Ton	
	High	Low	High	Low
1995	241.4064	235.1744	248.9760	248.8848
1994	255.6336	247.9120	246.6656	246.2400
1993	233.8672	229.9152	230.9792	230.6752
1992	266.8512	260.6800		
1991	284.7872	275.0288		
1990	363.0064	348.4752		
1989	295.9744	287.2800		
1988	275.4544	268.7968		
1987	239.4608	235.9344		
1986	213.7120	209.4560		
1985	333.7920	328.4720		
1984	311.9952	306.8880		
1983	330.8736	327.3776		
1982	369.9376	363.6752		
1981	402.0096	396.8416		
1980	389.6976	383.7392		
1979	359.0544	345.2832		
1978	176.6240	173.6448		
1977	154.9488	150.8752		
1976	167.0784	162.8832		

[a]Ref. 39. Courtesy of Dewitt & Co., Inc.

Table 15. United States Toluene, Commercial-Grade DCI Historical Prices[a]

Year	Spot, $/Metric Ton		Contract, $/Metric Ton	
	High	Low	High	Low
1995	225.9936	225.0208		
1994	247.6080	245.8144	235.5088	236.1168
1993	229.2160	226.9664	228.9728	228.6384
1992	257.5184	253.9312	233.5632	233.5632
1991	278.5248	272.1712		
1990	348.2624	334.4304		
1989	286.2768	280.5616		
1988	272.9008	264.2976		
1987	231.6480	227.2096		
1986	208.7568	203.6800		
1985	336.0416	330.9040		
1984	309.1680	304.2432		
1983	330.4784	326.3136		
1982	369.9376	363.6752		
1981	402.0096	396.8416		
1980	389.6976	383.7392		
1979	359.0544	345.2832		
1978	176.6240	173.6448		
1977	154.9488	150.8752		
1976	167.0784	162.8832		

[a]Ref. 39. Courtesy of Dewitt & Co., Inc.

Table 16. Specifications for Nitration-Grade Toluene, ASTM D841-80[a]

Property	Specification	ASTM test method
sp gr, 20°/20°C	0.8690–0.8730	D891
color	no darker than 20 (max) on Pt–Co scale	D1209
distillation range at 101.3 kPa	no more than 1°C including 110.6°C for any one sample	D850
paraffins	no more than 1.5 wt %	D851
acid-wash color	no darker than no. 2 color standard	D848
acidity	no free acid, no evidence of acidity	D847
sulfur compound	free of H_sS and SO_2	D853
copper corrosion	copper strip shows no iridescence, gray or black deposit, or discoloration	D849

[a]Ref. 40.

Table 17. Specifications for Industrial-Grade Toluene, ASTM D362-80[a]

Property	Specification	ASTM test method
sp gr, 20°/20°C	0.860–0.874	D891
color	no darker than 20 (max) on Pt–Co scale	D1209
distillation range at 101.3 kPa	no more than 2°C from initial boiling point to dry point, including 110.6°C	D850, D1078
odor	characteristic aromatic hydrocarbon odor as agreed on by buyer and seller	D1296
water	insufficient to show turbidity at 20°C	
acidity	no more than 0.005 wt % (free acid calculated as acetic acid) equivalent to 0.047 mg KOH (0.033 mg NaOH) per g of sample or no free acid; that is, no evidence of acidity	D847
acid-wash color	no darker than no. 4 color standard	D848
sulfur compounds	free of H_2 and SO_2	D853
corrosion 1/2 h at 100°C	copper strip shows no greater discoloration than Class 2 in Method D1616	D1616
solvent power	100 min kauri–butanol value	D1113

[a]Ref. 40.

Although the actual concentration of toluene in samples is not stipulated by these specifications, the purity is in fact controlled by the specific gravity and the boiling-range requirements of the method.

Purity of toluene samples as well as the number, concentration, and identity of other components can be readily determined using standard gas chromatography techniques (40–42). Toluene content of high purity samples can also be accurately measured by freezing point, as outlined in ASTM D1016. Toluene exhibits characteristic uv, ir, nmr, and mass spectra, which are useful in many specific control and analytical problems (2,43–45).

Analytical and Test Methods, Handling, Storage

Tables 16 and 17 list the analytical test methods for different properties of interest. The Manufacturing Chemists' Association, Inc. (MCA) has published the Chemical Safety Data Sheet SD 63, which describes in detail procedures for safe handling of use of toluene (46). The Interstate Commerce Commission classifies toluene as a flammable liquid. Accordingly, it must be packaged in authorized containers, and shipping must comply with ICC regulations. Properties related to safe handling are autoignition temperature, 536°C; explosive limits, 1.27–7.0 vol % in air; and flash point 4.4°C, closed cup.

Health and Safety Factors

Permissible exposure limits established by the U.S. Department of Health and Human Services and the U.S. Department of Labor are summarized below, with the more restrictive levels proposed by NIOSH (47).

	OSHA, mg/m^3 (ppm)	NIOSH, mg/m^3 (ppm)
average during 8-h shift (TWA)	752 (200)	376 (100)
not to exceed	1129 (300)	
except for 10-min average (TLV)	1181 (500)	752 (200)

Toluene generally resembles benzene closely in its toxicological properties; however, it is devoid of benzene's chronic negative effects on blood formation (48). General effects of inhalation are summarized in Table 18. A detailed discussion of physiological response may be found in Reference 48. The odor threshold for toluene has been determined to be ca 9.5 mg/m^3 (2.5 ppm) (49). In the human system, toluene is oxidized to benzoic acid, which in turn reacts with glycine to form hippuric acid, N-benzoylglycine, which is excreted in the urine. However, in animals the development of inflammatory and ulcerous lesions of the penis, prepuce, and scrotum have been demonstrated after inhalation of toluene. This implies that the metabolic processing of toluene produces irritating metabolites.

Table 18. Physiological Response to Inhaled Toluene[a]

Level, mg/L (ppm)	Result
0.38 (100)	transient irritation, psychological effects
0.76 (200)	transitory mild upper respiratory-tract irritation
1.52 (400)	mild eye irritations, lacrimation, hilarity
2.28 (600)	lassitude,, hilarity, slight nausea
3.03 (800)	rapid irritation, nasal secretion, metallic taste, drowsiness, and impaired balance

[a]Ref. 48.

Uses

It is difficult to estimate the total actual production capacity for toluene because it is dependent on the feedstocks used, the number of units operated, and the operating conditions of the units. About 85–90% of toluene produced annually in the United States is not isolated but is blended directly into the gasoline pool as a component of reformate and of pyrolysis gasoline. Capacity exists to isolate ca 6.1×10^6 t $(1.9 \times 10^9$ gal) per year, of which about 75–80% is used for chemicals and solvents or exported. The remainder is blended into gasoline to increase octane number of premium fuels. The largest use of toluene for chemicals (ca 40% in 1994) is in hydrodealkylation (HDA) operations to produce benzene. However, the period 1994–1995 witnessed a significant increase in the use of toluene as a result of major demand pull from *para*-xylene production in the MSTDP units, especially in the second quarter of 1995. This trend continued in 1996 and is expected to change the face of toluene, benzene, *para*-xylene, and mixed xylene industries as new and retrofit disproportionation plants with high solectivity to paraxylene come on stream. Unlike that for chemical-grade toluene, the level of use of toluene for gasoline blending is beginning to decline after the introduction of Reformulated Gasoline (RFG) in the fall of 1994 as a result of the Clean Air Act. Methyl *tert*-butyl ether (MTBE) and other oxygenates are beginning to take over the supply of octane and oxygen requirements in Reformulated Gasoline. However, toluene will still be used in special cases for vapor-pressure fine-tuning.

Automotive Fuels. About 90% of the toluene generated by catalytic reforming is blended into gasoline as a component of $>C_5$ reformate. The octane number $(R + M/2)$ of such reformates is typically in the range of 88.9–94.5, depending on severity of the reforming operation. Toluene itself has a blending octane number of 103–106, which, as shown in Table 19, is exceeded only by oxygenated compounds such as methyl *tert*-butyl ether, ethanol, and methanol.

Toluene is, therefore, a valuable blending component, particularly in unleaded premium gasolines. Although reformates are not extracted solely for the purpose of generating a high octane blending stock, the toluene that is coproduced when xylenes and benzene are extracted for use in chemicals, and that

Table 19. Blending Octane Number, $(R + M)/2^a$, for Selected Components in Unleaded Gasoline

Component	$(R + M)/2$	References
methanol	120, 117	46,47
ethanol	119, 113, 117	46–48
methyl *tert*-butyl ether	108, 106, 111	46–48
tert-butyl alcohol	97.5, 94.5, 96	46–48
toluene	106, 103.5, 102.9	47–49
C_8 aromatics	105.5	49
unleaded regular	88.0	50
unleaded premium	93.0	50

$^a R$ = Research method octane rating, ASTM D2699; M = motor method octane rating, ASTM D2700.

exceeds demands for use in chemicals, has a ready market as a blending component for gasoline.

As a blending component in automotive fuels, toluene has several advantages. First, as shown in Table 19, it has a high octane number compared to regular and premium unleaded gasoline. Second, its relatively low volatility permits incorporation into gasoline blends of other available and less expensive materials, eg, *n*-butane, with relatively high volatility. Since the principal use of toluene is in gasoline, with only ca 9% or less used in chemicals, there will always be an available supply for chemicals manufacture at a price essentially fixed by the value of toluene as a blending component in gasoline.

Manufacture of Benzene. Toluene is converted to benzene by hydrodemethylation either under thermal or catalytic conditions. Benzene produced from this source generally supplies 25–30% of the total benzene demand. Reaction conditions generally range from 600–800°C at 3.55–7.00 MPa (500–1000 psi), and the reaction is exothermic. Conversion per pass is 60–90% with selectivities to benzene >95%. With catalysts, typically supported Cr_2O_3, Mo_2O_3, and CoO, operating temperatures are lower than in the thermal process and selectivities are higher. These gains, however, are offset by the need to decoke the catalyst periodically. Losses to by-product formation, particularly biphenyls, are controlled by recycle of these materials to the reaction zone (50,51). A flow scheme for a typical catalytic process (51) is shown in Figure 12.

The feedstock is usually extracted toluene, but some reformers are operated under sufficiently severe conditions or with selected feedstocks to provide toluene pure enough to be fed directly to the dealkylation unit without extraction. In addition to toluene, xylenes can also be fed to a dealkylation unit to produce benzene. Table 20 lists the producers and their capacities for manufacture of benzene by hydrodealkylation of toluene. Additional information on hydrodealkylation is available in References 50 and 52.

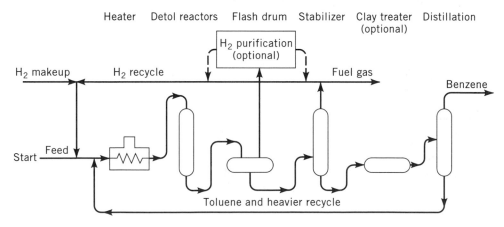

Fig. 12. Air Products and Chemicals toluene dealkylation (Detol) process (51). Courtesy of Gulf Publishing Co.

Table 20. U.S. Producers of Benzene by Hydrodealkylation of Toluene and Their Annual Capacities[a]

Company/Location	1992	1993	1994	1995	1996	1997	1998	1999	2000	Remarks
American Petrofina (Cosden)										
Big Spring	(89)	(89)	(89)	(89)	(89)	(89)	(89)	(89)	(89)	shut down late 1981
Arochem										
Penuelas	(124)	(124)	(124)	(124)	(124)	(124)	(124)	(124)	(124)	
Ashland Chemical										
Catlettsburg	(92)	(92)	(92)	(92)	(92)	(92)	(92)	(92)	(92)	
BP-Standard										
Alliance	157	157	157	157	157	157	157	157	157	
Lima	257	257	257	257	257	257	257	257	257	started up during first quarter of 1985
Chevron										
Philadelphia	(70)	(70)	(70)	(70)	(70)	(70)	(70)	(70)	(70)	
Port Arthur	134	134	134	134	134	134	134	134	134	
Coastal States										
Corpus Christi	261	261	261	261	261	261	261	261	261	
Crown Central Petroleum										
Houston	(60)	(60)	(60)	(60)	(60)	(60)	(60)	(60)	(60)	
Dow Chemical										
Freeport	(134)	0	0	0	0	0	0	0	0	
Plaquemine	401	401	401	401	401	401	401	401	401	

Table 20. (*Continued*)

Company/Location	1992	1993	1994	1995	1996	1997	1998	1999	2000	Remarks
Huntsman										
Bayport	50	50	50	50	50	50	50	50	50	
Koch Chemical										
Corpus Christi	204	204	204	204	204	204	204	204	204	
Oxychem										
Chocolate Bayou	134	134	134	134	134	134	134	134	134	
Corpus Christi	150	150	150	150	150	150	150	150	150	
Phillips										
Guayama	257	257	257	257	0	0	0	0	0	
Quintana Petrochemical										
Corpus Christi	(329)	(329)	(329)	(329)	(329)	(329)	(329)	(329)	(329)	
Shell Chemical										
Odessa	(20)	(20)	(20)	(20)	(20)	(20)	(20)	(20)	(20)	
Sun Petrochemical										
Toledo	(194)	(194)	(194)	(194)	(194)	(194)	(194)	(194)	(194)	
Tulsa	67	(67)	(67)	(67)	(67)	(67)	(67)	(67)	(67)	
Total	*2072*	*2005*	*2005*	*2005*	*1748*	*1748*	*1748*	*1748*	*1748*	

[a]Ref. 39. Courtesy of Dewitt & Co., Inc.

380

Use as Solvent. Toluene is more important as a solvent than either benzene or xylene. Solvent use accounts for ca 14% of the total U.S. toluene demand for chemicals. About two-thirds of the solvent use is in paints and coatings; the remainder is in adhesives, inks, pharmaceuticals, and other formulated products utilizing a solvent carrier. Use of toluene as solvent in surface coatings has been declining, primarily because of various environmental and health regulations. It is being replaced by other solvents, such as esters and ketones, and by changing the product formulation to use either fully solid systems or water-based emulsion systems.

Potential Uses of Toluene

Because much toluene is demethylated for use as benzene, considerable effort has been expended on developing processes in which toluene can be used in place of benzene to make directly from toluene the same products that are derived from benzene. Such processes both save the cost of demethylation and utilize the methyl group already on toluene. Most of this effort has been directed toward manufacture of styrene. An alternative approach is the manufacture of *para*-methylstyrene by selective ethylation of toluene, followed by dehydrogenation. Resins from this monomer are expected to displace polystyrene because of price and performance advantages. Another approach to developing large-scale uses of toluene is to find a reagent that reacts selectively in the para position to yield a derivative readily converted to a carboxylic acid. Such a process would provide a feedstock for manufacture of terephthalic acid and eliminate the need for separation of *para*-xylene, the traditional feedstock, from mixed C_8 aromatics.

Styrene from Toluene. Processes for forming styrene by reaction of methanol with toluene have been reported in both the Japanese and former USSR literature, and in the United States a patent has been issued to Monsanto (53–56). In the latter case, an X-type faujasite aluminosilicate, exchanged with cesium and promoted with either boron or phosphorus, was used as the catalyst. Toluene and methanol at a 5:1 mol ratio react at 400–475°C. About half the methanol is converted to an ethylbenzene–styrene mixture and about half is converted to carbon monoxide and hydrogen. Yields of toluene are very high. Provision must still be made for dehydrogenation of the ethylbenzene in the mixture. The product stream is quite dilute, ca 5%, necessitating large recycles. Because of the generation of carbon monoxide and hydrogen, such a plant would need to operate in conjunction with a methanol synthesis plant, a significant process disadvantage which may possibly be overcome by catalyst development.

Chem Systems Inc. proposed a process in which benzyl alcohol obtained by an undisclosed direct oxidation of toluene is homologated with synthesis gas to yield 2-phenylethyl alcohol, which is then readily dehydrated to styrene (57). This process eliminates the intermediate formation of methanol from synthesis gas but does require the independent production of benzyl alcohol.

A different approach, taken by both Monsanto (58) and Gulf Research and Development Company (59), involved the oxidative coupling of two molecules of toluene to yield stilbene. The stilbene is then subjected to a metathesis reaction with ethylene to yield two molecules of styrene.

$$2 \bigcirc\!\!-CH_3 \xrightarrow[\text{catalyst}]{\text{air}} \bigcirc\!\!-CH{=}CH{-}\!\bigcirc + 2\,H_2O$$

$$\bigcirc\!\!-CH{=}CH{-}\!\bigcirc + CH_2{=}CH_2 \longrightarrow 2 \bigcirc\!\!-CH{=}CH_2$$

A significant problem is the dehydrocoupling reaction, which proceeds only at low yields per pass and is accompanied by rapid deactivation of the catalyst. The metathesis step, although chemically feasible, requires that polar contaminants resulting from partial oxidation be removed so that they will not deactivate the metathesis catalyst. In addition, apparently both *cis*- and *trans*-stilbenes are obtained; consequently, a means of converting the unreactive *cis*-stilbene to the more reactive trans isomer must also be provided, thus complicating the process.

None of these potential toluene processes appears to be simple enough to compete with existing technology.

para-**Methylstyrene.**   Mobil Chemical has a process for the manufacture of *para*-methylstyrene from toluene (60,61). This monomer is produced by alkylating toluene with ethylene, using the Mobil ZSM-5 zeolite catalyst. The alkylation is highly selective, reportedly producing the para isomer with 97% selectivity. Conventional technology, employing a special catalyst to minimize by-products and optimize conversion, is used to dehydrogenate the *para*-ethyltoluene to *para*-methylstyrene.

Vinyltoluene, comprising a mixture of ca 33% *para*- and 67% *meta*-methylstyrene, has been marketed for ca 45 yr by Dow Chemical Company and also by Cosden. However, the performance properties of the polymers prepared from the para isomer are not only superior to those of the polymer prepared from the typical mixed isomers, but are generally superior to those of polystyrene (60). This advantage, coupled with a raw material cost advantage over styrene, suggests that *para*-methylstyrene may displace significant amounts of styrene, currently a 3.2×10^6 t/yr domestic market.

Terephthalic Acid from Toluene.   Both carbon monoxide and methanol can react with toluene to yield intermediates that can be oxidized to terephthalic acid. In work conducted mainly by Mitsubishi Gas Chemical Company (62,63), toluene reacts with carbon monoxide and molar excesses of HF and BF_3 to yield a *para*-tolualdehyde–HF–BF_3 complex. Decomposition of this complex under carefully controlled conditions recovers HF and BF_3 for recycle and *para*-tolualdehyde, which can be oxidized in place of *para*-xylene to yield terephthalic acid. One drawback of the process is the energy-intensive, and therefore high cost, de-complexing step. The need for corrosion-resistant materials for construction and the need for extra design features to handle the relatively hazardous HF and BF_3 also add to the cost. This process can be advantageous where toluene is available and xylenes are in short supply.

A second approach is the selective alkylation of toluene with methanol to yield C_8 aromatic mixtures containing 70–90% *para*-xylene and generally <1% of ethylbenzene (64). Such C_8 aromatic mixtures are excellent feedstocks for recovery of high purity *para*-xylene. The high selectivity to *para*-xylene is achieved by modifying typical HZMS-5 silica aluminate zeolites (65) with phosphorus and boron. To date, ca 1997, this process is not used commercially, probably because current feedstock needs for manufacture of terephthalic acid are met by *para*-xylene from typical reformate, and because conversions of toluene in the process are relatively low (ca 20% per pass) and significant amounts of methanol are converted to by-products such as HCHO, CO, CO_2, CH_4, C_2H_4, and C_3H_6 where stoichiometric quantities of methanol are used. Best results are obtained at 4:1 and higher mole ratios of toluene to methanol. Improved selectivity of the catalyst to permit better utilization of the methanol would enhance the economics of this process.

Derivatives

Toluene Diisocyanate. Toluene diisocyanate is the basic raw material for production of flexible polyurethane foams. It is produced by the reaction sequence shown below, in which toluene is dinitrated, the dinitrotoluene is hydrogenated to yield 2,4-diaminotoluene, and this diamine in turn is treated with phosgene to yield toluene 2,4-diisocyanate.

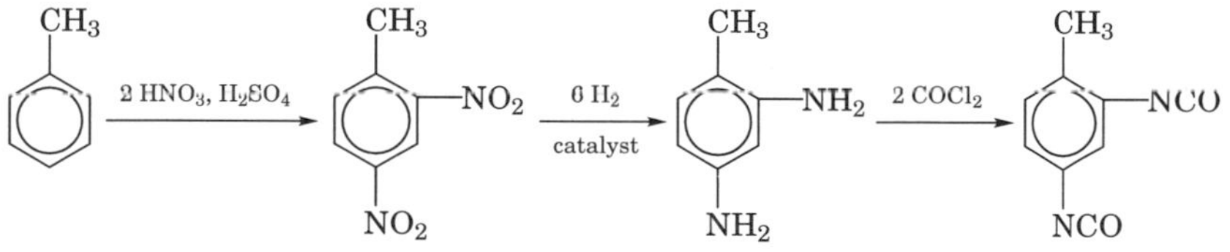

The nitration step produces two isomers, 2,4-dinitrotoluene and 2,6-dinitrotoluene, the former predominating. Mixtures of the two isomers are frequently used, but if single isomers are desired, particularly the 2,4-dinitrotoluene, nitration is stopped at the mono stage and pure *para*-nitrotoluene is obtained by crystallization. Subsequent nitration of this material yields only 2,4-dinitrotoluene for conversion to the diisocyanate.

Polyurethane foams are formed by reaction with glycerol; with poly-(propylene oxide), sometimes capped with poly(ethylene oxide) groups; with a reaction product of trimethylolpropane and propylene oxide; or with other appropriate polyols. A typical reaction sequence is shown below, in which HO–R–OH represents the diol. If a triol is used, a cross-linked product is obtained.

Water, in small amount, reacts with the diisocyanate to generate carbon dioxide, and amine and is used most frequently as the foaming agent. Polyurethanes have been treated in detail in the literature (66–68).

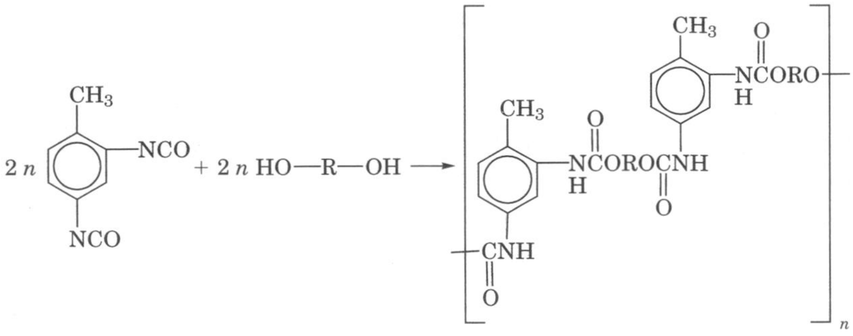

Benzoic Acid. Benzoic acid is manufactured from toluene by oxidation in the liquid phase using air and a cobalt catalyst. Typical conditions are 308–790 kPa (30–100 psi) and 130–160°C. The crude product is purified by distillation, crystallization, or both. Yields are generally >90 mol %, and product purity is generally >99%. Kalama Chemical Company, the largest producer, converts about half of its production to phenol, but most producers consider the most economic process for phenol to be peroxidation of cumene. Other uses of benzoic acid are for the manufacture of benzoyl chloride, of plasticizers such as butyl benzoate, and of sodium benzoate for use in preservatives. In Italy, Snia Viscosa uses benzoic acid as raw material for the production of caprolactam, and subsequently nylon-6, by the sequence shown below.

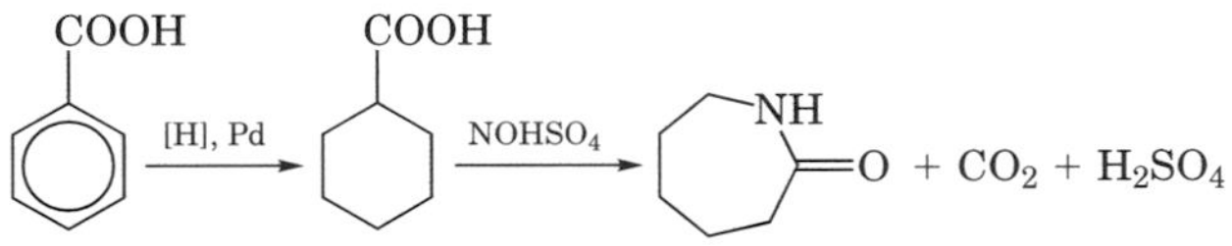

The Henkel process employing potassium benzoate has been used in Japan for the manufacture of terephthalic acid by the following scheme:

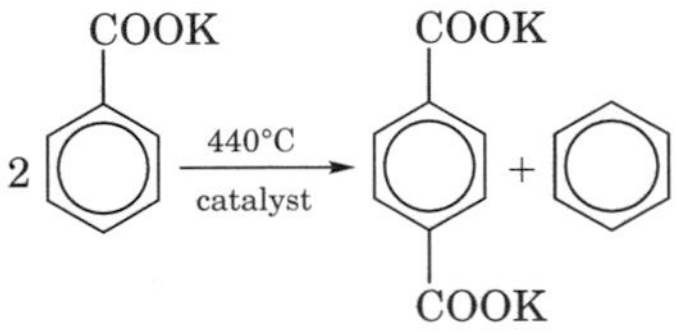

The Henkel process provides a means to convert toluene to benzene and at the same time makes use of the methyl group. Neither of these two processes is economically attractive for use in the United States.

Benzyl Chloride. Benzyl chloride is manufactured by high temperature free-radical chlorination of toluene. The yield of benzyl chloride is maximized by use of excess toluene in the feed. More than half of the benzyl chloride produced is converted by butyl benzyl phthalate by reaction with monosodium butyl phthalate. The remainder is hydrolyzed to benzyl alcohol, which is converted to aliphatic esters for use in soaps, perfume, and flavors. Benzyl salicylate is used as a sunscreen in lotions and creams. By-product benzal chloride can be converted to benzaldehyde, which is also produced directly by oxidation of toluene and as a by-product during formation of benzoic acid. By-product benzotrichloride is not hydrolyzed to make benzoic acid but is allowed to react with benzoic acid to yield benzoyl chloride.

Disproportionation to Benzene and Xylenes. With acidic catalysts, toluene can transfer a methyl group to a second molecule of toluene to yield one molecule of benzene and one molecule of mixed isomers of xylene.

mixed isomers

This disproportionation is an equilibrium reaction for which typical distributions are shown in Table 21. Disproportionation generates benzene from toluene and at the same time takes full advantage of the methyl group to generate a valuable product, ie, xylene. Economic utility of the process is strongly dependent on the relative values of toluene, benzene, and the xylenes. This xylene,

Table 21. Equilibrium Distribution for Toluene Disproportionation[a], Mol %

Temperature, K	Benzene	Toluene	Xylenes
300	31.3	37.4	31.3
400	28.8	42.4	28.8
500	26.5	47.0	26.5
600	23.9	52.2	23.9
700	21.4	57.2	21.4
800	18.7	62.6	18.7
900	16.2	67.6	16.2
1000	13.6	72.8	13.6

[a]Ref. 69.

which contains little or no ethylbenzene, at one time would have commanded a premium as a feed for *para*-xylene units. However, ethylbenzene-free feeds offer little advantage because the zeolite-based isomerization catalysts, eg, the Mobil ZSM-5 and the Amoco AMS-1B molecular sieves, very selectively hydrodeethylate and disproportionate ethylbenzene as they isomerize xylenes. Accordingly, toluene disproportionation processes have little advantage over catalytic reforming for production of xylenes. Two companies, Atlantic Richfield Company and Sun Company, operate disproportionation plants in the United States. Operation of such plants can be justified only where there is an excess of toluene and where both xylenes and benzene are a desired product. The disproportionation plant can then replace a toluene demethylation unit. By proper selection of catalysts, the xylene production can be controlled to give high selectivity to the para isomer; however, in order to accomplish this, catalyst reactivity is greatly diminished (70).

Vinyltoluene.   Vinyltoluene is produced by Dow Chemical Company and is used as a resin modifier in unsaturated polyester resins. Its manufacture is similar to that of styrene; toluene is alkylated with ethylene, and the resulting ethyltoluene is dehydrogenated to yield vinyltoluene. Annual production is in the range of 18,000–23,000 t/yr requiring 20,000–25,000 t ($6–7.5 \times 10^6$ gal) of toluene.

Toluenesulfonic Acid.   Toluene reacts readily with fuming sulfuric acid to yield toluene–sulfonic acid. By proper control of conditions, *para*-toluenesulfonic acid is obtained. The primary use is for conversion, by fusion with NaOH, to *para*-cresol. The resulting high purity *para*-cresol is then alkylated with isobutylene to produce 2,6-di-*tert*-butyl-*para*-cresol (BHT), which is used as an antioxidant in foods, gasoline, and rubber. Mixed cresols can be obtained by alkylation of phenol and by isolation from certain petroleum and coal–tar process streams.

The toluenesulfonic acid prepared as an intermediate in the preparation of *para*-cresol also has a modest use as a catalyst for various esterifications and condensations. Sodium salts of the toluenesulfonic acids are also used in surfactant formulations. Annual use of toluene for sulfonation is ca 100,000–150,000 t ($30–45 \times 10^6$ gal).

Benzaldehyde.   Annual   production   of   benzaldehyde requires   ca 6,500–10,000 t ($2–3 \times 10^6$ gal) of toluene. It is produced mainly as by-product during oxidation of toluene to benzoic acid, but some is produced by hydrolysis of benzal chloride. The main use of benzaldehyde is as a chemical intermediate for production of fine chemicals used for food flavoring, pharmaceuticals, herbicides, and dyestuffs.

Toluenesulfonyl Chloride.   Toluene reacts with chlorosulfonic acid to yield both *ortho*- and *para*-toluenesulfonyl chlorides, of which Monsanto is the only producer. The ortho isomer is converted to saccharin.

The para isomer is used for preparation of specialty chemicals. Annual toluene requirements are ca 6500 t (2×10^6 gal).

Miscellaneous Derivatives. Other derivatives of toluene, none of which is estimated to consume more than ca 3000 t (10^6 gal) of toluene annually, are mono- and dinitrotoluene hydrogenated to amines; benzotrichloride and chlorotoluene, both used as dye intermediates; *tert*-butylbenzoic acid from *tert*-butyltoluene, used as a resin modifier; dodecyltoluene converted to a benzyl quaternary ammonium salt for use as a germicide; and biphenyl, obtained as by-product during demethylation, used in specialty chemicals. Toluene is also used as a denaturant in specially denatured alcohol (SDA) formulas 2-B and 12-A.

ACKNOWLEDGMENT

To R. A. Wilsak and M. E. Carrera (Amoco Chemical Co.) and O. C. Okoroafor (Cooper Union for the Advancement of Science and Arts).

BIBLIOGRAPHY

"Toluene" in *ECT* 1st ed., Vol. 14, pp. 262–273, by M. Lapeyrouse, Esso Research and Engineering Co.; in *ECT* 2nd ed., Vol. 20, pp. 527–565, by H. E. Cier, Esso Research and Engineering Co; in *ECT* 3rd ed., Vol. 23, pp. 246–273, by M. C. Hoff, Amoco Chemicals Corp.

1. Design Institute for Physical Properties Research, *Project 801, Data Compilation*, 1995; *Beilstein Online*, Beilstein Institute for Organic Chemistry, Springer-Verlag, Heidelberg, Germany, 1995.
2. *American Petroleum Institute Research Project 44*, Thermodynamics Research Center, Texas Engineering Experiment Station, Texas A&M University, College Station, Tex., 1976.
3. *Technical Data Book—Petroleum Refining*, 4th ed., Refining Department, American Petroleum Institute, 1983.
4. R. C. Reid, J. M. Prausnitz, and B. E. Poling, *The Properties of Gases and Liquids*, 4th ed., McGraw-Hill Book Co., Inc., New York, 1987.
5. D. T. Jamieson, J. B. Irving, and J. S. Tudhope, *Liquid Thermal Conductivity: A Data Survey to 1973*, National Engineering Laboratory, Edinburgh, Scotland, 1975.
6. D. R. Stull, E. F. Westrum, Jr., and G. C. Sinke, *The Chemical Thermodynamics of Organic Compounds*, Robert E. Krieger Publishing Co., Malabar, Fla., 1987.
7. B. D. Smith and R. Srivastava, *Thermodynamic Data for Pure Compounds: Part A. Hydrocarbons and Ketones*, Elsevier, Amsterdam, the Netherlands, 1986.
8. T. Boublik, V. Fried, and E. Hala, *The Vapour Pressures of Pure Substances*, 2nd ed., Elsevier, Amsterdam, the Netherlands, 1984.
9. H. Knapp, R. Doring, L. Oellrich, U. Plocker, and J. M. Prausnitz, *Vapor–Liquid Equilibria for Mixtures of Low Boiling Substances*, Chemistry Data Series, Vol. VI, Dechema, 1982.
10. I. Wichterle, J. Linek, and E. Hala, *Vapor–Liquid Equilibrium Data Bibliography*, Elsevier Scientific Publishing Co., Amsterdam, the Netherlands, 1973, with Suppls. I, II, III, and IV, 1976, 1979, 1982, and 1985, respectively.
11. J. Wisniak and A. Tamir, *Liquid–Liquid Equilibrium and Extraction*, Elsevier Scientific Publishing Co., Amsterdam, Pt. A, 1980; Pt. B, 1981; Suppl. 1, 1985; Suppl. 2, 1987.

12. L. H. Horsley, *Azeotropic Data-III*, Advances in Chemistry Series, No. 116, American Chemical Society, Washington, D.C., 1973.
13. J. Gmehling, J. Menke, K. Fischer, J. Krafczyk, *Azeotropic Data: Part II*, VCH Publishers, Inc., New York, 1994.
14. L. J. Andrews and R. M. Keefer, *J. Am. Chem. Soc.* **71**, 3644 (1949); **72**, 3113 (1950).
15. H. C. Brown and J. D. Brady, *J. Am. Chem. Soc.* **74**, 3570 (1952).
16. R. E. Merrifield and W. D. Phillips, *J. Am. Chem. Soc.* **80**, 2778 (1958).
17. D. A. McCaulay and A. P. Lien, *J. Am. Chem. Soc.* **73**, 2013 (1951).
18. D. A. McCaulay and co-workers, *Ind. Eng. Chem.* **42**, 2103 (1950).
19. C. K. Ingold and co-workers, *J. Chem. Soc.*, 1959 (1931).
20. F. E. Condon, *J. Am. Chem. Soc.* **70**, 1963 (1948).
21. H. Pines and co-workers, *J. Chem. Soc.* **77**, 554 (1955).
22. G. G. Eberhardt and H. J. Peterson, *J. Org. Chem.* **30**, 82 (1965).
23. G. G. Eberhardt and W. A. Butte, *J. Org. Chem.* **29**, 2928 (1964).
24. H. Pines and L. A. Schaap in *Advances in Catalysis*, Vol. XII, Academic Press, New York, 1960, p. 117.
25. B. T. Brooks and co-workers, *The Chemistry of Petroleum Hydrocarbons*, Van Nostrand Reinhold Company, New York, 1955, Chapt. 56.
26. M. S. Kharasch and M. J. Berkman, *J. Org. Chem.* **6**, 810 (1941).
27. H. Pines, *The Chemistry of Catalysis Hydrocarbon Conversions*, Academic Press, Inc., New York, 1981, pp. 173–174.
28. *Hydrocarbon Process.* **60**(11), 147 (1981).
29. R. T. Morrison and R. N. Boyd, *Organic Chemistry*, 3rd ed., Allyn & Bacon, Inc., Boston, Mass., 1973, p. 323.
30. *Hydrocarbon Process.* **55**(5), 75 (May 1976).
31. R. Coates and co-workers, *Proc. API Div. Refin.* **53**, 251 (1973).
32. E. A. Sutton, A. R. Greenwood, and F. H. Adams, *Oil Gas J.* **70**(21), 52 (May 22, 1972).
33. Ref. 17, pp. 101–110.
34. J. H. Gary and G. E. Handwark, *Petroleum Refining Technology and Economics*, Marcel Dekker, Inc., New York, 1975, pp. 65–85.
35. J. E. Germain, *Catalytic Conversion of Hydrocarbons*, Academic Press, Inc., New York, 1969.
36. D. B. Broughton and G. F. Asselin, *Proc. Second World Pet. Congr.* **4**, 65 (1967).
37. R. C. Schucker, *AIChE National Meeting Topical Conference Preprints*, Vol. 1, pp. 357–379, Miami Beach, Fla., Nov. 12–17, 1995.
38. *1995 PERP Report No. 93-7 on Benzene/Toluene*, Chem Systems Inc., Tarrytown, N.Y., 1995.
39. *1995–96 Toluene–Xylenes Annuals*, Dewitt & Company Inc.; Houston, Tex., Nov. 1995.
40. *1992 Annual Book of ASTM Standards*, Sect. 6, Vol. 06.03, American Society for Testing and Materials, Philadelphia, Pa., 1992.
41. C. L. Stucky, *J. Chromatogr. Sci.* **7**, 177 (1969).
42. H. M. McNaire and E. J. Bonelli, *Basic Gas Chromatography*, Varian Aerograph, Walnut Creek, Calif.; R. R. Freeman, ed., *High Resolution Gas Chromatography*, 2nd ed., Hewlett-Packard Co., Palo Alto, Calif., 1981.
43. *The Sadtler Standard Spectra*, Sadtler Research Laboratories, Philadelphia, Pa., 1971.
44. C. J. Pouchert, *The Aldrich Library of Infrared Spectra*, 2nd ed., Aldrich Chemical Co., 1975.
45. J. G. Grasselli, *Atlas of Spectral Data and Physical Constants for Organic Compounds*, CRC Press, Cleveland, Ohio, 1973.
46. *Chemical Safety Data Sheet SD-63, Toluene*, Manufacturing Chemists' Association, Washington, D.C., 1956.

47. U.S. Department of Health and Human Services, *Occupational Health Guidelines for Chemical Hazards*, DHHS (NIOSH) Pub. No. 81-123, U.S. Government Printing Office, Washington, D.C., Jan. 1981.
48. G. D. Clayton and F. E. Clayton, eds., *Patty's Industrial Hygiene and Toxicology*, 3rd ed., Vol. 2B, John Wiley & Sons, Inc., New York, 1981.
49. C. P. Carpenter and co-workers, *Toxicol. Appl. Pharmacol.* **26**, 473 (1976).
50. K. Weissermel and H. Arpe, *Industrial Organic Chemistry*, Verlag Chemie, New York, 1978, pp. 288–289.
51. *Hydrocarbon Process.* (11), 138 (Nov. 1981).
52. A. L. Waddams, *Chemicals from Petroleum*, 4th ed., John Murray, London, 1978, Chapt. 13.
53. T. Yashima and co-workers, *J. Catal.* **26**, 303 (1972).
54. H. Itoh and co-workers, *J. Catal.* **64**, 284 (1980).
55. USSR Pat. 188,958 (Oct. 11, 1965), (to USSR).
56. U.S. Pat. 4,115,424 (Dec. 22, 1976), M. L. Unland and G. E. Barker (to Monsanto Co.).
57. A. P. Gelbein, *Toluene–Synthesis Gas Based Routes to Styrene, An Assessment*, ACS Petro Chemical Division, New York, Aug. 23–28, 1981.
58. U.S. Pat. 3,965,206 (Dec. 12, 1974), P. D. Montgomery, R. N. Moore, and W. R. Knox (to Monsanto Company).
59. R. A. Innes and H. E. Swift, *Chemtech*, 244 (Apr. 1981).
60. *Chem. Week*, 42 (Feb. 17, 1982).
61. *Chem. Eng. News*, 20 (May 31, 1982).
62. S. Fujiyama and T. Kashara, *Hydrocarbon Process.* **11**, 147 (1978).
63. A. Mitsutani, *Terephthalic Acid from Toluene*, R&D Review Report No. 8, Nippon Chemtec Consulting, Inc., Feb. 1978.
64. W. W. Kaeding and co-workers, *J. Catal.* **67**, 159 (1981).
65. U.S. Pat. 3,702,886 (Oct. 19, 1969), R. J. Arganer and R. G. Landolt (to Mobil Oil Co.).
66. H. J. Saunders and co-workers, *Polyurethanes: Chemistry and Technology*, Pt. 1, John Wiley & Sons, Inc., 1962, pp. 273–314.
67. P. F. Bruins, *Polyurethane Technology*, Wiley-Interscience, New York, 1969, pp. 1–37.
68. E. N. Doyle, *The Development and Use of Polyurethane Products*, McGraw-Hill Book Co., Inc., New York, 1971, pp. 233–255.
69. D. R. Stull and co-workers, *Chemical Thermodynamics of Hydrocarbon Compounds*, John Wiley & Sons, Inc., New York, 1969, p. 368.
70. W. W. Kaeding and co-workers, *J. Catal.* **69**, 392 (1981).
71. *1996–97 Toluene–Xylenes Annuals*, Dewitt & Co., Inc., Houston, Tex., Jan. 1997.

E. DICKSON OZOKWELU
Amoco Chemical Company

TOLUENEDIAMINES. See AMINES, AROMATIC–DIAMINOTOLUENES.

TOOL MATERIALS

Machining of materials using a cutting tool harder than the work material is a common manufacturing operation occurring in the production of a variety of parts. Geometrically defined, single- or multiple-point cutting tools are used to remove the unwanted material from the work material in the form of chips. Generally a numerically controlled (NC) machine tool is employed to provide the required relative motions to produce parts of a given shape, size, and accuracy. A trained operator can produce parts to specifications consistently and economically on a routine basis. Machining processes include turning, drilling, milling, boring, threading, tapping, and broaching. Each operates under a different set of machining conditions. Consequently, the requirements of the tool material differ from one operation to another. The tool materials presented herein are for the most part concerned with cutting operations involving metals and their alloys. Materials used for grinding, polishing, lapping, etc, that use abrasives where the cutting edges are not geometrically defined, ie, have random geometry, and where the geometry changes continuously as the process progresses are outside the scope of this article.

As of the mid-1990s, some estimated 300×10^9/yr was spent on labor and overhead costs alone for machining in the United States (1). This sum does not include the cost of the machine tools and the associated equipment, the cutting tools, the work material, etc. The total cost of the cutting tools used is only a small (ca 1–2%) fraction of this sum and is negligible when compared to the cost of a machine tool. The cutting tool insert, the lowest priced single unit in the machine tool system, however, offers the greatest opportunity for productivity improvement and cost reduction (2).

Machining of metals involves extensive plastic deformation (shear strain of ca 2–8) of the work material in a narrow region ahead of the tool. High tool temperatures (ca 1000°C) and freshly generated, chemically active surfaces (underside of the chip and the machined surface) that interact extensively with the tool material, result in tool wear. There are also high mechanical and thermal stresses (often cyclic) on the tool (3).

Modes of Tool Wear

The performance and life of a cutting tool depend on the cutting conditions as well as the combination of tool material, work material, and the lubricant used. Wear on a tool can be in any one of four areas: crater wear on the rake face, flank wear on the clearance face, flank wear on the nose of the tool, and depth-of-cut line (DCL) notch wear in the machining of certain difficult-to-machine materials such as superalloys using ceramic tools (Fig. 1) (5). In addition, part of the tool, eg, the nose, may be deformed plastically owing to inadequate strength at high operating temperatures. Moreover, cracks may be generated on the tool owing to thermal or mechanical cyclic stresses induced during interrupted cutting. A rapid cratering on the rake face of the tool can result either from high temperatures generated at cutting speeds much higher than recommended ones or from high chemical reactivity between the tool material and the work material. Flank wear

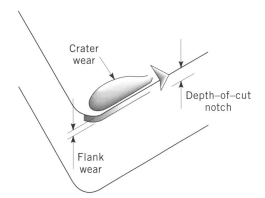

Fig. 1. Schematic showing typical wear modes on a cutting tool.

on the clearance face and on the nose is generally a result of inadequate abrasion resistance of the tool material.

The rapid cratering of the straight cemented tungsten carbide in machining steels in the late 1930s led to the development of cemented carbide tools containing solid solutions of multicarbide material of W, Ti, and Ta (Nb) for machining steels. Whereas Si_3N_4 tool can machine gray cast iron at very high speed (1500 m/min) with very little wear, and that only on the flank face, the same material when machining steels wears rapidly on the rake face, forming a deep crater. Detailed knowledge of the mechanism of wear, in this case chemical interactions between steel and Si_3N_4, can result in significant improvements in productivity. For example, some success in high speed machining of low carbon steels, malleable cast iron, or nodular cast iron using a ceramic tool containing about 70% Al_2O_3 and the remaining Si_3N_4 and minor amounts of sintering aids have been reported (6). This ceramic tool material was described as a mechanical mixture of Al_2O_3 and Si_3N_4 and not a single phase. The microstructure of Si_3N_4, which consists of elongated grains of β-Si_3N_4 that form an interlocking grain structure, is expected to provide additional toughness to this material. Similarly, SiC whisker-reinforced Al_2O_3 was found to be an excellent tool material for machining nickel-base superalloys, but this material wears rapidly when used in machining steels. Reaction of the micrometer-size SiC whiskers with the steel was postulated because in the case of Al_2O_3–Si_3N_4, the tool did not wear when used in machining steels. Thus where Si is present in the form of Si_3N_4 or SiO_2, wear does not occur, but where Si is present in the form of SiC whiskers the tool wears rapidly. Substitution of other whisker materials might then lead to an appropriate tool for machining steels at high cutting speeds.

Tool Materials

A wide range of cutting-tool materials is available. Properties, performance capabilities, and cost vary widely (2,7). Various steels (see STEEL); cast cobalt alloys (see COBALT AND COBALT ALLOYS); cemented, cast, and coated carbides (qv);

ceramics (qv), sintered polycrystalline cubic boron nitride (cBN) (see BORON COM-POUNDS) and sintered polycrystalline diamond; thin diamond coatings on cemented carbides and ceramics; and single-crystal natural diamond (see CARBON) are all used as tool materials. Most tool materials used in the 1990s were developed during the twentieth century. The tool materials of the 1990s will likely become the work materials of the twenty-first century.

The properties affecting performance of a cutting tool in machining a given material and a given cutting process can be described as mechanical, thermal, physical, or chemical. Chemical properties control the chemical interaction between the tool, the work material, and the environment. Mechanical properties control the wear, deformation, and fracture resistance. Thermal properties control the heat partition and thermal shock resistance of the tool. Thus, the hot hardness determines the abrasion resistance as well as hot deformation resistance. Transverse rupture strength (TRS) determines the toughness of the materials and the ability to withstand the loads applied. Thermal conductivity determines how much of the heat generated at the chip–tool interface is conducted into the tool versus how much goes into the chip. The product ($K \times$ TRS/α) of thermal conductivity, K, and TRS over the thermal expansion coefficient, α, is termed the thermal shock parameter and determines the tool's ability to withstand the thermal shock experienced during interrupted cutting. The fracture toughness of the tool determines the impact and fracture-resistance of the tool material. The various properties of cutting tool materials are summarized in Table 1.

The cutting tool is an important component of the machining system. Consequently, tool materials significantly affect machining operation productivity. Other elements include cutting conditions, tool geometry, and the characteristics of the work material, nature of parts produced, machine tool, and support system.

The methodology for tool selection is illustrated in Figure 2 (8). Whereas the selection of a particular class of tool material for a given application is relatively simple, selection of a precise tool grade, shape, geometry, chip groove profile, and size is much more difficult. Many times extensive machining tests are conducted in-house before any implementation on the shop floor. General guidelines for the selection of tool materials for different work materials and different machining operations are given in Tables 2 and 3, respectively.

Measurement of hardness (qv) at room temperature is relatively easy; however, it is the hot hardness at the temperature of cutting that is of importance for tool materials. Figure 3 shows the variation of hot (microindentation) hardness of various tool materials measured at different temperatures. The various suppliers of tool materials can be found in References 11–13 and other trade literature.

Carbon Steels and Low–Medium Alloy Steels.   Plain carbon steels, the most common cutting tool materials of the nineteenth century, were replaced by low–medium alloy steels at the turn of that century because of the need for increased machining productivity in many applications. Low–medium carbon steels have since then been largely superseded by other tool materials, except for some low speed applications.

Table 1. Summary of Properties for Cutting Tool Materials[a]

Parameter	Carbon and low–medium alloy steels	High speed steels	Cast cobalt alloys	Carbides		Ceramics	Polycrystalline	
				Cemented	Coated		cBN	Diamond
hot hardness			increasing →					
toughness			decreasing →					
impact strength			decreasing →					
wear resistance			increasing →					
chipping resistance			decreasing →					
cutting speed			increasing →					
depth of cut	light to medium	light to heavy	light to heavy	light to heavy	light to heavy	light to heavy	light to heavy	very light for single crystal diamond
finish obtainable	rough	rough	rough	good	good	very good	very good	excellent
method of processing	wrought	wrought, cast, HIP sintering	cast and HIP sintering	cold pressing and sintering	CVD[b] PVD[c]	cold pressing and sinter-ing or HIP	high pressure–high tempera-ture sintering	high pressure–high tempera-ture sintering
fabrication	machining and grinding	machining and grinding	grinding	grinding or as molded		grinding	grinding and polishing	grinding and polishing
thermal shock resistance			←— decreasing					
tool material cost			increasing —→					

[a]Overlapping of characteristics exists in many cases. Exceptions to the rule are common. In many classes of tool materials, a wide range of composition and properties are obtainable.
[b]CVD = chemical vapor deposition.
[c]PVD = physical vapor deposition.

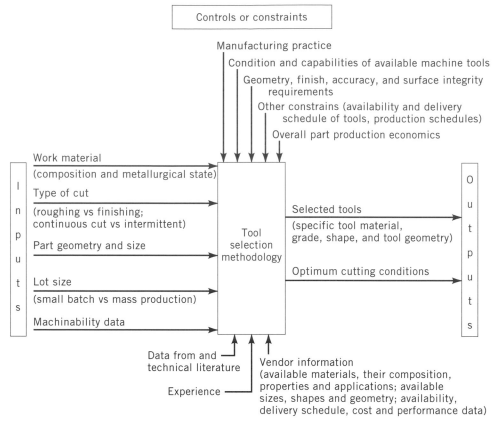

Fig. 2. Methodology for the selection of tool material, grade, shape, size, and geometry, and cutting conditions for a given application (8).

Low–medium alloy steels contain elements such as Mo and Cr for hardenability, and W and Mo for wear resistance (Table 4) (7,16,17) (see STEEL). These alloy steels, however, lose their hardness rapidly when heated above 150–340°C (see Fig. 3). Furthermore, because of the low volume fraction of hard, refractory carbide phase present in these alloys, their abrasion resistance is limited. Hence, low–medium alloy steels are used in relatively inexpensive tools for certain low speed cutting applications where the heat generated is not high enough to reduce their hardness significantly.

Low–medium alloy steels are relatively inexpensive and readily available on short notice or for a short run of parts. They can be heat-treated by simple hardening and tempering using relatively inexpensive equipment. They are easily formed and ground, and are processed in many job shops fabricating their own tools. However, these alloys have the following limitations in addition to low hot hardness (see Fig. 3): low wear resistance, poor hardenability, susceptibility to forming quench cracks and grinding cracks, and poor dimensional stability. Choice of a given grade depends on the tool requirement, availability, cost, and other factors.

Table 2. Guidelines for Tool Materials

Tool materials[a]	Work materials	Machining operation and cutting-speed range	Modes of tool wear or failure[b]	Limitations
carbon steels	low strength, softer materials, nonferrous alloys, plastics	tapping, drilling, reaming; low speed	buildup, plastic deformation, abrasive wear, microchipping	low hot hardness, limited hardenability and wear resistance, low cutting speed, low-strength materials
low–medium alloy steels	low strength–soft materials, nonferrous alloys, plastics	tapping, drilling, reaming; low speed	buildup, plastic deformation, abrasive wear, microchipping	low hot hardness, limited hardenability and wear resistance, low cutting speed, low-strength materials
HSS and TiN-coated HSS	all materials of low–medium strength and hardness	turning, drilling, milling, broaching; medium speed	flank wear, crater wear	low hot hardness, limited hardenability and wear resistance, low to medium cutting speed, low- to medium-strength materials
cemented carbide	all materials up to medium strength and hardness	turning, boring, drilling, milling, broaching; medium speed	flank wear, crater wear, nose wear thermal, cracks, deformation, fracture	not for low speed because of cold welding of chips and microchipping, not suitable for low speed application
coated carbides	cast iron, alloy steels, stainless steels, superalloys	turning; medium to high speed, boring, drilling, milling, threading, grooving, parting	flank wear, crater wear nose wear thermal, cracks, deformation, fracture	not for low speed because of cold welding of chips and microchipping, not for titanium alloys, not for nonferrous alloys since the coated grades do not offer additional benefits over uncoated

Table 2. (Continued)

Tool materials[a]	Work materials	Machining operation and cutting-speed range	Modes of tool wear or failure[b]	Limitations
ceramics	cast iron, Ni-base superalloys, nonferrous alloys, plastics	turning; high speed to very high speed	DCL notching, micro-chipping, gross fracture	low strength and thermo-mechanical fatigue strength, not for low speed operations or interrupted cutting, not for machining Al, Ti alloys
cBN	hardened alloy steels, HSS, Ni-base super-alloys, hardened chill-cast iron, commercially pure nickel	turning, milling; medium to high speed	DCL notching, chipping, oxidation, graphitization	low strength and chemical stability at higher temperature, but high strength, hard materials otherwise
diamond	pure copper, pure aluminum, aluminum-Si alloys, cold-pressed cemented carbides, rock, cement, plastics, glass–epoxy composites, nonferrous alloys, hardened high carbon alloy steels (for burnishing only), fibrous composites	turning, milling; high to very high speed	chipping, oxidation, graphitization	low strength and chemical stability at higher temperature, not for machining low carbon steels, Co, Ni, Ti, Zr

[a]HSS = high speed steel; cBN = cubic boron nitride.
[b]DCL = depth of cut line.

Table 3. Tool Materials for Cutting Operations[a]

Operation	Tool materials[b]	Speed range[c]
single-point turning	low–medium alloy steels, HSS, cemented carbide, coated carbide, ceramics, cBN, diamond	low to very high
drilling	low–medium alloy steels, HSS, solid cemented carbide <2.54 cm	low
tapping	carbon steels, low–medium alloy steels, cemented carbides[d]	low
reaming	HSS, cemented carbides, diamond	low to medium
broaching	HSS, cemented carbide	low to medium
end milling	HSS, solid cemented carbide <2.54 cm, brazed carbides >2.54 cm	low to medium
face milling	HSS, brazed carbides, cemented carbide inserts, diamond, cBN	medium to very high

[a]Refs. 9 and 10. [b]cBN = cubic boron nitride; HSS = high speed steel. [c]Low: 30 m/min; medium: 30–150 m/min; high: 150–300 m/min; very high: 300 m/min. [d]Limited application.

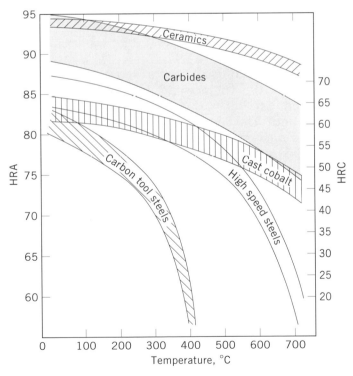

Fig. 3. Hot-hardness of tool materials as a function of temperatures (7). HRA and HRC are Rockwell A and Rockwell C hardness, respectively (see HARDNESS).

Table 4. Compositions of Carbon and Low–Medium Alloy Steels,[a] Wt %[b]

Type[c]	C	Mn	Si	Cr	W	Mo
		Carbon steels[d]				
W1	0.6–1.4					
W2	0.6–1.4[e]					
W3	0.6–1.4			0.5		
		Low–medium alloy steels[f]				
O1	0.9	1.00		0.5	0.5	
O2	0.9	1.60				
O6	1.45		1.00			0.25
O7	1.20			0.75	1.75	0.25

[a]Refs. 8, 14, and 15.
[b]Remainder Fe in all cases.
[c]W = water-hardening grade.
[d]Available in ranges of 0.1 wt % of carbon content.
[e]Also contains 0.25 wt % V.
[f]Cold worked.

High Speed Steels. Toward the latter part of the nineteenth century, a new heat-treatment technique for tool steels was developed in the United States (3,17) that enabled increased metal removal rates and cutting speeds. This material was termed high speed steel (HSS) because it nearly doubled the then maximum cutting speeds of carbon–low alloy steels. Cemented carbides and ceramics have since surpassed the cutting speed capabilities of HSS by 5–15 times.

High speed steels contain significant amounts of W, Mo, Co, V, and Cr in addition to Fe and C (18,19). The presence of these alloying elements strengthens the matrix beyond the tempering temperature, increasing the hot hardness and wear resistance. The materials are readily available at reasonable cost and exhibit the following desirable features: through hardenability; higher hardness than carbon steel and low–medium alloy steels; good wear resistance; high toughness (a feature especially desirable in intermittent cutting); and the ability to alter hardness appropriately by suitable heat treatment. This last facilitates manufacturing complex tools in the soft annealed condition followed by suitable heat treatment for hardening and grinding of tools and cutters to final shape. Associated with the advantages are the following limitations: hardness decreases sharply beyond 540°C, limiting these tools to low speed cutting operations (<30 m/min); wear resistance, chemical stability, and propensity to interact chemically with the chip and the machined surface are limited; and the chips tend to adhere to the tool.

Tool steels are broadly classified as T-type or M-type depending on whether W or Mo is the principal alloying element (Table 5), (Fig. 4), (7,16–19). The original HSS were T-type but concern for the shortage of tungsten, a strategic material, lead to an extensive search for its replacement. Molybdenum was found to serve as an equivalent with additional features in its favor, namely, Mo has half the atomic weight of W, therefore only half as much Mo is required. The two

Table 5. Chemical Composition of High Speed Steels

AISI tool steel type	Chemical composition[a], nominal %						
	C	Cr	V	W	Mo	Co	W_{eq}[b]
Tungsten high speed steel							
T1[c]	0.70	4.0	1.0	18.0			18.0
T2[c]	0.85	4.0	2.0	18.0			18.0
T3	1.00	4.0	3.0	18.0	0.60		19.2
T4	0.75	4.0	1.0	18.0	0.60	5.0	19.2
T5	0.80	4.25	1.0	18.0	0.90	8.0	19.8
T6	0.80	4.25	1.5	20.0	0.90	12.0	21.8
T7	0.80	4.0	2.0	14.0			14.0
T8	0.80	4.0	2.0	14.0	0.90	5.0	15.8
T9	1.20	4.0	4.0	18.0			18.0
T15	1.55	4.50	5.0	12.0	0.60	5.0	13.2
Molybdenum high speed steels							
M1[c]	0.80	4.0	1.00	1.5	8.0		17.5
M2[c]	0.85	4.0	2.00	6.0	5.0		16.0
M3	1.00	4.0	2.75	6.0	5.0		16.0
M4	1.30	4.0	4.00	5.5	4.5		14.5
M6	0.80	4.0	1.50	4.0	5.0	12.0	14.0
M7[c]	1.00	4.0	2.00	1.75	8.75		19.25
M8[d]	0.80	4.0	1.50	5.0	5.0		15.0
M10	0.85	4.0	2.00		8.0		16.0
High hardness (molybdenum base) cobalt high speed steels							
M30	0.85	4.0	1.25	2.0	8.0	5.0	18.0
M34	0.85	4.0	2.00	2.0	8.0	8.0	18.0
M35	0.85	4.0	2.00	6.0	5.0	5.0	16.0
M36	0.85	4.0	2.00	6.0	5.0	8.0	16.0
M41	1.10	4.25	2.00	6.75	3.75	5.0	14.25
M42	1.10	3.75	1.15	1.50	9.50	8.25	20.5
M43	1.20	3.75	1.60	2.75	8.00	8.25	18.75
M44	1.15	4.25	2.00	5.25	6.50	12.00	18.25
M45	1.25	4.25	1.60	8.25	5.0	5.50	18.25
M46	1.25	4.00	3.20	2.00	8.25	8.25	18.0

[a]Normal ranges of manganese, silicon, phosphorus, and sulfur are assumed (see STEEL). The balance is Fe in all cases.
[b]$W_{eq} = 2 (\% Mo) + \% W$.
[c]Widely available.
[d]Also contains 1.25% columbium.

types T- and M- can be used interchangeably because they possess more or less the same properties and have comparable cutting performance. However, M-type steels tend to decarburize more during heat treatment, for which the temperature range is narrow, and hence, care should be exercised during this treatment. In general, M-type tool steels are more popular, representing ca 85% of all tool steels, because they are less expensive by ca 30% than the corresponding T-type steels.

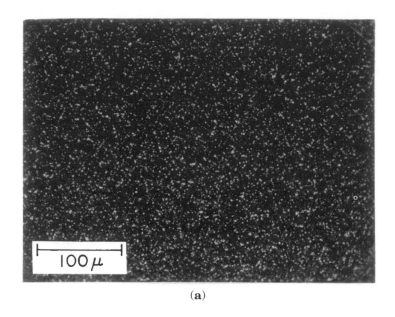

(a)

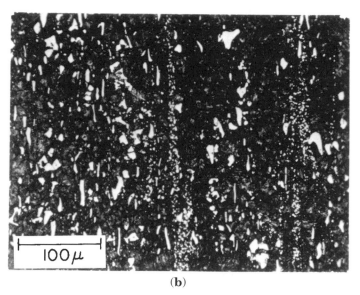

(b)

Fig. 4. Microstructure of AISI T15 tool steel (quenched and tempered) produced (a) from particles and (b) by the conventional technique (picral etch). In (a), the median and maximum carbide sizes are 1.3 and 3.5 mm, respectively; in (b), 6.2 and 34 mm, respectively. Courtesy of Crucible Steel Company.

High speed steel tools are available in cast, wrought, and sintered forms. Improper processing of both cast and wrought products can lead to undesirable microstructure, carbide segregation, formation of large carbide particles, significant variation of carbide size, and nonuniform distribution of carbides in the matrix. Such a material is difficult to shape by grinding and causes wide fluctuations of properties, inconsistent performance, distortion, and cracking.

A processing technique introduced in the late 1960s involves atomization of the prealloyed molten tool steel alloy into fine powder, followed by consolidation under hot isostatic pressure (HIP) (20–23). This technique, termed consolidation by powder metallurgy (CPM), when combined with suitable hardening and tempering, provides a microstructure consisting of a uniform and fine dispersion of carbides in a fine-grained, tempered, martensite matrix. For example, a mean of 1.3 μm and a maximum of 3.5 μm for carbide grain size results from CPM compared to a mean of 6.2 μm and a maximum of 34 μm by conventional cast and wrought processes. Tool steels made in this manner grind more easily, especially the highly alloyed tool steels, with grinding ratios two to three times better; exhibit more uniform properties; and perform more consistently (24). Also, highly alloyed tool steels that can attain HRC 70 cannot be made by the conventional casting or hot forming processes but can be made by CPM. Because of the fine size of the carbides present in tool steels made by CPM, tools made of this material have significant edge strength and provide edge sharpness during cutting, such as in end milling. Consequently, material made by this process is extensively used to produce relatively inexpensive tools, such as drills, milling cutters, and taps, as well as expensive form tools such as broaches, shaper cutters for gears, and various dies for metal forming applications. Tool steels up to HRA of 70 can be obtained using high Co (up to 20%) and high vanadium carbide (VC) (also up to 20%). This would, however, be at the expense of significant loss of toughness. Tool steel technology has matured. Improvements in cleanliness of tool steels, ie, control of the composition of tramp elements, and tighter tolerances on the chemical composition, etc, are underway, as is improvement in the overall quality of the product.

The heat-treatment procedure generally consists of first preheating the HSS tool steel to 730–840°C, then heating rapidly to 1177–1220°C for 2–5 min to fully austenitize the steel, followed by quenching, initially in a suitable molten salt bath to a certain intermediate temperature (ca 600°C), and then cooling in air (16,17,19–23,25). This treatment is followed by single or double tempering, where the steel is heated to 540–590°C for ca 1 h and then air-cooled to produce a tempered martensite structure containing unreacted larger carbides, and to relieve residual stresses.

Shortages and escalating costs of Co in the 1970s prompted tool-steel producers to seek an appropriate substitute. Hot hardness can be maintained without Co by appropriate increases of Mo–W or V content, or both (26). Higher concentrations of these latter elements in the matrix provide equivalent solid–solution strengthening at elevated temperatures. The compositions of steel grades with and without Co, yielding similar performance, are given in Table 6 (26). Micrographs of heat-treated (quenched and tempered) AISI M-42 tool steels with and without Co are shown in Figure 5. Despite heavy competition from

Table 6. Chemical Compositions of Equivalent Grades With and Without Cobalt

HSS Type	Chemical composition,[a] nominal %						
	C	Cr	V	W	Mo	Co	W_{eq}[b]
T 15	1.55	4	5	12.25		5	12.25
Co-less T 15	1.08	4	5	12.5	6.5		25.5
M 42	1.1	3.75	1.1	1.5	9.5	8	20.5
Co-less M 42	1.3	3.75	2.0	6.25	10.5		27.25

[a]Balance is Fe in all cases.
[b]$W_{eq} = 2$ (% Mo) + % W.

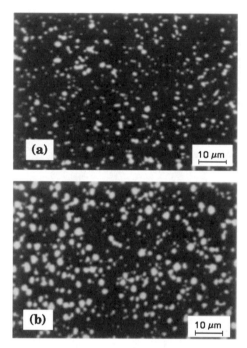

Fig. 5. Micrographs of the microstructure of fully hardened and tempered tool steels produced by the powder metallurgy technique, showing uniform distribution and fine carbide particles in the matrix. (**a**) M-42 (see Table 6) and (**b**) cobalt-free AISI T-15 having a higher concentration of fine carbide particles in the matrix.

cemented carbide, coated carbide, and ceramic tool materials, as of this writing (ca 1997) HSS accounts for the largest tonnage of tool materials used because of its unique properties (chiefly the toughness and the fracture resistance), flexibility in fabrication, and the fact that many cutting operations have to be conducted at a low enough speed range for HSS to perform efficiently and economically.

HSS tools are used mostly for low speed, heavy-duty applications. Thus built-up edge, adhesion of the chips to the tool, and high friction are the primary concerns for these tools rather than high tool temperatures. Consequently, when thin coatings on cemented carbides were developed in the late 1960s, the possible application of such coatings for HSS were considered. In fact, coatings for HSS tools preceded the use of coatings on cemented carbides (27,28). The chemical vapor deposition (CVD) technique used for coatings on cemented carbide requires the tools to be heated to temperatures from 950–1050°C. Application of such temperatures would be unsuitable for HSS tools, altering the metallurgical structure and consequently the properties of the HSS substrate material. Thus a low (450°C) temperature technique, known as physical vapor deposition (PVD) is used to provide a thin (ca 5-μm) coating of TiN on HSS. This coating more or less provides both the needed protection against metal buildup on the tool and low frictional conditions. Consequently, TiN-coated HSS tools are gaining in popularity in large part because of the improvement in tool life from three to ten times when the tool is used in the speed range capability of HSS tools. The substrate material of TiN-coated HSS tools will be affected (softened) when the tool is operated at higher cutting speeds than those recommended. TiN coating also has a golden color that is aesthetically attractive and enables determination of the extent of tool wear. Thus TiN coatings are considered for wear applications where decorative value is important such as for watch cases. Uncoated HSS tools are used widely but upward of 40% of HSS tools are coated and this percentage is expected to increase.

A newer tool steel material having a fine grain size of TiC (40–55%) in a steel matrix (45–60%) with several unique characteristics was developed. Additional Cr (3–17.5%), Mo (0.5–4%), Ni (0.5–12%), Co (5–5.7%), Ti (0.5–0.7%), and C (0.4–0.85%) are made to provide solid solution strengthening as well as hot hardness of the matrix material (29–31). This material, which combines the hardness (consequently, the wear resistance) of cemented carbides with the heat treatability of HSS, responds to heat treatment, such as annealing and quench hardening, and can be machined in the annealed condition. The material is produced by initially compacting TiC powder in a steel die. The resulting porous compact is sintered at high temperature and subsequently infiltrated with molten steel under vacuum. The upper limit of TiC content is determined by the degree of machinability desired in the annealed condition. The microstructure of this material in the annealed condition shows well rounded carbide grains in a spheroidite steel matrix and in the quenched condition in a fine martensite matrix. The relatively wide separation between carbide particles in the annealed condition accounts for its good machinability.

Hardness in the annealed condition of the TiC in a steel matrix material is ca 69 HRA and after heat treatment ca 86.5 HRA. By compacting the TiC powder at higher pressure prior to infiltration, closer spacing and a harder microstructure result. Similarly, higher TiC content also leads to closer spacing. The transverse rupture strength (TRS) is ca 2068 MPa (300 ksi). The modulus of elasticity is 303 GPa (44×10^6 psi). TiC is also chemically more stable when machining steels at medium speeds. In the annealed condition this material is soft enough that it can be machined to shape. After heat treatment, such as austenitizing followed by oil quenching and tempering, tempered martensite structure is formed in the binder phase, resulting in a significant increase in the hardness of this material. The additional feature of this material is that owing to comparable thermal expansion coefficients, it can be easily brazed or welded to the steel substrate without any danger of cracking.

Another HSS tool material, similar to the TiC in a steel matrix, is comprised of 30–60% of submicrometer (ca 0.1 μm) TiN hard phase dispersed in a heat treatable steel (Coronite) (32). It can be seen that the percentage of hard phase, TiN in this alloy is higher than in HSS but less than the lowest limit of cemented carbide. It is thus harder than any conventional HSS, but tougher than most cemented carbides. At the same time, the fine grain size of TiN ensures excellent edge strength especially for milling cutters, drills, etc used in the machining of steels. Because TiN is also chemically more stable when machining steels, this combined material should fill the gap between HSS and cemented carbide. This material can be heat-treated and ground more easily than the cemented TiC counterpart. Tools consist of a steel (HSS or spring steel) core on which the TiN–HSS material is pressed using powder metallurgy technology to comprise about 15% of the diameter. The outer surface can then be coated with TiCN or TiN by PVD.

Cast-Cobalt Alloys.   Cast-cobalt alloys were introduced about the same time as HSS for cutting tool applications. Popularly known as Stellite tools, these materials are Co-rich Cr–W–C cast alloys having properties and applications in the intermediate range between HSS and cemented carbides. Although comparable in room-temperature hardness to HSS tools, cast-cobalt alloy tools retain their hardness to a much higher temperature (see Fig. 2) and hence can be used at higher (25%) cutting speeds than HSS tools. Cast-cobalt alloys contain a primary phase of Co-rich solid solution (instead of Fe in HSS) strengthened by Cr and W, and dispersion-hardened by complex, hard, refractory carbides of W and Cr (33,34). Unlike HSS, cast-cobalt alloys are hard as cast, and cannot be softened or hardened by heat treatment. Cast-cobalt alloys have, however, been phased out owing to the high cost of Co, safety in handling Co-base alloys, and availability difficulties.

Cemented Carbides.   Tungsten carbide was first synthesized in the 1890s, but satisfactory methods for fabricating this material in bulk form with adequate strength in the form of cutting tools, dies, or wear parts were not developed until many years later (35). The main impetus for such a development was to replace the expensive diamond dies used in the manufacture of tungsten wire for lighting. The first cemented-carbide tool material, introduced in Germany in the mid-1920s, was an unalloyed tungsten carbide, WC, in a Co binder (36,37). The material was called Widia for *wie diamente*, like a diamond, and introduced at the Leipzig Trade Fair in 1927 (38). There are some 200 cemented carbide manufacturers worldwide as of the mid-1990s. Prominent among them are Sandvik, Kennametal, Velerite, Krupp Widia, Carboloy, Carmet, Teledyne Firth Sterling, Stellram, Mitsubishi, Sumitomo, Toshiba Tungaloy, and Iscar (12).

Cemented carbides are a class of tool material containing a large-volume fraction ($\geq$90%) of fine-grain, refractory carbides (WC or solid solutions of carbides of W, Ti, and Ta or Nb, and TiC) in a metal binder (Co for the first two types and Ni–Mo for TiC) produced by cold pressing followed by liquid-phase sintering (39–44) (see CARBIDES, INDUSTRIAL HARD CARBIDES). The binder material is chosen so that it wets the carbide to form a good bond thus enabling the carbide to be sintered into a dense mass. This introduces a ductile component into the microstructure, thereby increasing the material toughness. Moreover, by varying the amount of the binder phase, cemented carbides tool materials of different toughness values can be obtained. For metal-forming dies and wear parts the percentage of the binder can be quite high, as much as 25%. In cutting tool applications, the binder is also expected to provide refractoriness and chemical stability at high temperatures. Some tungsten is deliberately allowed to alloy with Co to provide these features, in addition to the solid solution strengthening. Where other properties, such as higher conductivity, or more chemical stability, or higher strength are required, suitable alloying elements can be added to the binder phase (see CARBIDES, CEMENTED CARBIDES; REFRACTORIES).

The function of Co in cemented WC is to act as a medium in which carbide grains can grow together to form a skeletal structure (39), not merely act as a binder. The carbide particles tend to grow together and may actually bond at several locations in the grain, forming a skeletal structure. Thus the continuous phase in cemented WC or multicarbide is the carbide phase which accounts for the high modulus, which is significantly higher (ca three times) than that of HSS.

Cemented carbides differ from HSS in many important respects. They are much harder, chemically more stable, and superior in hot hardness. They are also generally lower in toughness than HSS. They can be used at cutting speeds three to six times higher than HSS. Carbide is the continuous phase in cemented carbides just as the metallic phase (FC) is in HSS. As a result, the Young's modulus, E, of cemented carbide is two to three times that of HSS (414–689 GPa $(60–100 \times 10^6$ psi)). Consequently, cemented carbide is two to three times stiffer than HSS. Furthermore, a specific grade of cemented carbide can be used to machine a specific work material, thus minimizing chemical interaction between the tool and the work material. This is possible in cemented carbides because the chemistry of the primary (carbide) phase can be altered to provide the needed stability. Cemented carbides have a lesser tendency for adhesion, except at low speed and heavy loads, but are more brittle and expensive to fabricate and shape than HSS. A wide range of hard refractory coatings (qv) can be deposited with some reduction in TRS of the substrate in the CVD process. The strategic metals W, Co, and Ta are used extensively in cemented carbides.

Most cemented-carbide tools are WC-based and have Co as the binder. Other carbide tool materials based on TiC having a Ni–Mo binder were developed primarily for high (>300–500 m/min) speed finish machining of steels and gray cast irons for automotive applications.

In the machining of cast iron, cemented straight tungsten carbide tool material exhibits a long tool life even at three to six times the cutting speeds used with HSS. When machining steels, however, cemented WC develops a deep crater on the tool face owing to chemical interactions, thus leading to rapid wear. Improved stability of solid solutions of multicarbides of W–Ti or W–Ti–Ta over a mechanical mixture of WC, TiC, and TaC or unalloyed WC in providing considerable resistance to crater wear when machining steels has been observed (45,46). A unique process for the production of solid solution carbides of two or more refractory carbides has been developed (47–51). Different grades of cemented carbides were obtained by varying the Co content, the amount of different carbides, and the carbide grain size (see Table 7). The higher Co grades or coarser carbide size grades are tougher but less hard; the more complex carbides are harder, and chemically more resistant (especially to steels), but weaker than WC–Co alloys. Production of cemented tungsten carbide is a mature technology. Good control on quality is maintained irrespective of the tool manufacturer.

Figure 6 contains micrographs of representative nonsteel machining grades of cemented carbides (roughing, general-purpose, and finish-machining grades, respectively) containing unalloyed WC with decreasing grain size or Co content. Figure 7 contains micrographs of similar grades for steels containing different amounts of complex multicarbides in a Co binder. Progressing from a roughing to a finishing grade, the hardness increases, toughness decreases, and resistance to high temperature deformation and wear resistance increases. The variation of hardness, transverse rupture strength, impact strength, and elastic modulus with percent Co binder content for straight cemented-WC grades are shown in Figure 8 (52). Hardness and elastic modulus decrease with an increase in Co content; the impact strength and transverse rupture strength increase.

Submicrometer grain size (<1-μm and typically in the range of 0.1–0.5-μm) cemented tungsten carbide tool materials were developed in the late 1960s to increase toughness and edge strength (53,54). Fine dispersions of small amounts (0.5%) of submicrometer chromium carbide restrict grain growth of

Table 7. Composition and Properties of Some Representative Grades of Cemented-Carbide Tools[a]

Grade	Composition, wt %				Grain size	Density, g/cm^3	HRA[b]	TRS[c], MPa[d]
	WC	TiC	TaC	Co				
Nonsteel grades[g]								
roughing	94			6	coarse	15.0	91	2210
general purpose	94			6	medium	15.0	92	2000
finishing	97			3	fine		92.8	1790
Steel grades[h]								
roughing	72	8	11.5	8.5	coarse	12.6	91.1	1720
general purpose	71	12.5	12	4.5	medium	12.0	92.4	690
finishing	64	25.5	4.5	6	medium	9.9	93.0	130

[a]Ref. 8.
[b]Rockwell hardness A scale.
[c]Transverse rupture strength.
[d]To convert MPa to psi, multiply by 145.
[e]To convert GPa to psi, multiply by 145,000.
[f]To convert J to ft·lbf, divide by 1.356.
[g]C-1 to C-4.
[h]C-5 to C-8.

WC in this alloy. Because of the fine grain size the grain boundary area of this material is significantly higher than a similar material having larger grain size. Consequently, higher binder content can be used without sacrificing hardness, but increasing the toughness. The transverse rupture strength (TRS) of the sub-micrometer cemented tungsten carbide can be up to 2757 MPa (400 ksi) which is close to that of HSS. But, the hardness of this carbide is significantly higher, 91.5 R_A, compared to 70 for HSS (53,54).

Sometimes cemented carbide tools are used not only for hardness and wear resistance but also for high modulus or stiffness. For example, in end mills used in high speed machining of aluminum alloys, the deflection of the tool can affect the performance of the tool considerably. This includes chatter or vibrations of the tool, tolerance, and finish requirements. In such circumstances, a solid carbide provides nearly three times the stiffness of HSS as well as providing the wear resistance required at the cutting edge, thus overcoming some of the problems experienced in the shop floor and at the same time increasing the productivity significantly. A similar situation involves long-boring bars used in steel matching.

There are at least four different classification systems for cemented carbides (7,12). The U.S. system is based on relative performance; the U.K. system is based on properties, and the former USSR system on composition; the fourth system, widely used in Europe and supported by the ISO, is based on application and chip form. In this article, the U.S. system and the ISO system are briefly reviewed.

In the United States, the C-classification (C-1 to C-8) for cemented carbide tools, used unofficially for machining applications, was originally developed by the automobile industry to obtain a relative performance index of tools made by different tool producers. This is by far the simplest system. The grades are broadly divided into two classes (C-1 to C-4 and C-5 to C-8), according to the type of work material to be machined. Grades C-1 to C-4 are recommended for

Table 7. (Continued)

				Properties		
Elastic modulus E, GPa[e]	Impact strength, J[f]	Compressive strength, MPa[d]	Tensile strength, MPa[d]	Relative abrasion resistance, vol loss/cm^3	Thermal conductivity, W/(m·K)	Thermal expansion, per (°C^{-1} × 10^{-6})
				*Nonsteel grades[g]*		
640	16	5170	1520	15	120	4.3
650	16	5450	1950	35	100	4.5
610	12	5930	1790	60		4.3
				Steel grades[h]		
560	11	5170		8	50	5.8
570	9	5790		7	35	5.2
460	5	4900	480	5		5.9

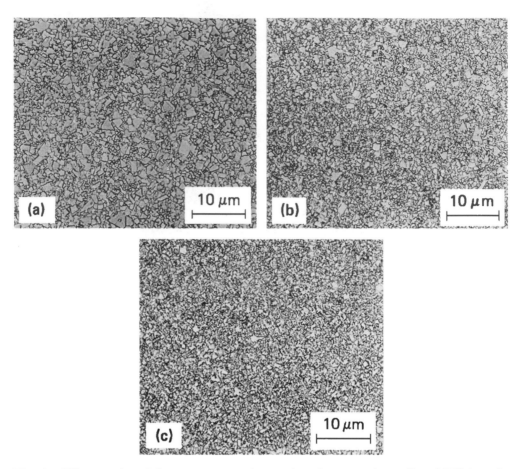

Fig. 6. Micrographs of three representative grades of cemented, unalloyed WC in a Co binder for (**a**) roughing, (**b**) general purpose, and (**c**) finish-machining of materials other than steels (9).

machining nonsteels, ie, cast iron, nonferrous alloys, and nonmetallics (nonsteel workmaterials), whereas C-5 to C-8 are recommended for machining carbon steels and alloy steels. Although the grades to be used for machining other difficult-to-machine materials, eg, the titanium alloys and Ni-base and Co-base superalloys, have not been specified explicitly in this classification, the nonsteel grades C-1 to C-4 are applicable for machining nonferrous alloys. Many users of this classification system are not familiar with this, as they consider grades C-1 to C-4 as cast-iron grades, and not as grades for machining materials other than steels. In general, the nonsteel grades are straight WC in a Co binder, whereas the steel grades are solid solutions of multicarbides in a Co binder.

Within each class, ie, C-1 to C-4 and C-5 to C-8, each grade is distinguished by the type of machining operation: C-1 and C-5 for roughing, C-2 and C-6 for general purpose, C-3 and C-7 for semifinishing, and C-4 and C-8 for precision-finishing operations. In general, from grades C-1 to C-4 or C-5 to C-8 within

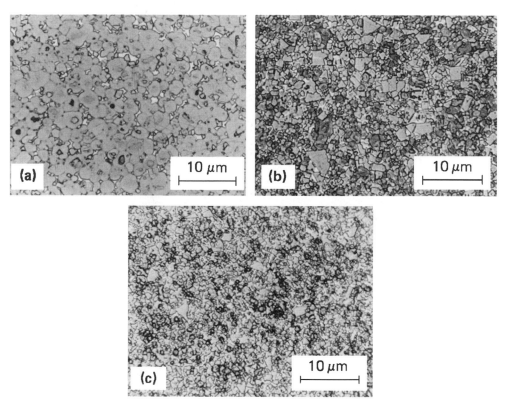

Fig. 7. Micrographs of representative grades of cemented carbides for steels containing different amounts of solid-solution multicarbides in a Co binder (9). (**a**) Roughing, (**b**) general purpose, and (**c**) finish-machining grades.

each class, the shock resistance decreases, hardness increases, high temperature deformation resistance and wear resistance increase, and the Co content and carbide grain size decrease. Roughing and general-purpose grades require more toughness to withstand heavy loads, whereas finishing and semifinishing grades require a high temperature deformation-resistant and a wear-resistant sharp edge. At one time, each tool producer associated one or more carbide grades with the eight grades in the C-classification. However, this comparison involves competition only of grades identified within each class, eg, C-1 or C-5 by different manufacturers, and hence there is a trend to disassociate from this classification. Individual cemented-carbide producers deviate from this rule slightly (in the carbide grain size and Co content) to gain a competitive edge.

The ISO classification for cemented-carbide cutting tools is given in Table 8 (12). Some prior knowledge of machining is expected in consulting this table, which is broadly divided into three categories and, for convenience as well as easy identification, color coded when used on the shop floor. P-grades (blue) are highly alloyed multicarbides used mainly for machining hard steels and steel castings; M-grades (yellow) are low alloy multicarbide alloys which are multipurpose

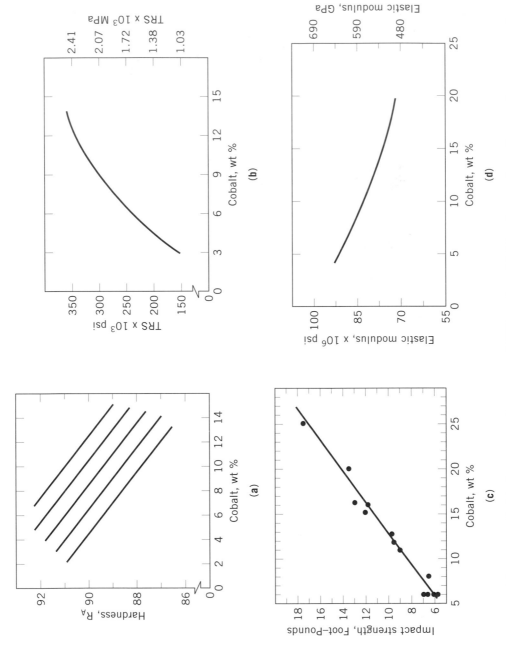

Fig. 8. Variation with percent Co binder content for cemented WC grades of (**a**) hardness, where the numbers represent carbide grain size in μm; (**b**) transverse rupture strength (TRS); (**c**) impact strength; and (**d**) elastic modulus (52).

Table 8. ISO Classification of Cemented Carbide Tools According to Use[a]

Main groups of chip removal		Designa-tion	Groups of application		Increase or decrease in characteristic	
Symbol (color)	Categories of material to be machined		Material to be machined[b]	Use and working conditions	Of cut	Of carbide
P (blue)	ferrous metals with long chips	P 01	steel, steel castings	finish turning and boring; high cutting speeds; small chip section; accuracy of dimensions and fine finish; vibration-free operation		
		P 10	steel, steel castings	turning, copying, threading, milling; high cutting speeds; small or medium chip sections		
		P 20	steel, steel castings, malleable cast iron with long chips	turning, copying, milling; medium cutting speeds; medium chip sections; planing with small chip sections		
		P 30	steel, steel castings, malleable cast iron with long chips	turning, milling, planing; medium or low cutting speeds; medium or large chip sections; machining in unfavorable conditions[b]		
		P 40	steel, steel castings with sand inclusion and cavities	turning, planing, slotting; low cutting speeds; large chip sections; possibility of large cutting angles for machining in unfavorable conditions[b] and work on automatic machines		
		P 50	steel, steel castings of medium or low tensile strength, with sand inclusion and cavities	for operations demanding very tough carbides; turning, planing, slotting; low cutting speeds; large chip sections; possibility of large cutting angles for machining in unfavorable conditions[b] and work on automatic machines		

Increasing speed — Decreasing feed → Wear resistance — Decreasing toughness

411

Table 8. (Continued)

	Main groups of chip removal		Groups of application[b]		Increase or decrease in characteristic	
Symbol (color)	Categories of material to be machined	Designation	Material to be machined	Use and working conditions	Of cut	Of carbide
M (yellow)	ferrous metals with long or short chips and nonferrous metals	M 10	steel, steel castings, manganese steel, gray cast iron, alloy cast iron	turning; medium or high cutting speeds; small or medium chip sections		
		M 20	steel, steel castings, austenitic or manganese steel, gray cast iron	turning, milling; medium cutting speeds; medium chip sections		
		M 30	steel, steel castings, austenitic steel, gray cast iron, high temperature resistant alloys	turning, milling, planing; medium cutting speeds; medium or large chip sections		
		M 40	mild free cutting steel, low tensile steel, nonferrous metals and light alloys	turning, parting off, particularly on automatic machines		
K (red)	ferrous metals with short chips, nonferrous metals, and nonmetallic materials	K 01	very hard gray cast iron, chilled castings of over 85 Shore, high silicon–aluminum alloys, hardened steel, highly abrasive plastics, hard cardboard, ceramics	turning, finish turning, boring, milling, scraping		

Increasing speed → Decreasing feed → Wear resistance → Decreasing toughness

Increasing speed
Decreasing feed
Wear resistance
Decreasing toughness

←

K 10	gray cast iron over 220 Brinell, malleable cast iron with short chips, hardened steel, silicon aluminum alloys, copper alloys, plastics, glass, hard rubber, hard cardboard, porcelain, stone	turning, milling, drilling, boring, broaching, scraping
K 20	gray cast iron up to 220 Brinell, nonferrous metals (copper, brass, aluminum)	turning, milling, planing, boring, broaching, demanding very tough carbide
K 30	low hardness gray cast iron, low tensile steel, compressed wood	turning, milling, planing, slotting, for machining in unfavorable conditions[b] and with the possibility of large cutting angles
K 40	soft wood or hard wood, nonferrous metals	turning, milling, planing, slotting, for machining in unfavorable conditions[b] and with the possibility of large cutting angles

[a]Ref. 12.
[b]Raw material or components in shapes which are awkward to machine: casting or forging skins, variable hardness, variable depth of cut, interrupted cut, work subject to vibrations, etc.

nonsteel grades used for machining high temperature alloys, low strength steels, gray cast iron, free machining steels, and nonferrous metals and their alloys; and K-grades (red) are straight WC grades for machining very hard gray cast iron, chilled castings, nonferrous metals and their alloys, and nonmetallics such as plastics, glass, glass–epoxy composites, hard rubber, and cardboard. Details of the work material to be machined, the type of cutting operation, eg, continuous vs intermittent or roughing vs finishing, and the type of chip formed are included in this classification. There are six basic categories in the P-type, four in the M-type, and five in the K-type. Carbide grades differing from these basic categories can be designated by appropriate in-between numbers within each class, eg, between P 01 and P 10. Thus, the coated grades that do not fall under the basic categories can be placed in these categories. The lower numbers, eg, P 01, M 10, or K 10, are for higher speed, finishing (lighter cut) applications (harder with low Co–finer carbide grain size), and the higher numbers (P 50, M 40, or K 40) are for lower speed, roughing (heavier cut) applications (tougher with higher Co–coarser carbide grain size).

The original objective of the ISO classification was to issue detailed standards for cemented carbides in terms of microstructure, composition, and properties for quality control and performance reliability. This objective, however, is yet to be realized. Increased emphasis on worldwide implementation of ISO 9000 standards and globalization of manufacturing, may lead the industry-at-large to adopt the ISO classification.

In selecting a carbide grade for a given application, the following general guidelines should be followed: the grade with the lowest Co content and the finest grain size consistent with adequate strength to eliminate chipping should be chosen; straight WC grades can be employed if cratering, seizure, or galling is not experienced and for work materials other than steels; to reduce cratering and abrasive wear when machining steels, TiC grades are preferred; and for heavy cuts in steel where high temperature and high pressure deform the cutting edge plastically, a multicarbide grade containing W–Ti–Ta(Nb) with low binder content should be used.

Composition, microstructure, and performance of cemented carbides depend on Co binder content, carbide grain size, and type and composition of various carbides. With increasing Co content, toughness as reflected by the transverse rupture strength (TRS) and impact strength increase, whereas hardness, Young's modulus, and thermal conductivity decrease (see Fig. 8) (52). Finer grain size gives a harder product than coarser grain-size carbides. Multicarbides increase chemical stability and both room-temperature and hot hardness, whereas TiC addition to WC controls crater wear, especially for machining steels. The proper grade of cemented carbide for a given work material should provide adequate crater-wear resistance, abrasion resistance, and toughness to prevent microchipping of the cutting edge.

Another type of carbide tool material, developed by the Ford Motor Company in the late 1950s for high speed (>300–400 m/min) finish machining (low feed) of steels, is based on cemented TiC in a Ni–Mo binder (55–59). According to the ASTM definition, cermet (*ceramic–meta*l) is an acronym to designate a heterogeneous combination of metal(s) or alloy(s) with one or more ceramic phases in which the latter constitutes approximately 15–85% by volume and in which there is relatively little solubility between the metallic and ceramic phases at the preparation temperature. Whereas cemented tungsten carbide also comes under this category, the common practice is to consider only TiC materials as cermets. Actually, TiC-based cermets with a Ni binder were explored in Germany in the 1930s (40,45,46). Because of poor wetting of TiC by Ni, this material was not as strong as cemented WC and hence not as effective as a cutting tool material.

A breakthrough occurred in the development of cermets when in 1956 (60) additions of Mo to TiC–Ni cermet were shown to improve wetting of the carbide by forming a mixed carbide shell (Ti,Mo)C around the TiC grains, thereby inhibiting carbide coalescence and grain growth. The resulting microstructure gives improved hardness and impact resistance. However, this material was not as hard and tough as the cemented multicarbide counterpart. Because of the wide popularity enjoyed by cemented carbide tools and the initial rather negative reputation gained by cemented TiC for being relatively brittle and easy to chip or fracture, there was a reluctance to change from the cemented WC tools to cemented TiC tools.

Further advancement of this cermet is based on the following: (1) improvements through the additions of other carbides, such as MoC, TaC, and WC as well as Co binder; (2) significant additions of TiN to TiC either as separate phases or as titanium carbonitride, TiCN, resulted in TiC–TiN cermets being widely used commercially, especially in Japan; (3) modification of the composition of the cermet by adding Al to the alloy, which precipitates fine Ni_3Al particles in the binder phase for improving the elevated temperature strength, similar to the strengthening effect found in nickel-base superalloys; and (4) TiN coating, preferably by physical vapor deposition (PVD), on the cermet tools. With these additions the hot hardness, transverse rupture strength (TRS), oxidation resistance, and thermal conductivity are significantly increased. Table 9 shows a comparison between the original TiC–Ni–Mo cermet and the cermet containing TiN, WC, TaC, and Co (61). The higher TRS provides better edge strength and chipping resistance, whereas higher thermal conductivity provides thermal shock resistance, both of which have limited the application of this material for a long time.

Compositions and properties of C-5 to C-8 steel grades of cemented TiC are given in Table 10. The TRS of cemented TiC is higher than that of most ceramics, but lower than that of cemented WC. Furthermore, the Young's modulus of

Table 9. A Comparison of Properties of Cermets

Composition of cermet	HV,[a] at 1000°C	TRS, 900°C, N/mm²	Oxidation resistance wt gain after 1 h at 1000°C, mg/cm²	Thermal conductivity at 1000°C, W/(K·m)
TiC–16.5% Ni–9% Mo	500	1050	11.8	24.7
TiC–20% TiN–15% WC– 10% TaC–5.5% Ni– 11% Co–9% Mo	650	1360	1.66	42.3

[a]HV = Vicker's hardness.

Table 10. Composition and Properties of Steel Grades of Cemented Titanium Carbide[a]

Grade[b]	Composition, wt %			Properties			
	TiC	Ni	Mo	HRA[c]	TRS, MPa[d]	Young's modulus, GPa[e]	Density, g/cm³
roughing	67–69	22	9–11	91	1900	413	5.8
general purpose	72–74	17	9–11	92	1620	431	5.6
finishing	77–79	12	9–11	92.8	1380	440	5.5

[a]Ref. 9.
[b]C-5 to C-8.
[c]Rockwell hardness A scale.
[d]To convert MPa to psi, multiply by 145.
[e]To convert GPa to psi, multiply by 145,000.

cemented TiC, although double that of HSS, is ca 25% less than that of cemented WC. Like the cemented WC tools, cemented TiC is processed by cold pressing followed by liquid-phase vacuum sintering. Finishing steel-grade cemented TiC was originally used for high speed finish turning and boring of steels. The improved grades are used in the high speed milling of steels and malleable cast irons. Coatings on cemented tungsten carbide are used extensively; those on cemented titanium carbide have begun to be used more recently (62). Refractory, hard fiber-reinforced, and multilayer-coated cemented TiC are expected to be available for a range of steel and cast iron machining applications in the future.

The TiC–TiN cermets or titanium carbonitride cermets, especially those having coatings, cover the machining range between cemented carbides at one end and ceramics on the other. These have attained a significant degree of prominence in Japan, accounting for some 30% of the cutting tool market, mostly for high speed finishing (light cuts) of near net-shaped parts (63). This success is mainly the result of their high hot hardness (and consequently high deformation resistance), high wear resistance (both flank and crater wear), high resistance to metal buildup, and higher thermal conductivity and thermal shock resistance, compared to cemented TiC (64). Consequently, improved surface finish on the part, consistent part tolerances, high speed capability, and longer tool life result. Different grades of TiC–TiN were developed to cover roughing (ca 12% binder), general purpose (ca 10% binder), and finishing (ca 9% binder). In addition, coated grades having high binder content (ca 18% binder) were developed for interrupted cutting, such as milling. In Europe, the use of TiC–TiN cermets is growing steadily (ca 5%). In the U.S., however, cermets are a niche market as in the auto industry where productivity gains, superior part finish, and tolerance are critical and specialized high-speed, high-power precision machine tools are available.

Although cemented-carbide tools can be brazed, most of the carbide tools are available in insert form such as squares, triangles, diamonds, and rounds. These can be easily clamped on to the tool shank, thereby avoiding the problems and complexities associated with brazing. It is this feature that widely extended the applications of cemented carbide tools. Many cemented carbide grades are relatively less strong compared to HSS, especially when machining high strength materials. Thus these grades are used in such a manner that a square insert cuts with all its eight corners successively.

Chip breaking and disposal during cutting was not as serious a problem when manually operated machine tools were used. The clamped chip breakers, which were adequate to control the chip flow, are not acceptable for numerically controlled (NC) machine tools. Thus built-in (mould-in) chip grooves of various shapes and complexities have been developed for various applications, such as different work materials, different machining processes, and for roughing, semifinishing, and finishing. A myriad of chip groove geometries are available. These built-in chip breakers not only break chips but also facilitate chip disposal during machining. They also reduce the forces involved in machining, thereby improving the efficiency of cutting.

Cemented carbides are not generally recommended for low speed cutting operations because the chips tend to weld to the tool face and cause microchipping and there is no economic incentive to use them at lower speeds. However, for applications requiring higher stiffness, and higher wear resistance, such as broaches and shaper cutters, they are used extensively at lower speeds. Thin coatings of TiN made this application even more attractive. Cemented carbides are especially effective at higher speeds, generally in the 45–180 m/min range. This speed can be much higher (>300 m/min) for materials that are easier to machine, eg, Al alloys, and much lower (ca 30 m/min) for materials more difficult to machine, eg, Ti alloys. In interrupted cutting applications, edge chipping is prevented by appropriate choice of cutter geometry and cutter position with respect to the workpiece in such a way as to transfer the point of application of the load away from the tool tip. Finer grain size and higher Co content improve toughness in straight WC–Co grades and are considered desirable in materials used for interrupted cutting. Because of the high hardness of cemented carbides, they can be finished only by diamond grinding. Abusive grinding can lead to thermal cracks and poor performance.

To conserve the strategic materials (W, Co, and Ta) and reduce costs, recycling of used cemented-carbide inserts (so-called disposable or throwaway inserts) is growing steadily (65). Cobalt can be removed either by chemical leaching or by heating to high temperature (ca 1700–1800°C) in a vacuum to vaporize some of the Co and embrittle the rest of the material, leaving the carbide particles intact. The mass is then pulverized and screened to produce a fine powder. Other separation techniques include the zinc reclaim process commercialized in the 1970s (43). Alternatively, the cemented carbide inserts can be reground for applications where the actual size of the insert is not of critical concern. Several commercial fabricators provide regrinding services on a regular basis. In the extreme case, to conserve these materials economically, new techniques could be developed wherein the cemented carbide is used only at and near the cutting tips (65).

Coated Tools

The difficulty of machining many advanced materials presented challenges to the cutting tool industry, leading to the introduction of coated cemented carbides in the late 1960s (66–71) and coated HSS in the late 1970s. The technique commonly used for coated cemented carbides, chemical vapor deposition (CVD), requires that the tools be heated to ca 950–1050°C. At these temperatures, the metallurgical structure of HSS can be altered significantly. Thus, only coatings requiring substrate heating below the HSS transformation temperature (ca 450–500°C) can be applied for HSS. This is accomplished by a thin coating of TiN on HSS using physical vapor deposition (PVD) processes which are operated at lower temperatures (ca 400–450°C). Because most HSS tools are used for low speed applications, material buildup and friction on the tool face are the main considerations. TiN coating provides an acceptable solution to these problems because of its significantly higher hardness than HSS. Although rapid advances in coated cemented-carbide technology have taken place, coating technology for HSS is still limited to coating of TiN by PVD.

An analysis of the cutting process indicates that the material requirements at or near the surface of the tool are different from those of the tool body. The clearance surface, to be abrasion-resistant, has to be hard, and the rake face, to prevent chemical interaction, has to be chemically inert. The weakest link in the case of cemented carbide is the Co binder which is a soft metal and has a lower melting temperature than the carbide. A thin, chemically stable, hard, refractory binderless coating often satisfies these requirements. If the coating is too thick, it exhibits its bulk qualities, principally the brittleness. The tool body, by contrast, should have adequate deformation resistance to withstand high temperature plastic deformation of the nose and the body of the cutting tool under the conditions of cutting. These requirements are somewhat conflicting. Coated tool design is thus considered in terms of engineered materials. This methodology is also termed surface engineering. A thin (ca 5-μm) coating of TiC was developed for cemented-carbide tools in the mid-1960s. Patents were issued in the early 1970s (72–82). Over 200 U.S. patents have been issued to various tool manufacturers worldwide on the development of coatings on cutting tools as of the mid-1990s.

An effective coating should be hard; refractory; chemically stable; chemically inert to shield the constituents of the tool and the work-material from interacting chemically under the conditions of cutting; binder free; of fine grain size with no porosity; metallurgically bonded to the substrate with a graded interface to match the properties of the coating and the substrate; thick enough to prolong tool life but thin enough to prevent brittleness; free of the tendency of metal chips to adhere to or seize to the tool face; able to provide residual compressive stress; easy to deposit in bulk quantities; and inexpensive. In addition, coatings should have low friction and exhibit no detrimental effects on the substrate or bulk properties of the tool.

In order for the coating to adhere strongly to the substrate, several factors should be considered (83). These include mechanical, physical, and chemical compatibilities between the coating and the substrate. Because the tools are subjected to a high intensity of loading during cutting, the substrate must have adequate hardness and deformation resistance to support the coating without deformation. Otherwise, the coating becomes delaminated from the substrate owing to the development of interfacial tensile stresses. The stress level is intensified in the coating where the coating is stiffer than the substrate. These stresses increase with increasing load and with increasing mismatch in the modulus of elasticity. The relative in-plane normal stress levels in the coating and the substrate are proportional to the ratio of the elastic moduli. Thermal expansion mismatch between the coating and the substrate is another factor responsible for the incompatibility. Tensile stresses in the coating are most damaging. The stress becomes more tensile with increasing temperature when the thermal expansion of the substrate is greater than the coating. Alternatively, if the thermal expansion coefficient of the substrate is lower, then residual compressive stresses are induced which are beneficial to the adhesion of the coating to the substrate.

To avoid failure of the coating at the interface, the strength of the interface should be very high. Where the properties of the coating and the substrate are significantly different, as in the case of diamond and carbide substrate or alumina and carbide substrate, it is preferable to develop a graded interface to take into account factors promoting strong bonding. If chemical bonding is not feasible, then mechanical bonding involving interlocking of asperities should be considered prior to coating deposition. This may be accomplished by mechanical action or chemical etching. To improve adhesion, many manufacturers develop a graded interface between the coating and the substrate. For example, a graded interface with a carbon-rich layer of TiC or TiCN adjacent to the substrate followed by a series of varying layers of coating material wherein the content of carbon progressively decreases and nitrogen progressively increases with the surface layer rich in nitrogen content, TiN, has been developed (84).

Several refractory coatings (qv) have been developed including single coatings of TiC, TiN, Al_2O_3, HfN, or HfC, and multiple coatings of Al_2O_3 or TiN on top of TiC, generally deposited by CVD. TiC is used as a hard wear-resistant coating at low speeds. Multiple coatings prolong tool life, as the thickness of the coating can be increased from ca 5 to 10 μm without inducing brittleness, provide a strong metallurgical bond between the coating and the substrate by choosing appropriate coatings that would provide graded interface(s), and provide protection for machining a range of work materials. Figure 9 shows representative micrographs of single and double coatings on cemented tungsten carbide. The dark regions between the coating and the substrate in these figures are the brittle, η-phase, which should be avoided to improve adhesion and increase the transverse rupture strength (TRS). A very thin (ca 5-μm) coating effectively reduces crater formation on the tool face by one or two orders of magnitude relative to the uncoated tools. At higher speeds, TiC oxidizes and loses its effectiveness. Hence Al_2O_3, which has good wear resistance (both flank and crater wear) at high temperatures, is used. Because TiN is known to provide low friction, it is generally used as the topmost layer. In addition, its lustrous gold color enhances marketability as well as ready recognition of tool wear. Most of the refractory hard coatings (either single or multiple) including carbides, borides, nitrides, oxides, or their combination, were patented. The trend is toward multiple coatings of TiC, TiCN, Al_2O_3, and TiN by a combination of high temperature CVD, medium temperature CVD, and PVD. For example, a multilayer consisting of 10 layers of TiC, TiCN, and TiN and four layers of alumina separated by three

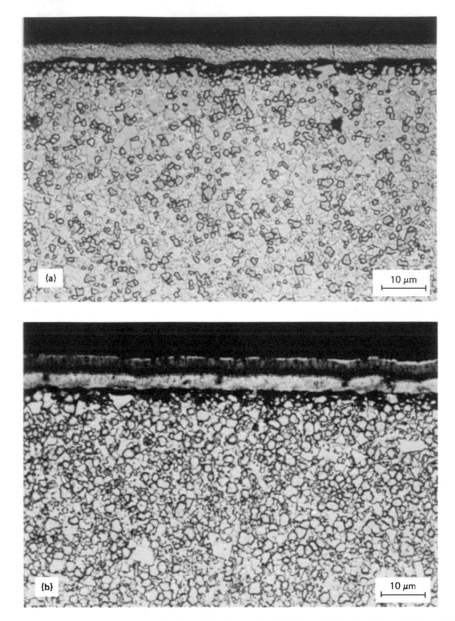

Fig. 9. Representative micrographs of (**a**) single (TiC) and (**b**) double (TiN and TiC) coatings on cemented tungsten carbide (9).

layers of TiN has been used (85). This tool showed considerable improvement in flank wear resistance when machining hot-rolled steel, chilled cast iron (54 R_C), and a nickel-base superalloy, Inconel 718, over commercial multilayer alumina-coated tool. Similarly an alumina-coated tool having an initial layer of 3-μm TiC followed by 19 layers of alumina and 19 layers of TiN to a total thickness of 6 μm, performed with improved crater and flank wear when machining an AISI 1060 steel and AISI 1045 steel in both continuous and interrupted cutting (86).

In the CVD process, several thousand tools are loaded in a vacuum chamber and are initially heated to a temperature of ca 950–1050°C for a coating of either TiC or Al_2O_3, or a combination of the two. This initial temperature is generally <900°C for TiN. The high temperatures ensure good interfacial bonding between the coating and the substrate, provided the material of the substrate is properly tailored for the coating. For a TiC coating, $TiCl_4$ and CH_4 are passed through a reaction chamber containing the tools to be coated in a hydrogen atmosphere at a pressure of ca 101 kPa (1 atm) or less. TiC deposits on the tool face from the vapor phase and establishes a metallurgical bond with the substrate at high temperatures. Alternatively, $TiCl_4$ and carbon (from the cemented carbide substrate) can react in a hydrogen atmosphere. A coating of ca 5–10 μm thickness is deposited for optimum performance, and the process requires ca 8–24 h. In the case of TiC coating, depletion of carbon from the substrate results in the formation of a brittle η-phase, $Co_xW_yC_z$, and associated microporosity at the coating–substrate interface. This brittle η-phase can reduce the transverse rupture strength (TRS) of coated tools by as much as 30% of the substrate's TRS (87), resulting in premature tool failure, especially in interrupted cutting or in roughing. This latter can be a serious limitation, in that interrupted cutting and hence methods to overcome or to augment the strength through other methods become essential for the success of the coated tools. Co-enrichment at and near the rake face of the tool, use of a lower temperature PVD technique which effectively eliminates the formation of the brittle η-phase, and use of a topmost coating of TiN by PVD to introduce residual compressive stresses are some of the attempts to overcome this deficiency. CVD coating technology has reached a stage wherein turnkey systems are available for coating a variety of materials.

Some of the early coated tools were notorious in regard to lack of adhesion of the coating to the substrate, owing to the presence of η-phase at the interface, and consequently exhibited inconsistent performance. This problem has been largely eliminated by a number of process and metallurgical innovations and technological advances resulting in more uniform coatings, better adhesion of the coating to the substrate with minimal interfacial η-phase and associated brittleness. For example, to minimize the affect of η-phase at the interface on coated tool performance, medium (700–900°C) temperature CVD coating technology was developed (70,71). Using a mixture of $TiCl_4$, H_2, and an organic C–N compound such as acetonitrile, TiCN coatings were developed. Similarly, PVD technology wherein the substrate is not heated above 450°C is applied to cemented carbides to eliminate the interfacial η-phase. Because the coating is extremely thin, the edge of a coated tool should be prepared, eg, by honing a radius or providing a small negative rake land, prior to coating and should not be altered subsequently. This ensures uniform coating around the edge. Reasons for honing include (1) honing restores some mechanical strength lost during the CVD process, and (2) having minimizes the tendency of extensive η-phase formation. Alternatively, application of coating by the PVD technique overcomes this problem without the need for honing (43). Whenever a sharp cutting edge is required, the latter technique is used, preferably with a submicrometer carbide substrate.

For a coating of Al_2O_3 on cemented multicarbide tools by the CVD process, a gaseous mixture of hydrogen, water vapor, and an aluminum halide such as aluminum trichloride is used in the temperature range of 900–1250°C (79). The water vapor is most conveniently formed by reacting hydrogen with CO_2 in the deposition chamber to form CO and water vapor. H_2 is found to be necessary to ensure oxidation of aluminum at the carbide interface and to form a dense, adherent coating. To enhance adhesion of the coating to the substrate an interlayer of a transition metal which is both a carbide former and an oxide former, such as Ti, Ta, Hf, Zr, or Nb, is selected (88). For example, the tool is first coated with a thin layer of TiC by the CVD process. It is then oxidized to form TiO at the surface, to which Al_2O_3 coating is deposited. Interlayers are thus provided for compatibility between oxides and carbides by a gradual transition or by forming oxi-carbides. In this manner, thermal expansion mismatches and chemical incompatibility are minimized. Similarly, TiN coating is deposited by reacting $TiCl_4$ and N_2 or NH_3 in a hydrogen atmosphere, and TiCN coating is deposited by reacting $TiCl_4$, methane, and N_2 in a hydrogen atmosphere.

To take full advantage of the coating potential, substrates are carefully matched or appropriately altered to optimize properties, resulting in significant gains in productivity. For example, various coatings of Al_2O_3 or TiC, or TiN–TiC or Al_2O_3–TiC, combined with different cemented-carbide substrate materials, provide a range of combination properties of the substrate-coating tailored for different applications. Coated tools with steel-grade substrates (multiple carbides) are still recommended for machining steels, whereas tools with nonsteel grade substrates are recommended for other materials. The coated tools can be used at higher speeds or higher removal rates, or for longer life at the prevailing speeds. However, as multilayer coating process technology advances, the need for large amounts of carbides for steel machining will decrease.

The modes of wear are different on the rake and the clearance faces. Thus coated tool technology has advanced in the direction of selective compositions and modifications of the substrate in strategic areas to prolong the tool life and to make the coated tool more versatile (89,90). This is especially true in interrupted cutting or heavy-duty cutting, where wear resistance, edge strength, and deformation resistance are all required. An example of the approach in this direction is the development of a multiphase (Ti–C–N or Ti–Al–N)-coated tools with a straight WC–Co-enriched layer (ca 25 μm thick) on the rake face and a Co-depleted layer on the clearance face of a multicarbide (86% WC, 8% (Ti,Ta,Nb)C, and 6% Co) substrate. The Co-enriched zone provides impact resistance characteristic of a high Co-cemented WC, while the flank face depleted of Co provides the superior deformation resistance and high abrasion resistance. This concept was further extended to coatings of TiC–TiCN–TiN on multicarbide with a Co-enriched zone at the rake face, and Co-depleted zone at the flank face provided superior edge while maintaining the edge and crater wear resistance of the coated layers (Fig. 10). The technique used for the Co-enrichment of the tool shown in

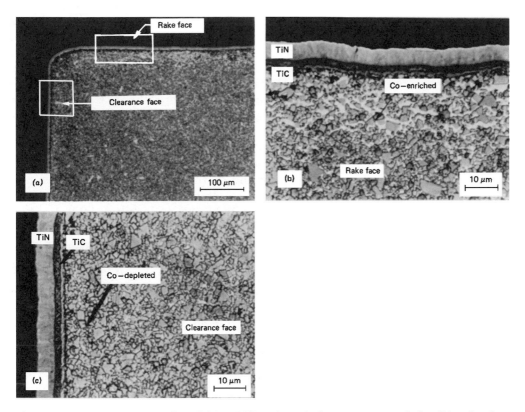

Fig. 10. Optical micrographs of (**a**) a WC tool, and the areas around the (**b**) rake face and (**c**) flank face, showing details of Co enrichment near the rake face and Co depletion near the flank face for the first generation Co enrichment technique.

Figure 10 resulted in a zone near the surface having Co content much higher than desired (200–300% of the bulk binder content instead of 150–200%).

The Co enrichment of the surface layers (25–40 μm) can be achieved either during vacuum sintering or by subsequent heat treatment of the cemented carbides in certain substrates. The Co-enriched zones are characterized by Type A (diameter of the pores ca 10 μm) or Type B (diameter of the pores <40 μm) porosity in the cemented carbide (35,89,90). Co-enrichment via heat treatment occurs more readily when the alloy contains a W-lean Co binder with a magnetic saturation in the range of 0.0145–0.0157 T·cm^3/g (145–157 G·cm^3/g). Additions of small amounts of TiN along with the necessary carbon to the base powder mix promotes the formation of the required magnetic saturation of the Co binder alloy, which would otherwise be difficult to achieve. This is accomplished by adding ca 0.5–2% of fine-grain TiN or TiCN to the WC–Co alloys. Because TiN is not completely stable during vacuum sintering, partial volatilization of nitrogen takes place, resulting in fine A- or B-type porosity in the material. Typically a layer of Co or C is formed on the surface of the substrate during this process which is removed by grinding prior to coating in order to obtain adherent bonding between the coating and the substrate. This also ensures the appropriate thickness of the Co-enrichment layer (ca 25 μm) on the rake face. As this is a surface-enrichment process all around the insert, the Co enrichment on the flank face of the surface has to be removed by grinding, resulting in the required Co-depleted zone on the flank face, leaving the Co enrichment zone on the rake face. Using this technique, the peripheral Co-enriched zone (ca 10–25 μm wide) is completely devoid of solid solution cubic carbides and the distribution of Co is homogeneous and nonstratified, unlike the first-generation coatings. This development, followed by multilayer coatings of Ti–C–N, or TiC–TiCN–TiN, has enabled users to conduct heavy interrupted machining, such as that encountered in scaled forgings and castings at low cutting speeds (89–91). Further, the introduction of TiC–Al_2O_3–TiN coating on this substrate material has enabled improved impact resistance as well as bulk deformation resistance at high temperatures, thus facilitating high speed interrupted machining (92).

A second technique involves heating the cemented tungsten carbide to the solidus–liquidus temperature region of the binder phase in a decarburizing atmosphere, such as CO_2 gas (93). Decarburization occurs at the surface whereby the carbon concentration at the surface is reduced to reach the solidus line of the binder phase, and the liquid phase solidifies. As a result, the liquid phase is supplied to the inner portion, and this also reaches near the surface where it is decarburized to reach the solidus line and this again solidifies. This procedure is repeated until Co is enriched in the zone near the rake face.

A third method is similar to the first one, in that additions of aluminum nitride in sufficient amounts (5–10%) are made, instead of TiN as in the case of the first method, to the cemented carbide mix and vacuum sintered containing Type B1 porosity. This enhances the surface toughness of the cemented carbide by promoting binder enrichment and depletion of aluminum nitride near the peripheral surface, obtained by decomposition of aluminum nitride during sintering (94). The Co-enriched zone is also characterized by the presence of straight tungsten carbide and depletion of multicarbides. Thus this zone consists of Co-enriched tungsten carbide, and after appropriate coating, the tool would provide the required toughness in interrupted or heavy-duty cutting.

Physical vapor deposition (PVD) technology includes vapor deposition, various types of sputtering, and ion plating (95) (see THIN FILMS). The mean free path in PVD is large, and the vapor species arrive at the substrate without extensive gas collision. Consequently, this is a low temperature, line-of-sight process. Because of lower substrate temperatures, adhesion of the coating to the substrate can be a problem. Also, the rate of deposition, especially using straight sputtering, can be low (on the order of nm/min). However, unlike in CVD, the formation of η-phase is practically eliminated owing to lower substrate temperatures (91). Also, the PVD process, eg, iron plating, can introduce compressive residual stresses beneficial to the adhesion of coating to the substrate and the overall performance of the coated tools (87).

Although coated tools have demonstrated significant performance gains over comparable uncoated tools, up to 3–10 times in certain cases, in the initial stages of their introduction several factors contributed to a less-than-complete acceptance. These include inadequate machine-tool systems (the significant performance gains possible with coated tools are accomplished at higher speeds, at higher removal rates, and with more rigid, high power machine tools; most older machine tools are somewhat limited in this respect); nonuniform and inconsistent performance of some earlier coated tools owing to quality-control problems; limited user knowledge, partly because many small-scale users have less knowledge on the performance and application of these coated tools, and partly because the technology is advancing rather rapidly; slightly higher cost, which should not be a consideration because the cost of coated tools is only fractionally higher than that of uncoated tools; larger inventory of different tool grades; and slightly lower toughness reported with some coated tools.

Coated tools, originally developed exclusively for machining steels, are used for a wide range of materials, including various types of steels, cast iron, stainless steel, nickel and Co-base superalloys, and titanium alloys. The tool manufacturers often encounter a challenge between developing a general-purpose grade that covers a wide range of work materials, cutting conditions, and machining operations to reduce the number of grades the user can stock vs the coatings targeted for niche areas, eg, coatings for machining titanium alloys and coatings for interrupted cutting. Generally, a coating for a niche area is developed first because of the specific need in that area. Subsequently its scope is broadened to cover either a class of work materials, a class of manufacturing operations, or a range of machining conditions. The selection of an appropriate tool material and proper tool geometry, chip groove geometry, coating, and substrate properties for machining a given material is not a simple selection based on rule of thumb, but rather is a more sophisticated decision-making process based on detailed knowledge and experimentation. Of the carbide tools used as of the mid-1990s, some 65% are coated, especially those in the United States, Western Europe, and Japan. This percentage is expected to increase considerably (as high as ca 80%) into the twenty-first century.

A thin (ca 5-μm) coating effectively reduces crater formation on the tool face by a factor of two to three relative to uncoated tools. Multiple coatings enable greater (ca 10–15 μm) thickness, create a strong metallurgical bond between the coating and the substrate, provide protection for different work materials, and prolong tool life. Single or multiple coatings of different thicknesses of TiN, TiN–TiCN–TiN, TiN–TiCN, TiN–Al_2O_3–TiC–TiCN, TiN–TiCN–TiC, TiN–Al_2O_3–TiCN, Al_2O_3–TiC, Al_2O_3–TiC–TiCN, TiN–Al_2O_3–TiC, and TiN–Al_2-O_3–TiN–Al_2O_3–TiN–Al_2O_3–TiCN engineered on selected cemented carbide (different Co content and straight WC or multicarbide) as well as Co-enriched substrates exist. In addition, one grade of TiN–Al_2O_3 on a Si_3N_4 substrate is also available. These coatings can be combined with suitable chip-groove geometry to result in an optimum level of performance. Coatings on cutting tools have become customized for various work-materials and cutting applications. The trend is toward coating the first layer of TiN or TiCN on the substrate by PVD to minimize or eliminate the formation of η-phase ($Co_xW_yC_z$), as well as to coat the last layer by PVD to induce residual compressive stresses on carbide substrates. Although PVD and medium temperature CVD coatings are beginning to appear, the majority of the coatings on cemented carbides are made by the CVD technology. Technology to coat crystalline Al_2O_3 by PVD is not available as of this writing (ca 1997).

Other coatings, such as TiAlN (96), TiCN, ZrO_2, and ZrN (97), and CrN (98) were developed for special applications. The last was developed for higher speed machining of titanium alloys. Sometimes a coating is developed not for its wear-resistance but for its heat insulation. The case in point is alumina coating of cBN to reduce the heat conductivity at the surface so that the cBN performance can be enhanced (99).

Multiple Nanolayered Coatings

The refractory hard materials used for coatings on cutting tools are generally brittle and hence not tough. The fracture mechanism consists of crack initiation at stress concentrations and its rapid propagation to failure (see FRACTURE MECHANICS). By arresting the propagation of the cracks, it is possible to increase the toughness of these hard coatings significantly without compromising on hardness. This is accomplished by applying multiple nanolayer coatings of alternating hard and tough materials (see NANOTECHNOLOGY (see SUPPLEMENT)). Investigations have been directed at improving the properties of materials significantly by reducing the microstructural or spatial scale of a material system to nanometer dimensions (100,101). In this approach, a crack initiated in any hard layer is stopped when it reaches the tough layer, facilitating higher toughness (100–103). The number of nanolayers can be several hundred in contrast to the few layers used on cutting tools prepared by CVD techniques.

Nanolayer coatings are generally expected to be harder, tougher, and chemically more stable than coatings of several micrometers or of bulk materials. For example, B_4C or SiC ceramics have high hardness but are not used as cutting tools either in monolithic form or as micrometer-thick coatings because of the ease of oxidation, reaction with most ferrous materials, and more importantly their inherent brittleness. However, in nanolayer coatings in multiples (literally hundreds) of layers having alternating hard material and tough ductile metal, the material is no longer brittle and even if a layer oxidizes after performing its cutting action, the next layer is ready to take its place during subsequent contact with the work-material. This case is somewhat similar to the self-sharpening action of abrasives in grinding, where the abrasives (qv) release new cutting edges by micro cleavage once the previous edges are worn out.

In addition to coatings of alternating hard and tough materials, multiple nanocoatings can also be developed based on alternating hard and lubricating materials, such as a carbide/oxide/boride/nitride–MoS_2 sequence to improve tool life and at the same time decrease friction at the chip–tool interface. Multiple coatings of alternate WC (1.2-nm) and Co (0.8-nm) (104), nanolayered coatings of Al–Al_2O_3 (105,106), and Ti–TiN coatings (107) have been developed, as have multinanolayer composite coatings consisting of alternate solid lubricant and a metal (108,109). Solid lubricant–metal multilayer nanocoatings have been developed for tribological applications (110,111). TiN-based coatings have been introduced to investigate the *in situ* solid lubrication of TiO_2 layers, believed to form by the reaction of TiN with water vapor in the air (112), Ti–TiN, Al–AlN nanolayers for tribological and corrosion protection have been investigated (113). The role of tribology in metal cutting has been discussed (114).

The number of material systems that can be used for nanolayer coatings is virtually unlimited. Any refractory hard material can be used as the hard material; compatible metal can be used as the tough material. Examples of material systems for nanocoatings include the following:

In situ formation of oxidation-protective and low friction layers

hard carbide/hard carbide systems $B_4C-SiC, HfC-B_4C, HfC-SiC$

hard carbide/metal co-sputtered or B_4C (HfC, SiC)$-W$ (Al, Cr, Ti, Si, Mg,
   layered systems   Zr)

Low friction, low stress coatings

layered lattice (solid lubricant)/metal $MoS_2-Mo, MoS_2-Ag-Mo, TaS_2-Ta,$
   systems WS_2-W

Low friction, hard coatings

hard carbide, oxide, nitride, boride/layered lattice (solid lubricant) systems

Multiple nanolayer coatings are deposited by PVD; chiefly magnetron sputtering using multiple targets is employed. The tool is rotated with respect to the targets. The tool then sees different sputtering targets alternately. Consequently, the coatings are endowed with the benefits associated with the PVD technique, ie, lower substrate temperatures, virtual elimination of the η-phase in the case of carbide substrates, and residual compressive stresses. The sputtering rate and the rotational speed of the tool (or duration during which the tool is exposed to a given target material) determines the thickness of each coating layer. Whereas each layer in the coating is only a few nanometers, the total thickness can be in the range of $2-5$ μm. As of this writing (ca 1997), no commercial nanocoatings for cutting tool applications are available in the marketplace. This technology is expected to become widely used in the twenty-first century.

Ceramics

Ceramics (qv), one of the newest classes of advanced tool materials, are used on the one hand for high speed finishing operations involving light feeds and on the other for high removal-rate machining involving low speeds and large depths of cut of some difficult-to-machine steels and cast irons (115–119). The ceramics used initially were predominantly alumina based, although silicon nitride-based materials (also called nitrogen ceramics) have been found to be very attractive for high speed machining of gray cast iron (1500 m/min) and nickel-base super-alloys (200 m/min or higher) (see ADVANCED CERAMICS). Ceramics, in general, are harder, more wear-resistant, more highly refractory, and chemically more stable than cemented carbides and HSS. Notching at the depth of cut line (DCL notching), microchipping of the tool edge, chipping owing to thermal or mechanical cyclic stresses during interrupted cutting, and gross fracture of the tool are the predominant modes of wear experienced with ceramic tools. Because of poor thermal and mechanical shock resistance, interrupted cutting is especially severe on ceramic tools, owing to repeated entry and exit of the cut. Both the tool and the part to be machined must be fully supported, and to prolong life, the machine tool must be extremely rigid. Ceramics are machined either dry or with a heavy stream of coolant, because intermittent application of coolant can cause thermal shock leading to fracture. High speed, high power precision machine tools are desirable to take full advantage of the potential of ceramic tools.

Although ceramic tools were considered for certain machining applications as early as 1905, transverse rupture strength (TRS) under the conditions of cutting was inadequate and the performance inconsistent (115,116). In the mid-1950s, ceramic tools were reintroduced for high speed machining of steels and gray cast iron for the automobile industry, and slow speed, high removal-rate machining of extremely hard (and difficult to machine) chilled cast iron or forged steel rolls used in the steel industry. These were basically fine-grain (<5 μm), alumina-based materials, alloyed with suboxides of Ti or Cr to form solid solutions, and contained small amounts of magnesia as a sintering aid (115). Carboloy in the United States developed an alumina–TiO ceramic (Grade O-30) that is characterized by a grain size of ca 3 μm and reasonably uniform microstructure. The TiO constitutes ca 10%. A density of ca 90–95% theoretical, a hardness of ca 93–94 HRA and a TRS >550 MPa (ca 80,000 psi) were achieved with cold-pressing followed by sintering (120,121). Similarly, the Carborundum Company developed a nearly pure alumina (Stupalox or CCT 707) having minor additions of MgO as a sintering aid and a grain-growth inhibitor (115). The Vascoloy Ramet/Wesson Corporation manufactured a similar material (VR 97), originally developed by the Norton Company. Extremely rigid, high powered (up to 450 kW (600 hp)) machine tools were specially designed having high stiffness and high precision, enabling material removal rates of two to three orders of magnitude higher than for conventional machine tools (122). Similarly, high (up to 5000 rpm) speed machine tools were specially built for machining gray cast iron using full automation to take advantage of the potential of this tool material.

Several factors have rejuvenated interest in the development and application of ceramic cutting tools (115–119). Applications of advanced ceramics for structural applications, advances in the ceramic-processing technology; progress in the understanding of the toughening mechanisms in ceramics; rapidly rising manufacturing costs; the need to use materials that are increasingly more difficult to machine; rapidly increasing costs and decreasing availability of W, Ta, and Co, which are the principal and strategic raw materials in the manufacture of cemented-carbide tools; and advances in machining science and technology have all played a role.

A comparison of the physical properties of ceramic tools and carbide tools is given in Table 11. Ceramics are harder (hence, more abrasion-resistant), have a higher melting temperature (thus are more refractory), and are chemically more stable up to their melting temperatures. Ceramics are, however, less dense and less tough (lower TRS), have lower thermal conductivity, and have lower thermal expansion coefficients than cemented carbides. Toughness of ceramics having smaller grain size can be improved by the introduction of a more ductile second phase. Because of lower TRS and high refractory characteristics, ceramics are generally recommended for higher cutting speed ($\geq$300-m/min), a lower rate of material removal, ie, high speed finish machining, and continuous-cutting applications. Lower fracture toughness (the value for alumina is ca 2.3 MPa·m$^{1/2}$) is the main limitation in the application of ceramics for heavy or interrupted cutting.

The next advancement in alumina-based ceramics is the development of pure alumina and alumina–TiC dispersion-strengthened ceramics (117). Alumina–TiC-based ceramics contain ca 30 wt % TiC and small amounts of yttria as a sintering agent, resulting in a density close to 99.50% theoretical. High purity, fine grain size, and elimination of porosity are the principal reasons for the high TRS (700–900 MPa (ca 100–130 ksi)). The slightly more expensive hot-pressed materials usually have higher TRS and more consistent performance than cold-pressed materials. The fracture toughness of this material is ca 3.3 (MPa·m)$^{1/2}$. The fracture toughness and TRS are improved through obstruction of the cracks, crack deflection, or crack branching caused by the dispersed hard TiC particles. The lower thermal expansion coefficient and higher thermal conductivity of the composite compared to straight alumina increases the thermal shock resistance and thermal shock cycling experienced in interrupted

Table 11. Physical Properties of Ceramic and Cemented-Carbide Cutting Tools[a]

Property[b]	Ceramics	Cemented carbide[c]
hardness, HRA[d]	91–95	90–93
TRS[e] for alumina-based ceramics, MPa[f]	690–930	1590–2760
melting range, °C	ca 2000	ca 1350
density, g/cm^3	3.9–4.5	12.0–15.3
modulus of elasticity, E, GPa[g]	410	70–648
grain size, μm	1–3	0.1–6
compressive strength, MPa[f]	2760	3720–5860
tensile strength, MPa[f]	240	1100–1860
thermal conductivity, W/(m·K)		41.8–125.5
thermal expansion coefficient, 10^{-6}/°C	7.8	4–6.5

[a] Ref. 9.
[b] The exact properties depend upon the materials used, grain size, binder content, volume fraction of each constituent, and processing method.
[c] Coated carbides are not included.
[d] Rockwell hardness A scale.
[e] Transverse rupture strength.
[f] To convert MPa to psi, multiply by 145.
[g] To convert GPa to psi, multiply by 145,000.

cutting. At temperatures exceeding 800°C, TiC, and TiN begin to oxidize and lose their strengthening properties. Figure 11a shows a micrograph of a hot-pressed alumina–TiC ceramic. Other Al_2O_3-based ceramic tools include Al_2O_3–TiB_2, Al_2O_3–ZrO_2, and silicon–aluminum–oxygen–nitrogen (Si–Al–O–N). The latter two materials are slightly less hard than alumina or Al_2O_3–TiC, but significantly tougher. When machining superalloys, hard chill-cast irons, and high strength steels in the medium speed range (ca 150-m/min), longer tool life results because of lower flank wear and, more important, lower DCL notching.

An Al_2O_3–ZrO_2 ceramic (Cer Max 460) was introduced in the United States by Carboloy. A similar material performs exceptionally well in the grinding industry as a tough abrasive in heavy-stock grinding operations, such as cut-off and snagging. The high toughness and superior grinding performance of this

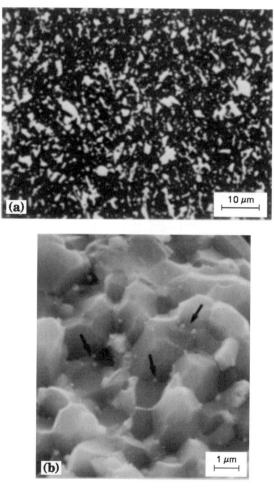

Fig. 11. Micrographs of (**a**) a hot-pressed alumina–TiC ceramic showing a white TiC phase and a dark alumina phase (3) and (**b**) a fracture surface of an alumina–zirconia alloy precipitation of zirconia at the alumina grain boundaries (3).

alloy are attributed to the rapid freezing of the alloy from the melt, which results in a dendritic freezing of the eutectic structure. Alternatively, a fine dispersion of unstabilized ZrO_2 in the matrix of Al_2O_3 can give rise to stress-induced transformation of tetragonal zirconia particles, inducing microcracks which absorb elastic energy and prevent cracks from propagating, thereby increasing the fracture toughness. The three popular compositions contain 10%, 25%, and 40% ZrO_2, respectively; the remainder in each case is alumina. The 40% ZrO_2–Al_2O_3 composition is close to the eutectic. The higher ZrO_2 compositions are less hard, but tougher.

The micrograph of a fracture surface of an alumina–zirconia alloy is shown in Figure 11**b**. The zirconia particles are concentrated predominantly at the alumina–grain boundaries. Although the fracture is intergranular, the presence of these particles is believed to provide additional toughness before failure can occur by fracture. In some machining tests on a tough chill-cast iron (HRC 42–44) used for steel rolls, this material performed exceptionally well, showing very little wear in plunge cuts at 150 m/min cutting speed, a feed rate of 0.4 mm per revolution, and a width of cut of 25.4 mm over a straight-alumina tool. Chipping was the predominant mode of wear in the latter case. Similarly, when machining solution-treated and aged Inconel 718, a nickel–iron base superalloy (HRC 42–44), at a cutting speed of 150 m/min, a feed rate of 0.3 mm per revolution, and a depth of cut of 0.3175 mm, this material (2.54-cm round) gave a tool life of >8 min and yielded an excellent finish (1–2.5 mm) on the machined surface. These tests were conducted dry. Based on other successful high speed machining tests (ca 300 m/min) where a coolant-lubricant was used, higher tool life at the same speed or increased cutting speed for the same tool life was obtained.

The second interesting class of ceramic tool material under development is based on Si_3N_4, either nearly pure Si_3N_4 (except for some minor additions of sintering aids) or having various additions of aluminum oxide, yttrium oxide, and TiC (123–134). It is a spin-off of the high temperature–structural ceramics technology developed in the 1970s for automotive gas turbines and other high temperature applications. Ford Motor Co. developed a ceramic tool of Si_3N_4 having additions of ca 12% yttria (Grade S 8). Norton Co. developed a Si_3N_4 ceramic based on MgO but has never commercialized it for cutting tool applications. Instead, it concentrated on advanced structural applications, including hybrid ceramic bearings. Similarly, General Electric and Westinghouse also developed Si_3N_4-based ceramics for high temperature–structural applications but did not extend them to cutting tools.

Si_3N_4 is a covalently bonded material that exists in two phases, α and β. The structure of these phases is derived from the basic Si_3N_4 tetrahedra joined in a three-dimensional network by sharing the corners, with each nitrogen corner being common to the three tetrahedra. This material provides a number of favorable properties, including higher elevated temperature strength, thermal stability, low thermal expansion coefficient, higher thermal conductivity, and higher fracture toughness than alumina. But because of the low self-diffusion coefficient, it is virtually impossible to fabricate pure Si_3N_4 into a dense body with no porosity by conventional sintering or hot pressing techniques without the sintering aids. The predominant impurity in Si_3N_4 is SiO_2, which is present on the surfaces of the Si_3N_4 particles. In the synthesis of Si_3N_4, sintering aids, such as MgO, Y_2O_3, TiO, and Cr_2O_3, are combined with fine powder of α-Si_3N_4, ball-milled, and used as the starting material for consolidation. It is cold-pressed to shape and hot-pressed in a N_2 atmosphere at temperatures ca 1600°C. At the densification temperature, the sintering aids combine with SiO_2 to form a glassy liquid phase. α-Si_3N_4 particles dissolve in this liquid phase and precipitate out as β-Si_3N_4. The β-Si_3N_4 nuclei grow as elongated grains and form an interlocked grain structure. The higher fracture toughness (4–6 MPa·m$^{1/2}$) and higher strength of Si_3N_4 are attributed to this elongated grain structure. Commercial Si_3N_4 is characterized by a two-phase structure, consisting of β-Si_3N_4 crystallites and an intergranular bonding phase.

Si_3N_4 is marketed by most tool manufacturers and used extensively in high speed machining of gray cast iron. However, to take full advantage of this tool material at high speeds, more rigid, high speed, high power machine tools are required. Many automotive and other industries are working with the machine-tool builders of these special machine tools. The fracture toughness of pure Si_3N_4, higher than alumina, is not adequate for rough machining, interrupted cutting, or in machining of castings with irregular surfaces or with a scale. To address this problem, GTE developed a ceramic tool of Si_3N_4 having additions of yttria (ca 6%), alumina (ca 2%), and a fine dispersed phase of TiC (ca 30%) (Grade Quantum 5000). Other additions of the dispersed phase include TiN, HfC, or a combination of TiN and TiC, which increases the hardness and fracture toughness (via crack interactions with the dispersoid and crack deflection) of the composite material. Many tool manufacturers have similar grades. Similarly, SiC or Si_3N_4 whisker-reinforced Si_3N_4 was developed to increase the fracture toughness of the base material for interrupted cutting applications.

In the late 1970s Lucas Industries Ltd. of the United Kingdom developed a ceramic tool of Si–Al–O–N with additions of yttria, and marketed under the trademark SYALON (in the United States this material was marketed initially as KYON by Kennametal Inc. under license; Al_2O_3-based materials are sold as KYON as of 1997). Oxygen (O^{2-}) can be substituted for nitrogen (N^{3-}) in the β-Si_3N_4 crystal provided Al^{3+} is simultaneously substituted for Si^{4+} to maintain charge neutrality. SiAlON tools are produced by sintering. The powder charge consists of a mixture of Si_3N_4, AlN, Al_2O_3, and Y_2O_3. The last is added as a sintering aid for full densification. The powder mix is first ball-milled, dried, pressed to shape by cold-pressing, and subsequently sintered at a temperature of ca 1800°C under isothermal conditions for ca 1 h before it is allowed to cool slowly. Yttria reacts with β-Si_3N_4 to form a silicate which is a liquid at the sintering temperature. The resulting Si_3N_4 material thus has a glassy intergranular phase. Some properties of the Sialon material are given in Table 12 (135).

Further developments in microstructure and composition may yield an even more refractory material consisting of β-Si_3N_4 and an intergranular phase of yttrium aluminum garnet (YAG) without an intergranular glassy phase. Similar to the alumina–zirconia ceramic, this material offers significant improvements in tool life, consistency in tool performance (more reliability), and higher removal

Table 12. Composition and Properties of Sialon Material[a]

Parameter	Value
Composition	
Si_3N_4, wt %	77
Al_2O_3, wt %	13
Y_2O_3, wt %	10
Properties	
density, g/cm^3	3.2–3.4
hardness, GPa[b,c] (kgf/mm^2)	17.65
Young's modulus, GPa[b]	300
compressive strength, MPa[d]	>3500
thermal conductivity, $W/(m \cdot K)$	20–25
thermal expansion coefficient, $10^{-6}/°C$	32

[a]Ref. 135.
[b]To convert GPa to psi, multiply by 145,000.
[c]Value is equal to 1800 kgf/mm^2.
[d]To convert MPa to psi, multiply by 145.

rates possible at reasonable cutting speeds (90–125 m/min) when machining nickel-base superalloys. With the increasing trend toward computer-controlled machining, consistency and reliability of tool performance are crucial. Furthermore, the trend toward more than one machine tool per operator is resulting in lower and more manageable but reliable cutting speeds. SiAlON, alumina–TiC, alumina–zirconia, and straight alumina are some of the tool materials that might meet the needs of these trends.

Because of its high toughness and good thermal shock resistance, test results indicate the possibility of using square-, triangular-, and diamond-shaped tools of SiAlON for machining superalloys in the intermediate speed range (ca 150 m/min) where only round tools are used currently with other ceramics.

Even though Si_3N_4 and SiAlON tool materials are used extensively for high speed machining of cast iron and machining of nickel-base superalloys, respectively, they could not be used for machining steels. To take advantage of their improved fracture toughness as well as their ability to machine ductile C-1, ceramic coatings on Si_3N_4, SiAlON, and modified compositions of the two were developed. Thin (2–5-μm) coatings on monolithic ceramic substrates have been developed mainly to limit chemical interactions between the tool and the steel work materials. Also, to take advantage of the high temperature deformation resistance of this material and to minimize chemical interactions when machining steels at high speeds, single (TiC, TiN, AlN, Al_2O_3) or multiple coatings of TiC–TiN, or Al_2O_3–TiC on silicon nitride, SiAlON, and SiAlON dispersed with TiC substrates were developed that are similar to the coatings on cemented carbides (136–140) (Fig. 12). Whereas attempts are continuously being made to improve the adhesion between the coating and the substrate, the problem remains. Even if this problem were to be solved, the extent to which such coated ceramic tools are used vs competing materials for high speed machining of steels

(a)

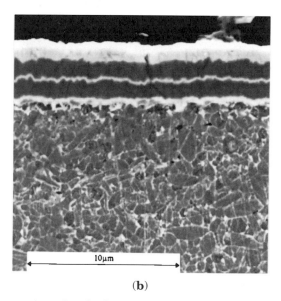

(b)

Fig. 12. (a) A cross section of multiple coatings of TiN on TiC on a silicon nitride-based tool material; (b) multicoatings on a SiAlON-based tool material.

and other materials would depend on the need for them and the economics of machining. The coated ceramics are still in the experimental stage and (ca 1997) are not available commercially.

A newer whisker-reinforced ceramic composite (SiC whisker-reinforced alumina) material possessing improved fracture toughness (K_{1C} ca 8.5–9 $(MPa \cdot m)^{1/2}$) has been introduced as a cutting tool for machining of nickel–iron-base superalloys used in aircraft engines (141–154). Single crystal whiskers of SiC possess very high tensile strength (7 GPa (1×10^6 psi)). SiC also has higher thermal conductivity and higher coefficient of thermal expansion than alumina. Consequently, the composite exhibits higher strength, fracture toughness, and thermal shock resistance.

SiC whiskers are commonly made from rice hulls, a waste product of agriculture (150). Rice hulls have a high (15–20%) ash content relatively high in silica (>95%) and cellulose. Thermal decomposition of this very high surface area material provides intimate contact in the rice hulls and SiC whiskers are readily formed. In the preparation of the SiC whisker-reinforced composite material, SiC whiskers (0.5–1 μm in diameter and 10–80 μm long) about 20% volume fraction are mixed homogeneously with micrometer-sized fine powder of alumina. The mixture is then hot-pressed to over 99% of the theoretical density at a pressure in the range of 28–70 MPa and temperature in the range of 1600–1950°C for pressing times varying from about 0.75 to 2.5 h. The fracture toughness of this SiC whisker-reinforced alumina material is by far the highest among ceramic cutting tools, nearly twice that of its closest ceramics, Si_3N_4 and Sialon. Details of the micromechanisms for improved fracture toughness of this ceramic composite tool material have not been clearly established (151,152).

Although monolithic Si_3N_4 is a reasonably tough ceramic tool material (K_{1C} ca 4.7 $MPa \cdot m^{1/2}$) compared to alumina, its fracture toughness is not high enough for interrupted cutting (milling or rough turning) of cast iron. Thus a tougher silicon nitride tool reinforced with SiC or Si_3N_4 whiskers was developed for this application. For example, a 30 vol % SiC whisker-reinforced Si_3N_4 showed an increase of 40% in the fracture toughness (K_{1C} of ca 4.7 $MPa \cdot m^{1/2}$ for Si_3N_4, compared to ca 6.4 $(MPa \cdot m)^{1/2}$ for SiC whisker-reinforced material) and 25% increase in strength (TRS) (151). Similarly, to increase the toughness of the matrix material even further, SiC whisker-reinforced Al_2O_3–ZrO_2 material was developed (145).

The SiC whisker-reinforced alumina composite, a model for engineered materials, has opened new vistas for tool material development. Whereas SiC whisker-reinforced alumina is used extensively for the machining of nickel-base superalloys, SiC whiskers react chemically with steel, causing rapid wear on the rake face. Attempts are underway to replace SiC whiskers with less reactive whiskers such as TiC or TiN.

Ceramic tools are inherently more brittle than cemented carbides, and a tool geometry of −10 deg rake and +10 deg clearance is recommended instead of −5 deg rake and +5 deg clearance (for cemented carbides). In interrupted cutting, attempts should be made to shift the point of application of the load away from the cutting edge to minimize chipping. Suitable edge preparation involving honing a small radius or a small negative land on the rake face is also recommended. The work materials recommended include hardened steels, chill-cast iron, and superalloys (Ni-base and Co-base).

Certain ceramic tools, especially those based on alumina, are not suitable for machining aluminum, titanium, and similar materials because of a strong tendency to react chemically. They are also not generally suited for low speed and intermittent cutting operations because of failure by chipping, unless they are used on extremely rigid high precision machine tools. Poor thermal shock resistance prevents the intermittent application of cutting fluids. Hence either heavy flooding or no coolant at all is recommended for machining with ceramic tools.

Diamond

Diamond is the hardest (Knoop hardness ca 78.5 GPa (ca 8000 kgf/mm^2)) of all known materials. Both the natural (single-crystal) and synthetic (polycrystalline sintered body) forms can be used for cutting-tool applications. Diamond tools exhibit high hardness, good thermal conductivity, ability to form a sharp edge by cleavage (especially the single-crystal natural diamond), low friction, nonadherence to most work materials, ability to maintain a sharp edge for a long period of time, especially when machining soft materials like copper and aluminum; and high wear resistance. Sometimes if the surface of a tool material is somewhat rough, metal may be stuck in the valleys of the tool surface and subsequent buildup can occur between this metal and the chips. Disadvantages of diamond tools include extensive chemical interaction with metallic elements of Groups (4–10) (IVB–VIII) of the Periodic Table (diamond wears rapidly when machining or grinding mild steel; it wear less rapidly with high carbon alloy steels than with low carbon steel and is occasionally employed to machine gray cast iron (high carbon content) with long life); a tendency to revert at higher (ca 700°C) temperatures to graphite and oxidize in air; extreme brittleness (single-crystal diamond cleaves easily); difficulty in shaping and reshaping after use; and high cost.

Steels account for a significant fraction of the work materials machined. Thus the inability to use diamond for machining steels is a significant limitation where general machining is concerned. There are, however, other applications where diamond is the ideal material, such as for ultraprecision machining of aluminum and copper for laser mirror applications where it is almost impossible to use other tool materials to produce such a surface to provide a long tool life economically. Use of a single-crystal diamond as microtome knives is another unique application. High quality, single-crystal industrial diamonds are the tools of choice for these applications because of long life and ability to machine accurately (with an extremely sharp edge formed by cleavage). Lower quality industrial diamonds are extensively used in high speed machining of aluminum–silicon alloys in the automobile industry; in polymers and glass–epoxy composites in the aircraft industry; in copper commutators in the electrical industry; for machining nonferrous (brass, bronze) and nonmetallic materials; for cold-pressed sintered–carbide performs for the metal-cutting and metal-forming industries; to shape and cut stone and concrete; and as dressing tools for alumina grinding wheels. Industrial quality natural diamonds give unreliable performance, caused by easy cleavage and unknown amounts of impurities and imperfections. Regrinding of these tools is difficult and expensive.

Limited supply, increasing demand, and high cost have led to an intense search for an alternative, dependable source of diamond. This search led to the high pressure (ca 5 GPa (0.5×10^6 psi)), high temperature (ca 1500°C) (HP–HT) synthesis of diamond from graphite in the mid-1950s (153–155) in the presence of a catalyst–solvent material, eg, Ni or Fe, and the subsequent development of polycrystalline sintered diamond tools in the late 1960s (156).

The polycrystalline diamond tools consist of a thin layer (ca 0.5–1.5 mm) of fine grain-size particles sintered together and metallurgically bonded to a cemented-carbide base. The cemented carbide provides the necessary elastic support base for the hard and brittle diamond layer above it (Fig. 13**a**). These tools are formed by a HP–HT process at conditions close to those used for the synthesis of diamond from graphite. Fine diamond powder (1–30-μm) is first packed on a support base of cemented carbide in the press. At the appropriate sintering conditions of pressure and temperature (in the diamond-stable region), complete consolidation and extensive diamond-to-diamond bonding takes place (Fig. 13**b**). Stress concentration at the sharp corners of the diamond crystals during sintering subjects these areas to local stresses of perhaps an order of magnitude higher than nominal (ca 50 GPa (5×10^6 psi)). As a result, individual diamond crystals are work-hardened (154), resulting in a sintered diamond compact which is probably much harder than an undeformed diamond, and consequently the abrasion resistance of the tool is increased. In addition to diamond-to-diamond bonding, good metallurgical bonding is established between the diamond layer and the cemented-carbide support base in this process. However, some of the binder phase and other impurities from the underlying cemented carbide can diffuse into the diamond layer above it. This can affect the high temperature performance of the polycrystalline diamond. In order to eliminate this effect,

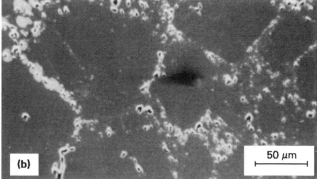

Fig. 13. (**a**) Photograph of a polycrystalline diamond tool showing a thin layer (ca 0.5–1.5 mm) of fine-grain size diamond particles sintered together and metallurgically bonded to a cemented carbide base; (**b**) micrograph of the polycrystalline diamond tool showing extensive diamond-to-diamond bonding.

the metallic impurities in the polycrystalline diamond are leached and the resulting diamond layer is subjected to HP–HT conditions to obtain a much denser, impurity-free material with superior performance (157). The polycrystalline diamond tools are then finished to shape by laser cutting or electrodischarge machining (EDM), followed by grinding, polishing, and lapping to size, finish, and accuracy.

Sintered polycrystalline diamond tools of various grain sizes are fabricated in an assortment of shapes (squares, rounds, triangles, and sectors of a circle of different included angles) and sizes from round blanks. The main advantages of sintered polycrystalline tools over natural single-crystal tools are better control over inclusions and imperfections, higher quality, and greater toughness and wear resistance (resulting from the random orientation of the diamond grains and the corresponding lack of simple cleavage planes). In addition, the availability of sintered diamond tools is not dictated by nature or some artificial control; thus, such tools can be manufactured to meet strategic needs.

Sintered polycrystalline diamond tools are much more expensive than conventional cemented-carbide or ceramic tools because of the high cost of the processing technique and the finishing methods used. Diamond tools, however, are economical on an overall-cost-per-part basis for certain applications because of long life and increased productivity.

Sintered diamond tools are used for applications similar to the lower quality industrial diamonds. Because of high reactivity, they are not recommended for machining soft low carbon steels, titanium, nickel, cobalt, or zirconium. Because they are inherently brittle, they are used with a negative rake ($-5°$) geometry with suitable edge preparation on materials that are difficult to machine, such as pressed and sintered cemented tungsten carbide, stone, and concrete. For softer materials, eg, Al–Si alloys, aluminum- or copper-front surface mirrors, and motor commutators, a high positive rake ($+15°$) geometry is used. Positive rake inserts with polycrystalline diamond tips are among the most commonly used tools for this application. The tips can be resharpened and are available in cartridges.

Low Pressure Diamond Coatings. In the early 1980s Japanese researchers (158) took earlier Russian findings (159,160) seriously and developed microwave-assisted chemical vapor deposition of diamond having growth rates reaching nearly 1 μm/h. They used a 2.45-GHz microwave plasma for the production of diamond films. This method has been used since that time to coat carbide tools and for optical applications, as the films grown have good quality. Other advantages of this process are its stability, large deposition area, and no metal contamination of the film.

Around the same time a similar technique was independently developed whereby micrometer sized diamond crystallites were grown (161). What is required in essence for the low pressure diamond synthesis is a source of carbon (typically a hydrocarbon gas), hydrogen, and a temperature above 2000°C to convert molecular hydrogen to its atomic state.

In 1988 a technique for diamond synthesis was announced based on an oxy-acetylene combustion flame with a slightly fuel-rich mixture (162). It uses an oxy-acetylene welding torch yielding high (50–150-μm/h) growth rates. This technique offers a very simple and economical means for diamond synthesis at growth rates one to two orders of magnitude higher than microwave or hot-filament-assisted CVD diamond techniques, which are typically $1-2$-μm/h. Because the diamond growth takes place under atmospheric conditions, expensive vacuum chambers and associated equipment are not needed. The flame provides its own environment for diamond growth and the quality of the film is dependent on such process variables as the gas flow rates, gas flow ratios, substrate temperature and its distribution, purity of the gases, distance from the flame to the substrate, etc.

Plasma-jet diamond techniques yield growth rates of about 980 μm/h (163,164). However, the rate of diamond deposition is still one to two orders of magnitude lower than the HP–HT technology, which is about 10,000 μm/h (165). Diamond deposition rates of ca 1 μm/s have been reported using laser-assisted techniques (166). This rate is comparable to the HP–HT synthesis.

In the microwave-assisted or hot-filament-assisted CVD of diamond, methane and hydrogen gases (CH_4 ca $1-5\%$ and H_2 ca $95-99\%$) are used. In addition, oxygen is used at times to produce improved diamond coatings. Methane provides the source of carbon. The microwave unit generates plasma in a stainless steel chamber. The microwave energy is coupled by the Symmetric Plasma Coupler to produce a uniform ball of plasma at, or slightly above, the substrate surface. The plasma in the case of microwave and hot tungsten filament at temperatures in excess of 2000°C in the case of hot filament ensures hydrogen in the atomic state. The purpose of atomic hydrogen is to etch away any graphite or nondiamond carbon from the coating. The CVD chamber is kept at a pressure of about 6.7 kPa (50 torr), and the substrate temperature is maintained at 950°C. Typical distance between the substrate and a fixed point of the cavity is ca $20-25$ mm. Some of the variables used include the type of substrate material, cg, cutting tool material, substrate preparation, substrate temperature, microwave power, distance between the substrate and a fixed point of the cavity, pressure inside the reactor, composition of the gases, their flow rates and flow ratios, and duration of the test. The deposition rate by this technique using CH_4–H_2 is ca 1 μm/h. Use of hot filament or microwave CVD has enabled the formation of crystalline diamond coatings with relative ease (Fig. 14). Depending on the processing conditions used, octahedral, cubic, or cauliflower-like structures can be formed. Whereas the rough surface topography can reduce the friction between the tool rake face and the sliding chip, it can be the base to which metal may be anchored during cutting of soft material and subsequent metal buildup. In such cases it may be necessary to polish the rake face. This, of course, adds to the cost of the cutting tool.

Figure 15 shows the variation of diamond deposition rates by various activated CVD techniques as well as the HP–HT technique (165). It can be seen that the highest growth rate of activated CVD diamond synthesis is still an order of magnitude lower than the HP–HT technique. However, CVD has the potential to become an alternative for diamond growth in view of the significantly lower cost of activated CVD equipment and lower running and maintenance costs.

The activated CVD diamond techniques can be more attractive in cases where the huge capital investment (several hundred million dollars) required

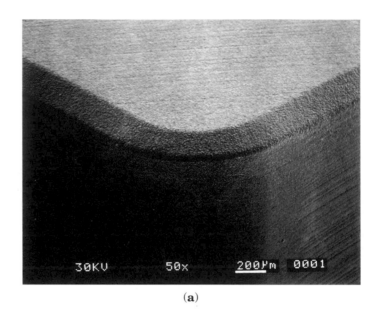

(a)

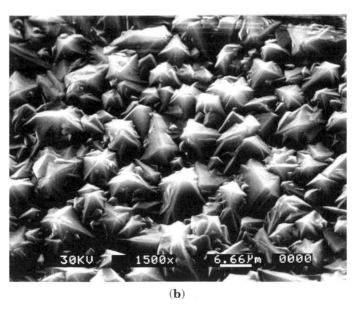

(b)

Fig. 14. A scanning electron micrograph of a diamond coating on a silicon nitride cutting tool; (**b**) at higher magnification, the octahedral growth of diamond.

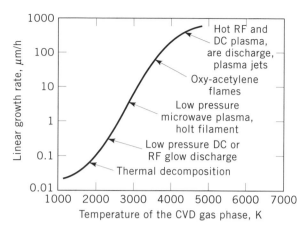

Fig. 15. Variation of diamond deposition rates by various activated CVD techniques as well as the HP–HT technique (165).

for the HP–HT technology is not available or where the high level of technical knowledge required for HP–HT synthesis is not available. In addition, most wear-resistant applications require diamond coatings only of the order of a few micrometers thick. Such coatings can be deposited directly on the finished product without the need for further finishing if CVD techniques are employed.

The low pressure CVD diamond data from various researchers has been summarized (167). Figure **16a** is the atomic C–H–O diamond deposition phase diagram for all diamond CVD methods used. Most of the combustion synthesis experiments were conducted along the acetylene line. Most of the plasma and hot filament experiments were conducted using highly diluted mixtures of hydrocarbon and hydrogen, sometimes with additional oxygen. The diamond region is very narow in the hydrogen-rich end of the phase diagram and broadens considerably on the C–O line. This diagram indicates that the low pressure diamond synthesis is feasible only within a well-defined field of the phase diagram.

The effect of the substrate temperature can also be considered (Fig. **16c**). As the substrate temperature increases, the triangular diamond domain region in the C–H–O equilibrium diagram shrinks to almost a line at the highest temperature.

In order to improve adhesion between diamond and the cutting tool substrate, various approaches are being taken. One reason advanced for the poor bonding in the case of cemented carbide substrates is the presence of Co (168). To overcome this problem, either a low Co cemented carbide grade is chosen or the surface Co is etched away before diamond coating takes place (169). One of the desirable characteristics for good bonding between the substrate and the coating is chemical compatibility and good match in crystal structure and lattice parameters between them. An element that comes closest to meeting these requirements in the case of diamond is silicon. Bonding between polycrystalline diamond films grown on a silicon substrate using the microwave CVD is very strong. It is therefore possible to deposit a thin film of silicon on the tool substrate

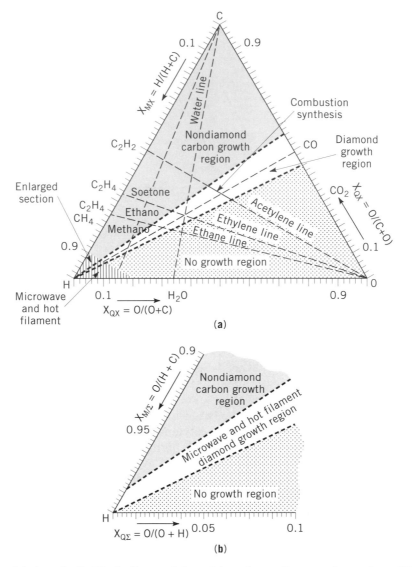

Fig. 16. (**a**) Atomic C–H–O diamond deposition phase diagram for various CVD diamond methods used showing a definite diamond growth region (adapted from Ref. 167); (**b**) enlarged hydrogen-rich region of the phase diagram shown in (**a**) where most of the plasma and hot filament experiments were conducted (adapted from Ref. 167); (**c**) effect of the substrate temperature on the diamond domain in the C–H–O diamond deposition phase diagram (adapted from Ref. 167).

before coating with diamond. Another factor in good bonding is good thermal expansion match between the coating and the substrate. Diamond with the highest thermal conductivity is difficult to match with most tool materials. However, an interlayer of high conductivity material, in which diamond particles

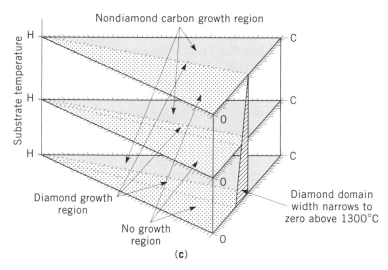

Fig. 16. *Continued*

can be cemented may be appropriate. The coating can, therefore, adhere strongly to the substrate without spalling.

Another approach is to coat the cutting tool material with a carbide former, such as titanium or silicon or their respective carbides by CVD and deposit diamond on top of it. The carbide layer may serve as an interface between diamond and the cemented carbide, thus promoting good bonding. Yet another method to obtain adherent diamond coatings is laser-induced microwave CVD. By ablating the surface of the substrate with a laser (typically, ArF excimer laser) and coating this surface with diamond by microwave CVD, it is possible to improve the adhesion between the tool and the substrate. Partial success has been achieved in this direction by many of these techniques.

Efforts to deposit diamond coatings on cutting tools are concentrated in the United States (chiefly Crystallume, Norton, Kennametal, and General Electric), Japan (Sumitomo Electric, Toshiba Tungaloy, Mitsubishi Metal Corporation, and Asahi Diamond Company), and Europe (Sandvik). Two approaches are generally taken for CVD diamond coatings on cutting tools. One is to grow thick (1–1.5-mm), free-standing polycrystalline diamond slabs cut to size, and braze them onto a cemented carbide substrate. However, the tools have to be finished before use. Most companies are attempting to commercialize this type of CVD diamond tool. As of this writing (ca 1997), such tools are available commercially on a limited basis. This tool type, in concept, is not much different from the polycrystalline diamond made by the HP–HT process.

The other approach is to develop thin (2–5-μm) coatings on cutting tools. This requires use of the CVD diamond process. Microwave CVD, hot filament CVD, plasma torch, and combustion synthesis are some of the techniques used either individually or in combination to deposit diamond coatings on cutting tools. The advantage of this technique is that no subsequent finishing of the tool is required, thus saving considerably on finishing costs. It is this aspect that made hard, refractory thin $(1-10\text{-}\mu\text{m})$ coatings on cemented carbide tools attractive and economical.

Adhesion of the diamond coating to the tool substrate is the main problem challenging researchers. Many difficulties have to be overcome before a good metallurgical bond between diamond coating and the substrate can be developed on cutting tools. Diamond is the most difficult material to bond with most materials because of its unique characteristics. There is therefore a lack of suitable substrate materials having similar properties. To form a good metallurgical bond at the interface in a coated tool, the characteristics of the coating and the substrate should be matched as closely as possible. This includes matching of thermal expansion coefficients, lattice parameters, chemistry, etc. A graded interface having properties of the coating near the substrate closer to the substrate properties and properties near the coating closer to the coating properties must be used in this case.

Cubic Boron Nitride (cBN)

Cubic boron nitride (cBN), next only to diamond in hardness (Knoop hardness 46.1 GPa (ca 4700 kgf/mm^2)), was developed in the late 1960s (153–155). It is a remarkable material in that it does not exist in nature and is produced by high temperature–high pressure (HP–HT) synthesis in a process similar to that used to produce diamond from graphite. Hexagonal boron nitride (hBN) is used as the starting material. Alkaline-earth metals and their compounds (instead of Ni in the case of diamond) are found to be the suitable catalyst–solvent for the production of cBN by the HP–HT process. Cubic boron nitride, although not as hard as diamond, is less reactive with ferrous materials like hardened steels, hard chill-cast iron, and nickel-base and Co-base superalloys. It can be used efficiently and economically at higher speed (ca five times), with a higher removal rate (ca five times) than cemented carbide, and with superior accuracy, finish, and surface integrity. Sintered cBN tools are fabricated in the same manner as sintered diamond tools and are available in the same sizes and shapes. Their costs are significantly higher than those of either cemented-carbide or ceramic tools because of higher processing and shaping costs. Like the sintered polycrystalline diamond tools, cBN tools are held on standard tool holders. In order to gain full potential of this material, very rigid precision machine tools with adequate speed and power capabilities are recommended.

Polycrystalline cBN is used extensively for machining of high hardness steels (Rc > 45), nickel-base superalloys, and alloyed cast iron. However, the development of other, less expensive tool materials, chiefly ceramics (SiC whisker-reinforced alumina and SiAlON) for machining of nickel-base alloys is challenging the use of this material. To compete with lower cost advanced tool materials, such as ceramics, coated carbides, newer fabrication technology is under development. For example, instead of fabricating sintered cBN tools on a cemented–carbide base, tools of ca 1.5 mm thickness are fabricated without the cemented carbide support base and the tool faces are ground on either side (170). Such a tool can be used on both sides, which roughly doubles its life. The tool, however, has to be properly supported and clamped during use to prevent premature failure. Polycrystalline cBN tools are very hard and consequently somewhat brittle. Also, cBN tools can be cut into segments of a pie and several tools can be made instead of one round tool.

In order to extend applications of cBN to include machining of medium-hardness steels, modifications of the cBN were introduced. An example is the fabrication of sintered cBN tools by the same HP–HT process, but using binder and second phase (either metallic or nonmetallic) such as TiN or TiC to increase toughness (171). In regard to phase distribution, cBN tools resemble cemented-carbide or alumina–TiC ceramic tools, but are tougher and have greater chemical stability.

The two predominant wear modes of cBN tools are DCL notching and microchipping. Polycrystalline cBN tools exhibit flank wear where alumina ceramic tools fail catastrophically. These tools have been used successfully for heavy interrupted cutting and for milling white cast iron and hardened steels. Negative lands and honed cutting edges were used. Like diamond, cBN is thermally unstable at elevated temperatures. The reaction products, however, when machining materials like steel- or nickel-base alloys, are generally not damaging to the process. cBN tools are not recommended for very low or very high speed cutting applications. Nevertheless, these tools are capable of very high removal rates when used with machine tools of adequate power and stiffness. In fact, they perform better with heavier cuts than with lighter cuts. Because they are inherently brittle, cBN tools are used with a negative rake ($-5°$) geometry. Suitable edge preparation, consisting of honing a small radius or a small negative land on the rake face, is also recommended.

Diamond and cBN tools provide significantly higher performance capability, and demands are being placed on the machine tools and manufacturing practice in order to take full advantage of the potential of these materials. Being extremely hard but brittle, the rigid machine tools must be used with gentle entry and exit of the cut in order to prevent microchipping by cleavage. High precision machine tools offer the advantage of producing high finish and accuracy. Use of machine tools with higher power and rigidity enables higher removal rates. A more recent application of cBN is the finish machining of hardened steels, such as in bearing races. Typically these materials are machined in the annealed condition to remove much of the unwanted material and subsequently heat-treated to the desired hardness and ground to the required size, accuracy, and finish. This can be a time-consuming and expensive operation. Using cBN tools, the bearing steels are obtained in their final hardness condition and are machined to the required size, accuracy, and finish without the need for subsequent grinding on rigid, high precision machine tools.

Economic Aspects

Machining costs (labor and overhead) in the United States have an estimated value of $> \$300 \times 10^9$/yr. The cost of labor and overhead for machining is based on the estimated number of total metal-cutting machine tools in various metal-cutting industries (1). This value does not take into account the cost of raw stock (work material), cutting tools, and many other support facilities. An estimated breakdown of cutting tool costs is given in Table 13 (172). Because of the competitive nature of these industries, most prefer to keep cost information proprietary.

In 1981, the value of disposable metal-cutting tools shipped to various U.S. manufacturing plants was estimated to be 2.13×10^9, only 1% of the total estimated U.S. manufacturing costs. Thus, the cost of cutting tools per se is only a small fraction of the total costs, although the tooling costs may be significant in a large manufacturing facility. The costs associated with the use of cutting fluids is estimated to be about 16% of the manufacturing costs (173–175).

High speed steels (HSS) and cemented carbides were the most extensively used tool materials in 1996, accounting for ca 2×10^9 in sales. From $1–1.25 \times 10^9$ were for HSS and the remaining portion for carbides. The market for ceramics is ca 25×10^6. Although uncoated tools of the HSS and cemented carbide materials are still used, the trend is toward more extensive (ca 60% in the United States and western Europe) use of coated tools. For HSS coated tools this percentage has not quite reached 50%. This is partly because HSS tools are relatively inexpensive and hence some of the customers have not fully appreciated the tangible benefits of coatings, especially when used in small batches as in small job shops, and also partly because many HSS tools are reground, ie, they have to be recoated for use as coated tools. Diamond and cubic boron nitride (cBN) are used for special applications where despite high cost, use is justified because of high hardness. The market for polycrystalline diamond and cBN is in the range of $25–50 \times 10^6$. Most of the cost information is kept confidential by individual companies and not disclosed; the costs given here are, therefore, estimates based on indirect information. Cast-cobalt alloys are presently phased out because of the high cost of raw materials (Co, Cr, and W), and because of safety problems encountered in the handling of Co and the increasing availability of alternative materials with superior performance at reduced cost. New ceramics will have significant impact on future manufacturing productivity, especially as improved fracture toughness and strength and hence the reliability of these materials occur.

Table 13. Estimated Breakdown of Cutting Tool Costs

Material	Cost, $
high carbon, low alloy, HSS	$(1–125) \times 10^9$
cemented carbides	750×10^{6a}
ceramics	25×10^6
diamond, cBN	$(25–50) \times 10^6$

^{a}About half is for coated grades.

Health and Safety Factors

Threshold limit values for the components of cemented carbides and tool steels are given in Table 14 (176). There is generally no fire or explosion hazard involved with tool steels, cemented carbides, or other tool materials. Fires can be handled as metal fires, eg, with Type D fire extinguishers. Most constituents of tool materials do not polymerize.

Table 14. Threshold Limit Values (TLV)

Constituent	TLV, mg/m^3
tungsten carbide	5
titanium carbide	na
tantalum carbide	5
chromium carbide	0.5
cobalt	0.1
nickel	1
iron	na
tungsten	5
copper	1

During machining operations, eye protection is recommended; during grinding operations, NIOSH-approved respirators for metal fumes and dust are recommended (177). Fine powder of Co is known to cause dermatitis and pulmonary disorders in humans. Most manufacturers supply safety information with their products (178). These should be followed strictly for the welfare of the personnel on the shop floor.

Safety is of particular concern in metal-cutting and metal-forming operations (178). Precautions should be taken to ensure protection of personnel and equipment from potential flying fragments and sharp edges as well as the large volumes of chips produced. Safety devices and protective shields or screens must be installed on metal-cutting machines. Chips should be handled with some mechanical device, never by hand. In automated machining, chip handling by effective chip groove geometry should be practiced. Tool overhang must be as short as possible to avoid instances of deflection, resulting in breakage or chatter. Noise caused by chatter or vibration can be highly objectionable to personnel nearby, in addition to the operator. Corrective measures should be taken wherever chatter prevails. These include change in the cutting conditions, modification of the tool–work support system to increase its stiffness, or operating at conditions below or above the natural frequencies of parts of the machine tool that cause chatter.

Some cutting fluids, eg, oils, may present a fire hazard. Some work materials, eg, magnesium, aluminum, titanium (under certain conditions), and uranium, in finely divided form, also present fire hazards. Very small metal chips or dust may ignite.

Adequate ventilation of grinding operations should be established to comply with existing government regulations, and management should remain alert to the possibility of symptoms, even in grinders working within established government standards.

The high temperatures generated in machining, especially at high cutting speeds, necessitate the use of a refractory cutting tool that can withstand the high temperature and provide long tool life. Cutting fluids are needed to absorb the high heat, cool the cutting tool at higher speeds, lubricate at low speeds and high loads, increase the tool life, improve the surface finish, reduce the cutting forces and power consumption, etc. The physiological effects of cutting fluids on the operator must be considered. Toxic vapors, unpleasant odors, smoke fumes, skin irritations (dermatitis), or effects from bacteria cultures from the cutting fluid are all factors. Consumption of cutting fluids has been reduced drastically by using mist lubrication. However, mist in the industrial environment can have a serious respiratory effect on the operator. Consequently, standards are being set to minimize this effect. Many industrialized nations, including Germany, have made commitment to provide a safer working environment and the United States has no option but to provide a similar manufacturing environment.

Future Outlook

The raw materials used in the cutting tools are currently (ca 1997) made by melting and subsequent size reduction by milling to fine size. Many defects such as microcracks, voids, etc, are generated during solidification and the subsequent size-reduction process. The comminution process limits the size of the crystallites to ca 1-μm. Newer technologies are being developed based on chemical routes, such as the sol–gel technique for the production of ultrafine materials in nanocrystalline size. These materials are also relatively free from defects and hence tool materials based on nanocrystalline materials may increasingly be used, especially for lower speed, roughing, or intermittent cutting operations or operations where edge strength is an essential requirement, as in milling.

In many of the thin coatings on cemented carbide, either single or multiple coatings of single phase materials, such as TiC, are used. It would appear that extending the use of solid solutions of multicarbides of W, Ti, and Ta or Nb for coatings may further enhance the performance of the coated carbides. It would not be difficult to accomplish this either by CVD or PVD techniques.

Even though TiC is much harder than WC at room temperature (3200 kg/mm^2 for TiC, vs 1800 kg/mm^2 for WC), at higher temperatures, TiC oxidizes and loses its hardness rapidly. Figure 17 is a plot of the variation of hardness of single crystals of various monocarbides with temperature (44). No similar data is available for multicarbides or other refractory hard materials, such as nitrides, borides, oxides, or any combination of them.

It is known that fracture toughness of materials can be increased significantly by transformation toughening as in the case of ZrO$_2$, by applying crack deflection methods involving fine dispersion of second-phase particles, as in the case of Al$_2$O$_3$ + TiC, and whisker pullout in as in the case of SiC-whisker-reinforced alumina. This approach is expected to find greater use for a range of tool materials where toughness is an important consideration, eg, in ceramic tools.

Metal cutting research in the 1950s (179) clearly showed that significant reductions in forces can result using an increase in the rake angle ($\leq 45°$). Consequently, the energy requirements and heat generated, etc, would also be reduced. For the most part as of 1996 $-5°$ rake and 5° clearance inserts were used. This is true partly owing to the application of the same insert for a range of work-materials and partly owing to concerns that high positive geometry might render the tool weak, especially when machining high strength materials.

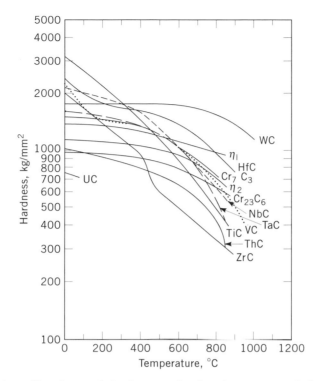

Fig. 17. Variation of hardness of single crystals of various monocarbides with temperature (44).

Use of higher rake angles (ca 45°) would permit higher cutting speed for the same tool life or longer tool life for the same cutting speed, improved surface finish, lower cutting forces resulting in lower cutting energy and power requirements or higher removal rates, lower thrust forces and consequently lower deflections, and reduced tool wear owing to lower interface temperature. However, strength of the cutting edge is rather critical owing to low included angle of the tool. Submicrometer-grain cemented carbide material or submicrometer-grain TiC or TiN in HSS steels are recommended for such applications, especially at low cutting speeds. All these approaches lead to increased efficiency and longer tool life.

The implementation of coated tool technology on the shop floor is proceeding at a significant pace. New coating combinations, tailored substrates, CVD/PVD technologies for different types of engineered coatings for different workmaterials, cutting conditions, machining operations are being addressed effectively. This effort is expected to continue. Multiple nanolayer coatings which have more recently been developed have improved hardness, strength, and chemical stability. The practically unlimited choice of coating combinations, ie, alternate hard and tough materials, or alternate hard material and solid lubricant, etc, should lead to numerous multiple nanolayer coating applications. Nanolayer coatings may be ideal for hard refractory coatings which have difficulty in bonding with the substrate or other coatings.

SiC whisker-reinforced alumina is a major advance in tool material development, as it provides a means to increase the fracture toughness of the material via the composite material approach. It is entirely possible that in the next century many new whiskers of refractory, hard materials will be made available economically for application to cutting tools. One may even consider SiC whisker-reinforced alumina as a model material on the basis of which many new tool materials may be developed. Tool material for high speed machining of titanium alloys may evolve from this concept as most tool materials are very reactive with respect to titanium. Some of the intermetallics may be candidate materials for this application.

Another interesting concept, one used in the development of superalloys, is the strengthening of the matrix by orderly precipitation of the second-phase materials and strengthening by dispersoids, such as Al_2O_3. Strengthening by dispersoid is to some extent already practiced, as in the cases of TiC in Al_2O_3, and TiN in cemented TiC. However, orderly precipitation by solution treatment and aging can be an attractive approach. For example, one can consider extensive dispersion of fine precipitates of TiC, TiN, or TiB_2 in a titanium matrix; ZrB_2 in a zirconium matrix; or B_4C in a boron matrix; or hot-pressing of B_4C in titanium matrix.

Cemented carbides are fairly expensive owing to the use of hard, refractory materials. This is expected to become even more the case as some of the strategic materials used in these tools become more expensive or newer but more expensive materials such as HfC or HfN come into more common use. It may be economical, therefore, to use these materials at or near the cutting edge instead of as the whole insert. The development of tools of TiC (40–55%) or TiN (30–60%) in a steel matrix on a steel core using powder metallurgy technology suggests a similar approach for cemented carbides as the need arises.

BIBLIOGRAPHY

"Tool Materials for Machining" in *ECT* 1st ed., Suppl. 2, pp. 873–882, by Roland B. Fischer, Battelle Memorial Institute; "Tool Materials for Machining" in *ECT* 2nd ed., Vol. 20, pp. 566–578, by Roland B. Fischer, The Dow Chemical Company; "Tool Materials" in *ECT* 3rd ed., Vol. 23, pp. 273–309, by R. Komanduri and J. D. Desai, General Electric Co.

1. *Machinability Data Handbook*, 3rd ed., Vols. 1 and 2, Machinability Data Center, Metcut Research Associates, Cincinnati, Ohio, 1980.
2. E. Dow Whitney, ed., *Ceramic Cutting Tools*, Noyes Publications, Park Ridge, N.J., 1994.
3. R. Komanduri, *ASME Appl. Mechan. Rev.* **46**(3), 80–132 (Mar. 1993).
4. E. D. Doyle, personal communication, 1981.
5. V. A. Tipnis, in M. B. Peterson and W. O. Winer, eds., *Wear Control Handbook*, ASME, New York, 1980, pp. 891–930.
6. U.S. Pat. 4,286,905 (Sept. 1, 1981), S. K. Samanta (to Ford Motor Co.).
7. H. G. Swinehart, ed., *Cutting Tool Materials Selection*, American Society of Tool and Manufacturing Engineers (now the Society of Manufacturing Engineers (SME)), Dearborn, Mich., 1968.

8. R. Komanduri, "Cutting Tool Materials" in M. B. Bever, ed., *Encyclopedia of Materials Science and Engineering*, Pergamon Press, Oxford, U.K., 1983; *General Electric TIS report no. 82CRD176*, Schenectady, N.Y., June 1982.

9. R. Komanduri and J. D. Desai, *Tool Materials for Machining*, General Electric TIS report no. 82CRD220, Schenectady, N.Y., Aug. 1982.

10. *Machinability Data Handbook*, 3rd ed., Vols. 1 and 2, Machinability Data Center, Metcut Research Associates, Cincinnati, Ohio, 1980.

11. *Manuf. Eng.*, 49 (Oct. 1977).

12. K. J. A. Brooks, *World Directory and Handbook of Hard Metals*, 5th ed., An Engineer's Digest Ltd., U.K., 1992.

13. R. L. Hatschek, *Am. Machinist*, 165–176 (May 1981).

14. *Properties and Selection of Tool Materials*, American Society for Metals, Metals Park, Ohio, 1975.

15. *Properties and Selection: Stainless Steels, Tool Materials and Special-Purpose Metals*, Vol. 3 of *Metals Handbook*, 9th ed., American Society for Metals, Metals Park, Ohio, 1979, pp. 421–488.

16. J. T. Berry, "*Recent Developments in the Processing of High-Speed Steels*," Climax Molybdenum Co., Greenwich, Conn., 1970.

17. *Properties and Selection of Tool Materials*, American Society for Metals, Metals Park, Ohio, 1962.

18. G. A. Roberts, J. C. Hamaker, and A. R. Johnson, *Tool Steels*, American Society for Metals, Metals Park, Ohio, 1962.

19. A. M. Bayer and B. A. Beecherer, in *ASM Handbook*, 9th ed., Vol. 16, *Machining*, American Society for Metals, Metals Park, Ohio, 1989, pp. 51–59.

20. E. J. Dulis and T. A. Neumeyer, *Materials for Metal Cutting*, ISI Publication P216, The Iron and Steel Institute, U.K., 1970, pp. 112–118.

21. M. G. H. Wells and L. W. Lherbier, eds., *Processing and Properties of High Speed Tool Steels*, Metallurgical Society of AIME, Warrendale, Pa., 1980.

22. R. W. Stevenson, in *ASM Handbook*, 9th ed., Vol. 7, *Powder Metallurgy*, 1984, pp. 784–793.

23. K. E. Pinnow and W. Stasko, in Ref. 19, pp. 60–68.

24. R. Komanduri and M. C. Shaw, *Proceedings of the Third North American Metal Working Research Conference (NAMRC-III), May 5–7, 1975*, Carnegie Press, Pittsburgh, Pa., 1975.

25. R. Wilson, *Metallurgy and Heat Treatment of Tool Steels*, McGraw-Hill Book Co., Inc., New York, 1975.

26. W. T. Haswell, W. Stasko, and F. R. Dax in Ref. 20, pp. 147–158.

27. S. J. Whalen, *Vapor Deposition of Titanium Carbide*, ASTME Technical Paper No. 690, American Society of Tool and Manufacturing Engineers (now Society of Manufacturing Engineers (SMF)), Dearborn, Mich., 1965.

28. U.S. Pats. 2,962,388 and 2,962,399 (Nov. 29, 1960), W. Ruppert and G. Schwedler (to Metallgesellschaft Aktinengesellschaft).

29. M. Epner and E. Gregory, *Cermets*, Reinhold Publishing Co., New York, 1960, pp. 146–149.

30. J. L. Ellis, *Tool Eng.* **38**, 103–105 (1957).

31. S. E. Tarkan and M. K. Mal, *Metal Progr.* **105**, 99–102 (1974).

32. *Modern Metal Cutting—A Practical Handbook*, Sandvik Coromant, Sweden, 1994, pp. III-41 to III-44.

33. *Cobalt Monograph*, prep. by Battelle Memorial Institute, Columbus, Ohio for Centre d'Information du Cobalt, Brussels, Belgium, 1960.

34. Technical information brochures, Stellite Division of Cabot Corp., Kokomo, Ind.

35. H. Moissan, *The Electrical Furnace*, tr. V. Lenher, Chemical Publishing Co., 1904.

36. U.S. Pat. 1,549,615 (Aug. 11, 1925), K. Schroter (to GE).

37. U.S. Pat. 1,721,416 (July 16, 1929), K. Schroter (to GE).
38. W. J. Loach, MR 71-901 in Ref. 67.
39. W. Dawihl, *Handbook of Hard Metals*, English transl., Her Majesty's Stationery Office, London, 1955.
40. P. Schwarzkopf and R. Keiffer, *Cemented Carbides*, The Macmillan Company, New York, 1960, pp. 116–120.
41. H. E. Exner, *Int. Met. Rev.*, (243), 149 (1979).
42. R. W. Stevenson, "Cemented Carbides," in Ref. 22, pp. 773–783.
43. A. T. Santhanam, P. Tierney, and J. L. Hunt, in *ASM Handbook*, 10th ed., Vol. 2, *Properties and Selection*, American Society for Metals, Metals Park, Ohio, 1990, pp. 950–977.
44. L. E. Toth, *Transition Metal Carbides and Nitrides*, Academic Press, New York, 1971.
45. U.S. Pat. 2,015,536 (Sept. 24, 1935), K. Schroter (to GE).
46. U.S. Pat. 1,925,910 (Sept. 5, 1933), P. Schwarzkopf and I. Hirschel (to GE).
47. U.S. Pat. 2,113,353 (Apr. 5, 1938), P. M. McKenna (to Kennametal).
48. U.S. Pat. 2,113,354 (Apr. 5, 1938), P. M. McKenna (to Kennametal).
49. U.S. Pat. 2,113,355 (Apr. 5, 1938), P. M. McKenna (to Kennametal).
50. U.S. Pat. 2,113,356 (Apr. 5, 1938), P. M. McKenna (to Kennametal).
51. U.S. Pat. 2,124,509 (July 19, 1938), P. M. McKenna (to Kennametal).
52. J. Gurland and P. Bardzil, *Trans. AIME*, 311 (Feb. 1955).
53. M. J. Kuderko, (EM 71–937) in Ref. 67.
54. O. Kasukawa, (EM 71-938) in Ref. 67.
55. J. E. Mayer, D. Moskowicz, and M. Humenik, *Materials for Metal Cutting*, ISI Special Report No. P126, The Iron and Steel Institute, U.K., 1970, pp. 143–151.
56. U.S. Pat. 2,967,349 (Jan. 10, 1961), M. Humenik and D. Moskowicz (to Ford Motor Co.).
57. D. Moskowitz and M. Humenik, Jr., *Proceedings of the 1975 International Powder Metallurgy Conference*, New York.
58. W. W. Gruss, in Ref. 19, pp. 90–97.
59. C. G. Goetzel, in Ref. 22, pp. 798–815.
60. M. Humenik and N. Parikh, *J. Am. Cer. Soc.* **39**, 60 (1956).
61. H. Doi, in *Science of Hard Materials*, E. A. Almond, C. A. Brookes, and R. Warren, eds., 1984, pp. 489–523.
62. U.S. Pat. 4,902,395 (Feb. 20, 1990), H. Yoshimura (to Mitsubishi).
63. C. W. Beeghly, R. V. Godse, and F. B. Battaglia, SME Technical Paper No. MR 90-247, *Proceedings of the 3rd Conference on Adv. Machining Technology*, Chicago, Ill., Sept. 4–6, 1990, p. 247.
64. H. Tanaka, in *Cutting Tool Materials: Proceedings of the ASM International Conference, Ft. Mitchell, Ky.*, ASM International, Metals Park, Ohio, Sept. 1980, pp. 349–361.
65. R. Komanduri, *Carbide Tool J.* **18**(4), 33–38 (July/Aug. 1986).
66. R. Komanduri, *Advances in Hard Materials Tool Technology*, Carnegie Press, Pittsburgh, Pa., 1976.
67. *Proceedings from the 1st International Cemented Carbide Conference, Chicago, Ill., Feb. 1–3, 1971*, Vols. 1 and 2, Cemented Carbide Producers Association and the Society of Manufacturing Engineers, Dearborn, Mich.
68. N. P. Suh, *Wear* **62**, 1–20 (1980).
69. N. A. Horlin, *Prod. Eng. (London)*, 153 (1971).
70. A. T. Santhanam and D. T. Quinto, in *ASM Handbook*, American Society for Metals, Metals Park, Ohio, 10th ed., Vol. 5, *Surface Engineering*, 1994, pp. 900–908.
71. H. G. Prengel, W. R. Pfouts, and A. T. Santhanam, *SME Manuf. Eng.*, 82–88 (July 1996).
72. G. Schuhmacher, "TiC-Coated Hard Metal," MR71-930 in Ref. 67.

73. U.S. Pat. 3,977,061 (Aug. 31, 1976), J. N. Lindstrom and F. J. O. E. Ohlsson (to Sandvik).

74. U.S. Pat. 3,616,506 (Nov. 2, 1971), C. S. G. Ekemar (to Sandvik).

75. U.S. Pat. 3,836,392 (Sept. 17, 1973), B. Lux, R. Funk, H. Schachner, and C. Triquet (to Sandvik AB).

76. W. Scott Buist, MR71-931 in Ref. 67.

77. U.S. Pat. 3,642,522 (Feb. 15, 1972), H. Gass and H. E. Hintermann (to Laboratoire Suisse De Recherches Horlogeres).

78. Y. Ohtsu, FC71-932 in Ref. 67.

79. U.S. Pat. 3,736,107 (May 29, 1973), T. E. Hale (to Carboloy).

80. M. Lee, R. H. Richman, and J. Stanislao, FC 71-928 in Ref. 67.

81. U.S. Pat. 3,717,496 (Feb. 20, 1973), R. Kieffer.

82. U.S. Pat. 3,744,979 (July 10, 1973), H. S. Kalish (to Adamas Carbide).

83. B. M. Kramer, *Thin Solid Films* **108**, 117–125 (1983).

84. U.S. Pat. 4,101,703 (July 18, 1978), W. Schintlmeister (to Schwarzkopf Development Corp.).

85. W. Schintlmeister, W. Wallgram, J. Kanz, and K. Gigl, *Wear* **100**, 153–169 (1984).

86. Dreyer and co-workers, *Development and Tool Life Behavior of Super-Wear Resistant Multilayer Coatings on Hard Metals*, Book No. 278, Metals Society (London), 1982, pp. 112–117.

87. D. T. Quinto, A. T. Santhanam, and P. C. Jindal, *Int. J. Refrac. Metals Hard Mater.* **8**(2), 95–101 (June 1989).

88. U.S. Pat. 4,463,062 (July 31, 1984), T. E. Hale (to Carboloy).

89. B. J. Nemeth, A. T. Santhanam, and G. P. Grab in H. M. Ortner, ed., *Proceedings of the 10th Plansee Seminar on Trends in Refractory Metals and Special Materials and Their Technology*, Metallwerk Plansee, Ruette, Austria, 1981, pp. 613–627.

90. U.S. Pat. 4,610,931 (Sept. 9, 1986), B. J. Nemeth (to Kennametal).

91. A. T. Santhanam, G. P. Grab, G. A. Rolka, and P. Tierney, in V. K. Sarin, ed., *Proceedings of the High Productivity Machining*, American Society for Metals, Metals Park, Ohio, 1985, pp. 105–112.

92. U.S. Pat. 4,828,612 (May 9, 1989), W. C. Yohe (to Carboloy).

93. U.S. Pat. 4,830,930 (May 16, 1989), Y. Taniguchi, K. Sasaki, M. Ueki, and K. Kobori (to Toshiba Tungaloy).

94. U.S. Pat. 5,372,873 (Dec. 13, 1994), H. Yoshimura, T. Tanaka, A. Osada, and T. Sudo (to Mitsubishi Materials).

95. S.. Ramalingam, in Ref. 5, pp. 385–411.

96. R. Horsfall and R. Fontana, *Cutting Tool Eng.*, 37–42 (Feb. 1993).

97. J. von Stebul and co-workers, *Surf. Coating Technol.* **68/69**, 762–769 (1994).

98. *Catalog*, Balzers Tool Coatings, Inc., North Tonawanda, N.Y., 1996.

99. U.S. Pat. 5,503,913 (Apr. 2, 1996), U. Konig and R. Tabersky (to Widia GmbH).

100. W. J. Clegg, K. Kendall, N. McN. Alford, T. W. Burton, and J. D. Birchall, *Nature (London)* **347**, 455–457 (1991).

101. P. Anderson, *Scripta Metall. Mater.* **27**, 687–692 (1992).

102. M. C. Shaw, D. B. Marshall, M. S. Dadkhah, and A. G. Evans, *Acta Metall. Mater.* **41**(11), 3311–3322 (1993).

103. C. A. Folsom, F. W. Zok, and F. F. Lange, *J. Am. Ceramic Soc.* **77**(3), 689–696 (1994).

104. U.S. Pat. 4,804,583 (Feb. 14, 1989), T. D. Moustakas (to Exxon R&E Co.).

105. A. T. Alpas, J. D. Embury, D. A. Hardwick, and R. W. Springer, *J. Mater. Sci.* **25**, 1603–1609 (1990).

106. Y. Ding, Z. Farhat, D. O. Northwood, and A. T. Alpas, *Surf. Coatings Technol.* **62**, 448 (1993).

107. Y. Ding, Z. Farhat, D. O. Northwood, and A. T. Alpas, *Surf. Coatings Technol.* **68/69**, 459–467 (1994).

108. U.S. Pat. 4,643,951 (Feb. 17, 1987), J. E. Keem and J. D. Flasck (to Ovonic Synthetic Materials).

109. U.S. Pat. 5,268,216 (Dec. 7, 1993), J. E. Keem and B. M. Kramer (to Ovonic Synthetic Materials).

110. M. R. Hilton and co-workers, *Surf. Coating Technol.* **53**, 13–23 (1992).

111. G. Jayaram, L. D. Marks, and M. R. Hilton, *Surf. Coating Technol.* **76–77**, 393–399 (1995).

112. E. Santner, D. Klaffke, and G. Meier zu Kocher, *Wear* **190**, 204–211 (1995).

113. R. Hubler and co-workers, *Surf. Coating Technol.* **60**, 561–565 (1993).

114. R. Komanduri and J. Larsen-Basse, *ASME Mechan. Eng.* **111**, 74–79 (Jan. 1989).

115. A. G. King and W. M. Wheildon, *Ceramics in Machining Processes*, Academic Press, Inc., New York, 1966.

116. F. W. Wilson, ed., *Machining with Carbides and Oxides*, McGraw-Hill Book Co., Inc., 1962.

117. E. D. Whitney, SAE Technical Paper No. 810319, Society of Automotive Engineers, International Congress and Exposition, Feb. 23–27, 1981.

118. R. Komanduri and S. K. Samanta, in Ref. 15, pp. 98–104.

119. R. Komanduri, *Int. J. Refract. Metals Hard Mater.* **8**(2), 125–132 (June 1989).

120. U.S. Pat. 2,873,198 (Feb. 10, 1959), E. W. Goliber (to GE/Carboloy).

121. E. W. Golliber, in Proceedings of the 66th Annual Meeting of the American Ceramic Society (ACS), Westerville, Ohio, 1960.

122. J. Binns, ASTME (now SME) Paper No. 633, Society of Manufacturing Engineers, Dearborn, Mich., 1963, pp. 1–15.

123. K. H. Jack, *Metals Technol.* **9**, 297–301 (July 1982).

124. Y. Oyama and O. Kamigaito, *Jpn. J. App. Phys.* **10**, 1637 (1971).

125. K. H. Jack and W. I. Wilson, *Nature (London) Phys. Sci.* **238**, 28–29 (1972).

126. U.S. Pat. 3,992,166 (Nov. 9, 1976), K. H. Jack and W. I. Wilson (to Joseph Lucas Industries Ltd.).

127. U.S. Pat. 3,991,148 (Nov. 9, 1976), R. J. Lumby, R. R. Wills, and R. F. Horsley (to Joseph Lucas Industries, Ltd.).

128. J. G. Baldoni and S. T. Buljan, in E. Dow Whitney, ed., *Ceramic Cutting Tools*, Noyes Publications, Park Ridge, N.J., 1994.

129. S. K. Bhattacharyya and A. Jawaid, *Int. J. Prod. Res.* **19**(5), 589–594 (1981).

130. S. K. Bhattacharyya, A. Jawaid, and J. Wallbank, in R. Komanduri, K. Subramanian, and B. F. von Turkovich, *Proceedings of the ASME High-Speed Machining Symposium*, New Orleans, La., PED Vol. 12, American Society of Mechanical Engineers, pp. 245–262.

131. U.S. Pat. 4,227,842 (Oct. 4, 1980), S. K. Samanta, S. Subramanian, and A. Ezis (to Ford Motor Co.).

132. U.S. Pat. 4,401,617 (1983), A. Ezia, S. K. Samanta, and K. Subramanian (to Ford Motor Co.).

133. U.S. Pat. 4,401,238 (Feb. 28, 1984), A. Ezia, S. K. Samanta, and K. Subramanian (to Ford Motor Co.).

134. S. K. Samanta and K. Subramanian, Proceedings of the North American Manufacturing Research Conference (NAMRC), University of California, Berkeley, Calif., 1985.

135. *Engineering*, 1009 (Sept. 1980).

136. V. K. Sarin and S. T. Buljan, in *High Productiv. Machin. Proc. Int. Conf. High Productivity Machining*, New Orleans, La., May 7–9, 1985, ASM International, Metals Park, Ohio, pp. 113–120.

137. U.S. Pat. 4,440,547 (Apr. 1984), V. K. Sarin and S. J. Buljan (to GTE).

138. U.S. Pat. 4,406,668 (Sept. 27, 1983), V. K. Sarin and S. J. Buljan (to GTE).

139. U.S. Pat. 4,406,669 (Sept. 27, 1983), V. K. Sarin and S. J. Buljan (to GTE).
140. U.S. Pat. 4,406,670 (Sept. 27, 1983), V. K. Sarin and S. J. Buljan (to GTE).
141. U.S. Pat. 4,543,345 (Sept. 24, 1985), G. C. Wei.
142. K. H. Smith, *Carbide Tool J.* **18**(5), 8–10 (Sept.–Oct. 1986).
143. U.S. Pat. 4,543,345 (Sept. 24, 1985), G. C. Wei (to Oakridge National Laboratories, DOE).
144. U.S. Pat. 4,749,667 (June 7, 1988), C. Jun (to Carboloy).
145. S. T. Buljan, J. G. Baldoni, and M. L. Juckabee, *Ceram. Bull.* **66**(2), 347–352 (1987).
146. P. K. Mehrotra, *M.P.R.*, 506–510 (July–Aug. 1987).
147. E. R. Billman, P. K. Mehrotra, L. F. Shuster, and C. W. Deeghly, *Ceram. Bull.* **67**(6), 1016–1019 (1988).
148. S. A. Buljan and S. F. Wayne, *Adv. Ceram. Mater.* **2**(4), 813–816 (1987).
149. J. R. Baldoni and S. T. Buljan, *Ceram. Bull.* **67**(2), 381–387 (1988).
150. J. G. Lee and I. B. Cutler, *Ceram. Bull.* **54**(2) (1975).
151. C. K. Jun and K. H. Smith, in Ref. 3, pp. 86–111.
152. C. F. Lewis, *Mater. Eng.*, 31–35 (July 1986).
153. F. P. Bundy, H. T. Hall, H. M. Strong, and R. H. Wentorf, Jr., *Nature* **176**, 51 (1955).
154. F. P. Bundy, *Sci. Am.* **231**, 62 (Aug. 1974).
155. R. H. Wentorf, Jr., R. C. DeVries, and F. P. Bundy, *Science* **208**, 873 (May 23, 1980).
156. L. E. Hibbs, Jr. and R. H. Wentorf, Jr., *High Temp. High Pressure* **6**, 409 (1974).
157. U.S. Pat. 4,224,380 (Sept. 23, 1980), H. P. Bovenkerk and P. D. Gigl (to GE).
158. S. Matsumoto, Y. Sato, M. Tsutsumi, and N. Setaka, *J. Mater. Sci.* **17**, 3106 (1982).
159. B. V. Derjaguin and D. V. Fedoseev, *Jzd. Nauka* (*Moscow*) (1977).
160. B. V. Derjaguin, L. L. Bouilov, and B. V. Spitsyn, *Arch. Nouki Mater.* **7**, 111 (1986).
161. T. Anthony, personal communication, 1982.
162. Y. Hirose, S. Amanoma, N. Okoda, and K. Komaki, Abstract, *First International Conference on New Diamond Science and Technology*, Tokyo, Japan, 1988, p. 38.
163. T. P. Ong and R. P. H. Chang, *Appl. Phys. Lett.* **55**(20), 2063–2065 (1989).
164. J. Suzuki, H. Kawarada, K. Mar, J. Wei, Y. Yokota, and A. Hiraki, *Jpn. J. App. Phys.* **28**(2), L281–L283 (1989).
165. R. Komanduri and S. Nandyal, *Int. J. Mech. Tools Manuf.* **33**(2), 285–296 (1993).
166. E. E. Sprow, *Manuf. Eng.*, 41–46 (Feb. 1995).
167. P. K. Bechmann, D. Leers, and H. Lydtin, *Diamond Relat. Mater.* **1**, 1–12 (1991).
168. R. Haubner and B. Lux, *J. de Phys.* **C5**(5), C5169–C5176 (May 1989).
169. U.S. Pat. 5,236,740 (Aug. 17, 1993), M. G. Peters and R. H. Cummings (to National Center for Manufacturing Sciences (NCMS)).
170. K. S. Reckling, *Tool Prod.*, 74 (Dec. 1981).
171. N. Tabuchi and co-workers, *Sumitomo Elec. Tech. Rev.* **18**, 57 (Dec. 1978).
172. P. M. Klutznick, *1981 U.S. Industrial Outlook for 200 Industries with Projections for 1985*, U.S. Department of Commerce, Washington, D.C., 1981.
173. G. Byrne and E. Scholta, *Ann. CIRP* **42**(1), 471–474 (1993).
174. *Machinery Prod. Eng.*, 14–20 (June 3, 1994).
175. R. B. Aronson, *Manuf. Eng.* **114**(1), 33–36 (Jan. 1995).
176. Material safety data sheets, Carboloy Company, Detroit, Mich.
177. M. E. Lichtenstein, F. Bartl, and R. T. Pierce, *Am. Ind. Hyg. Assoc. J.*, 879 (Dec. 1975).
178. *Turning Handbook of High Efficiency Cutting*, GT9-262, Carboloy Systems Business Department, General Electric Company, Detroit, Mich., 1980.
179. J. H. Crawford and M. E. Merchant, *Trans. ASME* **75**, 561–566 (May 1953).

RANGA KOMANDURI
Oklahoma State University

TOOTHPASTE.   See Dentifrices.

TORPEX.   See Explosives and propellants.

TOXAPHENE.   See Insect control technology.

TOXICOLOGY

Natural and synthetic chemicals affect every phase of our daily lives in both good and noxious manners. The noxious effects of certain substances have been appreciated since the time of the ancient Greeks. However, it was not until the sixteenth century that certain principles of toxicology became formulated as a result of the thoughts of Philippus Aureolus Theophrastus Bombastus von Hohenheim-Paracelsus (1493–1541). Among a variety of other achievements, he embodied the basis for contemporary appreciation of dose–response relationships in his often paraphrased dictum: "Only the dose makes a poison."

Subsequently, further concepts of toxicology came from the work and writings of Bonaventura Orfila (1787–1853), a Spanish physician. Among his most important accomplishments were the delineation of the discipline of toxicology, attempts to correlate chemical nature and toxic effect, and the establishment of bases for the specialization of forensic toxicology (see Forensic chemistry). The twentieth century has seen significant development of the discipline of toxicology, and its establishment as a professional activity. This has been at least in part the result of a phenomenal increase in the number of synthetic industrial and agricultural chemicals and pharmaceutical preparations. Relying particularly on developments in the chemical, physical, and general biological sciences, toxicologists have made substantial advances in defining the nature and mechanisms of toxic injury, in determining factors which may influence the expression of a toxic effect, in developing methodologies, and in obtaining information on the toxicity of a multitude of discrete or combined substances. The historical development of toxicology has been reviewed (1).

There are about as many definitions of toxicology as there exist textbooks on the subject. Although they differ in detail, all good definitions embrace the concept the toxicology is concerned with the potential of chemicals, or mixtures of them, to produce harmful effects in living organisms. Toxicology is a study of the interactions between chemicals and biological systems in order to determine quantitatively the potential for such chemicals to produce injury that results in adverse health effects in intact living organisms, and to investigate the nature, incidence, mechanism of production, and reversibility of such adverse effects.

In the context of the above definition, adverse health effects are taken to mean those which are detrimental either to the survival or to the normal functioning of the individual. This definition is intended to highlight the following points with respect to phenomena investigated in toxicology.

Materials that produce harmful effects must come into close structural or functional relationship with the tissue or organ they may affect. As a result, they can physically or chemically interact with particular biological components in order to effect the toxic response.

Investigations are carried out using a variety of biological systems, including observations on exposed whole animals (*in vivo* studies) or on appropriately treated isolated tissues and cells, homogenates of tissues, or cultured lower organisms (*in vitro* studies).

If possible, there should be measurement of the toxic effect in order quantitatively to relate the observations made to the degree of exposure (exposure dose). Ideally, there is a need to determine quantitatively the toxic response to several differing exposure doses, in order to determine the relationship, if any, between exposure dose and the nature and magnitude of any effect. Such dose–response relationship studies are of considerable value in determining whether an effect is causally related to the exposure material, in assessing the possible practical (in-use) relevance of the exposure conditions, and to allow the most reasonable estimates of hazard.

A prime consideration in toxicology, and a principal reason behind the need for dose–response relationship information, is the potential for a material to produce harmful effects. Thus, different materials may produce similar toxic effects, but the exposure doses to produce a just-detectable (threshold) effect, or a given degree of injury, may vary significantly among the materials. Hence, the magnitude of the injury produced varies for similar exposure doses to different materials. Also, with a particular material that is known to cause a variety of toxic effects, the individual effects may be produced only by differing exposure conditions. In assessing the relation of any induced toxic effects to practical situations, and in comparing and contrasting different materials, potency is clearly a central issue. Toxicity describes the nature of harmful effects produced with respect to the conditions necessary for their induction; ie, the toxicity of a material is its potential to produce biological injury.

Although studies are carried out by *in vitro* and *in vivo* procedures, it is the primary aim of toxicology to determine the potential for harmful effects in the intact living organism, usually with an ultimate goal of assessing the significance of the findings with respect to humans.

There are four components to a complete risk assessment (2):

(*1*) Hazard identification involves gathering and evaluating data on the types of health injury or disease that may be produced by a chemical and on the conditions of exposure under which injury or disease is produced. It may also involve characterization of the behavior of a chemical within the body and the interactions it undergoes with organs, cells, or even parts of cells. Hazard identification is not risk assessment. It is a scientific determination of whether observed toxic effects in one setting will occur in other settings.

(*2*) Dose–response evaluation is used in describing the quantitative relationship between the amount of exposure to a substance and the extent of toxic injury or disease. Data may be derived from animal studies or from studies in exposed human populations. Dose–response toxicity relationship for a substance varies under different exposure conditions. The risk of a substance can not be ascertained with any degree of confidence unless dose–response relations are described.

(3) Human exposure evaluation is used in describing the nature and size of the population exposed to a substance and the magnitude and duration of their exposure. The evaluation could concern past or current exposures, or exposure anticipated in the future.

(4) Risk characterization is defined as the integration of the data and analysis of the above three components to determine the likelihood that humans will experience any of the various forms of toxicity associated with a substance. When the exposure data are not available, hypothetical risk is characterized by the integration of hazard identification and dose–response evaluation data.

Ideally, any series of toxicological investigations should attempt to define the following: the nature of the harmful effects, ie, the basic injury produced; the incidence and severity of the effects as functions of the exposure dose; the mechanisms by which the effects are produced, ie, the fundamental biological interactions and consequent biochemical and biophysical aberrations which are responsible for the initiation and maintenance of the toxic responses; the detection of the effects, ie, the development of methodologies for the specific recognition and quantitation of the toxic effects; and whether there is reversibility of the toxic injury. This may involve a determination of whether spontaneous resolution of injury, ie, healing, occurs after cessation of exposure, or if it is possible to induce reversibility of toxic injury by antidotal or other measures, ie, treatment.

Toxicity, the potential to produce harmful effects, is to be clearly differentiated from hazard, which is the likelihood that a particular material will exhibit its known toxicity under specific conditions of use.

Classification of Toxic Effects

The diagram shown in Figure 1 gives a basis for the classification of toxic effects according to site and degree of exposure. In order to cause tissue injury, a substance must come into contact with an exposed body surface; this may be skin, eye, or the lining membranes of the respiratory and alimentary tracts. Toxic effects may be produced where a material comes into contact with a body surface, these being referred to as local effects. However, material may be absorbed from a contaminated site and disseminated by the circulatory system to various body organs and tissues. As a consequence, toxic injury may be produced in tissues and organs remote from the site of primary contamination; these are referred to as systemic effects. Systemic effects may be produced by the parent material that is absorbed, or by conversion products following absorption. They may be restricted to one organ or tissue system, or affect multiple organs and tissues. Many materials may cause both local and systemic toxicity.

The nature of a toxic effect and the probability of its occurring are often related to the number of exposures. The classification of toxic effects, and descriptions of toxicology tests, may be dictated by the number of exposures that elicit toxic effects. The following terms are convenient in this respect.

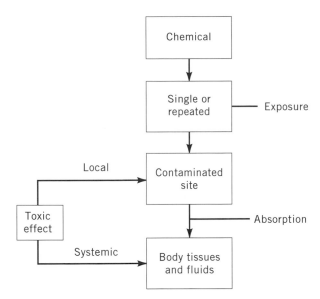

Fig. 1. Schematic representation showing the basis for classification of toxic effects into local and systemic by single or repeated exposures.

Acute exposures involve a single exposure to the test chemical in order to determine if this is effective in producing immediate, delayed, or persistent effects.

Short-term repeated exposures involve consecutive daily exposures to the test chemical which are continued over a period of a few days to a few weeks but usually not more than 5% of the lifespan of the animal. These test conditions are sometimes referred to as subacute, but this is a misleading term which should be avoided in order to prevent confusion between single and repetitive exposure toxicity.

Subchronic exposures involve consecutive daily exposures to the test material for a period amounting to usually no more than 10–15% of the lifespan of the test species.

Chronic exposures involve consecutive daily exposure to the test material over the lifespan of the test species, or a great portion of it.

The above terminology is useful in classifying toxic effects with respect to their development as a function of the number of exposures. For example, acute toxic effects are those resulting from a sufficiently large single challenge to the test chemical, and developing within a short time of that exposure; in contrast, chronic toxic effects are those resulting from repetitive exposures over a significant proportion of the lifespan.

It is important to remember that some materials of low acute toxicity may have a significant potential for producing harmful effects by repeated exposure, and vice versa. This stresses the need for a complete overview of the toxicity of a chemical by acute and repeated exposure in the process of hazard evaluation.

The following additional descriptive terms are also useful for the classification of toxic effects.

Latent effects occur either only after there has been a significant period free of toxic signs following exposure, or after resolution of acutely toxic effects which appeared immediately following exposure. They are also referred to as delayed-onset effects.

Persistent effects do not resolve, and may even become more severe after removal from the source of exposure. They can occur as a consequence of acute or repeated-exposure conditions. Thus, the use of the term persistent should be clearly differentiated from the implication of the use of the description of an effect as chronic. It should be noted, however, that some chronic effects may be persistent; an example is malignant neoplasia.

Cumulative effects are those where there is progressive injury and worsening of the toxic effect as a result of repeated-exposure conditions. Each exposure produces a further increment of injury adding to that already existing. Many materials known to induce a particular type of toxic effect by acute exposure can also elicit the same effect by a cumulative procedure from repetitive exposure to a dose less than that causing threshold injury by acute exposure.

Transient effects are those where there is repair of toxic physical injury or the reversal of induced biochemical aberrations.

Some examples of toxic effects produced by different chemicals, and classified according to the preceding guidelines, are shown in Table 1.

Depending on the circumstances of exposure, any given material may produce more than one type of toxic effect. Therefore, when describing toxicity for a particular material, it is necessary to define whether the effect is local, systemic, or mixed; the nature of the injury; the organs and tissues affected; and the conditions of exposure, including route of exposure, number of exposures, and magnitude of exposure.

Table 1. Examples of Differing Types of Toxic Effect Classified According to Time Scale for Development and Site Affected

Time scale	Site	Effect	Chemical	Reference
acute	local	lung damage	hydrogen chloride	3
	systemic	hemolysis	arsine	4
	mixed	lung damage methemoglobinemia	oxides of nitrogen	5
short-term	local	sensitization	ethylenediamine	6
	systemic	peripheral neuropathy	methyl-n-butyl ketone	7
	mixed	respiratory irritation kidney injury	pyridine	8
chronic	local	bronchitis	sulfur dioxide	9
	systemic	liver angiosarcoma	vinyl chloride	10
	mixed	emphysema kidney damage	cadmium	11
latent	local	pulmonary edema	phosgene	12
	systemic	neuropathy	organophosphates	13
		lung fibrosis	paraquat	14

The Nature of Toxic Effects

The biological response to chemical insult may take numerous forms, depending on the physicochemical properties of the material and the conditions of exposure. Listed below are some of the more significant and frequently encountered types of injury or toxic response; they may be defined in terms of tissue pathology, altered or aberrant biochemical processes, or extreme physiological responses.

Inflammation. This describes the local and immediate biological response to tissue injury (15). There is increased blood flow, leak of blood plasma into the tissues, and migration of particular blood cells to the affected area; these have protective functions. A process of repair follows. Depending on the duration of the inflammatory response and the type of cells in the affected tissue, inflammation may be described as acute or chronic. Acute inflammation is rapid in onset with early and complete healing of the injured area, and is produced by locally irritant chemicals. In chronic inflammation, there is persistence of the aggravating agent, such as insoluble particles, or continual repetitive exposure to an irritant material. A characteristic of chronic inflammation is that tissue destruction and the inflammatory process continue at the same time that healing processes are in operation. This may cause the development of excessive amounts of fibrous tissue (scar tissue), which may be sufficient to impair organ or tissue function; eg, in the lung there may be chronic progressive fibrotic disease.

Degeneration. This is a generic description for a variety of abnormal changes, visible on microscopy, that occur in tissue cells as a response to toxic injury. Acutely induced degenerative changes may be reversible, but repetitive exposure can cause progression of the degenerative changes, resulting in cell malfunction and, ultimately, cell death.

Necrosis. This term is used to describe the circumscribed death of tissue, and may be a consequence of many pathological processes induced by chemical injury.

Immune-Mediated Hypersensitivity Reaction. The immune system, as one of its primary functions, protects against invasion by foreign biological and other materials of potential harm. Such materials (antigens) stimulate various immune mechanisms in the host which cause functional elimination of the antigenic material. In some instances, there is an excess biological reactivity to the antigen, and a state of hypersensitivity develops (16). In the context of toxicology, the most important of such immune-mediated hypersensitivity reactions occur in skin and lungs. In skin, and following an appropriate period for the induction of immune defense mechanisms, the hypersensitivity reaction is recognized as an exaggerated inflammatory response at the site of application of the material; such materials are causes of allergic contact dermatitis (17). There is now an increasing awareness of the potential for immune-mediated hypersensitivity reactions by the inhalation of antigenic materials. Inhaling such materials results in the induction of a state of immunity against the antigen, which exhibits itself as a hypersensitivity reaction affecting the respiratory tract, and is clinically recognized as asthma (18,19). A classic cause of an immune-mediated hypersensitivity reaction affecting the respiratory tract caused by industrial chemicals is toluene diisocyanate (20).

Immunosuppression. Because a primary function of the immune system is protection against pathogenic foreign materials, any substance capable of producing a suppression of immune function will have a deleterious effect on such protective mechanisms, including defense against infective agents. Also, in view of the possible role of immunologic factors in neoplastic cell growth rate and other aspects of carcinogenesis, the influence of immunosuppressive chemicals in tumorigenesis is always a factor to be considered.

Neoplasia. Neoplasms are abnormal masses of cells in which growth control and divisional mechanism are impaired, resulting in aberrant proliferation and growth. Neoplasms are basically classified as benign or malignant. Benign neoplasms grow locally and without erosion of surrounding tissues. Adverse effects produced by benign neoplasms are due either to mechanical compressive effects, or to the liberation of biologically active materials from the tumor cells. Malignant neoplasms (cancers) may erosively invade surrounding tissues and become disseminated throughout the body, setting up secondary deposits of malignant-cell proliferation (metastasis). Induction of neoplasia is referred to as tumorigenesis or oncogenesis; the term carcinogenesis is used to describe the development of malignant neoplasms. The mechanisms of carcinogenesis, as yet incompletely understood, have been reviewed exhaustively (21–25).

Mutagenesis. Chemically induced mutagenesis involves an interaction between the causative agent and cellular constituents, which produces deoxyribonucleic acid (DNA) damage that is heritable. Chemically induced mutations can occur in somatic or germ cells, and may be reflected in altered structure or function of the cell. DNA damage can be classified broadly into that which can be visualized by light microscopic examination of the chromosomes (cytogenetics), and that which occurs at a strictly molecular level and is not visible by microscopy. The former may be visible as breaks, loss of chromosomal material, or rearrangement of segments of the chromosomes; this is frequently referred to as clastogenesis. DNA damage which is restricted to focal molecular lesions is often specifically referred to as mutagenesis. The implications for harmful effects from mutagenic events are multiple. If mutagenesis occurs in rapidly proliferating tissue, there may be abnormalities in the differentiation and proliferation of cells; should this occur in the embryo, a teratogenic effect may result. However, a variety of mechanisms may be involved in teratogenesis from differing materials, and a material which is devoid of mutagenic potential cannot necessarily be regarded as being devoid of teratogenic potential.

It is generally conceded where genotoxic carcinogens are concerned that an early irreversible stage in the complex process of carcinogenesis is likely to involve a mutagenic event. This has resulted in the use of mutagenicity testing procedures as screening methods for the detection of potentially carcinogenic materials. With appropriate test procedures, there is usually a reasonable correlation between the mutagenic potential of a material and its tumorigenicity as demonstrated in conventional chronic toxicity studies. Further, and of current concern, is the possibility that chemically induced mutations in germ cells could result in heritable alteration of cellular function, some of which might be deleterious to health or survival. Numerous excellent texts on mutagenesis are available (26–28).

Enzyme Inhibition. Some materials produce toxic effects by inhibition of biologically vital enzyme systems, leading to an impairment of normal biochemical pathways. The toxic organophosphates, for example, inhibit the cholinesterase group of enzymes. An important factor in their acute toxicity is the inhibition of acetylocholinesterase at neuromuscular junctions, resulting in an accumulation of the neurotransmitter material acetylcholine and causing muscle paralysis (29) (see NEUROREGULATORS).

Biochemical Uncoupling. The energy liberated by normal biochemical processes is stored in high energy phosphate molecules, eg, adenosine triphosphate. Uncoupling agents, such as dinitrophenol, interfere with the synthesis of these high energy phosphate molecules, resulting in the continual excess liberation of energy as heat.

Lethal Synthesis. This is a process in which the toxic substance has a close structural similarity to normal substrates in biochemical reactions. As a result, the material may be incorporated into the biochemical pathway and metabolized to an abnormal and toxic product. A classic example is fluoroacetic acid, which is accepted in place of acetic acid in the Krebs tricarboxylic acid cycle. The result is formation of fluorocitric acid, which is an inhibitor of aconitase and thus blocks energy production in the citric acid cycle.

Teratogenesis. Teratogenic effects are those resulting in the development of a structural or functional abnormality in the fetus or embryo. Depending on the nature of the material, teratogenic effects may be produced by a variety of mechanisms; these include mutagenesis, induction of chromosomal aberrations, interference with nucleic acid and protein synthesis, substrate deficiencies, and enzyme inhibition. With respect to the induction of structural abnormalities in development, the most critical time for exposure is during the early stage of gestation when the greatest degree of cell differentiation and definitive organ formation are occurring. However, there is increasing interest and concern about the effects of exposure to foreign chemicals in the later stages of gestation, which may induce functional, including behavioral, abnormalities. A number of excellent reviews on teratogenesis are available (30–33).

Sensory Irritation. Although not strictly a toxic effect, peripheral sensory irritation is important in many occupational health considerations. Materials described as peripheral sensory irritants are capable of interacting with sensory nerve receptors in body surfaces, producing local discomfort and related reflex effects. For example, with the eye there is pain, excess lachrymation, and involuntary closure of the eyelids (blepharospasm); inhaled sensory-irritant materials cause respiratory tract discomfort, increased secretions, and cough. Although these effects may be regarded as protective since they warn of exposure to a potentially harmful material, they are also distracting and thus likely to predispose to accidents. For this reason, information on sensory-irritant effects may be used extensively in assessing the suitability of exposure guidelines for workplace environments. A number of excellent review papers dealing with sensory irritation are available (34,35).

Endocrine System Disruption. The disruption of the endocrine system by certain substances mimicking or blocking the effects of estrogen, a hormone generated by the ovaries, testes, and adrenal that plays many roles in the body, such as ovulation, blood clotting, bone growth, and modulating the immune system, is still another toxic response (36–39). In the fetus, estrogen plays a principal role in organ development, including a part in determining whether the fetus develops male or female sex organs. Thus, endocrine disruption can cause disruption of various reproductive and development processes and increase the possibility of certain kinds of carcinogenesis.

Endocrine disrupters provide a uniquely different view of toxic substances and pose a great challenge to toxicologists. Dose–response toxicity evaluations of traditional substances do not hold true for endocrine disrupters. The greatest effects of endocrine disrupters are often observed at the smallest doses. Current methodology is not adequate to separate the effects caused by synthetic estrogen mimics from the effects caused by estrogen generated by the body. The Food Quality Protection Act of 1996 mandates development in two years by the U.S. Environmental Protection Agency (EPA) of a comprehensive strategy to screen and test common chemicals for endocrine disrupter effects, and to implement the testing program a year later.

Factors Influencing Toxicity

During the design, conducting, and evaluation of toxicology studies, there is a constant need to be aware of the numerous factors that may influence the nature, severity, and probability of induction of toxic injury. Some of the more important are listed below.

Number of Exposures.   Some toxic effects are produced in response to a single exposure of sufficient magnitude, while others require multiple exposures for their development (see Table 1).

Magnitude of Exposure.   As discussed in detail later, the magnitude of the exposure will influence both the likelihood of an effect being produced and its severity.

Species Tested.   In addition to the variation in susceptibility to chemically induced toxicity among members within a given population, there may be marked differences between species with respect to the relative potency of a given material to produce toxic injury. These species differences may reflect variations in physiological and biochemical systems, differences in distribution and metabolism, and differences in uptake and excretory capacity.

Route of Exposure.   As discussed below, the route of uptake may have a significant influence on the metabolism and distribution of a material. Differences in route of exposure may influence the amount of material absorbed and its subsequent fate. These differences may be reflected in variation in the nature and magnitude of the toxic effect.

Time of Dosing.   The time of day, or day of the year, may influence the toxic response. These changes reflect diurnal and seasonal variations in biochemical and physiological profiles, which may influence toxicity through a variety of mechanisms.

Formulation.   The formulation of a material may have a significant influence on its potential to cause toxic injury. For example, solvents may facilitate or retard the penetration and absorption of a chemical, resulting in enhancement or suppression of a toxic response, respectively.

Impurities. The presence of impurities may modify the toxic response, particularly if they have high toxicity.

The above are given as but a few examples of the factors which may influence the expression toxicity. The subject has been reviewed in detail (40,41).

Routes of Exposure

In order to induce a toxic effect, local or systemic, the causative material must first come into contact with an exposed body surface; these are the routes of exposure. In normal circumstances, and depending on the nature of the material, the practical routes of exposure are by swallowing, inhalation, and skin and eye contact. In addition, and for therapeutic purposes, it may be necessary to consider intramuscular, intravenous, and subcutaneous injections as routes of administration.

Swallowing. If it is sufficiently irritant or caustic, a swallowed material may cause local effects on the mouth, pharynx, esophagus, and stomach. Additionally, carcinogenic materials may induce tumor formation in the alimentary tract. Also, the gastrointestinal tract is an important route by which toxic materials are absorbed. The sites of absorption and factors regulating absorption have been reviewed (42,43).

Dietary Exposure Estimate. This estimate refers to exposure at the reference dose, ie, an estimate having an uncertainty of one order of magnitude or more of a lifetime daily dose of a chemical that is likely to be without significant risk to the human population, usually categorized in subgroups, such as adults, children, pregnant or nursing women, ethnic groups, people in certain geographical sections of the country, etc, of a given substance in typical human daily diets (44,45).

Skin. The skin may become contaminated accidentally or, in some cases, materials may be deliberately applied. Skin is a principal route of exposure in the industrial environment. Local effects that are produced include acute or chronic inflammation, allergic reactions, and neoplasia. The skin may also act as a significant route for the absorption of systemically toxic materials. Factors influencing the amount of material absorbed include the site of contamination, integrity of the skin, temperature, formulation of the material, and physicochemical characteristics, including charge, molecular weight, and hydrophilic and lipophilic characteristics. Determinants of percutaneous absorption and toxicity have been reviewed (32–35,42,43,46–49).

Inhalation. The potential for adverse effects from materials dispersed in the atmosphere depends on a variety of factors, including physical state, concentration, and time and frequency of exposure. Gases and vapors reach the alveoli. However, the solubility in water of a gas or vapor influences the depth of its penetration into the respiratory tract. Thus, the differences in solubility of chlorine and hydrogen chloride influence the depth of penetration or location and the irritant action of the two gases. The distribution of particles and fibers is determined by their size. In general, particles of mass median aerodynamic diameter greater than 50 μm do not enter the respiratory system; those greater than 10 μm are deposited in the upper respiratory tract; those in the range of 2–10 μm are deposited progressively in the trachea, bronchi, and bronchioles;

and only particles of $\leq$(1–2) μm reach the alveoli. It follows that larger respirable particles are more likely to cause local reactions in the upper airway than in the gas-exchanging tissues. The potential for alveolar involvement is greater with small-diameter particles. Factors governing the deposition of particles in the lung have been reviewed extensively (50–53). The aerodynamic behavior of fibers is such that those having diameters >3 μm are unlikely to penetrate into the lung. In general, fibers having a diameter of $\leq$3 μm and length not greater than 200 μm gain access to the lung. Fibers longer than 10 μm may not be readily removed by the normal pulmonary clearance mechanisms. Several studies indicate that maximum biological activity is associated with fibers less than 1.5 μm dia and more than 8 μm in length (54–57).

The likelihood that materials will produce local effects in the respiratory tract depends on their physical and chemical properties, solubility, reactivity with fluid-lining layers of the respiratory tract, reactivity with local tissue components, and (in the case of particulates) the site of deposition. Depending on the nature of the material, and the conditions of the exposure, the types of local response produced include acute inflammation and damage, chronic inflammation, immune-mediated hypersensitivity reactions, and neoplasia.

The degree to which inhaled gases, vapors, and particulates are absorbed, and hence their potential to produce systemic toxicity, depends on their solubility in tissue fluids, any metabolism by lung tissue, diffusion rates, and equilibrium state.

Eye. Adverse effects may be produced by splashes of liquids or solids, and by materials dispersed in the atmosphere. The eye is particularly sensitive to peripheral sensory irritants in the atmosphere. Toxic effects that may be induced include transient acute inflammation, persistent damage, and, occasionally, sensitivity reactions. Toxicologically significant amounts of material may be absorbed by the periocular blood vessels in cases of splash contamination of the eye with materials of high acute toxicity (58).

Multiple Exposures

Although toxicology testing is often performed with only a single material or a material in a relatively inert solvent, in most practical situations there is simultaneous exposure to multiple chemicals and thus a potential for complex biological interactions. The following descriptive terms are useful in classifying such effects.

Independent is an effect in which each material exerts its own effect irrespective of the presence of another.

Additive effects involve materials producing similar toxic effects where the magnitude of the response is numerically equal to the sum of the effect produced by each individual material.

Antagonism is applied to a situation where two chemicals, given together, interfere with each other's action or where one interferes with the action of the other. The result will usually be a decrease in toxic injury. A special case of antagonism is in studies on antidotal action.

Potentiation is applied to a condition where one material, of relatively low toxicity, enhances the expression of toxicity by another chemical. The result may be a larger response or more severe injury than that produced by the toxic chemical alone. A particular example is an enhancement of the absorption of a material of known toxicity by a surface-active material.

Synergism is applied to a situation where the effect of two or more chemicals that have common mechanism of toxicity, given together, is significantly greater than that expected from considerations on the toxicity of each material alone. This differs from potentiation in that both materials contribute to the toxic injury, and the net effect is always greater than additive.

Exposure to combinations of chemicals does not always necessarily produce clearly distinguishable interactions. Each situation must be considered in detail with due regard to all the factors that are required to be analyzed in the process of hazard evaluation (59,60).

Fate of Absorbed Chemicals Relative to Toxicity

The induction of systemic toxicity may involve a variety of complex interrelationships between the absorbed parent material, any conversion products, and their concentration and distribution in body tissues and fluids. The general pathway that a material may follow after its absorption is shown schematically in Figure 2.

Materials may be absorbed by a variety of mechanisms. Depending on the nature of the material and the site of absorption, there may be passive diffusion, filtration processes, facilitated diffusion, active transport and the formation of microvesicles for the cell membrane (pinocytosis) (61). Following absorption, materials are transported in the circulation either free or bound to constituents such as plasma proteins or blood cells. The degree of binding of

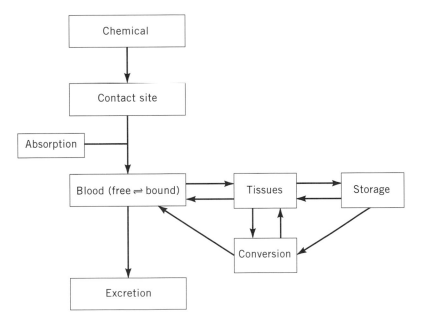

Fig. 2. Schematic representation of the possible fate of a chemical absorbed from a primary contact site.

the absorbed material may influence the availability of the material to tissue, or limit its elimination from the body (excretion). After passing from plasma to tissues, materials may have a variety of effects and fates, including no effect on the tissue, production of injury, biochemical conversion (metabolized or biotransformed), or excretion (eg, from liver and kidney).

The metabolism of a material may result in the formation of a transformation product of lower intrinsic toxicity than the parent molecule: ie, a process of detoxification has occurred. In other cases, the end result is a metabolite, or metabolites, of intrinsically greater toxicity than the parent molecule, ie, metabolic activation has occurred. Some examples of detoxification and metabolic-activation processes are given in Table 2.

The kidney is an important organ for the excretion of toxic materials and their metabolites, and measurement of these substances in urine may provide a convenient basis for monitoring the exposure of an individual to the parent compound in his or her immediate environment. The liver has as one of its functions the metabolism of foreign compounds; some pathways result in detoxification and others in metabolic activation. Also, the liver may serve as a route of elimination of toxic materials by excretion in bile. In addition to the liver (bile) and kidney (urine) as routes of excretion, the lung may act as a route of elimination for volatile compounds. The excretion of materials in sweat, hair, and nails is usually insignificant.

Parent substances and metabolites may be stored in tissues, such as fat, from which they continue to be released following cessation of exposure to the parent material. In this way, potentially toxic levels of a material or metabolite

Table 2. Examples of Metabolic Detoxification and Metabolic Activation of Chemicals by Biological Systems

Chemical	Transformation	Conversion	Ref.
cyanide, CN^-	detoxification	enzyme conversion to less acutely toxic thiocyanate	62
benzoic acid, $C_6H_5CO_2H$	detoxification	conjugation with glycine to produce less toxic hippuric acid	63
isoniazid,	detoxification	N-acetylation to less toxic acetyl derivative	64
parathion,	activation	converted by oxidative desulfuration to paraoxon, a potent cholinesterase inhibitor	65
carbon tetrachloride, CCl_4	activation	microsomal enzyme-mediated metabolic activation to hepatotoxic trichloromethyl radical	66
2-acetylaminofluorene,	activation	N-hydroxylation to the more potent carcinogen N-hydroxyacetylaminofluorene	67

may be maintained in the body. However, the relationship between uptake and release, and the quantitative aspects of partitioning, may be complex and vary between different materials. For example, volatile lipophilic materials are generally more rapidly cleared than nonvolatile substances, and the half-lives may differ by orders of magnitude. This is exemplified by comparing halothane and DDT (see ANESTHETICS; INSECT CONTROL TECHNOLOGY).

Both the metabolism of a material and its potential to cause toxic injury may vary with the route of exposure, although the magnitude of the dose and duration of dosing may influence this relationship. For example, materials that are metabolically activated by the liver are likely to exhibit a comparatively greater degree of toxicity when given perorally than when absorbed in the lung or across the skin. This is largely related to the anatomical routes of transport. Thus, the greatest proportion of material absorbed from the gastrointestinal tract passes via the portal vein directly to the liver. In contrast, materials absorbed as a result of respiratory exposure or skin contact initially pass to the lung and then into the systemic circulation, with only a small fraction of the cardiac output being delivered to the liver through the hepatic artery (Fig. 3). By similar reasoning, materials that are detoxified by the liver may be significantly less toxic by swallowing than by either inhalation or penetration across the skin. An example of the influence of route on toxicity is presented in Table 3. When assessing the relevance of metabolism in acute toxicity testing, and particularly when comparing toxicity by different routes of exposure, both the magnitude of the dose and the time period over which it is given must be considered. For example, when a single large dose (a bolus) of a metabolically activated material is given by gavage, it may be almost completely metabolized, resulting in the rapid development of acute toxic injury. When the same material is given orally at a slower rate, eg, by continuous inclusion in the diet, then there is a slow and continual absorption and metabolism of the material, and in these circumstances the rate of generation of the toxic species may approach that which occurs from the continuous absorption resulting from persistent exposure to an atmosphere of the material. The influence of dose magnitude–time relationships also apply to the interpretation of results with materials detoxified by the liver. With such materials, slow continuous peroral administration of the material results in slow titration to the liver and a high proportion of the material being detoxified. In this instance, the anticipated differential toxic effect between the oral and inhalation routes of exposure occurs. However, if a bolus of the material is introduced into the stomach, then the endogenous hepatic detoxification mechanisms may be exceeded, and unmetabolized material may enter the systemic circulation and initiate toxic injury.

Principal factors that determine the likelihood of toxic effects being produced and their severity include the rates at which the causative substances (parent molecule, toxic metabolite, or both) reach the tissues and the absolute amounts of materials to which the tissues are exposed. These determine the dose of material received at the target tissues.

With respect to environmental exposure conditions, the probability of adverse effects occurring depends on many factors; the more important of these include magnitude, duration, and number of exposures. These conditions determine the amount of material to which an organism is exposed (the

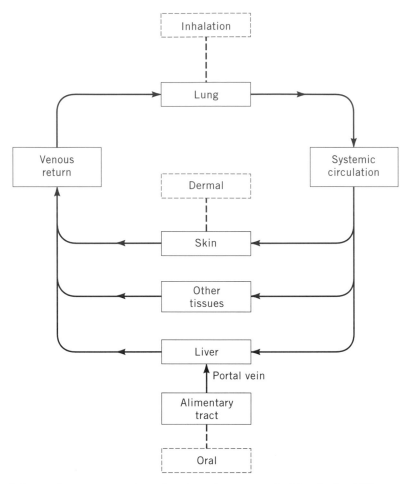

Fig. 3. Schematic representation showing the anatomical basis for differences in the quantitative supply of absorbed material to the liver. By swallowing (oral route), the main fraction of the absorbed dose is transported directly to the liver. Following inhalation or dermal exposure, the material passes to the pulmonary circulation and thence to the systemic circulation, from which only a portion passes to the liver. This discrepancy in the amount of absorbed material passing to the liver may account for differences in toxicity of a material by inhalation and skin contact, compared with its toxicity by swallowing, if metabolism of the material in the liver is significant in its detoxification or metabolic activation.

environmental exposure dose), and thus relates to the amount of material available for absorption at the contact site (the absorbed dose). The absorbed dose is an important factor that determines the amount of material available for distribution to body tissues, the amount of toxic metabolites formed, and thus the likelihood of inducing a toxic effect. Opposed to the influence of absorption and buildup of metabolites is the elimination of these materials from the body. Thus, for any given environmental exposure situation, the probability of inducing a toxic effect depends primarily on the dynamic equilibrium between the rate of absorption of the environmental chemical and the rate of excretion of the ab-

Table 3. Example of the Influence of Route on
the Acute Toxicity of Potassium Cyanide to the
Rabbit (Female)

Route	LD$_{50}$[a], mg/kg
intravenous	1.89 (1.66–2.13)
intraperitoneal	3.99 (3.40–4.60)
oral	5.82 (5.50–6.31)
percutaneous	22.3 (20.4–24.0)

[a]Confidence limits of 95%.

sorbed material and its metabolites. This relationship clearly determines the residence time for materials and metabolites in various body tissues and fluids, their fluctuation in concentration with time, the potential for storage, and availability for binding with macromolecules and structural cellular components.

The amount of material in contact with the absorbing surface is a principal determinant of the absorbed dose. In general, the higher the concentration, the greater the absorbed dose. However, when mechanisms other than simple diffusion across a concentration gradient are operating, a simple proportionate relationship between concentration and absorbed dose may not exist. In such cases, a rate-limiting factor could result in proportionately smaller increases in absorbed dose for incremental increases in concentration at the contact site. Also, and particularly when an active transport mechanism is involved in the absorptive process, there may be saturation and a ceiling value for absorption.

It is important to appreciate that the magnitude of the absorbed dose, the relative amounts of biotransformation product, and the distribution and elimination of metabolites and parent compound seen with a single exposure, may be modified by repeated exposures. For example, repeated exposure may enhance mechanisms responsible for biotransformation of the absorbed material, and thus modify the relative proportions of the metabolites and parent molecule, and thus the retention pattern of these materials. Clearly, this could influence the likelihood for target organ toxicity. Additionally, and particularly when there is a slow excretion rate, repeated exposures may increase the possibility for progressive loading of tissues and body fluids, and hence the potential for cumulative toxicity.

It is clear from the above considerations that the absorbed dose, and the distribution, excretion, and relative amounts of the absorbed material and its metabolites may be quantitatively different for acute and repeated exposures. This modifies the potential for the absorbed material to produce adverse effects by a given route of exposure.

Dose–Response Relationships and Their Toxicological Significance

The importance of determining a relationship between the magnitude of the exposure and the frequency of occurrence of a toxic effect is considered in detail below.

An observation which is fundamental to the interpretation of toxicology information is the variation in susceptibility to potentially harmful chemicals of individual members within a given population. Thus, if a group of animals of stated species and strains are exposed to a particular material by a given route of exposure, then as the exposure dose is increased so does the proportion of animals exhibiting a toxic effect. However, the biological variability with respect to individual susceptibility of animals to toxic materials is not a simple linear relationship. The most typical response is represented by a sigmoid curve (Fig. 4), indicating that a very small proportion of the exposed population is more susceptible (hypersensitive) and a few more resistant (hyposensitive) to the chemical. The majority of the population, however, responds over a defined exposure dose range around an average. This variability also exists among human populations, and the magnitude of the average response and the existence of a hypersensitive group clearly influences judgments on the hazards of materials studied.

In addition to the effect of biological variability in group response for a given exposure dose, the magnitude of the dose for any given individual also determines the severity of the toxic injury. In general, the considerations for dose−response relationship with respect to both the proportion of a population responding and the severity of the response are similar for local and systemic effects. However, if metabolic activation is a factor in toxicity, then a saturation level may be reached.

Dose−response relationships are useful for many purposes; in particular, the following: if a positive dose−response relationship exists, then this is good evidence that exposure to the material under test is causally related to the response; the quantitative information obtained gives an indication of the spread of sensitivity of the population at risk, and hence influences hazard evaluation; the data may allow assessments of no effects and minimum effects doses, and hence may be valuable in assessing hazard; and by appropriate considerations of the dose−response data, it is possible to make quantitative comparisons and contrasts between materials or between species.

A simplistic approach is illustrated in Figure 5 (68). When dosage is plotted against response, a response above the zero dose is considered to be an adverse effect, whereas a response below the zero level is considered to be beneficial.

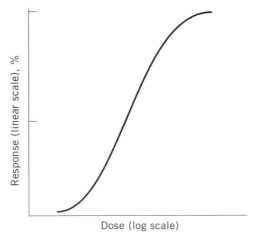

Fig. 4.   Typical sigmoid curve for the response of a biological system to chemical injury.

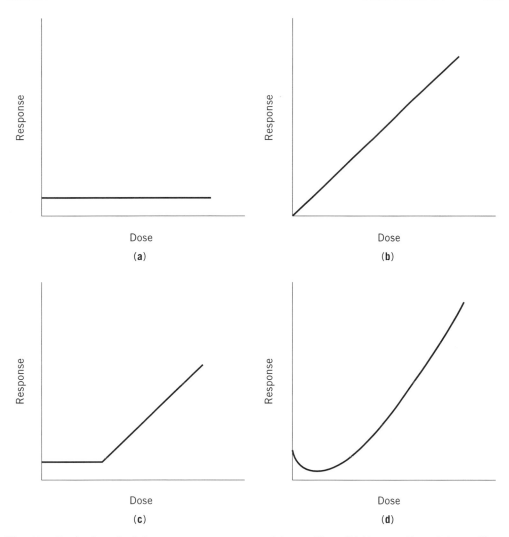

Fig. 5. Toxic chemical dose–response curves: (**a**) no effect; (**b**) linear effect; (**c**) no effect at low dose; and (**d**) beneficial at low dose.

Considerable caution is necessary in making quantitative comparisons between different materials, even when considering the same toxic end point. This can be conveniently illustrated using, as an example, death in response to a single exposure, ie, acute lethal toxicity. Studies to determine acute lethal toxicity by a particular route are usually conducted as described below.

Several groups of animals of a particular species and strain are given different doses of the test material; members within the same group receive similar doses. The animals are subsequently observed, usually for two weeks, and the number of mortalities noted. In most cases, the dose–mortality curve has a typical sigmoid form. This may be conveniently converted to a linear form by log–probit plot (Fig. 6). A frequently used numerical means to allow comparison of the lethal potential of different materials is to quote a particular level of

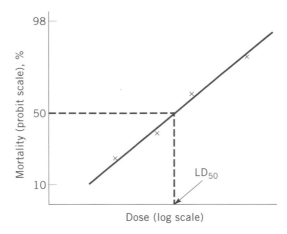

Fig. 6. Dose–response regression line for mortality data (represented by ×) expressed by log-probit plot.

mortality. Since the largest proportion of deaths is distributed around the 50% mortality level, this forms a convenient reference point, and is referred to as the median lethal dose $_{50}$ (LD_{50}). The LD_{50} may, therefore, be defined as the dose, calculated from the dose–mortality data, which will cause the death of half of the population exposed. In view of the multiplicity of factors that influence toxicity, the LD_{50} is valid only for the specific conditions of the test. It is, of course, possible to compare other levels of lethality, such as the LD_{10} and the LD_{90}; the difference between these limits will give an indication of the range of doses causing lethal toxicity to the majority of the population studied. They also give an indication of the relationship of hypersensitive and hyposensitive groups to the average response.

The LD_{50} is calculated from data obtained by using small groups of animals and usually for only a few dose levels. Therefore, there is an uncertainty factor associated with the calculation. This can be defined by determining the 95% confidence limits for the particular levels of mortality of interest (Fig. 7). The

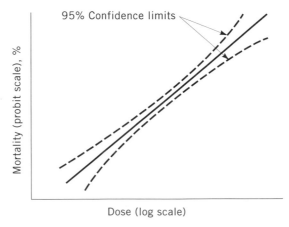

Fig. 7. Typical 95% confidence limits for dose–mortality regression data.

95% confidence limits give the dose range for which there is only a 5% chance that the LD_{50} will be outside.

In defining acute level toxicity for the purposes of comparing different materials, the LD_{50} itself is not sufficient; but the LD_{50} and the 95% confidence limits should be quoted as a minimum. For example, and as demonstrated in Figure 8, two materials (A and B) with different LD_{50} values, but overlapping 95% confidence limits, are to be considered not statistically significantly different with respect to mortality at the 50% level; this it based on the fact that there is a statistical probability that the LD_{50} of one material could lie in the 95% confidence limits of the other, and vice versa. Conversely, when there is no overlap in 95% confidence limits, as shown with material C, it may be concluded that the LD_{50} values are statistically significantly different.

A more complete comparison of the lethal potential of two or more materials requires that attention be paid to the slopes of the dose−response regression lines. For example, two materials having statistically significantly similar LD_{50} values (based on 95% confidence limit comparisons) and parallel dose−response regression lines also have statistically similar LD_{10} and LD_{90} values (Fig. 9). In contrast, materials having similar LD_{50} values (based on overlapping 95% confidence limits), but differing slopes for the regression lines, may have widely differing and separated 95% confidence limits at the LD_{10} and LD_{90} levels (Fig. 10). Such a significant difference in lethal toxicity at the hypersensitive and hyposensitive regions may markedly influence considerations with respect to the relative hazards of the two materials. For example, as shown in Figure 10, with the material having the steeper slope it is clear that once a lethal dose is reached, only a small increase in dose is necessary to affect the whole population. Thus, this material may present significant problems with respect to acute overexposure situations. In contrast, the material having a shallow dose−response regression

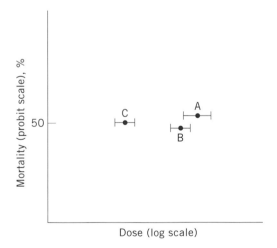

Fig. 8. Comparison of three materials on the basis of LD_{50} and 95% confidence internal data. Materials A and B are not statistically significantly different. Material C, however, has 95% confidence limits at the LD_{50} level which do not overlap those of A or B; it is statistically significantly more lethally toxic than either.

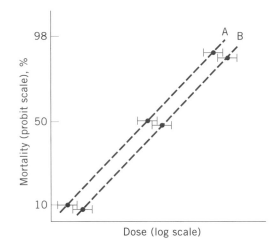

Fig. 9. The two materials, A and B, have overlapping 95% confidence limits at the LD_{50} level. Because the slopes of the dose–mortality regression lines for both materials are similar, there is no statistically significant difference in mortality at the LD_{10} and LD_{90} levels. Both materials may be assumed to be lethally equitoxic over a wide range of doses, under the specific conditions of the test.

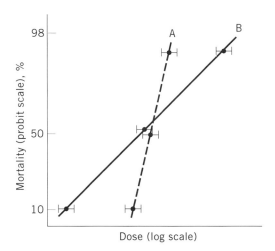

Fig. 10. Two materials, A and B, have statistically similar LD_{50} values but, because of differences in the slopes of the dose–mortality regression lines, there are significant differences in mortality at the LD_{10} and LD_{90} levels. Material A is likely to present problems with acute overexposure to large numbers of individuals in an exposed population when lethal levels are reached. With Material B, because of the shallow slope, problems may be encountered at low doses with hypersensitive individuals in the population.

line may present problems with respect to the development of toxic effects in a small hypersensitive group.

When making comparisons of lethal toxicity, it must be remembered that different mechanisms may be involved with different materials, and these need to be taken into account. Also, comparisons of acute toxicity should take note of differences in time to death, since marked differences in times between dosing and death may influence hazard evaluation procedures and their implications. In a few instances, it may be possible to calculate two LD_{50} values for mortality; one based on early death due to one mechanism, and a second based on delayed deaths due to a different mechanism (69).

Although acute lethal toxicity has been used as an example, the principles discussed apply in general to other forms of toxicity capable of being quantitated in terms of dose–response relationships.

For carcinogen pesticides (70,71), animal testings are subject to maximum tolerated doses (MTD). MTD is the maximum amount of a substance that can be administered to an experimental animal without causing extreme health consequences, such as death, to occur but while continuing to produce some measurable toxic effects. Current regulatory theory holds that carcinogen effects do not have a threshold and cannot be related to reference doses.

A substance's carcinogen potency is expressed quantitatively as a Q-star or Q^*. The Q^* represents the slope of the dose–response curve from animal studies yielding a positive oncogenic response, expressed in the units of excess tumor incidence/mg of substance exposure/kg of body weight/d. Mathematical extrapolation from tumor incidents resulting from very high doses administered in animal tests is used to predict the human carcinogenic potency of a substance in very low doses in human diets. The potency factor does not consider the type, site, or diversity of tumors observed. The EPA often combines malignant and benign tumors for the potency factor. A high Q^* indicates a strong carcinogen response to the administered dose; consequently, a low Q^* indicates a low response. The values of Q^*s represent the 95% upper-bound confidence limit of tumor induction likely to occur from a given dose and is considered to be a highly conversative model for quantifying carcinogen potency. Carcinogen risks below 10^{-6} (one in a million) are considered to be negligible risks.

Testing Procedures

For descriptive purposes, toxicology testing procedures can be conveniently subdivided into general and specific forms. General toxicology studies are those in which animals are exposed to a test material under appropriate conditions and then examined for all types of toxic effects that the monitoring procedures employed allow. Specific toxicological studies are those in which exposed animals are monitored specifically for a defined toxic end point or effect.

There are many guidelines that need to be followed and which are common to all types of toxicity testing, the most important of which are as follows:

There should be sufficiently large numbers of animals to allow a quantitative determination of the average response and the range of responses, including the demonstration of hypersensitive populations. When objective procedures are undertaken, these should be sufficient to allow valid statistical comparison to be made between treated and control groups.

Sufficient numbers of control animals should be employed. The use of such controls allows a determination of normal values for features monitored in the study and background incidence of pathology in the population studied; detection of the onset of adverse conditions, eg, infection, which are unrelated to, and detrimental to, the conduct of the study; and deviation of monitored features between controls and exposed animals, which may indicate a treatment-related effect.

Vehicle control animals may be necessary to allow an assessment of the possible contribution of the vehicle to any effects observed in exposed animals.

Exposure should be by the practical route. Other conditions, such as number and magnitude of exposures, should include at least one level representative of the practical situation; monitoring should be appropriate to the needs for conducting the study; and when practically and economically possible, pharmacokinetic observations should be undertaken in order to better define the relationship of dose to metabolic thresholds.

Factors in the design and conduct of studies have been published (72–74).

General Toxicology Studies.   Studies may be conducted in live specimens (*in vivo*), or in test tubes (*in vitro*). For reasons inherent in both the toxicity assessment procedure and the design of studies, it is usual to proceed in sequence from acute to the various stages of multiple-exposure studies. Acute studies give information on the type of toxic injury produced by a single exposure, including the effects of massive overexposure. The fact that a particular type of toxic injury is not produced by an acute exposure does not necessarily imply the absence of potential for that type of injury by the chemical, since multiple exposures may be necessary to induce the effect. However, effects produced by acute or relatively short-term repeated exposure may also be produced by longer-term repeated exposures, and at lower concentrations. Hence, in addition to giving information on potential for toxicity, the acute and short-term repeated studies are used to give guidance on exposure conditions to be followed in longer-term repeated exposure studies. The type of monitoring to be employed will depend on a variety of considerations, including the chemistry of the material, its known or suspected toxicology, the degree of exposure, and the reason for conducting the test. In general, since the multiple exposure studies are more likely to produce the widest spectrum of toxic effects, it is usual to employ the most extensive monitoring in these studies.

The types of monitoring employed to assess the functional status of the living animal and for the detection of injury in dead animals may include the following.

General Observations.   Animals are inspected at frequent intervals in order to discover any departure from normal appearance and function, the presence of abnormal patterns of behavior, and any other differences from the control animals. Simple observation of the animals may give information of considerable importance in assessing potential for toxicity and giving preliminary guidance on the nature of any injury.

Body Weight.   The detection of a decrease in the rate of gain in body weight, in comparison with controls, may be one of the earliest indications of the onset of toxic effects, particularly if it follows a dose–response relationship. Rarely, a relative increase in body weight may indicate a metabolic or endocrinological effect.

Food and Water Consumption.   Measurement of changes in food and water consumption may indicate a toxic potential, and can give guidance on the reason for abnormal body weight gains.

Hematology. The functional status of blood and of the blood-forming tissues can be assessed by tests which include red and white blood cell counts, platelet counts, clotting time, coagulation tests, and examination of bone marrow. Such tests, in addition to detecting abnormalities, may also allow differentiation between primary and secondary effects on blood and blood-forming tissues (75).

Chemical Pathology. Also referred to as clinical chemistry, this monitoring procedure involves the measurement of the concentration of certain materials in the blood, or of certain enzyme activities in serum or plasma. A variety of methods exist that allow (to variable degrees of specificity) the definition of a particular organ or tissue injury, the nature of the injurious process, and the severity of the effect (76).

Urinalysis. Urine is collected at various times and examined with respect to its volume, specific gravity, and the presence of abnormal constituents. The results may indicate kidney damage or suggest tissue injury at other sites (77).

Pathology. Animals are examined macroscopically at autopsy following death during the study or at planned sacrifice. This may show features apparent to the naked eye which are abnormal and suggestive of tissue damage (gross pathology). Sections of tissue examined under the light microscope allow a detailed evaluation of the interrelationships and structural integrity, or otherwise, of cells and intercellular materials. In this way, normality of tissue may be confirmed, or a specific pathological diagnosis attached to induced or coincidental tissue injury (histopathology) (78).

Organ Weight Determinations. Measurement of the weight of organs removed at autopsy is an integral part of most toxicology studies. This information may provide an indication of changes in these organs, although they have to be carefully related to the state of hydration and nutrition of the animal.

Special Investigations. Existing information on toxicology, or suspicions based on the chemistry of the material, may indicate the possibility for a particular type of organ toxicity. In such cases, it may be appropriate to incorporate certain special investigational procedures into the general toxicology study. For example, if there is evidence or suspicion that a material is capable of inducing progressive lung dysfunction, it is of importance that there be periodic monitoring of animals by respiratory function testing in any subchronic or chronic inhalation study with the material.

Types of Studies. Studies may be conducted in live specimens (*in vivo*), or in test tubes (*in vitro*). Studies may be carried out by single exposure or by repeated exposure over variable periods of time. The design of any one study, including the monitoring procedures, is determined by a large number of factors, including the nature of the test material, route of exposure, known or suspected toxicity, practical use of the material, and the reason for conducting the study.

Acute Toxicity Studies. These studies should provide the following information: the nature of any local or systemic adverse effects occurring as a consequence of a single exposure to the test material; an indication of the exposure conditions producing the adverse effects, in particular, information on dose–response relationships, including minimum and no-effects exposure levels; and data of use in the design of short-term repeated exposure studies.

Acute toxicity studies are often dominated by consideration of lethality, including calculation of the median lethal dose. By routes other than inhalation, this is expressed as the LD_{50} with 95% confidence limits. For inhalation experiments, it is convenient to calculate the atmospheric concentration of test material producing a 50% mortality over a specified period of time, usually 4 h; ie, the 4-h LC_{50}. It is desirable to know the nature, time to onset, dose–related severity, and reversibility of sublethal toxic effects.

Short-Term Repeated Studies.   These studies should give information about the potential for cumulative toxicity and allow the detection of toxicity, other than neoplasia, not detected in acute studies. Studies are generally carried out by exposing animals by an appropriate route for 5 d/wk for 1–4 wk. At a minimum, the conduct of these tests should include observations for signs of toxicity; measurement of body weight, food intake, and water consumption; autopsy; and gross pathology. Other monitoring requirements are dictated by the reasons for conducting the test.

Subchronic Studies.   Although short-term repeated exposure studies provide valuable information about toxicity over this time span, they may not be relevant for assessment of hazard over a longer time period. For example, the minimum and no-effects levels determined by short-term exposure may be significantly lower if exposure to the test material is extended over several months. Also, certain toxic effects may have a latency which does not allow their expression or detection over a short-term repeated-exposure period; for example, kidney dysfunction or disturbances of the blood-forming tissues may not become apparent until subchronic exposure studies are undertaken.

Typically, subchronic inhalation studies involve exposing the animals for 6 h/d, 5 d/wk for about 3 months. For feeding studies, the material is frequently included in the diet, provided unpalatability is not a problem. As with the shorter-term studies, several dose levels are used, together with a control group. Because of the potential for a wide spectrum of effects and the cost of conducting the basic test, a significant amount of relevant monitoring is employed in order to detect the nature, onset, progression, and severity of any toxic effects. Ideally, a small proportion of animals should be kept for several weeks after the end of the exposure period in order to determine the reversibility of any induced toxic effects. Subchronic exposure conditions usually detect all potential long-term repeated exposure toxicity, except for neoplasia.

Chronic Toxicity Studies.   With the exception of tumorigenesis, most types of repeated exposure toxicity are detected by subchronic exposure conditions. Therefore, chronic exposure conditions are usually conducted for the following reasons: if there is a need to investigate the tumorigenic potential of a material; if it is necessary to determine a no-effects or threshold level of toxicity for lifetime exposure to a material; and if there is reason to suspect that particular forms of toxicity are exhibited only under chronic exposure conditions.

For the above reasons, chronic exposure studies are frequently designed in such a way that it is possible to combine observations for tumorigenesis and nonneoplastic tissue injury. Chronic studies are usually extensively monitored. It is common practice to sacrifice animals at intervals during the study in order to detect the onset of any tissue injury. For two-year exposure studies, it is most meaningful to have interim sacrifices at 12 and 18 months.

Guidelines are available for conducting acute or repeated exposure studies by inhalation (79–81), application to the skin (82,83), or perorally (84,85).

Specific Toxicology Studies.   Many procedures, both *in vivo* and *in vitro*, are available to detect specific organ toxicity or quantitatively monitor for particular end points or effects. Although many of these studies are directed at measuring a particular toxic effect for hazard-evaluation purposes, some are employed as screening or short-term tests to determine the potential of a material to induce chronic toxic effects or those with a long latency period. In this context, screening means an experimental approach that allows the rapid and cost-effective prediction of the likelihood that a material exerts a particular type of adverse biological activity. Such approaches should be based on studies showing the method gives a high degree of correlation with conventional and credible methods for detecting the particular toxic end point. Some of the most commonly employed special toxicology methods and approaches are listed below.

Primary Irritancy Studies.   These studies are employed to determine the potential of materials to cause local inflammatory effects in exposed body surfaces, notably skin and eye, following acute or short-term repeated exposure. In general, the approach involves applying the test material to the surface of the skin or eye, and observing for signs of inflammation, their duration, and resolution. Reviews have been written about the conduct of primary eye irritation (58,86,87) and primary skin irritation studies (88,89).

Studies for Immune-Mediated Hypersensitivity.   Allergenic materials may produce hypersensitivity reactions by skin contact or by inhalation. In conventional tests for determining allergenic potential by skin contact, the basic approach involves repeatedly applying test material to the skin, or under the skin, in order to induce a state of hypersensitivity. After a latent period, the skin is challenged with test material to determine if an exaggerated local response, typical of delayed hypersensitivity contact dermatitis, has been produced. Details of the test procedures are available (90–92).

Experimental methods for determining the potential of materials to produce hypersensitivity reactions by inhalation use procedures to detect hyperreactivity of the airways as demonstrated by marked changes in resistance to air flow, and the detection of antibodies in blood serum (93).

Neurological and Behavioral Toxicology.   Observations on animals in general toxicology studies may indicate a potential for injury to the nervous system, particularly if there are abnormalities of movement, gait, and reaction to the environment. Where it is known or suspected that a material may produce structural or functional damage to the nervous system, special methods should be incorporated into general toxicology studies in order to determine the nature and extent of any neurological injury. This may include the use of simple observational test batteries in order to better assess the clinical status of the animal (94,95), more detailed examination of the potentially affected areas of the nervous system by light and electron microscopy (96,97), and the use of selective biochemical procedures (98).

However, in order to precisely define the nature of a neurotoxic process, its mechanism of production, and the quantitative determinants for the effect, it may be necessary to conduct specific studies. These may involve the use of electrophysiological, pharmacological, tissue culture, and metabolism techniques (99–102). Special observational methods are available for behavioral studies (103,104).

Sensory Irritation. In view of the ability of peripheral sensory irritants to produce local discomfort with related reflex effects, they may cause harassment and be a predisposing factor in accidents. Methods for quantitatively assessing the potential of materials to produce sensory irritation by inhalation (34,105) and contamination of the eye (106,107) are available. Techniques for assessing peripheral sensory irritant potential by inhalation are based on measuring a reflexly induced decrease in breathing rate. Those for assessing eye irritation are based on blepharospasm (involuntary closure of eyelids) in animals, and blepharospasm and discomfort in human subjects. The information so obtained may be useful considerations in assigning suitable exposure guidelines for materials in the workplace atmosphere (108).

Teratology. At present, most studies that are conducted to determine the teratogenicity of materials are aimed primarily at assessing structural defects of development. Basically, these studies involve administering the test material to the pregnant animal during the period of maximum organogenesis; for rats, it is usual to expose on days 6–15, and for rabbits on days 6–18. The day before anticipated normal parturition, fetuses are delivered by Caesarian section, to prevent the cannibalization of any deformed fetuses by the mother. Resorbed and dead fetuses are counted. The viable fetuses are sexed, weighed, measured (crown–rump length); some are used for examination of the integrity of the skeleton; and the remainder are dissected to determine the presence of any soft-tissue abnormalities. Additionally, observations are made for pathology in the maternal reproductive system. To allow for dose–response considerations, several exposure levels are used. Control groups should include animals that are untreated and others given the vehicle alone. The design and conduct of conventional teratology studies for the detection of structural abnormalities of development have been extensively reviewed (109–111).

Recently, there has been increasing interest in the development of test methods to assess the possible adverse functional effects of exposing the fetus, both early and late in gestation (112,113).

Reproductive Toxicology Tests. In contrast to teratology studies, which are aimed at assessing adverse effects on the developing fetus, reproductive studies cover a much wider spectrum of developmental biology. They are designed to assess the potential for adverse effects on gonads, fertility, gestation, fetuses, lactation, and general reproductive performance. Exposure to the chemical may be over one or several generations. Because of the necessarily comparatively low doses used over these long-term studies, there may be insufficient sensitivity to detect most potentially teratogenic materials in conventional multigeneration reproduction studies. Tests for reproductive toxicity have been reviewed (114–116).

Metabolism and Pharmacokinetics. Since the potential for systemic toxicity of a material may be highly dependent on its distribution, residence time, and bioconversion, studies on metabolism and pharmacokinetics can be of fundamental importance with respect to interpretation of the significance of conventional toxicology studies, determination of mechanisms of toxicity, relation of environmental exposure conditions to target organ toxicity, selection of dosages, and the design of further toxicology studies (see PHARMACODYNAMICS).

Metabolism is concerned with a determination of the biotransformation of the parent material, the sites at which this occurs, and the mechanism of the biotransformation.

Pharmacokinetic studies are designed to measure quantitatively the rate of uptake and metabolism of a material and determine the absorbed dose; to determine the distribution of absorbed material and its metabolites among body fluids and tissues, and their rate of accumulation and efflux from the tissues and body fluids; to determine the routes and relative rates of excretion of test material and metabolites; and to determine the potential for binding to macromolecular and cellular structures.

Pharmacokinetic studies should allow an assessment of the relationship between the environmental-exposure conditions and the absorbed dose, and how these influence the doses of test material and metabolites received by various body tissues and fluids, and the potential for storage. Numerous texts are available on the design and conduct of metabolism and pharmacokinetic studies (117–119).

Mutagenicity. Studies to determine the potential for materials to produce mutagenic events may be conducted *in vitro* and *in vivo*. The most widely used test system for mutagenic potential has been the Ames procedure (120). This test is based on the ability of mutagenic chemicals to cause certain bacteria to regain their ability to grow in media deficient in an essential amino acid. Other tests of the *in vitro* type make use of various end points, indicating a mutagenic event, in mammalian cells grown in culture. *In vivo* studies involve the exposure of animals to the test chemical after which cells are removed, usually from blood or bone marrow, and examined for chromosomal abnormalities or for focal mutagenic events using a biochemical or morphological marker. The tests for assessing mutagenic potential of chemicals have been extensively reviewed (121–124).

A positive result in a mutagenicity test system is not necessarily a directly usable end point in toxicity evaluation. There is general agreement that materials exhibiting a mutagenic potential, particularly by an *in vivo* approach, need to be reviewed in particular with respect to their possible genetic, teratogenic, and direct carcinogenic activity. The relationship between mutagenicity tests, both *in vivo* and *in vitro*, and the ability of a chemical to produce genetically transmitted adverse effects in the progeny of exposed individuals is unclear and the subject of much debate and research. A variety of mechanisms may be concerned in the induction of teratogenic effects by differing materials, of which one is mutagenicity. Thus, a material that exhibits a mutagenic potential in appropriate tests should be suspected of being teratogenic, and appropriate laboratory studies may be required. However, it is of the utmost importance to be aware that many other mechanisms for teratogenesis exist, and a material devoid of mutagenic potential may not necessarily be devoid of teratogenic potential. Perhaps the most common application of mutagenicity studies is to assess the carcinogenic potential of materials. There are correlative relations between mammalian carcinogens and mutagens (125). The latest thought is that cancer may result from the sequential or simultaneous genetic damage to several genes which govern cell growth, death, and maturation (126). Also, there is a considerable body of evidence to indicate that an early critical and irreversible stage in carcinogenesis is the induction of a mutagenic event in the affected cell line. For these reasons, materials which have been shown to be mutagens are suspect of being chemical carcinogens and this may necessitate, and assign a priority for, chronic exposure studies. However, this facet of mutagenicity testing requires considerable caution in its application. There is a need to look at the effect of metabolic activation, both *in vivo* and *in vitro*; the nature of the end point indicating a mutagenic potential; the potency of a material (or metabolite) in inducing events characteristic of chemical carcinogenesis; the possible influence of route on *in vivo* tests; and to determine that any response was not due to the presence of contaminants.

Another challenge to the toxicologists is the existence of significant toxicological differences between adults, and infants and children. A comprehensive study report (127) reveals some of the differences.

Human infants and children differ from human adults not only in size and body weight, but also in relative immaturity of biochemical and physiological functions in major body systems; in body composition in terms of proportions of water, fat, protein, and mineral mass, as well as the chemical constituents of these body components; in the anatomic structure of organs; and in the relative proportions of muscle, bone, solid organ, and brain. These structural and functional differences between neonates and adults can potentially influence the toxicity of substances, owing to qualitative and quantitative alteration in the magnitude of systemic adsorption, distribution, metabolism, interaction of the chemical with cellular components of target organs, and excretion. Without detailed knowledge of all age-related physiological changes and their potential interactions, it is not possible to extrapolate the impact of different biotic effects from mature to young animals. Therefore, in the experimental setting, the understanding of stage of growth and development of laboratory animals is essential in evaluating the toxicological effects of substances in infants and children.

Review of Toxicology Studies

The review and interpretation of toxicology studies is a professional matter, requiring experience in both the laboratory conduct of such studies and the practice of applied toxicology. Although all studies should be reviewed on a case-by-case basis, there are some general considerations to be kept in mind during the review process, described below.

The reviewer should establish that the laboratory reporting the study has the necessary professional reputation, scientific experience, and expertise in the area investigated. It should be confirmed that adequate quality-control facilities are in place and good laboratory practices and procedures followed.

The objectives of the study should be precisely stated and the work presented in a clear and coherent matter, with all the detail necessary to allow the reviewer to make his or her own assessment of the study. It should be confirmed that the overall design of the protocol satisfies the needs of the objectives of the study.

The material tested should be specified, including nature, relative proportions of any impurities, and stability over the test period. All details of the conduct of the study should be presented. It must be established that the methods employed for exposing and monitoring the animals are appropriate and sufficiently specific for the end points or effects planned to be studied.

Attention should be paid to the sufficiency of the study with respect to determining significance and assessing hazard, eg, whether the number of control and test animals is sufficient to allow detection of biological variability in response and for comparative statistical procedures.

There should be sufficient dose–response information to allow decisions on causal relationships and relevance.

The results of the study should allow decisions on whether injury is a direct result of toxicity or secondary to other events. In addition to confirming a causal relationship between exposure to the test material and development of an injury, the study should be reviewed in order to assess whether information is available to determine if the effect is traceable to parent material or metabolite.

In evaluating numerical information, it is important to remember that, although an effect may be statistically significant, this does not necessarily imply that the effect is of adverse biological significance. Conversely, a change or trend which is determined not to be statistically significant may be of biological consequence. Quantitative information, particularly when this involves does–response considerations, should be reviewed against the background of the study as a whole and the perspective of normal biological variations.

Hazard Evaluation Procedures and the Role of Toxicology

Hazard is the likelihood that the known toxicity of a material will be exhibited under specific conditions of use. It follows that the toxicity of a material, ie, its potential to produce injury, is but one of many considerations to be taken into account in assessment procedures with respect to defining hazard. The following are equally important factors that need to be considered: physicochemical properties of the material; use pattern of the material and characteristics of the environment where the material is handled; source of exposure, normal and accidental; control measures used to regulate exposure; the duration, magnitude, and frequency of exposure; route of exposure and physical nature of exposure conditions, eg, gas, aerosol, or liquid; population exposed and variability in exposure conditions; and experience with exposed human populations.

Consideration of the above information allows the exposure conditions to be defined and reviewed in the light of the known toxicity of the material being examined.

Relevance of Toxicology in Hazard Evaluation

Ideally, available information on the toxicology of a material should allow the following to be determined as part of a hazard evaluation procedure: nature of potential adverse effects; relevance of the conditions of the toxicology studies to the practical in-use situation; the average response, range of responses, the presence of a hypersensitive group, and an indication of minimal or no-effects levels; identification of factors likely to modify the toxic response; effects of acute gross overexposure, ie, accident situations; effects of repeated exposures; recognition of adverse effects; assistance in the definition of allowable and nonallowable exposure conditions; assistance in the definition of monitoring requirements; guidance on the need for personal and collective protection measures; guidance on first-aid, antidotal, and medical support needs; relevance of toxicity to coincidental disease; and definition of "at risk" individuals, eg, pregnant and fertile females; genetically susceptible individuals.

Information of the above type can only be obtained from carefully designed studies. In some instances, it may be economically impossible to conduct a complete spectrum of toxicology testing and, in such circumstances, it is necessary to carefully consider the most appropriate investigational approaches and their order of conduct for a hazard evaluation of a particular material under its anticipated conditions of use. The relevance and credibility of a toxicological study can be no better than its study design and conduct allow. It is of the utmost importance that meticulous detail be given in the planning of a study and preparing the protocol for that study (81).

The cost of toxicology studies varies, for example, from tens of thousands of dollars for simple acute toxicity tests to perhaps several million dollars for an extensive study. However, the precise costs for particular studies vary with the design and content of the protocol and the laboratory conducting the study. Those needing toxicology testing of their materials should understand the requirements of the end user and arrange for a careful and critical independent opinion on the nature of the testing required, and then obtain estimates for the timing and costing of these from several laboratories. Advice should be sought about the reputation of these laboratories and their ability to conduct particular studies. Sponsors should arrange for an independent audit of the conduct and reporting of their studies.

BIBLIOGRAPHY

"Industrial Hygiene and Toxicology" in *ECT* 1st ed., Vol. 7, pp. 847–870, by C. H. Hine, University of California, and L. Lewis, Industrial and Hygiene Associates; "Industrial Toxicology" in *ECT* 2nd ed., Vol. 11, pp. 595–610, by D. W. Fassett, Eastman Kodak Co.; "Industrial Hygiene and Toxicology" in *ECT* 3rd ed., Vol. 13, pp. 253–277, by G. D. Clayton, Clayton Environmental Consultants, Inc.; "Toxicology" in *ECT* 3rd ed., Supplement Vol., pp. 894–924, by B. Ballantyne, Union Carbide Corp.

1. L. J. Casarett and M. C. Bruce in J. Doull, C. D. Klaassen, and M. O. Amdur, eds., *Casarett and Doull's Toxicology*, Macmillan Publishing Co., Inc., New York, 1980, Chapt. 1.
2. P. A. Fenner-Crisp, "Risk Assessment Methods for Pesticides in Food and Drinking Water," Office of Pesticide Programs, U.S. Environmental Protection Agency, presented at the Florida Pesticide Review Council Meeting, July 7, 1989.
3. *Medical and Biological Effects of Environmental Pollutants, Chlorine and Hydrogen Chloride*, National Academy of Sciences, Washington, D.C., 1976.
4. M. Sittig, *Hazardous and Toxic Effects of Industrial Chemicals*, Noyes Data Corporation, N.J., 1979, p. 39.
5. R. Morley and S. J. Silk, *Ann. Occup. Hyg.* **13**, 101 (1970).
6. R. L. Baer, D. L. Ramsey, and E. Biondi, *Arch. Dermatol.* **108**, 74 (1973).

7. *Occupational Exposure to Ketones*, Publication No. 78-173, Department of Health, Education, and Welfare, National Institute for Occupational Safety and Health, Center for Disease Control, Washington, D.C., 1978.

8. E. Browning, *Toxicity and Metabolism of Industrial Solvents*, Elsevier Publishing Co., Amsterdam, the Netherlands, 1965, p. 304.

9. C. Zenz, ed., *Developments in Occupational Medicine*, Year Book Medical Publishers, Inc., Chicago, Ill., 1980, p. 403.

10. C. Maltoni, G. Lefemine, A. Ciliberti, G. Cotti, and D. Carretti, *Environ. Health Persp.* **41**, 3 (1981).

11. *Ann. Occup. Hygiene* **20**, 215 (1977).

12. *Occupational Exposure to Phosgene*, Publication No. 76–137, U.S. Department of Health, Education, and Welfare, National Institute for Occupational Safety and Health, Center for Disease Control, Washington, D.C., 1976.

13. C. S. Davis and R. J. Richardson in P. S. Spencer and H. H. Schaumburg, eds., *Experimental and Clinical Neurotoxicology*, Williams and Wilkins, Baltimore, Md., 1980, Chapt. 36.

14. K. Fletcher in B. Ballantyne, ed., *Forensic Toxicology*, John Wright & Sons, Ltd., Bristol, U.K., 1974, p. 86.

15. S. L. Robbins and R. S. Cotran, *Pathologic Basis of Disease*, W. B. Saunders Co., Philadelphia, Pa., 1979, Chapt. 3.

16. J. B. Walter and M. S. Israel, *General Pathology*, Churchill-Livingstone, Edinburgh, U.K., 1979, Chapt. 14.

17. R. Pittelkow in C. Zenz, ed., *Occupational Medicine*, Year Book Medical Publishers, Inc., Chicago, Ill., 1975, Chapt. 12.

18. W. R. Parkes, *Occupational Lung Disorders*, Butterworths, London, 1982, Chapt. 12.

19. A. Seaton in W. K. C. Morgan and A. Seaton, eds., *Occupational Lung Diseases*, W. B. Saunders Co., Philadelphia, Pa., 1975, Chapt. 12.

20. *Occupational Exposure to Diisocyanates*, Publication No. 78-215, Department of Health, Education, and Welfare, National Institute for Occupational Safety and Health, Center for Disease Control, Washington, D.C., Sept. 1978.

21. R. E. Kouri, *Genetic Differences in Chemical Carcinogenesis*, CRC Press Inc., West Palm Beach, Fla., 1980.

22. M. Sorsa in H. Vainio, M. Sorra, and K. Hemminki, eds., *Occupational Cancer and Carcinogenesis*, Hemisphere Publishing Corporation, Washington, D.C., 1979, p. 57.

23. J. H. Weisburger and G. M. Williams in Ref. 1, Chapt. 6.

24. S. H. Yuspa in D. Schottenfeld and J. F. Fraumeni, eds., *Cancer Epidemiology and Prevention*, W. B. Saunders Co., Philadelphia, Pa., 1982, Chapt. 3.

25. *Science* **218**, 975 (Dec. 3, 1982).

26. D. Brusick, *Principles of Genetic Toxicology*, Plenum Press, New York, 1980.

27. W. G. Flamm and M. A. Mehlman, eds., *Mutagenesis*, Hemisphere Publishing Corporation, Washington, D.C., 1978.

28. V. W. Mayer and W. G. Flamm in A. L. Reeves, ed., *Toxicology: Principles and Practice*, Vol. 1, John Wiley & Sons, Inc., New York, 1981.

29. J. M. Arena, *Poisoning*, Charles C. Thomas, Springfield, Ill., 1976, Chapt. 2.

30. R. H. Schwarz and S. J. Yaffe, eds., *Drug and Chemical Risks to the Fetus and Newborn*, Alan R. Liss, Inc., New York, 1980.

31. M. R. Juchau, ed., *The Biochemical Basis of Chemical Teratogenesis*, Elsevier/North Holland, New York, 1981.

32. C. A. Kimmel and J. Beulke-Sam, eds., *Developmental Toxicology*, Raven Press, New York, 1981.

33. K. S. Khera, *Fund. Appl. Tox.* **1**, 13 (1981).

34. Y. Alarie, L. Kane, and C. Barrow in Ref. 28, Chapt. 3.

35. Y. Alarie in B. K. J. Leong, ed., *Inhalation Toxicology and Technology*, Ann Arbor Science Publishers, Inc., Ann Arbor, Mich., 1981, p. 207.

36. *Food Quality Protection Act of 1996* (July 23, 1996), 104th Congress, 2nd Session Report 104–669 part 2, Sec. 408 (b)(2)(c) Safety to Infants and Children, Sec. 408 (p) Estrgenic Substance Screen Program.

37. *Environ. Sci. Technol./News* **30**(12), 540–544 1996.

38. T. Colborn, D. Dumanoski, and J. P. Myers, *Our Stolen Future*, E. P. Dutton, New York, 1996.

39. M. Hill, *Environ. Solutions*, 9 (Nov. 1996).

40. G. Zbinden and M. Flury-Roversi, *Arch. Toxicol.* **47**, 77 (1981).

41. J. Doull in Ref. 1, Chapt. 5.

42. C. D. Klaassen in Ref. 1, Chapt. 3.

43. R. S. Chhabra, *Environ. Health Persp.* **33**, 61 (1979).

44. W. G. Fong, A. Moye, J. Seiber, and J. Toth, *Pesticide Residue Methodology in Foods: Methods, Techniques and Regulations*, in progress.

45. F. J. Francis, *Food Safety: The Interpretation of Risk*, No. CC-1992-1, Council for Agricultural Science and Technology, Ames, Iowa, April, 1992.

46. T. A. Loomis in V. A. Drill and P. Lazar, eds., *Current Concepts in Cutaneous Toxicity*, Academic Press, New York, 1980, p. 153.

47. D. J. Birmingham in V. A. Drill and P. Lazar, eds., *Cutaneous Toxicity*, Academic Press, New York, 1977, p. 53.

48. P. H. Dugard in F. N. Marzulli and H. I. Maibach, eds., *Dermatotoxicology and Pharmacology*, Hemisphere Publishing Corp., Washington, D.C., 1977, Chapt. 22.

49. M. G. Bird, *Ann. Occup. Hyg.* **24**, 235 (1981).

50. C. N. Davies in B. Ballantyne, ed., *Respiratory Protection*, Year Book Medical Publishers, Inc., Chicago, Ill., 1981, Chapt. 4.

51. *Airborne Particles*, Committee on Medical and Biological Effects of Environmental Pollutants, National Research Council, University Book Press, Baltimore, Md., 1979.

52. W. K. C. Morgan in Ref. 19, Chapt. 3.

53. W. R. Parkes in Ref. 18, Chapt. 3.

54. I. M. Asher and P. P. McGrath, eds., *Symposium on Electron Microscopy of Microfibers, Proceedings of the First FDA Office of Science Symposium*, Stock No. 017-012-00244-7, Superintendent of Documents, U.S. Government Printing Office, Washington, D.C., 1976.

55. M. S. Stanton, *J. Natl. Cancer Inst.* **48**, 797 (1972).

56. M. S. Stanton, *J. Natl. Cancer Inst.* **67**, 965 (1981).

57. J. S. Harington, *J. Natl. Cancer Inst.* **67**, 977 (1981).

58. B. Ballantyne in B. Ballantyne, ed., *Current Approaches in Toxicology*, John Wright and Sons, Ltd., Briston, U.K., 1977, Chapt. 12.

59. K. J. Freundt, *Occup. Health Safety*, 10 (Aug. 1982).

60. E. J. Ariens, A. M. Simonis, and J. Offermeir, *Introduction to General Toxicology*, Academic Press, New York, 1976, Chapt. 7.

61. F. E. Guthrie in E. Hodgson and F. E. Guthrie, eds., *Introduction to Biochemical Toxicology*, Elsevier Publishing Co., New York, 1980, Chapt. 2.

62. W. B. Jakoby, R. D. Sekura, E. S. Lyon, C. J. Marcus, and J.-L. Wang in W. B. Jakoby, ed., *Enzymatic Basis of Detoxification*, Vol. 2, Academic Press, New York, 1980, Chapt. 11.

63. R. G. Killenberg and L. T. Webster, Jr., in Ref. 62, Chapt. 8.

64. W. W. Weber and I. B. Glowinski in Ref. 62, Chapt. 9.

65. A. de Bruin, *Biochemical Toxicology of Environmental Agents*, Elsevier Publishing Co., Amsterdam, 1976, p. 185.

66. R. O. Recknagel and L. A. Glende, *Crit. Rev. Toxicol.* **2**, 265 (1973).

67. J. A. Timbrell, *Principles of Biochemical Toxicology*, Taylor and Francis, Ltd., London, 1982, Chapt. 7.

68. F. J. Francis, *J. Sci. Food Agric.* **4**(1), 10–13 (1986).

69. R. P. Maickel and D. P. McFadden, *Res. Commun. Chem. Pathol. Pharmacol.* **26**, 75 (1979).

70. "Testing for Carcinogen," F. J. Francis, *J. Sci. Food Agric.* **4**(2) (1986).

71. C. K. Winter, *Reg. Toxicol. Pharmacol.* **15**, 137–150 (1992).

72. C. L. Galli, S. D. Murphy, and R. Paoletti, eds., *The Principles and Methods in Modern Toxicology*, Elsevier/North Holland, Amsterdam, 1980.

73. E. J. Gralla, ed., *Scientific Considerations in Monitoring and Evaluating Toxicological Research*, Hemisphere Publishing Corp., Washington, D.C., 1981.

74. *Chem. Eng. News* **61**(4), 42 (1983).

75. S. R. M. Bushby in G. E. Paget, ed., *Methods in Toxicology*, Blackwell Scientific Publications, Oxford, U.K., 1970, Chapt. 13.

76. A. E. Street in Ref. 75, Chapt. 12.

77. D. T. Plummer in J. W. Gorrod, ed., *Testing for Toxicity*, Taylor and Francis, London, 1981, Chapt. 12.

78. *Principles and Methods for Evaluating the Toxicity of Chemicals, Part 1, Environmental Health Criteria No. 6*, World Health Organization, Geneva, Switzerland, 1978, Chapt. 5.

79. D. Poynter in Ref. 58, Chapt. 8.

80. B. Ballantyne in B. Ballantyne, ed., *Respiratory Protection*, Chapman and Hall, London, 1981, Chapt. 5.

81. H. N. MacFarland in W. J. Hayes, ed., *Essays in Toxicology*, Vol. 7, 1976, p. 121.

82. B. P. McNamara in M. A. Mehlman, R. E. Shapiro, and H. Blumenthal, *New Concepts in Safety Evaluations*, Vol. 1, Hemisphere Publishing Corp., Washington, D.C., 1976, Part 1, Chapt. 4.

83. R. P. Giovacchini in Ref. 47, p. 31.

84. P. S. Elias in Ref. 72, p. 169.

85. *Principles and Procedures for Evaluating the Toxicity of Household Substances*, National Academy of Sciences, Washington, D.C., 1977, Chapt. 2.

86. R. Heywood in Ref. 77, Chapt. 17.

87. T. O. McDonald and J. A. Shadduck in Ref. 48, Chapt. 4.

88. A. M. McCreesh and M. Steinberg in Ref. 48, Chapt. 5.

89. A. B. Lansdown, *J. Soc. Cosmet. Chem.* **23**, 739 (1972).

90. W. E. Parish in Ref. 77, Chapt. 20.

91. F. Marzulli and H. C. Maguire, Jr., *Fd. Chem. Toxic.* **20**, 67 (1982).

92. G. Klecak in Ref. 48, Chapt. 9.

93. M. H. Karol in Ref. 35, p. 233.

94. S. C. Gad, *J. Toxicol. Environ. Health* **9**, 691 (1982).

95. H. A. Tilson, P. A. Cabe, and T. A. Burne, in Ref. 13, Chapt. 51.

96. S. Norton, *Environ. Health Persp.* **26**, 21 (1978).

97. P. S. Spencer, M. C. Bischoff, and H. H. Schaumburg in Ref. 13, Chapt. 50.

98. T. Damstra and S. C. Bondy in Ref. 13, Chapt. 56.

99. A. J. Dewar in Ref. 67, Chapt. 15.

100. C. L. Mitchell, ed., *Nervous System Toxicology*, Raven Press, New York, 1982.

101. L. Manzo, ed., *Advances in Neurotoxicology*, Pergamon Press, Oxford, U.K., 1980.

102. K. Presad and A. Vernadakis, eds., *Mechanisms of Action of Neurotoxic Substances*, Raven Press, New York, 1982.

103. I. Geller, W. C. Stebbins, and M. J. Wayner, eds., *Test Methods for Definition of Effects of Toxic Substances on Behavior and Neuromotor Function, Neurobehavioral Toxicology*, Vol. 1, Suppl. 1, 1979.

104. B. Weiss and V. G. Laties, eds., *Behavioural Toxicology*, Plenum Press, New York, 1975.

105. B. Ballantyne, M. F. Gazzard, and D. W. Swanston in Ref. 58, Chapt. 11.

106. B. Ballantyne and D. W. Swanston, *Acta Pharmacol. Toxicol.* **35**, 412 (1974).

107. E. J. Owens and C. L. Punte, *Am. Ind. Hyg. Assoc. J.* **24**, 262 (1963).

108. *TLVs for Chemical Substances in Workroom Air Adopted by ACGIH for 1981*, Publications Office, American Conference of Governmental Industrial Hygienists, Cincinnati, Ohio.

109. T. F. X. Collins and E. V. Collins in Ref. 82.

110. J. L. Schardein, *Drugs as Teratogens*, CRC Press, Cleveland, Ohio, 1976.

111. D. Neubert, H.-J. Merker, and T. E. Kwasigroch, eds., *Methods in Prenatal Toxicology*, Georg Thieme Publishers, Stutgart, Germany, 1977.

112. H. B. Pace, W. M. Davis, and L. A. Borgen, *Ann. N.Y. Acad. Sci.* **191**, 123 (1971).

113. I. Coyle, M. J. Wagner, and G. Singer, *Pharmacol. Biochem. Behav.* **4**, 191 (1976).

114. K. S. Rau and B. A. Schwetz in L. Breslow, ed., *Annual Review of Public Health*, Vol. 3, 1982, p. 1.

115. J. P. Griffin in Ref. 58, Chapt. 4.

116. I. C. Munro in Ref. 72, p. 125.

117. *"Chemobiokinetics and Metabolism," Principles and Methods for Evaluating the Toxicity of Chemicals*, Part 1, Environmental Health Criteria No. 6, World Health Organization, Geneva, 1978, Chapt. 4.

118. D. B. Tuey in E. Hodgson and F. E. Guthrie, eds., *Introduction to Biochemical Toxicology*, Elsevier Publishing Co., New York, 1980, Chapt. 3.

119. P. J. Gehring, P. G. Watenabe, and G. E. Blau in Ref. 82, Chapt. 8.

120. J. McCann and B. N. Ames in Ref. 27, Chapt. 5.

121. D. Brusick, *Principles of Genetic Toxicology*, Plenum Press, New York, 1980.

122. F. Vogel and G. Rohrborn, eds., *Chemical Mutagenesis in Mammals and Man*, Springer-Verlag, Berlin, 1970.

123. B. J. Kilbey, M. Legator, W. Nicols, and C. Ramel, eds., *Handbook of Mutagenicity Test Procedures*, Elsevier Scientific Publishing Company, Amsterdam, 1977.

124. A. W. Hsie, J. P. O'Neill, and V. K. McElheney, eds., *Mammalian Cell Mutagenesis: The Maturation of Test Systems*, Banbury Report No. 2, Cold Spring Harbor Laboratory, Cold Spring Harbor, New York, 1979.

125. A. Wallace Hays, ed., *Principles and Methods of Toxicology*, 3rd ed., 1994, pp. 545–546.

126. M. Patlak, *Env. Sci. Technol. News* **31**(4), 190A (1997).

127. *Pesticides in the Diets of Infants and Children*, National Research Council, National Academy Press, Washington, D.C., 1993.

General References

References 1, 4, 67, 77, 78, and 118 are also general references.

T. A. Loomis, *Essentials of Toxicology*, Lea and Febiger, Philadelphia, Pa., 1974.

V. K. Brown, *Acute Toxicity in Theory and Practice*, John Wiley & Sons, Inc., Chichester, U.K., 1980.

E. Boyland in R. S. F. Schilling, ed., *Occupational Health Practice*, Butterworths, London, 1981, Chapt. 24.

M. A. Cooke in A. W. Gardner, ed., *Current Approaches to Occupational Medicine*, John Wright and Sons, Ltd., Bristol, U.K., 1979, Chapt. 9.

C. R. Richmond, P. J. Walsh, and E. C. Copenhaver, eds., *Health Risk Analysis*, The Franklin Institute Press, Philadelphia, Pa., 1980.

N. H. Proctor and J. P. Hughes, *Chemical Hazards of the Workplace*, J. B. Lippincott Co., Philadelphia, Pa., 1978.

Registry of the Toxic Effects of Chemical Substances, Publication No. 81-116, Vols. 1 and 2, National Institute for Occupational Safety and Health, Superintendent of Documents, U.S. Government Printing Office, Washington, D.C., 1980.

G. D. Clayton and F. E. Clayton, eds., *Patty's Industrial Hygiene and Toxicology*, Vols. 2A, 2B, and 2C, John Wiley & Sons, Inc., New York, 1981.

W. M. Grant, *Toxicology of the Eye*, Charles C. Thomas, Springfield, Ill., 1974.

E. Cronin, *Contact Dermatitis*, Churchill Livingstone, Edinburgh, U.K., 1980.

I. Fishbein, *Potential Industrial Carcinogens and Mutagens*, Elsevier Scientific Publishing Co., Amsterdam, 1979.

Toxicology Abstracts, Cambridge Scientific Abstracts, Bethesda, Md. (monthly pub.).

Industrial Hygiene Digest, Industrial Health Foundation, Pittsburgh, Pa. (monthly pub.).

V. M. Traina, *Medicinal Research Reviews* **3**(1), 43 (1983).

BRYAN BALLANTYNE
Union Carbide Corporation

WILLIAM GEORGE FONG
Florida Department of Agriculture
and Consumer Services

TRACE AND RESIDUE ANALYSIS

Trace analysis is the detection of minute quantities of organic and inorganic materials. The definition of trace analysis continues to evolve. In the 1960s, it implied determinations under 0.01% of a sample. As of the mid-1990s, trace analysis is generally recognized as those determinations that represent around 0.0001%, ie, at the parts per million (ppm) level, where 1 ppm is equivalent to 1 μg/g (1). Ultratrace analysis, ie, determination below trace analysis, corresponds to levels below ppm or $<\mu$g/g (2–4). Residue analysis is the analysis of material left from an operation, ie, residual. Examples are solvents left in pharmaceuticals (qv) or pesticides (qv) left in fruits. The nature of the sample and the type of analysis to be performed, dictate methodology used. Analyses performed at or below ppm level, or those analyses where the actual analyte is at low micrograms level, are discussed herein, as are a variety of methodologies and some criteria for selection.

There are numerous applications for trace or ultratrace analyses in the chemical process industry. The following two examples highlight the need for such analyses. Although much controversy still surrounds the nature of its toxicity and possible safe levels, dioxin (2,3,7,8-tetrachlorodibenzo-*p*-dioxin (TCDD)) is frequently described as the worst poison known. It has been found to cause abortion in monkeys, even at a level of 200 parts per trillion (ppt) (5). Allowing for a hundredfold margin of safety for human exposure, the safe food level for TCDD would have to be less than 2 ppt. Polychlorinated biphenyls (PCBs) at 0.43 parts per billion (ppb) in water have been found to weaken the backbones of trout by interfering in collagen synthesis (6). The analysis of fish backbones from such water revealed excess calcium levels and a deficiency in collagen and phosphorus. The fish were also deficient in vitamin C, a cofactor in collagen synthesis. Thus it was concluded that the trout used vitamin C for detoxification of PCBs instead of for skeletal development.

The U.S. FDA monitors foods for half of the approximately 300 pesticides having official EPA tolerances as well as a number of other pesticides that have no official tolerances. Multiresidue methods, most of which are based on chromatography protocols, are employed (7). Not all pesticides are monitored on all foods and sampling (qv) is purposely biased to catch possible problems. The overall incidence of illegal pesticide residue is, however, quite small: 1% for domestic surveillance samples and 3% for imported foods. The methods employed can usually quantify residues present at 0.01 ppm. Quantitation limits range from 0.005 to 1 ppm.

Detectability limits are generally either not given in the scientific literature, or when limits are given, the units vary (8). It has been recommended (1) that grams/grams be used as the unit for data of ultratrace analyses. This would permit comparison of various detectabilities and allow easy calculation into ppm, ppb, etc. When this information is provided in molar units, the molecular weight should also be given. Moreover, the following important analytical parameters have been recommended to be reported for each method: amount present in grams in the original sample (APIOS) per mL or g; minimum amount detected (MAD) in g; minimum amount quantitated (MAQ) in g.

Frontiers of Low Level Detection

Extremely low level detection work is being performed in analytical chemistry laboratories. Detection of rhodamine 6G at 50 yoctomole (50×10^{-24} mol) has been reported using a sheath flow cuvette for fluorescence detection following capillary electrophoresis (9). This represents 30 molecules of rhodamine, a highly fluorescent molecule (see ELECTROSEPARATIONS, ELECTROPHORESIS; SPECTROSCOPY, OPTICAL).

Claims of single molecule detection in liquid samples have been made by combining the high sensitivity of laser-induced fluorescence (lif) and the spatial localization and imaging capabilities of optical microscopy (qv) (10). This technique combines confocal microscopy, diffraction-limited laser excitation, and a high efficiency detector. The probe volume is defined latitudinally by optical diffraction and longitudinally by spherical aberration. Using an unlimited excitation throughout and a low background level, this technique allows fluorescence detection of single rhodamine molecules at a signal-to-noise (S/N) ratio of approximately 10 in 1 ms. The use of confocal fluorescence microscopy can be extended to individual, fluorescently tagged biomolecules, including deoxynucleotides, whether single-stranded primers or double-stranded deoxyribonucleic acid (DNA).

Analysis of single mammalian cells by capillary electrophoresis has been reported using on-column derivatization and laser-induced fluorescence detection (11). Dopamine and five amino acids were determined in individual rat pheochromocytoma cells after on-column derivatization.

Radioactive tracers (qv) are powerful tools for trace detection. A method of labeling proteins using ^{99m}Tc has been described. The immunoreactivity of monoclonal antibodies after radiolabeling was demonstrated by radioimmunoimaging of thrombi using a ^{99m}Tc-labeled antifibrin monoclonal antibody (12). Radiotracer imaging agents have been used for mapping sympathetic nerves of the heart (13). The radioiodination of analogues of a calichemicin constituent have been employed as a possible brain-imaging agent (14) (see MEDICAL IMAGING TECHNOLOGY).

Samples

Sampling. A sample used for trace or ultratrace analysis should always be representative of the bulk material. The principal considerations are determination of population or the whole from which the sample is to be drawn, procurement of a valid gross sample, and reduction of the gross sample to a suitable sample for analysis (15) (see SAMPLING).

The analytical uncertainty should be reduced to one-third or less of sampling uncertainty (16). Poor results obtained because of reagent contamination, operator errors in procedure or data handling, biased methods, and so on, can be controlled by proper use of blanks, standards, and reference samples.

Sample Preparation. Sample contamination must be prevented throughout the sampling procedures. Factors that can influence sampling of an analyte include impinged material, residual solvents, sample preserving method, analyte absorption, and potential contamination from the environment. The contamination from sample holders or loss to them have to be considered. No significant changes should occur in the sample when it is being held for analysis. Sample stabilization generally includes storage at low temperatures; however, any stabilization step should be validated. A review of sample composition and properties is advised. This would include number of compounds present, chemical structures (functionality) of compounds, molecular weights of compounds, pK_a values of compounds, uv spectra of compounds, nature of sample matrix (solvent, fillers, etc), concentration range of compounds in samples of interest, and sample solubility. These properties provide the bases for the selection of an extraction solvent, or a disposable cartridge for sample-extract cleanup such as Supelclean, Quick-Sep, Sep-Pak, or Bond-Elut can be used (2). A detailed discussion of sample preparation methods is available (2). Table 1 lists some samples analyzed by solid-phase extractions from different matrices.

Frequently, preconcentration of an analyte is necessary because the detector used for quantitation may not have the necessary detectability, selectivity, or freedom from matrix interferences (32). Significant sample losses can occur during this step because of very small volume losses to glass walls of the recovery containers, pipets, and other glassware.

Solid-Phase Microextraction. Solid-phase microextraction (SPME), used as a sample introduction technique for high speed gc, utilizes small-diameter

Table 1. Samples Analyzed by Solid-Phase Extraction

Sample	Matrix	Solid-phase column	Detectability, ng/mL	Ref.
3-methoxy-4-hydroxyphenyl glycol	plasma	alumina	1−10	17
oxytetracycline	fish tissue	C-8	5−10[a]	18
doxefazepam	plasma	C-18	0.1[b]	19
cortisol	urine	C-18		20
basic drugs	plasma	CN	therapeutic levels	21
Δ^9-tetrahydrocannabinol	plasma	C-18	2, 100 pg	22
ibuprofen	plasma	C-2	1.3[b]	23
ranitidine	plasma, other fluids	CN	2	24
chlorpromazine and metabolites	plasma	C-8		25
cyclotrimethylenetrinitramine	biological fluids	C-18		26
sotalol	plasma, urine	C-8	10	27
cyclosporin A	serum, urine	cyanopropyl		28
cyclosporine	blood	C-18	10	29
growth factors	urine	C-1	200−1400-fold enrichment	30
carbamazine and metabolites	plasma	C-18	50	31

[a]Value is in ng/g.
[b]Value is in μm/mL.

fused-silica fibers coated with polymeric stationary phase for sample extraction and concentration (33). The trapped analyte can be liberated by thermal desorption. By using a specially designed dedicated injector, the desorption process can be shortened to a fraction of a second, producing an injection band narrow enough for high speed gc. A modified system has been investigated for the analysis of volatile compounds listed in EPA Method 624. Separation of all 28 compounds by ion trap mass spectrometric detector is achieved in less than 150 seconds.

SPME has been utilized for determination of pollutants in aqueous solution by the adsorption of analyte onto stationary-phase coated fused-silica fibers, followed by thermal desorption in the injection system of a capillary gas chromatograph (34). Full automation can be achieved using an autosampler. Fiber coated with 7- and 100-μm film thickness and a nitrogen−phosphorus flame thermionic detector were used to evaluate the adsorption and desorption of four s-triazines. The gc peaks resulting from desorption of fibers were shown to be comparable to those obtained using manual injection. The 7-μm fiber, designed for the analysis of semivolatile analytes, was used to investigate the effect of desorption temperature and on-column focusing temperature on peak response. The desorption temperature was found to be noncritical. An optimum focusing temperature of 40°C was used. Evaluation of 100-μm film fiber demonstrated its potential to adsorb greater quantities of analyte from solution. An absorption time of 15 minutes gave an equilibrium distribution of the solutes between the stationary and liquid phases. For the thicker film fiber the effectiveness of the desorption process was reduced at temperatures below 140°C. The linear dynamic range of the technique was evaluated over three orders of magnitude. To enhance method sensitivity, the fiber was used to extract 0.1 ppb solution of herbicide by repeatedly adsorbing and desorbing from the same solution and focusing the combined solutes at the front of the analytical column prior to elution and analysis.

Supercritical Fluid Extraction. Polycyclic aromatic hydrocarbons (PAHs) have been extracted from contaminated land samples by supercritical fluid extraction (SFE) with both pure and modified carbon dioxide (35) (see SUPERCRITICAL FLUIDS). An experimental design approach, based on central composite design, was used to determine which SFE variable affects the total recovery of 16 PAHs. Four parameters were chosen for evaluation: pressure, temperature, extraction time, and percentage of methanol modifier addition. Accessible levels of each parameter were dependent on instrumental constraints. A statistical treatment of the results indicated that extraction time and percentage of modifier addition were the only variables to affect PAH recovery significantly. The levels of these variables were set at the maximum values while the pressure and temperature were maintained at their midpoint value in design. These conditions were used in a repeatability study ($n = 7$), which extracted an average of 458.0 mg/kg total PAHs from the contaminated land sample with a relative standard deviation (RSD) of 3.1%. Sequential extractions on three of these samples, using identical operating conditions, did not show the presence of PAHs. The results were compared with the Soxhlet extraction and microwave-assisted (MAE) extraction of the sample, which recovered an average of 297.4 (RSD 10.0%) and 422.9 (RSD 2.4%) mg/kg, respectively.

Removing an analyte from a matrix using supercritical fluid extraction (SFE) requires knowledge about the solubility of the solute, the rate of transfer of the solute from the solid to the solvent phase, and interaction of the solvent phase with the matrix (36). These factors collectively control the effectiveness of the SFE process, if not of the extraction process in general. The range of samples for which SFE has been applied continues to broaden. Applications have been in the environment, food, and polymers (37).

Microwave-Assisted Extraction. Sample preparation techniques that prevent or minimize pollution in analytical laboratories, improve target analyte recoveries, and reduce sample preparation costs were evaluated with regard to the microwave-assisted extraction (MAE) procedure for 187 compounds and four Aroclors listed in EPA Methods 8250, 8081, and 8141A (38) (see MICROWAVE TECHNOLOGY). The results indicate that most of these compounds can be recovered in good yields from the matrices investigated. For example, recoveries ranged from 80 to 120% for 79 of the 95 compounds listed in Method 8250; 38 of the 45 organochlorine pesticides listed in Method 8081; and 34 of 47 organophosphorus pesticides listed in Method 8141A. When recoveries from freshly spiked oil samples were compared with those of aged samples, it was found that recoveries usually decreased in the aged samples. There was more spread in recoveries with increased aging time. For 15 compounds in a reference soil, the recoveries of 14 compounds by MAE were equal to or better than recoveries obtained by Soxhlet extraction (naphthalene was an exception).

For selected organochlorine pesticides, recoveries from spiked oil samples were at least 7% higher for MAE than for either Soxhlet or sonication extraction.

Comparative studies were performed to evaluate microwave digestion with conventional sample destruction procedures. These included the analysis of shellfish, meats, rocks, and soils. Generally, comparable accuracy at much shorter digestion time was found for the MAE vs the classical digestion method (39).

Sample Cleanup.   The recoveries from a quick cleanup method for waste solvents based on sample filtration through a Florisil and sodium sulfate column are given in Table 2 (40). This method offers an alternative for analysts who need to confirm the presence or absence of pesticides or PCBs.

Synthetic organic chemicals have been isolated by either resin adsorption or direct methylene chloride liquid–liquid extraction. Analyses for 48 distinct chemical entities in river water from a river located in North Carolina's Piedmont area were carried out. The river was sampled at three locations several times during a 13-month period (41). Most frequently included among the 48 chemicals found were atrazine, methyl atraton (triazine herbicides), dimethyl dioxane, 1,2,4-trichlorobenzene, tributylphosphate, triethylphosphate, trimethylindolinone, and tris(chloropropyl) phosphate. Many of these chemicals are indigenous to industrial and agricultural activities in Piedmont. The concentrations were in the ng/L to mg/L range.

Table 2. Florisil Filtration Recovery Efficiency[a]

Compound	Sample concentration, ppm	Recovery, %[b]	Standard deviation
aldrin	2	93	9.5
BHC	2	86	6.1
lindane	2	92	13.5
chlordane	2	79	4.7
DDD	2	69	16.0
DDE	2	92	11.0
DDT	2	89	7.2
dieldrin	2	88	3.5
endosulfan	2	91	7.6
endrin	2	98	7.6
heptachor	2	97	2.9
toxaphene	3	90	16.5
Arochlor 1016	20	93	2.0
Arochlor 1221	20	95	6.4
Arochlor 1232	20	100	8.2
Arochlor 1242	20	93	8.3
Arochlor 1248	10	95	9.1
Arochlor 1254	10	86	9.7
Arochlor 1260	10	87	12.2

[a]Ref. 40.
[b]Average recovery of triplicate analysis.

Method Validation

Statistically designed studies should be performed to determine accuracy, precision, and selectivity of the methodology used for trace or ultratrace analyses. The reliability requirements for these studies are that the data generated withstand interlaboratory comparisons.

The following principles should be used to establish a valid analytical method (42).

1. A specific detailed description and protocol should be written (standard operating procedure (SOP)).

2. Each step in the method should be investigated to determine the extent to which environmental, matrix, material, or procedural variables, from time of collection of material until the time of analysis and including the time of analysis, may affect the estimation of analyte in the matrix. Variability of the matrix owing to its physiological nature should be considered.

3. A method should be validated for its intended use with an acceptable protocol. All experiments conducted to make claims or draw conclusions about the validation of the method should be documented in a method validation report.

4. Wherever possible, the same matrix should be used for validation purposes. The stability of the analyte in the matrix during the collection process and the sample-storage period should be assessed, preferably before sample analysis. Accuracy, precision, reproducibility, response function, and specificity of the method, with respect to endogenous substances, metabolites, and known degradation products, should be established with reference to the biological matrix. With regard to specificity, there should be evidence that the substance being quantitated is the intended analyte.

5. The concentration range over which the analyte will be determined must be defined in the method, on the basis of actual standard samples over the range (standard curve).

6. It is necessary to use a sufficient number of standards to adequately define the relationship between concentration and response.

7. Determination of accuracy and precision should be made by analysis of replicate sets of analyte samples of known concentration from equivalent matrix. At least three concentrations representing the entire range of the calibration should be studied: one near the minimum (MAQ), one near the middle, and one near the upper limit of the standard curve.

Methodologies

The commonly used methods for ultratrace analyses together with the accepted detection limits are given in Table 3. Mass spectrometry (qv), which provides coverage for all elements at sensitivities of 10^{-12} to 10^{-15} g, is a sensitive detector for organic molecules and fragments. This technique is generally used in combination with gas or liquid chromatography for ultratrace analysis (see ANALYTICAL METHODS, HYPHENATED INSTRUMENTS). Derivatization chromatography can further help improve selectivity and detectability of a number of compounds (3).

Table 3. Ultratrace Analyses Methods[a]

Method	Minimum amount detected, g
mass spectrometry	
electron impact	10^{-12}
spark source	10^{-13}
ion scattering	10^{-15}
flame emission spectrometry	10^{-12}
liquid chromatography	
ultraviolet detection	10^{-11}
fluorescence detection	10^{-12}
gas chromatography	
flame ionization	10^{-12} to 10^{-14}
electron capture	10^{-13}
combination techniques	
liquid chromatography/mass spectrometry	10^{-12}
gas chromatography/mass spectrometry	10^{-12}
electron capture (negative)/ionization mass spectrometry	10^{-15}

[a]Ref. 3.

Atomic Absorption/Emission Spectrometry. Atomic absorption or emission spectrometric methods are commonly used for inorganic elements in a variety of matrices. The general principles and applications have been reviewed (43). Flame-emission spectrometry allows detection at low levels (10^{-12} g). It has been claimed that flame methods give better reproducibility than electrical excitation methods, owing to better control of several variables involved in flame excitation. Detection limits for selected elements by flame-emission spectrometry given in Table 4. Inductively coupled plasma emission spectrometry may also be employed.

Neutron Activation Analysis. A radiochemical neutron activation analysis technique for determination of 26 elements, including the emitting elements Th and U and Cu, Fe, K, Na, Ni, and Zn, has been developed (44). The radiochemical separation was performed by anion exchange on Dowex 1×8 column from HF and HF$-$NH$_4$F medium, leading to selective removal of the matrix-produced radionuclides ^{46}Sc, ^{47}Sc, and ^{48}Sc, and nearly selective isolation of ^{239}Np and ^{233}Pa, the indicator radionuclides of U and Th, respectively. For K, Na, Th, and U, a limit of detection of 30, 0.05, 0.03, and 0.07 ng/g, respectively, was achieved. For the other elements, the detection limits were between 0.002 ng/g for Ir and 45 ng/g for Zr.

Thin-Layer Chromatography. The most commonly used approach in thin-layer chromatography (tlc) entails separations on a silica (qv) gel plate where the silica gel is coated as a thin layer on a glass plate. The plate is developed using the mobile phase of choice after a sample has been applied to the starting line of the plate. Quantification is achieved directly by scanning the plate or indirectly by scraping and eluting the sample. A tlc assay, performed for the determination of rifampicin and its degradation products in drug-excipients interaction studies (45), involved a mobile phase consisting of chloroform$-$methanol$-$water (80:20:2.5). The peaks were quantified by den-

Table 4. Elemental Detection Limits by Flame Emission Spectrometry[a]

Elements	Emission lines, nm	Atomizer		
		Nebulizer flame, ng/mL	Rod-in-flame, pg	Particles-in-flame, pg
Ag	328.1	3	5	0.2
Ba	553.6	1	3	0.2
Ca	422.7	1	10	1
Cr	425.4	2	2	0.2
Cs	852.1	5×10^{-3}	0.03	10^{-4}
Cu	327.4	1	3	0.2
Eu	459.4	0.3	1	0.1
Ga	417.2	2	3	0.2
In	451.1	0.5	3	0.2
K	766.5	0.1	1	0.01
Li	670.8	10^{-4}	0.003	10^{-4}
Mg	285.2	3	10	1
Mn	403.1	1	1	0.05
Na	589.0	0.1	1	0.01
Rb	794.8	0.005	0.03	0.001
Sr	460.7	0.03	0.1	0.01
Tl	535.0	1	10	0.5
Yb	398.8	0.3	1	0.1

[a]Ref. 3.

sitometric evaluation of the chromatograms. The method gave a limit of detection of 10 ng per band and good precision and linearity in the range of 50–3000 ng per band for rifampicin, 3-formylrifamycin SV, rifampicin *N*-oxide and 25-desacetylrifampicin, and 100–350 ng per band of rifampicin quinone.

A rapid tlc immunoaffinity chromatographic method has been reported for quantitation in serum of an acute phase reactant, C-reactive protein (CRP), which can differentiate between viral and bacterial infections (46). The analysis is based on the sandwich assay format using monoclonal antibodies directed against two sites of CRP (see IMMUNOASSAYS). One of the antibodies is covalently bound to defined zones on a thin-layer immunoaffinity chromatography membrane, while the other antibody is covalently bound to deeply dyed blue latex particles. After incubation (CRP sample and latex particles), the CRP–latex immunocomplex is allowed to migrate along the immunoaffinity chromatography membrane. In the presence of antigen, a sandwich is formed between the CRP–latex immunocomplex and membrane-bound antibodies, resulting in the appearance of blue lines on the membrane. Antibody immobilization on the tlc membrane is made with a redesigned piezoelectric-driven ink-jet printer. The time required for analysis is less than 10 minutes. Quantitation is achieved either by counting the lines visually, using scanning reflectrometry, or using a modified bar-code reader. The limit of detection was estimated to be in the low femtomolar range by visual detection.

A number of compounds have been quantified by tlc or high performance thin-layer chromatography (hptlc) using absorption or fluorescence scanning densitometry (47). An example of trace determination relates gentiopicroside in various biological matrices with 40 ng/spot sensitivity by scanning at 270 nm (48). Using hptlc gave more consistent data and better precision when compared to hplc and a commercial elisa kit for determination of aflatoxins in peanut butter (49). Hptlc was combined with a new immunoblot approach called elisagram, to detect and quantify zearalenone and aflatoxin families at the picogram level (50). Aflatoxins and trichothecenes were separated and identified in submicrogram quantity by 2-D tlc/FAB-ms (51). Vitamin B_1 was quantified in pharmaceutical products by silica gel hptlc involving post-chromatography derivatization with potassium hexacyanoferrate(III)–sodium hydroxide reagent (500 pg/spot sensitivity) and fluorodensitometry (52).

Gas Chromatography. Gas chromatography is a technique utilized for separating volatile substances (or those that can be made volatile) between two phases, one of which is a gas. Purge-and-trap methods are frequently used for trace analysis. Various detectors have been employed in trace analysis, the most commonly used being flame ionization and electron capture detectors.

On-column gas chromatographic detection of nicotine at low picogram levels has been reported. Nicotine is first subjected to chemical derivatization with heptafluorobutyric anhydride in the presence of pyridine (53). The high yield reaction results in the opening of the N-methylpyrrolidine ring of nicotine to concomitant formation of a highly electrophilic N,O-diheptafluorobutyryl derivative. After the extraction of nicotine derivative into isooctane, it is subjected to splitless capillary gas chromatographic analysis using a ^{61}Ni electron-capture detector and moderately polar fused-silica capillary column. The nicotine derivative can be detected on-column at levels below 5 pg.

Gas chromatography/tandem mass spectrometry (gc/ms/ms) using selected reaction monitoring was applied to the analysis of urinary metabolites of sulfur mustard, derived from lyase pathway and from hydrolysis (54). In the case of lyase metabolites, a limit of detection of 0.1 ng/mL was obtained, compared to 2–5 ng/mL using single-stage gc/ms and selected ion monitoring. The gc/ms/ms methodology was less useful when applied to the analysis of thiodiglycol bis(pentafluorobenzoate) using negative ion chemical ionization although selected reaction chromatograms were cleaner than selected ion chromatograms. The advantage of using gc/ms/ms was demonstrated by the detection of low levels of -lyase metabolites in the urine of casualties who had been exposed to sulfur mustard.

The composition of technical DDT was investigated using achiral and chiral high resolution gas chromatography (hrgc) and electron-ionization mass spectrometry (ei/ms). 2,4'-DDT and 2,4'-DDD, two important components of technical DDT, were enantiomerically resolved by chiral hrgc using silylated β-cyclodextrin and by chiral hplc with permethylated γ-cyclodextrin as chiral selectors (55) (see CHIRAL SEPARATIONS (SUPPLEMENT)). The (+)- and (−)-enantiomers were assigned by chiral hplc using chiroptical measurements. Enantiopure isolates were then used to identify these enantiomers in chiral hrgc analyses. Previous data indicated (+)- and (−)-2,4'-DDT to have (S)- and (R)-configuration, respectively, but the absolute configurations for (+)- and (−)-2,4-DDD were hitherto unknown. These have been assigned via the reductive dechlorination of individual 2,4'-DDT enantiomers, which proceeded stereoselectively to the corresponding 2,4-DDD enantiomers. The results showed (+)- and (−)-2,4'-DDD to have (R)- and (S)-configurations, respectively. The enantiomers of 2,4'-DDD thus have reverse signs of rotation for polarized light compared to the 2,4'-DDT enantiomers with the same configuration. The method should be useful for the analysis of environmental and biological samples.

High Pressure Liquid Chromatography. High pressure liquid chromatography (hplc), frequently referred to as simply lc or as high performance liquid chromatography, is used in virtually all fields of chemistry. Nonvolatile or thermally labile compounds are best separated by hplc. Although techniques such as adsorption and ion-exchange chromatography have been used, the technique of choice is reversed-phase liquid chromatography (rplc). In rplc the stationary phase is nonpolar and the mobile phase is polar and its polarity can be suitably changed.

In hplc, detection and quantitation have been limited by availability of detectors. Using a uv detector set at 254 nm, the lower limit of detection is 3.5×10^{-11} g/mL for a compound such as phenanthrene. A fluorescence detector can increase the detectability to 8×10^{-12} g/mL. The same order of detectability can be achieved using amperometric, electron-capture, or photoionization detectors.

Hplc is capable of routine determination at the nanogram range (3). Using special techniques it is possible to perform analysis even when only a few nanograms of the analyte is available. Detection limits of a picogram or less have been demonstrated with the state-of-the-art capabilities. Achieving these low limits depends on the equipment, chromatographic conditions, special techniques, and the individual sample. The use of special techniques such as on-line derivatization assumes a great significance when only a small quantity of sample is available, since this includes sample enrichment as a means to improve detectability. To optimize detectability at trace or ultratrace levels, the user needs to have a thorough understanding of separation processes and various factors that affect them (56).

Equipment can play a significant role. Peak broadening arises from dispersion and mixing phenomena that occur in the injector, column connecting tubes, and detector cell, as well as from electronic constraints that govern the response speed of the detector and recorder. The quality of an instrument may therefore be judged by its ability to minimize extra-column band broadening and reproduce retention volumes. The constancy of retention volumes is primarily a function of solvent delivery system. The independent factors that contribute to extra-column band broadening can be treated as additive in their second moments or variances (2), according to the following relationship:

Total variance measured from chromatogram

= column variance + variance due to instrument volumes

+ variance due to electronic response time

The assumption that these individual contributions are independent of one another may not be true in practice. For an accurate calculation of instrument variance, it may be necessary to couple some of the individual contributions.

For an analyte of molecular weight 5000 and good chromatographic conditions, most photometric detectors can be expected to provide detection limits of 2–5 ng. Improvement into the mid-picogram or lower range normally requires the use of more sensitive detection means such as fluorescence or electrochemical detectors.

A study was conducted to measure the concentration of D-fenfluramine HCl (desired product) and L-fenfluramine HCl (enantiomeric impurity) in the final pharmaceutical product, in the possible presence of its isomeric variants (57). Sensitivity, stability, and specificity were enhanced by derivatizing the analyte with 3,5-dinitrophenylisocyanate using a Pirkle chiral recognition approach. Analysis of the calibration curve data and quality assurance samples showed an overall assay precision of 1.78 and 2.52%, for D-fenfluramine HCl and L-fenfluramine, with an overall intra-assay precision of 4.75 and 3.67%, respectively. The minimum quantitation limit was 50 ng/mL, having a minimum signal-to-noise ratio of 10, with relative standard deviations of 2.39 and 3.62% for D-fenfluramine and L-fenfluramine.

A reversed-phase isocratic hplc method has been developed for the determination of AG-331, a novel thymidylate synthase inhibitor, in human serum and urine (58). The method involves a solid-phase extraction from C-18 cartridges without addition of an internal standard. The methanol eluent is evaporated under nitrogen at 40°C, and reconstituted in mobile phase, acetonitrile–water (35:65) containing 25 mM ammonium phosphate. Separation of AG-331 was obtained on a C-18 column at a flow rate of 1 mL/min. Chromatographic signals were monitored by a photodiode array detector at a primary wavelength of 457 nm with a bandwidth of 4.8 mm. Standard curves are linear in the range of 22–175 ng/mL in plasma and 44–2175 ng/mL in urine, respectively. The extraction recovery ranged from 92.9 to 102.4%. Intraday coefficient of variation was less than 9.5%, and interday coefficient of variation was less than 14.3% for AG-331 concentration of 44 ng/mL. This method can be used to characterize the pharmacokinetics of AG-331 in cancer patients as part of ongoing phase I trials.

An hplc assay was developed suitable for the analysis of enantiomers of ketoprofen (KT), a 2-arylpropionic acid nonsteroidal antiinflammatory drug (NSAID), in plasma and urine (59). Following the addition of racemic fenprofen as internal standard (IS), plasma containing the KT enantiomers and IS was extracted by liquid–liquid extraction at an acidic pH. After evaporation of the organic layer, the drug and IS were reconstituted in the mobile phase and injected onto the hplc column. The enantiomers were separated at ambient temperature on a commercially available 250 × 4.6 mm amylose carbamate-packed chiral column (chiral AD) with hexane–isopropyl alcohol–trifluoroacetic acid (80:19.9:0.1) as the mobile phase pumped at 1.0 mL/min. The enantiomers of KT were quantified by uv detection with the wavelength set at 254 nm. The assay allows direct quantitation of KT enantiomers in clinical studies in human plasma and urine after administration of therapeutic doses.

Post-column in-line photochemical derivatization permits fluorescence detection of the common aflatoxins B1, B2, G1, and G2 (60). Chromatographic evidence indicates that photolysis causes the hydration of the nonfluorescent B1 and G1 components to B2a and G2a components, respectively. Analysis of naturally contaminated corn samples show no interfering peaks and permits the determination of 1 and 0.25 ppb for B1 and B2, respectively.

Derivatization is useful for detection of compounds such as amino acids and amines that lack easily detectable groups. For similar reasons, saccharides, as a class of compound, elicit much interest. Two derivatization schemes have been reported using benzamide (61) and FMOC–hydrazine (62) to produce fluorescent products.

An on-line concentration, isolation, and liquid chromatographic separation method for the analysis of trace organics in natural waters has been described (63). Concentration and isolation are accomplished with two precolumns connected in series: the first acts as a filter for removal of interferences; the second actually concentrates target solutes. The technique is applicable even if no selective sorbent is available for the specific analyte of interest. Detection limits of less than 0.1 ppb were achieved for polar herbicides (qv) in the chlorotriazine and phenylurea classes. A novel method for determination of tetracyclines in animal tissues and fluids was developed with sample extraction and cleanup based on tendency of tetracyclines to chelate with divalent metal ions (64). The metal chelate affinity precolumn was connected on-line to reversed-phase hplc column, and detection limits for several different tetracyclines in a variety of matrices were in the 10–50 ppb range.

A new cyanide dye for derivatizing thiols has been reported (65). This thiol label can be used with a visible diode laser and provide a detection limit of 8×10^{-6} M of the tested thiol. A highly sensitive laser-induced fluorescence detector for analysis of biogenic amines has been developed that employs a He–Cd laser (66). The amines are derivatized by naphthalenedicarboxaldehyde in the presence of cyanide ion to produce a cyanobenz[f]isoindole which absorbs radiation at the output of He–Cd laser (441.6 nm). Optimization of the detection system yielded a detection limit of 2×10^{-12} M.

Primary and secondary alcohols can be tagged using a fluorescence derivative (67). The detection limit for 1-propanol was 70 fmol for a 10-mL injection volume. Disodium EDTA and calcium chloride have been used as fluorescence intensity for determination of tetracycline antibiotics (68). The largest fluorescence increasing reagents for tetracycline was produced in a mobile phase when concentrations of EDTA and $CaCl_2$ were 25 and 35 mM, respectively, and pH was 6.5. The detection limit of the method ranged from 49 to 190 pg for three different tetracycline compounds. A post-column on-line immunochemical detection system was utilized for a very selective and sensitive method for determination of digoxin and digoxigenin (69). Fluorecine-labeled antibodies are used to target the chosen analytes, and the fluorescence detection system provides detection limits of 200 and 50 fmol, respectively, for digoxin and digoxigenin.

The limits of lifetime detection and resolution in on-the-flight fluorescence lifetime detection in hplc were evaluated for simple, binary systems of polycyclic hydrocarbons (70). Peak homogeneity owing to coelution was clearly indicated for two compounds having fluorescence lifetime ratios as small as 1.2 and the individual peaks could be recovered using predetermined lifetimes of the compounds. Limits of lifetime detection were determined to be 6 and 0.3 pmol for benzo[b]fluoranthene and benzo[k]fluoranthene, respectively.

A protein-binding assay (BA) coupled with hplc provided a highly sensitive post-column reaction detection system for the biologically important molecule biotin and its derivative biocytin, biotin ethylenediamine, 6-(biotinoylamino) caproic acid, and 6-(biotinoylamino)caproic acid hydrazide (71). This detection system is selective for the biotin moiety and responds only to the class of compounds that contain biotin in their molecules. In this assay a conjugate of streptavidin with fluorescamine isothiocyanate (streptavidin–FITC) was employed. Upon binding of the analyte (biotin or biotin derivative) to streptavidin–FITC, an enhancement in fluorescence intensity results. This enhancement in fluorescence intensity can be directly related to the concentration of the analyte and thus serves as the analytical signal. The hplc/BA system is more sensitive and selective than either the BA or hplc alone. With the described system, the detection limits for biotin and biocytin were found to be 97 and 149 pg, respectively.

An ion chromatographic system that included column switching and gradient analysis was used for the determination of cations such as Na^+, Ca^{2+}, Mg^{2+}, K^+, and NH_4^+ and anions such as Cl^-, NO_2^-, NO_3^-, and SO_4^{2-} in fog water samples (72). Ion-exchange chromatography compares very well with more generally used spectroscopic techniques for cation determinations. Determination limits range from 6 μg/L for Na^+ to 40 μg/L for K^+. Analysis of natural fog water samples can be performed at cost and time savings as compared to other techniques. Ion chromatography with self-regenerating suppression and conductivity detection was used for the simultaneous determination of Na^+, NH_4^+, K^+, Mg^{2+}, and Ca^{2+} in melted snow samples from high alpine sites (73). In order to determine low concentrations of winter snow samples (<2 g/kg), preconcentration of the sample was applied, resulting in better detection limits of 0.1–0.4 g/kg, depending on the cation.

Liquid Chromatography/Mass Spectrometry. Increased use of liquid chromatography/mass spectrometry (lc/ms) for structural identification and trace analysis has become apparent. Thermospray lc/ms has been used to identify by-products in phenyl isocyanate precolumn derivatization reactions (74). Five compounds resulting from the reaction of phenylisocyanate and the reaction medium were identified: two from a reaction between phenyl isocyanate and methanol, two from the reaction between phenyl isocyanate and water, and one from the polymerization of phenyl isocyanate. There were also two reports of derivatization to enhance either the response or structural information from thermospray lc/ms for linoleic acid lipoxygenase metabolites (75) and for cortisol (76).

A method for lc/ms of carotenoids has been reported that uses a C-30 reversed-phase column and a gradient solvent system containing methanol–methyl *tert*-butyl ether–ammonium acetate at a flow rate of 1.0 mL/min (77). The entire hplc column effluent passes through a photodiode array absorbance detector and then into the electrospray lc/ms interface without solvent splitting. In this way maximum sensitivity is achieved for both the diode array detector, which records the uv/visible spectra of each carotenoid, and mass spectrometer, which measures the molecular ion of each carotenoid. Molecular ions without evidence of any fragmentation were observed in the electrospray mass spectra of both xanthophylls and carotenes. In order to enhance the formation of molecular ions, solution-phase carotenoids oxidation was carried out by means of post-column addition of halogenated solvents to the hplc effluent. Several different halogenated solvents were evaluated, including chloroform, 2,2,3,3,4,4,4-heptafluoro-1-butanol, 2,2,3,3,4,4,4-heptafluorobutyric acid, 1,1,1,1,3,3,3-hexafluoro-2-propanol, and trifluoroacetic acid. Among these halogenated solvents, 2,2,3,3,4,4,4-heptafluor-1-butanol at a concentration of 0.1% was found to produce the best combination of carotenoid molecular ion abundance and reproducibility. The limits of detection for lutein and β-carotene were between 1 and 2 pmol each, which was a hundredfold lower than the detection limit of the photodiode array detector signal.

The development of methods of analysis of triazines and their hydroxy metabolites in humic soil samples with combined chromatographic and ms techniques has been described (78). A two-way approach was used for separating interfering humic substances and for performing structural elucidation of the herbicide traces. Humic samples were extracted by supercritical fluid extraction and analyzed by both hplc/particle beam ms and a new ms/ms method. The new ms/ms unit was of the tandem sector field-time-of-flight/ms type.

Liquid chromatography/thermospray mass spectrometric characterization of chemical adducts of DNA formed during *in vitro* reaction has been proposed as an analytical technique to detect and identify those contaminants in aqueous environmental samples which have a propensity to be genotoxic, ie, to covalently bond to DNA (79). The approach for direct acting chemicals includes the *in vitro* incubation of DNA with contaminated aqueous samples at 37°C, pH 7.0 for 0.5–6 h, followed by enzymatic hydrolysis of the DNA to deoxynucleosides and lc/ms of the resulting solution. A series of allylic reagents was used to model reactive electrophiles in synthetic aqueous samples to demonstrate that adduct formation was linear with both contaminant concentration and electrophilic activity potential. The characterization can also estimate the proportion of bonding to different sites on a base, for instance, the ratio of oxygen- vs nitrogen-bonding products, which is an important parameter in assessing the genotoxicology of chemicals. Electrospray lc/ms interface has been used to study various positive ions, including NH_4^+, Na^+, K^+, Cs^+, and Ca^+, and the protonated BH^+ ions of 30 organic nitrogen bases (B) (80). Detection limits in the subfemtomole to attomole range were achieved. The sensitivity for organic bases is pH-dependent and increase as the concentration of BH in solution increases. On-line hplc, electrospray ionization (es), and ms have been applied to the separation and identification of brevotoxins associated with red tide algae (81). Brevotoxins, toxic polyethers produced by the marine dinoflagellate *Gymnodinium breve*, are responsible for fish kills, and pose certain health risks to humans. The lc/ms method employs reversed-phase microbore hplc on a C-18 column with a mobile phase consisting of 85:15 methanol–water, a flow rate of 8 μL/min, and a post-column split ratio of 3:1 (uv absorbance detector/ms). A brevotoxin culture sample was found to contain at least six components, including two well-separated peaks corresponding to brevotoxin PbTx- and PbTx-1, as well as unknown compounds, including one with a molecular mass of 899 DA (possibly an isomer of PbTx-9). The brevotoxin molecules exhibited a high tendency to bind to alkali cations in the positive ion es/ms mode. For standard PbTX-9, PbTx-2, and PbTx-1, brevotoxins analyzed on the lc/ms system, the detection limits (employing ms scans of

100 m/z units) were determined to be less than 600 fmol, 1 pmol, and 50 fmol, respectively (S/N = 3).

Using capillary hplc, femtomole amounts of recombinant DNA-derived human growth hormone (rhgh) have been successfully detected from solutions at nanomolar concentrations (82). A sample of rhgh that was recovered from rat serum was analyzed by capillary reversed-phase hplc, using both acidic- and neutral-pH mobile phases, as well as by capillary ion-exchange chromatography. Submicrogram amounts of rhgh were also analyzed by tryptic mapping, using capillary hplc, and the resulting peptides were identified by capillary lc/ms.

Capillary Electrophoresis. Capillary electrophoresis (ce) is an analytical technique that can achieve rapid high resolution separation of water-soluble components present in small sample volumes. The separations are generally based on the principle of electrically driven ions in solution. Selectivity can be varied by the alteration of pH, ionic strength, electrolyte composition, or by incorporation of additives. Typical examples of additives include organic solvents, surfactants (qv), and complexation agents (see CHELATING AGENTS).

The inorganic composition of rat airway surface has been analyzed by ce (83). The pulmonary airways are covered by a layer of airway surface fluid (ASF), which is typically <30 μm in thickness. ASF composition is an important factor in the pathogenesis of several lung diseases, including cystic fibrosis. Because of the very small volume of ASF, it is difficult to determine its composition, particularly for inorganic ions, since sampling by lavage is not suitable. Using nanoliter injected volumes, capillary electrophoresis is ideally suited to ASF analysis. A novel technique has been developed using separate sampling and injection capillaries whereby microliter volumes of ASF (typically 100 nL) can be collected from airways and then analyzed by ce. Cations (Na^+, K^+, Ca_2^+, Mg_2^+), and anions including Cl^- are quantitated (RSD <10%) using indirect uv detection. In healthy rat lungs, ASF was found to be hypotonic, consistent with observations made in human airways.

A newer precolumn reagent for amino acid determination, 2-(9-anthryl)-ethyl chloroformate (AEOC), was introduced to obtain higher sensitivity in two capillary separation techniques, lc and ce (84). Replacement of the chromophore in the (9-fluorenyl)-methyl chloroformate (FMOC) reagent by anthracene resulted in a reagent having very high molar absorptivity (ϵ_{256} = 180,000 L/(mol·cm)) permitting AEOC-tagged species to be detected at nanomolar levels using a uv absorbance detection in standard 50 mm ID fused-silica capillaries. Weaker absorption bands match the uv argon laser lines of 351 and 368 nm, which allows for convenient lif detection. In this mode picomolar limits of detection are obtained.

Two different mixtures of peptides and alkaloids (qv) have been analyzed by ce/uv/ms using sims to determine whether this technique can detect trace impurities in mixtures (85). The first mixture consisted of two bioactive peptide analogues, which included Lys-bradykinin (kallidin) and Met-Lys-bradykinin. The presence of 0.1% Lys-bradykinin was detected by sim ce/ms but not by ce/uv at 0.1% level as it migrated from the capillary column prior to the main component, Met-Lys-bradykinin. The second mixture consisted of two antibacterial alkaloids, berberine and palmitine. The presence of 0.15% palmitine was detected by ce/uv and sim ce/ms at 0.15% level as it migrated from the capillary column, following the main component berberine. This technique can provide a complementary technique for trace components in such sample mixtures.

Determination of catecholamines by capillary zone electrophoresis using laser-induced fluorescence detection was performed on low concentration samples derivatized with naphthalene-2,3-dicarboxaldehyde to give highly fluorescent compounds (86). When the borate concentration in the derivatization medium was decreased from 130 to 13 mM, sensitivity for noradrenaline and dopamine was greatly enhanced, whereas resolution between the two compounds decreased. At 50 mM borate concentration, optimal resolution having high separation efficiency (3.1 million theoretical plates/m for dopamine) was achieved. The injection of 2.4 nL of a noradrenaline and dopamine solution derivatized at 10^{-9} M produced peaks with signal-to-noise ratio of 8:1 and 3:1, respectively, corresponding to 1.8 amol of each catecholamine. This method was used to determine noradrenaline in brain extracellular fluid: a peak corresponding to a basal level of 5×10^{-9} M of endogenous noradrenaline was observed in microdialysates from the medial frontal cortex of the rat, and its nature was confirmed by electrophoretic and pharmacological validations.

Methylmalonic acid (MMA) in serum is an established marker of cobalamine deficiency. MMA and other short-chain dicarboxylic acids react with L-pyrenyldiazomethane to form stable, highly fluorescent L-pyrenylmethyl monoesters (87). These esters have been analyzed in human blood by ce combined with lif detection. To miminize solute adsorption to the capillary wall, they were coated with polyacrylamide, and hydroxypropyl methylcellulose and dimethylformamide were used as buffer additives to achieve reliable separations. Separation was performed in tris-citrate buffer, pH 6.4, under reversed polarity conditions. The assay was linear for serum MMA concentrations in the range of 0.1–200 μmol/L.

A novel interface to connect a ce system with an inductively coupled plasma mass spectrometric (icpms) detector has been developed (88). The interface was built using a direct injection nebulizer (din) system. The ce/din/icpms system was evaluated using samples containing selected alkali, alkaline earths, and heavy-metal ions, as well as selenium (Se(IV) and Se(VI)), and various inorganic and organic arsenic species. The preliminary results show that the system can be used to determine metal species at ppt to ppb level.

Electrochromatography (ec) has been utilized to separate mixtures of 16 different polycyclic aromatic hydrocarbons (PAHs). Fused-silica capillary columns ranging in size from 50 to 150 mm ID were packed (20–40 cm sections) with 3 mm octadecylsilica particles (89). A potential of 15–30 kV was applied across the 30–50 cm total length capillary column to generate electro-osmotic flow that carries the PAHs through the stationary phase. An intracavity-doubled argon ion laser operating at 257 nm was used to detect PAHs by lif. The limits of detection for individual PAHs range between 10^{-17} and 10^{-20} mol.

Supercritical Fluid Chromatography. Supercritical fluid chromatography (sfc) combines the advantages of gc and hplc in that it allows the use of gc-type detectors when supercritical fluids are used instead of the solvents normally used in hplc. Carbon dioxide, n-petane, and ammonia are common supercritical fluids (qv). For example, carbon dioxide (qv) employed at 7.38 MPa (72.9 atm) and 31.3°C has a density of 448 g/mL.

Derivatization of primary and secondary amines using 9-fluorenylmethyl chloroformate to form a nonpolar, uv-absorbing derivative has been reported (90,91). Amphetamine and catecholamine were used as probes to evaluate this procedure. The derivatives were well behaved and allowed separation in a short time.

The analysis of mefloquine in blood, using packed-column sfc, a mobile phase consisting of n-pentane modified with 1% methanol and 0.15% n-butylamine, and electron capture detection has been reported (92). The method compares favorably to a previously published hplc-based procedure having a detection limit of 7.5 ng/mL in 0.1 mL blood sample.

The sfe of chlorpyrifos methyl from wheat followed by on-line lc/gc/ecd has been investigated (93). Extraction profiles were generated to determine the maximum analyte recovery and the minimum extraction time. Using pure CO_2, a 65% recovery of chlorpyrifos methyl spiked onto wheat at 50 ppb was reported. When 2% methanol was added to the CO_2, the recovery from a one gram sample averaged 97.8% ($n = 10$, 4.0% RSD).

Immunoassays. Immunoassays (qv) may be simply defined as analytical techniques that use antibodies or antibody-related reagents for selective determination of sample components (94). These make up some of the most powerful and widespread techniques used in clinical chemistry. The main advantages of immunoassays are high selectivity, low limits of detection, and adaptibility for use in detecting most compounds of clinical interest. Because of their high selectivity, immunoassays can often be used even for complex samples such as urine or blood, with little or no sample preparation.

A two-site immunometric assay of undecapeptide substance P (SP) has been developed. This assay is based on the use of two different antibodies specifically directed against the N- and C-terminal parts of the peptide (95). Affinity-purified polyclonal antibodies raised against the six amino-terminal residues of the molecule were used as capture antibodies. A monoclonal antibody directed against the carboxy terminal part of substance P (SP), covalently coupled to the enzyme acetylcholinesterase, was used as the tracer antibody. The assay is very sensitive, having a detection limit close to 3 pg/mL. The assay is fully specific for SP because cross-reactivity coefficients between 0.01% were observed with other tachykinins, SP derivatives, and SP fragments. The assay can be used to measure the SP content of rat brain extracts.

A number of solid-phase automated immunoassay analyzers have been used for performing immunoassays. Table 5 (96) provides useful information on maximum tests that can be run per hour, as well as the maximum number of analytes per sample. A number of immunoassay methods have been found useful for environmental analysis (see AUTOMATED INSTRUMENTATION).

Applications

Trace or ultratrace and residue analyses are widely used throughout chemical technology. Areas of environmental investigations, explosives, food, pharmaceuticals, and biotechnology rely particularly on these methodologies.

Table 5. Solid-Phase Automated Immunoassay Analyzers[a]

Instrument	Manufacturer	Tests/h	Phase/Sepn	Detection[b]	Analytes/run
ACS-180	Ciba-Corning	90–180	magnetic	chemi	13
Access	Sanofi Diagnostics	100	magnetic	chemi	24
Affinity	Becton-Dickinson	20–30	coated cuvet	color	[c]
AIA-600	Tosoh-Medics	60	magnetic	fluor	[c]
AIA1200DX	Tosoh Medics	120	magnetic	fluor	21
Cobas Core	Roche Diagnostics	100–150	magnetic	color	10
ES300	Boehringer Man.	120	coated tube	color	12
Immulite	Diagnostic Products	120	centrifugation	chemi	12
Immuno 1	Miles/Technicon	120	magnetic	color	16
IMX	Abbot Diagnostics	32–48	glass fiber	fluor	1
Luminomaster	Sankyo Co.	120	coated tube	chemi	20
Opus	PB Diagnostics	50–190	dry reagent	fluor	[c]
Opus Magnum	PB Diagnostics	50–190	dry reagent	fluor	>20
Radius	Bio-Rad	80–125	coated well	color	12
SRI	Serono-Baker	31–60	magnetic	color	[c]
Stratus Intelect	Baxter Diagnostics	45–72	radial	fluor	1
System 7000	Biotrol	100	magnetic	color	20
Vidas	Bio Merieux	36–45	coated tube	fluor	[c]
Vista	Syva	38–76	magnetic	fluor	15

[a]Ref. 96.
[b]Methods: chemi = chemiluminescence; color = colorimetry; and fluor = fluorescence.
[c]Manual loading.

Environment. Detection of environmental degradation products of nerve agents directly from the surface of plant leaves using static secondary ion mass spectrometry (sims) has been demonstrated (97). Pinacolylmethylphosphonic acid (PMPA), isopropylmethylphosphonic acid (IMPA), and ethylmethylphosphonic acid (EMPA) were spiked from aqueous samples onto philodendron leaves prior to analysis by static sims. The minimum detection limits on philodendron leaves were estimated to be between 40 and 0.4 ng/mm^2 for PMPA and IMPA and between 40 and 4 ng/mm^2 for EMPA. Sims analyses of IMPA adsorbed on 10 different crop leaves were also performed in order to investigate general applicability of static sims for detection of alkylmethylphosphoric acids (AMPAs) on a wide variety of leaves. Interference owing to organophosphorous pesticides were investigated as well as the effect of AMPA solution pH. The results suggest that static sims is a promising approach for rapid screening of organic surface contaminants on plant leaves.

The development of new immunosorbents for the selective solid-phase extraction of phenylurea and triazine herbicides from environmental waters has been presented (98). Silica-based sorbents were shown to be more effective than rigid hydrophilic polymers for covalent coupling of antibodies because no nonspecific interactions occurred between analytes and silica matrix. Off-line enrichment using a cartridge shows that immunosorbents can trap several analytes in the same group (9 from a mixture containing 13 phenylurea herbicides; 6 from a mixture containing 9 triazine herbicides). A stepwise elution using increasing amounts of methanol showed that the analytes had different affinities for the immunosorbents. High selectivity was illustrated by analyzing Seine River samples spiked with mixtures of phenylurea or triazine herbicides. Isoproturon was easily detected in surface waters at the range of 0.1 μg/L level with no cleanup.

A new chemical sensor based on surface transverse device has been developed (99) (see SENSORS). It resembles a surface acoustic wave sensor with the addition of a metal grating between the tranducer and a different crystal orientation. This sensor operates at 250 mHz and is ideally suited to measurements of surface-attached mass under fluid immersion. By immobilizing atrazine to the surface of the sensor device, the detection of atrazine in the range of 0.06 ppb to 10 ppm was demonstrated.

Hplc and high flow pneumatically assisted electrospray mass spectrometry using negative ionization was used for determination of several acidic herbicides (qv) (100). To achieve good separation, acidification of hplc eluent and subsequent post-column addition of a neutralization buffer are needed to avoid signal suppression. A combination of an on-line automated liquid extraction step using OSP-2 autosampler containing C-18 cartridges can be used for the trace determination of herbicides in environmental waters. Only 50 mL of water are required and the limit of detection is between 0.01 and 0.03 μg/L.

The monocarboxylic dicarboxylic metabolites of the herbicide Dacthal were quantitatively determined in groundwater samples by concentrating the analytes onto strong anion exchange disks (101) (see GROUNDWATER MONITORING). The metabolites are then eluted and derivatized to their ethyl esters. The concentrations of monocarboxylic metabolites ranged from 0.02 to 1.04 μg/L in a survey of groundwater wells from eastern Oregon, whereas the concentration of the dicarboxylic acid metabolites ranged from 0.60 to 18.70 μg/L.

Dissolved nitrite in natural waters was determined at subnanomolar levels by derivatization with 2,4-dinitro-phenylhydrazine followed by hplc (102). The method has a detection limit of 0.1 nM, with an average precision of 4% RSD at ambient natural water concentrations. Irgasan 300 ((2,4-dichlorophenoxy)-phenol) has been determined in the wastewater of a slaughterhouse, using gc/ecd after derivatization with diazomethane (103). The detection limit was found to be 0.2 ng/L.

Simultaneous quantification of the herbicides atrazine, simazine, terbutylazine, propazine, and prometryne and their principal metabolites has been reported in natural waters at 3–1500 ng/L concentration (104). The compounds were enriched on graphitized carbon black and analyzed with hplc and a diode array uv detector.

A multiresidue analytical method based on solid-phase extraction enrichment combined with ce has been reported to isolate, recover, and quantitate three sulfonylurea herbicides (chlorsulfuron, chlorimuron, and metasulfuron) from soil samples (105). Optimization for ce separation was achieved using an overlapping resolution map scheme. The recovery of each herbicide was >80% and the limit of detection was 10 ppb (see SOIL CHEMISTRY OF PESTICIDES).

High efficiency denuders that concentrate atmospheric SO_2 were coupled to an ion chromatograph to yield detection limits on the order of 0.5 ppt (106). A newer approach has been introduced for the quantitative collection of aerosol particles to the submicrometer size (107). When interfaced to an inexpensive ion chromatograph for downstream analysis, the detection limit of the overall system for particulate sulfate, nitrite, and nitrate are 2.2, 0.6, and 5.1 ng/m^3, respectively, for an 8-min sample. A two-stage membrane sampling system coupled with an ion trap spectrometer has been utilized for the direct analysis of volatile compounds in air, with quantitation limits to low ppt levels (108). Toluene, carbon tetrachloride, tricholoroethane, and benzene were used in these studies. The measurement of nitrogen dioxide at ppb level in a liquid film droplet has been described (109) (see AIR POLLUTION). A number of elements in environmental samples have been determined by thermal ionization ms (Table 6). The detection limit for Pu was as low as 4 fg.

Table 6. Thermal Ionization ms

Sample	Elements	Reference
precipitates	Fe, Mn, Zn, Cu, Ni, Pb	110
aerosol samples	Cd, Cu, Pb, Zn	111
lake sediments	^{240}Pu/^{239}Pu	112
marine sediments, tissues	Cu, Zn, Cd, Pb	113
aerosol particulates	Cr, Fe, Ni, Cu, Zn, Cd, Ti, Pb	114
marine sediments SRM 1941	S	115
reference materials	Tl	116
surface waters	Cu, Zn, Cd, Pb	117

Some of the methods used for determination of organic pollutants in the environment follow (118). The most notable are polyaromatic hydrocarbons (PAHs) and volatile organic compounds (VOCs).

Analytes	Methods
n-alkanes	gc/fid
VOCs	prefractionation and gc
PAHs	sample clean up and gc
bromoform and other bromine compounds	gc for flame-retardant compounds in air
nitrodibenzopyranone	isomeric separations based on gc
PCDDs and PCDFs	gc/ms
PAH	on-line lc/gc for PAH in air
chlorinated PAH	lc/gc/ms
nitrated PAH	hplc with electrochemical detection
azaarenes	column chromatography, cleanup, and tlc
toluene diisocyanate	micro lc
dimethyl sulfoxide or sulfone	atmospheric pressure chemical ionization (APCI) ms
VOCs	ion trap monitoring
volatile chlorine compounds	fiber-optic emission sensor-based on AA of chlorine
formaldehyde	monitoring tape: hydroxylamine sulfate, methyl yellow

methyl nitrite	nitrogen oxide-indicating tubes with ir detection
VOCs	ir-based methods for on-site analysis
isocyanate species	chemiluminescent techniques
benzene	photoionization detector
hydrazine	coulometric methods
nitrobenzene	piezoelectric sensors
fluorocarbon	metal oxide sensors
perchloroethylene	quartz balance and calorimetric transducer

A variety of organic and inorganic analytes have been analyzed in air and in water. The organic compounds in air include the following (119):

Analyte	Detection
nonmethane HC	gc
polar and nonpolar HC	gc/fid
alkadienes, alkenes	gc/pid and fid
benzene, toluene	differential optical absorption spectroscopy
organobromines	gc/ms
halocarbons	gc/ecd
glycol ethers	multidimensional gc/ms
organic acids	gc/fid or -npd
carboxylic acids	ion chromatography
formaldehyde	lc
carbonyl compounds	lc/fluorescence
PAH, nitro-PAH	gc/hrms
polyhalogenated dioxins and furans	gc/hrms
aliphatic amines	gc-*N*-selective; lc/fluorescence
monoethanolamine	gc/tsd
aliphatic polyamines	lc/fluorescence
alkyl nitrates, halocarbons	gc/ecd
nitrofluoranthene	lc/fluorescence
dimethyl sulfate	gc/flame photometric

The inorganic compounds in air include the following (120):

Analyte	Method
mercury	AA; ligand exchange
phosphine, arsine	sensor conductivity, potentiometric
plutonium 239	α-spectra
lead	AA; x-ray fluorescence
indium	epithermal neutron
Zn, Cd, Pb, Cu	anodic stripping voltammetry
trace metals	AA (flame, furnace, and electrothermal)

chlorine, bromine, bromide ion mobility	ion chromatography
sulfur	scanning electron micoscopy
hydrogen peroxide	chemiluminescence

In water the inorganic analytes include the following (121):

Analyte	Method
Al	spectrofluorometric, neutron activation
As	hplc coupled to icp/aes
Be	AAS
Co, Mo, V	icp/aes
Cu, Pb	potentiometric
ClO_2	ion chromatography
Ge	preconcentration
Pb	reaction followed by spectrophotometry
trace metals	preconcentrated as various complexes
U, Th, Po, Ra	α-spectroscopy and liquid scintillation

Organic compounds in water include the following (122):

Sample/technique	Method
aromatic hydrocarbons in presence of OH-containing species	electrochemiluminescent sensing
coated fibers for chlorinated HC	evanescent wave spectroscopy
PAHs	fluorescence
17 pesticides	atmospheric ionization
benzene, chlorobenzene, and dichloroethane	direct insertion membrane
ms/ms for chlorodinitrophenol isomers	electrospray ms
ms/ms used for pesticides and dyes	electrospray ion trap ms
amines	electrospray ion mobility
surfactants, atrazine	fast-atom bombardment
phenylurea pesticides, acidic, basic and neutral pesticides; C-18 disks used for isolation	particle beam ms

Explosives. Explosives can be detected using either radiation- or vapor-based detection. The aim of both methods is to respond specifically to the properties of the energetic material that distinguish it from harmless material of similar composition. A summary of techniques used is given in Table 7. These techniques are useful for detecting organic as well as inorganic explosives (see EXPLOSIVES AND PROPELLANTS).

Procedures for trapping accelerant vapors in the headspace of a closed container on charcoal that is either encased in a porous pouch or impregnated into

Table 7. Detection Techniques for the Explosives[a]

Technique	Type of explosive	Basis of detection
	Radiation measurements	
nmr, esr	organic	structural information
x-ray absorption	inorganic, organic	heavy metals, density
x-ray emission	inorganic	heavy metals
x-ray diffraction	solid, organic/inorganic	crystallinity
γ-ray absorption	organic nitro, inorganic	nitrogen density
thermal neutron activation	organic nitro, inorganic	nitrogen density
fast neutron activation	organic nitro, inorganic	C, O, N densities
	Vapor measurements	
ms	organic	molecular structure
gc/ec	organic	chromatographic properties
gc/chemluminescence	organic, nitro	chromatographic properties

[a]Ref. 123.

a flexible membrane have been described (124). Trace amounts of explosive compounds can be trapped from hplc effluents onto a porous polymer microcolumn for confirmatory gc examination (125).

An enzyme-linked immunosorbent assay (elisa) has been developed for the detection of residues on hands. As little as 50 pg of TNT can be detected (126). Liquid chromatography/thermospray negative-ion tandem ms has been successfully used to detect picogram levels of explosives in post-blast debris (127).

Food. Laws and regulations controlling contamination of food were once the province of religious organizations. As far back as the Dark Ages, government set standards. Significant changes have occurred only relatively recently, since the time analytical chemists could characterize most of the substances that comprise food and thus more effectively control contaminants in it. The reliability of data regarding medicated feeds (see FEEDS AND FEED ADDITIVES), dairy products (see MILK AND MILK PRODUCTS), seafood, meat products (qv), fruits and vegetables, and beverages has been reviewed (2).

The development of analytical strategies for the regulatory control of drug residues in food-producing animals has also been reviewed (128). Because of the complexity of biological matrices such as eggs (qv), milk, meat, and drug feeds, well-designed off-line or on-line sample treatment procedures are essential.

Methylmercury in fish was extracted and cleaned using column chromatography and then determined using flameless atomic absorption (129). The method gave a reproducibility of 18.2% (RSD) at 0.1 ppm level of Hg. Many trace elements, particularly the heavy metals, can be determined in food. For example, electrothermal atomic absorption has been applied for the determination of Cd, Co, and Pb in foods (130). Electrothermal AA has been compared to flame AA spectrophotometry for determination of cobalt (131). Electrothermal AA was more accurate, however, flame AA was more precise. The detection limits were 1.8 and 2.27 ppb by the electrothermal flame methods, respectively. A rapid and sensitive electrothermal AA spectrophotometric method using a combination of microwave digestion and palladium as a stabilizer was used for detecting Cd and Pb at sub ppm levels in vegetables and protein in foodstuffs (132).

The efficient recovery of volatile nitrosamines from frankfurters, followed by gc with chemiluminescence detection, has been described (133). Recoveries ranged from 84.3 to 104.8% for samples spiked at the 20 ppb level. Methods for herbicide residues and other contaminants that may also relate to food have been discussed. Inorganic elements in food can be determined by atomic absorption (AA) methods. These methods have been extensively reviewed. Table 8 lists methods for the analysis of elements in foods (134).

Pharmaceuticals. Examples of trace and ultratrace analyses of various drugs and pharmaceuticals have been provided throughout. The purity of the active ingredient, its content and availability in dosage form, therapeutic blood levels, delivery to target areas, elimination (urine, feces, and metabolites), and toxicity are always of importance.

Pharmaceutical Purity. A safety profile of a generic drug can differ from that of the brand-name product because different impurities may be present in each of the drugs (154). Impurities can arise out of the manufacturing processes and may be responsible for adverse interactions that can occur. For example, serious adverse reactions (Lyell syndrome) were observed upon the use of isoxicam in 1985. These seemed to have resulted from trace elements of a manufacturing by-product that was within the manufacturing quality control specifications.

Table 8. Analysis of Elements in Foods[a,b]

Element	Food	Reference
Icp atomic emission spectrometry		
As, Pb, Cd	seafood	136
various elements	tea	137, 138
Al	canned cherries	139
Fe, Cu, Zn	tinned mussels	140
Al, Ba, Mg, Mn	tea leaves	141
various elements	honey	142
Co, Cr, Cu, Fe, Pb, Zn	Brazilian beer	143
Ca, P, Mg, Fe, Cu, Zn	maternal milk	144
Icp mass spectrometry		
As, Cd, Pb	seafood	145
various elements	maternal milk	146
various elements	foods	147
Neutron activation analysis		
minor and trace	agricultural products	148
Mo	edible oils, margarine	149
I	human, animal, commercial milk	150
I	dates	151
As, Hg, Se, Zn	seafood	152
H, B, Cl, K, Na, S, Ca, Cd	orange juice	152
various	tea	153

[a]Refs. 134 and 135.
[b]Icp = inductively coupled plasma.

The subject of impurity analysis of pharmaceutical compounds has been insufficiently addressed in the scientific literature (3,4). Many monographs in the *United States Pharmacopeia* have nonspecific assay methods. An attempt has been made to address this problem by focusing on specific methodologies and delineating origination and concentration of impurities found in pharmaceutical compounds (2). A capsule review of methodologies used for the following classes of compounds is available (2,3).

alkaloids (qv)
antineoplastic agents
carboxylic acids
local anesthetics
steroids (qv)
vitamins (qv)
analgesics (see ANALGESICS,
    ANTIPYRETICS, AND ANTIINFLAM-
    MATORY AGENTS)

antibiotics (qv)
amines (qv)
carbonyl compounds
hydroxy compounds
prostaglandins (qv)
sulfonamides (see ANTIBACTERIAL
    AGENTS, SYNTHETIC)
amino acids/peptides
antidepressants, tranquilizers (see
    PSYCHOPHARMACOLOGICAL AGENTS)

Chiral separations have become of significant importance because the optical isomer of an active component can be considered an impurity. Optical isomers can have potentially different therapeutic or toxicological activities. The pharmaceutical literature is trying to address the issues pertaining to these compounds (155). Frequently separations can be accomplished by glc, hplc, or ce. For example, separation of $R(+)$ and $S(-)$ pindolol was accomplished on a reversed-phase cellulose-based chiral column with fluorescence emission (156). The limits of detection were 1.2 ng/mL of $R(+)$ and 4.3 ng/mL of $S(-)$ pindolol in serum, and 21 and 76 ng/mL in urine, respectively.

Biotechnology.   Particular attention must be paid to the detection of DNA in all finished biotechnology products because of the possibility that such DNA could be incorporated into the human genome and thus become a potential oncogene. The absence of DNA at the picogram-per-dose level should be demonstrated in order to assure the safety of biotechnology products (157).

The isolation and purification of DNA and ribonucleic acid (RNA) restriction fragments are of great importance in the area of molecular biology (158). These fragments are the product of site-specific digestion of large pieces of DNA and RNA with enzymes called restriction endonucleases. The fragments may range in size from a few base pairs to tens of thousands of base pairs. An ion-exchange column can provide DNA and RNA separations within one hour, giving resolution equivalent to that obtained with gel electrophoresis. Nucleic acid fragments are then visualized, using on-line uv detection and sample loading from 500 ng to 50 mg.

Molecular biologists are utilizing hplc for characterization and purification of proteins (qv), peptides, and antibodies (159). In fact, rplc is rapidly becoming the method of choice for resolving complex peptide mixtures from protein cleavage reactions (for example, peptide mapping of CNBr and tryptic digests), discrimination of homologous proteins from different species, and separation of synthetic diastereioisomeric peptides.

An electrospray ionization source interfaced to an ion trap storage/reflectron time-of-flight mass spectrometer was used as a sensitive detector for microbore hplc (160). Using the total ion-storage capabilities of the trap over a broad mass range, total ion chromatograms of tryptic digest bovine cytochrome c and bovine B-casein are obtained, following hplc separations with samples in the low picomole range.

Amino acids (qv) and peptides from both standard solutions and biological samples derivatized with 3-(4-carboxybenzoyl)-2-quinolinecarboxaldehyde at a low sample concentration to form highly fluorescent isoindoles gave minimum detectabilities using ce and laser-induced fluorescence detection in the low attomole range (161). Indirect fluorescence detection was used to monitor subfemtomolar components in tryptic digest by ce (162). Dideoxycytidine chain-terminated fragments were separated in gel-filled capillaries (163) and a post-column laser-induced fluorescence detector provided a mass detection limit below attomole range for fluorescamine-labeled DNA fragments.

Miscellaneous. Trace analyses have been performed for a variety of other materials. Table 9 lists some uses of electrothermal atomic absorption spectrometry (etaas) for determination of trace amounts of elements in a variety of matrices. The applications of icp/ms to geological and biological materials include the following (165):

Sample	Analyte
muscovites, granites	rare-earth elements
rocks, minerals	rare-earth elements
geological reference materials	Os, Ir, Pt, Au
coals	PGE
U.S. Geological Survey reference standard	28 elements
calcites	La, Ce, Eu, other REEs
high purity quartz sand	14 elements
electronic-grade quartz	U, Th, others
automotive catalysts	Pd, Rh, Ce, Ni, Fe, Ba
hplc components	Fe, Cr, Ni, Cd, Mn, Mo, Zn, Pb
blood serum	various elements
blood plasma	B
urine	Sb, Hg, Cd
rat hair	^{63}Cu, ^{65}Cu
animal tissue	Pt
brain tissue	39 elements
liver tissue	Th
dietary intake/uptake	Zn ratios/Sr ratios
milk powder	I
soil ingestion studies	Al, Ba, Mn, Si, Ti, V, Y, Zn
bone ash	U, Th

Table 9. Etaas Analysis of Solids and Slurries[a]

Matrix	Element	Comment
airborne aerosols	Pb	Zeeman BC
aluminum alloys	Ga	graphite powder
biological materials	Pd, Pb, Zn, Mn, Cu	D2 BC, autoprobe
bird feathers	Cd, Pb	Zeeman BC
nuclear waste	Mo, Ru, Rh, Pd	graphite modifier
urinary calculi	Cd, PB, Cr, Ni, Hg	Zeeman BC
rocks, minerals	alkali metals	FAES
soils	Pb	fast temperature program

[a]Ref. 164.

Thermal neutron activation analysis has been used for archeological samples, such as amber, coins, ceramics, and glass; biological samples; and forensic samples (see FORENSIC CHEMISTRY); as well as human tissues, including bile, blood, bone, teeth, and urine; laboratory animals; geological samples, such as meteorites and ores; and a variety of industrial products (166).

A discussion of methods and applications for trace analysis of cosmetics is available (167). Analyses of elements from Al to Zn by a variety of methods has also been described recently (168). Detection techniques for some of the elements of interest follow:

Analyte	Technique
Al	solid-phase spectrofluorometry
As	fluorescence
B	azomethine method
Be	photothermal spectrometry, fluorescence
Cd	fluorescence, neutron activation
Cl	on-wafer fabricated sensor
Cn	AAS
Cs	flame emission
Cu	energy-dispersive x-ray fluorescence; ion selective electrode
Fe	hplc
Hg	icpms, AAS, fluorescence, photoacoustic spectrometry
In	neutron activation
Li	neutron activation
Mg	amperometry
Mn	fluorescence, galvanostatic stripping
Os	kinetic spectrophotometry
Pb	anodic stripping, AA
Po	α-spectrometry
^{226}Ra	Cerenkov counting
Sb	neutron activation, AA
Se	gc/ec, spectrophotometry
Tb	fluorescence
^{230}Th	α-spectrometry
Zn	fluorescence

BIBLIOGRAPHY

1. S. Ahuja, *CHEMTECH*, **11**, 702 (1980).
2. S. Ahuja, *Trace and Ultratrace Analysis by HPLC*, John Wiley & Sons, Inc., New York, 1992.
3. S. Ahuja, *Ultratrace Analysis of Pharmaceuticals and Other Compounds of Interest*, John Wiley & Sons, Inc., New York, 1986.
4. S. Ahuja, *Chromatography of Pharmaceuticals*, ACS Symposium Series #512, American Chemical Society, Washington, D.C., 1992.
5. *Chem. Eng. News* (Aug. 7, 1978).
6. *Chem. Eng. News* (Sept. 25, 1978).
7. D. Noble, *Anal. Chem.* **67**, 375 R (1995).
8. E. Johnson, A. Abu-Shumay, and S. R. Abbot, *J. Chromatogr.* **134**, 107 (1977).
9. D. Chen and N. J. Dovichi, *J. Chromatogr.* **657**, 265 (1994).
10. S. Nie, D. T. Chiu, and R. N. Zaire, *Anal. Chem.* **67**, 2849 (1995).
11. S. D. Gilman and A. G. Ewing, *Anal. Chem.* **67**, 58 (1995).
12. D. Blok, R. Feltsma, M. N. Wasser, W. Nieuwenhuizen, and E. K. Pauwels, *Int. J. Rad. Appl. Instrum.* (*B*) **16**, 11 (1989).
13. K. C. Rosenspire, M. S. Haka, M. E. van Dort, D. M. Jewett, D. L. Gildersleeve, M. Scwaiger, and D. M. Wetland, *J. Nucl. Med.* **31**, 1328 (1990).
14. H. B. Patel, F. Hossain, R. P. Spenser, H. D. Scharf, M. S. Skulski, and A. Jansujwicz, *Nucl. Med. Biol.* **18**, 445 (1991).
15. B. Kratochvil and J. K. Taylor, *Anal. Chem.* **53**, 924A (1981).
16. W. J. Youden, *J. Assoc. Off. Anal. Chem.* **50**, 1007 (1967).
17. R. Yang, G. Campbell, H. Cheng, G. K. Tsuboyama, and K. L. Davis, *J. Liq. Chromatogr.* **11**, 3223 (1988).
18. A. Rogstad, V. Hormazabal, and M. Yndestad, *J. Liq. Chromatogr.* **11**, 2337 (1988).
19. G. Carlucci, *J. Liq. Chromatogr.* **11**, 1559 (1988).
20. E. P. Diamandis and M. D'Costa, *J. Liq. Chromatogr.* **426**, 25 (1988).
21. G. Musch and D. L. Massart, *J. Liq. Chromatogr.* **432**, 209 (1988).
22. P. G. M. Zweipfenning, J. A. Lisman, A. Y. N. Van Haren, G. R. Dijkston, and J. J. M. Holthius, *J. Liq. Chromatogr.* **456**, 83 (1988).
23. H. T. Karnes, K. Rajasekharaih, R. E. Small, and D. Farthing, *J. Liq. Chromatogr.* **11**, 489 (1988).
24. H. T. Karnes, K. Opong-Mensah, D. Farthing, and L. A. Beightol, *J. Liq. Chromatogr.* **422**, 165 (1987).
25. C. S. Smith, S. L. Morgan, S. V. Greene, and R. K. Abramson, *J. Liq. Chromatogr.* **423**, 207 (1987).
26. C. P. Turley and M. A. Brewster, *J. Liq. Chromatogr.* **421**, 430 (1987).
27. M. J. Bartek, M. Vekshteyn, M. P. Boarmand, and D. G. Gallo, *J. Liq. Chromatogr.* **421**, 309 (1987).
28. B. Brossat and co-workers, *J. Liq. Chromatogr.* **413**, 141 (1987).
29. P. M. Kabra and J. H. Wall, *J. Liq. Chromatogr.* **10**, 477 (1987).
30. W. R. Hudgins and K. Stromberg, *J. Liq. Chromatogr.* **10**, 3329 (1987).
31. R. Hartley, M. Lucock, W. I. Forsythe, and R. W. Smithells, *J. Liq. Chromatogr.* **10**, 2393 (1987).
32. F. W. Karasek, R. E. Clement, and J. A. Sweetman, *Anal. Chem.* **53**, 1050A (1981).
33. T. Gorecky and J. Pawliszyn, *Anal. Chem.* **67**, 3265 (1995).
34. I. J. Barnbas, J. R. Dean, I. A. Fowlis, and S. P. Owen, *J. Chromatogr.* **705**, 305 (1995).
35. I. J. Barnbas, J. R. Dean, I. A. Fowlis, W. R. Tomlinson, and S. P. Owen, *Anal. Chem.* **67**, 2064 (1995).

36. M. E. P. McNally, *Anal. Chem.* **67**, 308A (1995).
37. H. Engelhardt, J. Zapp, and P. Kolla, *Chromatographia*, **32**, 527 (1991).
38. V. Lopez-Avila, R. Young, J. Benedicto, P. Ho, R. Kim, and W. F. Beckert, *Anal. Chem.* **67**, 2096 (1995).
39. K. W. Jackson and H. Qiao, *Anal. Chem.* **64**, 50R (1992).
40. B. A. Pedersen and G. M. Higgins, *LC-GC*, **6**, 1016 (1988).
41. A. M. Dietrich, D. S. Millington, and Y. Seo, *J. Chromatogr.* **436**, 229 (1988).
42. V. P. Shah and co-workers, *J. Pharm. Sci.* **81**, 309 (1992).
43. K. W. Jackson and G. Chen, *Anal. Chem.* **68**, 231R (1996).
44. D. Wildhagen and V. Krivan, *Anal. Chem.* **67**, 2842 (1995).
45. K. C. Jindal, R. S. Chaudhary, S. S. Gangwal, A. K. Singlaand, S. Khanna, *J. Chromatogr.* **685**, 195 (1994).
46. S. Nilsson, C. Lager, T. Laurell, and S. Birnbaum, *Anal. Chem.* **67**, 3051 (1995).
47. J. Sherma, *Anal. Chem.* **64**, 134R (1992).
48. M. Du, C. Su, G. Han, S. Dai, and L. Lin, *J. Planar Chromatogr.-Mod. Tlc*, **3**, 407 (1990).
49. M. P. K. Dell, S. J. Haswell, O. G. Roch, R. D. Coker, V. F. Maedlock, and K. Tomlins, *Analyst*, **115**, 1435 (1990).
50. J. J. Pestka, *J. Immunol. Methods*, **136**, 177 (1991).
51. D. N. Tripathi, L. R. Chuhan, and A. Bhattacharya, *Anal. Sci.* **7**, 423 (1991).
52. W. Funk and P. Durr, *J. Planar Chromatogr.-Mod. Tlc*, **3**, 149 (1990).
53. J. M. Moore, D. A. Cooper, T. C. Kram, and R. F. X. Klein, *J. Chromatogr.* **645**, 273 (1993).
54. R. M. Black and R. W. Read, *J. Chromatogr.* **665**, 97 (1995).
55. H. R. Buser and M. D. Muller, *Anal. Chem.* **67**, 2791 (1995).
56. S. Ahuja, *Selectivity and Detectability Optimization in hplc*, John Wiley & Sons, Inc., New York, 1989.
57. L. Dou, J. N. Zeng, D. D. Gerochi, M. P. Duda, and H. H. Stating, *J. Chromatogr.* **679**, 367 (1994).
58. M. Qiau, P. J. O'Dwyer, and J. M. Gallo, *J. Chromatogr.* **666**, 307 (1995).
59. R. A. Carr, G. Colle, A. H. Nagoe, and R. T. Foster, *J. Chromatogr.* **668**, 175 (1995).
60. H. Joshua, *J. Chromatogr.* **654**, 247 (1993).
61. A. Coquet, J. L. Veuthy, and W. Herdi, *J. Chromatogr.* **553**, 255 (1991).
62. R. Zhang, Y. Cao, and M. W. Hearn, *Anal. Biochem.* **195**, 160 (1991).
63. V. Coquart and M. C. Henion, *J. Chromatogr.* **553**, 329 (1991).
64. W. H. H. Farrington, J. Tarbin, J. Bygrave, and G. Shearer, *Food Addit. Contam.* **8**, 55 (1991).
65. A. J. G. Mank, E. J. Molenaar, H. Lingeman, C. Gooijer, U. A. T. Brinkmann, and N. H. Velthorst, *Anal. Chem.* **65**, 2197 (1993).
66. J. M. Bostick, J. W. Strojek, T. Metcalf, and T. Kwana, *Appl. Spectrosc.* **46**, 1532 (1992).
67. T. Yoshida, Y. Moriyama, and H. Higuchi, *Anal. Sci.* **8**, 355 (1992).
68. K. Iwaki, N. Okumura, and M. Yamazaki, *J. Chromatogr.* **623**, 153 (1992).
69. H. Irth, A. J. Oosterkamp, W. van der Welle, U. R. Tjaden, and J. van der Greef, *J. Chromatogr.* **633**, 65 (1993).
70. M. B. Smalley and L. B. McGown, *Anal. Chem.* **67**, 1371 (1995).
71. N. G. Hentz and L. G. Bachas, *Anal. Chem.* **67**, 1014 (1995).
72. M. A. Chilli, L. Romele, W. Martinotti, and G. Sommariva, *J. Chromatogr.* **706**, 241 (1995).
73. A. Doscher, M. Schwikowski, and H. W. Gaggler, *J. Chromatogr.* **706**, 249 (1995).
74. S. Rakotomanga, A. Ballet, F. Pellerin, and D. Bayloc-Ferrier, *Chromatographia*, **32**, 125 (1991).

75. J. Paulson and C. Lindberg, *J. Chromatogr.* **554**, 149 (1991).
76. C. M. John and B. W. Gibson, *Anal. Biochem.* **187**, 281 (1990).
77. R. B. van Breemen, *Anal. Chem.* **67**, 2004 (1995).
78. S. Schutz, H. E. Hummel, A. Duhr, and H. Wolnik, *J. Chromatogr.* **683**, 141 (1994).
79. D. W. Kuehl, J. Serrano, and S. Naumann, *J. Chromatogr.* **683**, 113 (1994).
80. M. G. Ikonomou, A. T. Blades, and P. Kebarle, *Anal. Chem.* **62**, 957 (1990).
81. Y. Hua, W. Lu, M. S. Henry, R. H. Pierce, and R. B. Cole, *Anal. Chem.* **67**, 1815 (1995).
82. J. E. Battersby, A. W. Guzetta, and W. S. Hancock, *J. Chromatogr.* **662**, 335 (1995).
83. J. C. Transfiguracion, C. Dolman, D. H. Eideelman, and D. K. Lloyd, *Anal. Chem.* **67**, 2937 (1995).
84. A. Engstrom, P. E. Anderson, B. Josefesson, and W. D. Pfeffer, *Anal. Chem.* **67**, 3018 (1995).
85. F. Y. L. Hsieh, J. Cai, and J. Henion, *J. Chromatogr.* **679**, 206 (1994).
86. F. Robert, L. Bert, L. Denoroy, and B. Renaud, *Anal. Chem.* **67**, 1838 (1995).
87. J. Schneede and P. M. Ueland, *Anal. Chem.* **67**, 812 (1995).
88. Y. Liu, V. Lopez-Avila, J. J. Zhu, D. R. Wiedrin, and W. F. Beckert, *Anal. Chem.* **67**, 2020 (1995).
89. C. Yau, R. Dadoo, H. Zhao, R. Zare, and D. J. Rakestraw, *Anal. Chem.* **67**, 2026 (1995).
90. J. L. Veuthey and W. J. Haerdi, *J. Chromatogr.* **515**, 385 (1990).
91. A. A. Descombe, J. L. Veuthy, and W. Haerdi, *J. Anal. Chem.* **339**, 480 (1991).
92. D. L. Mount, L. C. Patchen, and F. C. Churchill, *J. Chromatogr.* **527**, 51 (1990).
93. R. M. Campbell, D. M. Meunier, and H. J. Cortes, *Microcolumn*, 302 (Sept. 1, 1989).
94. D. S. Hage, *Anal. Chem.* **65**, 420R (1993).
95. C. Creminon, O. Dery, Y. Frobert, J-Y. Couraud, P. Pradelles, and J. Grassi, *Anal. Chem.* **67**, 1617 (1995).
96. R. Haas, *Anal. Chem.* **65**, 444R (1993).
97. J. C. Ingram, G. S. Groenewald, A. D. Applehans, J. E. Delmore, and D. A. Dahl, *Anal. Chem.* **67**, 187 (1995).
98. V. Pichon, L. Chen, M. C. Henion, R. Daniel, A. Martel, F. le Goffic, J. Abian, and D. Barcelo, *Anal. Chem.* **67**, 2451 (1995).
99. M. Tom-Moy, R. L. Baer, D. Spira-Solomon, and T. P. Doherty, *Anal. Chem.* **67**, 1510 (1995).
100. S. Chiron, S. Pappiloud, W. Haerdi, and D. Barcelo, *Anal. Chem.* **67**, 1637 (1995).
101. J. A. Field and K. Monohan, *Anal. Chem.* **67**, 3357 (1995).
102. R. J. Kieber and P. J. Seaton, *Anal. Chem.* **67**, 3261 (1995).
103. M. Graovac, M. Todorovic, M. I. Tratanj, M. J. Kopecini, and J. J. Comor, *J. Chromatogr.* **705**, 313 (1995).
104. M. Berg, S. R. Muller, and R. P. Schwarzenbach, *Anal. Chem.* **67**, 1860 (1995).
105. G. Dinelli, A. Vicari, and V. Branolini, *J. Chromatogr.* **700**, 201 (1995).
106. P. K. Simon and P. K. Das Gupta, *Anal. Chem.* **65**, 1134 (1993).
107. *Ibid.*, **67**, 71 (1995).
108. M. E. Cisper, C. G. Gill, L. E. Townsend, and P. H. Hemberger, *Anal. Chem.* **67**, 1413 (1995).
109. A. A. Cardoso and P. K. Das Gupta, *Anal. Chem.* **67**, 71 (1995).
110. T. M. Church, A. Veron, C. C. Patterson, D. Settle, Y. Erel, H. R. Maring, and H. R. Flegal, *Global Biogeochem. Cycles*, **4**, 431 (1990).
111. A. L. Dick, *Geochim. Cosmochim. Acta*, **55**, 1827 (1991).
112. L. W. Green, F. C. Miller, J. A. Sparling, and S. R. Joshi, *J. Am. Soc. Mass Spectrom.* **2**, 240 (1991).

113. K. J. R. Rosman and N. K. Kempt, *Geostand. Newslett.* **15**, 117 (1991).

114. J. Voelkening and K. G. Herman, *J. Geophys. Res.* **95**, 20623 (1990).

115. M. M. Schantz and co-workers, *Fresenius J. Anal. Chem.* **338**, 501 (1990).

116. E. Waldman, K. Hilpert, and M. Stoeppler, *Fresenius J. Anal. Chem.* **338**, 572 (1990).

117. K. H. Coale and A. R. Flegal, *Sci. Total Environ.* **87–88**, 297–304 (1989).

118. R. E. Clement, G. A. Eiceman, and C. J. Koester, *Anal. Chem.* **67**, 227R (1995).

119. *Ibid.*, **65**, 90R (1993).

120. *Ibid.*, p. 92R.

121. *Ibid.*, **67**, 232R (1995).

122. *Ibid.*, p. 237R.

123. P. Kolla, *Anal. Chem.* **67**, 184A (1995).

124. W. R. Dietz, *J. Forensic Sci.* **36**, 111 (1991).

125. J. B. F. Lloyd, *J. Energ. Mater.* **9**, 1 (1991).

126. D. D. Fetteroff, J. L. Mudd, and K. Teten, *J. Forensic Sci.* **36**, 343 (1991).

127. A. M. A. Verweij, P. C. A. M. deBreyn, C. Choufoer, P. J. L. Lipman, *Forensic Sci. Int.* **60**, 7 (1993).

128. M. M. L. Aerts, A. C. Hoogenboom, U. A. Th. Brinkman, *J. Chromatogr.* **667**, 1 (1995).

129. W. Holak, *J. Assoc. Off. Anal. Chem.* **72**, 926 (1989).

130. R. Berbers, R. Farre, and D. Messado, *Nahrung*, **35**, 683 (1991).

131. R. Berbers, R. Farre, and D. Messado, *At. Spectrosc.* **9**, 6 (1988).

132. D. Littlejohn, J. N. Eglia, R. M. Gosland, U. K. Kunwar, C. Smith, and X. Shan, *Anal. Chim. Acta*, **250**, 71 (1991).

133. R. J. Maxwell, J. W. Pensabene, and W. Fiddler, *J. Chromatogr. Sci.* **31**, 212 (1993).

134. S. K. C. Chang, E. Holm, J. Schwarz, and P. Rayas-Duarte, *Anal. Chem.* **67**, 127R (1995).

135. Y. H. M. A., S. R. Ecknagel and co-workers, *Z. Lebensm. Unters. Forsch.* **197**, 444 (1993).

136. B. S. Sheppard, D. T. Heitkemper, and C. M. Gaston, *Analyst*, **119**, 1683 (1994).

137. A. M. Syed, M. Qadiruddin, and M. Pak, *J. Sci. Ind. Res.* **36**, 325 (1993).

138. C. F. Wang, C. H. Ke, and J. Y. Yang, *J. Radioanal. Nucl. Chem.* **173**, 195 (1993).

139. T. Maitani, D. R. Xing, C. Ikeda, Y. Goda, M. Takeda, and K. Yoshihira, *Shokuhin Eiseigaku Zasshi*, **35**, 201 (1994).

140. F. J. Copa-Rodriguez and M. I. Basadre-Pampin, *Fresenius J. Anal. Chem.* **348**, 390 (1994).

141. C. K. Manickum and A. A. Verbeek, *J. Anal. At. Spectrom.* **9**, 227 (1994).

142. P. Fodor and E. Molnar, *Mikrchim Acta*, **112**, 113 (1993).

143. I. Matsuhige and E. Oliveira, *Food Chem.* **47**, 205 (1993).

144. N. Carrion, A. Itriago, M. Murrilo, E. Eljuri, and A. Fernandez, *J. Anal. At. Spectrom.* **9**, 205 (1994).

145. B. S. Shepard, D. T. Heitkemper, and C. M. Gaston, *Analyst*, **119**, 1683 (1994).

146. T. Alkanani, J. K. Friel, S. E. Jackson, and H. P. Longerich, *J. Agric. Food Chem.* **42**, 1965 (1994).

147. G. Steffes, J. Luck, and W. Bloedron, *Alimenta*, **32**, 8 (1993).

148. A. Moauro, L. Triolo, P. Avino, and L. Ferrandi, *Dev. Plant Soil Sci.* **53**, 13 (1993).

149. S. Bajo, L. Tobler, and A. Wyttenbach, *Fresenius J. Anal. Chem.* **347**, 344 (1993).

150. R. R. Rao and A. Chatt, *Analyst*, **118**, 1247 (1993).

151. M. Dermelj, V. Sitbij, J. Stekar, and P. Stegnar, *Trace Elem. Man. Anim.* **7**, 22/1–22/2 (1990).

152. D. L. Anderson, W. C. Cunningham, and G. H. Alvarez, *J. Radioanal. Nucl. Chem.* **167**, 139 (1992).

153. C. F. Wang, C. H. Ke, and J. Y. Yang, *Radioanal. Nucl. Chem.* **173**, 195 (1993).

154. *FDC Report*, 8 (Oct. 23, 1968).
155. S. Ahuja, *Chiral Separations by hplc*, ACS Symposium Series, #471, American Chemical Society, Washington, D.C., 1991.
156. H. Zhang, J. T. Stewart, and M. J. Ujhelyi, *J. Chromatogr.* **668**, 309 (1995).
157. F. Bogdansky, *Pharm. Technol.*, 72 (Sept. 1987).
158. M. Merrim, W. Warren, C. Stacy, and M. E. Dwyer, *Waters Bull.*
159. R. H. Guenther, J. Coccuzza, H. D. Gopal, and P. F. Agris, *Am. Biotechnol.*, 22 (Sept.–Oct. 1987).
160. M. G. Qian and D. M. Lubman, *Anal. Chem.* **67**, 2870 (1995).
161. J. Liu, Y. Z. Hsieh, D. Weisler, and M. Novotny, *Anal. Chem.* **63**, 408 (1991).
162. B. L. Hogan and E. S. Yeung, *J. Chromatogr. Sci.* **28**, 15 (1990).
163. H. Swerdlow, S. Wu, H. Harke, and N. J. Dovichi, *J. Chromatogr.* **516**, 61 (1990).
164. K. W. Jackson and H. Qiao, *Anal. Chem.* **64**, 58R (1992).
165. D. W. Koppenaal, *Anal. Chem.* **64**, 326R (1992).
166. W. D. Ehmann, J. D. Robertson, and S. W. Yates, *Anal. Chem.* **64**, 4R (1992).
167. Ref. 2, p. 398.
168. R. E. Clement, G. A. Eiceman, and C. J. Koester, *Anal. Chem.* **67**, 236R (1995).

Satinder Ahuja
Ahuja Consulting

TRACERS.   See Imaging technology; Radioactive tracers.

TRAGACANTH.   See Gums.

TRANQUILIZERS.   See Psychopharmacological agents.

TRANSISTORS.   See Semiconductors.

TRANSPORTATION

The transportation of chemicals and related products is unusual in that substantial quantities are moved in packages as well as in bulk. Other materials, such as coal (qv), grain, and ore, are transported in bulk but seldom in packaged form. Moreover, most other bulk commodities, including petroleum and its products, are limited in the diversity of their chemical and physical characteristics and, therefore, do not require as wide a variety of packaging and bulk conveyances as is necessary for the movement of chemicals. Virtually all railroad tank cars are supplied by chemical producers rather than railroad companies, which furnish at least a portion of most other types of equipment used in rail transportation. The multiplicity of chemical and physical characteristics, as well as resulting variations in product value, density, volume of movement, and other factors, including the type and supply of packaging and conveyances, tend to complicate transport pricing and relations between chemicals shippers and the many transportation carriers they employ.

Since the late nineteenth century, the U.S. federal government and almost all states have regulated both the supply and pricing of transportation service, and such regulation has had a profound effect on the chemical industry. Although economic regulation was substantially relaxed by legislation enacted in the 1970s, 1980s, and early 1990s, such relaxation was accompanied by more intense regulation of hazardous materials transportation. Because most of the volume that the chemicals industry produces annually is classified as hazardous, transportation safety has become increasingly important to shippers and carriers of chemicals seeking to comply with a growing body of federal, state, and even local regulations in an effort to avoid civil and criminal penalties. In an era of increasing litigation and sustained public interest in environmental safety, an additional incentive to such shippers and carriers is the avoidance of civil liability and more burdensome regulation.

Table 1. Domestic Intercity t-km Carried by Mode, $t \times 10^6$ [a]

| | Rail | | | Truck | | | | | | Oil pipeline | | |
| | | | | ICC truck | | | Non-ICC truck | | | | | |
Year	Amt	%	Index	Amt	%	Index	Amt	%	Index	Amt	%	Index
1980	2,318	28.7	100	1,242	15.4	100	1,687	20.9	100	1,402	17.3	60
1983	2,009	26.7	87	1,150	15.3	93	1,646	21.9	98	1,434	19.0	62
1984	2,221	27.1	96	1,265	15.5	102	1,835	22.4	109	1,469	18.0	63
1985	2,116	26.2	91	1,240	15.3	100	1,869	23.1	111	1,487	18.4	64
1986	2,117	25.5	91	1,288	15.5	104	1,974	23.8	117	1,510	18.2	65
1987	2,250	25.9	97	1,360	15.7	110	2,089	24.1	124	1,525	17.6	66
1988	2,369	26.4	102	1,423	15.9	115	2,111	23.6	125	1,557	17.4	67
1989	2,352	25.9	101	1,466	16.1	118	2,244	24.7	133	1,536	16.9	66
1990	2,418	26.1	104	1,501	16.2	121	2,276	24.6	135	1,542	16.7	67
1991	2,374	25.6	102	1,547	16.7	125	2,369	25.5	140	1,529	16.5	66
1992	2,402	25.0	104	1,643	17.1	132	2,498	26.0	148	1,560	16.3	67
1993	2,403	24.3	104	1,819	18.4	147	2,647	26.7	157	1,557	15.7	67

Some of the more significant operational aspects of chemicals transportation in the context of the changing climate of both economic and safety regulation are discussed herein. The technical nature of safety regulations, especially in connection with hazardous materials and wastes, has necessitated frequent consultation between industrial distribution and technical personnel in the chemical industry, one of the largest users of commercial transportation services provided by rail, motor, water, air, and pipeline carriers, including combinations of such carriers as well as proprietary transportation. Shipments of chemicals are made in a wide variety of containers, such as tank cars and tank trucks, barges, self-propelled vessels, drums, barrels, cylinders, bags, and even small bottles for samples and laboratory specimens.

The cost of transportation has an important effect on the marketability of chemicals. For that reason, transportation, along with numerous other factors, is often a significant consideration in determining the location of chemical production facilities. In addition, convenient and economical access to water and rail transportation and the interstate highway system, as well as proximity to raw materials and markets, may influence the choice of warehouse and terminal sites for storage and redistribution of chemical products (see PLANT LOCATION).

Since the 1970s, the concept of transportation management in the chemicals industry has been broadened to include such functions as packaging (qv), order processing, sales service, warehousing, and scheduling of inbound raw materials. The expanded concept, commonly referred to as logistics or distribution (or physical distribution, as distinguished from market distribution), is in part a consequence of information technology, which has made it possible to relate total distribution costs to individual products, customers, and movements. Transportation, however, continues to be a central concern of most distribution managers in the chemical industry. Table 1 indicates the relative shares of U.S. intercity tonnage, including chemicals, carried by various modes of transportation.

Table 1. (Continued)

| | Water | | | | | | | | Air | | | Total | |
| | Great Lakes | | | River/canals | | | Coastwise | | | | | | | |
Year	Amt	%	Index	Amt	%	Index	Amt	%	Index	Amt	%	Index	Amt	Index
1980	168	2.1	100	781	9.7	100	481	6.0	100	6.7	0.1	100	8,086	100
1983	121	1.6	72	711	9.4	91	452	6.0	94	6.9	0.1	102	7,530	93
1984	143	1.7	85	792	9.7	101	449	5.5	93	7.9	0.1	117	8,182	101
1985	134	1.7	80	781	9.7	100	452	5.6	94	8.5	0.1	126	8,087	100
1986	127	1.5	76	817	9.9	105	449	5.4	93	9.2	0.1	137	8,291	103
1987	140	1.6	83	832	9.6	107	473	5.4	98	9.9	0.1	148	8,678	107
1988	160	1.8	96	858	9.6	110	474	5.3	98	10.9	0.1	163	8,964	111
1989	159	1.7	95	884	9.7	113	441	4.8	92	10.7	0.1	159	9,092	112
1990	160	1.7	96	909	9.8	116	436	4.7	91	10.9	0.1	163	9,254	114
1991	150	1.6	90	875	9.4	112	430	4.6	89	10.8	0.1	161	9,285	115
1992	156	1.6	93	906	9.4	116	416	4.3	86	11.5	0.1	172	9,591	119
1993	153	1.5	91	902	9.1	116	403	4.1	84	12.1	0.1	180	9,895	122

[a]Ref. 1.

Transportation Modes

Railroads. Until the 1980s, railroads were almost exclusively common carriers that offered their services to the public as transporters of virtually all commodities between all points on their lines, which are privately owned, maintained, and operated. In the 1990s, railroads provide a significant volume of transportation services under privately negotiated contracts with individual shippers, although such contracts were illegal prior to enactment of the Staggers Rail Act of 1980 (2).

Rail service may be single-line or joint-line. The former refers to movements that originate and terminate on a single railroad, without intermediate transportation by another rail carrier. Joint-line service occurs when more than one railroad participates in transportation from origin to destination, generally under agreements among the railroads involved for interchange at specified locations. Even where a single railroad serves both the origin and destination cities, joint-line service may be necessary or desirable in some situations, as, for example, where one carrier serves the shipper and another the consignee. In most cases where joint-line service is offered and appropriate routing restrictions are observed by the shipper, joint-line and single-line movements of the same goods between the same points are charged identically, and the carriers who participate in the joint-line service share in divisions of the total revenue. In recent years, the trend toward mergers of large railroads serving broad geographical areas has resulted in fewer joint-line movements.

Traditionally, railroads have furnished boxcars of varying sizes and capacities for general-purpose rail movement of packaged freight, including chemicals and related products, in drums, barrels, bags, and other containers. More recently, the introduction of energy-absorbing underframes and other technical innovations has enhanced the ability of boxcars to carry greatly increased loads without excessive damage to the lading.

Railroads generally do not supply tank cars and other special-purpose rail cars such as covered hopper cars for the movement of bulk plastic materials. Rather, shippers or receivers must furnish such equipment, usually through a purchase or lease arrangement with car manufacturers or lessors. Car manufacturers and intermediaries offer various forms of rail-car leases, ranging from short-term, full-maintenance rentals to long-term leases requiring outside financing (3). Many chemical shippers have substantial investments or lease commitments in tank cars and similar rail equipment, including cars constructed of or lined with special materials for particular products. Other cars may be thermally insulated to prevent excessive heat buildup in transit or for protection against fire.

At many chemical plants, as well as other manufacturing or receiving facilities dependent on rail transportation, railroad tracks are constructed within the plant to permit the shipment or receipt of rail cars. Such tracks, usually called industry or private tracks or sidetracks, connect directly with the tracks of the railroad(s) serving the plant. Because the sidetrack must be compatible with the railroad track to permit railroad switch engines and crews to enter the plant, the industry and railroad enter into a written sidetrack agreement (4), which defines their respective rights and obligations with regard to track construction, maintenance, and operation. Included in such agreements are provisions pertaining to required lateral and overhead track clearances and maintenance of hoppers, pits, or other loading or unloading devices.

When private tracks have insufficient capacity for the number of freight cars required to be stored, shippers or receivers of freight may lease additional trackage from a railroad in the vicinity of the plant (5). Because leased tracks are considered private during the term of the lease, demurrage is not payable on private cars held on such tracks, although a reasonable rental for the track lease must be paid. Frequently, tracks located at strategic places remote from a plant facility may be leased for storage of loaded cars in order to have them available for prompt delivery to customers or distributors in the vicinity of the track.

When a railroad provides transportation services under a continuing contract, many of these matters are addressed in the contract. Demurrage charges, in particular, are inapplicable, except to the extent the railroad and shipper agree on the terms of such charges, and the parties may negotiate all other terms and conditions in the same manner as any other contract would be negotiated. Transportation service provided pursuant to a contract is not subject to further regulation under the ICC Termination Act of 1995 (6).

Motor Carriage. Since the 1930s, motor carriage has been an essential part of the U.S. transportation system. Initially confined to movements of small shipments or over short distances, motor carriers took advantage of improved public highways, including the interstate system, to develop a network of transportation competitive with railroads in both rates and service. Less capital-intensive and, therefore, more numerous than railroads, and unconfined by the rigidity of tracks, motor carriers demonstrated a flexibility that broke historic patterns of industrial concentration in transportation centers, thereby contributing to the dispersion of manufacturing and other commercial enterprises to suburban and rural areas.

For both economic and legal reasons, individual motor carriers traditionally have specialized in the type of services offered, either in terms of commodities carried, areas or locations served, or type of equipment provided. Some truckers restrict services to particular categories of materials, such as "Chemicals, in bulk," "Acids, in packages," or single commodities, such as "Acetylene, in cylinders," as well as to specified cities, towns, counties, or states. As federal regulations were relaxed in the 1980s and early 1990s, most legal obstacles disappeared. In the 1990s, if a motor carrier restricts its service to particular commodities or geographic areas, it is usually an economic choice rather than a limitation imposed under a governmental franchise.

Motor carriage may fall within one of three different categories. Until the 1980s, most motor carriers offered services to the public as common carriers. All rates, terms, and conditions of common carriers were required to be published in tariffs filed with the Interstate Commerce Commission (ICC), and carriers could not deviate from these provisions. In 1994, U.S. Congress eliminated most tariff filing requirements for common carriers, allowing shippers and common carriers to negotiate individually determined rates, terms, and conditions. Tariffs are required for chemicals traffic only if the shipment moves in noncontiguous domestic trade, ie, transportation from or to Alaska, Hawaii, or a territory or possession of the United States (7). Some limited volume of traffic also may be subject to certain rates, terms, and conditions collectively established by a group of carriers pursuant to an agreement between such carriers, if approved and exempted from the antitrust laws by the Surface Transportation Board (STB) (8). Whenever any rate arranged between a shipper and carrier incorporates provisions of such an agreement, a shipper should request confirmation from the carrier that the carrier is a party to the agreement.

In addition to common carriers, motor contract carriers have been widely used in the chemical industry, in part because relaxation of federal regulations governing contract carriage resulted in a proliferation of such service in recent years. In practice, the distinction between common and contract motor carriers has been largely obscured. This has been particularly true since 1994 when tariff filing requirements were virtually eliminated for common carriers, leaving little to distinguish an agreement with a common carrier from a contract with a contract carrier. Indeed, the law no longer defines motor common and motor contract carriers separately and all laws previously applicable to common carriage are applicable to contract carriage (9). However, a carrier and shipper may waive application of any provision of the ICC Termination Act of 1995 by contract between them, except for provisions governing carrier registration, insurance, and safety fitness (10).

A third type of motor carriage of considerable importance to many industries, including the chemicals industry, is proprietary or private carriage. Such transportation is conducted in furtherance of a primary business other than transportation (11). Thus, manufacturers transporting goods that they have manufactured or processed or that they will use in such manufacturing or processing, or for purposes of bona fide sale or purchase, are engaged in private carriage. Contrary to common misconception, it is usually immaterial whether a private carrier does or does not have legal title to the transported goods. Indeed, when title to goods is acquired solely in an effort to create an appearance of private carriage, the transportation of such goods is considered to be for-hire carriage and, therefore, subject to governmental regulation (12). Furthermore, it is generally not required that a company use only vehicles that it owns rather than leases, or that it directly employ the drivers of such vehicles, provided that such company actually controls the transport operation and bears its characteristic burdens and financial risks (13).

Corporate members of a single group of corporations may lawfully perform such transportation for a parent or subsidiary, or for a sister subsidiary, provided that the one corporation wholly owns the other or that both are wholly owned by a common parent (14). Although compliance with certain minimal STB regulations is required (15), such transportation does not require a franchise and is not subject to other regulatory requirements. As a result, it is possible for many corporations to combine in a single vehicle their freight with that of other members of the same group of corporations, thereby improving equipment and labor utilization in consolidated private trucking operations.

Motor carriers use a wide variety of highway vehicles, including trucks, tractors, trailers, tank vehicles, hopper vehicles, low-boys, vans, and others. Unlike railroads, commercial motor carriers of bulk liquids or solids in tank or hopper trucks usually offer shippers both power equipment (tractor) and freight-carrying trailers, although shippers frequently supply such trailers under special arrangements. Highway tractors used for long, continuous journeys are usually equipped with sleeper-cabs to allow one driver to rest while a second driver operates the tractor-trailer.

The development of the interstate highway system and more permissive federal and state legislation have allowed the use of vehicular equipment of increased length and other dimensions, as well as higher weight-carrying capacity, thereby contributing to more economical motor transportation. Such legislation, however, has in turn given rise to disputes concerning highway tolls, fuel taxes, registration fees, and similar assessments, against both for-hire and proprietary truck operators, to permit adequate maintenance of highways. Because both passenger and freight-carrying vehicles use the highways, such disputes are not readily resolved.

Waterborne Transport.   Despite natural limitations, the transportation of chemicals by water has enjoyed substantial growth, especially since the end of World War II. Assisted by governmental development of the inland waterways system, including locks and other navigational aids, water carriers transport large quantities of bulk chemicals in barges between inland ports or between such ports and coastal ports. In addition, bulk chemicals are transported by self-propelled tank vessels between U.S. coastal points, and between U.S. ports and overseas destinations. In 1989, 56.1 million metric tons (61.9 million short tons) of chemicals were transported in the U.S. domestic waterborne commerce (16).

Although water carriers are sometimes classified as common or contract carriers, such distinctions are frequently insignificant, because water carriage of bulk chemicals in the United States is essentially unregulated. In conformity with long-standing practice in the maritime field, such transportation is often provided under various forms of agreement, such as bareboat charters, time charters, or voyage charters. In a bareboat charter, the owner of a vessel charters (leases) the vessel without crew; in a time or voyage charter, the vessel is leased with crew for a specified period or for a particular voyage. On U.S. inland waterways, chemical shippers sometimes engage towboat operators to tow barges that such shippers either own or charter (lease) from others. In the United States, little remains of a once flourishing liner trade in the transportation of packaged freight, although such liners are still engaged in such transportation to and from foreign ports.

Barges, like other transportation vehicles, are available in a variety of types, sizes, and capacities. On the inland waterways, barges are usually crewless and without power independent of towboats, which push several barges in a group. In deepwater, or ocean, transportation, barges sometimes carry a crew and are capable of self-propulsion. Deepwater barges, whether self-propelled or pulled by a hawser (cable) between the barge and towboat, are generally larger than river barges. Deepwater tows rarely consist of more than one or two barges.

As in the case of highways, considerable contention results from public maintenance of the inland waterways for recreation, flood control, and other purposes, as well as for the transportation of barges and other freight-carrying vessels. Because barge transportation of chemicals is considered essential to economical distribution, governmental tolls assessed for such maintenance are of critical interest to the chemicals industry.

Most oceangoing vessels, particularly those used between North America and other continents, are self-propelled. For the movement of packaged freight in international commerce, ocean transportation in recent years has been dominated by container ships designed to load and carry large, trailer-sized containers. Because such ships can be loaded and unloaded more quickly than traditional freight-carrying vessels, the amount of time these ships are docked at port has been greatly reduced, thereby increasing the number of voyages possible in a given period and reducing operating costs. Other types of ocean vessels include tankers and dry-bulk ships for the transportation of a wide variety of liquid hydrocarbons, chemicals, and materials such as coal, coke, and ores in large quantities. Chemical tankers tend to be smaller in size than petroleum tankers and usually have several compartments, each designed to carry one or more products.

Pipelines.   The feasibility of pipeline transportation depends on the availability of very large quantities of compatible materials between locations with sufficient storage facilities. Thus, pipeline transportation is predominantly, but not exclusively, limited to the movement of hydrocarbons, many of which are raw materials in the production of petrochemicals. Although proprietary pipelines (qv), generally of short distances, are not unusual, commercial petroleum pipelines are considered to be common carriers available to serve all customers who can tender sufficient quantities of acceptable liquids for transportation between terminals.

Air Transport.   Relatively small quantities of chemicals are transported by air, although availability of such service for the movement of samples, emergency shipments, and radioactive chemicals with a short half-life is important. Both economic and safety considerations impede the development of air carriage as a significant means of transporting a substantial volume of chemicals.

Other Services.   Domestic freight forwarders, although sometimes treated as common carriers, do not provide any physical transportation service. Instead, they arrange transportation services for their customers, usually the underlying shipper of goods, and perform related functions such as the booking of space with a carrier and preparing necessary documentation. One important function commonly performed by freight forwarders is the consolidation of multiple small shipments into carload or truckload lots, which are forwarded to a central location for subsequent distribution to individual destinations. In the export and import trade, where transportation is provided in whole or in part by a water carrier, similar services are provided by commercial operators known as nonvessel operating common carriers (NVOCC). Shipper cooperatives, more commonly known as shippers' associations, also provide consolidation and distribution services for the purpose of passing on the resulting savings in freight charges to their members. More recently, with the proliferation of contract carriage, particularly in the domestic transportation markets, many shippers' associations have undertaken the function of negotiating transportation agreements for the benefit of the members.

The diversity and flexibility of a highly developed transportation structure is demonstrated by intermodal transportation, ie, the combination of two or more transportation modes. Traditional combinations, such as rail and water or truck and water, are essentially end-to-end arrangements. However, since the late 1980s there has been substantial growth in combinations of the various transportation modes, such as the piggyback transportation of trucks or trailers on railroad flat cars, and similar loadings of trucks or containers on ships or barges. These methods of transportation are largely deregulated and have led to substantial economies for both carriers and shippers (17).

Warehouses and Terminals. Warehousing constitutes an integral part of the distribution system of the United States. Although employed primarily to store inventory, warehouses are also used to assure timely deliveries to customers remote from a production facility. Additionally, warehousing may facilitate the aggregation of large shipments, thus reducing transportation costs. Warehouses may be owned and operated by individual companies for their own purposes or they may be available to the public for storage of goods. The chemicals industry makes extensive use of bulk terminals for storage of liquid- and dry-bulk materials in a wide variety of sizes and types of tanks, silos, bins, and other facilities.

Warehouse and terminal operators, who offer their facilities and services for compensation, are liable for goods in their custody if they are negligent. Many operators limit the amount of their liability by provisions in the warehouse receipt, the customary document issued as evidence of goods held in storage. Warehouse charges are generally determined by the amount of space occupied by the stored goods and the period of storage, as well as the ease of handling, hazardous characteristics, and similar considerations.

Shipping

Shipping Terms. Although frequently referred to as shipping terms, fob, fas, and cif are actually terms of sale because they pertain to the relationship between vendor and vendee, rather than between shipper and carrier. The term fob, for example, means free on board and usually indicates that delivery of the goods to the vendee will occur when the goods, packaged in accordance with the terms of the sales agreement, are delivered aboard a vehicle of the type agreed upon at the fob point named. The risk of loss in transit is usually transferred at the point of delivery. Fob origin means that the vendee or consignee assumes such risk, whereas fob destination means that the vendor or consignor assumes it. In the absence of a contrary agreement between vendor and vendee, freight charges are payable by and are for the account of the party who bears the risk of transit loss.

The selection of shipping terms has a material effect on the sales contract. The party with the risk of loss must decide whether or not to insure against such risk and must prepare and file a claim against the transportation carrier when goods are lost or damaged in transit. Unless otherwise agreed, that party must also pay transportation charges and file any claims for freight overcharges. In export or import transactions, shipping terms such as fas (free alongside ship) or cif (cost, insurance, freight) may also determine the party responsible for preparing required documents, obtaining customs clearances, and similar matters (18).

Shipping Documents. The document most commonly used in both domestic and international transportation is the bill of lading which serves as a receipt for goods delivered to a carrier as well as a contract of carriage. Bills of lading may be negotiable documents and as such constitute evidence of title or the right to possession of the goods described in the document. The face of the bill of lading identifies the shipper, origin, consignee, destination, vehicle or car number, routing, commodity, containers, quantity shipped, and other information required for the carrier to properly transport and invoice the freight. The contract terms, usually on the reverse side of the bill of lading, specify rights and obligations of the shipper, consignee, and carrier, including most importantly, limitations on carrier liability, methods and time limits for submitting damage claims, payment of freight charges, and disposition of the goods in case of nondelivery. The short-form bill of lading does not reproduce all contract terms on the reverse side, but refers to such terms as published elsewhere.

In international trade, the ocean bill of lading serves essentially the same purposes, although it may differ in form and content and is frequently negotiated in such a manner that payment by the foreign consignee is required before delivery of goods by the carrier. Where a shipper's freight occupies the whole or a substantial portion of a particular vessel, the document used may be a voyage charter, which provides for use of the vessel for a single voyage. For shipments of bulk chemicals by tanker, a shipper may use a specified tank on a particular vessel under an arrangement referred to as a parcel charter. Another document commonly used in ocean shipping is a dock receipt, which is evidence that the goods have been delivered to a dock pending arrival of the vessel on which they will be loaded for transportation overseas.

A freight bill is an invoice issued by a carrier requesting payment for transportation services. Generally, the freight bill contains the information shown on the face of the bill of lading, together with the freight rate and charges and the carrier's invoice (pro) number. Carriers usually require submission of the original paid freight bill as part of a shipper's claim for freight loss, damage or delay, or for overcharge. A paid freight bill may also be required to prove that the vendor or vendee has paid freight charges, especially in cases where freight is added to or deducted from the merchandise invoice or is equalized with freight charges from competing shipping points.

A delivery receipt is a document, frequently a copy of the freight bill, which has been signed by the consignee as evidence of delivery of goods by the carrier. Where no exceptions have been noted on the delivery receipt, it constitutes prima facie proof of delivery in full, and in apparent good order and condition.

Interstate and Intrastate Commerce

The applicability of various federal and state transportation laws and regulations depends on whether transportation constitutes interstate or intrastate commerce. The transportation laws and regulations that may apply are both economic (ie, rates and routes) and safety related. Beginning in January 1995, however, Congress preempted all state economic regulation of intrastate motor carriage, but did not substitute federal regulation for the preempted state regulations (19). As a result, although limited economic regulation of interstate commerce by motor carrier remains, there is no economic regulation of intrastate commerce by motor carrier.

Except in rare instances, it can be assumed that transportation requiring physical movement across state boundaries is interstate commerce. On the other hand, transportation that takes place wholly within the confines of a single state is not necessarily intrastate commerce, because such transportation may be a portion of a continuous movement in interstate commerce. As a general rule, where there is a "fixed and persisting intent" that transportation be provided from a point in one state to a point in another state or in another country without coming to rest at an intermediate location, all portions of such transportation are considered interstate, even though one or more portions may be performed wholly within a single state (20). It is immaterial that different carriers or even different modes of for-hire transportation may be employed for each portion, or that new bills of lading are issued or separate freight bills rendered.

Thus, for example, where freight is transported by motor carrier from Springfield, Illinois, to a railroad piggyback ramp in Chicago for movement in railroad service to a place in New York, the truck service is in interstate commerce although neither the vehicle nor its driver physically leave the state of Illinois in the course of such transportation. Consequently, the motor carrier would be subject to federal franchise requirements. Similarly, a shipment from Albany, New York, to a New York City pier for export to Europe is considered interstate (or foreign) transportation and, therefore, subject to federal regulation, despite the issuance of a new, export bill of lading at the pier.

However, a shipment transported by motor carrier within New York City to a New York City pier for export, although likewise in interstate commerce, would generally not be subject to such regulation because a provision of the ICC Termination Act of 1995 exempts interstate motor carriage within a single municipality or commercial zone (21). An additional variation of the general rule occurs when freight is transported by a private carrier from one state to a second state, where it is given to a for-hire carrier for final delivery within the second state. In such cases, the ICC and the courts have concluded (22) that the for-hire portion of such movement is not subject to federal regulation even though the freight actually crossed a state line in the course of the through transportation. Neither, however, is the movement subject to state regulation because the movement is considered to be in interstate commerce (23).

Economic Regulation

In the United States, transportation has long been subjected to regulation by both federal and state governments. Generally, such regulation has been directed at operational safety or toward economic concerns such as discrimination in rates and services or excessive competition. In addition, regulatory statutes have provided for control of entry into the transportation business, regulation of freight rates and charges, and various finance, accounting, and insurance requirements, although there are numerous exceptions. Among the exceptions of particular importance to the chemical industry is that afforded to water carriage of liquid and dry-bulk commodities.

At the federal level, the STB, the Federal Maritime Commission (FMC) the Department of Transportation (DOT), and the Federal Energy Regulatory Commission (FERC) are all concerned with economic regulation of various modes of transportation. The STB regulates interstate railroads, motor and water carriers, and pipelines (other than water, gas, and oil). The DOT regulates motor carriers and international airlines, and the FMC regulates water transportation in foreign commerce. The FERC regulates pipeline transportation of oil, natural gas, water, and other energy resources.

There has been a significant trend toward relaxation of many economic regulatory controls since the 1970s which has resulted in substantial change in the transportation industry. Thus, entry into the motor-carrier business has been greatly liberalized resulting in more available carriers and, consequently, increased competition and reduction in freight rates. In addition, motor carriers are no longer required to publish rates and terms of transportation in public tariffs for most transportation. Similarly, railroads are no longer required to file tariffs for most transportation and are permitted to enter into contracts with individual shippers, providing for guaranteed volumes of movement at reduced rates, improved services or car supply, discounts for routing via specified railroads, and other flexible arrangements previously considered unlawful. Antitrust immunity, which most carriers formerly enjoyed in the collective establishment of rates, has been removed to a substantial extent and, in general, more reliance has been placed on market competition to achieve the objectives of economic regulation.

Regulation, however, has not been entirely abandoned and in many respects at least the form of railroad regulation remains essentially intact. Thus, for example, railroads must obtain STB authority to extend their lines (although few railroads have undertaken to do so) or to abandon existing service, although such abandonments are more freely allowed. Similarly, but to a much lesser extent, motor carrier regulation also remains intact. For example, motor carriers must be qualified to provide service by the DOT which issues motor carrier registrations. Registration, however, is based only on safety and financial fitness and is granted more freely than the operating franchises that were previously issued by the ICC. In most cases, both railroad and motor-carrier consolidations or mergers require STB approval.

With respect to freight rates, historic rules requiring that rates be reasonable and prohibiting discrimination or preference as between particular shippers or geographic areas have been phased out to varying degrees among different modes of transportation. To the extent these historic rules remain at all, their impact has been largely dissipated by provisions placing increased reliance on competitive forces. In connection with railroad rates, for example, the STB has lost virtually all of its powers to prescribe maximum reasonable rates, except in cases where railroads exercise market dominance with regard to particular movements (24). In most cases, however, the STB has tended to make it extremely difficult to prove the existence of such market dominance, much to the consternation of transportation managers in the chemical industry who contend that a substantial volume of chemicals railroad traffic is captive to that form of transport. With regard to air freight rates, the DOT has exempted most carriers and forwarders from rate regulations (25). Laws governing the rates of most motor carriers have been repealed in their entirety.

A significant result of regulatory relaxation is an increase in the authority of the STB to grant administrative exemptions from railroad and motor-carrier regulation (26). Such authority has been exercised in a variety of ways, including the virtually complete deregulation of most piggyback transportation (27). Thus, railroads are not treated as regulated carriers when performing piggyback service.

Regulated railroads, when not operating under contracts with shippers, are required to provide a shipper, upon request, its rates and terms for transportation as a common carrier. A rail common carrier cannot increase a rate or change the terms of service for a shipper who has requested this information within the previous 12 months unless the carrier provides 20 days prior notice (28). This is a significant change from previous law which required rail carriers to publish and file with the ICC tariffs or schedules of their rates and charges. Rail carriers were required to strictly adhere to these tariffs regardless of errors, conflicting promises or agreements, contrary intent, or other circumstances.

Shippers must bring a lawsuit to recover any overcharges by a railroad and carriers must bring a lawsuit to recover any undercharges within three years after the claim accrues (29). Transportation performed pursuant to a contract between a shipper and rail carrier is deemed unregulated and the parties must resolve any contractual disputes among themselves or in the courts, without the aid of the STB. With regard to motor carriers, a claim by a shipper against a motor carrier or a claim by a motor carrier against a shipper for recovery of freight charges, must be initiated within 18 months of delivery of the shipments (30).

Among other aspects of STB regulation that have survived reform are requirements pertaining to time periods for the collection of freight charges by common carriers pursuant to the extension of credit (31) and insurance requirements applicable to motor carriers (32). STB credit regulations require railroads to collect all freight charges within five days after issuance of a freight bill, and motor carriers must collect charges within seven days after billing. Motor carriers must also maintain certain minimal liability insurance coverage for the protection of the public, as well as insurance covering carrier liability for loss of or damage to freight. The latter type of insurance is frequently beneficial to shippers who have difficulty in collecting damage claims because a carrier has become bankrupt.

In the past, many state governments regulated economic activity in intrastate transportation in a manner similar to federal regulation. However, beginning in 1995, the federal government preempted all state economic regulation of intrastate motor carriage (33), although the federal government did not supplant state regulation. Instead, the market for intrastate motor carriage was left open to competitive market forces. With respect to railroads, federal legislation has effectively compelled the states to adopt federal regulatory requirements or to abandon railroad regulation entirely (34).

Freight Rates and Allowances. The establishment of freight rates, ie, a transportation price structure, embraces virtually all articles of commerce in a multitude of packages and quantities, via numerous routes and between innumerable locations. A variety of intermediate services are also often included, such as storage or reconsignment in transit and stop-offs to partially load or unload. The classification of freight into various categories or classes is the result of an effort to systemize the various factors considered in fixing a particular rate.

Freight classifications are generally based on freight density, susceptibility to damage or theft, value of the goods, etc.

Freight classification has established a more or less standard nomenclature to identify the numerous products shipped in commerce. This standardization facilitates preparation of shipping documents, determination of freight rates, and free interchange of freight between connecting or competing carriers. Chemical and freight nomenclatures, however, are frequently different. Thus, for example, many chemicals may be grouped under the single freight description "Chemicals, NOIBN," referring to chemicals that are not otherwise indexed (in the freight classification) by name. On the other hand, a particular product such as acetone may be specifically listed and, therefore, would not qualify for inclusion in the NOIBN category. Misdescription of freight resulting in misclassification is a frequent source of freight overcharges and undercharges.

An example of a few listings in the Uniform Freight Classification for railroads is given in Table 2 (35).

In the 1990s, the variety of possible transportation arrangements is virtually without limit, but it may be useful to generally describe the most common types of freight rates and charges. (1) Less-than-carload or less than-truckload rates are applicable to quantities of particular commodities less than a specified volume considered to constitute a carload or truckload quantity of such commodities. In most cases, small shipments are also subject to a minimum charge per shipment. However, almost all railroads have abandoned the transportation of less-than-carload freight. (2) Carload or truckload rates are applicable to quantities of a commodity sufficient to constitute a specified minimum carload or truckload volume. Such rates, of course, are substantially lower than less-than-carload or less-than-truckload rates. Freight rates are usually stated in dollars per 100 pounds, although rates on some materials such as coal or gravel may be stated per short ton or similar unit of weight. (3) Multiple car rates are applicable only when a specified number of carloads is tendered to a railroad for transportation in a single shipment. For commodities such as coal, which move in large volumes, trainload rates may be provided. (4) Annual (or periodic) volume rates are applicable to individual shipments that are part of an aggregate tonnage of a particular commodity or commodities that a shipper has agreed to ship between specified points in a specified period. (5) Accessorial charges

Table 2. Uniform Freight Classification 6000-K

Item	Article	Carload minimum, kg	Carload rating
23260	carbon tetrachloride		
	in carboys	13,608	45
	in containers in barrels or boxes		
	in metal cans completely jacketed, or in bulk		
	in steel barrels, or carload in tank cars	16,330	30
23315	chemicals, phosphoric in boxes: apply only on		
	separately packaged components of 1 liquid		
	and 1 solid chemical, nonexplosive prior to		
	mixing, which when in separate shipping		
	boxes are not subject to DOT regulations	18,144	35

are for services that are ancillary to line-haul transportation, such as switching, demurrage, storage or stopping in transit, reconsignment, and similar services.

For those regulated carriers required to publish tariffs, ie, motor carriage of household goods and domestic offshore water carriers, any payment by the carrier to the shipper could be construed as a reduction of tariff charges and, therefore, an illegal rebate. To avoid such rebates, carriers may publish allowances in their tariffs which they are willing to pay to the shipper, or deduct from the freight bill, for services performed by the shipper in lieu of the carrier. For example, a carrier who includes loading as part of transportation service may publish an allowance to shippers that perform such loading. Similarly, motor carriers may publish allowances for shippers who deliver freight to the carrier's terminal rather than requesting pick-up by the carrier.

The mileage (distance) allowances established by railroads for use of tank cars, hopper cars, and other railroad equipment furnished by shippers for transportation of products are of great importance to the chemical industry. Empty return movements are usually made without charge, provided that aggregate loaded and empty distances on each railroad are maintained in equilibrium. Such allowances, paid for loaded miles of car movement, represent large revenues for the industry and are an important consideration in calculating the actual (net) cost of transporting a given shipment. The amount of the per-mile allowance varies with the fair market value and age of a car, and may be eliminated altogether when freight rates are contractually established "net" of such allowances.

Freight Loss and Damage. Under the common law of the United States, common carriers by land were liable for loss of or damage to goods in their custody, except loss or damage resulting from an act of God, the act of a public enemy (revolution or hostility between governments), an act of governmental authority (eg, quarantine), inherent vice or defect of the goods, or the fault or negligence of the shipper. In the early 1900s, such liability was codified in the Interstate Commerce Act for railroads, motor carriers, and freight forwarders and provided that such carriers were liable for the full, actual loss of or damage to the goods. This codification is commonly referred to as the Carmack Amendment.

In general, common carriers could not limit their liability for loss or damage except for consideration in the form of a reduced freight rate if the shipper retains the right to select either full or limited liability. Historically, liability limitations generally required the approval of the ICC, which was sparingly given. However, because of the enactment of regulatory reform legislation in the early 1980s, approval is no longer required and shippers and carriers may agree on liability limitations. Freight rates applicable to shipments subject to limited carrier liability are known as released rates.

Recent amendments to the Carmack Amendment, however, arguably allow rail and motor carriers to unilaterally limit their liability by the establishment of released value rates, without the shipper's knowledge (36). Thus, under the current law, shippers are well advised to request from the rail or motor carrier a copy of the applicable rates, terms, and conditions, including liability limitations, that apply to the transportation service to be provided. The Act also provides that claims for loss or damage of goods transported via joint through routes may be filed against either the originating or destination carrier or against an intermediate carrier on whose line(s) the goods are known to have been damaged (36). Additionally, the Act provides a minimum time limit for filing a claim (nine months) and commencing suit (two years) (37). Despite the nine-month minimum for filing claims, shippers are well advised, in the case of damage that is not apparent at the time of delivery, to promptly notify the delivering carrier and afford an opportunity for inspection of the damaged goods.

Carriers generally are liable only for full actual loss. Thus, for example, a vendor who has prepaid freight charges that are not separately invoiced to the vendee may not recover both the invoice value of the goods and the freight paid, because such recovery would put the vendor in a more advantageous position than if the goods had been delivered undamaged and the vendee had paid the invoice as rendered.

When goods consigned to a shipper's warehouse or terminal are damaged, disputes frequently arise as to their value. Usually, the carrier contends that shippers should not earn profit on sales not made, and the shipper contends that it should not be required to produce goods merely to recover its costs. Such disputes are sometimes resolved by payment of the sales price less costs not incurred, such as the cost of delivery from the warehouse to the consignee.

Contract carriers generally are not held to the same standard of liability as common carriers because they are considered ordinary for-hire bailees and, therefore, are liable only for their failure to exercise a reasonable degree of care for goods in their custody or possession, although such liability may be varied by the contract. However, motor carriers providing service under contract are held to the same liability standard applicable to common carriers unless the statutory provisions imposing the standard are waived in the contract.

The liability of water carriers is established under principles of traditional admiralty law, which generally reflect the fundamental concept of liability for negligence, modified to accommodate risks peculiar to the long and dangerous voyages in ancient times. Thus, for example, ship owners may limit their liability to the value of the vessel and cargo after an accident, and cargo jettisoned at sea to save the venture may be compensated by general average charges against owners of the remaining cargo. Many other variations of water-carrier liability for cargo damage can be found upon examination of the multiplicity of charters, bills of lading, tariffs, and contracts employed in connection with such transportation.

The liability of common carriers by land for loss of or damage to freight is sometimes referred to as that of an insurer. This characterization is technically incorrect, however, because carriers are not liable for the fault or negligence of the shipper as, for example, in using faulty or defective packaging or in improper loading. Nevertheless, most transit loss or damage is recoverable from the carrier and, as a result, many shippers find it unnecessary to insure freight transported by land carriers, unless carrier liability has been limited in accordance with the legal principles discussed above. On the other hand, it is common practice for shippers or receivers to insure cargo transported by water, unless the carrier has contractually agreed to purchase such insurance. In some cases, as in transportation by air or in highway carriage of household goods, shippers may be afforded an opportunity to purchase insurance directly from the carrier.

Safety Regulation

Before the creation of the U.S. DOT in 1967, the now defunct ICC was authorized to prescribe rules and regulations for rail, truck, and pipeline safety. The Federal Aviation Administration (FAA) was responsible for air safety, and the U.S. Coast Guard for safety on the inland and coastal waterways. Upon establishment of DOT in 1967, the FAA and Coast Guard were transferred to the DOT, which assumed the safety functions the ICC formerly administered.

In general, DOT safety regulations fall into two categories. The first pertains to qualifications and hours of service of carrier employees and the safety of transport operations and equipment. The second, of special concern to the chemical industry, pertains to the transportation of hazardous materials and related commodities.

In connection with motor-carrier safety, DOT has assigned such responsibility to the Federal Highway Administration (FHA), which has prescribed extensive regulations regarding drivers' qualifications, the maintenance of drivers' logs, required vehicle equipment and inspections, and accident records and reports (38). Such regulations are applicable to all interstate carriers by highway, and violations are subject to criminal and civil penalties. The National Highway Traffic Safety Administration (NHTSA), a part of DOT, issues motor vehicle safety standards including, among others, standards for tires, brakes and brake fluids, bumper protection, and passenger restraint systems.

DOT's Federal Railroad Administration prescribes similar regulations for railroads, including requirements pertaining to the quality of tracks, train speeds, and freight-car construction. The FAA and Coast Guard continue to regulate safety by air and water, respectively, including aircraft and ship construction, maintenance, and operation. A separate, independent federal agency, the National Transportation Safety Board, is responsible for investigating serious accidents by all modes of transportation, including pipeline and passenger transportation. This agency makes recommendations to U.S. Congress and the regulatory agencies with respect to safe transportation practices and regulatory requirements.

Since the end of World War II there has been a substantial increase in the volume of hazardous materials transported domestically and internationally by all modes of carriage (estimated at 3.628 billion metric tons annually in nearly one-half million daily shipments). According to the Association of American Railroads, about 1.1 million carloads of hazardous materials were transported by U.S. railroads in 1988 (39). Prompted in part by such tragic occurrences as those at Bhopal and Chernobyl, increased concern with environmental protection and public safety has generated widespread interest in the safe transportation of explosives, toxic and radioactive materials, and other products with dangerous potential. Responsive to such concerns, the Chemical Manufacturers Association and other industrial organizations adopted a broad series of initiatives collectively known as Responsible Care, one part of which is the Distribution Code of Management Practices.

Federal legislation pertaining to the movement of hazardous materials, however, precedes such concerns. As early as 1866, Congress restricted the transportation of such products as nitroglycerin and blasting oil. In 1908 and 1909 the

ICC was authorized to issue regulations governing railroad transportation of explosives, and in 1921 such authority was extended to the regulation of flammable liquids and solids, oxidizing materials, corrosive liquids, compressed gases, and poisonous substances (40). Subsequent amendments to the 1921 legislation further extended ICC jurisdiction to transportation by contract and private carriers by motor vehicle and certain liquid pipelines, and added etiologic agents and radioactive substances to the categories of hazardous materials subject to such jurisdiction (see also EXPLOSIVES AND PROPELLANTS).

Following the initial transfer of safety functions from the ICC, FAA, and Coast Guard to DOT, it became evident that more comprehensive legislation and better coordinated regulation of all modes was required. Accordingly, in 1974, Congress enacted the Hazardous Materials Transportation Act (HMTA), and in 1990 amended HMTA with the Hazardous Materials Transportation Uniform Safety Act (HMTUSA) (41). This statute consolidated the authority of DOT with respect to safety regulation of the various modes, extended its jurisdiction to include manufacturers of containers used for transportation of hazardous materials, greatly increased penalties for violation and provided other enforcement mechanisms, and authorized the regulation of any substance or material that could create an unreasonable risk to health, safety, or property. Under HMTA, as amended by HMTUSA, authority has been assigned to DOT's Research and Special Programs Administration (RSPA), except for bulk water movements, which remain subject to the authority of the Coast Guard. In 1992 an RSPA official asserted that the Administration exercised jurisdiction over 200,000 carriers, shippers, and manufacturers of hazardous materials (42).

Generally, the purposes of hazardous materials regulation are to assure adequate containment of such goods in transit and to inform carrier personnel, cargo handlers, police, fire, other emergency personnel, and bystanders of immediate possible hazards in the event of containment failure. Thus, DOT hazardous materials regulations, published at 49 CFR Parts 171 to 180, prescribe in considerable detail specifications pertaining to the design of shipping containers, tank cars, tank trucks, and intermediate bulk containers. Also significant is Part 130, pertaining to the prevention of oil spills. To assure the proper transmission of hazard information, the regulations also require that specified information warning of potential dangers be shown on various prescribed labels, placards, and shipping documents. The diamond-shaped "red label" signifying the presence of a flammable liquid, for example, is almost universally recognized as indicating the presence of a possible fire hazard (Fig. 1) (43).

Many private industrial or professional organizations, eg, ASME, the Compressed Gas Association, the CMA, and the Bureau of Explosives, publish standards for containers, materials of construction, and tests. Such standards are frequently incorporated by reference in the DOT regulations. In 1990 and 1991, however, the DOT issued comprehensive regulations (44) adopting international performance-oriented packaging standards for nonbulk packaging (drums, boxes, etc). Contrary to prior packaging requirements, which included the prescription of package designs in considerable detail, performance standards rely primarily on package tests intended to simulate the transportation environment, such as drop, stacking, vibration, or pressure testing to demonstrate package integrity. Simultaneously, the DOT adopted the international system of identifying

Fig. 1. Label signifying the presence of a flammable liquid. The numeral in the lower corner represents the primary hazard class of the material in the labeled package.

hazard classes by class or division numbers. Flammable liquids, for example, are referred to as Class 3 materials, whereas poisonous gases are assigned to Division 2.3. Radioactive materials are identified as Class 7 materials, and infectious substances (etiologic agents) are in Division 6.2.

Another innovation derived from international regulatory systems was the DOT assignment of Packing Group (PG) designations to many individual materials, with PG I indicating the highest degree of danger and PG III, the lowest. Thus, for example, a poisonous (toxic) material in Division 6.1, PG I could be lethal when inhaled, and a Division 6.1, PG III material would require only a "Keep away from food" warning while in transportation. Owing to the extensive nature of revised regulations, the DOT allowed such regulations to be implemented in stages over a five-year period, with the last significant phase concluded on October 1, 1996 (45).

Although the hazardous materials regulations are frequently characterized as highly complex, the application is greatly facilitated through the use of the Hazardous Materials Table at 49 CFR §172.101, which alphabetically lists hundreds of hazardous material descriptions by proper shipping names as prescribed by DOT. Some materials are grouped into not otherwise specified (NOS) categories, which incorporate a variety of products. In addition, the Table contains the hazard class and a standard four-digit identification number for each listed material, the type or types of label prescribed for application to packages, the Packing Group number (if any), the required packaging and exceptions thereto, and certain restrictions applicable to shipments via cargo or passenger aircraft, railcar, and water vessels. International shipments of hazardous materials by water and air are governed by the regulations of the International Maritime Organization (46) and the International Civil Aviation Organization (47), respectively.

The proper selection of chemical shipping descriptions, and the determination of the hazard class, require chemical expertise and familiarity with DOT definitions of such classes, which are provided in Part 173 of the hazardous materials regulations. In some cases such definitions are sufficiently precise to permit objective determination of the classification of a material under consideration. Other definitions, however, may lead to disagreements among qualified experts as to the appropriate classification of particular products. Proper identification and classification of hazardous materials are critical prerequisites to compliance with the packaging and communications requirements of DOT regulations. Other materials subject to DOT regulations include hazardous wastes as defined by the EPA; hazardous substances which, if released in certain quantities, must be reported to the U.S. Coast Guard National Response Center; marine pollutants; and materials transported at elevated temperatures.

In addition to the package labels specified in the Hazardous Materials Table and in Subpart E of Part 172 of the regulations, packages and vehicles containing hazardous materials must be marked in accordance with Subpart D of Part 172 of the regulations and vehicles must be placarded in accordance with Subpart F of Part 172. Subpart D requires that certain packages be marked with the proper shipping name prescribed by DOT, the four-digit identification number, the name and address of the consignee or consignor, and package orientation markings in the case of certain liquid materials. Other markings are prescribed for specified materials or vehicles.

Although similar to labels in appearance, placards (Fig. 2) (48), larger and more durable than labels, are affixed to the exterior of rail cars and other

Fig. 2. Placard signifying the presence of a flammable gas. The numeral in the lower corner represents the primary hazard class of the material in the transport vehicle.

transport vehicles carrying hazardous materials. The regulations require that certain bulk packaging (tank cars, tank trucks, portable tanks) display the four-digit identification number, which may, except in certain cases, be substituted for the word or words, eg, Flammable or Poison, on a placard (Fig. 3) (49). In some cases, the identification number is displayed on an orange panel or a plain square-on-point configuration of prescribed specifications. Generally, bulk and nonbulk packagings that contain the residue of a hazardous material are regulated as if full (50).

Similar information is provided in a shipping paper prepared by the shipper, which must also contain a certification that the shipment has been properly classified, described, packaged, marked and labeled in accordance with DOT regulations. Most shippers of hazardous materials have incorporated DOT documentary requirements into their standard bill of lading. In addition to specific detailed information concerning each hazardous material (hazmat) in a shipment, the bill of lading or other shipping paper for each hazmat shipment must include a 24-hour emergency response telephone number and either contain or be accompanied by certain emergency response information (51). For hazardous waste, a hazardous waste manifest must be prepared by the waste generator (shipper) in accordance with EPA regulations (52).

Except for nonbulk packaging subject to performance-oriented packaging standards, DOT packaging requirements are extremely detailed and include precise specifications for shipping containers and materials of construction, tests, test reports, and manufacturer's marks (53). 49 CFR Part 178 also contains specifications for tank trucks and certain other motor vehicles as well as portable tanks and intermediate bulk containers. Tank car specifications are published

Fig. 3. Illustration of an identification number on a placard for acetone. The numeral in the lower corner represents the primary hazard class of the material.

in Part 179, and Part 180 provides for continuing qualification and maintenance of certain bulk packaging. The shipper is responsible under the regulations for selecting a type of container authorized for shipment of the hazardous materials to be transported. Generally, in using individual containers of the type selected, shippers may accept the package manufacturer's certification of compliance or identification of the package specification as evidence that such containers conform to DOT requirements for that specification, except for functions, such as closure, required to be completed by the shipper (54). For tank trucks supplied by a carrier, shippers may rely on the manufacturer's identification plate or certification by the carrier.

Parts 174, 175, 176, and 177 contain loading, handling, and operating requirements for carriage of hazardous materials by rail, air, water, and highway, respectively. Carriers are also subject to the provisions of Part 171 requiring them to report to DOT or other agencies any incidents involving a release or discharge of hazardous materials or wastes during transportation, or resulting in death, injury, or evacuation. Parts 106 and 107 of 49 CFR contain the procedural rules of the RSPA, including provisions pertaining to enforcement, preemption, and regulatory exemptions. Enforcement may result in compliance orders, injunctions, or the assessment of civil or criminal penalties as high as $500,000 and imprisonment for five years for each violation of the hazardous materials regulations (55). Exemptions are issued for a maximum of two years, but are renewable, and authorize specified deviations from prescribed packaging, tests, or other requirements, provided that equivalent safety can be demonstrated (see also PACKAGING, CONTAINERS FOR INDUSTRIAL MATERIALS).

The RPSA adoption of the four-digit United Nations' numbering system for identification of hazardous materials reflects a worldwide effort to improve response to transportation emergencies. Although accidental release of hazardous materials in transit is relatively rare (56), the potential for significant harm is of constant concern to the public and industry and is magnified by the fact that many public emergency response agencies have had little, if any, training or experience in dealing with chemical emergencies. In an effort to provide immediate and reliable information to carriers and public officials at the scene of an emergency, the Chemical Manufacturers' Association established the Chemical Transportation Emergency Center (CHEMTREC) in Washington, D.C. Since its formation in 1971, CHEMTREC has responded to hundreds of thousands of emergency calls, providing information from its files containing data on nearly 1.5 million chemical products. In 1994 alone, CHEMTREC responded to approximately 217,000 total (emergency and nonemergency) calls. Similarly, the Chlorine Institute has organized a mutual aid program, called CHLOREP, which offers assistance at the scene of emergencies involving chlorine. Industrial response teams are usually available for assistance in connection with cleanup of spills which may be hazardous to the public or environment (see ALKALI AND CHLORINE PRODUCTS).

Despite its traditional emphasis on product-based regulation, various DOT rules adopted in the early 1990s were more concerned with regulated persons than with regulated products. Thus, as mandated by statute, DOT issued new rules in 1992 requiring shippers and carriers of certain materials or types of shipment to register with DOT and pay an annual registration fee (57). Similarly,

DOT adopted regulations requiring all hazmat employees to be periodically trained and tested and to maintain certified records of such training (58).

Outlook

Transportation and distribution costs constitute a substantial portion of the total cost of the chemical industry. Most chemical producers, therefore, can be expected to pay continuing attention to the control of such costs and to the maintenance and development of more sophisticated distribution methods. Significant recent changes in the nature and extent of economic regulation in the transportation field promise new challenges to industrial distribution managers, especially in the areas of railroad and motor-carrier transportation. Rail carriers, armed with the freedom to price their services, to abandon unprofitable lines, and to merge with other railroads, have already demonstrated a tendency to increase rates on captive chemicals traffic. At the same time, however, the removal of regulatory restraints on contracts between shippers and railroads has generated a revolution in transport pricing.

In the motor-carrier field, increased competition resulting from the virtual elimination of economic regulatory controls has given motor carriers a degree of efficiency enabling them to challenge proprietary transportation in both cost and service. The availability of energy resources and the adequacy of the highway infrastructure may impose the most substantial constraints on the continued growth of motor transportation.

Without a breakthrough in energy usage, the technology of transportation is not expected to change dramatically in the foreseeable future. Improvement is likely to be concentrated on the transport infrastructure including intelligent transport systems, using electronic technology to increase the efficient use of highways and vehicles and to enhance traffic safety. Transportation safety will continue to be of concern to government at all levels, but such concern may be directed less at new regulations and restraints and more on the application of existing computer and communications technology to more effective emergency response.

BIBLIOGRAPHY

1. *Domestic Intercity Tonnage Carried By Mode, Transportation In America*, 12th ed., 1st Suppl., ENO Transportation Foundation, Nov., 1994, p. 7. ICC and Non-ICC refer to transportation formerly or not regulated by the ICC, respectively.
2. Pub. Law 96-448, 94 Stat. 1895.
3. S. Hoffman, *Model Legal Forms for Shippers*, Transport Law Research, Inc., Mamaroneck, N.Y., 1970, several forms of car leases are shown.
4. Ref. 3, Forms 3–13.
5. Ref. 3, Form 16.
6. 49 U.S.C. §10709(c)(1) (1996).
7. 49 U.S.C. §§13702, 13102(15) (1996).
8. 49 U.S.C. §13703 (1996); in 1996, the Surface Transportation Board (U.S. Dept. of Transportation) replaced the former Interstate Commerce Commission under provisions of *ICC Termination Act of 1995*, P.L. 104-88.

9. 49 U.S.C. §13102(4) (1996) for definition of "contract carriage".

10. 49 U.S.C. §14101(b) (1996).

11. 49 U.S.C. §13505 (1996).

12. *Wilson-Investigation of Operations*, 82 M.C.C. 651, 14 Fed. Carrier Cases, ¶34,886 (1960); *Utley Lumber*, 94 M.C.C. (motor-carrier cases) 458, 16 Fed. Carrier Cases ¶35,723 (1964).

13. *United States v. Drum*, 368 U.S. 370 (U.S. Supreme Ct. 1962).

14. 49 U.S.C. §13505(b) (1996).

15. *Code of Federal Regulations*, Title 49, Part 1167 (1995).

16. *Statistical Abstract of the United States 1994*, 114th ed., U.S. Dept. of Commerce, Corps of Engineers, Waterborne Commerce No. 1065, Washington, D.C., Sept. 14, 1994.

17. *Code of Federal Regulations*, Title 49, §1090.2 (1995).

18. *International Rules for the Interpretation of Trade Terms*, ICC Services SARL, Paris, France, 1990, available in the U.S. from ICC Publishing Corp., Inc., New York, for a comprehensive listing of shipping terms and explanations.

19. 49 U.S.C. §14501(c) (1996).

20. *Policy Statement—Motor Carrier Interstate Transportation: From Out-of-State Through Warehouses to Points in Same State*, Ex Parte No. MC-207, 8 I.C.C. 2d 470 (1992).

21. 49 U.S.C. §13506(b)(1) (1996).

22. *Pennsylvania R.R. v. Ohio P.U.C.*, 298 U.S. 170 (1936); *Motor Trans.*, 94 M.C.C. 541 (1964), affd., 382 U.S. 373 (1966).

23. *United States Dept. of Transportation—Petition for Rulemaking: Single-State Transportation in Interstate or Foreign Commerce*, Ex Parte No. MC-182, served Feb. 12, 1987.

24. 49 U.S.C. §10701(d) (1996).

25. 49 C.F.R. §§291.31, 296.10 (1995).

26. 49 U.S.C. §§10502 and 13541 (1996).

27. *1990 ICC LEXIS 1*, Ex Parte No. 230 (Sub. 7), Interstate Commerce Commission, Jan. 12, 1990.

28. 49 U.S.C. §11101(c) (1996).

29. 49 U.S.C. §11705(a), (b) (1996).

30. 49 U.S.C. §14705(a) and (b) (1996).

31. *Code of Federal Regulations*, Title 49, Part 1320 (1995).

32. *Code of Federal Regulations*, Title 49, Part 1043 (1995).

33. 49 U.S.C. §14501 (1996).

34. 49 U.S.C. §11501 (1996).

35. P. M. Boyle, Tariff Publishing Officer, *Uniform Freight Classification 6000-K*, ICC UFC 6000-K, Chicago, Ill., Mar. 7, 1994.

36. 49 U.S.C. §§11706(a)(1) and 14706(a) (1996).

37. 49 U.S.C. §§11706(e) and 14706(e) (1996).

38. *Code of Federal Regulations*, Title 49, Parts 390–397 (1995).

39. *Regulatory Review and Development Plan and Schedule of Rulemaking Actions*, Materials Transportation Bureau, Washington, D.C., Jan. 1979–Jan. 1980, p. 4; U.S. Congress, Office of Technology Assessment, Transportation of Hazardous Materials, OTA-SET-304, U.S. Government Printing Office, Washington, D.C., July, 1986; *Traffic World*, 35 (Apr. 30, 1990).

40. *Act of March 4, 1921*, c. 172, 41 Stat. 1444, 1445, later codified to 18 U.S.C. §§831-35 (1995).

41. 49 U.S.C. §1801 *et seq.*; recodified to 49 U.S.C. §5101, *et seq.*

42. *Transport Topics*, 14 (Feb. 17, 1992).

43. *Code of Federal Regulations*, Title 49, §172.419 (1995).

44. *Fed. Reg.* **55**, 52402 (Dec. 21, 1990), *Fed. Reg.* **56**, 66124 (Dec. 20, 1991), *inter alia*, DOT Docket No. HM-181, Washington, D.C.

45. *Code of Federal Regulations*, Title 49, §171.14 (1995).

46. *International Maritime Dangerous Goods Code*, 4 Vols., consol. ed., London, 1994; *Code of Federal Regulations*, Title 49, §171.12(b) (1995).

47. *Technical Instructions for the Safe Transport of Dangerous Goods by Air*, 1995–1996 ed., Doc. 9284-AN/905, International Civil Aviation Organization (ICAO), Montreal, Quebec, Canada; *Dangerous Goods Regulations*, 37th ed., IATA, Montreal, Quebec, Canada, effective Jan. 1, 1996; *Code of Federal Regulations*, Title 49, §171.11 (1995).

48. *Code of Federal Regulations*, Title 49, §172.532 (1995).

49. *Code of Federal Regulations*, Title 49, §172.332(d) (1995).

50. *Fed. Reg.* **61**, 28666 (June 5, 1996).

51. *Fed. Reg.* **54**, 27138 (June 27, 1989); rev. *Fed. Reg.* **55**, 870, 20796, 33707 (effective Dec. 31, 1990); codified at 49 U.S.C. §5110 (1996).

52. *Code of Federal Regulations*, Title 40, §262 (1995).

53. *Code of Federal Regulations*, Title 49, §178.37 (1995); see also Subpart B of §173.

54. *Code of Federal Regulations*, Title 49, §173.22(a)(3), (4) (1995).

55. 49 U.S.C. §§5123, 5124 (1996); 18 U.S.C. §3571 (1996).

56. *CMA News*, 5 (Sept. 1, 1982).

57. *Fed. Reg.* **57**, 30620, 33416, 37900; 49 U.S.C. §5108 (1996).

58. *Fed. Reg.* **57**, 20944, 22181; 49 U.S.C. §5107 (1996).

General References

United States Code Annotated, West Publishing Co., St. Paul, Minn. A compilation of U.S. laws of a general and permanent nature consisting of 50 Titles. Although many provisions of various Titles affect transportation, Titles 49 (Transportation) and 46 (Shipping) are of particular interest. Among other important statutes included in Title 49 are the *ICC Termination Act of 1995* (§10101 *et seq.*), the *Department of Transportation Act* (§101 *et seq.*), and the *Hazardous Materials Transportation Act* (§5101 *et seq.*). Title 46 collects various statutes pertaining primarily to water transportation.

Code of Federal Regulations, Office of the Federal Register, National Archives and Records Service, General Services Administration, Washington, D.C. A codification of rules and regulations published in *Fed. Reg.* by departments and agencies of the U.S. federal government. The Code is divided into 50 titles divided into chapters, subchapters, parts, and subparts. Title 49 contains the rules and regulations of the Surface Transportation Board, the Department of Transportation, and other federal agencies concerned with transportation.

Federal Register, Office of the Federal Register, National Archives and Records Service, General Services Administration, Washington, D.C., published daily, Monday through Friday (except official holidays). Provides uniform system for making available to the public U.S. federal regulations, proposed regulations, and other information.

Interstate Commerce Commission Reports, Superintendent of Documents, U.S. Government Printing Office, Washington, D.C., cited (Volume) ICC (Page). A series of decisions of the now defunct Interstate Commerce Commission; Vol. 1 was published 1887.

Motor Carrier Cases, Superintendent of Documents, U.S. Government Printing Office, Washington, D.C., cited (Volume) MCC (Page).

Federal Carriers Reporter, Commerce Clearing House, Inc., Chicago, Ill. Loose-leaf service, 4 Vols. of statutes, regulations, forms, and current court and administrative decisions pertaining to motor and water carriers.

Federal Carrier Cases, Commerce Clearing House, Inc., Chicago, Ill. Continuing series of volumes selectively reporting decisions of ICC and U.S. federal and state courts pertaining to motor carrier, water carrier, and domestic freight forwarder regulation.

Interstate Commerce Acts Annotated, Interstate Commerce Commission, Washington, D.C.; 22 Vols. plus *Advance Bulletins* (last published Aug. 1981). Compilation of *Interstate Commerce Act* and related U.S. federal laws, with digests of decisions of federal courts and ICC relating to each section of such Acts.

Hawkins Index-Digest-Analysis of Decisions Under the Interstate Commerce Act, Hawkins Publishing Co., Washington, D.C. Loose-leaf service, 10 Vols. collecting and digesting decisions under *Interstate Commerce Act* pertaining primarily to railroads.

Hawkins Index-Digest-Analysis of Decisions Under Part II and Part IV of the Interstate Commerce Act, Hawkins Publishing Co., Washington, D.C. Loose-leaf service, 4 Vols. collecting and digesting decisions under the *Interstate Commerce Act* pertaining to motor carriers and freight forwarders.

Hawkins Index-Digest-Analysis of Federal Maritime Commission Reports, Hawkins Publishing Co., Washington, D.C.

J. Guandolo, *Transportation Law*, 3rd ed., Wm. C. Brown Co., Dubuque, Iowa, 1979. Comprehensive review of transportation law and practice, with emphasis on railroads. Regulatory and legislative developments since 1979 suggest caution in volume use.

M. L. Fair and J. Guandolo, *Transportation Regulation*, 7th ed., Wm. C. Brown Co., Dubuque, Iowa, 1972. Comprehensive discussion of transport regulation prior to recent legislative reforms.

G. L. Shinn, *Freight Rate Application*, Simmons-Boardman Publishing Corp., New York, 1948.

J. C. Colquitt, *The Art and Development of Freight Classification*, National Motor Freight Traffic Association, Inc., Washington, D.C., 1956.

J. M. Miller, in R. R. Sigmon, ed., *Law of Freight Loss and Damage Claims*, 3rd ed., Wm. C. Brown Co., Dubuque, Iowa, 1967.

W. J. Augello, *Freight Claims in Plain English*, Shippers National Freight Claim Council, Inc., Huntington, N.Y., 1979.

S. Sorkin, *How to Recover for Loss or Damage to Goods in Transit*, Matthew Bender, New York. Loose-leaf service, 2 Vols.

G. Gilmore and C. L. Black, Jr., *The Law of Admiralty*, 2nd ed., The Foundation Press, Inc., Mineola, N.Y., 1975.

W. Poor, *American Law of Charter Parties @ Ocean Bills of Lading*, 5th ed., Matthew Bender, New York, 1968.

Journal of Law, Logistics and Policy (previously *ICC Practitioners' Journal*), Association for Transportation Law, Logistics and Policy, Washington, D.C., published quarterly.

Traffic World, The Journal of Commerce, Inc., Washington, D.C., published weekly. Widely read news magazine for the transportation industry.

I. L. Sharfman, *The Interstate Commerce Commission*, 5 Vols., The Commenwealth Fund, New York, 1931–1937. Classic and scholarly study, largely of historical interest.

T. G. Bugan, *When Does Title Pass*, Wm. C. Brown Co., Dubuque, Iowa, 1951. Valuable and useful discussion of shipping terms.

R. C. Colton and E. S. Ward, *Practical Handbook of Industrial Traffic Management*, 5th ed., The Traffic Service Corp., Washington, D.C.; C. H. Wager, rev. ed. 1973.

K. R. Feinberg, *Deregulation of the Transportation Industry*, Practicing Law Institute, Washington, D.C., Mar. 30–31, 1981. Course handbook; compilation of papers on transport deregulation and its effects.

1993 Emergency Response Guidebook, U.S. Dept. of Transportation, Washington, D.C. Guidebook for first responders during initial phase of hazardous materials incident (multilingual ed. scheduled 1996 publication).

Courier, periodic newsletter of the Hazardous Materials Advisory Council, Washington, D.C.

The Official Railway Guide, North American Freight Service ed., K-III Directory Corp., New York, published bimonthly. Maps of railroads, railroad officers and executives, lists of railroad stations, interchange points, and other useful railroad information.

United Nations Recommendations on the Transport of Dangerous Goods, 9th rev. ed., 1995. Recommendations of United Nations Committee of Experts on transport of dangerous goods.

United Nations Recommendations on the Transport of Dangerous Goods, Tests and Criteria, 2nd rev. ed., 1995.

Transport Topics, weekly newspaper of the American Trucking Associations, Inc., Washington, D.C.

Federal Register Extract Service, Hazardous Materials Advisory Council, Washington, D.C. Extracts of materials originally published in *Fed. Reg.* pertaining to hazardous materials.

D. P. Locklin, *Economics of Transportation*, 7th ed., Richard D. Irwin, Inc., Homewood, Ill., 1972. Classic textbook on transportation economics, with strong emphasis on regulation and regulatory history.

C. L. Dearing and W. Owen, *National Transportation Policy*, The Brookings Institution, Washington, D.C., 1949. Early postwar study of national transportation policy calling for reform of regulation, which ultimately led to revisions of recent years: the *Motor Carrier Act of 1980*, Public Law 96-296 (94 Stat. 793), and the *Staggers Rail Act of 1980*, Public Law 96-448 (94 Stat. 1895).

Hazardous Materials: A Guide for State and Local Officials, U.S. Dept. of Transportation, Washington, D.C., Feb. 1982. Comprehensive guide to DOT hazardous materials regulations, including discussions of legislative history and regulatory process enforcement.

W. J. Augello and S. Hoffman, *Transportation Contracts In Plain English*, Transportation Claims and Prevention Council, Inc., Huntington, N.Y., 1991. Review of laws and regulations pertaining to transportation contracts, with examination of numerous specific contract provisions.

Hazmat Transport News, biweekly newsletter, Business Publishers, Inc., Silver Spring, Md.

S. Hoffman, W. Wellman, and W. Kahler, *The Complete Shipping Papers Rules*, 2nd ed., Chilton Co., Radnor, Pa., 1995.

S. Hoffman, *Lines on the Law*, compilation of articles originally published in *Hazmat Transport News* (1987–1992), Business Publishers, Inc., Silver Spring, Md., 1992.

C. R. Bigelow, *Hazardous Materials Management in Physical Distribution*, Van Nostrand Reinhold, New York, 1994.

J. E. Tyworth, J. L. Cavinato, and C. John Langley, Jr., *Traffic Management: Planning, Operations, and Control*, Addison-Wesley Publishing Co., Inc., Reading, Mass., 1987.

S. Hoffman, W. Wellman, and W. G. Kahler, *HazMat Shipping Plain English, Road and Rail Edition*, Chilton Co., Radnor, Pa., 1995.

L. N. Moses and D. Lindstrom, eds., *Transportation of Hazardous Materials*, Kluwer Academic Publishers, Boston, Mass., 1993.

STANLEY HOFFMAN
Consultant

JEFFREY O. MORENO
KARYN A. BOOTH
ANTOINE P. COBB
RICHARD D. FORTIN
Donelan, Cleary, Wood & Maser PC

TRANSURANIUM ELEMENTS.   See ACTINIDES AND TRANSACTINIDES.

TRIAZINETRIOL.   See CYANURIC AND ISOCYANURIC ACIDS.

TRICRESYL PHOSPHATE.   See PLASTICIZERS; PHOSPHORUS COMPOUNDS.

TRIETHANOLAMINE.   See ALKANOLAMINES.

TRIETHYLENE GLYCOL DINITRATE.   See EXPLOSIVES AND PROPELLANTS.

TRIMENE BASE.   See RUBBER CHEMICALS.

TRIMETHYLOLETHANE.   See ALCOHOLS, POLYHYDRIC.

TRIMETHYLOLETHANE TRINITRATE.   See EXPLOSIVES AND PROPELLANTS.

TRIMETHYLOLPROPANE.   See ALCOHOLS, POLYHYDRIC.

TRIOXANE.   See FORMALDEHYDE.

TRIPENTAERYTHRITOL.   See ALCOHOLS, POLYHYDRIC.

TRIPHENYLMETHANE AND RELATED DYES

Triphenylmethane dyes comprise one of the oldest classes of synthetic dyes. They are of brilliant hue, exhibit high tinctorial strength, are relatively inexpensive, and may be applied to a wide range of substrates. However, they are seriously deficient in fastness properties, especially fastness to light and washing. Consequently, the use of triphenylmethane dyes on textiles such as wool, silk, and cotton has decreased as dyes from other classes with superior lightfastness and washfastness properties have become available (DYES AND DYE INTERMEDIATES). Interest in this class of dyes was revived with the introduction of polyacrylonitrile fibers (see ACRYLONITRILE POLYMERS; FIBERS, ACRYLIC). Triphenylmethane dyes are readily adsorbed on this fiber and show surprisingly high lightfastness and washfastness properties, compared with the same dyes on natural fibers. However, the durability of acrylic fibers created an even greater demand for fastness properties. Modifications of the classical triphenylmethane dyes intended to improve these properties met with limited success because they were generally accompanied by a reduction in tinctorial strength. Substitution of one of the aryl groups with a heteroaryl residue or the introduction of two or four cyanoethyl groups on the amine functionalities generally increases the lightfastness (1). Research has also led to the development of novel dye types from other chemical classes, such as the pendent cationic dyes, in which the localized positive charge is isolated from the chromophoric system, dyes which were specially designed to give high lightfastness. Similarly, the diazahemicyanine dyes (2), which offer both brightness and fastness, gradually replaced triphenylmethane dyes for incorporation onto acrylic fibers (see CYANINE DYES; POLYMETHINE DYES). Consequently, triphenylmethane dyes are being used in markets where brightness and cost effectiveness, rather than lightfastness and washfastness, are considered to be more important. An example is the coloration of paper. However, triarylmethane dyes such as malachite green [569-64-2] and fuchsine [632-99-5], owing to their high tinctorial strength and low cost, are still used on acrylic fibers as mixtures to produce deep colors, eg, black and navy blue.

The triarylmethane dyes are broadly classified into the triphenylmethanes (CI 42000–43875), diphenylnaphthylmethanes (CI 44000–44100), and miscellaneous triphenylmethane derivatives (CI 44500–44535). The triphenylmethanes are classified further on the basis of substitution in the aromatic nuclei, as follows: (1) diamino derivatives of triphenylmethane, ie, dyes of the malachite green series (CI 42000–42175); (2) triamino derivatives of triphenylmethane, ie, dyes of the fuchsine, rosaniline, or magenta series (CI 42500–42800); (3) aminohydroxy derivatives of triphenylmethane (CI 43500–43570); and (4) hydroxy derivatives of triphenylmethane, ie, dyes of the rosolic acid series (CI 43800–43875). Monoaminotriphenylmethanes are known but they are not included in the classification because they have little value as dyes.

Chemically, the triarylmethane dyes are monomethine dyes with three terminal aryl systems of which one or more are substituted with primary, secondary, or tertiary amino groups or hydroxyl groups in the para position to the methine carbon atom. Additional substituents such as carboxyl, sulfonic acid, halogen, alkyl, and alkoxy groups may be present on the aromatic rings. The number, nature, and position of these substituents determine both the hue or color of the dye and the application class to which the dye belongs. For instance, the triarylmethane dyes that have one amine substituent on only one of the terminal aryl groups give pale yellow dyeings on polyacrylonitrile fibers (3). The introduction of alkyl or alkoxy groups into the other unsubstituted aryl systems can give dyes which produce an orange or red dyeing on polyacrylonitrile fibers (4). Other colors, obtained from the more important commercial triarylmethane dyes, include vivid reds, violets, blues, greens, and even blacks when used in combination with a red azo dye, eg, CI Basic Red 18 [14097-03-1] (CI 11085). The application classes include pigments (qv) and basic (cationic), acidic (anionic), solvent, and mordant dyes. If no acidic groups are present, the dye is cationic or basic. If sulfonic acid groups are present, the dye is anionic or acidic. Carboxylic groups adjacent to hydroxyl groups in the dye confer mordant-dyeing properties. The free bases or the fatty acid salts of the triarylmethane dyes are used as solvent dyes. Pigments can be made by combining a cationic triarylmethane dye with an ion of opposite charge derived from the heteropoly acids of phosphorus, silicon, molybdenum, and tungsten. Pigments can also be made from sulfonated triarylmethane dyes (anionic); the cation is usually barium or calcium.

Structure

The first triarylmethane dyes were synthesized on a strictly empirical basis in the late 1850s; an example is fuchsine, which was prepared from the reaction of vinyl chloride with aniline. Their structural relationship to triphenylmethane was established by Otto and Emil Fischer (5) with the identification of pararosaniline [569-61-9] as 4,4',4''-triaminotriphenylmethane and the structural elucidation of fuchsine. Several different structures have been assigned to the triarylmethane dyes (6–8), but none accounts precisely for the observed spectral characteristics. The triarylmethane dyes are therefore generally considered to be resonance hybrids. However, for convenience, usually only one hybrid is indicated, as shown for crystal violet [548-62-9], CI Basic Violet 3 (**1**), for which $\lambda_{\max} = 589$ nm.

(**1**)   (**2**)

The ortho hydrogen atoms surrounding the central carbon atom show considerable steric overlap. Therefore, it can be assumed that the three aryl groups in the dye are not coplanar, but are twisted in such a fashion that the shape of the dye resembles that of a three-bladed propeller (9). Substitution in the para position of the three aryl groups determines the hue of the dye. When only one amino group is present, as in fuchsonimine hydrochloride [84215-84-9], $\lambda_{max} = 440$ nm (**2**), the shade is a weak orange-yellow.

However, when at least two or more amino groups are present in different rings, the resonance possibility is greatly increased, resulting in a much greater intensity of absorption and in a strong bathochromic shift to longer wavelengths, eg, Doebner's violet [3442-83-9] (**3**), $\lambda_{max} = 562$ nm, which is a reddish violet, and pararosaniline (**4**), $\lambda_{max} = 538$ nm, which is a bluish violet. The amino derivatives of commercial value contain two or three amino groups.

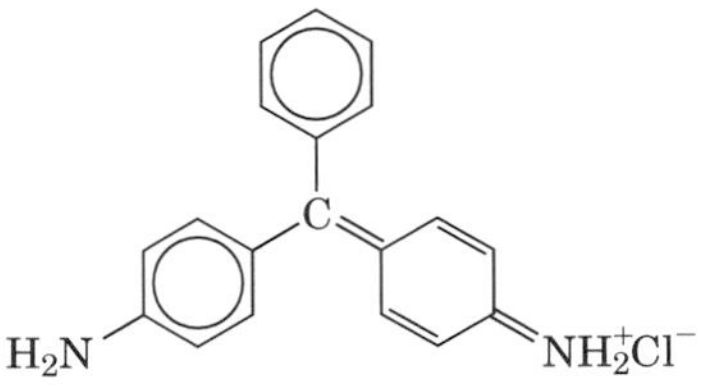

(**3**) 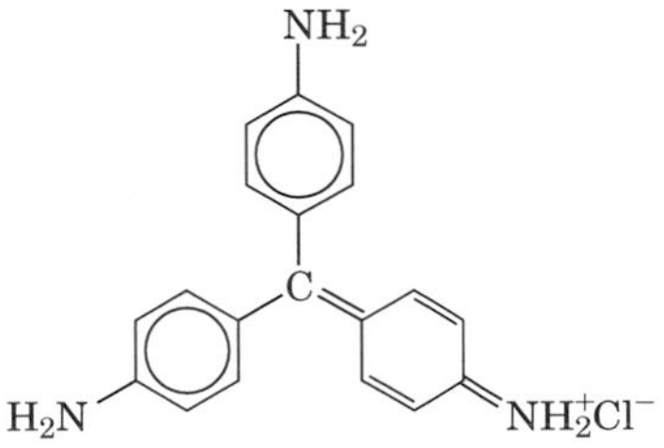   (**4**)

A further strong bathochromic shift is observed as the basicity of the primary amines is increased by N-alkylation, eg, malachite green [569-64-2], CI Basic Green 4, $\lambda_{max} = 621$ nm (**5**).

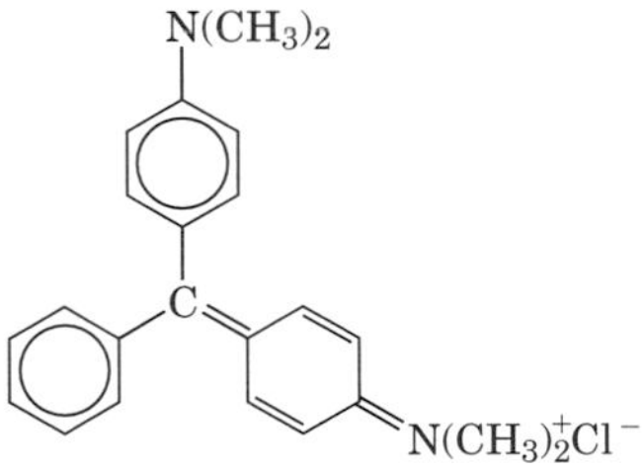

(**5**)

Phenylation of the primary amino groups also produces an increased bathochromic shift in the wavelength of absorption with increasing degree of phenylation. Only monophenylation of each amino group is possible, eg, as in (**6**) and (**7**).

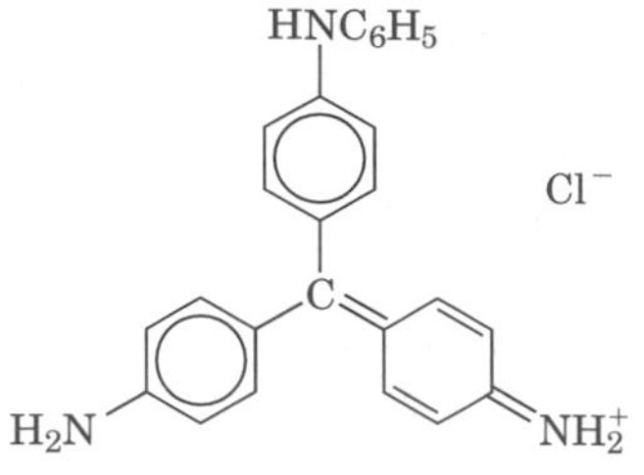

(6) 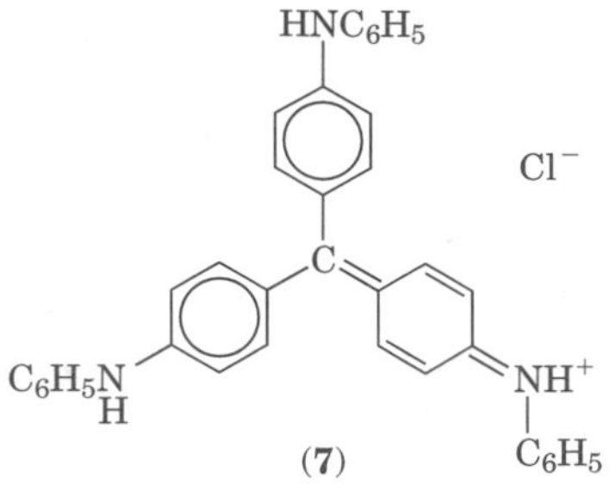　　　　　(7)

The steric effects of substituents on the color and constitution of triaryl-methane dyes have been studied extensively (10–19). Replacement of the hydrogen atoms ortho to the central carbon atom in crystal violet (λ_{max} = 589 nm) by methyl groups results in a uniform bathochromic shift (ca 8 nm per methyl group) to the 2,2′,2″-trimethyl derivative [84282-50-8] (λ_{max} = 614 nm) and reduced absorptivity values (20). These phenomena suggest that the axial rotational adjustment needed to accommodate the o-methyl groups is shared uniformly by the three phenyl rings. The 2,6-dimethyl derivative [117071-61-1] (λ_{max} = 635 nm) (**8**), however, shows a much larger bathochromic shift per methyl group, and it has been suggested that the dimethylaminoxylyl ring undergoes most of the rotational twist, relieving steric strain by twisting around the central bond in such a way that the charge is localized on the other two dimethylaminophenyl rings. Steric hindrance at the central carbon atom of the 2,6-dimethyl derivative of crystal violet is evident from the fact that the fuchsone derivative [85294-29-7] (λ_{max} = 562 nm) (**9**) and not the corresponding carbinol is formed by the action of a base on the dye (**8**). Thus, the dimethylamino group

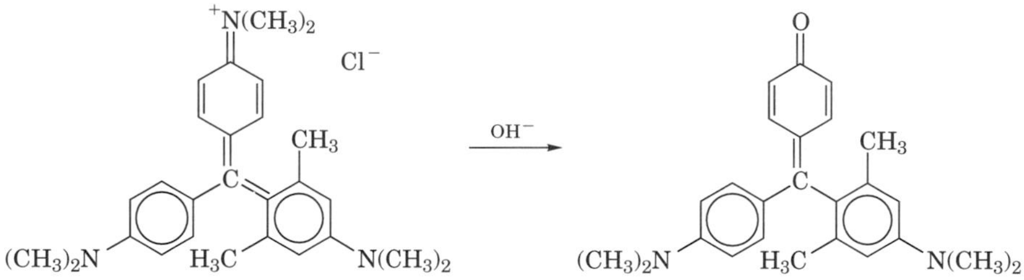

(8)　　　　　(9)

on the xylyl ring is partially deconjugated from the central carbon atom, making it less susceptible to nucleophilic displacement; consequently, the base replaces a terminal dimethylamino group on a phenyl ring with hydroxy. Therefore, steric hindrance facilitates the nucleophilic replacement of the terminal dimethylamino group by the hydroxyl group of the base (21).

Chemical Properties

Dyes in general and triarylmethane dyes in particular are rarely subjected to chemical processing once they have been formed. The introduction of substituents is usually carried out during the manufacture of the intermediates where the position and number of the groups introduced may be more precisely controlled. Dyes are sometimes exposed to oxidizing and reducing conditions during application and afterward.

Oxidation. Although many triarylmethane dyes are prepared by the oxidation of leuco bases, they are usually destroyed by strong oxidizing agents. Careful choice of both the oxidant and the reaction conditions is required to prevent loss of product during this stage of the manufacture. Overoxidation of malachite green (**5**) gives a quinone imine (**10**) identical to that obtained by oxidizing tetramethylbenzidine (**11**) or Michler's hydrol (**12**). Overoxidation may

also result in the oxidative cleavage of alkyl groups from the amino substituents. Thus, triarylmethane dyes are destroyed by sodium hypochlorite, further limiting their use as textile dyes.

The triarylmethane dyes are extremely sensitive to photochemical oxidation, a fact which accounts for their poor lightfastness on natural fibers (22–31). There are many factors which affect the rate of fading (degradation) of the triarylmethane dyes on natural and synthetic fibers. They include the type of substrate fiber, the nature of the fiber binding site (eg, a sulfonic acid or a carboxylic acid group) to which the dye is attached, the action of oxygen, and, to a lesser degree, water. The photodegradation products of malachite green on cellulosic substrates were identified as benzophenone and 4-dimethylaminobenzophenone (22). It has been proposed that decomposition of triarylmethane dyes occurs upon absorption of ultraviolet radiation by the carbinol form of the dye generated at the dye binding site. The excited carbinol form either undergoes radical fragmentation followed by reaction with water and oxygen or reacts directly with water and oxygen to give the products mentioned above.

Similar degradation products have been identified from the photo-oxidation of crystal violet using singlet oxygen sensitizers (25). In a proposed mechanism, the attack of singlet oxygen on the dye results in the formation of an unstable dioxetane intermediate (a four-membered ring containing two adjacent oxygen atoms). Several studies have revealed that *N*-dealkylation occurs simultaneously with the cleavage and contributes to the photodegradation. Introduction of substituents into the phenyl ring of malachite green produced no marked improvement in the lightfastness because of the presence of the N-alkyl groups in the molecule (31). Replacement with N-aryl groups in the analogues of the indolyldiphenylmethane dye, Wool Fast Blue FBL [6661-40-1], however, raised the lightfastness grade by 1–2 points on the 1–8 Gray scale for evaluating color change. Irradiation of N-alkyl groups in the presence of air and moisture converts them to aldehydes, eg, formaldehyde or acetaldehyde. Such N-dealkylation is a general phenomenon in dye photochemistry, insofar as it has been observed with thiazine dyes (32), rhodamine dyes (33), and N-methylaminoanthraquinones (34).

Reduction.   Triarylmethane dyes are reduced readily to leuco bases with a variety of reagents, including sodium hydrosulfite, zinc and acid (hydrochloric, acetic), zinc dust and ammonia, and titanous chloride in concentrated hydrochloric acid. Reduction with titanium trichloride (Knecht method) is used for rapidly assaying triarylmethane dyes. The $TiCl_3$ titration is carried out to a colorless end point which is usually very sharp (see TITANIUM COMPOUNDS, INORGANIC).

$$Ar_3COH + 2\ TiCl_3 + 2\ HCl \longrightarrow Ar_3CH + 2\ TiCl_4 + H_2O$$

Sulfonation.   The direct sulfonation of alkylaminotriphenylmethane dyes gives mixtures of substituted products. Although dyes containing anilino or benzylamino groups give more selective substitution, a sulfonated intermediate such as 3[(N-ethyl-N-phenylamino)methyl]benzenesulfonic acid (ethylbenzyl-anilinesulfonic acid) is the preferred starting material. However, Patent Blue V [3546-49-0], CI Acid Blue 3, was made from 3-hydroxybenzaldehyde and two moles of diethylaniline, followed by sulfonation of the leuco base and oxidation to the dye. FD&C Green 2 [5141-20-8], CI Acid Green 5, is still made by trisulfonation of the leuco base using ethylbenzylaniline and benzaldehyde as starting materials.

Spirit Blue [2152-64-9], CI Solvent Blue 23 (CI 42760), is one of the few dyes sulfonated as the leuco base. The degree of sulfonation depends on the conditions. Monosulfonated derivatives, commonly referred to as alkali blues, eg, CI Acid Blue 119 [1324-76-1], are used as their barium or calcium salts in printing inks. Disulfonated compounds, eg, CI Acid Blue 48 [1324-77-2], are employed as their sodium or ammonium salts for blueing paper, whereas the trisulfonic derivatives or ink blues, eg, CI Acid Blue 93 [28983-56-4] are used in writing inks (qv).

N-Alkylation and N-Arylation.   Dyes containing highly alkylated amino groups are prepared from highly alkylated intermediates and not by direct alkylation of dyes carrying primary amino groups. 4,4′,4″-Triaminotriphenylmethane (pararosaniline) may, however, be N-phenylated with excess aniline and benzoic acid to give the greenish blue, $N,N′,N″$-triphenylaminotriphenylmethane hydrochloride [2152-64-9], CI Solvent Blue 23 (**7**), $\lambda_{max} = 586$ nm. Shorter reaction times and use of less benzoic acid give more of a mixture of the reddish blue, mono- and diarylated products.

Pigment Formation.   Triarylmethane dyes can be converted into two types of insoluble compounds (35), which are used industrially as pigments (qv). Both are salts of triarylmethane dyes. Water-soluble cationic dyes are combined with phosphomolybdic acid, phosphotungstomolybdic acid, copper ferricyanide, and occasionally silicomolybdic acid and phosphotungstic acid to form insoluble complexes. Known as pigment lakes, these complexes provide clean, brilliant red and violet shades. These pigments are used in printing inks, especially packaging and special printing inks, but their use, ca 1996, is in decline because of higher production costs compared to other organic pigments that duplicate their color shades. The second type of pigments derived from triarylmethane dyes are known as alkali blues. These are inner salts of sulfonic acids, and the commercially important pigments are derived from either diarylated or triarylated rosaniline, eg, Pigment Blue 61 [1324-76-1] (13), CI 42765.1. They can

(13)

be prepared by either the diphenylmethane base method or the benzotrichloride method. Concentrated sulfuric acid is required to make the water-insoluble monosulfonates. The main use of the alkali blues is as shading pigments in inks based on carbon black, where an inexpensive blue component is needed to correct the natural brown tone of the base pigment. The main area of application is in printing inks, particularly offset, letterpress, and to a lesser extent in aqueous flexographic inks. They are used to color ribbons for typewriters and also to blue copy paper.

Manufacture

The preparation of triarylmethane dyes proceeds through several stages: formation of the colorless leuco base in acid media, conversion to the colorless carbinol base by using an oxidizing agent, eg, lead dioxide, manganese dioxide, or alkali dichromates, and formation of the dye by treatment with acid (Fig. 1). The oxidation of the leuco base can also be accomplished with atmospheric oxygen in the presence of catalysts.

The major products have been available since the late 1890s, and manufacturing processes have remained unchanged for much of this period. Older syntheses of triarylmethane dyes generally employed the isolation of the leuco base as a filter cake which could be easily washed free of contaminants such as

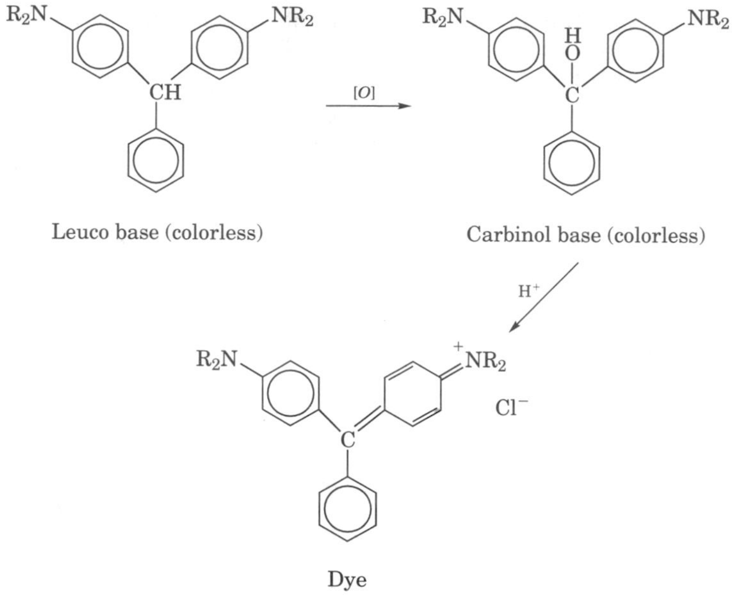

Fig. 1.   Preparation of triarylmethane dyes through the colorless leuco base.

unreacted aromatic amines. The more modern processes, especially the diphenyl-methane base and benzotrichloride methods, utilize excess aromatic amines as a convenient solvent to prepare the triarylmethane dyes without the need to isolate any of the intermediates. Consequently, the triarylmethane dyes are isolated as impure materials. The intermediates, by-products, and aromatic amine remain in the dye, reducing the yield and the tinctorial strength of the dye. Purification of the dyes usually involves physical processes, eg, membrane filtration, or chemical processes, eg, salting the mother liquors. These processes reduce the amount of dye isolated, thus adding to the overall cost. Increasingly, there has been a requirement to control effluents and air pollution in most developed countries as environmental controls have become more stringent. The limits for discharging color, acid, alkali, toxic heavy metals (eg, mercury, cadmium, lead, and chromium), lipophilic aromatic amines, or polychlorobiphenyls (PCBs) varies from country to country, but manufacturers have begun to minimize these losses in order to increase the yield of the dye, control the costs, and limit the effluent discharged to the drain. Manufacturing processes to prepare and purify triarylmethane dyes using alternative reagents and technology have appeared regularly in the patent literature since the early 1970s to meet the environmental concerns and lower costs (36–49).

Aldehyde Method.   This method is generally used for the preparation of diaminotriphenylmethane dyes or hydroxytriphenylmethane dyes. The central carbon atom is derived from an aromatic aldehyde or a substance capable of generating an aldehyde during the course of the condensation. Malachite green is prepared by heating benzaldehyde under reflux with a slight excess of dimethyl-aniline in aqueous acid (Fig. 2). The reaction mass is made alkaline and the

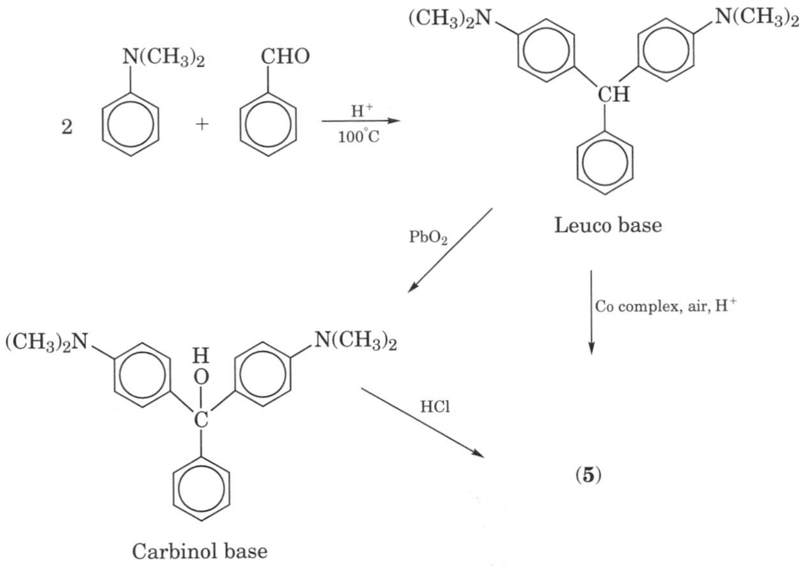

Fig. 2. Preparation of malachite green (**5**).

excess dimethylaniline is removed by steam distillation. The resulting leuco base is oxidized with freshly prepared lead dioxide to the carbinol base, and the lead is removed by precipitation as the sulfate. Subsequent treatment of the carbinol base with acid produces the dye, which can be isolated as the chloride, the oxalate [2437-29-8], or the zinc chloride double salt [79118-82-4].

The leuco base of malachite green has also been oxidized by air using a cobalt complex (36) or an iron complex (37) with chloranil in a mixture of glacial acetic acid and chloroform. Palladium (38), copper (39), vanadium, and molybdenum catalysts (40) have also been used to oxidize the leuco bases to the triphenylmethane dyes. Oxidation of the leuco base to the triarylmethane dye in glacial acetic acid can also be accomplished by treatment with air in the presence of chloranil and gaseous nitrogen oxides (41), eg, nitrogen monoxide and nitrogen dioxide (but not dinitrogen oxide). Under the acidic reaction conditions, other catalysts which release these nitrogen oxides can also be used, eg, iron(III) nitrate and sodium nitrite. Oxidative electrolysis has also been used to convert the leuco base to triarylmethane dyes (48).

The starting materials of the aldehyde method may be sulfonated. For example, CI Acid Blue 9 [2650-18-2], CI Food Blue 2 (CI 42090), is manufactured by condensing α-(N-ethylanilino)-m-toluenesulfonic acid with o-sulfobenzaldehyde. The leuco base is oxidized with sodium dichromate to the dye, which is usually isolated as the ammonium salt. In this case, the removal of the excess amine is not necessary. However, this color cannot be used in the food sector because separation of the chromium compounds from the dye is difficult. An alternative method which gives food-grade CI Acid Blue 9 (**14**) and dispenses with the use of sodium dichromate employs oxidative electrolysis of the leuco base (49).

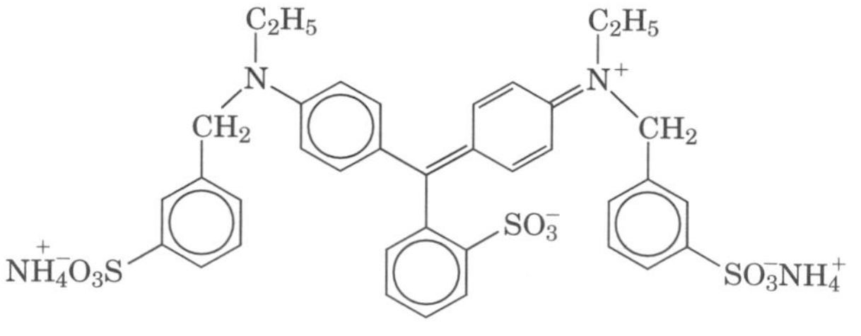

(14)

Ketone Method

In the ketone method, the central carbon atom is derived from phosgene (qv). A diarylketone is prepared from phosgene and a tertiary arylamine and then condenses with another mole of a tertiary arylamine (same or different) in the presence of phosphorus oxychloride or zinc chloride. The dye is produced directly without an oxidation step. Thus, ethyl violet [2390-59-2], CI Basic Violet 4 (**15**), is prepared from 4,4'-bis(diethylamino)benzophenone with diethylaniline in

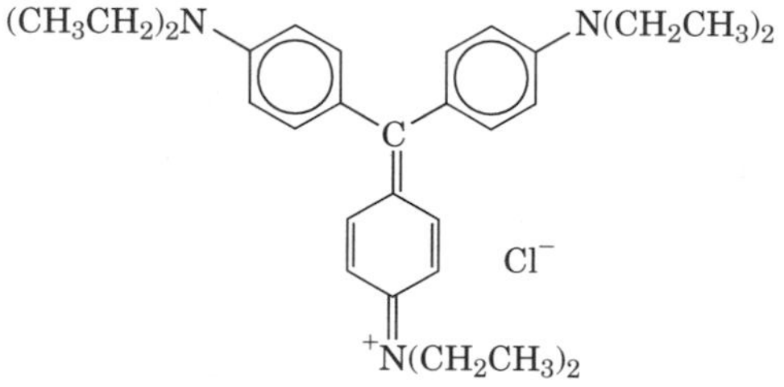

(15)

the presence of phosphorus oxychloride. This reaction is very useful for the preparation of unsymmetrical dyes. Condensation of 4,4'-bis(dimethylamino)-benzophenone [90-94-8] (Michler's ketone) with N-phenyl-1-naphthylamine gives the Victoria Blue B [2580-56-5], CI Basic Blue 26, which is used for coloring paper and producing ballpoint pen pastes and inks.

The manufacture of crystal violet (**1**), however, is a special case which does not involve the isolation of the intermediate Michler's ketone (Fig. 3). Thus, phosgene is treated with excess dimethylaniline in the presence of zinc chloride. Under these conditions, the highly reactive intermediate "ketone dichloride" is formed in good yield; this intermediate further condenses with another mole of dimethylaniline to give the dye.

Diphenylmethane Base Method.   In this method, the central carbon atom is derived from formaldehyde, which condenses with two moles of an arylamine to give a substituted diphenylmethane derivative. The methane base is oxidized with lead dioxide or manganese dioxide to the benzhydrol derivative. The reactive hydrols condense fairly easily with arylamines, sulfonated arylamines, and sulfonated naphthalenes. The resulting leuco base is oxidized in the presence of acid (Fig. 4).

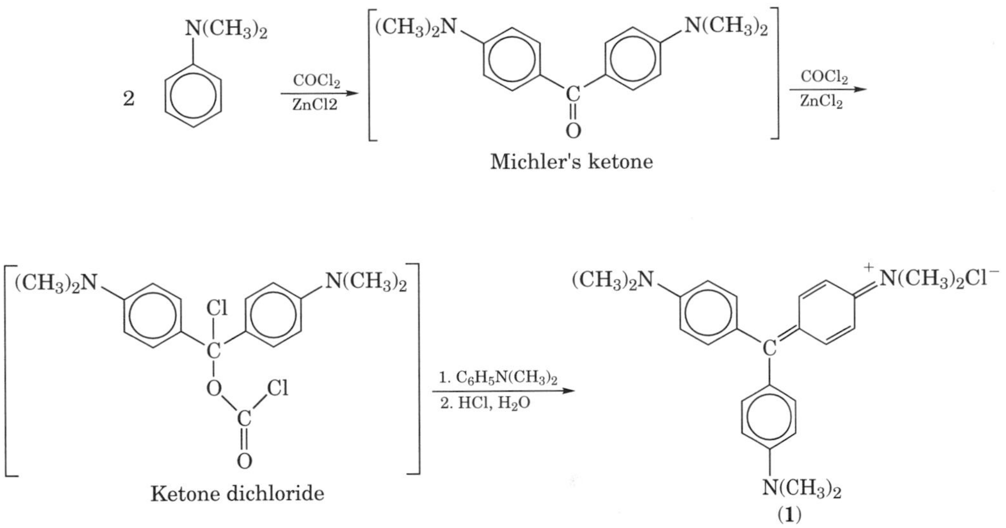

Fig. 3. Manufacture of crystal violet (**7**).

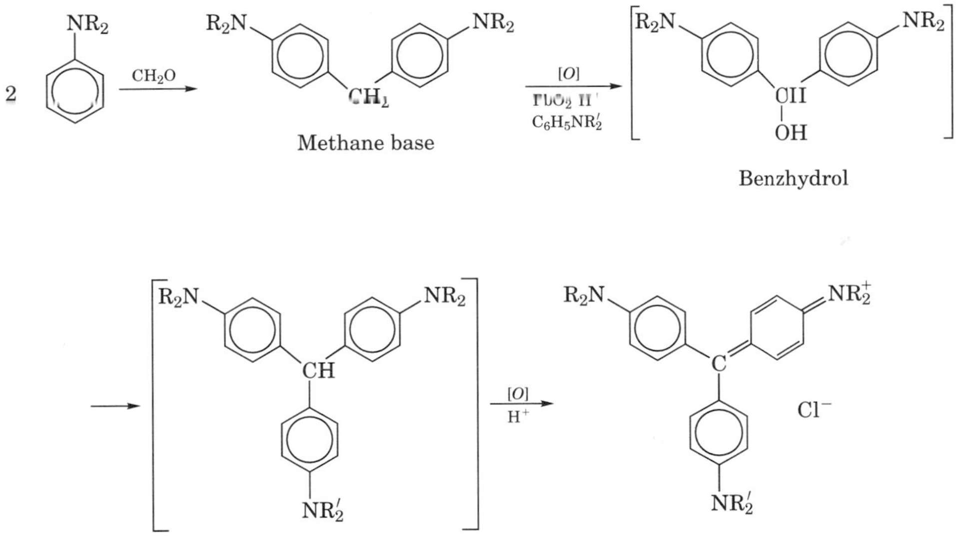

Fig. 4. Diphenylmethane base route to triarylmethane dyes.

In a variation of this method, isolation of the benzhydrol derivative is not required. The methane base undergoes oxidative condensation in the presence of acid with the same or a different arylamine directly to the dye. New fuchsine [*3248-91-7*], CI Basic Violet 2 (**16**), is prepared by condensation of two moles of *o*-toluidine with formaldehyde in nitrobenzene in the presence of iron salts to give the corresponding substituted diphenylmethane base. This base is also not isolated, but undergoes an oxidative condensation with another mole of *o*-toluidine to produce the dye.

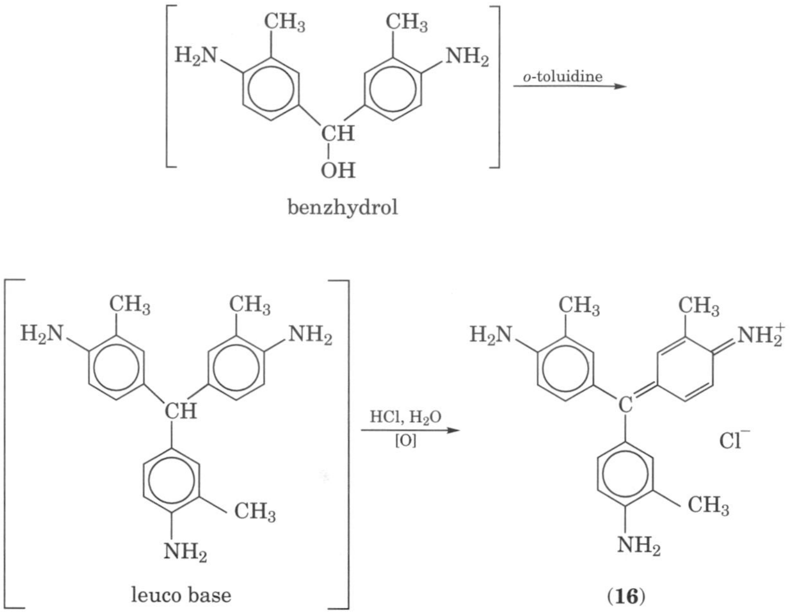

benzhydrol

leuco base   (16)

Methyl violet [8004-87-3], CI Basic Violet 1 (17), is made by the air oxidation of dimethylaniline in the presence of salt, phenol, and a copper sulfate catalyst. Initially, some of the dimethylaniline is oxidized to formaldehyde and N-methylaniline under those conditions. The formaldehyde then reacts with dimethylaniline to produce N,N,N',N'-tetramethyldiaminodiphenylmethane, which is oxidized to Michler's hydrol [119-58-4]. The hydrol condenses with N-methylaniline formed in the initial step to give the leuco base of methyl violet. Treatment with aqueous acid produces the dye. Because Michler's hydrol may also react with dimethylaniline instead of the N-methylaniline to give crystal violet, commercial-grade methyl violet is usually a mixture. A cobalt complex

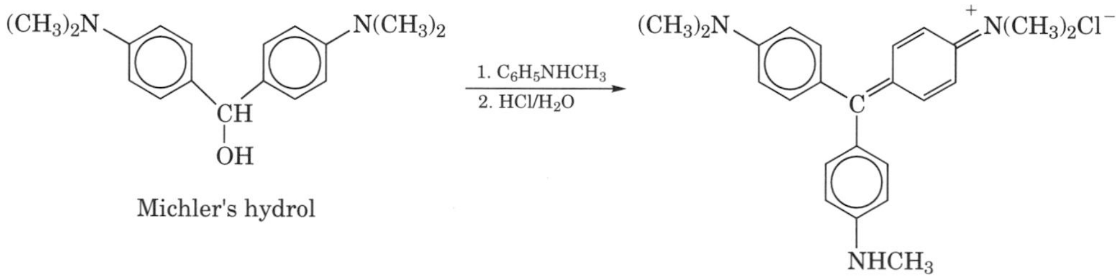

Michler's hydrol

(17)

has converted 4,4'-dimethylaminodiphenylmethane and dimethylaniline in the presence of atmospheric oxygen to crystal violet in one step (50).

Benzotrichloride Method. The central carbon atom of the dye is supplied by the trichloromethyl group from p-chlorobenzotrichloride. Both symmetrical and unsymmetrical triphenylmethane dyes suitable for acrylic fibers are prepared by this method. 4-Chlorobenzotrichloride is condensed with excess chlorobenzene in the presence of a Lewis acid such as aluminium chloride to produce the intermediate aluminium chloride complex of 4,4′,4″-trichlorotriphenylmethyl chloride (**18**). Stepwise nucleophilic substitution of the chlorine atoms of this intermediate is achieved by successive reactions with different arylamines to give both symmetrical (**51**) and unsymmetrical dyes (**52**), eg, *N*-(2-chlorophenyl)-4-[(4-chlorophenyl)[4-[(3-methylphenyl)imino]-2,5-cyclohexadien-1-ylidene]methyl]benzenamine monohydrochloride [*85356-86-1*] (**19**) from *m*-toluidine and *o*-chloroaniline.

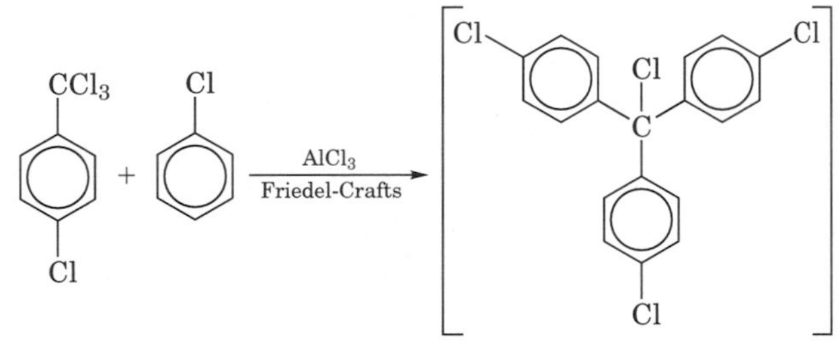

(**18**)

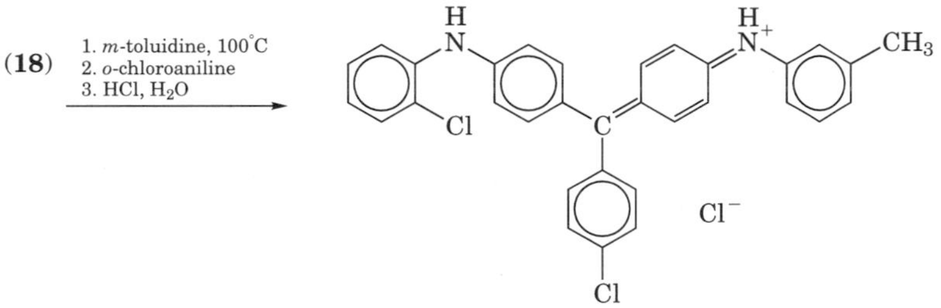

(**19**)

Economic Aspects

Since 1973, the U.S. International Trade Commission has reported the manufacture and sales of dyes by application class only. In 1972, the last year for which statistics are available by chemical class, 3900 metric tons of triarylmethane dyes were manufactured, which represents approximately 4% of total dyestuff production in the United States. At that time, there were 185 triarylmethane dyes listed in the *Colour Index*. From the latter half of the 1970s through the 1980s, annual dye production in the United States, including triarylmethane dyes, changed very little. In 1981, methyl violet, with an annual production of 725 t, was the only triarylmethane dye for which production statistics were available. Some triarylmethane dyes were imported, eg, malachite green (163 t in 1981), methyl violet (40 t), new fuchsine (30 t), and other dyes totalling less than 15 t.

Statistics for the production of basic dyes include those products listed as cationic dyes, eg, cyanines, for dyeing polyacrylonitrile fibers and the classical triarylmethane dyes, eg, malachite green, for coloring paper and other office applications (2,53). Moreover, statistics for triarylmethane dyes are also hidden in the production figures for acid, solvent, mordant, and food dyes, and also organic pigments. Between 1975 and 1984, the annual production of basic dyes in the United States varied from 5000–7700 t. However, from 1985–1990, annual production of basic dyes varied from 5000–5700 t, and the annual sales value increased from \$56 to \$73 million per year.

Health, Safety, and Environmental Information

In the 1960s, problems were encountered with the interpretation of toxicological studies on animals given triarylmethane dyes used as food colorants (54). The disagreement between experts largely persuaded certain authorities (U.K. and EEC) to remove Brilliant Blue FCF, CI Food Blue 2 (CI 42090), from the permitted lists. Although Brilliant Blue FCF was reinstated on the U.K. list, as of 1997 the problem still exists of correlating laboratory tests and actual human exposure. Conflicting test data have been obtained on various triarylmethane dyes (55). Positive and negative results for the same dye in different assays have only led to more genotoxic studies. For example, both Fast Green FCF [2353-45-9], CI Food Green 3 (CI 42053), and Green S [3087-16-9], CI Food Green 4 (CI 44090), have been subjected to many screenings and are permitted in food in several countries. However, other triarylmethane dyes have been proved genotoxic, eg, Acid Violet 6B [1694-09-3], CI Food Violet 2 (CI 42640), Methyl Violet 2B, CI Basic Violet 1 (CI 42535), and Victoria Blue B, CI Basic Blue 26 (CI 44045). These dyes are approved for limited use in food in some countries, however in the United States, CI Food Violet 2 was delisted for use in food, drugs, and cosmetics in 1973. There are triarylmethane dyes which have also been delisted worldwide for use in food, eg, Guinea Green B [4680-78-8], CI Food Green 1 (CI 42085), and Violet BNP [80539-34-0], CI Food Violet 3.

The triarylmethane dyes of the rosaniline family, eg, fuchsine and crystal violet, show similar toxic responses in assays. They are moderately toxic after acute exposure, but the effects usually pass within a couple of days. These effects are no cause for alarm as long as the dyes are not permitted for food use or contact (56). There is evidence that the toxicological effects might be the result of impurities, eg, aromatic amines, or of certain functional groups, notably amino substituents found in the dyes (55). The metabolic fate of the triarylmethane dyes in humans or animals has also been studied (54,57). Most dyes show almost quantitative excretion and no detectable coloring in the urine or metabolites in the body fluids. There have been many studies on the incidence of cancers among dye workers and users. Cancers among workers have declined since the use of the dye intermediates benzidine and β-naphthylamine were restricted in many developed countries. By the 1970s, it was thought that there was no evidence that human cancers were caused by dyes or pigments even if they showed animal carcinogenicity (58). There is very little information about the effects of triarylmethane dyes in products used by the consumer.

The toxicity of dyes to aquatic organisms has also been investigated (59), with most of the work done on fish. In these studies, over 3000 dyes in common use were tested, of which 27 dyes had a LC_{50} around 0.05 mg/L. Ten of these cases had triarylmethane structures. The American Dye Manufacturers Institute (AMDI) has also actively investigated this area. In their studies, the dye of highest toxicity was CI Basic Violet 1 with a LC_{50} of 0.05 mg/L (60). Environmental regulations have made it necessary to examine whether the dyes are toxic, biodegradable, accumulate in fish, or persist in such a way that they could prove hazardous for any downstream uses (see DYES, ENVIRONMENTAL CHEMISTRY). The Ecological and Toxicological Association of the Dyestuffs Manufacturing Industry (ETAD) and AMDI have built up a body of knowledge concerning the toxicity and environmental impact of many dyestuffs, including triarylmethane dyes.

Environmental Concerns

The main route by which dyes enter the environment is via wastewater, both from their manufacture and their use. Accurate data on dyes released into the environment are not available, although lists of materials released to the environment from the processes operated for the production of some triarylmethane dyes have been reported (61). However, various estimates and calculations have suggested that 1–2% of a dye is lost at the manufacturing stage and 1–10% is lost at the user stage (62). One estimate revealed that 2–3% of a basic dye is lost in exhaust and wash liquors. Manufacturers of dyes have increasingly investigated various technologies for effluent treatment, and some discussion of the treatment of triarylmethane dye waste has appeared in the literature (63,64).

Uses

Present usage of triarylmethane dyes, ca 1996, is confined mainly to nontextile applications. Substantial quantities are used in the preparation of organic pigments for printing inks, pastes, and for the paper printing trade, where cost and brilliance of shade are more important than lightfastness. Triarylmethane dyes and their colorless precursors, eg, carbinols and lactones, are used extensively in heat-, light-, and pressure-sensitive recording materials for high speed photoduplicating and photoimaging systems and for the production of printing plates and integrated circuits. They are also used for specialty applications such as tinting automobile antifreeze solutions and toilet sanitary preparations, in the manufacture of carbon paper, in ink for typewriter ribbons, and ink jet printing for high speed computer printers.

In addition to the dyeing and printing of natural and acrylic fibers, triarylmethane dyes are suitable for the coloration of other substrates such as paper, ceramics, leather, fur, anodized aluminium, waxes, polishes, soaps, plastics, drugs, and cosmetics. Several triarylmethane dyes are used as food colorants and are manufactured under stringent processing controls (see COLORANTS FOR FOODS, DRUGS, COSMETICS, AND MEDICAL DEVICES). They are usually bright green and blue, but red and violet shades are available for food coloring. Triphenylmethane dyes are also used extensively as microbiological stains. Some triarylmethane derivatives are very effective mothproofing agents for wool. Their use as antihalation dyes for photographic materials and as indicators is mentioned in the literature.

Triarylmethane dyes can be used for the coloration of glass. Using water-soluble dyes, eg, CI Acid Blue 83 [*6104-59-2*] (CI 42660), it is possible to prepare color filters by photomicrolithography. A red, green, and blue matrix of the three primary colors can be built up on the glass to produce the color filters. Several Japanese companies are investigating this technology to produce flat screen televisions. Other high technology applications using triarylmethane dyes include electrophotography and optical data storage. As the number of photocopying machines using selenium has decreased, organic photoconductors, especially positive charge control agents (CCAs), have become more important. Triarylmethane dyes such as *N*-{4-(bis[4-(phenylamino)phenyl]methylene)-2,5-cyclohexadien-1-ylidene}-3-methyl-benzeneamine sulfate [*57877-94-8*] (**20**) have been claimed as positive CCAs (65). The absorption spectra of the triarylmethane dyes can be extended into the near-infrared region. The use of triarylmethane dyes as infrared absorbers for optical information recording media (66) and as infrared color formers in carbonless copy paper has been claimed.

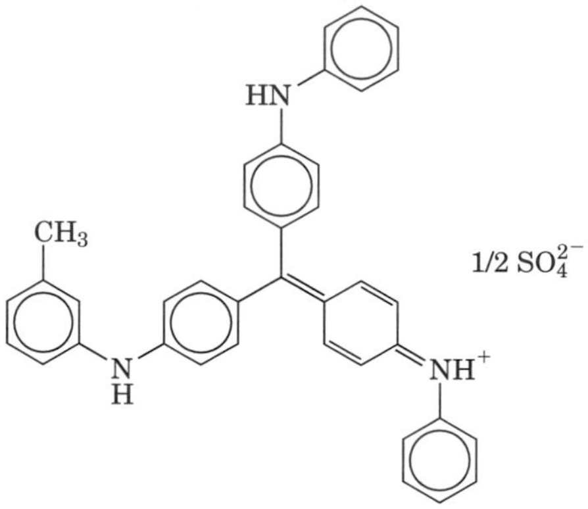

(**20**)

Related Dyes

Diphenylmethane Dyes.   The diphenylmethane dyes are usually classed with the triarylmethane dyes. The dyes of this subclass are ketoimine derivatives, and only three such dyes are registered in the *Colour Index*. They are Auramine O [*2465-27-2*] CI Basic Yellow 2 (CI 41000) (**21**, R = CH$_3$), Auramine G [*2151-60-2*] CI Basic Yellow 3 (CI 41005) (**22**), and CI Basic Yellow 37 [*6358-36-7*] (CI 41001) (**21**, R = C$_2$H$_5$). These dyes are still used extensively for the coloration of paper and in the preparation of pigment lakes.

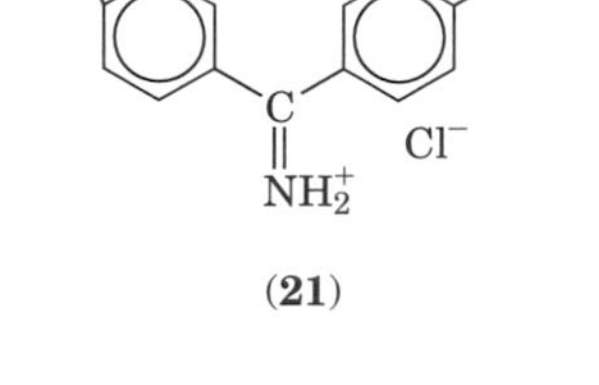

(**21**) 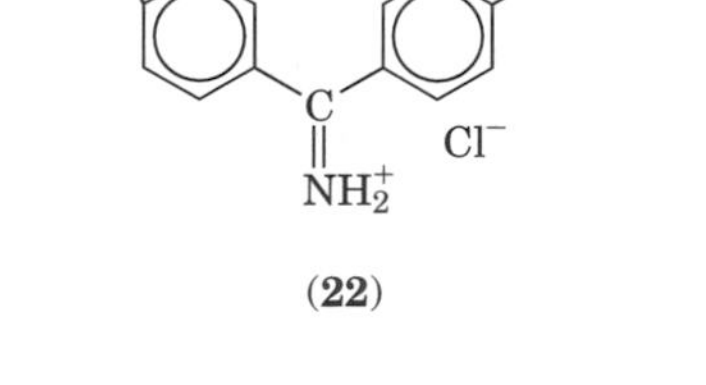   (**22**)

Auramine O is manufactured by heating 4,4'-bis(dimethylaminodiphenyl)-methane with a mixture of urea, sulfamic acid, and sulfur in ammonia at 175°C. The auramine sulfate [52497-46-8] formed in the reaction may be used directly in the dyeing process or can be converted into auramine base [492-80-8]. Highly concentrated solutions for use in the paper industry can be prepared by dissolving auramine base in formamide containing sodium bisulfate. The nitrate and nitrite salts exhibit excellent solubility in alcohols, which facilitates their use in lacquers and flexographic printing colors. Alkyl and halogen derivatives of N-phenyl(leucauramine) are colorless, stable, crystalline compounds that turn dark blue when in contact with acidic inorganic compounds such as aluminium sulfate, zinc sulfate, bentonite, or kaolin. They are useful in the production of colorless transfer sheets that on contact with an acidic copying sheet yield blue prints.

Phthaleins. Dyes of this class are usually considered to be triarylmethane derivatives. Phenolphthalein [77-09-8] (**23**, R = CO) and phenol red [143-74-8] (**23**, R = SO$_2$) are used extensively as indicators in colorimetric and titrimetric determinations (see HYDROGEN-ION ACTIVITY). These compounds are prepared by the condensation of phenol with phthalic anhydride or o-sulfobenzoic anhydride, respectively, in the presence of a dehydrating agent.

(**23**)

Heteroarylmethane Dyes. Dyes of this class usually have either one or two heteroaryl groups attached to the methane carbon atom. Trihetarylmethane dyes are known and have been investigated for their pharmacological activity (67–69) as well as their color characteristics. Dyes with only one heteroaryl substituent are prepared by the condensation of a diarylketone with the heteroaryl, eg, indole or carbazole. These dyes produce blue, green, and black dyeings on polyacrylonitrile fibers, whereas the dyes with two heteroaryl substituents produce red, violet, and blue dyeings. Dyes with two heteroaryl groups are prepared similarly, by condensation of an arylhetarylketone with a heteroaryl. These types of triarylmethane dyes and their derivatives are used as color formers in thermoreactive and pressure-sensitive recording materials (70–73).

Triarylmethane Dyes with Near-Infrared Absorption. The long wavelength absorption bands of triarylmethane dyes can be shifted into the near-infrared region, but the dyes still remain colored because other absorption bands are shifted to or stay in the visible region. There are two methods for shifting the absorption into the near-infrared region (74–76). The first method is to extend the chromophore by inserting extra conjugation, eg, carbon–carbon double (**24**) or triple bonds (**25**), or p-phenylene groups, between the methane carbon atom and the aryl groups. The second method involves linking two of the aryl rings at their 2- and 2′-positions with a direct bond to give a 9-arylfluorene derivative (**26**, **27**). These types of triarylmethane dyes and their derivatives have been claimed as infrared absorbers for optical information recording media (66) and security devices (77), and as organic photoconductors for use in lithographic plate production (78) (Fig. 5).

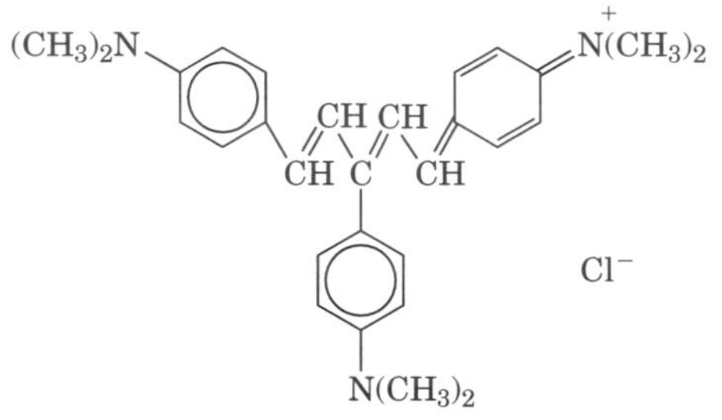

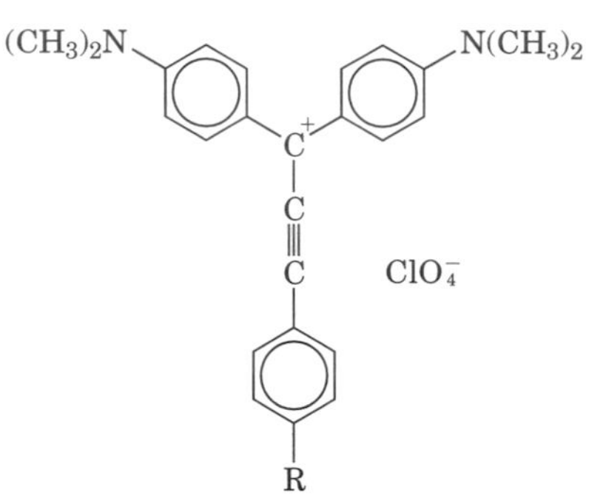

Fig. 5. Triarylmethane dyes with near-infrared absorption. For (**24**) [76438-66-9] in acetic acid, $\lambda_{max} = 770$ nm. For (**25**), if R = H [47544-33-2], $\lambda_{max}(CH_2Cl_2) = 688$ nm and $\log E = 5.00$; if R = N(CH_3)_2 [118751-95-4], $\lambda_{max}(CH_2Cl_2) = 663$ nm and $\log E = 5.12$. For the 9-arylfluorenone (**26**) [35324-95-9], $\lambda_{max} = 647$ and 850 nm; for (**27**) [127877-25-2], $\lambda_{max}(CH_2Cl_2) = 956$ nm and $\log E = 4.18$.

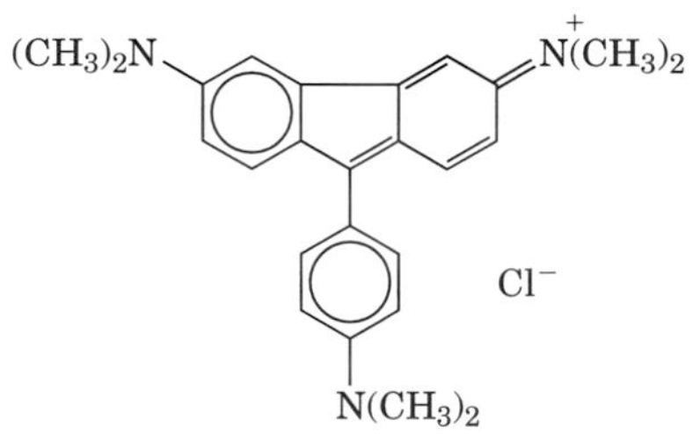

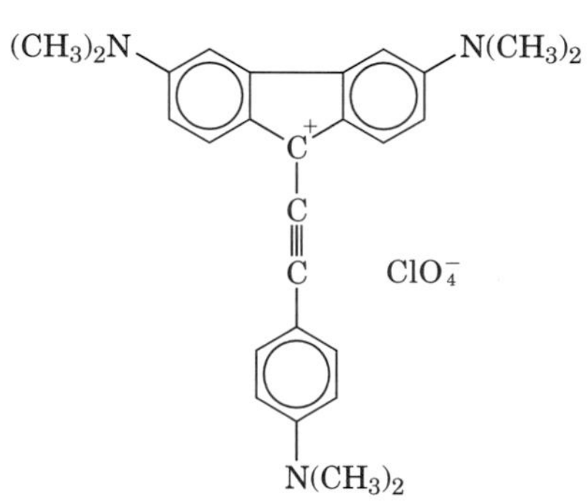

BIBLIOGRAPHY

"Triphenylmethane and Diphenylnaphthylmethane Dyes" in *ECT* 1st ed., Vol. 14, pp. 302–329, by A. J. Cofrancesco, General Aniline & Film Corp.; "Triphenylmethane and Related Dyes" in *ECT* 2nd ed., Vol. 20, pp. 672–737, by V. G. Witterholt, E. I. du Pont de Nemours & Co., Inc.; in *ECT* 3rd ed., Vol. 23, pp. 399–412, by D. Bannister and J. Elliott, Ciba-Geigy Corp.

1. Brit. Pat. 1,195,004 (Jan. 12, 1968), H. P. Kuelthau and R. Raue (to Bayer); Brit. Pat. 1,191,190 (Mar. 10, 1967), U. Blass (to Sandoz).
2. R. Raue, *Rev. Prog. Coloration*, **14**, 187–203 (1984).
3. Fr. Pat. 1,098,497 (Apr. 24, 1953) (to Ciba).
4. Ger. Pat. 1,811,652 (Nov. 29, 1968), R. Raue, H. P. Kuehlthau, and W. Eifler (to Bayer); see also U.S. Pat. 3,647,349 and U.S. Pat. 4,115,413.
5. O. Fischer and E. Fischer, *Ber.* **11**, 1079 (1878).
6. R. Wizinger, *Ber.* **60**, 1377 (1927).
7. C. R. Bury, *J. Am. Chem. Soc.* **57**, 2115 (1935).
8. L. Pauling, *Proc. Nat. Acad. Sci. USA*, **25**, 577 (1939).
9. W. Klyne and P. B. D. de la Mare, *Progress in Stereochemistry*, Vol. 2, Academic Press, Inc., New York, 1958, p. 42.
10. C. C. Barker, in G. W. Gray, ed., *Steric Effects in Conjugated Systems*, Butterworths, London, 1958, p. 34.
11. C. C. Barker, G. Hallas, and A. Stamp, *J. Chem. Soc.* **82**, 3790 (1960); C. C. Barker and G. Hallas, *J. Chem. Soc.* **83**, 1529 (1961).
12. G. Hallas, *J. Soc. Dyers Colour.* **83**, 368 (1968).
13. D. E. Grocock, G. Hallas, and J. D. Hepworth, *J. Soc. Dyers Colour.* **86**, 200 (1970); A. S. Ferguson and G. Hallas, *J. Soc. Dyers Colour.* **87**, 187 (1971).
14. A. S. Ferguson and G. Hallas, *J. Soc. Dyers Colour.* **89**, 22 (1973).
15. G. Hallas, K. N. Paskins, D. R. Waring, J. R. Humpston, and A. M. Jones, *J. Chem. Soc., Perkin Trans. II*, 450 (1977).
16. B. M. Fox, J. D. Hepworth, D. Mason, J. Sawyer, and G. Hallas, *J. Soc. Dyers Colour.* **98**, 10 (1982).
17. G. Hallas and M. M. Mitchell, *J. Soc. Dyers Colour.* **102**, 15 (1986).
18. S. F. Beach and co-workers, *J. Chem. Soc. Perkin Trans. II*, 1087 (1989).
19. H. Nakazumi, T. Kuriyama, and T. Kitao, *Shikizai Kyokaishi*, **59**, 747 (1986); H. Nakazumi, T. Kuriyama, Y. Shiraishi, and T. Kitao, *Shikizai Kyokaishi*, **60**, 420 (1987).
20. S. S. Ghandi, G. Hallas, and J. Thomasson, *J. Soc. Dyers Colour.* **93**, 451 (1977).
21. C. C. Barker and G. Hallas, *J. Chem. Soc.* **83**, 2642 (1961).
22. J. J. Porter and S. B. Spears, *Text. Chem. Color.* **2**, 191 (1970).
23. N. S. Allen, J. F. McKellar, and B. Mohajerani, *Dyes Pigm.* **1**, 49 (1980); N. S. Allen, *Rev. Prog. Color.* **17**, 61 (1987).
24. I. H. Leaver, *Photochem. Photobiol.* **16**, 189 (1972).
25. N. Kuramato and T. Kitao, *Dyes Pigm.* **3**, 49 (1982).
26. K. Iwamoto, *Bull. Chem. Soc. Jpn.* **10**, 420 (1935).
27. D. Bitzer and H. J. Brielmaier, *Melliand Textilber.* **41**, 62 (1960).
28. J. Wegmann, *Melliand Textilber.* **39**, 408 (1958).
29. E. D. Owen and R. T. Allen, *J. Appl. Chem. Biotechnol.* **22**, 799 (1972).
30. C. H. Giles, C. D. Shah, W. E. Watts, and R. S. Sinclair, *J. Soc. Dyers Colour.* **88**, 433 (1972).
31. N. A. Evans and I. W. Stapleton, *J. Soc. Dyers Colour.* **89**, 208 (1973).
32. H. Obata, Y. Usui, and M. Koizumi, *Bull. Chem. Soc. Jpn.* **34**, 1049 (1961).
33. N. A. Evans, *J. Soc. Dyers Colour.* **89**, 332 (1973).
34. C. H. Giles and R. S. Sinclair, *J. Soc. Dyers Colour.* **88**, 109 (1972).
35. W. Herbst and K. Hunger, *Industrial Organic Pigments, Production, Properties, Applications*, VCH, Weinheim, Germany, 1993, pp. 521–546.
36. Ger. Pat. 2,138,931 (Aug. 4, 1971), H. Kast, H. Baumann, U. Mayer, and A. Oberlinner (to BASF); see also U.S. Pat. 3,828,071.
37. Ger. Pat. 2,736,679 (Aug. 16, 1977), H. Kast and U. Mayer (to BASF); see also GB Pat. 2,003,096.

38. T. Sakakibara, J. Kotobuki, Y. Dogomori, *Chem. Letts.*, 25 (1977).

39. Jpn. Pat. 56,057,848 (May 20, 1981), K. Takeo, S. Nobuo, and Y. Hiroyoshi (to Hodogaya Chemical Co).

40. U.S. Pat. 4,321,207 (Mar. 23, 1982), F. F. Cesark (to American Cyanamid).

41. U.S. Pat. 4,330,476 (May 18, 1982), K. H. Hermann (to Bayer).

42. U.S. Pat. 4,477,381 (Oct. 16, 1984), U. Mayer, E. Hahn, and J. Jesse (to BASF).

43. U.S. Pat. 4,566,999 (Jan. 28, 1986), A. Engelmann (to Hoechst and Cassella).

44. U.S. Pat. 4,678,613 (July 7, 1987), R. J. Flores (to PMC Specialties Group Inc.).

45. U.S. Pat. 4,824,610 (Apr. 25, 1989), R. J. Sappok, L. R. de Alvare, and F. G. Spence (to BASF).

46. U.S. Pat. 5,013,857 (May 7, 1991), H. Berneth and R. Raue (to Bayer).

47. U.S. Pat. 5,198,558 (Mar. 30, 1993), S. P. Rines and C. J. Zullig (to BASF).

48. Jpn. Pat. 2,194,188 (July 31, 1990), T. Aizawa (to Nippon Kayaku KK).

49. U.S. Pat. 4,775,451 (Oct. 4, 1988), W. Habermann, U. Mayer, P. Hammes, and B. Landmann (to BASF).

50. Ger. Pat. 2,152,703 (Oct. 22, 1971), H. Kast, H. Baumann, U. Mayer, and A. Oberlinner (to BASF); see also U.S. Pat. 3,828,071.

51. Ger. Pat. 1,098,652 (Feb. 2, 1961), G. Schafer and F. Quint (to Hoechst).

52. Ger. Pat. 2,753,072 (Nov. 29, 1977), M. Haehnke and E. Mundlos (to Hoechst).

53. E. Fox, ed., *Amer. Dystuff Rep.*, 141–154 (1992).

54. J. J.-P. Drake, *Toxicology* **5**, 3 (1975).

55. R. D. Combes and R. B. Haveland-Smith, *Mutat. Res.* **98**, 101–248 (1982).

56. S. Clemmensen, J. C. Jensen, N. J. Jensen, O. Meyer, P. Olsen, and G. Wurtzen, *Arch. Toxicol.* **56**, 43 (1984).

57. MAFF (1979), *Interim Report on the Review of the Colouring Matter in Food Regulations*, FAC/REP/29, Ministry of Agriculture, Fisheries and Food, Food Additives and Contaminants Committee, HMSO, London, 1973.

58. T. Gadian, *Rev. Prog. Coloration* **7**, 85 (1976).

59. E. A. Clarke and R. Anliker, *Rev. Prog. Coloration* **14**, 84 (1984). M. L. Richardson and A. Waggott, *Ecotoxicol. Environmental Safety* **5**, 424 (1981).

60. *Dyes and the Environment*, ADMI Reports on Selected Dyes and their Effects, Vol. 1 (Sept. 1973) and Vol. 2 (Sept. 1974), American Dye Manufacturers Institute.

61. J. E. Gwinn and D. C. Bomberger, *Wastes from Manufacture of Dyes and Pigments*, Vol. 5: *Diphenylmethane and Triarylmethane Dyes and Pigments*, Report No. EPA-600/2-84-111E, SRI International, Menlo Park, Calif., 1984.

62. I. G. Laing, *Rev. Prog. Coloration* **21**, 56 (1991).

63. C. Yatome, T. Ogawa, and E. Idaka, Sen-I Gakkaishi **40**, T344 (1984).

64. V. V. Solodovnikov, V. M. Zadorskii, G. P. Fedorchenko, and Z. I. Selemeneva, *Vopr. Khim. Khim. Tekhnol.* **72**, 117 (1983).

65. Ger. Pat. 3,641,525 (Dec. 5, 1986), H.-T. Macholdt and A. Sieber (to Hoechst); see also U.S. Pat. 5,061,585.

66. U.S. Pat. 4,832,992 (May 23, 1989), M. Yabe and Y. Inagaki (to Fuji Photo Film KK).

67. E. Akgun and M. Tunali, *Arch. Pharm.* **321**, 921 (1988).

68. C. de Diego, C. Avendano, and J. Elguero, *Chem. Scr.* **28**, 403 (1988).

69. R. Naef, *Dyes Pigm.* **6**, 233 (1985).

70. U.S. Pat. 4,720,449 (Jan. 19, 1988), A. L. Borror and E. W. Ellis (to Polaroid Corp.).

71. U.S. Pat. 5,028,725 (July 2, 1991), P. F. King (to Polaroid Corp.).

72. U.S. Pat. 5,094,688 (Mar. 10, 1992), U. Eckstein, H. Psaar, and G. Jabs (to Bayer).

73. U.S. Pat. 5,233,048 (Aug. 3, 1993), U. Eckstein and R. Raue (to Bayer).

74. M. Sumitana, *Kagaku Kogyo*, 379 (1986).

75. S. Akiyama, S. Nakatsuji, K. Nakashima, and M. Watanabe, *J. Chem. Soc., Chem. Commun.*, 710 (1987); S. Akiyama, S. Nakatsuji, K. Nakashima, and S. Yamasaki, *Dyes Pigm.* **9**, 459 (1988).

76. S. Akiyama, S. Nakatsuji, K. Nakashima, M. Watanabe, and H. Nakazumi, *J. Chem. Soc., Perkin Trans.* **I**, 3155 (1988); S. Nakatsuji, K. Nakashima, S. Akiyama, and H. Nakazumi, *Dyes Pigm.* **24**, 37 (1994).

77. U.K. Pat. 2,173,914 (Oct. 22, 1986), P. F. Gordon (to ICI).

78. Jpn. Pat. 63,226,667 (Sept. 21, 1988), J. Yamada and T. Baba (to Mitsubishi Paper Mills).

General References

D. R. Waring and G. Hallas, eds., *The Chemistry and Applications of Dyes*, Plenum Publishing Corp., New York, 1990.

P. Gregory, *High-Technology Applications of Organic Colorants*, Plenum Publishing Corp., New York, 1990.

G. Booth, *The Manufacture of Organic Colorants and Intermediates*, The Society of Dyers and Colourists, Bradford, U.K., 1988.

P. F. Gordon, P. Gregory, *Organic Chemistry in Colour*, Springer-Verlag, Berlin, 1983.

G. Hallas, in J. Shore, ed., *Colorants and Auxiliaries*, Vol. 1, The Society of Dyers and Colourists, Bradford, U.K., 1990, pp. 279–294.

E. N. Abrahart, *Dyes and their Intermediates*, 2nd ed., Chemical Publishing, New York, 1977.

R. L. M. Allen, *Color Chemistry*, Appleton-Century-Crofts, New York, 1971.

K. Ventkataraman, ed., *The Chemistry of Synthetic Dyes*, Vol. 2, Academic Press Inc., New York, 1952; N. R. Ayyanger and B. D. Tilak, and D. R. Baer in Vol. 4, 1971; J. Lenoir in Vol. 5, 1971; E. Gurr, N. Anand, M. K. Unni, and N. R. Ayyanger in Vol. 7, 1974; N. A. Evans and I. W. Stapleton in Vol. 8, 1978.

P. Bentley and co-workers, *Review in Progress of Color and Related Topics*, Vol. 5, The Society of Dyers and Colourists, Bradford, U.K., 1974.

Manufacture of Triphenylmethane Dyestuffs and Intermediates at Ludwigshafen and Hoechst, BIOS Final Report 959 (1945); *Manufacture of Triphenylmethane Dyestuffs at Hoechst, Ludwigshafen, and Leverkusen*, BIOS Final Report 1433, British Intelligence Objectives Subcommittee (1946); German Dyestuffs and Dyestuff Intermediates (Field Information Agency, Technical, FIAT) Final Report 1313 (1946).

Colour Index, 3rd ed., The Society of Dyers and Colourists, Bradford, U.K. and the American Association of Textile Chemists and Colorists, Research Triangle Park, N.C., Vols. 1–6, 1971.

P. Rys and H. Zollinger, *Fundamentals of the Chemistry and Applications of Dyes*, Wiley-Interscience, New York, 1972.

H. A. Lubs, ed., *The Chemistry of Synthetic Dyes and Pigments*, American Chemical Society Monograph Series, Reinhold Publishing Corp., New York, 1955.

K. Venkataraman, ed., *The Analytical Chemistry of Synthetic Dyes*, Wiley-Interscience, New York, 1977.

J. Fabian and H. Hartmann, *Light Absorption of Organic Colorants*, Springer-Verlag, Berlin, 1980.

T. E. Furia, ed., *Handbook of Food Additives*, 2nd ed., CRC Press, Cleveland, Ohio, 1972.

D. M. Marmion, *Handbook of U.S. Colorants for Food, Drugs, and Cosmetics*, John Wiley & Sons, New York, 1979.

E. A. Clarke and R. Anliker, in O. Hutzinger, ed., *The Handbook of Environmental Chemistry*, Vol. 3, Part A, pp. 181–215, Springer-Verlag, Berlin, 1980.

E. Gurr, *Synthetic Dyes in Biology, Medicine and Chemistry*, Academic Press, London, 1971.

O. Valcl, I. Nemcova, and V. Suk, *Handbook of Triarylmethane and Xanthene Dyes: Spectrophotometric Determination of Metals*, CRC Press, Inc., Boca Raton, Fla., 1985.

DEAN THETFORD
Zeneca Specialties

TRYPSIN.   See ENZYME APPLICATIONS.

TRYPTOPHAN.   See AMINO ACIDS.

TUADS.   See RUBBER CHEMICALS.

TUNG OIL.   See FATS AND FATY OILS; DRYING OILS.

TUNGSTEN AND TUNGSTEN ALLOYS

Tungsten [*7440-33-7*] (wolfram), atomic number 74, atomic weight 183.85, is a silver-gray metallic element that appears in Group VIB of the Periodic Table, below chromium and molybdenum. There are 31 isotopes of this element, ranging from 160 to 190; the abundance of the five stable isotopes is given in Table 1. Tungsten has a very low vapor pressure, the highest melting point (3695 K) of any metal, as well as the highest tensile strength of any metal above 1650°C.

The name tungsten, meaning in Swedish heavy (*tung*) stone (*sten*), was first applied to a tungsten-containing mineral in 1755. The mineral was subsequently

Table 1. Isotopes of Stable Tungsten[a]

Isotope	CAS Registry Number	Abundance, %
^{180}W	[*14265-79-3*]	0.13
^{182}W	[*14265-80-6*]	26.30
^{183}W	[*14265-81-7*]	14.30
^{184}W	[*14265-82-8*]	30.67
^{186}W	[*14265-83-9*]	28.60

[a] Ref. 1.

identified by Scheele in 1781 as containing lime and a then-unknown acid which he called tungstic acid. The mineral was then named scheelite. Metallic tungsten was first produced in 1783 in Spain by the carbon reduction of tungstic acid and termed wolfram (W). This designation became common usage in Germany. In 1957, the IUPAC chose the English name tungsten and the French name tungstene, reserving wolfram as an alternative. However, W is still used as the chemical symbol.

During the nineteenth century, tungsten remained a laboratory material. The latter half of the century saw the development of high speed tool steels containing tungsten, which became the primary use for the metal in the first half of the twentieth century. The pure metal itself was first used as a filament for electric lamps at the beginning of the twentieth century. After some limited success with paste-extruded tungsten powder, the Coolidge process was developed in 1908, by which a pressed and sintered tungsten ingot could be worked at high temperatures by swaging and drawing to form a fine-wire filament. This was a landmark in the development of the incandescent-lamp industry and, later, in the use of tungsten as a welding electrode. In the 1920s, the search for an alternative to expensive diamond dies required for the drawing of the tungsten wire led to the manufacture of cemented carbides, which by the end of the twentieth century accounted for over half of the tungsten consumption in the world.

Tungsten is the eighteenth most abundant metal, having an estimated concentration in the earth's crust of 1–1.3 ppm. Of the more than 20 tungsten-bearing minerals, only four are of commercial importance: ferberite (iron tungstate), huebnerite (manganese tungstate), wolframite (iron–manganese tungstate containing ca 20–80% of each of the pure components), and scheelite (calcium tungstate).

The WO_3 content of wolframite minerals varies from 76.3% in $FeWO_4$ to 76.6% in $MnWO_4$. These minerals, commonly called black ores because their colors range from black to brown, occur as well-defined crystals to irregular masses of bladed crystals. They have a Mohs hardness of 5.0–5.5, a specific gravity of 7.0–7.5, and tend to be very brittle as well as weakly magnetic. Scheelite, containing 80.6% WO_3, is white to brown and strongly fluorescent in short-wave ultraviolet radiation. Occurring as massive crystals and small grains, Scheelite has a Mohs hardness of 4.5–5.0, a specific gravity of 5.6–6.1, and is very brittle.

Tungsten deposits occur in association with metamorphic rocks and granitic igneous rocks throughout the world (Table 2). Deposits in China constitute over half of the world reserves and over five times the reserves of the second largest source, Canada.

Physical Properties

Some of the physical properties of tungsten are given in Table 3; further property data are available (12–14). For thermodynamic values, References 5,15, and 16 should be consulted. Two values are given for the melting point. The value of 3660 K was selected as a secondary reference for the 1968 international practical temperature scale. However, since 1961, the four values that have been reported ranged from 3680 to 3695 and averaged 3688 K.

Table 2. World Tungsten Resources[a], 10^3 t

Country	Reserves	Other[b]	Total
North America			
United States	125	325	450
Canada	270	320	590
Mexico	20	5	25
other	1	2	3
Total[c]	*420*	*650*	*1070*
South America			
Bolivia	39	86	125
Brazil	18	40	58
other	2	2	4
Total[c]	*60*	*130*	*190*
Europe			
Austria	18	55	73
CIS	210	320	530
France	16	2	18
Portugal	24	30	54
United Kingdom	0.5	65	65
other	30	9	39
Total[c]	*300*	*480*	*780*
Africa			
Zimbabwe	5	5	10
other	5	14	19
Total[c]	*10*	*18*	*28*
Asia			
Burma	30	75	105
China	1400	2300	3700
North Korea	110	140	250
South Korea	80	80	160
Malaysia	15	30	45
Thailand	20	20	40
Turkey	75	14	89
other	5	5	10
Total[c]	*1700*	*2620*	*4320*
Oceania			
Australia	110	260	370
other	0.5	2	3
Total[c]	*110*	*260*	*370*
World total	*2600*	*4200*	*6800*

[a]Ref. 2.
[b]Derived in collaboration with the U.S. Geological Survey.
[c]Data may not add to totals shown because of independent rounding.

Chemical Properties

The oxidation states of tungsten range from $+2$ to $+6$; compounds that have zero oxidation state also exist. Above 400°C, tungsten is very susceptible to oxidation.

Table 3. Physical Properties of Tungsten

Property	Value	Reference
crystal structure	bcc	
lattice constant at 298 K, nm	0.316524	
shortest interatomic distance at 298 K, nm	0.2741	
density[a] at 298 K, g/cm^3	19.254	
melting point, K	3660	3
	3695 +/− 15	4
boiling point, K	5936	5
linear expansion per K		6
293–1395 K	$4.226 \times 10^{-6} (T - 293) + 8.479 \times 10^{-10} (T - 293)^2 - 1.974 \times 10^{-13} (T - 293)^3$	
1395–2495 K	$0.00548 + 5.146 \times 10^{-6} (T - 1395) + 1.952 \times 10^{-10} (T - 1395)^2 + 4.422 \times 10^{-13} (T - 1395)^3$	
2495–3600 K	$0.01226 + 7.451 \times 10^{-6} (T - 2495) + 1.654 \times 15^{-9} (T - 2495)^2 + 7.568 \times 10^{-14} (T - 2495)^3$	
specific heat, C_p, at 273–3300 K, J/(mol·K)[b]	$24.94\left(1 - \dfrac{4805}{T^2}\right) + 1.674 \times 10^{-3} T + 4.25 \times 10^{-10} T^3$	6
enthalpy, $H_T - H_{298}$, J/mol[b]	$24.94\left(T + \dfrac{4805}{T}\right) + 8.372 \times 10^{-4} T^2 + 1.062 \times 10^{-10} T^4 - 7917.8$	6
entropy at 298 K, J/mol[b]	32.66	5
heat of fusion, kJ/mol[b]	46.0	7
heat of sublimation at 298.13 K kJ/mol[b]	859.8	8
vapor pressure at 2600–3100 K, Pa[c]	$\log P_{pa} = -\dfrac{45395}{T} + 12.8767$[d]	9
thermal conductivity at K, W/(cm·K)		
0	0.0	
10	97.1	
50	4.28	
100	2.08	
500	1.46	
1000	1.18	
2000	1.00	
3400	0.90	
electrical resistivity, ρ, at 4–3000 K, nΩ·m	$\dfrac{0.04535\, T^{1.2472} - 2.90 \times 10^{-9}\, T^3}{1 + \frac{3.442 \times 10^5}{T^{2.98}}} + \rho_0$	10
total emissivity, ϵ_H, 1600–2800 K	$-2.685790 \times 10^{-2} + 1.819696 \times 10^{-4}\, T^4 - 2.194616 \times 10^{-8}\, T^2$	11

[a]Determined by x-ray. [b]To convert J to cal, divide by 4.184. [c]To convert Pa to mm H$_g$, multiply by 0.0075. [d]To convert $\log P_{pa}$ to $\log P_{mm\,H_g}$, subtract 2.1225.

At 800°C, sublimation of the oxide becomes significant and the oxidation is destructive. Very fine powders are pyrophoric. Above 600°C, the metal reacts vigorously with water to form oxides. In lamps, in the presence of water vapor, a phenomenon called water cycle occurs, in which the tungsten is oxidized in the hottest part of the filament and then reduced and deposited on the cooler portion of the filament. Tungsten is stable in nitrogen to over 2300°C. In ammonia, nitrides form at 700°C. Carbon monoxide and hydrocarbons react with tungsten to give tungsten carbide at 900°C. Carbon dioxide oxidizes tungsten at 1200°C. Fluorine is the most reactive halogen gas and attacks tungsten at room temperature. Chlorine reacts at 250°C, whereas bromine and iodine require higher temperatures.

Tungsten is resistant to many chemicals. At room temperature, it is rapidly attacked only by a mixture of hydrofluoric and nitric acids. Attack by aqua regia is slow. Hot sulfuric, nitric, and phosphoric acids also react slowly. Sodium, potassium, and ammonium hydroxide solutions slowly attack tungsten at room temperature in the presence of an oxidizing agent such as potassium ferricyanide or hydrogen peroxide. Molten sodium and potassium hydroxide attack tungsten only moderately. The attack is accelerated by the addition of an oxidizer. Tungsten resists attack by many molten metals. The maximum temperature of stability for various metals is as follows:

metal	Mg	Hg	Al	Zn	Na	Bi	Li
temperature, °C	600	600	680	750	900	980	1620

In contact with various refractory materials (qv), tungsten is stable in vacuum. In reducing atmospheres, however, the temperatures are lower.

refractory	Al_2O_3	BeO	MgO	ThO_2	ZrO_2
temperature, °C	1900	1500	2000	2200	1600

Manufacture

Mining and Beneficiation. Tungsten mines are generally small, producing less than 200 metric tons of raw ore per day. Worldwide, there are only about 20 mines producing over 300 t/d. Many small mines, limited by the nature of the ore body, are inactive at times, depending on the market price of tungsten. Mining is almost exclusively by underground methods. Where open-pit mining has been employed, underground methods are used as the deposit diminishes. Ore deposits usually range from 0.3 to 1.5% WO_3, but in exceptional cases can be as high as 4% WO_3. Because of the low tungsten content of the deposits, all mines have beneficiation facilities that produce a concentrate containing 60–75% WO_3.

Because scheelite and wolframite are both friable, care must be taken to avoid overgrinding, which can lead to sliming problems. The ores are crushed and ground in stages, and the fines are removed after each stage. Jawcrushers are employed for the first stage because the tonnages are low. Either jaw- or cone-type crushers are used for the second stage. In some cases, sizes are further reduced by rod milling. After each stage, the fines are removed and the coarse fraction is recirculated. Screening is the preferred method for fairly large particle sizes. Mechanical and hydraulic classifiers can also be used.

Because tungsten minerals have a high specific gravity, they can be beneficiated by gravity separation, usually by tabling. Flotation is used for many scheelites having a fine liberation size, but not for wolframite ores. Magnetic separators can be used for concentrating wolframite ores or cleaning scheelites.

Extractive Metallurgy. In extractive metallurgy, a relatively impure ore concentrate is converted into a high purity tungsten compound that can subsequently be reduced to metal powder. This is a particularly important step because high purity is required for all uses of tungsten except as a steel-alloying additive. The two most common intermediate tungsten compounds are tungstic acid, H_2WO_4, and ammonium paratungstate (APT), $(NH_4)_{10}W_{12}O_{41}\cdot5H_2O$ (Fig. 1). Most commercial processes in the 1990s use APT. Depending on the source, the impurities in ore concentrates can vary considerably, but those of most concern are sulfur, phosphorus, arsenic, silicon, tin, lead, boron, and molybdenum compounds.

The concentrate may first be pretreated by leaching or roasting. In scheelite concentrates, hydrochloric acid leaching reduces phosphorus, arsenic, and sulfur contents. Roasting of either scheelite or wolframites eliminates sulfur, arsenic, and organic residues left from the flotation process. Next, the concentrate is digested to extract the tungsten. For lower grade scheelites, the high pressure soda process is commonly employed. The concentrate is first ground to <100 μm (−150 mesh) size, and then digested in an autoclave with an aqueous sodium carbonate solution at ca 200°C and a pressure of >1.2 MPa (ca 11.9 atm).

$$NaCO_3 + CaWO_4 \longrightarrow Na_2WO_4 + CaCO_3$$

The sodium tungstate solution is filtered from the resulting slurry. Similarly, in the alkali roasting process, the concentrate, either scheelite or wolframite, is

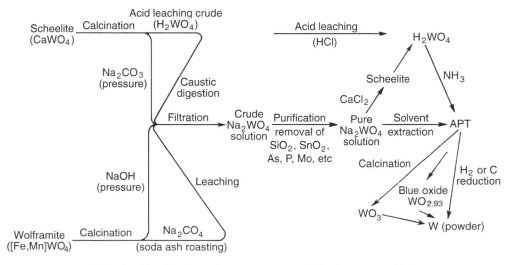

Fig. 1. Production of tungsten materials from ores (17).

heated with sodium carbonate in a rotary kiln at 800°C and then leached using hot water to remove the sodium tungstate. Wolframite ores are also decomposed by reaction with a sodium hydroxide solution at 100°C.

$$(Fe, Mn)WO_4 + 2\, NaOH \longrightarrow (Fe, Mn)(OH)_2 + Na_2WO_4$$

The insoluble hydroxides are removed by filtration. In another process, scheelite is leached using hydrochloric acid:

$$CaWO_4 + 2\, HCl \longrightarrow CaCl_2 + H_2WO_4$$

In this case, the tungstic acid is insoluble and is removed by filtration and washed. For purification, it is digested in aqueous ammonia to give an ammonium tungstate solution. Magnesium oxide is added to precipitate phosphorus and arsenic as magnesium ammonium phosphate and arsenate. Addition of activated carbon removes colloidal hydroxides and silica, which are filtered. Evaporation of the ammonium tungstate solution gives crystals of APT. The evaporation is not carried to completion so that most of the impurities still present remain in the mother liquor and are removed.

The sodium tungstate from the soda and caustic processes is purified by first adding aluminum and magnesium sulfates to remove silicon, phosphorous, and arsenic, then by adding sodium hydrogen sulfide to remove molybdenum and other heavy metals. The pH is controlled and the impurities are removed by filtration. The sodium tungstate is subsequently converted to ammonium tungstate by means of a liquid ion-exchange process. Because this involves the exchange of anions, most of the impurities present as cations are left behind. The ammonium tungstate solution is then evaporated to a fixed specific gravity under constant agitation to give APT crystals.

Reduction to Metal Powder.   The metal powder is obtained from APT by stepwise reduction with carbon or hydrogen. The intermediate products are the yellow oxide, WO_3; blue oxide, W_4O_{11} (see TUNGSTEN COMPOUNDS); and brown oxide, WO_2. Because carbon introduces impurities, hydrogen is preferred. The reduction is carried out in either tube furnaces or rotary furnaces, heated by gas or electricity and having three separately controlled heat zones. A tube furnace consists of multiple tubes 7–9-m long 80–150 mm in diameter. The boats containing the oxide have rectangular or semicircular cross sections and are 375–450-mm long. Both tubes and boats are made of Inconel or a similar heat-resistant alloy (see HIGH TEMPERATURE ALLOYS). The boats are either manually or automatically stoked through the furnaces. Hydrogen is fed through each tube countercurrent to material flow. The hydrogen is recirculated and is scrubbed, purified, and dried, and new hydrogen is added in each cycle. To avoid a buildup of nitrogen in the recirculating system, the ammonia present in the APT must be removed. This is achieved in the calcining to WO_3 and also in the preparation of the blue oxide. For environmental and economic reasons, however, it is easier to collect the ammonia in the closed blue oxide furnace. The bulk of tungsten is produced in the 1990s via the blue-oxide process. A rotary furnace consists of a large tube 3 to >10-m long. It is partitioned into three sections to restrict powder movement down the tube and has longitudinal vanes to carry the powder. The tube is tilted at a small angle and rotated to provide continuous flow of powder through the furnace.

Ammonium paratungstate is decomposed to yellow oxide by heating in air at 800–900°C. The blue oxide is obtained by heating APT to about 900°C in a self-generated atmosphere in the absence of air. Metal powder can be made directly from APT, but particle size is better controlled by using a two-stage reduction and regulating temperature, bed depth, and hydrogen flow. Because rotary furnaces have effectively a shallow bed, they tend to produce fine powders, and tube furnaces are preferred for the final reduction. Temperatures of 600–900°C are used to produce particle sizes of 1–8 μm.

For the production of lamp-filament wire, aluminum, potassium, and silicon dopants are added to the blue oxide. Some dopants are trapped in the tungsten particles upon reduction. Excess dopants are then removed by washing the powder in hydroflouric acid. For welding electrodes and some other applications, thorium nitrate is added to the blue oxide. After reduction, the thorium is present as a finely dispersed thorium oxide.

Consolidation. Because of its high melting point, tungsten is usually processed by powder metallurgy techniques (see POWDER METALLURGY). Small quantities of rod are produced by arc or electron-beam melting.

For rod and wire production, ingots ranging in sizes of 12–25-mm^2 s by 600–900 mm are mechanically pressed at ca 200 MPa (30,000 psi). The bars, which are very fragile, are presintered at 1200°C in hydrogen to increase their strength. Sintering is achieved by electric-resistance heating. The ingot is mounted between two water-cooled contacts inside a water-jacketed vessel containing a hydrogen atmosphere. A current is passed through the ingot, heating it to about 2900°C. This not only sinters the bar to a density of 17.2–18.1 g/cm^3, but also results in considerable purification by volatilization of the impurities. Larger billets for forging or rolling are isostatically pressed. The powder is placed in a plastisol bag, sealed tightly, and placed in a fluid in a high pressure chamber at 200–300 MPa (30,000–45,000 psi). Sintering takes place in an electric-resistance or induction-heated furnace at 2200°C in a hydrogen atmosphere. Densities after sintering are 17.9–18.1 g/cm^3. Small parts are made by mechanically pressing powder to which a lubricant has been added, followed by sintering at 1800–2100°C. For some applications, the sintering temperature can be lowered to 1500°C by the addition of small amounts of nickel or palladium. These, however, embrittle the tungsten.

Metalworking. Tungsten is unusual in that its ductility increases with working. As-sintered or after a full recrystallization anneal, it is as brittle as glass at room temperature. For this reason, tungsten is initially worked at very high temperatures, and large reductions are required to achieve ductility. Furthermore, the low specific heat of tungsten causes it to cool very rapidly. A working operation also requires rapid transfer from furnace to working equipment and frequent reheating during the working operation.

Swaging, the oldest process used for the metalworking of tungsten, is the method used first for the manufacture of lamp wire (Coolidge process). Swaging temperatures start at 1500–1600°C and decrease to ca 1200°C as the bar is worked. These temperatures are just below the recrystallization temperature and the working is, therefore, technically cold working. Reductions per pass start as low as 5% but then increase to as high as 40% as the size decreases. Total reductions of ca 60–80% are typical between anneals. For rod and wire production, rod rolling or, more recently, Kocks rolling is also applied, at least in the initial breakdown stages. For rod rolling, oval-to-square sequences are used with reductions of 15–25% per pass. A Kocks mill consists of 8–12 roll stands in sequence. Each stand consists of three rolls at 120° to each other, producing a hexagonal cross section. The material is rolled through the stands at very high speeds to avoid cooling problems. Wire drawing starts at ca 4 mm and at temperatures as high as 1000°C, decreasing to 500°C for fine wire. The graphite lubricant required must be replaced after each draw pass. Reduction per pass is 35–40% at the start of drawing and drops to 7–10% in fine wire. For drawing dies, tungsten carbide is used to a diameter of 0.25 mm; for smaller diameters, diamonds are used.

For larger-diameter rod or plate, rolling is also employed, with temperatures starting at 1600°C. However, when rolling large cross sections, most equipment is not powerful enough to handle large reductions and, as a result, large center-to-edge variations develop. These variations lead to center bursting or nonuniform structures after annealing. However, large-total-size reductions can overcome this effect. As the material is worked to larger reductions, the temperature is gradually lowered to avoid recrystallization during reheating. Once reductions of >95% are achieved, temperatures can be as low as 300°C.

Forging is also used on tungsten. Hammer forging is generally preferred to press forging because the temperature is better maintained for the higher rate of deformation. Temperature control is also critical here and even more critical for extrusion. Conventional extrusion of glass-coated billets is employed. A rather delicate balance between heat loss in the billet and heat generation in the die is required for good results.

Economic Aspects

The price of tungsten has fluctuated over a wide range since the 1950s, essentially driven by a supply-and-demand relationship, with extraneous factors exaggerating the variations (Table 4). Trading in tungsten ores and intermediates is conducted in units. A unit is one-hundredth of a ton and in countries using the metric system is known as one metric ton unit (MTU). In the United States this is the short ton unit (STU), to distinguish it from the English long

Table 4. Tungsten Prices[a], $/kg[b]

Year	WO$_3$	W
late 1950s	2.42	3.05
1962	0.88	1.10
1977	17.80	22.19
1982	10.63	13.40
1994	4.54	5.65

[a]Ref. 18.
[b]To convert $/kg to STU, multiply by 9.07. The STU contains 9.07 kg WO$_3$ or 7.19 kg W.

ton unit (LTU). The weight of a unit refers to that amount of material that contains the equivalent of 1/100 of a ton of tungsten measured as WO_3, regardless of the actual chemical form. Thus, one STU indicates that amount of a material that contains the equivalent of 20 pounds of WO_3 (15.86-lb W); for one MTU, it would be 10 kilograms of WO_3 (7.928-Kg W). There is a market for tungsten in England and current prices are quoted in the *London Metal Bulletin* (LMB).

Changes in the stockpiling policies of the General Services Administration (GSA) had a considerable influence on tungsten prices in the past, but the GSA has essentially reached its objectives with a stockpile level equivalent to about a three-year supply for the United States, so this factor is no longer a dynamic variable. As shown in Figure 2, there is a wide variation in the price of tungsten over the years. The single most important factor has been the actions of the Chinese producers and the policies of the Chinese government (18). There was a large increase in the export of concentrates from China, which depressed world prices. This was followed by marketing of high quality intermediates at very low prices, comparable to the price for concentrates. The U.S. government reacted by imposing higher import tariffs to control dumping and the Chinese authorities for their part introduced a licensing system to control mine output. These low prices caused the shutdown of almost all mining activity in North America. More recently, the countries of the CIS are exporting intermediates. This trend is helping to stabilize the market, which is still at a level where mines in the United States will remain on a care-and-maintenance status for the foreseeable future.

World production of tungsten is given in Table 5 and salient tungsten statistics in Table 6. The forms of tungsten produced, and the distribution by industry, are given in Table 7. Tungsten carbide products accounting for >65%

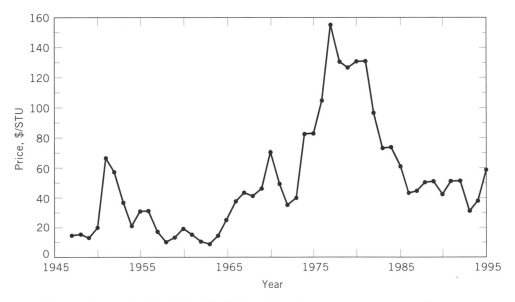

Fig. 2. Mean *London Metal Bulletin* price of tungsten concentrate by year.

Table 5. World Tungsten Concentrate Production by Country[a,b]

Country	Tungsten content[c], t				
	1990	1991	1992	1993	1994
Argentina	6	5 R			
Australia	1,090	237	159 R	23 R	11 E
Austria	1,380	1,310	1,490 R	104 R	
Bolivia	1,010	1,070	851	262 R	450 E
Brazil	316	223	205	245 R	250 E
Burma[d]	443 R	356 R	531 R	524 R	580 E
China[e]	32,000 E	31,800 E	25,000 E	21,600 R	16,500 E
CIS[f]	8,800 E	8,000 E			
Czechoslovakia[g]	84 E	13 E			
India	10 E	11	11	1 R	1 E
Japan	260	279	347	200	
Kazakstan			200 R	150 R	100 E
North Korea	1,000 E	1,000 E	1,000 E	1,000 E	900 E
South Korea	1,360	780	247	200	
Malaysia		2	3	2	
Mexico	183	194	162	160 E	150 E
Mongolia	500 E	300 E	260 E	250 E	250 E
Peru	1,540	1,230	802	398 R	800 E
Portugal	1,410 R	971 R	1,870 R	1,280 RE	1,000 E
Russia			6,500 R	5,000 R	4,000 E
Rwanda	156[h] E	175 E	175 E	175 E	30 E
Spain	10				
Tajikastan			200 E	150 E	100 E
Thailand	290	230	70	18 RE	20 E
Uganda	4 E	4 E	66[i] E	60 E	60 E
United Kingdom	42	9			
Uzekistan			300 E	300 E	300 E
Zaire	17	15 E			
Zimbabwe	1 E	1 E			
Total	*51,900 R*	*48,200 R*	*40,400 R*	*32,000 R*	*25,500*

[a]Ref. 18.

[b]Previously published and 1994 data are rounded off by the U.S. Bureau of Mines to three significant digits and may not add to totals shown. Table includes data available through July 5, 1995.

[c]E = estimated; R = revised.

[d]Includes content of tin–tungsten concentrate.

[e]Based on data in the *Yearbook of Nonferrous Metals Industry of China* in 1992 and 1993.

[f]Dissolved in December 1991.

[g]Dissolved in December 1992.

[h]Estimate based on reported gross weight having a content of 54% W and 68% WO_3.

[i]Reported figure.

of tungsten usage are predicted for a 3.2% annual growth. The main area of growth is expected to be in cutting, wear-resistant, and hard-facing applications.

Analytical Methods

Tungsten is usually identified by atomic spectroscopy. Using optical emission spectroscopy, tungsten in ores can be detected at concentrations of 0.05–0.1%,

Table 6. Salient Tungsten Statistics[a,b]

Use	Tungsten content, t				
	1990	1991	1992	1993	1994
U.S. concentrate					
consumption	5,880[c]	5,310[c]	4,310	2,870[d]	3,630[d]
exports	139	21	38	63	44
imports for consumption	6,420	7,840	2,480	1,720	2,960
stocks[e]					
producer	16	26	44	44	44
consumer	1,080	1,780	702	592	756
APT					
production	6,330[c]	5,860[c]	5,760	4,730[c]	536[f]
consumption	8,790	8,900	7,010	6,970	7,080
stocks[e], consumer and producer	896	578	333	420	82
Primary products					
net production	4,680[g]	8,980	8,450	9,410	7,410
consumption	8,500	7,980	6,910	7,580	8,110
stocks[e]					
producer	1,460[h]	1,670[h]	1,510[h]	1,480[h]	1,160[h]
consumer	793	796	601	716	849
World concentrate[i]					
production	51,900 R	48,200 R	40,400 R	32,000 R	25,500 E
consumption[j]	45,300	41,900 R	38,500 R	31,800 R	29,800 E

[a]Ref. 18.

[b]Previously published and 1994 data are rounded off by the U.S. Bureau of Mines to three significant digits.

[c]Excludes two months of withheld data.

[d]Excludes three months of withheld data.

[e]December 31.

[f]Excludes eleven months of withheld data.

[g]Includes only hydrogen-reduced metal powder and chemicals.

[h]Excludes tungsten carbide, cast and crystalline.

[i]E = estimated; R = revised.

[j]Based on data received from United Nations Conference on Trade and Development in April 1995.

whereas x-ray spectroscopy detects 0.5–1.0%. Scheelite in rock formations can be identified by its luminescence under ultraviolet excitation. In a wet-chemical identification method, the ore is fired with sodium carbonate and then treated with hydrochloric acid; addition of zinc, aluminum, or tin produces a beautiful blue color if tungsten is present.

In the classical method for the quantitative analysis of tungsten in ore concentrates, the ore is digested with acid, the tungsten is complexed with cinchonine, purified, ignited, and weighed. More commonly, x-ray spectrometry is used and its accuracy is enhanced by using tantalum as an internal standard. Plasma spectroscopy determines concentrations as low as 0.1 ppm in solutions

Table 7. Reported Consumption and Stocks of Tungsten Products by End Use in the United States in 1994[a–c]

End use	Ferro tungsten[d]	Tungsten metal powder	Tungsten carbide powder	Tungsten scrap[e]	Other tungsten materials[f]	Total
steel						
stainless and heat resistance	30			5	.	20
alloy	18					19
tool	529			W	W	529
superalloys	W	W	33	199	W	300
alloys[g]						
cutting and wear-resistant materials		97	5640	W	W	5920
other alloys[h]	W	W		W		
mill products made from powder		1200				1200
chemical and ceramic uses		W	i		105	108
miscellaneous					W	W
Total	582	1330	5670[j]	388	141	8110
consumer stocks[k]	20	22	767[j]	40	l	849

[a]Ref. 18.

[b]W = withheld to avoid disclosing company proprietary data, but included in totals.

[c]Previously published; 1994 data are rounded off by the U.S. Bureau of Mines to three significant digits and may not add totals shown.

[d]Includes scheelite, natural and synthetic.

[e]Does not include those used in making primary tungsten products.

[f]Includes tungsten chemicals and others.

[g]Excludes steels and superalloys.

[h]Includes welding and hard-facing rods as well as materials and nonferrous alloys.

[i]Included in tungsten carbide powder cutting and wear-resistant materials.

[j]Based on reported consumption plus information from a secondary sources on companies not canvassed, including estimates.

[k]December 31, 1994.

[l]Included in tungsten scrap.

and 10 ppm in solids. Measurement by atomic absorption is a very rapid method, but not accurate enough for assay-grade analyses. The thiocyanate–tungsten color complex is specific for tungsten and used for colorometric analyses for concentrations in the 0.1–1% range. However, it is not accurate enough for assaying concentrates.

The industry operates by using standardized procedures for testing and characterizing materials. These procedures are published and updated by the American Society for Testing Materials in consultation with interested parties in the industry.

Health and Safety Factors

There are no documented cases of tungsten poisoning in humans. However, numerous cases of pneumoconiosis have been reported in the cemented-carbide industry, but its cause, ie, WC or cobalt, has not been determined. It has been stated that the principal health hazards from tungsten arise from inhalation of aerosols during mining and milling operations. The principal compounds of tungsten to which workers are exposed are ammonium paratungstate, oxides of tungsten (WO_3, W_2O_5, WO_2), metallic tungsten, and tungsten carbide. In the production and use of tungsten carbide tools for machining, exposure to the cobalt used as a binder or cementing substance may be the most important hazard to the health of the employees. Because the cemented tungsten carbide industry uses other metals such as tantalum, titanium, niobium, nickel, chromium, and vanadium in the manufacturing process, the occupational exposures are generally to mixed dust.

Potential occupational exposures to sodium tungstate are found in the textile industry, where the compound is used as a mordant and fireproofing agent, and in the production of tungsten from some of its ores, where sodium tungstate in an intermediate product. Potential exposures to tungsten and its compounds are also found in the ceramics, lubricants, printing inks, paint, and photographic industries.

Permissible Exposure Limits in Air. Occupational exposure to insoluble tungsten needs to be controlled so that employees are not exposed to insoluble tungsten at a concentration greater than 5 mg tungsten/m^3 air, determined as a TWA concentration for up to a 10-h workshift in a 40-h workweek. An STEL value of 10 mg/m^3 has been set by ACGIH in 1983.

Occupational exposure to soluble tungsten must be controlled so that employees are not exposed to soluble tungsten at a concentration greater than 1 mg tungsten/m^3 air, determined as a TWA concentration for up to a 10-h workshift in a 40-h workweek. An STEL value of 3 mg/m^3 has been set by ACGIH in 1983.

Occupational exposure to dust of cemented carbide that contains more than 2% cobalt must be controlled so that employees are not exposed at a concentration greater than 0.1 mg cobalt/m^3 air, determined as a TWA concentration for up to a 10-h workshift in a 40-h workweek.

Occupational exposure to dust of cemented carbide that contains more than 0.3% nickel needs to be controlled so that employees are not exposed at a concentration greater than 15 μg nickel/m^3 air, determined as TWA concentration for up to a 10-h workshift in a 40-h workweek.

Hazards encountered with tungsten may be caused by substances associated with the production and use of tungsten, eg, As, Sb, Pb, and other impurities in tungsten ores, Co aerosols and dust in the carbide industry, and thoria used in welding electrodes. Lanthanum is being promoted as a substitute for thoria in this application.

Uses

Tungsten is used in four forms: as tungsten carbide, as an alloy additive, as essentially pure tungsten, and as tungsten chemicals. Tungsten carbide, because of its high hardness at high temperatures, is used for cutting tools, abrasion-resistant surfaces, and forming tools. This application accounts for ca 65% of tungsten usage, mostly in the form of cemented carbides. Tungsten carbide is produced by the reaction of tungsten powder with carbon black at 1500°C. The tungsten carbide is then milled and blended with 3–25% cobalt and pressed and sintered at ca 1400°C. Addition of tantalum and titanium carbides improves hardness and cratering resistance in steel-cutting applications. Cemented carbides are used for cutting tools, mining and drilling tools, forming and drawing dies, bearings, and numerous other wear-resistant applications.

About 16% of tungsten usage is as an alloy additive. Tungsten added to steels forms a dispersed tungsten carbide phase that imparts a finer-grain structure and increases the high temperature hardness. The finer-grain size improves toughness and produces a more durable cutting edge. For this type of application, up to 3% tungsten is used, usually with 1–4% chromium. For hot-work tool steels, up to 18% tungsten is added. Such steel, when quenched from a very high temperature and tempered, retains its hardness up to red heat. Tungsten is also used as an additive to nickel and cobalt-base superalloys. Here again, tungsten imparts high temperature strength and wear resistance (see HIGH TEMPERATURE ALLOYS).

Metallic tungsten accounts for 16% of tungsten consumption. Frequently, tungsten is used because of its high melting point and low vapor pressure. The best known use is the manufacture of lamp filaments, where potassium, silicon, and aluminum dopants are added to the oxide. After sintering, about 60-ppm potassium is retained in small voids. During working, these voids are stretched into long stringers. After heating, the voids form rows of tiny bubbles that control grain boundary movement, resulting in grains that are much longer than the wire diameter and whose boundaries are at a small angle to the wire axis. This structure is very creep-resistant and allows the coil to maintain its shape throughout life. Tungsten is widely employed as an electron emitter because it can be used at very high temperatures. Thoria is added to reduce the work function and improve emission. It also improves arc stability and gives longer life to welding electrodes. Tungsten is used as the target in high intensity x-ray tubes. Other high temperature applications include furnace elements, heat shields, vacuum metallizing coils and boats, glass-melting equipment, and arc-lamp electrodes. Because of its high elastic modulus and wear resistance, tungsten is used in high modulus and wear-resistance applications as well as in high speed impact printers. Its low thermal expansion makes tungsten ideal for glass-to-metal seals and as a base for silicon semiconductors. There are few tungsten-base alloys. W–Re alloys are used for thermocouple wire and for shock-resistant lamp filaments. Alloys of tungsten with various combinations of iron, nickel, and copper are called heavy alloys, which have the high density of tungsten and are easier to machine. These are used as counterweights, armor-piercing penetrator cores, x-ray shielding, gyroscope rotors, dart bodies, and other high density applications. A similar material is made by infiltrating porous tungsten with copper or silver. Such alloys are used as electrical contact materials and rocket nozzles. Composite materials with barium and strontium compounds are used in electron-emitting devices.

Nonmetallurgical uses include brilliant organic tungsten dyes and pigments that can be used in a variety of materials. Tungstates are used as phosphors in fluorescent lights, cathode-ray tubes, and x-ray screens, whereas tungsten compounds are used as catalysts in petroleum refining.

Recycling. In more recent years, processes that can convert used carbide cutting tools and used tungsten alloy penetrators back into powdered form that can be used directly into new products have been developed. It is estimated that in 1996 ca 25% of cutting inserts used in the United States were recycled in this way.

Recovery of used carbide tools by immersing them in molten zinc has been proposed in the United Kingdom (19). Cobalt and zinc are mutually soluble. By penetration of the zinc into cemented carbides, there is an increase in volume. On cooling, the carbides are left dispersed in the Co–Zn alloy. The carbides are recovered by acid-leaching the Co–Zn. The U.S. Bureau of Mines has demonstrated an improvement on this process by removing the zinc using vacuum distillation (20). This allows recovery of both carbides and cobalt as fine powders. Initially, the driving incentive for this process was recovery of high tantalum-containing materials, but later cobalt and tungsten prices increased such that they were of equal importance for recovery. Residual traces of zinc are released during vacuum sintering of cemented carbides made with recovered powders. This can be troublesome when a buildup of zinc occurs in the furnace. Teledyne Advanced Materials further developed this process on a commercial basis by achieving zinc levels in the low ppm range (<30 ppm). The fact that the materials were vacuum-sintered in their original form where certain impurities are removed leads to lower impurity levels in the recovered powders. There is a slight oxidation or loss of carbon that must be compensated, otherwise the recycled powder is not in any way inferior to the original.

High density tungsten alloy machine chips are recovered by oxidation at about 850°C, followed by reduction in hydrogen at 700–900°C. Typically, the resultant powders are about 3-μm grain size and resinter readily. There can be some pickup of refractory materials used in furnace construction, which must be controlled. This process is important commercially. For materials that may be contaminated with other metals or impurities, the preferred recovery process is the wet chemical conversion process used for recovery of tungsten from ores and process wastes. Materials can always be considered for use as additions in alloy steel melting.

BIBLIOGRAPHY

"Tungsten and Tungsten Alloys" in *ECT* 1st ed., Vol. 14, pp. 353–362, by B. Kopelman, Sylvania Electric Products, Inc.; "Wolfram" in *ECT* 2nd ed., Vol. 22, pp. 334–346, by B. Kopelman and J. S. Smith, Sylvania Electric Products, Inc.; "Tungsten and Tungsten Alloys" in *ECT* 3rd ed., Vol. 23, pp. 413–425, by J. A. Mullendore, GTE Product Corp.

1. *Handbook of Chemistry and Physics*, 67th ed., CRC Press, Inc., Boca Raton, Fla., 1987.
2. P. T. Stafford, *Tungsten*, U.S. Bureau of Mines, Washington, D.C., 1980.
3. Comité International des Poids et Mesures, *Metrologia*, **5**(2), 35 (1969).

4. A. Cezairhyan, *High Temp. Sci.* **4**, 248 (1972).

5. D. R. Stull and H. Prophet, *JANAF Thermochemical Tables*, 2nd ed., NBS, Washington, D.C., 1971.

6. *Thermophysical Properties of Matter*, Vol. 4, *Thermal Expansion*, Plenum Publishing Corp., New York, 1970.

7. W. Shaner, G. R. Gathers, and C. Minichino, *High Temp. High Pressures*, **8**, 425 (1976).

8. E. R. Plante and A. B. Sessions, *J. Res. Nat. Bur. Stand.* **77A**, 237 (1973).

9. C. Y. Ho, R. W. Powell, and P. E. Liley, *J. Phys. Chem. Ref. Data*, **3**, 1 (1974).

10. J. G. Hust, *High Temp. High Pressures*, **8**, 377 (1976).

11. R. F. Taylor, *High Temp. High Pressures*, **4**, 59 (1972).

12. G. D. Rieck, *Tungsten and Its Compounds*, Pergamon Press, London, U.K., 1967.

13. C. J. Smithells, *Tungsten*, Chemical Publishing Co., New York, 1953.

14. S. W. H. Yih and C. T. Wang, *Tungsten*, Plenum Publishing Corp., New York, 1979.

15. I. Barin and O. Knacke, *Thermochemical Properties of Inorganic Substances*, Springer-Verlag, New York, 1973.

16. O. Kubaschewski and C. B. Alcock, *Metallurgical Thermochemistry*, 5th ed., Pergamon Press, New York, 1979.

17. M. J. Hudson, *Chem. Br.* **18**, 438 (1982).

18. *Mineral Industries Surveys, Tungsten Annual Review*, U.S. Department of Interior Bureau of Mines, Washington, D.C., 1994.

19. U.S. Pat. 2,407,752 (1946) (to Trent Powder Alloys Ltd.).

20. U.S. Pat. 3,595,484 (1971) (to Barnard Bureau of Mines).

General References

M. Hoch. *High Temp. High Pressures*, **1**, 531 (1969).

Mineral Commodity Summaries 1981, U.S. Bureau of Mines, Washington, D.C., 1981.

C. C. Clark and J. B. Sutliff, *Am. Metal Market*, (Jan. 23, 1981).

Handbook of Toxic and Hazardous Chemicals and Carcinogens, 2nd ed., M. Sittig Marshall Publishers, Park Ridge, N.J.

THOMAS W. PENRICE
Consultant

TUNGSTEN COMPOUNDS

Tungsten is a Group VIB transition element having an atomic number of 74 and a valence state of 0, +2, +3, +4, +5, or +6 in compounds. However, tungsten alone has not been observed as a cation. Its most stable, and therefore most common, valence state is +6. Tungsten complexes vary widely in stereochemistry and oxidation states. Complex formation is exemplified by the large number of polytungstates. Simple tungsten compounds, such as the halides, are also known.

The chemical uses of tungsten have increased substantially in more recent years. Catalysis (qv) of photochemical reactions and newer types of soluble organometallic complexes for industrially important organic reactions are among the areas of these new applications.

Tungsten Hexacarbonyl. Tungsten hexacarbonyl [14040-11-0], W(CO)$_6$, may be prepared in yields >90% by the aluminum reduction of tungsten hexachloride [1283-01-7] in anhydrous ether under a pressure of 0.1 MPa (ca 1 atm) of carbon monoxide at 70°C. It is purified by sublimation or steam distillation. A colorless to white solid, tungsten hexacarbonyl decomposes without melting at ca 150°C, although it sublimes *in vacuo*. It is a zero-valent monomeric compound, having a relatively low vapor pressure of 13.3 Pa (0.1 mm Hg) at 20°C and 160 Pa (1.20 mm Hg) at 67°C, and is fairly stable in air, water, or acid, but is decomposed by strong bases and attacked by halogens. Tungsten carbonyl is slightly soluble in organic solvents but insoluble in water (see CARBONYLS).

Various applications such as lubricant additives, dyes, pigments, and catalysts are under investigation. Tungsten can be deposited from tungsten hexacarbonyl, but carbide formation and gas-phase nucleation present serious problems (1,2). As a result, tungsten halides are the preferred starting material.

Tungsten Halides and Oxyhalides. Tungsten forms binary halides for all oxidation states between +2 and +6; oxyhalides are only known for oxidation states +5 and +6. In general, tungsten halogen compounds are reactive toward water and oxygen in the air and must therefore be handled in an inert atmosphere. These are all solid, colored compounds at room temperature, except the fluorides, and many decompose on heating before melting. The hexachloride and hexafluoride [7783-82-6] are commercially available and are particularly suitable starting materials for the chemical vapor deposition of tungsten, which is an important process technique for coatings and free-standing parts such as thin-walled tubing. The resulting structure is generally columnar, but more recently a method has been described for obtaining a fine-grained, noncolumnar tungsten structure (3).

Fluorides. Tungsten hexafluoride [7783-82-6], WF$_6$, is a colorless gas at room temperature, sp gr 12.9 with respect to air. At 17.5°C, it condenses into a pale-yellow liquid, and at 2.5°C, a white solid is formed. It may be prepared by treating hydrogen fluoride, arsenic trifluoride, or antimony pentafluoride with tungsten hexachloride or by direct fluoridation of tungsten:

$$WCl_6 + 6\,HF \longrightarrow WF_6 + 6\,HCl$$

$$WCl_6 + 2\,AsF_3 \longrightarrow WF_6 + 2\,AsCl_3$$

$$WCl_6 + 3\,SbF_5 \longrightarrow WF_6 + 3\,SbF_3Cl_2$$

$$W + 3\,F_2 \longrightarrow WF_6$$

Direct fluoridation of pure tungsten in a flow system at atmospheric pressure at 350–400°C is the most convenient procedure (4). Tungsten hexafluoride is extremely unstable in the presence of moisture and hydrolyzes completely to tungstic acid [7783-03-1]:

$$WF_6 + 4\,H_2O \longrightarrow H_2WO_4 + 6\,HF$$

Tungsten hexafluoride dissolves in benzene or cyclohexane to give a bright red color, in dioxane a pale red, and in ether a violet-brown.

Tungsten pentafluoride [*19357-83-6*], WF_5, is prepared by the reduction of the hexafluoride on a hot tungsten filament in almost quantitative yield (5).

Tungsten tetrafluoride [*13766-47-7*], WF_4, is a nonvolatile, hygroscopic, reddish-brown solid. It has been prepared in low yields by the reduction of the hexafluoride with phosphorus trifluoride in the presence of liquid anhydrous hydrogen fluoride at room temperature (6).

Tungsten oxytetrafluoride [*13520-79-1*], WOF_4, mp 110°C, bp 187.5°C, forms colorless plates. It is prepared by the action of an oxygen–fluorine mixture on the metal at elevated temperatures (7). The compound is extremely hygroscopic and decomposes to tungstic acid in the presence of water.

Tungsten oxydifluoride [*14118-73-1*], , WO_2F_2, is a white solid prepared by the hydrolysis of WOF_4 (8). Its chemistry has not been investigated.

Chlorides. Tungsten hexachloride [*13283-01-7*], WCl_6, mp 275°C, bp 346.7°C, is a blue-black crystalline solid. It is prepared by the direct chlorination of pure tungsten in a flow system at atmospheric pressure at 600°C. Solidification usually occurs without incident, but further cooling may result in a violent, explosion-like expansion of the solid mass at 168–170°C. This phenomenon may be associated with an $\alpha_2 \rightarrow \alpha_1$ transition. However, tungsten hexachloride may be safely cooled if it occupies not more than one-half of the containing vessel. In the presence of moisture or oxygen, some $WOCl_4$ is formed as an impurity. Tungsten hexachloride is very soluble in carbon disulfide but decomposes in water to form tungstic acid. The hexachloride is easily reduced by hydrogen to the lower halides and finally to the metal itself (2).

Tungsten pentachloride [*13470-13-8*], WCl_5, mp 243°C, bp 275.6°C, is a black, crystalline, deliquescent solid. It is only slightly soluble in carbon disulfide and decomposes in water to the blue oxide, $W_{20}O_{58}$. Magnetic properties suggest that tungsten pentachloride may contain trinuclear clusters in the solid state, but this structure has not been defined. Tungsten pentachloride may be prepared by the reduction of the hexachloride with red phosphorus (9).

Tungsten tetrachloride [*13470-14-9*], WCl_4, is obtained as a coarse, crystalline, deliquescent solid that decomposes upon heating. It is diamagnetic and may be prepared by the thermal-gradient reduction of WCl_6 with aluminum (10).

Tungsten dichloride [*13470-12-7*], WCl_2, is an amorphous powder. It is a cluster compound and may be prepared by the reduction of the hexachloride with aluminum in a sodium tetrachloroaluminate melt (11).

Tungsten oxytetrachloride [*13520-78-0*], $WOCl_4$, mp 211°C, bp 327°C, is a red crystalline solid. It is soluble in carbon disulfide and benzene and is decomposed to tungstic acid by water. It may be prepared by refluxing sulfurous oxychloride, $SOCl_2$, on tungsten trioxide (12) and purified after evaporation by sublimation.

Tungsten oxydichloride [*13520-76-8*], WO_2Cl_2, a pale-yellow crystalline solid having an mp of 266°C, is soluble in cold water and in alkaline solution, although partly decomposed by hot water. It is prepared by the action of carbon tetrachloride on tungsten dioxide at 250°C in a bomb (13).

Tungsten oxytrichloride [*14249-98-0*], $WOCl_3$, a green solid, is prepared by the aluminum reduction of $WOCl_4$ in a sealed tube at 100–140°C (14).

Bromides. Tungsten hexabromide [13701-86-5], WBr_6, bluish-black crystals having an mp of 232°C, is formed by metathetical exchange reaction of BBr_3 with tungsten hexachloride (15).

Tungsten pentabromide [13470-11-6], WBr_5, violet-brown crystals having an mp of 276°C and a bp of 333°C, is extremely sensitive to moisture. It is prepared by the action of bromine vapor on tungsten at 450–500°C (16).

Tungsten tetrabromide [12045-94-2], WBr_4, black orthorhombic crystals, is formed by the thermal-gradient reduction of WBr_5 with aluminum, similar to the reduction of WCl_4 (10).

Tungsten tribromide [15163-24-3], WBr_3, prepared by the action of bromine on WBr_2, in a sealed tube at 50°C (17), is a thermally unstable black powder that is insoluble in water.

Tungsten dibromide [13470-10-5], WBr_2, formed by the partial reduction of the pentabromide with hydrogen, is a black powder that decomposes at 400°C.

Tungsten oxytetrabromide [13520-77-9], $WOBr_4$, black, deliquescent needles having an mp of 277°C and a bp of 327°C, is formed by the action of carbon tetrabromide on tungsten dioxide at 250°C (13).

Tungsten oxydibromide [13520-75-7], WO_2Br_2, light-red crystals, is formed by passing a mixture of oxygen and bromine over tungsten at 300°C.

Iodides. Tungsten tetraiodide [14055-84-6], WI_4, is a black powder that is decomposed by air. It is prepared by the action of concentrated hydriodic acid on tungsten hexachloride at 100°C.

Tungsten triiodide [15513-69-6], WI_3, is prepared by the action of iodine on tungsten hexacarbonyl in a sealed tube at 120°C (18).

Tungsten diiodide [13470-17-2], W_6I_2, sp gr 6.79, is a brownish crystalline substance. It can be prepared by the reaction between iodine and $W(CO)_6$ in a nitrogen atmosphere. In this reaction, $W(CO)_6$ and iodine gas first form WI_x, where x is ~3.2 at 250°C. At 600°C, WI_x and iodine gas form W_6I_{12} (19).

Tungsten oxydiiodide [14447-89-3], WO_2I_2, is prepared by heating a mixture of tungsten and tungsten trioxide with excess iodine in a 500–700°C temperature gradient for 36 h (20,21).

Oxides, Acids, and Salts. Tungsten oxides form a series of well-defined ordered phases to which precise stoichiometric formulas can be assigned (22–29) (Table 1). The composition of the tungsten oxides may vary over a fixed range without change in crystalline structure. Thus, the homogeneity ranges are represented by $WO_{2.95-3.0}$, $WO_{2.88-2.92}$ [12165-57-0], $WO_{2.664-2.776}$, and $WO_{1.99-2.02}$. Each tungsten atom is octahedrally surrounded by six oxygen atoms. In WO_3, these WO_6 units are joined through sharing of corner oxygen atoms only. As the

Table 1. Tungsten Oxides

Oxide	CAS Registry Number	Phase	O:W, average	Theoretical density, gm/cm³	Color
WO_3	[1314-35-8]	α	3.00	7.29	yellow
$W_{20}O_{58}$	[12037-58-0]	β	2.90	7.16	blue-violet
$W_{18}O_{49}$	[12037-57-9]	γ	2.72	7.78	reddish-violet
WO_2	[12036-22-5]	δ	2.00	10.82	brown
W_3O	[39368-90-6]	(β-W)	0.33	14.4	gray

oxygen-to-tungsten ratio decreases, however, the WO_6 units become more intricately joined in combinations of corners, edges, and faces to form chains and slabs. The loss of each oxygen atom from the oxide lattice means that two electrons are added to the conduction band of the lattice, and it is meaningless to speak of pentavalent and tetravalent tungsten atoms in such a lattice.

Tungsten trioxide [1314-35-8], WO_3, is a yellow powder. However, the smallest diminution of oxygen brings about a change in color. Tungsten trioxide, which is pseudorhombic at room temperature but tetragonal above 700°C, is usually prepared from tungstic acid or tungstates. It is the most important tungsten oxide and is the starting material for the production of tungsten powder. Tungsten trioxide is reduced to the metal by carbon above 1050°C and by hydrogen as low as 650°C. At lower temperatures, intermediate oxides are formed. Tungsten trioxide is insoluble in water and in all acid solutions except hydrofluoric, but gives tungstate with strong alkali.

$$2\,NaOH + WO_3 \longrightarrow Na_2WO_4 + H_2O$$

When heated in a hydrogen chloride atmosphere, WO_3 is completely volatilized at ca 500°C, forming the oxydichloride, WO_2Cl_2.

Tungsten dioxide [12036-22-5], WO_2, is a brown powder formed by the reduction of WO_3 with hydrogen at 575–600°C. Generally, this oxide is obtained as an intermediate in the hydrogen reduction of the trioxide to the metal. On reduction, first a blue oxide, then a brown oxide (WO_2), is formed. The composition of the blue oxide was in doubt for a long time. However, it has since been resolved that $W_{20}O_{58}$ and W are formed as intermediates, which may also be prepared by the reaction of tungsten with WO_3.

The oxide W_3O is regarded as both an oxide and a metal phase. It is gray and has a density of 14.4 g/cm^3, and is prepared by the electrolysis of fused mixtures of WO_3 and alkali-metal phosphates. At ca 700°C, it decomposes into W and WO_2; β-tungsten is W_3O.

Tungsten Bronze. Tungsten bronzes (30,31) constitute a series of well-defined nonstoichiometric compounds of the general formula $M_{1-x}WO_3$, where x is a variable between 0 and 1, and M is some other metal, generally an alkali metal, although many other metals can also be substituted.

The systems most extensively investigated are the sodium tungsten bronzes. These compounds are intensely colored, ranging from golden-yellow to bluish-black, depending on the value of x, and, in crystalline form, exhibit a metallic sheen. The compounds have a positive temperature coefficient of resistance for Na:WO_3 ratios of >0.3, and a negative temperature coefficient of resistance at lower ratios. Sodium tungsten bronzes are inert to chemical attack by most acids, but may be dissolved by basic reagents. Sodium tungsten bronzes serve as promoters for the catalytic oxidation of carbon monoxide and reformer gas in fuel cells (32) (see BATTERIES, SECONDARY CELLS). In general, these bronzes form cubic or tetragonal crystals, the lattice constants of which increase with sodium concentration. They are prepared by electrolytic reduction, vapor-phase deposition, fusion, or solid-state reaction (33,34). The latter method is the most versatile, in which the reagents are finely ground and heated at 500–850°C in vacuum for prolonged periods of time.

Tungsten Blue. The mild reduction, eg, by Sn (11), of acidified solutions of tungstates, tungsten trioxide, or tungstic acid in solutions gives intense blue products, which are referred to by the general name of tungsten blues, and which resemble molybdenum blues in many respects. Tungsten trioxide acquires a bluish tint merely on exposure to underwater ultraviolet radiation. If hydrogen is produced in a tungstate solution by means of zinc and hydrochloric acid, blue precipitates that are stable in air can form. These are believed to be hydrogen analogues of the tungsten bronzes. These blue hydrogen tungsten bronzes, $H_{1-x}WO_3$, are prepared by the wet reduction of tungstic acid and are structurally related to the alkali tungsten bronzes (35–37). Tungsten blues have a strong tendency to form colloids.

Tungstic Acid and Tungstates. Tungstic acid [7783-03-1], H_2WO_4 or $WO_3 \cdot H_2O$, is an amorphous yellow powder that is practically insoluble in water or acid solution, but dissolves readily in a strongly alkaline medium. It may be precipitated from hot tungstate solutions with strong acids. However, if the tungstate solution is acidified in the cold, a white voluminous precipitate of hydrated tungstic acid forms, which has the formula $WO_3 \cdot xH_2O$, where x is ca 2. This is converted to the yellow form by boiling in an acid medium. Both the yellow and white forms tend to become colloidal on washing. Tungstic acid forms a series of stable salts of the types $M(I)_2WO_4$, $M(II)WO_4$, and $M(III)_2(WO_4)_3$, of which some also exist in the hydrated form. Except for tungstates of the alkali metals and magnesium, these salts are generally sparingly soluble in water. They are decomposed by hot mineral acids (except phosphoric) to tungstic acid. The insoluble tungstates are prepared by adding a sodium tungstate solution to a solution of the appropriate salt. Some properties of these tungstates are given in Table 2.

Ammonium tungstate [11140-77-5], $(NH_4)_2WO_4$, cannot be obtained from an aqueous solution because it decomposes when such a solution is concentrated. It is prepared by the addition of hydrated tungstic acid to liquid ammonia.

Table 2. Properties of Normal Tungstates

Compound	CAS Registry Number	Properties	Specific gravity
$BaWO_4$	[7787-42-0]	colorless, tetragonal, $a = 0.564$ nm, $c = 1.270$ nm	5.04
$CdWO_4$	[7790-85-4]	yellow, rhombic	
$CaWO_4$	[7790-75-2]	white, tetragonal, $a = 0.524$ nm, $c = 1.138$ nm, $n_D^{20} = 1.9263$	6.06
$Ce_2(WO_4)_3$	[52345-28-5]	yellow, monoclinic, $a = 1.151$ nm, $b = 1.172$ nm, $c = 0.782$ nm, $\beta = 109° 48'$, mp 1089°C	6.77
$PbWO_4$	[7759-01-5]	colorless, monoclinic, mp 1123°C	8.46
Ag_2WO_4	[13465-93-5]	pale yellow	
Na_2WO_4	[13472-45-2]	white, rhombic, mp 698°C	4.179
$Na_2WO_4 \cdot 2H_2O$	[10213-10-2]	white, rhombic, loses 2 H_2O at 100°C	3.245
$SrWO_4$	[13451-05-3]	white, tetragonal, $a = 0.540$ nm, $c = 1.190$ nm	6.187

Anhydrous sodium tungstate, Na_2WO_4, is prepared by fusing tungsten trioxide in the proper proportion with sodium hydroxide or sodium carbonate:

$$WO_3 + 2\,NaOH \longrightarrow Na_2WO_4 + H_2O$$

$$WO_3 + Na_2CO_3 \longrightarrow Na_2WO_4 + CO_2$$

On crystallization from aqueous solution, the dihydrate is generally obtained. The tungstates are of particular interest in electronic and optical applications, but are also used for ceramics, catalysts, pigments, corrosion, and as fire inhibitors, etc.

Polytungstates. An important and characteristic feature of the tungstate ion is its ability to form condensed complex ions of isopolytungstates in acid solution (38). As the acidity increases, the molecular weight of the isopolyanions increases until tungstic acid precipitates. However, the extensive investigations on these systems have been hampered by lack of well-defined solid derivatives.

The chemistry of tungsten in solution has been studied by chromatography and spectroscopy (39,40). Much of the reported work concerns the existence of tungstate species in acid solutions, with particular reference to the molar ratio of soluble tungstate species.

If polytungstates are considered as formed by the addition of acid to WO_4^{2-}, then a series of isopolytungstates appears, in which the degree of aggregation in solution increases with decreasing pH. The relationships of the species, in order of increasing $H_3O^+:WO_4^{2-}$ ratio, are shown in Table 3.

Metatungstates of the alkali, alkaline-earth, rare-earth, and transition metals have been reported. However, classical synthesis rarely gives high yields of the pure compounds. The rare-earth tungstates, eg, $La_2(H_2W_{12}O_{40})\cdot xH_2O$, may be prepared by the action of lanthanide carbonates on metatungstic acid, $H_6(H_2W_{12}O_{40})$ [12299-86-4]. Other salts are prepared by the action of carbonates or sulfates of the corresponding metal on metatungstic acid or metatungstates. Generally, these compounds are heat-sensitive and should be recovered by freeze drying. Alkali metal and ammonium metatungstates, $M_6(H_2(W_3O_{10})_4)\cdot xH_2O$, may be prepared by the digestion of hydrated tungsten trioxide with the corresponding base (42–47). These salts are generally known for their high solubility in water; the most important is ammonium metatungstate [12028-48-7], $(NH_4)_6(H_2W_{12}O_{40})$.

Table 3. Polytungstates in Order of Increasing Ratio of $H_3O^+:WO_4^{2-a}$

$H_3O^+:WO_{2-4}^-$	Polytungstate	CAS Registry Number	Common name
0.333	$W_{12}O_{46}^{20-}$		para Z
0.667	$W_3O_{11}^{4-}$	[39898-14-1]	tritungstate
	$H_4W_3O_{13}^{4-}$		
1.167	$H_{10}W_{12}O_{46}^{10-}$	[12401-49-9]	para B
	$HW_6O_{21}^{5-}$	[11080-77-6]	para A
1.33	$W_{12}O_{40}^{8-}$		
1.50	$H_2W_{12}O_{40}^{6-}$	[12207-61-3]	meta
	$H_3W_6O_{21}^{3-}$	[12273-48-2]	pseudo meta
2.0	$WO_3\cdot H_2O(H_2WO_4)$	[7783-03-1]	tungstic acid

^{a}Ref. 41.

The paratungstates are generally crystallized from slightly basic solutions. By far the most important salt is ammonium paratungstate [1311-93-9], $(NH_4)_{10}W_{12}O_{40}\cdot 5H_2O$, usually known as the heavy form of commercial ammonium paratungstate. It is usually formed by crystallization from a boiling solution. However, if crystallization is allowed to take place slowly at room temperature, an undecahydrate [12383-34-5], $(NH_4)_{10}W_{12}O_{40}\cdot 11H_2O$, is formed. This hydrate is known as the light form of ammonium paratungstate. Both forms are insoluble in water and decompose in acid or alkali. They are reduced to the metal by heating in a hydrogen atmosphere. Ammonium paratungstate is widely used as a catalyst. Peroxytungstic acid [41486-83-3], $H_2WO_2(O_2)_2$, and peroxytungstates are known but tend to be unstable; the instability increases with increasing ratio of oxygen to tungsten (48).

Heteropolyanions are closely related to the isopolyanions. Over thirty elements are known to function as the heteroatom, with many stoichiometric ratios between the heteroatom and the anion. Both the acids and the salts are known and are usually hydrated when crystallized from aqueous solutions. As a class, heteropoly compounds are characterized by a number of properties independent of the heteroatom and the metallic component. Typically heteropoly tungsten compounds show the following characteristics: high molecular weight, usually >3000; high degree of hydration; unusually high solubility in water and some organic solvents; strong oxidizing action in aqueous solution; strong acidity in free acid form; decomposition in strongly basic aqueous solutions to give normal tungstate solutions; and highly colored anions or colored reaction products. Heteropoly anions may be classified according to the ratio of the number of central atoms to tungsten, as shown in Table 4.

Structures of heteropolytungstate and isopolytungstate compounds have been determined by x-ray diffraction. The anion structures are represented by polyhedra that share corners and edges with one another. Each W is at the center of an octahedron, and an O atom is located in each vertex of the octahedron. The central atom is similarly located at the center of an XO_4 tetrahedron or XO_6

Table 4. Principal Species of Heteropolytungstates

Ratio of hetero atoms to W atoms	Principal central atoms, X	Typical formulas	Structure by x-ray
1:12	P^{5+}, As^{5+}, Si^{4+}, Ge^{4+}, Ti^{4+}, Co^{3+}, Fe^{3+}, Al^{3+}, Cr^{3+}, Ga^{3+}, Te^{4+}, B^{3+}	$[X^{n+}(W_{12}O_{40})]^{(8-n)-}$	known
1:10	Si^{4+}, Pt^{4+}	$[X^{n+}(W_{10}O_x)]^{(2x-60-n)-}$	unknown
1:9	Be^{2+}	$[X^{2+}(W_9O_{31})]^{6-}$	unknown
1:6	series A: Te^{6+}, I^{7+}	$[X^{n+}(W_6O_{24})]^{(12-n)-}$	isomorphous with six molybdates
	series B: Ni^{2+}, Ga^{3+}	$[X^{n+}(W_6O_{24}H_6)]^{(6-n)-}$	known
2:18	P^{5+}, As^{5+}	$[X_2^{n+}(W_{18}O_{62})]^{(12-n)-}$	known
2:17	P^{5+}, As^{5+}	$[X_2^{n+}(W_{17}O_x)]^{(2x-102-2n)-}$	unknown
$1m:6m^a$	As^{3+}, P^{3+}	$[X^{n+}(W_6O_x)]_m^{m(2x-36-n)-}$	unknown

$^a m$ = unknown.

octahedron. Each such polyhedron containing the central atom is generally surrounded by WO_6 octahedra, which share corners, edges, or both with it and with one another. Thus, the correct total number of oxygen atoms is utilized. Each WO_6 octahedron is directly attached to a central atom through a shared oxygen atom. In the actual structures, the octahedra are frequently distorted. The oxygens are relatively large spheres, and practically all space within the anion structure is taken up by the bulky oxygens that are close-packed or nearly so. When the large heteropolytungstate anions are packed together as units in a crystal, the interstices between the anions are very large compared to water molecules or most simple cations. In most compounds, there is apparently no direct linkage between the individual heteropoly anions, eg, in the structures of $K_6CoW_{12}O_{40}\cdot20H_2O$ [37346-54-6] and $K_6P_2W_{18}O_{62}$ [60748-58-5]. Instead, the complexes are joined by hydrogen bonding through some molecules of water of hydration. These principles are illustrated in the crystal structure of $H_3PW_{12}O_{40}\cdot xH_2O$ [12501-23-4], as determined by x-ray diffraction (49,50).

Heteropoly salts of large cations, eg, cesium, frequently crystallize as acid salts regardless of the ratio of cations to anions in the mother liquor. Furthermore, salts of these cations are frequently less highly hydrated than salts of smaller cations. Apparently, the larger cations take up so much of the space between the heteropoly anions that there is less room for water. There is often not enough room for the large cations required to form a normal salt. Instead, solvated hydrogen ions fill in to balance the negative charge of the anions, and a crystalline acid salt results.

Commercially, heteropolytungstates, particularly the heteropolytungstates, are produced in large quantities as precipitants for basic dyes, with which they form colored lakes or toners (see also DYES AND DYE INTERMEDIATES). They are also used in catalysis, passivation of steel, etc.

Sulfides.   Tungsten disulfide [12138-09-9], WS_2, although found in nature, is usually prepared by heating tungsten powder with sulfur at 900°C. It is a soft, grayish-black powder, relatively inert and unreactive, with sp gr of 7.5. This disulfide is insoluble in water, hydrochloric acid, alkali, and organic solvents or oils, and decomposes in hot, strong oxidizing agents, eg, aqua regia, concentrated sulfuric acid, and nitric acid. Heating in air or in the presence of oxygen yields WO_3. However, its thermal stability in air is ca 90°C higher than that of MoS_2.

Tungsten disulfide forms adherent, soft, continuous films on a variety of surfaces and exhibits good lubricating properties similar to molybdenum disulfide and graphite (51) (see also LUBRICATION AND LUBRICANTS). It is also reported to be a semiconductor (qv).

Tungsten trisulfide [12125-19-8], WS_3, is a chocolate-brown powder, slightly soluble in cold water, but readily forming a colloidal solution in hot water. It is prepared by treating an alkali-metal thiotungstate with HCl (52). Tungsten trisulfide is soluble in alkali carbonates and hydroxides.

Tungsten forms thiotungstates corresponding to the tungstates, but one, two, three, or all of the oxygen atoms are replaced by sulfur. These compounds form with solutions of the alkali or alkaline-earth tungstates saturated with hydrogen sulfide. They vary in color from pale yellow to yellowish brown and, in general, crystallize well. Acidifying a solution of these salts precipitates tungsten trisulfide.

Potassium tetrathiotungstate [14293-75-5], K_2WS_4, forms yellow rhombic crystals that are soluble in water. Ammonium tetrathiotungstate [13862-78-7], $(NH_4)_2WS_4$, forms bright orange crystals that exhibit a metallic iridescence. These crystals are stable in dry air and soluble in water. Ammonium tetrathiotungstate is generally prepared by treating a solution of tungstic acid with excess ammonia and saturating with hydrogen sulfide. It is readily decomposed in a nonoxidizing atmosphere to WS_2, for which it is a convenient source.

Interstitial Compounds. Tungsten forms hard, refractory, and chemically stable interstitial compounds with nonmetals, particularly C, N, B, and Si. These compounds are used in cutting tools, structural elements of kilns, gas turbines, jet engines, sandblast nozzles, protective coatings, etc (see also REFRACTORIES; REFRACTORY COATINGS).

Carbides. Tungsten and carbon form two binary compounds, tungsten carbide [12070-12-1], WC, sp gr 15.63, and ditungsten carbide [12070-13-2], W_2C, sp gr 17.15. Both are prepared by heating tungsten and carbon at high temperatures. The presence of hydrogen or a hydrocarbon gas promotes the reaction. The relative quantities of the reactants and the temperature determine the phase formed. Tungsten carbide may also be prepared from oxygen-containing compounds of tungsten, but because of the tendency to form oxycarbides, a final heating in vacuum above 1500°C is necessary. Both carbides melt at ca 2800°C and have a hardness approaching that of diamond. Tungsten carbides are insoluble in water, but are readily attacked by HNO_3–HF. The most important commercial application is in hard metals. Tungsten carbides are brittle, but combination with, for example, cobalt decreases the brittleness. Approximately 67% of tungsten production is for the manufacture of WC (see CARBIDES).

Nitrides. The nitrides of tungsten are quite similar to the carbides. Although nitrogen does not react directly with tungsten, the nitrides can be prepared by heating tungsten in ammonia. The two phases, ditungsten nitride [12033-72-6], W_2N, and tungsten nitride [12058-38-7], WN, have been extensively studied (see NITRIDES) (53,54).

Borides. Ditungsten boride [12007-09-9], W_2B, and tungsten boride, WB, are prepared by hot-pressing tungsten and boron. Ditungsten pentaboride [12007-98-6], W_2B_5, is prepared by heating tungsten trioxide, graphite, and boron carbide *in uacuo*. Tungsten borides are extremely hard and exhibit almost metallic electrical conductivity. More recently, the formation of tungsten boride phases in the manufacture of boron filaments for structural composites for space vehicles and aircraft has been reported (55) (see BORON COMPOUNDS, REFRACTORY BORON COMPOUNDS).

Silicides. Tungsten silicides form a protective oxide layer over tungsten to prevent destructive oxidation at elevated temperatures. The layer fails to protect at lower temperatures, a behavior referred to as disilicide pest. This failure can be explained by the silicon being initially oxidized to SiO_2 at the surface and depleting the surface of Si, forming pentatungsten trisilicide [12039-95-i], W_5Si_3. At high temperatures, a uniform layer of W_5Si_3 is formed, but at lower (pest) temperatures, the attack is not uniform and seems to follow grain boundaries or subgrain boundaries in the disilicide. The next stage is the rapid growth and penetration of the complex oxide into the disilicide layer. This process ultimately consumes the disilicide, causing oxidation of the tungsten substrate. The existence of tritungsten disilicide [12509-47-6], W_3Si_2, and ditungsten silicide [56730-24-6], W_2Si, has been reported (56).

Ditungsten trisilicide [12138-30-6], W_2Si_3, gray in color and having an sp gr of 10.9, is insoluble in water, acid, or alkaline solutions. It is readily attacked by HNO_3-HF and fused alkali-metal carbonates and hydroxides.

Tungsten disilicide [12039-88-2], WSi_2, forms bluish-gray tetragonal crystals ($a = 0.3212$ nm, $c = 0.7880$ nm). It is insoluble in water and melts at 2160°C. The compound is attacked by fluorine, chlorine, fused alkalies, and HNO_3-HF. It may be used for high temperature thermocouples in combination with $MOSi_2$ in an oxidizing atmosphere.

Anionic Complexes. Compounds of tungsten with acid anions other than halides and oxyhalides are relatively few in number, and are known only in the form of complex salts. A number of salts containing hexavalent tungsten are known. Potassium octafluorotungstate [57300-87-5], K_2WF_8, can be prepared by the action of KI on $W(CO)_6$ in an IF_5 medium. The addition of tungstates to aqueous hydrofluoric acid gives salts that are mostly of the type $M(I)_2(W_2F_4)$. Similarly, double salts of tungsten oxydichloride are known.

Salts containing pentavalent tungsten may be obtained by the reduction of alkali tungstate in concentrated hydrochloric acid. Salts of types $M(I)_2(WOCl_5)$ (green), $M(I)(WOCl_4)$ (brown-yellow), and $M(I)(WOCl_4\cdot H_2O)$ (blue) have been isolated. Thiocyanato and bromo salts are also known.

Salts containing tetravalent tungsten have been prepared by various methods. The most important are the octacyanides, $M(I)_4(W(CN)_8)$, which form yellow crystals and are very stable. They are isolated as salts or free acids and can be oxidized by $KMnO_4$ in H_2SO_4 to compounds containing pentavalent tungsten, $M(I)_3(W(CN)_8)$ (yellow).

The only known trivalent tungsten complex is of the type $M(I)_3(W_2Cl_9)$. It is prepared by the reduction of strong hydrochloric acid solutions of K_2WO_4 with tin. If the reduction is not sufficient, a compound containing tetravalent tungsten, $K_2(WCl_5(OH))$ [84238-10-0], is formed (57).

Toxicity

A considerable difference in the toxicity of soluble and insoluble compounds of tungsten has been reported (58). For soluble sodium tungstate, $Na_2WO_4\cdot 2H_2O$, injected subcutaneously in adult rats, LD_{50} is 140–160 mg W/kg. Death results from generalized cellular asphyxiation. Guinea pigs treated orally or intravenously with $Na_2WO_4\cdot 2H_2O$ suffered anorexia, colic, incoordination of movement, trembling, and dyspnea.

Orally in rats, the toxicity of sodium tungstate was highest, tungsten trioxide was intermediate, and ammonium tungstate [15855-70-6] lowest (59,60). In view of the degree of systemic toxicity of soluble compounds of tungsten, a threshold limit of 1 mg of tungsten per m^3 of air is recommended. A threshold limit of 5 mg of tungsten per m^3 of air is recommended for insoluble compounds (61).

Uses

Tungsten compounds, especially the oxides, sulfides, and heteropoly complexes, form stable catalysts for a variety of commercial chemical processes, eg, petroleum processing (62). The tungsten compounds may function as principal catalysts or as promoters of other catalysts. The blue oxide, $W_{20}O_{58}$, is an important catalyst in industrial chemical synthesis involving hydration, dehydration, hydroxylation, and epoxidation (63). The application of tungsten catalysts is expected to increase greatly because many tungsten compounds are commercially available.

Tungsten hexachloride is used for preparing tungsten metathesis catalysts, which are very interesting because they form double and triple bonds with carbon. It is claimed that these catalysts permit the systematic control of a class of compounds used in the production of petroleum, plastics, synthetic fibers, and detergents. The improved control is expected to cut costs by providing more efficient use of raw materials (64). Films of tungsten deposited on various substrates improve the electrical conductivity of transparent tin oxide coatings on aircraft windows and windshields. Other uses include fire-retardant catalysts and as a fluxing agent in welding.

Tungsten disulfide forms adherent, soft, continuous films on a variety of substrates and exhibits good lubrication under extreme conditions of temperature, load, and vacuum. Applied as a dry powder, suspension, bonded film, or aerosol, it can be an effective lubricant in wire drawing, metal forming, valves, gears, bearings, packing materials, etc. Oil-soluble tungsten compounds, such as the ammonium salts of tungstate or tetrathiotungstate, are reported to be effective lubricating-oil additives.

Sodium tungstate is used in the manufacture of heteropolyacid color lakes, which are used in printing inks, plants, waxes, glasses, and textiles. It is also used as a fuel-cell electrode material and in cigarette filters. Other uses include the manufacture of tungsten-based catalysts, for fireproofing of textiles, and as an analytical reagent for the determination of uric acid.

Calcium tungstate is fluorescent when exposed to ultraviolet radiation and is therefore widely used in the manufacture of phosphors. It is also used in lasers, fluorescent lamps, high voltage sign tubes, and oscilloscopes for high speed photographic processes. Small crystals have been used for injection into malignant tumors, thus affording by transillumination a means of x-ray treatment. Other uses include screens for x-ray observations and photographs, luminous paints, and scintillation counters.

Ammonium paratungstate is commercially significant because it is the precursor of high purity tungsten oxides, tungsten, and tungsten carbide powders. It is slightly soluble in water but reacts with hydrogen peroxide to produce soluble peroxytungsten compounds. Ammonium metatungstate, on the other hand, is commercially significant because of its high solubility in water. This property as well as its acid characteristics makes it a very desirable starting material for catalysts and the impregnation of catalyst carriers with alkali-free solutions of tungsten. Other uses include nuclear shielding, corrosion inhibitors, and the preparation of other tungsten chemicals.

Tungsten trioxide is a principal source of tungsten metal and tungsten carbide powders. Because of its bright yellow color, it is used as a pigment in oil and water colors (see PIGMENTS). It is also used in a wide variety of catalysts and, more recently, in the control of air pollution and industrial hygiene. Tungsten carbides, on the other hand, are widely used in the manufacture of hard carbides for high speed machining tools, wire-drawing dies, wear surfacing, drills, etc.

Heteropoly tungstic acids are useful in analytical chemistry and biochemistry as reagents; in atomic-energy work as precipitants and inorganic ion exchangers; in photographic processes as fixing agents and oxidizing agents; in plating processes as additives; in plastics, adhesives, and cements for imparting water resistance; and in plastics and plastic films as curing or drying agents. An important use for 12-tungstophosphoric acid [*12067-99-1*] and its sodium salts is the manufacture of organic pigments. These compounds are also extensively used for the surface treatment of furs. In the textile industry, the salts are useful as antistatic agents. The acids are used in diverse applications, eg, printing inks, paper coloring, nontoxic paints, and wax pigmentation. The tungstates and molybdates are good corrosion inhibitors and have been used for some time in antifreeze solutions. In addition, they are used as laser-host materials, phosphors, and for the flameproofing of textiles.

BIBLIOGRAPHY

"Tungsten Compounds" in *ECT* 1st ed., Vol. 14, pp. 363–372, by B. Kopelman, Sylvania Electric Products, Inc.; "Wolfram Compounds" in *ECT* 2nd ed., Vol. 22, pp. 346–358, by M. B. MacInnis, Sylvania Electric Products, Inc.; "Tungsten Compounds" in *ECT* 3rd ed., Vol. 23, pp. 426–438, by M. B. MacInnis and T. K. Kim, GTE Products Corp.

1. J. J. Lander and L. H. Germer, *Am. Inst. Mining Met. Eng. Inst. Met. Div. Met. Technol.* **14**(6), 2259 (1947).
2. C. F. Powell, J. H. Oxley, and J. M. Blocher, Jr., *Vapor Deposition*, John Wiley & Sons, Inc., New York, 1966.
3. R. L. Landingham and J. H. Austin, *I. Less-Common Met.* **18**(3), 229 (1969).
4. E. J. Barber and G. H. Cady, *J. Phys. Chem.* **60**, 505 (1956).
5. A. R. D. Peacock, *J. Inorg. Nuel. Chem.* **35**(3), 751 (1973).
6. T. A. O'Donnell and D. F. Stewart, *Inorg. Chem.* **5**, 1434 (1966).
7. H. Cady and G. B. Hargreaves, *J. Chem. Soc.* 1568 (1961).
8. O. Ruff, F. Eisner, and W. Heller, *Z. Anorg. Allgem. Chem.* **52**, 256 (1907).
9. G. I. Novikoy, N. Y. Andreeva, and O. G. Polyachenok, *Russ. J. Inorg. Chem.* **6**, 1019 (1961).
10. R. E. McCarley and T. M. Brown, *Inorg. Chem.* **3**, 1232 (1964).
11. W. C. Dorman, *IS-T-510*, National Technical Information Service, Department of Commerce, Washington, D.C., 1972.
12. R. Colton and I. B. Tomkins, *Aust. J. Chem.* **18**, 447 (1965).
13. E. R. Epperson and H. Frye, *Inorg. Noel. Chem. Lett.* **2**, 223 (1966).
14. G. W. Fowles and J. L. Frost, *Chem. Common.*, 252 (1966).
15. P. M. Druce and M. F. Lappert, *J. Chem. Soc.* **A22**, 3595 (1971).
16. R. Colton and I. B. Tomkins, *Aust. J. Chem.* **19**, 759 (1966).
17. R. E. McCarley and T. M. Brown, *J. Am. Chem. Soc.* **84**, 3216 (1962).
18. C. Djordjevic and co-workers, *J. Chem. Soc. Inorg. Phys. Theor.* **1**, 16 (1966).

19. Schulz and co-workers, *J. Less Common Met.* **22**, 136–138 (1970).
20. J. Tillack, P. Eckerlin, and J. H. Dettingmeijer, *Angew. Chem.* **78**, 451 (1966).
21. A. Bartecki, M. Cieslak, and S. Weglowski, *J. Less-Common Met.* **26**(3), 411 (1972).
22. E. Gebert and R. J. Ackermann, *Inorg. Chem.* **5**(1), 136 (Jan. 1966).
23. J. Neugebauer, T. Miller, and L. Imre Tungsram, *Techn. Mitteil.* (2), (Mar. 1961).
24. G. Hagg and A. Magneli, *Rev. Pure Appl. Chem.* **4**, 235 (1954).
25. O. Glemser and H. Sauer, *Z. Anorg. Chem.* **252**, 144 (1943).
26. L. L. Y. Chang and B. Phillips, *J. Am. Cer. Soc.* **52**(10), 527 (1969).
27. G. Hagg and N. Schonberg, *Acta Cryst.* **7**, 351 (1954).
28. A. Magneli, *Ark. Kemi*, **1**, 513 (1950).
29. A. Magneli, *J. Inorg. Noel. Chem.* **2**, 330 (1956).
30. P. G. Dickens and M. S. Whittingham, *Q. Rev. Chem. Soc.* **22**(1), 30 (1968).
31. M. J. Sienko, *Adv. Chem. Ser.* **39**, 224 (1963).
32. L. W. Niedrach and H. I. Zeliger, *J. Electrochem. Soc.* **116**(1), 152 (1969).
33. J. P. Randin, *J. Electrochem. Soc.* **120**(3), 378 (1973).
34. V. I. Spitsyn and T. I. Drobasheva, *Zh. Inorg. Khim.* **21**(7), 1787 (1976).
35. Glemser and C. Naumann, *Z. Anorg. Chem.* **265**, 288 (1951).
36. P. G. Dickens and R. J. Hurditch, *Nature*, **215**, 1266 (1967).
37. E. Schwarzmann and R. Birkenberg, *Z. Naturforsek.* **B26**(10), 1069 (1971).
38. D. L. Kepert, *Progr. Inorg. Chem.* **4**, 199 (1962).
39. P. Tekula-Buxbaum, *Acta Tech. Acad. Sci. Hung.* **78**(3,4), 325 (1974).
40. H. M. Ortner, *Anal. Chem.* **47**(1), 162 (1975).
41. T. K. Kim, R. W. Mooney, and V. Chiola, *Sep. Sci.* **3**(5), 467 (1968).
42. U.S. Pat. 3,175,881 (Mar. 30, 1965), V. Chiola, J. M. Lafferty, Jr., and C. D. Vanderpool (to Sylvania Electric Products, Inc.).
43. U.S. Pat. 3,591,331 (July 6, 1971), V. Chiola and co-workers (to Sylvania Electric Products, Inc.).
44. U.S. Pat. 3,857,928 (Dec. 31, 1974), T. K. Kim and co-workers (to GTE Sylvania Inc.).
45. U.S. Pat. 3,857,929 (Dec. 31, 1974), L. R. Quatrini and co-workers (to GTE Sylvania Inc.).
46. U.S. Pat. 3,936,362 (Feb. 3, 1976), C. D. Vanderpool, M. B. MacInnis, and J. C. Patton, Jr. (to GTE Sylvania Inc.).
47. U.S. Pat. 3,956,474 (May 11, 1976), J. E. Ritsko (to GTE Sylvania Inc.).
48. A. Chretien and D. Helgorsky, *C. R. Acad. Sci. Paris*, **252**, 742 (1961).
49. A. J. Bradley and J. W. Illingworth, *Proc. Roy. Soc. London Ser.* **A49**(157), 113 (1936).
50. R. Signer and H. Gross, *Helv. Chim. Acta*, **17**, 1076 (1934).
51. V. R. Johnson, M. T. Lavik, and E. E. Vaughn, *J. App. Phys.* **28**, 821 (1957).
52. O. Glemser, H. Saver, and P. Konig, *Z. Inorg. Chem.* **257**, 241 (1948).
53. A. G. Mattock and co-workers, *J. Chem. Soc. Dalton Trans.* **12**, 1314 (1974).
54. L. A. Cherezova and B. P. Kryzhanovskii, *Opt. Spektrosk*, **34**(2), 414 (1973).
55. A. L. Buryking, Y. V. Dzyrdykevich, and V. V. Gorskii, *Poroshk Metall.* **2**, 74 (1973).
56. N. N. Matynshenko, L. N. Efimenko, and D. N. Solonikin, *Fiz. Met. Metalloved.* **8**, 878 (1959).
57. E. Konig, *Inorg. Chem.* **2**, 1238 (1963).
58. U.S. DHEW (NIOSH) Publication No. 77-127, U.S. Dept. of Health, Education and Welfare, Washington, D.C., 1977.
59. F. W. Kinard and J. Van de Erve, *Am. J. Med. Sci.* **199**, 668 (1940).
60. V. G. Nadeenko, *Hyg. Sanit.* **31**, 197 (1966).
61. *Documentation of TLV*, American Conference of Industrial Hygienists, Cincinnati, Ohio, 1966, Appendix C.
62. C. H. Kline and V. Kollonitsch, *Ind. Eng. Chem.* **57**(7), 53 (1965).

63. C. H. Kline and V. Kollonitsch, *Ind. Eng. Chem.* **57**(9), 53 (1965).
64. D. N. Clark and R. R. Schrock, *J. Am. Chem. Soc.* **100**, 6774 (1978).

General References

S. W. H. Yih and C. T. Wang, *Tungsten*, Plenum Press, New York, 1979.
G. D. Rieck, *Tungsten and Its Compounds*, Pergamon Press, London, 1967.
K. C. Li and C. Y. Wang, *Tungsten*, Reinhold Publishing Corp., New York, 1955.
C. J. Smithells, *Tungsten*, Chapman and Hall, London, 1952.
J. H. Canterford and R. Colton, *Halides of the Transition Elements*, John Wiley & Sons, Inc., New York, 1978.

THOMAS W. PENRICE
Consultant

TURBIDITY AND NEPHELOMETRY.   See ANALYTICAL METHODS.

TURKEY RED OIL.   See CASTOR OIL.

TURPENTINE.   See TERPENOIDS.

TYPE METAL.   See LEAD ALLOYS.

TYROCIDINE.   See ANTIBIOTICS, PEPTIDES.

TYROTHRICIN.   See ANTIBIOTICS, PEPTIDES.

ULTRAFILTRATION

Ultrafiltration is a pressure-driven filtration separation occurring on a molecular scale (see DIALYSIS; FILTRATION; HOLLOW-FIBER MEMBRANES; MEMBRANE TECHNOLOGY; REVERSE OSMOSIS). Typically, a liquid including small dissolved molecules is forced through a porous membrane. Large dissolved molecules, colloids, and suspended solids that cannot pass through the pores are retained.

Ultrafiltration separations range from ca 1 to 100 nm. Above ca 50 nm, the process is often known as microfiltration. Transport through ultrafiltration and microfiltration membranes is described by pore-flow models. Below ca 2 nm, interactions between the membrane material and the solute and solvent become significant. That process, called reverse osmosis or hyperfiltration, is best described by solution–diffusion mechanisms.

Membrane-retained components are collectively called concentrate or retentate. Materials permeating the membrane are called filtrate, ultrafiltrate, or permeate. It is the objective of ultrafiltration to recover or concentrate particular species in the retentate (eg, latex concentration, pigment recovery, protein recovery from cheese and casein wheys, and concentration of proteins for biopharmaceuticals) or to produce a purified permeate (eg, sewage treatment, production of sterile water or antibiotics, etc). Diafiltration is a specific ultrafiltration process in which the retentate is further purified or the permeable solids are extracted further by the addition of water or, in the case of proteins, buffer to the retentate.

Membrane filtration has been used in the laboratory for over a century. The earliest membranes were homogeneous structures of purified collagen or zein. The first synthetic membranes were nitrocellulose (collodion) cast from ether in the 1850s. By the early 1900s, standard graded nitrocellulose membranes were commercially available (1). Their utility was limited to laboratory research because of low transport rates and susceptibility to internal plugging. They did, however, serve a useful role in the separation and purification of colloids, proteins, blood sera, enzymes, toxins, bacteria, and viruses (2).

In the late 1950s and 1960s, a technique was developed that produced highly anisotropic or asymmetric structures, ie, membranes constructed of a very thin, tight surface skin having a porous substructure. The substructure provided the necessary mechanical support for the skin without the hydraulic resistance of previous isotropic structures. Flux rates improved by orders of magnitude, and inherent resistance to plugging increased. A molecule entering a pore through the skin traverses a channel of increasing diameter. Both high flux and plugging resistance are important for achieving an economical membrane performance in industrial applications.

The subsequent improvement of the physical and chemical characteristics of these membranes, their incorporation into machines, and the development of procedures to prevent or clean surface-fouling films were the principal areas of significant advancement. By 1990, the industrial ultrafiltration market had grown to an estimated $(90-100) \times 10^6$.

Media

Most ultrafiltration membranes are porous, asymmetric, polymeric structures produced by phase inversion, ie, the gelation or precipitation of a species from a soluble phase (see MEMBRANE TECHNOLOGY).

Typically, a polymer is first dissolved in a mixture of miscible solvents and nonsolvents. This mixture (lacquer solution) is frequently a better polymer solvent than any of the components (3,4). The lacquer solution is dearated and spread as a thin film on a suitable support. The surface of the film is then placed in contact with a nonsolvent diluent (precipitant) miscible with the solvent. This precipitates or gels the surface almost instantaneously, forming a membrane skin.

Macroscopically, the solvent and precipitant are no longer discontinuous at the polymer surface, but diffuse through it. The polymer film is a continuum with a surface rich in precipitant and poor in solvent. Microscopically, as the precipitant concentration increases, the polymer solution separates into two interspersed liquid phases: one rich in polymer and the other poor. The polymer concentration must be high enough to allow a continuous polymer-rich phase but not so high as to preclude a continuous polymer-poor phase.

The skin is highly stressed because of the polymer consolidation. The surface tears at polymer-poor sites, forming cracks or pores that expose a more fluid internal polymer layer to the precipitant—solvent mixture (5). The pores propagate into so-called fingers by drawing the precipitating polymer from the bottom to the side of the pore (Fig. 1). Because this process proceeds along a moving boundary into the polymer film, additional pores do not form on the walls. The polymer solution behind these precipitated walls gels into an open-sponge structure (Fig. 2). The capillary stresses (surface activity) must be low enough to avoid collapsing the structure. Polymers with high elastic moduli and solvents that do not plasticize the polymer are preferred.

Membrane structure is a function of the materials used (polymer composition, molecular weight distribution, solvent system, etc) and the mode of preparation (solution viscosity, evaporation time, humidity, etc). Commonly used polymers include cellulose acetates, polyamides, polysulfones, dynels (vinyl chloride—acrylonitrile copolymers) and poly(vinylidene fluoride).

Modification of the membranes affects the properties. Cross-linking improves mechanical properties and chemical resistivity. Fixed-charge membranes

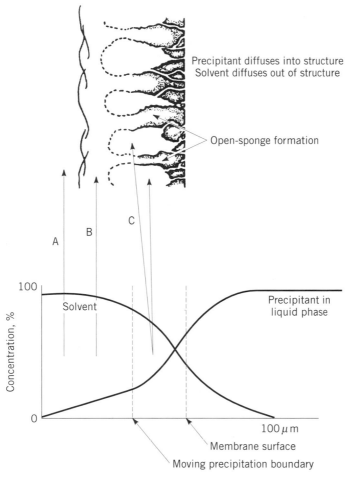

Fig. 1. Formation of an ultrafiltration membrane: A, unprecipitated polymer solution; B, polymer solution separating into two phases; C, pore fingers with precipitant–solvent mixture.

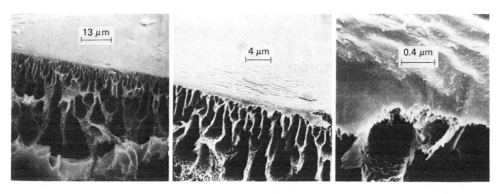

Fig. 2. A series of progressively closer (scanning electron microscope) SEM photographs of the same membrane cross section, clearly showing skin and substructure.

are formed by incorporating polyelectrolytes into polymer solution and cross-linking after the membrane is precipitated (6), or by substituting ionic species onto the polymer chain (eg, sulfonation). Polymer grafting alters surface properties (7). Enzymes are added to react with permeable species (8–11) and reduce fouling (12,13).

Polyelectrolyte complex membranes are phase-inversion membranes where polymeric anions and cations react during the gelation. The reaction is suppressed before gelation by incorporating low molecular weight electrolytes or counterions in the solvent system. Both neutral and charged membranes are formed in this manner (14,15). These membranes have not been exploited commercially because of their lack of resistance to chemicals.

Inorganic ultrafiltration membranes are formed by depositing particles on a porous substrate (16,17). In one form, inorganic particles (alumina, Zr_2SiO_2) of two discrete sizes are deposited. The smaller size can pass through the porous support whereas the larger size cannot. The mixture forms a controlled porosity film at the entrance of the support's pores. These membranes can be removed and regenerated *in situ*. Alternatively, inorganic or organic binders can be added as stabilizers. Inorganic membranes exhibit good thermal and chemical stability.

Dynamic membranes are concentration–polarization layers formed *in situ* from the ultrafiltration of colloidal material analogous to a precoat in conventional filter operations. Hydrous zirconia has been thoroughly investigated; other materials include bentonite, poly(acrylic acid), and films deposited from the materials to be separated (18).

Track-etched membranes are made by exposing thin films (mica, polycarbonate, etc) to fission fragments from a radiation source. The high energy particles chemically alter material in their path. The material is then dissolved by suitable reagents, leaving nearly cylindrical holes (19) (see PARTICLE-TRACK ETCHING (SUPPLEMENT)).

Process

Pore-flow models most accurately describe ultrafiltration processes. Other membrane transport mechanisms, which may occur simultaneously although generally at a much lower rate, include dialysis (diffusion), osmosis (solvent by osmotic gradient), anomalous osmosis (osmosis with a charged membrane), reverse osmosis (solvent by pressure gradient larger and opposite to osmotic gradient), electrodialysis (solute ions by electric field), piezodialysis (solute by pressure gradient), electroosmosis (solvent in electric field), Donnan effects, Knudsen flow, thermal effects, chemical reactions (including facilitated diffusion), and active transport.

When pure water is forced through a porous ultrafiltration membrane, Darcy's law states that the flow rate is directly proportional to the pressure gradient:

$$J = \frac{V}{A \cdot t} = \frac{K_m \Delta P}{\mu} \tag{1}$$

where J is permeate flux in units of volume V per membrane area A, at time t, K_m is the membrane hydraulic permeability, μ is the fluid viscosity, and ΔP is the membrane pressure drop between the retentate and permeate.

The membrane hydraulic permeability K_m is a function of the pore size, tortuosity, and length, and any resistance in the substructure. Because ultrafiltration membranes are plastic and can yield (compact) or creep under pressure, K_m is also a function of the pressure history. Dynamic pressure drops from flow through a membrane and static pressure drops from a force applied on a membrane surface (eg, across a fouling film) can both cause compaction. Initial compaction occurs rapidly during startup, whereas long-term compaction occurs slowly over the operating life of the membrane. Swelling agents can sometimes (partially) reverse compaction.

Initial membrane compaction is illustrated by Figure 3. Equation 1 predicts a straight-line response of J to ΔP, or J_3 at P_1. Owing to the compaction, a lower flux J_2 is observed. Once a membrane has been subjected to some pressure (P_1), equation 1 is valid for predicting flux up to that pressure (Fig. 3, curve B). If the membrane is subsequently subjected to higher pressure (P_2), the hydraulic permeability constant is changed (Fig. 3, curve D).

The addition of small membrane-permeable solutes to the water affects permeate transport in the following ways. (1) Solute–solvent interactions change the permeating fluid viscosity. (2) Solute adsorption reduces the apparent membrane-pore diameter (20). Because of high interfacial tension between water and certain materials, the water phase in the pores can be replaced. Dynel and polysulfone membranes, for example, preferentially extract partially soluble alcohols from water. Surfactants suppress hydrophobic adsorption. Adsorption of permeate species is characterized by a lag in permeate concentration as a

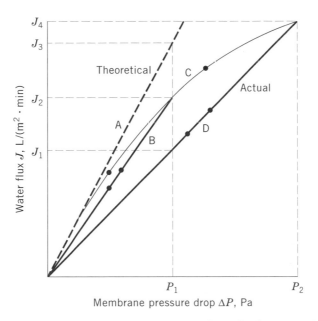

Fig. 3. Water flux versus pressure. Equation 1 predicts the line (———) having flux J_3 at P_1. Actual initial water flux follows curve C to flux J_2 at P_1. Subsequent operation at pressure drops less than P_1 follows curve B (eq. 1). If pressure is increased above P_1, flux follows curve C (additional compaction) to P_2. A new value of K_m is used in equation 1. Operation at pressure drops less than P_2 follows curve D. Flux at P_1 is lowered to J_1.

function of time. (3) The interfacial charge between the membrane-pore wall and the liquid affects permeate transport when the Debye screening length approaches (ca 10%) the membrane-pore size. Flux declines, rejection increases, and electrolyte is retained. Other electrokinetic phenomena become pronounced and may influence fouling (21). (4) High surface tension on hydrophobic membranes forces water molecules to form large clusters in the pores. Water-structuring ions (eg, Na^+, Mg^{2+}, and OH^-) tend to decrease permeability and increase rejection; destructuring ions (eg, Cl^-, NO_3^-, and ClO_4^-) have the opposite effect (22). (5) Solvents, swelling agents, and plasticizers that diffuse into the polymer structure can change the apparent pore size (K_m in eq. 1) or increase the rate of long-term compaction. The rejection R of a solute is defined as:

$$R = 1 - \frac{C_{pi}}{C_{bi}} \tag{2}$$

where C_p is the permeate concentration of species i and C_b is the concentration of that species in the retentate. There are two components of rejection. Observed rejection, R_o, is based on the concentration of the solute in the bulk solution, C_b. The intrinsic rejection, R_i, is based on the concentration of the solute on the surface of the membrane, C_w.

$$R_o = 1 - \frac{C_p}{C_b} \qquad R_i = 1 - \frac{C_p}{C_w}$$

If the solute size is approximately the (apparent) membrane-pore size, it interferes with the pore dimensions. The solute concentration in the permeate first increases, then decreases with time. The point of maximum interference is further characterized as a minimum flux. Figure 4 is a plot of retention and flux versus molecular weight. It shows the minimum flux at ca 60–90% retention.

If the solute size is greater than the pore dimensions, the solute is retained by mechanical sieving.

Membrane pores are not of uniform size (23). They are not cylindrical, but rather resemble fissures (24) or cracks (5). Similarly, molecules are not spherical.

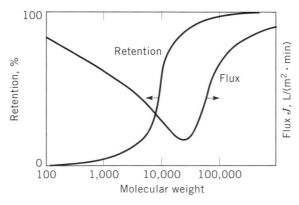

Fig. 4.   Retention and flux versus molecular weight.

A long chain of 100,000 mol wt (eg, dextran) may readily pass through a pore which retains a globular protein of 20,000 mol wt. Branching chains may block or plug pores. Frequently, macromolecules change shape as a function of solution pH or ionic strength. The transition between solute molecular weight and rejection is therefore gradual and involves conformational considerations (25,26). The slope of the retention curve of Figure 4 is a measure of the interaction between the pore-size and the solute-size distributions.

Retained species are transported to the membrane surface at the rate:

$$J_i = JC_{bi} \tag{3}$$

where J is the permeate flux and C_{bi} is the bulk concentration of the retained species i. They accumulate in a boundary layer at the membrane surface (Fig. 5). This deposit is composed of suspended particles similar to conventional filter cakes, and more importantly, a slime that forms as retained solutes exceed their solubility. The gel concentration C_g is a function of the feed composition and the membrane-pore size. The gel usually has a much lower hydraulic permeability and smaller apparent pore size than the underlying membrane (27). The gel layer and the concentration gradient between the gel layer and the bulk concentration are called the gel-polarization layer.

The concentration boundary layer forms because of the convective transport of solutes toward the membrane due to the viscous drag exerted by the flux. A diffusive back-transport is produced by the concentration gradient between the membranes surface and the bulk. At equilibrium the two transport mechanisms are equal to each other. Solving the equations leads to an expression of the flux:

$$J = K \ln\left(\frac{C_w}{C_b}\right)$$

where K is the mass-transfer coefficient, C_w is the concentration of the solute at the surface of the membrane, and C_b is the solute concentration at the bulk.

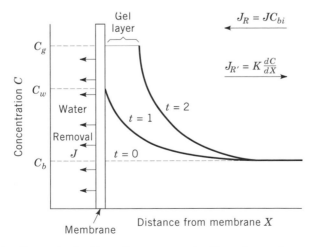

Fig. 5. Concentration polarization: C_w = concentration at membrane wall, C_b = bulk concentration, C_{bi} = bulk concentration of species i, J = flux, and C_g = gel concentration.

The concentration boundary layer can form a resistance to the flux owing to the formation of a gel, or to the osmotic pressure created by the layer.

Feed–constituent interactions further affect retention (28,29). Dispersing agents and emulsifiers are partially retained because they attach to the dispersed phase. Small molecules may similarly adsorb onto larger particles.

The gel-layer thickness is limited by mass transport back into the solution bulk at the rate:

$$J_i = K \frac{dC_i}{dX} \tag{4}$$

where the mass-transfer coefficient K is multiplied by the concentration gradient.

At steady state,

$$JC_b = K \frac{dC}{dX} \tag{5}$$

where C_b is the bulk concentration of all retained species. Integration gives

$$J = K \cdot \ln \frac{C_g}{C_b} \tag{6}$$

In a static system, the gel-layer thickness rapidly increases and flux drops to uneconomically low values. In equation 6, however, K is a function of the system hydrodynamics. Typically, high flux is sustained by moving the solution bulk tangentially to the membrane surface. This action decreases the gel thickness and increases the overall hydraulic permeability. For any given channel dimension, there is an optimum velocity which maximizes productivity (flux per energy input).

A number of analytical solutions have been derived for K as a function of channel dimensions and fluid velocity (30). In practice, the fit between theory and data for K is poor except in idealized cases. Most processes exhibit either higher fluxes, presumably caused by physical disruption of the gel layer from the nonideal hydrodynamic conditions, or lower fluxes caused by fouling (31). In addition, K is a function of the fluid composition.

Ultrafiltration equipment suppliers derive K empirically for their equipment on specific process fluids. Flux J is plotted versus log C_b for a set of operation conditions in Figure 6; K is the slope, and C_w is found by extrapolating to zero flux. Operating at different hydrodynamic conditions yields differently sloped curves through C_w.

The gel-polarization layer has an hydraulic permeability of K_g. Equation 6 states that flux is independent of pressure, and K_g must therefore decrease with increasing pressure. Equation 1 becomes

$$J = \frac{\Delta P}{\mu \left(\dfrac{1}{K_m} + \dfrac{1}{K_g} \right)} = \frac{\Delta P}{\mu (R_m + R_g)} \tag{7}$$

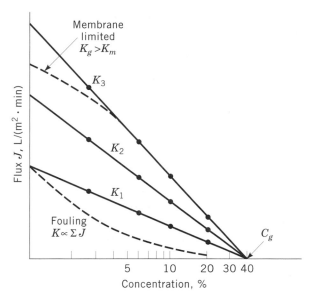

Fig. 6. Flux versus concentration, illustrating the effect of operating conditions on K and deviations from equation 6.

where R_m and R_g are the hydraulic resistances of the membrane gel.

Flux is independent of pressure when the process flux is much less than the water flux ($K_g \ll K_m$). If $K_g > K_m$, the process is limited by the membrane water flux and flux would flatten out at low concentrations of solids (see Fig. 6).

For very small ΔP, flux is linear with pressure. Figure 7 shows a graph of flux versus pressure. Curve A is the pure water flux from equation 1, curve B is the theoretical permeate flux (TPE) for a typical process. As the gel layer forms, the flux deviates from the TPF following equation 7 and curve D results. Changing the hydrodynamic conditions changes K_g and results in a different operating curve, curve C.

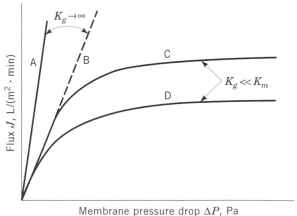

Fig. 7. Flux versus membrane ΔP.

Fouling.   If the gel-polarization layer is not in hydrodynamic equilibrium with the fluid bulk, the membrane may be fouled. Fouling is caused either by adsorption of species on the membrane or on the surface of the pores, or by deposition of particles on the membrane or within the pores. Fouled systems are characterized as follows: flux is a function of total permeate production when hydrodynamic conditions are constant (see Fig. 6); if hydrodynamic conditions are changed, hydraulic permeability response of the gel layer is not reversible; and theoretical permeate flux (TPF) changes with time. A sensitive test for predicting fouling or process instability is to measure change in TPF after subjecting the system to process extremes (eg. high pressure with no flow).

Fouling is controlled by selection of proper membrane materials, pretreatment of feed and membrane, and operating conditions. Control and removal of fouling films is essential for industrial ultrafiltration processes.

Suspensions of oil in water (32), such as lanolin in wool (qv) scouring effluents, are stabilized with emulsifiers to prevent the oil phase from adsorbing onto the membrane. Polymer latices and electrophoretic paint dispersions are stabilized using surface-active agents to reduce particle agglomeration in the gel-polarization layer.

Dairy wheys containing complex mixtures of proteins, salts, and microorganisms rapidly foul membranes. Heat treatment and pH adjustment accelerate the aggregation of β-lactoglobulin with other whey components (33,34). Otherwise, they would interact within the polarization layer (35,36), forming sheet-like fouling gels. These methods also reduce microbial fouling and the formation of apatite gels. Other whey pretreatment methods include demineralization, clarification, and centrifugation (37,38).

Pretreatment of membranes with dynamically formed polarization layers and enzyme precoats have been effective (12,13,39). Pretreatment with synthetic permeates prevents startup instability with some feed dispersions.

When fouling is present or possible, ultrafiltration is usually operated at high liquid shear rates and low pressure to minimize the thickness of the gel polarization layer.

Cleaning.   Fouling films are removed from the membrane surface by chemical and mechanical methods. Chemicals and procedures vary with the process, membrane type, system configuration, and materials of construction. The equipment manufacturer recommends cleaning methods for specific applications. A system is considered clean when it has returned to >75% of its original water flux.

In order to develop an effective cleaning method, it is essential to know the fouling constituents and whether the cleaning agents solubilize or disperse the foulants. Detergents emulsify oils, fats, and grease (40), whereas protein films are dispersed by proteolytic enzymes and alkaline detergents (38). Acids or alkalies solubilize inorganic salts; sodium hypochlorite is used as a cleaning agent for organics. If the feed contains a mixture of different components, several cleaners may be needed. Depending on the process, cleaning agents may be used in combination or sequentially, separated by rinses.

Dissolved fouling material may pass into the membrane pores. Reprecipitation upon rinsing must be avoided. Membrane-swelling agents, such as hypochlorites, flushout material which may be lodged in the pores.

Cleaning is frequently aided mechanically. Foam balls scour the center of tubes, and hollow-filter systems can be back-flushed. Hollow fibers and membranes attached to rigid supports can be back-pressured, thereby eliminating the pressure drop that holds redispersed films on the membrane surface.

Unless redispersed foulants are completely flushed away before using membrane swelling agents (for sanitizing), they may become entrapped in the membrane structure. Water flux does not recover and the subsequent process fouls faster than usual. This phenomenon is discontinuous and differs from a steady reduction in water flux over many cleaning cycles, which indicates a gradual buildup of a fouling component not attacked by the cleaning composition.

Certain applications require that the equipment meet FDA and USDA sanitary requirements. These requirements ensure that the products are not contaminated by extractables or microorganisms from the equipment. Special considerations are given to the design of such equipment (41–44) (see STERILIZATION TECHNIQUES).

Practical Aspects

The theoretical models cannot predict flux rates. Plant-design parameters must be obtained from laboratory testing, pilot-plant data, or in the case of established applications, performance of operating plants.

Flux response to concentration, cross flow or shear rate, pressure, and temperature should be determined for the allowable plant excursions. Fouling must be quantified and cleaning procedures proven. The final design flux should reflect long-range variables such as feed-composition changes, reduction of membrane performance, long-term compaction, new foulants, and viscosity shifts.

Flux is maximized when the upstream concentration is minimized. For any specific task, therefore, the most efficient (minimum membrane area) configuration is an open-loop system where retentate is returned to the feed tank (Fig. 8). When the objective is concentration (eg, enzyme), a batch system is employed. If the object is to produce a constant stream of uniform-quality permeate, the system may be operated continuously (eg, electrocoating).

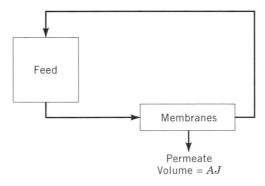

Fig. 8. Open-loop system.

The upstream concentration C_b starts at C_o and ends at C_f, as described by the following relationship:

$$C_b = C_o \frac{(V_o)^R}{V_b} \tag{8}$$

$$V_b = V_o - AJt \tag{9}$$

where V_o is the original volume, R is rejection, A is the membrane area, and t is time. Because J is a function of C_b (eq. 6), the solution can only be approximated.

Open-loop systems have inherently long residence times which may be detrimental if the retentate is susceptible to degradation by shear or microbiological contamination. A feed-bleed or closed-loop configuration is a one-stage continuous membrane system. At steady state, the upstream concentration is constant at C_f (Fig. 9). For concentration, a single-stage continuous system is the least efficient (maximum membrane area).

The single-pass system and the staged cascade (Figs. 10 and 11) have high flux at low residence time. Both trade the concentration dependence of the batch system on time for concentration dependence on position in the system. Thus, a uniform flux is maintained (assuming no fouling) allowing continuous process integration. In practice, the single-pass system is difficult to implement, and therefore most commercial systems are multistaged cascade. The more stages used, the closer the average flux approaches the batch flux. Table 1 compares the flux for batch and staged systems operating on cheese whey.

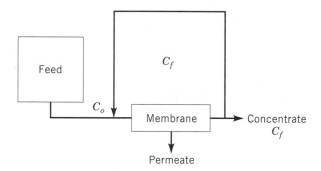

Fig. 9. Closed-loop system (feed bleed).

Electroultrafiltration

Electroultrafiltration (EUF) combines forced-flow electrophoresis (see ELECTRO-SEPARATIONS, ELECTROPHORESIS) with ultrafiltration to control or eliminate the gel-polarization layer (45–47). Suspended colloidal particles have electrophoretic mobilities measured by a zeta potential (see COLLOIDS; FLOTATION). Most naturally occurring suspensoids (eg, clay, PVC latex, and biological systems), emulsions, and protein solutes are negatively charged. Placing an electric field across an ultrafiltration membrane facilitates transport of retained species away from

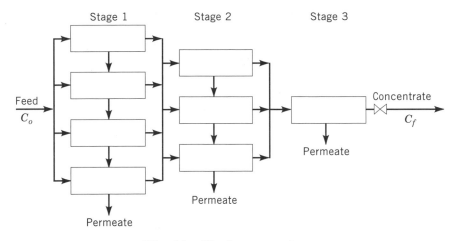

Fig. 10.   Single-pass system.

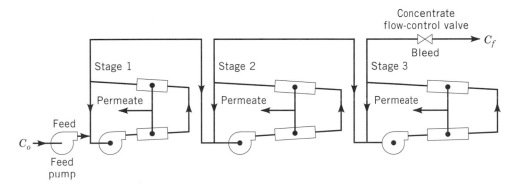

Fig. 11.   Continuous multistage (cascade system).

Table 1. Flux Comparisons Between Batch and Staged Systems Operating on Cheese Whey

Configuration	Relative flux, %
batch open loop	100
single-stage feed bleed	67
two-stage cascade	82
three-stage cascade	87
four-stage cascade	89

the membrane surface. Thus, the retention of partially rejected solutes can be dramatically improved (see ELECTRODIALYSIS).

Electroultrafiltration has been demonstrated on clay suspensions, electrophoretic paints, protein solutions, oil–water emulsions, and a variety of other materials. Flux improvement is proportional to the applied electric field E up to some field strength E_c, where particle movement away from the membrane is equal to the liquid flow toward the membrane. There is no gel-polarization layer and (in theory) flux equals the theoretical permeate flux. It follows, therefore, that E_c is proportional to ΔP.

At electric-field strengths greater than E_c, flux is proportional to ΔP up to the critical pressure P_c where E becomes E_c (Fig. 12).

Anodic deposition is controlled by either fluid shear (cross-flow filtration) (48), similar to gel-polarization control, or by continual anode replacement (electrodeposited paints) (46). High fluid shear rates can cause deviations from theory when $E > E_c$ (49). The EUF efficiency drops rapidly with increased fluid conductivity.

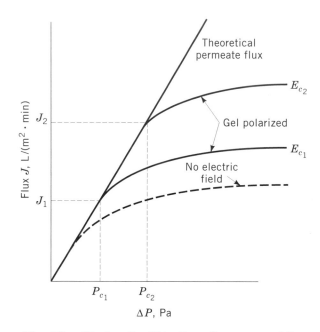

Fig. 12. Electroultrafiltration, flux versus ΔP.

Diafiltration

Diafiltration is an ultrafiltration process where water or an aqueous buffer is added to the concentrate and permeate is removed (50). The two steps may be sequential or simultaneous. Diafiltration improves the degree of separation between retained and permeable species.

Constant-volume batch diafiltration is the most efficient process mode. For species that freely permeate the membrane,

$$\ln\!\left(\frac{C_o}{C_t}\right) = \frac{V_p}{V_o} = N \equiv \text{turnover ratio} \qquad (10)$$

where C_o is the permeate concentration at the start of diafiltration, C_t is the instantaneous permeate concentration at time t, V_o is the constant retentate volume, and V_p is the total permeate volume at t, which also equals the added water volume.

The fractional recovery of permeable solids in the retentate is

$$Y_r = \left(\frac{C_t}{C_o}\right) = \exp(-N) = 1 - Y_p \tag{11}$$

where Y_p is the fractional recovery in the permeate.

For partially retained solutes, equation 10 becomes

$$\ln\left(\frac{C_o}{C_t}\right) = 1 - \delta\left(\frac{V_p}{V_o}\right) \tag{12}$$

Area–time requirements for a specific diafiltration mission are defined as

$$A \cdot t = \frac{V_p}{J} = \frac{N V_o}{J} \tag{13}$$

When flux is independent of C_p:

$$A \cdot t = \frac{K}{C_b \ln\left(\dfrac{C_g}{C_b}\right)} \tag{14}$$

The optimum concentration for any diafiltration (minimum area time) is the minimum of the plot:

$$\frac{1}{J \cdot C_b} \text{ versus } C_b \tag{15}$$

When fouling is absent, the optimum concentration is 0.37 C_g. If the permeate solids are of primary value, it is usually preferable to diafilter at the minimum retentate volume to minimize permeate dilution.

Sequential batch diafiltration is a series of dilution–concentration steps. The concentration of membrane-permeable species is

$$\frac{C_o}{C_t} = \left(1 + \frac{V_p}{V_o}\right)^{n(1-\delta)} \tag{16}$$

where V_p is the permeate volume produced in each of n equal operations, and δ is the rejection of solids. As $n \to \infty$, equation 16 approaches equation 10.

Continuous diafiltration practiced in one or more stages of a cascade system has the same volume turnover relationship for overall recoveries as sequential batch diafiltration. The residence time however is dramatically reduced. If recovery of permeable solids is of primary importance, the permeate from the last stage may be used as diafiltration fluid for the previous stage. This countercurrent diafiltration arrangement results in higher permeate solids at the expense of increased membrane area.

Membrane Equipment

Commercial industrial ultrafiltration equipment first became available in the late 1960s. Since that time, the industry has focused on five different configurations.

Parallel-Leaf Cartridge.   A parallel-leaf cartridge consists of several flat plates, each having membrane sealed to both sides (Fig. 13). The plates have raised (2–3 mm) rails along the sides in such a way that, when they are stacked, the feed can flow between them. They are clamped between two stainless-steel plates with a central tie rod. Permeate from each leaf drains into an annular channel surrounding the tie rod (33).

Another type has several flat plates manifolded into a plastic header. The surface of the laminate is suitable for dip-casting membranes, whereas the interior is several orders of magnitude more porous. Permeate collects in the center of the laminate and drains into the header.

Cartridges are inserted in series into plastic or stainless-steel tubular pressure housings of square cross section (Fig. 14). Feed flows parallel to the leaf surface. A permeate fitting secures each cartridge to the housing wall, which allows permeate egress and facilitates sealing between concentrate, atmosphere, and permeate channels.

Plate and Frame.   Plate-and-frame systems consist of plates (Fig. 15) each with a membrane on both sides. The plates have a frame around their perimeter which forms flow channels ca 1 mm wide between the plates when they are stacked. The stack is clamped between two end plates, sealing the frames together.

At least one hole near the perimeter of each plate connects the flow channels from one side of the plate to the other. The membrane is sealed around the hole to isolate the permeate from the concentrate. Permeate collects in a drain grid behind the membrane and exits from a withdrawal port on the frame perimeter.

Spiral Wound.   A spiral-wound cartridge has two flat membrane sheets (skin side out) separated by a flexible, porous permeate drainage material. The membrane sandwich is adhesively sealed on three sides. The fourth side of one or more sandwiches is separately sealed to a porous or perforated permeate

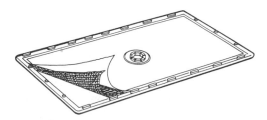

Fig. 13.   Flat-plate membrane element.

Fig. 14. Flat plate cartridge and housing.

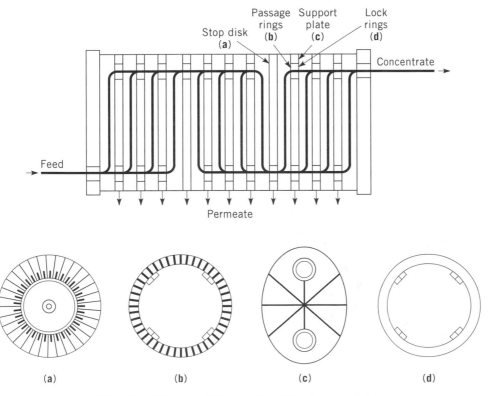

Fig. 15. Plate-and-frame ultrafiltration module.

withdrawal tube. An open-mesh spacer is placed on top of the membrane, and both the mesh and the membrane are wrapped spirally around the tube (Fig. 16).

Spiral-wound cartridges are inserted in series into cylindrical pressure vessels. Feed flows parallel to the membrane surfaces in the channel defined by the mesh spacer which acts as a turbulence promoter. Permeate flows into the center permeate-withdrawal tube which is sealed through the housing end caps.

Supported Tube.   There are three types of supported tubular membranes: cast in place (integral with the support tube), cast externally and inserted into the tube (disposable linings), and dynamically formed membranes.

The most common supported tubes are those with membranes cast in place (Fig. 17). These porous tubes are made of resin-impregnated fiber glass, sintered

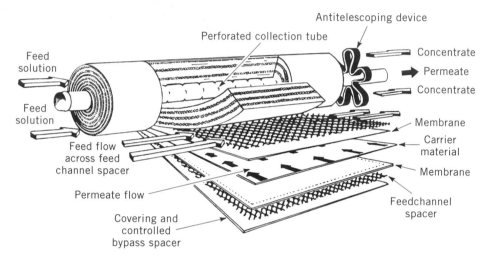

Fig. 16. Spiral-wound membrane configuration (51).

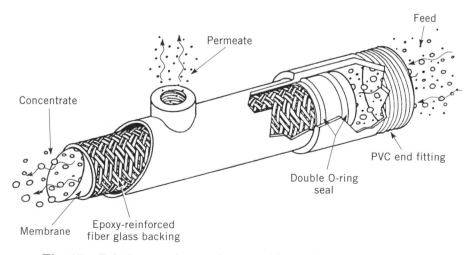

Fig. 17. Tubular membrane element with membrane cast in place.

polyolefins, and similar materials. Typical inside diameters are ca 25 mm. The tubes are most often shrouded to aid in permeate collection and reduce airborne contamination.

External cast membranes are first formed on the inside of paper, polyester, or polyolefin tubes. These are then inserted into reusable porous stainless-steel support tubes; inside diameters are ca 12 mm. The tubes are generally shrouded in bundles to aid in permeate collection.

Tubes for dynamic membranes are usually smaller (ca 6-mm ID). Typically, the tubes are porous carbon or stainless steel with inorganic membranes (silica, zirconium oxide, etc) formed in place.

Self-Supporting Tubes. Depending on the membrane material and operating pressure, self-supporting tubes are less than 2-mm ID; inside diameters as small as 0.04 mm are commercially available. Hollow fibers with the skin on the inside are extruded from a set of concentric nozzles. Membrane casting solution is forced through the outer annulus while diluent nonsolvent is pumped through the center (52).

A large number of fibers are cut to length, and potted in epoxy resin at each end (see EMBEDDING). The fiber bundle is shrouded in a cylinder which aids in permeate collection, reduces airborne contamination, and allows back pressing of the membrane. Hollow-fiber membranes (qv) have also found use in ultrafiltration.

Systemization

Each of the membrane devices may be assembled by connecting the modules into combinations of series, parallel-flow paths, or both. These assemblies are connected to pumps, valves, tanks, heat exchangers, instrumentation, and controls to provide complete systems.

Because of the broad differences between ultrafiltration equipment, the performance of one device cannot be used to predict the performance of another. Comparisons can only be made on an economic basis and only when the performance of each is known.

Uses

Applications of ultrafiltration are summarized in Table 2.

Table 2. Ultrafiltration Applications

Application	Process	Kirk-Othmer article	Refs.
electrophoretic paint	control of properties, recovery of solids from rinse systems	PAINT	52–56
dairy wheys	protein recovery, concentration, purification, diafiltration	MILK AND MILK PRODUCTS	33–38, 57–59

Table 2. (*Continued*)

Application	Process	Kirk-Othmer article	Refs.
milk	cheese and yogurt mfg, 15–20% yield improvement, standardization	MILK AND MILK PRODUCTS	38, 60–68
oil–water emulsions	concentration	EMULSIONS	32, 40, 69, 70
effluents of wool and yarn scouring	lanolin recovery, pollution abatement	TEXTILES; WOOL	71, 72
enzymes	concentration, purification	ENZYME APPLICATIONS	9–12, 73–78
biological reactors	antibiotic mfg, alcohol fermentation, sewage treatment	ANTIBIOTICS; WATER, SEWAGE	74, 79–82
vegetable proteins		FERMENTATION; FOODS, NONCONVENTIONAL; PROTEINS	75, 83–85
latex concentration		ELASTOMERS, SYNTHETIC; LATEX TECHNOLOGY	67
production of pure[a] water		WATER	51, 86
pulp and paper	lignosulfonate separation from spent liquor	PAPER; PULP; WOOD	
blood and blood products	fractionation, purification	FRACTIONATION, BLOOD	
vaccines	concentration, purification	VACCINE TECHNOLOGY	
biotechnology products	concentration, purification	GENETIC ENGINEERING; HORMONES; INSULIN AND OTHER ANTIDIABETIC DRUGS	

[a]Virus-free.

Nomenclature

Symbol	Definition
A	membrane area
C	concentration
C_b	bulk concentration of all retained species
C_{bi}	concentration of species i in retentate
C_{bR}	concentration of bulk of the retentate
C_f	final concentration
C_g	gel concentration
C_o	initial concentration
C_{pi}	concentration of species i in permeate
C_t	concentration at time t
C_w	concentration at membrane wall

E	electric field
E_c	critical field strength
i	species
J	permeate flux on membrane filtration rate
J_R	flux of retentate toward membrane surface
K	mass-transfer coefficient
K_g	gel hydraulic permeability coefficient
K_m	membrane hydraulic permeability coefficient
P_c	critical pressure
R, R_i, R_o	rejection
R_m, R_g	hydraulic resistances of membrane gel
t	time
V	volume
V_o	constant retentate volume
V_p	total permeate volume
X	distance from membrane
Y_p, Y_r	fractional recovery of permeable solids in permeate
μ	fluid viscosity
ΔP	membrane pressure drop
δ	rejection of solute

BIBLIOGRAPHY

"Ultrafiltration" in *ECT* 3rd ed., Vol. 23, pp. 439–461, by P. R. Klinkowski, Dorr-Oliver, Inc.

1. M. C. Porter, *AIChE Symp. Ser.* **73**, 83 (1977).
2. R. E. Kesting, *Synthetic Polymeric Membranes*, McGraw-Hill, New York, 1971, pp. 8–9.
3. C. M. Hansen, *I&EC Product Research and Development* **8**(1), (Mar. 1969).
4. J. D. Crowley, G. S. Teague, and J. W. Lowe, *J. Paint Technol.* **38**, 270 (May 1966).
5. H. Strathman, K. Kock, P. Amar, and R. W. Baker, *Desalination* **16**, 179 (1975).
6. H. P. Gregor, in E. Selegny, ed., *Charged Gels and Membranes*, D. Reidel Publishing Co., Holland, 1976.
7. R. E. Kesting, in Ref. 2, p. 140.
8. Brit. Pat. 147,4594 (May 25, 1977), P. Klinkowski and E. Ondera (to Dorr-Oliver Inc.).
9. H. P. Gregor and P. W. Rauf, *Enzyme Coupled Ultrafiltration Membranes*, work supported by National Science Foundation Grant GI-32497.
10. R. A. Korus and A. C. Olson, *J. Food Sci.* **42**, 258 (1977).
11. U.S. Pat. 4,033,822 (July 5, 1977), H. P. Gregor.
12. O. Velicangil and J. A. Howell, *Biotechnology and Bioengineering*, Vol. 23, John Wiley & Sons, Inc., New York, 1981, pp. 843–854.
13. S. S. Wang, B. Davidson, C. Gillespie, L. R. Harris, and D. S. Lent, *J. Food Sci.* **45**, 700 (1980).
14. A. Michaels, H. Bixler, R. Hausslein, and S. Flemming, *OSW Research Development*, Report No. 149, U.S. Government Printing Office, Washington, D.C., Dec. 1965; *Ind. Eng. Chem.* **57**, 32 (1945).
15. A. Michaels, *Ind. Eng. Chem.* **57**(10), 32 (1965).
16. U.S. Pat. 3,977,967 (Aug. 31, 1976), O. C. Trulson and L. M. Litz (to Union Carbide Corp.).
17. U.S. Pat. 4,060,488 (Nov. 28, 1977), F. W. Hoover and R. K. Iler (to E. I. du Pont de Nemours & Co., Inc.).

18. H. Z. Friedlander and L. M. Litz in M. Bier, ed., *Membrane Processes in Industry and Biomedicine*, Plenum Press, New York, 1971.
19. R. Fleischer, P. B. Price, and E. Symes, *Science* **143**, 249 (1964).
20. C. R. Bennet, R. S. King, and P. J. Petersen, *Pressure Effects on Macromolecular—Water Interactions with Synthetic Membranes*, ACS 181st National Meeting, Atlanta, Ga., Mar. 1981.
21. G. B. Westermann-Clark, Ph.D. dissertation, Carnegie-Mellon University, Pittsburgh, Pa., 1981.
22. S. Sourirajan, *Ind. Eng. Chem. Fundam.* **2**(1), 51 (1963).
23. S. Jacobs, *Filtr. Sep.*, 525 (Sept.–Oct. 1972).
24. M. P. Freeman, *Colloid Interface Sci.* **5**, 133 (1976).
25. L. Zeman and M. Wales, in A. F. Turbak, ed., *Synthetic Membranes*, Vol. 2, ACS Symposium Series 154, American Chemical Society, Washington, D.C., 1980.
26. A. S. Michaels, *Sep. Sci. Technol.* **15**, 1305 (1980).
27. P. Dejmek, "Permeability of the Concentration Polarization Layer in Ultrafiltration of Macro Molecules," *Proceedings of the International Symposium, Separation Processes by Membranes*, Paris, Mar. 13–14, 1975.
28. Q. T. Nguyen, P. Aptel, and J. Neel, *J. Membr. Sci.* **6**(1), 71 (1980).
29. P. S. Leung, in A. R. Cooper, ed., *Ultrafiltration Membranes and Applications*, Plenum Press, New York, 1980, pp. 415–421.
30. W. F. Blatt, A. Dravid, A. S. Michaels, and L. Nelson, in J. E. Finn, ed., *Membrane Science and Technology*, Plenum Press, New York, 1970, pp. 47–97.
31. R. F. Madsen and W. K. Nielsen, in Ref. 29, pp. 423–438.
32. Brit. Pat. 1,456,304 (Nov. 24, 1976) (to Abcor, Inc.).
33. A. C. Epstein and S. R. Korchin, paper presented at *91st National Meeting, AIChE*, Detroit, Mich., Aug. 1981.
34. M. W. Hickey, R. D. Hill, and B. R. Smith, *N.Z. J. Dairy Sci. Technol.* **15**(2), 109 (1980).
35. D. N. Lee and R. L. Merson, *J. Dairy Sci.* **58**, 1423 (1975).
36. D. N. Lee and R. L. Merson, *J. Food Sci.* **41**, 403 (1976); **41**, 778 (1976).
37. L. L. Muller and W. J. Harper, *J. Agric. Food Chem.* **27**, 662 (1979).
38. W. J. Harper, in Ref. 29, pp. 321–347.
39. J. A. Howell and O. Velicangil, in Ref. 29, pp. 217–229.
40. P. A. Bailey, *Filtr. Sep.* **14**(1), 53 (1977).
41. F. E. McDonough and R. E. Hargrove, *J. Milk Food Technol.* **35**(2), 102 (1972).
42. R. G. Semerad, "Sanitary Considerations Involved with Membrane Equipment," *Proceedings of the Whey Product Conference*, Atlantic City, N.J., 1976.
43. N. C. Beaton, *J. Food Protection*, **42**, 584 (July 1979).
44. B. S. Horton, *N. Z. J. Dairy Sci. Technol.* **14**(2), 93 (1979).
45. M. Bier, in Ref. 18.
46. U.S. Pat. 3,945,900 (Mar. 23, 1976), P. R. Klinkowski (to Dorr-Oliver, Inc.).
47. J. M. Radovich and R. E. Sparks, in Ref. 29, pp. 249–268.
48. J. D. Henry, Jr., L. Lawler, and C. H. A. Kuo, *AIChE J.* **23**, 851 (1977).
49. J. D. Henry, Jr., and C. H. A. Kuo, paper presented at *The Symposium on Recent Developments in Colloidal Phenomena*, AIChE National Meeting, New Orleans, La., Nov. 1981.
50. P. R. Klinkowski and N. C. Beaton, *J. Sep. Proc. Tech.* (1991).
51. *Abcor Sanitary Spiral-Wound Ultrafiltration Modules*, Product Bulletin, Abcor, Inc., Wilmington, Mass., 1981.
52. B. R. Breslau, A. J. Testa, B. A. Milnes, G. Medjanis, in Ref. 29, pp. 109–127.
53. R. A. Scaddan, *Ind. Finish. Surf. Coat.* **27**, 326 (1975).
54. G. E. F. Brewer, paper presented at *Electrocoat 72*, Electrocoating Committee and the National Paint and Coatings Association, Chicago, Ill., Oct. 2–4, 1972.
55. B. J. Weissman, in Ref. 54.

56. U.S. Pat. 3,663,397 (May 16, 1972), L. R. LeBras and J. Ostrowski (to PPG Industries); U.S. Pats. 3,663,398 and 3,663,402–403 (May 16, 1972), R. Christenson and R. R. Twack (to PPG Industries); U.S. Pats. 3,663,399 and 3,663,404 (May 16, 1972), F. M. Loop (to PPG Industries); U.S. Pats. 3,663,400–401 and 3,663,405 (May 16, 1972), R. Christenson and L. R. LeBras (to PPG Industries); U.S. Pat. 3,633,406 (May 16, 1972), L. R. LeBras and R. R. Twack (to PPG Industries).

57. F. E. McDonough, W. A. Mattingly, and J. H. Vestal, *J. Dairy Sci.* **54**, 1406 (1971).

58. B. S. Horton, R. L. Goldsmith, and R. R. Zall, *Food Technol.* **26**(2) (1972).

59. M. E. Matthews, *N. Z. J. Dairy Sci. Tech.* **14**(2), 86 (1979).

60. F. Lang and A. Lang, *Milk Ind.* **78**(9), 16 (1976).

61. O. J. Olsen, in Ref. 6, Vol. 2.

62. J. L. Maubois, "Application of Ultrafiltration to Milk Treatment for Cheese Making," in Ref. 27.

63. H. R. Covacevich and F. V. Kosikowski, *J. Food Sci.* **42**, 362 (1977).

64. J. E. Ford, F. A. Glover, and K. J. Scott, *Int. Dairy Congress*, 1068 (1978).

65. F. A. Glover, P. J. Skudder, P. H. Stuthart, and E. W. Evans, *J. Dairy Res.* **45**, 291 (1978).

66. S. Jepsen, *J. Cultured Dairy Products* **14**(1), 5 (1979).

67. J. L. Maubois, in Ref. 29, pp. 305–318.

68. P. R. Poulsen, *J. Dairy Sci.* **61**(6), 807 (1978).

69. R. L. Goldsmith, D. A. Roberts, and D. L. Burre, *J. Water Pollut. Control Fed.* **46**, 2183 (1974).

70. I. K. Bansal, *Ind. Water Eng.* **13**, 6 (Oct.–Nov. 1976).

71. N. C. Beaton, *Textile Institute and Industry* **13**, 361 (Nov. 1975).

72. J. A. C. Pearson, C. A. Anderson, and G. F. Wood, *J. Water Pollut. Control Fed.* **48**, 945 (1976).

73. D. I. C. Wang, A. J. Sinskey, and T. A. Butterworth, in Ref. 30.

74. N. C. Beaton, in Ref. 29, pp. 373–404.

75. H. S. Olsen and J. Adler-Nissen, in Ref. 25, Vol. 2.

76. U.S. Pat. 3,720,583 (Mar. 13, 1973), E. E. Fisher (to A. E. Staley).

77. J. Hong, G. T. Tsao, P. C. Wankat, in Ref. 12, pp. 1501–1516.

78. W. D. Deeslie and M. Cheryan, *J. Food Sci.* **46** 1035 (1981).

79. J. Gawel and F. V. Kosikowski, *J. Food Sci.* **43**, 1717 (1978).

80. U.S. Pat. 3,472,765 (Oct. 14, 1969), W. E. Budd and R. W. Okey (to Dorr-Oliver, Inc.).

81. A. G. Fane, C. J. D. Fell, and M. T. Nor, in Ref. 29, pp. 631–658.

82. A. S. Michaels, C. R. Robertson, and S. N. Cohen, paper presented at *The 180th National Meeting ACS*, San Francisco, Calif., Aug. 1980.

83. G. Jackson, M. M. Stawiarski, E. T. Wilhelm, R. L. Goldsmith, and W. Eykamp, *AIChE Symp. Ser.* **70**, 514 (1974).

84. M. Cheryan, in Ref. 29, pp. 343–325.

85. J. T. Lawhon, L. J. Manak, and W. E. Lusas, in Ref. 29.

86. G. Belfort, T. F. Baltutis, and W. F. Blatt, in Ref. 29, pp. 439–474.

General References

Reference 30 is also a general reference.

N. C. Beaton, "Advances in Enzyme and Membrane Technology," *Institute of Chemical Engineers Symposium Series No. 51*, Institute of Chemical Engineers, London, 1977, pp. 59–70.

N. C. Beaton and H. Steadly, in N. Li, ed., *Recent Developments in Separation Science*, Vol. 7, CRC Press, Boca Raton, Fla., 1982, pp. 2–25.

A. R. Cooper, ed., *Ultrafiltration Membranes and Applications*, *Proceedings of 178th National ACS Meeting*, Washington, D.C., 1979, Plenum Press, New York, 1980.

R. P. deFillipi and R. L. Goldsmith, in J. E. Flinn, ed., *Membrane Science and Technology*, Plenum Press, New York, 1970, pp. 33–46.

R. J. Gross and J. F. Osterle, *J. Chem. Phys.* **49**, 228 (1968).

S. Hwang and K. Kammermeyer, *Membranes in Separations*, John Wiley & Sons, Inc., New York, 1975; good study of membrane transport phenomenon.

R. E. Kesting, *Synthetic Polymeric Membranes*, McGraw-Hill, New York, 1971; good bibliographies.

A. S. Michaels, L. Nelson, and M. C. Porter, in M. Bier, ed., *Membrane Processes in Industry and Biomedicine*, Plenum Press, New York, 1971, pp. 197–232.

U.S. Pat. 3,615,024 (Oct. 26, 1971), A. S. Michaels (to Amicon Corp.).

A. S. Michaels, paper presented at *Clemson University Membrane Technology Conference*, Mar. 1979; OWRT Contract No. 14-34-0001-8548.

A. S. Michaels, *Chem. Technol.* **2**(1), 37 (Jan. 1981).

A. S. Michaels, *Desalination* **35**, 329 (1980).

L. Mir, W. Eykamp, and R. L. Goldsmith, *Ind. Water Eng.* **14**, (May–June 1977).

M. C. Porter and L. Nelson, in N. N. Li, ed., *Recent Developments in Separation Science*, Vol. 2, CRC Press, Cleveland, Ohio, 1972, pp. 227–267.

M. C. Porter, *Ind. Eng. Chem. Prod. Res. Dev.* **2**, 234 (1972).

M. C. Porter, in P. A. Schweitzer, ed., *Handbook of Separation Techniques for Chemical Engineers*, McGraw-Hill, New York, 1979.

R. N. Rickles, *Membranes, Technology and Economics*, Noyes Development Co., Park Ridge, N.J., 1967.

H. Strathman, K. Kock, P. Amar, and R. W. Baker, *Desalination* **16**, 179 (1975).

A. F. Turbak, ed., *Synthetic Membranes*, Vols. 1 and 2, ACS Symposium Series 153 and 154, ACS, Washington, D.C., 1981.

Equipment Available for Membrane Processes, Bulletin No. 115, International Dairy Federation, Brussels, p. 1979 (available through Library of Congress or Harold Wainess and Associates, Northfield, Ill.).

L. J. Zeman and A. L. Sydney, *Microfiltration and Ultrafiltration*, Marcel Dekker, Inc., New York, 1996.

RALF KURIYEL
Millipore Corporation

ULTRAMARINE. See PIGMENTS, INORGANIC.

ULTRAVIOLET ABSORBERS. See UV STABILIZERS.

UMBER. See PIGMENTS, INORGANIC.

UNADS. See RUBBER CHEMICALS.

UNDERGROUND STORAGE TANKS. See TANKS AND PRESSURE VESSELS.

UNITS AND CONVERSION FACTORS

The barleycorn, inch, foot, yard, rod, furlong, mile, league, ell, fathom, and chain are units of length that the North American colonies inherited from the British. The inch was originally the length of three barleycorns; the yard was the distance from the tip of the nose to the tip of the middle finger on the outstretched arm of a British king (Henry I); the acre was the amount of land plowed by a yoke of oxen in a day. This system of units is very old and may be traced back to ancient Egypt.

A jumble of units existed throughout the world even until the late eighteenth century. In 1790, the French National Assembly requested of the French Academy of Sciences that it work out a system of units suitable for adoption by the whole world. This system was based on the meter as a unit of length and the gram as a unit of mass. Industry, commerce, and especially the scientific community benefited greatly. In 1893, the United States actually adopted the meter and the kilogram as the fundamental standards of length and mass. Although the spellings metre and litre are preferred by the author and ASTM, meter and liter are used in the *Encyclopedia*.

The foundation to international standardization of units was laid with an international treaty, the Meter Convention, which was signed by 17 countries, including the United States, in 1875. This treaty established a permanent International Bureau of Weights and Measures and defined the meter and the kilogram, from which evolved a set of units for the measurement of length, area, volume, capacity, and mass. Also established was the General Conference on Weights and Measures (CGPM), which was to meet at regular intervals to consider any needed improvements in the standards. The National Institute of Standards and Technology (NIST) represents the United States in these activities.

From these early beginnings, several variants of the metric system evolved. With the addition of the second as a unit of time, the centimeter–gram–second (cgs) system was adopted in 1881. In the early 1900s, practical measurements in metric units began to be based on the meter–kilogram–second (mks) system. In 1935, the International Electrotechnical Commission adopted a proposal to link the mks system of mechanics with the electromagnetic system of units by adding the ampere as a base unit and forming the mksA (meter–kilogram–second–ampere) system.

In 1954, the 10th CGPM added the degree Kelvin as the unit of temperature and the candela as the unit of luminous intensity. At the time of the 11th CGPM in 1960, this new system with six base units was formalized with the title International System of Units. Its abbreviation in all languages is SI, from the French *Le Système International d'Unités*.

Since 1960, various refinements to the system have been made, including redefinition of the second based on the atomic frequency of cesium; change of the name of the unit of temperature from degree Kelvin to the kelvin (symbol K); redefinition of the candela (all in 1967); addition of a seventh base unit, the mole (mol), as the unit of amount of substance; the pascal (Pa) as a special name for the SI unit of pressure or stress, equal to a newton per square meter; the siemens (S) as a special name for the unit of electric conductance, equal to the ampere per volt (all in 1971); addition of two SI units for ionizing radiation, the becquerel (Bq) as the unit of activity, equal to one reciprocal second, and the gray (Gy) as the unit of absorbed dose, equal to one joule per kilogram; prefixes for 10^{18}, exa (E), and 10^{15}, peta (P) (all in 1975); addition of the sievert (Sv) as the unit of dose equivalent, equal to one joule per kilogram; further redefinition of the candela; recognition of both l and L as symbols for liter (all in 1979); and interpretation of the radian and the steradian as dimensionless derived units for which the CGPM allows the freedom of use or nonuse in expressions for SI-derived units (1980).

In order to increase the precision of realization of the base unit meter, the definition based on the wavelength of a krypton-86 radiation was replaced in 1983 by one based on the speed of light. Also added were the prefixes zetta (Z) for 10^{21}, zepto (z) for 10^{-21}, yotta (Y) for 10^{24}, and yocto (y) for 10^{-24}.

In 1995 the 20th CGPM approved eliminating the class of supplementary units as a separate class in SI. Thus the new SI consists of only two classes of units: base units and derived units, with the radian and steradian subsumed into the class of derived units of the SI.

Advantages of SI

SI is a decimal system. Fractions have been eliminated, and multiples and submultiples are formed by a system of prefixes ranging from yotta, for 10^{24}, to yocto, for 10^{-24}. Calculations, therefore, are greatly simplified.

Each physical quantity is expressed in one and only one unit, eg, the meter for length, the kilogram for mass, and the second for time. Derived units are defined by simple equations relating two or more base units. Some are given special names, such as newton for force and joule for work and energy.

In an energy-conscious world, SI provides a direct relationship among mechanical, electric, chemical, thermodynamic, molecular, and solar forms of energy. All power ratings are given in watts.

The system is coherent. There is no duplication of units for a quantity, and all derived units are obtained by a direct one-to-one relation of base units or derived units; eg, one newton is the force required to accelerate one kilogram at the rate of one meter per second squared; one joule is the energy involved when a force of one newton is displaced one meter in the direction of the force; and one watt is the power that in one second gives rise to the production of energy of one joule.

The same simplified system of units can be used by the research scientist, the technician, the practicing engineer, and by members of the lay public.

The International System of Units

SI rests on seven base units and a number of derived units, some of which have special names. A list of these units is given in the introduction to this volume.

Base Units.   *Meter.*   The meter is the length of the path traveled by light in a vacuum during a time interval of 1/299 792 458 of a second.

This definition, adopted in 1983 by the 17th CGPM, superseded the definition based on the wavelength of a krypton-86 radiation.

Kilogram. The kilogram is the unit of mass; it is equal to the mass of the international prototype of the kilogram.

This international prototype, adopted by the 1st and 3rd CGPM in 1889 and 1901, is a particular cylinder of platinum–iridium kept at the International Bureau of Weights and Measures near Paris. It is the only base unit still defined by an artifact.

Second. The second is the duration of 9 192 631 770 periods of the radiation corresponding to the transition between the two hyperfine levels of the ground state of the cesium-133 atom.

This definition was adopted by the 13th CGPM in 1967 to replace previous definitions based on the mean solar day and, later, the tropical year.

Ampere. The ampere is that constant current which, if maintained in two straight, parallel conductors of infinite length, of negligible circular cross section, and placed one meter apart in a vacuum, would produce between these conductors a force equal to 2×10^{-7} newton per meter of length.

This definition was adopted by the 9th CGPM in 1948. The electrical units for current and resistance had been first introduced by the International Electrical Congress in 1893. These international units were replaced officially by so-called absolute units by the 9th CGPM.

Kelvin. The kelvin unit of thermodynamic temperature is the fraction 1/273.16 of the thermodynamic temperature of the triple point of water (0.01°C).

Before the 13th CGPM in 1967, when this definition was adopted, the unit was called the degree Kelvin (symbol °K, now K).

Mole. The mole is the amount of substance of a system that contains as many elementary entities as there are atoms in 0.012 kilogram of carbon-12.

When the mole is used, the elementary entities must be specified and may be atoms, molecules, ions, electrons, other particles, or specified groups of such particles.

This definition was adopted by the 14th CGPM in 1971. Previously, physicists and chemists had based the amount of substance, then called gram–atom or gram–molecule, on the atomic weight of oxygen (by general agreement taken as 16), but with slight differences depending on the isotope used. The 1971 agreement assigned the value of 12 to the isotope 12 of carbon to give a unified scale. At its 1980 meeting, the International Committee for Weights and Measures (CIPM), under the authority of the CGPM, specified that in this definition "it is understood that unbound atoms of carbon-12, at rest and in their ground state, are referred to."

Candela. The candela is the luminous intensity, in a given direction, of a source that emits monochromatic radiation of frequency 540×10^{12} hertz and that has a radiant intensity in that direction of 1/683 watt per steradian.

This unit, most recently defined by the 16th CGPM in 1979, replaced the candle and, later, the new candle and a definition of the candela based on the luminous intensity of a specified projected area of a blackbody emitter at the temperature of freezing platinum.

Derived Units. The largest class of SI units, the derived units, consists of a combination of base and derived units according to the algebraic relations linking the corresponding quantities. When two or more units expressed in base units are multiplied or divided to obtain derived quantities, the result is a unit value. The fact that no numerical constant is introduced maintains this coherent system. Special names have been given to 21 derived units. For example, the joule is the name given to the product of a newton and a meter; the siemens is the name given to the quotient of an ampere divided by a volt. A list of derived SI units is given in the introduction to this volume (pp. xiv–xvi). The SI units with special names and their definitions are given in Table 1.

Prefixes. In SI, 20 prefixes are used and are directly attached to form decimal multiples and submultiples of the units (see the introduction to this volume, p. xvi). Prefixes indicate the order of magnitude, thus eliminating nonsignificant digits and providing an alternative to powers of 10; eg, 45 300 kPa becomes 45.3 MPa and 0.0043 m becomes 4.3 mm.

Preferably, the prefix should be selected in such a way that the resulting value lies between 0.1 and 1000. To minimize variety, it is recommended that prefixes representing 1000 raised to an integral power be used. For example, lengths can be expressed in micrometers, millimeters, meters, or kilometers and still meet the 0.1-to-1000 limits. There are three exceptions to these rules: (*1*) In expressing area and volume, the intermediate prefixes may be required, eg, hm^2, dL, and cm^3. (*2*) In tables of values, for comparison purposes it is generally preferable to use the same multiple throughout, and one particular multiple is also used in some applications. For example, millimeter is used for linear dimensions in mechanical engineering drawings even when the values are far outside the range 0.1 to 1000 mm. (*3*) The centimeter is often used for body-related measurements, eg, clothing.

Compound Units. It is usually recommended that only one prefix be used in forming a multiple of a compound unit, and that it should be attached to the numerator. An exception is the base unit kilogram, where it appears in the denominator. Multiples of kilogram are formed by attaching the prefix to the word gram (g). Compound prefixes are not used; eg, 1 pF is correct, not 1 $\mu\mu$F.

Units Used with SI. A number of non-SI units are used in SI (Table 2).

Time. Although the SI unit of time is the second, the minute, hour, day, and other calendar units may be necessary where time relates to calendar cycles. Automobile velocity is, for example, expressed in kilometers per hour.

Plane Angle. The radian, although the preferred SI unit, is not always convenient, and the use of the degree is permissible. Minute and second should be reserved for special fields such as cartography.

Volume. The special name liter (L) has been approved for the cubic decimeter, but its use is restricted to volumetric capacity, dry measure, and measure of fluids (both gases and liquids).

Mass. The metric ton (symbol t), equal to 1000 kg, is used widely in commerce, although the megagram (Mg) is the appropriate SI unit.

Units Used Temporarily with SI. Additional non-SI units are used with SI units until the CIPM considers their use no longer necessary (Table 3).

Length. The nautical mile is a special unit employed for marine and aerial navigation to express distances. The conventional value was adopted by the 1st International Extraordinary Hydrographic Conference, Monaco, 1929, under the name International nautical mile.

Table 1. SI Derived Units with Special Names

Quantity	Name	Symbol	Formula	Definition
absorbed dose	gray[a]	Gy	J/kg	absorbed dose when energy per unit mass imparted to matter by ionizing radiation is one joule per kilogram
activity	becquerel	Bq	1/s	activity of radionuclide decaying at rate of one spontaneous nuclear transition per second
angle				
plane	radian	rad		plane angle between two radii of a circle that cut off on the circumference an arc equal in length to the radius
solid	steradian	sr		solid angle that, having its vertex in the center of a spere, cuts off an area of the surface of the sphere equal to that of a square with sides of length equal to the radius of the sphere
Celsius temperature	degree Celsius	°C		equal to kelvin and used in place of kelvin for expressing Celsius temperature, t, defined by equation $t = T - T_0$, where T is the thermodynamic temperature and $T_0 = 273.15$ K by definition
dose equivalent	sievert	Sv	J/kg	dose equivalent when absorbed dose of ionizing radiation multiplied by dimensionless factors Q (quality factor) and N (product of any other multiplying factors) stipulated by the International Commission on Radiological Protection is one joule per kilogram
electric capacitance	farad	F	C/V	capacitance of a capacitor between plates of which there appears a difference of potential of one volt when charged by a quantity of electricity equal to one coulomb
electric charge, quantity of electricity	coulomb	C	A·s	quantity of electricity transported in one second by current of one ampere
electric conductance	siemens	S	A/V	electric conductance of a conductor in which current of one ampere is produced by electric potential difference of one volt
electric inductance	henry	H	Wb/A	inductance of a closed circuit in which electromotive force of one volt is produced when electric current in circuit varies uniformly at rate of one ampere per second

631

Table 1. (Continued)

Quantity	Name	Symbol	Formula	Definition
electric potential, potential difference, electromotive force	volt	V	W/A	difference of electric potential between two points of conductor carrying constant current of one ampere, when power dissipated between these points is equal to one watt
electric resistance	ohm	Ω	V/A	electric resistance between two points of conductor when constant difference of potential of one volt, applied between these two points, produces in this conductor a current of one ampere, this conductor not being the source of any electromotive force
energy, work, quantity of heat	joule	J	N·m	work done when point of application of force of one newton is displaced a distance of one meter in the direction of force
force	newton	N	kg·m/s²	that force which, when applied to a body having mass of one kilogram, gives it acceleration of one meter per second squared
frequency	hertz	Hz	1/s	frequency of periodic phenomenon of which period is one second
illuminance	lux	lx	lm/m²	illuminance produced by luminous flux of one lumen uniformly distributed over surface of one square meter
luminous flux	lumen	lm	cd·sr	luminous flux emitted in a solid angle of one steradian by point source having uniform intensity of one candela
magnetic flux	weber	Wb	V·s	magnetic flux which, linking a circuit of one turn, produces in it an electromotive force of one volt as it is reduced to zero at uniform rate in one second
magnetic flux density	tesla	T	Wb/m²	magnetic flux density given by magnetic flux of one weber per square meter
power, radiant flux	watt	W	J/s	power which gives rise to the production of energy at rate of one joule per second
pressure or stress	pascal	Pa	N/m²	pressure or stress of one newton per square meter

[a]The gray is also used for the ionizing radiation quantities, specific energy imparted, kerma, and absorbed dose index, which have the SI unit joule per kilogram.

632

Table 2. Units in Use with SI

Unit	Symbol	Value in SI units
minute	min	1 min = 60 s
hour	h	1 h = 60 min = 3600 s
day	d	1 d = 24 h = 86 400 s
degree	°	1° = $(\pi/180)$ rad
minute	′	1′ = $(1/60)°$ = $(\pi/10\ 800)$ rad
second	″	1″ = $(1/60)′$ = $(\pi/648\ 000)$ rad
liter	L	1 L = 1 dm^3 = 10^{-3} m^3
metric ton	t	1 t = 10^3 kg

Table 3. Units Temporarily in Use with SI

Unit	Symbol	Value in SI units
nautical mile		1 nautical mile = 1852 m
knot		1 nautical mile per hour = (1852/3600) m/s
hectare	ha	1 ha = 1 hm^2 = 10^4 m^2
kilowatt-hour	kWh	1 kWh = 3.6 MJ
barn	b	1 b = 10^{-28} m^2
bar	bar	1 bar = 10^5 Pa
curie	Ci	1 Ci = 3.7×10^{10} Bq
roentgen	R	1 R = 2.58×10^{-4} C/kg
rad	rd	1 rd = 0.01 Gy
rem	rem	1 rem = 0.01 Sv = 10 mSv

Area.   The SI unit of area is the square meter (m^2). The hectare (ha) is a special name for the square hectometer (hm^2). Large land or water areas are generally expressed in hectares or in square kilometers (km^2).

Energy.   The kilowatthour (kWh) is widely used as a measure of electric energy, but it should eventually be replaced by the megajoule (MJ) (1 kWh = 3.6 MJ).

Pressure.   Although both bar and torr are widely used for pressure, the use of the torr is strongly discouraged in favor of the pascal and its multiples. The bar, however, is still approved for temporary use. The millibar is widely used in meteorology (1 mbar = 100 Pa).

Radiation Units.   Units in use for activity of a radionuclide, ie, the curie, the roentgen (exposure to x and gamma rays), the rad (absorbed dose), and the rem (dose equivalent), should eventually be replaced by the becquerel (Bq), coulomb per kilogram (C/kg), gray (Gy), and the sievert (Sv), respectively.

Units to Be Abandoned.   Except for the non-SI units referred to in the two preceding sections, a great many other metric units should be avoided in order to maintain the advantages of using one common coherent system of units, eg, units of the cgs system with special names such as the erg, dyne, poise, stokes, gauss, oersted, maxwell, stilb, phot, and angstrom. Other unit names to be deprecated are the kilogram-force, calorie, torr, millimeter of mercury, and the mho (see also pp. xxvi–xxvii in this *Encyclopedia*).

Mass, Force, and Weight.   Weight is a force: the weight of a body is the product of its mass and the acceleration due to gravity.

The use of the same term for units of force and mass causes confusion. When the non-SI units are used, a distinction should be made between force and mass, eg, lbf to denote force in gravimetric engineering units, and lb for mass.

The term load means either mass or force, depending on its use. A load that produces a vertically downward force because of the influence of gravity acting on a mass may be expressed in mass units. Any other load is expressed in force units.

Temperature.   The kelvin is the SI unit of thermodynamic temperature, and is generally used in scientific calculations. Wide use is made of the degree Celsius (°C) for both temperature and temperature interval. The temperature interval 1°C equals 1 K exactly. Celsius temperature, t, is related to thermodynamic temperature, T, by the following equation:

$$t = T - 273.15$$

The name degree centigrade was dropped in 1948 in favor of the degree Celsius because in some countries the grade has been used as a unit of angular measure.

Pressure and Vacuum.   Pressure is usually designated as gauge pressure, absolute pressure, or, if below ambient, vacuum. Pressures are expressed in pascals with appropriate prefixes. When the term vacuum is used, it should be made clear whether negative gauge pressure or absolute pressure is meant. The correct way to express pressure readings is "at a gauge pressure of 13 kPa" or "at an absolute pressure of 13 kPa."

Quantities and Units Used in Rotational Mechanics.   *Angle, Angular Velocity, and Angular Acceleration.*   Their SI units are rad, rad/s, and rad/s², respectively. Because the radian is here taken to be dimensionless, the units 1, 1/s, and 1/s² are also used where appropriate.

Moment of Force (Torque or Bending Moment).   Moment of force is force times moment arm (lever arm). Its SI unit is N·m.

Moment of Inertia.   Moment of inertia, I, is a property of the mass distribution of a body around an axis ($I = \Sigma mr^2$). Its SI unit is kg·m².

Angular Momentum (Moment of Momentum).   Angular momentum is linear momentum (kg·m/s) times moment arm (m). Its SI unit is kg·m²/s. For a rotating body the total angular momentum is equal to the moment of inertia I (kg·m²) times the angular velocity ω (rad/s or 1/s).

Rotational Kinetic Energy.   Rotational kinetic energy of a rotating body is equal to 1/2 $I\omega^2$. Its SI unit is J.

Rotational Work.   Rotational work is equal to torque (N·m) times angle of rotation (rad). Its SI unit is J.

Torsional Stiffness (Torsion Constant).   Torsional stiffness of a body is applied torque (N·m) divided by angle of twist (rad). Its SI unit is N·m/rad.

Centripetal Acceleration.   Centripetal acceleration, v^2/r or $\omega^2 r$, where v is the tangential linear velocity (m/s), r the radius (m), and ω the angular velocity (rad/s), is, like any other linear acceleration, measured in SI units m/s². Centripetal force, equal to mass times centripetal acceleration, is, like any force in SI, measured in newtons.

Impact Energy Absorption.   This quantity, often incorrectly called impact resistance or impact strength, is measured in terms of the work required to break a standard specimen; the proper unit is joule.

Nominal Dimensions. Some dimensions do not have an SI equivalent because their values are nominal, that is, a value is assigned for the purpose of convenient designation. For example, a 1-in. pipe has no dimension that is 25.4 mm. Another common example is the 2-by-4 piece of lumber, which is considerably smaller than 50.8 by 101.6 mm in its finished form.

Dimensionless Quantities. Certain quantities, eg, refractive index and relative density (formerly specific gravity), are expressed by pure numbers. In these cases, the corresponding SI unit is the ratio of the same two SI units, which cancel each other, leaving a dimensionless unit. The SI unit of dimensionless quantities may be expressed as 1. Units for dimensionless quantities such as percent and parts per million (ppm) may also be used with SI; in the latter case, it is important to indicate whether the parts per million are by volume or by mass.

Density and Relative Density. Density is mass per unit volume and in SI is normally expressed as kilograms per cubic meter (density of water = 1000 kg/m^3 or 1 g/cm^3). The term specific gravity was formerly the accepted dimensionless value describing the ratio of the density of solids and liquids to the density of water at 4°C or for gases to the density of air at standard conditions. The term specific gravity is being replaced by relative mass density, a more descriptive term.

Style and Usage. If the advantages of SI are to be realized, everyone must use the system in the same manner. Listed below are a number of editorial rules that must be followed:

(*1*) SI symbols are always in roman type, not italics.

(*2*) A space is required between the number and the unit, eg, 150 mm, not 150mm.

(*3*) A period is not placed after a symbol unless the symbol is at the end of a sentence.

(*4*) The plural form of a symbol is the same as the singular. Plurals of unit names are formed by adding an "s", except in henries; hertz, lux, and siemens are not changed.

(*5*) Y, Z, E, P, T, G, and M, the prefixes for 10^6 and above, are capitalized, as are the symbols whose unit names have been derived from proper names, eg, N for newton (Sir Isaac Newton) and Pa for pascal (Blaise Pascal); an exception is the use of L for liter.

(*6*) The product of two or more symbols is indicated by a centered dot and the product of unit names preferably by just a space, eg, N·m for newton meter.

(*7*) A solidus indicates the quotient of two unit symbols and the word per the division of two unit names: m/s for meter per second. The horizontal line or negative powers are also permissible. The solidus or the word per is not repeated in the same expression, eg, acceleration as m/s^2 for meter per second squared and thermal conductivity as W/(m·K) for watt per meter kelvin.

(*8*) An exponent attached to a symbol containing a prefix indicates that the multiple of the unit is raised to the power expressed by the exponent, eg, 1 cm^3 = (10^{-2} m)^3 = 10^{-6} m^3.

(*9*) Compound prefixes are not used, eg, pF, not $\mu\mu$F.

(*10*) Because the comma is used as a decimal marker in many countries, a comma should not be used to separate groups of digits. The digits can be separated into groups of three to the left and right of the decimal point, and a space separates the groups, eg, 1 234 567 or 0.123 456. If there are only four digits, the space can be deleted; eg, 1.1234.

(*11*) Because of the difference in the meaning of the word billion in the United States and most other countries, this term must be avoided; the prefix giga is unambiguous.

(*12*) When using powers with a unit name, the modifier squared or cubed is used after the unit name, except for areas and volumes, eg, second squared, gram cubed, but square millimeter, cubic meter.

Conversion and Rounding. Conversion of quantities should be handled with careful regard to the implied correspondence between the accuracy of the data and the number of digits. In all soft conversions (a soft conversion being defined as the conversion of an existing non-SI measurements to acceptable SI units without a significant change in size or magnitude), the number of significant digits retained should be such that accuracy is neither sacrificed nor exaggerated. Following are some examples.

A length is reported as 75 ft. The exact metric conversion is 22.86 m. If the reported length is a value rounded to the nearest 1 ft, it would be more appropriate to round the metric value to the nearest 0.1 m, ie, 22.9 m. If the 75-ft length, however, was rounded to the nearest 5 ft, then the appropriate rounding would be to the nearest 1 m, or 23 m.

Significant Digits. Any digit that is necessary to define the specific value or quantity is said to be significant. A problem arises, however, when a value of, eg, 4 in. is given. This may be intended to represent 4, 4.0, 4.00, 4.000 or even more accuracy with a corresponding increase in significant digits (equivalent to 102, 101.6, 101.60, and 101.600 mm, respectively).

Tolerances. Linear Units. The following procedure is used for converting linear units to the proper number of significant places: the maximum and minimum limits in inches are calculated. The corresponding two values are converted exactly into millimeters by multiplying each by the conversion factor 1 in. = 25.4 mm. The results are rounded in accordance with Table 4.

Temperature. General guidance for converting tolerances from degrees Fahrenheit to kelvins or degrees Celsius is given in Table 5.

Table 4. Rounding of Linear Units

Original tolerance, inches		Fineness of rounding, mm
At least	Less than	
0.000 04	0.0004	0.0001
0.000 4	0.004	0.001
0.004	0.04	0.01
0.04	0.4	0.1
0.4		1

Table 5. Temperature Conversion Tolerances

°F	K or °C
2 ± 1	1 ± 0.5
4 ± 2	2 ± 1
10 ± 5	6 ± 3
20 ± 10	11 ± 5.5
30 ± 15	17 ± 8.5
40 ± 20	22 ± 11
50 ± 25	28 ± 14

Pressure or Stress. Values with an uncertainty of more than 2% may be converted without rounding by using the approximate factor 1 lbf/in.^{2} = 7 kPa.

Conversion Factors. Excellent tables of conversion factors are available (1–3), in which the conversion factors are listed both alphabetically and classified by physical quantity.

The conversion factors are presented for ready adaptation to computer readout and electronic data transmission. The factors are written as a number equal to or greater than one and less than 10, with six or fewer decimal places. The number is followed by E (for exponent), a plus or minus symbol, and two digits which indicate the power of 10 by which the number must be multiplied to obtain the correct value. For example:

$$3.523\ 907\ \text{E-}02 = 3.523\ 907 \times 10^{-2} = 0.035\ 239\ 07$$

An asterisk (*) after the sixth decimal place indicates that the conversion factor is exact and that all subsequent digits are zero. Where fewer than six decimal places are shown, more precision is not warranted.

The conversion factors for other compound units not listed can easily be generated from numbers given in the alphabetical list by the substitution of the converted units; eg, to find the conversion factor from lb·ft/s to kg·m/s:

$$1\ \text{lb} = 0.453\ 592\ 4\ \text{kg}$$

$$1\ \text{ft} = 0.3048\ \text{m (exactly)}$$

Substituting,

$$(0.453\ 592\ 4\ \text{kg}) \times (0.3048\ \text{m})/\text{s} = 0.138\ 255\ 0\ \text{kg·m/s}$$

Thus, the factor is 1.382 550 E-01.

To find the conversion factor from oz·in.^{2} to kg·m^2,

$$1\ \text{oz} = 0.028\ 349\ 52\ \text{kg}$$

$$1\ \text{in.}^2 = 0.000\ 645\ 16\ \text{m}^2\ \text{(exactly)}$$

Substituting,

$$(0.028\ 349\ 52\ \text{kg}) \times (0.000\ 645\ 16\ \text{m}^2) = 0.000\ 018\ 289\ 98\ \text{kg·m}^2$$

Thus, the factor is 1.828 998 E-05.

BIBLIOGRAPHY

"Units" in *ECT* 2nd ed., Suppl. Vol., pp. 984–1007, by M. L. McGlashan, The University, Exeter, U.K.; "Units and Conversion Factors" in *ECT* 3rd ed., Vol. 23, pp. 491–502, by R. P. Lukens, American Society for Testing and Materials.

1. *Standard Practice for Use of the International System of Units (SI) E 380-93*, American Society for Testing and Materials, West Conshohocken, Pa., 1993.
2. *American National Standard for Metric Practice*, ANSI/IEEE Std 268-1992, Institute of Electrical and Electronics Engineers, Inc., New York, 1992.
3. B. N. Taylor, *Guide for the Use of the International System of Units (SI)*, NIST Special Publication 811, Superintendent of Documents, U.S. Government Printing Office, Washington, D.C., 1995.

General References

References 1, 2, and 3 are also General References.

The International System of Units (SI), NIST Special Publication 330, Superintendent of Documents, U.S. Government Printing Office, Washington, D.C., 1991.

F. D. Rossini, *Fundamental Measures and Constants for Science and Technology*, CRC Press, Boca Raton, Fla., 1974.

T. Wildi, *Units and Conversion Charts: A Handbook for Engineers and Scientists*, Sperika Enterprises, Ltd., Quebec, Canada, 1988.

ROBERT P. LUKENS
American Society for Testing and Materials
Committee E-43 on SI Practice

UNSATURATED POLYESTERS.   See POLYESTERS, UNSATURATED.

URANIUM AND URANIUM COMPOUNDS

Uranium [7440-61-1] is a naturally occurring radioactive element with atomic number 92 and atomic mass 238.03. Uranium was discovered in a pitchblende [1317-75-5] specimen in 1789 by M. H. Klaproth (1) who named the element *uranit* after the planet Uranus, which had been recently discovered. For 50 years the material discovered by Klaproth was thought to be metallic uranium. Péligot showed that the uranit discovered by Klaproth was really uranium dioxide [1344-57-6], UO_2, and obtained the true elemental uranium as a black powder in 1841 by reduction of UCl_4 [10026-10-5] with potassium (2).

In 1896, Becquerel discovered that uranium was radioactive (3). Becquerel was studying the fluorescence behavior of potassium uranyl sulfate, and observed that a photographic plate had been darkened by exposure to the uranyl salt. Further investigation showed that all uranium minerals and metallic uranium behaved in this same manner, suggesting that this new radioactivity was a property of uranium itself. In 1934, Fermi bombarded uranium with neutrons to produce new radioactive elements (4).

Prior to 1939, uranium played no significant role in technical processes, and was only used as a pigment in glass and ceramics (qv). In 1939, Hahn and Strassman reported their discovery of nuclear fission which announced the dawn of the nuclear age (5). Following this report, uranium gained importance as fuel for nuclear reactors and as starting material for the synthesis of plutonium. There are 19 isotopes of uranium with masses 218, 222, 225–240, and 242 and radioactive half-lives ranging from 1 μs (^{222}U) to 4.468×10^9 yr, the latter for the main naturally occurring (99.27%) uranium isotope, ^{238}U (6).

Uranium is the fourth element of the actinide ($5f$) series. In the actinide series the $5f$ electrons are more effectively shielded by the $7s$ and $7p$ electrons relative to the $4f$ electrons (shielded by $6s$, $6p$) in the lanthanide ($4f$) series. Thus, there is a greater spatial extension of $5f$ orbitals for actinides than $4f$ orbitals for lanthanides. This results in a small energy difference between $5f^n 7s^2$ and $5f^{n-1} 6d 7s^2$ electronic configurations, and a wider range of oxidation states is therefore accessible to the early members of the actinide series (U–Am). Uranium has four common oxidation states, III, IV, V, and VI.

Of the four oxidation states (III, IV, V, and VI), only the IV and VI states are stable enough to be of general importance. Aqueous solutions of uranium(III) may be prepared, but they are readily oxidized to the IV state with evolution of hydrogen; and the V state disproportionates into the IV and VI states in the presence of water or hydrolytic compounds. The ease of alternation between IV and VI states has economic significance. The highly stable and disseminated grains of uraninite in igneous rock formations are in the IV state, but when altered to the VI state, they are soluble enough to dissolve in circulating groundwater. The solubility of uranium in the VI state accounts for its wide distribution in seawater, fresh water, and hydrothermal deposits.

Isotopes

Natural uranium is a mixture of three α-emitting isotopes: ^{238}U (99.274%, half-life = 4.47×10^9 yr, 4.15 MeV α), ^{235}U (0.7202%, half-life = 7.08×10^8 yr, 4.29 MeV α), and ^{234}U (0.0057%, half-life of 2.45×10^5 yr, 4.78 MeV α). Uranium is the progenitor of two naturally occurring decay series, ^{238}U ($4n + 2$), shown in Figure 1 and ^{235}U ($4n + 3$) which terminates at stable ^{207}Pb. The man-made Np series, which ends in ^{209}Bi, includes ^{233}U (7). Two of the isotopes of the ^{238}U($4n + 2$) chain, ^{226}Ra and ^{222}Rn, have significant historical and radiological implications. Natural uranium is not highly radioactive in a relative sense; for example, 2800 kg of natural uranium have a radioactivity equivalent to that of approximately one g of radium-226 (8).

Uranium isotopes and their radioactive decay products, from thorium to lead, are used extensively in determining the geochronology and geochemistry of a wide variety of minerals, rocks, and geologic formations. In the uranium decay series (^{238}U parent) eight α-particles are emitted in the decay from ^{238}U to ^{206}Pb.

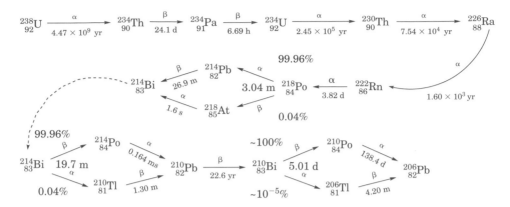

Fig. 1. Uranium decay scheme (6).

A mineral can thus be dated once the concentration ratios of ^{238}U and He are known. Another common method of dating U-minerals is based on considering the distribution of lead isotopes. Lead has four stable isotopes of which three are end products of radioactive decay series. The fourth isotope, ^{204}Pb, is found in lead minerals in about 1.4% isotopic abundance and has no radiogenic origin (9).

The U/Pb decay schemes have been used to date the oldest known terrestrial rocks (10). The types of samples studied using these methods are extensive, including but not limited to archaeological deposits, carbonates and other sediments, pebble conglomerates, zircons, volcanic and pyroclastic rocks, granites, basalts, and uranium ores (11–16). For example, uranium-bearing quartz conglomerates have been used to study uranium ore mineralization (13). The Greenbushes pegmatite is a giant Archean pegmatite dike with substantial Li–Sn–Ta mineralization, including half the world's Ta resource, and was dated utilizing imprecise whole-rock Pb–Pb and precise U–Pb zircon techniques (14). Uranium oxide ages have been used to study ore formations and uranium mineralization (11,12).

Ratios of ^{234}U and ^{238}U to ^{230}Th and ^{226}Ra daughters, combined with differences in chemical reactivity have been used to investigate the formation and weathering of limestone in karst soils of the Jura Mountains, and of the mountains in the central part of Switzerland. Uranium contained within calcite is released during weathering, and migrates as stable uranyl(VI) carbonato complexes through the soil. In contrast, the uranium decay products, Th and Ra, hydrolyze and are strongly sorbed to soil particles and/or form insoluble compounds that become more and more enriched in the soil as a function of time (17). It is interesting to note that the ratios of ^{234}U and ^{238}U provide information on weathering, because ^{234}U, as the product of a lattice damaging α-decay of ^{238}U, is preferentially leached (18).

In addition, uranium and lead transport mechanisms in radioactive minerals have been studied in order to evaluate the suitability of mineral phases as hosts for radioactive wastes. Zircon is one of the most commonly used geochronometers, as well as a proposed nuclear waste matrix material, and there are many mechanisms by which uranium and lead can migrate through its structure (19).

Occurrence in Nature

Uranium is widely distributed in nature (20). It is found in significant concentrations in rocks, oceans, lunar rocks, and meteorites. Estimates of the concentrations of uranium in various geological matrices are given in Table 1. It can be seen in Table 1 that uranium is present at about 2 ppm in the earth's crust, and is thus more abundant than many other common elements, such as Cd, Ag, Hg, etc. In general, igneous rocks with a high silicate content, such as granite, contain an above average uranium concentration; whereas basic rocks, such as basalts contain a below average uranium content. Sedimentary rocks also generally contain below average uranium concentrations. Despite this low uranium content, sedimentary rocks like sandstones and conglomerates contain approximately 90% of the world's uranium resources.

Uranium resources can be assigned on the basis of their geological setting to fifteen main categories of uranium ore deposit types arranged according to their approximate economic significance: (*1*) unconformity-related deposits; (*2*) sandstone deposits; (*3*) quartz pebble conglomerate deposits; (*4*) vein deposits; (*5*) breccia complex deposits; (*6*) intrusive deposits; (*7*) phosphorite deposits; (*8*) collapse breccia pipe deposits; (*9*) volcanic deposits; (*10*) surficial deposits; (*11*) metasomatite deposits; (*12*) metamorphite deposits; (*13*) lignite; (*14*) black shale deposits; and (*15*) other deposits (1).

Unconformity-related deposits are found near principal unconformities. Examples include the ore bodies at Cluff Lake, Key Lake, and Rabbit Lake in northern Saskatchewan, Canada, and in the Alligator Rivers area in northern Australia. Sandstone deposits are contained in rocks that were deposited under fluvial or marginal marine conditions. The host rocks nearly always contain

Table 1. Occurrence of Uranium in Nature

Location	U concentration, ppm	Reference
igneous rocks		
basalts	0.6	21
granites (normal)	4.8	21
ultrabasic rocks	0.03	22
sandstones, shales, limestones	1.2–1.3	23
earth's crust	2.1	21
oceanic	0.64	21
continental	2.8	21
earth's mantle	~0.01	21
seawater	0.002–0.003	23
meteorites	0.05	21
chondrites	0.011	21
uraniferous materials		
high grade veins	$(3–8.5) \times 10^5$	
vein ores	$(2–10) \times 10^3$	23
sandstone ores	$(0.5–4) \times 10^3$	23
gold ores (South Africa)	150–600	23
uraniferous phosphates	50–300	23
uraniferous granites	15–100	23
Chattanooga shale (Tenn., U.S.)	60	23

pyrite and organic plant matter. The sediments are commonly associated with tuffs. Unoxidized deposits of this type consist of pitchblende and coffinite in arkasoic and quartzitic sandstones. Upon weathering, secondary minerals such as carnotite, tuyamunite, and uranophane are formed. More information on these and other uranium deposit types is available (1).

Approximately 155 minerals are known that contain uranium as an important, or major constituent, and another 60 that contain minor amounts of uranium, or contain uranium as an impurity (24). A listing of several representative uranium minerals is given in Table 2. Uranium minerals can be divided into two mineral classes, primary and secondary. Primary uranium minerals are those that formed during the last stages of magma crystallization, and are rich in silicates such as quartz and feldspar. Primary uranium minerals include uraninite [*1317-99-3*], pitchblende [*1317-75-5*], and a large number of complex multiple oxides such as uranium-containing lanthanide niobates, tantalates, and titanates (25). Uraninite and pitchblende are very important uranium minerals with a composition that varies from UO_2 to $UO_{2.67}$ and are found in veins, pegmatites, and unweathered portions of conglomerate and sandstone ores which contain the bulk of the world's economic uranium deposits. Secondary uranium minerals are produced by hydration, metathesis, oxidation or possibly transport and redeposition. Primary minerals are generally black and contain uranium in an average oxidation state less than VI, while secondary minerals are generally yellow, green, or orange, and contain uranium in the hexavalent state. Uraninite can be considered both a primary and secondary mineral, and there are a wide variety of theories regarding the mechanism of formation of uraninite veins (22,26,27).

There is extensive literature regarding all aspects of uranium geology, mineralogy, and mining. There are comprehensive descriptions of uranium minerals including elemental compositions (24,28), mineralogical properties (28), ore distributions (23,29), and typical uranium contents (23). A 1970 symposium discussed the geology of known uranium deposits, theories of the genesis of ore deposits and uranium mineralization, and means of predicting where further deposits may be found (30). Excellent reviews of these data have been provided (20,31). The most comprehensive coverage of the literature of ore deposits, mineralogy and geochemistry is still the *Gmelin Handbook*.

The Oklo Phenomenon.   Naturally occurring uranium consists mainly of ^{238}U and fissionable ^{235}U. The isotopic ratio can be calculated from the relative decay rates of the two isotopes. Because ^{235}U decays faster than ^{238}U, the isotopic ratio decreases with time. In 1997, the isotopic abundance of ^{235}U in natural uranium is $0.7202 \pm 0.006\%$. In 1972, uranium samples from the Oklo open-pit uranium mine in southeastern Gabon Republic were found to be depleted in ^{235}U, relative to the expected natural isotopic composition. The levels of ^{235}U depletion were inhomogeneous throughout the ore body, with the lowest isotopic ratio 0.296%. After much study, it was determined that the uranium ore deposit at Oklo was the site of at least six natural nuclear reactors (qv). Geochemical reactions (weathering) and geological changes created different regions enriched and depleted in uranium, and at that time in geologic history the ^{235}U enrichment was estimated to be approximately 3%, the same as used in commercial nuclear reactors. Under these conditions, a critical mass could

Table 2. Selected Uranium Minerals[a]

Name	Chemical composition	CAS Registry Number	Name	Chemical composition	CAS Registry Number
oxides			silicates		
uraninite	UO_2–$UO_{2.67}$	[1317-99-3]	uranophane	$Ca(UO_2)_2(Si_2O_7)\cdot6H_2O$	[12195-76-5]
schoepite	$UO_3\cdot2H_2O$	[22972-07-2]	coffinite	$U(SiO_4)_{1-x}(OH)_{4x}$	[14485-40-6]
Becquerelite	$7UO_3\cdot11H_2O$	[12378-67-5]	Soddyite	$(UO_2)_5(SiO_4)_4(OH)_2\cdot5H_2O$	[12196-99-5]
phosphates			arsenates		
autunite	$Ca(UO_2)_2(PO_4)_2\cdot10$–$12H_2O$	[16390-74-2]	Abernathyite	$K_2(UO_2)_2(AsO_4)_2\cdot6H_2O$	[12005-93-5]
torbernite	$Cu(UO_2)_2(PO_4)_2\cdot12H_2O$	[26283-21-6]	Metakahlerite	$Fe(UO_2)_2(AsO_4)_2\cdot nH_2O$	[12255-22-0]
natroautunite	$Na_2(UO_2)_2(PO_4)_2\cdot8H_2O$	[161334-19-6]	Novacekite	$Mg(UO_2)(AsO_4)_2\cdot8$–$10H_2O$	[12255-29-1]
meta-ankoleite	$K_2(UO_2)_2(PO_4)_2\cdot6H_2O$	[12169-00-5]			
parsonsite	$Pb_2(UO_2)(PO_4)_2$	[12137-57-4]			
carbonates			molybdates		
andersonite	$Na_2Ca(UO_2)(CO_3)_3\cdot6H_2O$	[12202-87-8]	cousinite	$MgO\cdot2MoO_3\cdot UO_2\cdot6H_2O$	
liebigite	$Ca_2(UO_2)(CO_3)_3\cdot10H_2O$	[14831-68-6]	Wulfenite	$Pb(Mo,U)O_4$	[14913-82-7]
bayleyite	$Mg_2(UO_2)(CO_3)_3\cdot18H_2O$	[19530-04-2]	Irginite	$UO_3\cdot2MoO_3\cdot3H_2O$	
rutherfordine	UO_2CO_3	[12202-79-8]			
vanadates					
carnotite	$K_2(UO_2)_2(VO_4)_2\cdot1$–$3H_2O$	[60182-49-2]			
tyuyamunite	$Ca(UO_2)_2(VO_4)_2\cdot5$–$8H_2O$	[12196-95-1]			

[a]Compiled from Ref. 24.

be attained and a nuclear fission reaction initiated with groundwater or water in the clays as a neutron moderator. The chain reaction is thought to have cycled. As the reaction heated up, water would be driven away thereby slowing the reaction. As the reaction cooled, water could return, thereby slowing the neutrons and starting the chain reaction all over again. The identification of fission products in the proper ratios, and differing from that expected for natural occurrence, gave unequivocal evidence that a nuclear chain reaction had taken place (32,33). In addition to the purely scientific interest, the Oklo phenomenon has also led to a better understanding of environmental migration of radioactive materials (34). As such, the information attained at Oklo is of great importance in understanding the aqueous transport and redistribution of uranium as it pertains to the safe disposal of radioactive waste (35,36).

Resources

Resource estimates are divided into separate categories reflecting different levels of confidence in the quantities reported, and further separated into categories based on the cost of production. A listing of uranium resources by country is given in Table 3.

Reasonably Assured Resources (RAR) refers to uranium in known mineral deposits of size, grade, and configuration such that recovery is within the given production cost ranges with currently proven mining and processing technology. The majority of these resources are found in Australia, Brazil, Canada, Namibia, Niger, South Africa, and the United States (see Table 3).

Estimated additional resources (EAR) is a term that applies to resources that are inferred to occur as extensions of well-explored deposits, little-explored deposits, or undiscovered deposits believed to exist along a well-defined geological continuity with known deposits. There are two types of EAR, EAR-I and EAR-II, which are inferred based on direct or indirect evidence of existence, respectively.

In January 1993, RAR recoverable at costs of $\leq$\$130/kg U, for selected countries, were estimated at 2.093×10^6 t of uranium, excluding the Commonwealth of Independent States (CIS), most European countries, and China (1). Estimates of total RAR recoverable at costs between \$80–\$130/kg U accounted for 660,000 t. Total RAR recoverable at costs of $\leq$\$80/kg U were estimated at 1.424×10^6 t uranium (1), and RAR at costs <\$80/kg were estimated at 670,000 t. This represents a decrease of ~2.4% from the 1991 values and is related to mine closures in traditional supplier countries (Canada, France, South Africa, and the United States). Significant uranium resources are known to exist in a number of Asian and Eastern European countries including Bulgaria, China, India, Mongolia, Romania, and the CIS. Kazakhstan, Mongolia, Russian Federation, and Ukraine have reported known uranium resources of 1041×10^6 t uranium in *in situ* resources with an estimated production cost of $\leq$\$130/kg U. Uzbekistan reports known resources of 230,000 tons uranium with no assigned cost. There are also significant quantities of unconventional uranium resources, primarily associated with marine phosphate deposits. Production from these resources is declining.

Table 3. Uranium Resources by Country, January 1993, 10^3 t[a]

Country	Reasonably assured resources			Estimated additional resources		
	$80/kg	$80–$130/kg	$130/kg	$80/kg	$80–$130/kg	$130/kg
Algeria	26.00	0.00	26.00			
Argentina	4.60	2.70	7.30	2.30	0.30	2.60
Australia	462.00	55.00	517.00	272.00	122.00	394.00
Brazil	162.00	0.00	162.00	94.00	0.00	94.00
Canada	277.00	120.00	397.00	31.00	43.00	74.00
Central African Republic	8.00	8.00	16.00			
Czech Republic	15.85	6.40	22.25	1.35	20.00	21.35
Denmark	0.00	27.00	27.00	0.00	16.00	16.00
Finland	0.00	1.50	1.50			
France	19.85	13.80	33.65	3.55	3.18	6.73
Gabon	9.78	4.65	14.43	1.30	8.30	9.60
Germany	0.00	3.00	3.00	0.00	4.00	4.00
Greece	0.30	0.00	0.30	6.00	0.00	6.00
Hungary	0.62	0.51	1.13	1.32	15.34	16.66
Indonesia	0.00	5.42	5.42	0.00	2.15	2.15
Italy	4.80	0.00	4.80	0.00	1.30	1.30
Japan	0.00	6.60	6.60			
Korea, Republic of	0.00	11.80	11.80	0.00	3.00	3.00
Mexico	0.00	1.70	1.70	0.00	0.70	0.70
Namibia	80.62	16.00	96.64	30.00	23.00	53.00
Niger	159.17	6.65	165.82	295.77	10.00	305.77
Peru	1.79	0.00	1.79	1.72	0.14	1.86
Portugal	7.30	1.40	8.70	1.45	0.00	1.45
Slovenia	0.00	1.80	1.80	5.00	0.00	5.00
Somalia	0.00	6.60	6.60	0.00	3.40	3.40
South Africa	144.40	96.44	240.84	34.72	19.70	54.42
Spain	17.85	21.15	39.00	4.20	0.00	4.20
Sweden	2.00	2.00	4.00	1.00	5.30	6.30
Thailand	0.00	0.01	0.01			
Turkey	9.13	0.00	9.13			
United Kingdom	0.00	0.00	0.00	0.00	0.00	0.00
United States	114.00	255.00	369.00	0.00	0.00	0.00
Zaire	1.80	0.00	1.80	1.70	0.00	1.70
Zimbabwe	1.80	0.00	1.80			
Total (rounded)	*1531*	*675*	*2206*	*789*	*302*	*1091*
Total (adjusted)	*1424*	*659*	*2093*	*670*	*292*	*966*

[a]Data compiled from the OECD Nuclear Enery Agency and the IAEA (1).

Uranium exploration has decreased significantly since 1990, primarily due to decreased expenditures in France and the United States. However, exploration programs are still being conducted in Australia, Canada, France, India, and the United States. A listing of total uranium production by country in 1994 is given in Table 4. In 1992, production of 36,246 t of uranium accounted for only 63% of world reactor requirements of 57,182 t. It is anticipated that world reactor

Table 4. World Uranium Production by Country, 1994[a]

Country	U, t	Country	U, t
Argentina	64	India	200
Australia	2,183	Kazakhstan	2,240
Belgium	40	Namibia	1,901
Brazil	50	Niger	2,975
Bulgaria	50	Portugal	23
Canada	9,694	Romania	120
China	780	Russia	2,968
Czech Republic	541	South Africa	1,690
France	1,028	Spain	255
Gabon	650	Ukraine	500
Germany	395	United States	1,400
Hungary	402	Uzbekistan	2,015
		Totals	32,188

[a]Data compiled from Ref. 38.

requirements will reach 76,673 t by the year 2010 (1). The shortfall of uranium production in selected countries is estimated to reach 27,300 t by the year 2000, and over 50,000 t by the year 2010. The shortfall between fresh production and reactor requirement is expected to be filled by several sources. These sources include excess inventory drawdown and production from planned or prospective centers. They could also include development of low cost resources that could soon become available as a result of technical developments, or perhaps more significantly, new policy changes affecting recycled material such as spent fuel, and low enriched uranium (LEU) converted from the high enriched uranium (HEU) found in warheads. If the technical and political problems can be overcome, reactor-grade uranium produced from HEU warhead material could contribute significantly to meeting the anticipated fresh uranium production shortfall. Natural uranium demands will also depend on the utilization of warhead plutonium and low enrichment mixed oxide fuels in commercial reactors.

The demand for uranium in the commercial sector is primarily determined by the consumption and inventory requirements of nuclear power reactors. In March 1997, there were 433 nuclear power plants operating worldwide with a combined capacity of about 345 GWe (net gigawatts electric) (34). Projections for the 1990s show steady growth in nuclear capacity to 446 GWe by the year 2010, representing a 30% increase.

During the past few years, governments and governmental agencies have become involved in regulating international trade in uranium involving supply from the CIS republics. In early 1993, a purchase agreement was reached between the United States and Russia for the highly enriched uranium contained in the warheads of dismantled weapons. It is expected that the potential introduction of military material will have a dominant influence on the nuclear fuel supply and demand balance in years to come (37).

Recovery From Ores

The extractive metallurgy of uranium has been discussed in detail in various older books (23,29,39–41) and in a number of more recent papers (42–48). A comprehensive discussion and overview is provided in the *Gmelin Handbook*. The extraction of uranium from ores varies widely, and depends on the nature of the ore involved. The ore may vary from hard, igneous rock to soft, weakly cemented sedimentary rock. The principal gangue mineral may be quartz, which is chemically inactive, or an acid-consuming mineral, such as calcite. Some ores are highly refractory and require intensive processing, whereas others break down between the mine and the mill. In order to recover the uranium from ores, a series of steps is often required including crushing and concentrating by conventional physical means; roasting and leaching the ore with acid in the presence of an oxidant to ensure conversion to UO_2^{2+}; recovery of the uranium from the leach solution, and refining to a high purity product.

Preconcentration.   Preconcentration enriches low grade ores to the point where they can be processed economically. In general, conventional ore-dressing techniques have not been successful in the preconcentration of uranium minerals. However, it is usually possible to obtain acceptable concentration ratios. Where the uranium values occur as masses in pegmatitic rock with large areas of unmineralized pegmatite separating the ore minerals, they are preconcentrated by electronic sorting devices. Gravity separations are sometimes possible owing to the high density of uranium minerals relative to most gangue components. However, uranium minerals tend to concentrate with the fines in the grinding and crushing process of some ores. Electrostatic methods generally give low recoveries at low concentrations. Magnetic gangue minerals, eg, magnetite, ilmenite, and garnet, may be separated by magnetic methods, which do not affect the nonmagnetic uranium component. Jaw crushers are employed for coarse crushing; smaller jaw crushers, gyratories, or hammer mills are used for secondary crushing. Rod mills, ball mills, and hammer mills are used for grinding. Uranium is concentrated in the cementing material and in the coating of the sand grains that are separated from the barren sand during the grinding action. In many cases, the ore is so poorly consolidated that there is no need to close the grinding circuit with screens or classifiers.

Roasting and Calcining.   It is usually desirable to subject the ores to a high temperature calcination prior to leaching. Carbonaceous material can be removed by an oxidizing roast, which at the same time converts the uranium to a soluble form. An oxidizing roast converts sulfides or other sulfur compounds to sulfates, in order to avoid poisoning of ion-exchange resins in subsequent treatments, and removes other reductants that might interfere in the leaching step. Roasting also improves the characteristics of many ores. Clays of the montmorillonite type, for instance, cause thixotropic slurries, and thus interfere with leaching, settling, and filtering. Vanadium-containing ores are roasted with sodium chloride to convert vanadium into a soluble sodium vanadate, which in turn is believed to form soluble uranyl vanadates (23). Sodium chloride roasting also converts silver to silver chloride, rendering the silver insoluble for easier separation.

Leaching.   Treatment with suitable solvents (acids or alkalies) converts uranium contained in the ore to water-soluble species. The uranium is separated by chemical processing, including at least one digestion step with acid or alkaline solution. Most mills use acid leaching, which completely extracts uranium. Because of its low cost, sulfuric acid is preferred, however, the more corrosive hydrochloric acid is used where it is a by-product of salt roasting. As a general rule, only uranium(VI) compounds dissolve readily in H_2SO_4, whereas minerals such as uraninite, pitchblende, or others in which the uranium has a lower valence, do not. For minerals containing uranium in the lower oxidation state, oxidizing conditions must be provided by addition of suitable oxidants, such as manganese dioxide or sodium chlorate. Iron must be present in solution as a catalyst in order for either reaction to be effective. Typical leach reactions are listed in equations 1–5.

$$2\,H_2SO_4 + MnO_2 + UO_2 \longrightarrow UO_2SO_4 + MnSO_4 + 2\,H_2O \tag{1}$$

$$3\,H_2SO_4 + NaClO_3 + 3\,UO_2 \longrightarrow 3\,UO_2SO_4 + NaCl + 3\,H_2O \tag{2}$$

$$UO_2 + 2\,Fe^{3+} \longrightarrow UO_2^{2+} + 2\,Fe^{2+} \tag{3}$$

$$2\,Fe^{2+} + MnO_2 + 4\,H^+ \longrightarrow 2\,Fe^{3+} + Mn^{2+} + 2\,H_2O \tag{4}$$

$$6\,Fe^{2+} + ClO_3^- + 6\,H^+ \longrightarrow 6\,Fe^{3+} + Cl^- + 3\,H_2O \tag{5}$$

In most ores, sufficient Fe is already present. For some ores, it is necessary to add metallic iron. In practice, the oxidation potential of the solution can be monitored and controlled using the Fe^{2+}/Fe^{3+} ratio. Very high leaching efficiencies with H_2SO_4 are common, eg, 95–98% dissolution yield of uranium (39). If acid consumption exceeds 68 kg/t of ore treated, alkaline leaching is preferred. The comparative costs of acid, sodium hydroxide, and sodium carbonate differ widely in different areas and are the determining factor.

Carbonate leaching is usually carried out using sodium carbonate. The utility of the carbonate process arises owing to the high thermodynamic stability and solubility of the $UO_2(CO_3)_3^{4-}$ [24646-13-1] ion in aqueous media at low hydroxide ion concentration. This method takes advantage of the fact that U(VI) is very soluble in carbonate solution, unlike the majority of other metal ions that form insoluble carbonates and hydroxides under similar solution conditions. The sodium carbonate leach is therefore inherently more selective than the sulfuric acid leach process. Minerals containing U(IV) require the addition of an oxidant to generate the more soluble VI state. Oxygen (as air) or permanganate is typically used to provide the needed oxygen, and the dissolution of simple uranium oxide follows the reactions shown in equations 6–8. Bicarbonate is typically employed to keep the hydroxide concentration low and avoid precipitation of uranates according to equation 9. Under proper oxidizing conditions, carbonate extractions yield 90–95% of the uranium (23). For this process, fine ores are required.

$$2\,UO_2 + O_2 \longrightarrow 2\,UO_3 \tag{6}$$

$$UO_3 + 3\,CO_3^{2-} + H_2O \longrightarrow UO_2(CO_3)_3^{4-} + 2\,OH^- \tag{7}$$

$$OH^- + 2\,HCO_3^- \longrightarrow CO_3^{2-} + H_2O \tag{8}$$

$$2\,UO_2(CO_3)_3^{4-} + 6\,OH^- + 2\,Na^+ \longrightarrow Na_2U_2O_7 + 6\,CO_3^{2-} + 3\,H_2O \tag{9}$$

Carbonate leaching under ambient conditions is extremely slow with poor recoveries. Therefore, the ore is typically leached in an autoclave with air providing most of the needed oxygen. The leach liquor is separated from the solid in a countercurrent–decantation system of thickeners, and the uranium is precipitated from the clarified sodium carbonate solution with addition of sodium hydroxide (eq. 9) (23).

Recovery of Uranium from Leach Solutions. The uranium can be recovered from leach solutions using a variety of approaches including ion exchange (qv), solvent extraction, and chemical precipitation. The most common methods in practice are ion exchange and solvent extraction to purify and concentrate the uranium prior to final product precipitation.

Ion Exchange. The recovery of uranium from leach solutions using ion exchange is a very important process (42). The uranium(VI) is selectively adsorbed to an anion-exchange resin as either the anionic sulfato or carbonato complexes. In carbonate solutions, the uranyl species is thought to be the tris carbonato complex, $UO_2(CO_3)_3^{4-}$ [24646-13-7], and from sulfate solutions the anion is likely to be $UO_2(SO_4)_n^{2-2n}$, where n is 3 [56959-61-6] or 2 [27190-85-8]. The uranium is eluted from the resin with a salt or acid solution of 1 M MCl or MNO_3 (M = H^+, Na^+, NH_4^+). The sulfate solution is acidified and the carbonate solution is kept slightly basic with addition of bicarbonate (23). From this solution, the uranium is precipitated and recovered as a fairly pure uranium concentrate. The uranium ion-exchange process has been extensively reviewed, and specific flow sheets, processing rates, recycle methods for reagent conservation, and process equipment are available elsewhere (23,39,42).

Solvent Extraction. Solvent extraction has widespread application for uranium recovery from ores. In contrast to ion exchange, which is a batch process, solvent extraction can be operated in a continuous countercurrent-flow manner. However, solvent extraction has a large disadvantage, owing to incomplete phase separation because of solubility and the formation of emulsions. These effects, as well as solvent losses, result in financial losses and a potential pollution problem inherent in the disposal of spent leach solutions. For leach solutions with a concentration greater than 1 g U/L, solvent extraction is preferred. For low grade solutions with <1 g U/L and carbonate leach solutions, ion exchange is preferred (23). Solvent extraction has not proven economically useful for carbonate solutions.

For extraction of uranium from sulfate leach liquors, alkyl phosphoric acids, alkyl phosphates, and secondary and tertiary alkyl amines are used in an inert diluent such as kerosene. The formation of a third phase is suppressed by addition of modifiers such as long-chain alcohols or neutral phosphate esters. Such compounds also increase the solubility of the amine salt in the diluent and improve phase separation.

Amine extraction from sulfate solutions is mechanistically similar to anion-exchange separation of uranium from uranyl sulfate solution. Uranyl(VI) sulfato complexes are extracted by the alkyl ammonium cations at pH 1–2, and the $UO_2(SO_4)_3^{4-}$ [56959-61-6] complex is the predominant solution species extracted. The amine structure affects selectivity and affinity. Other anions, such as nitrate or chloride, may interfere with the uranium extraction. Nitrate interferes with secondary amines and chloride with tertiary amines. The choice of suitable stripping agents depends on such factors, as does the recycling of solutions. Molybdenum (present in the ore) is extracted more readily than uranium. It builds up as a poison in the amine, and inhibits the process by precipitation at the aqueous–organic interface. The problem may be solved by including one or more specific molybdenum stripping steps in the process (23). Vanadium is also extracted to some extent (see VANADIUM AND VANADIUM ALLOYS). Various other ions function as salting-out agents for uranium, which is stripped from the organic solvent in their presence. The affinity of nitrate to the amine is so high that the latter has to be scrubbed by means of a hydroxide or carbonate wash before it can be recycled to another extraction run. Chloride does not give this complication, except with secondary amines, which have a high chloride affinity.

Monoalkyl phosphate extractants exhibit good efficiency in the presence of dilute nitrate, sulfate, or chloride, and cause fewer phase-separation problems. However, they are less selective, and other cations present in the leach liquor are co-extracted with the uranium from which they must be separated. The most widely used extractants are di(2-ethylhexyl) phosphate (D2EHPA) and dodecylphosphate (DDPA). The selectivity for uranium is about equal. Fe(III) interferes and has to be quantitatively reduced to Fe(II) in the feed liquor prior to extraction. DPPA has substantially higher solubility losses than D2EHPA, and strong acids are required for back-extraction from DPPA. For D2EHPA, sodium carbonate solution is used as the stripping agent; for DPPA, hydrochloric or hydrofluoric acids are used. The D2EHPA solvent extraction process is generally referred to the DAPEX process (dialkyl phosphate extraction).

Chemical Precipitation.   The product of the extraction processes, whether derived from acid or carbonate leach, is a purified uranium solution that may or may not have been upgraded by ion exchange or solvent extraction. The uranium in such a solution is concentrated by precipitation and must be dried before shipment. Solutions resulting from carbonate leaching are usually precipitated directly from clarified leach liquors with caustic soda without a concentration step, as shown in equation 9.

Losses are kept to a minimum by carbonation of the mother liquor with CO_2 and recycle of the carbonated product back to the leach system. From acid solutions, uranium is usually precipitated by neutralization with ammonia or magnesia. Ammonia gives an acceptable precipitate, for which compositions such as $(NH_4)_2(UO_2)_2SO_4(OH)_4 \cdot nH_2O$ were calculated. The ammonium salt is preferred if the product is to be used in the manufacture of fuel–element material.

A higher uranium content can be obtained by precipitation with magnesia (MgO) to yield yellowcake. The magnesium sulfate formed is water-soluble, and the uranium compound can be separated by filtration. Yellowcake consists of either ammonium diuranate [7783-22-4] or magnesium diuranate [13568-61-1]. The ammonium diuranate in yellowcake is not a stoichiometric compound, but a mixture of compounds ranging in composition from $(NH_4)_2UO_4$ [13597-77-8] to $(NH_4)_2U_8O_{25}$, and having the approximate composition $(NH_4)_2U_2O_7$ (49). Sodium uranate [13721-31-4] may be the product from carbonate leach plants.

Refining to a High Purity Product.   The normal yellowcake product of uranium milling operations is not generally pure enough for use in most nuclear applications. Many additional methods have been used to refine the yellowcake into a product of sufficient purity for use in the nuclear industry. The two most common methods for refining uranium to a high purity product are tributyl phosphate (TBP) extraction from HNO_3 solutions, or distillation of UF_6, since this is the feedstock for uranium enrichment plants.

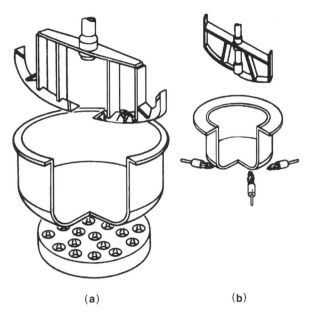

(a)   (b)

Fig. 2.   Gas-fired denitration pots for denitration of $UO_2(NO_3)_2 \cdot 6H_2O$. (**a**) The large pot (1.68 M ID, 0.81 m height) is heated by three concentric rings of small radiant gas burners. (**b**) The small pot (76 cm ID, 46 cm height) is heated by four gas burners inside a ceramic furnace (40).

In TBP extraction, the yellowcake is dissolved in nitric acid and extracted with tributyl phosphate in a kerosene or hexane diluent. The uranyl ion forms the mixed complex $UO_2(NO_3)_2(TBP)_2$ which is extracted into the diluent. The purified uranium is then back-extracted into nitric acid or water, and concentrated. The uranyl nitrate solution is evaporated to uranyl nitrate hexahydrate [*13520-83-7*], $UO_2(NO_3)_2 \cdot 6H_2O$. The uranyl nitrate hexahydrate is dehydrated and denitrated during a pyrolysis step to form uranium trioxide [*1344-58-7*], UO_3, as shown in equation 10. The pyrolysis is most often carried out in either a batch reactor (Fig. 2) or a fluidized-bed denitrator (Fig. 3). The UO_3 is reduced with hydrogen to uranium dioxide [*1344-57-6*], UO_2 (eq. 11), and converted to uranium tetrafluoride [*10049-14-6*], UF_4, with HF at elevated temperatures (eq. 12). The UF_4 can be either reduced to uranium metal or fluorinated to uranium hexafluoride [*7783-81-5*], UF_6, for isotope enrichment. The chemistry and operating conditions of the TBP refining process, and conversion to UO_3, UO_2, and ultimately UF_4 have been discussed in detail (40).

$$UO_2(NO_3)_2 \cdot 6H_2O \longrightarrow UO_3 + 2\,NO_2 + 1/2\,O_2 + 6\,H_2O \qquad (10)$$

$$UO_3 + H_2 \longrightarrow UO_2 + H_2O \qquad (11)$$

$$UO_2 + 4HF \longrightarrow UF_4 + 2\,H_2O \qquad (12)$$

Uranium Metal

Properties.   Uranium metal is a dense, bright silvery, ductile, and malleable metal. Uranium is highly electropositive, resembling magnesium, and tarnishes rapidly on exposure to air. Even a polished surface becomes coated

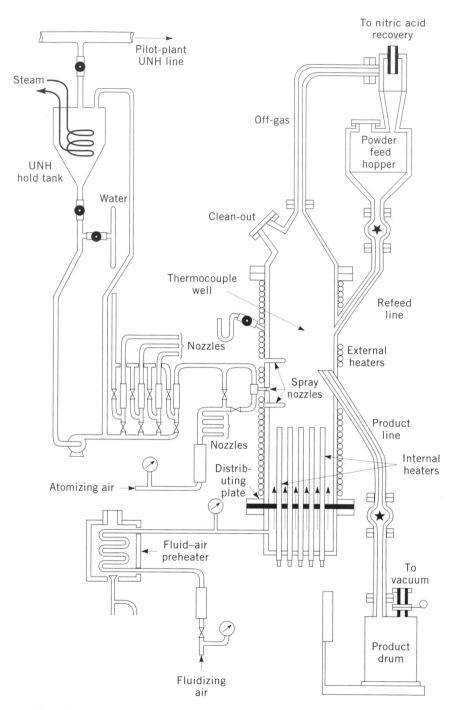

Fig. 3. Pilot-plant fluidized-bed denitrator for denitration of $UO_2(NO_3)_2 \cdot 6H_2O$ where UNH = uranyl nitrate hexahydrate (40).

with a dark-colored oxide layer in a short time upon exposure to air. At elevated temperatures, uranium metal reacts with most common metals and refractories. Finely divided uranium reacts, even at room temperature, with all components of the atmosphere except the noble gases. The silvery luster of freshly cleaned uranium metal is rapidly converted first to a golden yellow, and then to a black oxide–nitride film within three to four days. Powdered uranium is usually pyrophoric, an important safety consideration in the machining of uranium parts. The corrosion characteristics of uranium have been discussed in detail (28).

In the solid state, uranium metal exists in three allotropic modifications. The transformation temperatures and the enthalpies of transformation are given in Table 5. The thermodynamic properties of uranium metal have been determined with great accuracy and have been discussed (50).

Uranium metal is weakly paramagnetic, with a magnetic susceptibility of 1.740×10^{-5} A/g at 20°C, and 1.804×10^{-5} A/g (A = 10 emu) at 350°C (51). Uranium is a relatively poor electrical conductor. Superconductivity has been observed in α-uranium, with the value of the superconducting temperature, T_c, being pressure-dependent. This was shown to be a result of the fact that there are actually three transformations within α-uranium (37,52).

Uranium metal exhibits three crystalline forms before finally melting at 1132.4°C. The crystallographic parameters have been determined by various workers, and the relevant values are listed in Table 5. The α-phase exists at room temperature and consists of corrugated sheets of atoms. The β-phase exists between 668 and 775°C, and the γ-phase is formed at temperatures above 775°C. The α- and β-phases of uranium have rather unique crystalline structures, and therefore do not form solid solutions with other metals to any great extent. Both phases do form a wide variety of alloys and intermetallic compounds (U_6Fe, UPt_2, U_2Ga, etc) with other metals, and these have been described in detail (31). Uranium alloys are of interest for many reasons. Uranium is a relatively weak and extremely reactive metal, and alloys provide improved strength and corrosion resistance. A comprehensive listing and description of uranium alloys, including phase diagrams is available (28).

Preparation of Uranium Metal. Uranium is a highly electropositive element, and extremely difficult to reduce. As such, elemental uranium cannot be prepared by reduction with hydrogen. Instead, uranium metal must be prepared using a number of rather forcing conditions. Uranium metal can be prepared by reduction of uranium oxides (UO_2 [1344-59-8] or UO_3 [1344-58-7] with strongly electropositive elements (Ca, Mg, Na), reduction of uranium halides (UCl_3 [10025-93-1], UCl_4 [10026-10-5], UF_4 [10049-14-6] with electropositive metals (Li, Na, Mg, Ca, Ba), electrodeposition from molten salt baths, and decomposition of uranium halides (the van Arkel-de Boer method). Typical reaction stoichiometries are given in equations 13–18. There are several comprehensive treatments of developments in this field (28,53).

$$UO_3 + 6\,Na \longrightarrow U(0) + 3\,Na_2O \tag{13}$$

$$UO_3 + 3\,Mg \longrightarrow U(0) + 3\,MgO \tag{14}$$

$$UO_2 + 2\,Ca \longrightarrow U(0) + 2\,CaO \tag{15}$$

Table 5. Some Physical Properties of Uranium Metal[a]

Property	Value
crystallographic properties	
α, orthorhombic, nm,[b] 298°C	$a = 0.28537$
	$b = 0.58695$
	$c = 0.49584$
density, g/cm^3	19.07
atoms per unit cell, Z	4
β, tetragonal, nm,[b] 720°C	$a = 1.0763$
	$c = 0.5652$
density, g/cm^3	18.11
atoms per unit cell, Z	30
γ, body-centered cubic, nm,[b] 805°C	$a = 0.3524$
density, g/cm^3	18.06
atoms per unit cell, Z	2
melting point, °C	1132.4 ± 0.8
enthalpy of vaporization, kJ/mol,[c] at 25°C	446.7
enthalpy of fusion, kJ/mol[c]	19.7
enthalpy of sublimation, kJ/mol[c]	487.9
vapor pressure	
1720–2340 K	$\log(P/\text{atm}) = -(26{,}210 \pm 270)\,(T)^{-1}$
	$+(5.920 \pm 0.135)^d$
1480–2420 K	$\log(P/\text{atm}) = -(25{,}230 \pm 370)\,(T)^{-1}$
	$+(5.71 \pm 0.17)^d$
phase-transformation temperature, °C	
$\alpha \to \beta$	667.8 ± 1.3
$\beta \to \gamma$	774.9 ± 1.6
$\gamma \to$ liquid	1132.4 ± 0.8
enthalpy of phase transformation, kJ/mol[c]	
$\Delta H_L(\alpha \to \beta)$	2.791
$\Delta H_L(\beta \to \gamma)$	4.757
$\Delta H_L(\gamma \to$ liquid)	9.142
thermal conductivity, W/(cm·K) at 100°C	0.263
elastic constants	
elastic modulus, kPa[e]	1758×10^6
shear modulus, kPa[e]	73.1×10^6
bulk modulus, kPa[e]	97.9×10^6
Poisson's ratio	0.20

[a]Data compiled from Ref. 50.
[b]To convert from nm to angstroms, multiply by 10.
[c]To convert from J to cal, divide by 4.184.
[d]T = temperature, P = pressure. 1 atm = 101.3 kPa[e].
[e]To convert kPa to mm Hg, multiply by 7.5.

$$UO_2 + 4\,Na \longrightarrow U(0) + 2\,Na_2O \tag{16}$$

$$UCl_4 + 2\,Ca \longrightarrow U(0) + 2\,CaCl_2 \tag{17}$$

$$UF_4 + 2\,Ca \longrightarrow U(0) + 2\,CaF_2 \tag{18}$$

A combination of technical considerations makes the reduction of UF_4 by Mg or Ca the preferred method for the preparation of uranium metal. Most important is that the reaction mixture must be fluid for the molten uranium metal to collect into an ingot at the bottom of the reaction vessel. This is an important safety consideration because finely divided uranium metal is pyrophoric. Because MgF_2 and CaF_2 have low melting points relative to MgO and CaO, the reduction of uranium halides generates low melting reaction products, and therefore Mg or Ca reduction of a uranium fluoride is preferred. The availability of large quantities of magnesium in high purity make it the reagent of choice for most applications. In addition, UF_4 is the starting material of choice because of its greater air and moisture stability.

In practice, uranium ore concentrates are first purified by solvent extraction with tributyl phosphate in kerosene to give uranyl nitrate hexahydrate. The purified uranyl nitrate is then decomposed thermally to UO_3 (eq. 10), which is reduced with H_2 to UO_2 (eq. 11), which in turn is converted o UF_4 by high temperature hydrofluorination (eq. 12). The UF_4 is then converted to uranium metal with Mg (eq. 19).

$$UF_4 + 2\,Mg \longrightarrow U(0) + 2\,MgF_2 \tag{19}$$

Reduction of uranium tetrafluoride by magnesium metal has been described in detail (40,53). It is often referred to as the Ames process, since it was demonstrated at the Ames Laboratory in early 1942. The reaction is very exothermic and the reduction process is carried out in a sealed bomb due to volatility at the temperatures reached in the reaction (Fig. 4). To avoid reaction between molten uranium and the steel container, and to prevent undue heat loss from the bomb, a refractory liner of CaO or MgO is placed inside the reactor. The reactor is then lined with MgF_2, and filled with a thoroughly mixed charge of anhydrous UF_4 powder and Mg chips. A typical reaction would employ 202 kg UF_4, and 32.1 kg of Mg chips. The charge is covered with MgF_2 powder and the bomb sealed.

The heat produced by the reaction is not sufficient to maintain a temperature high enough to ensure fluidity of the mixture. Therefore, the bomb is typically heated to 700°C to provide sufficient heat to completely liquefy the reaction mixture. Alternatively, another oxidant or reaction booster can be added to the reaction mixture. During the reaction, uranium settles to the bottom of the bomb to form a metal button or ingot, which can be recast into a shape suitable for machining and fuel element fabrication. The average reaction time for a bomb containing a charge of the size given above is approximately 4.5 h, depending on the purity of UF_4. If 98% UF_4 is used, the yield may be 97% pure metal, which corresponds to 140 kg of uranium for the reaction size given above.

A direct ingot (dingot) method has been applied to charges of uranium up to 1540 kg of metal (40). The liquid uranium generated in the reduction collects as a pool in the bottom of the bomb and solidifies to an ingot with a diameter of about 25 cm and a height of 25 cm. Dingot metal is of high quality and can be fed directly to a milling machine or to an extrusion press without intermediate recasting. Magnesium is used as the reductant around 1900°C.

A unique problem arises when reducing the fissile isotope ^{235}U. The amount of ^{235}U that can be reduced is limited by its critical mass. In these cases, where

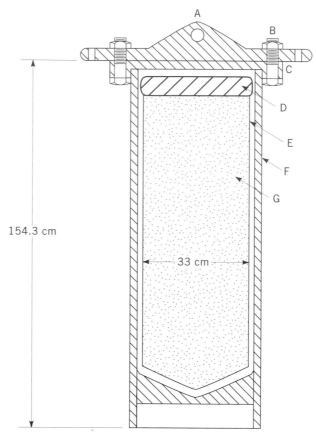

Fig. 4. Bomb reactor for the reduction of UF_4 with Mg by the Ames process (capacity 144.2 kg uranium metal): A, steel cover flange with lifting eye; B, bolt and nut; C, top flange of bomb; D, graphite cover; E, liner of fused dolomitic oxide; F, steel bomb, and G, charge, where ▨ represents steel, □ liner, ▨ graphite, and ▨ charge.

the charge must be kept relatively small, calcium becomes the preferred reductant, and iodine is often used as a reaction booster. This method was introduced by Baker in 1946 (54). Researchers at Los Alamos National Laboratory have recently introduced a laser-initiated modification to this reduction process that offers several advantages (55). A carbon dioxide laser is used to initiate the reaction between UF_4 and calcium metal. This new method does not require induction heating in a closed bomb, nor does it utilize iodine as a booster. This promising technology has been demonstrated on a 200 g scale.

Isotope Enrichment

Uranium-235 Enrichment. The enrichment of uranium is expressed as the weight percent of ^{235}U in uranium. For natural uranium the enrichment level is 0.72%. Many applications of uranium require enrichment levels above 0.72%, such as nuclear reactor fuel (56,57). Normally for lightwater nuclear reactors (LWR), the 0.72% natural abundance of ^{235}U is enriched to 2–5% (9,58). There are special cases such as materials-testing reactors, high flux isotope reactors, compact naval reactors, or nuclear weapons where ^{235}U enrichment of 96–97% is used.

Uranium isotope enrichment can be achieved in a number of ways. At the time of writing, the methods that have been, or are currently in use include gaseous diffusion, gaseous centrifugation, electromagnetic separation, chemical exchange, laser photoionization and photodissociation, separation nozzle, and cyclotronic resonance isotope separation. Most of these processes are of historical significance, and have been described (56). The gaseous diffusion and centrifugation processes (GDP and GDC) are the only methods employed on an industrial scale in the United States and Europe. Vigorous research programs are under way in the United States (52), France (59), and Japan (60), for a new industrial technique known as atomic-vapor laser isotope separation (AVLIS, or SILVA in France). The AVLIS process is expected to have a lower cost per separative work unit (SWU) than the diffusion process for uranium enrichment (57).

Gaseous-Diffusion Process. This process is used for the separation of ^{235}U and ^{238}U on an industrial scale (57,61–64). It is based on the fact that molecular transport through small pores takes place via Knudsen diffusion, where the speed of transport is inversely proportional to the square root of the mass of the molecules (65). Highly purified gaseous UF_6 is pumped through a barrier tube with porous walls. $^{235}UF_6$ and $^{238}UF_6$ diffuse through the barrier tubes at slightly different rates, based on the mass difference between $^{235}UF_6$ and $^{238}UF_6$ molecules. The separation efficiency for this process is very small because of the small mass difference (theoretical separation factor of 1.0043) between these uranium isotopes. To obtain an enrichment of the natural ^{235}U from 0.72% to 3% necessary for power generation, more than 1000 separation steps are needed. For many applications, the process must be repeated hundreds of thousands of times to obtain high enrichments of ^{235}U. This is accomplished by coupling many diffuser units in a series referred to as a cascade or stage (Fig. 5). Approximately 12 stages make up a larger unit known as a cell, and several cells make up a single GDP unit. Gaseous diffusion plants are enormous in size, often covering hundreds of acres, and requiring huge amounts of electric power to operate. The three U.S. gaseous-diffusion plants are in Oak Ridge, Tennessee, Paducah, Kentucky, and Portsmouth, Ohio. The Oak Ridge Plant began operation in 1945 during the Manhattan Engineering District Project, and had the capability to enrich uranium greater than 90%. The Oak Ridge Plant was permanently closed in 1987 and the Portsmouth high enrichment section has been closed since 1992. An excellent discussion of U.S. gas enrichment plants, the plans for their decontamination and decommissioning, and the ultimate disposition of natural reserves of UF_6 has been published by the National Research Council (57).

Centrifugal Isotope Separation. The high capital cost and large power requirements of gaseous-diffusion plants led to an extensive investigation into centrifugal separation of ^{235}U and ^{238}U by a number of countries (66–70). The advantage of centrifugal separation over gaseous diffusion is that the separation factor is not proportional to the square root of the ratio of the masses of ^{235}U and ^{238}U, but instead to the mass difference of isotopes. Therefore, centrifugal separation is more efficient than gaseous diffusion. However, the atmosphere is highly corrosive, and careful maintenance is required.

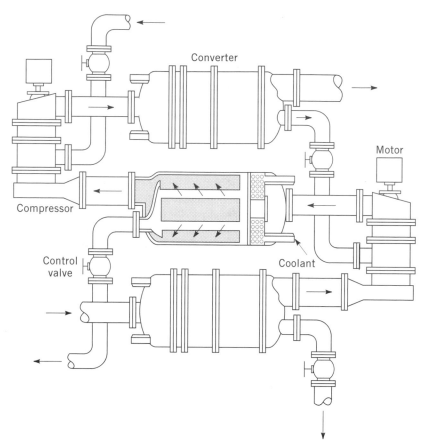

Fig. 5. Schematic of a gaseous-diffusion stage showing a single converter in center.

Evaporative current and countercurrent centrifuges with long rotors and ultrahigh speed capability have been extensively studied (66). In these centrifuges, a hollow rotor is partially filled with the isotope mixture in liquid form. High velocity is attained under temperature and pressure conditions that favor evaporation. The lighter isotope may be drawn axially from the rotor, whereas the depleted heavier isotope remains in the rotor. A concurrent type of centrifuge, which does not employ the evaporative principle, has been used experimentally. It is claimed that this device is more efficient than the evaporative centrifuge. Increases in the abundance ratio of ^{235}U to ^{238}U of as much as 5.6% have been reported with a single centrifugation. An analysis of market needs for gas centrifuge technology has been reported (71) and centrifuges have been successfully deployed commercially in Europe.

Electromagnetic Separation. The electromagnetic separation method was developed at the University of California (Berkeley) Radiation Laboratory, and employed on an industrial scale in the electromagnetic separation plant at Oak Ridge (72–74). Because the prototype machines were tested in the Berkeley cyclotron magnets, the Oak Ridge separators were called calutrons (California University Cyclotron). A comprehensive description of the calutron process has been given (75). A calutron separator is essentially a 180° mass spectrograph designed for a large throughput of ions. The uranium to be separated is converted to the tetrachloride and loaded into the charge bottle of the calutron, or is generated in the charge bottle from UO_2 [1344-57-6] and CCl_4 vapor. The ions generated from the UCl_4 vapor in the ion source are accelerated into the magnetic field and deflected through an angle of 180°. The radius of curvature of the heavy beam (^{238}U) is larger than that of the light beam (^{235}U). Thus, the two beams are focused on two different locations on the receiver. The receiver, built of graphite, is equipped with pockets in which the beams of the two isotopes are collected. After each run, the receiver is dismantled, and the individual isotopes are worked up separately. Electromagnetic separation is therefore an incredibly labor-intensive process.

The electromagnetic separation plant built during World War II at Oak Ridge, involved two types of calutrons, alpha and beta. The larger alpha calutrons were used for the enrichment of natural uranium, and the beta calutrons were used for the final separation of ^{235}U from the pre-enriched alpha product. For the electromagnetic separation process, UO_3 was converted into UCl_4 [10026-10-5] with CCl_4. The UCl_4 was fed into the calutron for separation. The calutron technique has been used to separate pure samples of ^{234}U, ^{236}U, and stable isotopes of many other elements. The Y-12 calutron plant at Oak Ridge was shut down in 1980. There remains a great deal of interest in electromagnetic separation as evidenced by the frequency of new patents at the time of writing. Electromagnetic separation is used in Russia to produce small quantities of research-grade isotopes of extremely high purity (76,77), and the International Atomic Energy Agency (IAEA) found evidence that the Iraqi government was using electromagnetic separation of uranium prior to the 1991 Gulf War (78).

Atomic-Vapor Laser Isotope-Separation. Although the technology has been around since the 1970s, laser isotope separation has only recently matured to the point of industrialization. In particular, laser isotope separation for the production of fuel and moderators for nuclear power generation is on the threshold of pilot-plant demonstrations in several countries. In the atomic vapor laser isotope-separation (AVLIS) process, vibrationally cooled ^{235}U metal atoms are selectively ionized by means of a high power (1–2 kW) tunable copper vapor or dye laser operated at high (kHz) repetition rates (51,59,60).

In the U-AVLIS process under development at Lawrence Livermore National Laboratory in California, an electron beam gun produces high energy electrons that melt and vaporize the uranium metal feed. The electron beam is focused onto the uranium melt using magnetic fields. The uranium melt is contained in a water-cooled copper crucible, and uranium feed is introduced into the melt in the form of a bar. The uranium atomic vapor produced by e-beam melting expands into a vacuum and becomes collimated upon cooling. Adiabatic cooling puts most of the uranium atoms in the ground electronic state. At this point, a high power laser is tuned to ionize ^{235}U atoms selectively, leaving the ^{238}U atoms unaffected. An electromagnetic field is used to strip the ^{235}U ions from the vapor stream and the unionized ^{238}U atomic vapor stream flows to the roof of the chamber (52).

This process has been proposed to solve the problem of the disposition of 500,000 metric tons of depleted UF_6, stored at Paducah, Portsmouth, and Oak

Ridge (57). In 1992 the DOE announced their intention to build a demonstration enrichment plant, however, as of 1997 no construction has begun. In France, a pilot plant has been built and is in operation (79).

Other Isotope Separation Methods. A number of other methods for separating uranium isotopes have been developed, but none of these has been advanced beyond the pilot-plant stage, and many have received little attention due to the improved economics of gaseous diffusion and centrifugation, and AVLIS. The liquid thermal-diffusion process was installed in a pilot plant in Oak Ridge at the time of the Manhattan Engineering District Project (80). Uranium hexafluoride, kept liquid at temperatures above its triple point, is subjected to diffusion in a thermal-diffusion column, the center of which is kept at 188–286°C, whereas the wall is maintained at 65°C. By thermal diffusion, the ^{235}U is enriched at the top, while ^{238}U migrates to the bottom of the column. The S-50 plant provided slightly enriched feed material for the calutron plant at Y-12. The S-50 operations were terminated in Sept. 1945, when the K-25 gaseous-diffusion plant went into operation. In the separation-nozzle method, uranium hexafluoride, UF_6, vapor is effused out of a nozzle (66). During the effusion, the light isotope is enriched in certain parts of the gas jet and may be enriched by stripping those parts away from the outer parts of the jet. The gas is an H_2-UF_6 mixture containing 5% UF_6. The process has been demonstrated on the pilot-plant scale at the Nuclear Engineering Institute of the Karlsruhe Nuclear Research Center in Germany.

Analytical Methods for Uranium Determination

The uranium content of a sample can be determined by fluorimetry, α-spectrometry, neutron activation analysis, x-ray microanalysis with a scanning-transmission electron (sem) microscope, mass spectrometry, and by cathodic stripping voltammetry (8). In most cases, measurements of environmental or biological materials require preliminary sample preparations such as ashing and dissolution in acid, followed by either solvent extraction or ion exchange. For uranium isotope analysis, inductively coupled plasma–mass spectrometry may also be used (81). Another uranium detection technique that has become very popular within the last few years is x-ray absorption near edge structure (xanes) spectroscopy. This method can provide information about the oxidation state or local structure of uranium in solution or in the solid state. The approach has recently been used to show that U(VI) was reduced to U(IV) by bacteria in uranium wastes (82), to determine the uranium speciation in soils from former U.S. DOE uranium processing facilities (83,84), and the mode of U(VI) binding to montmorillonite clays (85,86).

Uses and Economic Aspects

Uranium is a synthetic precursor of transuranium elements and the source of the light isotope, ^{235}U. The primary use of ^{235}U, as a source of nuclear energy for nuclear power generators and nuclear weapons, is well known. The thermal energy generated by fission of 1 g of ^{235}U is equivalent to that released by burning 2200 L of crude oil or 2.7 t of coal. The predominate species utilized as nuclear fuel in power stations is UO_2. However, other uranium compounds and alloys are being considered for use as nuclear fuels in alternative reactors. Uranium carbide (UC) has been utilized in sodium or lead cooled reactors, whereas uranium silicides have been proposed as a fuel source in lightwater reactors. Uranium–zirconium and uranium–aluminum alloys are used in materials and research testing reactors. Uranium–zirconium alloys are also widely used in marine reactors, while the hydrogenated U–Zr alloys ($UZrH_x$) are fuels for spacecraft reactors. The full details of the role of uranium in the nuclear fuel cycle have been presented in previous editions of this *Encyclopedia*, the *Gmelin Handbook*, and *Ullmann's Encyclopedia of Industrial Chemistry*.

Depleted uranium (^{238}U), which is about 0.2% ^{235}U, has a density more than twice that of steel. This property has been utilized for military purposes in the production of armor and armor-piercing projectiles, also known as kinetic energy penetrators (87). Depleted uranium and uranium alloys such as $UTi_{0.75}$ [*39460-95-2*] are very useful as armor-piercing projectiles (88), and the penetration mechanism has been determined (89). The superior penetration behavior of depleted uranium penetrators is presently attributed to the ability to self-sharpen during armor penetration by failure along adiabatic shear bands. This is in contrast to conventional tungsten heavy alloy (WHA) penetrators which form a "mushroom" head that decreases the energy delivered to the target. The radiological hazard of depleted uranium combined with chemical corrosion during storage has resulted in significant research into improving the penetration behavior of tungsten alloys (90,91).

The high density of uranium makes it attractive for flywheels, and its density and effectiveness at absorbing gamma-rays also suggests a possible use for shielding of spent nuclear fuels. Two types of storage canister have been proposed; one made of metallic uranium and one made of UO_2-containing concrete. One of the difficulties in designing metallic uranium shields is the tendency of uranium to corrode in air, forming oxide surfaces. A corrosion-resistant material can be obtained by alloying uranium with 2–8 wt % molybdenum, and therefore, can be used in medical radiation equipment shielding and in aircraft trimming weights.

Uranium Compounds

Oxides. Oxides of uranium are some of the most prevalent and technologically important binary uranium compounds known. Numerous oxide phases have been observed and characterized, including uranium oxide [*12035-97-1*], UO, UO_2 [*1344-57-6*]; U_4O_9 [*12037-15-9*]; U_3O_7 [*1203-04-6*]; U_3O_8 [*1344-59-8*]; UO_3 [*1344-58-7*]; hydrated species such as $UO_3 \cdot xH_2O$ and the peroxo complex, $UO_4 \cdot xH_2O$; and anionic uranates including $[U_2O_7]^{2-}$ [*85096-44-2*] and $[U_4O_{13}]^{2-}$ [*128085-85-8*]. Of these oxide phases, UO_2, U_3O_8, and UO_3 are extremely important both industrially and in the nuclear energy cycle. The preparation of some of these most important oxides is given in equations 20–26.

$$UO_3 + H_2 \xrightarrow{700^\circ C} UO_2 + H_2O \tag{20}$$

$$UO_3 + CO \xrightarrow{350^\circ C} UO_2 + CO_2 \tag{21}$$

$$3\ UO_2 + O_2 \xrightarrow{600°C} U_3O_8 \tag{22}$$

$$3\ UO_3 \xrightarrow{700°C} U_3O_8 + 1/2\ O_2 \tag{23}$$

$$U_3O_8 + 5\ UO_2 \xrightarrow{\text{sealed tube}} 2\ U_4O_9 \tag{24}$$

$$4\ UO_2 + 1/2\ O_2 \xrightarrow[900°C]{0.1\ \text{mm air}} U_4O_9 \tag{25}$$

$$3\ UO_2 + 1/2\ O_2 \xrightarrow{<200°C} U_3O_7 \tag{26}$$

Uranium dioxide [*1344-57-6*], UO_2, is found in nature as the mineral pitchblende and as a component in uraninite. The crystalline solid melts at 2878°C and is paramagnetic with a room temperature magnetic moment of 3.2 μ_B. The density has been found to range from 10.79 to 10.95 g/cm^3, lower values are observed for hyperstoichiometric complexes, UO_{2+x}. Industrially, UO_2 is prepared by the decomposition of ammonium uranyl carbonate on the scale of 10 kg/d, using a fluidized-bed furnace (92). In addition to the industrial process, pure UO_2 has been synthesized by (*1*) oxidation of uranium metal, (*2*) reduction of higher valent oxides, (*3*) thermal decomposition of uranyl uranates, (*4*) oxidation or reduction of uranium halides, (*5*) decarboxylation of uranium compounds of carbonic acids, (*6*) hydrometallurgical preparation, and (*7*) electrolysis of uranium halides. Single crystals of UO_2 have been grown by a variety of techniques, including vapor or electrolytic deposition from salt melts or vapor deposition on ionic substrates. The nature of the bonding in UO_2 is best described as ionic. The x-ray analysis of the stoichiometric complex reveals a face-centered cubic CaF_2 type structure, with the uranium atoms occupying the face-centered sites. Hyper- and hypo-stoichiometric, UO_{2+x} and UO_{2-x}, respectively, are also known and have been analyzed by x-ray crystallography. In the case of the UO_{2+x}, extra oxygen atoms occupy central lattice holes in the normal UO_2 structure. For the hypostoichio-metric complex, the structure indicates the presence of layers of UO_2 and UO.

The main technological uses for UO_2 are found in the nuclear fuel cycle as the principal component for light and heavy water reactor fuels. Uranium dioxide is also a starting material for the synthesis of UF_4 [*10049-14-6*], UF_6 [*7783-81-5*] (both critical for the production of pure uranium metal and isotopic enrichment), UCl_4 [*10026-10-5*], and $UO_2(NO_3)_2 \cdot 6H_2O$ [*10102-06-4*]. In order to be useful as a nuclear fuel, the material must have certain physical and chemical properties in the reactor temperature range, ie, small coefficients of linear and volume expansion, reasonable heat conductivity, and chemical stability. Uranium dioxide has been found to exhibit a majority of these desirable properties with an average thermal coefficient of expansion of 10.8×10^{-6} (20–946°C), specific heat from 0.237 to 0.338 J/gK (300–1773°C), and a thermal conductivity of 8.281 to 2.353 W/mK (300–1773°C) at 0 atomic % burnup UO_2. The sintered complex has also been found to be chemically stable toward air and H_2O up to 300°C. The thermodynamic and transport properties of UO_2, including hypo/hyperstoichiometric compounds as well as doped species under reactor conditions, are under investigation (93–101).

For most nuclear applications, UO_2 must be produced as uniform spheres and pellets. Three techniques utilized for microsphere fabrication are sol-gel, gel-precipitation, and plasma spheroidization (102). Details on the sol-gel and gel-purification processes, the two most popular, can be obtained from the *Gmelin Handbook*. The common method for producing UO_2 pellets consists of pressing granules in the presence of binding agents and lubricants with a subsequent sintering, after the organics have been removed. High density spheres without open porosity have been fabricated from soft UO_2 microspheres, using a gel pelletization technique (103).

Triuranium octaoxide [*1344-59-8*], U_3O_8, is a greenish black material which is also a constituent of pitchblende. This complex has been identified with a number of different oxygen deficiencies, depending mostly on the temperature and partial pressure of O_2 used in the preparation. From 900 to 1500°C, the oxide decomposes prior to melting or subliming to form gaseous UO_3. The material is paramagnetic and epr-active with a room temperature μ_{eff} of 1.32 μ_B. XPS studies of U_3O_8 have indicated the presence of two valences, U(IV) and U(VI), in a 1:2 ratio (104). The density of the stoichiometric complex has been found to range from 8.16 to 8.41 g/cm^3. This density for U_3O_8 is significantly smaller than that of UO_2, a property that is problematic for nuclear fuel cells (*vide infra*). The preparation of U_3O_8 has been accomplished by a variety of means, including thermal decomposition of $(NH_4)_2U_4O_{13}$ [*129002-73-9*], $UO_4 \cdot xH_2O$ [*12036-71-4*], and $UO_2(NO_3)_2 \cdot 6H_2O$ [*10102-06-4*] at 600°C, the oxidation of UO_2 under streaming O_2 at 600°C (eq. 22), and the reduction of UO_3 at high temperatures (600–800°C) under streaming oxygen (eq. 23). One of the problems associated with synthesizing stoichiometric U_3O_8 is the propensity of oxygen loss at elevated temperatures (500–700°C), producing oxygen deficient complexes U_3O_{8-x}. Depending on the partial pressure of O_2 the oxygen deficiency can go as low as $UO_{2.62}$. In addition to the different U/O ratios, U_3O_8 has been found to exist in at least five different crystalline modifications. In the α-form, the uranium atom displays a pentagonal bipyramidal coordination geometry, whereas the β-modification has two different coordination environments, one pseudo-octahedral and one pentagonal bipyramidal. A full description of the anionic sheets has been reported (105,106).

Industrially, U_3O_8 has been shown to be active in the decomposition of organics, including benzene and butanes (107,108) and as supports for methane steam reforming catalysts (109). In the nuclear fuel industry, U_3O_8 is an oxidation product of UO_2 (SIMFUEL), and thus, a large component of spent fuel rods (110,111). As previously mentioned, U_3O_8 is less dense than UO_2 and as a result, the production of U_3O_8 in nuclear fuel can lead to the destruction of the UO_2 pellet by pulverization. It is for this reason that many studies of the formation kinetics (111) and the thermal and mechanical properties of U_3O_8 have been reported. Triuranium octaoxide is not always a destructive force in the fuel cycle, it is actually quite useful in the initial production of UO_2 pellets for fuel (103,112), in the manufacturing of mixed oxide (MOX) pellets (113), as well as being a dispersive nuclear fuel itself (114).

Uranium trioxide [*1344-58-7*], UO_3, is a versatile solid that also has important applications in the nuclear fuel cycle. The trioxide has been isolated in six well-defined stoichiometric modifications as well as a hypostoichiometric modification, $UO_{2.9}$. Similar to U_3O_8, the trioxide decomposes into lower oxides prior to melting or subliming. Even though UO_3 is formally U(VI), a small temperature-dependent paramagnetism exists with molar magnetic susceptibility values ranging from 128 to 157×10^{-6} cm^3/mol. A general trend has been observed for the densities of uranium oxides; an inverse proportionality between the O/U ratio and the density of the material. The trioxide does not deviate from this trend with density values ranging from 6.99 to 8.54 g/cm^3, depending on the modification. The preparation of UO$_3$ has been accomplished by a variety of means. Industrially, the complex is prepared by three main routes, thermal decomposition of UO$_4$·xH$_2$O [12036-71-4], (NH$_4$)$_2$U$_4$O$_{13}$ [128085-85-8], or UO$_2$(NO$_3$)$_2$·6H$_2$O [10102-06-4] under O$_2$. For the latter complex, the techniques utilized to acomplish the decomposition include batch decomposition, continuous stirred-bed, fluidized bed, and spray decomposition. The trioxide can also be synthesized by the oxidation of lower oxides, UI$_3$ [13775-18-3] (low temperature), UI$_4$ [13470-22-9] (low temperature) (115), UC [12070-09-6], or UN [25658-43-9] (116) with O$_2$, and by the calcination of (NH$_4$)$_4$UO$_2$(CO$_3$)$_3$ [17872-00-3] (117,118).

As mentioned above, uranium trioxide exists in six well-defined modifications with colors ranging from yellow to brick-red. Of these phases, the γ-phase has been found to be the most stable, however, other phases, especially α and β, are also frequently used and studied. The structure of the α-modification is based on sheets of hexagons, whereas the β-, γ-, and δ-modifications contain an infinite framework. All of these topologies have been fully described (105,106). They are α-brown, hexagonal; β-orange, monoclinic; γ-yellow, rhombic; δ-red, cubic; ϵ-brick red, triclinic; and η-rhombic.

The most important role of UO$_3$ is in the production of UF$_4$ [10049-14-6] and UF$_6$ [7783-81-5], which are used in the isotopic enrichment of uranium for use in nuclear fuels (119–121). The trioxide also plays a part in the production of UO$_2$ for fuel pellets (122). In addition to these important synthetic applications, microspheres of UO$_3$ can themselves be used as nuclear fuel. Fabrication of UO$_3$ microspheres has been accomplished using sol-gel or internal gelation processes (19,123–125). Finally, UO$_3$ is also a support for destructive oxidation catalysts of organics (126,127).

Nitrides. Uranium nitrides are well known and are used in the nuclear fuel cycle. There are three nitrides of exact stoichiometry, uranium nitride [25658-43-9], UN; U$_2$N$_3$ [12033-85-1]; and U$_4$N$_7$ [12266-20-5]. In addition to these, nonstoichiometric complexes, U$_2$N$_{3+x}$, where the N/U ratio ranges from 1.64 to 1.84, have been identified (128). The brown mononitride, which is the only nitride complex stable above 1300°C, melts at 2600°C. Uranium mononitride is the most dense of the nitrides with a density of 14.31 g/cm^3. The magnetic properties of the nitrides are extremely dependent on the phase and stoichiometry of the complex. The mononitride, α-U$_2$N$_3$, and β-U$_2$N$_3$ are paramagnetic at room temperature with the former becoming antiferromagnetic and the latter becoming ferromagnetic at low temperatures. Classically, the different nitrides have been prepared from direct interaction of the elements under the appropriate conditions. A number of alternatives to this preparation have been investigated, including uranium metal under static NH$_3$ at 300–350°C to yield U$_2$N$_3$ (129,130), uranium metal or uranium carbides with NH$_3$ or N$_2$ at 600–900°C to produce U$_2$N$_{3+x}$ (131,132), uranium carbide fuels reacted with N$_2$/H$_2$ to form UN (133), and a self-propagating metathetical reaction, thermolysis at 500°C of UCl$_4$ [10026-10-5] with Li$_3$N, yielding UN and U$_2$N$_3$ (134,135).

The structures of some of the nitrides have been determined. The mononitride has a face-centered cubic NaCl type structure. The sesquinitride complex has two modifications: the α-phase is found with a body-centered cubic Mn_2O_3-type structure (128), while the high temperature β-phase crystallizes in a hexagonal Mn_2O_3-type structure. The substoichiometric nitride complex, $UN_{1.9}$, crystallizes in the hexagonal CaF_2-type lattice.

Uranium and mixed uranium–plutonium nitrides have a potential use as nuclear fuels for lead cooled fast reactors (136–139). Reactors of this type have been proposed for use in deep-sea research vehicles (136). However, similar to the oxides, in order for these materials to be useful as fuels, the nitrides must have an appropriate size and shape, ie, spheres. Microspheres of uranium nitrides have been fabricated by internal gelation and carbothermic reduction (140,141). Another use for uranium nitrides is as a catalyst for the cracking of NH_3 at 550°C, which results in high yields of H_2 (142).

Carbides. Uranium carbides, UC [12070-09-6], U_2C_3 [12076-62-9], and UC_2 [12071-33-9] are all dark gray solids with a metallic luster. In addition to these binary materials, numerous mixed uranium–plutonium and uranium–transition-metal carbides have been prepared and are mainly utilized in nuclear fuel. The melting points of UC and U_2C_3 are 2400°C and 2417°C, respectively, and the dicarbide melts at 2475°C and boils at 4370°C (760 mm Hg). The monocarbide is the most dense of the carbide series with a room temperature density of 13.60 g/cm^3, whereas U_2C_3 and UC_2 have densities of 12.85 g/cm^3 and 11.69 g/cm^3, respectively. The magnetic properties of the carbides are extremely composition dependent. All three materials are paramagnetic at room temperature, with U_2C_3 becoming antiferromagnetic at low temperatures. The typical techniques involved in the synthesis of the carbides include the reaction of carbon or hydrocarbons with uranium metal or UH_3 [13598-56-6] at elevated temperature, precipitation from metal melts, and reduction of uranium halides. Techniques for the synthesis include the carbothermic reduction of UO_2 (143) and the direct interaction of uranium and carbon under highly exothermic conditions (144). The crystal structure of UC is a face-centered cubic NaCl-type lattice, identical to that of UN [25658-43-9]. The sesquicarbide crystallized with a body-centered cubic Pu_2C_3-type structure, while two modifications exist for UC_2. The α-phase of UC_2 is a body-centered tetragonal CaC_2-type structure and the β-phase crystallizes in a face-centered cubic KCN-type lattice. It should also be noted that the α-modification undergoes a phase transition from a tetragonal to a hexagonal lattice under increased pressure, ie, 17.6 GPa.

As previously stated, uranium carbides are used as nuclear fuel (145). Two of the typical reactors fueled by uranium and mixed metal carbides are thermionic, which are continually being developed for space power and propulsion systems, and high temperature gas-cooled reactors (83,146,147). In order to be used as nuclear fuel, carbide microspheres are required. These microspheres have been fabricated by a carbothermic reduction of UO_3 and elemental carbon to form UC (148,149). In addition to these uses, the carbides are also precursors for uranium nitride based fuels.

Oxo Ion Salts. Salts of oxo anions, such as nitrate, sulfate, perchlorate, iodate, hydroxide, carbonate, phosphate, oxalate, etc, are important for the separation and reprocessing of uranium, hydroxide, carbonate, and phosphate ions are important for the chemical behavior of uranium in the environment (150–153).

Nitrate complexes are very weak, and the determination of the formation constants for aqueous nitrate solution species is extremely difficult. There appears to be reliable thermodynamic data only for the formation of $UO_2(NO_3)^+$ (154). Although determination of the formation constants is complicated, there is little doubt that under high nitric acid conditions, $UO_2(NO_3)_2$, and perhaps $UO_2(NO_3)_3^-$ are formed, at least to some extent. Solid uranyl nitrate [*10102-06-4*], $UO_2(NO_3)_2 \cdot xH_2O$, is obtained as the orthorhombic hexahydrate from dilute nitric acid solutions, and as the trihydrate from concentrated acid. The melting point of the hexahydrate is at 118°C. Uranyl nitrate plays an important role in the reprocessing of uranium in spent fuel and in uranium extraction from aqueous solutions. The preparation of the anhydrous uranyl nitrate by dehydration is extremely difficult. Several molecular structures of uranyl nitrate complexes have been reported, and all show the common formula unit of $UO_2(NO_3)_2(OH_2)_2$ with a local hexagonal bipyramidal coordination about the central uranyl ion (**1**) (155–157). The technologically important $UO_2(NO_3)_2(TBP)_2$ complex also displays trans nitrate ligands with the TBP ligands occupying the same coordination sides as H_2O in (**1**).

(**1**)　　　　　(**2**)　　　　　(**3**)

There is reasonable evidence for the formation of aqueous U(IV) nitrate complexes of general formula $U(NO_3)_n^{4-n}$ where $n = 1–4$. However, owing to the inherent weakness of the complexes, quantitative data on the formation constants is only available for $U(NO_3)^{3+}$ and $U(NO_3)_2^{2+}$ (154). No neutral U(IV) nitrates have been obtained from aqueous solution, but a number of anionic complexes of general formula $M_2[U(NO_3)_6]$, where M = NH_4, Rb, Cs, and $M[U(NO_3)_6] \cdot 8H_2O$, where M = Mg, Zn have been isolated and characterized. These solids contain the 12 coordinate anionic U(IV) center shown in (**2**) (158). Neutral, U(IV) nitrate complexes of formula $U(NO_3)_4L_2$ (**3**) (L = $OP(C_6H_5)_3$, $OP(NC_4H_8)_3$) have also been isolated from aqueous solutions and structurally characterized (159).

The aqueous uranyl sulfate system has been extensively studied and complexes of formula $UO_2(SO_4)_n^{2n-2}$, where $n = 0$, 1, and 2, are likely to be formed in solution. Quantitative data only exists for the formation of complexes with

$n = 0$ and 1 (154). $UO_2(SO_4)_2 \cdot xH_2O$ ($x = 1$, 2, 2.5, 3, or 4) can be precipitated from aqueous solutions as the trihydrate. The monohydrate and the anhydrous salt can be obtained by dehydration of the trihydrate. The fluorosulfate, $UO_2(SO_3F)_2$ [75357-79-8], is obtained by treating $UO_2(MeCO_2)_2$ [541-09-3] with HSO_3F (160). A large number of ternary U(VI) sulfates of the general formula $(M)_k(UO_2)_m(SO_4)_n \cdot xH_2O$, where M = monovalent cation, ie, NH_4 or alkali metals, $M(UO_2)_m(SO_4)_n \cdot xH_2O$, where M = bivalent cation, such as alkaline-earth or transition metals (Mn, Cd, Hg), have been reported. A layered structure is observed for $(NH_4)_2UO_2(SO_4)_2 \cdot 2H_2O$ [12357-71-0] with local pentagonal biypyramidal coordination around the U atom, and bridging sulfate groups joining the uranyl polyhedra (161). In $K_4UO_2(SO_4)_3$ [69567-87-9] each uranium in the pentagonal bipyramid is coordinated to five oxygen atoms from four sulfate groups in the equatorial plane (162).

The U(IV) sulfate system has also been studied in strong acid solutions, and quantitative data are only available for $U(SO_4)^{2+}$ and $U(SO_4)_2$ (154). Solids of composition $U(SO_4)_2 \cdot 8H_2O$ [14355-39-6] and $U(SO_4)_2 \cdot 4H_2O$ can be precipitated from weak and concentrated sulfuric acid solution, respectively. In neutral solution the basic salt, $UOSO_4 \cdot 2H_2O$ [18902-45-9], is formed. The octahydrate looses four hydration waters at 70°C, the remaining four molecules of water can be removed at temperatures over 400°C. In $U(SO_4)_2 \cdot 4H_2O$ the uranium atoms are surrounded by a square antiprism of O atoms, with each U bonded to four molecules of water and linked by bridging sulfate groups to other uranium atoms (163). Several U(IV) fluorosulfates have been obtained involving mono and bidentate SO_3F^- groups as reported for $U(SO_3F)_4$ (160).

Crystals of uranyl perchlorate, $UO_2(ClO_4)_2 \cdot xH_2O$ [13093-00-0], have been obtained with six and seven hydration water molecules. The uranyl ion is coordinated with five water molecules (**4**) in the equatorial plane with a U−O(aquo) distance of 245 nm (2.45 Å). The perchlorate anion does not complex the uranyl center. The unit cells contain two $[ClO_4]^-$ and one or two molecules of hydration water held together by hydrogen bonding (164).

Hydroxides. The hydrolysis of uranium has been recently reviewed (154,165,166), yet as noted in these compilations, studies are ongoing to continue identifying all of the numerous solution species and solid phases. The very hard uranium(IV) ion hydrolyzes even in fairly strong acid ($\sim 0.1\ M$) and the hydrolysis is complicated by the precipitation of insoluble hydroxides or oxides. There is reasonably good experimental evidence for the formation of the initial hydrolysis product, $U(OH)^{3+}$; however, there is no direct evidence for other hydrolysis products such as $U(OH)_2^{2+}$, $U(OH)_3^+$, and $U(OH)_4$ (or $UO_2 \cdot 2H_2O$). There are substantial amounts of data, particularly from solubility experiments, which are consistent with the neutral species $U(OH)_4$ (154,167). It is unknown whether this species is monomeric or polymeric. A new study under reducing conditions in NaCl solution confirms its importance and reports that it is monomeric (168). Solubility studies indicate that the anionic species $U(OH)_5^-$, if it exists, is only of minor importance (169). There is limited evidence for polymeric species such as $U_6(OH)_{15}^{9+}$ (154).

The hydrolysis of the uranyl(VI) ion, UO_2^{2+}, has been studied extensively and begins at about pH 3. In solutions containing less than $10^{-4}M$ uranium, the first hydrolysis product is the monomeric $UO_2(OH)^+$, as confirmed using time-resolved laser induced fluorescence spectroscopy. At higher uranium concentrations, it is accepted that polymeric U(VI) species are predominant in solution, and the first hydrolysis product is then the dimer, $(UO_2)_2(OH)_2^{2+}$ (154,170). Further hydrolysis products include the trimeric uranyl hydroxide complexes $(UO_2)_3(OH)_4^{2+}$ and $(UO_2)_3(OH)_5^+$ (154). At higher pH, hydrous uranyl hydroxide precipitate is the stable species (171). In studying the sol-gel UO_2-ceramic fuel process, ^{17}O nmr was used to observe the formation of a trimeric hydrolysis product, $((UO_2)_3(\mu_3\text{-}O)(\mu_2\text{-}OH)_3)^+$ which then condenses into polymeric $UO_2O_{6/3}$, layers of a gel based on the hexagonal structure of α-$UO_2(OH)_2$. In the same process there is a second pathway where a uranyl derivative is treated with excess hydroxide in the absence of a metal or H-bonding ammonium cations which form insoluble solid uranates. Condensation of the resulting solution of $UO_2(OH)_n^{2-n}$ anions can then lead to a similar $UO_2O_{6/3}$ gel (172,173).

A study performed in the nonstandard electrolyte, tetramethylammonium trifluoromethanesulfonate, provided data on additional species, $(UO_2)_3(OH)_7^-$, $(UO_2)_3(OH)_8^{2-}$, and $(UO_2)_3(OH)_{10}^{4-}$ (174). Solid-state structures of uranyl hydroxides are limited, but are known for the important cations of formula $(UO_2)_2(OH)_2(OH_2)_6^{2+}$ (**5**) (175), and $(UO_2)_3O(OH)_3(OH_2)_6^+$, (**6**) (176).

(**4**)   (**5**)

(**6**)

Carbonates.   Actinide carbonate complexes are of interest not only because of their fundamental chemistry and environmental behavior (150), but also because of extensive industrial applications, primarily in uranium recovery from ores and nuclear fuel reprocessing.

The aqueous U(VI) carbonate system has been very thoroughly studied, and there is little doubt about the compositions of the three monomeric complexes

$UO_2(CO_3)$, $UO_2(CO_3)_2^{2-}$, and $UO_2(CO_3)_3^{4-}$ present under the appropriate conditions (154). There is also a great deal of evidence from emf, solubility, and spectroscopic data supporting the existence of polymeric solution species of formulas $(UO_2)_3(CO_3)_6^{6-}$, $(UO_2)_2(CO_3)(OH)_3^-$, $(UO_2)_3O(OH)_2(HCO_3)^+$, and $(UO_2)_{11}(CO_3)_6$-$(OH)_{12}^{2-}$ which form only under conditions of high metal ion concentration or high ionic strength (154,177). Determining the formation constant for the triscarbonato uranyl monomer, $UO_2(CO_3)_3^{4-}$, was complicated because this species is in equilibrium with the hexakiscarbonato uranyl trimer, $(UO_2)_3(CO_3)_6^{6-}$. Thermal lensing spectroscopy, which is sensitive enough to allow the study of relatively dilute solutions where the trimer is not favored, has been used to determine the equilibrium constant for the addition of one carbonate to $UO_2(CO_3)_2^{2-}$ to form $UO_2(CO_3)_3^{4-}$, and used to calculate the formation constant, β_{13} (178).

The trimetallic uranyl cluster $(UO_2)_3(CO_3)_6^{6-}$ has been the subject of a good deal of study, including nmr spectroscopy (179–182) solution x-ray diffraction (182), potentiometric titration (177,183,184), single crystal x-ray diffraction (180), and exafs spectroscopy in both the solid and solution states (180). The data in this area have consistently led to the proposal and verification of a trimeric $(UO_2)_3(CO_3)_6^{6-}$ cluster (181,182,185).

The known uranium(VI) carbonate solids have empirical formulas, $UO_2(CO_3)$, $M_2UO_2(CO_3)_2$, and $M_4UO_2(CO_3)_3$. The solid of composition $UO_2(CO_3)$ is a well-known mineral, rutherfordine, and its structure has been determined from crystals of both the natural mineral and synthetic samples. Rutherfordine is a layered solid in which the local coordination environment of the uranyl ion consists of a hexagonal bipyramidal arrangement of oxygen atoms with the uranyl units perpendicular to the orthorhombic plane. Each uranium atom forms six equatorial bonds with the oxygen atoms of four carbonate ligands, two in a bidentate manner and two in a monodentate manner.

Biscarbonato complexes of uranium(VI) are well-established in solution (154) and there are many reports dating from the late 1940s through the 1960s of solid phases with the general stoichiometry $M_2UO_2(CO_3)_2$, where M is a monovalent cation (Na^+, K^+, Rb^+, Cs^+, NH_4^+, etc). A summary of the preparative details is available (186), as is a listing of the compounds (187). A careful examination of the more recent literature and a detailed understanding of the solution chemistry suggests that the claims of some of these early reports on solid $M_2UO_2(CO_3)_2$ compounds should be reinterpreted. It is known that solids of general composition $M_2UO_2(CO_3)_2$ form trimetallic clusters of molecular formula $M_6(UO_2)_3(CO_3)_6$. The trimetallic cluster forms in solution when high metal ion concentrations are present, and these solutions are relatively unstable unless the pH is kept near 6 and a CO_2 atmosphere is maintained over the solution. A recent single-crystal x-ray diffraction study of $[C(NH_2)_3]_6[(UO_2)_3(CO_3)_6]\cdot6.5\ H_2O$ revealed that the central $(UO_2)_3(CO_3)_6^{6-}$ anion (**7**) possesses a D_{3h} planar structure in which all six carbonate ligands and the three uranium atoms lie within the molecular plane (180). The six uranyl oxygen atoms are perpendicular to the plane, with three above, and three below the plane. The local coordination geometry about each uranium is hexagonal bipyramidal.

The triscarbonato solids, $M_4UO_2(CO_3)_3$ (M = monovalent cation) are the most thoroughly studied uranium(VI) carbonate solids. These solid phases are generally prepared by evaporation of an aqueous solution of the components, or by precipitation of the UO_2^{2+} ion with an excess of carbonate. Some of these salts can be further purified by dissolution in water and recrystallization by evaporation. Single-crystal x-ray diffraction studies have been reported for a large number of these uranyl complexes. Detailed lists of complexes that have been characterized by x-ray diffraction are available (187,188). In the solid state, all monomeric $M_4AnO_2(CO_3)_3$ complexes show the same basic structural features: a hexagonal bipyramidal coordination geometry where three bidentate carbonate ligands lie in a hexagonal plane (**8**), and the trans oxo ligands occupy coordination sites above and below the plane. Typical metrical parameters for these structures have $An = O$ bond distances within the relatively narrow range of 1.7–1.9 Å (0.17–0.19 nm), and $An-O$ bonds to the carbonate ligands in the range 2.4–2.6 Å (0.24–0.26 nm).

(7)   (8)

Although there is a great deal of qualitative information regarding anionic carbonato complexes of the tetravalent actinides, reliable quantitative data are rare (154). Quantitative data exist only for $U(CO_3)_5^{6-}$ and $U(CO_3)_4^{4-}$ (189,190). Tetracarbonato uranium salts of composition $[C(NH_2)_3]_4[U(CO_3)_4]$ and $[C(NH_2)_3]_3(NH_4)[U(CO_3)_4]$ have been reported (191). The pentacarbonato salts of formula $M_6U(CO_3)_5 \cdot nH_2O$ (M_6 = Na_6, K_6, Tl_6, $(Co(NH_3)_6)_2$, $[C(NH_2)_3]_3[(NH_4)]_3$, $[C(NH_2)_3]_6$; $n = 4-12$) have been reported (186,192,193). The sodium salt can be prepared by chemical or electrochemical reduction of $Na_4UO_2(CO_3)_3$, followed by the addition of Na_2CO_3 to form a precipitate. The potassium salt, $K_6U(CO_3)_5 \cdot 6H_2O$ can be prepared by dissolution of freshly prepared U(IV) hydroxide in K_2CO_3 solution in the presence of CO_2; and the guanidinium salt can be prepared by addition of guanidinium carbonate to a warm $U(SO_4)_2$ solution, followed by cooling (191). All of the uranium(IV) complexes are readily air-oxidized to uranium(VI) complexes, and therefore there is no structural information for the uranium(IV) analogues.

Phosphates. Inorganic phosphate ligands are important with respect to the behavior of uranium in the environment and as potential waste forms. There have been a number of experimental studies to determine the equilibrium constants in the uranium–phosphoric acid system, but they have been complicated by the formation of relatively insoluble solid phases and the formation of ternary uranium complexes in solution (154). In acidic solution (hydrogen-ion concentration range 0.25–2.00 M) H_3PO_4 and $H_2PO_4^-$ are potential ligands, whereas in neutral to basic solution, HPO_4^{2-} and PO_4^{3-} ligands are predominant. Numerous U(VI) phosphate complexes have been identified and their formation constants determined. However, relatively little thermodynamic data have been recommended with confidence. There is good evidence for the formation of $UO_2(PO_4)^-$, $UO_2(HPO_4)$, $UO_2(H_2PO_4)^+$, $UO_2(H_2PO_4)_2$, $UO_2(H_3PO_4)^{2+}$, and $UO_2(H_3PO_4)(H_2PO_4)^+$ complexes in solution. There are only a few studies on the U(IV) phosphate system, and recent reviews on the thermodynamics of uranium have not recommended any thermodynamic data for the U(IV) phosphate system (154).

Solid uranium–phosphate complexes have been reported for the IV and VI oxidation states, as well as for compounds containing mixed oxidation states of U(IV) and U(VI). Only a few solid state structures of U(IV) phosphates have been reported, including the metaphosphate $U(PO_3)_4$, the pyrophosphate $U(P_2O_7)$, and the orthophosphate, $CaU(PO_4)_2$. The crystal structure of orthorhombic $CaU(PO_4)_2$ is similar to anhydrite (194). Compounds of the general formula $MU_2(PO_4)_3$ have been reported for M = Li, Na, and K, but could not be obtained with the larger Rb and Cs ions (195). In the solid state, uranium(IV) forms the triclinic metaphosphate, $U(PO_3)_4$. Each uranium atom is eight-coordinate with square antiprismatic UO_8 units bridged by $(P_4O_{12})^{4-}$ rings (196,197). The pyrophosphate of uranium(IV) belongs to the family of ZrP_2O_7-type structures (198). The dissolution of U chips in HCl and H_3PO_4 results in the formation of orthorhombic $U_2O(PO_4)_2$ (199). Each uranium atom is seven-coordinate, with a local edge-sharing pentagonal bipyramidal coordination geometry, and linear U–O–U and bridging bidentate phosphate units making up a three-dimensional structure. Pyrophosphates of composition $U_2O_3P_2O_7$ and $U_3O_5P_2O_7$ have been synthesized containing uranium in oxidation states IV and VI in a ratio 1:1 and 2:1, respectively. $U_2O_3P_2O_7$ melts at 1442°C, is stable under oxic conditions up to 1250°C, but decomposes in nitrogen and argon at 950°C. $U_3O_5P_2O_7$ oxidizes in air above 500°C to form $(UO_2)_3(PO_4)_2$ at 950°C (200). $M_2UO_2P_2O_7$, with M = alkali, have been synthesized by heating uranyl nitrate in the presence of alkali metal pyrophosphates and ammonium dihydrogenphosphate. The new mixed valence $U(UO_2)(PO_4)_2$ has been synthesized and characterized spectroscopically showing the absence of pyrophosphate and the existence of the dioxo cation unit, UO_2^{2+}, as one of the two independent U atoms. Bidentate phosphates ligands connect the chains generating a three-dimensional network (201).

Uranium(VI) phosphates have been widely investigated and can be divided in several structure types: orthophosphates $M(UO_2)_n(PO_4)_m \cdot xH_2O$, hydrogenphosphates $M(UO_2)_n(H_kPO_4)_m \cdot xH_2O$, pyrophosphates $U_mO_nP_2O_7$, metaphosphates $(UO_2)_n(PO_3)_m \cdot xH_2O$, and polyphosphates $(UO_2)_n(P_aO_b)_m \cdot xH_2O$ (188).

A few uranyl metaphosphates have been described in the literature. $UO_2(PO_3)_2$ is formed in 85% H_3PO_4 at 300–350°C or by thermal decomposition of $UO_2(H_2PO_4)_2$ at 800 to 850°C. At 600–700°C $UO_2(PO_3)_2$ forms uranium(IV) pyrophosphate, UP_2O_7. Uranium(VI) orthophosphates of the general formula $M(UO_2)_n(PO_4)_m \cdot xH_2O$, where M = H^+, M^+, or M^{2+}, are readily prepared by reaction of uranyl nitrates or perchlorates with phosphoric acid. Several hydrates are known for the uranyl phosphates. Hydrogen uranyl phosphates readily exchange the hydrogen with alkali or alkaline-earth metals. Some of the latter compounds are identical with natural minerals. The tetrahydrate, $H(UO_2)(PO_4)\cdot 4H_2O$, is reported to form three different polymorphic modifications at room temperature (202). The geometry about the uranyl ion in $K_4(UO_2)(PO_4)_2$ is tetragonal bipyramidal with four oxygen atoms in the equatorial plane from four tetrahedral phosphate groups, making up a $[UO_2(PO_4)_2]_n^{4n-}$ layer (203). The neutral compound, $(UO_2)_3(PO_4)_2\cdot xH_2O$, has been synthesized as mono-, tetra-, and hexahydrate. $(UO_2)_3(PO_4)_2\cdot 4.8H_2O$ was prepared by addition of 0.5 M uranyl nitrate to 0.36 M H_3PO_4 at 60°C and pH 1. Orthorhombic crystals formed with lower symmetry than the natural analogue troegerite (204). The anhydrous salt can be obtained from the hydrates upon heating to 250–500°C. The tetrahydrate is the stable form under normal conditions and precipitates at $U:PO_4^{3-}$ ratios ≤1:3. The hexahydrate has been found to precipitate from a solution low in phosphate (<0.014 M).

The trihydrate of uranyl dihydrogenphosphate, $UO_2(H_2PO_4)_2\cdot 3H_2O$, has been obtained from a suspension of $HUO_2PO_4\cdot 4H_2O$ in 85% phosphoric acid after stirring for several days. Uranium(VI) polyphosphates, $(UO_2)_2(P_3O_{10})_2\cdot xH_2O$, were obtained by precipitation of a uranyl solution with $Na_5P_3O_{10}$. The dodecahydrate and the hydrate with 21 hydration water molecules were precipitated from a solution 0.06–0.02 M and 0.15–0.25 M, respectively, in uranyl ion. The U(VI) ultraphosphate, $(UO_2)_2(P_6O_{17})_2$, was obtained from a phosphate melt with a P:U = ratio of 8:1 at 400–420°C.

Coordination Complexes

The coordination chemistry of uranium continues to be of great interest, and has expanded greatly since the 1970s (205–207). Considered "hard" metal ions, U(III to VI) have the greatest affinity for hard donor atoms such as N, O, and the light halides. Tetravalent and hexavalent uranium coordination complexes are the most common, however trivalent and pentavalent complexes have been identified with increasing frequency. As with all of the actinides, the ionic radius of any uranium ion is significantly larger compared to a transition metal ion in an identical oxidation state. The result of this increased ionic radius is an expansion of the possible coordination environments (3- to 14-coordinate) and electron counts (up to 24 electrons).

Nitrogen Donors. There are numerous N-donating ligands which have been complexed with uranium. Classic examples range from neutral mono-, bi-, and polydentate ligands, ie, ammonia, primary, secondary, and tertiary amines, alkyl–aryldiamines (en = ethylenediamine, 1,4-diaminobenzene), N-heterocycles (py = pyridine, bipy = bipyridine, terp = terpyridyl), nitriles (CH_3CN), to anionic amides $[N(C_2H_5)_2, N(Si(CH_3)_3)_2^-]$, thiocyanates, and polypyrazolylborates. A complete listing of ligands can be found in the general references, *Gmelin Handbook* and *Comprehensive Coordination Chemistry*.

Very few U(III) coordination complexes with neutral N-donor ligands have been identified due in part to the ease of oxidation, examples include $UX_3(NH_3)_n$ (X = Cl, n = 1, 3 (**9**); X = Br, n = 3, 4, 6), UI_3py_4 (**10**) $UI_3(tmed)_2$, and $UCl_3(CH_3CN)$ [27459-32-1]. Structural details of these complexes are also limited, however the acetonitrile adduct of UCl_3 is believed to be polymeric. A tris-silylamido complex, $U[N(Si(CH_3)_3)_2]_3$ (**11**) [110970-66-6] has been synthesized and shown to be a useful starting material for the synthesis of other trivalent uranium compounds (208). Polypyrazol-1-yl borates of the form $U(HBpz_3)_m I_{3-m}(THF)_n$ ($m = 1$, $n = 2$ [*159438-24-1*]; $m = 2$, $n = 0$ [*159438-25-2*]) have been synthesized, which are also becoming useful synthons for stabilizing U(III) organometallic complexes (209). Complexed to U(III), the pyrazolyl borate ligand displays an unusual side-on interaction between the pyrazolyl borate and the U(III) center in $UI[HB(Me_2pz)_3]_2$ (**12**) (210).

(**9**)   (**10**)   (**11**)

(**12**)

N-Donor coordination complexes of U(IV) are numerous and have been well characterized. Adducts of UX_4 (X = halogen, alkoxide) have been isolated with all of the ligand types described above. The most common coordination environments for U(IV) are 8–12, as exemplified by UX_4L_n [X = Cl, L = NH_3 ($n = 1$–10), en ($n = 4$), bipy ($n = 2$); X = Br, I L = NH_3 ($n = 4$–6), 1,4-diaminobenzene ($n = 4$); X = alkyl, L = CH_3CN ($x = 4$)]. Lower coordination environments can be obtained by increasing the steric bulk of the anionic ligand, ie, $U(OR)_4(NH_3)_n$ (R = alkyl, aryl; $n = 1$, 2). Amido complexes of U(IV) are important starting materials owing to their highly reactive nature toward insertion reactions and protonation. For example, $[U(NR_2)_4]_n$ (R = alkyl, aryl) undergoes insertion reactions with CO_2, COS, CS_2, and CSe_2 to form carbamate complexes and reacts with alcohols to form alkoxide complexes. The structural motif of the amido complexes is highly dependent on the steric bulk of the R group. When R = phenyl, the complex is monomeric with a pseudotetrahedral geometry (**13**), however, when R = ethyl a dimeric species with an equatorial-bridged trigonal bipyramidal geometry is obtained (**14**). A tripodal amido complex has also been recently synthesized, $[UCl(N(CH_2CH_2NSi(CH_3)_3)_3)]_2$ (**15**) [*157342-43-3*] with which numerous metathetical reactions may be performed (211). Cationic complexes have been stablized with amido ligands, ie, $[U(N(C_2H_5)_2)_3(THF)_3]B(C_6H_5)_4$ [*109622-17-7*] (212). The structure of this complex showed a pseudo-octahedral geometry with the amido and THF ligands in a facial configuration. Polypyrazol-1-yl borate complexes of uranium are important in stabilizing organometallic uranium complexes, however, purely inorganic materials have been isolated and fully characterized. Examples of these include UCl_2L_2 (L = $HBPz_3$ [*55914-06-2*], $(C_6H_5)_2BPz_2$ [*60459-89-4*], BPz_4 [*60459-92-9*]) and UL_4 (L = H_2BPz_2 [*55914-08-4*], $HBPz_3$ [*55914-07-3*]). Another *N*-donor ligand set which has been complexed to U(IV) are porphyrins yielding bisporphyrin complexes, $U(P)_2$ (P = octaethylporphyrin [*149214-28-8*], tetra-*p*-tolylporphyrin [*514923-00-6*]). These complexes are useful for studying the electronic structure and activities of interacting porphyrins (213).

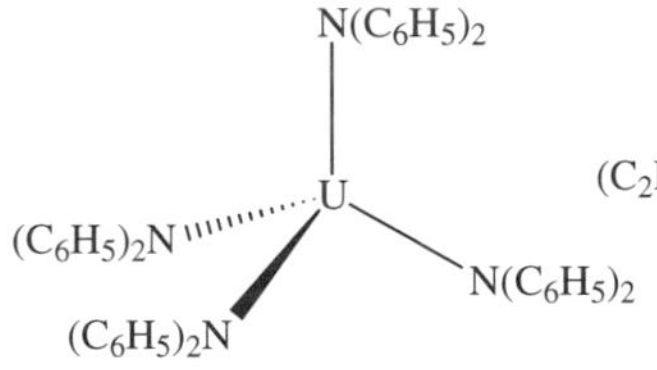

(**13**)

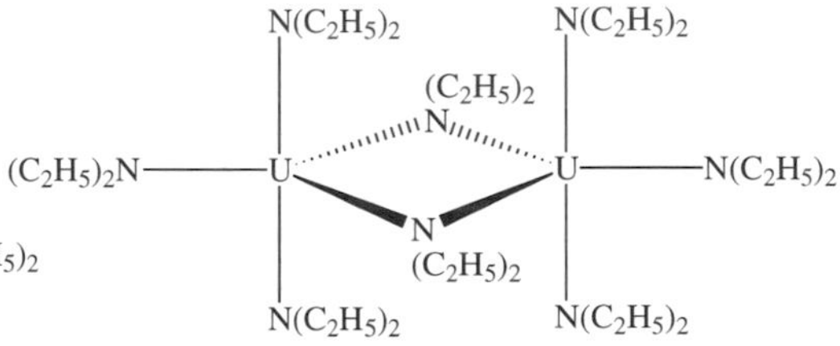

(**14**)

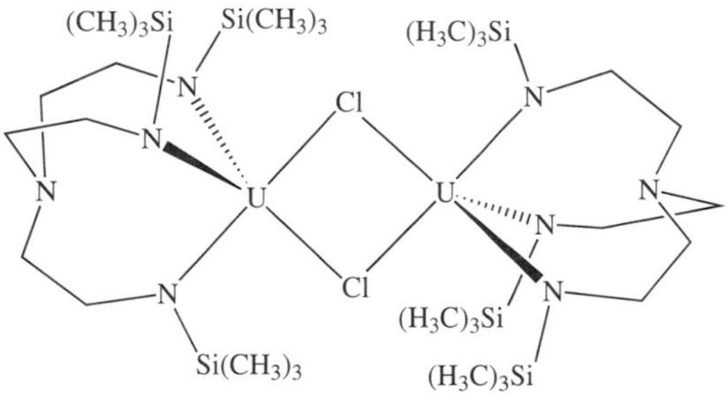

(**15**)

As in the case of U(III), coordination chemistry of U(V) *N*-donor complexes are relatively unexplored owing to disproportionation which yields U(IV) and U(VI) complexes. Typically, ammonia, secondary amines, pyridines, pyrazines, and nitrile adducts of $U(OR)_5$ and UX_5 have been isolated with coordination numbers ranging from 6–8. Examples of these complexes are given by the following, $U(OCH_2CF_3)_5(NH_3)_n$ (n = 6–12), $U(O-t-Bu)_5(py)$ [*104577-09-5*], $[UCl_4(bipy)]Cl$ [*30370-04-8*], and $UBr_5(CH_3CN)_n$ (n = 2, 3). Structural details of these complexes are scarce.

The majority of U(VI) coordination chemistry has been explored with the *trans*-dioxo uranyl cation, UO_2^{2+}. The simplest complexes are ammonia adducts, of importance because of the ease of their synthesis and their versatility as starting materials for other complexes. In addition to ammonia, many of the ligand types mentioned in the introduction have been complexed with U(VI) and usually have coordination numbers of either 6 or 8. As a result of these coordination environments a majority of the complexes have an octahedral or hexagonal bipyramidal coordination environment. Examples include $UO_2X_2L_n$ (X = halide, OR, NO_3, RCO_2, L = NH_3, primary, secondary, and tertiary amines, py; $n = 2-4$), $UO_2(NO_3)_2L_n$ (L = en, diaminobenzene; $n = 1, 2$). The use of thiocyanates has lead to the isolation of typically 6 or 8 coordinate neutral and anionic species, ie, $[UO_2(NCS)_x]^{2-x} \cdot yH_2O$ ($x = 2-5$).

Phosphorus Donors. Phosphine coordination complexes of uranium are rare owing to the preference of uranium for hard donor atoms. The majority of the stable uranium–phosphine coordination complexes isolated contain the chelating ligand, dmpe = 1,2-(bis-dimethylphosphino)ethane, ie, $U(BH_4)_3$(dmpe) and UX_4(dmpe)$_2$ (X = Cl [80290-55-7], Br [80290-57-9], CH_3 [80290-60-4]); however, complexes with monodentate phosphines, ie, $P(CH_3)_3$, have been identified. The benefit of these complexes is their versatility in synthetic uranium organometallic chemistry. Uranium(V) phosphine complexes have been synthesized, using amido ligands with a phosphine appendage, such as $UCl_2[N(CH_2CH_2PPri_2)_2]_3$ [158845-39-7] (214,215).

The phosphido complex, $U(PPP)_4$ [165825-64-9] (PPP = $P(CH_2CH_2P-(CH_3)_2)$, was prepared and fully characterized (216). This complex was one of the first actinide complexes containing exclusively metal-phosphorus bonds. The x-ray structural analysis indicated a distorted bicapped triganol prism with 3–3-electron donor phosphides and 1–1-electron phosphide, suggesting a formally 24-electron complex. Similar to the amido system, this phosphido compound is also reactive toward insertion reactions, especially with CO (216).

Oxygen Donors. A wide variety of O-donors have been used to complex uranium. The predominate oxidation states are IV and VI; however, complexes with U(III) and U(V) are also known. The majority of the complexes have coordination numbers of 6 to 12, depending mostly on the steric bulk of the ancillary ligands. Owing to the prevalence of O-donating ligands in natural systems, ie, aquo, hydroxide, carbonate, phosphate, carboxylate, and catecholate, understanding the complexation of uranium and other radioactive nuclides is important to environmental, waste processing and storage, and bioinorganic chemistry. Some of the other O-donating ligands which have been studied are crown ethers (217), Schiff bases (218,219), polyglycols, and cryptands (220). These ligands have been proposed as actinide sequestering agents (221). A complete listing of the O-donating ligands complexed with uranium can be found in the general references, *Gmelin Handbook* and *Comprehensive Coordination Chemistry*.

Oxygen-Containing Organics. Neutral and anionic oxygen-containing organic molecules form a wide variety of complexes with uranium. Much work has focused on alkoxides (222), aryloxides and carboxylates; complexes with alcohols, ethers, esters, ketones, aldehydes, ketoenolates, and carbamates are also well known.

Alkoxides and Aryloxides. Alkoxide and aryloxide ligands have been studied extensively in transition-metal chemistry, and more recently these studies have been extended to uranium. These studies have focused on determining which ligand systems yield crystalline compounds and provide useful starting materials. Oligomerization in uranium alkoxide complexes, as well as many of their solution properties, are highly dependent on the steric requirements of the alkoxide ligands. In the case of the sterically demanding ligand O-2,6-t-$Bu_2C_6H_3$, monomeric $U(O$-2,6-t-$(C_4H_9)_2C_6H_3)_4$ (**16**) can be readily isolated (223). As the steric bulk of the alkoxide ligand decreases, dimers, $U_2(O$-t-$(C_4H_9)_2)_8(HO$-t-$C_4H_9)$, (**17**) or $U_2(O$-i-$C_3H_7)_{10}$, (**18**) and trimers, $U_3O(O$-t-$C_4H_9)_{10}$ (**19**) are usually observed (222).

(16)   (17)   (18)

(19)

Halides. Uranium halide complexes can be found in all four of the available metal oxidation states, III, IV, V, and VI. In general, fluoride ligands tend to favor higher oxidation states, and iodide ligands tend to favor the lower oxidation states. As a result of the important industrial applications of binary fluorides and chlorides (*vide infra*), the majority of the halide discussion focuses on the binary systems. A selected listing of physical constants for the binary uranium halides is provided in Table 6. Halide complexes are extremely useful as starting materials for coordination and organometallic complexes. Common starting species for these systems are the hydrocarbon soluble $UX_m(THF)_n$ ($m = 3, 4, 5$). The oxohalide complexes of the form, UOX_n ($n = 2, 3$) and $[UO_2X_n]^{4-n}$ ($n = 1$ to 4), have been isolated and fully characterized. The latter U(VI) dioxohalide complexes have usually been isolated with ancillary ligands, ie, H_2O, phosphine oxides when $x < 4$. Mixed halide systems are also known, ie, UClBrI [*84370-90-1*]. Ternary uranium halides are well known and are usually isolated with alkali or alkaline-earth metal ions. These halide complexes can be described by the general formulas, M_xUF_y ($x = 1, y = 5, 6, 7; x = 2, y = 6, 7, 8; x = 3, y = 7, 8, 9; x = 4, y = 8$), M_xUCl_y ($x = 1, y = 6; x = 2, y = 5, 6, 7; x = 3, y = 6, 7, 8$), M_xUBr_y ($x = 1, 2, y = 6$), and MUI_6.

Table 6. Physical Constants for Selected Uranium Halides

Compound	Density, g/mL	Melting point, °C	Boiling point, °C
UF_6	4.68	64.5–64.8	56.2[765]
UF_4	6.70	960	
UF_3		>1000 dec	
UCl_5	3.81	>300 dec	
UCl_4	4.87	590	792[760]
UCl_3	5.44	842	
UBr_4	5.35	516	792[760]
UBr_3	6.53	730	volatile
UI_4	5.6	506	759[760]

Fluorides. Uranium fluorides play an important role in the nuclear fuel cycle as well as in the production of uranium metal. The dark purple UF_3 [13775-06-9] has been prepared by two different methods neither of which neither have been improved. The first involves a direct reaction of UF_4 [10049-14-6] and uranium metal under elevated temperatures, while the second consists of the reduction of UF_4 [10049-14-6] by UH_3 [13598-56-6]. The local coordination environment of uranium in the trifluoride is pentacapped trigonal prismatic with an 11-coordinate uranium atom. The trifluoride is insoluble in H_2O but is soluble in strong acids, ie, nitric, hot sulfuric and perchloric.

The tetrafluoride [10049-14-6], UF_4, is a green solid, which can be isolated with high purity and has industrially important properties, ie, high stability and low volatility. As a result of these properties UF_4 is widely used as a starting material in uranium production processes. The solid-state structure of the tetrafluoride indicates halogen-bridged polymers with the metal center in a distorted square antiprismatic geometry. The preparation of UF_4 has been accomplished by reaction of HF with UO_2 [1344-57-6] at elevated temperatures (eq. 12) or by electrolytic reduction of uranyl fluoride in aqueous HF. The latter process leads to a hydrated species, which may be dehydrated by heating under reduced pressure or flowing N_2. Another preparation of UF_4 is the direct fluorination of β-UO_3 [1344-58-7] at 753 K by flowing Freon-12. It has been found that impurities of nitrates or ammonium ions facilitate the conversion to the tetrafluoride (119). The solubility properties of the tetrafluoride are the following: insoluble in dilute acids and bases, soluble in strong acids and bases, and very slightly soluble in cold H_2O.

Uranium pentafluoride [13775-07-0], UF_5, has been isolated under different conditions, leading to two different modifications, α and β. The former is a grayish white solid, which is synthesized from the interaction of UF_6 [7783-81-5] and HBr or by heating UF_4 [10049-14-6] and UF_6 to 80–100°C. The yellowish white β-modification is also obtained by reacting UF_4 and UF_6, but at higher temperatures (150–200°C). The two different modifications of UF_5 have both been structurally characterized. The α-form consists of infinite chains of octahedral UF_6 units. The β-form has eight-coordinate uranium atoms with the fluorides in a geometry between dodecahedral and square antiprismatic.

Uranium hexafluoride [7783-81-5], UF_6, is an extremely corrosive, colorless, crystalline solid, which sublimes with ease at room temperature and atmospheric pressure. The complex can be obtained by multiple routes, ie, fluorination of UF_4 [10049-14-6] with F_2, oxidation of UF_4 with O_2, or fluorination of UO_3 [1344-58-7] by F_2. The hexafluoride is monomeric in nature having an octahedral geometry. UF_6 is soluble in H_2O, CCl_4 and other chlorinated hydrocarbons, is insoluble in CS_2, and decomposes in alcohols and ethers. The importance of UF_6 in isotopic enrichment and the subsequent applications of uranium metal cannot be overstated. The U.S. government has approximately 500,000 t of UF_6 stockpiled for enrichment or quick conversion into nuclear weapons had the need arisen (57). With the change in political tides and the downsizing of the nation's nuclear arsenal, debates over releasing the stockpiles for use in the production of fuel for civilian nuclear reactors continue.

Chlorides. The olive-green trichloride [10025-93-1], UCl_3, has been synthesized by chlorination of UH_3 [13598-56-6] with HCl. This reaction is driven by the formation of gaseous H_2 as a reaction by-product. The structure of the trichloride has been determined and the central uranium atom possesses a nine-coordinate tricapped trigonal prismatic coordination geometry. The solubility properties of UCl_3 are as follows: soluble in H_2O, methanol, glacial acetic acid; insoluble in ethers.

Uranium tetrachloride [10026-10-5], UCl_4, has been prepared by several methods. The first method, which is probably the best, involves the reduction/chlorination of UO_3 [1344-58-7] with boiling hexachloropropene. The second consists of heating UO_2 [1344-57-6] under flowing CCl_4 or $SOCl_2$. The structure of the dark green tetrachloride is identical to that of Th, Pa, and Np, which all show a dodecahedral geometry of the chlorine atoms about a central actinide metal atom. The tetrachloride is soluble in H_2O, alcohol, and acetic acid, but insoluble in ether, and chloroform. Industrially the tetrachloride has been used as a charge for calutrons.

The reddish brown pentachloride, uranium pentachloride [13470-21-8], UCl_5, has been prepared in a similar fashion to UCl_4 [10026-10-5] by reduction–chlorination of UO_3 [1344-58-7] under flowing CCl_4, but at a lower temperature. Another synthetic approach which has been used is the oxidation of UCl_4 by Cl_2. The pentachloride has been structurally characterized and consists of an edge-sharing bioctahedral dimer, U_2Cl_{10}. The pentachloride decomposes in H_2O and acid, is soluble in anhydrous alcohols, and insoluble in benzene and ethers.

The hexachloride, uranium hexachloride [13763-23-0], UCl_6, is best prepared by chlorination of UCl_4 [10026-10-5] with $SbCl_5$. An alternative preparative approach is the disproportionation UCl_5 [13470-21-8] to UCl_4 and UCl_6 under reduced pressure. The obvious disadvantage of the second method is contamination by UCl_4, however, sublimation is a possible purification technique. Isostructural with the hexafluoride, the hexachloride is monomeric with an octahedral arrangement of the chlorine atoms around the uranium center.

Bromides and Iodides. The red-brown tribromide, UBr_3 [13470-19-4], and the black triiodide, UI_3 [13775-18-3], may both be prepared by direct interaction of the elements, ie, uranium metal with X_2 (X = Br, I). The tribromide has also been prepared by interaction of UH_3 and HBr, producing H_2 as a reaction product. The tribromide and triiodide complexes are both polymeric solids with a local bicapped trigonal prismatic coordination geometry. The tribromide is soluble in H_2O and decomposes in alcohols.

The best synthetic approach to isolate UBr$_4$ [13470-20-7] and uranium tetraiodide [13470-22-9], UI$_4$, is by direct interaction of the elements. This is typically accomplished by heating uranium turnings under flowing nitrogen–halogen gas. The tetrabromide is dark brown and hygroscopic. The black tetraiodide is unstable, undergoing reduction to uranium triiodide [13775-18-3], UI$_3$, and I$_2$. Structural details of the tetrabromide and the tetraiodide are not available. The tetrabromide is soluble in H$_2$O and liquid NH$_3$, but decomposes in alcohols, whereas the tetraiodide is soluble in cold H$_2$O and acetonitrile, and decomposes in hot H$_2$O.

Uranium pentabromide [13775-16-1], UBr$_5$, is unstable toward reduction and the pentaiodide is unknown. Two synthetic methods utilized for the production of UBr$_5$ involve the oxidation of uranium tetrabromide [13470-20-7], UBr$_4$, by Br$_2$ or by bromination of uranium turnings with Br$_2$ in acetonitrile. The metastable pentabromide is isostructural with the pentachloride, being dimeric with edge-sharing octahedra U$_2$Br$_{10}$.

Organometallic Complexes

The organometallic chemistry of uranium has grown rapidly since the 1970s. The majority of the organouranium complexes are found with U(IV) centers; however, there are some examples of higher and lower valent species being isolated. Uranium organometallic compounds have potential uses in homogeneous and heterogeneous catalysis (206–208) with activities ranging from the hydrogenation and polymerization of olefins to the selective activation of alkanes (224). In addition to these potentially important industrial uses, uranium complexes are also used as innocuous models for other more radioactive actinides. A wide range of organic molecules have been complexed with uranium including: hydrocarbyl, allyl, arene, cyclooctatetraenyl, and a host of cyclopentadienyl-based ligands. More detailed discussion, background material, and extensive references to the primary literature can be found in excellent reviews in References 206, 225–233.

Cyclopentadienyl and Substituted Cyclopentadienyl Complexes. Uranium complexes containing cyclopentadienyl rings, Cp (C$_5$H$_5^-$), and its modified analogues, Cp* (C$_5$(CH$_3$)$_5^-$), Cp† ((CH$_3$)$_3$SiC$_5$H$_4^-$), Cp‡ [((CH$_3$)$_3$Si)$_2$C$_5$H$_3^-$], Cp$'$ (CH$_3$C$_5$H$_4^-$), are among the most common organouranium complexes known. Electron-deficient mono-ring U(IV) compounds of general formula (η-ring)UX$_3$ (X = halogen) are unstable, but can be isolated as Lewis base adducts of general formula CpUX$_3$L$_n$ (**20**) (n = 2, X = Cl, Br, L = THF, DME, amides, organonitriles, R$_3$PO; n = 3, X = NCS, L = R$_3$PO), or with the more sterically demanding permethylcyclopentadienyl ligand as in Cp*UX$_3$L$_2$ (X = Cl, Br; L = R$_3$PO, THF, and organonitriles). The Lewis base serves the important role of saturating the coordination environment of the U(IV) ion. This is a fairly common theme in mono-ring U(IV) chemistry. Other examples of this synthetic strategy make use of ligands capable of high uranium coordination numbers such as CpU(HBpz$_3$)Cl$_2$L (L = THF, R$_3$PO). The trispyrazolylborate ligand (HBpz$_3$) is not only sterically demanding, but occupies three coordination sites, making this system formally nine-coordinate. Borohydrides, which generally coordinate in a tripodal fashion, have also found use in stabilizing monocyclopentadienyl–uranium systems, ie, $(\eta\text{-ring})U(BH_4)_3$ (ring = Cp [103948-79-4], Cp′ [120628-95-7], $C_5H_4P(C_6H_5)_2$ [157620-00-3] (234). The borohydride complexes still show a tendency to add additional ligands to form adducts of formula $CpU(BH_4)_3L_2$ (L = THF [122651-47-2], 0.5 DME [122651-48-3], $(C_6H_5)_3PO$ [122651-49-4]). Three-legged piano stool complexes (**21**) of formula $(\eta\text{-ring})UR_3$ can also be stabilized using ligands capable of π-donation or multihapto-coordination to the metal center. Examples of this type of ligand include amides, allyl or benzyl ligands which form stable complexes of the type $CpU(N(C_4H_9)_2)_3$ [88898-38-8], $[Cp^*U(N(C_2H_5)_2)_2(THF)_2]^+$ [171975-70-5] (212), $Cp^*U(C_3H_5)_3$ [84895-69-2], or $(\eta\text{-ring})U(CH_2C_6H_5)_3$ (ring = Cp* [82511-74-8], $C_4((CH_3)_4)P$ [155706-47-1] (235). The benzyl analogue has been shown to exist in a rapid $\eta^1-\eta^3$ equilibrium in solution, indicating the preference for higher coordination environments. Mono-ring complexes can also be stabilized through the use of sterically demanding ancillary ligands, as in the case of $(\eta\text{-ring})UCl(OR)_2$ (ring = Cp, Cp′) and $CpUX_3$ (X = diketo-enolate, alkoxide) (236). Trivalent organouranium complexes are relatively rare; however, the trivalent species $CpU(BH_4)_3^-$, $Cp^*U(BH_4)_3^-$, $Cp^*UI_2(THF)_3$ (**22**) [120410-81-3] have been isolated and fully characterized. A pseudo-octahedral coordination geometry was revealed for the latter compound with a trans-mer arrangement of iodide and THF ligands (208).

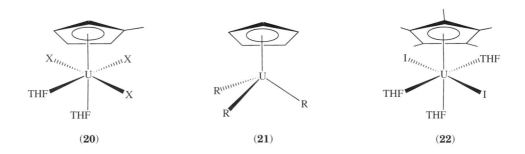

(**20**)   (**21**)   (**22**)

Both U(III) and U(IV) bis-Cp complexes have been isolated, with the former exhibiting slightly lower stabilities. Dimeric uranium(III) bis-ring complexes $[Cp_2^{\ddagger}UX]_2$ (X = Cl [109144-35-6], Br [109192-52-1], I [109168-46-9], aryloxide) are readily cleaved in THF solution to give $Cp_2^{\ddagger}UX(THF)$, further indicating the tendency for high coordination numbers to stabilize this oxidation state. There are several other examples of monomeric U(III) complexes stabilized by Lewis bases or by anionic ligands possible of multiple bonding. Examples include $Cp_2U(CN)$ [54006-98-3], $Cp_2'U(ER)$ (E= O, S; R = alkyl), and $Cp_2^*UH(dmpe)$ [80602-96-6].

Although Cp_2MX_2 compounds are ubiquitous in early transition-metal organometallic chemistry, the uranium analogues are prone to ligand redistribution reactions. Thus, attempts to prepare Cp_2UCl_2 in dme solution actually provide a mixture of Cp_3UCl and $CpUCl_3(dme)$. In some cases, these complexes have been successfully stabilized against ligand redistribution by adding chelating ancillary ligands. Cp_2UCl_2 has been stabilized by the chelating phosphate 1,2-bis-(diphenylphosphineoxide)ethane to form $Cp_2UCl_2((C_6H_5)_2P(O)CH_2CH_2P-(O)(C_6H_5)_2)$ [67588-78-7]. Stabilization against redistribution has also been achieved through the use of sterically demanding X groups, or ligands capable of both σ- and π-donation to the actinide metal center, thereby saturating the metal coordination sphere. Treatment of $Cp_2U(N(C_6H_5)_2)_2$ [54068-37-0] with bulky acidic ligands gives a variety of stable $Cp_2U(ER)_2$ (E = O, S; R = alkyl, aryl) complexes. The amido complex has been shown to react with isocyanides, which insert into the U–N bond. Sterically demanding ring systems can also help in affording stability. Complexes of the form, $Cp_2^*UR_2$ (R = halide, alkoxide, thiolate, amide, hydrocarbyl, etc) have been synthesized and fully characterized all with a monomeric pseudotetrahedral geometry (**23**). Many of these compounds undergo insertion reactions, especially with CO. The amides insert CO into the U–N bond to give carbamoyls, whereas the alkyl, aryl, and benzyl analogues insert CO to produce acyl complexes. High valent uranium organometallic complexes with organoimido and oxo groups have been synthesized, ie, $[Cp_2^*U(NR)_xCl_{2-x}$ (R = alkyl, x = 1, 2) (237). Bis-Cp hydrides and borohydrides have also been stabilized, ie, $(\eta\text{-ring})_2UH_2$ (ring = Cp, Cp*) and $Cp_2U(BH_4)_2$ [65888-45-1]. The hydrides behave like typical transition-metal hydrides, inserting CO and catalyzing homogenous alkene hydrogenation.

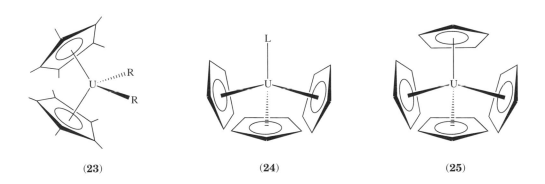

(23)   (24)   (25)

Metallacycle formation has also been observed in bis-Cp complexes. Heating $Cp_2^*UR[P(Si(CH_3)_3)_2]$ (R = Cl [146840-37-1], CH_3 [146840-39-3]) results in the metallation of the phosphido ligand. These complexes are structurally similar to the group 4 and 6 transition-metal metallacycle complexes, but show a dramatically reduced reactivity.

Tris-Cp uranium complexes have been isolated with U(III, IV, V) metal centers. The $(\eta\text{-ring})_3U$ system has been studied with a variety of substituted ring systems. These complexes form 1:1 adducts with Lewis bases, ie, THF, nitriles, isonitriles, phosphine oxides, etc, all of which show a pseudotetrahedral geometry (**24**). The relatively high solubility of the THF adducts make these complexes excellent starting materials. Anionic U(III) complexes have also been isolated of the general form, $[(\eta\text{-ring})_3UR]^-$ (R = alkyl, BH_4, H). The alkyl derivatives react with terminal alkenes and H_2. The hydrides are dimeric in nature with bridging hydride ligands.

Uranium(IV) ring systems of general formula (η-ring)$_3$UX (X = halide, CN, PPh$_2$, N(C$_2$H$_5$)$_2$, BH$_4$, OR, SR, C(CN)$_3$, NO$_3$, ClO$_4$) are well known and have been prepared by metathesis with (η-ring)$_3$UCl. The crystal structure of Cp$_3$U(SCH$_3$) [174576-68-2] has been determined, showing the common pseudotetrahedral geometry illustrated in (**24**) (238). These complexes are highly Lewis acidic and tend to react with neutral ligands to afford adducts of the form (η-ring)$_3$UXL (X = anionic ligand, L = neutral 2-electron donor). An oxo-bridged dimer of Cp$_3$U has been isolated, [Cp$_3$U]$_2$O [175608-90-9]. This complex exhibits a pseudotetrahedral geometry with a linear bridging oxo group (239). The Cp ligand has been found to stabilize the U–C bond in complexes of the form, Cp$_3$UR (R = alkyl, aryl, vinyl, and allyl). These complexes are all air-sensitive and engage in migratory insertion reactions with CO (240). Hydride complexes with a tris-Cp uranium backbone are also known, ie, Cp$_3$UH [134097-41-9] and Cp$_3$U(AlH$_4$) [107633-87-4]. The hydride complex inserts CO_2 and CO into the U–H bond.

Organometallic U(V) complexes are extremely rare. Examples include Cp$_3'$U(NR) (R = C$_6$H$_5$ [94161-46-3], Si(CH$_3$)$_3$ [94202-28-5]) and Cp$_3$U[CHP(CH$_3$(C$_6$H$_5$)$_2$] [77357-86-9]. In all cases multiple bonding characteristics have been found (U=NR, U=CHR), and they can be considered as U(V) members of the class of pseudotetrahedral Cp$_3$UX compounds.

Tetrakis-Cp uranium complexes are readily prepared via metathesis of UCl$_4$ and KCp in refluxing benzene. These complexes are a relatively rare example of a pseudotetrahedral complex with four η^5-Cp rings, (η-ring)$_4$U (**25**). The Cp derivative has been shown to react with CO and CO_2 to give acyl and carboxylato complexes. This complex also reacts with alkyl halides to afford the U(IV) complex, Cp$_3$UX (X = halide).

Cyclooctatetraenyl Compounds. Sandwich-type complexes of uranium with the cyclooctatetraenyl anion (COT = C$_8$H$_8^{2-}$) are significant in the history of organouranium chemistry with the synthesis of uranocene, U(COT)$_2$, the first example of an organouranium bis-ring sandwich complex (**26**). Uranium(III) complexes of the type [K(solvent)] [(COT)$_2$U] have been isolated as either a lithium or potassium salt with COT and alkyl-substituted COT rings. The structure of the potassium–diglyme complex has a methylCOT ring bridging the U and potassium atoms (**27**). Uranocene, (COT)$_2$U [11079-26-8] (**26**), the simplest and most prominent of the COT derivatives has been prepared by the interaction of UCl$_4$ [10026-10-5] with two equivalents of K$_2$C$_8$H$_8$. These compounds are thermally stable, but exceeding sensitive to oxygen. Mössbauer data on the Np analogue suggests appreciable covalency in the Np(IV)–COT bond.

Mono-COT compounds have been studied, and these are usually stabilized by Lewis basis, ancillary ligands capable of π-bonding, or both. Examples include (COT)UX$_2$(THF)$_2$ (**28**) (X = Cl [117097-69-5], BH$_4$), [(COT)U(N(C$_2$H$_5$)$_2$)$_x$]$^{2-x}$ (x = 2 [152249-41-7], 3 [152249-43-9]), (COT)U(CH$_2$Si(CH$_3$)$_3$)$_2$(HMPA) [136937-74-1], and [(COT)U(CH$_2$Si(CH$_3$)$_3$)$_3$]$^-$ [130950-94-6]. The chloro and borohydride complexes have been used in metathetical reactions with a host of anionic ligands, ie, β-diketonates, amides, cyclopentadienyl ring systems, etc, to produce stable compounds. Cationic uranium COT species have also been generated by protonolysis of amides to form [(COT)U(N(C$_2$H$_5$)$_2$)(THF$_2$)]B(C$_6$H$_5$)$_4$ [152249-47-3] and

[Cp*(COT)U(THF)$_2$]B(C$_6$H$_5$)$_4$ [*171975-76-1*] (212). The only successful preparation of a mixed ring U(III) complex is illustrated by [((CH$_3$)$_3$Si)COT]UCp*(THF)
(**29**) and the bipyridine analogue (COT)UCp*(4,4′-(CH$_3$)$_2$-2,2′-bipy) (241).

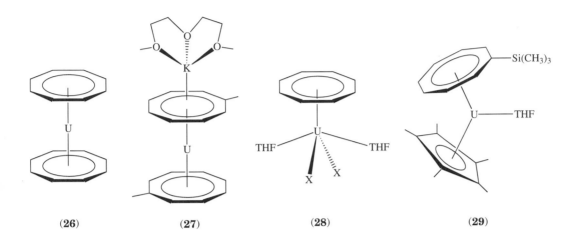

(**26**)   (**27**)   (**28**)   (**29**)

Allyl Complexes. Allyl complexes of uranium are known and are usually
stabilized by cyclopentadienyl ligands. Allyl complexes can be accessed via the
interaction of a uranium halide and an allyl grignard reagent. This synthetic
method was utilized to obtain a rare example of a "naked" homoleptic allyl
complex, U(η^3-C$_3$H$_5$)$_4$ [*12701-96-1*], which decomposes at 0°C. Other examples,
which are more stable than the homoleptic allyl complex have been synthesized,
ie, U(allyl)$_2$(OR)$_2$ (R = alkyl), U(allyl)$_3$X (X = halide), and U(allyl)(bipy)$_3$.

Hydrocarbyl Complexes. Stable homoleptic and heteroleptic uranium hydrocarbyl complexes have been synthesized. Unlike the thorium analogues,
uranium alkyl complexes are generally thermally unstable due to β-hydride
elimination or reductive elimination processes. A rare example of a homoleptic
uranium complex is U(CH(Si(CH$_3$)$_3$)$_2$)$_3$, the first stable U(III) homoleptic complex
to have been isolated. A structural study indicated a triganol pyramidal geometry with stabilizing γ-agostic U–H interactions.

Heteroleptic complexes of uranium can be stabilized by the presence of the
ancillary ligands; however, the chemistry is dominated by methyl and benzyl
ligands. Examples of these materials include UR$_4$(dmpe) (R = alkyl, benzyl) and
U(benzyl)$_4$MgCl$_2$. The former compounds coordinate "soft" chelating phosphine
ligands, a rarity for the hard U(IV) atom.

Another class of heteroleptic alkyl complexes contains π-donating ancillary
ligands such as RU[N(Si(CH$_3$)$_3$)$_2$]$_3$ (R = CH$_3$, H, BH$_4$). The hydride species can
be converted into the methyl species via reaction with BuLi and CH$_3$Br. The
methyl compound has exhibited insertion chemistry with small molecules including aldehydes, ketones, nitriles, and isocyanides (206). Stable metallacycle
compounds are also known, ie, [((CH$_3$)$_3$Si)$_2$N]$_2$U(CH$_2$Si((CH$_3$)$_2$)N(Si(CH$_3$)$_3$)). Generally, uranium metallacycles are quite reactive inserting a host of organics, ie,
CO, secondary amines, nitriles, isonitriles, aldehydes, ketones, and alcohols.

Bimetallic Complexes. There are two types of bimetallic organometallic uranium complexes, those with, and those without metal—metal interactions. Examples of species containing metal—metal bonds are complexes with Mo, W, Fe, or Ru carbonyl fragments. Examples of these include (η-ring)$_3$U(μ-OC)(M(CO)$_2$(η-ring)) M = Mo, W), [N(Si(CH$_3$)$_3$)$_2$]$_3$U(μ-OC){M(CO)$_2$(η-ring)} (M = Mo, W), and (η-ring)$_3$U(CpM(CO)$_2$) (M = Fe, Ru).

The chemistry of nonmetal—metal bonded species is more extensive. Diphenylphosphinocyclopentadienyl (CpP) complexes of uranium have been used to coordinate molybdenum carbonyl fragments, ie, R$_2$U(μ-CpP)$_2$(Mo(CO)$_4$) (R = N(C$_2$H$_5$)$_2$, CpP). Hydrides have also been utilized as a bridging ligand to rhenium in Cp$_3$UReH$_6$L$_2$ (L = P(C$_6$H$_5$)$_3$, PC$_6$H$_4$F-p). In this case, the rhenium hydride acts like a borohydride coordinated in a tridentate fashion. The polyoxoanions are represented by [N-n-(C$_4$H$_9$)$_4$] [Cp$_3$U(MW$_5$O$_{19}$)$_2$] (M = Nb, Ta); compounds of this type are also known for thorium. All of the Cp ligands are bound in an η^5-fashion with two polyoxoanions in a trans geometry bound through the oxygen atoms. A single uranium σ-bonded ferrocenyl complex, UO$_2$(acac)(FcN) [143687-88-1] FcN = (dimethylaminomethyl)ferrocenyl, has been isolated. This complex is eight-coordinate with two oxo groups, four oxygens from the acac ligands and two nitrogens from the ferrocenyl group.

Health and Safety Factors

Exposure and Health Effects. Uranium is a general cellular poison which can potentially affect any organ or tissue. Uranium and its compounds can be damaging due to chemical toxicity and by the injury caused by ionizing radiation. The chemical toxicity of uranium compounds depends on their solubility in biological media. Highly soluble and therefore highly transportable and toxic compounds include fluorides, chlorides, nitrates, and carbonates of uranium(VI); moderately transportable compounds include corresponding uranium(IV) compounds; slightly transportable compounds include oxides, hydrides, and carbides. In experiments where uranium was administered to laboratory animals, the dose after which 50% of the animals died on days 14 to 21 (LD$_{50}$) were as follows: for uranyl fluoride, 2.5 mg/kg for male rats and 1 mg/kg for female rats; and for uranyl nitrate 2.0 and 1.0 mg/kg for male and female rats, respectively (8).

Uranium can enter the human body orally, by inhalation, and through the skin and mucous membranes. Uranium compounds, both soluble and insoluble, are absorbed most readily from the lungs. In the blood of exposed animals, uranium occurs in two forms in equilibrium with each other: as a nondiffusible complex with plasma proteins and as a diffusible bicarbonate complex (242).

Studies show that the main sites of uranium deposition are the renal cortex and the liver (8). Uranium is also stored in bones; deposition in soft tissues is almost negligible. Uranium(VI) is deposited mostly in the kidneys and eliminated with the urine; whereas, tetravalent uranium is preferentially deposited in the liver and eliminated in the feces. The elimination of uranium absorbed into the blood occurs via the kidneys in urine, and most, ~84%, of it is cleared within 4 to 24 hours (8).

The critical organs for the chemical toxicity of uranium and its soluble compounds are the kidney and the liver. In acute uranium poisoning, kidney lesions, renal hemorrhage, and liver cell changes were observed. Occupational exposure to uranium usually involves inhalation of aerosols carrying particles varying in size and density and containing a mixture of uranium compounds with different solubilities. Insoluble particles are deposited in the lungs, retained there for a long time, and can cause radiation damage of varying degree or silicosis (8).

In the 1940s, an accident was caused by the sudden rupture of a tank containing gaseous UF_6 and resulted in the death of two persons, the serious injury of three others, and minor injury (dispensary treatment) of 13 others. Health effects of the exposed personnel included renal lesions and diffuse inflammatory changes in the lungs, and gastrointestinal tract effects. The renal damage was thought to be caused by absorbed uranium; skin, eye, and respiratory effects were presumed to be due to direct action of fluorine.

A study performed in the 1940s, where 31 workers were examined after year-long inhalation exposure to dusts of uranium(VI) oxide, uranium peroxide, and uranium chlorides (at concentrations that at times reached 155 mg/m^3 in terms of uranium), did not reveal any symptoms or signs of chronic poisoning (243). However, examination of 237 uranium mine workers revealed anemia in 31% of them, marked leukopenia in 23%, and lymphocytosis in 14%. In another similar group of uranium workers employed for about 10 years, reduced body weight and pathological changes in the lungs, kidneys, and blood were observed and shown to have resulted from radiation exposure. On the other hand, none of the 100 individuals who had been exposed for five years to slightly soluble uranium compounds exhibited any signs of renal damage nor any pulmonary or blood changes (8).

In another study of workers exposed to UF_6, the review of two years of follow-up medical data on 31 workers who had been exposed to uranium(VI) fluoride and its hydrolysis products following the accidental rupture of a 14-t shipping cylinder in early 1986 indicated that none of the 31 workers sustained any observable health effects from exposure to U; even though an exposure limit of 9.6 mg was exceeded by eight of the workers (244).

Results of extensive studies have demonstrated that the consumption of drinking water containing uranium at elevated levels of 0.04–0.05 mg/L is not detrimental to human health (8).

Occupational Protection and Radiation Considerations. The main adverse factor during the mining and processing of uranium and uranium-containing minerals is airborne dust. Personal protection should include respirators, protective clothing, surgical gloves, suitable footwear, use of wet processes wherever possible, and in operations involving dust formation, face masks, constant ventilation, and glove boxes. Local exhaust ventilation is necessary since one of the daughter products of U is radioactive radon. Environmental protection measures against radioactive contamination in the mining and processing of uranium ores has been described in detail (245). In the United States, the hygienic standards for both soluble and insoluble compounds of natural uranium are, expressed as threshold limit values, 0.2 mg/m^3 for an 8-h time weighted average and 0.6 mg/m^3 for a short-term exposure (8). Finely divided uranium metal, some alloys, and uranium hydride are pyrophoric, therefore such materials should be handled in an inert atmosphere glovebox.

The toxicity of uranium caused by its radiation depends on the isotopes present. Natural uranium does not constitute an external radiation hazard since it emits mainly low energy α-radiation. It does, however present an internal radiation hazard if it enters the body by inhalation or ingestion. The concentration of 1 mg U/g biological tissue corresponds to an absorbed dose of 0.006 Sv per year. Radiation exposure may occur in the mining of uranium ores, although the inhalation of particulates to which miners are exposed is a greater hazard (8). Isotopes such as ^{232}U, which emit fairly strong γ-radiation, should be handled in a hot cell; ^{233}U, ^{234}U, and ^{236}U should be handled in glove boxes; ^{235}U and ^{238}U because of their soft radiation, can be handled on an open laboratory bench or in a fume hood. The laboratory should be equipped as an α-laboratory. In the handling of uranium, as in the case of all radionuclides, radioactivity due to the progeny, such as Th, Ra, Rn, etc, should be considered.

Large quantities of fissile isotopes, ^{233}U and ^{235}U, should be handled and stored appropriately to avoid a criticality hazard. Clear and relatively simple precautions, such as dividing quantities so that the minimum critical mass is avoided, following administrative controls, using neutron poisons, and avoiding critical configurations (or shapes), must be followed to prevent an extremely treacherous explosion (246).

BIBLIOGRAPHY

"Uranium and Uranium Compounds" in *ECT* 1st ed., Vol. 14, pp. 432–458, by J. J. Katz, Argonne National Laboratory; in *ECT* 2nd ed., Vol. 21, pp. 1–36, by V. L. Mattson, Kerr-McGee Corp.; in *ECT* 3rd ed., Vol. 23, pp. 502–547, by F. Weigel, University of Munich.

1. *Uranium, Resources, Production and Demand: a joint report by the OECD Nuclear Energy Agency and the International Atomic Energy Agency*, Organisation for Economic Cooperation and Development, Nuclear Energy Agency, Paris, France, 1993.
2. E. Péligot, *Compt. Rend.* **13**, 417 (1842).
3. A. H. Becquerel, *C. R. Acad. Sci.* **128**, 771 (1896).
4. E. Fermi, E. Amaldi, O. D'Agostino, R. Rasetti, and E. Segré, *Proc. R. Soc. London* **A146**, 483 (1934).
5. O. Hahn and F. Strassman, *Naturwissenschaften* **27**, 11 (1939).
6. N. E. Holden, in D. R. Lide, ed., *Handbook of Chemistry and Physics*, 76th ed., Chemical Rubber Publishing Co., Boca Raton, Fla., 1997.
7. N. N. Greenwood and A. Earnshaw, *Chemistry of the Elements*, Pergamon Press, Oxford, U.K., 1988.
8. V. A. Filov and A. L. Ivin, in *Harmful Chemical Substances, Vol. 1: Elements in Group I–IV of the Periodic Table and their Inorganic Compounds*, Ellis Horwood, New York, 1993, pp. 351–373.
9. G. Choppin, J. Rydberg, and J. O. Liljenzin, *Radiochemistry and Nuclear Chemistry*, 2nd ed., Butterworth-Heinemann Ltd., Oxford, U.K., 1995.
10. S. Moorbath, P. N. Taylor, and N. W. Jones, *Chem. Geol.* **57**(1–2), 63–86 (1986).
11. D. S. Bhattacharyya, *Precambrian Res.* **58**(1–4), 71–83 (1992).
12. K. Bell, *Can. Mineral Mag.* **44**(336), 371–381 (1981).
13. G. R. N. Das, R. V. Viswanath, and S. A. Pandit, *Mem.—Geol. Soc. India* **9**, 29–31 (1988).
14. G. A. Partington, N. J. McNaughton, and I. S. Williams, *Econ. Geol.* **90**(3), 616–635 (1995).

15. R. D. Tucker and W. S. McKerrow, *Can. J. Earth Sci.* **32**(4), 368–379 (1995).
16. *Proceedings of the Mineralogical Association of Canada Short Course on Radiogenic Isotope Systems to Problems in Geology*, Toronto, 1991.
17. H. R. von Gunten, H. Surbeck, and E. Roessler, *Environ. Sci. Technol.* **30**(4), 1268–1274 (1996).
18. G. Faure, *Principles of Isotope Geology*, 2nd ed., John Wiley & Sons, Inc., New York, 1986, pp. 370–372.
19. J. K. W. Lee, *Australia. Dev. Petrol.* **14**, 423–446 (1993).
20. B. DeVito, F. Ippolitio, G. Capoldi, and P. R. Simpson, eds., *Uranium Geochemistry, Mineralogy, Geology, Exploration and Resources*, Institution of Mining and Metallurgy, London, 1984.
21. K. S. Heier and J. J. W. Rogers, *Geochim. Cosmochim. Acta* **27**, 137 (1963).
22. S. H. U. Bowie, in *IAEA*, Vienna, STI/PUB/277, 1970, pp. 285–300.
23. R. C. Merritt, *The Extractive Metallurgy of Uranium*, Colorado School of Mines Research Institute and U.S. AEC, Golden, Colo., 1971.
24. J. W. Frondel, M. Fleischer, and R. S. Jones, *Glossary of Uranium and Thorium-Bearing Minerals*, 4th ed., U.S. Geological Survey Bulletin 1250, Washington, D.C., 1967.
25. C. Frondel, *Systematic Mineralogy of Uranium and Thorium*, U.S. Geological Survey Bulletin 1064, Washington, D.C., 1958.
26. J. W. Gabelman, in STI/PUB/277, IAEA, Vienna, 1970, pp. 315–330.
27. D. S. Robertson, in STI/PUB/277, IAEA, Vienna, 1970, pp. 267–284.
28. W. D. Wilkinson, *Uranium Metallurgy*, Vol. I, *Process Metallurgy*, Vol. II, *Uranium Corrosion and Alloys*, Wiley-Interscience, New York, 1962.
29. J. H. Gittus, *Uranium*, Butterworths, Washington, 1963.
30. *IAEA, Vienna*, STI/PUB/277, in Vienna, April 13–17, 1970, pp. 285–300.
31. F. Weigel, in J. J. Katz, G. T. Seaborg, and L. R. Morss, eds., *The Chemistry of the Actinide Elements*, Chapman and Hall, London, 1986.
32. G. A. Cowan, *Scientific American* **235**, 36 (1976).
33. R. West, *J. Chem. Ed.* **53**, 336 (1976).
34. A. Berzero and M. D'Alessandro, *Comm. Eur. Communities* [*Rep.*] *Eur.* (1990).
35. F. Gauthier-Lafay, P. Holliger, and P.-L. Blanc, *Geochim. Cosmochim. Acta* **60**, 4831–4851 (1996).
36. L. Raimbault, H. Peycelon, and P.-L. Blanc, *Radiochim. Acta*, **74**, 283–287 (1996).
37. R. Bhar, in T. Brewis, J. Chadwick, A. Kennedy, R. Morgan, G. Pearse, J. Schooling, J. Spooner, and M. West, eds., *Metals and Minerals Annual Review*, Mining Journal Ltd., London, 1994.
38. S. Kidd, in T. Brewis, J. Chadwick, D. Clifford, R. Ellis, M. Forest, A. Kennedy, R. Morgan, G. Pearse, N. Rosin, and M. West, eds., *Metals and Minerals Annual Review*, Mining Journal Ltd., London, 1996.
39. J. W. Clegg and D. D. Foley, *Uranium Ore Processing*, Addison-Wesley Publishing Co., Reading, Mass., 1958.
40. C. D. Harrington and A. E. Ruehle, *Uranium Production Technology*, Van Nostrand, Princeton, N.J., 1959.
41. R. G. Bellamy and N. A. Hill, *The Extraction and Metallurgy of Uranium, Thorium, and Beryllium*, Pergamon Press, Oxford, U.K., 1963.
42. J. P. Bibler, in *Recent Dev. Exch. 2*, [*Proc. Int. Conf. Ion Exch. Processes*], *2nd*, Elsevier, London, held at Westinghouse Savannah River Co., Aiken, S.C., 1990, pp. 121–133.
43. J. G. H. Du Preez, *Radiat. Prot. Dosim.* **26**, 7 (1989).
44. N. Petrescu, S. G. Choi, and L. Ganovici, *Metalurgia* (*Bucharest*) **40**, 88 (1988).
45. J. H. Cavendish, *Sci. Technol. Tributyl Phosphate* **2**(PtA), 1–41 (1987).

46. D. R. Weir, *Can. Metall. Q.* **23**, 353 (1984).

47. H. Geier, in C. Keller and H. Moellinger, eds., *Kernbrennstoffkreislauf*, Huethig, Heidelberg, Germany, Vol. 1, 1978, pp. 109–130.

48. S. J. Browning, *Aust. Mining* **64**, 48 (1972).

49. E. H. P. Cordfunke, *J. Inorg. Nucl. Chem.* **24**, 303 (1962).

50. F. L. Oetting, M. H. Rand, and R. J. Ackermann, *The Chemical Thermodynamics of Actinide Elements and Compounds, Part 1, The Actinide Elements*, IAEA, Vienna, STI/PUB/424/1, 1976.

51. R. H. Hackel and B. E. Warner, *Laser Isotope Separation, SPIE Proceedings Series*, Vol. 1859, Society of Photo-Optical Instrumentation Engineers, Bellingham, Wash., 1993.

52. T. M. Anklam, L. V. Berzins, K. G. Hagans, G. W. Kamin, M. A. McClelland, R. D. Scarpetti, and D. W. Shimer, in Ref. 51, p. 277.

53. J. C. Warner, *Metallurgy of Uranium*, Oak Ridge, Tennessee, Natural Nuclear Energy Series, Div. IV, U.S. AEC Technical Information Service, 1953.

54. R. D. Baker, B. R. Hayward, C. Hull, H. Raich, and A. R. Weiss, *Preparation of Uranium Metal by the Bomb Method*, LA-472, Los Alamos National Laboratory, N.M., 1946.

55. M. H. West, M. M. Martinez, J. B. Nielson, D. C. Court, and Q. D. Appert, *Synthesis of Uranium Metal Using Laser-Initiated Reduction of Uranium Tetrafluoride by Calcium Metal*, LA-12996-MS, Los Alamos National Laboratory, N.M., 1995.

56. S. Villani, *Uranium Enrichment*, Springer-Verlag, New York, 1979.

57. National Research Council (U.S.). Committee on Decontamination and Decommissioning of Uranium Enrichment Facilities *Affordable Cleanup?: Opportunities for Cost Reduction in the Decontamination and Decommissioning of the Nation's Uranium Enrichment Facilities*, National Academy Press, Washington, D.C., 1996.

58. M. Taylor, ed., *Uranium and Nuclear Energy: 1993*; Proceedings of the Eighteenth International Symposium held by the Uranium Institute, The Uranium Institute, London, 1993.

59. N. Camarcat, A. Lafon, J. P. Pervés, A. Rosengard, and G. Sauzay, in Ref. 51, p. 14.

60. N. Morioka, in Ref. 51, p. 2.

61. *Data on New Gaseous Diffusion Plants*, U.S. DOE Oak Ridge Operations Office, ORO-685, 1972.

62. D. Massignon, *Top. Appl. Phys.* **35**(Uranium Enrich.), 55 (1979).

63. E. Von Halle, *AIChE Symp. Ser.* **76**, 82 (1980).

64. J. F. Petit, *Bull. Inf. Sci. Tech., Commis. Energ. At.* (Fr.) **223**, 19 (1977).

65. K. Keizer and H. Verweij, *Chemtech*, 37 (1996).

66. S. Villani, *Isotope Separation*, American Nuclear Society, La Grange Park, Ill., 1976.

67. E. I. Abbakumov, V. A. Bazhenov, Y. V. Verbin, A. A. Vlasov, A. S. Dorogobed, A. K. Kaliteevskii, V. F. Kornilov, D. M. Levin, E. I. Mikerin, and co-workers, *USSR. At. Energ.* **67**, 255 (1989).

68. J. Buist, E. Coester, P. G. T. de Jong, H. Rakhorst, P. H. M. te Riele, and K. Schneider, *Synth. Appl. Isot. Labelled Compd. 1994, Proc. Int. Symp.*, Wiley, 1995, pp. 263–268.

69. E. M. Iisen, V. D. Borisevich, and E. V. Levin, *At. Energ.* **72**, 44 (1992).

70. H. Makihara and T. Ito, *J. Nucl. Sci. Technol.* **26**, 1023 (1989).

71. P. H. Readle and P. Wilcox, *Nucl. Eng.* (Inst. Nucl. Eng.) **28**, 170 (1987).

72. H. W. Savage, *Separation of Isotopes in Calutron Units*, Oak Ridge, Tennessee, National Nuclear Energy Series Div. I, Vol. 7, TID-5233, U.S. AEC Technical Information Service, 1951.

73. G. A. Akin, H. P. Kackenmaster, R. J. Schrader, J. W. Strohecker, and R. E. Tate, *Chemical Processing Plant Equipment: Electromagnetic Separation Process*, Oak Ridge, Tennessee, National Nuclear Energy Series, Div. I, Vol. 12, TID-5232, U.S. AEC Technical Information Service, 1950.

74. I. Alexeff, *Current Limitations in Calutrons*, Oak Ridge National Laboratory, ORNL-TM-3722, 10 pp.; Avail. Dep. NTIS, 1972.

75. *Oak Ridge, Tennessee, National Nuclear Energy Series Div. I*, Vols. 1–13, McGraw-Hill Book Co., Inc., New York, and U.S. AEC Technical Information Service, 1949–1952.

76. S. P. Vesnovskii and V. N. Polynov, *Nucl. Instrum. Methods Phys. Res., Sect. B* **B70**, 9 (1992).

77. S. M. Abramychev, N. V. Balashov, S. P. Vesnovskii, V. N. Vyachin, V. G. Lapin, E. A. Nikitin, and V. N. Polynov, *Nucl. Instrum. Methods Phys. Res., Sect. B* **B70**, 5 (1992).

78. D. L. Donohue and R. Zeisler, *Anal. Chem.* **65**, 359A–360A, 364A–368A (1993).

79. B. Lorrain and R. Sobrero, in Ref. 51, p. 154.

80. P. A. Abelson, N. Rosen, and J. I. Hoover, *Liquid Thermal Diffusion*, Oak Ridge, Tennessee, National Nuclear Energy Series, Div. IX, Vol. 1, TID-5229, U.S. AEC Technical Information Service, 1951.

81. W. Burkart, in E. Merian, ed., *Metals and Their Compounds in the Environment, Occurrence, Analysis, and Biological Relevance*, VCH, New York, 1991, pp. 1275–1287.

82. A. J. Francis, C. J. Dodge, F. Lu, G. P. Halada, and C. R. Clayton, *Environ. Sci. Technol.* **28**, 636 (1994).

83. D. E. Morris, P. G. Allen, J. M. Berg, C. J. Chisholm-Brause, S. D. Conradson, R. J. Donohoe, N. J. Hess, J. A. Musgrave, and C. D. Tait, *Environ. Sci. Technol.* **30**, 2322–2331 (1996).

84. P. M. Bertsch, D. B. Hunter, S. R. Sutton, S. Bajt, and M. L. Rivers, *Environ. Sci. Technol.* **28**, 980 (1994).

85. C. Chisholm-Brause, S. D. Conradson, P. G. Eller, and D. E. Morris, *Mat. Res. Soc., Sci. Basis for Nucl. Waste Management* **XV**, 315 (1992).

86. C. Chisholm-Brause, S. D. Conradson, C. T. Buscher, P. G. Eller, and D. E. Morris, *Geochim. Cosmochim. Acta* **58**, 3625 (1994).

87. D. J. Sandstrom, *Los Alamos Sci.* **17**, 36–50 (1989).

88. J. W. Hopson, L. W. Hantel, and D. J. Sandstrom, *Evaluation of Depleted-Uranium Alloys for Use in Armor-Piercing Projectiles (U)*, LA-5238, Los Alamos Scientific Laboratory, N.M., 1973.

89. P. S. Dunn and B. K. Damkroger, in *Tungsten Refract. Met.–1994, Proc. Int. Conf., 2nd (1995), Meeting Date 1994, 199–211*; Metal Powder Industries Federation, Princeton, N.J., 1995.

90. S. Guha, C. Kyriacou, J. C. Withers, and R. O. Loutfy, *Development of a Tungsten Heavy Alloy that Falls by an Adiabatic Shear Mechanism*, Phase 1, Material Electrochemical Research Corp., Tucson, Ariz., ARL-CR-56, Order No. AD-A265867, 120 pp; Avail., NTIS, 1993.

91. A. Bose, J. Lankford, and H. Couque, *Development and Characterization of Adiabatic Shear Prone Tungsten Heavy Alloys*, Wyman-Gordon Co., Worcester, Mass., SWRI-06-4601, Order No. AD-A270 477, 81 pp., Avail., NTIS, 1993.

92. U. C. Gupta, M. Anuradha, and R. Meena, *Adv. Chem. Eng. Nucl. Process Inc.* (1994).

93. J. K. Fink, *Tables of Thermodynamic and Transport Properties of Uranium Dioxide*, 1982.

94. Y. Arita and T. Matsui, *Thermochim. Acta* **267**, 389–396 (1995).

95. Y. Arita, S. Hamada, and T. Matsui, *Thermochim. Acta* **247**, 225–236 (1994).

96. M. Amaya, T. Kubo, and Y. Korei, *J. Nucl. Sci. Technol.* **33**, 636–640 (1996).

97. M. Hoch, *J. Nucl. Mater.* **130**, 94–101 (1985).

98. P. G. Lucuta, H. J. Matzke, and R. A. Verrall, *J. Nucl. Mater.* **223**, 51–60 (1995).

99. L. G. Nikiforov, *Izv. Akad. Nauk Sssr, Neorg. Mater.*, 22 (1986).

100. S. Peng and G. Grimvall, *J. Nucl. Mater.* **210**, 115–122 (1994).

101. R. A. Verrall and P. G. Lucuta, *J. Nucl. Mater.* **228**, 251–253 (1996).

102. I. Amato, P. G. Cappelli, and M. Ravizza, *Metall. Ital.* **22**, 323–327 (1967).

103. S. Suryanarayana, N. Kumar, Y. R. Bamankar, V. N. Vaidya, and D. D. Sood, *J. Nucl. Mater.* **230**, 140–147 (1996).

104. S. Liu, K. Guo, Y. Hu, Q. Wang, D. Gu, and Z. Shen, *Z. Fenxi Huaxue* **22**, 984–988 (1994).

105. P. C. Burns, M. L. Miller, and R. C. Ewing, *The Canadian Mineralogist* **34**, 845–880 (1996).

106. M. L. Miller, R. J. Finch, P. C. Burns, and R. C. Ewing, *Mater. Res. Soc. Symp. Proc.* (1996).

107. G. J. Hutchings, C. S. Heneghan, I. D. Hudson, and S. H. Taylor, *ACS Symp. Ser.* **638**, 58–75 (1996).

108. *Pct Int. Appl.* U.S. Pat. WO 9630085 A1 961003, S. H. Taylor, I. Hudson, and G. J. Hutchings.

109. L. G. Gordeeva, Y. I. Aristov, E. M. Moroz, N. A. Rudina, V. I. Zaikovskii, Y. Y. Tanashev, and V. N. Parmon, *J. Nucl. Mater.* **218**, 202–209 (1995).

110. G. S. You, K. S. Kim, D. K. Min, S. G. Ro, and E. K. Kim, *J. Korean Nucl. Soc.* **27**, 67–73 (1995).

111. J. W. Choi, R. J. McEachern, P. Taylor, and D. D. Wood, *J. Nucl. Mater.* **230**, 250–258 (1996).

112. B. G. Kim, K. W. Song, J. W. Lee, K. K. Bae, M. S. Yang, and H. S. Park, *Yoop Hakhoechi* **32**, 471–481 (1995).

113. Jpn. Kokai Tokkyo Koho, JP 08062363 A2 960308, K. Tokai and A. Ooe.

114. G. L. Hofman and J. L. Snelgrove, *Mater. Sci. Technol.* (1994).

115. M. Handa, *Bull. Chem. Soc. Jpn.* **39** (1966).

116. R. M. Dell, V. J. Wheeler, and E. J. McIver, *Trans. Faraday Soc.* **62**, 3591–3606 (1966).

117. E. H. Kim, C. S. Choi, J. H. Park, and I. S. Chang, *Yoop Hakhoechi* **30**, 279–288 (1993).

118. E. H. Kim, C. S. Choi, J. H. Park, and I. S. Chang, *Yoop Hakhoechi* **30**, 289–298 (1993).

119. B. S. Girgis and N. H. Rofail, *Radiochim. Acta* **57**, 41–44 (1992).

120. Fr. Demande, FR 264447 A1 900921, H. R. Cartmell and J. F. Ellis.

121. J. H. Pashley, *Radiochim. Acta* **25**, 135–138 (1978).

122. Jpn. Kokai Tokkyo Koho, JP 63279195 A2 881116, T. Ozawa.

123. A. F. Bishay, H. A. S. Abdel, F. H. Hammad, and A. M. Elaslaby, *J. Therm. Anal.* **32**, 1415–1420 (1987).

124. A. F. Bishay, H. A. S. Abdel, F. H. Hammad, M. F. Abadir, and A. M. Elaslaby, *J. Therm. Anal.* **35**, 1405–1412 (1989).

125. S. Yamagishi and Y. Takahashi, *J. Nucl. Sci. Technol.* **23**, 711–721 (1986).

126. S. Mori and M. Uchiyama, *Sekiyu Gakkai Shi* **19**, 758–762 (1976).

127. C. V. Cortes, G. Kremenic, and T. L. Gonzalez, *React. Kinet. Catal. Lett.* **36**, 235–240 (1988).

128. H. Serizawa, K. Fukuda, Y. Ishii, Y. Morii, and M. Katsura, *J. Nucl. Mater.* **208**, 128–134 (1994).

129. H. Serizawa, K. Fukuda, and M. Katsura, *J. Alloys Compd.* **223**, 39–44 (1995).

130. M. Miyake, M. Hirota, S. Matsuyama, and M. Katsura, *J. Alloys Compd.* **213**, 444–446 (1994).

131. M. Katsura, M. Hirota, and M. Miyake, *J. Alloys Compd.* **213**, 440–443 (1994).

132. M. Katsura, *Seisan to Gijutsu* **47**, 37–39 (1995).

133. K. Ananthasivan, S. Anthonysamy, V. Chandramouli, I. Kaliappan, and R. P. R. Vasudeva, *J. Nucl. Mater.* **228**, 18–23 (1996).

134. J. C. Fitzmaurice and I. P. Parkin, *New J. Chem.* **18**, 825–832 (1994).

135. I. P. Parkin and J. C. Fitzmaurice, *J. Mater. Sci. Lett.* **13**, 1185–1186 (1994).

136. A. Otsubo and K. Haga, *Emerging Nucl. Energy Syst., Int. Conf., 7th*, 1994.

137. H. Sekimoto and Z. Su'ud, *Nucl. Technol.* **109**, 307–313 (1995).

138. Z. Su'ud and H. Sekimoto, *J. Nucl. Sci. Technol.* **32**, 834–845 (1995).

139. S. Zaki and H. Sekimoto, *Ann. Nucl. Energy* **22**, 711–722 (1995).

140. F. Ingold, S. Daumas, S. Pillon, M. Baumann, and G. Ledergerber, *Psi Proc.* (1995).

141. G. Ledergerber, Z. Kopajtic, F. Ingold, and R. W. Stratton, *J. Nucl. Mater.* **188**, 28–35 (1992).

142. d. R. S. M. Rizzo, d. A. A. Rodrigues, and A. Abrao, *An. Assoc. Bras. Quim.* **44**, 33–40 (1995).

143. R. S. Mehrotra, *Mater. Res. Bull.* **28**, 1193–1199 (1993).

144. L. L. Wang, H. G. Moore, and J. W. Gladson, *Aip Conf. Proc.* 1994.

145. A. S. Gontar, R. Y. Kucherov, N. V. Lapochkin, and Y. V. Nikolaev, *Aip Conf. Proc.* (1994).

146. A. G. Lanin, P. V. Zubarev, and K. P. Vlasov, *At. Energ.* **74**, 42–47 (1993).

147. S. Sahin and E. B. Kennel, *Nucl. Technol.* **107**, 155–184 (1994).

148. S. K. Mukerjee, G. A. R. Rao, J. V. Dehadraya, V. N. Vaidya, V. Venugopal, D. D. Sood, *J. Nucl. Mater.* **199**, 247–257 (1993).

149. S. K. Mukerjee, J. V. Dehadraya, V. N. Vaidya, and D. D. Sood, *J. Nucl. Mater.* **210**, 107–114 (1994).

150. D. L. Clark, D. E. Hobart, and M. P. Neu, *Chem. Rev.* **95**, 25–48 (1995).

151. R. J. Silva and H. Nitsche, *Radiochim. Acta* **70/71**, 377 (1995).

152. J. I. Kim, *Mat. Res. Soc. Symp. Proc.* **294**, 3 (1993).

153. J. I. Kim, in A. J. Freeman and C. Keller, eds., *Handbook on the Physics and Chemistry of the Actinides*, Elsevier Science Publishers, B.V., 1986; Chapt. 8, p. 413.

154. I. Grenthe, J. Fuger, R. J. M. Konigs, R. J. Lemire, A. B. Muller, C. Nguyen-Trung, and H. Wanner, *Chemical Thermodynamics of Uranium*, Vol. 1, Elsevier Science Publishing Company, Inc., New York, 1992.

155. J. H. Burns and P. L. Ritger, *Am. Cryst. Assoc. Ser. 2* **11**, 27 (1983).

156. A. Elbasyouny, H. J. Brugge, K. von Deuten, M. Dickel, A. Knochel, K. U. Koch, J. Kopf, D. Melzer, and G. Rudolph, *J. Am. Chem. Soc.* **105**, 6568 (1983).

157. P. G. Eller and R. A. Penneman, *Inorg. Chem.* **15**, 2439 (1976).

158. J. Rebizant, C. Apostolidas, M. R. Spirlet, G. D. Andreeti, and B. Kanellakopulos, *Acta Cryst. C* **44**, 2098 (1988).

159. J. L. M. Dillen, C. A. Strydom, C. P. J. van Vuuren, and P. H. van Rooyen, *Acta Cryst. C* **44**, 1921 (1988).

160. R. C. Paul, S. Singh, and R. D. Verma, *J. Fluorine Chem.* **16**, 153 (1980).

161. L. Niinisto, J. Toivonen, and J. Valkonen, *J. Acta. Chem. Scand., Ser. A* **33**, 621 (1977).

162. Y. N. Mikhailov, L. A. Kokh, V. G. Kutznetsov, T. G. Grevtseva, S. K. Sokol, and G. V. Ellert, *Sov. J. Coord. Chem. (Engl. Transl.)* **3**, 388 (1977).

163. P. Kierkegaard, *Acta. Chem. Scand.* **10**, 599 (1956).

164. N. W. Alcock and S. Esperas, *J. Chem. Soc., Dalton Trans.* **9**, 893–896 (1977).

165. E. I. Sergeyeva, O. A. Devina, I. L. Khodakovsky, and J. Vernadsky, *J. Alloys Compd.* **213/214**, 125–131 (1994).

166. S. Ahrland in Ref. 153, Chapt. 9.

167. I. Engkvist and Y. Albinsson, *Radiochim. Acta.* **58/59**, 109 (1992).

168. T. Yajima, Y. Kawamura, and S. Ueta, *Mater. Res. Soc. Symp. Proc.* **353**, 1137–1142 (1995).

169. D. Rai, A. R. Felmy, and J. L. Ryan, *Inorg. Chem.* **29**, 260 (1990).

170. Y. Kato, G. Meinrath, T. Kimura, and A. Yoshida, *Radiochim. Acta.* **64**(2), 107–111 (1994).

171. C. F. Baes and R. E. Mesmer, *The Hydrolysis of Cations*, John Wiley & Sons, Inc., New York, 1976.

172. C. M. King, R. B. King, A. R. Garber, M. C. Thompson, and B. R. Buchanan, *Mater. Res. Soc. Symp. Proc.* **180**, 1075–1082 (1990).

173. C. M. King, R. B. King, and A. R. Garber, *Mater. Res. Soc. Symp. Proc.* **180**, 1083–1085 (1990).

174. D. A. Palmer and C. Nguyen-Trung, *J. Solution Chem.* **24**(12), 1281–1291 (1995).

175. A. Navaza and F. Villain, *Polyhedron* **3**, 143 (1984).

176. M. Aberg, *Acta Chem. Scand., Ser. A* **A32**, 101–107 (1978).

177. I. Grenthe and B. Lagerman, *Acta Chem. Scand.* **45**, 122 (1991).

178. G. Bidoglio, P. Cavalli, I. Grenthe, N. Omenetto, P. Qi, and G. Tanet, *Talanta* **38**, 433 (1991).

179. I. Bányai, J. Glaser, K. Micskei, I. Tóth, and L. Zékány, *Inorg. Chem.* **34**, 3785–3796 (1995).

180. P. G. Allen, J. J. Bucher, D. L. Clark, N. M. Edelstein, S. A. Ekberg, J. W. Gohdes, E. A. Hudson, N. Kaltsoyannis, W. W. Lukens, M. N. Neu, P. D. Palmer, T. Reich, D. K. Shuh, C. D. Tait, and B. D. Zwick, *Inorg. Chem.* **34**, 4797–4807 (1995).

181. D. Ferri, J. Glaser, and I. Grenthe, *Inorg. Chim. Acta* **148**, 133–134 (1988).

182. M. Aberg, D. Ferri, J. Glaser, and I. Grenthe, *Inorg. Chem.* **22**, 3981 (1983).

183. I. Grenthe, D. Ferri, F. Salvatore, and G. Riccio, *J. Chem. Soc. Dalton Trans.*, 2439 (1984).

184. L. Ciavatta, D. Ferri, M. Grimaldi, R. Palombari, and F. Salvatore, *Inorg. Nucl. Chem.* **41**, 1175 (1979).

185. L. Ciavatta, D. Ferri, I. Grenthe, and F. Salvatore, *Inorg. Chem.* **20**, 463–467 (1981).

186. I. I. Chernyaev, V. A. Golovnya, and A. K. Molodkin, *Russ. J. Inorg. Chem.* **3**, 100 (1958).

187. K. W. Bagnall, in *Gmelin Handbook of Inorganic Chemistry*, Supplement Vol. C7, 1988, p. 1.

188. F. Weigel, in Ref. 153.

189. L. Ciavatta, D. Ferri, I. Grenthe, F. Salvatore, and K. Spahiu, *Inorg. Chem.* **22**, 2088 (1983).

190. J. Bruno, I. Grenthe, and P. Robouch, *Inorg. Chim. Acta* **158**, 221 (1989).

191. V. A. Golovnya and G. T. Bolotova, *Russ. J. Inorg. Chem.* **6**, 1256 (1961).

192. J. Dervin, J. Faucherre, and P. Herpin, *Bull. Soc. Chim. France* **7**, 2634 (1973).

193. J. Dervin and J. Faucherre, *Bull. Soc. Chim. France* **3**, 2930 (1973).

194. Y. Dusausoy, N. E. Ghermani, R. Podor, and M. Cuney, *Eur. J. Mineral.* **8**(4), 667–673 (1996).

195. A. A. Burnaeva, Y. F. Volkov, A. I. Kryukova, O. V. Skiba, V. I. Spiryakov, I. A. Korshunov, and T. K. Samoilova, *Radiokhim.* **29**(1), 3–7 (1987).

196. R. Masse and J. C. Grenier, *Fr. Bull. Soc. Fr. Mineral Cryst.* **95**(1), 136–142 (1972).

197. S. A. Linde, Y. E. Gorbunovaz, and A. V. Lavrov, *Zh. Neorg. Khim.* **28**(6), 1391–1395 (1983).

198. A. Cabeza, M. A. G. Aranda, F. M. Cantero, D. Lozano, M. Martinez-Lara, and S. Bruque, *J. Solid State Chem.*, 181–189 (1996).

199. P. Benard, D. Loueur, N. Dacheux, V. Brandel, and M. Genet, *An. Quim. Int. Ed.* **92**(2), 79–87 (1996).

200. J. M. Schaekers and W. G. Greybe, *J. Appl. Cryst.* **6**(3), 249–250 (1973).

201. N. Dacheux, V. Brandel, and M. Genet, *New J. Chem.* **19**(1), 15–25 (1995).

202. M. G. Shilton and A. T. Howe, *J. Solid State Chem.* **34**(2), 137–147 (1980).

203. S. A. Linde, Y. E. Gorbunova, and A. V. Lavrov, *Russ. J. Inorg. Chem. (Engl. Transl.)* **25**, 1105 (1980).

204. G. A. Sidorenko, I. G. Zhil'tsova, I. K. Moroz, and A. Valueva, *Dokl. Akad. Nauk SSSR* **222**(2), 444–447 (1975).

205. D. L. Clark, A. P. Sattelberger, W. G. Van der Sluys, and J. G. Watkin, *J. Alloys Compd.* **180**, 303–315 (1992).
206. M. Ephritikhine, *New J. Chem.* **16**, 451–469 (1992).
207. G. Folcher, *J. Less-Comm. Met.* **122**, 139–151 (1986).
208. D. L. Clark and A. P. Sattelberger, in R. B. King, ed., *Encyclopedia of Inorganic Chemistry*, Wiley-Interscience, New York, 1994, Vol. 1, p. 24.
209. R. McDonald, Y. Sun, J. Takats, V. W. Day, and T. A. Eberspracher, *J. Alloys Compd.* **213**, 8–10 (1994).
210. Y. Sun, R. McDonald, J. Takats, V. W. Day, and T. A. Eberspacher, *Inorg. Chem.* **33**, 4433 (1994).
211. P. Scott and P. B. Hitchcock, *J. Chem. Soc., Dalton Trans.* **4**, 603–609 (1995).
212. J. C. Berthet, C. Boisson, M. Lance, J. Vigner, M. Nierlich, and M. Ephritikhine, *J. Chem. Soc., Dalton Trans.* **18**, 3019–3025 (1995).
213. K. M. Kadish, G. Moninot, Y. Hu, D. Dubois, A. Ibnlfassi, J. M. Barbe, and R. Guilard, *J. Am. Chem. Soc.* **115**, 8153–8166 (1993).
214. S. J. Coles, A. A. Danopoulos, P. G. Edwards, M. B. Hursthouse, and P. W. Read, *J. Chem. Soc., Dalton Trans.* **20**, 3401–3408 (1995).
215. S. J. Coles, P. G. Edwards, M. B. Hursthouse, and P. W. Read, *J. Chem. Soc., Chem. Commun.* **17**, 4922–4936 (1994).
216. P. G. Edwards, J. S. Parry, and P. W. Read, *Organometallics* **14**, 3649–3658 (1995).
217. R. D. Rogers, C. B. Bauer, and A. H. Bond, *J. Alloys Compds.* **213/214**, 305–312 (1994).
218. B. T. Thaker, A. Patel, J. Lekhadia, and P. Thaker, *Indian J. Chem., Sect. A: Inorg., Bio. Inorg., Phys., Theor. Anal. Chem.* **6**, 483–488 (1996).
219. A. Aguiari, N. Brianese, S. Tamburini, and P. A. Vigato, *New J. Chem.* **19**, 627–639 (1995).
220. P. Thuery, N. Keller, M. Lance, J. D. Vigner, and M. Nierlich, *New J. Chem.* **19**, 619–625 (1995).
221. D. W. Whisenhunt, Jr., M. P. Neu, Z. Hou, J. Xu, D. C. Hoffman, and K. N. Raymond, *Inorg. Chem.* **35**, 4128–4136 (1996).
222. A. P. Sattelbergcr and W. G. Van der Sluys, *Chem. Rev.* **90**, 1027 (1990).
223. J. M. Berg, D. L. Clark, J. C. Huffman, D. E. Morris, A. P. Sattelberger, W. E. Streib, W. G. Van der Shuys, and J. G. Watkin, *J. Am. Chem. Soc.* **114**, 10811–10821 (1992).
224. M. S. Eisen and T. J. Marks, *Organometallics* **11**, 3939–3941 (1992).
225. T. J. Marks, *Science* **217**, 989 (1982).
226. T. J. Marks and R. D. Ernst, in G. Wikinson, F. G. A. Stone, and E. W. Abel, eds., *Comprehensive Organometallic Chemistry*, Vol. 3, Pergamon Press, Oxford, 1982, Chapt. 21.
227. T. J. Marks, in J. J. Katz, G. T. Seaborg, and L. R. Morss, eds., *The Chemistry of the Actinide Elements*, Chapman and Hall, New York, 1986, Chapt. 23.
228. T. J. Marks, *Acc. Chem. Res.* **25**, 57 (1992).
229. T. J. Marks and A. J. Streitwieser, in Ref. 227, Chapt. 22.
230. J. Takats, in T. J. Marks and I. Fragala, eds., *Fundamental and Technological Aspects of Organo-f-Element Chemistry*, Reidel, Dordrecht, 1985, p. 159.
231. I. Santos, P. D. Matos, and A. G. Maddock, *Adv. Organomet. Chem.* **34**, 65 (1989).
232. B. E. Bursten and R. J. Strittmatter, *Agnew. Chem. Int. Ed. Engl.* **30**, 1069 (1991).
233. C. J. Burns and B. E. Bursten, *Comments Inorg. Chem.* **9**, 61 (1989).
234. D. Baudry, A. Dormond, A. Hafid, and C. Raillard, *J. Organomet. Chem.* **511**, 37–45 (1996).
235. P. Gradoz, D. Baudry, M. Ephritikhine, M. Lance, M. Nierlich, and J. Vigner, *J. Organomet. Chem.* **466**, 107–118 (1994).
236. M. S. Gill and V. S. Sagoria, *Indian J. Chem., Sect. A: Inorg. Bio. Inorg. Phys. Theor. Anal. Chem.* **12**, 997–999 (1995).

237. D. S. J. Arney and C. J. Burns, *J. Am. Chem. Soc.* **117**, 9448–9460 (1995).
238. P. C. Leverd, M. Ephritikhine, M. Lance, J. Vigner, and M. Nierlich, *J. Organomet. Chem.* **507**, 229–237 (1996).
239. M. R. Spirlet, J. Rebizant, C. Apostolidis, E. Dornberger, B. Kanellakopulos, and B. Powietzka, *Polyhedron* **15**, 1503–1508 (1996).
240. M. Weydert, J. G. Brennan, R. A. Andersen, and R. G. Bergman, *Organometallics* **14**, 3942–3951 (1995).
241. A. R. Schake, L. R. Avens, C. J. Burns, D. L. Clark, A. P. Sattelberger, and W. H. Smith, *Organometallics* **12**, 1497 (1993).
242. M. Berlin and B. Rudell, in L. Friberg and co-eds., *Handbook on the Toxicology of Metals*, Elsevier Science, Amsterdam, 1979, pp. 647–658.
243. J. W. Howland, in C. Voegtlin and H. C. Hodge, eds., *Pharmacology and Toxicology of Uranium Compounds of Uranium Compounds*, McGraw-Hill, New York, 1949.
244. D. R. Fisher, M. J. Swint, and R. L. Kathre, Pacific Northwest National Laboratory, PNL-7328, 1990.
245. O. S. Andreyeva and co-workers, *Natural and Enriched Uranium*, Atomizdat, Moscow, Russia, 1979.
246. R. D. Carter, G. R. Kiel, and K. R. Ridgway, in *Criticality Handbook*, 1968, pp. ARH-600, June 30, 1968 and ARH-600, May 23, 1969.

General References

References 1, 6–9, 18, 20, 28, 31, 39–41, 45, 50, 57, 58, 81, 150, 171, 187, 188, 208, 222, 226, 227, 229, 230, 232, 243, 246 are good general references.
K. W. Bagnal, in Sir G. Wilkinson, R. D. Gillard, and J. A. McCleverty, eds., *Comprehensive Coordination Chemistry: The Synthesis, Reactions, Properties, and Applications of Coordination Compounds*, 1st ed., Pergamon Press, New York, 1987, pp. 1120–1130.
F. T. Edelman, in E. W. Abel, F. G. A. Stone, Sir G. Wilkinson, eds., *Comprehensive Organometallic Chemistry II: A Review of the Literature 1982–1994*, 1st ed., Pergamon Press, New York, 1995, pp. 12–192.
M. Peehs, T. Walter, and S. Walter, in B. Elvers and S. Hawkins, eds., *Ullmann's Encyclopedia of Industrial Chemistry*, VCH, Weinheim, Germany, 1996, p. 281.

DAVID L. CLARK
D. WEBSTER KEOGH
MARY P. NEU
WOLFGANG RUNDE
Glenn T. Seaborg Institute for Transactinium Science
Los Alamos National Laboratory

UREA–FORMALDEHYDE RESINS. See AMINO RESINS AND PLASTICS.

URETHANE POLYMERS

The addition polymerization of diisocyanates with macroglycols to produce urethane polymers was pioneered in 1937 (1). The rapid formation of high molecular weight urethane polymers from liquid monomers, which occurs even at ambient temperature, is a unique feature of the polyaddition process, yielding products that range from cross-linked networks to linear fibers and elastomers. The enormous versatility of the polyaddition process allowed the manufacture of a myriad of products for a wide variety of applications.

Polyurethanes contain carbamate groups, $-NHCOO-$, also referred to as urethane groups, in their backbone structure. They are formed in the reaction of a diisocyanate with a macroglycol, a so-called polyol, or with a combination of a macroglycol and a short-chain diol extender. In the latter case, segmented block copolymers are generally produced. The macroglycols are based on polyethers (qv), polyesters, or a combination of both. A linear polyurethane polymer has the structure of (1), whereas a linear segmented copolymer obtained from a diisocyanate, a macroglycol, and a diol extender, $HO(CH_2)_xOH$, has the structure of (2).

$$\left(RO-\overset{O}{\overset{\|}{C}}-NH-R'-NH-\overset{O}{\overset{\|}{C}}-O\right)_{n}$$

(1)

$$\left[RO-\overset{O}{\overset{\|}{C}}-NH-R'-NH-\overset{O}{\overset{\|}{C}}-O-(CH_2)_x-O-\overset{O}{\overset{\|}{C}}-NH-R'-NH-\overset{O}{\overset{\|}{C}}-O\right]_{n}$$

(2)

In addition to the linear thermoplastic polyurethanes obtained from difunctional monomers, branched or cross-linked thermoset polymers are made with higher functional monomers. Linear polymers have good impact strength, good physical properties, and excellent processibility, but, owing to their thermoplasticity, limited thermal stability. Thermoset polymers, on the other hand, have higher thermal stability but sometimes lower impact strength (rigid foams). The higher functionality is obtained with higher functional isocyanates (polymeric isocyanates), or with higher functional polyols (see ISOCYANATES; GLYCOLS; POLYESTERS). Cross-linking is also achieved by secondary reactions. For example, urea groups are generated in the formation of water-blown flexible foams. An isocyanato group reacts with water to form a carbamic acid, which dissociates into an amine and carbon dioxide, with the latter acting as a blowing agent. The amine reacts with another isocyanate to form a urea linkage. Further reaction of the urea group with the isocyanate leads to cross-linking via a biuret group. Water-blown flexible foams contain urethane, urea, and some biuret groups in their network structure (see FOAMED PLASTICS). Urea-modified segmented polyurethanes are manufactured from diisocyanates, macroglycols, and diamine extenders.

Urethane network polymers are also formed by trimerization of part of the isocyanate groups. This approach is used in the formation of rigid polyurethane-modified isocyanurate (PUIR) foams (**3**).

(**3**)

History

The early German polyurethane products were based on toluene diisocyanate (TDI) and polyester polyols. In addition, a linear fiber, Perlon U, was produced from the aliphatic 1,6-hexamethylene diisocyanate (HDI) and 1,4-butanediol. Commercial production of flexible polyurethane foam in the United States began in 1953. In Germany a toluene diisocyanate consisting of an isomeric mixture of 65% 2,4-isomer and 35% 2,6-isomer was used in the manufacture of flexible foam, whereas in the United States the less expensive 80:20 isomer mixture was used. In 1956, Du Pont introduced poly(tetramethylene glycol) (PTMG), the first commercial polyether polyol; the less expensive polyalkylene glycols appeared by 1957. The availability of the lower cost polyether polyols based on both ethylene and propylene oxides provided the foam manufacturers with a broad choice of suitable raw materials, which in turn afforded flexible foams with a wide range of physical propeties. Polyether polyols provide foams with better hydrolytic stability, whereas polyester polyols give superior tensile and tear strength. The development of new and superior catalysts, such as Dabco (triethylenediamine) and organotin compounds, has led to the so-called one-shot process in 1958, which eliminated the need for an intermediate prepolymer step. Prior to this development, part of the polyol was treated with excess isocyanate to give an isocyanate-terminated prepolymer. Further reaction with water produced a flexible foam.

The late 1950s saw the emergence of cast elastomers, which led to the development of reaction injection molding (RIM) at Bayer AG in Leverkusen, Germany, in 1964 (see PLASTICS PROCESSING). Also, thermoplastic polyurethane elastomers (TPUs) and Spandex fibers (see FIBERS, ELASTOMERIC) were introduced during this time. In addition, urethane-based synthetic leather (see LEATHER-LIKE MATERIALS) was introduced by Du Pont under the trade name Corfam in 1963.

The late 1950s also witnessed the emergence of a new polymeric isocyanate (PMDI) based on the condensation of aniline with formaldehyde. This product was introduced by the Carwin Company (later Upjohn and Dow) in 1960 under the trade name PAPI. Similar products were introduced by Bayer and ICI in Europe in the early 1960s. The superior heat resistance of rigid foams derived from PMDI prompted its exclusive use in rigid polyurethane foams. The large-scale production of PMDI made the coproduct 4,4'-methylenebis(phenyl isocyanate) (MDI) readily available, which has since been used almost exclusively in polyurethane elastomer applications. Liquid derivatives of MDI are used in RIM applications, and work has been done since the 1990s to reinforce polyurethane elastomers with glass, graphite, boron, and aramid fibers, or mica flakes, to increase stiffness and reduce thermal expansion. The higher modulus thermoset elastomers produced by reinforced reaction injection molding (RRIM) are also used in the automotive industry. In 1969 Bayer pioneered an all-plastic car having RIM-molded bumpers and fascia; in 1983 the first plastic-body commercial automobile (Pontiac Fiero) was produced in the United States.

The availability of PMDI also led to the development of polyurethane-modified isocyanurate (PUIR) foams by 1967. The PUIR foams have superior thermal stability and combustibility characteristics, which extend the use temperature of insulation foams well above 150°C. The PUIR foams are used in pipe, vessel, and solar panel insulation; glass-fiber-reinforced PUIR roofing panels having superior dimensional stability have also been developed. More recently, inexpensive polyester polyols based on residues obtained in the production of dimethyl terephthalate (DMT) have been used in the formulation of rigid polyurethane and PUIR foams.

One of the trends in polyurethanes is the gradual replacement of TDI by the less volatile PMDI or MDI in many applications. Elimination of chlorinated fluorocarbon (CFC) blowing agents and the reduction of emission of volatile organic compounds (VOCs) have been ongoing. Flexible foam producers have eliminated auxiliary blowing agents, and the rigid foam producers use water-blown formulations in combination with hydrochlorofluorocarbons (HCFCs), hydrofluorocarbons (HFCs), or hydrocarbons. Adhesives and sealants are reformulated to 100% solid and water-based systems.

Formation and Properties

Polyurethane Formation.   The key to the manufacture of polyurethanes is the unique reactivity of the heterocumulene groups in diisocyanates toward nucleophilic additions. The polarization of the isocyanate group enhances the addition across the carbon–nitrogen double bond, which allows rapid formation of addition polymers from diisocyanates and macroglycols.

$$R-\overset{..}{\underset{..}{N}}-\overset{+}{C}=\overset{..}{O} \rightleftarrows RNCO \rightleftarrows R\overset{+}{N}=\overset{..}{C}-\overset{..}{\underset{..}{O}}:$$

The liquid monomers are suitable for bulk polymerization processes. The reaction can be conducted in a mold (casting, reaction injection molding), continuously on a conveyor (block and panel foam production), or in an extruder (thermoplastic polyurethane elastomers and engineering thermoplastics). Also, spraying of the monomers onto the surface of suitable substrates provides insulation barriers or cross-linked coatings.

The polyaddition reaction is influenced by the structure and functionality of the monomers, including the location of substituents in proximity to the reactive isocyanate group (steric hindrance) and the nature of the hydroxyl group (primary or secondary). Impurities also influence the reactivity of the system; for example, acid impurities in PMDI require partial neutralization or larger amounts of the basic catalysts. The acidity in PMDI can be reduced by heat or epoxy treatment, which is best conducted in the plant. Addition of small amounts of carboxylic acid chlorides lowers the reactivity of PMDI or stabilizes isocyanate terminated prepolymers.

The steric effects in isocyanates are best demonstrated by the formation of flexible foams from TDI. In the 2,4-isomer (**4**), the initial reaction occurs at the nonhindered isocyanate group in the 4-position. The unsymmetrically substituted ureas formed in the subsequent reaction with water are more soluble in the developing polymer matrix. Low density flexible foams are not readily produced from MDI or PMDI; enrichment of PMDI with the 2,4′-isomer of MDI (**5**) affords a steric environment similar to the one in TDI, which allows the production of low density flexible foams that have good physical properties. The use of high performance polyols based on a copolymer polyol allows production of high resiliency (HR) slabstock foam from either TDI or MDI (2).

(**4**)   (**5**)

The uncatalyzed reaction of diisocyanates with macroglycols is of no significance in the formation of polyurethanes. Tailoring of performance characteristics to improve processing and properties of polyurethane products requires the selection of efficient catalysts. In flexible foam manufacturing a combination of tin and tertiary amine catalysts are used in order to balance the gelation reaction (urethane formation) and the blowing reaction (urea formation). The tin catalysts used include dibutyltin dilaurate, dibutylbis(laurylthio)stannate, dibutyltinbis(isooctylmercapto acetate), and dibutyltinbis(isooctylmaleate). The principal tertiary amines used are listed in Table 1.

Strong bases, such as potassium acetate, potassium 2-ethylhexoate, or amine−epoxide combinations are the most useful trimerization catalysts. Also, some special tertiary amines, such as 2,4,6-tris(N,N-dimethylaminomethyl)-phenol (DMT-30) (**6**), 1,3,5-tris(3-dimethylaminopropyl)hexahydro-s-triazine (**7**), and ammonium salts (Dabco TMR) (**8**) are good trimerization catalysts.

Table 1. Tertiary Amine Catalysts for Flexible Foams

Name	CAS Registry Number	Structure	Activity	Application
Dabco		(bicyclic diamine structure)	gelation catalysts	flexible foams
pentamethyldipropylene-triamine		$(CH_3)_2N(CH_2)_3N(CH_2)_3N(CH_3)_2$ with CH_3 on middle N	balanced blow and gelation catalyst	cold-cure HR foams
bis(dimethylamino ethyl ether)		$(CH_3)_2N(CH_2)_2O(CH_2)_2N(CH_3)_2$	blowing catalyst	slabstock foam
pentamethyldiethylene-triamine		$(CH_3)_2N-(CH_2)_2-N-(CH_2)_2-N(CH_3)_2$ with $-N(CH_3)_2$ and CH_3 on middle N	blowing catalyst	semiflexible foam
DBU[a]		(bicyclic amidine structure)	heat-activated catalyst	molded foam
dimethylcyclohexylamine	[98-96-2]	(cyclohexyl)$-N(CH_3)_2$	balanced gelation and blowing catalyst	slabstock foam

[a]Phenol salt.

(**6**)   (**7**)   (**8**)

Hydroxy group containing tertiary amines are also used because they become incorporated into the polymer structure, which eliminates odor formation in the foam (3). Delayed-action or heat-activated catalysts are of particular interest in molded foam applications. These catalysts show low activity at room temperature but become active when the exotherm builds up. In addition to the phenol salt of DBU (4), benzoic acid salts of Dabco are also used (5).

Amine catalysts for polyurethane applications are sold by Air Products (Dabco), Abbott (Polycat), Kao Corporation (Kaolizer), Tosoh Corporation (Toyocat), and Union Carbide (Niax).

For the reaction of TDI with a polyether triol, bismuth or lead compounds can also be used. However, tin catalysts are preferred mainly because of their slight odor and the low amounts required to achieve high reaction rates. Carboxylic acid salts of calcium, cobalt, lead, manganese, zinc, and zirconium are employed as cocatalysts with tertiary amines, tin compounds, and tin–amine combinations. Carboxylic acid salts reduce cure time of rigid foam products. Organic mercury compounds are used in cast elastomers and in RIM systems to extend cream time, ie, the time between mixing of all ingredients and the onset of creamy appearance.

The formation of cellular products also requires surfactants to facilitate the formation of small bubbles necessary for a fine-cell structure. The most effective surfactants are polyoxyalkylene–polysiloxane copolymers. The length and ethylene oxide/propylene oxide (EO/PO) ratio of the pendant polyether chains determine the emulsification and stabilizing properties. In view of the complexity of the interaction of surfactant molecules with the growing polymer chains in foam production, it is essential to design optimal surfactants for each application. Flexible polyurethane foams require surfactants that promote improved cell-wall drainage. This allows the cell walls to become more open during the foaming reaction. Also the shift away from TDI to MDI in molded high resiliency foams adds new demands on foam surfactants (6).

The physical properties of polyurethanes are derived from their molecular structure and determined by the choice of building blocks as well as the supramolecular structures caused by atomic interaction between chains. The ability to crystallize, the flexibility of the chains, and spacing of polar groups are of considerable importance, especially in linear thermoplastic materials. In rigid cross-linked systems, eg, polyurethane foams, other factors such as density determine the final properties.

Thermoplastic Polyurethanes. The unique properties of polyurethanes are attributed to their long-chain structure. In segmented polyether- and polyesterurethane elastomers, hydrogen bonds form between $-NH-$ groups (proton donor) and the urethane carbonyl, polyether oxygen, or polyester carbonyl groups. The symmetrical MDI is more suitable for the preparation of segmented polyurethane elastomers having excellent physical properties. Segmented polyurethanes are also obtained from 2,6-TDI, but an economically attractive separation process for the TDI isomers has yet to be developed.

The melt viscosity of a thermoplastic polyurethane (TPU) depends on the weight-average molecular weight and is influenced by chain length and branching. TPUs are viscoelastic materials, which behave like a glassy, brittle solid, an elastic rubber, or a viscous liquid, depending on temperature and time scale of measurement. With increasing temperature, the material becomes rubbery because of the onset of molecular motion. At higher temperatures a free-flowing liquid forms.

The melt temperature of a polyurethane is important for processibility. Melting should occur well below the decomposition temperature. Below the glass-transition temperature (T_g), the molecular motion is frozen, and the material is only able to undergo small-scale elastic deformations. For amorphous polyurethane elastomers, the T_g of the soft segment is ca -50 to $-60°C$, whereas for the amorphous hard segment, T_g is in the $20-100°C$ range. The T_g and T_m of the more common macrodiols used in the manufacture of TPU are listed in Table 2.

The choice of macrodiol influences the low temperature performance, whereas the modulus, ie, hardness, stiffness, and load-bearing properties, increases with increasing hard-segment content.

The pseudocross-links, generated by the hard-segment interactions, are reversed by heating or dissolution. Without the domain crystallinity, thermoplastic polyurethanes would lack elastic character and be more gum-like in nature. In view of the outlined morphology, it is not surprising that many products develop their ultimate properties only on curing at elevated temperature, which allows the soft- and hard-phase segments to separate.

Mesogenic diols, such as 4,4'-bis(ω-hydroxyalkoxy)biphenyls, are used with 2,4-TDI or 1,4-diisocyanatobenzene (PPDI) to construct liquid crystalline polyurethanes (7). Partial replacement of the mesogenic diols by PTMG shows that the use of lower molecular weight flexible spacers form polymers that have a more stable mesophase and exhibit higher crystallinity (8). Another approach to liquid crystal polyurethanes involves the attachment of cholesterol to the polyurethane chain utilizing the dual reactivity in 2,4-TDI (9).

Polyurethane ionomers are segmented polymers in which ionic groups are separated by long-chain apolar segments (10). In the presence of water the ionic

Table 2. Macrodiols for Thermoplastic Polymethane[a]

Polyol	T_g, °C	T_m, °C
poly(propylene glycol) (PPG)	-73	
poly(tetramethylene glycol) (PTMG)	-100	32
poly(1,4-butanediol adipate)	-71	56
poly(ethanediol–1,4-butanediol adipate)	-60	17(37)
polycaprolactone	-72	59
poly(1,6-hexanediol carbonate)	-62	49

[a]Mol wt = 2000.

centers are hydrated. This effect enables ionomers to form stable dispersions in water, and solventless polyurethane coatings are formulated in this manner. The use of N-alkyldiethanolamines or dimethylolpropionic acid as extenders allows incorporation of ionic groups into the polymer backbone. Also, reaction of $-NH-$ group containing polyurethane chains having 1,3-propane sultone affords ionomers. If the ionic centers are located in the hard segment, these then align to form a domain morphology. Anionic dispersions have greater stability than cationic dispersions, but cationic polymers show better adhesion to glass.

Thermoset Polyurethanes. The physical properties of rigid urethane foams are usually a function of foam density. A change in strength properties requires a change in density. Rigid polyurethane foams that have densities of <0.064 g/cm^3, used primarily for thermal insulation, are expanded with HCFCs, HFCs, or hydrocarbons (see INSULATION, THERMAL). Often water or a carbodiimide catalyst is added to the formulation to generate carbon dioxide as a coblowing agent. High density foams are often water-blown. In addition to density, the strength of a rigid foam is influenced by the catalyst, surfactant, polyol, isocyanate, and the type of mixing. By changing the ingredients, foams can be made that have high modulus, low elongation, and some brittleness (friability), or relative flexibility and low modulus (see FOAMED PLASTICS).

Rigid polyurethane foams generally have an elastic region in which stress is nearly proportional to strain. If a foam is compressed beyond the yield point, the cell structure is crushed. Compressive strength values of 10 to 280 kPa (1–14 psi) can be obtained using rigid polyurethane foams of 0.032-g/cm^3 density. In addition, the elastic modulus, shear strength, flexural strength, and tensile strength all increase with density.

Most low density rigid polyurethane foams have a closed-cell content of $>90\%$. Above 0.032 g/cm^3, closed-cell content increases rapidly and is generally $>99\%$ above 0.192 g/cm^3. Bun foam, produced under controlled conditions, has a very fine-cell structure, with cell sizes of 150–200 μm.

The properties of thermoset flexible polyurethane foams are also related to density; load-bearing properties are likewise important. Under normal service temperatures, flexible foams exhibit rubber-like elasticity to deformations of short duration, but creep under long-term stress. Maximum tensile strength is obtained at densities of ca 0.024–0.030 g/cm^3. The densities are controlled by the amount of water in the formulation and may range from 0.045 to 0.020 g/cm^3 by raising the amount of water from 2 to 5%. Auxiliary blowing agents are also used to reduce density and control hardness. The size and uniformity of the cells are controlled by the efficiency of mixing and the nucleation of the foam mix. Flexible foams are anisotropic and the load-bearing properties are best when measured in the direction of foam rise. Some of the polyurethane machine suppliers, eg, Bayer/Hennecke and Cannon, have developed machinery to use liquid carbon dioxide as coblowing agent in the manufacture of water-blown flexible foams.

Hyperbranched polyurethanes are constructed using phenol-blocked trifunctional monomers in combination with 4-methylbenzyl alcohol for end capping (11). Polyurethane interpenetrating polymer networks (IPNs) are mixtures of two cross-linked polymer networks, prepared by latex blending, sequential polymerization, or simultaneous polymerization. IPNs have improved mechanical properties, as well as thermal stabilities, compared to the single cross-linked polymers.

In pseudo-IPNs, only one of the involved polymers is cross-linked. Numerous polymers are involved in the formation of polyurethane-derived IPNs (12).

Raw Materials

Isocyanates. The commodity isocyanates TDI and PMDI are most widely used in the manufacture of urethane polymers (see also ISOCYANATES, ORGANIC). The former is an 80:20 mixture of 2,4- and 2,6-isomers, respectively; the latter a polymeric isocyanate obtained by phosgenation of aniline–formaldehyde-derived polyamines. A coproduct in the manufacture of PMDI is 4,4′-methylenebis(phenyl isocyanate) (MDI). A 65:35 mixture of 2,4- and 2,6-TDI, pure 2,4-TDI and MDI enriched in the 2,4′-isomer are also available. The manufacture of TDI involves the dinitration of toluene, catalytic hydrogenation to the diamines, and phosgenation. Separation of the undesired 2,3-isomer is necessary because its presence interferes with polymerization (13).

Polymeric isocyanates or PMDI are crude products that vary in exact composition. The main constituents are 40–60% 4,4′-MDI; the remainder is the other isomers of MDI, trimeric species, and higher molecular weight oligomers. Important product variables are functionality and acidity. Rigid polyurethane foams are mainly manufactured from PMDI. The so-called pure MDI is a low melting solid that is used for high performance polyurethane elastomers and spandex fibers. Liquid MDI products are used in RIM polyurethane elastomers.

The basic raw materials for the manufacture of PMDI and its coproduct MDI is benzene. Nitration and hydrogenation affords aniline (see AMINES, AROMATIC). Reaction of aniline with formaldehyde in the presence of hydrochloric acid gives rise to the formation of a mixture of oligomeric amines, which are phosgenated to yield PMDI. The coproduct, MDI, is obtained by continuous thin-film vacuum distillation. Liquid MDI (Isonate 143-L) is produced by converting some of the isocyanate groups into carbodiimide groups, which react with the excess isocyanate present to form a small amount of the trifunctional four-membered ring cycloadduct (9). The presence of (9) lowers the melting point of MDI to give a liquid product.

Liquid MDI products are also made by reaction of the diisocyanate with small amounts of glycols. These products are called prepolymers.

Several higher priced specialty aromatic diisocyanates are also available, including 1,5-naphthalene diisocyanate (NDI), *p*-phenylene diisocyanate (PPDI), and bitolylene diisocyanate (TODI). These symmetrical, high melting diisocyanates give high melting hard segments in polyurethane elastomers.

Urethanes obtained from aromatic diisocyanates undergo slow oxidation in the presence of air and light, causing discoloration, which is unacceptable in some applications. Polyurethanes obtained from aliphatic diisocyanates are color-stable, although it is necessary to add antioxidants (qv) and uv-stabilizers (qv) to the formulation to maintain the physical properties with time. The least costly aliphatic diisocyanate is hexamethylene diisocyanate (HDI), which is obtained by phosgenating the nylon intermediate hexamethylenediamine. Because of its low boiling point, HDI is mostly used in form of its derivatives, such as biurets, allophanates, dimers, or trimers (14). Isophorone diisocyanate (IPDI) and its derivatives are also used in the formulation of rigid coatings; hydrogenated MDI (HMDI) and cyclohexane diisocyanate (CHDI) are used in the formulation of flexible coatings and polyurethane elastomers.

HMDI was originally produced by Du Pont as a coproduct in the manufacture of Qiana fiber. Du Pont subsequently sold the product to Bayer. In the 1990s MDA is hydrogenated by Air Products for Bayer (see AMINES, AROMATIC–METHYLENEDIANILINE). Commercial HMDI is a mixture of three stereoisomers. Semicommercial aliphatic diisocyanates include *trans*-cyclohexane-1,4-diisocyanate (CHDI) and *m*-tetramethylxylylene diisocyanate (TMXDI). A coproduct in the production of TMXDI is *m*-isopropenyl-α, α-dimethylbenzyl isocyanate (TMI), which can be copolymerized with other olefins to give aliphatic polyisocyanates.

Masked or blocked diisocyanates are used in coatings applications. The blocked diisocyanates are storage-stable, nonvolatile, and easy to use in powder coatings. Blocked isocyanates are produced by reaction of the diisocyanate with blocking agents such as caprolactam, 3,5-dimethylpyrazole, phenols, oximes, acetoacetates, or malonates. Upon heating at 120–60°C, the blocked isocyanates dissociate and the generated free isocyanate reacts with hydroxyl groups available in the formulation to give high molecular weight polyurethanes. In the case of acetoacetates and malonates, the free isocyanates are not regenerated, but the adducts undergo transesterification reactions with the present polyol upon heating (15). A phenol-blocked methylene diisocyanate (**10**) is obtained in the reaction of phenyl carbamate with formaldehyde (16).

$$2 \; \langle\!\!\langle \rangle\!\!\rangle\text{---OCONH}_2 + \text{CH}_2\text{O} \longrightarrow \; \langle\!\!\langle \rangle\!\!\rangle\text{---OCONHCH}_2\text{NHCO O} \langle\!\!\langle \rangle\!\!\rangle$$

$$(\mathbf{10})$$

Since the early 1970s, attempts have been made by the principal global producers of isocyanates to avoid use of the toxic phosgene (qv) in the manufacture of isocyanates. Attempts to produce TDI and PMDI by nonphosgene processes have failed, but several aliphatic diisocyanates, eg, CHDI and TMXDI, have been manufactured using nonphosgene processes. Hüls and BASF have announced plans to use nonphosgene processes for the manufacture of IPDI in their new plants under construction. In the new, nonphosgene chemistry, isocyanic acid, generated by thermolysis of urea, reacts with diamines to give a bis-urea derivative. Subsequent reaction with diethylamine affords tri-substituted urea derivatives, which are thermolyzed in an inert solvent in the presence of an acidic catalyst to give the diisocyanate (17). Gaseous ammonia is the only by-product in this process. Also, reaction of aliphatic diamines with carbon dioxide, in the presence of triethylamine, affords bis-carbamate salts, which can be dehydrated with phosphoryl chloride to give the diisocyanate (18). Sterically hindered aromatic diamines can be converted in high yield into diisocyanates at room temperature using di-*t*-butyl dicarbonate in the presence of dimethylaminopyridine (19).

The properties and manufacturers of the commercial isocyanates are presented in Table 3. Several TDI producers, ie, Allied, Du Pont, Union Carbide, ICI, have stopped the production of TDI. Although Rhone-Poulenc has sold its TDI plants in France to SNPE, the company has made arrangements with Arco to market TDI.

Polyether Polyols.   Polyether polyols are addition products derived from cyclic ethers (Table 4). The alkylene oxide polymerization is usually initiated by alkali hydroxides, especially potassium hydroxide. In the base-catalyzed polymerization of propylene oxide, some rearrangement occurs to give allyl alcohol. Further reaction of allyl alcohol with propylene oxide produces a monofunctional alcohol. Therefore, polyether polyols derived from propylene oxide are not truly difunctional. By using zinc hexacyano cobaltate as catalyst, a more difunctional polyol is obtained (20). Olin has introduced the difunctional polyether polyols under the trade name POLY-L. Trichlorobutylene oxide-derived polyether polyols are useful as reactive fire retardants. Poly(tetramethylene glycol) (PTMG) is produced in the acid-catalyzed homopolymerization of tetrahydrofuran. Copolymers derived from tetrahydrofuran and ethylene oxide are also produced.

Polyether polyols are high molecular weight polymers that range from viscous liquids to waxy solids, depending on structure and molecular weight. Most commercial polyether polyols are based on the less expensive ethylene or propylene oxide or on a combination of the two. Block copolymers are manufactured first by the reaction of propylene glycol with propylene oxide to form a homopolymer. This polymer upon further reaction with ethylene oxide affords the block copolymer. Because primary hydroxyl groups, resulting from the polymerization of ethylene oxide, are more reactive than secondary hydroxyl groups, the polyols produced in this manner are more reactive. Random copolymers are obtained by polymerizing mixtures of propylene oxide and ethylene oxide. The viscosity of polyether polyols increases with hydroxyl equivalent weight. The higher molecular weight polyether polyols are soluble in organic solvents. Poly(propylene oxide) is soluble in water up to a molecular weight of 760, and copolymerization with ethylene oxide expands the range of water solubility.

With amine initiators the so-called self-catalyzed polyols are obtained, which are used in the formulation of rigid spray foam systems. The rigidity or stiffness of a foam is increased by aromatic initiators, such as Mannich

Table 3. Properties of Commercial Diisocyanates

Name	Acronym	CAS Registry Number	Structure	Bp, °C$_{kPa}$[a]	Mp, °C	Producer
p-phenylene diisocyanate	PPDI	[104-49-4]		$110-112_{1.6}$	94–96	Akzo, Du Pont
toluene diisocyanate	TDI	[1321-38-6]		$121_{1.33}$	14^b	BASF, Bayer, Enichem, Dow, Olin, Rhone-Poulenc, Mitsui
4,4′-methylenebis-(phenylisocyanate)	MDI	[101-68-8]		$171_{0.13}$	39.5	BASF, Bayer, Dow, Enichem, ICI, Mitsui
polymethylene polyphenyl isocyanate	PMDI	[9016-87-9]				BASF, Bayer, Dow, Enichem, ICI, Mitsui
1,5-naphthalene diisocyanate	NDI	[3173-72-6]		$244_{0.017}$	130–132	Bayer, Mitsui

706

Name	Abbreviation	CAS No.	Structure	bp	mp	Producer
bitolylene diisocyanate	TODI	[91-97-4]	see structure	$160–170_{0.066}$	71–72	Nippon-Soda
m-xylylene diisocyanate	XDI	[3634-83-1]	see structure	$159–162_{1.6}$		Takeda
m-tetramethyl-xylylene	TMXDI	[58067-42-8]	see structure	$150_{0.4}$		American Cyanamid
hexamethylene diisocyanate	HDI	[822-06-0]	$OCN(CH_2)_6NCO$	$130_{1.73}$		Bayer, Olin
1,6-diisocyanato-2,2,4-tetra-methylhexane	TMDI	[83748-30-5]	see structure	$149_{1.33}$		Rhône Poulenc, Mitsui, Hüls

Table 3. (*Continued*)

Name	Acronym	CAS Registry Number	Structure	Bp, °C$_{kPa}$[a]	Mp, °C	Producer
1,6-diisocyanato-2,4,4-trimethyl-hexane	TMDI	[15646-96-5]	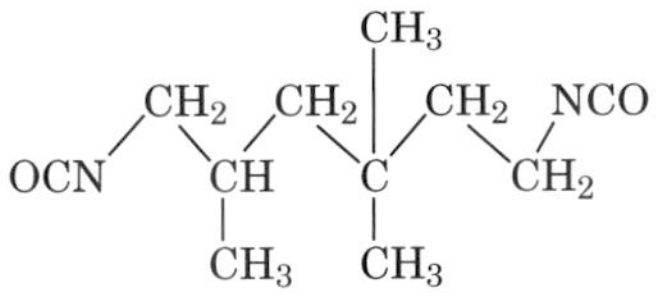 (structure)			
trans-cyclohexane-1,4-diisocyanate	CHDI	[2556-36-7]	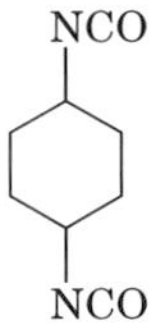 (structure)	122–124$_{1.6}$		Akzo
1,3-bis(isocyanato-methyl)cyclohexane	HXDI[c]	[38661-72-2]	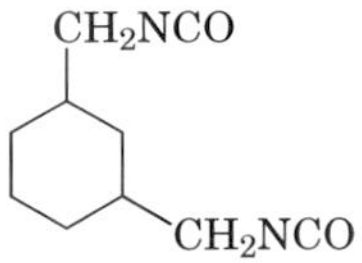 (structure)	98$_{0.053}$		Takeda
3-isocyanato-methyl-3,5,5-trimethylcyclo-hexyl isocyanate	IPDI	[4098-71-9]	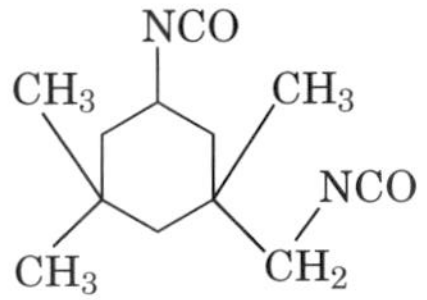 (structure)	153$_{1.33}$		BASF, Bayer, Hüls, Olin
dicyclohexylmethane diisocyanate	HMDI[c]	[5124-30-1]	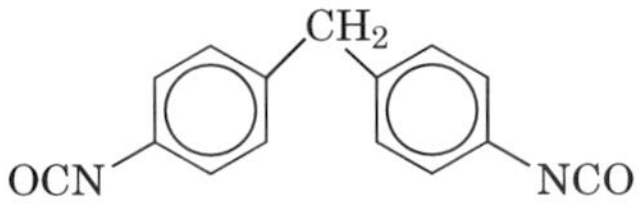 (structure)	179$_{0.12}$		Bayer

[a] To convert kPa to mm Hg, multiply by 7.5.
[b] Mixture of 80% 2,4-isomer [584-84-9] and 20% 2,6-isomer [91-08-7].
[c] Mixture of stereoisomers.

708

Table 4. Commercial Polyether Polyols

Product	Nominal functionality	Initiator[a]	Cyclic ether[b]
poly(ethylene glycol) (PEG)	2	water or EG	EO
poly(propylene glycol) (PPG)	2	water or PG	PO
PPG/PEG[c]	2	water or PG	PO/EO
poly(tetramethylene glycol)	2	water	THF
glycerol adduct	3	glycerol	PO
trimethylolpropane adduct	3	TMP	PO
pentaerythritol adduct	4	pentaerythritol	PO
ethylenediamine adduct	4	ethylenediamine	PO
phenolic resin adduct	4	phenolic resin	PO
diethylenetriamine adduct	5	diethylenetriamine	PO
sorbitol adduct	6	sorbitol	PO/EO
sucrose adduct	8	sucrose	PO

[a]EG = ethylene glycol; PG = propylene glycol.
[b]EO = ethylene oxide; PO = propylene oxide; THF = tetrahydrofuran.
[c]Random or block copolymer.

bases derived from phenol, phenolic resins, toluenediamine, or methylenedianiline (MDA).

In the manufacture of highly resilient flexible foams and thermoset RIM elastomers, graft or polymer polyols are used. Graft polyols are dispersions of free-radical-polymerized mixtures of acrylonitrile and styrene partially grafted to a polyol. Polymer polyols are available from BASF, Dow, and Union Carbide. *In situ* polyaddition reaction of isocyanates with amines in a polyol substrate produces PHD (polyharnstoff dispersion) polyols, which are marketed by Bayer (21). In addition, blending of polyether polyols with diethanolamine, followed by reaction with TDI, also affords a urethane/urea dispersion. The polymer or PHD-type polyols increase the load bearing properties and stiffness of flexible foams. Interreactive dispersion polyols are also used in RIM applications where elastomers of high modulus, low thermal coefficient of expansion, and improved paintability are needed.

Polyester Polyols. Initially polyester polyols were the preferred raw materials for polyurethanes, but in the 1990s the less expensive polyether polyols dominate the polyurethane market. Inexpensive aromatic polyester polyols have been introduced for rigid foam applications. These are obtained from residues of terephthalic acid production or by transesterification of dimethyl terephthalate (DMT) or poly(ethylene terephthalate) (PET) scrap with glycols.

Polyester polyols are based on saturated aliphatic or aromatic carboxylic acids and diols or mixtures of diols. The carboxylic acid of choice is adipic acid (qv) because of its favorable cost/performance ratio. For elastomers, linear polyester polyols of ca 2000 mol wt are preferred. Branched polyester polyols, formulated from higher functional glycols, are used for foam and coatings applications. Phthalates and terephthalates are also used.

In addition, polyester polyols are made by the reaction of caprolactone with diols. Poly(caprolactone diols) are used in the manufacture of thermoplastic polyurethane elastomers with improved hydrolytic stability (22). The hydrolytic stability of the poly(caprolactone diol)-derived TPUs is comparable to TPUs based on the more expensive long-chain diol adipates (23). Polyether/polyester polyol hybrids are synthesized from low molecular weight polyester diols, which are extended with propylene oxide.

Uses

Flexible Foam.   Flexible slab or bun foam is poured by multicomponent machines at rates of >45 kg/min. One-shot pouring from traversing mixing heads is generally used. A typical formulation for furniture-grade foam having a density of 0.024 g/cm^3 includes a polyether triol, mol wt 3000; TDI; water; catalysts, ie, stannous octoate in combination with a tertiary amine; and surfactant. Coblowing agents are often used to lower the density of the foam and to achieve a softer hand. Coblowing agents are methylene chloride, methyl chloroform, acetone, and CFC 11, but the last has been eliminated because of its ozone-depletion potential. Additive systems (24) and new polyols (25) are being developed to achieve softer low density foams. Higher density (0.045 g/cm^3) slab or bun foam, also called high resiliency (HR) foam, is similarly produced, using polyether triols having molecular weight of 6000. The use of polymer polyols improves the load-bearing properties.

Flexible foams are three-dimensional agglomerations of gas bubbles separated from each other by thin sections of polyurethanes and polyureas. The microstructures observed in TDI- and MDI-based flexible foams are different. In TDI foams monodentate urea segments form after 40% conversion, followed by a bidentate urea phase, which is insoluble in the soft segment. As the foam cures, annealing of the precipitated discontinuous urea phase occurs to optimize alignment through hydrogen bonding (26).

Flame retardants (qv) are incorporated into the formulations in amounts necessary to satisfy existing requirements. Reactive-type diols, such as N,N-bis(2-hydroxyethyl)aminomethyl phosphonate (Fyrol 6), are preferred, but nonreactive phosphates (Fyrol CEF, Fyrol PCF) are also used. Often, the necessary results are achieved using mineral fillers, such as alumina trihydrate or melamine. Melamine melts away from the flame and forms both a nonflammable gaseous environment and a molten barrier that helps to isolate the combustible polyurethane foam from the flame. Alumina trihydrate releases water of hydration to cool the flame, forming a noncombustible inorganic protective char at the flame front. Flame-resistant upholstery fabric or liners are also used (27).

There are four main types of flexible slabstock foam: conventional, high resiliency, filled, and high load-bearing foam. Filled slabstock foams contain inorganic fillers to increase the foam density and improve the load-bearing characteristics. High load-bearing formulations incorporate a polymer polyol. Slabstock flexible foam is produced on continuous bun lines. The bun forms while the material moves down a long conveyor. In flat-top bun lines, the liquid chemicals are dispensed from a stationary mixing head to a manifold at the bottom of a trough. More rectangular foams are produced by several newer processes such as Draka, Petzetakis, Hennecke, Planiblock, and Econo Foam. However, the most popular rectangular block foam process is the Maxfoam process. The high outputs require faster and longer conveyors. An exception is the Vertifoam process, in which the reaction mixture is introduced at the bottom of an enclosed expansion chamber. The chamber is lined with paper or polyethylene film, which is drawn upward at a controlled rate. Because the Vertifoam machine is much smaller than the horizontal machines, operational savings can be achieved (28). Two newer slabstock foam manufacturing processes have been developed. The Cannon CarDio process injects liquid carbon dioxide into the foam mix to reduce the density, whereas the Bayer/Hennecke NovaFlex process is conducted under reduced pressure to lower the density.

A high rate of block foam production (150–220 kg/min) is required in order to obtain large slabs to minimize cutting waste. Bun widths range from ca 1.43 to 2.2 m, and typical bun heights are 0.77–1.25 m. In a flexible foam plant, scrap can amount to as much as 20%. Most of it is used as carpet underlay and in pillows and packaging (see PACKAGING MATERIALS). The finished foam blocks are stored in a cooling area for at least 12 h before being passed to a storage area or to slitters where the blocks are cut into sheets. In the production plant the fire risk must be minimized. Temperatures of up to 150°C can be reached in the interior of the foam blocks. Blowing of ambient air through the porous foam allows dissipation of the heat generated in the exothermic reaction (29).

Most flexible foams produced are based on polyether polyols; ca 8–10% (15–20% in Europe) of the total production is based on polyester polyols. Flexible polyether foams have excellent cushioning properties, are flexible over a wide range of temperatures, and can resist fatigue, aging, chemicals, and mold growth. Polyester-based foams are superior in resistance to dry cleaning and can be flame-bonded to textiles.

In more recent years, molded flexible foam products are becoming more popular. The bulk of the molded flexible urethane foam is employed in the transportation industry, where it is highly suitable for the manufacture of seat cushions, back cushions, and bucket-seat padding. TDI prepolymers were used in flexible foam molding in conjunction with polyether polyols. The introduction of organotin catalysts and efficient silicone surfactants facilitates one-shot foam molding, which is the most economical production method.

The need for heat curing has been eliminated by the development of cold-molded or high resiliency foams. These molded HR foams are produced from highly reactive polyols and are cured under ambient conditions. The polyether triols used are 4500–6500 mol wt and are high in ethylene oxide (usually >50% primary hydroxyl content). Reactivity is further enhanced by triethanolamine, liquid aromatic diamines, and aromatic diols. Generally, PMDI, TDI, or blends of PMDI–TDI are used. Load-bearing characteristics are improved by using polymer polyol. High resiliency foams exhibit relatively high SAC (support) factors, ie, load ratio; excellent resiliency (ball rebound >60%); and improved flammability properties.

Semiflexible molded polyurethane foams are used in other automotive applications, such as instrument panels, dashboards, arm rests, head rests, door liners, and vibrational control devices. An important property of semiflexible foam is low resiliency and low elasticity, which results in a slow rate of recovery after deflection. The isocyanate used in the manufacture of semiflexible foams is PMDI, sometimes used in combination with TDI or TDI prepolymers. Both polyester as well as polyether polyols are used in the production of these water-blown foams. Sometimes integral skin molded foams are produced.

Semirigid foams are also manufactured. These foams do not fully recover after deformation; they are used in the construction of energy-absorbing automobile bumpers. Integral skin molded foams have an attached densified water skin, which is produced during manufacture. The preferred isocyanate for integral skin foams is carbodiimide-modified liquid MDI, which is used with ethylene oxide-capped polyols or polymer polyols. Thicker skins are obtained by lowering mold temperatures and increasing the percentage of overpack.

Rigid Foams.   Rigid polyurethane foam is mainly used for insulation (qv). The configuration of the product determines the method of production. Rigid polyurethane foam is produced in slab or bun form on continuous lines (Fig. 1), or it is continuously laminated between either asphalt or tar paper, or aluminum, steel, and fiberboard, or gypsum facings (Fig. 2). Rigid polyurethane products, for the most part, are self-supporting, which makes them useful as construction insulation panels and as structural elements in construction applications. Polyurethane can also be poured or frothed into suitable cavities, ie, pour-in-place applications, or be sprayed on suitable surfaces. Spray-applied polyurethane foams are produced in densities ranging from 0.021 to 0.048 g/cm^3. The lower density foams are used primarily in nonload-bearing applications, eg, cavity walls and residential stud-wall insulation, whereas the higher density foams are used in roofing applications. Applicators can buy formulated systems consisting of the isocyanate component, as well as the polyol side containing the catalysts, surfactants, and blowing agent.

Almost all rigid polyurethanes are produced from PMDI. Some formulations, particularly those for refrigerator and freezer insulation, are based on modified TDI (golden TDI) or TDI prepolymers, but these are being replaced by PMDI formulations. The polyols used include propylene oxide adducts of polyfunctional hydroxy compounds or amines (see Table 4). The amine-derived polyols are used in spray foam formulations where high reaction rates are required. Crude aromatic polyester diols are often used in combination with the multifunctional polyether polyols. Blending of polyols of different functionality,

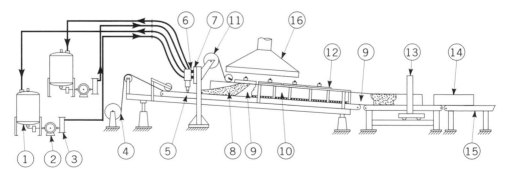

Fig. 1.   Rigid bun foam line: 1, material tank with agitators; 2, metering pump; 3, heat exchanger; 4, bottom paper roll; 5, conveyor; 6, mixing head; 7, traverse assembly; 8, rising foam; 9, side paper; 10, adjustable side panels; 11, top paper roll; 12, top panels with adjustable height; 13, cutoff saw (traversing); 14, cut foam bun; 15, roller conveyor; 16, exhaust hood.

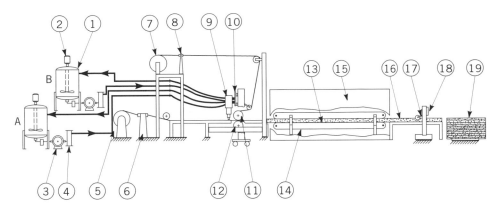

Fig. 2.   Rigid foam laminating line: 1, material tank; 2, agitator; 3, metering pump; 4, heat exchanger; 5, bottom facer roll; 6, bottom facer alignment device; 7, top facer roll; 8, top facer alignment device; 9, mixing head; 10, traverse assembly; 11, top nip roll; 12, bottom nip roll; 13, take-up conveyor top belt with adjustable height; 14, take-up conveyor bottom belt; 15, curing oven; 16, laminate; 17, side-trim saws; 18, cutoff saw (traversing); 19, laminated-panel stack and packaging.

molecular weight, and reactivity is used to tailor a polyol for a specific application. Polyether–polyester polyol hybrids are also synthesized from low mol wt polyesters, which are subsequently propoxylated. The high functionality of the polyether polyols combined with the higher functionality of PMDI contributes to the rapid network formation required for rigid polyurethane foams. Reactive or nonreactive fire retardants, containing halogen and phosphorous, are often added to meet the existing building code requirements. The most commonly used reactive fire retardants are Fyrol 6, chlorendic anhydride-derived diols, and tetrabromophthalate ester diols (PHT 4-Diol). There is a synergistic effect of nitrogen and phosphorus observed in P–N compounds. Phosphonates, such as Fryol 6, are effective in char formation, whereas phosphine oxide-derived fire retardants are reactive in the gas phase. Because the reactive fire retardants are combined with the polyol component, storage stability is important. Nonreactive fire retardants include halogenated phosphate esters, such as tris(chloroisopropyl) phosphate (TMCP) and tris(chloroethyl) phosphate (TCEP), and phosphonates, such as dimethyl methylphosphonate (DMMP). Highly halogenated aromatic compounds, borax, and melamine are also used as fire retardants in rigid foams.

Insulation foams are halocarbon-blown. Chlorofluorocarbons, eg, CFC-11 (bp 23.8°C) and CFC-12 (bp −30°C), were used extensively as blowing agents in the manufacture of rigid insulation foam. Because of the mandatory phaseout of CFCs by January 1, 1996, it had become necessary to develop blowing agents that have a minimal effect on the ozone layer. As a short-term solution, two classes of blowing agents are considered: hydrochlorofluorocarbons (HCFCs) and hydrofluorocarbons (HFCs). For example, HCFC 141b, CH_3CCl_2F (bp 32°C), is a drop-in replacement for CFC-11, and HFC 134a, CF_3CH_2F (bp −26.5°C), was developed to replace CFC-12. HCFC 142b, CH_3CClF_2 (bp −9.2°C), is the blowing agent used in the 1990s. Addition of water or carbodiimide catalysts to the formulation generates carbon dioxide as a coblowing agent. Longer-range environmental considerations have prompted the use of hydrocarbons such as pentanes and cyclopentane as blowing agents. Pentane blown foams have already been used in the appliance industry in Europe. Pentane-based formulations are typically used in conjunction with water. Because rigid foams blown with alternative blowing agents have λ-values (m·W/m·K) of about 19.5, as compared to 18.0 for CFC-11-blown foams, they are thus less efficient in their insulation performance. In addition, because rigid polyurethane foams at a density of 0.032 g/cm^3 are ca 97% gas, the blowing agents determine the k-factor (insulation value).

From the onset of creaming to the end of the rise during the expansion process, the gas must be retained completely in the form of bubbles, which ultimately result in the closed-cell structure. Addition of surfactants facilitates the production of very small uniform bubbles necessary for a fine-cell structure.

The catalysts used in the manufacture of rigid polyurethane foams include tin and tertiary amine catalysts. Combinations of catalysts are often used to achieve the necessary balance of reaction rates. This is especially necessary if part of the blowing agent is carbon dioxide, generated in the reaction of the isocyanate with added water. New surfactants are required for the emerging water-coblown formulations, using pentanes as the main blowing agent (30). A typical water-coblown rigid polyurethane formulation is shown in Table 5.

During the molding of high density rigid foam parts, the dispensed chemicals have to flow a considerable distance to fill the cavities of the mold. In the filling period, the viscosity of the reacting mixture increases markedly from the initial low value of the liquid mixture to the high value of the polymerized foam. If the viscosity increases rapidly, incomplete filling results. Chemical factors that influence flow properties are differential reactivity in the polyol components and the addition of water to the formulation. Because venting holes allow the escape of air displaced by the rising foam, a moderate degree of overpacking is often advantageous. Newer high pressure RIM machines have simplified the mold-filling procedure, so filling of intricate molds is no longer a problem.

Many of the rigid insulation foams produced in the 1990s are urethane-modified isocyanurate (PUIR) foams. In the formulation of poly(urethane isocyanurate) foams an excess of PMDI is used. The isocyanate index can range from 105 to 300 and higher. PUIR foams have a better thermal stability than polyurethane foams (32). The cyclotrimerization of the excess isocyanate groups produces heterocyclic triisocyanurate groups, which do not revert to the starting materials, but rather decompose at much higher temperatures. In the decompo-

Table 5. Typical Rigid Polyurethane Panel Formulation[a]

Ingredients	Parts
PMDI	135.0
polyol	98.0
water	1.9
catalyst	2.0
surfactant	2.0
HFC 134a	17.0

[a]Ref. 31.

sition of the PUIR foams a char is formed, which protects the foam underneath the char.

The formation of isocyanurates in the presence of polyols occurs via intermediate allophanate formation, ie, the urethane group acts as a cocatalyst in the trimerization reaction. By combining cyclotrimerization with polyurethane formation, processibility is improved, and the friability of the derived foams is reduced. The trimerization reaction proceeds best at 90–100°C. These temperatures can be achieved using a heated conveyor or a RIM machine. The key to the formation of PUIR foams is catalysis. Strong bases, such as potassium acetate, potassium 2-ethylhexoate, and tertiary amine combinations, are the most useful trimerization catalyst. A review on the trimerization of isocyanates is available (33).

Modification of cellular polymers by incorporating amide, imide, oxazolidinone, or carbodiimide groups has been attempted but only the urethane-modified isocyanurate foams are produced in the 1990s. PUIR foams often do not require added fire retardants to meet most regulatory requirements (34). A typical PUIR foam formulation is shown in Table 6.

CASE Polyurethanes. CASE is the acronym for coatings, adhesives, sealants, and elastomers. Polyurethane coatings are mainly based on aliphatic isocyanates and acrylic or polyester polyols because of their outstanding weatherability. For flexible elastomeric coatings, HMDI and IPDI are used with polyester polyols, whereas higher functional derivatives of HDI and IPDI with acrylic polyols are mainly used in the formulation of rigid coatings. Plastics coatings, textile coatings, and artificial leather are based on either aliphatic or aromatic isocyanates. For light-stable textile coatings, combinations of IPDI and IPDA (as chain extender) are used. The poly(urethane urea) coatings are applied either directly to the fabric or using transfer coating techniques. The direct-coating method is applied for rainwear because the microporous coating is permeable to air and water vapor, but not to liquid water. Microporous polyurethane sheets (poromerics) are used for shoe and textile applications. Alcantara is an artificial velour leather used in the fabrication of fashion wear. Polyurethane binder resins are also used to upgrade natural leather.

Blocked aliphatic isocyanates or their derivatives are used for one-component coating systems. Masked polyols are also used for this application. For example, polyols capped with vinyl or isopropenyl ethers produce polyacetals, which do not react with isocyanates. Hydrolysis of the acetals with moist

Table 6. Typical PUIR Foam Formulation[a]

Ingredients	Parts
PMDI (250 index)	208.7
Terate 203[b]	100.0
Dabco K-15	5.2
Dabco TMR 30	1.2
surfactant	2.0
HCFC 141b	35.0

[a]Ref. 35.
[b]Crude aromatic polyester diol.

air regenerate the hydroxyl groups, which undergo polyurethane reaction with isocyanate-terminated prepolymers. In addition, substituted oxazolines are used as masked cross-linkers (36). Ketimine cross-linkers are also utilized in the formulation of one-component coating systems (37). Hydrolysis of ketimines produces diamines, which undergo a very fast reaction with isocyanate-terminated prepolymers. Blocked isocyanates are also used in the cross-linking of acrylic resins for automotive coatings. Incorporation of masked diisocyanates into epoxy resins lowers the moisture absorption in the derived coatings (38).

Powder coatings are formulated from the reaction product of trimethylolpropane and IPDI, blocked with caprolactam, and polyester polyols. The saturated polyester polyols are based on aromatic acid diols, neopentyl glycol, and trimellitic anhydride for further branching. To avoid the release of caprolactam in the curing reaction, systems based on IPDI dimer diols are used.

Water-borne polyurethane coatings are formulated by incorporating ionic groups into the polymer backbone. These ionomers are dispersed in water through neutralization. The experimental 1,12-dodecane diisocyanate (C12DI; Du Pont) is especially well suited for the formation of water-borne polyurethanes because of its hydrophobicity (39). Cationomers are formed from IPDI, N-methyldiethanolamine, and poly(tetramethylene adipate diol) (40); anionic dispersions are obtained from IPDI, PTMG, poly(propylene glycol) (PPG), and dimethylol propionic acid (41). The ionic groups can also be introduced in the polyol segment. For example, reaction of diesterdiol, obtained from maleic anhydride and 1,4-butanediol, with sodium bisulfite produces the ionic building block, which on reaction with HDI gives a polyurethane ionomer (42). The weatherability of aliphatic polyurethane coatings is related to their structures. Polyester polyol-based polyurethanes are more uv-resistant than polyether polyol-based polyurethanes, but the latter offer better hydrolytic stability.

Ionic polymers are also formulated from TDI and MDI (43). Poly(urethane urea) and polyurea ionomers are obtained from divalent metal salts of p-aminobenzoic acid, MDA, dialkylene glycol, and 2,4-TDI (44). In the case of polyureas, the glycol extender is omitted. If TDI is used in coatings applications, it is usually converted to a derivative to lower the vapor pressure. A typical TDI prepolymer is the adduct of TDI with trimethylolpropane (Desmodur L). Carbodiimide-modified MDI offers advantages in polyester-based systems because of improved hydrolytic stability (45). Moisture cure systems based on aromatic isocyanates are also available.

Polyurethane adhesives are known for excellent adhesion, flexibility, toughness, high cohesive strength, and fast cure rates. Polyurethane adhesives rely on the curing of multifunctional isocyanate-terminated prepolymers with moisture or on the reaction with the substrate, eg, wood and cellulosic fibers. Two-component adhesives consist of an isocyanate prepolymer, which is cured with low equivalent weight diols, polyols, diamines, or polyamines. Such systems can be used neat or as solution. The two components are kept separately before application. Two-component polyurethane systems are also used as hot-melt adhesives.

Water-borne adhesives are preferred because of restrictions on the use of solvents. Low viscosity prepolymers are emulsified in water, followed by chain extension with water-soluble glycols or diamines. As cross-linker PMDI can be used, which has a shelf life of 5 to 6 h in water. Water-borne polyurethane coatings are used for vacuum forming of PVC sheeting to ABS shells in automotive interior door panels, for the lamination of ABS/PVC film to treated polypropylene foam for use in automotive instrument panels, as metal primers for steering wheels, in flexible packaging lamination, as shoe sole adhesive, and as tie coats for polyurethane-coated fabrics. PMDI is also used as a binder for reconstituted wood products and as a foundry core binder.

Polyurethane sealant formulations use TDI or MDI prepolymers made from polyether polyols. The sealants contain 30–50% of the prepolymer; the remainder consists of pigments, fillers, plasticizers, adhesion promoters, and other additives. The curing of the sealant is conducted with atmospheric moisture. One-component windshield sealants utilize diethyl malonate-blocked MDI prepolymers (46). Several polyurethane hybrid systems, containing epoxies, silicones, or polysulfide, are also used.

The largest segment of the CASE family of polyurethanes are elastomers. Cast polyurethane elastomers reached a new dimension when high pressure impingement mixing led to reaction injection molding (RIM). This technology is used widely in the automotive industry, and reinforced versions (RRIM) and structural molded parts (SRIM) have been added in more recent years.

Polyurethane elastomers are either thermoplastic or thermoset polymers, depending on the functionality of the monomers used. Thermoplastic polyurethane elastomers are segmented block copolymers, comprising of hard- and soft-segment blocks. The soft-segment blocks are formed from long-chain polyester or polyether polyols and MDI; the hard segments are formed from short-chain diols, mainly 1,4-butanediol, and MDI. Under ambient conditions the higher melting hard segments are incompatible with the soft segments, and microphase separation occurs. The hard segments aggregate into crystalline domains, in which hydrogen bonding of the $-NH-$ groups of the urethane chain bond to neighboring carbonyl groups. Upon melting, the crystalline domains are disrupted and the polymer can be processed.

Polyester and polyether diols are used with MDI in the manufacture of thermoplastic polyurethane elastomers (TPU). The polyester diols are obtained from adipic acid and diols, such as ethylene glycol, 1,4-butanediol, or 1,6-hexanediol. The preferred molecular weights are 1,000 to 2,000, and low acid numbers are essential to ensure optimal hydrolytic stability. Also, caprolactone-derived diols and polycarbonate diols are used. Polyether diols are mainly poly(tetramethylene glycol) (PTMG), but polyalkylene oxide-derived diols are also used. TPUs contain wax to aid in mold release and diatomaceous silica for added slip and as antiblocking agents in films. Antioxidants (hindered phenols or hindered amines) and uv-stabilizers (benzotriazoles) are also added to improve the environmental resistance.

Thermoset polyurethanes are cross-linked polymers, which are produced by casting or reaction injection molding (RIM). For cast elastomers, TDI in combination with 3,3'-dichloro-4,4'-diphenylmethanediamine (MOCA) are often used. In the RIM technology, aromatic diamine chain extenders, such as diethyltoluenediamine (DETDA), are used to produce poly(urethane ureas) (47), and replacement of the polyether polyols with amine-terminated polyols produces polyureas (48). The aromatic diamines are soluble in the polyol and provide fast reaction rates.

In 1985, internal mold release agents based on zinc stearate compatibilized with primary amines were introduced to the RIM process to minimize mold preparation and scrap from parts torn at demold. Some physical properties of RIM systems are listed in Table 7.

Polyurethane engineering thermoplastics are also manufactured from MDI and short-chain glycols (49). These polymers were introduced by Upjohn/Dow under the trade name Isoplast. The glycols used are 1,6-hexanediol and cyclohexanedimethanol. 1,4-Butanediol is too volatile at the high processing temperatures used in the reaction extrusion process. Blends of engineering thermoplastics with TPU are also finding uses in many applications (50).

Segmented elastomeric polyurethane fibers (Spandex fibers) based on MDI have also been developed. Du Pont introduced Lycra in 1962. The generic name Spandex fibers designates elastomeric fibers, in which the fiber-forming substance is a long-chain polymer consisting of >85% of polyurethane (see FIBERS, ELASTOMERIC). Extenders used in Spandex fibers include hydrazine and ethylenediamine. Du Pont uses a dry-spinning process, in which the polymer solution in dimethylformamide (DMF) is extruded through a spinerette into a column of circulating hot air. Other producers that use the dry-spinning process include Bayer (Dorlastan) and Asahi Chemical Industry (Asahi Kasei Spandex). Wet-spinning processes are also used, in which isocyanate-terminated prepolymers are extruded into a nonaqueous diamine bath. Globe manufacturing is producing Glospan using this process. Nishin Spinning Company uses a melt extrusion process to produce Mobilon.

Recycling. The methods proposed for the recycling of polyurethanes include pyrolysis, hydrolysis, and glycolysis. For example, introducing pyrolysis products obtained from a mixture of plastic products into a refinery could be a viable method for the recycling of mixtures of plastics. Energy recovery from scrap polyurethanes or mixed plastic waste products by incineration is another useful recycling method. Polyurethane regrind is used extensively as fillers in a variety of polyurethane applications. For example, flexible foam powder can be used as filler (15–25%) in the molding of automotive car seats, and up to 20% of reground flexible foam can be added to virgin polyol in the manufacture of slabstock foam. Regrind from polyurethane RIM elastomers is used as filler in some RIM as well as compression molding applications. The RIM chips are also used in combination with rubber chips in the construction of athletic fields, tennis courts, and pavement of working roads of golf courses.

Table 7. Properties of RIM Systems

Properties	Flexural modulus, MPa[a]		
	0.137–0.517	0.517–1.03	1.37–2.75
elongation at break, %	100–300	50–200	<50
Izod impact, J/m[b]	534–801	267–801	<267
impact strength	high	medium high	low
material description	elastomer	pseudo-plastic	plastic
automotive application	fascia	fender	hood or deck lid

[a]To convert MPa to psi, multiply by 145.
[b]To convert J/m to ft·lb/in., divide by 53.38.

The use of rebound flexible foam for carpet underlay and for high load-bearing padding for furniture or for gymnasium mats is already a reality. Rebound flexible foam can also be used for sound dampening in cars. Rebounding of rigid foam particles with PMDI produces polyurethane particle boards. These boards are unaffected by water and are therefore used in furniture aboard ships. Rigid foam scrap is also used as filler in the manufacture of building products.

The most convenient chemical recycling process, pioneered by Upjohn in the 1970s, consists of glycolysis of solid polyurethane products (51). Heating of polyurethane scrap in a mixture of glycols and diethanolamine converts the cross-linked polymers into linear soluble oligomers via a transesterification process. Replacement of 10–30% of virgin polyol by glycolysate has been achieved in rigid foam production. In RIM applications only 10–15% of recycled polyol can be used (52).

Economic Aspects

In 1993, a total of over 6×10^6 t of polyurethanes were consumed worldwide (Table 8). The flexible foam market in the United States totaled 932,000 tons in 1994. Flexible slab foam is used predominantly in furniture, carpet underlay, and bedding; molded foam is used extensively in transportation. Carpet underlay is manufactured from either virgin or scrap polyurethane foam, which is combined with a binder adhesive. The consumption of flexible polyurethane foam in the various U.S. markets in 1994 is shown in Table 9.

More than 50% of the rigid polyurethane foams manufactured in the United States in 1994 are used in the construction industry (Table 10). About 60% of the total rigid foam is used in laminated boards and insulation panels; about

Table 8. 1993 Total World Consumption of Polyurethanes

Application	Consumption, 10^3 t
flexible foam	3000
rigid foam	1550
RIM elastomers and integral skin foams	450
TPU	200
others	900
Total	*6100*

Table 9. 1994 U.S. Flexible Foam Use[a]

Applications	Consumption, 10^3 t
furniture	272
transportation	236
carpet underlay	210
bedding	98
packaging	89
textile laminates	12
miscellaneous	15
Total	*932*

[a]Ref. 53.

Table 10. 1994 U.S. Rigid Foam Consumption[a]

Application	Consumption, 10^3 t
building and construction	227
refrigeration	113
tank and pipe insulation	48
packaging	30
transportation	16
furniture	6
other	18
Total	*458*

[a]Ref. 53.

30% is poured in place. Insulated appliances is another important use of rigid insulation foam.

The total elastomer consumption in the United States in 1994 accounted for 145,000 t; automotive RIM applications account for 63% of the total. Coatings consumption in the United States in 1994 was 95,000 t, and powder coatings accounted for about 20% of the total. Adhesives and sealants consumption amounted to another 95,000 t.

The polyurethane industry is dominated by the multinational isocyanate producers. Several of the principal isocyanate producers, including BASF, Bayer, Dow, ICI, and Olin, also manufacture polyols, the other significant building block for polyurethanes. Annual production capacities of the global aromatic isocyanate producers are listed in Table 11. Polyols, mainly used for flexible foam production, account for 65 wt % in a flexible foam formulation, 35% in rigid polyurethane foams, and even less in PUIR foams.

Health and Safety Factors

Fully cured polyurethanes present no health hazard; they are chemically inert and insoluble in water and most organic solvents. However, dust can be generated in fabrication, and inhalation of the dust should be avoided. Polyether-based polyurethanes are not degraded in the human body, and are therefore used in biomedical applications.

Some of the chemicals used in the production of polyurethanes, such as the highly reactive isocyanates and tertiary amine catalysts, must be handled with caution. The other polyurethane ingredients, polyols and surfactants, are relatively inert materials having low toxicity.

Isocyanates.   Isocyanates in general are toxic chemicals and require great care in handling. Oral ingestion of substantial quantities of isocyanates can be tolerated by the human body, but acute symptoms may develop from the inhalation of much smaller amounts. The inhalation of isocyanates presents a hazard for the people who work with them as well as the people who live in the proximity of an isocyanate plant. Adequate control of exposure is necessary to achieve a safe working environment. The suppliers Material Safety Data Sheets (MSDS) have to be consulted for the most current information on the safe handling of isocyanates.

Table 11. 1995 Worldwide Aromatic Isocyanate Capacities, 10^3 t

Company	MDI	TDI
Western Hemisphere		
BASF	90	166
Bayer	145[a]	110
Bayer de Brasil		20
Bayer, Mexico		12
Dow	163	63[b]
ICI	136	
Olin		110[c]
Pronor, Brazil		53
Total	*514*	*534*
Europe		
BASF	193	35
Bayer	257	189
Borsodchem, Hungary	25	
Dow	115	
Enichem	55	90
ICI	105[d]	
Poland (Bydgoszi)		15
Rhone-Poulenc[e]		120
Ukraine (Dzerzhinsk)	22	13
Total	*710*	*432*
Asia		
BASF/Hanwha, Korea	40	
Baiyin, China		20
Cangzhou, China		20
Chin Yang, Korea		10
Korean Fine Chem.		25
Mitsui-Toatsu, Japan	55	35[f]
Nippon PU, Japan	31	15
Sumitomo Bayer, Japan	40	13
Taiyuan, China		20
Takeda/BASF, Japan	30[g]	22
Yantai, China	10	
Total	*206*	*180*
Total world	*1512*	*1176*

[a]A 136,000-t expansion is planned for 1998.
[b]A doubling of capacity is planned for 1997.
[c]A 136,000-t expansion is planned for 1997.
[d]A 90,000-t expansion is planned by 1997.
[e]Isocyanates are now produced for Rhone-Poulenc by SNPE.
[f]A 27,000-t expansion is planned.
[g]Distillation plant.

Respiratory effects are the primary toxicological manifestations of repeated overexposure to diisocyanates (54). Once a person is sensitized to isocyanates, lower concentrations can trigger a response (55). Most of the industrial diisocyanates are also eye and skin irritants. Controlling dermal exposure is good industrial hygiene practice. The 1997 American Conference of Governmental Industrial Hygienists (ACGIH) exposure guideline for TDI is 0.005 ppm as a TWA-TLV (an eight-hour time-weighted average concentration); the 1997 TLV for TDI in Japan is 20 ppb.

Overexposure to TDI can cause chemical bronchitis (isocyanate asthma) in sensitized individuals. Transient acute asymptomatic changes in respiratory function and deterioration of lung function following long-term repeated exposure have also been encountered. Allergic sensitization may occur within months or after years of exposure to isocyanates. Animal studies using TDI showed no teratologic response at exposure concentrations up to 0.5 ppm. A chronic gavage study indicated tumor formation in the animals, but the study was found to be of doubtful toxicological relevance because of the method used and the excessively high dose levels. Vapor exposure to MDI is limited by the low vapor pressure, corresponding to a saturated atmosphere of 0.1 mg/m^3 at 25°C. An acute aerosol inhalation study on PMDI using rats indicated that the 4-h LC$_{50}$ is 490 mg/m^3 (56). The current ACGIH TLV for MDI is 0.051 mg/m^3 (0.005 ppm) as a TWA. The OSHA PEL is 0.02 ppm as a ceiling limit.

The toxicity of aliphatic diisocyanates also warrants monitoring exposure to its vapors. HDI has a moderate potential for acute systemic dermal toxicity; rabbit dermal LD$_{50}$ is 570 ml/Kg (57). However, HDI is severely irritating to the skin and eyes. Irritation, lacrimation, rhinitis, burning sensation to throat and chest, and coughing have all been reported in humans following acute inhalation exposure to HDI. HMDI has a low eye and dermal irritation potential, as well as a low potential for acute toxicity. Exposure to HMDI aerosol can cause dermal sensitization of laboratory animals. IPDI can cause skin sensitization reactions as well as eye irritation. The acute toxicity of diisocyanates in rats is shown in Table 12.

There are a multitude of governmental requirements for the manufacture and handling of isocyanates. The U.S. Environmental Protection Agency (EPA) mandates testing and risk management for TDI and MDI under Toxic Substance Control Administration (TSCA). Annual reports on emissions of both isocyanates are required by the EPA under SARA 313.

Thermal degradation of isocyanates occurs on heating above 100–120°C. This reaction is exothermic, and a runaway reaction can occur at temperatures >175°C. In view of the heat sensitivity of isocyanates, it is necessary to melt MDI with caution and to follow suppliers' recommendation. Disposal of empty contain-

Table 12. Acute Toxicity of Diisocyanates in Rats[a]

Isocyanate	LC$_{50}$, mg/kg	LC$_{50}$, 1 h, mg/m^3	Concentration,[b] ppm
HDI	710	310[c]	6.8
IPDI	>2,500	260	0.34
TDI	5,800	58–66	19.6
MDI	>31,600		0.1
NDI	>10,000		0.02[d]

[a]Ref. 58.
[b]Vapor pressure at STP.
[c]4 h.
[d]Vapor pressure at 50°C.

ers, isocyanate waste materials, and decontamination of spilled isocyanates are best conducted using water or alcohols containing small amounts of ammonia or detergent. For example, a mixture of 50% ethanol, 2-propanol, or butanol; 45% water, and 5% ammonia can be used to neutralize isocyanate waste and spills. Spills and leaks of isocyanates should be contained immediately, ie, by dyking with an absorbent material, such as saw dust.

The total U.S. airborne emission of volatile TDI is estimated by the International Isocyanate Institute (III) to be <25 t, or less than 0.005% of the annual U.S. production. Published data show that TDI has a 1/3 life of 8 s in air at 25°C and 50% rh, and a 0.5 s to 3 d half-life in water, depending on pH and agitation. Without agitation, isocyanates sink to the bottom of the water and react slowly at the interface. Because of this reactivity, there is no chance of bioaccumulation.

Tertiary Amine Catalysts. The liquid tertiary aliphatic amines used as catalysts in the manufacture of polyurethanes can cause contact dermatitis and severe damage to the eye. Inhalation can produce moderate to severe irritation of the upper respiratory tracts and the lungs. Ventilation, protective clothing, and safety glasses are mandatory when handling these chemicals.

Polyurethanes. These polymers can be considered safe for human use. However, exposure to dust, generated in finishing operations, should be avoided. Ventilation, dust masks, and eye protection are recommended in foam fabrication operations. Polyurethane or polyisocyanurate dust may present an explosion risk under certain conditions. Airborne concentrations of $25-30$ g/m^3 are required before an explosion occurs. Inhalation of thermal decomposition products of polyurethanes should be avoided because carbon monoxide and hydrogen cyanide are among the many products present.

Because polyurethanes are combustible, they have to be applied in a safe and responsible manner. At no time should exposed foam be used in building construction. An approved fire-resistive thermal barrier must be applied over foam insulation on interior walls and ceilings. Model U.S. building codes specify that foam plastic used on interior walls and ceilings must have a flame-spread rating, determined by ASTM E84, of <75, and smoke generation of <450. The foam plastic must be covered with a fire-resistive thermal barrier either having a finish rating of not less than 15 min or equivalent to 12.7-mm gypsum board, or having a flame-spread rating of <25, smoke generation of <450 (if covered with approved metal facing), and protection by automatic sprinklers. Under no circumstances should direct flame or excessive heat be allowed to contact polyurethane or polyisocyanurate foam. The ASTM numerical flame-spread rating is not intended to reflect hazards presented under actual fire conditions.

Commercial Applications

Flexible Foam. The largest markets for flexible polyurethane foam are in the furniture, transportation, and bedding industries. Most furniture cushioning is made of polyurethane foam, predominantly cut from slabs or buns having a density of $0.0192-0.0288$ g/cm^3. Polyurethane foam core mattresses are used increasingly in bedding. High resiliency flexible foam having a density of 0.040 g/cm^3 is used for seat cushions in higher priced furniture. Molded flexible polyurethane foam is used in the automotive industry for seating, instrument panels, head rests, and arm rests applications. Semiflexible molded polyurethane foams are used in dashboards and door liners. Semiflexible foams are also formulated for sound and vibrational control in automotive applications. Other foam uses include carpet underlay, packaging, textile laminates, and interior padding. Specialty applications include reticulated foams for filtration and foams for such consumer products as sponges, scrubbers, squeegees, and paint applicators.

Rigid Foam.   The bulk of the rigid polyurethane and polyisocyanurate foam is used in insulation (qv). More than half (60%) of the rigid foam consumed in 1994 was in the form of board or laminate; the remainder was used in pour-in-place and spray foam applications. Laminates are used for residential sheathing and board for flat-deck commercial roofing. Commercial buildings are often covered with polyurethane spray foam. Pour-in-place foam is typically integrated in large-scale assembly operations, such as aircraft carriers. Insulation of truck trailers, truck bodies, railroad freight cars, and cargo containers are some of the other spray or pour-in-place applications. Tank and pipe insulation is either spray-applied or cut from bun stock.

Ships transporting liquid natural gas (LNG) are usually insulated with rigid PUIR foam laminates, which provide temperature stabilities from −180 to 150°C. The main fuel tank of the National Aeronautics and Space Administration (NASA) space shuttles is also insulated with PUIR foam. Rigid polyurethane foam is used in engineered foamed-in-place packaging of industrial or scientific equipment and in the molding of furniture, simulated-wood ceiling beams, and a variety of decorative and structural furniture components. Rigid foam is also used in movie props, for the repair of river barges, and in boat flotation applications.

Polyurethane Coatings.   Polyurethane surface coatings are used wherever applications require abrasion resistance, skin flexibility, fast curing, good adhesion, and chemical resistance (see COATINGS). The polyaddition process allows formulation of solvent-based or solventless liquid two-component systems, water-based dispersions, or powder coatings. Aliphatic isocyanates or TDI are mainly used in the formulation of coatings. Water-based coatings, formulated from polyurethane ionomers, are used for aircraft, appliances, as well as automotive, industrial, and farm machinery applications. For baking enamels, wire and powder coatings blocked isocyanates are used. Blocking agents include caprolactam, phenol, oximes, malonates, and 3,5-dimethylpyrazole. Unlike the other polyurethane coatings, blocked urethane coatings do not cure below a certain threshold baking temperature. Catalysts, such as tertiary amines or organometallic tin compounds, are used to lower the curing temperature. Blocked isocyanates are also used in the formulation of one-component systems.

Synthetic leather products are also produced using a urethane binder. These poromeric materials are produced from textile-length fiber mats impregnated with DMF solutions of polyurethanes. Permeability to moisture vapor is the key property needed in synthetic leather. In addition to shoe applications, poromerics are used for handbags, luggage, and apparel (see LEATHER-LIKE MATERIALS). Polyurethane films having oxygen and water permeability are applied in bandages and wound dressings and as artificial skin for burn victims.

Polyurethane Elastomers.   Polyurethane elastomers are used in applications where toughness, flexibility, strength, abrasion resistance, and shock-absorbing qualities are required. Thermoplastic polyurethane elastomers and polyurethane engineering thermoplastics are molded or extruded to produce elastomeric products used as automobile parts, shoe soles, ski boots, roller skate and skateboard wheels, pond liners, cable jackets, and mechanical goods. Cast and RIM elastomers are used in auto fascia, bumper and fender extensions, printing and industrial rolls, industrial tires, and industrial and agricultural parts, such as oil well plugs and grain buckets. Elastomeric spandex fibers are used in hosiery and sock tops, girdles, brassieres, support hose, and swim wear. The use of spandex fibers in sport clothing is increasing.

BIBLIOGRAPHY

"Urethanes" in *ECT* 1st ed., Vol. 14, pp. 473–480, by J. A. Garman, Food Machinery and Chemical Corp.; "Urethane Polymers" in *ECT* 1st ed., Suppl. Vol., pp. 888–908, by J. H. Saunders and E. E. Hardy, Mobay Chemical Co.; in *ECT* 2nd ed., Vol. 21, pp. 56–106, by K. A. Pigott, Mobay Chemical Co.; in *ECT* 3rd ed., Vol. 23, pp. 576–608, by H. Ulrich, The Upjohn Co.

1. O. Bayer, *Angew. Chem.* **A59**, 257 (1947).
2. H. Mispreuve and P. Knaub, *Proceedings of Polyurethane World Congress*, Vancouver, Canada, 1993, p. 297.
3. N. Malwitz and co-workers, *J. Cell. Plast.* **23**, 461 (1987).
4. F. M. H. Casati, F. W. Arbir, and D. S. Raden, *J. Cell. Plast.* **19**, 11 (1983).
5. K. Diblitz and C. Diblitz, in Ref. 2, p. 619.
6. D. R. Battice and W. J. Lopes, *J. Cell. Plast.* **23**, 158 (1987).
7. J. B. Lee and co-workers, *Macrom.* **26**, 4989 (1993).
8. B. Szczepaniak and co-workers, *J. Polym. Sci.* **A31**, 3223 (1993).
9. M. Tanaka and T. Nakaya, *J. Macromol. Sci. Chem.* **A26**, 1655 (1989).
10. D. Dietrich, W. Keberle, and H. Witt, *Angew. Chem.* **82**, 53 (1970).
11. R. Spindler and J. M. J. Frechet, *Macrom.* **26**, 4809 (1993).
12. L. H. Sperling, D. Klempner, and L. A. Utracki, *Interpenetrating Polymer Networks*, American Chemical Society, Washington, D.C., 1994.
13. W. J. Schnabel and E. Kober, *J. Org. Chem.* **34**, 1162 (1969).
14. H. J. Laas, R. Halpaap, and J. Pedain, *J. Prakt. Chem.* **336**, 185 (1994).
15. Z. W. Wicks, Jr., *Progr. Org. Coat.* **9**, 3 (1981).
16. Z. Wirpsza, *Proceedings of UTECH*, 1994.
17. Eur. Pat. Appl. EP 408,277 (1991), E. T. Shawl, J. G. Zajazek, and H. S. Kessling, Jr.
18. T. E. Waldman and W. D. McGhee, *J. Chem. Soc. Chem. Commun.* 957 (1994).
19. H. J. Knölker, T. Braxmeier, and G. Schlechtingen, *Angew. Chem. Int. Ed.* **34**, 2497 (1995).
20. C. P. Smith, J. W. Reisch, and J. M. O'Connor, *J. Elastom. Plast.* **24**, 306 (1992).
21. K. G. Spitler and J. J. Lindsey, *J. Cell. Plast.* **17**, 43 (1981).
22. H. W. Bonk and co-workers, *J. Elastom. Plast.* **3**, 157 (1971).
23. C. S. Schollenberger and F. D. Stewart, *Advances in Urethane Science and Technology*, Vol. 1, Technomics, Stamford, Conn., 1971, p. 65.
24. R. M. Hennington, V. Zellmer, and M. Klincke, *Proceedings of SPI 33rd Annual Conference*, 1990, p. 492.
25. J. S. Hicks and A. K. Schrock, in Ref. 24, p. 348.
26. J. V. McClusky and co-workers, in Ref. 2, p. 507.
27. J. Schuhmann and G. Hartzell, *J. Fire Sci.*, 386 (1989).
28. A. C. M. Griffiths and P. Shreeve, *Cell. Polym.*, 195 (1984).
29. C. D. Mcaffe and co-workers, in Ref. 2, p. 279.

30. J. Grimminger and K. Muha, in Ref. 2, p. 609.
31. C. Cecchini, V. Cancellier, and B. Cellarusi, in Ref. 2, p. 354.
32. H. Ulrich, *J. Cell. Plast.* **17**, 31 (1981).
33. A. K. Zhitinkina, N. A. Shibanova, and O. G. Tarakanov, *Rus. Chem. Rev.* **54**, 1104 (1985).
34. H. E. Reymore, R. J. Lockwood, and H. Ulrich, *J. Cell. Plast.* **14**, 95 (1978).
35. T. W. Bodnar, J. J. Koch, and J. D. Thornsberry, *Proceedings of Polyurethane World Congress*, Nice, France, 1991, p. 191.
36. D. C. Scholl, *Proceedings of Water Borne and Higher Solids Conference*, New Orleans, La., 1985, p. 120.
37. M. Bock and R. Halpaap, *J. Coat. Tech.* **59**, 755 (1987).
38. S. V. Lonikar and co-workers, *J. Appl. Polym. Sci.* **28**, 759 (1990).
39. R. A. Smiley and co-workers, *Proceedings of UTECH*, 1990, p. 205.
40. J. C. Lee and B. Kim, *J. Polym. Sci.* **A32**, 1983 (1994).
41. B. K. Kim, J. C. Lee, and K. H. Lee, *J. Macromol. Sci. Chem.* **A31**, 1241 (1994).
42. D. Dietrich, *Prog. Org. Coat.* **9**, 281 (1981).
43. H. A. Al-Salah and co-workers, *J. Polym. Sci.* **B24**, 2681 (1986).
44. H. Matsuda and S. Takechi, *J. Polym. Sci.* **A29**, 83 (1991); **A28**, 1895 (1990).
45. H. W. Bonk, H. Ulrich, and A. A. R. Sayigh, *J. Elastom. Plast.* **4**, 259 (1979).
46. Z. Wicks and B. Kostyk, *J. Coat. Tech.* **49**, 634 (1977).
47. U.S. Pat. 4,219,502 (1980), K. G. Ihrman and M. Brandt.
48. H. G. Schmelzer and co-workers, *J. Prakt. Chem.* **336**, 483 (1994).
49. U.S. Pat. 4,567,236 (1986), D. J. Goldwasser and R. Oertel.
50. H. W. Bonk and co-workers, *Proceedings of 43rd SPE conference*, New York, 1985, p. 1300.
51. H. Ulrich and co-workers, *Polym. Eng. Sci.* **18**, 844 (1978).
52. H. Ulrich and co-workers, *J. Elastom. Plast.* **11**, 208 (1979).
53. End-use market survey on the polyurethane industry, compiled by the Polyurethane Division of SPI, SPI Conference, Chicago, Ill., 1995.
54. A. W. Musk, J. M. Peters, and D. H. Wegman, *Am. J. Ind. Medicine*, **13**, 331 (1988).
55. M. H. Karol, *CRC Crit. Rev. Toxicol.* **16**, 349 (1986).
56. P. G. J. Reuzel, Report to the International Isocyanate Institute, III, New York, 1987 and 1988.
57. R. J. Davies, *Clin. Immonol. All.* **4**, 103 (1984).
58. I. F. Carney, *Toxicology of Isocyanates*, International Isocyanate Institute, Inc., New Canaan, Conn., 1980.

General References

G. Oertel, *Polyurethane Handbook*, 2nd ed., Carl Hanser Publishers, Munich, Germany, 1993.

J. K. Backus and co-workers, in H. F. Mark and co-workers, eds., *Encyclopedia of Polymer Science and Engineering*, 2nd ed., Vol. 13, John Wiley & Sons, Inc., New York, 1988.

G. Woods, *The ICI Polyurethanes Book*, John Wiley & Sons, Inc., New York, 1987.

R. Herrington and K. Hook, eds., *Flexible Polyurethane Foams*, Dow Chemical Company, Midland, Mich., 1991.

W. F. Gum, W. Riese, and H. Ulrich, eds., *Reaction Polymers*, Hanser Gardner Publications, Cincinnati, Ohio, 1992.

H. Ulrich, *The Chemistry and Technology of Isocyanates*, John Wiley & Sons, Inc., New York, 1996.

HENRI ULRICH
Consultant

VACCINE TECHNOLOGY

A vaccine is a preparation used to prevent a specific infectious disease by inducing immunity in the host against the pathogenic microorganism. The practice is also called immunization. The first human immunization was performed in 1796 by Edward Jenner in England which led to the discovery of smallpox vaccine. However, classical vaccinology developed 100 years later, after the work by Louis Pasteur demonstrated that microorganisms are causes of diseases.

During the early 1900s, vaccines against major human epidemic diseases such as pertussis, diphtheria, tetanus, and tuberculosis were developed. Vaccines for many animal diseases were also available. In the early 1950s, the development of cell culture techniques by J. E. Enders at Harvard was followed by another series of major advances in vaccine development. Vaccines against polio, mumps, measles, and rubella were licensed during the 1960s.

However, with the discovery and widespread use of antibiotics, beginning in the 1950s, the interest in vaccine research disappeared. It was anticipated that infectious diseases would no longer be a threat to human health. In fact, since the licensure of polio in 1963, no other vaccines were licensed for infant primary series immunization until the introduction of *Haemophilus influenza* type b conjugate vaccine in 1990. The development of biotechnology and modern immunology created new opportunities for producing new antigens and vaccine research has become a primary focus in recent years. As a result, several vaccines such as Hepatitis B, Hepatitis A, *H. influenza*, and Varicella have been approved. A new vaccine against pertussis has been recently approved in the United States.

Preventive medicine through vaccination continues to be the most cost-effective public health practice, even with the drastic advance in modern medicine. Mass vaccination programs have eradicated smallpox from the earth. The World Health Organization (WHO) has a major campaign underway to eradicate polio by the year 2000. The development of vaccines has saved millions of lives and prevented many more from suffering. However, there are still many diseases without effective vaccines, such as malaria. With the recent emergence of antibiotic-resistance strains and exotic viruses, an effective vaccine development program becomes a top priority of public health policy.

Commercial Vaccines

Vaccines can be roughly categorized into killed vaccines and live vaccines. A killed vaccine can be (*1*) an inactivated, whole microorganism such as pertussis, (*2*) an inactivated toxin, called toxoid, such as diphtheria toxoid, or (*3*) one or more components of the microorganism commonly referred to as subunit vaccines. The examples are capsular polysaccharide of *Streptococcus pneumoniae* and the surface antigen protein for Hepatitis B virus vaccine.

Live vaccines are normally weakened strains that do not cause diseases in the host, but still can stimulate the immune response. A typical example is the polio vaccine. The weakening of microorganisms or attenuation of the virus or bacteria can be accomplished by passage through different substrates and/or at different temperatures. Modern genetic engineering techniques can also be used to attenuate a virus or bacterium.

Vaccines for human use are regulated by the Food and Drug Administration in the United States and Boards of Health in other countries. The manufacturing of vaccines requires adherence to strict current good manufacturing practices (cGMPs) and in the United States licenses for both the process and the facility where the vaccine is produced are required. The Center for Biologics Evaluation and Research is the branch of the FDA that regulates vaccines. Basic requirements are described in the *Code of Federal Regulations* (CFR) (1).

Vaccines are used in either the general population of children or adults or for special groups. Recommendations for vaccine usage are made by the Advisory Committee on Immunization Practices (ACIP) of the Centers for Disease Control. The Committee on Infectious Diseases of the American Academy of Pediatrics (Redbook Committee) also makes recommendations for infants through adolescents, and the American Academy of Family Physicians makes recommendations for adults. An excellent review of vaccine history, development, usage, and related regulatory issues is available (2).

Vaccines for the General Population

Vaccines in this category protect children and adults from polio, diphtheria, tetanus, pertussis (whooping cough), measles (rubeola), mumps, rubella (German measles), hepatitis B, and haemophilus disease (meningitis, epiglotitis). The basic schedule is given in Table 1 (3).

Poliomyelitis.   Two vaccines are licensed for the control of poliomyelitis in the United States. The live, attenuated oral polio virus (OPV) vaccine can be used for the immunization of normal children. The killed or inactivated vaccine is recommended for immunization of adults at increased risk of exposure to poliomyelitis and of immunodeficient patients and their household contacts. Both vaccines protect against the three serotypes of poliomyelitis that cause disease. A mixed schedule, or immunization with inactivated poliovirus vaccine (IPV) followed by OPV has most recently been accepted as an immunization regimen for children.

Table 1. Recommended U.S. Childhood Immunization Schedule, January 1995[a]

Vaccine	Birth	2 Months	4 Months	6 Months	12 Months	15 Months	18 Months	4–6 Years	11–12 Years	14–16 Years
Hepatitis B	Hep 1 →	Hep 2 →			Hep 2 →					
diphtheria–tetanus–pertussis (DTP)		DTP[b]	DTP[b] →	DTP[b] →				DTP or DTaP	Td →	
Haemophilus influenzae type b (Hib)		Hib	Hib	Hib →		→				
poliovirus		OPV[c]	OPV[c] →			→		OPV		
measles–mumps–rubella (MMR)						MMR →		MMR →	MMR	

[a]Ref. 3.
[b]As of mid-1996, DTaP can be used.
[c]As of 1997, IPV can be used.

729

Composition and Methods of Manufacture. The live vaccine is a mixture of three types of the Sabin-attenuated polioviruses that have been propagated in cultures of monkey kidney cells, human diploid cell lines, or a monkey kidney heteroploid line. The cells are grown in the presence of a nutrient medium consisting of salts, amino acids, vitamins, dextrose, and sodium bicarbonate. Fetal calf serum is also used to promote growth. Additional supplements can include a pH indicator such as phenol red and antibiotics. The final vaccine is prepared by diluting the three serotypes to the appropriate levels in a modified cell culture medium and adding a stabilizer such as sorbitol or magnesium chloride (4).

Preparation of the killed vaccine involves the additional steps of concentrating and purifying the viruses, followed by inactivation with formaldehyde. The final vaccine is a combination of the three serotypes with a stabilizer such as human serum albumen and possibly a preservative (5). The development of bioreactor technology suitable for preparing the cell substrates for poliovirus propagation has allowed for greater ease in preparing more potent and additionally purified inactivated poliovirus vaccines (6).

Standardization and Testing. Requirements for the licensure and release testing in the United States are described in the CFR (1). After August 1996, requirements were given within the manufacturer's license. Worldwide requirements are described by World Health Organization's Biologicals Standards Documents or in documents specific to individual BOH country requirements. In general both *in vivo* (small animals, monkeys) and *in vitro* (cell culture) tests are performed at several stages during manufacture of the vaccine to control for safety, potency, and the absence of adventitious agents, eg, other viruses or microorganisms. Potency of the live vaccine is standardized by calculating the tissue culture infective doses while that of the killed vaccine is standardized by determining the mass of the active antigenic component (4,5). Passage levels of the attenuated vaccine strains are strictly controlled to ensure that the vaccine maintains its attenuated characteristics. The poliovirus genome has been cloned and sequenced, such that is possible to maintain a cDNA repository, and develop a better understanding of the biology of the vaccine viruses (7,8).

Diphtheria, Tetanus, and Pertussis. These vaccines in combination (DTP) have been routinely used for active immunization of infants and young children since the 1940s. The recommended schedule calls for immunizations at 2, 4, and 6 months of age with boosters at 18 months and 4–5 years of age. Since 1993 these vaccines have been available in combination with a vaccine that protects against *Haemophilus* disease, thus providing protection against four bacterial diseases in one preparation. A booster immunization with diphtheria and tetanus only is recommended once every 10 years after the fifth dose.

Composition and Methods of Manufacture. The diseases of diphtheria and tetanus are caused by toxins synthesized by the organisms *Corynebacterium diphtheriae* and *Clostridium tetani*, respectively. Diphtheria and tetanus vaccines contain purified toxins that have been inactivated by formaldehyde to form toxoids.

Tetanus toxin can be obtained by growing *Cl. tetani* in a complex medium specially formulated for production of high yields of toxin. The medium contains complex nutrients such as an enzymatic digest of casein, dextrose, sodium chloride, and other essential nutrients (9,10). The medium for growth of the diphtheria organism is also a complex liquid (11). The toxins of these bacteria are liberated into the growth medium and harvested away from intact cells and debris by filtration. The toxins may then be further purified, then toxoided (or toxoided first) followed by purification. Toxoiding involves incubation with formaldehyde at specified conditions of time, temperature, and pH. Purification can be accomplished by alcohol (methanol, acid pH) fractionation (12) or separation by serial ammonium sulfate fractionation from the culture fluid, or by column chromatography (12,13).

The pertussis vaccine consists either of whole, killed bacteria, or selected subunits of the bacteria. Until 1991, only the whole cell vaccine was available. In 1991, FDA (Center for Biological Evaluation and Research) licensed the first vaccine that contained pertussis subunits (referred to as acellular pertussis) for use in the last two doses of the five-dose series. In 1996, an acellular vaccine was licensed for use in the first three doses. It is expected that acellular vaccine will be licensed for all five doses in the near future.

The whole cell pertussis vaccine is prepared by first growing the organism in a complex medium containing casein, minerals, and other growth factors. The bacteria are collected and concentrated by centrifugation and killed and detoxified by heat and chemicals such as thimerosal, formaldehyde, or a combination of both (14). The acellular vaccines are prepared by a variety of methods that begin with growth of the bacteria in a complex medium, followed by chemical extraction and purification of the desired components away from the medium and cell debris. Formaldehyde treatment is used to inactivate any remaining toxic activity. Acellular pertussis vaccines contain one or more in varying proportions of the following proteins: fimbrial hemagglutinin, pertussis toxin, pertactin, and agglutinogens 1, 2, and 3 (15,16).

The final vaccine contains the two toxoids, as well as pertussis (whole cell or acellular), a buffer, and an adjuvant, ie, a substance that increases the response to an antigen when combined with the antigen, eg, aluminum. As noted above, the final vaccine can also contain a component that protects against *Haemophilus* disease.

Standardization and Testing. Requirements for DTP have been described (17). Standardization of potency for the toxoids relies on antigenic and flocculation tests. In principle, the antigenic tests are conducted to measure the ability of the vaccine to induce specific antibodies in guinea pigs. The flocculation test provides a quantitative estimate of the amount of toxoid in the vaccine.

The U.S. standard pertussis vaccine is used to standardize the potency of the whole cell pertussis vaccine. The number of protective units in the vaccine is estimated for each lot from the results of simultaneous intracerebral mouse-protection tests of the vaccine being studied and the U.S. reference standard (14,17). The potency of the acellular vaccines is estimated by their ability to produce antibodies to the proteins in the vaccine in a mouse model. These vaccines also undergo a series of animal safety tests to ensure that the inactivation and toxoiding steps were carried out correctly (14,17).

Measles, Mumps, Rubella. Live, attenuated vaccines are used for simultaneous or separate immunization against measles, mumps, and rubella in children from around 15 months of age to puberty. Two doses, one at 12–15 months of age and the second at 4–6 or 11–12 years are recommended in the United States.

Composition and Methods of Manufacture. The combined vaccine for simultaneous immunization in the United States is a mixture of the three live, attenuated viruses: measles (Moraten strain), mumps (Jeryl Lynn strain), and rubella (RA27/3 strain). Other strains of each of these viruses are used throughout the world. The measles and mumps viruses are propagated in cultures of primary chick embryo cells, whereas the rubella is propagated in the WI-38 strain or the MRC-5 strain of human diploid cells. In either case the cells are propagated and the viruses grown in a tissue culture medium. After sufficient viral replication, the fluids are collected, clarified, and mixed together in the proper proportions along with stabilizers such as gelatin, sorbitol, and amino acids. The vaccine is presented in freeze-dried vials and must be reconstituted with sterile distilled water before injection (18,19).

Standardization and Testing. Potency is determined by titration of the amount of live virus in susceptible tissue culture and is run in parallel with a U.S. standard. Both *in vivo* and *in vitro* tests are used to assess safety (17).

Haemophilus influenza serotype b. Three vaccines are available for immunizing infants. Two of these vaccines are administered at 2, 4, and 6 months of age with a booster given at 12–15 months of age, and the third vaccine is administered at 2 and 4 months of age with a booster at 12–15 months of age.

Composition and Methods of Manufacture. The vaccines suitable for immunization of infants are all forms of the capsular polysaccharide of the *H. influenza* b strain conjugated to a carrier protein. The antibodies generated to the capsule are protective. The carrier proteins are either CRM197 (a naturally nontoxic variant of diphtheria toxin), tetanus toxoid, or an outer membrane protein from the *Neisseria meningitis* bacterium. Manufacturing is accomplished by separate process streams for fermenting, purifying, and inactivating, if necessary, the carrier protein and the capsular polysaccharide. The polysaccharide may be cleaved into smaller units before conjugation to the carrier protein. Conjugation is accomplished by activating either the protein or the saccharide and joining with or without a linking agent (20–23). The vaccine can be presented in liquid or freeze-dried form and can be combined with DTP as a complete liquid vaccine in one case or rehydrated just before injection with a DTP liquid vaccine.

Standardization and Testing. Requirements are generally specified within licenses in the United States, and include a variety of in-process tests to assess purity, safety, and potency of the individual components and potency and safety of the final product. Potency is standardized by determining the size of the conjugate and the quantitative amount of saccharide that is bound to the carrier protein. General safety and immunogenicity is assessed in animals.

Hepatitis B. Although Hepatitis B (Hep B) is not an infant disease, it is recommended for infant immunization to better control spread, because compliance with vaccine immunization programs is easier to achieve in an infant population. Infants receive immunizations at birth, 1–2 months, and a third dose at 6 months. Other schedules are available for immunization of adolescents and adults who have not previously received the vaccine.

Composition and Methods of Manufacture. Hepatitis B vaccines consist predominantly of 22-nm particles of the S antigen of the Hepatitis B surface antigen. Some vaccines also include varying amounts of pre-S1 and/or pre-S2 antigens that are precursors to the fully matured surface antigen of this virus. The antigen can be derived from the plasma of chronic carriers using plasma fractionation techniques that ensure purity and inactivation of any unwanted live agents, or more commonly from recombinant organisms. There are systems using recombinants of yeasts or Chinese hamster ovary (CHO) cells that can be used to produce the surface antigen on a large scale (24–26).

Standardization and Testing. Potency is determined by quantitating the Hepatitis B antigen by an antibody-binding assay combined with a determination of the amount of protein. Safety testing typical for cell culture-derived products is also performed, and includes assuring the absence of live virus.

Varicella. The varicella (chicken pox) vaccine was approved in April 1995 for immunization of children. A single dose at one year of age is recommended. In the future it may be combined with measles, mumps, and rubella.

Composition and Methods of Manufacture. Vaccine is produced from the Oka attenuated strain. Vaccine is produced in human diploid cells such as MRC-5. After growth in the cell substrate, the cells themselves are harvested into the growth medium and sonicated to release the cell-associated virus. Sucrose and buffering salts are generally in the medium to help stabilize the virus. The vaccine is presented in a freeze-dried vial to be reconstituted with sterile distilled water before injection (27).

Standardization and Testing. Potency is determined by titrating the amount of live virus using a suitable cell substrate. Safety testing is also performed on seed lots to assure proper attenuation and on vaccine to assure absence of unwanted contaminants.

Vaccines for Special Populations

Vaccines for special populations are listed in Table 2. Two vaccines that are in fairly widespread use in the adult population are vaccines that prevent viral influenza and pneumococcal pneumonia.

Influenza. The ACIP recommends annual influenza vaccination for all persons who are at risk from infections of the lower respiratory tract and for all older persons. Influenza viruses types A and B are responsible for periodic outbreaks of febrile respiratory disease.

Composition and Methods of Manufacture. Two types of influenza viruses, A and B, are responsible for causing periodic outbreaks of febrile respiratory disease. The manufacture of an effective vaccine is complicated by antigenic variation or drift, which can occur from year to year within the two virus types, making the previous year's vaccine less effective. Each year, antigenic characterization is important for selecting the virus strains to be included in the vaccine.

Vaccines are prepared by growing high yielding strains of influenza viruses in embryonated chicken eggs. The viruses are harvested and purified from the allantoic fluid and are either inactivated (formalin treatment) as whole virions or split prior to inactivation. Splitting is generally accomplished by chemical agents to break the virus down into subvirion particles. Chemicals useful for splitting

Table 2. Selectively Used Vaccines

Type	Composition	Use	Reference
	Viral		
rabies	inactivated rabies grown in cultures of human diploid cells	post-exposure for treatment of animal bites, pre-exposure for those at high risk	28–30
yellow fever	live, attenuated virus grown in embryonated chicken eggs	international travel to high risk areas	31
adenovirus	live virus, types 4 and 7	prevention of acute respiratory disease in military recruits	32
hepatitis A	inactivated virus	high risk for exposure	33
	Bacterial		
meningitis	purified capsular polysaccharides of *Neisseria meningitidis* serogroups A, C, Y, W_{135}	control outbreaks and for military recruits	34, 35
cholera	inactivated *Vibrio* cholera strains Inaba and Ogawa	international travel to high risk areas	36
typhoid	inactivated whole cells, capsular polysaccharide, or live attenuated bacteria	international travel to high risk areas	37, 38
plague	inactivated Yersinia pestis	high risk of exposure	39
anthrax	inactivated cell culture filtrate of *Bacillus anthracis*	high risk of exposure	40
tuberculosis	mixture of live and killed bacteria, Bacille Calmette-Guerin	high risk groups	41

the virus include triton X-100, Triton N101, cetyltrimethylammonium bromide, sodium dodecyl sulfate, Tween 80, and tri-*n*-butyl phosphate (42–46).

Standardization and Testing. The final vaccine is tested for safety, potency, and residual chemicals. Safety includes testing for endotoxin and sterility. Potency is evaluated by quantitative determination of the amount of hemagglutinin in the vaccine. Antibody to this glycoprotein is associated with protection. The single radial immunodiffusion (SRID) technique is used to standardize the mass of this protein in comparison to a reference preparation.

Pneumococcal   Polysaccharide. *Indications   for   Use.* Pneumococcal polysaccharide vaccine may be used for immunization of persons two years of age or older who are at increased risk of pneumococcal disease.

Composition and Methods of Manufacture. The vaccine consists of a mixture of purified capsular polysaccharides from 23 pneumococcal types that are responsible for over 90% of the serious pneumococcal disease in the world (47,48). Each of the polysaccharide types is produced separately and treated to remove impurities. The latter is commonly achieved by alcohol fractionation, centrifugation, treatment with cationic detergents, proteolytic enzymes, nucleases or activated charcoal, diafiltration, and lyophilization (49,50). The vaccine contains 25 micrograms of each of the types of polysaccharide and a preservative such as phenol or thimerosal.

Standardization and Testing. The Center for Biologics Evaluation and Research has set guidelines for the vaccine which include standards for size of the individual polysaccharides and specifications for both purity (absence of protein and nucleic acid) and chemical and immunological identity.

Vaccines Being Developed

Despite the tremendous advances since the 1960s in the biomedical fields, including the total eradication of smallpox and reduction of mortality resulting from various diseases, there remains a large number of diseases that are endemic in many parts of the world. The Third World or developing countries bear the brunt of several of these, eg, malaria, trypanosomiasis, and schistosomiasis. In developed countries, diseases such as herpes and gonorrhea are becoming increasingly prevalent. Vaccines for many of the etiological agents that still cause disease have not been manufactured for several reasons. These include a lack of understanding of how immunity can be artificially induced and an inability to grow sufficient quantities of these agents to produce vaccines. In 1985, the Institute of Medicine identified the 10 most needed vaccines for the United States and the developing world (51). Some of these vaccines have been developed and licensed, whereas good progress is advancing in other areas. In the meantime, emerging exotic viruses such as HIV and drug-resistant pathogens continue to appear. There is an urgent need to expand vaccine R&D in order to reduce the risk of disease in the future.

Meningitis. *Haemophilus influenze*, type b (Hib), *Streptococcus pneumoniae*, and *Neisseria meningitidis* are the major cause of meningitis in infants. Vaccines against Hib disease prepared using conjugate technology have been in use worldwide, and have been efficacious in eliminating the disease from the population (52–54). This same technology is being applied to the development of vaccines for *S. pneumoniae* and *N. meningitidis* (55).

S. pneumoniae has more than 80 sero-types. The current polysaccharide vaccine consists of 23 serotypes and covers about 87% of all pneumococcal diseases in the United States. Current vaccine development is based on conjugate technology and concentrates on the most prevalent 7–9 serotypes. Three multivalent vaccine candidates are in clinical trials. All are based on conjugating the polysaccharide to a T-dependent protein carrier. The results of phase I and II trials in infants have demonstrated the safety and immunogenicity (56–58) of these vaccines. Phase III trials to demonstrate efficacy against systematic diseases and otitis media are in progress and final approval of this vaccine for infant immunization will be by the year 2000.

N. meningititidis also has several groups and serotypes. Most of the diseases are caused by groups A, B, and C. A multivalent polysaccharide vaccine consisting of types A, C, Y, and W_{135} is available. However, like other polysaccharide vaccines, it is not immunogenic in infants. Conjugate vaccines against groups A and C are being developed, using different protein carries and conjugate chemistries. Clinical trials of these vaccines are in progress (59–61).

The capsular polysaccharide of group B meningococcus is not immunogenic in humans. Thus, a conjugate vaccine of the group B polysaccharide will not improve its efficacy, and this remains a major challenge in developing the vaccine against group B organisms. One approach to improve the immunogenicity is to modify the polysaccharide (62). Another approach is to use the outer membrane proteins of the bacteria. Because of the different serotypes, a multivalent vaccine will be needed, if outer-membrane proteins are to be used as the vaccine (63). The outer-membrane protein preparation is normally a crude complex mixture or vesicle, or it can be a purified subunit protein. Both approaches have been in clinical trial (64,65). However, it is not anticipated that this vaccine will be in general use in the near future. Another approach is to use the lipooligosaccharide (LOS) of the bacteria as the antigen, which is a component of the bacterial endotoxin and conserved in all serotypes. A conjugate vaccine based on the LOS is being developed (66).

Rotavirus.   Rotavirus causes infant diarrhea, a disease which has major socio-economic impact. In developing countries it is the major cause of death in infants worldwide, causing up to 870,000 deaths per year. In the United States, diarrhea is still a primary cause of physician visits and hospitalization, although the mortality rate is relatively low. Studies have estimated a substantial cost benefit for a vaccination program in the United States (67–69). Two membrane proteins (VP4 and VP7) of the virus have been identified as protective epitopes and most vaccine development programs are based on these two proteins as antigens. Both live attenuated vaccines and subunit vaccines are being developed (68).

By using the technique of viral gene re-assortant, a multivalent live attenuated vaccine for rotavirus has been developed (68). The vaccine candidates are generated by transferring the VP7 gene from a human rotavirus to Rhesus monkey (or bovine) rotavirus. The re-assortant virus will express human VP7 protein without causing disease in humans. The expressed VP7 will be able to stimulate an immune response protective against human rotavirus. Similar approaches have been used for transfer of VP4 or both epitopes to different species of rotavirus. A multivalent Rhesus monkey re-assortant virus consisting of four serotypes has been in clinical trials. It has been demonstrated to be safe and efficacious. The efficacy of the vaccine is about 50% against all rotavirus diseases. The protection efficacy for infants from severe disease, especially hospitalization, is about 90% (69). The licensure of this vaccine for worldwide use will occur during the next few years. A subunit vaccine based on the VP4 and VP7 proteins is also being developed; however, it will be well into the twenty-first century before a vaccine will be available.

Respiratory Syncytial Virus.   Respiratory syncytial virus (RSV) causes severe lower respiratory tract disease in infants. It is the major cause of hospitalization in the United States (~90,000 events/yr) and it has a high mortality rate in neonates and other high risk populations, such as the geriatric population (51). Development of an RSV vaccine has always been a major priority, however, earlier attempts have mostly failed (70).

Both subunit and live, attenuated vaccine approaches are being developed for RSV. A candidate subunit vaccine based on the surface (F-) protein is being tested. The vaccine is prepared by infecting an appropriate cell substract, such as a vero cell, which then expresses the F-protein on its surface. After the cells lyse, the protein is purified by chromatography procedures. Clinical trials (71) for both infant and elderly are in progress. Another approach to protect the infant is passive immunization, through immunization of pregnant women. This maternal immunization of pregnant women with the subunit vaccine candidate is also being proposed.

Live attenuated vaccines for RSV are also being developed. Most of these vaccine candidates are derived from cold adaptation, by passing the virus at progressingly lower temperatures than human body temperatures. However, other means of mutagenesis have been used to generate vaccine candidates (72). Several clinical trials of these vaccines are also in progress (73,74).

Parainfluenza. Parainfluenza viruses (PIV) also causes viral pneumonia in infants. It is similar to RSV, therefore similar approaches are being used for developing a vaccine. A live attenuated PIV-3 vaccine has been in clinical trial (74).

Otitis Media. Otitis media is thought to be caused by several bacteria (75), significantly *S. pneumoniae*, nontypable *H. influenza*, and *Moraxella catarrhalis*. Viruses such as influenza, RSV, and PIV may also play a role in the disease. The use of a pneumococcal vaccine is the first step in the development of an otitis media vaccine (76). A clinical trial in Finland to compare two different types of pneumococcal conjugate vaccines will be used to demonstrate efficacy of these vaccine candidates against otitis media. In the meantime, vaccines against nontypable *H. influenza* and *Moraxella* are at the development stages. Both vaccine candidates are derived from the exposed proteins of the bacteria. The proteins are purified by extraction with different detergents followed by chromatography methods. Due to lack of animal models, efficacy of these subunit vaccines may be difficult to prove (77) without extensive clinical trials.

Herpes Simplex. There are two types of herpes simplex virus (HSV) that infect humans. Type I causes orofacial lesions and 30% of the U.S. population suffers from recurrent episodes. Type II is responsible for genital disease and anywhere from $3 \times 10^4 - 3 \times 10^7$ cases per year (including recurrent infections) occur. The primary source of neonatal herpes infections, which are severe and often fatal, is the mother infected with type II. In addition, there is evidence to suggest that cervical carcinoma may be associated with HSV-II infection (78–80).

Vaccine development is hampered by the fact that recurrent disease is common. Thus, natural infection does not provide immunity and the best method to induce immunity artificially is not clear. The genome of these viruses is also able to cause transformation of normal cells, thus conferring on them one of the properties attributed to cancerous cells. Vaccine made from herpes viruses must, therefore, be carefully purified and screened to eliminate the possibility of including any active genetic material.

Vaccine candidates are based on the two viral surface proteins, gD and gB (80). Recombinant methods are used to express the proteins, either in Chinese hamster ovary (CHO) cells or in baculovirus. The proteins are purified as subunits and formulated with different adjuvants. Clinical trials with these vaccine candidates have been performed, but the results to date have not been encouraging.

A much better understanding of the pathogenesis of the virus and virus-host interactions are required for the efficient development of the vaccine. Recently, DNA immunization is being proposed as a means to stimulate the appropriate Th 1 response which might provide long-term protection (81).

Influenza.   Although current influenza vaccine (subunit split vaccine) has been in use yearly for the elderly, it is not recommended for the general population or infants. Improvements to increase or prolong the immunogenicity, reduce the side-effects (due to egg production procedure), and provide mass protection are still being pursued. One approach is to use a live, attenuated virus though cold adaptation. A vaccine has been used in Russia and demonstrated to be safe and efficacious for infants (82). Clinical trials for a similar vaccine are being carried out in the United States (83).

Subunit vaccines based on the surface proteins of virus are also being explored. It has been demonstrated that the two major protective antigens are haemagglutinin (HA) and neuraminidase (NA). The genes for these antigens have been cloned and expressed in baculovirus in insect cell culture (84).

Production of the virus in a bioreactor reactor, using a continuous cell line, has also been studied (85,86). This will reduce production costs and side effects. Both Madin-Darby canine kedney (MDCK) and Vero cell lines are being developed for production of the vaccine.

Malaria.   Malaria infection occurs in over 30% of the world's population and almost exclusively in developing countries. Approximately 150×10^6 cases occur each year, with one million deaths occurring in African children (87). The majority of the disease in humans is caused by four different species of the malarial parasite. Vaccine development is problematic for several reasons. First, the parasites have a complex life cycle. They are spread by insect vectors and go through different stages and forms (intercellular and extracellular; sexual and asexual) as they grow in the blood and tissues (primarily liver) of their human hosts. In addition, malaria is difficult to grow in large quantities outside the natural host (88). Despite these difficulties, vaccine development has been pursued for many years. An overview of the state of the art is available (89).

One of the early vaccine candidates was directed against sporozoites, the form of the parasites that is first injected into the host by a mosquito. With recent development of recombinant techniques, several circumsporozoite proteins or its related peptides were proposed as the vaccine candidates. Clinical trials have been carried out. The vaccines were immunogenic, but did not provide sufficient protective efficacy (90,91).

Interest in vaccine development has centered around the asexual erothrocytic stage of the life cycle, especially the merozoite. Several proteins associated with these stages have been identified and produced by recombinant techniques (92,93). The most prominent is the MSA-1 protein of the merozoite. A clinical trial with this protein is being planned (93).

The most advanced of all the malaria vaccine candidates is SPf66 (94,95). It is a synthetic polypeptide. The peptide represents several protective epitopes correlated to several proteins of the pre-erythrocytic and asexual blood stage of *Plasmodian falciporum*. Extensive clinical trials with this vaccine have been carried out in South America and Africa (95–97). The efficacy of the vaccine varied in different regions, and generally lower than expected in a developed country. Consequently, the general application of the vaccine still generates much debate (98). However, this vaccine represents a major advance in the development of a malaria vaccine.

Gonorrhea.   Gonorrhea, caused by *Neisseria gonorrheae*, is the most commonly reported communicable disease in the United States. Approximately 10^6 cases were reported to the Center for Disease Control (CDC) in 1979, but actual cases could be two to three times higher (99,100). In addition, an increasing number of strains are becoming resistant to penicillin, the antibiotic that is usually used to treat this disease.

Development of a vaccine is problematic because natural infection does not necessarily provide immunity. Whether this results from a poor immunological response or to strain differences is not certain. Studies are being carried out on various structural components of the gonococcal bacterium, including pili, outer membrane proteins, lipopolysaccharide, and the outer capsule, in an effort to develop a vaccine (100). One of the more promising approaches involves a vaccine made with pili. These structures are responsible for attachment of the gonococci to mucosal surfaces, the first step necessary for infection to occur. Antisera against pili may prevent disease by preventing this attachment. One method for obtaining pili involves growth of the gonococci in liquid culture followed by mechanical shearing of the pili from the surface of the bacterium (101). Pili are further purified by differential centrifugation and ammonium sulfate precipitations. This type of preparation was shown to yield a protein pili vaccine that is immunogenic in human volunteers (102). Additional human studies indicate that a pili vaccine stimulates antibody formation that is 50–100 times the prevaccination level and is effective in preventing disease after challenge (103).

Human Immunodeficiency Virus.   Human immunodeficiency virus (HIV) causes Acquired Immunodeficiency Syndrome (AIDS), which has no cure. HIV infects the cells of the human immune system, such as T-lymphocytes, monocytes, and macrophages. After a long period of latency and persistent infection, it results in the progressive decline of the immune system, and leads to full-blown AIDS, resulting in death.

Since the discovery of HIV-1 as a causative agent for AIDS, the development of vaccine against HIV-1 has been a top priority of the national public health agencies and medical research institutes. After 20 years of extensive research, there is much better understanding of the physiology and pathogenesis of the virus and host-virus interactions and responses. However, the effort in developing a vaccine has not been as successful as expected. The main problem is the tremendous antigenic variability of the virus (104). An antigen derived from the cultured strain might not be the same as the clinical strain. Another problem is the fact that the virus infects the cells of the human immune system, making the design of the vaccine more complex. It will require certain combinations of immune responses to provide long-term protection or eliminate the virus from the host. So far, the proper immune mechanism for achieving this goal has not been identified, although it is generally agreed that a cell-mediated immune response (CMI) is essential.

Up to the 1990s, most of the vaccine candidates have been derived from the surface proteins of the virus. HIV-1 envelope glycoproteins gp120 and gp160 have been extensively studied. Peptides or polypeptides related to these proteins are also being studied. Although these candidates all show immunogenicity and are protective in animal models, clinical studies of these proteins have not been able to demonstrate protection against the disease (105–107). With the disappointing outcomes from several clinical trials, the National Institute of Allergy and Infection Diseases (NIAID) has decided to stop planning for any further Phase III trials of any candidate vaccines (108). However, efforts in development of the vaccine are being continued in the public and private research institutes.

Other Vaccines

There are many other diseases which do not have effective vaccines. These diseases are mostly regional in nature, epidemic in the developing world. Vaccines against parasites are also becoming critical to public health. Vaccines are being developed for Lyme disease, dengue, *Helicobacter pylori*, Japanese encephalitis, Equine encephalitis, Tick-borne encephalitis, cholera, shigellas, schistosomiasis, group B streptococcus, and other sexually transmitted diseases.

Future Technology

Vaccines for many diseases are unavailable because of an inability to determine the appropriate method for vaccination or difficulty in obtaining large quantities of antigens. Advances in medical science and immunology have substantially improved the understanding of the design and delivery of antigens. Genetic engineering offers further advances in providing the techniques for construction and production of large quantities of antigens. Development of these fields has been responsible for the rapid advances of vaccinology. Development of new vaccines also requires different process technology for the production of antigens and preparation of delivery system for vaccines.

Immunology

Immunology is the basis of vaccine technology. Only through the better understanding of the function of the human immune system can better antigens as vaccine candidates be designed. For example, the discovery of the functions of T- and B-lymphocytes led to the development of capsular saccharide–protein conjugate vaccines. Discovery of the different Th 1 and Th 2 immune responses also generated great interest in designing a vaccine that can stimulate a specific immune response, which may be critical for some viral vaccines. Cell-mediated immunity (CMI) has also been demonstrated to be critical for a successful vaccine. Several vaccine candidates, especially for viral vaccines, have been based on this approach. The mucosal and secretory immune system has also been studied extensively. This area will lead to the better design of vaccines for oral delivery or intranasal delivery of vaccine, which may be more efficacious for diseases originating in the mucosal system.

Genetic Engineering

Genetic engineering (recombinant-DNA technology) has been reviewed (62,109–113). It involves preparation of DNA fragments (passengers) coding for the substance of interest, inserting the DNA fragments into vectors (cloning vehicles), and introducing the recombinant vectors into living host cells where the passenger DNA fragments replicate and are expressed, ie, transcribed and translated, to yield the desired substance. The passenger DNA fragments can be obtained from natural DNA molecules by treatment with restriction endonucleases (enzymes that cut DNA at specific sites) or by mechanical shearing (114,115). They can also be synthesized either from messenger RNA (mRNA), through the actions of reverse transcriptase and DNA polymerase, or by pure chemical methods (116–122). The vectors are autonomously replicating DNA molecules (replicons), eg, plasmids, bacteriophages, and animal viruses. Small plasmids and bacteriophages are the most suitable vectors because their maintenance does not require integration into the host genome and their DNA can be isolated readily in an intact form (113). Many plasmid and bacteriophage vectors of improved qualities have been constructed by addition to and deletion of some of their genetic elements. Insertion of passenger DNA fragments into cloning vehicles can be carried out by one of three methods: ligation of cohesive ends produced by restriction endonuclease, homopolymer tailing, and blunt-end ligation. *Escherichia coli* has been exclusively used as the host cells for cloning. However, other microorganisms, eg, *Bacillus subtilis* and *Saccharomyces cerevisiae*, have also been used successfully (123,124). Introduction of the recombinant vectors into host cells (transformation) can be accomplished by different methods, depending on the vector–host cell system used. A calcium heat-shock treatment has been used exclusively in the plasmid–*E. coli* system (125). Successfully transformed host cells can be selected from the whole population using the drug resistance and nutritional markers carried by plasmid vectors, the plaque-forming abilities of phage vectors, immunochemical methods by means of antibodies directed to the substance of interest, or nucleic acid hybridization methods (126,127).

Since the 1970s, genetic engineering has evolved to become the most powerful and routine tool in the study of immunology and the development of new vaccines. It offers new, and in some instances safer and more effective methods for production of vaccines of higher quality. It has allowed an efficient way for the study of construction of new attenuated live viral or bacterial vaccines. It can also be used to study the pathogenicity and immunology of viruses or bacteria. Recently, direct injection of DNA as a means of vaccination is also being developed. Extensive reviews on the application of genetic engineering in vaccinology are available (128,129).

Recombinant hepatitis B vaccine is the first approved human vaccine based on a genetic engineering technique. The viral antigen is produced by a yeast-expressing system. A newer version of the vaccine is being tested. It is produced in cell culture, using chicken embryo cells as the substance. Almost all new protein antigens being developed such as HIV-1, herpes, rotavirus, and malaria are derived from genetic engineering techniques.

The genetic engineering techniques can also be used to reduce the virulence of a pathogen which can then be used to produce vaccines. Thus a mutant of pertussis has been constructed which will produce a cross-reactive material of the pertussis toxin (130). It has none of the toxicity associated with pertussis toxin, but still produces all the immunological properties of the antigen. This vaccine has been shown to be effective and is approved for use in several countries. A similar technique is being used for developing other vaccine candidates (131). Vaccine candidates for *Salmonella typhi* have been generated (131,132) by deleting the genes in the aromatic amino acid pathway. This vaccine candidate can be used to protect against typhoid. Similar vaccines are being developed for shigella and cholera (131).

Live vectors (131,133) are another application of genetic engineering. In this case, the genes from a pathogen are inserted into a vaccine vector, such as salmonella or vaccinia. In the case of salmonella, it will be possible to develop an oral vaccine. Vectors for this application include salmonella, BCG, polio, adenovirus, and vaccinia.

The use of naked DNA as a vaccine is the most recent development in this field. Since the demonstration of the possibility of genetic immune response by direct injection of DNA into muscle cells, the field is developing rapidly (134). Not only does it allow for large-scale production of vaccine, the use of naked DNA also has the advantage of stimulating the desirable Th1 response (135). The main obstacle for this approach is the low uptake rate of the injected DNA. Thus, the recent developments in this field are the different approaches in preparing DNA and injection techniques for improving the immune response (136). Clinical trials for influenza (137), hepatitis (138), HIV-1, and herpes simplex are being initiated.

Adjuvants

Adjuvants are substances which can modify the immune response of an antigen (139,140). With better understanding of the functions of different arms of the immune system, it is possible to explore the effects of an adjuvant, such that the protective efficacy of a vaccine can be improved. At present, aluminum salt is the only adjuvant approved for use in human vaccines. New adjuvants such as QS-21, 3D-MPL, MF-59, and other liposome preparations are being evaluated. Several of these adjuvants have been in clinical trial, but none have been approved for human use. IL-12 has been proposed as an adjuvant which can specifically promote T-helper 1 cell response, and can be a very promising adjuvant for future vaccine development.

Peptide Vaccines

Development of a peptide vaccine is derived from the identification of the immunodominant epitope of an antigen (141). A polypeptide based on the amino acid sequence of the epitope can then be synthesized. Preparation of a peptide vaccine has the advantage of allowing for large-scale production of a vaccine at relatively low cost. It also allows for selecting the appropriate T- or B-cell epitopes to be included in the vaccine, which may be advantageous in some cases.

Several vaccines based on peptide approaches, such as SPf66 (95) for malaria and an HIV-1 peptide (142) have been in clinical trials. No peptide vaccines are licensed as yet.

Other Developments

With the advance of immunology, the scope of vaccinology is also expanding. The technology can be applied into developing vaccines against cancer, allergies, and autoimmune diseases. Rapid progression in these areas have been documented in recent years.

Process Technology

In the preparation of classical killed or toxoid vaccines, simple process technology was used. With the advance of new vaccines, far more sophisticated process technologies are needed. The desire to reduce side effects of vaccination requires processes which will yield antigens of extreme purity. The new regulation in cGMP requires consistent production procedures, and global competition also demands that the most efficient process technology be applied.

The basic process technology in vaccine production consists of fermentation for the production of antigen, purification of antigen, and formulation of the final vaccine. In bacterial fermentation, technology is well established. For viral vaccines, cell culture is the standard procedure. Different variations of cell line and process system are in use. For most of the live viral vaccine and other subunit vaccines, production is by direct infection of a cell substrate with the virus.

Alternatively, some subunit viral vaccines can be generated by rDNA techniques and expressed in a continuous cell line or insect cells. Recent advances in bioreactor design and operation have improved the successful production of IPV in large-scale bioreactors. However, roller bottles or flasks are still used for most current vaccine production. Development of insect cell culture will allow for very large-scale liquid suspension culture (143). Several vaccine candidates such as gp160 for HIV and gD protein for herpes have been demonstrated in the insect cell culture system. However, no vaccine has been approved for human use.

The purification of an antigen normally utilizes the most advanced technology available, due to the high value of the product. The antigens are mostly protein or polysaccharide which can degrade easily. To avoid the degradation of these products, more drastic purification procedures cannot be used. The most common separation procedures are ultrafiltration (qv) and chromatography. The chromatography methods are ion exchange, gel permeation, and hydroxyapatite. Affinity chromatography can also be used, but requires extensive regulatory reviews. Classical separation procedures such as salting out or solvent precipitation are also used.

Development of conjugate and peptide vaccines requires the typical organic synthesis process and purification. This is a new area for vaccine technologists. Again, the main concern is to maintain the immunogenicity of the vaccine candidate during the chemical reaction and purification steps. Most of these procedures are proprietary. Formulation development is also becoming more complex for preparation and delivery of new vaccines. The classical vaccines are mostly prepared as injectable solutions. Aseptic techniques are required in the design and operation of the facilities.

To take advantage of the advance in immunology and adjuvants, future vaccines will be formulated to target a specific part of the immune system. The desire of combining several antigens to reduce the number of injections will require a detailed study of the vaccine formulation. Oral and intranasal delivery may also become common practice. All of these will need different technology for the preparation of the final vaccine dosage form and will present new challenges in vaccine technology.

Economic Aspects

The worldwide market is approximately $3.0 billion in sales, with the pediatric portion accounting for about 35%. Basic, required childhood vaccines (DTP, polio, measles/MMR, BCG, and TT) account for 3640×10^6 doses of this global market. In the United States doses distributed in the pediatric sector have risen from around 45×10^6 in 1982, covering basic, childhood vaccines, to around 75×10^6 in 1993 due primarily to the addition of vaccines for *Haemophilus* disease, hepatitis B, and a second dose of MMR to the recommended childhood series (144). The majority of vaccines for the U.S. market are produced by Merck, Wyeth-Lederle, Pasteur-Merieux Connaught, and SmithKline Beecham.

Costs of vaccine manufacture vary according to the type of vaccine produced and how it is supplied. Live virus vaccines are generally less expensive because the quantitative mass to be given to the recipient is less than an inactivated or subunit vaccine. The purification process and yield and the number of strains or components in any given vaccine also affect the cost of manufacture. New vaccines often have a royalty cost, in addition to manufacturing and testing costs. Because of the requirement for specialized facilities and training of personnel, there is a large fixed cost burden to the manufacturing process, making production volumes a key factor in overall cost per dose. Filling and packaging is often the most expensive part of the manufacturing process and the cost varies by how many doses are filled and packed into one unit. Average estimated costs for vaccines used and produced in the United States in the 1990s range from approximately $0.50 to $2.50 per dose (144).

In addition to covering the cost of goods, the price of vaccines to the consumer must also be able to cover sales, marketing, and distribution costs (including a cold chain, a vaccine shipping and storage temperature requirement), and research and facilities for new products. The setting for immunization will affect the cost to the consumer. Slightly more than half of the vaccines used in the United States for pediatric immunization are purchased in large volumes by the government and administered in the public sector environment, while the remainder are administered in the private practice sector. An excise tax on vaccines, used to cover adverse reaction events, is also included in the price. Full immunization in the public sector was estimated at $111 and in the private sector at $238 per child in 1993 (144).

Another important aspect of vaccine technology is the cost–benefit relationship between prevention vaccination and disease treatment. Generally the cost savings are high. For the early period of polio immunization (1955–1961), the net savings as a result of immunization were calculated to be 327×10^6. If loss of income were added the savings would amount to ~$1 billion. Measles vaccination was estimated to have saved 100×10^6 in medical and lost work costs from 1963–1967 (145). Studies of the cost effectiveness of immunization of children against diphtheria, tetanus, and pertussis disease have yielded a benefit-to-cost ratio of 6.2:1 for direct costs and 20.1:1 when indirect costs are included (146,147). Projected savings from a rotavirus immunization program (vaccine not yet licensed) have also been calculated. A partially protective vaccine would yield an average savings of 78×10^6 per year in the United States in health care costs, and 466×10^6 when overall costs to society were considered (67). Direct and indirect savings for commonly used childhood vaccines as studied by the CDC are given in Table 3.

Cost–benefit analyses for adult immunizations have also been performed. Influenza immunization during the period from 1971 to 1977 resulted in over 13 million more years of life at a cost of only $63 per year of life gained. Productivity gains were estimated to have a value of 250×10^6 (148). Projected costs of pneumonia have been calculated at 3.6 times the cost of vaccination, or a savings of $141 per person is achieved among those at risk for developing pneumonia or over the age of 50 years (149).

Liability for adverse reaction events associated in time with immunization have also played a principal role in vaccine economics. Prior to 1988, compensation for any adverse reaction associated in time with vaccination required that the vaccine recipient bring suit against the manufacturer or the health care provider that administered the vaccine. The uncertainty of numbers and costs associated with lawsuits contributed to the decline in the number of providers of routine childhood vaccines during the late 1970s and 1980s. Lawsuits peaked for the DTP vaccine in 1986 leading to the enactment in 1988 of the National Vaccine Injury Compensation Program. This program, as well as the National Childhood Vaccine Injury Act of 1986, was provided as a nonfault alternative to the tort system for resolving claims resulting from adverse reactions to mandated childhood vaccines, and has achieved its goal of providing compensation to those injured by rare adverse events associated with vaccination and providing some stability for the vaccine market.

Table 3. Benefit–Cost Analysis of Commonly Used Vaccines[a]

Vaccine	Direct medical savings, $	Direct and indirect[b] savings, $
DTP	6.0	29.1
MMR	16.3	21.3
OPV	3.4	6.1
integrated schedule (DTP, MMR, OPV combined)	7.4	25.5
H. influenzae type b	1.4	2.2
Varicella	0.9	5.4

[a]Savings per dollar invested.
[b]Indirect savings include work loss, death, and disability.

BIBLIOGRAPHY

"Vaccine Technology" in *ECT* 3rd ed., Vol. 23, pp. 628–643, V. A. Jegede and co-workers, Lederle Laboratories, American Cynamid Co.

1. *Code of Federal Regulations*, Title 21, Parts 200 and 600.
2. S. A. Plotkin and E. A. Mortimer, *Vaccines*, 2nd ed., W. B. Saunders Co., Philadelphia, Pa., 1994.
3. *Mobidity and Mortality Weekly Report (MMWR)*, Vol. 44, RR 5, Centers for Disease Control, Atlanta, Ga., June 1993.
4. M. B. Ritchey, *Nineteenth Immunization Conference Proceedings*, 1984, p. 75.
5. A. L. VanWezel, G. VanSteenis, P. VanderMarck, and A. Oslerhair, *Rev. Infect. Dis.* **6**(Suppl.), 5335 (1984).
6. B. Montagnon, B. Fanget, and J-C. Vincent-Falquet, *Rev. Infect. Dis.* **6**, 5341 (1984).
7. M. Koham, S. Abe, S. Kuge, and co-workers, *Virology*, **151**, 21 (1986).
8. V. R. Racaniello and D. Baltimore, *Science*, **214**, 915 (1981).
9. J. H. Mueller and P. A. Miller, *J. Immunol.* **56**, 143 (1947).
10. L. Weinstein, *N. Eng. J. Med.* **289**, 1293 (1973).
11. D. W. Stainer and M. J. Scholte, *Biotechnol. Bioeng. Symp.* **4**, 283 (1973).
12. L. Pillemer, D. B. Grossberg, and R. G. Wittler, *J. Immunol.* **54**, 213 (1946).
13. *Diphtheria and Tetanus Toxoids and Pertussis Vaccine Adsorbed*, USP Package Insert, Connaught Laboratories, Surftuster, Pa., revised 1994.
14. Ref. 4, p. 87.
15. E. M. Edwards, B. D. Meade, and M. D. Dechs, *Pediatr. Res.* **4**(12), 91A (1992).
16. H. S. Schmitt and S. Wagner, *Eu. J. Pediatr.* **162**, 402 (1993).
17. Ref. 1, Part 610.
18. A. Y. Elliott, *Nineteenth Immunization Conference Proceedings*, 1984, p. 79.
19. U.S. Pat. 4,147,772 (Apr. 3, 1979), U. J. McAleen and H. Z. Markus (to Merck and Co., Inc.).
20. L. K. Gorden, in R. M. Chanock and R. A. Learner, eds., *Modern Approaches to Vaccines*, Cold Spring Harbor, N.Y., 1984.
21. P. Anderson, *Infect. Immunol.* **39**, 233 (1983).
22. C. Y. Chu and co-workers, *Infect. Immunol.* **40**, 245 (1983).
23. PedVax Hib product circular, *PDR* **50** (1996).
24. P. Valenzucla, A. Medina, W. J. Rultz, and co-workers, *Nature*, **298**, 347 (1982).
25. W. J. McAleer, E. B. Bugnak, R. Z. Margeths, and co-workers, *Nature*, **307**, 178 (1984).
26. A. Siddiqui, in G. Papaevangelou and W. Hennersu, eds., *Developmental Biological Studies of Standardization in Immunophylaxis of Infections by Hepatic Viruses*, Vol. 54, S. Karger, Basel, 1983, p. 19.
27. Varivax Package Insert, Merck and Co., Rahway, N.J., 1995.
28. *Morbid. Mortal. Weekly Rep.* **29**, 265 (1980).
29. S. A. Plotkin and T. J. Wiktor, *Am. Rev. Med.* **29**, 583 (1978).
30. M. Bahmanyer, A. Fagaz, S. Nour-Saleki, M. Mahammadi, and H. Koperski, *JAMA*, **236**, 2751 (1976).
31. *United States Pharmacopeia XXII*, Mack Publishing Co., Easton, Pa., 1989, p. 1461.
32. F. H. Top, Jr., *Yale J. Biol. Med.* **48**, 185 (1975).
33. F. E. Andre, A. Hepburn, and E. D. Hondt, in J. L. Melnick, ed., *Prog. Med. Virol. Basel* (Karger) **72** (1990).
34. J. Armand, F. Arminym, M. C. Mynard, and C. Lafai, *J. Biol. Stand.* **10**, 335 (1982).
35. W. P. McKenny and G. P. Bannar, *JAMA*, **260**(5), 638 (1988).
36. *Cholera Vaccine*, USP Package Insert, Wyeth Laboratories, Marietta, Pa., 1994.
37. *Typhoid Vaccine*, USP Package Insert, Wyeth Laboratories, Marietta, Pa., 1994.

38. S. J. Cryz, Jr., E. Fust, L. S. Baron, K. F. Noon, F. A. Rubin, and D. J. Kopus, *Infect. Immun.* **57**, 3863 (1989).

39. *Morbid. Mortal. Weekly Rep.* **27**, 255 (1978).

40. B. E. Ivins and S. L. Welkes, *Eu. J. Epidemiol.* **4**, 12 (1988).

41. M. Gheorghiu, *J. Biol. Stand.* **15**, 15 (1988).

42. U.S. Pats. 4,064,232 (Dec. 20, 1977), and 4,140,762 (Feb. 20, 1979), H. Backmayer and G. Schmidt (to Sandoz, Ltd.).

43. U.S. Pat. 4,029,763 (June 14, 1977), E. D. Kilbourne (to Mt. Sinai School of Medicine of the City University of New York).

44. U.S. Pat. 4,158,054 (June 12, 1979), I. G. S. Furminger and M. I. Brady (to Duncan Flockart and Co., Ltd.).

45. U.S. Pat. 4,000,257 (Dec. 28, 1976), F. R. Cano (to American Cyanamid Co.).

46. U.S. Pat. 3,962,421 (June 8, 1976), A. R. Neurston (to American Home Products Corp.).

47. D. S. Fedson, S. A. Plotkin, and E. A. Mortimer, Jr., eds., *Vaccines*, 1st ed., W. B. Saunders Co., Philadelphia, Pa., 1988, p. 271.

48. J. B. Robbins, R. Austran, C. J. Lee, and co-workers, *J. Inf. Dis.* **148**, 1136 (1983).

49. U.S. Pat. 4,242,501 (Dec. 30, 1980), F. R. Cano and J. S. C. Kuo (to American Cyanamid Co.).

50. Eur. Pat. 0002,404A1 (June 13, 1971), D. C. Carlo, K. H. Nolstadt, T. H. Stoudt, R. B. Walton, and J. Y. Zeitner (to Merck and Co.).

51. P. J. Baker, ed., National Institute of Allergy and Infectious Diseases, *The Jordan Report, Accelerated Development of Vaccines, 1993*, NIH, Bethesda, Md.

52. H. R. Shinefiled and S. Black, *Pediatr. Infect. Dis. J.* **14**, 978 (1995).

53. CDC, "Progress Toward Elimination of *Haemophilus influenzae* Type b Disease Among Infants and Children—United States, 1993–1994," *MMWR*, **44**(29) (1995).

54. D. C. Madore, *Infect. Agents Dis.* **5**, 8 (1996).

55. R. Eby, in M. Powell and M. Newman, eds., *Vaccine Design: The Subunit and Adjuvant Approach*, Plenum Press, New York, 1995, Chapt. 31.

56. S. Black, H. Shinefield, P. Ray, and co-workers, *Pediatr. Res.* **34**, 39(4), 167A, Abs. 986 (1996).

57. K. Zangquill, D. Greenberg, K. Wong, and co-workers, *Pediatr. Res.* **39**(4), 188A, Abs. 1116 (1996).

58. E. L. Anderson, D. J. Kennedy, K. M. Geldmacher, J. Donnelly, and P. M. Mendelman, *J. Pediatr.* **128**(5), 649 (1996).

59. M. Rennals, K. Edwards, H. Keyserling, and co-workers, *Pediatr. Res.* **39**(4), 183A, Abs. 1083 (1996).

60. J. M. Lieberman and co-workers, *JAMA*, **275**(19), 1499 (1996).

61. P. Costantino, S. Viti, A. Podda, M. Velmonte, L. Nenicioni, and R. Rappuoli, *Vaccine*, **10**, 691 (1992).

62. R. P. Novick, *Sci. Am.* **243**(6), 102 (1980).

63. H. J. Jennis, R. Roy, and A. Bamian, *J. Immunol.* **37**, 1708 (1986).

64. J. T. Poolman, *Infect. Agents Dis.* **4**, 13 (1995); G. Bjune, E. Hoiby, J. Gronnesby, and O. Arnesen, *Lancet*, **338**, 1093 (1991).

65. L. Milagres, A. Lemos, and co-workers, *Brit. J. Med. Biol. Res.* **28**, 981 (1995).

66. X. Gu and C. Tsai, *Infect. Immun.* **61**, 1873 (1993).

67. J. Smith, A. Haddix, S. Teutsch, and R. Glass, *Pediatr.* **96**, 609 (1995).

68. K. Midthan and A. Z. Kapikian, *Clin. Microbiol. Rev.* **9**(3), 423 (1996).

69. M. B. Rennels, R. L. Ward, M. E. Mack, and E. T. Zito, *J. Infect. Dis.* **173**(2), 306 (1996).

70. G. L. Toms, *Archiv. Dis. Childhood*, **72**, 1 (1995).

71. P. A. Piedra and co-workers, *Pediatr. Infect. Dis. J.* **15**, 23 (1996).

72. J. Crowe, P. Bui, W. Lond, A. Davis, P. Hung, R. Chanock, and B. Murphy, *Vaccine*, **12** (1994).

73. B. R. Murphy, S. Hall, A. Kulkarni, J. Crowe, P. Collins, M. Connors, R. Karron, and R. Chanock, *Virus Res.* **32**, 13 (1994).

74. R. Karron, P. Wright, and co-workers, *J. Infect. Dis.* **172**, 1445 (1995).

75. P. H. Karma, H. Bakaletz, G. Giehink, B. Mugi, and B. Rynnel-Dag, *Int. J. Pediatr. Otorhinocoryngol.* **32**(Suppl.), S127 (1995).

76. S. Giebink, J. Meier, and co-workers, *J. Infect. Dis.* **173**, 119 (1996).

77. L. Alphen, *J. Infect. Dis.* **165**, 177 (1992).

78. A. L. Notkins, R. A. Bankowski, and S. Baron, *J. Infect. Dis.* **127**, 117 (1973).

79. F. Rapp, *Conn. Med.* **44**(3), 131 (1980).

80. R. L. Burke, *Virology* **4**, 187 (1993).

81. N. Bourne, L. Stanberry, D. Bernstein, and D. Lew, *J. Infect. Dis.* **173**, 800 (1996).

82. A. S. Skhan and co-workers, *J. Infect. Dis.* **173**, 453 (1996).

83. W. Gruber and co-workers, *J. Infect. Dis.* **173**, 1313 (1996).

84. J. Treanor, R. Betts, G. Smith, E. Anderson, C. Hackett, B. Wilkinson, B. Belshe, and D. Powers, *J. Infect. Dis.* **173**, 1467 (1996).

85. J. Robertson, P. Cook, A. Attwell, and S. Williams, *Vaccine* **13**, 1583 (1995).

86. E. Govorkova, N. Kaverin, V. Gubareua, M. Meignier, and R. Webster, *J. Infect. Dis.* **172**, 250 (1995).

87. D. Sturchler, *Parasitol. Today* **5**, 39 (1989).

88. A. Mizrah, I. Hertman, M. A. Klingberg, and A. Kohn, eds., *Progress in Clinical and Biological Research*, Vol. 47, Alan R. Liss, Inc., New York, 1980.

89. M. F. Good, A. Saul, and P. Graves, "Malaria Vaccines," in R. Ellis, ed., *Vaccines, New Approaches to Immunological Problems*, Butterworth-Heinmann, Boston, Mass., 1992.

90. J. A. Sherwood, R. Copeland, K. Taylor, and co-workers, *Vaccine* **14**, 817 (1996).

91. L. F. Fries, D. Schneider, and co-workers, *Infant Immun.* **60**, 1534 (1992).

92. T. Hodder, P. Crewther, M. Lett, and co-workers, "Apical Membrane Antigen 1: A potential Malaria Vaccine Candidate," *7th Malaria Meeting of British Society of Parasilolgy*, London, Sept. 19–21, 1995.

93. W. Huber, S. Steiger, and co-workers, "A Pilot Study of Genotyping of MSA1 and 2 During a Trial of Spf66 Vaccine in Tenzania," in Ref. 92.

94. G. J. Jennings, A. Belkum, L. van Boorn, and M. Wiser, "Analysis of Variability of the Merozoite Surface Protein of Plasmodian berghei," *44th Annual Meeting of the American Society of Tropical Medicine and Hygiene*, San Antonio, Tex., Nov. 17–21, 1995.

95. R. Amador and J. Patarroyo, *J. Clin. Immunol.* **16**, 183 (1996).

96. P. L. Alonso, T. Smith, and co-workers, *Lancet* **344**, 1175 (1994).

97. U. D'Alexandro, A. Leach, C. Drakeley, and co-workers, *Lancet* **346**, 462 (1995).

98. M. P. Tanner and P. Alonso, *Schqueizersche Medizinische Wochenschift*, **126**, 1210 (1996).

99. J. I. Ito, Jr., *Ariz. Med.* **38**, 626 (1981).

100. G. Schoolnik and T. Mietzner, "Vaccines Against Gonococcal Infection," in G. Woodrow and M. Levin, eds., *New Generation Vaccines*, Marcel Dekker, New York, 1990.

101. B. T. M. Buchanan, Jr., *Exper. Med.* **141**, 1470 (1975).

102. J. D. Nelson and C. Grasi, eds., *Current Chemotherapy and Infectious Disease*, ASM, Washington, D.C., 1980, p. 1239.

103. *New Drug Commentary* **8**, 17 (1981).

104. F. E. Mccutchan, A. Artenstein, E. Sandersbuell, and co-workers, *J. Virol.* **70**, 3331 (1996).

105. J. Mascola, S. Synder, O. Weislow, and co-workers, *J. Infect. Dis.* **173**, 340 (1996).

106. F. Valentine, S. Kundu, P. Haslett, and co-workers, *J. Infect. Dis.* **137**, 1136 (1996).

107. G. Gorse, M. Keefer, B. Belsha, and co-workers, *J. Infect. Dis.* **137**, 330 (1996).

108. *Science*, **271**, 1227 (Mar. 1, 1996).

109. S. N. Cohen, *Sci. Am.* **233**(1), 24 (July 1975).

110. R. L. Sinsheimer, *Annu. Rev. Biochem.* **46**, 415 (1977).

111. W. Gilbert and L. Villa-Komaroff, *Sci. Am.* **243**(4) (1980).

112. P. Chambon, *Sci. Am.* **244**(5), 60 (1981).

113. N. G. Carr and co-workers, eds., *Principles of Gene Manipulation: An Introduction to Genetic Engineering*, University of California Press, Berkeley, Calif., 1980.

114. P. C. Wensink and co-workers, *Cell* **3**, 315 (1974).

115. L. Clarke and J. Carbon, *Cell* **9**, 91 (1976).

116. E. Y. Friendman and M. Rosabash, *Nucleic Acids Res.* **4**, 3455 (1977).

117. G. N. Buell and co-workers, *J. Biol. Chem.* **253**, 471 (1978).

118. M. P. Wickens and co-workers, *J. Biol. Chem.* **253**, 2483 (1978).

119. K. Itakura and co-workers, *Science* **190**, 1056 (1977).

120. D. V. Goeddel and co-workers, *Proc. Natl. Acad. Sci. USA* **76**, 106 (1979).

121. H. G. Khorana, *Science* **203**, 614 (1979).

122. K. Itakura and A. D. Riggs, *Science* **209**, 1041 (1981).

123. P. S. Lovett and K. M. Keggins, *Methods Enzymol.* **68**, 342 (1979).

124. A. Hinnen and co-workers, *Proc. Natl. Acad. Sci. USA* **75**, 1929 (1978).

125. M. Mandel and A. Higa, *J. Mol. Biol.* **63**, 159 (1970).

126. S. Broome and W. Gilbert, *Proc. Natl. Acad. Sci. USA* **75**, 2246 (1978).

127. M. Grunstein and D. S. Hogness, *Proc. Natl. Acad. Sci. USA* **72**, 3961 (1975).

128. G. Woodrow, "An Overview of Biotechnology as Applied to Vaccine Development," in G. Woodrow and M. Levine, eds., *New Generation Vaccines*, Marcel Dekker, New York, 1990.

129. R. Ellis, "The Application of rDNA Technology to Vaccines," in S. A. Plotkin and B. Fantini, eds., *Vaccinia, Vaccination and Vaccinology*, Elsevier, Paris, 1996.

130. R. Rappuoli, G. Douce, and M. Pizza, *Int. Arch. Allergy Immunol.* **108**, 327 (1995).

131. A. Lindberg, *Dev. Biolog. Stand.* **84**, 211 (1995).

132. M. Levin, J. Galen, E. Barry, and co-workers, *J. Biotechnol.* **44**, 193 (1996).

133. F. Schodel and F. Curtiss, *Dev. Biolog. Stand.* **84**, 248 (1995).

134. J. Wolff, R. Malone, P. Williams, and co-workers, *Science* **247**, 1465 (1990).

135. M. Liu, J. Donnelly, T. Fu, M. Cawlfield, and J. Ulmer, "DNA Vaccines: Mechanism of Antigen Presentation from feneration of MHC Class I Restricted Cytotoxic T-Lymphocyte Responses," *Biomedicine '96: Medical Research from Bench to Bedside*, Washington, D.C., May 3–6, 1996.

136. H. Hofland and co-workers, *Proc. Nat. Acad. Sci.* **14**, 7305 (1996).

137. J. Donnelly, A. Friedman, D. Montgomery, and co-workers, "Polynucleotide Vaccination Against Influenza," in E. Nerrby, F. Brown, B. Chanock, and H. Ginsberg, eds., *Modern Approaches to New Vaccines including Preventing AIDS*, Lab Press, Cold Spring Harbor, New York, 1994, p. 55.

138. H. Davis, M. Mancin, M. Michel, and R. Whalen, *Vaccine* **14**, 910 (1996).

139. R. Gupta and G. Siber, *Vaccine* **13**, 1263 (1995).

140. J. Bliss, V. VanCleave, and co-workers, *J. Immunol.* **156**, 887 (1996).

141. A. Saul and H. Geysen, in G. Woodrow and M. Levine, eds., *Identification of Epitopes through Peptide Technology in New Generation Vaccines*, Marcel Dekker, New York, 1990.

142. World Pat. 9529700-A1 (1995), B. F. Haynes and T. J. Palker, "New Peptide Corresponding to HIV Sequences Used for Inducing Protective Immunity to HIV and in the Treatment of e.g. Auto-Immune Disease Infectious Diseases and Tumors" (to Duke University).

143. J. Vlak and R. Keus, "Baculovirus Expression Vector System for Production of Viral Vaccines," in A. Mizrahi, ed., *Viral Vaccines*, Wiley-Liss, New York, 1990.

144. *Mercer Management Consulting Report on the U.S. Vaccine Industry to the U.S. Dept. Health and Human Services*, Washington, D.C., June 14, 1995.

145. H. H. Fudenberg, *J. Lab. Clin. Med.* **79**, 353 (1972).

146. Battelle (MEDTAP) Final Report, *A Cost Benefit Analysis of DTP Vaccine to the U.S. Dept. Health and Human Services*, Washington, D.C., Mar. 4, 1994.

147. Technical data, Centers for Disease Control, Atlanta, Ga.

148. J. Perez Tirse and P. A. Gross, *Pharmaco. Economics*, **2**(3), 198 (1992).

149. C. B. Gable, S. S. Holzer, L. Engehardt, and co-workers, *JAMA*, **264**, 2910 (1990).

CHIA-LUNG HSIEH
MARY B. RITCHEY
Wyeth-Lederle Vaccine and Pediatrics

VACUUM TECHNOLOGY

Vacuum technology concerns the means to predict, effect, and control subatmospheric pressure environments (vacuum) (1). Increasingly, each vacuum environment must be not only safe and cost-, energy-, and materials-effective, but also tailored to serve each use profitably. However, as of this writing (1997), the likely interactions between a dynamic endeavor and a potentially profitably serving vacuum environment may still not be straightforward. A learning experience is likely. Adequate results can depend on luck in choosing and following recipes, cut-and-dry methods, guessing, and intuition, rather than understanding traceable to scientific first principles.

Vacuum production was essential in the early steam-actuated pumps used for pumping water (qv). In these engines, steam pressure raised a piston against atmospheric pressure. The work stroke in the engine was created by condensing the steam, which allows atmospheric pressure to perform the output of the engine. These vacuum-force-type engines were replaced by the steam engine of James Watt. Philosophical notions of void and vacuum developed over centuries turned out to be more realistic than the idea that subatmospheric pressure in a gas characterizes a vacuum.

The vacuum environment offers a great range and diversity of uses (Table 1). The host of parameters are set by the USE and may not include gas pressure except heuristically. Construction and control represent a challenge to the engineers and technicians. Vacuum environments can be grouped into the following operational levels: crude (CR), rough (R), controlled (C), highly controlled (HC), and ultracontrolled (UC) (2). In some instances, these correspond to the traditional categories where vacuum is referred to as low, medium, high, ultrahigh, and beyond ultrahigh. The traditional categorization focuses

Table 1. Vacuum Applications

Vacuum environment category[a]	Gas pressure, ~Pa[b]	Pump[c]	Use
C, HC	$<10^3$	A, B, D	aneroid barometers
all	all	all	annealing
HC, UC	$<10^{-3}$	D	arc circuit breakers and switches
CR, R, C	<1	A, B, D	arc furnaces
HC	$<10^{-4}$	D	betatrons
C	3	B	blood-plasma dehydration
CR, R, C	10^3	A, D	capacitors
CR, R	all	A, D	casting
R, C	60	A	citrus-juice dehydration
CR	5×10^4	A	cleaning
C		A	coffee packing
CR	5×10^4	A	concrete casting
all	all	all	cooling
R, C	10^{-5}	cryo	cryogenic wind tunnels
R, C	3	B	dehydration of antibiotics
R, C	3	B	distillation of plasticizers
C, HC	$<10^{-2}$	D	electron- and ion-beam lithography
C	$<10^{-2}$	D	electron-beam furnaces
C	$<10^5$	A, B, D	electron-beam welding
HC, UC	$<10^{-3}$	D	electron-diffraction cameras
HC, UC	$<10^{-3}$	D	electron and ion linear accelerators
C, HC, UC	$<10^{-2}$	D	electron microscopes
C	50	A	essential-oil distillation
CR, R, C	10^3	A	filtration
C	$<2 \times 10^{-2}$	D	fluorescent lights
R, C	2	A, B, D	freeze-drying foods and pharmaceuticals
C, HC, UC	$<10^{-2}$	A, B, D	fusing analysis
HC, UC	$<10^{-1}$	D	fusion power research
HC	$<5 \times 10^{-3}$	D	geiger-counter tubes
C, HC, UC	10^{-2}	D	helium-leak detectors
CR, R	20	A	impregnation of cables
CR, R	20	A	impregnation of capacitors
CR, R	50	A	impregnation of castings
CR, R	10	A	impregnation of wood
C	<1	B, D	incandescent lamps
all	$<10^{-1}$	B, D	induction melting
R, C	$<10^{-1}$	A, D	infrared spectrometers
C, HC, UC	$<10^{-4}$	D	isotope separators
C, HC, UC	$<10^{-2}$	D	mass spectrometers
HC, UC	2	A, B, D	mercury switches
C, HC	2	A, B	mercury thermometers
CR, R	10^{-1}	A, B	metalizing capacitor paper
all	$<10^{-2}$	D	metal evaporation
C, HC	$<10^{-1}$	B, D	metal sputtering (triode)
R, C, HC, UC	10^{-1}	A, B, D	molecular distillation
C, HC, UC	$<10^{-2}$	A, D	molecular, ion, and electron beams
CR	5×10^4	A	milking machines

751

Table 1. (Continued)

Vacuum environment category[a]	Gas pressure, ~Pa[b]	Pump[c]	Use
R, C	10^{-2}	D	neon signs
R, C	10^3	A	oil deodorizers
C	$<10^{-2}$	D	optics coating
CR	10^4	A	papermill equipment
CR, R	10^4	A	petroleum distillation
C, HC	10^{-4}	D	photoelectric cells
C, HC, UC	$<10^{-5}$	D	photomultiplier tubes
C, HC	2	B, D	radio-receiving tubes
C, HC, UC	$<10^{-3}$	D	ratio-transmitting tubes
C, HC	3	B, D	radiofrequency diode sputtering
R, C	3	A, B	refrigeration units
C, HC	1	B, D	relays
C, HC, UC	1	B, D	sealed resistors
C	5	B	serum ampules
all	$<10^{-1}$	D	sintering
R	10	A	solvent recovery
CR, R	5	B	smelting
all	all	all	space simulation
R, C	$<10^{-1}$	all	spectrophotometers
CR, R	2×10^3	A	steam-turbine exhaust
CR, R, C	50	A	steel degassing
HC, UC	$<10^{-8}$	D	storage-ring and colliding-beam machines
CR, R	10^4	A	sugar-evaporating pans
R, C	<50	A, cryo	supersonic wind tunnels
C, HC, UC	$<10^{-3}$	D	synchrotrons for electron, ion
R, C	2×10^{-3}	D	thermocouples
C	2×10^{-2}	D	thermos bottles
C	5×10^{-5}	D	television tubes
HC, UC	$<10^{-5}$	D	thin-film circuits
CR	50	A	transformer-oil drying
CR, R	$<10^{-2}$	D	ultracentrifuge
C, HC	$<10^{-2}$	D	ultraviolet instruments
C	2	B, D	vaccines
C, HC	<1	B, D	vapor lamps
CR	5×10^4	A	vehicular transportation, brakes, engines, etc
CR, R, C	$<10^4$	A, B, D	wind tunnels
HC, UC	$<10^{-4}$	D	x-ray tubes

[a]CR represents crude; R, rough; C, controlled; HC, highly controlled; UC, ultracontrolled.
[b]To convert Pa to torr, divide by 133.3.
[c]A represents mechanical pump or steam ejector; B, booster pump; D, cryo, turbomolecular, sorption, ion, or trapped diffusion pumps.

on the magnitude of pressure rather than on the parameters and their magnitudes that are essential to a given use. The degree of control of a vacuum environment demonstrated by the base pressure of the system is regarded as a balance between the rate at which molecules enter the gas phase and the rate at which molecules are pumped away. There is no explicit recognition of the types of molecules present, and the base pressure of a system at room temperature is not sufficient information for predicting the behavior of the system when a dynamic process or experiment is attempted in the chamber. Regarding vacuum as a molecular environment from the beginning can help to keep in view the importance of any possible consequence between the process and vacuum systems in part and/or as a whole.

Within vessels, vacuum environments comprise gaseous molecular phase(s) in contact but not necessarily in equilibrium with condensed molecular phases (2). The condensed phases consist of desirable, undesirable, and tolerable aggregations of molecules and structures such as grain boundaries, dislocations, vacancies, and defects. Typically, these are found in configurations that include undifferentiated bulk at interfacial, thin-film, and surface locations. The gaseous phase, especially under extreme dynamic conditions, may include neutral, excited, metastable, and electrically charged cluster, as well as particulate species that can consist of beneficial, tolerable, and deleterious molecular species, eg, contamination can be good or benign as well as bad.

Nonmolecular species, including radiant quanta, electrons, holes, and phonons, may interact with the molecular environment. In some cases, the electronic environment (3), in a film for example, may be improved by doping with impurities (4). Contamination by undesirable species must at the same time be limited. In general, depending primarily on temperature, molecular transport occurs in and between phases (5), but it is unlikely that the concentration ratios of molecular species is uniform from one phase to another or that, within one phase, all partial concentrations or their ratios are uniform. Molecular concentrations and species that are anathema in one application may be tolerable or even desirable in another. Toxic and other types of dangerous gases are handled or generated in vacuum systems. Safety procedures have been discussed (6,7).

Through its committees, divisions, and chapters, the American Vacuum Society has produced a nearly complete bibliography (to 1996) (8), a dictionary of terms (9), a monograph series, and a number of other useful publications (10). Another source of information is the Association of Vacuum Equipment Manufacturers. A history of vacuum ideas and technology development from the Middle Ages to Newton has been given (11).

Vacuum Dynamics

Units and Concentration. In the gaseous as well as the condensed phases, molecular concentration by molecular species is of prime importance. By convention, total pressure in a Maxwellian gas is used as though it indicates the quality of the vacuum and as though Maxwellian gases were the rule rather than the exception (12). In general, in dynamic systems, gas pressure (or its partial pressure components) is neither isotropic nor an adequate indicator of molecular significance.

For a Maxwellian gas in steady state, one standard atmosphere is defined as being equal to 101,323.2 N/m^2; 1 N/m^2 = 1 pascal (Pa) $\rightarrow$ 2.6 $\times$ 10^{20} molecules per cubic meter; 1 torr = 133.32 Pa $\cong$ 1 mm Hg = 1000 micrometers; 1 millibar

= 100 Pa; 1 in. Hg = 3386.33 Pa; and 1 lb/in.^{2} = 6895.3 Pa (see PRESSURE MEASUREMENT).

Condensed-Phase vs Surface-Phase Concentration. There are ca 4×10^{18} molecules/m^2 in a monolayer on the surface of a plane, depending on substrate and adsorbed species. At bulk impurities of 10^{-6} and a bulk thickness of 1 mm, ca 7×10^{19} molecules/m^2 can diffuse to the surface as a function of temperature and time. Thus, in a vessel of 1 m^3 at 100 μPa (7.5×10^{-7} mm Hg) and 300 K, the reservoir of molecules in the gas phase is 2.6×10^{16}, whereas surface and bulk are apt to hold ca 2×10^{19} and $>4 \times 10^{20}$ impurity molecules, respectively.

Interaction between Gaseous and Condensed Phases. In a closed vessel of volume V containing a nonionized, unexcited molecular gas having total number of molecules N, the change in the pressure P in the gas can often be predicted if the steady-state absolute temperature T is changed to another steady, constant level:

$$PV = NkT \tag{1}$$

where k is the Boltzmann constant, relating the steady-state absolute temperature T and the equilibrium pressure P in the gas.

However, it is not practical to set the gas temperature in steady state without equally setting the temperature of the surface and bulk phases bounding the gas. Consideration of the response of the system as a vacuum environment can then provide a sufficiently precise prediction of the pressure P and the surface coverage θ at temperature T for molecules of a known species in a known state on a known surface. For example, an isotherm is established between the surface of the condensed and the gaseous phases, depending, eg, on the heat of desorption Q. For submonolayer coverage on a known surface, the pressure P is likely to be an exponential function of T. Among several isotherms, the Temkin isotherm (13) may be used to predict a specific pressure:

$$P = C \cdot \exp -(Q_\theta/RT) \tag{2}$$

where C is a constant. Changing the temperature can follow or significantly depart from equation 1, the kinetic formula predicting the dependency of P on T.

An example would be a cubical chamber (1000 cm^3) at 298 K constructed of metal, ceramic, or glass that is evacuated (pumped down) and then filled with hydrogen to a pressure of ca 13 mPa (9.75×10^{-5} torr). On cooling to 77 K, the H$_2$ pressure drops to ca $77/298 \times 1.3 \times 10^{-2} = 3.4$ mPa (25.5×10^{-4} torr), in agreement with the kinetic theory. If a clean tungsten surface is deposited by evaporating tungsten, and then enough H$_2$ gas is admitted intermittently to give a nonvarying gas-phase pressure at 298 K of ca 13 mPa (9.75×10^{-5} torr), a concentration of $N_{H_2} = 3 \times 10^{15}$ is obtained. Then the tungsten surface (projected area only) would have absorbed ca 5.6×10^{18} hydrogen molecules/m^2, or $>3.4 \times 10^{17}$ molecules (100 times the total in the gas phase). If the chamber is again cooled to 77 K, the hydrogen partial pressure in the gas phase falls to <10 fPa ($<75 \times 10^{-18}$ torr), as predicted by the Temkin isotherm (13).

Kinetics Modified by Dynamic Interaction. The kinetic theory of gases is a valuable tool for vacuum technology. The unmodified kinetic theory must not be applied when the gas interacts significantly with itself or with the molecular phases that bound it. When interaction occurs, as it does for many molecular species in the systems considered here, the kinetic predictions must be modified by dynamic considerations. The condensed phase dominates the behavior of the gaseous phase in almost every respect under free molecular conditions. In general, measuring vacuum is not equivalent to measuring any single parameter (1).

Partial Concentration. The sum of the partial concentrations (pressures) in a free molecular gas is equal to the total concentration (pressure). However, all gaseous components, at the same partial pressure or absolute pressure or ratios thereof, are not likely to have the same significance to any or all vacuum applications. The significance of the condensed-phase concentrations must therefore be considered.

Essential Parameters. Traditionally, all vacuum environments are characterized in terms of one parameter, ie, pressure in the gaseous phase. However, when costs, energy, safety, hazardous wastes, and other requirements are taken into account, each system must be characterized by a host of parameters. Their magnitudes must be determined in order to judge system performance.

The role of a component as a function of position, use, history, and time may change. For example, a gas-pumping system is always a source of contamination, which may be negligible when the system is in good condition. The significance of a given component in a system at a given time must be taken into account. The following are examples of vacuum environments.

Electrical Breakdown. The electrical breakdown between parallel planar vacuum electrodes is seen to be a function of gas species and pressure (14,15). Markedly higher a-c and d-c voltages can be held off at gas pressures of 100 μPa to 1 Pa (7.5×10^{-7} to 7.5×10^{-3} torr) (Fig. 1). The composition of the surface molecular phase is a key factor.

Zinc Coating of Capacitors. In the zinc coating of paper strip for capacitors, the paper strip is fed from air through locks into a vacuum environment. There, it is coated by thermally evaporated zinc. The rate of evaporation is so high that contamination of the zinc vapor is excluded. The paper is fed at the maximum rate permitted by its own strength.

Electron Phenomena. Deleterious electron (r-f) phenomena are erased in electron linacs by an improved vacuum environment. Before the development of the 3000-m-long Stanford linear accelerator (SLAC), electron linacs were plagued by so-called multipactoring because secondary electrons were trapped in the r-f cavities of the accelerator when voltage was first applied. No beam could be accelerated through the linac until these cavities were conditioned. The vacuum achieved was thought not to contribute to multipactoring. The vacuum thought to be satisfactory for the accelerator was based on the prevention of beam interaction with the residual gas. This mean-free-path consideration is species-dependent, but by custom the vacuum was specified in terms of total pressure in the gas phase only. When the SLAC was planned, each klystron r-f power tube was to be open to the accelerator vacuum because reliable r-f windows transmitting the power required had not yet been developed. Thus, SLAC was built to have a vacuum good enough for a klystron tube. However, r-f windows

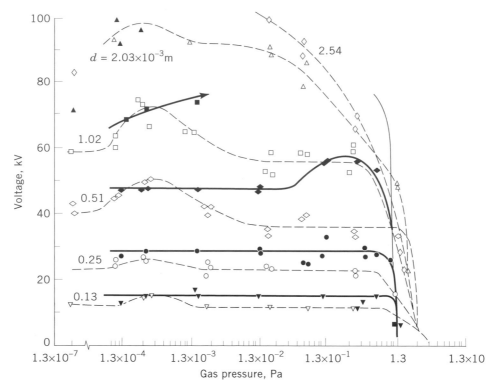

Fig. 1. Currents are d-c (dashed lines and empty symbols) and peak a-c (solid lines and solid symbols) breakdown voltage vs gas pressure in nickel for the various gaps, d, noted by values $\times 10^{-3}$ m on the curves (16). To convert Pa to torr, divide by 133.3.

were developed but the accelerator vacuum environment was kept clean as a klystron. When a beam was first attempted at SLAC, no multipactoring occurred, and the beam passed through from electron source to target the first time r-f voltage was applied.

Film Contamination from Bulk Phase. The contamination of an epitaxial film of GaAs from an oven charge can be corrected by doping (4).

Oil Contamination of Helium Gas. For more than 20 years, helium gas has been used in a variety of nuclear experiments to collect, carry, and concentrate fission-recoil fragments and other nuclear reaction products. Reaction products, often isotropically distributed, come to rest in helium at atmospheric concentration by collisional energy exchange. The helium is then allowed to flow through a capillary and then through a pinhole into a much higher vacuum. The helium thus collects, carries, and concentrates products that are much heavier than itself, electrically charged or neutral, onto a detector that may be a photographic emulsion. If the helium is contaminated with pump oil, the efficiency of delivery to the detector is markedly increased. Oil contamination, anathema in some systems, is desirable for this purpose.

Field Emission of Electrons. Nonthermionic emission from Spindt-type arrays of cold cathode tips is affected by gas-phase concentration and species (Fig. 2) (17). The time response of emission at constant voltage to changes, and

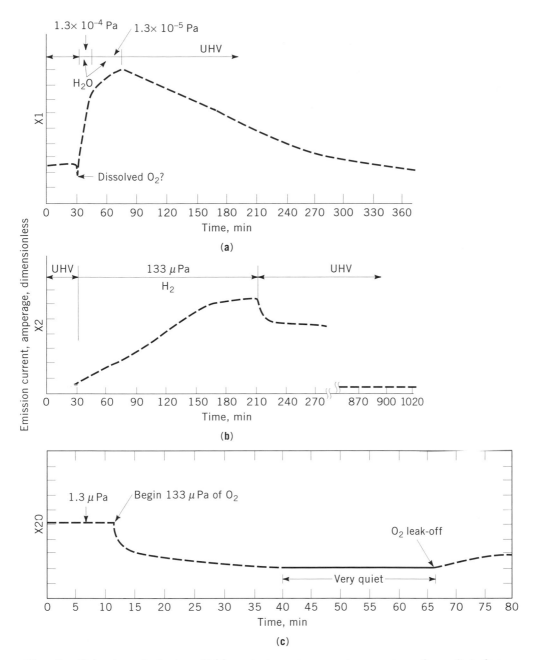

Fig. 2. Behavior of electron-field emission at room temperature from Spindt-type arrays of 5000 tips per mm^2, beginning and ending with ultrahigh vacuum (UHV), eg, ultracontrol (UC): (**a**) water; (**b**) hydrogen; and (**c**) oxygen, where the dashed line indicates noise. To convert Pa to torr, divide by 133.3.

reverse, from negligible to significant concentrations of water, hydrogen, and oxygen indicates that this emission may depend on diffusion rates into and out of the bulk phase in addition to cathode surface adsorption–desorption. Emission from Spindt-type tips was also increased by other active gases tested, eg, NH_3, CH_4, and H_2S (not O_2), but was unchanged by corresponding concentrations of He, Ne, or Ar. The bulk and surface phases of the anode receiving the electron emission must be degassed at ca 1200 K in order to avoid electrical breakdown.

Action of Vacuum on Spacecraft Materials.   For service beyond the atmosphere, the vacuum environment allows materials to evaporate or decompose under the action of various forces encountered (1,18,19). These forces include the photons from the sun, charged particles from solar wind, and dust. The action of space environment on materials and spacecraft can be simulated by a source–sink relationship in a vacuum environment. Thus, for example, the lifetime of a solar panel in space operation may be tested (see PHOTOVOLTAIC CELLS).

A vacuum system can be constructed that includes a solar panel, ie, a leak-tight, instrumented vessel having a hole through which a gas vacuum pump operates. An approximate steady-state base pressure is established without test parts. It is assumed that the vessel with the test parts can be pumped down to the base pressure. The chamber is said to have an altitude potential corresponding to the height from the surface of the earth where the gas concentration is estimated to have the same approximate value as the base pressure of the clean, dry, and empty vacuum vessel.

In general, the test object cannot be heated above its operating temperature in space. As free molecular conditions are obtained around the object, it outgases and, if solar-spectrum photons impinge on the object, increases the release of gas. Because the object is in a vessel and the area of the hole leading to the gas pump is small compared with the projected interior area of the vessel, molecules originating from the test object can return to the test object provided that they do not interact in some manner with the vessel walls and the other components of the molecular environment. The object inside the vessel establishes an entirely different system than the clean, dry, and empty vacuum vessel. The new system no longer has the capability to reach the clean, dry, and empty base pressure within a reasonable time.

This simulation can be achieved in terms of a source–sink relationship. Rather than use the gas concentration around the test object as a target parameter, the test object can be surrounded by a sink of ca 2-π solid angle. The solar panel is then maintained at its maximum operating temperature and irradiated by appropriate fluxes, such as those of photons. Molecules leaving the solar panel strike the sink and are not likely to come back to the panel. If some molecules return to the panel, proper instrumentation can determine this return as well as their departure rates from the panel as a function of location. The system may be considered in terms of sets of probabilities associated with rates of change on surfaces and in bulk materials.

Electronic Vacuum Tube.   In special electronic vacuum diode tubes, with spacing between the cathode and anode of 10 μm, high gas concentrations of some types are beneficial to the operation of the tube under proper control.

Pump Down

Many problems encountered in producing a highly controlled vacuum result from the system's design, history, contents, use, and maintenance.

Initially, the vessel is filled with ambient air. Any given macrosample of air may contain at least 1600 substances that have been identified (20). Among these are the gases and vapors listed in Table 2. Most important is usually water; others include viable and nonviable particulates (21), aerosols (21), and cluster species of molecules that can be in states of excitation or ionization. Aerosols originate from both anthropomorphic and natural sources; in urban air, the former far outweigh the latter. Typical contaminants include cigarette smoke, lead (qv), and asbestos (qv). Gasoline and diesel fuel contribute significantly to contamination. Effluents and exudates are present, including skin particles, hair, and innumerable other substances from human sources (see AIR POLLUTION).

Microstructure on Surfaces.   Gross cracks and voids are usually lined with microstructure, as indicated by Figure 3. As the depth–width, D/W, ratio of a crack is held constant but the dimensions approach molecular dimensions, the crack becomes more retentive. At room temperature, gaseous molecules can enter such a crack directly and by two-dimensional diffusion processes. The amount of work necessary to remove completely the water from the pores of an artificial zeolite can be as high as 400 kJ/mol (95.6 kcal/mol). The reason is that the water molecule can make up to six H-bond attachments to the walls

Table 2. Typical Average Diurnal Concentrations of Molecular Species in Nonpolluted Ambient Air[a]

Species	Concentration, molecules/m^3
More chemically reactive and surface adsorbing	
O_2	5×10^{24}
H_2O	ca 2.5×10^{23}
O_3	ca 1×10^{18}
NO	ca 1.5×10^{17}
HO_2	ca 6.5×10^{14}
HO	ca 4×10^{11}
Cl	ca 2×10^{10}
N_2O	ca 5×10^{19}
H_2	ca 5×10^{19}
Less chemically reactive and surface adsorbing	
N_2	2.1×10^{25}
CO_2	8.9×10^{21}
CH_4	ca 5×10^{19}
Chemically inert but slightly adsorbing	
Ar	2.5×10^{23}
Ne	4.9×10^{20}
He	1.4×10^{20}
Kr	3.0×10^{19}
Xe	2.3×10^{18}
Ra	2×10^4

[a] Ref. 20.

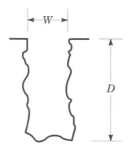

Fig. 3.   Crack of width W and depth D; ratio D/W = constant (22).

of a pore when the pore size is only slightly larger. In comparison, the heat of vaporization of bulk water is 42 kJ/mol (10 kcal/mol), and the heat of desorption of submonolayer water molecules on a plane, solid substrate is up to 59 kJ/mol (14.1 kcal/mol). The heat of desorption appears as a exponential in the equation correlating desorption rate and temperature (see MOLECULAR SIEVES).

Turbulent Gas Flow (Rough Pumping).   An oil-sealed mechanical pump in good condition, having vented or trapped exhaust, is gas purged by running for several hours. A liquid-nitrogen (LN) trap is between pump and vessel. It can consist of a U-shaped tube of thin-walled stainless steel clad externally with heavy-wall copper along its vertical legs (23). In each riser leg of the U-trap, a twisted piece of copper of width d, ie, the ID of tube, is inserted to ensure that under contaminated flow conditions, oil cannot pass without encountering a liquid-nitrogen-cooled surface. An in-line, hot (450 K), all-metal valve on the system side of the trap is connected in the line above the copper-clad leg by a 0.02-m-long stainless-steel neck (2.5×10^{-4}-m wall thickness). Thus, the rate of boil-off of the liquid nitrogen from the trap is kept reasonable. The trap can be filled automatically from a local reservoir, from a built-in LN supply line, or by hand. By keeping the all-metal valve always at 450 K, it is possible to close it, allow the trap to warm up when needed, refill the trap, and reopen the valve. This arrangement provides satisfactory control of contamination from an oil-sealed mechanical pump (23).

Using absorbent material is time-consuming and expensive, and can contribute minute, solid pieces of the sorbent into the system. Metallic bonded, high surface area materials can be used instead.

With valves open, the air in the vessel is exhausted through the U-trap by the oil-sealed pump. As the pressure falls, the composition of the gas in the vessel begins to change. At a pressure of ca 13 Pa (0.097 mm Hg), water is the dominant species in the gas phase. The surface phases then change appreciably, although initially water was the dominant species on the surfaces. The bulk phase is unlikely to contain any water molecules as such, except in voids and gross defects. Water is desorbed from glass as a result of OH radicals changing to H_2O at the surface.

Diffusion Pump System.   After the pump line and trap have been shut off, a large valve is opened slowly enough that the mass flow of gas from the chamber through the valve into the oil-diffusion pump system does not disrupt the top jet of the diffusion pump (DP) (Fig. 4). When the liquid nitrogen is replenished

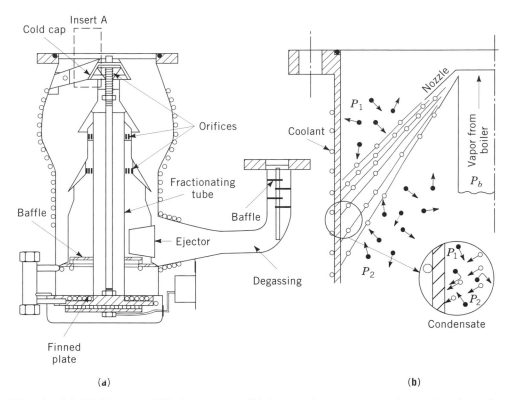

Fig. 4. (a) Multistage diffusion pump; (b) insert A, jet spray, where closed circles indicate gas molecules P_1 and open circles indicate vapor-jet molecules P_2. P_b is boiler pressure (23).

after the trap has been operated for some time, release of previously trapped gas must be avoided. The so-called ionization-gauge response pips at the start of the liquid-nitrogen replenishment are an indication of trap ineffectiveness.

The trap fill line is separated from the high vacuum region by an overhanging copper skirt, ie, the creep barrier, which also serves to keep the interior surface exposed to the working environment at a temperature independent of the liquid level in the trap. Oil creepage in two dimensions along surfaces is effectively inhibited at <200 K. Contribution to creepage by the liquid-nitrogen trap is usually small compared with the contamination delivered during the filling of the trap and reduction of LN level.

DP Speed Factor. Pumping-speed efficiency depends on trap, valve, and system design. For gases having velocities close to the molecular velocity of the DP top jet, system-area utilization factors of 0.24 are the maximum that can be anticipated: eg, less than one quarter of the molecules entering the system can be pumped away where the entrance area is the same as the cross-sectional area above the top jet (see Fig. 4). The system speed factor can be quoted together with the rate of contamination from the pump set. Utilization factors of <0.1 for N_2 are common.

The rate of contamination from the pump set is $<10^9$ molecule/(m²·s) for molecular weights >44 (23). This is the maximum contamination rate for routine service for a well-designed system that is used constantly and subject to automatic liquid-nitrogen filling and routine maintenance.

A fraction of gas, depending on species, is pumped by the diffusion pump or trapped on the cold surfaces of the LN trap; for example, 0.05 for H_2 and 0.9 for H_2O, respectively, have been measured. With enough flow, multilayers of gas can build up on the liquid-nitrogen-cooled surface, thereby evaporating at the bulk vapor pressure of the gas at LN temperature. To cope with such buildups, a well-designed valve is required immediately between the system and the LN-cooled surface. Designs incorporating this valve within the LN trap itself are available. The valve can be closed and the trap warmed sufficiently to purge accumulated gases such as carbon dioxide. Furthermore, the trap can be allowed to warm to room temperature without delivering contamination into crannies and surfaces of the valve exposed to the chamber when the valve is open.

Leaks

A vacuum system can be stalled by gas leaks (4,6,24). Traditionally, leaks are categorized as real or virtual. A real leak refers to permeation processes or cracks or holes that allow external gas (air) to seep into the vacuum environment. Atmospheric gases such as helium and hydrogen permeate glass equipment, especially at elevated temperature. The noble gases do not permeate metals, but hydrogen does. Virtual leaks refer to gases that originate from within, eg, from trapped volumes, the gauges, pumps and the bulk and surface-phase species. For example, carbon in bulk stainless steel may precipitate along grain boundaries and then combine with surface oxygen to give CO, which is then desorbed into the gas phase (25). Proper instruments readily distinguish real leaks from virtual leaks.

In practice, it is often necessary to take readings from hot-filament ionization gauges or other devices. Figure 5 gives pump-down curves for six different types of pumping equipment on the same vacuum chamber (23). The shape of curve 1 indicates that a real leak could be responsible for the zero slope demonstrated by the Bayard-Alpert gauge (BAG). The shape of the other curves could result from a combination of real and virtual leaks.

In fact, the leveling-off slope of curve 1 was entirely owing to gas issuing from the pumping equipment itself, and there were no other sources of leakage. Curves 2–6 all resulted from combinations of virtual and real leakage. Most of the leakage in curves 2–4 originated in the pumping equipment. This was also true for the early part of curve 5. The phenomenology of the early stage of curve 6 resulted from chamber-wall outgassing. The latter stages of curve 6 show a combination of wall outgassing and a small leakage from the pump itself.

Figure 6 is instructive regarding the magnitude of leakage from pump 5, shown in Figure 5, after ca 5 h of pumping. Pump 5 was a commercial Orbitron made by the National Research Corporation. After ca 4 h from the start of the pump down, the electric power to the pump was turned off, whereas the power to the BAG remained on. The total pressure in the system rose very rapidly, and this response is recorded in Figure 6. Within a short time, however, this rise reached a maximum and assumed a small slope downward. After ca 200 s, the reading on the BAG was close to the reading achieved before the Orbitron power

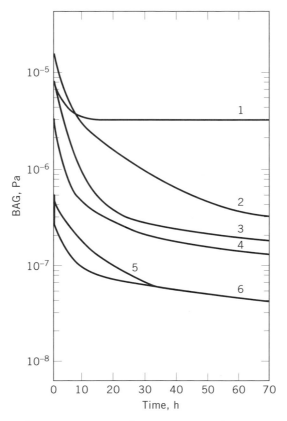

Fig. 5. Plots of pump-down performance for pumps operating on 0.1-m-dia × 0.43-m-long stainless-steel tubing. Curves 1–4 are sputter-ion pumps of different makes; curve 5 is Orbitron type; and curve 6, LN-trapped oil DP. Pressure is measured using Bayard-Alpert gauge (BAG) (26). To convert Pa to torr, divide by 133.3.

had been turned off. It remained essentially constant for two months after the initial pump down.

Figure 6 can be interpreted as follows. The Orbitron pump is delivering a virtual leak back into the vacuum environment. As with many other pumps, this quantity of gas is significant. The operational speed of the Orbitron just before the electric power was turned off was zero. Thus, an equilibrium was established between the source–sink properties of the vacuum environment. When the Orbitron power is turned off, a slug of electron-bombarded, evaporating titanium remains hot for some time and continues to outgas. As the slug cools down sufficiently, however, the BAG intrinsic speed is sufficient to balance the decreasing rate of gas evolved. The intrinsic steady speed for the BAG is known to be ca 100 cm^3/s. Thus, after ca 4 min from the time the electric power was switched off to the Orbitron, the operational speed of the system is provided by the ionization gauge and the sorption properties of the system. The ionization gauge is thus able to provide almost the same equilibrium base pressure provided by pump 5, which has a rated speed nearly 4000 times greater than that of the BAG. The significance of equilibrium base pressure is use-dependent. The

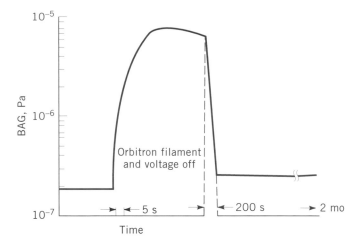

Fig. 6. Bayard-Alpert gauge response vs time upon cutting off power to Orbitron pump (27). To convert Pa to torr, divide by 133.3.

category virtual leak must include evolution from the pumping system and the condensed phases, as well as from atmospheric air and gas trapped in cracks and voids. The question of virtual leak vs real leak may be important in every system.

Molecular Transport

Molecular transport concerns the mass motion of molecules in condensed and gaseous phases. The mass motions are driven primarily by temperature. As time progresses, the initial mass motion results in concentration gradients. In the condensed phase, flow along concentration gradients is described by Fick's law.

Standard texts may be consulted on the topic of diffusion in solids (6,12,13). Some generalizations, however, are possible. No noble gas permeates a metal. Metals are, however, permeated readily by hydrogen. Stainless steel, for example, can be permeated by hydrogen from concentrations likely in air. The least permeable material for hydrogen is carbon. Glasses are permeable, especially by the light noble gases at elevated temperatures.

After a bake-out of 600–700 K, the bulk phase is likely to far exceed the surface phase as a source of atomic (molecular) impurities that desorb into the gas phase (28). Bake-out at 1300 K greatly reduces bulk-phase impurities.

Gas Transport. Initially, in a vessel containing air at atmospheric pressure, mass motion takes place when temperature differences exist and especially when a valve is opened to a gas pump. Initial flow in practical systems has been discussed (29), as have Monte Carlo methods to treat shockwave, turbulent, and viscous flow phenomena under transient and steady-state conditions (5).

Viscous Transport. Low velocity viscous laminar flow in gas pipes is commonplace. Practical gas flow can be based on pressure drops of <50% for low velocity laminar flow in pipes whose length-to-diameter ratio may be as high as several thousand. Under laminar flow, bends and fittings add to the frictional loss, as do abrupt transitions.

Free Molecular Transport. The free molecular gas regime is illustrated by Figure 7. A duct of maximum transverse dimension D and length L connects two chambers, each of minimum interior dimension $\gg D$. Free molecular transport (Knudsen flow) is often sufficiently approximate when $\lambda \geq D$. For a right circular cylinder of length $L = D$, the diameter, the internal pressure drop in free molecular flow is ca 50% (Figs. 8–10). In free molecular flow at steady state (Fig. 7), the temperature of the gas entering a duct determines the rate of passage through the duct, not the temperature (other than zero) of the duct itself. Volumetrically, Knudsen flow is proportional to gas entering velocity only; it is constant, independent of gas concentration or gradient. Free molecular flow can be described in terms of a statistically valid number of molecules interacting with the surfaces of a duct, provided that the entering spatial and ongoing reflection distributions from the walls by the molecules are known.

If a Maxwellian gas at steady state is entering end 1 in Figure 8, then the duct conductance, C, is

$$C = 1/4 \ (\overline{v} A_1 W_{1 \to 2}) \tag{3}$$

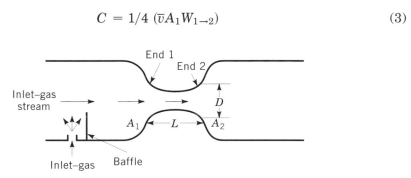

Fig. 7. Duct connecting two volumes of dimension $\gg D$, where A_1 is the area of end 1; A_2, that of end 2.

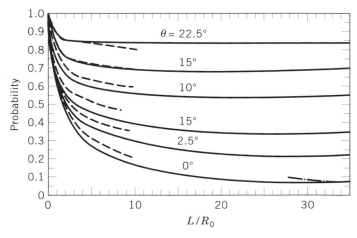

Fig. 8. Conical ducts having length L, small end of radius, R_0, and half angle θ (30). (———) represents Monte Carlo method (29,31); (-··-··-), long-tube asymptote. The divergence of solid and dashed lines results from an error in maintaining the sample size during the Monte Carlo method (30,31).

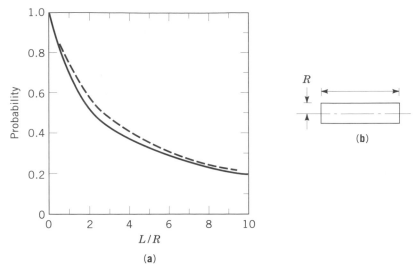

Fig. 9. Molecular-transmission probability: (**a**) for a cylindrical tube; and (**b**) as a function of the ratio of length to radius, L/R. The conductance is $C = 1/4 \ (\pi R^2 \bar{v} W_{1 \to 2})$. The solid curve is from Clausing's calculation; the dashed curve corresponds to the approximation $W = 1/(1 + (3L/8R))$ (32).

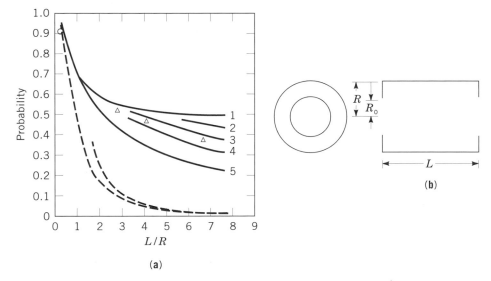

Fig. 10. Molecular-transmission probability (**a**) for circular cylinder (**b**) having two restricted ends where the solid line represents fraction transmitted without wall encounter; and the dashed line, prediction of this fraction by the formula R_c^2/L^2 vs L/R_0. $(R/R_0)^2$: 1, ∞; 2, 3; 3, 2; 4, 1.5; 5, 1.0. ($\bigcirc$), Argon, $(R/R_0)^2 = 2$; and ($\triangle$), nitrogen, $(R/R_0)^2 = 2$ (33).

where $\bar{v}$ is the average velocity of the Knudsen Maxwellian gas entering end 1 (before wall encounter); A_1 is the area of end 1; and the transmission probability of a Maxwellian gas (typically reflecting from the walls by a cosine distribution) is $W_{1 \to 2}$ = number in end 1 per number out through end 2.

Under isothermal conditions where energy is not added or removed from the system, the second law of thermodynamics obtains, and

$$A_1 W_{1\to 2} = A_2 W_{2\to 1} \tag{4}$$

Thus, the probability of transmission must be directional for $A_1 \neq A_2$, but the conductance cannot be directional (see THERMODYNAMICS). If C is directional, energy must be supplied from external sources and C is a pumping speed. Thus, work is performed by pumping but not by conductance. Under free molecular flow, the volumetric rates of transport in the gas phase are independent of the pumps being on or off; bends in a duct little alter the probability of passage over a straight duct having the same axial length. The intuitive idea that the walls of transition sections must be fared in (smooth flow) does not apply in free molecular flow. For example, conical transitions (see Fig. 9) (30) and the area above the top jet in Figure 4 always have smaller transmission probabilities than right circular cylinders having one restricted end where diameters and length are the same as in a given conic frustum. A right circular cylinder without a restricted end has a free molecular conductance that is greater than that of the conic frustum. However, there can be other reasons than conductance for using conical transitions in a given instance.

Wall Geometries. Rougher-than-rough wall geometries can reduce transmission probabilities in Knudsen flow by as much as 25% compared to the so-called rough-wall cosine reflection (34,35). For this and other reasons, conductance calculations that claim accuracy beyond a few percent may not be realistic.

In free molecular flow, if gaseous conductance were not independent of the flow direction, a perpetual-motion machine could be constructed by connecting two large volumes by a pair of identical ducts having a turbine in front of one of the ducts. A duct that has asymmetrically shaped grooves on its wall surface could alter the probability of molecular passage in such a way that for a tube of equal entrance and exit areas, the probability of passage would be made directional.

On purely kinetic grounds, however, the term random must be used carefully in describing a Maxwellian gas. The probability of a Maxwellian gas entering a duct is not a random function. This probability is proportional to the cosine of the angle between the molecular trajectory and the normal to the entrance plane of the duct. The latter assumption is consistent with the second law of thermodynamics, whereas assuming a random distribution entry is not.

The probability of passage is independent of the entrance velocity of free molecules and the subsequent velocity ($v \neq 0$) of these molecules within the tube, but depends on the entering angular and wall-reflection distributions of the molecules. It is difficult to predict the distribution functions of molecules reflecting from single-crystal surfaces (31,34–36) and thus also from engineering surfaces. For engineering surfaces and gases at room temperature, reasonable results within ± 10 percent are obtained by assuming that a statistical number of molecules impinging on a surface exhibits a cosine distribution upon reflection from the surface. Some engineering surfaces may be classed as rougher than rough, however, and the distribution of reflected molecules from an element of surface shows a maximum near the normal to the wall of the tube (34) rather than the spherical shape predicted by the cosine law. Thus, cooling or heating a structure such as a trap does not alter the probability of molecular passage through the structure. This statement assumes that the spatial distribution of molecules reflecting from the walls of the trap is not a function of temperature and that a Maxwellian gas in steady state is entering the trap. In actuality, arranging constant temperature differences in steady-state flow can work on a free molecular gas and cause a pumping effect. This pumping effect is distinguished from gaseous conductance because net work is performed on the free molecular gas. In other words, in special cases, series arrays of ducts can be caused to pump when temperature differences are maintained in the arrays. These constant temperature differences are maintained because energy is supplied to the system from external sources to maintain the temperature differences. This energy from the external sources keeps the system operating at steady state at an end-to-end pressure (concentration) difference.

In free molecular flow, all tubes of a geometrically similar shape have the same free molecular probability of passage for the same entering gas distribution. When dealing with Maxwellian gases, therefore, the probability of passage through a duct can be plotted and this probability can be used to calculate the conductance for all pipes of that shape (Figs. 8–12). Under free molecular flow, bends in the tube of equal entrance and exit areas little alter the probability of molecular passage compared to a straight tube having the same axial length. In a few cases, the probability of passage through a duct can be obtained by

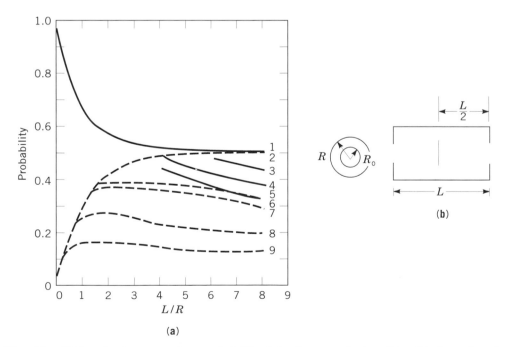

Fig. 11. Molecular-transmission probability (**a**) of a round pipe (**b**) with (———) and without (– – –) blocking plate and restricted entrance and exit apertures. $(R/R_0)^2$: 1, ∞; 2, ∞; 3, 3; 4, 2; 5, 2.25; 6, 1.5; 7, 2.0; 8, 1.5; and 9, 1.25 (37).

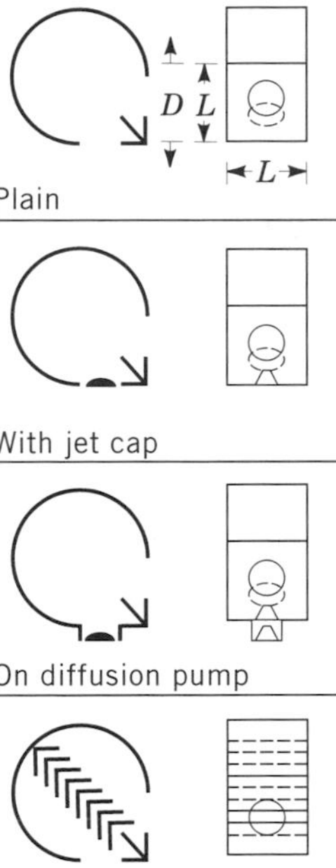

Bulged elbow	L/D	Probability, experimental
Plain	2.00	0.44
	1.33	0.39
	1.00	0.32
With jet cap	2.00	0.33
	1.33	0.30
On diffusion pump	2.00	0.32
	1.33	0.27
With Chevron (L/D = 5)	2.00	0.38
	1.66	0.35
	1.33	0.31

Fig. 12. Transmission probability for bulged-elbow geometries (31).

inspection. For example, through a thin orifice, the probability is 1; through a large box that has a pinhole entrance and a pinhole exit, it is 0.5. If a small plate is placed in the box and obscures the line of sight between the pinholes, the probability of passage is still 0.5.

Combining Conductances. Combining short conductances may be difficult because, if a free molecular gas that is Maxwellian in steady state enters conductance 1 (length ≠ 0), the gaseous distribution is no longer Maxwellian at exit 1. This corresponds to the so-called beaming effect. The overall conductance can be estimated if the probabilities of passage of the individual components are known and if the juxtaposed components do not vary more than about a factor of two in cross-sectional areas (6,33). In general, exact methods of calculating conductance depend on knowing the spatial distribution of molecules encountering and reflecting from real surfaces.

Pumping Speed. If the standard formulas for gas flow in vacuum are applied, eg, pumping speed $S = Q/P$, where Q = mass of gas transported and P = pressure, it is assumed that a Maxwellian free molecular gas is entering the pump. The pump must be used on a chamber having a cross-sectional area much larger than that of the cross-sectional entrance of the pump. The gas concentration is then sufficiently uniform in the chamber and across the entrance to the pump. As the chamber is made smaller, finally coinciding with the pumps' entrance area, the gaseous distribution entering the pump is increasingly non-Maxwellian. This does not present a problem, however, provided that this fact is taken into account when assessing the system. For example, the behavior of a free molecular gas from point to point can be predicted mathematically, whether it is Maxwellian or not (5). The calculations can be augmented by experimental measurement. When using only experimental measurements, the data are treated like those obtained with Maxwellian gases. Calculating the probability of passage and pumping speed for more complex shapes may require high speed computers, but once a code is obtained, only a limited amount of machine time may be needed (5). Similar codes have been worked out for neutron and radiation problems.

Instead of calculations, practical work can be done with scale models (33). In any case, calculations should be checked wherever possible by experimental methods. Using a Monte Carlo method, for example, on a shape that was not measured experimentally, the sample size in the computation was allowed to degrade in such a way that the results of the computation were inaccurate (see Fig. 8) (30,31). Reversing the computation or augmenting the sample size as the calculation proceeds can reveal or eliminate this source of error.

System Pumping Speed.   The operational speed of the pump is a systems effect. All current pumps perform more than one function in the system. Pumps are sinks and at the same time sources for molecules. Pumping speed relates to work on the gas phase. Each type of pumping system occupies a projected area on the vacuum vessel wall. This area may be a hole in the wall, a surface that getters gas, or a cold surface that condenses gas. In the case of cryogenic or gettering surfaces, there can be some advantage in making the actual area of the surface larger than its projected area. If the primary interest is to increase the pumping speed of the system as distinct from the molecular handling capacity, the speed can be increased for molecular species having sticking coefficients of <0.3.

Increasing the specific area of a solid surface that retains the molecules that strike it impedes a free molecular gas in reaching more deeply into the extended surface area. This conductance effect is geometrical and thus nearly independent of the dimensions of the increased area. However, depending on temperatures and species, when molecular-size pores such as encountered in sorbent material are used, molecules that arrive at the pores initially can plug the pores and prevent subsequent molecules from entering. Furthermore, depending on the temperature of the surface and the molecular species, some two-dimensional adjustment can take place where molecules can penetrate into higher specific-area regions by two-dimensional flow along surfaces. The system pumping speed for holes cut in the side of a vacuum system can be evaluated in terms of efficiency and cost for this pumping. For Maxwellian gases in free molecular flow, the chance of molecules being pumped through a hole is independent of the concentration. This pumping probability has been incorrectly called the Ho coefficient for the system (6). The probability of a molecule exiting from a system through a hole and not returning is equivalent to the Clausing factor for Maxwellian free molecular gases. Pumping systems, whether based on turbomolecular, ionic, getter, cryogenic, or diffusion-pump principles, have yet been unable to utilize holes fully and be of reasonable size and cost. If the area of a hole in the vessel is equal to the cross-sectional entrance area of a pump operating on that hole for the light gases hydrogen and helium, the system pumping probability is ca 1:10. Thus, theoretically, another pump could be placed on the same hole and deliver

10 times the pumping speed available. This inability to utilize holes and still keep the cost of a pump within reasonable bounds stems from several difficulties.

Diffusion pumps, whether of oil or mercury, require a trap to prevent the working fluid of the pump from contaminating the vacuum environment. Liquid-nitrogen traps are usually employed. Their temperature, however, is not low enough to condense any hydrogen or helium gas even on sorbents of large specific area. In addition to the trap, a valve is often essential in the system. Analytical estimates of the maximum system probability realizable for a diffusion pump showed it to be less than 3:10 for molecular species having velocities near to or less than the molecular velocity of the top jet of the diffusion pump (Fig. 4). It thus appears impractical to increase the molecular velocity of the top jet. The pumping of light gases such as hydrogen and helium is fundamentally limited by the velocity of working fluid molecules in the top jet. From a user's point of view, there should be a sizeable market for pumping equipment that could utilize the potential of wall area better without increasing the system cost.

Turbomolecular pumps operating on holes are faced with the same fundamental problem as the diffusion pump. The tip speed of the turbine rotor in commercial pumps, because of the strength-to-density ratio of the turbine blade material, is limited to $<6 \times 10^2$ m/s. The average velocity for helium at room temperature is ca 1×10^3 m/s; for hydrogen, 2×10^3 m/s. Patents have been filed related to a novel vacuum pump (38). Magnetically suspended self-balancing rotating arrays of fibers permitting tip speeds of ca 3×10^3 m/s are self-cutting and of low mass (light weight).

Instrumental Measurement

Especially important under dynamic conditions, the role of a system and of each component can be disclosed by appropriate measurements. Thus, it can be established when the system environment is ready for a dynamic use, eg, if the pump is likely to perform as a molecular sink, a source, or some combination of these.

Gauges. Because there is no way to measure and/or distinguish molecular vacuum environment except in terms of its use, readings related to gas-phase concentration are provided by diaphragm, McCleod, thermocouple, Pirani gauges, and hot and cold cathode ionization gauges (manometers).

Ionization gauges (IG) do not give pressure in the gaseous phase but are set to provide readings proportional to the concentration of molecules in the gaseous and, to a lesser extent, the condensed phases. These concentrations are translated to units of gaseous pressure. The hot and cold cathode ionization gauges, such as those shown in Figures 13 and 14, provide information about vacuum environment as a host of parameters. Perhaps largely by trial and error, selection of a gauge readout indicates when to begin the process or experiment. Turning the gauge off and on provides the so-called flash-filament gauge response. When the filament is turned off, it cools, and molecules from the gaseous environment impinge and stick to the surface of the cooling filament. When the filament is turned on again and its temperature rises to incandescence, these molecules are desorbed. Sufficient electrons are provided by the filament to indicate the gas-phase concentration increase from this desorbed material. An abruptly rising pip

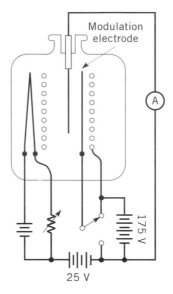

Fig. 13.   The introduction of a modulator electrode in a Bayard-Alpert gauge permits the pressure range to be lowered by a factor of 10^{-1} to 133 pPa (10^{-12} torr) (39).

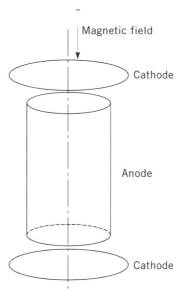

Fig. 14.   Geometry of a single cell of a penning gauge. Space charge of the trapped, circulating electrons equalizes the axis potential with that of the cathode. Thus, the electric field is radial. Electron density is at a maximum a short distance from the anode. Electrons progress radially toward the anode only as they lose kinetic energy, mainly through inelastic (ionizing) collisions with molecules (40).

having a longer decaying tail can be recorded from the output of the gauge. The area under this desorption peak is a measure of the integrated pumping effect of the cold bare filament. The filament is bombarded by the free molecular gas over a ca 2-π solid angle. The flux of particles striking the filament is given by

$n \cdot \bar{v}$, where n is the molecular concentration in the gas phase and $\bar{v}$ is the average velocity of this Maxwellian gas. By changing the power supplied to the filament in its heating phase and varying the time interval when the filament is left cold, approximate but useful information can be obtained (13).

Residual Gas Analyzers. A gaseous molecular phase is analyzed using a mass spectrometer (41) (see ANALYTICAL METHODS). The surface and bulk phases of the vacuum environment can be probed for appropriate uses. Desorbed species, neutrals, and metastable, positive, and negative ions can all be read by the residual gas analyzer (RGA) through its ion source and/or by its electron multiplier (13). If heat is delivered to the condensed phase or if electrons are caused to strike its surface, molecular desorption provides a signal for the mass spectrometer analysis. Direct electron desorption does not involve heating. For contaminated surfaces, the probability of a molecule being desorbed per incident electron can be as high as 10^{-3}. For submonolayer coverage of some species, this probability may decrease to $< 10^{-7}$. These probabilities are large enough to allow a significant signal from the mass spectrometer. The electron current delivered to the surface can be raised to deliver intense heat, providing information about the bulk phase near the surface. Photons desorb molecules directly from surfaces (except for water vapor).

Ultrasound frequencies can be introduced into the walls of the vacuum system. If a source of ultrasound is placed on the wall of an ultrahigh vacuum system, a large hydrogen peak is observed (see ULTRASONICS). Related phenomena, presumably from frictional effects, are observed if the side of a vacuum system is tapped with a hammer; a desorption peak can be seen. Mechanical scraping of one part on another also produces desorption.

Vacuum Systems and Equipment

Wall Materials. Glass (qv) is often the material of choice for small laboratory systems and sealed systems in commercial practice. Glass has a wide range of useful properties, including high compressive strength but relatively low tensile strength, requiring careful selection of the glass and design. Evacuated glass tubes, such as photomultiplier tubes, are exposed to temperatures as high as 720 K. In some applications, glass has been displaced by high alumina ceramic. Borosilicate glasses, in the form of industrial glass pipe, have been used in large experimental systems that frequently require demountable joints.

Industrial glass pipe and other wall materials are connected to other materials that have demountable seals made of elastomers or soft metal wire, eg, lead, indium, tin and their alloys. The soft-wire seals, however, have led to difficulties. High temperature bake-out must be avoided, and glow-discharge cleaning must not reach the gasket. Furthermore, the gasket material can creep under load and admit a leak, except in the case of indium, which creeps but does not usually leak (see also PACKING MATERIALS).

The choice of metals for vacuum walls is largely based on the ease of fabrication of the metal, machining, cleaning (26), welding, etc. Aluminum alloys are the material of choice for out-gassing at room temperature.

Demountable joints are commercially available in great variety in stainless steel, but less so in aluminum alloy or related materials. Experimental joints have been made in which aluminum flanges employ aluminum foil as gasket material.

Vacuum Brazing Furnace.   The cross sections of a vacuum brazing furnace of the bell-jar type are shown in Figures 15 and 16. Contamination is very low even when rapidly cycled from one work load to the next. The bell jar can serve on one hearth while another hearth is being loaded for the next operation.

Rapid cool-down by helium heat transfer is made possible at an interior of ca 100 Pa (0.1 atm). A convective fan transfers heat efficiently from the interior hot surfaces of the furnace to water-cooled base and wall parts.

During startup, the baseplate heat shields are lowered, thereby allowing pump-down through a large gap. The heat shields can then be raised to produce a narrow gap when maximum temperature is required.

Offset design gives ready access along the axis of the hot zone. This design permits routine operation and cycling of the furnace without sacrificing control of contamination, access, and speed for condensables or noncondensables.

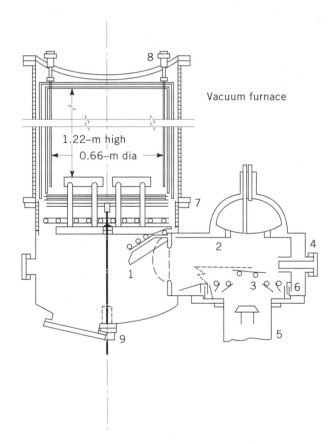

Fig. 15.   A low contamination vacuum brazing furnace for brazing large metal-to-ceramic joints: 1, valve, open (—) and shut (–––); 2, LN trap; 3, water-cooled baffle; 4, valve actuator feedthroughs for lowering the bottom heat shields; 5, diffusion pump oil, 1.25-cm dia; 6, creep barrier; 7, flange separation for opening the furnace; 8, electrical feedthroughs for lowering the bottom heat shields; 9, seal for raising and lowering of the furnace (42).

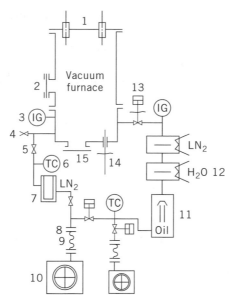

Fig. 16. Schematic representation of the vacuum furnace shown in Figure 15, where 1 is electrical feedthrough; 2, viewport; 3, vacuum-gauge hot-filament ionization; 4, air valve; 5, valve; 6, thermocouple vacuum gauge; 7, thimble trap; 8, demountable coupling; 9, flexible line; 10, two-stage liquid-sealed mechanical pump; 11, DP; 12, water-cooled baffle; 13, pneumatic valve having sealed bellows; 14, linear motion feedthrough having sealed bellows; and 15, blind flange port (43). The graphic symbols used are among those contained in the American Vacuum Society Standard 7.1. IG = ionization gauge.

An optimized relationship is obtained between the bell jar, 60° swing-leaf valve, LN trap, baffle for the oil, and the plane of action for the diffusion pump (DP) top jet. The valve open area equals 0.38 of the cross-sectional area of the inside diameter of the furnace. The volumetric speed factor for water vapor is thus $0.38 \times 0.9 \simeq 0.34$, where 0.9 is the Clausing factor.

No gas pips occur during the filling of the LN trap, and the temperature of the trap surfaces does not vary more than 0.01 K with the liquid-nitrogen level variation in the reservoir. The high conductance water-cooled baffle provides a minimum restriction of flow to the DP, yet returns back-streamed pump oil along the walls of the pump attached at its room temperature edge only. A thin, stainless-steel, anticreep barrier, cooled by radiation to the LN-chilled annulus, prevents surface migration of pump oil to the furnace. The bottom heat shields may be lowered by a rod during the warming of the furnace to provide high conductance from the hot zone to the trap inlet. A hot-zone temperature of >1500 K has been achieved.

Liquid-Nitrogen Traps. The principal reason that cold traps are frequently ineffective in preventing the passage of oil or mercury is the warming of the trap and its internal filling lines when LN is added and/or as LN depletes. However, some designs have eliminated this problem (Fig. 15). A liquid-nitrogen trap need not have an active chemical surface, but in some cases, it is advantageous to cool a chemically active surface to liquid-nitrogen temperature in order to obtain effective trapping of methane and other low molecular weight species. This method, however, has a predictable finite life determined by the formation of monolayers on the chemically active surface. A plain metal surface cooled to liquid-nitrogen temperature in a trap has two primary functions, namely, to act as a cryopump for water vapor and to prevent contamination from DP working fluid from reaching a given vacuum environment. A well-designed LN trap can provide a pumping speed of at least 10^2 m^3/s per m^2 pf of system entrance area for water vapor (at room temperature, under free molecular conditions) and confine oil contamination to negligible levels. Measurements have shown that oil molecules and cracked-oil products having molecular weights of >60 pass through a particular design at rate of $<10^9$ molecules/(m^2·s) (23). These rates correspond to a vapor pressure at room temperature of less than ca 0.1 pPa $(7.5 \times 10^{-16}$ mm Hg).

Typically, a baffle condenses a vapor flow to a liquid in such a way that liquid can drain off for recycle. A well-designed and useful trap, on the other hand, catches and retains condensables such as water and the higher vapor pressure fraction of the working fluid of a pump. Although the terms baffle and trap seem to describe the same type of function, namely, to slow down but not prevent the rate at which some species of molecules pass through, traps can completely stop some molecular species (23).

Molecules arrive at the surfaces of traps and baffles by volume flow and surface creep. Molecules are trapped in vacuum systems by binding with energies much greater than kT of the surface, where k is Boltzmann's constant and T the absolute temperature, or by lowering the temperature of the surface in such a way that kT is less than the heat of physisorption of a molecular species on a surface.

If a trap is placed directly over a DP top jet (Fig. 4) without baffling action, all of the working fluid is caught by the trap in the time predicted by the rate at which the fringe of the top jets feeds the working fluid to the trap. This rate may be many orders of magnitude greater than the rate predicted from the equilibrium vapor pressure of the working fluid at DP wall temperature.

Valve, trap, and baffle can be combined in such a manner that elastomers can be used in the valve and contamination is controlled from the valve-actuator mechanisms and from the gasket of the valve-plate seal (Fig. 15) (42).

The role, design, and maintenance of creepproof barriers in traps, especially those in oil DPs, remain to be fully explored. In general, uncracked oil from a DP is completely inhibited from creeping by a surface temperature of <223 K. On the other hand, a cold trap, to perform effectively in an ordinary vacuum system, must be <173 K because of the vapor pressure of water, and ≤ 78 K because of the vapor pressure of CO_2. For ultracontrolled vacuum environments, LN temperature or lower is required. CO_2 accumulation on the trap surface must be less than one monolayer. The effectiveness of a LN trap can be observed by the absence of pressure pips on an ionization gauge when LN is replenished in the reservoir.

DP systems can be shut down when not in use to conserve energy. If a liquid-nitrogen trap is incorporated, the manner in which this trap is warmed up and the DP is cooled down should be determined by the presence or absence of a valve between the chamber and the liquid-nitrogen trap. In critical systems, this head valve can be included in order to permit rapid shutdown and rapid return to operation. The assertion that dry nitrogen gas can be used to sweep contamination from traps and pumps in such manner that oil contamination is prevented from running counter to the nitrogen-sweeping flow direction is questionable. Proper placement of valves can eliminate the need of a sweep gas.

Process Equipment

A good survey of process equipment widely used in chemical engineering is available (44). The pressure in the gaseous phase characterizes the likely vacuum process environment. Rough vacuum ranges from 101 kPa to ca 100 Pa (760 to 0.75 torr); medium vacuum ranges from 100 Pa to ca 0.1 Pa (0.75 to 0.00075 torr). The chemical engineer is likely to work in the rough vacuum region in which distillation, evaporation, drying, and filtration are normally conducted. The medium vacuum range is employed in molten-metal degassing, molecular distillation, and freeze drying.

Vacuum equipment requires strength to withstand the pressure of the surrounding atmosphere. The full load is ca 101.3 kPa when the internal gas pressure in the system is sufficiently reduced (see PUMPS).

Steam Ejectors. Ejectors are simple vacuum pumps. They have no moving parts but can accomplish compression through fluid-momentum transfer. A high pressure motive fluid enters the ejector and expands through the converging and diverging section of the nozzle. The initial pressure energy is converted to velocity. This increase in velocity entrains the load to be handled, which enters through the suction inlet. The motive and suction fluids are then mixed and recompressed through the defuser, which discharges this mixture into a higher intermediate pressure. By convention, an ejector represents a single-point design and is most efficient at a single set of conditions. Ejectors are either single-stage or multistage, depending on the suction pressures required. A practical compression ratio for single-stage ejectors is ca 6:1 when discharging to atmospheric pressure. This compression ratio can be as high as $10^6:1$ for six-stage ejectors. Multistage steam ejectors are usually equipped with direct-contact or surface-type condensors between stages in order to prevent contamination of cooling fluid. When designing a multistage steam ejector, condensables and noncondensables in the intake must be distinguished. Booster stages followed by a condenser are used to handle large quantities of condensables. Intercondensers are sized for both the motive steam from the booster and the process vapors. Thus, the load to the ejector stages downstream is significantly reduced. Such an ejector system uses considerably less steam, and purchase and installation costs are lower than those of a system designed to handle process vapors as condensables.

Liquid-Ring Pumps. In a liquid-ring pump, the rotor is the only moving part. The liquid ring performs all the functions normally done by mechanical pistons or vanes. There are a number of variations of this design, but all operate on the principle that before startup the pump casing is partially filled with a sealant liquid. When the rotating impeller is turned on, the liquid is caused to contact the periphery of the casing centrifugally. Thus, a liquid ring is formed that seals off the cylindrical pump body. Because the rotor axis is offset from the body axis, a piston action is established as the liquid fills and then almost empties each of the chambers between the rotor blades. Compression ratios as high as 10:1 can be achieved when discharging to atmospheric pressure for a single-stage pump. Because of the apparent difficulty in rating a liquid-ring pump on the basis of swept volume, it is classed as an isothermal machine rather than a true positive displacement compressor. The liquid ring acts as a heat sink to maintain constant-temperature operation. Assuming a normal sealing-liquid flow, a temperature rise in compressing air from ca 26.6 to 101.3 kPa (ca 0.25 to 1 atm) is ca 3–6 K. For comparison, an uncooled adiabatic compressor can cause a temperature rise exceeding ca 90 K. In sizing the pump, both evaporative and condensing effects must be considered. Evaporative cooling takes place whenever dry gases are introduced at temperatures higher than those of the sealed liquid. Condensation occurs when pumping gas that is saturated with vapor, ie, when the pump behaves like a direct-contact condenser.

Rotary-Piston Pumps. Positive-displacement, oil-sealed machines that compress a specific volume of gas with each revolution, and compress and exhaust it to the atmosphere, are the rotary-piston pumps. The oil-sealed piston traps the aspirated gas ahead of it by closing the inlet port. The gas is compressed, the discharge valve opens, and the gas is exhausted to the atmosphere. Compression ratios can be as high as more than 10^6:1 for a single-stage pump. These pumps operate in an internal oil bath which lubricates the pump and seals against backstreaming from the exhaust into the intake. The rate of flow and the distribution of oil through the pump are the important features of the design. The piston must be sufficiently lubricated to prevent failure. Failures are often caused by breakdowns of the oil distribution systems.

Rotary-piston pumps are frequently stalled by condensation of process vapors in the lubricating fluid. In addition, condensate can accumulate in the pump oil, resulting in mechanical failure and permanent damage. Water vapor, a persistent source of oil contamination, is almost always present in vacuum-processing operations. Higher alcohols and other solvents normally encountered also have a tendency to condense in the lubricating oil during compression. A number of techniques have been developed to prevent condensation of process vapors in oil-sealed pumps. Gas ballast, the most common technique, involves drilling a hole in the head of the pump to admit air or other gas into the cylinder during the latter portion of the compression stroke. Ballasting takes effect while the gas being compressed is sealed off from the intake by the piston. The ballasting method reduces the partial pressure of the condensible vapor and thus reduces or eliminates condensation within the oil. The introduction of gas-ballast air into the pump increases the pressure differential across the seals between intake and exhaust. Ballasting results in increased leakage past the seals, which can significantly reduce the capacity of rotary-piston pumps operating below ca 100 Pa (ca 0.1 atm). However, the effect on pump capacity is negligible in most processing applications. Below the processing range, however, ballasting when handling the water in ordinary atmospheric air is not needed.

Rotary-Vane Pumps. Rotary-vane pumps are positive-displacement machines having spring-loaded vanes that contact the inside of the pump casing. Gas entering the pump is trapped between adjacent blades, compressed, and forced out to the atmosphere through the discharge point. Maintenance requirements have severely limited the use of these pumps in process applications. Rotary-vane pumps are still found in laboratory applications and can achieve compression ratios well above 10^6:1 for discharge directly into the atmosphere.

Oil contamination by process fluids causes a deterioration in the pump performance and can necessitate frequent cleaning and reassembly of the pump.

Rotary-Blower Pumps. This type of pump employs two interlocking rotors to trap and compress gases. The rotors are prevented from touching one another, and there is no sealing liquid in the pump. The gears and rotor bearings are lubricated with oil, but are external to the rotors. Clearance between the rotors is generally $25-100$ μm. Typically, these pumps operate at high speeds of $3000-4000$ rpm. Because there is no positive seal between the rotors, the rotary blower is limited to small compression ratios, though it can also be designed for higher throughput than any other mechanical pump. In process and most other applications, the rotary blower is limited to operation in conjunction with other mechanical pumps. Inherently, rotary blowers are potentially subject to overheating because of the lack of a discharge valve separating the heated gas. As a result, the compression ratio of single-stage blowers is limited to ca 2.3:1. However, if the rotary blower discharges into a rotary-piston, oil-sealed pump, the combination can exceed 10^6:1. When operating below 133 Pa (1 torr), overheating need not be a consideration because the work done in compressing the process load is small.

Economic Aspects

Vacuum systems, largely for the semiconductor industry, are the main source of sales (see SEMICONDUCTORS). The sales of all vacuum equipment, pumps (qv), valves, sensors (qv), etc, in the United States, including applications not in vacuum systems, generally exceed 500×10^6/yr. A reasonably comprehensive list of high vacuum manufacturers is supplied by the American Vacuum Society's exhibitor's list. In Europe, a special issue of the journal *Vacuum* serves similarly.

Capital investment, capital costs, operating costs, return on investment, and energy conservation have all been discussed (6). In the economic analysis, the speed of each type of pump considered is normalized to 1 m^3/s as a common basis.

With regard to fixed amortized investment (45), utilization of the wall area of the chamber to be evacuated using a given pumping method must be considered. Depending on the process, it may or may not be possible to expose gettering or cryogenic pumping surfaces directly. Many uses would contribute dynamically to a gettering or cryogenic surface, making it uneconomic to handle the energy flux, sputtering processes, etc, which direct exposure might entail (see also CRYOGENICS).

Energy costs are not directly related to the energy efficiency of the process (6,42). Even if the thermal efficiency of a steam ejector, for example, is less than that of mechanical equipment run by an electric motor, the overall cost of the energy to run the steam ejector may still be less.

BIBLIOGRAPHY

"Vacuum Technique" in *ECT* 1st ed., Vol. 14, pp. 503–536, by B. B. Dayton, Consolidated Vacuum Corp.; "Vacuum Technology" in *ECT* 2nd ed., Vol. 21, pp. 123–157, by B. B.

Dayton, The Bendix Corp.; in *ECT* 3rd ed., Vol. 23, pp. 644–673, N. Milleron, EMR Photoelectric.

1. N. Milleron, in F. J. Clauss, ed., *Surface Effects on Space Craft Materials, 1st Symposium 1959*, John Wiley & Sons, Inc., New York, 1960, pp. 260, 303, 325–342.
2. N. Milleron, in J. A. Dillon, Jr., and V. J. Harwood, eds., *Experimental Vacuum Science and Technology*, Marcel Dekker, Inc., New York, 1973.
3. J. D. Dow and co-workers, *J. Vac. Sci. Technol.* **19**, 502 (1981).
4. P. D. Kirchner and co-workers, *J. Vac. Sci. Technol.* **19**, 604 (1981).
5. G. A. Bird, *Molecular Gas Dynamics*, Oxford University Press, Oxford, U.K., 1976.
6. J. F. O'Hanlon, *A User's Guide to Vacuum Technology*, Wiley-Interscience, New York, 1980.
7. L. C. Beavis, V. J. Harwood, and M. T. Thomas, *Vacuum Hazards Manual*, American Vacuum Society, New York, 1975.
8. P. Holloway, *Vacuum Book Bibliography*, American Vacuum Society, New York, 1982.
9. M. S. Kaminsky and J. J. Lafferty, *Dictionary of Terms for Vacuum Science and Technology, Surface Science, Thin-Film Technology, Vacuum Metallurgy and Electronic Materials*, American Vacuum Society, New York, 1980.
10. J. L. Vossen, *Bibliography on Metallization Materials and Techniques for Silicon Devices*, Vols. 6 (1980), 7 (1981), and 8 (1982), American Vacuum Society, New York.
11. E. Grant, *Much Ado About Nothing: Theories of Space and Vacuum from the Middle Ages to the Scientific Revolution*, Cambridge University Press, New York, 1981.
12. G. L. Weissler and R. W. Carlson, eds., *Vacuum Physics and Technology*, Academic Press, Inc., New York, 1979, p. 4.
13. P. A. Redhead, J. P. Hobson, and E. V. Kornelsen, *The Physical Basis of Ultra High Vacuum*, Chapman and Hall, London, 1968.
14. R. V. Latham, *High Voltage Insulation*, Academic Press, Inc., New York, 1981.
15. J. M. Lafferty, ed., *Vacuum Arcs Theory and Application*, John Wiley & Sons, Inc., New York, 1980.
16. R. Hackam and L. Altcheh, *J. Appl. Phys.* **46**(2), 631 (1975).
17. C. A. Spindt, *Development Program on a Cold Cathode Electron Gun*, NASA CR 159570, Stanford Research Institute, Menlo Park, Calif., 1979, pp. 37–38.
18. I. J. Scialdone, *NASA TN D-7250*, Goddard Space Flight Center, Greenbelt, Md., 1972.
19. N. Milleron, *Res. Dev.* **11a**, 44 (Jan. 1964).
20. T. E. Graedel, *Chemical Compounds in the Atmosphere*, Academic Press, Inc., New York, 1978.
21. A. D. Zimon, *Adhesion of Dust and Powder*, Consultants Bureau, New York, 1982.
22. Ref. 12, p. 425.
23. N. Milleron, *IEEE Trans. Nucl. Sci.* **14**(3), 794 (1967).
24. N. G. Wilson and L. C. Beavis, *Handbook of Vacuum Leak Detection*, American Vacuum Society, New York, 1976.
25. D. J. Mattox, *Surface Cleaning in Thin-Film Technology*, American Vacuum Society, New York, 1975.
26. Ref. 23, p. 800.
27. Ref. 23, p. 801.
28. L. C. Beavis, *J. Vac. Sci. Technol.* **20**, 972 (1982).
29. D. S. Miller, *Internal Flow*, Cranfield British Hydro Mechanical Research Association, 1971.
30. E. M. Sparrow and V. K. Jansson, *AIAA J.* **1**(5), 1081 (1963).
31. D. H. Davis, L. L. Levenson, and N. Milleron, in L. Talbot, ed., *Rarified Gas Dynamics*, Academic Press, Inc., New York, 1961, p. 99.
32. Ref. 12, p. 19.
33. L. L. Levenson, N. Milleron, and D. H. Davis, *Transactions of the American Vacuum Society*, Pergamon Press, Inc., Elmsford, N.Y., 1961.

34. D. Davis, L. L. Levenson, and N. Milleron, *J. Appl. Phys.* **35**, 529 (1964).

35. F. O. Goodman and H. Y. Wachman, *Dynamics of Gas Surface Scattering*, Academic Press, Inc., New York, 1976.

36. R. H. Edwards, *Low Density Flows through Tubes and Nozzles*, Vol. 51, Pt. 1, American Institute of Aeronautics and Astronomy, New York, 1977.

37. D. H. Davis, *J. Appl. Phys.* **31**, 1169 (1960).

38. N. Milleron and D. N. Frank, *J. Vac. Sci. Technol.* **20**, 1052 (Apr. 1982).

39. Ref. 12, p. 71.

40. Ref. 12, p. 219.

41. M. J. Drinkwine and D. Lichtman, *Partial Pressure Analysis*, American Vacuum Society, New York, 1977.

42. Ref. 12, p. 277.

43. Ref. 12, p. 278.

44. J. L. Ryans and S. Croll, *Chem. Eng.* **73** (Dec. 14, 1981).

45. E. P. De Garmo, J. R. Canada, and W. G. Sullivan, *Engineering Economy*, 6th ed., Macmillan-Collier, New York, 1979.

General References

A. Berman, *Vacuum Engineering Calculations, Formulas, and Solved Exercises*, Academic Press, Inc., New York, 1992.

A Greenwood, *Vacuum Switchgear*, Institution of Electrical Engineers, 1994.

R. V. Latham, ed., *High Voltage Vacuum Insulation: Basic Concepts and Technological Practice*, Academic Press, Inc., New York, 1995.

D. Konig, *Proceedings of the 15th International Symposium on Discharges and Electrical Insulation in Vacuum*, Darmstadt, Germany, 1992.

G. A. Mesyats, *Proceedings of the 16th International Symposium on Discharges and Electrical Insulation in Vacuum*, Moscow, CIS, 1994.

Proceedings of the 6th International Vacuum Microelectronics Conference, Newport, R.I., 1993.

J. P. Looney and S. Tison, *J. Res.* **100**, 75 (Jan./Feb. 1995).

P. A. Redhead, *Vacuum Science and Technology: Pioneers of the 20th Century, History of Vacuum Science and Technology*, Vol. 2, AIP Press for the American Vacuum Society, New York, 1994.

G. L. Saksaganskii, *Molecular Flow in Complex Vacuum Systems*, Gordon & Breach, New York, 1988.

Norman Milleron
SEN Vac Services

VANADIUM AND VANADIUM ALLOYS

Vanadium

Vanadium [7440-62-2], V, (at. no. 23, at. wt 50.942) is a member of Group 5 (VB) of the Periodic Table. It is a gray body-centered-cubic metal in the first transition series (electronic configuration $4s^2 3d^3$). When highly pure, it is very soft and dutile. Because of its high melting point, vanadium is referred to as a refractory metal, as are niobium, tantalum, chromium, molybdenum, and tungsten (see REFRACTORIES). The principal use of vanadium is as an alloying addition to iron (qv) and steel (qv), particularly in high strength steels and, to a lesser extent, in tool steels and castings (see TOOL MATERIALS). Vanadium is also an important beta-stabilizer for titanium alloys (see TITANIUM AND TITANIUM ALLOYS). Interest has been shown in the intermetallic compound V_3Ga [12024-15-6] for superconductor applications (see SUPERCONDUCTING MATERIALS).

Vanadium was first discovered in 1801 by del Rio while he was examining a lead ore obtained from Zimapan, Mexico. The ore contained a new element and, because of the red color imparted to its salts on heating, it was named erythronium (redness). The identification of the element vanadium did not occur until 1830 when it was isolated from cast iron processed from an ore from mines near Taberg, Sweden. It was given the name vanadium after Vanadis, the Norse goddess of beauty. Shortly after this discovery, vanadium was shown to be identical to the erythronium that del Rio had found several years earlier.

Occurrence.   Vanadium is widely distributed throughout the earth but in low abundance, ranking 22 among the elements of the earth's crust. The lithosphere contains ca 0.07 wt % vanadium and few deposits contain more than 1–2 wt %. Vanadium occurs in uranium-bearing minerals of Colorado, in the copper, lead, and zinc vanadates of Africa, and with certain phosphatic shales and phosphate rocks in the western United States. It is a constituent of titaniferous magnetites, which are widely distributed with large deposits in Russia, South Africa, Finland, the People's Republic of China, eastern and western United States, and Australia. At one time, the largest and most important vanadium deposits were the sulfide and vanadate ores from the Peruvian Andes, but these are depleted. Most of the vanadium reserves are in deposits in which the vanadium would be a by-product or coproduct with other minerals, including iron, titanium, phosphate, and petroleum.

Trace amounts of vanadium have been found in meteorites and seawater, and it has been identified in the spectrum of many stars including the earth's sun. The occurrence of vanadium in oak and beech trees and some forms of aquatic sea life indicates its biological importance.

There are over 65 known vanadium-bearing minerals, some of the more important are listed in Table 1. Patronite, bravoite, sulvanite, davidite, and roscoelite are classified as primary minerals, whereas all of the others are secondary products which form in the oxidizing zone of the upper lithosphere. The carnotite and roscoelite ores in the sandstones of the Colorado Plateau have been important sources of vanadium as well as of uranium.

Table 1. Important Minerals of Vanadium

Mineral	CAS Registry Number	Color	Formula	Location
patronite	[12188-60-2]	greenish black	$V_2S + nS$	Peru
bravoite	[12172-92-8]	brass	$(Fe,Ni,V)S_2$	Peru
sulvanite	[15117-74-5]	bronze-yellow	$3Cu_2S \cdot V_2S_6$	Australia, United States (Utah)
davidite	[12173-20-5]	black	titanate of Fe, U, V, Cr, and rare earths	Australia
roscoelite	[12271-44-2]	brown	$2K_2O \cdot 2Al_2O_3(Mg,Fe)O \cdot 3V_2O_5 \cdot 10SiO_2 \cdot 4H_2O$	United States (Colorado, Utah)
carnotite	[1318-26-9]	yellow	$K_2O \cdot 2U_2O_3, V_2O_5 \cdot 3H_2O$	southwest United States
vanadinite	[1307-08-0]	reddish brown	$Pb_5(VO_4)_3Cl$	Mexico, United States, Argentina
descloizite	[19004-61-6]	cherry-red	$4(Cu,Pb,Zn)O \cdot V_2O_5 \cdot H_2O$	Namibia, Mexico, United States
cuprodescloizite	[12325-36-9]	greenish brown	$5(Cu,Pb)O \cdot (V,As)_2O_5 \cdot 2H_2O$	Namibia
vanadiferous phosphate rock			$Ca_5(PO_4)_3(F,Cl,OH)$; VO_4 ions replace some PO_4 ions	United States (Montana)
titaniferous magnetite			$FeO \cdot TiO_2 - FeO \cdot (Fe,V)O_2$	Russia, People's Republic of China, Finland, Union of South Africa

The metallic vanadates of lead, copper, and zinc, which occur in Namibia (southwest Africa) and Zambia, are also a large resource of vanadium-bearing ores, as are the phosphatic shales and rocks of the phosphoria formation in Idaho and Wyoming. Vanadium salts are obtained as by-products of the phosphoric acid and fertilizer industries (see FERTILIZERS). Large reserves of vanadium in Arkansas and Canada and the titaniferous magnetite ores will probably become increasingly significant as sources of production. Certain petroleum crude oils, especially those from South America, contain varying amounts of vanadium compounds. These accrue as fly ash or boiler residues upon combustion of the crude oils and they can be reclaimed (see AIR POLLUTION CONTROL METHODS).

Physical Properties. Vanadium is a soft, ductile metal in pure form, but it is hardened and embrittled by oxygen, nitrogen, carbon, and hydrogen (1,2). Selected metal additions lead to higher strength alloys which maintain a reasonable level of ductility (3). Its thermal conductivity is significantly lower than that of copper. Important physical properties of vanadium are listed in Table 2. Some of these properties depend on the purity of the material used for the determinations. Although a purity level of 99.99% has been achieved experimentally, such high purity material has not been used for all of the determinations listed (7).

Table 2. Physical Properties of Vanadium Metal[a]

Property	Value
melting point, °C	1890 ±10
boiling point, °C	3380
vapor pressure, from 1393–1609°C, kPa[b]	$R \ln P = \dfrac{121,950}{T} - 5.123 \times 10^{-4}\, T + 38.3$
crystal structure	bcc
lattice constant, nm	0.3026
density, g/cm^3	6.11
specific heat, 20–100°C, J/g[c]	0.50
latent heat of fusion, kJ/mol[c]	16.02
latent heat of vaporization, kJ/mol[c]	458.6
enthalpy, at 25°C, kJ/mol[c]	5.27
entropy, at 25°C, kJ/(mol·°C)[c]	29.5
thermal conductivity, at 100°C, W/(cm·K)	0.31
electrical resistance, at 20°C, $\mu\Omega$·cm	24.8–26.0
temperature coefficient of resistance, at 0–100°C, ($\mu\Omega$·cm)/°C	0.0034
magnetic susceptibility, m^3/mol[d]	0.11
superconductivity transition, K	5.13
coefficient of linear thermal expansion, °C^{-1}	
at 20–720°C (x-ray)	$(9.7 \pm 0.3) \times 10^{-6}$
200–1000°C (dilatometer)	8.95×10^{-6}
thermal expansion, at 23–100°C, $\times 10^{-6}$/°C	8.3
recrystallization temperature, °C	800–1000
modulus of elasticity, MPa[e]	$(1.2–1.3) \times 10^5$
shear modulus, MPa[e]	4.64×10^4
Poisson ratio	0.36
thermal neutron absorption, m^2/at.[f]	$(4.7 \pm 0.02) \times 10^{-28}$
capture cross section for fast (1 MeV) neutrons, m^2/at.[f]	3×10^{-31}

[a]Refs. 4–6.
[b]To convert kPa to atm, divide by 101.3. In this, the Antoine equation, R = gas constant, T is temperature in K, and P is pressure in kPa.
[c]To convert J to cal, divide by 4.184.
[d]To convert m^3/mol to cgs units, multiply by $4\,\pi \times 10^{-6}$.
[e]To convert MPa to psi, multiply by 145.
[f]To convert m^2 to barns, multiply by 1×10^{28}.

Chemical Properties. Vanadium has oxidation states of +2, +3, +4, and +5. When heated in air at different temperatures, it oxidizes to a brownish black trioxide, a blue-black tetroxide, or a reddish orange pentoxide. It reacts readily with chlorine at fairly low temperatures (180°C) forming VCl$_4$ and with carbon and nitrogen at high temperatures forming VC and VN, respectively. The pure metal in massive form is relatively inert toward oxygen, nitrogen, and hydrogen at room temperature.

Vanadium is resistant to attack by hydrochloric or dilute sulfuric acid and to alkali solutions. It is also quite resistant to corrosion by seawater but is reactive toward nitric, hydrofluoric, or concentrated sulfuric acids. Galvanic corrosion tests run in simulated seawater indicate that vanadium is anodic with respect to stainless steel and copper but cathodic to aluminum and magnesium. Vanadium exhibits corrosion resistance to liquid metals, eg, bismuth and low oxygen sodium.

Manufacture

Ore Processing. Vanadium is recovered domestically as a principal mine product, as a coproduct or by-product from uranium–vanadium ores, and from ferrophosphorus as a by-product in the production of elemental phosphorus. In Canada, it is recovered from crude-oil residues and in the Republic of South Africa as a by-product of titaniferous magnetite. Whatever the source, however, the first stage in ore processing is the production of an oxide concentrate.

The principal vanadium-bearing ores are generally crushed, ground, screened, and mixed with a sodium salt, eg, NaCl or Na_2CO_3. This mixture is roasted at ca 850°C and the oxides are converted to water-soluble sodium metavanadate, $NaVO_3$. The vanadium is extracted by leaching with water and precipitates at pH 2–3 as sodium hexavanadate, $Na_4V_6O_{17}$, a red cake, by the addition of sulfuric acid. This is then fused at 700°C to yield a dense black product which is sold as technical-grade vanadium pentoxide. This product contains a minimum of 86 wt % V_2O_5 and a maximum of 6–10 wt % Na_2O.

The red cake can be further purified by dissolving it in an aqueous solution of Na_2CO_3. The iron, aluminum, and silicon impurities precipitate from the solution upon pH adjustment. Ammonium metavanadate then precipitates upon the addition of NH_4Cl and is calcined to give vanadium pentoxide of greater than 99.8% purity.

Vanadium and uranium are extracted from carnotite by direct leaching of the raw ore with sulfuric acid. An alternative method is roasting the ore followed by successive leaching with H_2O and dilute HCl or H_2SO_4. In some cases, the first leach is with a Na_2CO_3 solution. The uranium and vanadium are then separated from the pregnant liquor by liquid–liquid extraction techniques involving careful control of the oxidation states and pH during extraction and stripping.

In the Republic of South Africa, the recovery of high vanadium slags from titaniferous magnetites has been achieved on a large scale (8). The ore, containing about 1.75 wt % V_2O_5, is partially reduced with coal in large rotary kilns. The hot ore is then fed to an enclosed, submerged-arc electric smelting furnace which produces a slag containing substantial amounts of titania and pig iron containing most of the vanadium that was in the ore. After tapping from the furnace and separation of the waste slag, the molten pig iron is blown with oxygen to form a slag containing up to 25 wt % V_2O_5. The slag is separated from the metal and may then be used as a high raw material in the usual roast–leach process.

Solvent extraction following roasting and leaching is a promising processing method for dolomitic shale from Nevada (9).

Ferrovanadium. The steel industry accounts for the majority of the world's consumption of vanadium as an additive to steel. It is added in the steelmaking process as a ferrovanadium alloy [12604-58-9], which is produced commercially by the reduction of vanadium ore, slag, or technical-grade oxide with carbon, ferrosilicon, or aluminum. The product grades, which may contain 35–80 wt %

vanadium, are classified according to their vanadium content. The consumer use and grade desired dictate the choice of reductant.

Carbon Reduction. The production of ferrovanadium by reduction of vanadium concentrates with carbon has been supplanted by other methods. An important development has been the use of vanadium carbide as a replacement for ferrovanadium as the vanadium additive in steelmaking. A product containing ca 85 wt % vanadium, 12 wt % carbon, and 2 wt % iron is produced by the solid-state reduction of vanadium oxide with carbon in a vacuum furnace.

Silicon Reduction. The preparation of ferrovanadium by the reduction of vanadium concentrates with ferrosilicon has been used but not extensively. It involves a two-stage process in which technical-grade vanadium pentoxide, ferrosilicon, lime, and fluorspar are heated in an electric furnace to reduce the oxide; an iron alloy containing ca 30 wt % vanadium but undesirable amounts of silicon is produced. The silicon content of the alloy is then decreased by the addition of more V_2O_5 and lime to effect the extraction of most of the silicon into the slag phase. An alternative process involves the formation of a vanadium–silicon alloy by the reaction of V_2O_5, silica, and coke in the presence of a flux in an arc furnace. The primary metal then reacts with V_2O_5 yielding ferrovanadium.

A silicon process has been developed by the Foote Mineral Company and has been used commercially to produce tonnage quantities of ferrovanadium (10). A vanadium silicide alloy containing less than 20 wt % silicon is produced in a submerged-arc electric furnace by reaction of vanadium-bearing slags with silica, flux, and a carbonaceous reducer followed by refinement with vanadium oxide. This then reacts with a molten vanadiferous slag in the presence of lime yielding a ferrovanadium alloy (Solvan) containing ca 28 wt % vanadium, 3.5 wt % silicon, 3.8 wt % manganese, 2.8 wt % chromium, 1.25 wt % nickel, 0.1 wt % carbon, and the remainder iron. A unique feature of this process is its applicability to the pyrometallurgical process of vanadium-bearing slags of the type described in the preceding section.

Aluminum Reduction. The aluminothermic process for preparing a ferrovanadium alloy differs from the carbon and silicon reduction processes in that the reaction is highly exothermic. A mixture of technical-grade vanadium oxide, aluminum, iron scrap, and a flux are charged into an electric furnace and the reaction between aluminum and vanadium pentoxide is initiated by the arc. The temperature of the reaction is controlled by adjusting the size of the particles and the feed rate of the charge by using partially reduced material or by replacing some of the aluminum with a milder reductant, eg, calcium carbide, silicon, or carbon. Ferrovanadium containing as much as 80 wt % vanadium is produced in this way.

Ferrovanadium can also be prepared by the thermite reaction, in which vanadium and iron oxides are co-reduced by aluminum granules in a magnesite-lined steel vessel or in a water-cooled copper crucible (11) (see ALUMINUM AND ALUMINUM ALLOYS). The reaction is initiated by a barium peroxide–aluminum ignition charge. This method is also used to prepare vanadium–aluminum master alloys for the titanium industry.

Pure Vanadium. Vanadium, like its sister Group 5 (VB) elements, dissolves significant quantities of oxygen, nitrogen, hydrogen, and carbon interstitially into its lattice. In so doing, a severe loss of ductility results. Formation of a ductile pure metal or alloy requires that contamination by these elements is carefully controlled. Generally, small quantities of some or all of these elements, particularly oxygen, are tolerated in a compromise between increased strength and loss of ductility. The production method for pure vanadium and vanadium-based alloys must be tailored with this need for high purity and for economic and engineering considerations in mind. Most reduction processes suffer from inabilities to reduce the amount of all impurities simultaneously to the desired level. Consequently, one or more purification methods are used to overcome the limitations of the original reduction.

Vanadium metal can be prepared either by the reduction of vanadium chloride with hydrogen or magnesium or by the reduction of vanadium oxide with calcium, aluminum, or carbon. The oldest and most commonly used method for producing vanadium metal on a commercial scale is the reduction of V_2O_5 with calcium. Recently, a two-step process involving the aluminothermic reduction of vanadium oxide combined with electron-beam melting has been developed. This method makes possible the production of a purer grade of vanadium metal, ie, of the quality required for nuclear reactors (qv).

Calcium Reduction.   High purity vanadium pentoxide is reduced with calcium to produce vanadium metal of ca 99.5% purity. The exothermic reaction is carried out adiabatically in a sealed vessel or bomb. In the original process, calcium chloride was added as a flux for the CaO slag (12). The vanadium metal was recovered in the form of droplets or beads. A massive ingot or regulus has been obtained by replacing the calcium chloride flux with iodine (13). This latter reaction became the basis of the first large-scale commercial process for producing vanadium. The reaction is initiated either by preheating the charged bomb or by internal heating with a fuse wire embedded in the charge. Calcium iodide formed by the reaction of calcium with iodine serves both as a flux and as a thermal booster. Thus, sufficient heat is generated by the combined reactions to yield liquid metal and slag products. The resulting metal contains ca 0.2 wt % carbon, 0.02–0.8 wt % oxygen, 0.01–0.05 wt % nitrogen, and 0.002–0.01 wt % hydrogen. Two factors that contribute to the relative inefficiency of this process are the rather low metal yields (75–80%) and the required amount of calcium reductant (50–60% excess of stoichiometric quantity).

Vanadium powder can be prepared by substituting V_2O_3 for the V_2O_5 as the vanadium source. The heat generated during the reduction of the trioxide is considerably less than for the pentoxide, so that only solid products are obtained. The powder is recovered from the product by leaching the slag with dilute acid.

Aluminothermic Process.   In the development of the liquid-metal fast-breeder reactor, vanadium has been considered for use as a fuel-element cladding material (see NUCLEAR REACTORS). Difficulty was encountered in the fabrication of alloys prepared from the calcium-reduced metal, a factor attributable to the high interstitial impurity content. An aluminothermic process was developed by the AEC (now the NRC) in order to meet the more stringent purity requirements for this application (14). In this process, vanadium pentoxide reacts with high purity aluminum in a bomb to form a massive vanadium–aluminum alloy. Use of proprietary additions to either increase the reaction temperature, decrease the melting point of the slag or metal, or increase the fluidity of the two phases leads to formation of a solid metallic regulus that is relatively free of slag. The alloyed aluminum and dissolved oxygen are subsequently removed in a high temperature, high vacuum processing step to yield metal of greater than 99.9% purity.

Purified V_2O_5 powder and high purity aluminum granules are charged into an alumina-lined steel crucible. The vessel is flushed with an inert gas to minimize atmospheric contamination and then is sealed. The reaction is initiated by a vanadium fuse wire. Sufficient heat is generated by the chemical reaction to produce a molten alloy of vanadium containing ca 15 wt % aluminum; a fused aluminum oxide slag also forms. The liquid alloy separates from the alumina slag and settles to the bottom of the crucible as a massive product. The feasibility of carrying this reaction out in a water-cooled copper crucible, thus eliminating the alumina liner which is a source of some contamination, has been demonstrated.

Examination of the metallic product (regulus) of such aluminothermically produced vanadium metal reveals the presence of oxide phases in the metal matrix. This suggests that there is a decreasing solubility for aluminum and oxygen below the melting point. To date, no purification processes have been developed that take advantage of the purification potential of this phenomenon.

The vanadium alloy is purified and consolidated by one of two procedures, as shown in the flow diagram of the entire aluminothermic reduction process presented in Figure 1. In one procedure, the brittle alloy is crushed and heated in a vacuum at 1790°C to sublime most of the aluminum, oxygen, and other impurities. The aluminum facilitates removal of the oxygen, which is the feature that makes this process superior to the calcium process. Further purification and consolidation of the metal is accomplished by electron-beam melting of pressed compacts of the vanadium sponge.

The alternative procedure involves direct electron-beam melting of the vanadium–aluminum alloy regulus. Two or more melting steps are required to achieve the desired levels of aluminum and oxygen in the final ingot. Chemical analyses of two ingots of vanadium metal prepared from the identical vanadium–aluminum alloy and processed by the two methods described above are presented in Table 3. Comparable purities are obtained by these procedures. Ingots weighing up to 454 kg have been prepared by this process involving direct electron-beam melting of the alloy.

Refining of Vanadium. In addition to the purification methods described above, vanadium can be purified by any of three methods: iodide refining (van Arkel-deBoer process), electrolytic refining in a fused salt, and electrotransport.

Metal of greater than 99.95% purity has been prepared by the iodide-refining method (15). In this process, an impure grade of vanadium metal reacts with iodine at 800–900°C forming vanadium diiodide and the volatilized iodide is thermally decomposed and deposited on a hot filament at ca 1300°C. The refining step is carried out in an evacuated and sealed tube. The main impurities removed in the process are the gaseous elements and those metals that form stable or nonvolatile iodides. Vanadium metal containing 5 ppm nitrogen, 150 ppm carbon, and 50 ppm oxygen has been prepared in this way.

An electrolytic process for purifying crude vanadium has been developed at the U.S. Bureau of Mines (16). It involves the cathodic deposition of vanadium from an electrolyte consisting of a solution of VCl_2 in a fused KCl–LiCl eutectic.

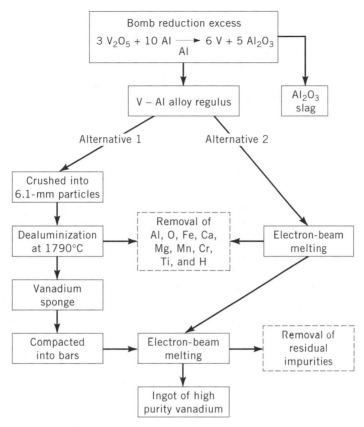

Fig. 1. Flow diagram for aluminothermic process showing alternative methods of aluminum removal from alloy regulus.

Table 3. Vanadium Prepared by Dealuminization with Vacuum Heating[a]

	By dealuminization, ppm		By direct melting, ppm	
Impurity	Sponge	Electron-beam melted ingot	First melt	Second melt
C	100	100	100	100
O	50	60	200	150
N	40	45	40	50
H	<1	<1	<1	<1
Al	>1000	300	ca 1000	100
Ca	<10	<10	<10	<10
Cu	<20	<20	<20	<20
Fe	70	70	60	60
Mg	<20	<20	<20	<20
Mn	<20	<20	<20	<20
Ni	60	60	60	60
Si	300	300	300	300

[a]Compared to that prepared by direct electron-beam melting of vanadium alloy containing 15 wt % aluminum.

The vanadium content of the mixture is 2–5 wt % and the operating temperature of the cell is 650–675°C. Metal crystals or flakes of up to 99.995% purity have been obtained by this method.

The highest purity vanadium reported has been purified by an electro-transport technique (17). A high density current is passed through a small rod of electrolytically refined metal, heating it to 1700–1850°C. Under these conditions, interstitial solute atoms, eg, carbon, oxygen, and nitrogen, migrate to the negative end of the bar, which results in a high degree of purification along the remainder of the rod. Small amounts of vanadium containing less than 10 ppm of carbon, oxygen, and nitrogen and having a resistance ratio, $R_{300 \text{ K}}$:$R_{4.2 \text{ K}}$, of greater than 1100 have been prepared by this technique.

Consolidation, Fabrication, and Properties

Because no process has been developed for selectively removing impurities in vanadium and vanadium alloys in the metallic state, it is essential that all starting materials, in aggregate, be pure enough to meet final product purity requirements. In addition, the consolidation method must be one that prevents contamination through reaction with air or with the mold or container material.

Consolidation. Consolidation by the consumable-electrode electric-arc melting technique is ideally suited for vanadium and is used extensively for preparation of ingots of most of the reactive and refractory metals (18,19). An electrode consisting of carefully weighed portions of each alloy constituent is prepared by combining various sizes and shapes of starting materials. For example, an alloy composed of vanadium plus 15 wt % titanium and 7.5 wt % chromium might consist of an electron-beam purified vanadium ingot to which has been welded high purity titanium plat and electrolytically produced chromium granules. Welding is performed either in a vacuum or under an inert gas, eg, argon, helium, or a mix of the two.

The electrode is attached mechanically and electrically to a system that feeds it downward as well as in all horizontal directions so as to center the electrode in the crucible. The electrode is then placed in a water-cooled copper crucible whose internal diameter is 2.5–7.6 cm greater than the circle circumscribing the electrode. Chips or blocks of vanadium or of the alloy to be melted are placed in the bottom of the crucible, forming a starter pad for the arc melting. The furnace assembly is sealed and the chamber evacuated and then backfilled with a pressure of helium and argon, which varies from between a few millipascals (10^{-5} mm Hg) to nearly 101.3 kPa (=1 atm). An arc is struck between the electrode and the starter pad; voltage and amperage are varied to control the energy input and, thus, the rate at which the electrode is melted and consumed to form the alloy ingot.

Multiple-arc melting for a minimum of two melts is conventionally used to ensure a homogeneous ingot. Although conventional arc-melt practice involves a negative electrode, improved alloying is achieved with a positive electrode for at least one of the several melts and usually the first melt.

The as-cast ingot generally exhibits a rough side wall because of entrapment of porosity as the molten metal contacts the water-cooled copper crucible. Also, a shrinkage cavity usually forms near the top of the ingot as a result of volumetric changes during solidification. Both of these artifacts are normally removed prior to fabrication, because most vanadium alloys exhibit only limited ductility in the as-cast condition, and the porosity, if not removed, would lead to cracking. Machining is the conventional surface-conditioning method.

Fabrication. Primary or initial fabrication is generally performed by either forging or extrusion at 1000–1200°C (19). Because the alloy oxidizes quite rapidly at elevated temperatures and because the pentoxide melts at 690°C, the machined ingot is clad and sealed in a mild-steel container. Following the initial hot-working sequences, subsequent working can be performed at RT to 500°C, depending on the alloy and the stage of processing. Intermediate and final recrystallization annealing are performed at 650–1000°C. Annealing is done in a vacuum or inert gas to reduce metal loss and contamination of the metal which would occur if heating were done in air.

The fabrication of most vanadium alloys is difficult because of increased strength and decreased ductility, especially at low temperatures. Generally, higher temperatures are used for each step of fabrication. Also, processes, eg, extrusion in which the forces are largely compressive, are used for the initial ingot breakdown.

Properties. Most of the alloys developed to date were intended for service as fuel cladding and other structural components in liquid-metal-cooled fast-breeder reactors. Alloy selection was based primarily on the following criteria: corrosion resistance in liquid metals, including lithium, sodium, and NaK, and a mixture of sodium and potassium; strength; ductility, including fabricability; and neutron considerations, including low absorption of fast neutrons as well as irradiation embrittlement and dimensional-variation effects. Alloys of greatest interest include V 80, Cr 15, Ti 5 [*39308-80-0*]; V 80, Ti 20 [*12611-15-3*]; V 77.5, Ti 15, Cr 7.5 [*51880-37-6*]; and V 86.65, Cr 9, Fe 3, Zr 1.3, C 0.05 [*84215-85-0*].

Based on considerations of economics, fabricability, and performance, reasonable allowable levels of interstitial elements are as follows:

Element	Allowable level, ppm wt
carbon	200
nitrogen	100
oxygen	300
hydrogen	100

Because of the effects of impurity content and processing history, the mechanical properties of vanadium and vanadium alloys vary widely. The typical RT properties for pure vanadium and some of its alloys are listed in Table 4. The effects of alloy additions on the mechanical properties of vanadium have been studied and some alloys that exhibit room-temperature tensile strengths of 1.2 GPa (175,000 psi) have strengths of up to ca 1000 MPa (145,000 psi) at 600°C. Beyond this temperature, most alloys lose tensile strength rapidly.

As in the case of many metal–alloy systems, weld ductility is not as good as that of the base metal. Satisfactory welds can be made in vanadium alloys provided the fusion zone and the heat-affected zone (HAZ) are protected from contamination during welding. Satisfactory welds can be made by a variety of

Table 4. Typical Room Temperature Properties of Vanadium and Vanadium Alloys[a]

Metal	Tensile strength, MPa[b]	Yield strength, MPa[b]	Elongation, %
pure vanadium			
annealed or hot-worked	380–550	410–480	20–27
cold-worked	910	760	2–7
alloys			
V 87.5, Ti 15, Cr 7.5 annealed	730	620	30
V 80, Cr 15, Ti 5 annealed	600	500	28
various high strength alloys, warm rolled	1210		2–5

[a]Refs. 3 and 20.
[b]To convert MPa to psi, multiply by 145.

weld methods, including electron-beam and tungsten-inert-gas (TIG) methods. It is also likely that satisfactory welds can be made by advanced methods, eg, laser and plasma techniques (see LASERS; PLASMA TECHNOLOGY).

Economic Aspects

The United States dominated world vanadium production for all uses until the late 1960s when several countries, notably the former USSR, expanded production significantly. At about the same time, the United States shifted from being a net exporter to a net importer; this situation continues. In 1978, the United States supplied 15% of the total world production but consumed 23%. World production values and anticipated capacities are shown in Table 5 (21); U.S. production and demand, as well as forecasts, are shown in Table 6 (21).

Usually vanadium is produced as a coproduct or by-product of other materials, including uranium, phosphorus, iron, crude oil, or tars. As such, pricing and availability depends on the aggregate supply–demand relationships of several other commodities. In 1979, the price for vanadium oxide was ca $13.90/kg of contained vanadium. The price for high purity vanadium metal ingot in 1997 is ca $661/kg, but this could decrease markedly if consumption were to increase. U.S. import duties vary from 3% for unwrought, ie, cast or unworked, alloys to 45% for wrought metals, depending on the most-favored-nation status.

Because of the strategic nature of many of the uses, vanadium is one of the materials designated in the National Defense Stockpile Inventory. The goals for 1980 for vanadium-containing materials was 907 metric tons of contained vanadium in ferrovanadium, and 6985 t of contained vanadium in vanadium pentoxide. As of March 1981, the inventory consisted of 491 t of contained vanadium in vanadium pentoxide; there was no ferrovanadium in the inventory (22).

Health and Safety Factors, Toxicology

In the consolidated form, vanadium metal and its alloys pose no particular health or safety hazard. However, they do react violently with certain materials, including BrF_3, chlorine, lithium, and some strong acids (23). As is true with

Table 5. World Vanadium Production (1978) and Capacity (1978, 1979, and 1985), t

Country	Production 1978	Capacity 1978	Capacity 1979	Capacity 1985
North America				
United States	4,721[a]	7,711	7,983	10,886
South America				
Chile	ca 689	1,087	1,087	1,087
Europe				
Finland	2,805	3,131	3,131	3,131
Norway	ca 463	1,179	1,179	1,179
Russia	ca 9,526	14,518	14,518	17,282
Total	*12,794*	*18,828*	*18,828*	*21,592*
Africa				
Republic of South Africa	ca 11,249	12,955	14,225	17,237
Nambia (South-west Africa)	440	726	726	726
Total	*11,689*	*13,681*	*14,951*	*17,963*
Asia				
People's Republic of China	ca 1,996	2,631	8,709	8,709
Oceania				
Australia			635	3,084
World Total	*31,889*	*43,938*	*52,193*	*63,321*

[a]Recovered vanadium.

Table 6. Comparison of U.S. Vanadium Production and Demand: 1958–2000, t

Year	U.S. primary demand	U.S. primary production
1958	1,270	2,532
1961	2,315	5,277
1964	4,280	4,580
1967	5,523	5,372
1970	6,410	5,075
1973	7,756	4,413
1976	8,872	5,622
1979	7,755	5,224
1990	11,521 estd	8,437 estd[a]
2000	16,692 estd	8,800 estd[a]

[a]21-year trend.

many metals, there is a moderate fire hazard in the form of dust or fine powder or when the metal is exposed to heat or flame. Since vanadium reacts with oxygen and nitrogen in air, control of such fires normally involves smothering the burning material with a salt.

Vanadium compounds, including those which may be involved in the production, processing, and use of vanadium and vanadium alloys, are irritants chiefly to the conjuctivae and respiratory tract. Prolonged exposure may lead to pulmonary complications. However, responses are acute, never chronic. Toxic effects vary with the vanadium compound involved. For example, LD_{50} (oral) of vanadium pentoxide dust in rats is 23 mg/kg of body weight (24).

The toxicity of vanadium alloys may depend on other components in the alloy. For example, the V_3Ga alloy requires precautions related to both vanadium and gallium, and gallium is highly toxic. Similarly, alloys with chromium may require precautions associated with that metal.

The adopted values for TWAs for airborne vanadium, including oxide and metal dusts of vanadium, is 0.5 mg/m^3; the values for fumes of vanadium compounds is 0.05 mg/m^3. These limits are for normal 8-h workday and 40-h workweek exposures. The short-term exposure limit (STEL) is 1.5 mg/m^3 for dusts (25). A description of health hazards, including symptoms, first aid, and organ involvement, personal protection, and respirator use has been published (26).

The ammonium salts of vanadic acid and vanadium pentoxide have been listed as toxic constituents in solid wastes under the Resource Conservation and Recovery Act (27).

Uses

The most important use of vanadium is as an alloying element in the steel industry where it is added to produce grain refinement and hardenability in steels. Vanadium is a strong carbide former, which causes carbide particles to form in the steel, thus restricting the movement of grain boundaries during heat treatment. This produces a fine-grained steel which exhibits greater toughness and impact resistance than a coarse-grained steel and which is more resistant to cracking during quenching. In addition, the carbide dispersion confers wear resistance, weldability, and good high temperature strength. Vanadium steels are used in dies or taps because of their deep-hardening characteristics and for cutting tools because of their wear resistance. They are also used as constructional steel in light and heavy sections; for heavy iron and steel castings; forged parts, eg, shafts and turbine motors; automobile parts, eg, gears and axles; and springs and ball bearings. Vanadium is an important component of ferrous alloys used in jet-aircraft engines and turbine blades where high temperature creep resistance is a basic requirement (see HIGH TEMPERATURE ALLOYS).

The principal application of vanadium in nonferrous alloys is the titanium 6–4 alloy (6 wt % Al–4 wt % V), which is becoming increasingly important in supersonic aircraft where strength-to-weight ratio is a primary consideration. Vanadium and aluminum impart high temperature strength to titanium, a property that is essential in jet engines, high speed air frames, and rocket-motor cases. Vanadium foil can be used as a bonding material in the cladding of titanium to steel. Vanadium is added to copper-based alloys to control gas content and microstructure. Small amounts of vanadium are added to aluminum alloys to be used in pistons of internal combustion engines to enhance the alloy's strength and reduce their thermal expansion coefficients. Because of its low capture cross section for fast neutrons as well as its resistance to corrosion by liquid sodium and its good high temperature creep strength, vanadium alloys are receiving considerable attention as a fuel-element cladding for fast-breeder reactors. Vanadium is a component in several permanent-magnet alloys containing cobalt, iron, sometimes nickel, and vanadium. The vanadium content in the most common of these alloys is 2–13 wt % (see MAGNETIC MATERIALS). Vanadium and several vanadium compounds are also used as catalysts in certain chemical and petrochemical reactions.

Liquid-Metal Systems. The liquid-metal fast-breeder reactor (LMFBR) program in the United States and corresponding fast-reactor programs in other countries have considered the use of vanadium as a fuel cladding since its neutron economy, high temperature strength, and corrosion resistance in liquid metals promises higher operating temperatures than stainless steel, which is the reference cladding material.

Interstitial mass transfer is the dominant mechanism of corrosion in vanadium-based alloy–liquid alkali metal systems, eg, lithium or sodium, at 500–650°C (28). Vanadium loses oxygen but absorbs carbon and nitrogen when in contact with lithium at these temperatures; in sodium, vanadium absorbs all three elements. Because varying absorption and loss mechanisms exist for the liquid metals in contact with other structural metals in a system, these effects can be quite severe. Since all three elements contribute to vanadium's strength and ductility, large changes in properties can occur. Alloy additions to vanadium do not change the direction of interstitial transfer but can alter the rate of transfer as well as the morphology of the resultant structure. Formation of surface layers or intermetallic compounds can lead to spalling.

Specific restraints of use in the LMFBR program precluded consideration of all alloy systems. Accordingly, it is reasonable to expect that a much broader range of property improvements is possible through a broader use of alloying additions. No significant programs aimed at achieving these properties have been reported, largely since no commercially attractive results are anticipated.

Superconductivity. One potential future use of vanadium is in the field of superconductivity. The compound V_3Ga exhibits a critical current at 20 T (20×10^4 G), which is one of the highest of any known material. Although niobium–zirconium and Nb_3Sn have received more attention, especially in the United States, the vanadium compound is being studied for possible future application in this field since V_3Ga exhibits a critical temperature of 15.4 K as opposed to 18.3 K for Nb_3Sn (see SUPERCONDUCTING MATERIALS).

Like Nb_3Sn, V_3Ga has a Type A15 structure, which exhibits low ductility. Consequently, efforts at producing the compound in fine-grained fibrous structures have involved unique processing techniques (29). One method consists of forming a billet consisting of vanadium rods surrounded by a Cu–Ga bronze alloy. The billet is fabricated to wire by conventional extrusion, rod-rolling, or swagging and wire-drawing techniques. Following production of a wire, a heat treatment is performed which leads to diffusion of Ga to the fine V filaments, thereby forming the V_3Ga intermetallic compound.

Another technique, developed originally for Cu–Nb alloys, begins by rapidly chilling a liquid Cu–V alloy. During solidification, a fine dispersion of primary vanadium grains form in the matrix of the Cu–V alloy. The cast billet is fabricated by the methods described above. Following formation of the wire, gallium is plated on it and thermally diffuses, forming the V_3Ga which is in long, thin platelets ca 0.01–0.1 μm dia.

Fusion Reactors. The development of fusion reactors requires a material exhibiting high temperature mechanical strength, resistance to radiation-induced swelling and embrittlement, and compatibility with hydrogen, lithium and various coolants. One alloy system that shows promise in this application, as well as for steam-turbine blades and other applications in nonoxidizing atmospheres, is based on the composition $(Fe,Co,Ni)_3V$ (30). Through control of an ordered–disordered transformation, the yield strength of these alloys increases at elevated temperatures above the room-temperature yield strength. One composition, for example, exhibits a yield strength of 480 MPa (70,000 psi) at ca 750°C compared to its room temperature value of 345 MPa (50,000 psi). The alloys also show good resistance to radiation-induced swelling (see FUSION ENERGY). These alloys can be fabricated, eg, by rolling, at 1000–1100°C and are then heat-treated at 800–1100°C.

BIBLIOGRAPHY

"Vanadium and Vanadium Alloys" in *ECT* 1st ed., Vol. 14, p. 583, by J. Strauss, Vanadium Corp. of America; in *ECT* 2nd ed., Vol. 21, pp. 157–167, by O. N. Carlson and E. R. Stevens, Ames Laboratory of the U.S. Atomic Energy Commission; in *ECT* 3rd ed., Vol. 23, pp. 673–687, E. F. Baroch, International Titanium, Inc.

1. S. A. Bradford and O. N. Carlson, *ASM Trans. Q.* **55**, 493 (1962).
2. R. W. Thompson and O. N. Carlson, *J. Less-Common Met.* **9**, 354 (1965).
3. D. L. Harrod and R. E. Gold, *International Metals Reviews*, 163 (1980).
4. C. A. Hampel, ed., *The Encyclopedia of the Chemical Elements*, Reinhold Publishing Corp., New York, 1968, p. 790.
5. C. A. Hampel, ed., *Rare Metals Handbook*, 2nd ed., Reinhold Publishing Corp., New York, 1961, p. 634.
6. *Metals Handbook*, 9th ed., Vol. 2, American Society for Metals, Metals Park, Ohio, 1979, p. 822.
7. O. N. Carlson in H. Y. Sohn, O. N. Carlson, and J. T. Smith, eds., *Extractive Metallurgy of Refractory Metals*, The Metallurgical Society of AIME, Warrendale, Pa., 1980, p. 191.
8. T. J. McLeer, Foote Mineral Co., Exton, Pa., personal communication, July 1969.
9. P. T. Brooks and G. M. Potter, *Recovering Vanadium from Dolomitic Nevada Shale*, Bureau of Mines RI 7932, U.S. Bureau of Mines, Washington, D.C., 1974, 20 pp.
10. U.S. Pat. 3,420,659 (Oct. 11, 1967), H. W. Rathmann and R. T. C. Rasmussen (to Foote Mineral Co.).
11. F. H. Perfect, *Trans. Metall. Soc. AIME* **239**, 1282 (1967).
12. J. W. Marden and M. N. Rich, *Ind. Eng. Chem.* **19**, 786 (1927).
13. R. K. McKechnie and A. U. Seybolt, *J. Electrochem. Soc.* **97**, 311 (1950).
14. O. N. Carlson, F. A. Schmidt, and W. E. Krupp, *J. Met.* **18**, 320 (1966).
15. O. N. Carlson and C. V. Owen, *J. Electrochem. Soc.* **108**, 88 (1961).
16. T. A. Sullivan, *J. Met.* **17**, 45 (1965).
17. F. A. Schmidt and J. C. Warner, *J. Less-Common Met.* **13**, 493 (1967).
18. R. W. Huber and I. R. Lane, Jr., *Consumable-Electrode Arc Melting of Titanium and Its Alloys*, Bureau of Mines R.I. 5311, U.S. Bureau of Mines, Washington, D.C., 1957, 36 pp.
19. R. W. Buchman, Jr., *International Metals Reviews*, 158 (1980).
20. Ref. 6, pp. 822–823.
21. G. A. Morgan, "Vanadium" in *Mineral Facts and Problems*, Bureau of Mines Bulletin 671, U.S. Bureau of Mines, Washington, D. C., 1980, 10 pp.

22. L. O. Giuffrida, *Stockpile Report to the Congress, October 1980–March 1981*, P&P-1, Federal Emergency Management Agency, Washington, D.C., Nov. 1981, 30 pp.

23. N. I. Sax, *Dangerous Properties of Industrial Materials*, 5th ed., Van Nostrand Reinhold Co., New York, 1979, p. 1082.

24. Ref. 23, p. 1083.

25. *Threshold Limit Values for Chemical Substances in Workroom Air adopted By ACGIH for 1981*, American Conference of Governmental Industrial Hygienists, Cincinnati, Ohio, 1981.

26. *NIOSH/OSHA Pocket Guide to Chemical Hazards*, American Optical Corporation, Southbridge, Mass., 1978, pp. 108–109.

27. *Fed. Reg.* **45**, 33121, 33133 (May 19, 1980).

28. R. L. Ammion, *International Metals Reviews*, 255 (1980).

29. B. N. Das, J. E. Cox, R. W. Huber, and P. A. Meussner, *Metall. Trans.* **8A**, 541 (1977).

30. C. T. Liu, *J. Nucl. Mat.* **85/86**, 907 (1979).

General References

R. Rostoker, *The Metallurgy of Vanadium*, John Wiley & Sons, Inc., New York, 1965.

T. E. Dietz and J. W. Wilson, *Behavior and Properties of Refractory Metals*, Stanford University Press, Stanford, Calif., 1965.

M. E. Weeks, *Discovery of the Elements*, 6th ed., Mack Printing Co., Easton, Pa., 1956.

C. A. Hampel, ed., *Rare Metals Handbook*, 2nd ed., Reinhold Publishing Corp., New York, 1961, p. 634.

Economic Analysis of the Vanadium Industries, U.S. Dept. of Commerce Document PB-176 471, U.S. Dept. of Commerce, Washington, D.C., June 1967.

D. R. Spink, G. L. Rempel, and C. O. Gomez-Bueno, in H. Y. Sohn, O. N. Carlson, and J. T. Smith, eds., *Extractive Metallurgy of Refractory Metals*, The Metallurgical Society of AIME, Warrendale, Pa., 1980, p. 147.

G. Gabra and I. Malinsky, in H. Y. Sohn, O. N. Carlson, and J. T. Smith, eds., *Extractive Metallurgy of Refractory Metals*, The Metallurgical Society of AIME, Warrendale, Pa., 1980, p. 167.

Economic Analysis of the Vanadium Industries, U.S. Dept. of Commerce Document PB-176 471, U.S. Dept. of Commerce, Washington, D.C., June 1967, p. 70.

R. C. Svedberg and R. W. Buchman, Jr., *International Metals Reviews*, 223 (1980).

R. E. Gold and D. L. Harrod, *Int. Metals Rev.*, 232 (1980).

U.S. Pat. 4,002,504 (Jan. 11, 1977), D. G. Howe.

D. G. Howe, T. L. Francavilla, and D. U. Gubser, *IEEE Trans. Magn.* **13**, 815 (Jan. 1977).

VANADIUM COMPOUNDS

Vanadium is widely dispersed in the earth's crust at an average concentration of ca 150 ppm. Deposits of ore-grade minable vanadium are rare. Vanadium is ordinarily recovered from its raw materials in the form of the pentoxide, but sometimes as sodium and ammonium vanadates. These initial compounds have catalytic and other chemical uses (see CATALYSIS). For such uses and for conversion to other vanadium chemicals, granular V_2O_5 usually is made by decomposing ammonium metavanadate. For metallurgical uses, which represent ca 90% of vanadium consumption, some oxides are prepared for conversion to master alloys by fusion and flaking to form glassy chips. The preparation and application of alloying materials are described elsewhere (see VANADIUM AND VANADIUM ALLOYS). Vanadium–aluminum master alloy is made for alloying with titanium metal. A part of such vanadium-rich raw materials as slags, ash, and residues is smelted directly to master alloys or to alloy steels. Figure 1 shows the prevailing process route (1).

Possibly because of price and performance competition from chromium, titanium, and other transition elements, only about a dozen vanadium compounds are commercially significant; of these, vanadium pentoxide is dominant.

Physical Properties

Some properties of selected vanadium compounds are listed in Table 1. Detailed solubility data are available (3), as are physical constants of other vanadium compounds (4). Included are the lattice energy of several metavanadates and the magnetic susceptibility of vanadium bromides, chlorides, fluorides, oxides, and sulfides (5).

Vanadium, a typical transition element, displays well-characterized valence states of 2–5 in solid compounds and in solutions. Valence states of -1 and 0 may occur in solid compounds, eg, the carbonyl and certain complexes. In oxidation state 5, vanadium is diamagnetic and forms colorless, pale yellow, or red compounds. In lower oxidation states, the presence of one or more $3d$ electrons, usually unpaired, results in paramagnetic and colored compounds. All compounds of vanadium having unpaired electrons are colored, but because the absorption spectra may be complex, a specific color does not necessarily correspond to a particular oxidation state. As an illustration, vanadium(IV) oxy salts are generally blue, whereas vanadium(IV) chloride is deep red. Differences over the valence range of 2–5 are shown in Table 2. The structure of vanadium compounds has been discussed (6,7).

Chemical Properties

The chemistry of vanadium compounds is related to the oxidation state of the vanadium. Thus, V_2O_5 is acidic and weakly basic, VO_2 is basic and weakly acidic, and V_2O_3 and VO are basic. Vanadium in an aqueous solution of vanadate salt occurs as the anion, eg, $(VO_3)^-$ or $(V_3O_9)^{3-}$, but in strongly acid solution, the cation $(VO_2)^+$ prevails. Vanadium(IV) forms both oxyanions $((V_4O_9)^{2-}$ and oxycations $(VO)^{2+}$. Compounds of vanadium(III) and (II) in solution contain the hydrated ions $[V(H_2O)_6]^{3+}$ and $[V(H_2O)_6]^{2+}$, respectively.

Coordination compounds of vanadium are mainly based on six coordination, in which vanadium has a pseudooctahedral structure. Coordination number four is typical of many vanadates. Coordination numbers five and eight also are known for vanadium compounds, but numbers less than four have not been reported. The coordination chemistry of vanadium has been extensively reviewed (8–12) (see COORDINATION COMPOUNDS).

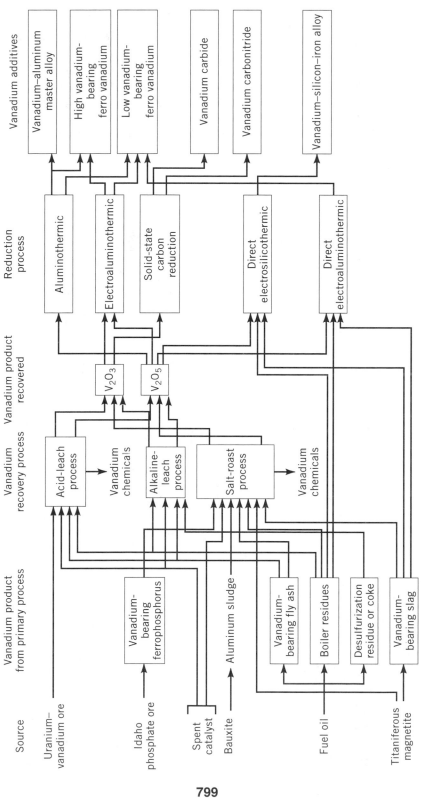

Fig. 1. Generalized flow sheet of minerals processing to vanadium products (1).

Table 1. Physical Properties of Some Industrial and Other Selected Vanadium Compounds[a]

Compound	CAS Registry Number	Formula	Appearance	Mol wt	Density, g/cm^3	Mp, °C	Bp, °C	Solubility
vanadic acid, meta	[13470-24-1]	HVO_3	yellow scales	99.95				soluble in acid and alkali
ammonium metavanadate	[7803-55-6]	NH_4VO_3	white–yellowish or colorless crystals	116.98	2.326	200 dec		slightly soluble in H_2O
potassium metavanadate	[13769-43-2]	KVO_3	colorless crystals	134.04				soluble in hot H_2O
sodium metavanadate	[13718-26-8]	$NaVO_3$	colorless, monoclinic prisms	121.93		630		soluble in H_2O
sodium orthovanadate	[13721-39-6]	Na_3VO_4	colorless, hexagonal prisms	183.94		850–856		soluble in H_2O
sodium pyrovanadate	[13517-26-5]	$Na_4V_2O_7$	colorless, hexagonal prisms	305.84		632–654		soluble in H_2O
vanadium carbide	[12070-10-9]	VC	black cubic	62.95	5.77	2810	3900	insoluble in H_2O; soluble in HNO_3 with decomposition
vanadium nitride	[24646-85-3]	VN	black cubic crystals	64.95	6.13	2320 dec		soluble in *aqua regia*
vanadium(III) trichloride	[7718-98-1]	VCl_3	pink crystals, deliquescent	157.301	3.00			soluble (deliquescent) in H_2O; soluble in methanol and ether
vanadium(IV) tetrachloride	[7632-51-1]	VCl_4	red-brown liquid	192.75	1.816	28 ± 2	148.5	soluble (deliquescent) in H_2O; soluble in methanol, ether, and chloroform
vanadium(V) oxytrichloride	[7727-18-6]	$VOCl_3$	yellow liquid	173.30	1.829	−77 ± 2	126.7	soluble (deliquescent) in H_2O; soluble in methanol, ether, acetone, and acid
vanadium(II) oxide	[12035-98-2]	VO	light green crystals	66.95	5.758	ignites		soluble in acid
vanadium(III) oxide	[1314-34-7]	V_2O_3	blue crystals	149.88	4.87	1970		soluble in HNO_3, HF, and alkali in presence of oxide
vanadium(IV) oxide	[12036-21-4]	VO_2	blue crystals	82.94	4.339	1967		soluble in acid and alkali
vanadium(V) oxide	[1314-62-1]	V_2O_5	yellow-red rhombohedra	181.88	3.357	690	1750 dec	slightly soluble in H_2O; soluble in acid and alkali
vanadium(IV) disilicide	[12039-87-1]	VSi_2	metallic prisms	107.11	4.42			soluble in HF
vanadium(III) acetylacetonate	[13476-99-8]	$V(C_5H_7O_2)_3$	brown crystals	348.27	0.9–1.2	178–190		soluble in methanol, acetone, benzene, and chloroform
biscyclopentadienyl-vanadium chloride	[12083-18-6]	$(C_5H_5)_2VCl_2$	pale green crystals	252.04		<250 dec		soluble in methanol and chloroform

[a]Ref. 2.

Table 2. Magnetism and Color of Vanadium Compounds

Compound	CAS Registry Number	Vanadium valence	No. of d electrons	Magnetic moment, $J/T \times 10^{-23a}$	Color
$VOCl_3$	[7727-18-6]	5	0	0	yellow
$VOSO_4 \cdot 5H_2O$	[12439-96-2]	4	1	1.60	blue
$(NH_4)V(SO_4)_2 \cdot 12H_2O$	[29932-01-2]	3	2	2.60	blue
$VSO_4 \cdot 7H_2O$	[36907-42-3]	2	3	3.47	violet

^{a}To convert J/T to μ_B, divide by 9.274×10^{-24}.

Aqueous pentavalent vanadium is readily reduced to the tetravalent state by iron powder or by SO_2 gas. A stronger reducing agent, eg, zinc amalgam, is needed to yield divalent vanadium. Divalent and trivalent vanadium compounds are reducing agents and require storage under an inert atmosphere to avoid oxidation by air.

Interstitial and Intermetallic Compounds. In common with certain other metals, eg, Hf, Nb, Ti, Zr, Mo, W, and Ta, vanadium is capable of taking atoms of nonmetals into its lattice. Such uptake is accompanied by a change in the packing pattern to a cubic close-packed structure. Carbides, hydrides, and nitrides so formed are called interstitial compounds. Their composition is determined by geometrical packing arrangements rather than by valence bonding. As all possible vacant lattice sites need not be filled, the compositions display a range of nonmetal content up to the theoretical limit. A range for the vanadium nitride compound is $VN_{0.71-1.00}$. Compounds corresponding to V_2C [12012-17-8], VC, and $VH_{1.8}$ [12713-06-3] have been formed. Vanadium borides also are known. In VB [12045-27-1], the boron atoms form a zigzag chain; whereas, in VB_2 [12007-37-3], the interstitial boron atoms are linked in a layer of hexagons.

Large-atomed nonmetals, eg, Si, Ge, P, As, Se, and Te, form compounds with vanadium that are intermediate between being interstitial and intermetallic. The interstitial and intermetallic vanadium compounds are very hard crystalline solids (9–10 on Mohs' scale), have high melting points (2000–3000°C), and generally are resistant to attack by mineral acids. Except for carbides that are used in impure form, eg, in making alloy steels, this group of compounds is little used. Vanadium disilicide, VSi_2, which is made by the reaction of vanadium pentoxide with silicon metal at ca 1200°C, has limited use as a refractory material (see REFRACTORIES). Preparation of vanadium nitride from the reaction of vanadium oxides with ammonia was investigated by the U.S. Bureau of Mines (13). In subsequent studies, the nitrides were to be converted to vanadium metal.

Vanadium Oxides. Vanadium pentoxide (V_2O_5) is intermediate in behavior and stability between the highest oxides of titanium, ie, TiO_2, and of chromium, ie, CrO_3. It is thus less stable to heat than TiO_2 and more heat-stable than CrO_3. Also, V_2O_5 is more acidic and a stronger oxidant than TiO_2, but less so than CrO_3. An excess of aluminum is capable of reducing V_2O_5 to a V–Al alloy, but only calcium can reduce the oxide to vanadium metal. Solubility in water is 0.1–0.8 g/100 g H_2O and is lowest for crystals solidified from the molten state. The pentoxide readily dissolves in both acids and alkalies. In strongly alkaline solutions (pH >13), simple mononuclear vanadate ions are present. In strongly acid solutions, the main species is the dioxyvanadium(V) cation $(VO_2)^+$. More complex ions are present in solutions of intermediate strength alkali or acid; these include $(VO_3)^-$, $(HVO_4)^{2-}$, $(V_3O_9)^{3-}$, $(V_4O_{12})^{4-}$, $(V_{10}O_{28})^{6-}$, and others.

Vanadium(IV) Oxide. Vanadium(IV) oxide (vanadium dioxide, VO_2) is a blue-black solid, having a distorted rutile (TiO_2) structure. It can be prepared from the reaction of V_2O_5 at the melting point with sulfur or carbonaceous reductants such as sugar or oxalic acid. The dioxide slowly oxidizes in air. Vanadium dioxide dissolves in acids to give the stable $(VO)^{2+}$ ions and in hot alkalies to yield vanadate(IV) species, eg, $(HV_2O_5)^-$.

Vanadium(III) Oxide. Vanadium(III) oxide (vanadium sesquioxide, V_2O_3) is a black solid, having the corundum (Al_2O_3) structure. It can be prepared by reduction of the pentoxide by hydrogen or carbon. Air oxidation proceeds slowly at ambient temperatures, but oxidation by chlorine at elevated temperatures to give $VOCl_3$ and V_2O_5 is rapid.

Vanadium(II) Oxide. Vanadium(II) oxide is a nonstoichiometric material with a gray-black color, metallic luster, and metallic-type electrical conductivity. Metal–metal bonding increases as the oxygen content decreases, until an essentially metal phase containing dissolved oxygen is obtained (14).

Vanadates. Ammonium metavanadate and, to a lesser extent, potassium and sodium metavanadates, are the main vanadates of commercial interest. The pure compounds are colorless crystals. Vanadates(V) are identified as the meta (MVO_3), ortho (M_3VO_4), and pyro ($M_4V_2O_7$) compounds, and the nomenclature commonly used is that applied to the phosphates (M is univalent). Orthovanadates contain discrete tetrahedral $(VO_4)^{3-}$ anions. The pyrovanadates also contain discrete ions $(V_2O_7)^{4-}$ and have two VO_4 tetrahedra sharing a corner. The metavanadate structure is different for the anhydrous and hydrated salts. In the anhydrous salts KVO_3 and NH_4VO_3, the vanadium atoms are four-coordinate, with VO_4 tetrahedra linking through two oxygen atoms. In the hydrated form, $KVO_3 \cdot H_2O$, the vanadium atoms are five-coordinate, with three shared oxygen atoms per vanadium and two terminal oxygen atoms.

Vanadium Halides and Oxyhalides. Known halides and oxyhalides of vanadium, their valences, and their colors are listed in Table 3. Only vanadium(V) oxytrichloride ($VOCl_3$) and the tetrachloride (VCl_4) have appreciable commercial importance. The trichloride (VCl_3) is of minor commercial interest. The absence of pentavalent vanadium halides, other than the fluoride, is attributed to the relative weakness of the fluoride–fluoride bond compared with analogous bonds in other halides. Even VCl_4 is somewhat unstable, and VBr_4 decomposes at $-23°C$. The halides and oxyhalides have been well-characterized (6).

Vanadium(V) Oxytrichloride. Vanadium(V) oxytrichloride ($VOCl_3$) is readily hydrolyzed and forms coordination compounds with simple donor molecules, eg, ethers, but is reduced by reaction with sulfur-containing ligands and molecules. It is completely miscible with many hydrocarbons and nonpolar metal halides, eg, $TiCl_4$, and it dissolves sulfur.

Vanadium(IV) Chloride. Vanadium(IV) chloride (vanadium tetrachloride, VCl_4) is a red-brown liquid, is readily hydrolyzed, forms addition compounds with donor solvents such as pyridine, and is reduced by such molecules to trivalent vanadium compounds. Vanadium tetrachloride dissociates slowly at room temperature and rapidly at higher temperatures, yielding VCl_3 and Cl_2.

Table 3. Vanadium Halides and Oxyhalides

Vanadium valence	Formula	CAS Registry Number	Appearance[a]
		Halides	
V	VF_5	[7783-72-4]	white
IV	VF_4	[10049-16-8]	green
	VCl_4	[7632-51-1]	red-brown liquid
	VBr_4[b]	[13595-30-7]	magenta
III	VF_3	[10049-12-4]	green
	VCl_3	[7718-98-1]	red-violet
	VBr_3	[13470-26-3]	gray
	VI_3	[15513-94-7]	dark brown
II	VF_2	[13842-80-3]	blue
	VCl_2	[10580-52-6]	pale green
	VBr_2	[14890-41-6]	orange-brown
	VI_2	[15513-84-5]	red
		Oxyhalides	
V	VOF_3	[13709-31-4]	pale yellow
	$VOCl_3$	[7729-18-6]	yellow liquid
	$VOBr_3$	[13520-90-6]	deep red liquid
	VO_2F	[14259-82-6]	brown
	VO_2Cl	[13759-30-3]	orange
IV	VOF_2	[13814-83-0]	yellow
	$VOCl_2$	[10213-09-9]	green
	$VOBr_2$	[13520-89-3]	yellow-brown
III	$VOCl$	[13520-87-1]	brown
	$VOBr$	[13520-88-2]	violet

[a] At room temperature; solid unless identified as liquid.
[b] Decomposes at $-23°C$.

Decomposition also is induced catalytically and photochemically. This instability reflects the difficulty in storing and transporting it for industrial use.

Vanadium(III) Chloride. Vanadium(III) chloride (vanadium trichloride, VCl_3) is a pink-violet solid, is readily hydrolyzed, and is insoluble in nonpolar solvents but dissolves in donor solvents, eg, acetonitrile, to form coordination compounds. Chemical behavior of the tribromide (VBr_3) is similar to that of VCl_3.

Vanadium Sulfates. Sulfate solutions derived from sulfuric acid leaching of vanadium ores are industrially important in the recovery of vanadium from its raw materials. Vanadium in quadrivalent form may be solvent-extracted from leach solutions as the oxycation complex $(VO)^{2+}$. Alternatively, the vanadium can be oxidized to the pentavalent form and solvent-extracted as an oxyanion, eg, $(V_3O_9)^{3-}$. Pentavalent vanadium does not form simple sulfate salts.

Vanadium(IV) Oxysulfate. Vanadium(IV) oxysulfate pentahydrate (vanadyl sulfate), $VOSO_4·5H_2O$) is an ethereal blue solid and is readily soluble in water. It forms from the reduction of V_2O_5 by SO_2 in sulfuric acid solution. Vanadium(III) sulfate [13701-70-7] ($V_2(SO_4)_3$) is a powerful reducing agent and has been prepared in both hydrated and anhydrous forms. The anhydrous form is insoluble in either water or sulfuric acid. Vanadium(II) sulfate heptahydrate ($VSO_4·7H_2O$)

is a light red-violet crystalline powder that can be prepared by electrolytic reduction of $VOSO_4$. The powder is oxidized by air and dissolves in water to give a red-violet solution.

Manufacture

Primary industrial compounds produced directly from vanadium raw materials are principally 98 wt % fused pentoxide, air-dried (technical-grade) pentoxide, and technical-grade ammonium metavanadate (NH_4VO_3). Much of the fused and air-dried pentoxides produced at the millsite is made by thermal decomposition of ammonium vanadates. Prior to 1960, the main vanadium mill products were fused technical-grade pentoxide (black cake) containing 86–92 wt % V_2O_5 and air-dried technical-grade pentoxide (red cake) containing 83–86 wt % V_2O_5, both being high in alkali content. An historical review of the manufacture of vanadium compounds until 1960 is available (15). Some milling practices for the production of primary vanadium compounds from ores have been reviewed (16).

Vanadium raw materials are processed to produce vanadium chemicals, eg, the pentoxide and ammonium metavanadate (AMV) primary compounds, by salt roasting or acid leaching. Interlocking circuits, in which unfinished or scavenged material from one process is diverted to the other, are sometimes used. Such interlocking to enhance vanadium recovery and product grade became more feasible in the late 1950s with the advent of solvent extraction.

Salt Roasting. Iron ore concentrate, uranium–vanadium ores, ferrophosphorus from manufacture of elemental phosphorus, vanadiferous shale, and assorted slag, ash, fumes, residues, and depleted catalysts, singly or in combination, are suitable feed for the salt-roast process. Sometimes, substitution of sodium carbonate for part or all of the salt results in improved vanadium recovery. The presence of calcium and magnesium carbonates is deleterious, because these form vanadates that are insoluble in water. Interference from calcium carbonate can be ameliorated by acidulating the ore with sulfuric acid before roasting or by adding pyrite to the charge, which results in formation of calcium sulfate which is innocuous in salt roasting. Calcium vanadates, present in a salt-roast calcine, can be dissolved by acid leaching the calcine. Limestone of low magnesium content must be added to the charge when salt roasting ferrophosphorus to combine with the phosphate that forms.

The ore is ordinarily ground to pass through a ca 1.2-mm (14-mesh) screen, mixed with 8–10 wt % NaCl and other reactants that may be needed, and roasted under oxidizing conditions in a multiple-hearth furnace or rotary kiln at 800–850°C for 1–2 h. Temperature control is critical because conversion of vanadium to vanadates slows markedly at ca 800°C, and the formation of liquid phases at ca 850°C interferes with access of air to the mineral particles. During roasting, a reaction of sodium chloride with hydrous silicates, which often are present in the ore feed, yields HCl gas. This is scrubbed from the roaster off-gas and neutralized for pollution control, or used in acid-leaching processes at the mill site.

Hot calcine from the kiln is water-quenched or cooled in air before being lightly ground and leached. Air cooling allows back reactions, which adversely affect vanadium extraction for some ores. Leaching and washing of the residue is by percolation in vats or by agitation and filtration. Extraction of vanadium is 65–85%. Vanadium solution from water leaching of the calcine has a pH of 7–8 and a vanadium content of ca 30–50 g V_2O_5/L. If the water-leach residue is leached subsequently with sulfuric acid solution, as for uranium extraction, as much as 10–15 wt % more vanadium may dissolve. Originally, such acid-leached vanadium precipitated from the uranium solvent extraction raffinate as an impure sludge that was recycled to the salt-roast kiln. The vanadium generally is recovered from the uranium raffinate by solvent extraction (see URANIUM AND URANIUM COMPOUNDS).

Recovery of Vanadium from Salt-Roast Leach Solution. When recovery of vanadium as sodium red cake was practiced, the leach solution pH was adjusted to 2.7 by adding sulfuric acid. Any vanadium not already in the pentavalent form was oxidized by addition of sodium chlorate, and the solution was boiled for 2–8 h to precipitate substantially all of the vanadium as shotlike particles of sodium red cake. Although melting and flaking converted the highly alkaline red cake to black cake suitable for most metallurgical uses, extensive purification, usually by conversion to ammonium metavanadate, was required to prepare compounds for catalytic and chemical use. The process routes favored for recovering vanadium from the leach solution are solvent extraction and precipitation of ammonium vanadates or vanadic acid. Sometimes both solvent extraction and direct precipitation are used in integrated circuits.

For solvent extraction of pentavalent vanadium as a decavanadate anion, the leach solution is acidified to ca pH 3 by addition of sulfuric acid. Vanadium is extracted in about four countercurrent mixer–settler stages by a 3–5 wt % solution of a tertiary alkyl amine in kerosene. The organic solvent is stripped by a soda-ash or ammonium hydroxide solution, and addition of ammoniacal salts to the rich vanadium strip liquor yields ammonium metavanadate. A small part of the metavanadate is marketed in that form and some is decomposed at a carefully controlled low temperature to make air-dried or fine granular pentoxide, but most is converted to fused pentoxide by thermal decomposition at ca 450°C, melting at 900°C, then chilling and flaking.

For solvent extraction of a tetravalent vanadium oxyvanadium cation, the leach solution is acidified to ca pH 1.6–2.0 by addition of sulfuric acid, and the redox potential is adjusted to −250 mV by heating and reaction with iron powder. Vanadium is extracted from the blue solution in ca six countercurrent mixer–settler stages by a kerosene solution of 5–6 wt % di-2-ethylhexyl phosphoric acid (EHPA) and 3 wt % tributyl phosphate (TBP). The organic solvent is stripped by a 15 wt % sulfuric acid solution. The rich strip liquor containing ca 50–65 g V_2O_5/L is oxidized batchwise initially at pH 0.3 by addition of sodium chlorate; then it is heated to 70°C and agitated during the addition of NH_3 to raise the pH to 0.6. Vanadium pentoxide of 98–99% grade precipitates, is removed by filtration, and then is fused and flaked.

For direct precipitation of vanadium from the salt-roast leach liquor, acidulation to ca pH 1 without the addition of ammonia salts yields an impure vanadic acid; when ammonium salts are added, ammonium polyvanadate precipitates. The impure vanadic acid ordinarily is redissolved in sodium carbonate solution, and ammonium metavanadate precipitates upon addition of ammonium salts. Fusion of the directly precipitated ammonium salts can yield high purity V_2O_5

for the chemical industry. Amine solvent extraction is sometimes used to recover 1–3 g/L of residual V_2O_5 from the directly precipitated tail liquors.

Other Developments.   Recovery of vanadium from a dolomitic shale by salt roasting, acid leaching, and amine solvent extraction has been reported (17). A patent was issued for recovery of vanadium-bearing solution from silica containing titaniferous magnetic ore by roasting a mixture of ore, a sodium salt, and cryolite at <1350°C for 30 min to 2 h and then leaching the calcine (18). Roasting of iron ores with limestone to form calcium vanadate [14100-64-2] and leaching of the vanadate with ammonium carbonate or bicarbonate solution have also been patented (19). A patent was granted for heating a mixture of slag and sodium carbonate in a converter at 600–800°C in the presence of oxygen to solubilize the vanadium, leach with water, and recover vanadium from the leach solution (20). Another patent for vanadium recovery from slag calls for an oxidizing roast, followed by leaching of the calcine with phosphoric acid to dissolve the vanadium, and recovery of vanadium from the leach solution (21).

Acid Leaching.   Direct acid leaching for vanadium recovery is used mainly for vanadium–uranium ores and less extensively for processing spent catalyst, fly ash, and boiler residues. Although V_2O_5 in spent catalysts dissolves readily in acid solutions, the dissolution of vanadium from ores and other feed materials requires leaching for ca 14–24 h in strong, hot, oxidizing sulfuric acid solutions. Ore is ground to ca 0.60 mm (28 mesh) and is leached at 50–55 wt % solids and 75°C in a series of about four agitated tanks. Enough sulfuric acid and sodium chlorate are added to the first tank in the series to maintain ca 70 g/L of free acid in the second tank and a terminal redox potential of at least −430 mV. Such intensive leaching conditions dissolve ca 75% of the vanadium and 95% of the uranium. Excess acid in the leach liquor, after liquid–solids separation, is neutralized by reaction with fresh ore. A second liquid–solids separation produces a clarified leach liquor of ca pH 1 for solvent extraction. Vanadium is recovered from the uranium solvent-extraction raffinate.

For vanadium solvent extraction, iron powder can be added to reduce pentavalent vanadium to quadrivalent and trivalent iron to divalent at a redox potential of −150 mV. The pH is adjusted to 2 by addition of NH_3, and an oxyvanadium cation is extracted in four countercurrent stages of mixer–settlers by a diesel oil solution of EHPA. Vanadium is stripped from the organic solvent with a 15 wt % sulfuric acid solution in four countercurrent stages. Addition of NH_3, steam, and sodium chlorate to the strip liquor results in the precipitation of vanadium oxides, which are filtered, dried, fused, and flaked (22). Vanadium can also be extracted from oxidized uranium raffinate by solvent extraction with a tertiary amine, and ammonium metavanadate is produced from the soda-ash strip liquor. Fused and flaked pentoxide is made from the ammonium metavanadate (23).

Australian Vanadium–Uranium Ore.   A calcareous carnotite ore at Yeelirrie, Australia, is ill-suited for salt roasting and acid leaching. Dissolution of vanadium and uranium by leaching in sodium carbonate solution at elevated temperature and pressure has been tested on a pilot-plant scale (24).

Halides and Oxyhalides.   Vanadium(V) oxytrichloride is prepared by chlorination of V_2O_5 mixed with charcoal at red heat. The tetrachloride (VCl_4) is prepared by chlorinating crude metal at 300°C and freeing the liquid from dissolved chlorine by repeated freezing and evacuation. It now is made by chlorinating V_2O_5 or $VOCl_3$ in the presence of carbon at ca 800°C. Vanadium trichloride (VCl_3) can be prepared by heating VCl_4 in a stream of CO_2 or by reaction of vanadium metal with HCl.

Production

The bulk of world vanadium production is derived as a by-product or coproduct in processing iron, titanium, and uranium ores, and, to a lesser extent, from phosphate, bauxite, and chromium ores and the ash, fume, or coke from burning or refining petroleum. Total world production of V_2O_5 was ca 131×10^6 lbs in 1996.

Most U.S. production (20×10^6 lbs in 1996) of primary vanadium compounds has been as by-products or coproducts of uranium and of ferrophosphorus derived from smelting Idaho phosphates. Most of this processing was from leaching acids, residues, and spent catalysts. The only domestic commercially mined ore, for its sole production of vanadium, is Arkansas brookite. It has contributed significantly to domestic supply since ca 1969, however, it has not been mined since 1992 (25).

Most foreign vanadium is obtained as a coproduct of iron and titanium. South Africa, Norway, and Finland are suppliers. Chile produces slag from an iron operation. Australia's first vanadium operation started producing fused pentoxide flake from a vanadium mine in 1980. Russia and the People's Republic of china produce slag and pentoxide from iron–titanium ores.

Economic Aspects and Specifications

Prices are only published in the *London Metal Bulletin*. In June 1997, the following equivalent prices per kilogram of FeV and V_2O_5, respectively, were reported (26): $18.40–18.90, on the basis of metals containing 70–80% V, and $3.75–$3.80, based on minimum oxide of 98% V_2O_5.

Product data issued by a producer of vanadium catalysts include the following specifications. Vanadium(III) acetylacetonate, $V(C_5H_7O_2)_3$ 98.0 wt % min, $VO(C_5H_7O_2)_2$ 2 wt % max, must be maintained under inert atmosphere (nitrogen containing <10 ppm O_2 is recommended). Vanadium oxytrichloride, $VOCl_3$ 95.0 wt % min, V^{5+} 28.5 wt % min, Cl 60.5–61.5 wt % Fe 0.06 wt % max, must be maintained under inert atmosphere (N_2 containing <10 ppm O_2); exposure to moisture in air results in formation of HCl and HVO_3. Vanadium tetrachloride, vanadium(IV) 22.5 wt % min, vanadium(V) 3.0 wt % max, free Cl_2 ca 3.0 wt %, chloride 73.0 wt %; store under inert atmosphere and away from heat; reacts vigorously with water.

Energy Use in the Manufacture of Fused Pentoxide. The energy required to produce fused pentoxide from a vanadium–uranium ore containing 1.3 wt % V_2O_5 and 0.20 wt % U_3O_8 is ca 360 MJ/kg (155,000 Btu/lb) (27). Treatment is assumed to be by salt roasting, water and acid leaching, uranium and vanadium solvent extraction, and production of fused pentoxide. Vanadium recovery from ore is ca 80%. The hypothetical carnotite ore selected for the energy estimation is relatively rich in vanadium and uranium. Processing of lower grade feed would require more energy per unit of product.

Imports and Exports.   The United States has long been a significant importer of vanadium slags, but imports of pentoxide were negligible until they rose quickly to 850 metric tons in 1974, and 2000 t in 1975 (mostly from the Republic of South Africa). Pentoxide imports then declined to 1400 t in 1980 with Finland being the main and South Africa the minor suppliers. In recent years, U.S. imports of ammonium and potassium vanadates and of other vanadium compounds have been 100–200 t/yr, mainly from the U.K., Germany, and the Republic of South Africa.

Annual U.S. exports of the pentoxide and other compounds were 1300–1400 t in 1978 and declined to 800–900 t in 1979 and 1980. The anhydrous pentoxide accounted for roughly three-fourths of the compounds exported.

U.S. Stockpile.   A U.S. government stockpile goal for vanadium pentoxide of 6985 t contained vanadium was announced on May 1, 1980. This is equivalent to 12,470 t of V_2O_5. At the time of the announcement, the stockpile contained only 491 t of vanadium in the form of the pentoxide (28). Physical requirements are that V_2O_5 be supplied as broken flake, all of a size to pass a 2.54-cm screen and not more than 5 wt % to pass a 4.7-mm screen. Packaging in polyethylene film inside 208-L steel drums and marking of the drums has been described in detail (29).

Analytical and Test Methods

A delicate qualitative test for the presence of vanadium is the formation of brownish red pervanadic acid upon addition of hydrogen peroxide to a solution of a vanadate. Although titanium reacts similarly, its color disappears when fluoride or phosphate ions are added (30). Quantitative determinations over a wide range of vanadium content are readily performed by atomic absorption spectroscopy. Acetylene or nitrous oxide flames are ordinarily used. A highly sensitive atomic-absorption technique involving a carbon-filament atom reservoir has a detection limit of 100 pg V in a 1-μL sample. Volumetric, colorimetric, and spectrographic methods for vanadium are well-developed (31). Conversely, gravimetric methods are seldom used. X-ray absorption spectroscopy is a convenient means for identifying traces of vanadium in coal (32).

Health, Safety, and Environmental Considerations

The effect of vanadium compounds in the workplace and in ambient air on human health and safety has been extensively reviewed (33). In humans, toxic effects have been observed from occupational exposure to airborne concentrations of vanadium compounds that were probably several milligrams or more per cubic meter of air. Direct irritation of the bronchial passageways results from such exposure and is accompanied by coughing, spitting, wheezing, and eye, nose, and throat irritation. Some workers exhibit weakness, neurasthenia, and slight anemia, which suggests chronic toxic effect from vanadium absorption. Threshold limits for V_2O_5 in the air of the workplace have been established by OSHA as 0.5 mg V/m^3 for dust and 0.05 mg V/m^3 for fumes.

Oral vanadium toxicity in humans is minimal. Ingestion of 4.5 mg/d of vanadium has been without effect, but higher doses produce gastrointestinal distress and the greentongue associated with excessive inhalation of vanadium (34). The concentration of vanadium in vegetation varies from undetectable to 4 ppm in alfalfa and in animal tissues from 0.25 to slightly over 1 ppm. Drinking water contains up to 220 ppb (parts per billion (10^9)) of vanadium. Although contamination of water supplies by seepage from vanadium processing wastes seems possible, evidence of such contamination has not been found. Postulated to be an essential trace element for human well-being, the function of vanadium and its limiting concentrations have yet to be established (35).

In the United States, the largest concentration of atmospheric vanadium occurs over Eastern seaboard cities where residual fuels of high vanadium content from Venezuela are burned in utility boilers. Coal ash in the atmosphere also contains vanadium (36). Ambient air samples from New York and Boston contain as much as 600–1300 ng V/m^3, whereas air samples from Los Angeles and Honolulu contained 1–12 ng V/m^3. Adverse public health effects attributable to vanadium in the ambient air have not been determined. Increased emphasis by industry on controlling all plant emissions may have resulted in more internal reclamation and recycle of vanadium catalysts. An apparent drop in consumption of vanadium chemicals in the United States since 1974 may be attributed, in part, to such reclamation activities.

Uses

Conversion of fused pentoxide to alloy additives is by far the largest use of vanadium compounds. Air-dried pentoxide, ammonium vanadate, and some fused pentoxide, representing ca 10% of primary vanadium production, are used as such, purified, or converted to other forms for catalytic, chemical, ceramic, or specialty applications. The dominant single use of vanadium chemicals is in catalysts (see CATALYSIS). Much less is consumed in ceramics and electronic gear, which are the other significant uses (see BATTERIES). Many of the numerous uses reported in the literature are speculative, proposed, obsolete, or in such small quantities as to be generally reported under such consolidated headings as miscellaneous or other.

The NRC Committee's estimates of catalyst use for 1972, 1974, and 1975 were 40–80% higher than the BOM's, and the former's estimates for overall use, including ceramics, electronics, and unspecified uses, were 3–39% higher. Data published by the BOM through 1980 show a precipitous decline in total annual use from ca 311 t in 1974 to ca 82 t in 1980. The decline appears to have been caused mainly by unfavorable business conditions and conservation efforts, but part of the apparent decline may stem from incomplete collection of data.

Catalytic uses result in little consumption or loss of vanadium. The need to increase conversion efficiency for pollution control from sulfuric acid plants, which require more catalyst, and expanded fertilizer needs, which require more acid plants, were factors in the growth of vanadium catalyst requirements during the mid-1970s. Use was about evenly divided between initial charges to new plants and replacements or addition to existing plants.

Minor uses of vanadium chemicals are preparation of vanadium metal from refined pentoxide or vanadium tetrachloride; liquid-phase organic oxidation reactions, eg, production of aniline black dyes for textile use and printing inks;

color modifiers in mercury-vapor lamps; vanadyl fatty acids as driers in paints and varnish; and ammonium or sodium vanadates as corrosion inhibitors in flue-gas scrubbers.

Uses reported in the early literature, but which were insignificant in recent years at least in the United States, include refractories. V_2O_3 for coloring glass green, and V_2O_5 for ultraviolet screening in glass. Other developments include an expanded role as catalyst in such applications as vapor-phase polymerizations of ethylene and propylene, ammoxidations to form acrylonitrile and terephthalonitrile, and hydrodesulfurization of crude oils; V_3Ga [12024-15-6] as a superconductor; VO_2 as a thermal or light-activated resistor–conductor; vanadate glasses as electrooptical switches; rare-earth vanadites as magnetic materials; and addition of ca 5 wt % V_2O_5 to silicon carbide refractories for increased oxidation resistance at high temperatures; the latter is reportedly being practiced in Germany (37).

BIBLIOGRAPHY

"Vanadium Compounds" in *ECT* 1st ed., Vol. 14, pp. 594–602, by H. E. Dunn and C. M. Cosman, Vanadium Corp. of America; in *ECT* 2nd ed., Vol. 21, pp. 167–180, by G. W. A. Fowles, University of Reading, Reading, England; in *ECT* 3rd ed., Vol. 23, pp. 688–704, by J. B. Rosenbaum, Consultant.

1. *Vanadium Supply and Demand Outlook*, Report No. NMAB-346, National Research Council, Washington, D.C., 1978, 125 pp.
2. R. C. Weast, ed., *Handbook of Chemistry and Physics*, 62nd ed., CRC Press, Inc., Boca Raton, Fla., 1981–1982.
3. Seidell, *Solubilities—Inorganic and Metal-Organic Compounds*, 4th ed., Vol. 2, American Chemical Society, Washington, D.C., 1965.
4. Ref. 2, pp. B-162 and C-689.
5. Ref. 2, pp. D-84 and D-123.
6. R. J. H. Clark, *The Chemistry of Titanium and Vanadium*, Elsevier Publishing Co., Amsterdam, the Netherlands, 1968.
7. *Transition Metal Compounds*, Vol. 4, of A. F. Trotman-Dickenson, ed., *Comprehensive Inorganic Chemistry*, Pergamon Press, Ltd., Oxford, U.K., 1973.
8. D. Nichols, *Chem. Rev.* **1**, 379 (1966).
9. S. F. Ashcroft and C. T. Mortimer, *Thermochemistry of Transition Metal Complexes*, Academic Press, Inc., New York, 1970.
10. A. E. Martel, ed., *Coordination Chemistry*, Vol. 1, ACS Monograph 168, Van Nostrand Reinhold Co., New York, 1971.
11. R. G. Wilkins, *The Study of Kinetics and Mechanisms of Reactions of Transition Metal Complexes*, Allyn and Bacon, Inc., Boston, Mass., 1974.
12. A. E. Martel, ed., *Coordination Chemistry*, vol. 2, ACS Monograph 174, American Chemical Society, Washington, D.C., 1978.
13. R. A. Guidotti, G. B. Atkinson, and D. G. Kesterke, *Nitride Intermediates in the Preparation of Columbium, Vanadium, and Tantalum Metals*, Pt. I, RI-8079, U.S. Bureau of Mines, Washington, D.C., 1975, 25 pp.
14. Ref. 7, p. 352.
15. P. M. Busch, *Vanadium*, IC-8060, U.S. Bureau of Mines, Washington, D.C., 1961, 95 pp.
16. J. B. Rosenbaum, *Vanadium Ore Processing, Meeting of High Temperature Metal Committee*, preprint A71-52, AIME, New York, 1971, 14 pp.

17. P. T. Brooks and G. M. Potter, *Recovering Vanadium from Dolomitic Nevada Shale*, RI-7932, U.S. Bureau of Mines, Washington, D.C., 1974, 20 pp.

18. U.S. Pat. 3,733,193 (May 15, 1973), J. S. Fox and W. H. Dresher (to Union Carbide Corp.).

19. U.S. Pat. 3,853,982 (Dec. 10, 1974), C. B. Bore and J. W. Pasquali (to Bethlehem Steel Corp.).

20. U.S. Pat. 3,929,460 (Dec. 30, 1975), F. J. W. M. Peters, S. Middelholk, and A. Rijkelboer (to Billiton Research, BV).

21. U.S. Pat. 4,039,614 (Aug. 2, 1977), N. P. Slotvinsky-Sidak and N. V. Grinberg.

22. L. White, *Eng. Min. J.*, 87 (Jan. 1976).

23. C. E. Baker and D. K. Sparling, *Min. Eng.*, 382 (Apr. 1981).

24. *Eng. Min. J.*, 105 (Feb. 1979).

25. I. R. Taylor, *Min. Eng.*, 82 (Apr. 1969).

26. *London Metal Bulletin*, 43 (June 16, 1997).

27. *Energy Use Patterns in Metallurgical and Non-Metallic Mineral Processing, Phase 6—Low Priority Commodities*, final report to U.S. Bureau of Mines, Battelle Columbus Laboratory, Columbus, Ohio, July 1976.

28. *Met. Week* **51**(18), 5, 10 (May 1980).

29. *National Stockpile Purchase Specifications, P-58-R-2*, U.S. Dept. of Commerce with approval of the Federal Emergency Management Agency, Washington, D.C., June 25, 1981, 4 pp.

30. H. H. Willard and H. Diehl, *Advance Quantitative Analysis*, D. Van Nostrand Company, New York, 1943, p. 239.

31. W. J. Williams, *Handbook of Anion Determination*, Butterworth & Co., London, 1979, pp. 251–261.

32. D. H. Maylotte, J. Wong, R. L. St. Peters, F. W. Lytle, and R. B. Greegor, *Science* **214**, 554 (Oct. 1981).

33. *Medical and Biological Effects of Environmental Pollutants—Vanadium*, National Academy of Sciences, Washington, D.C., 1974, 117 pp.

34. V. W. Oehme, *Toxicity of Heavy Metals in the Environment*, Pt. 2, Marcel Dekker, Inc., New York, 1979.

35. W. Mertz, *Science* **213**, 1332 (Sept. 18, 1981).

36. R. D. Smith, J. A. Campbell, and W. D. Felix, *Min. Eng.*, 1603 (Nov. 1980).

37. Ref. 1, p. 115.

General References

Vanadium, in *Mineral Facts and Problems*, Bull. 671, U.S. Bureau of Mines, Washington, D.C., 1980.

MIKE WOOLERY
U.S. Vanadium

VANILLIN

Vanillin [121-33-5], a natural product, can be found as a glucoside (glucovanillin) in vanilla beans, at concentrations of about 2%. It can be extracted with water, alcohol, or other organic solvents. Approximately 250 by-products have been identified in natural vanilla, out of which 26 are present at levels in excess of 1 ppm. The balance of all these products contributes to the subtle taste of vanilla beans. The vanilla bean contains about 2% vanillin, but the 10% extract prepared from beans has several times the strength of a solution of 2% vanillin. For this reason, the U.S. Food and Drug Administration (FDA) regulations state that one part of vanilla beans is equivalent to 0.07 parts vanillin in flavor strength. The best known natural source of vanillin is the vanilla plant, *Vanilla planifolia* A., which belongs to the orchid family. It is cultivated mainly in Mexico, Madagascar, Reunion, Java, and Tahiti.

The long and expensive process of extracting vanillin from vanilla beans yields a product that has an inconsistent quality. The demand for this universally popular flavoring cannot be satisfied by vanilla beans alone. For technological and economic reasons, the consumption of naturally occurring vanilla has gradually given way to synthetic vanillin. Synthetic vanillin is identical to that contained in the pod, but differs in smell and flavor from natural vanillin as a result of the various compounds in the natural extract that do not exist in artificial vanillin. These other compounds represent only 2% of the extract; the remaining 98% is vanillin. Vanillin is the common name for 3-methoxy-4-hydroxybenzaldehyde [121-33-5] (**1**).

(**1**)

Production

Vanillin was observed long before it was reported in chemical literature, as it crystallizes on the surface of vanilla beans after harvesting, processing, and storage. The first report in the literature was probably made by Bucholtz in 1816. Some years later, Bley referred to vanillin as vanilla camphor. In 1858, Gobley crystallized out vanillin from alcoholic solutions of vanilla bean extract and succeeded in obtaining it in a relatively pure form. He reported its composition to be $C_{10}H_6O_2$. Its correct analysis, $C_8H_8O_2$, was established in 1872 by Carles, who also gave its correct melting point, 81°C (177.8°F).

In 1874, Tiemann and Haarmann examined the structure of vanillin and reported it to be 3-methoxy-4-hydroxybenzaldehyde. This was not a difficult task because, on treatment with potassium hydroxide, vanillin (**1**) gave protocatechaic acid [99-50-3] (**2**), which, in turn, was decarboxylated to catechol [120-80-9] (**3**) by

$$(1)$$

dry distillation (eq. 1). As both compounds were known at that time, the position of the substituent groups in vanillin was established. Finally, Reimer synthesized vanillin from guaiacol [90-05-1] and thus proved the identity of its structure. In 1894 Rhône-Poulenc began producing vanillin on an industrial scale. Since then, many other producers have entered into vanillin production, often only to leave it behind.

The manufacture of vanillin shows the progress made in the chemistry and chemical engineering of the substance. Most commercial vanillin is synthesized from guaiacol; the remainder is obtained by processing waste sulfite liquors. Preparation by oxidation of isoeugenol is of historical interest only.

Preparation from Guaiacol and Glyoxylic Acid. Several methods can be used to introduce an aldehyde group into an aromatic ring. Condensation of guaiacol (**4**) with glyoxylic acid (**5**), followed by oxidation of the resulting mandelic acid (**6**) to the corresponding phenylglyoxylic acid (**7**) and decarboxylation continues to be a competitive industrial process for vanillin synthesis (eq. 2).

$$(2)$$

In the 1990s, guaiacol is synthesized from catechol, which is prepared by acid-catalyzed hydroxylation of phenol with hydrogen peroxide (see HYDROQUINONE, RESORCINOL, AND CATECHOL). Glyoxylic acid is obtained as a by-product in the synthesis of glyoxal from acetaldehyde (qv), and it can also be produced by oxidation of glyoxal with nitric acid. Condensation of guaiacol with glyoxylic acid proceeds smoothly in alkaline media. Crude vanillin is obtained by acidification and simultaneous decarboxylation of the 4-hydroxy-3-methoxyphenyl glyoxylic acid solution. Commercial grades are obtained by vacuum distillation and subsequent recrystallization.

This process has the advantage that, under the reaction conditions, the glyoxyl radical enters the aromatic guaiacol ring almost exclusively para to the phenolic hydroxyl group. Tedious separation procedures are thus avoided.

Preparation from Waste Sulfite Liquors. The starting material for vanillin production can also be the lignin (qv) present in sulfite wastes from the cellulose industry. The concentrated mother liquors are treated with alkali at elevated temperature and pressure in the presence of oxidants. The vanillin formed is separated from the by-products, particularly acetovanillone, 4-hydroxy-3-methoxyacetophenone, by extraction, distillation, and crystallization.

A large number of patents describe various procedures for the mainly continuous hydrolysis and oxidation processes, as well as for the purification steps required to obtain high grade vanillin. Lignin is degraded either with sodium hydroxide or with calcium hydroxide solution and simultaneously oxidized in air in the presence of catalysts. When the reaction is completed, the solid wastes are removed. Vanillin is extracted from the acidified solution with a solvent, eg, butanol or benzene, and reextracted with sodium hydrogen sulfite solution. Reacidification with sulfuric acid followed by vacuum distillation yields technical-grade vanillin, which must be recrystallized several times to obtain food-grade vanillin. Water, associated with small amounts of ethanol, is used as the solvent in the last crystallization step.

The process starting from lignin has faced serious problems, such as reduced availability and environmental impact. The availability is reduced because the new process for making paper paste yields less liquor. As a result, it is likely that the larger companies will not reinvest in new factories to process liquors to meet demand. The process's environmental impact is also problematic because over 160 t of caustic waste are produced for every ton of vanillin manufactured.

Many impurities are present in vanillin produced by the lignin process, principally 5-formylvanillin, *para*-coumaric acid, *para*-hydroxybenzaldehyde, syringic aldehyde and syringic acid, and acetovanillone (4-hydroxy-3-methoxyacetophenone). The last-mentioned is the main impurity present in vanillin from lignin. It can be found, although only rarely, at levels of up to 1000 ppm. Its relatively strong odor is responsible for the typical odor of the head space observed in lignin vanillin samples.

In contrast to vanillin from lignin, the principal impurity found in vanillin from guaiacol is 5-methyl vanillin, typically present at levels of about 100 ppm in Rhovanil Extra Pure (Rhône-Poulenc), although levels as high as 3000 ppm have been found in samples from other producers. This impurity is completely odorless.

No residual guaiacol can be found in vanillin produced by the guaiacol process. In contrast to vanillin from lignin, vanillin from guaiacol is extremely consistent in quality owing to the consistency of the supply source, and shows no variation in taste, odor, or color.

Specifications

The physical properties of Rhovanil Extra Pure vanillin of Rhône-Poulenc, the leading company in this area, are shown in Table 1.

Table 1. Physical Properties of Rhovanil[a] Extra Pure Vanillin

Property	Value
white to off-white nonhygroscopic crystalline powder	
melting point, capillary, °C	81–83
assay, %	99.96 min
bulk density	~0.6
flash point, °C	153
boiling point, °C	
at 101.3 kPa[b]	284–285
10 mm Hg	154
sublimation[c], °C	70

[a]Trademark of Rhône-Poulenc.
[b]To convert kPa to mm Hg, multiply by 7.5.
[c]At normal pressure.

Solubility.   Solubility in water is less than 2%; the solubility in ethanol is given by the ratio one part vanillin to two parts alcohol. Certain manufacturing processes require that the product be in liquid form. Depending on the application, the solvent must be chosen in accordance with the manufacturing process and regulation requirements.

The solubility of the Rhovanil vanillin in water–ethanol, water–propylene glycol, and water–glycerol solutions are shown in Figure 1. In addition, the influence of temperature and solvent concentration are important in maximizing the vanillin concentration.

Particle-Size Distribution.   Particle size, crystal shape, and distribution of vanillin are important and greatly affect parameters such as taste, flavor, solubility, ease of dispersion in solvent, flowability of the powder, caking effect, and production of dust (Fig. 2).

The particle size distribution of Rhovanil Extra Pure vanillin shows a less narrow profile than other standard mesh grades available on the market. The product shows an improved mixability in blending operations, allowing shorter blending time of compounds or food mixes, and better homogeneity of vanillin content, especially in low content vanillin blends.

Taste and Flavor.   The taste effect is generally sweet, but depends strongly on the base of preparation. For tasting purposes, vanillin is often evaluated in ice-cold milk with about 12% sugar. A concentration of 50 ppm in this medium is clearly perceptible. Vanilla is undoubtedly one of the most popular flavors; its consumption in the form of either vanilla extracts or vanillin is almost universal.

The food flavor industry is the largest user of vanillin, an indispensable ingredient in chocolate, candy, bakery products, and ice cream. Commercial vanilla extracts are made by macerating one part of vanilla beans with ten parts of 40–50% alcohol. Although vanillin is the primary active ingredient of vanilla beans, the full flavor of vanilla extract is the result of the presence of not only vanillin but also other ingredients, especially little-known resinous materials which contribute greatly to the quality of the flavor.

It is easy to smell a difference in the quality of vanillins from different origins, but it is normally difficult to taste the same difference, provided the

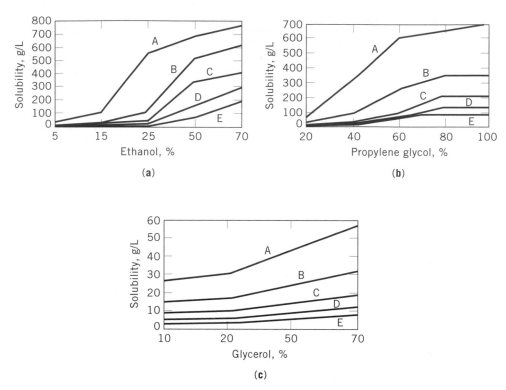

Fig. 1. Solubility of vanillin in (a) ethanol solutions, (b) propylene glycol solutions, and (c) glycerol solutions, where A is 40°C; B, 30°C; C, 20°C; D, 10°C; and E, 0°C.

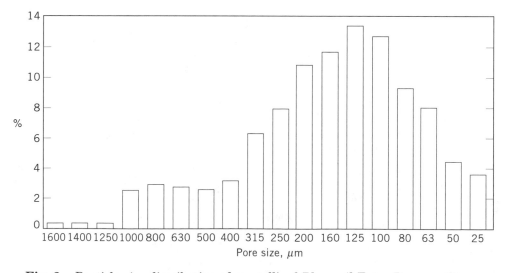

Fig. 2. Particle size distribution of crystallized Rhovanil Extra Pure vanillin.

various samples are of good quality. Vanillin is sensitive to contamination with other crystalline odor chemicals. Adherent or absorbed odor on the crystals perceptibly affects the odor of vanillin from a fiber-drum or other large containers.

So far, several blind-test studies have been conducted to detect possible differences in odor and taste (retroolfaction) of vanillin from lignin and guaiacol, respectively. In none of these trials did the taste panel observe significant differences between lignin and guaiacol vanillin. However, a difference in odor can be observed in the head space of pure vanillin containers, ie, top note. But no difference in top note is observed between vanillin from lignin and vanillin from guaiacol after acetovanillone is added to the latter, at the level present in the lignin vanillin. In fact, guaiacol-based vanillin has the real, unadulterated odor of pure vanillin, in contrast to lignin-based vanillin, whose odor is influenced by the presence of acetovanillone at variable levels.

Available Grades.   Rhovanil Extra Pure is the trade name of the food-grade vanillin of Rhône-Poulenc, worldwide leader in the diphenols area. The following grades are commercially available: Rhovanil Extra Pure crystallized, Rhovanil Fine Mesh, Rhovanil Free Flow, and Rhovanil Liquid.

Rhovanil Extra Pure is the standard mesh, multipurpose quality of food-grade extra pure vanillin. Its broad particle-size distribution shows a versatile granulometry, compatible with a wide range of granulometric profiles from any other ingredients, and allows a homogeneous powder mixability, even at low content in a given blend.

Rhovanil Fine Mesh, a specially calibrated extra pure vanillin that avoids demixing with other very fine dry ingredients such as sucrose, flour, and dextrose, provides a faster dissolution rate at lower stirring, at lower temperature, in low acidity medium, or in viscous liquids.

Rhovanil Free Flow is obtained by adding an anticaking agent (0.5% max) to the extra pure vanillin. The flowability is increased, making it particularly suitable for self-dispensing equipment (instant beverage), while both mixability and dispersion/dissolution ratios remain as good as the standard Rhovanil Extra Pure vanillin.

Two different grades of Rhovanil liquid are also available. They are prepared with Rhovanil Extra Pure crystallized vanillin diluted in food-grade ethanol or monopropylene glycol, acting both as carriers. There are other producers of vanillin from guaiacol, in Japan (UBE Industries), China, and Norway for example, but Rhône-Poulenc is the only company with a broad range of vanillin grades. Competitors of Rhône-Poulenc generally have only one standard quality at their disposal.

Chemical Properties

Vanillin is a compound that possesses both a phenolic and an aldehydic group. It is capable of undergoing a number of different types of chemical reactions. Addition reactions are possible owing to the reactivity of the aromatic nucleus.

On distillation at atmospheric pressure, vanillin undergoes partial decomposition with the formation of pyrocatechol. This reaction was one of the first to be studied and contributed to the elucidation of its structure. Exposure to air causes vanillin to oxidize slowly to vanillic acid. When vanillin is exposed to light in an alcoholic solution, a slow dimerization takes place with the formation of dehydrodivanillin. This compound is also formed in other solvents. When fused with alkali (eq. 3), vanillin (**1**) undergoes oxidation and/or demethylation,

$$\text{(3)}$$

       (**1**)   (**8**)   (**2**)

yielding vanillic acid [*121-34-6*] (**8**) and/or protocatechaic acid (**2**).

Reduction of vanillin by means of platinum black in the presence of ferric chloride gives vanillin alcohol in excellent yields. In 1875, Tiemann reported the reduction of vanillin to vanillin alcohol by using sodium amalgam in water. The yields were poor, however, and there were a number of by-products. High yields of vanillin alcohol have been obtained by electrolytic reduction.

Because vanillin is a phenol aldehyde, it is stable to autooxidation and does not undergo the Cannizzaro reaction. Numerous derivatives can be prepared by etherification or esterification of the hydroxy group and by aldol condensation at the aldehyde group. All three functional groups in vanillin are highly reactive. The hydroxy group can be methylated and acetylated. Acetals and mercaptals have also been prepared.

Applications

In flavor formulations, vanillin is used widely either as a sweetener or as a flavor enhancer, not only in imitation vanilla flavor, but also in butter, chocolate, and all types of fruit flavors, root beer, cream soda, etc. It is widely acceptable at different concentrations; 50–1000 ppm is quite normal in these types of finished products. Concentrations up to 20,000 ppm, ie, one part in fifty parts of finished goods, are also used for direct consumption such as toppings and icings. Ice cream and chocolate are among the largest outlets for vanillin in the food and confectionery industries, and their consumption is many times greater than that of the perfume and fragrance industry.

Vanillin, being an aldehyde, is able to form acetals and hemiacetals. Therefore, in flavor formulations using high concentrations of vanillin in conjunction with carriers such as propylene glycol, a glc analysis often shows a reduced vanillin peak after storage of the compounded flavor, and the presence of new peaks indicating acetal formation. Addition of about 0.5% of water to the formula reverses the reaction, ie, there is a reduction of acetal, and the reappearance of vanillin peaks.

Food Flavoring Compounds.   When vanillin is not used as a single flavoring ingredient, it is a key part of flavor compounding. At least 30% of food-grade vanillin consumed in the world is through flavoring compounds. Flavor compounding requires expertise to develop well-balanced, complex flavors, such as fruit flavors, by mergers of multiples and different single flavors. Single flavors can be classified according to the stage of flavor perception, incorporating both direct olfaction and retronasal olfaction at 35°C. Head note describes the first fugacious impression; intermediary note, a more sustained impression when the food is in front of the nose; body note, an impression more solid or dense when the food is in the mouth; and queue or long-lasting note, the feelings or sensations after deglutition. These four principal notes correspond to various components (Table 2).

The work of several flavors always corresponds to an optimized work of assembling these four notes into a given direction, starting with a basis or a body note. Vanillin and ethylvanillin belong to the category of body notes. Vanilla extracts, owing to more than 200 minor components in addition to vanillin, are considered a queue note.

Baking. In the industrial production of dry cookies, cakes, and pastries, the vanillin content ranges between 20 and 50 g per 100 kg of dough. Often, vanillin is added at the dry stage of dough preparation as the flour and sugar are being mixed. In fact, it is better to take advantage of the properties of fatty ingredients that are excellent at retaining flavor. Better results can be obtained by incorporating vanillin into the fatty ingredients, as is the case, for example, during creaming or topping.

In fat-free recipes where this method cannot be practiced, it is possible to add and mix vanillin powder with eggs. An alternative to vanillin powder for highly mechanized processes is to use a vanillin solution or liquid flavor; the solvent used is either ethanol- or monopropylene glycol-based.

Chocolate. Vanillin is added during the manufacturing process, in powder form, in average amounts of 20 g per 100 kg of the finished product. However, this amount varies according to the quality of the chocolate being made. Chocolate having a high melting point, eg, the type of chocolate used in icing, is more strongly flavored, around 25 g/100 kg of mixture. Chocolate with a low melting point is less strongly flavored, around 15 g/100 kg of mixture.

Vanillin can be added either in the first stage of production during crushing and before conching, or just before conching, or in the last steps of conching. In the first two cases, some flavor may be lost if the temperature is raised to over 80°C. An example of the last case is by dissolving the vanillin in cocoa butter, which is added to adjust the product's melting point.

Table 2. Classification of Food Flavors Depending on Perception of Main Aromatic Note[a]

Heat notes	Intermediary notes	Body notes	Queue notes
amyl acetate	aldehydes C_8, C_9, C_{11}, and C_{12}	aldehydes C_6 and C_8	amines
geranyl acetate	cinnamyl acetate	ethyl vanillin	cocoa infusion
lemon zest	citronellol	ionones	fruit juice concentrate
linalool	terpineol thyme	terpenic components vanillin	vanilla extract

[a]Ref. 1.

Although the vanillin concentration is a matter of taste depending on different factors in each individual case, the following concentrations are generally accepted:

Vanillin flavoring	g/100 kg
dark chocolate	15–60
milk chocolate	5–30

Confections.   Main applications are sugared almonds, caramel, nougat, and sweets. For sugared almonds and caramel, vanillin is mixed into the sugar in the dry phase of the recipe. For nougat, Vanillin is added during the liquid phase of manufacturing. In sweets, vanillin is added in the form of a 10% ethanol solution.

The appropriate concentration of vanillin depends on the nature of the finished product and the flavor desired. Therefore, in caramel for which vanillin is used to give a pleasant but not pronounced taste, 15 g of vanillin is sufficient for 100 kg of confectionery. However, if a pronounced taste is required, at least 40 g of vanillin is required for 100 kg of the finished product. The vanillin flavoring can be incorporated either in liquid extract form or in the form of vanillin-flavored sugar, or even by mixing the required amount of a special finer mesh grade such as Rhovanil Fine Mesh (Rhône-Poulenc) with a little sugar and adding this to the rest of the ingredients. If the process requires the mixture to be oven-baked, the vanillin should only be added at the end of the cooking to avoid losses caused by evaporation. Although the concentration may be different in individual cases, the following may serve as a rough guide:

Vanillin flavoring	g/100 kg
soft-center sweets	5–15
other sweets	15–30
caramel	15–55
chewing gum	15–45
nougat	40–55

Vanillin is used in flavored milk, desserts, yogurts, sorbets, and ice cream. Generally, vanillin is used in liquid form either in ethanol solution with a vanillin concentration up to 400 g/L or in monopropylene glycol with a vanillin concentration to 300 g/L. Both concentrations are given for a temperature of 20°C to avoid recrystallization problems.

The content of vanillin, in relation to the product to be flavored, is around 5 g/100 kg of the finished product. Flavoring is carried out by adding the appropriate quantity of vanillin solution during one of the product's mixing stages. To obtain an even distribution of the flavoring in the product, it is best to add the flavoring solution as early as possible during the manufacturing process.

Vanillin Sugar.   This product is prepared by dry mixing or impregnating the sugar with a vanillin alcohol solution and evaporating the alcohol. However, modern techniques increasingly involve grinding the sucrose and vanillin mixture very finely.

Vanillin sugar can also be prepared by mixing 20 g of vanillin with 980 g of sieved sugar to remove any lumps and then placing the mixture in an airtight container for the amount of time necessary for the synthetic vanillin aroma and taste to develop. This method of preparation is used for smaller volumes or household needs. Vanillin sugar can be used for flavoring all dry preparations. This method of dry flavoring can also be used for dispersing agents other than sugar, eg, flour, corn starch, or any other suitable ingredient used in the food industry.

Beverages. Vanillin confers a pleasant note to liqueur flavoring and improves the flavor of fortified wines by giving them a greatly enhanced bouquet. For example, vanillin is used for flavoring grenadine as well as chocolate-flavored drinks.

Animal Feed. Vanillin is used as a palatability enhancer to make animal feed more appetizing by flavor-masking minerals with off-taste. Approximately 5 g of vanillin/100 kg of feed is added when preparing feed for lambs and pigs in order to increase feed intake and stimulate the growth of the animals. Vanillin is added during the manufacturing process either by mixing into the dry ingredients or in its liquid form. Increasingly, vanillin is also used as a substitute for aniseed.

Perfumes and Cosmetics. Vanillin, a crystal, is the main constituent of the vanilla bean. Its importance can be illustrated by the fact that human preferences in fragrances and in flavors, as determined by various studies, comprise three main smells or tastes: rose, vanilla, and strawberry.

In addition, vanillin was among the three or four aroma chemicals that helped perfumers of the past to imagine a new generation of fragrance combinations. The work done in organic chemistry by chemical companies has helped perfumers who had been restrained in their creation by the availability of raw materials, which include natural oils and extracts made by the enfleurage process, ie, the property of perfumes to stick on fat and greases. Those extracts were weak, flat, and unpleasant.

The perfumers of the early twentieth century were delighted to be able to use a perfectly vanilla-like product in their compounds, but one that was at least a hundred times stronger than vanilla. Thus began what was subsequently called sophisticated perfumery. For many years a perfumery culture and language based on the experience and subjectivity of a few individual experts has developed and thrived (see PERFUMES).

The denomination of odors was schematically related to two separate domains, both related to the memory stimulus of an event concomitant with the perception of the odor. One domain was based on an actual reference point that contains the odor vectors; the other was associated with an odor stimulus based on imagination, ie, what image is evoked by the stimulus. With such a system, the final descriptive terminology used would more often than not be expressed in esoteric language, causing confusion and even communication breakdown. The work of Jaubert (1) was the origin of a more standardized descriptive system in the field of aroma description.

The model developed, which has been in use for many years both for the training of professionals and specialists and the preliminary education of nonspecialists, leads toward a universal language for odor relationships, and is named the spectrum or field of odor. This spatial model has been based on 42 reference odorants, including vanillin, and is becoming the methodological reference for describing odors (see ODOR MODIFICATION).

The uses of vanillin in international perfumery are many. In aldehydic perfumes, vanillin provides the powdery impression given by the background smell, usually up to 2% in the perfume concentrate. In fruity notes, vanillin enhances the various fruity constituents (0.1–0.5% in pears; up to 2% in peaches); for instance, a peach note is not fully peach without vanillin. When vanillin is combined with some floral notes, such as heliotrope and orchid, which actually contain strong vanilla impressions, amounts of 2–5% are possible. However, with notes such as rose, orange flower, and jonquil, the addition of 0.1–2% vanillin can bring warmth and elegance. In woody families such as fougère and chypre, and also in spicy perfumes, the harsh impression also needs the fine, smooth aroma provided by vanillin traces.

In some cases, especially when the family is directed more toward Oriental fragrances, the use of vanillin can be up to 5%. These types of perfumes include Spanish fougère, sweet or fruity chypres, woody Oriental, or spicy Oriental notes. All the perfumes based on the sweet, warm, and powdery impressions brought by vanillin belong to the great family of the Oriental notes and of the amber notes. Vanillin, when used together with coumarin and nitro-musks, can have a concentration of up to 10%.

In detergent perfumes, the stability of vanillin is not always certain. It depends on the association made with other raw materials, eg, with patchouli, frankincense, cloves, most of the animal notes, and such chemicals as amyl salicylate, methyl ionones, heliotropin, gamma undecalactone, linalool, methyl anthranilate, benzyl acetate, phenyl ethyl alcohol, cedar wood derivatives, oak mosses, coumarin, benzoin, Peru balsam, and cistus derivatives. In some cases, these mixtures can cause discoloration effects.

In cosmetics, as in bath products, most of the problems arising with the use of vanillin are related to the soap perfumery problems. However, because the amount of perfume concentrate used in bath products is usually lower than that used in fine fragrances, the problems should be studied separately.

The use of high concentrations of vanillin in soap perfumery can cause discoloring effects; over time, dark or black spots appear on the soap and foaming power is reduced. In some cases, however, the use of Rhodiarome ethylvanillin is possible, because ethylvanillin [121-32-4] does not cause the same discoloration problems and, being at least three times more powerful than vanillin, can be used alone. Some surprising cases show that with oak or tree mosses and large amounts of methyl ionones, the soap perfume may look fine and have a low discoloration, and yet over time vanillin crystals can appear on the soap itself.

Flavor-Masking Deodorant.   In addition to its use as a constituent of perfume compositions, vanillin is also useful as a deodorant to mask the unpleasant odor of many manufactured goods. As a masking agent for numerous types of ill-smelling mass-produced industrial products, particularly those of synthetic rubber, plastics, fiber glass, inks, etc, vanillin finds extensive use. It is often the most inexpensive material for the amount of masking effect it provides. Only traces are required for this purpose as the odor of vanillin is perceptible in dilutions of 2×10^{-7} mg/m^3 of air. Crude vanillin is acceptable for such purposes.

Pharmaceutical Products. Rhône-Poulenc offers a flaked technical-grade vanillin, Vaniltek, to be used in pharmaceutical applications. The single largest use for vanillin is as a starting material for the manufacture of an antihypertensive drug having the chemical name of Methyldopa or L-3-(3,4-dihydroxyphenyl)-2-methylalanine.

L-Dopa and Trimethoprim are two other drugs that can be made from vanillin. L-Dopa is used for the treatment of Parkinson's disease; Trimethoprim is an antiinfective agent used mainly for urinary tract infections and certain venereal diseases. Also, Mebeverine, an antispasmodic agent, and Verazide, a generic antitubercular agent, are drugs that can be made from vanillin or its derivatives.

Papaverine, used to treat heart diseases as a vasodilator, is a drug that was originally made from vanillin but has since been made from veratrole and *ortho*-1,2-dimethoxybenzene. Vanillin is also used as a pharmaceutical excipient.

Antimicrobial Effect. Vanillin itself has some bacteriostatic properties and therefore has been used in formulations to treat dermatitis. More specifically, it has been reported that, at dosages between 500 and 1000 ppm, vanillin and to a lesser extent ethyl vanillin, showed clear antimicrobial effects. Both compounds are more effective against fungi and nonlactic gram-positive than against gram-negative bacteria. Effectiveness was greater at pH 6 than at pH 8, which suggests that such antimicrobial activity is reinforced in foods with unfavorable pH conditions, osmotic pressure, and growth of temperature (cf acid foods).

Agrochemical Products. Hydrazones of vanillin have been shown to have a herbicidal action similar to that of 2,4-D, and the zinc salts of dithio-vanillic acid. Made by the reaction of vanillin and ammonium polysulfide in alcoholic hydrochloric acid, dithiovanillic acid is a vulcanization inhibitor. 5-Hydroxymercurivanillin, 5-acetoxymercurivanillin, and 5-chloromercurivanillin have been prepared and found to have disinfectant properties.

Ripening Agent. A new potential use for vanillin is as a ripening agent to increase the yield of sucrose in sugarcane by the treatment of the cane crop a few weeks before harvest.

Industrial Applications. The antiultraviolet protection properties of vanillin have been patented and look promising for the plastics and cosmetics (suncreams) industries.

Other uses for vanillin include the prevention of foaming in lubricating oils, as a brightener in zinc coating baths, as an activator for electroplating of zinc, as an aid to the oxidation of linseed oil, as an attractant in insecticides, as an agent to prevent mouth roughness caused by smoking tobacco, in the preparation of syntans for tanning, as a solubilizing agent for riboflavin, and as a catalyst to polymerize methyl methacrylate.

Analysis

Identification. When a solution of ferric chloride is added to a cold, saturated vanillin solution, a blue color appears that changes to brown upon warming to 20°C for a few minutes. On cooling, a white to off-white precipitate (dehydro-divanillin) of silky needles is formed. Vanillin can also be identified by the white to slightly yellow precipitate formed by the addition of lead acetate to a cold aqueous solution of vanillin.

Determination. Various classical techniques are used for the analysis of vanillin, including colorimetric, gravimetric, spectrophotometric, and chromatographic (tlc, gc, and hplc) methods. The *Food Chemical's Codex* (FCC) prescribes infrared spectrophotometry for identifying and testing vanillin. However, more vanillin analyses are made by either gc or hplc.

Gas chromatography is a widely used method for analyzing vanillin products, vanilla extracts, and compounded flavors. It can also be used to monitor the vanillin manufacturing process. It is ideal for detecting many trace impurities associated with manufacturing either lignin vanillin or guaiacol vanillin. It can determine the levels (ppm) of impurities in the vanillin finished product. Many commercial chromatographic column packings are available for analyzing vanillin.

In the 1990s hplc has become widely used in the flavor and fragrance industry to measure vanillin and other phenolic compounds. Routine methods have been developed that are particularly adapted to thermosensitive products, such as vanillin and its derivative products, with elution gradient and uv detection at given wavelengths. Certain critical impurities can thus routinely be traced to very low (10 ppm) concentrations.

Health and Safety Factors

Vanillin is listed in the *Code of Federal Regulations* by the FDA as a Generally Recognized As Safe (GRAS) substance. The Council of Europe and the FAO/WHO Joint Expert Committee on Food Additives have both given vanillin an unconditional Acceptable Daily Intake (ADI) of 10 mg/kg.

Vanillin has a low potential for acute and chronic toxicity, with a reported oral LD_{50} in rats of 1580–3300 mg/kg. Dietary doses up to 20,000 ppm administered to rats for two years resulted in no adverse toxicologic or carcinogenic effects. Vanillin is classified as a GRAS substance by FEMA. Consequently, at levels normally found in the human diet, vanillin would present no significant health or carcinogenic risk to humans.

Vanillin is known to cause allergic reactions in people previously sensitized to balsam of Peru, benzoic acid, orange peel, cinnamon, and clove, but vanillin itself is not an allergic sensitizer.

Vanillin has been reported to be a bioantimutagen, demonstrating the ability to protect against mutagenic effects by enhancement of an error-free postreplication repair pathway. Vanillin has been reported to be nonmutagenic in bacterial systems, but conflicting results in mammalian systems leave no clear indication of the SCE-inducing potential of vanillin.

Vanillin was reported to be nonteratogenic in chicks. No mammalian teratology studies were found in the literature. One study of the influence of vanillin on MNNG teratogenicity indicated that vanillin could enhance or reduce MNNG teratogenicity, depending on the specific endpoint examined and the teratogenic mechanism involved.

BIBLIOGRAPHY

"Vanillin" in *ECT* 1st ed., Vol. 14, pp. 603–611, by D. M. C. Reilly, Food Machinery and Chemical Corp.; in *ECT* 2nd ed., Vol. 21, pp. 180–196, by D. G. Diddams, Sterling Drug Inc., and J. K. Krum, The R. T. French Co.; in *ECT* 3rd ed., Vol. 23, pp. 704–717, by J. H. Van Ness, Monsanto Co.
1. J.-N. Jaubert, C. Tapiero, and J.-C. Dore, *Perfum. Flavor.* **20** (May–June 1995).

General References

K. Bauer, in D. Garbe, ed., *Common Fragrance and Flavor Materials: Preparation, Properties and Uses*, VCH, Weinheim, Germany, 1985.
P. Z. Bedoukian, ed., *Perfumery and Flavoring Synthetics*, Allured Publishing Corp.
S. Arctander, ed., *Perfume and Flavor Chemicals (Aroma Chemicals)*, Montclair, N.J., 1969.
Ullmann's Encyclopedia of Industrial Chemistry, 5th ed., Vol. A 11, VCH, Weinheim, Germany, 1988, pp. 199–200.
G. S. Clark, *Perfum. Flavor.* **15** (Mar.–Apr. 1990).
J. M. Jay and G. M. Rivers, *J. Food Safety* **6**, 129–139 (1984).

LAWRENCE J. ESPOSITO
K. FORMANEK
G. KIENTZ
F. MAUGER
V. MAUREAUX
G. ROBERT
F. TRUCHET
Rhône-Poulenc

VAPOR–LIQUID EQUILIBRIA. See ABSORPTION; DISTILLATION.

VARNISH. See INSULATION, ELECTRIC–PROPERTIES AND MATERIALS; RESINS, NATURAL.

VEGETABLE FIBERS. See FIBERS, VEGETABLE.

VELVETEX. See SURFACTANTS; DETERGENCY.

VERMICULITE. See INSULATION, THERMAL.

VERMILION. See PIGMENTS, INORGANIC.

VETERINARY DRUGS

The use of pharmaceuticals (qv) in the treatment and prevention of animal diseases has expanded greatly since the 1960s. The modern veterinarian, whether in companion-animal practice, equine service, feedlot medicine, swine and poultry dairy work, zoo management, or any animal medical or surgical specialty, has a wide range of products from which to choose. These products may generally be classified as one of the following: antimicrobial agents (see ANTIBACTERIAL AGENTS, SYNTHETIC; ANTIBIOTICS; GROWTH REGULATORS–ANIMAL); antiinflammatory agents (see ANALGESICS, ANTIPYRETICS, AND ANTIINFLAMMATORY AGENTS); parasiticides (see ANTIPARASITIC AGENTS); hormones (qv); anesthetics (qv) and tranquilizers (see HYPNOTICS, SEDATIVES, ANTICONVULSANTS, AND ANXIOLYTICS); cancer chemotherapeutics (see CHEMOTHERAPEUTIC AGENTS, ANTICANCER); or production enhancers (eg, BGT), reproductive hormones, and growth promoters (eg, steroids).

All drugs used for veterinary purposes are subject to governmental regulations. In the United States, pharmaceuticals are regulated by the Center for Veterinary Medicine of the FDA. Vaccines and many immunotherapeutics are classified as biologicals, and thus, are regulated by the USDA. Topical insecticides and growth regulators are regulated by the EPA.

Governmental Regulations

The exact requirements for regulatory approval of a given product for veterinary purposes, whether prescription or over-the-counter, vary from country to country. Product development often takes a minimum of eight to ten years and usually requires large ($>\$10 \times 10^6$) sums of money.

In general, the following information must be provided:

Acute toxicity in laboratory animals and target species, including eye and skin irritation and toxicity.

Subacute toxicity in laboratory animals by 28- and 90-day feedings.

Chronic toxicity in laboratory animals by two-year or lifetime feedings in two species of laboratory animals, multiple-generation teratogenicity, and, not always required, one-year feeding of dogs.

Specialized *in vitro* mutagenicity tests.

Overdose or extended treatment studies in the target species.

Efficacy to justify label claims.

Drug stability studies to determine rate of compound degradation and incompatibilities with other compounds or feed ingredients according to the anticipated conditions of use.

Metabolism studies to identify site of metabolism and principal metabolites.

Tissue residues in food-producing animals to document persistence in edible tissues.

Manufacturing methods to assure product consistency and the safety of personnel exposed to the drug or process intermediates.

Environmental effects, including effect on methanogenic and nitrifying bacteria, persistence in the environment, and projections of possible liability to relevant ecosystems.

Comprehensive labeling and directions.

Antimicrobial Agents

The use of drugs to control infection is considerably older than the recognition of the causes of infection or the complex physiological responses to the infectious agent. Historically such unlikely and diverse agents as vinegar (qv) (wine), copper salts, and honey, not to mention various natural plant products, were used to combat a recognized infection. The discoveries of sulfanilamide and penicillin in the 1930s and 1940s, however, ushered in the golden age of antimicrobial therapy, in which the use of antimicrobial agents came to be based on the knowledge of specific causative organisms and their corresponding activities.

The selection of the most appropriate antimicrobial agent depends on an accurate diagnosis and identification of the offending organism. *In vitro* culture and sensitivity testing of isolated organisms are routine methods for determining the antimicrobial of choice. The spectrum of activity of most antibiotics is broadly described in terms of activity against gram-positive or gram-negative organisms. This classification is based on staining characteristics with a blue primary stain of crystal violet with iodine and a red counterstain, usually safranin. The biochemical foundation for an organism retaining the blue, gram-positive color or not is related to the physical and chemical characteristics of the cell wall or cytoplasmic membrane. Empirically, the functional activity of many antibacterials correlates to some degree with the gram-staining reaction. The medical trend is toward use of relatively narrow-spectrum therapeutics having strong activity against specific organisms, as opposed to use of broad-spectrum agents without the supporting diagnostics. When the time factor is critical, as in life-threatening conditions, therapy using a broad-spectrum agent may be started before the culture and sensitivity testing. In addition to organism identification, consideration is given to whether an agent kills or inhibits the organism or its growth. The route of administration, dosage rate, frequency of treatment, and overall duration of treatment must also be considered. Microbes are constantly undergoing genetic change. Some variant strains have the ability to deactivate certain antimicrobials or grow in the presence of antimicrobials to which earlier generations were sensitive. Under antimicrobial therapy, the resistant strains may survive and render the agent less effective. Because of this phenomenon of selection, several agents representative of structurally different antimicrobial families may be used consecutively or, less often, concurrently for an evolving microbial population.

Antimicrobial agents are also used as prophylactics during surgery or at generally lower levels of administration to promote an animal's ability to withstand pathogenic challenge when under stress. In addition to the various families of drugs discussed herein, antimicrobial agents include carbadox [6804-07-5], the cephalosporins, nitrofurans, oxytetracycline [79-57-2], cefluofor, tilmecosen, trimethoprim sulfa, florfenicol [73231-34-2], [76639-94-6], and tylosin [1401-69-0].

Sulfonamides.   The sulfonamides (sulfas) are derivatives of *para*-amino-benzenesulfonamide [63-74-1]. These agents are active against a broad spectrum of gram-positive and gram-negative organisms. Their mode of action is by competitive antagonism of *para*-aminobenzoic acid (PABA), a folic acid precursor. Because mammalian cells do not synthesize folic acid, as do the bacteria that are sensitive to the sulfas, mammalian toxicity is low. The antibacterial activity of the sulfas can be augmented by concurrent use of trimethoprim [738-70-5], which blocks another step in folic acid synthesis.

Although the antibacterial spectrum is similar for many of the sulfas, chemical modifications of the parent molecule have produced compounds with a variety of absorption, metabolism, tissue distribution, and excretion characteristics. Administration is typically oral or by injection. When absorbed, they tend to distribute widely in the body, be metabolized by the liver, and excreted in the urine. Toxic reactions or untoward side effects have been characterized as blood dyscrasias; crystal deposition in the kidneys, especially with insufficient urinary output; and allergic sensitization. Selection of organisms resistant to the sulfonamides has been observed, but has not been correlated with cross-resistance to other antibiotic families (see ANTIBACTERIAL AGENTS, SYNTHETIC—SULFONAMIDES).

Penicillins.   Since the discovery of penicillin in 1928 as an antibacterial elaborated by a mold, *Penicillium notatum*, the global search for better antibiotic-producing organism species, radiation-induced mutation, and culture-media modifications have been used to maximize production of the compound. These efforts have resulted in the discovery of a variety of natural penicillins differing in side chains from the basic molecule, 6-aminopenicillanic acid [551-16-6]. These chemical variations have produced an assortment of drugs having diverse pharmacokinetic and antibacterial characteristics (see ANTIBIOTICS, β-LACTAMS).

The mechanism of antibacterial activity is through inhibition of gram-positive bacterial cell-wall synthesis; thus, the penicillins are most effective against actively multiplying organisms. Because mammalian cells do not have a definitive cell-wall structure as do bacteria, the mammalian toxicity of the penicillins is low. Allergic phenomena in patients following sensitization may occur.

The penicillins as natural and semisynthetic agents are used primarily against susceptible *Pasteurella* sp., staphylococci, streptococci, clostridia, and *Corynebacterium* sp. Penicillin is widely used for therapeutic purposes against these organisms and in animal feeds as a growth promoter. The latter effect is considered to be a result of subtle and reversible effects on the gastrointestinal microflora.

Aminoglycosides.   The aminoglycosides, such as streptomycin [128-46-1], neomycin [119-04-0], kanamycin [59-01-8], and gentamycin [1403-66-3], have a hexose nucleus joined to two or more amino sugars (see ANTIBIOTICS, AMINOGLYCOSIDES). These all tend to be poorly absorbed from the gastrointestinal tract but are absorbed well following parenteral administration. They are rapidly bactericidal by inhibiting intracellular protein synthesis. The active transport mechanism which allows intracellular access strongly depends on pH, divalent cations, osmolality, and oxygen tension. The latter renders many anaerobes resistant to the aminoglycosides, which typically have very broad activity spectra, with greater activity against gram-negative bacteria. For this reason, a penicillin may be complementary to an aminoglycoside and provide, overall, a broader spectrum. Toxicity following exaggerated or prolonged dosage schedules is characterized by renal failure or damage to the eighth cranial nerve (auditory) with auditory- or vestibular-balance dysfunction.

Tetracyclines. The tetracyclines, including chlortetracycline [57-62-5] and oxytetracycline [79-57-2], are produced as fermentation products (see ANTI-BIOTICS, TETRACYCLINES). These have a broad antibacterial spectrum including gram-positive and gram-negative organisms, rickettsiae, *Chlamydia* sp., and *Mycoplasma* sp. In addition, the tetracyclines are commonly employed at low dosages as growth promoters in the main food-producing species. Most tetracyclines are incompletely absorbed following oral administration and are antagonized through chelatin primarily with divalent cations. The mechanism of action is, like the aminoglycosides, by intracellular inhibition of protein synthesis. Toxicity is rare. Although seldom of clinical significance, the tetracyclines are incorporated into metabolically active calcified tissues, resulting in discoloration. This phenomenon is most prominent in rapidly growing bone, eg, fetal or juvenile skeletal and dental systems.

Growth Promoters. The tetracyclines and penicillin, when administered to food-producing species (poultry, swine, and cattle) during active growth, improve the rate of weight gain and efficiency of feed utilization significantly (see FEEDS AND FEED ADDITIVES). Other antibacterials are also used for this purpose, including avoparcin [37332-99-3], monensin [17090-79-8], bacitracins, virginiamycin [11006-76-1], lincomycin [154-21-2], tylosin, and flavomycin [11015-37-5]. These effects are not related specifically to prevention or treatment of bacterial diseases, but to subtle shifts in enteric processes (1) (see GROWTH REGULATORS, ANIMAL).

Antifungal Agents. Fungi and related organisms are encountered most commonly in superficial infections of the skin, and less commonly as systemic or deep mycoses affecting internal organs (see ANTIPARASITIC AGENTS—ANTIMYCOTICS). Superficial lesions may range from minor hair loss (ringworm) to severe, generalized hair loss and marked pathological changes in the skin, with secondary bacterial infections. The systemic diseases are typically refractory to most therapeutics and are frequently fatal. Many of the mycotic infections have, to some highly variable degree, zoonotic potential. Because they tend to be spore-forming organisms, mycoses also tend to be periodically recurrent or persistent in a given environment.

Favorable responses of the superficial infections have been observed following exposure to sunlight or administration of vitamin A [68-26-8] (qv). Some infections remit spontaneously as a young animal matures. More often, topical application of an antifungal such as nystatin [34786-70-4] or cuprimyxin [28069-65-0] or systemic griseofulvin [126-07-8] over six to 12 weeks is justified. The systemic mycoses are sensitive to very few therapeutic agents. Some, such as actinomycosis and actinobacillosis of cattle, respond to sulfa therapy, but others (cryptococcosis, blastomycosis, etc) may show response only to amphotericin B [1397-89-3], a relatively toxic antibiotic, or itraconazole [84625-61-6] (dogs and cats). Animals having systemic infections are frequently euthanized because of the history of limited therapeutic success and the zoonotic disease potential.

Parasiticides

Parasiticides can be roughly divided according to parasites, host species, or chemical classification (see ANTIPARASITIC AGENTS–ANTHELMINTICS; ANTIPARASITIC AGENTS–ANTIPROTOZOALS). By any classification, these are ubiquitous in the management and control of parasites of both companion and food-producing animals (2,3).

Organophosphates and Carbamates. The main pharmacologic action of organophosphates and carbamates is the inhibition of the cholinesterase enzymes, primarily acetylcholinesterase (AChE) (see ENZYME INHIBITORS). Generally, acetylcholine (ACh) is responsible for transmission of neutral impulses at voluntary neuromuscular junctions, at the sympathetic ganglia synapses, and throughout the parasympathetic system (see CHOLINE). Under normal conditions, it is rapidly hydrolyzed and inactivated by AChE. In the presence of AChE inhibitors, the enzyme is phosphorylated, with the consequent pharmacologic and toxic actions produced by excessive accumulations of ACh. Because ACh is an integral part of insect and helminth physiology, the antiparasitic utility spectrum of the drug family is immense.

Organophosphates and carbamates are typically lipid-soluble and are, as a consequence, rapidly absorbed following inhalation or oral, parenteral, or topical administration. Once absorbed, metabolism is primarily by hepatic hydrolysis or oxidation. Various organophosphates (O–Ps) and carbamates are used against virtually all animal parasites. The use of O–Ps and carbamates is widespread, both as animal antiparasiticides and as agricultural and home-use pesticides (see INSECT CONTROL TECHNOLOGY). In addition, concurrent exposure to more than one agent results in cumulative physiologic effect, and therefore the incidence of toxic effect is relatively high. Atropine is an excellent antidote by virtue of its blocking the action of ACh within the parasympathetic nervous system. It is neither a complete antagonist nor does it modify the rate at which the enzyme is regenerated. Pralidoxime chloride is another antidote frequently used as an adjunct to atropine specifically for O–P toxicity. It acts by regenerating the enzyme throughout the system, but may exacerbate toxicity in cases of reversible carbamate–esterase bonding.

Avermectins. The avermectins [65195-52-0, 65195-58-6] are fermentation products derived from *Streptomyces avermitilis* (see ANTIPARASITIC AGENTS, AVERMECTINS). They are macrocyclic lactones having a very broad spectrum of insecticidal and anthelmintic activity. First commercially available in 1981, abamectin [71751-41-2] (avermectin B_1) and ivermectin [70288-86-7] are highly active at doses of ca 0.2 mg/kg body weight against the internal parasites of cattle and horses, respectively, as well as against lice, internally migrating fly larvae (warbles), and mites. They are active in dogs against larval stages of heartworm disease and intestinal parasites, with the exception of tapeworms. Milbemycin, sulfar, and metronidazole [443-48-1] are used in small-animal practice.

Levamisole. The racemic mixture of the *d* and *l* isomers of tetramisole [6649-23-6] was first described in 1966. It is used as an anthelmintic against a wide variety of nematodes, including lungworms, of ruminants, swine, horses, dogs, and poultry. Anthelmintic activity resides in the *l*-isomer, levamisole [14769-73-4], the form used.

Benzimidazoles. The benzimidazoles include a large family of anthelmintics, eg, thiabendazole [148-79-8], albendazole [54965-21-8], cambendazole [26097-80-3], fenbendazole [43210-67-9], mebendazole [31431-39-7], oxfendazole [53716-50-0], and oxibendazole [20559-55-1]. Administration is oral, and the spectrum of activity is broad against nematode parasites of the intestinal tract. The usual dosage of thiabendazole is 50–110 mg/kg body weight. Dosage of other benzimidazoles are 2–30 mg/kg. The activity of the individual compound varies against specific parasite species; none, however, is effective against lungworms. Benzimidazoles have the advantage of a low mammalian toxicity, ca 10–30 times the recommended dosage. Absorption is rapid, parent compound and metabolites are excreted in the urine. There has been some indication of teratogenic effects with use of albendazole and cambedazole.

Other Parasiticides. The parasiticides described below have a relatively limited usage owing to a narrow spectrum of antiparasitic activity or because of the introduction of inherently safer or more effective products.

Immiticide (melarsonine hydrochloride) is now the drug of choice in dogs against the adult stage of heartworm infection.

Carbon disulfide (qv) is used, in combination with other orally administered anthelmintics, by stomach tube for bots (*Gastrophilus sp.* larvae) and ascarids (roundworms) of horses.

Coccidiosis. Coccidiosis, caused by protozoans of the genera Eimeria and Isospora, may be present in any domesticated animal species but is ubiquitous in the poultry industry, with serious consequences (see ANTIPARASITIC AGENTS, ANTIPROTOZOALS). The life cycle involves both asexual and sexual intracellular parasitic stages characterized by a rapid development and multiplication of infective stages with consequent destruction of, primarily, the intestinal lining of the host. This leads to growth retardation and, when severe, a high mortality. Anticoccidial agents are added routinely as feed components through the life of broiler chickens. A partial list of additives includes amprolium [121-25-5], ethopabate [59-06-3], robenidine [25875-50-7], arprinocid [55779-18-5], monensin, lasalocid [25999-31-9], chlortetracycline [57-62-5], and the sulfa compounds. The most widely used additives are representatives of the ionophore antibiotics, ie, monensin and lasalocid. Historically, the appearance of resistance by the coccidia to anticoccidial agents has been rapid, frustrating efforts aimed at control. This resistance has not yet been observed to any notable degree with the ionophores even after more than a decade of extensive use. A program of rotation, where anticoccidials with differing modes of action are used in succession minimizes the impact of resistance.

Diethylcarbamazine [98-89-1] is a piperazine derivative which is given daily as a prophylactic for canine heartworm disease (*Dirofilaria immitis*) or as a therapeutic for roundworms. *D. immitis* is transmitted only by mosquitoes, and therefore the period of administration varies geographically, depending on temperatures and humidity which regulate the mosquito life cycle.

Phenothiazine [58-37-7] (thiodiphenylamine) is used orally against intestinal nematodes of ruminants and horses. It is used with occasional gastrointestinal upset, hemolytic processes, and photosensitivity. It is used routinely at low concentrations on horse farms to suppress the egg production of intestinal parasites (strongyles) and thus limit pasture contamination and transmission (4).

Hexachloroethane [67-72-1] has, like carbon tetrachloride [56-23-5], been used to remove liver flukes from ruminants. Also used are albenzadole, previously mentioned as a benzimidazole, and clioxanide [144327-41-3], oxychozanide [2277-92-1], or rafoxanide [22662-39-1]. Ciba's Ivomect is used for fluke control in Europe.

Niclosamide [50-65-7] (2',5-dichloro-4'-nitrosalicylanilide) has been commonly used against tapeworms in small animals (5). Although tapeworms (cestodes) are frequently refractory to anthelmintics highly active against other intestinal parasites, they are sensitive to niclosamide as well as praziquantel [55268-74-1] and epsirantel. Pyrantel pamoate [22204-24-6] is probably the most common animal wormer used in the 1990s.

Antiinflammatory Agents

Inflammation is a defense mechanism of the body that plays a key role in righting disease and initiating wound healing. It is clinically characterized by local redness, swelling, pain, and heat. Because of these symptoms, inflammation can be more detrimental than beneficial to normal body function and at such times the practitioner chooses to slow it down. This group of compounds is used on an individual animal basis and tends to be used more in the companion-animal and equine specialities (4) (see ANALGESICS, ANTIPYRETICS, AND ANTIINFLAMMATORY AGENTS).

The classic example of an antiinflammatory drug is aspirin [50-78-2], acetosalicylic acid, an effective analgesic for many years. It is well tolerated by the dog and the horse, but is relatively toxic to cats. Under the proper clinical circumstances, it can be used for prolonged therapy in chronic inflammatory diseases such as arthritis. Rimadyl is presently used.

Pyrazolone derivatives, specifically phenylbutazone [50-33-9] and, for limited conditions, dipyrone [5907-38-0], are very popular with equine practitioners and are particularly useful in managing cases of lameness and controlling inflammation after trauma or surgery (4). Dipyrone is an analgesic for cases of equine colic. Phenylbutazone is an effective antiinflammatory drug, but is more toxic than the salicylates, which limits its long-term use. For short-term usage, its ease of administration as injectable or oral preparations makes it a popular product for the equine or small-animal specialist. However, ketoprofen [22071-15-4] is a new drug of choice.

The most widely used group of antiinflammatory drugs are the corticosteroids and their synthetic analogues. This group of compounds has several physiologic actions, including effects on sodium retention and liver glycogen deposition as well as inhibitory effects on wound healing and, more recently recognized, proliferation of cancer cells.

The natural compounds cortisol [50-23-7], cortisone [53-06-5], and corticosterone [50-22-6] vary only slightly in structures and pharmacologic properties (see STEROIDS). The synthetic analogues in more modern practice, prednisolone [52438-85-4], dexamethasone [50-02-2], triamcinolone [124-94-7], and betamethasone have greater antiinflammatory potency, and their effects on sodium retention tend to be less severe.

The uses of corticosteroid antiinflammatory drugs in veterinary medicine are many and varied. In the intact animal, the glucocorticoids and mineralcorticoids are produced in the adrenal glands. Exogenous compounds are, therefore, used for their glucogenic physiologic effect in cases where the animal is unable to produce sufficient quantities of these compounds. When given at pharmacologic dosage, their effects include antiinflammatory aspects useful in controlling healing and inflammation following trauma or surgery; controlling inflammation in severe dermatologic cases, thereby improving effective treatment of the cause of the problem; and in helping to control allergic reactions. On a cellular level, these compounds exert less well-defined effect in helping to preserve cell-membrane integrity and improve cellular metabolism. These effects, in addition to the effects on the microcirculation, are the basis for corticosteroid use in shock-syndrome therapy, which is, however, controversial.

Hormones

Hormones (qv) as naturally occurring, semisynthetic, or synthetic compounds are used to regulate reproductive cycles, gestation, and parturition. They are also used as therapeutics for hormonal imbalances and responsive physical or physiological abnormalities, or as growth promoters in ruminants (see GROWTH REGULATORS, ANIMAL). The application of other than sex-related hormones is not as complex as in human therapy because of the relatively short life spans of animals and high cost.

Hormones can either delay or induce estrus. The therapeutic manipulations of normal, or abnormal, estrus sequences are based on the intricate biological feedback relationship between the pituitary gland and the gonads. In broad terms, follicle-stimulating hormone [9034-38-2] releases estrogens. The estrogens cause a decrease in FSH and an increase in luteinizing hormone [9002-67-9] (LH) which causes ovulation and formation of an ovarian corpus luteum (CL). The CL releases natural progesterone, the level of which either helps maintain pregnancy or, at some point, reinitiates the cycle. The therapeutic applications are based on adjusting or creating the normal sequence of events. Administration of progesterone [57-83-0] or progestogens (mibolerone [3704-09-4] or megestrol acetate [3562-63-8]) simulates the hormonal action of the corpus luteum and, in so doing, delays the onset of an estrus. Estrus can be terminated with progesterone, LH, or some prostaglandins (qv), with a resulting fertile ovulation. Stilbestrol is used in female dogs for estrogen-responsive incontinence. An artificial estrus, which does not produce a fertile ovulation, can be induced by exogenous estrogens such as diethylstilbestrol [56-53-1]. Pregnant mare serum, a functional gonadotrophin, the chorionic gonadotrophins and FHS create superovulation with an increased number of developed ova. The estrogens are frequently used to prevent zygote implantation and thus pregnancy, when given shortly after an unintended mating. Estradiol cypronate is used for mismating in dogs. Oxytocin [50-56-6] of pituitary origin stimulates sensitive uterine muscle at parturition. Oxytocin and prolactin [12585-34-1] facilitate milk production and let-down in lactating animals. For fertility control and estrus synchronization, the following are used: prostaglandins as lukeolytic agent ($PGF_{2\alpha}$, both natural and synthetic) in cattle; GNRH (aptorellin and Factrel) in cattle; FSH for superovulation in cattle and sheep and in embryo-transfer work; and anabolic steriods, eg, Winstrol-V (stanozolal). Cases of enlarged prostate glands in males are frequently responsive to estrogen therapy. The progestins, notably megestrol acetate, have been used successfully in the management of a variety of dermatitides and behavioral problems in small animals, and discrete clinical syndromes are associated with estrogen and testosterone imbalances in both males and females. Medical therapy, in light of the multisystem effects of these compounds, is always conservative, and adverse effects related to feminization of males or aggressive masculinization of females is frequent. Therapy is often an adjunct to surgical ovariectomy or castration.

Estrogens, testosterone [58-22-0], or compounds such as zeranol [26538-44-3] or trenbolone [10161-33-8] which can mimic their effects, have shown utility in accelerating the rate of weight gains and decreasing the amount of feed required to produce these gains in food-producing animals (6). The potential for human consumption of these compounds via the food supply has come under severe regulatory scrutiny and most of the drugs used for this purpose are administered as an implant or pellet in a part of the body, usually the ear, which is discarded at the time of slaughter. Extended withdrawal periods between the time of administration and the allowable date of slaughter depend on the release characteristics of the implant and may range from zero to one year. Dosages are relatively low, allowing drug release of 2–5 mg/d (see CONTROLLED RELEASE TECHNOLOGY; DRUG DELIVERY SYSTEMS).

Tranquilizers and Anesthetics

Tranquilizers find their niche in veterinary medicine in the management of excitement in individual animals (7) (see PSYCHOPHARMACOLOGICAL AGENTS). This group of compounds allows the practitioner to examine the frightened or injured patient with less chance of further damage or injury to the animal, the owner, and the veterinarian. Tranquilizers are also useful in the management of stress to avoid injury during the shipping of animals (see also ANESTHETICS; HYPNOTICS, SEDATIVES, ANTICONVULSANTS AND ANXIOLYTICS).

Acepromazine [61-00-7], a phenothiazine, is used in most animal species in both oral and injectable forms. It can be used at varying dosages to provide the state of tranquilization desired by the veterinarian. The product has a good margin of safety and has been used successfully by the veterinary profession for many years. Xylazine hydrochloride [23076-35-9] is another product used for both large and small animals. A thiazine compound unrelated to the phenothiazines, it acts primarily as a sedative. This compound is especially useful in examining fractious horses under field conditions (4). Both acepromazine and xylazine can be combined with other anesthetics for varying degrees of anesthesia or tranquilization. Detonimide is more potent than tylazine.

Tranquilizers are employed for restraint in minor surgical procedures. The cardiovascular system, respiratory system, and blood chemistry are all greatly altered by general anesthesia. In larger animals, such as cattle and horses, the weight of the animal's body alone resting on its side can be a physiologically adverse stress on the heart and lower lung field that the veterinarian would prefer to avoid. If the surgery is of a confined or local nature, such as the repair of a superficial laceration, the animal can be tranquilized and analgesics provided to the wound area by use of a specific nerve block or by infiltrating the area around the site with a local anesthetic. The latter are synthetic nonnarcotic substitutes for cocaine, the first anesthetic ever used. The synthetic substitutes include procaine hydrochloride [51-05-8], tetracaine hydrochloride [136-47-0], lidocaine [137-58-6], and mepivacaine hydrochloride [1722-62-9]. Lidocaine is preferred in veterinary medicine.

These agents are often combined with a vasoconstrictant such as epinephrine [51-43-4]. By using such a combination, the local anesthetic is held in the area for a longer period of time and its effect extended; hemorrhage is minimized, blood loss prevented, and a better surgical repair obtained.

A drug combination in popular use in dogs is a mixture of fentanyl [437-38-7], a narcotic analgesic, and droperidol [548-73-2], a butyrophenone tranquilizer. This combination produces a state of neuroleptanalgesia in which sedation and analgesia are achieved. The mixture is sold commercially and can be administered by both subcutaneous and intramuscular injection. Because the combination contains a narcotic, it has the advantage of being rapidly reversible with narcotic antagonists such as naloxone [465-65-6] and nalorphine [62-67-9] once the effects are no longer needed.

Another injectable anesthetic widely used in feline and primate practice is ketamine hydrochloride [1867-66-9]. Ketamine, a derivative of phencyclidine, can be chemically classified as a cyclohexamine and pharmacologically as a dissociative agent. Analgesia is produced along with a state that resembles anesthesia but in humans has been associated with hallucinations and confusion. For these reasons, ketamine is often combined with a tranquilizer. The product is safe when used in accordance with label directions, but the recovery period may be as long as 12–24 h.

Another group of anesthetics is comprised of barbiturates. By substituting various side chains on the basic structure, anesthetic activity can be greatly altered with regard to onset and duration of action. Short-acting barbiturates, such as thiopental [77-27-0], often provide only a few minutes of sedation. These products are useful for induction to other types of general anesthesia, trachea intubation, and minor manipulations, examinations, and procedures. Pentobarbital [57-33-0] is longer-acting and can be useful in more extensive or time-consuming surgery or in procedures requiring an extended sedation. Long-acting barbiturates, such as barbital and phenobarbital, have a prolonged effect in the animal, but also have a delayed onset of activity. These have generally been replaced in veterinary medicine by inhalation anesthetics. Phenobarbital [50-06-6], phenytan [57-41-0], and primidone [125-33-7] are used as anticonvulsants.

In veterinary medicine, the list of inhalation anesthetics generally includes only two agents, halothane [151-67-7] and methoxyflurane [76-38-0]. Although ether (ethyl ether) is used extensively in experimental work with laboratory animals, the risks associated with its use and the advantages of halothane and methoxyflurane have removed ether from general use by the practitioner.

Halothane and methoxyflurane are volatile and are used in a vaporizer and delivered to the animal via an oxygen carrier. Both agents can be delivered with nitrous oxide [14522-82-8], a mild anesthetic that when combined with halothane or methoxyflurane can induce anesthesia faster than halothane or methoxyflurane alone. The recovery is faster because of the low solubility of nitrous oxide in the blood. Nitrous oxide can also be used alone, but must be supplemented with a barbiturate or a narcotic. At present, isoflurane [26675-46-7] is the most commonly used and preferred gas anesthetic in veterinary medicine.

It must be remembered that all anesthetics and tranquilizers are used by the practitioner following a risk–benefit evaluation. General anesthesia, even being administered by an experienced practitioner, can result in death through cardiac or respiratory depression. The veterinarian is acutely aware of these risks and chooses the drug and method of administration considering the patient's health status, the nature of and need for the procedure, and the likelihood of success.

Cancer Chemotherapy

In the veterinary as in the human patient, neoplasms are often metastatic and widely disseminated throughout the body. Surgery and irradiation are limited in use to well-defined neoplastic areas and, therefore, chemotherapy is becoming more prevalent in the management of the veterinary cancer victim (see CHEMOTHERAPEUTICS, ANTICANCER). Because of the expense and time involved, such management must be restricted to individual animals for which a favorable risk–benefit evaluation can be made and treatment seems appropriate to the practitioner and the owner. In general, treatment must be viewed not as curative, but as palliative.

The purpose of cancer chemotherapy, most briefly put, is to kill specific cells. The compounds are most active against rapidly growing and dividing cells, ideally the neoplastic cells, but all dividing cells can be attacked. For this reason, toxic signs such as alopecia, anemias and leukopenias, anorexia, vomiting, and other gastrointestinal signs may be indicative of undesirable effects of therapy. Periods of rest are often built into the treatment regimen to allow the animal's body a chance to recover and reestablish normal function.

Chemotherapeutic agents are grouped by cytotoxic mechanism. The alkylating agents, such as cyclophosphamide [50-18-0] and melphalan [148-82-3], interfere with normal cellular activity by alkylation deoxyribonucleic acid (DNA). Antimetabolites, interfering with complex metabolic pathways in the cell, include methotrexate [59-05-2], 5-fluorouracil [51-21-8], and cytosine arabinoside hydrochloride [69-74-9]. Antibiotics such as bleomycin [11056-06-7] and doxorubicin [23214-92-8] have been used, as have the plant alkaloids vincristine [57-22-7] and vinblastine [865-21-4].

These compounds vary in their specific mechanism of action and often have different effects on the individual patients. Thus, they are generally used in combinations, eg, corticosteroids with an alkylating agent, or an antimetabolite with a plant alkaloid in a rotating schedule.

Immunostimulation

The body's immune mechanism, both humoral and cell-mediated, affords a primary defense against invasion by foreign substances, ie, exogenous entities that the body may encounter, including viruses, bacteria, chemicals, drugs, grafts, and transplants. The reaction by the immune system kills, neutralizes, or rejects the entity. The mechanisms involved in this complex system are under intense investigation, and a better understanding of the immune system will, in the future, permit the control of disease by means only speculated about today.

Human and veterinary practitioners have been manipulating the immune system for many years with bacterins and virus vaccines, in order to induce a response in the immune system. The animal forms antibodies which destroy the antigen. When the same or similar antigen is encountered again, as during exposure to the disease organism, the immune system is activated more quickly through an anamnestic response, thereby preventing the disease. Vaccines and bacterins are widely used in veterinary medicine for most domestic and exotic species (8) (see IMMUNOTHERAPEUTIC AGENTS; VACCINE TECHNOLOGY).

The prophylactic stimulation of the immune system using vaccines and bacterins is time-consuming. Of even greater value would be the ability to activate the system to combat a disease attack already underway, or to be able to increase the response to abnormal cells and neutralize neoplasia in any organ of the body. Several compounds, some unique entities and some already in use for other purposes, have shown potential utility as such nonspecific immune stimulants.

In 1971, levamisole, an anthelmintic compound widely used in cattle and swine, was shown to improve the effects of an experimental *Brucella abortus* vaccine in mice. Since that time, the veterinarians and physicians have explored the effects of levamisole in such diverse areas as arthritis, lupus erythematosis, cancer therapy, respiratory diseases, Newcastle disease, foot-and-mouth disease, mastitis, and vaccine potentiation. Although the exact mechanism of action has as yet not been determined there is substantial evidence that, under defined circumstances, levamisole can augment the animal's natural immune response (9). New immunostimulants include *Staph Lysate acemannon, MAB-31.*

Discovered in 1957, a group of natural substances called interferons has been the subject of therapeutic interest. Interferons are glycoproteins synthesized by cells that are under attack by a virus. Interferon seems to be an integral part of the body's basic defense mechanism, but more recent work indicates broad therapeutic activity and the possibility of cross-species efficacy. Human interferons are used in cats with FIV and FetV. Feline interferon has been approved. In Japan, cyclosporine [59865-13-3] is used in organ transplantation. Prednisone [53-03-2] is approved for topical application in treatment of autoimmune ocular disease. The emergence of recombinant DNA technology might allow sufficient quantities of interferon to be produced at a reasonable cost and thus may make interferon therapy a practical reality (see GENETIC ENGINEERING, ANIMALS).

BIBLIOGRAPHY

"Veterinary Drugs" in *ECT* 2nd ed., Vol. 21, pp. 241–254, by A. L. Shor and R. J. Magee, American Cyanamid Co.; in *ECT* 3rd ed., Vol. 23, pp. 742–753, by D. M. Petrick and R. B. Dougherty, American Cyanamid Co.

1. *Feed Additive Compendium*, Miller Publishing Co., Minneapolis, Minn., 1982, published annually.

2. O. H. Siegmund, *The Merck Veterinary Manual*, 5th ed., Merck and Co., Inc., Rahway, N.J., 1979.

3. J. R. Georgi, *Parasitology for Veterinarians*, 3rd ed., W. B. Saunders Co., Philadelphia, Pa., 1980.

4. E. J. Catcott and J. F. Smithcors, *Equine Medicine and Surgery*, 2nd ed., American Veterinary Publications, Inc., Wheaton, Ill., 1972.

5. R. W. Kirk, *Current Veterinary Therapy, Small Animal Practice*, 7th ed., W. B. Saunders Co., Philadelphia, Pa., 1980.

6. J. L. Howard, *Current Veterinary Therapy, Food Animal Practice*, W. B. Saunders Co., Philadelphia, Pa., 1981.

7. L. R. Soma, *Textbook of Veterinary Anesthesia*, The Williams and Wilkins Co., Baltimore, Md., 1971.

8. S. Krakowka, *Mod. Vet. Pract.* **62**, 447 (1981).

9. J. Symoens and M. Rosenthal, *J. Reticuloendothel. Soc.* **21**, 175 (1977).

VETIVER.   See OILS, ESSENTIAL.

VINEGAR

Vinegar is the liquid condiment or food flavoring used to give a sharp or sour taste to foods. It is also used as a preservative in pickling and as the sour component in many different sauces, dressings, and gravies. Asian-style sweet–sour sauces and so-called health beverages are also based on vinegars (1,2). The word vinegar is derived from Latin via the old French *vinaigre*, meaning eager wine. In old English and old French the word eager (*aigre*) meant sour or sharp. Thus, vinegar is a sharp or sour wine (3) and, as such, consists principally of water, acetic acid, mineral salts, and the organic constituents of the natural organic starting material.

Vinegar results from the action of the enzymes of bacteria of the genus *Acetobacter* and some others on dilute solutions of ethyl alcohol such as cider, wine, beer, or diluted distilled alcohol (see ETHANOL). Most vinegars for table use, eg, in the dressing of salads, derive from the acetic acid–bacterial fermentation of wine or cider (see ACETIC ACID). These latter, in turn, are produced by alcoholic fermentation (see FERMENTATION) of dilute sugar solutions such as grape juice, apple juice, or malt. *Saccharomyces cerevisiae* is the yeast involved most frequently in alcoholic fermentation, ie, in the enzymatic conversion of fermentable sugars to dilute alcoholic solutions (see YEASTS). Although fruits and honey are used most frequently as sources of fermentable sugar for vinegar production, barley malt and, in the Orient, rice, after hydrolysis of starch, serve as primary sources (see FOOD PROCESSING; FRUIT JUICES). Some raw materials used for vinegar production are listed in Table 1.

In the United States, standards of identity for vinegar date back to the Federal Food and Drug Act of 1906, in which six types of vinegar are defined as follows: "Vinegar, cider vinegar, apple vinegar, is the product made by the alcoholic and subsequent acetous fermentations of the juice of apples, and contains, in 100 cubic centimeters (20°C), not less than 4 grams of acetic acid" (33). The other five types of vinegar are defined in the same terms, except that cider vinegar is replaced by wine vinegar, malt vinegar, sugar vinegar, glucose vinegar, or spirit vinegar. In the case of the malt vinegar, the sugar of a hydrolyzed starch solution is fermented to ethanol. This solution, known as dilute beer, is immediately oxidized by *Acetobacter* to vinegar (see BEER). The quantity of 4 g in 100 cm^3 of the quoted federal regulation (33) is equivalent to 40 g/L acetic acid, or 40-grain strength in the terms used by vinegar producers (see BEVERAGE SPIRITS, DISTILLED). The United States *Federal Register* carries regulatory announcements concerning vinegar production at frequent intervals. Table 2 lists published figures for vinegar production in the United States and for interstate trade among members of the European Economic Community and Asia.

A number of factors govern the composition of vinegar: the nature of the raw material, the substances added to promote alcoholic fermentation and the growth and activity of *Acetobacter*, the procedure used for the acetification, and finally the aging, stabilization, and bottling operations. Vinegars are made from natural solutions containing fermentable sugars, such as fruit juices and honey; solutions in which the sugars are produced by hydrolysis of starch, such as beers and sake; and solutions of distilled ethyl alcohol. Distilled alcohol, in turn, can be derived from the sugars of fruits or that from hydrolysis of starch, or from synthetic processes such as the hydration of ethylene from petroleum. Most countries distinguish between fermentation vinegars and synthetic vinegars. Mediterranean countries usually permit only fermentation vinegars (some only

Table 1. Raw Materials Used to Make Vinegar

Raw material	Reference	Raw material	Reference
Mainly sugary		*Mainly starchy*	
jujube	4	potato, corn flour	22
sweet potato	5	soybean	23
dates	6,7	seaweed	24
citrus	8,9	rice	25
persimmon	10,11	grain starch	26
pear	12		
sugar cane	13	*Various*	
plum	14	onions	27
tomato	15	bamboo grass	28
kiwi fruit	16	wood	29
pineapple	17	whey	30
molasses	18	coconut water	31
honey	19	vinasse (distillation residue)	32
palm sap	20		
muscavado (brown sugar)	21		

Table 2. Vinegar Production, 1000 m³[a]

Year	Fermented, 4%[b]	Distilled, 10%[b]	Total
United States[c]			
1972	202.9	282.4	485.3
1982	199.1	509.1	708.2
1987	126.4	581.4	707.8
1992	202.5	592.0	794.5
European Economic Community[d]			
1990	180	270	450
Japan[e]			
1989			381.1
1992			391.2
South Korea[f]			
1990	21.2	1.48	22.7
1994	33.6	0.58	34.2
Asia (other than Japan, Korea, and China)[e]			
1989			392.9
1992			482.5

[a]To convert m³ to U.S. gallons, multiply by 264.
[b]Percent acetic acid.
[c]Ref. 34.
[d]Ref. 35.
[e]Ref. 36.
[f]Ref. 37.

wine vinegar) as foods, whereas in some northern European countries even dilute solutions of synthetic acetic acid are acceptable.

Grape and apple juices usually contain all of the trace nutrients required by *Saccharomyces* for fermentation of sugars to alcohol. Other fruit and diluted honey, as well as barley malt and rice extract, frequently need additions of nitrogen, phosphorus, and potassium compounds, together with some autolyzed yeast to facilitate the yeast growth necessary for fermentation. Stimulation of *Acetobacter* growth frequently requires the addition of autolyzed yeast, vitamin B complex (see VITAMINS) and phosphates. The character and composition of vinegar is influenced greatly by the method used for the acetification and the subsequent processing steps.

Manufacture

Primitive people very likely encountered vinegar-like liquids in hollows in rocks or downed timber into which berries or fruit had fallen. Wild yeasts and bacteria would convert the natural sugars to alcohol and acetic acid. Later, when early peoples had learned to make wines and beers, they certainly would have found that these liquids, unprotected from air, would turn to vinegar. One can postulate that such early vinegars were frequently sweet, because the fruit sugars would have been acted on simultaneously by both bacteria and yeast. Only since the middle 1800s has it been known that yeast and bacteria are the cause of fermentation and vinegar formation.

Starch Hydrolysis and Alcoholic Fermentation. In general, because yeasts cannot utilize starch directly as a carbon source, the starch must first be hydrolyzed to sugar. Malt vinegars, commonly used as table vinegar in the United Kingdom, are made from malted barley or a mixture of malted barley with other starchy grains. Malt enzymes convert starch to sugars readily fermentable by *Saccharomyces* yeasts. In Japan, where vinegars are made from rice, a mixture of hydrolyzing enzymes produced by the fungus *Aspergillus oryzae* converts rice starches to sugars. A small amount of cooked rice cultured with the fungus is added to a larger quantity of cooled steamed rice. Frequently, the yeast that converts the sugar to alcohol is added at the same time, resulting in a dilute alcoholic solution rather than a dilute sugar solution. For the alcoholic fermentation itself, strains of *Saccharomyces cerevisiae* are most frequently used. The malt alcoholic solutions usually contain 5–10 vol % of ethanol; rice-derived alcoholic solutions, ie, sake, may contain 15–20%. These alcoholic solutions usually are converted immediately to vinegar but, if they are to be stored, the pH should be low enough to discourage growth of undesirable organisms. In particular, lactic bacteria can cause problems at higher pH values. Addition of 20–30 mg SO_2/L of vinegar helps to prevent undesirable bacterial activity during storage.

Alcoholic fermentation, according to the Gay-Lussac equation,

$$C_6H_{12}O_6 \longrightarrow 2\ CO_2 + 2\ C_2H_5OH$$

yields 0.5114 g alcohol for each 1 g of sugar used. The theoretical yield is never obtained because of side reactions, volatilization of alcohol with the evolving carbon dioxide, competition from other organisms, and other factors (38). Yields of 88–94% are considered good commercial practice (see FERMENTATION).

The alcoholic fermentation is frequently conducted in two phases, although in a modern vinegar plant it can be conducted in one. The first phase is a vigorous fermentation during which the rapid evolution of carbon dioxide protects the alcoholic solution from air. The second or slower phase is fermentation of the residual sugar at a lower rate, during which, again, protection from air is required. In the first phase, 50–100 mg SO_2/L of sugar-containing mash is added, followed after approximately 1 h by 1–3% of an actively fermenting, pure-culture starter of *Saccharomyces cerevisiae*. The fermentation process is monitored for disappearance of sugar and increase in temperature. Rates of alcoholic fermentation are highest at ca 25–30°C; higher temperatures tend to damage enzyme systems. The decrease in sugar content is measured hydrometrically and usually is expressed in degree Brix (the weight percentage of sucrose in a sucrose–water mixture). However, the sugar content of fruit juice of corresponding density is only slightly less than the equivalent percentage of glucose and fructose. Some error is introduced by the presence of nonsugar solids. Alcohol produced during the fermentation has a density less than that of water, causing the degree Brix reading to decrease more than it would from the loss of the sugar alone. Temperatures during the fermentation are controlled by cooling when necessary. Cooling jackets or internally mounted coils are used, although in older installations the fermenting medium is pumped from the tank through an external heat exchanger and returned to the tank. Near the end of the vigorous fermentation (about 0°Brix), the wine or beer is siphoned off and placed in another tank for final fermentation. Equipped to permit escape of CO_2 but prevent entry of air, this tank is used for fermentation of residual sugar. In a modern plant, the complete fermentation can be conducted in a single closed stainless steel tank. When all the sugar has been fermented (negative °Brix readings and <1 g/L reducing sugars), the wine or beer may be acetified immediately or stored protected from air. For prolonged storage, addition of 50 mg/L SO_2 helps to prevent growth of lactic organisms.

Both the fermentation of hexose sugars to ethanol and carbon dioxide and the oxidation of ethanol to acetic acid are exothermic (heat yielding) processes (see SUGAR). The first reaction is expressed as follows:

$$180 \text{ g } C_6H_{12}O_6 \longrightarrow 92 \text{ g } C_2H_5OH + 88 \text{ g } CO_2 + 234 \text{ kJ } (55.6 \text{ kcal})$$

The yeast enzymes capture ca 92 kJ (22 kcal) of this energy for the formation of adenosine triphosphate (ATP), and the actual waste heat per 180 g (mol wt) of sugar fermented is ca 142 kJ (33.9 kcal). Depending upon the size of the fermenter and the rates of fermentation and aeration, loss of waste heat is apportioned among radiation, conduction, and vaporization of water and ethanol plus carbon dioxide. Although small fermenters may require no cooling, large ones require more cooling than occurs through natural radiation and conductance.

Wines have a low sodium and high potassium content and also contain tartaric, malic, and succinic acids, a wide spectrum of amino acids, phenolic materials, and trace quantities of vitamins and growth factors. These substances are found also in wine vinegar. Vinegar materials from other fruits, honey, and sugar-containing natural materials also contain a wide spectrum of nutrients from the base material. Distilled alcohol solutions, in contrast, do not contain nonvolatiles. Growth of acetobacter in these substrates requires addition of nitrogen, potassium, and phosphorus salts, trace amounts of other elements, and some organic growth factors.

Acetic Acid. Ethyl alcohol is converted to acetic acid by air oxidation catalyzed by the enzymes within bacteria of the genus *Acetobacter*:

$$46 \text{ g } C_2H_5OH + 32 \text{ g } O_2 \longrightarrow 60 \text{ g } CH_3COOH + 18 \text{ g } H_2O + 487.2 \text{ kJ } (116.4 \text{ kcal})$$

One gram of ethanol should yield 1.304 g acetic acid. Practical yields are 77–85%. To avoid killing bacteria, excess heat must be removed during the course of the oxidation. For example, oxidation of 1 m^3 (264 gal) of a solution of ethanol (10% by volume) yields 836.7 MJ (ca 2×10^5 kcal). If this oxidation occurs over a 4-d period, and the heat is liberated at a uniform rate, heat production amounts of 2.42 kW.

In contrast with the well-known Embden-Meyerhof-Parnass glycolysis pathway for the conversion of hexose sugars to alcohol, the steps in conversion of ethanol to acetic acid remain in some doubt. Likely, ethanol is first oxidized to acetaldehyde and water (39). For further oxidation, two alternative routes are proposed: more likely, hydration of the acetaldehyde gives $CH_3CH(OH)_2$, which is oxidized to acetic acid. An alternative is the Cannizzaro-type disproportionation of two molecules of acetaldehyde to one molecule of ethanol and one molecule of acetic acid. Possibly *Acetobacter* initiates both reactions (40). The disproportionation reaction is favored by slightly alkaline pH values, whereas the dehydrogenation more readily occurs under acidic conditions. In slightly acid media such as wine and beer, dehydrogenation seems more likely than disproportionation.

Orleans Process. Early mention of vinegar is found in the Talmud, in accounts of wine and beer turning to vinegar (41). The Babylonians, ca 5000 BC, made and used vinegar as a flavor enhancer and as a pickling agent or preservative. Production of vinegar by the Greeks and Romans is described in numerous writings, but it was not until the fourteenth century that there are records of vinegar-making, in Modena, Italy, and near Orleans, France. In the Orleans process, the wine oxidizes slowly in a barrel where it is covered with a film of *Acetobacter*. Holes that are covered with screens to exclude insects are bored in each barrel head to permit access to air. Wine is added through the bung hole with a long-stemmed funnel below the surface of the bacterial film and without disturbing the film. In operation, vinegar is removed through a spigot mounted near the bottom of the barrel head and is replaced with an equivalent quantity of wine through the funnel. Wines with 10–12% ethanol give vinegars of 8–10% acetic acid concentration. Orleans vinegars are characterized by a relatively high concentration of ethyl acetate, detected by its pleasantly strong fruity odor. Although Orleans process wine vinegar is much preferred for table use, its production is slow and relatively costly and has been replaced for processing vinegars. Furthermore, the vinegar in the Orleans process barrels tends to become slimy from the production of exocellular bacterial cellulose generated by *Acetobacter xylinum*. This slimy cellulose, called mother of vinegar, encapsulates the bacterial cells and dramatically slows the rate of production. Significant amounts of red wine vinegar are produced in Europe by the Pasteur modification of the Orleans process. A wooden grating is placed at the surface of the liquid in the partially filled barrels or in shallow tanks to support the film of vinegar bacteria.

In 1973, a multistage surface-fermentation process was patented in Japan for the production of acetic acid (42); eight surface fermenters were connected in series and arranged in such a way that the mash passed slowly through the series without disturbing the film of *Acetobacter* on the surface of the medium. This equipment is reported to produce vinegar of 5% acidity and 0.22% alcohol with a mean residency time in the tanks of 22 h.

Modena-Style or Balsamic Vinegar Process. Balsamic vinegars are made in and around the city of Modena in central Italy and consist of two general types: "Aceto Balsamico Tradizionale di Modena o di Reggio Emilia" and "Aceto Balsamico di Modena" (43). The traditional product is made from juice of the Trebbiano grape concentrated to about 40% sugar by direct flame heating of the container. The concentrate is added to the first (youngest) barrel of a series which is operated as a fractional blending system, the product removed from the oldest barrel being replaced by that from the one next older, and so forth through the system. The barrels are made of many different woods and vary considerably in size. *Zygosaccharomyces*, *Saccharomyces cerevisiae*, and *Gluconobacter* in the barrels convert the sugars to ethyl alcohol and gluconic acid as principal products. The ethyl alcohol is converted to acetic acid by the *Acetobacter* and *Gluconobacter*.

Subsequent chemical and bacterial reactions result in the formation of many other flavor and odor substances, among the more important of which is the ester ethyl acetate. Years of age as the product passes through the blending system create a condiment exhibiting a harmonious balance of sweetness and tartness. Analysis of 99 aliquots from ten different Aceto Balsamico Traditionale systems of barrels showed increases of heavy metal cations (Cr, Ni, Cu, Zn, Cd, and Pb) with age (44). This indicates that evaporation of volatiles concentrated the non-volatile metal salts in the product as it progressed through the system. Another analysis of product from each barrel of 13 systems showed increasing concentrations of gluconic acid with barrel age in about half of the systems and that gluconic acid concentration was significantly greater in Aceto Balsamico Traditionale than in Aceto Balsamico di Modena, wine and cider vinegars (43). Aceto Balsamico di Modena differs from the traditional product in that wine vinegar is blended into the grape juice concentrate and the age of the product is usually less.

Generator Process.   References to quick or generator processes for vinegar production are found as early as the seventeenth century (45). Usually, generators are packed with shavings of beech wood, which tend to curl and thus provide packing that does not consolidate but allows open spaces for the free flow of liquid and air. In addition, beech wood does not contribute undesirable flavors or impurities to the vinegar. In the modern generator, a recirculating pump transfers the partially acetified alcoholic mixture from the bottom section of the generator to a distributing system at the top of the packed section. Air is measured into the upper part of the storage section below the packed section of the vat and exhaust air is vented from the top of the packed section to the outside, although in some generators it is recirculated (46). Cooling coils may be located in the packed section, but more frequently are placed at the bottom of the receiver section or are incorporated in the line for recirculating the liquid. Some packing materials other than the traditional beech wood shavings are coke, grape twigs, rattan bundles, corn cobs, or unglazed ceramic saddles. A pilot-plant Frings generator makes vinegar at a rate of 20.4 μgL^{-1}s^{-1} (47).

Various species and many strains of *Acetobacter* are used in vinegar production (48,49). Aeration rates, optimum temperatures and nutrient requirements vary with individual strains. In general, fermentation alcohol substrates require minimal nutrient supplementation while their addition is necessary for distilled alcohol substrates.

Submerged-Culture Generators.   Adaptation of the surface-film growth procedure for producing antibiotics to an aerated submerged-culture process has been successful in making vinegar. A mechanical system keeps the bacteria in suspension in the liquid in the tank, in intimate contact with fine bubbles of air. The excess heat must be removed and the foam, which accumulates at the top of the tank, must be destroyed. The most widely used submerged-culture oxidizer is the Frings acetator (50). It uses a bottom-driven hollow rotor turning in a field of stationary vanes arranged in such a way that the air which is drawn in is intimately mixed with the liquid throughout the whole bottom area of the tank (51,52). In the United States, continuous cavitator units are used widely for cider-vinegar production.

A strain of thermophilic *Acetobacter* was patented in Japan for oxidizing ethanol in a submerged culture oxidizer at temperatures as high as 37°C with considerable savings in cooling water. Another thermophilic strain of *Acetobacter* maintained full activity at 35°C, and 45% of its maximum activity at 38°C.

A Frings acetator consisting of a 48-m^3 (12,700-gal) tank produces 12 m^3 (3,200 gal) of 10% acetic acid vinegar/d. It required 2.2 L of cooling water/s at 15°C and an energy input of ca 36 MW (8600 kcal/s) (39). Thus, the submerged-culture oxidizer is capable of producing vinegar at nearly twice the rate of the best generator. Furthermore, submerged culture oxidizers are smaller for a given amount of production and, most important, they are more flexible in their operation. It is possible to change from one vinegar type to another with different feed-alcohol concentrations and nutrient requirements more quickly than with a generator.

Submerged-culture oxidizers are usually operated on a semicontinuous basis. In most cases, ca half the liquid in the tank is removed every 1–2 d, when the alcohol concentration has dropped to 0.1–0.2 vol %. The removed vinegar is replaced with wine or mash of richer ethanol and lower acetic acid concentration, giving a mixture in the tank of 5–6 vol % ethanol and 6–8 vol % of acetic acid. These are the optimum conditions for *Acetobacter* growth. Fermentation alcohol substrates do not require the addition of nutrients, but diluted distilled alcohol solutions need about 10–15 g of inorganic substances such as diammonium acid phosphate, potassium chloride, and traces of other metals and 30–50 g of organic materials such as glucose, autolyzed yeast, citric acid, and powdered whey per liter of alcohol. The pH of the fermenter mixture should be 3.9–5.0, the ideal temperature between 28 and 31°C. Since the new charge of mash or wine to the oxidizers lowers the temperature, cooling may be interrupted until the temperature again reaches 28–31°C. The rate of aeration depends on the surface of contact between air and liquid and is an inverse function of the bubble size. At optimum aeration rate, 50 mmol O_2/h is introduced into the solution per liter of mash, ie, 1870 cm^3 air/s/m^3 of mash. The maximum value for aeration recommended in Reference 39 is 1100 cm^3/s/m^3 mash.

Foam production is most troublesome under conditions adverse to bacterial growth and thus can be minimized by keeping nutrient, ethanol, and acetic acid concentrations in the optimum ranges (see DEFOAMERS). Temperature and aeration rate are also critical. Dead or dying cells seem to promote foam formation. Even under optimum conditions, some type of foam breaker mounted in the top of the oxidizer is needed; the foam is usually broken down by centrifugal force, but food-grade silicone antifoaming agents may be employed.

Submerged culture oxidizers can also be operated on a continuous basis. Continuous monitoring of ethanol and acetic acid concentrations, temperature, and aeration rates permit control of feed and withdrawal streams. Optimum production, however, is achieved by semicontinuous operation because the composition of vinegar desired in the withdrawal stream is so low in ethanol that vigorous bacterial growth is impeded. Bacterial concentrations up to 100×10^6 cells/cm^3 have been reported in generators making about 20% vinegars.

A submerged-culture oxidizer with instrumentation to control the oxygen concentration of the mash accurately and with a heat-transfer system that efficiently controls the temperature is described in Reference 40. Clear vinegar may be withdrawn from the oxidizer by use of tangential filters which retain the bacteria in the system. Blinding of the filter is precluded by the rapid flow of liquid across the filter surface (53,54). The clear vinegar is removed from the system and the bacterial cells are retained to continue their work. Glycerol catalyzes the production of vinegar from the alcoholic solution obtained from malt wort (55), and its degradation pathways have been elucidated. Certain strains of *Saccharomyces cerevisiae* produce enough SO_2 to slow the start of oxidation by *Acetobacter* (56). A scrubber has been patented which greatly increases the efficiency of vinegar production by recycling ethanol and acetic acid vapors normally lost with the exhaust air stream (57).

Fluidized-Bed Vinegar Reactors.   Intimate contact of air with *Acetobacter* cells is achieved in fluidized-bed or tower-type systems. Air introduced through perforations in the bottom of each unit suspends the mixture of liquid and microorganisms within the unit. Air bubbles penetrating the bottom plate keep *Acetobacter* in suspension and active for the ethanol oxidation in the liquid phase. Addition of a carrier for the bacterial cells to the liquid suspension is reported to improve the performance (58–60).

Vinegars with High Concentrations of Acetic Acid.   The U.S. regulations require at least 4 g acetic acid/100 cm^3 vinegar. Commercial vinegar and many quality table vinegars are significantly more concentrated. Submerged-culture oxidizers easily give acetic acid concentrations of 10–13 g/cm^3. Production rate is somewhat less than for lower acetic acid concentrations. Submerged-culture oxidizer techniques that produce vinegars with acetic acid concentrations ranging from 15–20% are now in commercial use (61–63). Continuous aeration, careful stepwise addition of ethanol as it is oxidized and careful control of temperature seem to be the keys to successful operation. The increased oxidation rate is the result of a greater cell mass per unit volume and of the selection of bacteria more tolerant of ethanol and of acetic acid (64–66).

In order to obtain ever higher acetic acid concentrations, water is removed after the generation step by freezing (67–70). Ice crystals are removed from the slush by filtration or centrifugation. Concentrated vinegars are of particular value in the pickling industry where dilution of the vinegary, spiced mixture by the water from the cucumbers is a serious and costly problem (40). Concentrated wine vinegar is stabilized with bentonite, silica gel, or $K_4Fe(CN)_6$. In another process, water is removed by formation of a hydrate of trichlorofluoromethane. The solid hydrate is separated from the concentrated vinegar and the fluorocarbon is recovered and recycled (71) (see INCLUSION COMPOUNDS). In another suggested process, vinegar acetic acid is neutralized with Na_2CO_3 and the water is removed by reduced-pressure distillation. H_2SO_4 is then added to liberate the acetic acid, which is removed by low pressure distillation (72). Production of a very high acetic acid (40%) vinegar by extraction of normal (15%) vinegar with liquid CO_2 under pressure is described in a Japanese patent application (73).

Vinegar Eels and Mother of Vinegar.   The nematode *Anguilla aceti* grows readily in packed-tank vinegar generators. Although it is esthetically undesirable, it is not harmful. These nematodes, known as vinegar eels, may actually be of some assistance in consuming dead bacteria from the surface of the packing material in the tank, and thus may aid in prolonging the operation of the system. Vinegar eels may also make nutrients more readily available to *Acetobacter* (74). Vinegar eels are removed from the raw vinegar by filtration and pasteurization before the vinegar is sold or used further in pickling or other processes.

Mother of vinegar is the term given to the cellulosic slime that coats the bacterial cells and is produced by a strain of *Acetobacter xylinium*. Different strains and different medium compositions result in different consistencies and crystalline forms of the cellulosic slime. Although it does not cause any problems in submerged-culture oxidizers, the slime can effectively block the passageways in packed-tank generators. High concentrations of acetic acid in the vinegar and the generator discourage the production of mother of vinegar slime.

Processing and Preparation for Marketing

Clarification. Raw vinegars as removed from the production unit vary widely in stability, depending upon the raw material and the type of generator or oxidizer employed. Table vinegars produced from wine, cider, malt, or other natural materials frequently contain unstable phenolic materials, pectins, and traces of proteins which form clouds or deposits. Vinegars from distilled alcohol are more stable, but still might contain traces of unstable materials. Generator vinegars are relatively free of *Acetobacter* cells, in contrast to the submerged-culture vinegars, which carry a high and cloudy suspension of bacterial cells. Clarification and stabilization of vinegars generally follow the standard practices of beverage industries. Bentonite is used as clarifier and, occasionally a proprietary formulation of potassium ferrocyanide is used to remove traces of heavy metals. Submerged-culture vinegar is clarified with mixed suspensions of bentonite and alginic acid (75); treatment with bentonite prepared with $NaHCO_3$ has been patented (76). Japanese patents describe vinegar stabilization with alumina or silica gels (77) or poly(vinylpyrrolidinone), cellulose, and Dowex A-1 for the same purpose. A Polish patent application describes vinegar clarification by foaming. Aeration is interrupted momentarily and then resumed vigorously for 5 min, creating a layer of foam amounting to ~1% of the tank volume. The clearer liquid from the bottom portion of the tank is reported to remain clear and stable (78). Activated carbon adsorbs some compounds causing clouding or precipitates in bottled vinegars. Vinegars are usually given a rough filtration on plate-and-frame or leaf filters with pads coated with diatomaceous earth. Immediately before bottling, the vinegar is filtered through more retentive pads or possibly membranes of pore size small enough to exclude all yeasts cells and bacteria. Membrane filtration can be combined with aseptic bottling to provide a vinegar free of all microorganisms, but the process is expensive and not essential to the stability of bottled vinegars. A membrane can be used for continuous microfiltration of cloudy vinegar (79,80). The surface is kept clean with a vigorous flow of liquid across the membrane parallel to its surface. This method has been successful in both laboratory and pilot plant.

Sterilizing and Packing. Many vinegars bottled for table use or pickling are pasteurized before shipment. In the low strength vinegars, the cellulose-producing acetic bacteria and certain strains of lactic bacteria may create problems. The former cause clouding, whereas the latter alter the flavor. Small amounts of sulfur dioxide are frequently added to minimize lactic-organisms' growth. Sterile filtration through very tight pads or through membranes followed by aseptic bottling is possible but difficult in the case of bacterial contaminants. For pasteurization, vinegar may be heated in bulk to 65–70°C, filled hot into bottles, sealed, and cooled slowly, or filled and sealed bottles may be pasteurized by heating to 65–70°C. Sterilization of many submerged-culture vinegars with high bacterial cell concentrations requires a pasteurization temperature of 77–80°C. The vinegar must be protected from exposure to bacterial or yeast contaminants and iron or copper in all processing steps following stabilization, ie, the filling equipment must be of stainless steel, plastic, or glass (see STERILIZATION TECHNIQUES).

Analyses of Vinegars

Because there is a considerable difference in cost between vinegar derived from petroleum-based or other synthetic ethanols and that derived from present-day biogenic sources such as grapes, apples, barley, or rice, there has always been need for analytical methods to detect blending. The presence of ^{14}C and ^{3}H in the methyl group of vinegar acetic acid is evidence of recent biogenic origin. Vinegar from petroleum alcohol has none of these unstable carbon and hydrogen isotopes (81,82). Some success has been attained in differentiating among biogenic vinegars on the basis of ^{13}C/^{12}C and ^{2}H/^{1}H isotopic ratios in the methyl group of the acetic acid of the vinegar (83–90) and by statistical studies of concentrations of minor compounds which differ with different sugar sources (91,92).

BIBLIOGRAPHY

"Vinegar" in *ECT* 1st ed., Vol. 14, pp. 675–686, by M. A. Joslyn, University of California; in *ECT* 2nd ed., Vol. 21, pp. 254–269, by M. A. Joslyn, University of California; in *ECT* 3rd ed., Vol. 23, pp. 753–763, by A. D. Webb, University of California, Davis.

1. K. Yamauchi, K. Sakata, C. Iwata, A. Yagi, and K. Ina, *Nippon Shokuhin Kogyo Gakkaishi* **41**(9), 600–605 (1994).
2. Chin. Pat. Appl. CN 93-105989 (May 20, 1993), K. Shan (to Aixin Health Beverage Factory).
3. *Oxford English Dictionary*, Compact Ed., Vol. II, Oxford University Press, Oxford, U.K., 1971, p. 3633.
4. Chin. Pat. Appl. CN 92-113596 (Nov. 28, 1992), Y. Fan, H. Cui, and H. Wang.
5. Jpn. Pat. Appl. JP 93-124951 (Apr. 27, 1993), Takayuki Nakano.
6. M. A. Qadeer, M. Y. Chaudhry, M. A. Shah, R. Ahmad, and F. H. Shah, *Pak. J. Biochem.* **25**(1,2), 77–83 (1992).
7. M. A. Mehaia and M. Cheryan, *Enzyme Microb. Technol.* **13**(3), 257–261 (1991).
8. M. Shiga, *Koryo* **179**, 105–109, (1993).
9. Jpn. Pat. Appl. JP 90-324931 (Nov. 26, 1990) Z. Seike, Akamatsu and Imai.
10. H. Suenaga and co-workers, *Nippon Shokuhin Kogyo Gakkaishi* **40**(4), 275–277 (1993).
11. H. Noda, K. Nakamichi, and M. Tada, *Kagawa-ken Nogyo Shikenjo Kenkyu Hokoku* **42**, 27–32 (1991).
12. Y. J. Oh, *Han'guk Yongyang Siklyong Hakhoechi* **21**(4), 377–380 (1992).
13. H. K. Tewari, S. S. Marwaha, A. Gupta, and P. K. Khanna, *J. Res (Punjab Agric. Univ.)* **28**(1), 77–84 (1991).
14. H. S. Grewal and H. K. Tewari, *J. Res. (Punjab Agric. Univ.)* **27**(2), 272–275 (1990).

15. Jpn. Pat. Appl. JP 88-94914 (Apr. 18, 1988), T. Kuroshima, N. Suzuki, and T. Kanamori.
16. Jpn. Pat. Appl. JP 87-85280 (Apr. 7, 1987), Iwao Kikuhara.
17. G. Joseph and M. Mahadeviah, *Indian Food Packer* **42**(1), 46–58 (1988).
18. Jpn. Pat. Appl. JP 86-187979 (Aug. 11, 1986), Y. Hayano.
19. Ind. Pat. Appl. IN 83-MA231 (Nov. 28, 1983), S. M. Chakalakkal and J. A. Chemmarappally.
20. J. A. Ekundayo, *Brit. Mycol. Soc. Symp. Ser. 3 (Fungal Biotechnol.*), 243–271, 1980.
21. Jpn. Pat. Appl. JP 78-78534 (June 30, 1978), Y. Awakuni.
22. L. Qui, H. Wu, S. Liu, *Zhongguo Niangzao* **6**, 19–21 (1993).
23. Jpn. Pat. Appl. JP 91-35594 (Feb. 4, 1991), M. Toda, S. Yamashoji, and S. Ooonishi.
24. Jpn. Pat. Appl. JP 88-106934 (Apr. 27, 1988), S. Kitahara.
25. A. Saeki, *Nippon Shokuhin Kogyo Gakkaishi* **36**(9), 726–31 (1989).
26. Jpn. Pat. Appl. JP 86-213963 (Sept. 12, 1986), J. Osuge, K. Umemoto, N. Nakamura, and A. Mori.
27. Jpn. Pat. Appl. JP 92-288979 (Oct. 27, 1992), M. Ukon.
28. Jpn. Pat. Appl. JP 92-91887 (Mar. 17, 1992), S. Itoku.
29. S. Jodai, S. Yano, and T. Uehara, *Mokuzai Gakkaishi* **35**(6), 555–563 (1989).
30. E. Sobczak and E. Konieczna, *Acta Aliment. Pol.* **13**(4), 351–358 (1987).
31. P. C. Sanchez, S. Lilial, C. L. Gerpacio, and H. Lapitan, *Philipp. Agric.* **68**(4), 439–448 (1985).
32. A. G. Korotaev, T. A. Nacheva, and N. K. Strel'nikova, *Vinodel. Vinograd. SSSR* **1**, 42–44 (1984).
33. "Pure Food and Drug Act," C. 3915, June 30, 1906, *52 United States Statutes at Large*, p. 1040.
34. *Census of Manufacturers, Industry Series*, U.S. Department of Commerce, Table 6a, 1992 and 1987.
35. A. Carnacini and V. Gerbi, *Wein-Wiss.* **47**(6), 216–225 (1992).
36. United Nations Industrial Commodity Statistics.
37. *Vinegar Institute 1995 Annual Meeting*, Ottogi Foods Co., Korea.
38. M. A. Amerine and co-workers, *Technology of Wine Making*, 4th ed., AVI, Westport, Conn., 1980.
39. G. Keszthelyi, *Mitt. Hoeher. Bundesl. Versuchsanst. Wein Obstbau Klosterneuburg* **24**, 445 (1974).
40. H. A. Conner and R. J. Allgeier, *Adv. Appl. Microbiol.* **20**, 81 (1976).
41. E. Huber, *Dtsch. Essigind.* **31**(1), 12 (1927); **31**(2), 28 (1927).
42. Brit. Pat. 1,305,868 (Feb. 7, 1973), (to Kewpie Jozo Kabushiki Kaisha).
43. P. Guidici, *Ind. Bevande*, **22**(124), 123–125 (1993).
44. F. Corradini, L. Marcheselli, A. Marchetti, and C. Prete, *J. AOAC Int.* **77**(3), 714–717 (1994).
45. C. A. Mitchell, *Vinegar: Its Manufacture and Examination*, 2nd ed., Griffin, London, 1926.
46. J. M. Gomez, L. E. Romero, I. Caro, and D. Cantero, *Biotechnol. Tech.* **8**(10), 711–716 (1994).
47. R. J. Allgeier, R. T. Wisthoff, and F. M. Hildebrandt, *Ind. Eng. Chem.* **44**, 669 (1952); **45**, 489 (1953); **46**, 2023 (1954).
48. J. B. Nickol in H. J. Peppler and D. Perlman, eds., *Microbial Technology*, 2nd ed., Vol. 2, Academic Press, New York, 1979, pp. 5–72.
49. M. Fukaya, *Bioprocess Technol.* **19**, 529–542 (1994).
50. O. Hromatka and H. Ebner, *Enzymologia* **13**, 369 (1949).
51. U.S. Pat. 2,997,424 (Aug. 22, 1961), J. E. Mayer (to Hunt Foods and Industries); E. Mayer, *Food Technol.* **17**, 582 (1963).

52. U.S. Pat. 2,913,343 (Nov. 17, 1959), A. C. Richardson (to California Packing Corp.).
53. H. Ebner, *Chem. Ing. Tech.* **53**(1), 25–31 (1981).
54. G. Meglioli, *Ind. Bevande* **23**(131), 243–246 (1994).
55. Ger. Offen. 2,215,456 (Oct. 4, 1973), R. N. Greenshields and D. D. Jones.
56. C. Zambonelli, M. E. Guerzoni, M. Nanni, and G. Gianstefani, *Riv. Vitic. Enol.* **25**, 214 (1972).
57. Jpn. Pat. Appl. JP 75-76508 (June 24, 1975), Hiroshi Masai, Hirotake Yamada, and Mikio Nishimura.
58. A. Mori, *Bioprocess Technol.* **16**, 291–313 (1993).
59. J. F. Kennedy, *Enzyme Eng.* **4**, 323 (1978).
60. J. F. Kennedy, J. D. Humphreys, S. A. Barker, and R. N. Greenshields, *Enzyme Microb. Technol.* **2**, 209 (1980).
61. Y. C. Lee and co-workers, *Sanop Misaengmul Hakhoechi* **21**(5), 511–512 (1993).
62. Y. C. Lee and co-workers, *Sanop Masaengmul Hakhoechi* **20**(6), 663–667 (1992).
63. A. Saeki, *Nippon Shokuhin Kagyo Gakkaishi* **37**(3), 191–198 (1990).
64. M. Kittelmann, W. W. Stamm, H. Follman, and H. G. Trueper, *Appl. Microbiol. Biotechnol.* **30**(1), 47–52 (1989).
65. M. Fukaya, *Bioprocess Technol.* **19**, 529–142 (1994).
66. M. Fukaya and co-workers, *Agric. Biol. Chem.* **53**(9), 2435–2440 (1989).
67. F. K. Lawler, *Food Eng.* **23**, 68, 82 (1961).
68. J. R. Dooley and D. D. Lineberry, in *Symposium on New Developments in Bioengineering—Minneapolis*, preprint, American Institute of Chemical Engineering, New York, 1965, 12 pp.
69. Jpn. Pat. Appl. JP 92-313034 (Nov. 24, 1992).
70. L. C. Dickey and J. C. Craig, Jr., *Phys. Chem. Food Processes* **2**, 542–553 (1993).
71. Brit. Pat. GB 75-36629 (Sept. 5, 1975).
72. Jpn. Pat. Appl. JP 86-285227 (Nov. 28, 1986).
73. Jpn. Pat. Appl. JP 85-44573 (Mar. 8, 1985).
74. R. C. Zalkan and F. W. Fabian, *Food Technol.* **7**, 453 (1953).
75. Jpn. Kokai Tokkyo Koho 74 108,295 (Oct. 15, 1974), H. Masai and K. Yamada (to Nakano Vinegar Co., Ltd.).
76. Czech. Pat. 151,118 (Nov. 15, 1973), J. Vyslouzil.
77. Jpn. Kokai Tokkyo Koho 81 11,430 and 81 11,431 (Mar. 14, 1981), (to Nisshin Flour Milling co., Ltd.).
78. Pol. Pat. Appl. PL 88-276725 (Dec. 22, 1988).
79. G. Meglioli, *Ind. Bevande* **23**(131), 243–246 (1994).
80. H. C. Van der Horst and J. H. Hanemaaiajer, *Desalination* **77** 235–258 (1990).
81. E. R. Schmid and I. Fogy, *Ernaehrung (Vienna)* **2**(4), 187–190 (1978).
82. G. Volonterio and P. Resmini, *Riv. Vitic. Enol.* **37**(12), 671–681 (1984).
83. E. R. Schmid and co-workers, *Biomed. Mass Spectrom.* **8**(10), 496–499 (1981).
84. D. A. Krueger and H. A. Krueger, *Biomed. Mass Spectrom.* **11**(9), 472–474 (1984).
85. D. A. Krueger and H. W. Krueger, *J. Assoc. Off. Anal. Chem.* **68**(3), 449–452 (1985).
86. K. Kanno, Y. Kawamura, and K. Kato, *Nippon Nogei Kagaku Kaishi* **63**(7), 1207–1211 (1989).
87. K. Kanno and co-workers, *Nippon Nogei Kagaku Kaishi* **64**(4), 897–899 (1990).
88. C. Nubling and co-workers, *Ann. Falsif. Expert Chim. Toxicol.* **82**(880), 385–392 (1989).
89. G. Remaud, *Fresenius J. Anal. Chem.* **342**(4–5), 457–461 (1992).
90. D. A. Krueger, *J. AOAC Int.* **75**(4), 725–728 (1992).
91. M. I. Guerro, F. J. Heredia, and A. M. Troncoso, *J. Sci. Food Agric.* **66**(2), 209–212 (1994).
92. K. Yamauchi and co-workers, *Nippon Shokuhin Kogyo Gakkaishi* **41**(9), 600–605 (1994).

General References

A. D. Webb, in J. Robinson, ed., *The Oxford Companion to Wine*, Oxford University Press, Oxford, U.K., 1994, p. 1032.

M. Fukaya, *Bioprocess Technol.* **19**, 529–542 (1994).

J. Nieto and co-workers, *Dev. Food Sci.* **32**, 469–500 (1993).

L. R. Partin and W. H. Heise, *Chem. Ind.* **49**, 3–13 (1993).

A. Carnacini and V. Gerbi, *Ind. Bevande* **21**(122), 465–477, 483 (1992).

B. Zhu, *Zhong-guo Tiaoweipin* **5**, 1–6 (1990) (Chinese).

M. Ameyama and S. Otsuka, eds., *Science of Vinegar*, Asakura Publishing Co., Ltd., Tokyo, 1990, 224 pp.

A. Kuriyama, *New Food Ind.* **32**(3), 17–26 (1990) (Japanese).

A. Mori, *Biseibutsu* **4**(4), 329–342 (1988) (Japanese).

H. Masai, *Kagaku Kogaku* **50**(12), 879–886 (1986) (Japanese).

H. Ebner and H. Follmann, in G. Reed, ed., *Biotechnology*, Vol. 5, Verlag Chemie, Weinheim, Germany, 1983, pp. 425–446.

E. Levonen and C. Llaguno, *Sem. Vitivinic.* **36**(1821–1822), 2525, 2527, 2529–2531, 2533–2535 (1981) (Spanish).

M.-H. Lai, W. T. H. Chang, and B. S. Luh, in B. S. Luh, eds., *Rice: Production and Utilization*, AVI, Westport, Conn., 1980, pp. 712–735.

A. Dinsmoor Webb
University of California, Davis

VINYLBENZENE. See Styrene.

VINYL CHLORIDE

Vinyl chloride [*75-01-4*], CH_2=CHCl, by virtue of the wide range of application for its polymers in both flexible and rigid forms, is one of the largest dollar-volume commodity chemicals in the United States and is an important item of international commerce. Growth in vinyl chloride production is directly related to demand for its polymers and, on an energy-equivalent basis, rigid poly(vinyl chloride) (PVC) [*9002-86-2*] is one of the most energy-efficient construction materials available. Initial development of the vinyl chloride industry in the 1930s stemmed from the discovery that, with plasticizers, PVC can be readily processed and converted into a rubbery product (1). However, it was not until after World War II that vinyl chloride production grew rapidly as a result of the increased volume of PVC products for the consumer market.

The early history of vinyl chloride has been documented (2–6). Justus von Liebig at the University of Giessen, Germany, won the distinction of being the first person to synthesize vinyl chloride when, in the 1830s, he reacted the so-called oil of the Dutch chemists, dichloroethane [*1300-21-6*], with alcoholic potash to make vinyl chloride. Liebig's student, Victor Regnault, confirmed his discovery and was allowed to publish it as sole author in 1835 (7). In 1872, E. Baumann observed that white flakes precipitated from vinyl chloride upon prolonged exposure to sunlight in a sealed tube (8). This material was further investigated in the early 1900s by Ivan Ostromislensky, who named it *Kauprenchlorid* (cauprene chloride), and gave it the empirical formula $(C_2H_3-Cl)_{16}$ (9). However, vinyl chloride was of little commercial interest until Waldo Semon's work with plasticized PVC for the B. F. Goodrich Company beginning in 1926 (10). Some years earlier, Fritz Klatte had developed the first practical route to vinyl chloride while looking to find uses for acetylene [*74-86-2*] for Chemische Fabrik Griesheim-Elektron. This process, in which hydrogen chloride [*7647-01-0*], HCl, is added to acetylene over a mercuric chloride [*7487-94-7*] catalyst, was patented in 1912 (11). By 1926, Griesheim-Elektron had concluded that the patent held no commercial value and allowed it to lapse. Klatte's process eventually formed the basis of the vinyl chloride industry for many years from its beginnings in the 1930s, but it was ultimately supplanted by a balanced process from ethylene [*74-85-1*] and chlorine [*7782-50-5*] in which vinyl chloride is made by pyrolysis of 1,2-dichloroethane [*107-06-2*] (ethylene dichloride (EDC)).

Vinyl chloride (also known as chloroethylene or chloroethene) is a colorless gas at normal temperature and pressure, but is typically handled as the liquid (bp $-13.4°C$). However, no human contact with the liquid is permissible. Vinyl chloride is an OSHA-regulated material.

Physical Properties

The physical properties of vinyl chloride are listed in Table 1 (12). Vinyl chloride and water [*7732-18-5*] are nearly immiscible. The equilibrium concentration of vinyl chloride at 1 atm partial pressure in water is 0.276 wt % at 25°C, whereas the solubility of water in vinyl chloride is 0.0983 wt % at 25°C and saturated pressure (13). Vinyl chloride is soluble in hydrocarbons, oil, alcohol, chlorinated solvents, and most common organic liquids.

Reactions

Polymerization.   The most important reaction of vinyl chloride is its polymerization and copolymerization in the presence of a radical-generating initiator.

Substitution at the Carbon–Chlorine Bond.   Vinyl chloride is generally considered inert to nucleophilic replacement compared to other alkyl halides. However, the chlorine atom can be exchanged under nucleophilic conditions in the presence of palladium [*7440-05-3*], Pd, and certain other metal chlorides and salts. Vinyl alcoholates, esters, and ethers can be readily produced from these reactions.

Use of alcohol as a solvent for carbonylation with reduced Pd catalysts gives vinyl esters. A variety of acrylamides can be made through oxidative addition of carbon monoxide [*630-08-0*], CO, and various amines to vinyl chloride in the presence of phosphine complexes of Pd or other precious metals as catalyst (14).

Table 1. Physical Properties of Vinyl Chloride[a]

Property	Value
molecular weight	62.4985
melting point (1 atm), K	119.36
boiling point (1 atm), K	259.25
heat capacity at constant pressure, J/(mol·K)[b]	
vapor at 20°C	53.1
liquid at 20°C	84.3
critical temperature, K	432
critical pressure, MPa[c]	5.67
critical volume, cm³/mol	179
critical compressibility	0.283
acentric factor	0.100107
dipole moment, C·m	4.84×10^{-30}
enthalpy of fusion (melting point), kJ/mol[b]	4.744
enthalpy of vaporization (298.15 K), kJ/mol[b]	20.11
enthalpy of formation (298.15 K), kJ/mol[b]	28.45
Gibbs energy of formation (298.15 K), kJ/mol[b]	41.95
vapor pressure, kPa[c]	
−30°C	49.3
−20°C	78.4
−10°C	119
0°C	175
viscosity, mPa·s	
−40°C	0.345
−30°C	0.305
−20°C	0.272
−10°C	0.244
explosive limits in air, vol %	
lower limit	3.6
upper limit	33
autoignition temperature, K	745

[a]Ref. 12.
[b]To convert J to cal, divide by 4.184.
[c]To convert MPa to psi, multiply by 145.

Reaction of vinyl chloride with butyllithium [109-72-8] and then with carbon dioxide [124-38-9], CO_2 in diethyl ether [60-29-7] at low temperatures gives high yields of α, β-unsaturated carboxylic acids.

Vinylmagnesium chloride [3536-96-7] (Grignard reagent) can be prepared from vinyl chloride (15) and then used to make a variety of useful end products or intermediates by adding a vinyl anion to organic functional groups (16–22). For instance, the vinylmagnesium compounds can be coupled with cuprous chloride [7758-89-6], CuCl, at −60°C to give 1,3-butadiene [106-99-0], while vinyl ketones and alcohols can be prepared by the addition of vinylmagnesium chloride to organic acids.

Vinyl chloride similarly undergoes Grignard reactions with other organomagnesium halide compounds. For example, cross-coupling with 1- or 2-phenylethylmagnesium bromide [*41745-02-2* or *3277-89-2*] yields 4- or 3-phenyl-1-butene [*768-56-9* or *934-10-1*], respectively (23,24), whereas 4-(3-methyl-2-butenyl)styrene [*85964-33-6*] (*p*-prenylstyrene) is obtained by coupling 4-(3-methyl-2-butenyl)phenylmagnesium chloride [*106364-41-4*] with vinyl chloride (25).

Vinyllithium [*917-57-7*] can be formed directly from vinyl chloride by means of a lithium [*7439-93-2*] dispersion containing 2 wt % sodium [*7440-23-5*] at 0–10°C. This compound is a reactive intermediate for the formation of vinyl alcohols from aldehydes, vinyl ketones from organic acids, vinyl sulfides from disulfides, and monosubstituted alkenes from organic halides. It can also be converted to vinylcopper [*37616-22-1*] or divinylcopper lithium [*22903-99-7*], which can then be used to introduce a vinyl group stereoselectively into a variety of α, β-unsaturated systems (26), or simply add a vinyl group to other α,β-unsaturated compounds to give γ, δ-unsaturated compounds. Vinyllithium reagents can also be converted to secondary alcohols with trialkylboranes.

Vinyl chloride reacts with sulfides, thiols, alcohols, and oximes in basic media. Reaction with hydrated sodium sulfide [*1313-82-2*] in a mixture of dimethyl sulfoxide [*67-68-5*] (DMSO) and potassium hydroxide [*1310-58-3*], KOH, yields divinyl sulfide [*627-51-0*] and sulfur-containing heterocycles (27). Various vinyl sulfides can be obtained by reacting vinyl chloride with thiols in the presence of base (28). Vinyl ethers are produced in similar fashion, from the reaction of vinyl chloride with alcohols in the presence of a strong base (29,30). A variety of pyrroles and indoles have also been prepared by reacting vinyl chloride with different ketoximes or oximes in a mixture of DMSO and KOH (31).

The carbon–chlorine bond can also be activated at high temperatures. Vinyl chloride reacts with alkanethiols, and with dialkyl sulfides and disulfides at 400°C to form vinyl sulfide compounds in low yields (32). Aryl and thienyl vinyl sulfides can be prepared in similar fashion, reacting aryl and thienyl thiols with vinyl chloride in a quartz tube at 380–440°C (33). Thiophene [*110-02-1*] and its substituted forms can be obtained by heterocyclization of vinyl chloride with hydrogen sulfide [*7783-06-4*] in the presence of an acetylenic reagent at 500–550°C (34) (see THIOPHENE AND THIOPHENE DERIVATIVES). Reaction of vinyl chloride with germanium tetrachloride [*10038-98-9*] and trichlorosilane [*10025-78-2*] in a tubular reactor at 600–700°C gives vinyltrichlorogermane [*4109-83-5*] in moderate yield (35). Vinyltrichlorosilane [*75-94-5*] is obtained at yields of up to 77% by reacting vinyl chloride with trichlorosilane at 400–750°C in a tubular reactor (36), while vinylsilanes can be made from mono-, di-, or trisilane [*7803-62-5*, *1590-87-0*, or *7783-26-8*] and vinyl chloride in a tubular reactor at 100–650°C (37).

Finally, the reaction of vinyl chloride with hydrogen fluoride [*7664-39-3*], HF, over a chromia [*1308-38-9*], Cr_2O_3, on-alumina [*1344-28-1*], Al_2O_3, catalyst at 380°C yields vinyl fluoride [*75-02-5*] (38).

Oxidation. The chlorine atom [*22537-15-1*]-initiated, gas-phase oxidation of vinyl chloride yields 74% formyl chloride [*2565-30-2*] and 25% CO at high oxygen [*7782-44-7*], O_2, to Cl_2 ratios; it is unique among the chloro olefin oxidations because CO is a major initial product and because the reaction proceeds by a nonchain path at high O_2/Cl_2 ratios. The rate of the gas-phase reaction of chlorine atoms with vinyl chloride has been measured (39).

The oxidation of vinyl chloride with oxygen in the gas phase proceeds by a nonradical path which, again, is unique among the chloro olefins. No C_2 carbonyl compounds are made; the main products are formyl chloride, CO, HCl, and formic acid [64-18-6]. Complete oxidation of vinyl chloride with oxygen in the gas phase can be achieved using a cobalt chromite [12016-69-2] catalyst (40). At −15 to −20°C, vinyl chloride reacts with oxygen, with ultraviolet (uv) light initiation, to give a peroxide, reported as [OCH$_2$CHClO]$_2$. On heating to 35°C, this peroxide decomposes to formaldehyde [50-00-0], CO, and HCl.

Reaction with triplet oxygen $O(^3P)$ atoms [17778-80-2] gives high yields of CO and chloroacetaldehyde [107-20-0], with smaller amounts of acetyl chloride [75-36-5], HCl, methane [74-82-8], and polymer. The rate of the gas-phase reaction of vinyl chloride with $O(^3P)$ atoms has also been reported (41).

Oxidation of vinyl chloride with ozone [10028-15-6] in either the liquid or the gas phase gives formic acid and formyl chloride. The ozone reaction with vinyl chloride can be used to remove it from gas streams in vinyl chloride production plants.

Vinyl chloride can be completely oxidized to CO_2 and HCl using potassium permanganate [7722-64-7] in an aqueous solution at pH 10. This reaction can be used for wastewater purification, as can ozonolysis, peroxide oxidation, and uv irradiation (42). The aqueous phase oxidation of vinyl chloride with chlorine yields chloroacetaldehyde (43).

The combustion of vinyl chloride in air at 510−795°C produces mainly CO_2 and HCl along with CO. A trace of phosgene [75-44-5] also forms.

Addition.　Chlorine adds to vinyl chloride to form 1,1,2-trichloroethane [79-00-5] (44−46). Chlorination can proceed by either an ionic or a radical path. In the liquid phase and in the dark, 1,1,2-trichloroethane forms by an ionic path when a transition-metal catalyst such as ferric chloride [7705-08-0], FeCl$_3$, is used. The same product forms in radical reactions up to 250°C. Photochemically initiated chlorination also produces 1,1,2-trichloroethane by a radical path (47). Above 250°C, the chlorination of vinyl chloride gives unsaturated chloroethylenes produced by dehydrochlorination of 1,1,2-trichloroethane. The presence of small amounts of oxygen greatly accelerates the rate of the radical-chain chlorination reaction at temperatures above 250−300°C (48). Other halogens can be added to vinyl chloride to form similar 1,2-addition products but these reactions have not been thoroughly studied. Vinyl chloride can be halofluorinated in the sulfur tetrafluoride [7783-60-0]−HF−Cl$_2$ system (49).

Hydrogen halide addition to vinyl chloride in general yields the 1,1-adduct (50−52). The reactions of HCl and hydrogen iodide [10034-85-2], HI, with vinyl chloride proceed by an ionic mechanism, while the addition of hydrogen bromide [10035-10-6], HBr, involves a chain reaction in which a bromine atom [10097-32-2] is the chain carrier (52). In the absence of a transition-metal catalyst or antioxidants, HBr forms the 1,2-adduct with vinyl chloride (52). HF reacts with vinyl chloride in the presence of stannic chloride [7646-78-8], SnCl$_4$, to form 1,1-difluoroethane [75-37-6] (53).

Various vinyl chloride adducts can be formed under acid-catalyzed Friedel-Crafts conditions. Vinyl chloride can add tertiary alkyl halides (54). It can be condensed with ethyl chloride [75-00-3] to yield 1,1,3-trichlorobutane [13279-87-3] and 1,1-dichloroethane [75-34-3] (55). The reaction of 2-chloropropane

[75-29-6] with vinyl chloride yields 1,1-dichloro-3-methylbutane [625-66-1] (55). At 0–5°C, vinyl chloride reacts with benzene [71-43-2], resulting in a mixture of 1-chloroethylbenzene [672-65-1] and 1,1-diphenylethane [612-00-0] (56). Reaction with toluene (qv) [108-88-3] leads to 1,1-ditolylethane [29036-13-3], whereas reaction with anisole [100-66-3] (methoxybenzene) gives 1,1-di-p-anisylethane [10543-21-2]. Phenol [108-95-2] also reacts to give p-vinylphenol [2628-17-3].

Vinyl chloride forms a photo [2 + 2] cycloadduct with 1-isoquinolone [491-30-5] and its N-methyl derivative (57).

Condensation of vinyl chloride with formaldehyde and HCl (Prins reaction) yields 3,3-dichloro-1-propanol [83682-72-8] and 2,3-dichloro-1-propanol [616-23-9]. The 1,1-addition of chloroform [67-66-3] as well as the addition of other polyhalogen compounds to vinyl chloride are catalyzed by transition-metal complexes (58). In the presence of iron pentacarbonyl [13463-40-6], both bromoform [75-25-2], $CHBr_3$, and iodoform [75-47-8], CHI_3, add to vinyl chloride (59,60). Other useful products of vinyl chloride addition reactions include 2,2-difluoro-4-chloro-1,3-dioxolane [162970-83-4] (61), 2-chloro-1-propanol [78-89-7] (62), 2-chloropropionaldehyde [683-50-1] (63), 4-nitrophenyl-β,β-dichloroethyl ketone [31689-13-1] (64), and β,β-dichloroethyl phenyl sulfone [3123-10-2] (65).

Sodium β-chloroethanesulfonate [15484-44-3] can be obtained by reacting vinyl chloride with sodium bisulfite [7631-90-5] (66). Reaction with nitronium tetrafluoroborate [13826-86-3] yields 1-chloro-1-fluoro-2-nitroethane [461-70-1] (67).

Vinyl chloride reacts with ammonium chloride [12125-02-9] and oxygen in the vapor phase at 325°C over a cupric chloride [7447-39-4], $CuCl_2$, catalyst to make 1,1,2-trichloroethane and ammonia (68).

Vinyl chloride can be hydrogenated over a 0.5% platinum [7440-06-4], Pt, on alumina catalyst to ethyl chloride and ethane [74-84-0]. This reaction is zero order in vinyl chloride and first order in hydrogen.

Photochemistry.   Vinyl chloride is subject to photodissociation. Photexcitation at 193 nm results in the elimination of HCl molecules and Cl atoms in an approximately 1.1:1 ratio (69). Both vinylidene ($^3B_2$) [2143-69-3] and acetylene have been observed as photolysis products (70), as have H_2 molecules (71) and H atoms [12385-13-6] (72). HCl and vinylidene appear to be formed via a concerted 1,1 elimination from excited vinyl chloride (70). An adiabatic recoil mechanism seems likely for Cl atom elimination (73). As expected from the relative stabilities of the 1- and 2-chlorovinyl radicals [50663-45-1 and 57095-76-8], H atoms are preferentially produced by detachment from the β carbon (72). Finally, a migration mechanism appears to play a significant role in H_2 elimination (71).

Pyrolysis.   Vinyl chloride is more stable than saturated chloroalkanes to thermal pyrolysis, which is why nearly all vinyl chloride made commercially comes from thermal dehydrochlorination of EDC. When vinyl chloride is heated to 450°C, only small amounts of acetylene form. Little conversion of vinyl chloride occurs, even at 525–575°C, and the main products are chloroprene [126-99-8] and acetylene. The presence of HCl lowers the amount of chloroprene formed.

Decomposition of vinyl chloride begins at approximately 550°C, and increases with increasing temperature. Acetylene, HCl, chloroprene, and vinylacetylene [689-97-4] are formed in about 35% total yield at 680°C (74). At higher temperatures, tar and soot formation becomes increasingly important. Vinyl chloride pyrolysis is a free-radical chain process, in which Cl atoms are important carriers. Abstraction of H atoms from vinyl chloride by Cl atoms leads to 2-chlorovinyl and 1-chlorovinyl radicals, respectively. The former lose Cl atoms to form acetylene, while the latter add to vinyl chloride, ultimately resulting in chloroprene. Addition of HCl increases the acetylene–chloroprene product ratio, probably owing to reversal of the reaction leading to the 1-chlorovinyl radical through H atom transfer from HCl (74).

When dry and in contact with metals, vinyl chloride does not decompose below 450°C. However, if water is present, vinyl chloride can corrode iron, steel, and aluminum because of the presence of trace amounts of HCl. This HCl may result from the hydrolysis of the peroxide formed between oxygen and vinyl chloride.

Manufacture

Vinyl chloride monomer was first produced commercially in the 1930s from the reaction of HCl with acetylene derived from calcium carbide [75-20-7]. As demand for vinyl chloride increased, more economical feedstocks were sought. After ethylene became plentiful in the early 1950s, commercial processes were developed to produce vinyl chloride from ethylene and chlorine. These processes included direct chlorination of ethylene to form EDC, followed by pyrolysis of EDC to make vinyl chloride. However, because the EDC cracking process also produced HCl as a co-product, the industry did not expand immediately, except in conjunction with acetylene-based technology. The development of ethylene oxychlorination technology in the late 1950s encouraged new growth in the vinyl chloride industry. In this process, ethylene reacts with HCl and oxygen to form EDC. Combining the component processes of direct chlorination, EDC pyrolysis, and oxychlorination provided the so-called balanced process for production of vinyl chloride from ethylene and chlorine, with no net consumption or production of HCl.

Although a small fraction of the world's vinyl chloride capacity is still based on acetylene or mixed actylene–ethylene feedstocks, nearly all production is conducted by the balanced process based on ethylene and chlorine (75). The reactions for each of the component processes are shown in equations 1–3 and the overall reaction is given by equation 4:

Direct chlorination $$CH_2 = CH_2 + Cl_2 \longrightarrow ClCH_2CH_2Cl \tag{1}$$

EDC pyrolysis $$2\ ClCH_2CH_2Cl \longrightarrow 2\ CH_2 = CHCl + 2\ HCl \tag{2}$$

Oxychlorination $$CH_2 = CH_2 + 2\ HCl + 1/2\ O_2 \longrightarrow ClCH_2CH_2Cl + H_2O \tag{3}$$

Overall reaction $$2\ CH_2 = CH_2 + Cl_2 + 1/2\ O_2 \longrightarrow 2\ CH_2 = CHCl + H_2O \tag{4}$$

In a typical balanced plant producing vinyl chloride from EDC, all the HCl produced in EDC pyrolysis is used as the feed for oxychlorination. On this basis, EDC production is about evenly split between direct chlorination and oxychlorination, and there is no net production or consumption of HCl. The three principal operating steps used in the balanced process for ethylene-based vinyl chloride production are shown in the block flow diagram in Figure 1, and a schematic of the overall process for a conventional plant is shown in Figure 2 (76). A typical material balance for this process is given in Table 2.

Direct Chlorination of Ethylene.   Direct chlorination of ethylene is generally conducted in liquid EDC in a bubble column reactor. Ethylene and chlorine dissolve in the liquid phase and combine in a homogeneous catalytic reaction to form EDC. Under typical process conditions, the reaction rate is controlled by mass transfer, with absorption of ethylene as the limiting factor (77). Ferric chloride is a highly selective and efficient catalyst for this reaction, and is widely used commercially (78). Ferric chloride and sodium chloride [7647-14-5] mixtures have also been utilized for the catalyst (79), as have tetrachloroferrate compounds, eg, ammonium tetrachloroferrate [24411-12-9], NH_4FeCl_4 (80). The reaction most likely proceeds through an electrophilic addition mechanism, in which the catalyst first polarizes chlorine, as shown in equation 5. The polarized chlorine molecule then acts as an electrophilic reagent to attack the double bond of ethylene, thereby facilitating chlorine addition (eq. 6):

$$FeCl_3 + Cl_2 \rightleftharpoons FeCl_4^- - Cl^+ \tag{5}$$

$$FeCl_4^- - Cl^+ + CH_2 \!=\! CH_2 \longrightarrow FeCl_3 + ClCH_2CH_2Cl \tag{6}$$

The direct chlorination process may be run with a slight excess of either ethylene or chlorine, depending on how effluent gases from the reactor are subsequently processed. For example, the noncondensables could easily be routed to an oxychlorination process in the case of excess ethylene (81). Conversion of the limiting component is essentially 100%, and selectivity to EDC is greater than 99% (78). The main by-product is 1,1,2-trichloroethane, which most likely forms through radical reactions beginning with homolytic dissociation of a small fraction of the chlorine. However, oxygen, which is frequently present as an impurity in chlorine, tends to increase selectivity to EDC by inhibition of free-radical reactions that produce, 1,1,2-trichloroethane. Consequently, oxygen is often added to a level of about 0.5% of the chlorine feed. Amides, eg, N,N-dimethylformamide [68-12-2], also increase selectivity to EDC, as do aromatic hydrocarbons and phenols (82).

The direct chlorination reaction is very exothermic ($\Delta H = -180$ kJ/mol for eq. 1, Ref. 83) and requires heat removal for temperature control. Early direct chlorination reactors were operated at moderate temperatures of 50–65°C to take advantage of lower by-product formation, and utilized conventional water cooling for heat removal. As energy costs became more significant, various schemes for recovering the heat of reaction were devised. A widely used method involves operating the reactor at the boiling point of EDC, allowing the pure product to vaporize, and then either recovering heat from the condensing vapor, or replacing one or more EDC fractionation column reboilers with the reactor itself (84–86). An alternative method entails operation of the reactor at higher pressure to raise the boiling point of EDC; in this case, the reactor operates

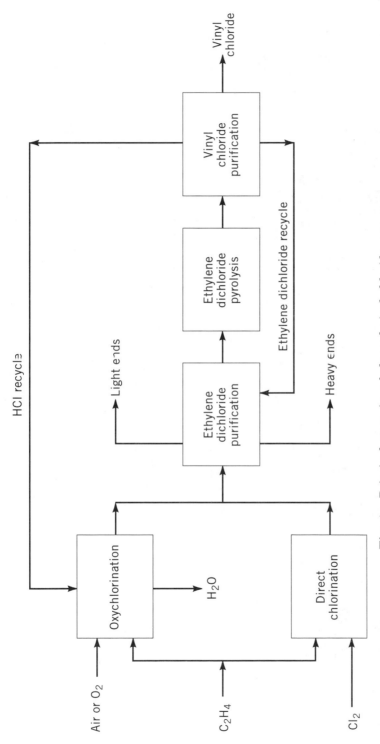

Fig. 1. Principal steps in a balanced vinyl chloride process.

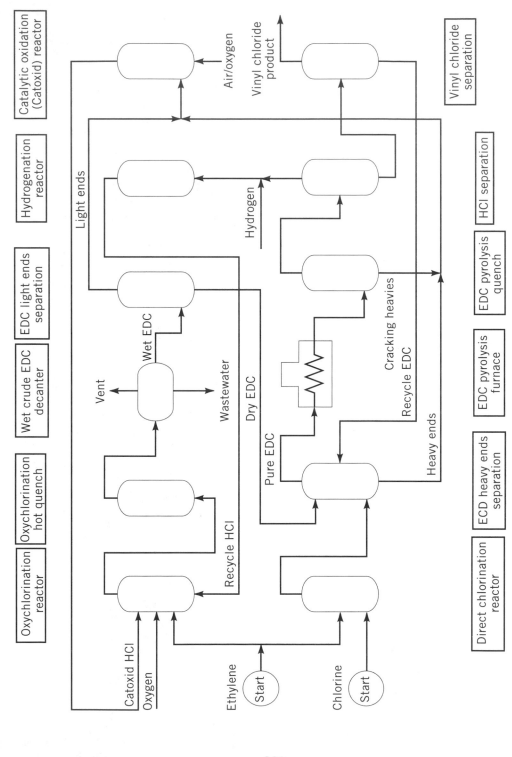

Fig. 2. Typical (Geon) balanced vinyl chloride process with oxygen-based oxychlorination (76).

860

Table 2. Typical Material Balance for Vinyl Chloride Production by the Air-Based Balanced Ethylene Process

Components, kg	Raw materials	Intermediates	By-products	Aqueous streams	Vent streams			Product
					Direct chlorination[a]	Oxychlorination	Distillation columns	
C_2H_4	0.4656				0.0025		0.0001	
Cl_2	0.5871				0.0001		0.0003	
N_2	0.5782					0.5779		
O_2	0.1537					0.0214		
CO_2	0.0003					0.0116		
CO						0.0032		
$ClCH_2CH_2Cl$		1.6370[b]	0.0029		0.0016	0.0017	0.0045	
HCl		0.6036						
H_2O	0.0171		0.1438	0.1196		0.0413		
NaOH				0.0008				
NaCl				0.0014				
lights			0.0029		0.0003	0.0025		
heavies			0.0023					
$CH_2{=}CHCl$			6.0008		0.0001	0.0012	0.0024	1.0000
Total, kg/kg vinyl chloride	*1.8020*	*2.2406*	*0.1527*	*0.1218*	*0.0046*	*0.6608*	*0.0073*	*1.0000*

[a]Inerts present in chlorine feed are emitted in this vent stream.
[b]Represents EDC necessary for a stoichiometric balance, including that converted to by-products, but no recycled EDC.

without boiling, but at higher temperatures (75–200°C) to allow more efficient heat transfer to some other part of the process (87,88). For reactors equipped with liquid product removal, the EDC is usually treated to remove ferric chloride. The latter, which would lead to rapid fouling of the EDC cracking reactor, can be removed by washing with water or by adsorption on a solid. With dry feedstocks (<10 ppm water) and good temperature control, carbon steel can be used in direct chlorination reactors operating at low temperature and in auxiliary equipment. Higher temperature operation generally requires materials that are more resistant to erosion–corrosion in the reactor, eg, hard alloy cladding below the liquid level and nickel alloy feed spargers.

Oxychlorination of Ethylene. When compared with direct chlorination, the oxychlorination process is characterized by higher capital investment, higher operating costs, and slightly less pure EDC product. However, use of the oxychlorination process is dictated by the need to consume the HCl generated in EDC pyrolysis.

In oxychlorination, ethylene reacts with dry HCl and either air or pure oxygen to produce EDC and water. Various commercial oxychlorination processes differ from one another to some extent because they were developed independently by several different vinyl chloride producers (78,83), but in each case the reaction is carried out in the vapor phase in either a fixed- or fluidized-bed reactor containing a modified Deacon catalyst. Unlike the Deacon process for chlorine production, oxychlorination of ethylene occurs readily at temperatures well below those required for HCl oxidation.

Oxychlorination catalysts typically contain cupric chloride as the primary active ingredient, impregnated on a porous support, eg, alumina, silica–alumina [37287-16-4], diatomaceous earth [7631-86-9], etc, and may also contain numerous additives (89–91). Although the detailed catalytic mechanism is not known, $CuCl_2$ is widely recognized as the active chlorinating agent. The CuCl produced during the ethylene chlorination step is rapidly reconverted to $CuCl_2$ under reaction conditions, and the presence of some CuCl is thought to be advantageous because it readily complexes with ethylene, bringing it into contact with $CuCl_2$ long enough for chlorination to occur (78). A very simple representation of this heterogeneous catalytic cycle is given in equation 7–9, and the overall, net reaction is given in equation 3.

$$CH_2 = CH_2 + 2\ CuCl_2 \longrightarrow 2\ CuCl + ClCH_2CH_2Cl \qquad (7)$$

$$1/2\ O_2 + 2\ CuCl \longrightarrow CuOCuCl_2 \qquad (8)$$

$$2\ HCl + CuOCuCl_2 \longrightarrow 2\ CuCl_2 + H_2O \qquad (9)$$

Other mechanisms, involving initial formation of ethylene oxide [75-21-8] as the possible rate-limiting step, complexation of $CuCl_2$ with HCl (92), and Cl_2 as the chlorinating agent (93) have been suggested.

Because commercial oxychlorination processes differ with respect to catalysts in terms of their composition, morphology, and physical properties, with respect to the catalyst contacting method (fluidized- or fixed-bed reactor), and with respect to oxygen source (air or pure oxygen feed), the operating conditions, feed ratios, conversions, and yields also vary, depending on the particular combination used and on the methods employed for secondary recovery of feed-stock and product. For any particular combination of reactor type and oxygen source, however, good temperatures control of this highly exothermic reaction ($\Delta H = -239$ kJ/mol for eq. 3, Ref. 83) is essential for efficient production of EDC. Increasing reactor temperatures lead to increased by-product formation, mainly through increased oxidation of ethylene to carbon oxides and increased cracking of EDC. Cracking, ie, dehydrochlorination of EDC, results in the formation of vinyl chloride, and subsequent oxychlorination and cracking steps lead progressively to by-products with higher levels of chlorine substitution. High temperatures (>300°C) can also cause catalyst deactivation through increased sublimation of $CuCl_2$.

Fluidized-bed reactors typically are vertical cylindrical vessels equipped with a support grid/feed sparger system for adequate fluidization and feed distribution, internal cooling coils for heat removal, and either external or internal cyclones to minimize catalyst carryover. Fluidization of the catalyst assures intimate contact between feed and product vapors, catalyst, and heat-transfer surfaces, and results in a uniform temperature within the reactor (78). Reaction heat can be removed by the generation of steam within the cooling coils or by some other heat-transfer medium. An operating temperature of 220–245°C and reactor gauge pressures of 150–500 kPa (22–73 psig) are typical for oxychlorination with a fluidized catalyst. Given these operating conditions, fluidized-bed reactors can be constructed with a carbon steel shell and with internal parts made from a corrosion-resistant alloy.

While fluidized-bed oxychlorination reactors are generally well-behaved and operate predictably, under certain (usually upset) conditions, they are subject to a phenomenon known, appropriately, as catalyst stickiness. This can be described as catalyst particle agglomeration, which is characterized by declining fluidization quality and, in severe cases, can result in a slumped or collapsed bed. Oxychlorination catalyst stickiness is brought on by adverse operating conditions that promote the formation of dendritic growths of cupric chloride on the surface of individual catalyst particles, which leads to increasing interparticle interactions and agglomeration. All fluidized-bed oxychlorination catalysts normally exhibit some level of catalyst particle agglomeration/deagglomeration dynamics, and the severity of stickiness depends on catalyst characteristics as well as process operating conditions. Stickiness can be largely avoided by using catalyst formulations that exhibit excellent fluidization characteristics over a wide range of operating conditions (91).

Fixed-bed reactors resemble multitube heat exchangers, with the catalyst packed in vertical tubes held in a tubesheet at top and bottom. Uniform packing of catalyst within the tubes is important to ensure uniform pressure drop, flow, and residence time through each tube. Reaction heat can be removed by the generation of steam on the shell side of the reactor or by some other heat-transfer fluid. However, temperature control is more difficult in a fixed-bed than in a fluidized-bed reactor because localized hot spots tend to develop in the tubes. The tendency to develop hot spots can be minimized by packing the reactor tubes with active catalyst and inert diluent mixtures in proportions that vary along the length of the tubes, so that there is low catalyst activity at the inlet, but the activity steadily increases to a maximum at the outlet (78). Another method for minimizing hot spots is to pack the tubes with catalysts having a progressively higher loading of $CuCl_2$ so as to provide an activity gradient along the length of the tubes. Multiple reactors are also used in fixed bed oxychlorination, primarily to control heat release by staging the air or oxygen feed. Each successive reactor may also contain catalyst with a progressively higher loading of $CuCl_2$. These methods of staging the air or oxygen feed and of grading the catalyst activity tend to flatten the temperature profile and allow improved temperature control. Compared with the fluidized-bed process, fixed-bed oxychlorination generally operates at higher temperatures (230–300°C) and gauge pressures (150–1400 kPa (22–203 psig)). Given these operating conditions, a corrosion-resistant alloy is needed for the reactor tubes, and steel tubesheets and reactor heads need to be clad with nickel, whereas the reactor shell itself can be constructed of carbon steel.

In the air-based oxychlorination process with either a fluidized- or fixed-bed reactor, ethylene and air are fed in slight excess of stoichiometric requirements to ensure high conversion of HCl and to minimize losses of excess ethylene that remains in the vent gas after product condensation. Under these conditions, typical feedstock conversions are 94–99% for ethylene and 98–99.5% for HCl with EDC selectivities of 94–97%. Downstream product recovery involves cooling the reactor exit gases by either direct quench or with a heat exchanger, and condensation of the EDC and water, which are then separated by decantation (see Fig. 2). The remaining gases still contain 1–5 vol % EDC and hence are further processed in a secondary recovery system involving either solvent absorption or a refrigerated condenser. In air-based processes operating at high ethylene conversion, the dilute ethylene remaining in the vent is generally incinerated, but in those operating at lower conversion, various schemes are first used to recover unconverted ethylene, usually by direct chlorination to EDC (94–96).

The use of oxygen instead of air in the oxychlorination process with either a fixed- or fluidized-bed reactor permits operation at lower temperatures and results in improved operating efficiency and product yield (97). Unlike the air-based process, ethylene is generally fed in somewhat larger excess over stoichiometric requirements. The reactor exit gas is cooled, purified from traces of unconverted HCl, separated from EDC and water by condensation, recompressed to the reactor inlet pressure, reheated, and recycled to the oxychlorination reactor. Recycle of the effluent gas permits lower ethylene conversion per pass through the reactor with minimal loss in overall ethylene yield. A small amount of reactor off-gas, typically 2–5 vol %, is continuously purged from the system to prevent accumulation of impurities, eg, carbon oxides, nitrogen, argon, and unreacted hydrocarbons, which either form in the oxychlorination reactor or enter the process as impurities in the feed streams. An important advantage of oxygen-based oxychlorination technology over air-based operation is the drastic reduction in volume of the vent gas discharge. Since nitrogen is no longer present in the reactor feed streams, only a small amount of purge gas is vented. On a volume comparison, the reduced purge gas stream typically amounts to only 2–5% of the vent gas volume for air-based operation. Air-based processes release significant quantities of vent gases to the atmosphere, generally after treatment by incineration and scrubbing. Typically, for every kilogram of EDC produced by oxychlorination, ca 0.7–1.0 kg of vent gas is emitted from the air-based process (see Table 2). Therefore, for an air-based, balanced vinyl chloride plant with a rated capacity of 450,000 metric tons of vinyl chloride per year, the total vent gas volume released to the atmosphere would be 70–100 m^3/s (2,450–3,530 ft^3/s). However, the vent gas consists mainly of nitrogen, some unconverted oxygen, and small amounts of carbon oxides. Depending on the type of oxychlorination process involved, however, there are differing levels of undesirable impurities, ethylene, and chlorinated hydrocarbons in the oxychlorination vent gas.

Chlorinated by-products of ethylene oxychlorination typically include 1,1,2-trichloroethane; chloral [75-87-6] (trichloroacetaldehyde); trichloroethylene [79-01-6]; 1,1-dichloroethane; cis- and trans-1,2-dichloroethylenes [156-59-2 and 156-60-5]; 1,1-dichloroethylene [75-35-4] (vinylidene chloride); 2-chloroethanol [107-07-3]; ethyl chloride; vinyl chloride; mono-, di-, tri-, and tetrachloro-methanes (methyl chloride [74-87-3], methylene chloride [75-09-2], chloroform, and carbon tetrachloride [56-23-5]); and higher boiling compounds. The production of these compounds should be minimized to lower raw material costs, lessen the task of EDC purification, prevent fouling in the pyrolysis reactor, and minimize by-product handling and disposal. Of particular concern is chloral, because it polymerizes in the presence of strong acids. Chloral must be removed to prevent the formation of solids which can foul and clog operating lines and controls (78).

Oxychlorination reactor feed purity can also contribute to by-product formation, although the problem usually is only with low levels of acetylene which are normally present in HCl from the EDC cracking process. Since any acetylene fed to the oxychlorination reactor will be converted to highly chlorinated C$_2$ by-products, selective hydrogenation of this acetylene to ethylene and ethane is widely used as a preventive measure (78,98–102).

Purification of Ethylene Dichloride for Pyrolysis. By-products contained in EDC from the three main processes must be largely removed prior to pyrolysis. These include by-products from direct chlorination and oxychlorination, and in the recovered, unreacted EDC from the cracking process. EDC used for pyrolysis to vinyl chloride must be of high purity, typically greater than 99.5 wt %, because the cracking process is highly susceptible to inhibition and fouling by trace quantities of impurities (78). It must also be dry (less than 10 ppm water) to prevent excessive corrosion downstream. Inadvertent moisture pickup, however, is always possible. In such cases, the corrosion of steel equipment tends to be greatest in reboilers, the bottom section of distillation columns, bubble caps, plates, condensers, water separators, valves, pumps, and fittings.

Direct chlorination usually produces EDC with a purity greater than 99.5 wt %, so that, except for removal of the FeCl$_3$, little further purification is necessary. Ferric chloride can be removed by adsorption of a solid, or the EDC can be distilled from the FeCl$_3$ in a boiling reactor, as noted above. Alternatively, the FeCl$_3$ can be removed by washing with water, usually in conjunction with EDC from the oxychlorination process.

EDC from the oxychlorination process is less pure than EDC from direct chlorination and requires purification by distillation. It is usually first washed with water and then with caustic solution to remove chloral and other water-extractable impurities (103). Subsequently, water and low boiling impurities are taken overhead in a first (light ends or heads) distillation column, and finally, pure, dry EDC is taken overhead in a second (heavy ends or product) column (see Fig. 2).

Unreacted EDC recovered from the pyrolysis product stream contains a variety of cracking by-products. A number of these, eg, trichloroethylene, chloroprene, and benzene, are not easily removed by simple distillation and require additional treatment (78). Chloroprene can build up in the light ends column where it can polymerize and cause serious fouling. Benzene boils very close to EDC, as does trichloroethylene, which also forms an azeotrope with EDC. If allowed to accumulate in the recovered EDC, these by-products can inhibit the cracking reaction and increase coking rates. Because they are all unsaturated, they can be converted to higher boiling compounds for easy separation by subjecting the recovered EDC stream to chlorination prior to distillation (104–107). Chloroprene can also be removed by treatment with HCl and by hydrogenation.

Ethylene Dichloride Pyrolysis to Vinyl Chloride. Thermal pyrolysis or cracking of EDC to vinyl chloride and HCl occurs as a homogenous, first-order, free-radical chain reaction. The accepted general mechanism involves the four steps shown in equations 10–13:

Initiation
$$ClCH_2CH_2Cl \longrightarrow ClCH_2C{\cdot}H_2 + Cl{\cdot} \qquad (10)$$

Propagation
$$Cl{\cdot} + ClCH_2CH_2Cl \longrightarrow ClCH_2C{\cdot}HCl + HCl \qquad (11)$$

$$ClCH_2C{\cdot}HCl \longrightarrow CH_2{=}CHCl + Cl{\cdot} \qquad (12)$$

Termination
$$Cl{\cdot} + ClCH_2C{\cdot}H_2 \longrightarrow ClCH{=}CH_2 + HCl \qquad (13)$$

The overall, net reaction is given in equation 2. Reactions 11 and 12 are the chain propagation steps, because each elementary step consumes one of the two chain carriers and simultaneously produces the other. The net effect of reactions 11 and 12 is continuation of the chain by conversion of EDC to vinyl chloride. Thus, the two chain carriers are chlorine atoms and 1,2-dichloroethyl radicals [23273-86-1]. In general, anything that consumes a chain carrier is an EDC cracking inhibitor, and anything that produces a chain carrier is a promoter. Therefore, any molecular or radical species that consumes a chain carrier without simultaneously producing either 1,2-dichloroethyl radicals or chlorine atoms is an EDC cracking inhibitor, eg, propylene [115-07-1]. The allylic hydrogen atoms of propylene can be easily abstracted by one of the chain carriers, either 1,2-dichloroethyl radicals or chlorine atoms. The resulting allyl radical can then combine with a chlorine atom forming allyl chloride [107-05-1]. Because the same sequence can occur two more times, one molecule of propylene can consume up to six chain carriers. Reaction initiators or accelerators include carbon tetrachloride, chlorine, bromine [7726-95-6], iodine [7553-56-2], or oxygen, although exclusion of oxygen is claimed to result in considerably less fouling on the pyrolysis tube walls. Carbon tetrachloride, a minor by-product of the oxychlorination process, is often controlled in the EDC purification step to allow a small amount to enter the EDC feed as a cracking promoter in the pyrolysis process.

The endothermic cracking of EDC ($\Delta H = 71$ kJ/mol EDC reacted for eq. 2, Ref. 83) is relatively clean at atmospheric pressure and at temperatures of 425–550°C. Commercial pyrolysis units, however, generally operate at gauge pressures of 1.4–3.0 MPa (200–435 psig) and at temperatures of 475–525°C to provide for better heat transfer and reduced equipment size, and to allow separation of HCl from vinyl chloride by fractional distillation at noncryogenic temperatures. EDC conversion per pass through the pyrolysis reactor is normally maintained at 53–63%, with a residence time of 2–30 s. Cracking reaction selectivity to vinyl chloride of >99% can be achieved at these conditions. Increasing cracking severity beyond this level gives progressively smaller increases in EDC conversion, with progressively lower selectivity to vinyl chloride, since some of the by-products generated during pyrolysis act as cracking inhibitors. Higher conversion also increases pyrolysis tube coking rates and causes problems with downstream product purification. To minimize coke formation, it is necessary to quench or cool the pyrolysis reactor effluent quickly. Substantial yield losses to heavy ends and tars can occur if cooling is done too slowly. Therefore, the hot effluent gases are normally quenched and partially condensed by direct contact with cold EDC in a quench tower. Alternatively, the pyrolysis effluent gases can first be cooled by heat exchange with cold liquid EDC furnace feed in a transfer line exchanger (TLE) prior to quenching in the quench tower. In this case, application of a TLE to preheat and vaporize incoming EDC furnace feed saves energy by decreasing the amount of fuel gas required to fire the cracking furnace and/or steam needed to vaporize the feed.

Although there are minor differences in the HCl–vinyl chloride recovery section from one vinyl chloride producer to another, in general, the quench column effluent is distilled to remove first HCl and then vinyl chloride (see Fig. 2). The vinyl chloride is usually further treated to produce specification product, recovered HCl is sent to the oxychlorination process, and unconverted EDC is purified for removal of light and heavy ends before it is recycled to the cracking furnace. The light and heavy ends are either further processed, disposed of by incineration or other methods, or completely recycled by catalytic oxidation with heat recovery followed by chlorine recovery as EDC (76).

By-products from EDC pyrolysis typically include acetylene, ethylene, methyl chloride, ethyl chloride, 1,3-butadiene, vinylacetylene, benzene, chloroprene, vinylidene chloride, 1,1-dichloroethane, chloroform, carbon tetrachloride, 1,1,1-trichloroethane [71-55-6], and other chlorinated hydrocarbons (78). Most of these impurities remain with the unconverted EDC, and are subsequently removed in EDC purification as light and heavy ends. The lightest compounds, ethylene and acetylene, are taken off with the HCl and end up in the oxychlorination reactor feed. The acetylene can be selectively hydrogenated to ethylene. The compounds that have boiling points near that of vinyl chloride, ie, methyl chloride and 1,3-butadiene, will codistill with the vinyl chloride product. Chlorine or carbon tetrachloride addition to the pyrolysis reactor feed has been used to suppress methyl chloride formation, whereas 1,3-butadiene, which interferes with PVC polymerization, can be removed by treatment with chlorine or HCl, or by selective hydrogenation.

By-Product Disposal. By-product disposal from vinyl chloride manufacturing plants is complicated by the need to process a variety of gaseous, organic liquid, aqueous, and solid streams, while ensuring that no chlorinated organic compounds are inadvertently released. Each class of by-product streams poses its own treatment and disposal challenges.

Gaseous vent streams from the different unit operations may contain traces (or more) of HCl, CO, methane, ethylene, chlorine, and vinyl chloride. These can sometimes be treated chemically, or a specific chemical value can be recovered by scrubbing, sorption, or other method when economically justified. For objectionable components in the vent streams, however, the common treatment method is either incineration or catalytic combustion, followed by removal of HCl from the effluent gas.

Organic liquid streams include the light and heavy ends from EDC purification (see Fig. 2 and Table 2). The light ends typically consist of ethyl chloride, *cis*- and *trans*-1,2-dichloroethylene, chloroform, and carbon tetrachloride. The heavy ends are mainly composed of 1,1,2-trichloroethane, lesser amounts of tetrachloroethanes, chlorinated butanes, and chlorinated aromatics, and many other chlorinated compounds present at small concentrations. If there is economic justification, these streams can be fractionated to recover specific, useful components, and the remainder subsequently incinerated and scrubbed to remove HCl. An alternative method involves combining all liquid by-product streams and passing them along with air or oxygen-enriched air into a fluidized bed, catalytic oxidation reactor (76). The resulting combustion product stream, consisting essentially of HCl, CO_2, H_2O, O_2, and N_2, is fed directly into an oxychlorination reactor where the HCl content is recovered as EDC. Furthermore, the heat of combustion is recovered as high pressure steam in a manner similar to that in fluidized-bed oxychlorination processes. In addition, there is no direct vent to the atmosphere, and any unconverted chlorinated organic material is recovered in the crude EDC from the oxychlorination process and ultimately recycled back to the catalytic oxidation unit. Minor drawbacks to this process include increased inerts loading and the potential for carryover of entrained oxidation catalyst into the oxychlorination reactor.

Process water streams from vinyl chloride manufacture are typically steam-stripped to remove volatile organics, neutralized, and then treated in an activated sludge system to remove any nonvolatile organics. If fluidized-bed oxychlorination is used, the process wastewater may also contain suspended catalyst fines and dissolved metals. The former can easily be removed by sedimentation, and the latter by precipitation. Depending on the specific catalyst formulation and outfall limitations, tertiary treatment may be needed to reduce dissolved metals to acceptable levels.

Solid by-products include sludge from wastewater treatment, spent catalyst, and coke from the EDC pyrolysis process. These need to be disposed of in an environmentally sound manner, eg, by sludge digestion, incineration, landfill, etc.

Economic Aspects

Yearly U.S. production volumes and prices of vinyl chloride are listed in Table 3 (75,108–110). The most recent increase in price (between 1992 and 1995) was brought on by a tight chlorine supply, which raised chlorine prices. Contract

Table 3. U.S. Vinyl Chloride Production and Prices[a]

Year	Production, 10^3 t	Contract price, ¢/kg
1955	240	22
1960	470	26
1965	907	18
1970	1,833	11
1975	1,903	24
1980	2,933	49
1985	3,527	36.5
1990	4,678	43.3
1991	5,031	32.6
1992	5,374	30.3
1993	5,496	36.7
1994	6,020	46.5
1995	6,316	52.0
1996	6,487	44.0

[a]Ref. 108 for 1955–1980, Ref. 109 for 1985, Ref. 75 for 1990–1994, and Ref. 110 for 1995–1996.

prices for vinyl chloride monomer in the United States are largely determined by the prices of the two raw materials, ethylene and chlorine, and by the price of the end product, PVC. The interplay of these components, which fluctuate monthly, results in relatively stable monomer prices. In general, the price is high enough to allow monomer producers to make a profit, yet low enough so as not to place nonintegrated PVC producers at a significant competitive disadvantage (75). The lower relative cost of PVC has allowed it to compete effectively with metals in the construction and automobile industries.

U.S production capacities are listed in Table 4 (110). Worldwide production capacity of each country for vinyl chloride is listed in Table 5 (110). The United States is the world's largest exporter of EDC, vinyl chloride, and PVC owing primarily to excess capacity relative to domestic demand, and regional shortages in the international market that keep export prices attractively high. The fastest-growing markets for vinyl products are in the Asia/Pacific region. Worldwide demand for PVC is expected to grow at nearly 5% annually, which will require significant increases in vinyl chloride production capacity (75).

Environmental Considerations

Since the early 1980s, there has been much debate among environmental activist organizations, industry, and government about the impact of chlorine chemistry on the environment (111,112). One aspect of this debate involves the incidental manufacture and release of trace amounts of hazardous compounds such as polychlorinated dibenzodioxins, dibenzofurans, and biphenyls (PCDDs, PCDFs, and PCBs, respectively, but often referred to collectively as dioxins) during the production of chlorinated compounds like vinyl chloride. Initial concerns were prompted by the acute animal toxicity of 2,3,7,8-tetrachlorodibenzodioxin [1746-01-6] (TCDD) and 2,3,7,8-tetrachlorodibenzofuran [51207-31-9] (TCDF). More recently, the focus has shifted to the reported estrogen mimic capabilities of dioxins (111). In 1994, the EPA released a review draft of its reassessment of

Table 4. U.S. Producers of Vinyl Chloride Monomer[a]

Company	Capacity, 10^3 t/yr[b]	Remarks
Borden, Inc. (Borden Chemicals & Plastics Limited Partnership)		
Geismar, La.	426	
Condea Vista Co.		name changed from Vista
Lake Charles, La.	413	Chemical Co., Inc. in 1996
The Dow Chemical Co., Inc.		
Oyster Creek, Tex.	862	
Plaquemine, La.	635	
Formosa Plastics Corp., USA		
Baton Rouge, La.	560	
Point Comfort, Tex.	363	
The Geon Company		
LaPorte, Tex.	938	formerly a division of BF Goodrich, separated from it in 1993
Georgia Gulf Corp.		
Plaquemine, La	574	
Oxymar		
Corpus Christi, Tex.	635	joint venture between Occidental Petroleum Corp. and Marubeni Corp. (Japan)
Oxy Chemical Corp.		
Deer Park, Tex.	510	
PPG Industries, Inc.		
Lake Charles, La.	349	
Westlake Chemical Corp. (Westlake Monomers Corp.)		
Calvert City, Ky.	476	purchased BF Goodrich Calvert City plant in 1990
Total	*6,741*	

[a]Ref. 110.
[b]Average annual capacity for calendar year 1996.

the impact of dioxins in the environment on human health, which prompted speculation as to the amount of dioxins that might be attributed to chlorine-based industrial processes (113). The U.S. vinyl industry responded to this document by committing to a voluntary characterization of dioxin levels in its products and in emissions from its facilities to the environment. The results of this study to date support the vinyl industry's position that it is a minor source of dioxins in the environment (114). Furthermore, a recent process model analysis of a vinyl chloride purification distillation unit demonstrated that, owing to the roughly 500°C boiling point difference between vinyl chloride and TCDD, any incidentally manufactured TCDD would be quantitatively removed from the vinyl chloride product (115). In addition, a global benchmark study released recently

Table 5. World Wide Vinyl Chloride Capacity as of January 1997[a]

Country	Capacity, 10^3 t/yr[b]
North America	
Canada	374
Mexico	270
United States	6,741
Total	*7,385*
Europe	
Western Europe	
Belgium	1,010
France	1,205
Germany	1,710
Italy	620
Netherlands	520
Norway	470
Spain	399
Sweden	115
United Kingdom	360
Total	*6,409*
Eastern Europe	
Commonwealth of Independent States	1,085
Czech Republic and Slovakia	217
former Yugoslavia	260
Hungary	185
Poland	325
Romania	270
Total	*2,342*
South America	
Argentina	160
Brazil	540
Venezuela	180
Total	*880*
Asia	
Far East	
China	1,356
India	610
Indonesia	150
Japan	2,965
Korea (North)	24
Korea (South)	755
Pakistan	5
Philippines	10
Taiwan	1,030
Thailand	240
Total	*7,145*
Middle East	
Iran	213
Israel	110
Saudi Arabia	360
Turkey	179
Total	*862*

Table 5. (*Continued*)

Country	Capacity, 10^3 t/yr[b]
Africa	
Algeria	40
Egypt	100
Libya	62
Morocco	38
South Africa	165
Total	*405*
Australia	36
Grand total	*25,464*

[a]Ref. 110.
[b]Average annual capacity for calendar year 1996.

by the American Society of Mechanical Engineers found no relationship between the chlorine content of waste and dioxin emissions from combustion processes (116). Reviews of this issue and other environmental considerations related to vinyl chloride production are available (111,112,117).

Because of the toxicity of vinyl chloride, the EPA in 1975 proposed the following emission standards for vinyl chloride manufacture: (*1*) emissions from all point sources except oxychlorination would be limited to 10 ppm vinyl chloride; (*2*) emissions from the oxychlorination process would be limited to 0.02 kg vinyl chloride per 100 kg EDC produced by oxychlorination; (*3*) no preventable relief valve discharges would be allowed; and (*4*) fugitive emissions would be minimized by enclosing emission sources and collecting all emissions (118). This proposal was subsequently enacted as EPA Regulation 40 CFR 61, Subpart F (119). Compliance testing began in 1978. Additional EPA and state actions were initiated in 1977 to reduce hydrocarbon emissions from vinyl chloride plants in nonattainment regions. These were aimed primarily at lowering the ethylene content of vent streams from air-based oxychlorination units (78).

Environmental concerns and government regulations have prompted a major increase in the amount of add-on technology used in U.S. vinyl chloride production plants. Primary and redundant incineration facilities for all vinyl chloride point source and collected fugitive emissions are needed to ensure compliance. The incinerators are typically equipped with HCl scrubbing and neutralization or recovery units. Process sewers and sewage collection systems are closed. Larger and/or redundant strippers are used to remove trace organics from wastewater. Dual mechanical seals on pumps and agitators are required. Other common emissions control and reduction measures include vinyl chloride and EDC leak-detection systems and portable monitors, enclosed sampling and analytical systems, and vapor recovery systems for vinyl chloride loading and unloading and equipment cleaning (78).

Technology Trends

The ethylene-based, balanced vinyl chloride process, which accounts for nearly all capacity worldwide, has been practiced by a variety of vinyl chloride producers since the mid-1950s. The technology is mature, so that the probability of significant changes is low. New developments in production technology will likely be based on incremental improvements in raw material and energy efficiency, environmental impact, safety, and process reliability.

More recent trends include widespread implementation of oxygen-based oxychlorination, further development of new catalyst formulations, a broader range of energy recovery applications, a continuing search for ways to improve conversion and minimize by-product formation during EDC pyrolysis, and chlorine source flexibility. In addition, the application of computer model-based process control and optimization is growing as a way to achieve even higher levels of feedstock and energy efficiency and plant process reliability.

Nearly all oxychlorination processes built since 1990 are oxygen-based, and many existing, air-based units are being retrofitted for pure oxygen feed. This is the result of significant advantages of oxygen- over air-based operation, described above. The greatest benefit is drastic reduction (over 95%) in the volume of vent gas discharged by the process, making destruction of any environmentally objectionable compounds in this stream more manageable. Savings in ethylene and chlorine feedstock, incineration, and air compression costs can more than offset the oxygen raw material cost. For existing, air-based plants, the decision whether to convert to oxygen depends on local emission standards, oxygen availability and cost, electrical energy cost, and the viability of alternative, add-on processes for cleaning the vent stream, eg, catalytic oxidation or sorption methods.

New catalyst developments have steadily progressed, especially for ethylene oxychlorination (91), so that the remaining potential increases in reactor productivity and feedstock efficiency keep shrinking. Nevertheless, there is still room for further catalyst improvements. Direct chlorination, for example, would benefit from a catalyst that provides increased reaction selectivity to EDC and minimizes by-product formation at the higher temperatures required for boiling reactors. A catalyst that is more resistant to poisoning by the metals content of by-product streams, eg, iron from direct chlorination, sodium from caustic washing, tramp metals from corrosion, etc, would improve the performance of catalytic hydrogenation and oxidation processes.

In order to take advantage of the heat released by the reaction, most direct chlorination processes built since the early 1980s use boiling reactors with energy recovery. Some alternative designs that also utilize the heat of reaction, but without boiling, have likewise been developed (87,88). More recently, EDC pyrolysis furnaces are being built with TLEs, which use the thermal energy of the product stream to vaporize the EDC furnace feed or make steam (120–122), and with air preheaters, which preheat the furnace air feed by exchange with the flue gas, to reduce the energy requirements for pyrolysis. Further design modifications to improve energy efficiency are likely, particularly if the cost of energy should increase significantly.

The search for new EDC cracking promoters and by-product inhibitors (123–126) and for improved feed purification methods is expected to continue. Since current cracking technology limits EDC conversion to 55–65%, considerable energy and cost savings would be realized if conversion could be increased without concurrent loss of EDC to undesirable side reactions and coking.

Laser-induced EDC cracking has been studied as a way to promote thermal cracking (127–130). At temperatures comparable to those used in commercial pyrolysis reactors, laser-induced cracking is claimed to increase conversion while decreasing by-product formation. However, commercial application at the current state of development appears unlikely. Periodic furnace decoking is still normally accomplished thermally by controlled air/stream oxidation. However, shot peen and catalytic decoking are two more recent alternative methods aimed at faster turnaround times and less thermal stress on the cracking furnaces.

The recent (since 1993) tightness of the chlorine market, which was accompanied by rising chlorine prices, sparked interest in alternative sources of chlorine, ie, HCl and EDC, where these are available. Some plants no longer operate in a strictly balanced mode, but instead operate with more than half of their EDC made from oxychlorination (owing to importation of HCl or EDC as a chlorine source, thus bypassing direct chlorination). The ideal situation is one in which the plant can adapt to any feed combination, allowing operation at the optimum mix of feedstocks as determined by minimization of the sum of raw material and operating costs.

Alternatives to oxychlorination have also been proposed as part of a balanced VCM plant. In the past, many vinyl chloride manufacturers used a balanced ethylene–acetylene process for a brief period prior to the commercialization of oxychlorination technology. Addition of HCl to acetylene was used instead of ethylene oxychlorination to consume the HCl made in EDC pyrolysis. Since the 1950s, the relative costs of ethylene and acetylene have made this route economically unattractive. Another alternative is HCl oxidation to chlorine, which can subsequently be used in direct chlorination (131). The Shell-Deacon (132), Kel-Chlor (133), and MT-Chlor (134) processes, as well as a process recently developed at the University of Southern California (135) are among the available commercial HCl oxidation technologies. Each has had very limited industrial application, perhaps because the equilibrium reaction is incomplete and the mixture of HCl, O_2, Cl_2, and water presents very challenging separation, purification, and handling requirements. HCl oxidation does not compare favorably with oxychlorination because it also requires twice the direct chlorination capacity for a balanced vinyl chloride plant. Consequently, it is doubtful that it will ever displace oxychlorination in the production of vinyl chloride by the balanced ethylene process.

If the production of vinyl chloride could be reduced to a single step, such as direct chlorine substitution for hydrogen in ethylene or oxychlorination/cracking of ethylene to vinyl chloride, a major improvement over the traditional balanced process would be realized. The literature is filled with a variety of catalysts and processes for single-step manufacture of vinyl chloride (136–138). None has been commercialized because of the high temperatures, corrosive environments, and insufficient reaction selectivities so far encountered. Substitution of lower cost ethane or methane for ethylene in the manufacture of vinyl chloride has also been investigated. The Lummus-Transcat process (139), for instance, proposes a molten oxychlorination catalyst at 450–500°C to react ethane with chlorine to make vinyl chloride directly. However, ethane conversion and selectivity to vinyl chloride are too low (30% and less than 40%, respectively) to make this process competitive. Numerous other catalysts and processes have been patented as well, but none has been commercialized owing to problems with temperature, corrosion, and/or product selectivity (140–144). Because of the potential payback, however, this is a very active area of research.

Specifications

Polymerization-grade vinyl chloride should not contain more than the amounts of impurities listed in Table 6 (145).

Table 6. Typical Impurity Levels in Monomer Grade Vinyl Chloride[a]

Impurity	Maximum level, ppm
acetylene	0.5–2.0
acidity, as HCl, by wt	0.1–1.0
acetaldehyde	0.4–1.0
alkalinity, as NaOH, by wt	0.25
1,3-butadiene	8–12
ethyl chloride	35
EDC	10
iron, as Fe, by wt	0.15–0.4
methyl chloride	60–75
vinyl acetylene	10
water	100
nonvolatiles	25–50
total C_4 unsaturates	40
oxygen in vapor space after loading	200–1,000
vinyl chloride, wt %	99.96–99.98[b]

[a]Ref. 145.
[b]Minimum vinyl chloride content.

Health and Safety Factors

Vinyl chloride is an OSHA-regulated substance (146). Current OSHA regulations impose a permissible exposure limit (PEL) to vinyl chloride vapors of no more than 1.0 ppm averaged over any 8-h period. Short-term exposure is limited to 5.0 ppm averaged over any 15-min period. Contact with liquid vinyl chloride is prohibited. Monitoring is required at all facilities where vinyl chloride is produced or PVC is processed. OSHA regulations also define an action level of 0.5 ppm, 8-h time-weighted average. Employers must demonstrate that monitoring results show exposure below the action level of 0.5 ppm on subsequent readings taken not less than five working days apart in order to discontinue monitoring. Where concentrations cannot be lowered below the 1.0-ppm PEL, the employer must establish a regulated area with controlled access, a respirator program conforming to paragraph g of the OSHA standard (146), and a written plan to reduce vinyl chloride levels. OSHA regulations require facilities that handle vinyl chloride to develop a medical surveillance program with annual physical examinations and blood serum analyses for all employees exposed at levels above the action level. Surveillance frequency increases to semiannually for these employees when they attain 10 or more years of service in the manufacture of vinyl chloride.

While the current OSHA limitations for EDC exposure are not as restrictive as those for vinyl chloride, a new regulation similarly limiting EDC exposure is expected to be issued in 1997.

Contact with liquid vinyl chloride can cause frostbite. Chronic exposure to vinyl chloride at concentrations of 100 ppm or more is reported to have produced Raynaud's syndrome, lysis of the distal bones of the fingers, and a fibrosing dermatitis. However, these effects are probably related to continuous intimate contact with the skin. Chronic exposure is also reported to have produced a rare cancer of the liver (angiosarcoma) in a small number of workers after continued exposure for many years to large amounts of vinyl chloride gas (147). Consequently, the vinyl industry worked in conjunction with government regulatory agencies to develop the much more stringent hygiene standards that exist. Toxicology data on vinyl chloride, eg, TC_{Lo} (human), TC_{Lo} (rat), LD_{50} (rat), and threshold limit values, are reported in Reference 148. Vinyl chloride monomer is listed as a cancer-suspect agent by OSHA (146). The American Conference of Governmental Industrial Hygienists (ACGIH) calls it a confirmed human carcinogen, while the National Toxicology Program (NTP) and the International Agency for Research on Cancer (IARC) both regard it as a human carcinogen (149).

Vinyl chloride also poses a significant fire and explosion hazard. It has a wide flammability range, from 3.6 to 33.0% by volume in air (150). Large fires of the compound are very difficult to extinguish, while vapors represent a severe explosion hazard. Vapors are more than twice as dense as air and tend to collect in low lying areas, increasing the risk of fire. Workers entering these low lying areas risk suffocation, which can occur at levels above 18,000 ppm. The mild, sweet odor of vinyl chloride becomes detectable around 260 ppm (151).

Vinyl chloride is generally transported via pipeline, and in railroad tank cars and tanker ships. Containers of vinyl chloride must be labeled "vinyl chloride," "extremely flammable gas under pressure," and "cancer-suspect agent" (146). Because hazardous peroxides can form on standing in air, especially in the presence of iron impurities, vinyl chloride should be handled and transported under an inert atmosphere. The presence of peroxide from vinyl chloride and air can initiate polymerization of stored vinyl chloride; however, stabilizer can be added to prevent polymerization. Inhibitors such as hydroquinone [123-31-9] are often added, particularly when shipping long distances in warmer climates.

Vinyl chloride is listed as "ethene, chloro-" on the Toxic Substances Control Act (TSCA) inventory and on the Canadian Domestic Substances List (DSL). It is listed as "chloroethylene" on the European Inventory of Existing Commercial Chemical Substances (EINECS), bearing the identification number 2008 310 (149).

Uses

Vinyl chloride has gained worldwide importance because of its industrial use as the precursor to PVC. It is also used in a wide variety of copolymers. The inherent flame-retardant properties, wide range of plasticized compounds, and low cost of polymers from vinyl chloride have made it a major industrial chemical. About 95% of current vinyl chloride production worldwide ends up in polymer or copolymer applications (83). Vinyl chloride also serves as a starting material for the synthesis of a variety of industrial compounds, as suggested by the number of reactions in which it can participate, although none of these applications will likely ever come anywhere near PVC in terms of volume. The primary nonpolymeric uses of vinyl chloride are in the manufacture of vinylidene chloride and tri- and tetrachloroethylene [127-18-4] (83).

BIBLIOGRAPHY

"Vinyl Chloride" under "Chlorine Compounds, Organic" in *ECT* 1st ed., Vol. 3, p. 786, by J. Werner, General Aniline & Film Corp., General Aniline Works Division; "Vinyl Chloride" under "Vinyl Compounds" in *ECT* 1st ed., Vol. 14, pp. 723–726, by C. H. Alexander and G. F. Cohan, B. F. Goodrich Chemical Co.; "Vinyl Chloride" under "Chlorocarbons and Chlorohydrocarbons" in *ECT* 2nd ed., Vol. 5, pp. 171–178, by D. W. F. Hardie, Imperial Chemical Industries, Ltd.; "Vinyl Chloride" under "Vinyl Polymers" in *ECT* 3rd ed., Vol. 23, pp. 865–885, by J. A. Cowfer and A. J. Magistro, B. F. Goodrich Co.

1. U. S. Pats. 2,188,396 (Jan. 30, 1940) and 1,929,453 (Oct. 10, 1933), W. L. Semon (to the B. F. Goodrich Co.).
2. C. E. Schildknecht, *Vinyl and Related Polymers: Their Preparations, Properties, and Applications in Rubbers, Plastics, Fibers, and in Medical and Industrial Arts*, John Wiley & Sons, Inc., New York, 1952, pp. 387–388.
3. M. Kaufman, *The Chemistry and Industrial Production of Polyvinyl Chloride: The History of PVC*, Gordon and Breach Science Publishers Inc., New York, 1969, pp. 387–388.
4. W. L. Semon and G. A. Stahl, *J. Macromol. Sci., Chem.* **A15**, 199 (1981).
5. *C & E News* **62**(25), 38 (1984).
6. R. B. Seymour, ed., *Pioneers in Polymer Science*, Kluwer Academic Publishers, Dordrecht, the Netherlands, 1989, p. 119.
7. V. Regnault, *Ann. Chim. Phys.* **58**, 301 (1835).
8. E. Baumann, *Ann. Chem. Pharm.* **163**, 308 (1872).
9. I. I. Ostromislensky, *J. Russ. Phys. Chem. Soc.* **48**, 1132 (1916).
10. E. M. Smith, ed., *Waldo Lonsbury Semon, A Man of Ideas: The Inventor of Plasticized Polyvinyl Chloride*, The Geon Co., Cleveland, Ohio, 1993.
11. Ger. Pat. 278,249 (Oct. 11, 1912) (to Chemische Fabrik Griesheim-Elektron).
12. DIPPR Data Compilation File (STN International online service), Design Institute for Physical Property Data, Pennsylvania State University, University Park, Pa., Aug. 1994.
13. A. L. Horvath, *Halogenated Hydrocarbons: Solubility-Miscibility With Water*, Marcel Dekker, Inc., New York, 1982, pp. 494, 550.
14. U.S. Pat. 5,312,984 (May 17, 1994) and Eur. Pat. 185,350 (June 25, 1986), P. P. Nicholas (to the B. F. Goodrich Co.); P. P. Nicholas, *J. Org. Chem.* **52**, 5266 (1987).
15. E. Ger. Pat. 260,276 (Sept. 21, 1988), U. Thust and co-workers, (to VEB Chemiekombinat Bitterfeld).
16. Jpn. Pat. 56 139,434 (Oct. 30, 1981), T. Keiichi and co-workers (to T. Hasegawa Co., Ltd.).
17. N. F. Cherepennikova and V. V. Semonov, *Zh. Obshch. Khim.* **59**, 965 (1989).
18. V. I. Zhun, M. K. Ten, and V. D. Sheludyakov, *Khim. Prom-st. (Moscow)*, 15 (1989).
19. Jpn. Pat. 58 103,328 (June 20, 1983), A. Takehiro and co-workers (to Taisho Pharmaceutical Co., Ltd.).
20. D. Arnould and co-workers, *Bull. Soc. Chim. Fr.*, 130 (1985).

21. Ger. Pat. 3,126,022 (Jan. 13, 1983), H. Sauter and co-workers (to BASF AG).
22. Ger. Pat. 2,918,801 (Nov. 20, 1980), B. Zeeh, E. Ammermann, and E. H. Pommer (to BASF AG).
23. A. Indolese and G. Consiglio, *J. Organomet. Chem.* **463**, 23 (1993).
24. S. Nunomoto, Y. Kawakami, and Y. Yamashita, *Toyama Kogyo Koto Senmon Gakko Kiyo* **20**, 45 (1986).
25. T. Amano and co-workers, *Bull. Chem. Soc. Jap.* **59**, 1656 (1986).
26. E. J. Corey and R. L. Carney, *J. Am. Chem. Soc.* **93**, 7318 (1971); E. J. Corey and co-workers, *J. Am. Chem. Soc.* **94**, 4395 (1972).
27. B. A. Trofimov and co-workers, *Zh. Org. Khim.* **21**, 2324 (1985); **17**, 1098 (1981), B. A. Trofimov and co-workers, *Sulfur Lett.* **2**, 99 (1984); Russ. Pat. 852,862 (Aug. 7, 1981), B. A. Trofimov and co-workers, (to Irkutsk Institute of Organic Chemistry and "Karbolit" Kemerovo Scientific-Industrial Enterprises).
28. Jpn. Pat. 03 287,572 (Dec. 18, 1991), T. Nishitake and S. Matsuoka (to Tokuyama Soda Co., Ltd.).
29. B. A. Trofimov and co-workers, *Zh. Org. Khim.* **26**, 725 (1990).
30. Brit. Pat. 332,605 (Mar. 22, 1929), (to I. G. Farbenindustrie AG).
31. I. A. Aliev and co-workers, *Khim. Geterotsikl. Soedin.* 1320 (1991); 1337 (1990); 1359 (1984); *Zh. Org. Khim.* **24**, 2436 (1988); **22**, 489 (1986); I. A. Aliev, A. I. Mikhaleva, and M. V. Sigalov, *Sulfur Lett.* **2**, 55 (1984); S. E. Korostova and co-workers, *Zh. Org. Khim.* **20**, 1960 (1984); A. I. Mikhaleva and co-workers, *Zh. Org. Khim.* **18**, 2229 (1982); U.S.S.R. Pat. 840,038 (June 23, 1981), B. A. Trofimov, A. I. Mikhaleva, and A. N. Vasil'ev (to Irkutsk Institute of Organic Chemistry).
32. M. A. Kuznetsova, E. N. Deryagina, and M. G. Voronkov, *Zh. Org. Khim.* **21**, 2331 (1985).
33. M. G. Voronkov, E. N. Deryagina, and M. A. Kuznetsova, *Zh. Org. Khim.* **16**, 1776 (1980).
34. Russ. Pat. 1,776,655 (Nov. 23, 1992), N. D. Ivanova and co-workers (to Irkutsk Institute of Organic Chemistry).
35. V. V. Shcherbinin and co-workers, *Zh. Obshch. Khim.* **63**, 1915 (1993).
36. Eur. Pat. 456,901 (Nov. 21, 1991), W. Hange and co-workers, (to Hüls AG).
37. Eur. Pat. 383,566 (Aug. 22, 1990), M. Itoh and co-workers (to Mitsui Toatsu Chemicals, Inc.).
38. A. Akramkhodzhaev, T. S. Sirlibaev, and Kh. U. Usmanov, *Uzb. Khim. Zh.*, 29 (1980).
39. R. Atkinson and S. M. Aschmann, *Int. J. Chem. Kinet.* **19**, 1097 (1987).
40. O. V. Konopatsky, O. G. Chernitsky, and V. M. Vlasenko, *Teor. Eksp. Khim.* **29**, 462 (1993); *Ekotekhnol. Resursosberezhenie* (5), 13 (1993); O. V. Konopatsky, V. M. Vlasenko, and O. G. Chernitsky, *Teor. Eksp. Khim.* **28**, 202 (1992).
41. J. Hranisavljevic and co-workers, *Combust. Sci. Technol.* **101**, 231 (1994).
42. U.S. Pat. 4,849,114 (July 18, 1989), J. D. Zeff and E. Leitis (to Ultrox International).
43. V. V. Popov and co-workers, *Khim.-Farm. Zh.* **23**, 629 (1989); **19**, 988 (1985).
44. U.S. Pat. 5,315,052 (May 24, 1994), T. G. Taylor and co-workers (to PPG Industries, Inc.).
45. E. Huang, Z. Xu, and S. Wang, *Huaxue Fanying Gongcheng Yu Gongyi* **4**, 60 (1988).
46. U.S.S.R. Pat. 910,573 (Mar. 7, 1982), O. A. Zaidman and co-workers, (to U.S.S.R.).
47. F. S. Dainton, D. A. Lomax, and M. Weston, *Trans. Faraday Soc.* **58**, 308 (1962); P. B. Ayscough and co-workers, *Trans. Faraday Soc.* **58**, 318 (1962).
48. G. D. Ivanyk and Yu. A. Pazderskii, *Khim. Prom-st.* (*Moscow*), 333 (1984).
49. V. B. Kunshenko and co-workers, *Zh. Org. Khim.* **28**, 672 (1992).
50. U.S. Pat. 5,345,018 (Sept. 6, 1994), P. L. Bak and co-workers, (to the Geon Co.).
51. Jpn. Pat. 59 044,290 (Oct. 29, 1984), (to Tokuyama Soda Co., Ltd.).
52. M. S. Kharasch and C. W. Hannum, *J. Am. Chem. Soc.* **56**, 712 (1934).

53. Chin. Pat. 1,069,019 (Feb. 17, 1993), G. Yan, X. Guo, and M. Liu (to Chemical Industry Institute, Zhejiang Province).
54. Eur. Pat. 380,052 (Aug. 1, 1990), M. Yamamoto and co-workers, (to Sumitomo Chemical Co., Ltd.).
55. L. Schmerling, *J. Am. Chem. Soc.* **68**, 1650 (1946).
56. J. M. Davidson and A. Lowy, *J. Am. Chem. Soc.* **51**, 2978 (1929).
57. T. Chiba and co-workers, *Chem. Pharm. Bull.* **38**, 3317 (1990).
58. M. Kotora and M. Hajek, *J. Mol. Catal.* **77**, 51 (1992); *React. Kinet. Catal. Lett.* **44**, 415 (1991).
59. R. A. Amriev and co-workers, *Izv. Akad. Nauk SSSR, Ser. Khim.*, 2626 (1984).
60. R. Kh. Freidlina, F. K. Velichko, and R. A. Amriev, *C1 Mol. Chem.* **1**, 193 (1985).
61. W. Navarrini and co-workers, *J. Fluorine Chem.* **71**, 111 (1995).
62. Jpn. Pat. 06 100,481 (Apr. 12, 1994), Y. Watabe and F. Hayakawa (to Mitsui Toatsu Chemicals, Inc.).
63. Jpn. Pat. 05 194,301 (Aug. 3, 1993), T. Saeki, N. Arashiba, and S. Kyono (to Mitsui Toatsu Chemicals, Inc.); Jpn. Pat. 01 121,233 (May 12, 1989), H. Ono, T. Kasuga, and S. Kyono (to Mitsui Toatsu Chemicals, Inc.).
64. Jpn. Pat. 63 303,969 (Dec. 12, 1988), M. Sasaki and co-workers, (to Dainippon Ink and Chemicals, Inc.).
65. Pol. Pat. 146,123 (Dec. 31, 1988), S. Stefaniak (to Zaklady Chemiczne "Organika-Zachem").
66. Jpn. Pat. 07 082,237 (Mar. 28, 1995), T. Matsuoka (to Tosoh Corp.).
67. A. G. Talybov et al., *Izv. Akad. Nauk SSSR, Ser. Khim.*, 654 (1982).
68. Jpn. Pat. 56 057,720 (May 20, 1981), (to Asahi-Dow Ltd.).
69. M. Umemoto and co-workers, *J. Chem. Phys.* **83**, 1657 (1985).
70. A. Fahr and A. H. Laufer, *J. Phys. Chem.* **89**, 2906 (1985).
71. G. He and co-workers, *J. Chem. Phys.* **103**, 5488 (1995).
72. Y. Huang and co-workers, *J. Chem. Phys.* **99**, 2752 (1993).
73. P. T. A. Reilly, Y. Xie, and R. J. Gordon, *Chem. Phys. Lett.* **178**, 511 (1991).
74. J. A. Manion and R. Louw, *Recl. Trav. Chim. Pays-Bas* **105**, 442 (1986).
75. *1995 World Vinyls Analysis*, Chemical Market Associates, Inc., Houston, Tex., Jan. 1996.
76. *Hydrocarbon Process.* **74**(3), 148 (1995).
77. S. Wachi and H. Morikawa, *J. Chem. Eng. Jpn.* **19**, 598 (1986).
78. R. W. McPherson, C. M. Starks, and G. J. Fryar, *Hydrocarbon Process.* **58**(3), 75 (1979).
79. Ger. Pat. 4,318,609 (July 28, 1994), J. Eichler and co-workers, (to Hoechst AG).
80. Ger. Pat. 3,245,366 (June 14, 1984), J. Hundeck, H. Scholz, and H. Hennen (to Hoechst AG).
81. Eur. Pat. 260,650 (Mar. 23, 1988), J. A. Cowfer (to the B. F. Goodrich Co.).
82. World Pat. 82/02,197 (July 8, 1982), T. Akiyama and co-workers (to Ryo-Nichi Co., Ltd.).
83. K. Weissermel and H.-J. Arpe, *Industrial Organic Chemistry*, 2nd ed., VCH Publishers, Inc., New York, 1993, pp. 215–218.
84. E. Lundberg, *Kem. Tidskr.* **96**, 34 (1984).
85. Ger. Pat. 3,604,968 (Aug. 21, 1986), S. Wachi, Y. Ariki, and H. Oshima (to Kanegafuchi Chemical Industry Co., Ltd.).
86. M. G. Avet'yan and co-workers, *Khim. Prom-st.* (*Moscow*), 323 (1991).
87. Eur. Pat. 471,987 (Feb. 26, 1992), G. Rechmeier (to Hoechst AG).
88. Ger. Pat. 4,133,810 (Apr. 15, 1993), H. Perkow, M. Winhold, and F. Seidelbach (to Hoechst AG).
89. Jpn. Pats. 56 158,148 (Dec. 5, 1981) and 57 2,224 (Jan. 7, 1982), T. Toshiyuki and M. Masataka (to Tokuyama Soda K.K.).

90. U.S. Pat. 5,166,120 (Nov. 24, 1992), K. Deller and co-workers, (to Degussa AG and Wacker-Chemie GmbH).

91. U.S. Pat. 4,446,249 (May 1, 1984), J. S. Eden (to the B. F. Goodrich Co.); U.S. Pat. 4,849,393 (July 18, 1989), J. S. Eden and J. A. Cowfer (to the B. F. Goodrich Co.); U.S. Pat. 5,292,703 (Mar. 8, 1994), G. H. Young, J. A. Cowfer, and V. J. Johnston (to the Geon Co.); U.S. Pat. 5,600,043 (Feb. 4, 1997); J. A. Cowfer and V. J. Johnston (to the Geon Co.).

92. Yu. M. Bakshi and co-workers, *Kinet. Katal.* **32**, 740 (1991).

93. A. J. Ruoco, *Applied Catal. A: Gen.* **117**, 139 (1994).

94. Jpn. Pat. 54 55,502 (May 2, 1979), Y. Yoshito and co-workers (to Mitsui Toatsu Chemicals, Inc.).

95. U.S. Pat. 4,754,088 (June 28, 1988), L. Schmidhammer and co-workers (to Wacker-Chemie GmbH).

96. Jpn. Pat. 63 201,136 (Aug. 19, 1988), H. Tejima and T. Kawaguchi (to Tosoh Corp.).

97. R. G. Markeloff, *Hydrocarbon Process.* **63**(11), 91 (1984).

98. Brit. Pat. 1,090,499 (Nov. 8, 1967), R. T. Carroll (to the B. F. Goodrich Co.).

99. Ger. Pat. 2,353,437 (May 15, 1975), P. R. Laurer, G. De Beuchelaer, and J. Langens (to BASF AG).

100. Ger. Pat. 2,438,153 (Feb. 19, 1976), G. Vollheim and co-workers (to Degussa AG).

101. U.S. Pat. 4,668,833 (May 26, 1987), H. Ohshima and O. Kakimoto (to Kanegafuchi Chemical Industry Co., Ltd.).

102. U.S. Pat. 4,839,153 (June 13, 1989), L. Schmidhammer and co-workers, (Wacker-Chemie GmbH).

103. U.S. Pat. 3,996,300 (Dec. 7, 1976), R. C. Ahlstrom, Jr. (to The Dow Chemical Co.).

104. Eur. Pat. 452,909 (Oct. 23, 1991). L. Schmidhammer and co-workers, (to Wacker-Chemie GmbH).

105. Ger. Pat. 3,441,045 (May 15, 1986), G. Dummer and co-workers, (to Wacker-Chemie GmbH).

106. Brit. Pat. 2,054,574 (Feb. 18, 1981), J. Riedl, W. Froehlich, and E. M. Maier (to Hoechst AG).

107. Ger. Pat. 1,917,933 (Dec. 3, 1970), A. Jacobowsky and co-workers, (to Knapsack AG).

108. *Chemical Economics Handbook, Marketing Research Report, Vinyl Chloride Monomer (VCM)*, SRI International, Menlo Park, Calif., Oct. 1990.

109. *1988 World Vinyls Analysis*, Chemical Market Associates, Inc., Houston, Tex., Oct. 1988.

110. *1996/97 World Vinyls Analysis*, Chemical Market Associates, Inc., Houston, Tex., Jan. 1997.

111. B. Hileman, J. R. Long, and E. M. Kirschner, *C&E News* **72**(47), 12 (1994).

112. B. Hileman, *C&E News* **71**(16), 11 (1993).

113. *U.S. Environmental Protection Agency Report Nos. EPA/600/6-88/005Ca, b, and c; EPA/600/BP-92/001a, b, and c*, U.S. Environmental Protection Agency, Washington, D.C., June 1994.

114. *The Vinyl Industry's Dioxin Testing Program: Background and Status Report*, The Vinyl Institute, Morristown, N.J., Dec. 1995; *1995 Annual Report*, The Geon Co., Avon Lake, Ohio, 1996.

115. Technical data, The Geon Co., Avon Lake, Ohio, 1997.

116. H. G. Rigo, A. J. Chandler, and W. S. Lanier, *The Relationship Between Chlorine in Waste Streams and Dioxin Emissions From Waste Combustor Stacks*, The American Society of Mechanical Engineers, New York, 1995.

117. N. R. Kamsvåg and J. Baldwin, *PVC and the Environment*, Norsk Hydro, as, Oslo, Norway, Sept. 1992.

118. *U.S. Environmental Protection Agency Report No. EPA-450/2-75-009*, Environmental Protection Agency, Research Triangle Park, N.C., Oct. 1975.

119. *EPA Regulations*, 40 CFR 61, Subpart F—National Emission Standard for Vinyl Chloride, Aug. 27, 1993.
120. K. Shirai, *Kagaku Kogaku* **59**, 268 (1995).
121. E. Ger. Pat. 298,236 (Feb. 13, 1992), W. Richert and co-workers, (to Buna AG).
122. Eur. Pat. 270,007 (June 8, 1988), Y. Teshima and S. Onishi (to Tosoh Corp.).
123. Jpn. Pats. 01 052,731 (Feb. 28, 1989) and 63 156,734 (June 29, 1988), Y. Murata and co-workers, (to Mitsubishi Kasei Corp.).
124. U.S. Pat. 4,590,318 (May 20, 1986), D. A. Longhini (to PPG Industries, Inc.).
125. Jpn. Pat. 58 110,528 (July 1, 1983), O. Hiroshi and T. Yuzuru (to Kanegafuchi Chemical Industry Co., Ltd.).
126. Brz. Pat. 81 03,277 (Feb. 16, 1982), Y. Uemura and co-workers, (to Mitsui Toatsu Chemicals, Inc.).
127. P. Ma and co-workers, *J. Chem. Soc., Faraday Trans.* **89**, 4171 (1993); J. Liu, P. Ma, and G. Chen, *Chin. Chem. Lett.* **1**, 29 (1990); P. Ma, J. Liu, and G. Chen, *Spectrochim. Acta, Part A* **46A**, 577 (1990).
128. P. E. Dyer and co-workers, *J. Chem. Soc., Faraday Trans.* **87**, 2151 (1991).
129. M. Schneider and J. Wolfrum, *Ber. Bunsen-Ges. Phys. Chem.* **90**, 1058 (1986); *Proc. SPIE-Int. Soc. Opt. Eng.* **669**, 110 (1986); J. Wolfrum, *Umschau* **84**, 480 (1984); J. Wolfrum and M. Schneider, *Proc. SPIE-Int. Soc. Opt. Eng.* **458**, 46 (1984); K. Kleinermanns and J. Wolfrum, *Laser Chem.* **2**, 339 (1983); M. Schneider, *Report No. MPIS-21/1981*, Max-Planck-Inst. Strömungsforsch., Göttingen, Germany, 1981; Ger. Pat. 2,938,353 (Apr. 2, 1981), J. Wolfrum, M. Kneba, and P. N. Clough (to Max-Planck-Gess. zur För. der Wiss.); Eur. Pat. 27,554 (Apr. 29, 1981), J. Wolfrum and co-workers, (to Max-Planck-Gess. zur För. der Wiss.).
130. J. B. Clark, J. C. Stevens, and D. J. Perettie, *Proc. SPIE-Int. Soc. Opt. Eng.* **458**, 82 (1984).
131. E. W. Wong and co-workers, *Hydrocarbon Process.* **71**(8), 129 (1992).
132. S. Polet, *Chem. Tech. (Amsterdam)* **23**, 617 (1968).
133. W. C. Schreiner and co-workers, *Hydrocarbon Process.* **53**(11), 151 (1974).
134. T. Mitani, *Kagaku Kogaku* **54**, 422 (1990).
135. *Chem. Eng. Progress* **89**(4), 16 (1993).
136. Brit. Pat. 2,036,718 (July 2, 1980), O. A. Zaidman and co-workers (to U.S.S.R.).
137. Yu. A. Treger and co-workers, *Khim. Prom-st. (Moscow)*, 67 (1988).
138. G. K. Shestakov, L. A. Zakharova, and O. N. Temkin, *Kinet. Katal.* **29**, 371 (1988).
139. H. D. Riegel, H. Schindler, and M. C. Sze, *AIChE Symp. Ser.* **69**(135), 96 (1973).
140. Yu. A. Treger and co-workers, *Khim. Prom-st. (Moscow)*, 12 (1985).
141. Yu. A. Treger and co-workers, *Khim. Prom-st. (Moscow)*, 3 (1988).
142. S. C. Che and co-workers, *Report No. GRI-87/0004* Kinetic Technology International Corp., Monrovia, Calif., 1987.
143. World Pat. 92/10,447 (June 25, 1992), K. Viswanathan, H. C. B. Chen, and S. W. Benson (to Occidental Chemical Corp.); World Pat. 92/12,946 (Aug. 6, 1992), S. W. Benson and M. A. Weissman (to Univ. of South. Calif.).
144. World Pats. 95/07,249, 95/07,251, and 95/07,252 (Mar. 16, 1995), I. M. Clegg and R. Hardman (to EVC Technology AG).
145. J. P. Allison, The Geon Co., personal communication, Apr. 1, 1996; generally accepted, industry-wide specifications.
146. *OSHA Regualtions*, 29 CFR 1910.1017, Mar. 11, 1983.
147. *Prudent Practices for Handling Hazardous Chemicals in Laboratories*, National Academy Press, Washington, D.C., 1981, pp. 150–152.
148. N. I. Sax, *Dangerous Properties of Industrial Chemicals*, CD-ROM ed., Van Nostrand Reinhold Co., New York, 1994.
149. *ICRMS (International Chemical Regulatory Monitoring System)*, Ariel Research Corp., Bethesda, Md., 1996.

150. *Vinyl Chloride*, Material safety data sheet, The Geon Co., June 1995.
151. P. C. Conlon, *Emergency Action Guides*, Association of American Railroads, 1984.

JOSEPH A. COWFER
MAXIMILIAN B. GORENSEK
The Geon Company

VINYL ETHER. See ANESTHETICS; VINYL POLYMERS.

VINYL FIBERS. See FIBERS, POLY(VINYL ALCOHOL).

VINYLIDENE CHLORIDE
MONOMER AND POLYMERS

Vinylidene chloride copolymers were among the first synthetic polymers to be commercialized. Their most valuable property is low permeability to a wide range of gases and vapors. From the beginning in 1939, the word Saran has been used for polymers with high vinylidene chloride content, and it is still a trademark of The Dow Chemical Company in some countries. Sometimes Saran and poly(vinylidene chloride) are used interchangeably in the literature. This can lead to confusion because, although Saran includes the homopolymer, only copolymers have commercial importance. The homopolymer, ie, poly(vinylidene chloride), is not commonly used because it is difficult to fabricate.

The principal solution to fabrication difficulties is copolymerization. Three types of comonomers are commercially important: vinyl chloride; acrylates, including alkyl acrylates and alkylmethacrylates; and acrylonitrile. When extrusion is the method of fabrication, other solutions include formulation with plasticizers, stabilizers, and extrusion aids plus applying improved extrusion techniques. The literature on vinylidene chloride copolymers through 1972 has been reviewed (1).

Monomer

Properties. Pure vinylidene chloride [75-35-4] (1,1-dichloroethylene) is a colorless, mobile liquid with a characteristic sweet odor. Its properties are summarized in Table 1. Vinylidene chloride is soluble in most polar and nonpolar organic solvents. Its solubility in water (0.25 wt %) is nearly independent of temperature at 16–90°C (4).

Table 1. Properties of Vinylidene Chloride Monomer [a,b]

Property	Value
molecular weight	96.944
odor	pleasant, sweet
appearance	clear, liquid
color (APHA)	0–10
solubility of monomer in water at 25°C, wt %	0.25
solubility of water in monomer at 25°C, wt %	0.035
normal boiling point, °C	31.56
freezing point, °C	−122.56
flash point, °C	
Tag closed cup	−28
Tag open cup	−16
flammable limits in air (ambient conditions), vol %	6.5–15.5
autoignition temperature, °C	513[b]
latent heat of vaporization, ΔH_v°, kJ/mol[c]	
at 25°C	26.48 ± 0.08
at normal boiling point	26.14 ± 0.08
latent heat of fusion at freezing point, ΔH_m, J/mol[c]	6514 ± 8
heat of polymerization at 25°C, ΔH_p, kJ/mol[c]	−75.3 ± 3.8
heat of combustion, liquid monomer at 25°C, ΔH_c, kJ/mol[c]	1095.9
heat of formation	
liquid monomer at 25°C, ΔH_f, kJ/mol[c]	−25.1 ± 1.3
gaseous monomer at 25°C, ΔH_f, kJ/mol[c]	1.26 ± 1.26
heat capacity	
liquid monomer at 25°C, C_p, J/(mol·K)[c]	111.27
gaseous monomer at 25°C, C_p, J/(mol·K)[c]	67.03
critical temperature, T_c, °C	220.8
critical pressure, P_c, MPa[d]	5.21
critical volume, V_c, cm³/mol	218
liquid density, g/cm³	
−20°C	1.2852
0°C	1.2499
20°C	1.2137
index of refraction, n_D	
10°C	1.43062
15°C	1.42777
20°C	1.42468
absolute viscosity, mPa·s(=cP)	
−20°C	0.4478
0°C	0.3939
20°C	0.3302
vapor pressure[e], T, °C	$\log P_{kPa} = 6.1070$ $- 1104.29/(T + 237.697)$

[a] Refs. 2 and 3.
[b] Inhibited with methyl ether of hydroquinone.
[c] To convert J to cal, divide by 4.184.
[d] To convert MPa to atm, divide by 0.101.
[e] P measured from 6.7 − 104.7 kPa. To convert kPa to mm Hg, multiply by 7.5 (add 0.875 to the constant to convert $\log_{kPa}$ to $\log_{mm\,Hg}$).

Manufacture. Vinylidene chloride monomer can be conveniently prepared in the laboratory by the reaction of 1,1,2-trichloroethane [79-00-5] with aqueous alkali:

$$2\ CH_2ClCHCl_2 + Ca(OH)_2 \longrightarrow 2\ CH_2 = CCl_2 + CaCl_2 + 2\ H_2O$$

Other methods are based on bromochloroethane [25620-54-6], trichloroethyl acetate [625-24-1], tetrachloroethane [79-34-5], and catalytic cracking of trichloroethane (5). Catalytic processes produce as by-product HCl, rather than less valuable salts, but yields of vinylidene chloride have been too low for commercial use of these processes. However, good results have been reported with metal-salt catalysts (6–8).

Vinylidene chloride (VDC) is prepared commercially by the dehydrochlorination of 1,1,2-trichloroethane with lime or caustic in slight excess (2–10%) (3,9). A continuous liquid-phase reaction at 98–99°C yields ~90% VDC. Caustic gives better results than lime. Vinylidene chloride is purified by washing with water, drying, and fractional distillation. It forms an azeotrope with 6 wt % methanol (10). Purification can be achieved by distillation of the azeotrope, followed by extraction of the methanol with water; an inhibitor is usually added at this point. Commercial grades contain 200 ppm of the monomethyl ether of hydroquinone (MEHQ). Many other inhibitors for the polymerization of vinylidene chloride have been described in patents, but MEHQ is the one most often used. The inhibitor can be removed by distillation or by washing with 25 wt % aqueous caustic under an inert atmosphere at low temperatures.

For many polymerizations, MEHQ need not be removed; instead, polymerization initiators are added. Vinylidene chloride from which the inhibitor has been removed should be refrigerated in the dark at $-10°C$, under a nitrogen atmosphere, and in a nickel-lined or baked phenolic-lined storage tank. If not used within one day, it should be reinhibited.

Health and Safety Factors. Vinylidene chloride is highly volatile and, when free of decomposition products, has a mild, sweet odor. Its warning properties are ordinarily inadequate to prevent excessive exposure. Inhalation of vapor presents a hazard, which is readily controlled by observance of precautions commonly taken in the chemical industry (2). A single, brief exposure to a high concentration of vinylidene chloride vapor, eg, 2000 ppm, rapidly causes intoxication, which may progress to unconsciousness on prolonged exposure. The $LC_{50}/4$ h in rats is 6350 ppm. However, prompt and complete recovery from the anesthetic effects occurs when the exposure is for short duration. A single, prolonged exposure and repeated short-term exposures can be dangerous, even when the concentrations of the vapor are too low to cause an anesthetic effect. They may produce organic injury to the kidneys and liver. For repeated exposures, the vapor concentration of vinylidene chloride should be much lower. The American Conference of Governmental Industrial Hygienists (ACGIH) threshold limit value (TLV) of 5 ppm and short-term exposure level (STEL) of 20 ppm have been established to provide an adequate margin of safety. Investigations of vinylidene chloride exposure have been reviewed (11,12).

Vinylidene chloride is hepatotoxic, but does not appear to be a carcinogen (13–18). Pharmacokinetic studies indicate that the behavior of vinyl chloride and vinylidene chloride in rats and mice is substantially different (19). No unusual health problems have been observed in workers exposed to vinylidene chloride monomer over varying periods (20). Because vinylidene chloride degrades rapidly in the atmosphere, air pollution is not likely to be a problem (21). Worker exposure is the main concern. Sampling techniques for monitoring worker exposure to vinylidene chloride vapor are being developed (22).

The liquid is irritating to the skin after only a few minutes of contact. The inhibitor MEHQ may be partly responsible for this irritation. Inhibited vinylidene chloride is moderately irritating to the eyes. Contact causes pain and conjunctival irritation, and possibly some transient corneal injury and iritis. Permanent damage, however, is not likely.

Peroxide Formation. In the presence of air or oxygen, uninhibited vinylidene chloride forms a violently explosive complex peroxide at temperatures as low as 40°C. Decomposition products of vinylidene chloride products are formaldehyde, phosgene, and hydrochloric acid. A sharp, acrid odor indicates oxygen exposure and probable presence of peroxides. This is confirmed by the liberation of iodine from a slightly acidified dilute potassium iodide solution. Formation of insoluble polymer may also indicate peroxide formation. The peroxide adsorbs on the precipitated polymer, and separation of the polymer may result in an explosive composition. Any dry composition containing more than ~15 wt % peroxide detonates from a slight mechanical shock or from heat. Vinylidene chloride containing peroxides may be purified by being washed several times, either with 10 wt % sodium hydroxide at 25°C or with a fresh 5 wt % sodium bisulfite solution. Residues in vinylidene chloride containers should be handled with great care, and the peroxides should be destroyed with water at room temperature.

Copper, aluminum, and their alloys should not be used in handling vinylidene chloride. Copper can react with acetylenic impurities to form copper acetylides, whereas aluminum can react with the vinylidene chloride to form aluminum chloralkyls. Both compounds are extremely reactive and potentially hazardous.

Polymerization

Vinylidene chloride polymerizes by both ionic and free-radical reactions. Processes based on the latter are far more common (23). Vinylidene chloride is of average reactivity when compared with other unsaturated monomers. The chlorine substituents stabilize radicals in the intermediate state of an addition reaction. Because they are also strongly electron-withdrawing, they polarize the double bond, making it susceptible to anionic attack. For the same reason, a carbonium ion intermediate is not favored.

The 1,1-disubstitution of chlorine atoms causes steric interactions in the polymer, as is evident from the heat of polymerization (see Table 1) (24). When corrected for the heat of fusion, it is significantly less than the theoretical value of -83.7 kJ/mol (-20 kcal/mol) for the process of converting a double bond to two single bonds. The steric strain apparently is not important in the addition step, because VDC polymerizes easily. Nor is it sufficient to favor depolymerization; the estimated ceiling temperature for poly(vinylidene chloride) (PVDC) is about 400°C.

Homopolymerization.   The free-radical polymerization of VDC has been carried out by solution, slurry, suspension, and emulsion methods.

Solution polymerization in a medium that dissolves both monomer and polymer has been investigated (25). The kinetic measurements lead to activation energies and frequency factors in the normal range for free-radical polymerizations of olefinic monomers. The kinetic behavior of VDC is abnormal when the polymerization is heterogeneous (26). Slurry polymerizations are usually used only in the laboratory. They can be carried out in bulk or in common solvents, eg, benzene. Poly(vinylidene chloride) is insoluble in these media and separates from the liquid phase as a crystalline powder. The heterogeneity of the reaction makes stirring and heat transfer difficult; consequently, these reactions cannot be easily controlled on a large scale. Aqueous emulsion or suspension reactions are preferred for large-scale operations. Slurry reactions are usually initiated by the thermal decomposition of organic peroxides or azo compounds. Purely thermal initiation can occur, but rates are very slow (27).

Bulk Polymerization.   The spontaneous polymerization of VDC, so often observed when the monomer is stored at room temperature, is caused by peroxides formed from the reaction of VDC with oxygen. Very pure monomer does not polymerize under these conditions. Irradiation by either uv or γ-rays (26,28) also induces polymerization of VDC.

The heterogeneous nature of the bulk polymerization of VDC is apparent from the rapid development of turbidity in the reaction medium following initiation. The turbidity results from the presence of minute PVDC crystals. As the reaction progresses, the crystalline phase grows and the liquid phase diminishes. Eventually, a point is reached where the liquid slurry solidifies into a solid mass. A typical conversion-time curve is shown in Figure 1 for a mass polymerization catalyzed by benzoyl peroxide. The first stage of the reaction is characterized by rapidly increasing rate, which levels off in the second stage to a fairly constant value. This is often called the steady-state region. Throughout the first two stages, monomer concentration remains constant because the polymer separates into another phase. In the third stage, there is a gradual decrease in rate to zero as the monomer supply is depleted. Because the mass solidifies while monomer is still present (usually at conversions below 20%), further polymerization generates void space. The final solid, therefore, is opaque and quite porous. A similar pattern of behavior is observed when vinylidene chloride is polymerized in solvents, eg, benzene, that do not dissolve or swell the polymer. In this case, however, the reaction mixture may not solidify if the monomer concentration is low.

Heterogeneous polymerization is characteristic of a number of monomers, including vinyl chloride and acrylonitrile. A completely satisfactory mechanism for these reactions has not been determined. This is true for VDC also. Earlier studies have not been broad enough to elucidate the mechanism (26,30,31). Morphologies of as-polymerized poly(vinyl chloride) (PVC) and polyacrylonitrile (PAN) are similar, suggesting a similar mechanism.

The morphology of as-polymerized PVDC is quite different (31). Nearly spherical aggregates form in the PVC and PAN systems, whereas anisotropic growth takes place in the PVDC case. The differences in morphologies may be a consequence of the relative rates of polymerization and crystallization. PVDC is

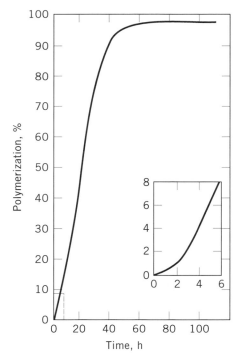

Fig. 1. Bulk polymerization of vinylidene chloride at 45°C, with 0.5 wt % benzoyl peroxide as initiator (29).

unique because polymerization and crystallization probably occur nearly simultaneously. It has been reported that the average lifetime of a growing radical (τ_s) is between 0.1 and 10 s (32). The half-times ($t_{1/2}$) for crystallization of PVDC copolymers in monomer were measured to be about 1 s at 60°C and about 0.01 s at 90°C (33). This information is important for developing an understanding of a mechanism that includes a contribution from a surface reaction which has the potential for autoacceleration.

Emulsion Polymerization. Emulsion and suspension reactions are doubly heterogeneous; the polymer is insoluble in the monomer and both are insoluble in water. Suspension reactions are similar in behavior to slurry reactors. Oil-soluble initiators are used, so the monomer–polymer droplet is like a small mass reaction. Emulsion polymerizations are more complex. Because the monomer is insoluble in the polymer particle, the simple Smith-Ewart theory does not apply (34).

A kinetic model for the particle growth stage for continuous-addition emulsion polymerization has been proposed (35). Below the monomer saturation point, the steady-state rate of polymerization, R_p, depends on the rate of monomer addition, R_a, according to the following reciprocal relationship:

$$1/R_p = 1/K + 1/R_a$$

where K depends on the number of particles and the propagation rate constant. A later study explored the kinetics of emulsion polymerization of nonswelling and swellable latex particles to define the locus of polymerization (36). There are no significant differences between the behavior of swelling and nonswelling emulsion particles and neither follows Smith-Ewart kinetics. The results indicate strongly that polymerization takes place at the particle–water interface or in a surface layer on the polymer particle.

Redox initiator systems are normally used in the emulsion polymerization of VDC to develop high rates at low temperatures. Reactions must be carried out below ~80°C to prevent degradation of the polymer. Poly(vinylidene chloride) in emulsion is also attacked by aqueous base. Therefore, reactions should be carried out at low pH.

Ionic Mechanisms. The instability of PVDC is one of the reasons why ionic initiation of VDC polymerization has not been used extensively. Many of the common catalysts either react with the polymer or catalyze its degradation. For example, butyllithium polymerizes VDC by an anionic mechanism, but the product is a low molecular weight, discolored polymer having a low chlorine content (37). Cationic polymerization of VDC seems unlikely in view of its structure (38). Some available data, however, suggest the possibility. In the low temperature, radiation-induced copolymerization of VDC with isobutylene, reactivity ratios vary markedly with temperature, indicating a change from a free-radical mechanism (39). Coordination complex catalysts may also polymerize VDC by a nonradical mechanism. Again, this speculation is based on copolymerization studies. Poly(vinylidene chloride) telomers can be prepared by using chlorine as the initiator and chain-transfer agent (40,41). Plasma polymerization of VDC in a radio-frequency glow discharge yields cross-linked polymer, which is partially degraded (42).

Copolymerization. The importance of VDC as a monomer results from its ability to copolymerize with other vinyl monomers. Its Q value equals 0.22 and its e value equals 0.36. It most easily copolymerizes with acrylates, but it also reacts, more slowly, with other monomers, eg, styrene, that form highly resonance-stabilized radicals. Reactivity ratios (r_1 and r_2) with various monomers are listed in Table 2. Many other copolymers have been prepared from monomers for which the reactivity ratios are not known. The commercially important copolymers include those with vinyl chloride (VC), acrylonitrile (AN), or various alkyl acrylates, but many commercial polymers contain three or more components, of which VDC is the principal one. Usually one component is introduced to improve the processability or solubility of the polymer; the others are added to modify specific

Table 2. Reactivity of Vinylidene Chloride (r_1) with Important Monomers (r_2)[a]

Monomer	r_1	r_2
styrene	0.14	2.0
vinyl chloride	3.2	0.3
acrylonitrile	0.37	0.91
methyl acrylate	1.0	1.0
methyl methacrylate	0.24	2.53
vinyl acetate	6	0.1

[a]Ref. 43.

use properties. Most of these compositions have been described in the patent literature, and a list of various combinations has been compiled (44). A typical terpolymer might contain 90 wt % VDC, with the remainder made up of acrylonitrile and an acrylate or methacrylate monomer.

Bulk copolymerizations yielding high VDC-content copolymers are normally heterogeneous. Two of the most important pairs, VDC–VC and VDC–AN, are heterogeneous over most of the composition range. In both cases and at either composition extreme, the product separates initially in a powdery form; however, for intermediate compositions, the reaction mixture may only gel. Copolymers in this composition range are swollen but not completely dissolved by the monomer mixture at normal polymerization temperatures. Copolymers containing more than 15 mol % acrylate are normally soluble in the monomers. These reactions are therefore homogeneous and, if carried to completion, yield clear, solid castings of the copolymer. Most copolymerizations can be carried out in solution because of the greater solubility of the copolymers in common solvents.

During copolymerization, one monomer may add to the copolymer more rapidly than the other. Except for the unusual case of equal reactivity ratios, batch reactions carried to completion yield polymers of broad composition distribution. More often than not, this is an undesirable result.

Vinylidene chloride copolymerizes randomly with methyl acrylate and nearly so with other acrylates. Very severe composition drift occurs, however, in copolymerizations with vinyl chloride or methacrylates. Several methods have been developed to produce homogeneous copolymers regardless of the reactivity ratio (43). These methods are applicable mainly to emulsion and suspension processes where adequate stirring can be maintained. Copolymerization rates of VDC with small amounts of a second monomer are normally lower than its rate of homopolymerization. The kinetics of the copolymerization of VDC and VC have been studied (45–48).

Studies of the copolymerization of VDC with methyl acrylate (MA) over a composition range of 0–16 wt % showed that near the intermediate composition (8 wt %), the polymerization rates nearly followed normal solution polymerization kinetics (49). However, at the two extremes (0 and 16 wt % MA), copolymerization showed significant autoacceleration. The observations are important because they show the significant complexities in these copolymerizations. The autoacceleration for the homopolymerization, ie, 0 wt % MA, is probably the result of a surface polymerization phenomenon. On the other hand, the autoacceleration for the 16 wt % MA copolymerization could be the result of Trommsdorff and Norrish-Smith effects.

Copolymers of VDC can also be prepared by methods other than conventional free-radical polymerization. Copolymers have been formed by irradiation and with various organometallic and coordination complex catalysts (28,44,50–53). Graft copolymers have also been described (54–58).

Polymer Structure and Properties

Chain Structure. The chemical composition of poly(vinylidene chloride) has been confirmed by various techniques, including elemental analysis, x-ray diffraction analysis, degradation studies, and ir, Raman, and nmr spectroscopy.

The polymer chain is made up of vinylidene chloride units added head-to-tail:

$$— CH_2CCl_2— CH_2CCl_2— CH_2CCl_2—$$

Because the repeat unit is symmetrical, no possibility exists for stereoisomerism. Variations in structure can occur only by head-to-head addition, branching, or degradation reactions that do not cause chain scission, including such reactions as thermal dehydrochlorination, which creates double bonds in the structure to give, for example, $— CH_2CCl_2CH = CClCH_2CCl_2—$ and a variety of ill-defined oxidation and hydrolysis reactions that generate carbonyl groups.

The infrared spectra of PVDC often show traces of unsaturation and carbonyl groups. The slightly yellow tinge of many of these polymers comes from the same source; the pure polymer is colorless. Elemental analyses for chlorine are normally slightly lower than the theoretical value of 73.2%.

The high crystallinity of PVDC indicates that no significant amounts of head-to-head addition or branching can be present. This has been confirmed by nmr spectra (59). Studies of well-characterized oligomers having degrees of polymerization (DP) of 2–10 offer further nmr evidence (40), ie, a single peak from the methylene hydrogens. Either branching or another mode of addition would produce nonequivalent hydrogens and a more complicated spectrum. However, nmr cannot detect small amounts of such structures. The ir and Raman spectra can also be interpreted in terms of the simple head-to-tail structure (60,61).

Molecular weights of PVDC can be determined directly by dilute solution measurements in good solvents (62). Viscosity studies indicate that polymers having degrees of polymerization from 100 to more than 10,000 are easily obtained. Dimers and polymers having DP < 100 can be prepared by special procedures (40). Copolymers can be more easily studied because of their solubility in common solvents. Gel-permeation chromatography studies indicate that molecular weight distributions are typical of vinyl copolymers.

Crystal Structure.   The crystal structure of PVDC is fairly well established. Several unit cells have been proposed (63). The unit cell contains four monomer units with two monomer units per repeat distance. The calculated density, 1.96 g/cm³, is higher than the experimental values, which are 1.80–1.94 g/cm³ at 25°C, depending on the sample. This is usually the case with crystalline polymers because samples of 100% crystallinity usually cannot be obtained. A direct calculation of the polymer density from volume changes during polymerization yields a value of 1.97 g/cm³ (64). If this value is correct, the unit cell densities may be low.

The repeat distance along the chain axis (0.468 nm) is significantly less than that calculated for a planar zigzag structure. Therefore, the polymer must be in some other conformation (65–67). Based on ir and Raman studies of PVDC single crystals and normal vibration analysis, the best conformation appears to be $qfqf'$, where q, the skeletal angle, is 120°, and the torsional angle f (f' of opposite sign) is 32.5°. This conformation is in agreement with theoretical predictions (68).

The melting temperature, T_m, of PVDC is independent of molecular weight above DP = 100. However, as shown in Figure 2, it drops sharply at lower molecular weights. Below the hexamer, the products are noncrystalline liquids.

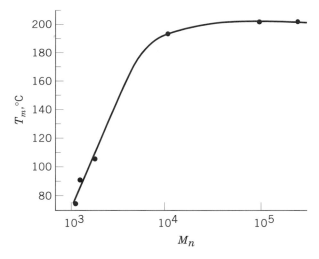

Fig. 2.   Crystalline melting temperatures of poly(vinylidene chloride) (40).

The properties of PVDC (Table 3) are usually modified by copolymerization. Copolymers of high VDC content have lower melting temperatures than PVDC. Copolymers containing more than ~15 mol % acrylate or methacrylate are amorphous. Substantially more acrylonitrile (25%) or vinyl chloride (45%) is required to destroy crystallinity completely.

The effect of different types of comonomers on T_m varies. VDC–MA copolymers more closely obey Flory's melting-point depression theory than do copolymers with VC or AN. Studies have shown that, for the copolymers of VDC with MA, Flory's theory needs modification to include both lamella thickness and surface free energy (69). The VDC–VC and VDC–AN copolymers typically have severe composition drift, therefore most of the comonomer units do not belong to crystallizing chains. Hence, they neither enter the crystal as defects nor cause lamellar thickness to decrease, so the depression of the melting temperature is less than expected.

The glass-transition temperatures, T_g, of vinylidene chloride copolymers have been studied extensively (70,71). The effect of various comonomers on the glass-transition temperature is shown in Figure 3. In every case, T_g increases

Table 3. Properties of Poly(vinylidene chloride)

Property	Best value	Reported values
T_m, °C	202	198–205
T_g, °C	−17	−19 to −11
transition between T_m and T_g, °C	80	
density at 25°C, g/cm³		
amorphous	1.775	1.67–1.775
unit cell	1.96	1.949–1.96
crystalline		1.80–1.97
refractive index (crystalline), n_D	1.63	
heat of fusion (ΔH_m), J/mol[a]	6275	4600–7950

[a]To convert J to cal, divide by 4.184.

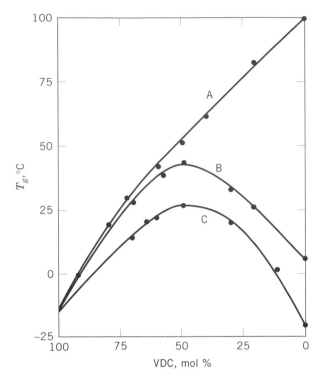

Fig. 3. Effect of comonomer structure on the glass-transition temperature of VDC copolymers (72), where A represents acrylonitrile; B, methyl acrylate; and C, ethyl acrylate.

with the comonomer content at low comonomer levels, even in cases where the T_g of the other homopolymer is lower. The phenomenon has been observed in several other copolymer systems as well (73). In these cases, a maximum T_g is observed at intermediate compositions. In others, where the T_g of the other homopolymer is much higher than the T_g of PVDC, the glass-transition temperatures of the copolymers increase over the entire composition range. The glass-transition temperature increases most rapidly at low acrylonitrile levels but changes the slowest at low vinyl chloride levels. This suggests that polar interactions affect the former, but the increase in T_g in the VDC–VC copolymers may simply result from loss of chain symmetry. Because of these effects, the temperature range in which copolymers can crystallize is drastically narrowed. Crystallization induction times are prolonged, and subsequent crystallization takes place at a low rate over a long period of time. Plasticization, which lowers T_g, decreases crystallization induction times significantly. Copolymers having lower glass-transition temperatures also tend to crystallize more rapidly (74).

Crystallization curves have been determined for 10 mol % acrylate copolymers of varying side-chain length. Among the acrylate copolymers, the butyl acrylate copolymer has a T_g of 8°C; the octyl acrylate, −3°C; and the octadecyl acrylate, −16°C. The rates of crystallization of these copolymers are inversely related to the glass-transition temperatures. Apparently, the long alkyl side chains act as internal plasticizers, lowering the melt viscosity of the copolymer even though the acrylate group acts to stiffen the backbone.

The maximum rates of crystallization of the more common crystalline copolymers occur at 80–120°C. In many cases, these copolymers have broad composition distributions containing both fractions of high VDC content that crystallize rapidly and other fractions that do not crystallize at all. Poly(vinylidene chloride) probably crystallizes at a maximum rate at 140–150°C, but the process is difficult to follow because of severe polymer degradation. The copolymers may remain amorphous for a considerable period of time if quenched to room temperature. The induction time before the onset of crystallization depends on both the type and amount of comonomer; PVDC crystallizes within minutes at 25°C.

Recrystallization of a copolymer having 15 wt % VC has been found to be nucleated by material that survives the melting process plus new nuclei (74). The maximum crystallization rate occurred at 373 K; the maximum nucleation rate at 283 K. Attempts to melt all the polymer led to degradation that interfered with recrystallization.

Orientation or mechanical working accelerates crystallization and has a pronounced effect on morphology. Crystals of uniaxially oriented filaments are oriented along the fiber axis (63). The long period (lamellar thickness), as determined by small-angle x-ray scattering, is 7.6 nm and decreases with comonomer content. The fiber is 43% crystalline and has a melting temperature of 195°C and an average crystal thickness of 4.5 nm. The crystal size is not greatly affected by comonomer content, but both crystallinity and melting temperature decrease with increasing comonomer.

Copolymerization also affects morphology under other crystallization conditions. Copolymers in the form of cast or molded sheets are much more transparent because of the small spherulite size. In extreme cases, crystallinity cannot be detected optically, but its effect on mechanical properties is pronounced. Before crystallization, films are soft and rubbery, with low modulus and high elongation. After crystallization, they are leathery and tough, with higher modulus and lower elongation.

Copolymers of vinylidene chloride and methyl acrylate have been studied by x-ray techniques (75). For example, the long period (lamellar thickness) for an 8.5 wt % methyl acrylate copolymer was found to be 9.2 nm by small-angle x-ray scattering. The unit cell is monoclinic, with $a = 0.686$ and $c = 1.247$ nm by wide-angle x-ray scattering.

Significant amounts of comonomer also reduce the ability of the polymer to form lamellar crystals from solution. In some cases, the polymer merely gels the solution as it precipitates rather than forming distinct crystals. At somewhat higher VDC content, it may precipitate in the form of aggregated, ill-defined particles and clusters.

Morphology and Transitions. The highly crystalline particles of PVDC precipitated during polymerization are aggregates of thin lamellar crystals (76). The substructures are 5–10-nm thick and 100 or more times larger in other dimensions. In some respects, they resemble the lamellar crystals grown from dilute solution (77–79). The single crystals are better characterized than the as-polymerized particles. They are highly branched, with branching angles of 65–70°; the angle appears to be associated with a twin plane in the crystal (80).

Melting temperatures of as-polymerized powders are high, ie, 198–205°C as measured by differential thermal analysis (dta) or hot-stage microscopy (76). Two peaks are usually observed in dta curves: a small lower temperature peak and the main melting peak. The small peak seems to be related to polymer crystallized by precipitation rather than during polymerization.

As-polymerized PVDC does not have a well-defined glass-transition temperature because of its high crystallinity. However, a sample can be melted at 210°C and quenched rapidly to an amorphous state at < -20°C. The amorphous polymer has a glass-transition temperature of -17°C as shown by dilatometry (70). Glass-transition temperature values of -19 to -11°C, depending on both method of measurement and sample preparation, have been determined.

Once melted, PVDC does not regain its as-polymerized morphology when subsequently crystallized. The polymer recrystallizes in a spherulitic habit. Spherulites between crossed polarizing plates show the usual Maltese cross and are positively birefringent. The size and number of spherulites can be controlled. Quenching and low temperature annealing generate many small nuclei that, on heating, grow rapidly into small spherulites. Slow crystallization at higher temperatures produces fewer but much larger spherulites. The melting temperature and degree of crystallinity of recrystallized PVDC also depend on crystallization conditions. The melting temperature increases with crystallization temperature, but the as-polymerized value cannot be achieved. There is no reason to believe that even these values indicate the true melting point of PVDC; it may be as high as 220°C. Slow, high temperature recrystallization and annealing experiments are not feasible because of the thermal instability of the polymer (81). Other transitions in PVDC have been observed by dynamic mechanical methods.

Solubility and Solution Properties.   Poly(vinylidene chloride), like many high melting polymers, does not dissolve in most common solvents at ambient temperatures. Copolymers, particularly those of low crystallinity, are much more soluble. However, one of the outstanding characteristics of vinylidene chloride polymers is resistance to a wide range of solvents and chemical reagents. The insolubility of PVDC results less from its polarity than from its high melting temperature. It dissolves readily in a wide variety of solvents above 130°C (81).

The polarity of the polymer is important only in mixtures having specific polar aprotic solvents. Many solvents of this general class solvate PVDC strongly enough to depress the melting temperature by more than 100°C. Solubility is normally correlated with cohesive energy densities or solubility parameters. For PVDC, a value of 20 ± 0.6 $(J/cm^3)^{1/2}$ $(10 \pm 0.3$ $(cal/cm^3)^{1/2})$ has been estimated from solubility studies in nonpolar solvents. The value calculated from Small's relationship is 20.96 $(J/cm^3)^{1/2}$ $(10.25$ $(cal/cm^3)^{1/2})$. The use of the solubility parameter scheme for polar crystalline polymers such as PVDC has limited value. A typical nonpolar solvent of matching solubility parameter is tetrahydronaphthalene. The lowest temperature at which PVDC dissolves in this solvent is 140°C. Specific solvents, however, dissolve PVDC at much lower temperatures. A list of good solvents is given in Table 4. The relative solvent activity is characterized by the temperature at which a 1 wt % mixture of polymer in solvent becomes homogeneous when heated rapidly.

Poly(vinylidene chloride) also dissolves readily in certain solvent mixtures (82). One component must be a sulfoxide or N,N-dialkylamide. Effective cosol-

Table 4. Solvents for Poly(vinylidene chloride)[a]

Solvents	T^b, °C
Nonpolar	
1,3-dibromopropane	126
bromobenzene	129
1-chloronaphthalene	134
2-methylnaphthalene	134
o-dichlorobenzene	135
Polar aprotic	
hexamethylphosphoramide	−7.2
tetramethylene sulfoxide	28
N-acetylpiperidine	34
N-methylpyrrolidinone	42
N-formylhexamethyleneimine	44
trimethylene sulfide	74
N-n-butylpyrrolidinone	75
diisopropyl sulfoxide	79
N-formylpiperidine	80
N-acetylpyrrolidinone	86
tetrahydrothiophene	87
N,N-dimethylacetamide	87
cyclooctanone	90
cycloheptanone	96
di-n-butyl sulfoxide	98

[a]Ref. 81.
[b]Temperature at which a 1 wt % mixture of polymer in solvent becomes homogeneous.

vents are less polar and have cyclic structures. They include aliphatic and aromatic hydrocarbons, ethers, sulfides, and ketones. Acidic or hydrogen-bonding solvents have an opposite effect, rendering the polar aprotic component less effective. Both hydrocarbons and strong hydrogen-bonding solvents are nonsolvents for PVDC.

As-polymerized PVDC is not in its most stable state; annealing and recrystallization can raise the temperature at which it dissolves (78). Low crystallinity polymers dissolve at a lower temperature, forming metastable solutions. However, on standing at the dissolving temperature, they gel or become turbid, indicating recrystallization into a more stable form.

Copolymers having enough vinylidene chloride content to be quite crystalline behave much like PVDC. They are more soluble, however, because of their lower melting temperatures. The solubility of amorphous copolymers is much higher. The selection of solvents in either case varies somewhat with the type of comonomer. Some of the more common types are listed in Table 5. Solvents that dissolve PVDC also dissolve the copolymers at lower temperatures. The identification of solvents that dissolve PVDC at low temperatures makes possible the study of dilute solution properties. Both light-scattering and intrinsic-viscosity studies have been reported (62). Intrinsic viscosity–molecular weight relationships for the three solvents investigated ($[\eta]$ in dL/g) are $[\eta] = 1.31 \times 10^{-4} M_w^{0.69}$

Table 5. Common Solvents for Vinylidene Chloride Copolymers

Solvents	Copolymer type	Temperature, °C
tetrahydrofuran	all	<60
2-butanone	low crystallinity	<80
1,4-dioxane	all	50–100
cyclohexanone	all	50–100
cyclopentanone	all	50–100
ethyl acetate	low crystallinity	<80
chlorobenzene	all	100–130
dichlorobenzene	all	100–140
dimethylformamide	high acrylonitrile	<100

for N-methylpyrrolidinone (MP); $1.39 \times 10^{-4} M_w^{0.69}$ for tetramethylene sulfoxide (TMSO); and $2.58 \times 10^{-4} M_w^{0.65}$ for hexamethylphosphoramide (HMPA). The relative solvent power (HMPA > TMSO > MP) agrees with solution-temperature measurements. The characteristic ratio C_∞ is about 8 ± 1, which is slightly larger than that of a similar polymer, polyisobutylene.

The dilute solution properties of copolymers are similar to those of the homopolymer. The intrinsic viscosity–molecular weight relationship for a VDC–AN copolymer (9 wt % AN) is $[\eta] = 1.06 \times 10^{-4} M_w^{0.72}$ (83). The characteristic ratio is 8.8 for this copolymer.

An extensive investigation of the dilute solution properties of several acrylate copolymers has been reported (80). The behavior is typical of flexible-backbone vinyl polymers. The length of the acrylate ester side chain has little effect on properties.

Intrinsic viscosity–molecular weight relationships have been obtained for copolymers in methyl ethyl ketone. The value for a 15 wt % ethyl acrylate (EA) copolymer is $[\eta] = 2.88 \times 10^{-4} M_w^{0.6}$.

In early literature, the molecular weights of PVDC and VDC copolymers were characterized by the absolute viscosity of a 2 wt % solution in o-dichlorobenzene at 140°C. The exact correlation between this viscosity value and molecular weight is not known. Gel-permeation chromatography is the preferred method for characterizing molecular weight; studies of copolymers have been reported (84,85).

Mechanical Properties. Because PVDC is difficult to fabricate into suitable test specimens, very few direct measurements of its mechanical properties have been made. In many cases, however, the properties of copolymers have been studied as functions of composition, and the properties of PVDC can be estimated by extrapolation. Some characteristic properties of high VDC content, unplasticized copolymers are listed in Table 6. The performance of a given specimen is sensitive to morphology, including the amount and kind of crystallinity, as well as orientation. Tensile strength increases with crystallinity, whereas toughness and elongation decrease. Orientation, however, improves all three properties. The effect of stretch ratio applied during orientation on properties of VDC–VC monofilaments is shown in Table 7.

The dynamic mechanical properties of VDC–VC copolymers have been studied in detail. The incorporation of VC units in the polymer results in a drop

Table 6. Mechanical Properties of High Vinylidene Chloride Copolymers

Property	Range
tensile strength, MPa[a]	
unoriented	34.5–69.0
oriented	207–414
elongation, %	
unoriented	10–20
oriented	15–40
softening range (heat distortion), °C	100–150
flow temperature, °C	>185
brittle temperature, °C	−10 to 10
impact strength, J/m[b]	26.7–53.4

[a]To convert MPa to psi, multiply by 145.
[b]To convert J/m to ft·lbf/in., divide by 53.38 (see ASTM D256).

Table 7. Effect of Stretch Ratio on Tensile Strength and Elongation of a VDC–VC Copolymer[a,b]

Stretch ratio	Tensile strength, MPa[c]	Elongation, %
2.50:1	235	23.2
2.75:1	234	21.7
3.00:1	303	26.3
3.25.1	268	33.1
3.50:1	316	19.2
3.75:1	330	21.8
4.00:1	320	19.7
4.19:1	314	16.2

[a]Ref. 86.
[b]Average of five determinations, using the Instron test at 5 cm/min.
[c]To convert MPa to psi, multiply by 145.

in dynamic modulus because of the reduction in crystallinity. However, the glass-transition temperature is raised; therefore, the softening effect observed at room temperature is accompanied by increased brittleness at lower temperatures. These copolymers are normally plasticized in order to avoid this. Small amounts of plasticizer (2–10 wt %) depress T_g significantly without loss of strength at room temperature. At higher levels of VC, the T_g of the copolymer is above room temperature and the modulus rises again. A minimum in modulus or maximum in softness is usually observed in copolymers in which T_g is above room temperature. A thermomechanical analysis of VDC–AN (acrylonitrile) and VDC–MMA (methyl methacrylate) copolymer systems shows a minimum in softening point at 79.4 and 68.1 mol % VDC, respectively (86).

In cases where the copolymers have substantially lower glass-transition temperatures, the modulus decreases with increasing comonomer content. This results from a drop in crystallinity and in glass-transition temperature. The loss in modulus in these systems is therefore accompanied by an improvement in low temperature performance. However, at low acrylate levels (<10 wt %), T_g

increases with comonomer content. The brittle points in this range may therefore be higher than that of PVDC.

The long side chains of the acrylate ester group can apparently act as internal plasticizers. Substitution of a carboxyl group on the polymer chain increases brittleness. A more polar substituent, eg, an N-alkyl amide group, is even less desirable. Copolymers of VDC with N-alkylacrylamides are more brittle than the corresponding acrylates even when the side chains are long (87). Side-chain crystallization may be a contributing factor.

Barrier Properties.   Vinylidene chloride polymers are more impermeable to a wider variety of gases and liquids than other polymers. This is a consequence of the combination of high density and high crystallinity in the polymer. An increase in either tends to reduce permeability. A more subtle factor may be the symmetry of the polymer structure. It has been shown that both polyisobutylene and PVDC have unusually low permeabilities to water compared to their monosubstituted counterparts, polypropylene and PVC (88). The values listed in Table 8 include estimates for the completely amorphous polymers. The estimated value for highly crystalline PVDC was obtained by extrapolating data for copolymers.

The effect of copolymer composition on gas permeability is shown in Table 9. The inherent barrier in VDC copolymers can best be exploited by using films containing little or no plasticizers and as much VDC as possible. However, the permeability of even completely amorphous copolymers, for example, 60% VDC–40% AN or 50% VDC–50% VC, is low compared to that of other polymers. The primary reason is that diffusion coefficients of molecules in VDC copolymers are very low. This factor, together with the low solubility of many gases in VDC copolymers and the high crystallinity, results in very low permeability. Permeability is affected by the kind and amounts of comonomer as well as crystallinity. A change from PVDC to 50 wt % VC or 40 wt % AN increases permeability 10-fold, but has little effect on the solubility coefficient.

A more polar comonomer, eg, an AN comonomer, increases the water-vapor transmission more than VC when other factors are constant. For the same reason, AN copolymers are more resistant to penetrants of low cohesive energy density. All VDC copolymers, however, are very impermeable to aliphatic hydrocarbons. Comonomers that lower T_g and increase the free volume in the amorphous phase

Table 8. Comparison of the Permeabilities of Various Polymers to Water Vapor[a]

Polymer	Density, g/mL		Permeability[b]	
	Amorphous	Crystalline	Amorphous	Crystalline
ethylene	0.85	1.00	200–220	10–40
propylene	0.85	0.94	420	
isobutylene	0.915	0.94	90	
vinyl chloride	1.41	1.52	300	90–115
vinylidene chloride	1.77	1.96	30	4–6

[a]Refs. 43, 88, and 89.
[b]In g/(h·100 m²) at 7.1 kPa (53 mm Hg) pressure differential and 39.5°C for a film 25.4-μm (1-mil) thick.

Table 9. Effect of Composition on the Permeability of Various Gases Through VDC Copolymers[a]

Polymer	Gas	T, °C	P, $\frac{nmol}{m \cdot s \cdot GPa}$[b]
PVDC	O_2	25	<0.04
	N_2	25	<0.02
	CO_2	25	<0.10
90/10 VC	He	25	2.23
	H_2	25	2.54
	O_2	25	0.14
	N_2	25	0.03
	CO_2	25	0.98
	H_2S	30	0.10
85/15 VC	He	34	10
	O_2	25	0.40
	CO_2	20	2.0
70/30 VC	O_2	25	0.36
50/50 VC	O_2	25	1.2
80/20 AN	O_2	25	0.14
	N_2	25	0.02
	CO_2	25	0.35
60/40 AN	O_2	25	0.71
	N_2	25	0.09
	CO_2	25	1.6

[a]Ref. 90.
[b]To convert $\frac{nmol}{m \cdot s \cdot GPa}$ to $\frac{cm^3 \cdot cm}{cm^2 \cdot s \cdot kPa}$, divide by 4.46×10^{12}.

increase permeability more than the polar comonomers; higher acrylates are an example. Plasticizers increase permeability for similar reasons.

The effect of plasticizers and temperature on the permeability of small molecules in a typical vinylidene chloride copolymer has been studied thoroughly. The oxygen permeability doubles with the addition of about 1.7 parts per hundred resin (phr) of common plasticizers, or a temperature increase of 8°C (91). The effects of temperature and plasticizer on the permeability are shown in Figure 4. The moisture (water) vapor transmission rate (MVTR or WVTR) doubles with the addition of about 3.5 phr of common plasticizers (92). The dependence of the WVTR on temperature is a little more complicated. WVTR is commonly reported at a constant difference in relative humidity and not at a constant partial pressure difference. WVTR is a mixed term that increases with increasing temperature because both the fundamental permeability and the fundamental partial pressure at constant relative humidity increase. Carbon dioxide permeability doubles with the addition of about 1.8 phr of common plasticizers, or a temperature increase of 7°C (93).

Table 10 compares the permeabilities of small molecules for several common polymers. The oxygen permeability is an important property for food packaging. In this application, humidity is an important consideration. The oxygen permeability is not affected by humidity for vinylidene chloride copolymers, nitrile barrier resins, poly(vinyl chloride), polystyrene, and polyolefins. Hence, for these polymers, the data in Table 10 are useful. The oxygen permeabilities of most

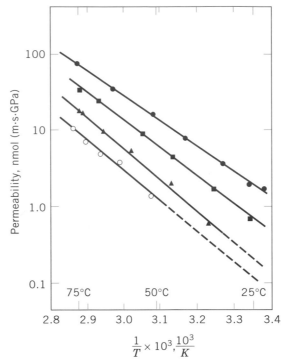

Fig. 4. Oxygen permeability in a vinylidene chloride copolymer film at selected levels of plasticizer (Citroflex A-4). Plasticizer level in parts per hundred resin (phr): ●, 7.2 phr; ■, 4.9 phr; ▲, 2.7 phr; and ○, 0.5 phr. To convert nmol/(m·s·GPa) to cm³·cm/(cm²·s·kPa), divide by 4.46×10^{12}.

nylons are increased modestly by increasing humidity, and the oxygen permeability of poly(ethylene terephthalate) is decreased modestly by increasing humidity. In contrast, the ethylene–vinyl alcohol copolymers (EVOH) are very sensitive to humidity. At low humidities, the oxygen permeabilities of EVOH are quite low. However, at the high humidities that contact food packages, the oxygen permeabilities of EVOH copolymers are much higher. Small changes in humidity can cause large changes in the oxygen permeabilities. Hence, data from tables are not as useful for EVOH.

The permeation of flavor/aroma compounds has become more important. This is a consequence of more sophisticated food packaging applications. The permeation of flavor/aroma molecules differs from the permeation of small molecules in some important ways. The diffusion coefficients, D, for flavor/aroma compounds are typically $10^2 – 10^5$ times smaller. This is a direct result of the larger permeant size. The solubility coefficients, S, for flavor/aroma compounds are typically $10^2 – 10^6$ times larger. This is partly related to higher boiling points. Table 11 compares the D and S of large and small molecules in several polymers. A low D and a high S mean that loss of flavor/aroma molecules into a polymer package wall can be more than the loss of flavor/aroma molecules through a polymer package.

Table 10. Barrier Properties of Polymers[a]

Polymer	Gas permeability at 23°C, $\frac{nmol}{m \cdot s \cdot GPa}$[b]			WVTR[c], $\frac{nmol}{m \cdot s}$[d]
	O_2	N_2	CO_2	
high barrier vinylidene chloride copolymers	0.04–0.3	0.01–0.1	0.1–0.5	0.02–0.1
nitrile barrier resin	1.6		6	1.0–1.2
nylon-6,6; nylon-6	2–5		3–9	1.5–5.5
polypropylene	300	60	1200	0.06–0.2
poly(ethylene terephthalate) (PET)	10–18	2–4	30–50	0.4–0.7
rigid poly(vinyl chloride)	10–40		40–100	0.2–1.3
high density polyethylene	300		1200	0.1
low density polyethylene	500–700	200–400	2000–4000	0.2–0.4
polystyrene	600–800	40–50	2000–3000	0.5–3.0
ethylene vinyl alcohol				
32 mol % ethylene				
0% rh	0.02	0.002	0.09	0.9[e]
100% rh	2.3			
44 mol % ethylene				
0% rh	0.18	0.015	0.8	0.3[e]
100% rh	1.3			

[a]Refs. 94 and 95.
[b]To convert $\frac{nmol}{m \cdot s \cdot GPa}$ to $\frac{cc \cdot mil}{100\ in.^2 \cdot d \cdot atm}$, divide by 2.
[c]WVTR = water vapor transmission rate at 90% rh and 38°C.
[d]To convert nmol/(m·s) to g·mil/(100 in.²·d), multiply by 4.
[e]40°C.

Table 11. Diffusion Coefficients and Solubility Coefficients of Selected Penetrants in Polymers at 25°C[a]

Penetrant	Polymer	D, m²/s	S, kg/(m³Pa)
oxygen	poly(ethylene terephthalate)	3×10^{-13}	9.8×10^{-7}
oxygen	high density polyethylene	1.7×10^{-11}	6.6×10^{-7}
oxygen	VDC copolymer	1.5×10^{-14}	3.5×10^{-7}
CO_2	acrylonitrile copolymer	1.0×10^{-13}	1.6×10^{-6}
CO_2	poly(vinyl chloride)	8.9×10^{-13}	3.4×10^{-6}
CO_2	VDC copolymer	1.4×10^{-14}	1.1×10^{-6}
d-limonene	high density polyethylene	7.0×10^{-14}	0.3
d-limonene	VDC copolymer	3.0×10^{-18}	0.6
methyl salicylate	nylon-6	2.1×10^{-17}	0.9
methyl salicylate	VDC copolymer	5.8×10^{-16}	0.3

[a]Ref. 96.

Humidity does not affect the permeability, diffusion coefficient, or solubility coefficient of flavor/aroma compounds in vinylidene chloride copolymer films. Studies based on *trans*-2-hexenal and D-limonene from 0 to 100% rh showed no difference in these transport properties (97,98). The permeabilities and diffusion coefficients of *trans*-2-hexenal in two barrier polymers are compared in Table 12. Humidity does not affect the vinylidene chloride copolymer. In contrast, transport in an EVOH film is strongly plasticized by humidity.

Table 12. Transport of *trans*-2-Hexenal in Barrier Films at 75°C

Film	Condition	Permeability, $\frac{nmol}{m \cdot s \cdot GPa}$ [a]	Diffusivity, 10^{-16} m^2/s
VDC copolymer[b]	dry	460	44
VDC copolymer	90/0[c]	420	39
EVOH[d]	dry	230	160
EVOH	90/0[c]	10,000	7,200

[a]To convert $\frac{nmol}{m \cdot s \cdot GPa}$ to $\frac{cm^3 \cdot cm}{cm^2 \cdot s \cdot kPa}$, divide by 4.46 × 10^{12}.
[b]Dow experimental resin XU32024.13.
[c]90% rh on the upstream side, 0% rh on the downstream side.
[d]44 mol % ethylene.

Degradation Chemistry

Vinylidene chloride polymers are highly resistant to oxidation, permeation of small molecules, and biodegradation, which makes them extremely durable under most use conditions. However, these materials are thermally unstable and, when heated above about 120°C, undergo degradative dehydrochlorination. For this reason, the superior characteristics of the homopolymer cannot be exploited. Furthermore, it degrades with rapid evolution of hydrogen chloride within a few degrees of its melting temperature (200°C). As a consequence, the copolymers of VDC with vinyl chloride, alkyl acrylates or methacrylates, acrylonitrile or methacrylonitrile, rather than the homopolymer, have come to commercial prominence. Such copolymers have often served as substrates for a study of the degradation reaction (99–102). The thermal degradation of vinylidene chloride copolymers occurs in two distinct steps. The first involves degradative dehydrochlorination via a chain process to generate poly(chloroacetylene) sequences (101,103). Subsequent Diels-Alder-type condensation between conjugated sequences affords a highly cross-linked network, which, upon further dehydrochlorination, leads to the formation of a large surface area, highly absorptive carbon (1). The initial dehydrochlorination occurs at moderate temperatures and is a typical chain process involving distinct initiation, propagation, and termination phases (104,105). Initiation is thought to occur via carbon–chlorine bond scission promoted by a defect structure within the polymer. An effective defect site in these polymers is unsaturation (100). Introduction of a random double bond produces an allylic dichloromethylene unit activated for carbon–chlorine bond scission. Initiation by the thermally induced cleavage of this bond, followed by propagation by successive dehydrochlorination along the chain, ie, the so-called unzipping reaction, can then proceed readily. The thermal stability of these polymers is decreased by pretreatment with ultraviolet irradiation (106), electron-beam irradiation (107), and basic solvents or reagents (108,109); by an atmosphere of oxygen (106,110) or nitric oxide (106); and by the presence of either peroxide linkages within the polymer (110,111), residues of emulsifying agents (110), organometallic initiator residues (110), ash from a previous decomposition (110), peroxide initiator residues (106–115), or metal ions (116). All the foregoing are sufficient to introduce random double bonds into the polymer structure. This can be demonstrated by examination of the treated sample by ultraviolet and infrared spectroscopic methods (107). Prolonged treatment of the polymer with basic reagents leads to more extensive dehydrohalogenation (116–121);

x- or γ-irradiation promotes carbon–carbon bond scission (106). The principal steps in the thermal degradation of VDC polymer are formation of a conjugated polyene sequence followed by carbonization.

$$(CH_2 - CCl_2)_n \xrightarrow[\Delta]{fast} (CH = CCl)_n + n\ HCl \xrightarrow[slow, \Delta]{carbonization} (CH = CCl)_n$$

$$\longrightarrow 2n\ C + n\ HCl$$

On being heated, the polymer gradually changes color from yellow to brown and finally to black. Early in the reaction, the polymer becomes insoluble, which indicates that cross-linking has occurred. The temperature of melting decreases, and the presence of unsaturation may be detected by spectroscopic (ultraviolet, infrared, nuclear magnetic resonance) methods. The polymer eventually becomes infusible, and the crystal structure as detected by x-ray diffraction disappears even though the gross morphology is retained (122). The presence of carbon radicals can be detected by electron spin resonance (esr) measurements. If the temperature is raised substantially above 200°C, aromatic structures are formed. Finally, at very high temperatures (>700°C), complete carbonization occurs.

The first of these reactions, ie, the loss of the first mole of hydrogen chloride, has had the greatest impact on the end use of VDC polymers and has been most studied and well characterized. The propensity of these polymers to undergo degradation is influenced by a wide variety of factors, including physical changes in the solid (annealing effects) and the method of preparation and purity of the polymer. The most stable polymers are those produced by bulk polymerization at low temperature using a nonoxygen initiator. In general, the stability of the polymer reflects the method of preparation, with bulk > solution > suspension ≫ emulsion. In the absence of elevated temperatures, suspending agents, polar solvents, redox initiators, etc, a more perfect polymer structure is formed, ie, one containing a minimum level of unsaturation.

The impact of a less defective structure may be seen in Figure 5, which depicts hydrogen chloride evolution for the thermal degradation of a typical vinylidene chloride polymer. Initiation of degradation occurs at activated allylic sites within the polymer, but initiation does not occur simultaneously at all sites.

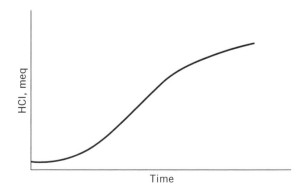

Fig. 5. Hydrogen chloride evolution for the thermal degradation of a typical vinylidene chloride polymer.

Therefore, early in the reaction, hydrogen chloride evolution increases as a function of time as initiation occurs at more and more sites. In other words, unzipping is started in an increasing number of chain segments. This gives rise to the acceleratory induction period characteristic of VDC copolymer degradation. When a greater number of initiation sites are present within the polymer, ie, at higher levels of unsaturation, the rate of hydrogen chloride evolution during the initiation phase of the degradation is greater. After unzipping has begun in all chains containing defect structures, hydrogen chloride production is essentially first order until termination by completion or other means becomes a prominent reaction. Undoubtedly, some random double-bond initiation continues to occur during propagation. However, during this period termination is roughly in balance with initiation. As the rate of termination significantly exceeds that of initiation, deceleration of degradation is observed (103,105,106). An accurate representation of the dehydrochlorination reaction over the entire range of degradation may be achieved using a kinetic expression containing two constants (104,123).

Much evidence has been accumulated to establish the radical nature of the degradation reaction (1). Prominent components of this include slight inhibition of the reaction by certain radical scavengers and changes in the electron spin resonance spectrum of a sample undergoing degradation (124,125). Both suggest that radical intermediates are generated during the degradation. The exact nature of the chain-carrying species is made more apparent from the results of degradation in bibenzyl solution (101,103). Bibenzyl is an efficient radical scavenger that is converted to stilbene on interaction with a radical. Stilbene can be readily quantitated by gas–liquid partition chromatography. For the degradation of typical VDC polymers, the ratio of hydrogen chloride evolved to stilbene produced is approximately 35:1 (101,103). This is in sharp contrast to the 2:1 ratio expected for trapping of chlorine atoms with perfect efficiency and suggests that the propagating species is a radical pair that does not dissociate appreciably. Thus propagation most probably occurs by a radical chain process in which the chain-carrying species is a radical pair that decomposes to alkene and hydrogen chloride without dissociation.

To some extent, the stability of VDC polymers is dependent on the nature of the comonomer present. Copolymers with acrylates degrade slowly. Apparent degradation propagation rates decrease somewhat as the acrylate content of the copolymer increases (126). The polyene sequences generated by dehydrochlorination are limited in size by the level of acrylate incorporation; that is, the acrylate molecules act as stopper units for the unzipping reaction. The impact of this chain-stopping is that the termination rate for higher acrylate content polymers is greater than for those containing smaller amounts of acrylate. Therefore initiation and termination rates are in balance for a shorter portion of the overall reaction period.

Copolymers with acrylonitrile or methacrylate undergo degradation more readily. In addition, the degradation is more complex than that observed for acrylate copolymers. Acrylonitrile copolymers release hydrogen cyanide as well as hydrogen chloride; products of thermal degradation of methyl methacrylate copolymers contain methyl chloride in addition to hydrogen chloride (100,102,127,128). In both cases, degradation apparently begins in VDC units adjacent to comonomer units (127,128).

The degradation of VDC polymers in nonpolar solvents is comparable to degradation in the solid state (101,125,129,130). However, these polymers are unstable in many polar solvents (131). The rate of dehydrochlorination increases markedly with solvent polarity. In strongly polar aprotic solvents, eg, hexamethylphosphoramide, dehydrochlorination proceeds readily (129,132). This reaction is clearly unlike thermal degradation and may well involve the generation of ionic species as intermediates.

Polymers of high VDC content are reactive toward strong bases to yield elimination products and toward nucleophiles to yield substitution products. Agents capable of functioning as both a base and a nucleophile react with these polymers to generate a mixture of products (119,133,134). Weakly basic agents such as ammonia, amines, or polar aprotic solvents accelerate the decomposition of VDC copolymers. Amines function as bases to remove hydrogen chloride and introduce unsaturation along the polymer main chain which may serve as initiation sites for thermal degradation. The overall effectiveness of a particular amine for dehydrohalogenation may be dependent on several factors, including inherent basicity, degree of steric hindrance at nitrogen, and compatibility with the polymer (135). Phosphines are more nucleophilic but less basic than amines. However, phosphines also promote dehydrohalogenation rather than displacement of allylic chlorine (136).

Amines can also swell the polymer, leading to very rapid reactions. Pyridine, for example, would be a fairly good solvent for a VDC copolymer if it did not attack the polymer chemically. However, when pyridine is part of a solvent mixture that does not dissolve the polymer, pyridine does not penetrate into the polymer phase (108). Studies of single crystals indicate that pyridine removes hydrogen chloride only from the surface. Kinetic studies and product characterizations suggest that the reaction of two units in each chain-fold can easily take place; further reaction is greatly retarded either by the inability of pyridine to diffuse into the crystal or by steric factors.

Aqueous bases or nucleophiles have little impact on VDC polymers, primarily because the polymer is not wetted or swollen by water. However, these polymers do slowly degrade in hot concentrated aqueous sodium hydroxide solution (116).

Lewis acids, particularly transition-metal salts, strongly promote the thermal degradation of VDC polymer (116,125,137–147). The rate of initiation of degradation is greatly enhanced in the presence of metal ions (140). The metal ion (or other Lewis acid) coordinates chlorine atoms, making them much better leaving groups. This facilitates the introduction of initial double bonds, which act as defect sites from which degradative dehydrohalogenation may propagate. Care must be taken to avoid metal ions, particularly precursors of iron chloride, during the preparation and processing of VDC polymers.

Copolymers of VDC that are free of impurities do not degrade at an appreciable rate in the absence of light below 100°C. However, when exposed to ultraviolet light, these polymers discolor (148). Again, the primary reaction seems to be dehydrochlorination. Hydrogen chloride is evolved and cross-linking occurs (106). Polyene sequences of narrow sequence length distribution are formed (149,150). These function as initiation sites for subsequent thermal degradation (106). Other photodegradation processes, eg, hydroperoxide formation at the methylene groups, probably also occur but are less important for these polymers than is polyene formation (139).

Stabilization. The stabilization of VDC polymers toward degradation is a highly developed art and is responsible for the widespread commercial use of these materials. Although the mode of action is often not understood, some general principles of effective stabilization have been established (151). The ideal stabilizer system should (1) absorb or combine with evolved hydrogen chloride irreversibly under conditions of use, but not strip hydrogen chloride from the polymer chain; (2) act as a selective uv absorber; (3) contain a reactive dienophilic moiety capable of preventing discoloration by reacting with and disrupting the color-producing conjugated polymer sequences; (4) possess nucleophilicity sufficient for reaction with allylic dichloromethylene units; (5) possess antioxidant activity so as to prevent the formation of carbonyl groups and other chlorine-labilizing structures; (6) be able to scavenge chlorine atoms and other free radicals efficiently; and (7) chelate metals, eg, iron, to prevent chlorine coordination and the formation of metal chlorides.

Acid acceptors are of two general types: alkaline-earth and heavy-metal oxides or salts of weak acids, such as barium or calcium fatty acid salts; and epoxy compounds, such as epoxidized soybean oil or glycidyl ethers and esters. Epoxidized oils are less effective for the stabilization of VDC polymers than for other halogenated polymers in similar processes (141). The function of these materials as plasticizers and processing lubricants is probably responsible for modest improvements in processing stability. Effective light stabilizers have a chemical configuration that leads to hydrogen bonding and chelation, showing exceptional conjugative stability and very good uv absorption properties. The principal compounds of commercial interest are derivatives of salicylic acid, resorcylic acid, benzophenone, and benzotriazole. Examples of dienophiles that have been used are maleic anhydride and dibasic lead maleate.

Antioxidants are generally of two types: those that react with a free radical to stop a radical chain, that is, to scavenge chlorine atoms or peroxy radicals; and those that reduce hydroperoxides to alcohols. Phenolic antioxidants, eg, 2,6-di-*tert*-butyl-4-methylphenol and substituted bisphenols, are of the first type. Because the chain-carrying species for the degradative dehydrochlorination is a tight chlorine-atom carbon-radical pair that does not dissociate appreciably during the reaction, the effectiveness of these agents is limited (142). The second type is exemplified by organic sulfur compounds and organic phosphites. The phosphites, ethylenediaminetetraacetic acid [60-00-4] (EDTA), citric acid [77-92-9], and citrates, can chelate metals. The ability of organic phosphites to function as antioxidants and as chelating agents illustrates the dual role of many stabilizer compounds. It is common practice to use a combination of stabilizing compounds to achieve optimum results (148). In addition, stabilization packages usually contain lubricants and other processing aids that enhance the effectiveness of the stabilizing compounds. The presence of these agents is particularly important to minimize the shearing component of degradation during extrusion and other processing steps (143).

Metal carboxylates have been considered as nucleophilic agents capable of removing allylic chlorine and thereby affording stabilization (143). Typical PVC stabilizers, eg, tin, lead, or cadmium esters, actually promote the degradation of VDC polymers. The metal cations in these compounds are much too acidic to be used with VDC polymers. An effective carboxylate stabilizer must contain a metal cation sufficiently acidic to interact with allylic chlorine and to facilitate its displacement by the carboxylate anion, but at the same time not acidic enough to strip chlorine from the polymer main chain (144).

Vinylidene chloride polymers containing stabilizing features have been prepared. In general, these have been polymers containing comonomer units with functionality that can consume evolved hydrogen chloride and do so in such a manner that good radical scavenging sites are exposed (145,146).

Commercial Methods of Polymerization and Processing

Processes that are essentially modifications of laboratory methods and that allow operation on a larger scale are used for commercial preparation of vinylidene chloride polymers. The intended use dictates the polymer characteristics and, to some extent, the method of manufacture. Emulsion polymerization and suspension polymerization are the preferred industrial processes. Either process is carried out in a closed, stirred reactor, which should be glass-lined and jacketed for heating and cooling. The reactor must be purged of oxygen, and the water and monomer must be free of metallic impurities to prevent an adverse effect on the thermal stability of the polymer.

Emulsion Polymerization. Emulsion polymerization is used commercially to make vinylidene chloride copolymers. In some applications, the resulting latex is used directly, usually with additional stabilizing ingredients, as a coating vehicle to apply the polymer to various substrates. In other applications, the polymer is first isolated from the latex before use. When the polymer is not used in latex form, the emulsion/coagulation process is chosen over alternative methods. The polymer is recovered in dry powder form, usually by coagulating the latex with an electrolyte, followed by washing and drying. The principal advantages of emulsion polymerization are twofold. First, high molecular weight polymers can be produced in reasonable reaction times, especially copolymers with vinyl chloride. The initiation and propagation steps can be controlled more independently than in the suspension process. Second, monomer can be added during the polymerization to maintain copolymer composition control. The disadvantages of emulsion polymerization result from the relatively high concentration of additives in the recipe. The water-soluble initiators, activators, and surface-active agents generally cause the polymer to have greater water sensitivity, poorer electrical properties, and poorer heat and light stability. Recovery of the polymer by coagulation, washing, and drying to some extent improves these properties over those of the polymer deposited in latex form.

A typical recipe for batch emulsion polymerization is shown in Table 13. A reaction time of 7–8 h at 30°C is required for 95–98% conversion. A latex is produced with an average particle diameter of 100–150 nm. Other modifying ingredients may be present, eg, other colloidal protective agents such as gelatin or carboxymethylcellulose, initiator activators such as redox types, chelates, plasticizers, stabilizers, and chain-transfer agents.

Commercial surfactants are generally anionic emulsifiers, alone or in combination with nonionic types. Representative anionic emulsifiers are the sodium

Table 13. Recipe for Batch Emulsion Polymerization[a]

Ingredient	Parts by wt
vinylidene chloride	78
vinyl chloride	22
water	180
potassium peroxysulfate	0.22
sodium bisulfite	0.11
Aerosol MA[b], 80 wt %	3.58
nitric acid, 69 wt %	0.07

[a]Ref. 152.
[b]Aerosol MA (American Cyanamid Co.) = dihexyl sodium sulfosuccinate.

alkylaryl sulfonates, the alkyl esters of sodium sulfosuccinic acid, and the sodium salts of fatty alcohol sulfates. Nonionic emulsifiers are of the ethoxylated alkylphenol type. Free-radical sources other than peroxysulfates may be used, eg, hydrogen peroxide, organic hydroperoxides, peroxyborates, and peroxycarbonates. Many of these are used in redox pairs, in which an activator promotes the decomposition of the peroxy compound. Examples are peroxysulfate or perchlorate activated with bisulfite, hydrogen peroxide with metallic ions, and organic hydroperoxides with sodium formaldehyde sulfoxylate. The use of activators causes the decomposition of the initiator to occur at lower reaction temperatures, which allows the preparation of a higher molecular weight polymer within reasonable reaction times. This is an advantage, particularly for copolymers of vinylidene chloride with vinyl chloride. Oil-soluble initiators are usually effective only when activated by water-soluble activators or reducing agents.

To ensure constant composition, the method of emulsion polymerization by continuous addition is employed. One or more components are metered continuously into the reaction. If the system is properly balanced, a steady state is reached in which a copolymer of uniform composition is produced (153). A process of this type can be used for the copolymerization of VDC with a variety of monomers. A flow diagram of the apparatus is shown in Figure 6; a typical recipe is shown in Table 14. The monomers are charged to the weigh tank A, which is kept under a nitrogen blanket. The emulsifiers, initiator, and part of the water are charged to tank B; the reducing agent and some water to tank C. The remaining water is charged to the reactor D, and the system is sealed and purged. The temperature is raised to 40°C and one-tenth of the monomer and initiator charges is added, then one-tenth of the activator is pumped in. Once the reaction begins, as indicated by an exotherm and pressure drop, feeds of A, B, and C are started at programmed rates that begin slowly and gradually increase. The emulsion is maintained at a constant temperature during the run by cooling water that is pumped through the jacket. When all components are in the reactor and the exotherm begins to subside, a final addition of initiator and reducing agent completes the reaction.

Suspension Polymerization. Suspension polymerization of vinylidene chloride is used commercially to make molding and extrusion resins. The principal advantage of the suspension process over the emulsion process is the use of fewer ingredients that might detract from the polymer properties. Stability

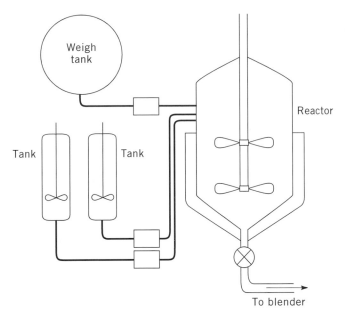

Fig. 6. Apparatus for continuous-addition emulsion polymerization of a VDC-acrylate mixture (153).

Table 14. Recipe for Emulsion Polymerization by Continuous Addition[a]

Ingredient	Parts by wt
vinylidene chloride	468
comonomer	52
emulsifiers	
Tergitol[b] NP 35	12
sodium lauryl sulfate, 25 wt %	12
initiator ammonium peroxysulfate	10
sodium metabisulfite ($Na_2S_2O_5$), 5 wt %	10
water	436

[a]Ref. 153.
[b]Nonionic wetting agent produced by Union Carbide.

is improved and water sensitivity is decreased. Extended reaction times and the difficult preparation of higher molecular weight polymers are disadvantages of the suspension process compared to the emulsion process, particularly for copolymers containing vinyl chloride.

A typical recipe for suspension polymerization is shown in Table 15. At a reaction temperature of 60°C, the polymerization proceeds to 85–90% conversion in 30–60 h. Unreacted monomer is removed by vacuum stripping, then it is condensed and reused after processing. The polymer is obtained in the form of small (150–600 μm (30–100 mesh)) beads, which are dewatered by filtration or centrifugation and then dried in a flash dryer or fluid-bed dryer. Suspension polymerization involves monomer-soluble initiators, and polymerization occurs inside suspended monomer droplets, which form by the shearing action of the

Table 15. Recipe for Suspension Polymerization[a]

Ingredient	Parts by wt
vinylidene chloride	85
vinyl chloride	15
deionized water	200
400 mPa·s(=cP) methyl hydroxypropylcellulose	0.05
lauroyl peroxide	0.3

[a]Ref. 154.

agitator and are prevented from coalescence by the protective colloid. It is important that the initiator be uniformly dissolved in the monomer before droplet formation. Unequal distribution of initiator causes some droplets to polymerize faster than others, leading to monomer diffusion from slow-polymerizing to fast-polymerizing droplets. The fast-polymerizing droplets form polymer beads that are dense, hard, glassy, and extremely difficult to fabricate because of their inability to accept stabilizers and plasticizers. Common protective colloids that prevent droplet coalescence and control particle size are poly(vinyl alcohol), gelatin, and methylcellulose. Organic peroxides, peroxycarbonates, and azo compounds are used as initiators for vinylidene chloride suspension polymerization.

The batch-suspension process does not compensate for composition drift, whereas constant-composition processes have been designed for emulsion or suspension reactions. It is more difficult to design controlled-composition processes by suspension methods. In one approach (155), the less reactive component is removed continuously from the reaction to keep the unreacted monomer composition constant. This method has been used effectively in VDC–VC copolymerization, where the slower reacting component is a volatile and can be released during the reaction to maintain constant pressure. In many other cases, no practical way is known for removing the slower reacting component.

Economic Aspects

Vinylidene chloride monomer is produced commercially in the United States by The Dow Chemical Company and PPG Industries. The monomer is produced in Europe by Imperial Chemical Industries, Ltd., in the United Kingdom; Badische Anilin und Soda Fabrik and Chemische Werk Hüls in Germany; Solvay S.A. and Amaco et Compagnie in France; and The Dow Chemical Company in the Netherlands. The monomer is produced in Japan by the Asahi Chemical Company, Kureha Chemical Industries, and Kanto Denka Kogyo Company.

Although Saran is a generic name for VDC copolymers in the United States, it is a Dow trademark in most foreign countries. Other trade names include Daran (Hampshire Chemical Corporation) and Serfene (Morton Chemical) in the United States, and Haloflex (Zeneca Resins), Diofan (BASF), Ixan (Solvay SA), and Polyidene (Scott-Bader) in Europe. The monomer is of particular economic interest because it is only 27 wt % hydrocarbon. In addition, B. F. Goodrich Chemicals (GEON) supply non-barrier VDC copolymers.

Applications

Molding Resins. Vinylidene chloride–vinyl chloride copolymers were originally developed for thermoplastic molding applications, and small amounts are still used for this purpose. The resins, when properly formulated with plasticizers and heat stabilizers, can be fabricated by common methods, eg, injection, compression, or transfer molding. Conventional or dielectric heating can be used to melt the polymers. Rapid hardening is achieved by forming in heated molds to induce rapid crystallization. Cold molds result in supercooling of the polymer. Because the interior of the molded part remains soft and amorphous, the part cannot be easily removed from the mold without distortion. Mold temperatures of up to 100°C allow rapid removal of dimensionally stable parts. The range of molding temperatures is rather narrow because of the crystalline nature of the resin and thermal sensitivity. All crystallites must be melted to obtain low polymer melt viscosity, but prolonged or excessive heating must be avoided to prevent dehydrochlorination.

Thermal degradation is a problem even when the resin is formulated with the very best stabilizers. Molding equipment is designed to alleviate this problem by having all passages through the heating cylinder streamlined to prevent plastic buildup. Any plastic that remains in the heating cylinder for longer than a few minutes decomposes, releasing HCl and forming carbon. The carbon may build up or break off and contaminate molded parts with black specks. It is especially important that an injection-molding heating cylinder not be shut down when loaded with molten resin. The cylinder must be purged with a more stable resin, eg, polystyrene.

The metal parts of the injection molder, ie, the liner, torpedo, and nozzle, that contact the hot molten resin must be of the noncatalytic type to prevent accelerated decomposition of the polymer. In addition, they must be resistant to corrosion by HCl. Iron, copper, and zinc are catalytic to the decomposition and cannot be used, even as components of alloys. Magnesium is noncatalytic but is subject to corrosive attack, as is chromium when used as plating. Nickel alloys such as Duranickel, Hastelloy B, and Hastelloy C are recommended as construction materials for injection-molding metal parts. These and pure nickel are noncatalytic and corrosion-resistant; however, pure nickel is rather soft and is not recommended.

The injection mold need not be made of noncatalytic metals; any high grade tool steel may be used because the plastic cools in the mold and undergoes little decomposition. However, the mold requires good venting to allow the passage of small amounts of acid gas as well as air. Vents tend to become clogged by corrosion and must be cleaned periodically.

Molded parts of vinylidene chloride copolymer plastics are used to satisfy the industrial requirements of chemical resistance and extended service life. They are used in such items as gasoline filters, valves, pipe fittings, containers, and chemical process equipment. Complex articles are constructed from molded parts by welding; hot-air welding at 200–260°C is a suitable method. Molded parts have good physical properties but lower tensile strength than films or fibers, because crystallization is random in molded parts. Higher strength is developed by orientation in films and fibers. Physical properties of a typical molded vinylidene chloride copolymer plastic are listed in Table 16.

Extrusion Resins. Extrusion of VDC–VC copolymers is the main fabrication technique for filaments, films, rods, and tubing or pipe, and involves the same concerns for thermal degradation, streamlined flow, and noncatalytic materials of construction as described for injection-molding resins (84,122). The plastic leaves the extrusion die in a completely amorphous condition and is maintained in this state by quenching in a water bath to about 10°C, thereby inhibiting recrystallization. In this state, the plastic is soft, weak, and pliable. If it is allowed to remain at room temperature, it hardens gradually and recrystallizes partially at a slow rate with a random crystal arrangement. Heat treatment can be used to recrystallize at controlled rates.

Crystal orientation is developed in the supercooled extrudate by plastic deformation and heat treatment. In the manufacture of filaments, stretching produces orientation in a single direction and develops unidirectional properties of high tensile strength, flexibility, long fatigue life, and good elasticity. The filaments are removed from the supercooling tank, wrapped several times around smooth takeoff rolls, and then wrapped several times around orienting rolls, which have a linear speed about four times that of the takeoff rolls. The difference in roll speeds produces mechanical stretching and causes orientation of crystallites along the longitudinal axis while the polymer is crystallizing. Heat treatment may be used during or after stretching to affect the degree of crystallization and control the physical properties of the oriented filaments.

A variation of the preceding process is used to produce oriented vinylidene chloride copolymer films. The plastic is extruded into tube form and then is supercooled and subsequently biaxially oriented in a continuous bubble process. The supercooled tube is flattened and passed through two sets of pinch rolls, which are arranged so that the second set of rolls travels faster than the first set. Between the two sets, air is injected into the tube to create a bubble that

Table 16. Properties of Resin for Injection-Molding Applications[a]

Typical resin properties	Test method	Value
ultimate tensile strength, MPa[b]	ASTM D638	24.1–34.5
yield tensile strength, MPa[b]	ASTM D638	19.3–26.2
ultimate elongation, %	ASTM D638	160–240
modulus of elasticity in tension, MPa[b]		345–552
Izod impact strength, J/m[c] of notch	ASTM D256	21.35–53.38
density, g/cm^3	ASTM D792	1.65–1.72
hardness, Rockwell M	ASTM D785	50–65
water absorption, % in 24 h	ASTM D570	0.1
mold shrinkage, cm/cm (injection-molded)	ASTM D955	0.005–0.025
limiting oxygen index, %	ASTM D2863	60.0[d]
UL 94	UL 94 Test	V–O[d]

[a]Ref. 156.
[b]To convert MPa to psi, multiply by 145.
[c]To convert J/m to ft·lbf/in. of notch, divide by 53.38.
[d]The results of small-scale flammability tests are not intended to reflect the hazards of this or any other material under actual fire conditions.

is entrapped by the pinch rolls. The entrapped air bubble remains stationary while the extruded tube is oriented as it passes around the bubble. Orientation is produced in the transverse and the longitudinal directions, creating excellent tensile strength, elongation, and flexibility in the film. The commercial procedure has been described (157).

Unoriented film can be formed by extruding through a slit die. The temperature must be controlled to promote crystallization before winding on a roll. Extruded monofilaments in diameters of 0.13–0.38 mm have been widely used in the textile field as furniture and automobile upholstery, drapery fabric, outdoor furniture, venetian blind tape, filter cloths, etc. Chemically resistant tubing and pipe liners are extruded. The pipe liner is inserted into an oversized steel pipe, which is swaged to size, and lengths are connected by flanged joints and vinylidene chloride copolymer gaskets. Pipe fittings are lined with injection-molded liners.

The biaxially oriented extruded films are used in packaging applications where their excellent resistance to water vapor and most gases makes them ideal transparent barriers. Because they are highly oriented, these films exhibit some shrinkage when exposed to higher than normal temperatures. Preshrinking or heat-setting can be performed to minimize residual shrink, or the shrinkage may be used to advantage in the heat-shrinking of overwraps on packaged items. Shrink bags made from high shrink film are used in the packaging of cuts of fresh red meat. Films for packaging are used in tube form or as flat film for overwraps or conversion to film bags on modified bag making machinery. The electronic or dielectric seal is the most satisfactory type for sealing the film to itself, although hot-plate sealers or cement-type seals produce an air-tight seal on overwraps without fusing of the material.

Multilayer Film. A significant application for vinylidene chloride copolymer resins is in the construction of multilayer film and sheet (158,159). This permits the design of a packaging material with a combination of properties not obtainable in any single material. A VDC copolymer layer is incorporated into multilayer film for perishable food packaging because it provides a barrier to oxygen. A special high barrier resin is supplied specifically for this application. Typically, multilayer packaging films contain outer layers of a tough, low cost polymer such as high density polyethylene (HDPE) with VDC copolymer as the core layer. The film is made in a special coextrusion process. The properties of a 0.05-mm multilayer film are listed in Table 17.

Rigid Barrier Containers. Rigid containers for food packaging can be made from coextruded sheet that contains a layer of a barrier polymer. A simple example is a sheet with five layers that has a total thickness of about 1.3 mm (50 mil). The outermost layers might be polypropylene, polyethylene, polystyrene, high impact polystyrene (HIPS), or other nonbarrier polymer having good mechanical properties. The innermost barrier layer is about 125 μm (5 mil) of a vinylidene chloride copolymer. Adhesive layers connect the outer layers and the barrier layer. This coextruded sheet can be formed into containers by any of several techniques, including solid-phase pressure forming (SPPF) and melt-phase forming (MPF). The final container has a total wall thickness of about 500 mm and a barrier layer that is about 50 mm thick. Such a container is capable of protecting oxygen-sensitive foods at ambient temperatures for a year or more.

Table 17. Physical Properties[a] of a Multilayer Barrier Film[b] and a Polyethylene Film[c]

Property[d]	Multilayer film	Polyethylene film	Test method
yield tensile strength, MPa[e]			ASTM D882-61T
MD	14	12.1	
TD	13	9.7	
ultimate tensile strength, MPa[e]			ASTM D882-61T
MD	24	20.0	
TD	17	17	
tensile modulus, MPa[e]			ASTM D882-61T
MD	170	180	
TD	150	180	
elongation, %			ASTM D882-61T
MD	400	325	
TD	400	550	
Elmenford tear strength, g			ASTM D1922
MD	800	325	
TD	650	250	
gas transmission at 24°C, $\frac{nmol}{m^2 \cdot s \cdot GPa}$[f]			ASTM D1434-63
oxygen	3.8×10^4	1.6×10^7	
carbon dioxide	12.0×10^4	10.0×10^7	
nitrogen	0.8×10^4	0.8×10^7	
water vapor transmission at 95% rh and 38°C, $\frac{nmol}{m \cdot s}$[g]	3200	6400	ASTM E96-63T

[a]Ref. 158.
[b]0.05-mm total thickness with layers of polyethylene, adhesive, and vinylidene chloride copolymer.
[c]0.05-mm polyethylene, 0.921 g/cm^3 density.
[d]MD is machine direction; TD is transverse direction.
[e]To convert MPa to psi, multiply by 145.
[f]To convert $\frac{nmol}{m^2 \cdot s \cdot GPa}$ to $\frac{cc}{100 \, in.^2 d \cdot atm}$, divide by 7.9×10^4.
[g]To convert $\frac{nmol}{m^2 \cdot s}$ to $\frac{g}{100 \, in.^2 \cdot d}$, multiply by 1×10^4.

These containers are lightweight, microwavable, nonbreakable, and attractive. More sophisticated containers may have more than five layers and improved economics by including a layer of scrap or recycled polymers in the structure.

Lacquer Resins. Vinylidene chloride polymers have several properties that are valuable in the coatings industry: excellent resistance to gas and moisture vapor transmission, good resistance to attack by solvents and by fats and oils, high strength, and the ability to be heat-sealed (160,161). These characteristics result from the highly crystalline nature of the very high vinylidene chloride content of the polymer, which ranges from ~80 to 90 wt %. Minor constituents in these copolymers generally are vinyl chloride, alkyl acrylates, alkyl methacrylates, acrylonitrile, methacrylonitrile, and vinyl acetate. Small concentrations of vinylcarboxylic acids, eg, acrylic acid, methacrylic acid, or itaconic acid, are sometimes included to enhance adhesion of the polymer to the substrate. The ability to crystallize and the extent of crystallization are reduced with increasing concentration of the comonomers; some commercial polymers do not crystallize. The most common lacquer resins are terpolymers of VDC–methyl methacrylate–acrylonitrile (162,163). The VDC level and the methyl methacrylate–acrylonitrile ratio are adjusted for the best balance of solubility and permeability. These polymers exhibit a unique combination of high solubility, low permeability, and rapid crystallization (164).

Acetone, methyl ethyl ketone, methyl isobutyl ketone, dimethylformamide, ethyl acetate, and tetrahydrofuran are solvents for vinylidene chloride polymers used in lacquer coatings; methyl ethyl ketone and tetrahydrofuran are most extensively employed. Toluene is used as a diluent for either. Lacquers prepared at 10–20 wt % polymer solids in a solvent blend of two parts ketone and one part toluene have a viscosity of 20–1000 mPa·s(=cP). Lacquers can be prepared from polymers of very high vinylidene chloride content in tetrahydrofuran–toluene mixtures and stored at room temperature. Methyl ethyl ketone lacquers must be prepared and maintained at 60–70°C or the lacquer forms a solid gel. It is critical in the manufacture of polymers for a lacquer application to maintain a fairly narrow compositional distribution in the polymer to achieve good dissolution properties.

The lacquers are applied commercially by roller coating, doctor and dip coating, knife coating, and spraying. Spraying is useful only with lower viscosity lacquers, and solvent balance is important to avoid webbing from the spray gun. Solvent removal is difficult from heavy coatings, and multiple coatings are recommended where a heavy film is desired. Sufficient time must be allowed between coats to avoid lifting of the previous coat by the solvent. In the machine coating of flexible substrates, eg, paper and plastic films, the solvent is removed by infrared heating or forced-air drying at 90–140°C. Temperatures of 60–95°C promote the recrystallization of the polymer after the solvent has been removed. Failure to recrystallize the polymer leaves a soft, amorphous coating that blocks or adheres between concentric layers in a rewound roll. A recrystallized coating can be rewound without blocking. Handling properties of the coated film are improved with small additions of wax as a slip agent and of talc or silica as an antiblock agent to the lacquer system. The concentration of additives is kept low to prevent any serious detraction from the vapor transmission properties of the vinylidene chloride copolymer coating. For this reason, plasticizers are seldom, if ever, used.

A primary use of vinylidene chloride copolymer lacquers is the coating of films made from regenerated cellulose or of board or paper coated with polyamide, polyester, polypropylene, poly(vinyl chloride), and polyethylene. The lacquer imparts resistance to fats, oils, oxygen, and water vapor (165). These coated products are used mainly in the packaging of foodstuffs, where the additional features of inertness, lack of odor or taste, and nontoxicity are required. Vinylidene chloride copolymers have been used extensively as interior coatings for ship tanks, railroad tank cars, and fuel storage tanks, and for coating of steel piles and structures (166,167). The excellent chemical resistance and good adhesion result in excellent long-term performance of the coating. Brushing and spraying are suitable methods of application.

The excellent adhesion to primed films of polyester combined with good dielectric properties and good surface properties makes the vinylidene chloride copolymers very suitable as binders for iron oxide pigmented coatings for magnetic tapes (168–170). They perform very well in audio, video, and computer tapes.

Vinylidene Chloride Copolymer Latex.   Vinylidene chloride polymers are often made in emulsion, but usually are isolated, dried, and used as conventional resins. Stable latices have been prepared and can be used directly for coatings (171–176). The principal applications for these materials are as barrier coatings on paper products and, more recently, on plastic films. The heat-seal characteristics of VDC copolymer coatings are equally valuable in many applications. They are also used as binders for paints and nonwoven fabrics (177). The use of special VDC copolymer latices for barrier laminating adhesives is growing, and the use of vinylidene chloride copolymers in flame-resistant carpet backing is well known (178–181). VDC latices can also be used to coat poly(ethylene terephthalate) (PET) bottles to retain carbon dioxide (182).

Poly(vinylidene chloride) latices can be easily prepared by the same methods but have few uses because they do not form films. Copolymers of high VDC content are film-forming when freshly prepared but soon crystallize and lose this desirable characteristic. Because crystallinity in the final product is very often desirable, eg, in barrier coatings, a significant developmental problem has been to prevent crystallization in the latex during storage and to induce rapid crystallization of the polymer after coating. This has been accomplished by using the proper combination of comonomers with VDC.

Most vinylidene chloride copolymer latices are made with varying amounts of acrylates, methacrylates, and acrylonitrile, as well as minor amounts of vinylcarboxylic acids, eg, itaconic and acrylic acids. Low foam latices having high surface tension are prepared with copolymerizable sulfonate monomers (179,181–183). The total amount of comonomer ranges from about 8 wt % for barrier latices to as high as 60 wt % for binder and paint latices. The properties of a typical barrier latex used for paper coating are listed in Table 18. Barrier latices are usually formulated with antiblock, slip, and wetting agents. They can be deposited by conventional coating processes, eg, with an air knife (184,185).

Coating speeds in excess of 305 m/min can be attained. The latex coating can be dried in forced-air or radiant-heat ovens (186,187). Multiple coats are applied, particularly in paper coating, to reduce pinholing (188). A precoat is often used on porous substances to reduce the quantity of the more expensive VDC copolymer latex needed for covering (189). The properties of a typical coating are listed in Table 19.

Vinylidene Chloride Copolymer Foams.   Low density, fine-celled VDC copolymer foams can be made by extrusion of a mixture of vinylidene chloride

Table 18. Properties of a Typical Barrier Latex

Properties	Value
total solids, wt %	54–56
viscosity at 25°C, mPa·s(=cP)	25
pH	2
color	creamy white
particle size, nm	ca 250
density, g/cm^3	1.30
mechanical stability	excellent
storage stability	excellent
chemical stability	not stable to di- or trivalent ions

Table 19. Film Properties of Vinylidene Chloride Copolymer Latex

Property	Value
water vapor transmission at 38°C and 95% rh, $\frac{nmol}{m \cdot s}$[a]	0.012[b]
grease resistance	excellent
scorability and fold resistance	moderate
oxygen permeability at 25°C, $\frac{nmol}{m \cdot s \cdot GPa}$[c]	0.07
heat sealability[d]	good
light stability	fair
density, g/cm^3	1.60
color	watery white
clarity	excellent
gloss	excellent
odor	none

[a]To convert $\frac{nmol}{m \cdot s}$ to $\frac{g \cdot mil}{100 \text{ in.}^2 \cdot d}$, multiply by 4.
[b]Values 0.37 g/(24 h·100 in.^{2}) for 0.5 mils.
[c]To convert $\frac{nmol}{m \cdot s \cdot GPa}$ to $\frac{cc \cdot mil}{100 \text{ in.}^2 \cdot d \cdot atm}$, divide by 2.
[d]Face-to-face.

copolymer and a blowing agent at 120–150°C (190). The formulation must contain heat stabilizers, and the extrusion equipment must be made of noncatalytic metals to prevent accelerated decomposition of the polymer. The low melt viscosity of the VDC copolymer formulation limits the size of the foam sheet that can be extruded.

Expandable VDC copolymer microspheres are prepared by a microsuspension process (191). The expanded microspheres are used in reinforced polyesters, blocking multipair cable, and in composites for furniture, marble, and marine applications (192–195). Vinylidene chloride copolymer microspheres are also used in printing inks and paper manufacture (196).

Vinylidene Chloride Copolymer Ignition Resistance. The role of halogen-containing compounds in ignition and flame suppression has been studied for many years (197–202). Vinylidene chloride copolymers are an abundant source of organic chlorine, eg, often above 70 wt %. Vinylidene chloride emulsion copolymers are used in a variety of ignition-resistant binding applications (203,204). Powders dispersible in nonsolvent organic polymer intermediates, eg, polyols, are used for both reinforcement and ignition resistance in polyurethane foams. VDC copolymer powder is also used as an ignition-resistant binder for cotton batt (205–207).

The halogenated polymers generate significantly more smoke than polymers that have aliphatic backbones, even though the presence of the halogen does increase the limiting oxygen index. Heavy-metal salts retard smoke generation in halogenated polymers (207,208). A VDC emulsion copolymer having a high acrylonitrile graft can be used to make ignition-resistant acrylic fibers (209). A rubber-modified VDC copolymer combines good ignition resistance with good low temperature flexibility (210,211). The rubber-modified VDC copolymer has been evaluated in wire coating where better ignition resistance and lower smoke generation are needed.

Materials are also blended with VDC copolymers to improve toughness (211–214). Vinylidene chloride copolymer blended with ethylene–vinyl acetate copolymers improves toughness and lowers heat-seal temperatures (215,216). Adhesion of a VDC copolymer coating to polyester can be achieved by blending the copolymer with a linear polyester resin (217).

BIBLIOGRAPHY

"Vinylidene Polymers (Chloride)" in *ECT* 2nd ed., Vol. 21, pp. 275–303, by R. Wessling and F. G. Edwards, The Dow Chemical Co.; "Vinylidene Chloride and Poly(Vinylidene Chloride)" in *ECT* 3rd ed., Vol. 23, pp. 764–798, by D. S. Gibbs and R. A. Wessling, Dow Chemical U.S.A.

1. R. A. Wessling, *Polyvinylidene Chloride*, Gordon & Breach, New York, 1977.
2. *Vinylidene Chloride Monomer Safe Handling Guide*, No. 00-6339-88-SAI, The Dow Chemical Co., Midland, Mich., 1988.
3. L. G. Shelton, D. E. Hamilton, and R. H. Fisackerly, in E. C. Leonard, ed., *Vinyl and Diene Monomers, High Polymers*, Vol. 24, Wiley-Interscience, New York, 1971, pp. 1205–1282.
4. P. T. DeLassus and D. D. Schmidt, *J. Chem. Eng. Data*, **26**, 274 (1981).
5. U.S. Pat. 2,238,020 (Apr. 8, 1947), A. W. Hanson and W. C. Goggin (to The Dow Chemical Co.).
6. U.S. Pat. 3,760,015 (Sept. 18, 1973), S. Berkowitz (to FMC Corp.).
7. U.S. Pat. 3,870,762 (Mar. 11, 1975), M. H. Stacey and T. D. Tribbeck (to Imperial Chemical Industries, Ltd.).
8. U.S. Pat. 4,225,519 (Sept. 30, 1980), A. E. Reinhardt III (to PPG Industries).
9. P. W. Sherwood, *Ind. Eng. Chem.* **54**, 29 (1962).
10. U.S. Pat. 2,293,317 (Aug. 18, 1942), F. L. Taylor and L. H. Horsley (to The Dow Chemical Co.).
11. T. J. Haley, *Clin. Toxicol.* **8**, 633 (1975).
12. H. S. Warren and B. E. Ricci, *Oak Ridge National Labl Tox Information Response Center Report*, No. 77/3, NITS, Washington, D.C., 1978.
13. P. L. Viola and A. Caputo, *Environ. Health Perspect.* **21**, 45 (1977).
14. C. C. Lee and co-workers, *J. Toxicol. Environ. Health*, **4**, 15 (1978).
15. V. Ponomarkov and L. Tomatis, *Oncology*, **37**, 136 (1980).
16. R. D. Short and co-workers, *EPA Report*, No. PB281713, Environmental Protection Agency, Washington, D.C., 1977.
17. T. R. Blackwood, D. R. Tierney, and M. R. Piana, *EPA Report*, No. PB80-146442, Environmental Protection Agency, Washington, D.C., 1979.
18. J. M. Norris, personal communication, The Dow Chemical Co., Midland, Mich., 1982.
19. M. J. McKenna, P. G. Watanabe, and P. J. Gehring, *Environ. Health Perspect.* **21**, 99 (1977).
20. M. G. Ott and co-workers, *J. Occup. Med.* **18**, 735 (1976).
21. J. Hushon and M. Kornreich, *EPA Report*, No. PB280624, Environmental Protection Agency, Washington, D.C., 1978.
22. D. Foerst, *Am. Ind. Hyg. Assoc. J.* **40**, 888 (1979).
23. G. Talamini and E. Peggion, in G. E. Ham, ed., *Vinyl Polymerization*, Vol. 1, Marcel Dekker, Inc., New York, 1967, Part 1, Chapt. 5.
24. P. J. Flory, *Principles of Polymer Chemistry*, Cornell University Press, Ithaca, N.Y., 1953, Chapt. 6.
25. W. H. Stockmayer, K. Matsuo, and G. W. Nelb, *Macromolecules*, **10**, 654 (1977).
26. J. D. Burnett and H. W. Melville, *Trans. Faraday Soc.* **46**, 976 (1950).
27. C. E. Bawn, T. P. Hobin, and W. J. McGarry, *J. Chem. Phys.* **56**, 791 (1959).
28. W. J. Burlant and D. H. Green, *J. Polym. Sci.* **31**, 227 (1958).

29. R. C. Reinhardt, *Ind. Eng. Chem.* **35**, 422 (1943).

30. W. I. Bengough and R. G. W. Norrish, *Proc. R. Soc. London Ser. A*, **218**, 149 (1953).

31. R. A. Wessling and I. R. Harrison, *J. Polym. Sci. Part A-l*, **9**, 3471 (1971).

32. G. Odian, *Principles of Polymerization*, John Wiley & Sons, Inc., New York, 1981, Chapt. 3.

33. B. E. Obi and co-workers, *J. Polym. Sci: Part B: Polym. Phys.* **33**, 2019–2032 (1995).

34. J. L. Gardon, in C. E. Schildknecht, ed., *Polymerization Processes*, John Wiley & Sons, Inc., New York, 1977, Chapt. 6.

35. R. A. Wessling, *J. Appl. Polym. Sci.* **12**, 309 (1968).

36. R. A. Wessling and D. S. Gibbs, *J. Macromol. Sci. Chem.* **A7**, 647 (1973).

37. A. Konishi, *Bull. Chem. Soc. Jpn.* **35**, 197 (1962).

38. *Ibid.*, p. 193.

39. A. P. Sheniker and co-workers, *Dokl. Akad. Nauk SSSR*, **124**, 632 (1959).

40. D. R. Roberts and R. H. Beaver, *J. Polym. Sci. Polym. Lett. Ed.* **17**(3), 155 (1979).

41. B. A. Howell, A. M. Kelly-Rowley, and P. B. Smith, *J. Vinyl. Tech.* **11**, 159 (1989).

42. A. R. Westwood, *Eur. Polym. J.* **7**, 377 (1971).

43. J. Brandrup and E. H. Immergut, eds., *Polymer Handbook*, 2nd ed., John Wiley & Sons, Inc., New York, 1975.

44. J. F. Gabbett and W. Mayo Smith, in G. E. Ham, ed., *Copolymerization*, John Wiley & Sons, Inc., New York, 1964, Chapt. 10.

45. K. Matsuo and W. H. Stockmayer, *Macromolecules*, **10**, 658 (1977).

46. W. I. Bengough and R. G. W. Norrish, *Proc. R. Soc. London Ser. A*, **218**, 155 (1953).

47. C. Pichot, Q. T. Pham, and J. Guillot, *J. Macromol. Sci. Chem.* **12**, 1211 (1978).

48. C. Pichot and Q. T. Pham, *Eur. Polym. J.* **15**, 833 (1979).

49. Technical data, B. E. Obi, The Dow Chemical Company, Midland, Mich., 1995.

50. N. Yamazaki and co-workers, *Polym. Prepr. Am. Chem. Soc. Div. Polym. Chem.* **5**, 667 (1964).

51. A. Konishi, *Bull. Chem. Soc. Jpn.* **35**, 395 (1962).

52. B. L. Erusalimskii and co-workers, *Dokl. Akad. Nauk SSSR*, **169**, 114 (1966).

53. Brit. Pat. 1,119,746 (July 10, 1968) (to Chisso Corp.).

54. Can. Pat. 798,905 (Nov. 12, 1968), R. Buning and W. Pungs (to Dynamit Nobel Corp.).

55. U.S. Pat. 3,366,709 (Jan. 30, 1968), M. Baer (to Monsanto Co.).

56. U.S. Pat. 3,509,236 (Apr. 28, 1970), H. G. Siegler, R. B. Oberlar, and W. Pungs (to Dynamit Nobel Aktiengeselfichaft Co.).

57. U.S. Pat. 3,655,553 (Apr. 11, 1972), R. C. DeWald (to Firestone Tire and Rubber Co.).

58. M. Pegoraro, E. Beati, and J. Bilalov, *Chim. Ind. Milan*, **54**, 18 (1972).

59. T. Fisher, J. B. Kinsinger, and C. W. Wilson, *Polym. Lett.* **5**, 285 (1967).

60. M. Meeks and J. L. Koenig, *J. Polym. Sci. Part A-1*, **9**, 717 (1971).

61. S. Krimm, *Fortschr. Hochpolym. Forsch.* **2**, 51 (1960).

62. K. Matsuo and W. H. Stockmayer, *Macromolecules*, **8**, 660 (1975).

63. K. Okuda, *J. Polym. Sci. Part A*, **2**, 1749 (1964).

64. E. J. Arlman and W. M. Wagner, *Trans. Faraday Soc.* **49**, 832 (1953).

65. M. M. Coleman and co-workers, *J. Macromol. Sci. Phys.* **15**, 463 (1987).

66. M. S. Wu and co-workers, *J. Polym. Sci. Polym. Phys. Ed.* **18**, 95 (1980).

67. *Ibid.*, p. 111.

68. R. H. Boyd and L. Kesner, *J. Polym. Sci. Polym. Phys. Ed.* **19**, 393 (1981).

69. B. E. Obi, P. T. DeLassus, and E. A. Grulke, *Macromolecules*, **27**, 5491–5497 (1994).

70. R. F. Boyer and R. S. Spencer, *J. Appl. Phys.* **15**, 398 (1944).

71. K. H. Illers, *Kolloid Z.* **190**, 16 (1963).

72. R. A. Wessling and co-workers, *Appl. Polym. Symp.* **25**, 83 (1974).

73. N. W. Johnston, *Rev. Macromol. Chem.* **C14**, 215 (1976).

74. G. R. Riser and L. P. Witnauer, *Polym. Prepr. Am. Chem. Soc. Div. Polym. Chem.* **2**, 218 (1961).

75. B. G. Landes, P. T. DeLassus, and I. R. Harrison, *J. Macromol. Sci.* **B22**, 735 (1983–1984).

76. R. A. Wessling, J. H. Oswald, and I. R. Harrison, *J. Polym. Sci. Polym. Phys. Ed.* **11**, 875 (1973).

77. I. R. Harrison and E. Baer, *J. Colloid Interface Sci.* **31**, 176 (1969).

78. R. A. Wessling, D. R. Carter, and D. L. Ahr, *J. Appl. Polym. Sci.* **17**, 737 (1973).

79. A. F. Burmester and R. A. Wessling, *Bull. Am. Phys. Soc.* **18**, 317 (1973).

80. M. Asahina, M. Sato, and T. Kobayashi, *Bull. Chem. Soc. Jpn.* **35**, 630 (1962).

81. R. A. Wessling, *J. Appl. Polym. Sci.* **14**, 1531 (1970).

82. *Ibid.*, p. 2263.

83. M. L. Wallach, *Polym. Prepr. Am. Chem. Soc. Div. Polym. Chem.* **10**, 1248 (1969).

84. A. Revillion, B. Dumont, and A. Guyot, *J. Polym. Sci. Polym. Chem. Ed.* **14**, 2263 (1976).

85. A. Revillion, *J. Liq. Chromatogr.* **3**, 1137 (1980).

86. G. S. Kolesnikov and co-workers, *Izu. Akad. Nauk SSSR Otd. Khim. Nauk*, 731 (1959).

87. E. F. Jordan and co-workers, *J. Appl. Polym. Sci.* **13**, 1777 (1969).

88. S. W. Lasoski, *J. Appl. Polym. Sci.* **4**, 118 (1960).

89. S. W. Lasoski and W. H. Cobbs, *J. Polym. Sci.* **36**, 21 (1959).

90. H. J. Bixler and O. S. Sweeting, in O. J. Sweeting, ed., *The Science and Technology of Polymer Films*, Vol. 2, Wiley-Interscience, New York, 1971, Chapt. 1.

91. P. T. DeLassus, *J. Vinyl Technol.* **1**, 14 (1979).

92. P. T. DeLassus and D. J. Grieser, *J. Vinyl Technol.* **2**, 195 (1980).

93. P. T. DeLassus, *J. Vinyl Technol.* **3**, 240 (1981).

94. *Introduction to Barrier Polymer Performance*, No. 190-333-1084, The Dow Chemical Co., Midland, Mich., 1984.

95. *Kuraray EVAL Resin*, No. 6-1000-605, Kuraray Co., Ltd., Osaka, Japan, 1986.

96. P. T. DeLassus, *Proceedings of COEX America*, Scotland, 1986, p. 187.

97. P. T. DeLassus and co-workers, in J. H. Hotchkiss, ed., *Food and Packaging Interactions*, American Chemical Society, Washington, D.C., 1987, Chapt. 2.

98. P. T. DeLassus, G. Strandburg, and B. A. Howell, *Tappi J.* **71**(12), 152 (1988).

99. R. F. Boyer, *J. Phys. Colloid Chem.* **51**, 80 (1947).

100. G. M. Burnett, R. A. Haldon, and J. N. Hay, *Eur. Poly. J.* **4**, 83 (1968).

101. B. A. Howell and P. T. DeLassus, *J. Polym. Sci. Polymer. Chem. Ed.* **25**, 1967 (1987).

102. U.S. Pat. 3,321,417 (May 23, 1967) (to Union Carbide Corp.); N. L. Zutty and F. J. Welch, *J. Polym. Sci. Part A*, **1**, 2289 (1963).

103. B. A. Howell, *J. Polym. Sci. Polym. Chem. Ed.* **25**, 1981 (1987), and references cited therein.

104. J. D. Danforth, *Polym. Prepr.* **21**, 140 (1980).

105. J. D. Danforth, in P. O. Klemchuk, ed., *Polymer Stabilization and Degradation*, American Chemical Society, Washington, D.C., 1985, Chapt. 20, and references cited therein.

106. D. H. Everett and D. J. Taylor, *Trans. Faraday Soc.* **67**, 402 (1971).

107. D. Vesely, *Ultramicroscopy*, **14**, 279 (1984).

108. I. R. Harrison and E. Baer, *J. Colloid Interface Sci.* **31**, 176 (1969).

109. D. R. Roberts and A. L. Gatzke, *J. Polym. Sci. Polym. Chem. Ed.* **16**, 1211 (1978).

110. B. Dolezel, M. Pegoraro, and E. Beati, *Eur. Polym. J.* **6**, 1411 (1970).

111. P. Pendleton, B. Vincent, and M. L. Hair, *J. Colloid Interface Sci.* **80**, 512 (1981).

112. R. D. Bohme and R. A. Wessling, *J. Appl. Polym. Sci.* **16**, 1961 (1972).

113. A. Crovato-Arnaldi and co-workers, *J. Appl. Polym. Sci.* **8**, 747 (1964).

114. G. M. Burnett, R. A. Haldon, and J. N. Hay, *Eur. Polym. J.* **3**, 449 (1967).

115. D. H. Davies, D. H. Everett, and D. J. Taylor, *Trans. Faraday Soc.* **67**, 382 (1971).

116. R. A. Wessling, *Am. Chem. Soc. Div. Org. Coat. Plast. Chem. Pap.* **34**, 380 (1976).
117. F. F. He and H. Kise, *J. Polym. Sci. Polym. Chem. Ed.* **21**, 1972 (1983).
118. D. H. Davies, *J. Chem. Soc. Faraday Trans.* **1**, 72, 2390 (1976).
119. E. Tsuchido and co-workers, *J. Polym. Sci. Part A*, **2**, 3347 (1964).
120. S. S. Barton and co-workers, *Trans. Faraday Soc.* **67**, 3534 (1971).
121. S. S. Barton, J. R. Dacey, and B. H. Harrison, *Am. Chem. Soc. Div. Org. Coat. Plast. Chem. Pap.* **31**, 768 (1971).
122. A. Bailey and D. H. Everett, *J. Polym. Sci. Part A-2*, **7**, 87 (1969).
123. R. Simon, *Polym. Degr. Stab.* **43**, 125 (1994).
124. G. M. Burnett, R. A. Haldon, and J. N. Hay, *Eur. Polym. J.* **3**, 449 (1967).
125. D. E. Agostini and A. L. Gatzke, *J. Polym. Sci. Polym. Chem. Ed.* **11**, 649 (1973).
126. B. A. Howell, P. T. DeLassus, and C. Gerig, *Polym. Prepr.* **28**(1), 278 (1987).
127. P. L. Kumler and co-workers, *Macromolecules*, **22**, 2994 (1989).
128. V. Rossbach and co-workers, *Angew. Makromol. Chem.* **40–41**, 291 (1974).
129. D. H. Grant, *Polymer*, **11**, 581 (1970).
130. D. L. C. Jackson and W. S. Reid, *Nature*, **162**, 29 (1948).
131. B. A. Howell and P. B. Smith, *J. Polym. Sci. Polym. Phys. Ed.* **26**, 1287 (1988).
132. D. H. Davies and P. M. Henheffer, *Trans. Faraday Soc.* **66**, 2329 (1970).
133. E. Tsuchida and co-workers, *J. Polym. Sci. Part A*, **2**, 3347 (1964).
134. S. S. Barton and co-workers, *Trans. Faraday Soc.* **67**, 3534 (1971).
135. B. A. Howell and H. Liu, *Thermochim. Acta*, **212**, 1 (1992).
136. B. A. Howell and B. B. S. Sastry, *Proceedings of 22nd North American Thermal Analysis Society Meeting*, 1993, pp. 122–127.
137. U.S. Pat. 3,852,223 (Dec. 3, 1974), R. D. Bohme and R. A. Wessling (to The Dow Chemical Co.).
138. A. Ballistreri and co-workers, *Polymer*, **22**, 131 (1981).
139. S. Gopalkrishnan and W. H. Starnes, Jr., *Polym. Prepr.* **30**(2), 201 (1989).
140. B. A. Howell and J. R. Keeley, *Proceedings of 23rd North American Thermal Analysis Society Meeting*, 1994, pp. 672–677.
141. B. A. Howell and co-workers, *Thermochim. Acta*, **166**, 207 (1990).
142. B. A. Howell, M. F. Debney, and C. V. Rajaram, *Thermochim. Acta*, **212**, 115 (1992).
143. S. R. Betso and co-workers, *J. Appl. Polym. Sci.* **51**, 781 (1994).
144. B. A. Howell and C. V. Rajaram, *J. Vinyl Tech.* **15**, 202 (1993).
145. B. A. Howell and co-workers, *Polym. Adv. Tech.* **5**, 485 (1994).
146. B. A. Howell and co-workers, in Ref. 140, pp. 185–190.
147. J. Ozaki, T. Watanabe, and Y. Nishiyama, *J. Phys. Chem.* **97**, 1400 (1983).
148. L. A. Matheson and R. F. Boyer, *Ind. Eng. Chem.* **44**, 867 (1952).
149. G. Oster, G. K. Oster, and M. Kryszewski, *J. Polym. Sci.* **57**, 937 (1962).
150. M. Kryszewski and M. Mucha, *Bull. Acad. Pol. Sci. Ser. Sci. Chim. Geol. Geogr.* **13**, 53 (1965).
151. U.S. Pat. 4,418,168 (Nov. 29, 1983), E. H. Johnson (to the Dow Chemical Co.).
152. U.S. Pat. 3,033,812 (May 8, 1962), P. K. Isacs and A. Trofimow (to W. R. Grace & Co.).
153. D. M. Woodford, *Chem. Ind. (London)*, **8**, 316 (1966).
154. U.S. Pat. 2,968,651 (Jan. 17, 1961), L. C. Friedrich, Jr., J. W. Peters, and M. R. Rector (to The Dow Chemical Co.).
155. U.S. Pat. 2,482,771 (Sept. 27, 1944), J. Heerema (to The Dow Chemical Co.).
156. *Saran Resins*, No. 190-289-79, The Dow Chemical Co., Midland, Mich., 1979.
157. Ref. 90, Chapt. 6.
158. D. L. Roodvoets, in P. F. Bruin, ed., *Packaging with Plastics*, Gordon & Breach, New York, 1974, p. 85.
159. *SARANEX Films*, The Dow Chemical Co., Midland, Mich., 1979.

160. S. F. Roth, *American Chemical Society Chemical Marketing Economic Div. Symposium (N.Y.)*, American Chemical Society, Washington, D.C., 1976, p. 29.

161. *Saran F Resin*, Technical Bulletin, The Dow Chemical Co., Horgen, Switzerland, 1969.

162. U.S. Pat. 3,817,780 (June 18, 1974), P. E. Hinkamp and D. F. Foye (to The Dow Chemical Co.).

163. U.S. Pat. 3,879,359 (Apr. 22, 1975), P. E. Hinkamp and D. F. Foye (to The Dow Chemical Co.).

164. U.S. Pat. 4,097,433 (June 27, 1978), W. P. Kane (to E. I. du Pont de Nemours & Co., Inc.).

165. U.S. Pat. 2,462,185 (Feb. 22, 1949), P. M. Hauser (to E. I. du Pont de Nemours & Co., Inc.).

166. W. W. Cranmer, *Corrosion Houston*, **8**(6), 195 (1952).

167. R. L. Alumbaugh, *Mater. Prot.* **3**(7), 34, 39 (1964).

168. U.S. Pat. 3,144,352 (Aug. 11, 1964), J. P. Talley (to Ampex Corp.).

169. U.S. Pat. 3,865,741 (Feb. 11, 1975), F. J. Sischka (to Memorex Corp.).

170. U.S. Pat. 3,894,306 (July 1, 1975), F. J. Sischka (to Memorex Corp.).

171. L. J. Wood, *Mod. Packag.* **33**, 125 (1960).

172. R. F. Avery, *Tappi J.* **45**, 356 (1962).

173. A. D. Jordan, *Tappi J.* **45**, 865 (1962).

174. B. J. Sauntson and G. Brown, *Rep. Prog. Appl. Chem.* **56**, 55 (1972).

175. P. S. Bryant, *European FlexographicTechnical Association Barrier Coatings and Laminations Seminar*, Vol. 1, Manchester, U.K., 1977, p. 7.

176. G. H. Elschnig and co-workers, *Pop. Plast.* **17**(2), 19 (1972).

177. U.S. Pat. 3,787,232 (Jan. 22, 1974), B. K. Mikofalvy and D. P. Knechtges (to B. F. Goodrich Co.).

178. R. G. Jahn, *Adhes. Age*, **20**(6), 37 (1977).

179. U.S. Pat. 3,946,139 (Mar. 23, 1976), M. Bleyle and co-workers (to W. R. Grace & Co.).

180. U.S. Pat. 3,850,726 (Nov. 26, 1974), D. R. Smith and H. Peterson (to A. E. Staley Co.).

181. U.S. Pat. 3,617,368 (Nov. 2, 1971), D. S. Gibbs and R. A. Wessling (to The Dow Chemical Co.).

182. P. T. DeLassus, D. L. Clarke, and T. Cosse, *Mod. Plast.* 86 (Jan. 1983).

183. Brit. Pat. 1,233,078 (May 26, 1971), H. Gould and J. A. Zaslowsky (to Alcolac Chemical Co.).

184. G. H. Elschnig and A. F. Schmid, *Pop. Plast.* **17**(3), 36 (1972).

185. *Ibid.*, (4), p. 17.

186. *Idib.*, (6), p. 17.

187. G. H. Elschnig and A. F. Schmid, *Paintindia*, **22**(6), 22 (1972).

188. F. C. Caruso, *Proceedings of Test. Pap. Synth. Conference*, TAPPI, Atlanta, Ga., 1974, p. 167.

189. E. A. Chirokas, *TAPPI J.* **50**, 59A (1967).

190. U.S. Pat. 3,983,080 (Sept. 28, 1976), K. S. Suh, R. E. Skochdopole, and M. E. Luduc (to The Dow Chemical Co.).

191. U.S. Pat. 3,615,972 (Oct. 26, 1971), D. S. Morehouse and R. J. Tetreault (to The Dow Chemical Co.).

192. D. S. Morehouse and H. A. Walters, *SPE J.* **25**, 45 (1969).

193. T. E. Cravens, *Am. Chem. Soc. Div. Org. Coat. Plast. Chem. Pap.* **33**, 74 (1973).

194. R. C. Mildner and co-workers, *Mod. Plast.* **47**(5), 98 (1970).

195. T. F. Anderson, H. A. Walters, and C. W. Glesner, *J. Cell. Plast.* **6**(4), 171 (1970).

196. *Mater. Plast. Elastomer.* **10**, 468 (Oct. 1980).

197. D. L. Chamberlain, in W. C. Kuryla and A. J. Papa, eds., *Flame Retardancy of Polymeric Materials*, Vol. 2, Marcel Dekker, Inc., New York, 1973, pp. 109–168.

198. D. W. Van Krevelen, *Polymer*, **16**, 615 (1975).
199. L. G. Imhoff and K. C. Stueben, *Polym. Eng. Sci.* **13**, 146 (1973).
200. E. R. Larsen, *J. Fire Flamm. Fire Ret. Chem.* **1**, 4 (1974).
201. *Ibid.*, pp. 2 and 5.
202. R. C. Kidder, *Proceedings of Fire Retardant Chemicals Association Semi-Annual Meeting*, 1977, pp. 45–51.
203. J. Knightly and J. C. Bax, *Proceedings of the European Conference on Flammability Fire Retardance*, 1979, pp. 75–83.
204. J. R. Goots and D. P. Knechtges, *Polym. Plast. Technol. Eng.* **5**, 131 (1975).
205. C. V. Neywick, R. E. Yoerger, and R. F. Peterson, *J. Cell. Plast.* **16**, 171 (1980).
206. U.S. Pat. 4,232,129 (Nov. 4, 1980), D. S. Gibbs, J. H. Benson, and R. T. Fernandez (to The Dow Chemical Co.).
207. U.S. Pat. 4,002,597 (Jan. 11, 1977), E. D. Dickens (to B. F. Goodrich Co.).
208. U.S. Pat. 4,055,538 (Oct. 25, 1977), W. J. Kronke (to B. F. Goodrich Co.).
209. U.S. Pat. 4,186,156 (Jan. 29, 1980), D. S. Gibbs (to The Dow Chemical Co.).
210. *Plast. Technol.* **26**(1), 13 (1980).
211. U.S. Pat. 4,206,105 (June 3, 1980), O. L. Stafford (to The Dow Chemical Co.).
212. U.S. Pat. 4,239,799 (Dec. 16, 1980), A. S. Weinberg (to W. R. Grace & Co.).
213. U.S. Pat. 3,840,620 (Oct. 8, 1974), R. Gallagher (to Stauffer Chemical Co.).
214. U.S. Pat. 3,513,226 (May 19, 1970), T. Hotta (to Kureha Kagaku Kogyo Kabushiki Kalsha Co.).
215. U.S. Pat. 3,565,975 (Feb. 23, 1971), F. V. Goff, F. Stevenson, and W. H. Wineland (to The Dow Chemical Co.).
216. U.S. Pat. 3,558,542 (Jan. 26, 1971), J. W. McDonald (to E. I. du Pont de Nemours & Co., Inc.).
217. U.S. Pat. 3,896,066 (July 22, 1975), R. O. Ranck (to E. I. du Pont de Nemours & Co., Inc.).

R. A. WESSLING
D. S. GIBBS
P. T. DeLASSUS
B. E. OBI
The Dow Chemical Company

B. A. HOWELL
Central Michigan University

VINYLIDENE POLYMERS, POLY(VINYLIDENE FLUORIDE) ELASTOMERS. See FLUORINE COMPOUNDS, ORGANIC.

VINYL POLYMERS

VINYL ACETAL POLYMERS

Vinyl acetal polymers are made by the acid-catalyzed acetalization of poly(vinyl alcohol) [*9002-89-5*] with aldehydes (1).

Analogously, poly(vinyl ketals) can be prepared from ketones, but since poly(vinyl ketals) are not commercially important, they are not discussed here. The acetalization reaction strongly favors formation of the 1,3-dioxane ring, which is a characteristic feature of this class of resins. The first of this family, poly(vinyl benzal), was prepared in 1924 by the reaction of poly(vinyl alcohol) with benzaldehyde in concentrated hydrochloric acid (2). Although many members of this class of resins have been made since then, only poly(vinyl formal) [*9003-33-2*] (PVF) and poly(vinyl butyral) [*63148-65-2*] (PVB) continue to be made in significant commercial quantities.

Commercialization of PVF and PVB began during the 1930s and 1940s following development efforts by a number of companies, including Union Carbide, DuPont, Shawinigan Chemicals (now Monsanto Chemical), Wacker-Chemie, and I.G. Farben Industrie (now BASF) (3–12). One incentive for this activity was the discovery that safety windshields made with plasticized PVB interlayer offered significant advantages over windshields made with plasticized cellulose acetate, which was then in general use (13,14).

PVB accounts for about 90% of the poly(vinyl acetal) resin made in the 1990s. Most PVB is plasticized and made into interlayer for vehicle and architectural safety glass, and the remainder is used as an ingredient in a variety of coating, binding, printing, and adhesive applications. PVF accounts for the remaining 10% of the poly(vinyl acetal) made. PVF's primary use is for wire and cable insulation. In this application, PVF is combined with other reactive resins and cured to form tough, chemical- and abrasion-resistant coatings (see

INSULATION, ELECTRIC). Applications for PVF and PVB resins make use of the toughness, resilience, optical clarity, high pigment/filler binding capacity, and high adhesion the resins can provide when appropriately formulated.

The poly(vinyl acetal) prepared from acetaldehyde was developed in the early 1940s by Shawinigan Chemicals, Ltd., of Canada and sold under the trade name Alvar. Early uses included injection-molded articles, coatings for paper and textiles, and replacement for shellac. Production peaked in the early 1950s and then decreased as a result of competition from less expensive resins such as poly(vinyl chloride) (see VINYL POLYMERS, POLY(VINYL CHLORIDE)). Alvar is no longer manufactured.

Synthesis and Structure

Poly(vinyl alcohol) used to manufacture the poly(vinyl acetal)s is made from poly(vinyl acetate) homopolymer (see VINYL POLYMERS, VINYL ALCOHOL POLY-MERS; VINYL POLYMERS, VINYL ACETATE POLYMERS). Hydrolysis of poly(vinyl acetate) homopolymer produces a polyol with predominantly 1,3-glycol units. The polyol also contains up to 2 wt % 1,2-glycol units that come from head-to-head bonding during the polymerization of vinyl acetate monomer. Poly(vinyl acetate) hydrolysis is seldom complete, and for some applications, not desired. For example, commercial PVF resins may contain up to 13 wt % unhydrolyzed poly(vinyl acetate). Residual vinyl acetate units on the polymer help improve resin solubility and processibility (15). On the other hand, the poly(vinyl alcohol) preferred for commercial PVB resins has less than 3 wt % residual poly(vinyl acetate) units on the polymer chain.

Poly(vinyl acetals) are made from poly(vinyl alcohol) and aldehydes by acid-catalyzed addition–dehydration. The mechanism of acetalization has been proposed (16,17). The degree of acetalization and the conditions used during the reaction significantly affect product properties. Batch and continuous processes in both aqueous and organic media are used during manufacturing. In single-stage batch processes, hydrolysis of poly(vinyl acetate) and acetalization of the poly(vinyl alcohol) hydrolysis product are carried out in the same kettle at the same time. In two-stage batch processes, hydrolysis and acetalization take place in separate kettles.

Assuming that acetalization is irreversible and that both 1,3- and 1,2-glycol units are present in poly(vinyl alcohol), the highest degree of acetalization that can be expected on a statistical basis is 81.6% (18). However, acetalization is reversible (19), and even higher degrees of acetalization are possible (7,20,21). For most applications, complete acetalization is not needed, and would be difficult to achieve on a commercial scale without a significant amount of intermolecular acetalization.

Commercial poly(vinyl acetal)s are terpolymers with varying amounts of vinyl acetate and vinyl alcohol units remaining on the backbone after acetalization. The class can be represented by the following structure, showing acetal (**1**), vinyl alcohol (**2**), and vinyl acetate (**3**) units.

The physical properties of the resin can be modified over a wide range of values because resin properties are a function of the relative amounts of the three monomeric units, average molecular weight, molecular weight distribution, aldehyde chain length, and the stereochemistry of the backbone pendent groups, all of which can be manipulated during manufacturing. Aldehyde blends have also been used to achieve unique properties (22).

Both intramolecular and intermolecular acetalization can occur, although intramolecular acetalization predominates during early stages of the reaction. Late in the reaction intermolecular acetalization begins to take place when isolated hydroxyl groups from two different polymer chains form acetal linkages. As the level of intermolecular acetalization increases, the resin becomes more difficult to process and gel particles form as cross-linked networks begin to build.

For commercial applications, a small degree of intermolecular acetalization is tolerated and, to a limited extent, it can be manipulated to control the molecular weight distribution of the resin. This in turn affects the solubility and rheological properties of the resin. Other than the possibility of cross-linking, hydrolysis and acetalization do not significantly alter the resin's chain length. Thus the molecular weight of an acetylated resin is largely determined by that of the poly(vinyl acetate) from which it is derived.

The poly(vinyl alcohol) made for commercial acetalization processes is atactic and a mixture of *cis*- and *trans*-1,3-dioxane stereoisomers is formed during acetalization. The precise cis/trans ratio depends strongly on process kinetics (16,17) and small quantities of other system components (23). During formylation of poly(vinyl alcohol), for example, *cis*-acetalization is more rapid than *trans*-acetalization (24). In addition, the rate of hydrolysis of the *trans*-acetal is faster than for the *cis*-acetal (25). Because hydrolysis competes with acetalization during acetal synthesis, a high cis/trans ratio is favored. The stereochemistry of PVF and PVB resins has been studied by proton and carbon nmr spectroscopy (26–29).

The resin's unacetalated hydroxyl groups take part in both intramolecular and intermolecular hydrogen bonding. Intermolecular hydrogen bonding helps bind individual polymer molecules together, making it more difficult for polymer chains to untangle and slip by each other. Resin glass-transition temperature (T_g), viscosity, modulus, and tensile strength increase as the hydroxyl level increases. Simultaneously, processability and resiliency are reduced. Backbone hydroxyl groups also can form hydrogen and covalent bonds to the surface of polar substrates and are largely responsible for the adhesion characteristics of this resin class (30).

Physical Properties

The thermal glass-transition temperatures of poly(vinyl acetal)s can be determined by dynamic mechanical analysis, differential scanning calorimetry, and nmr techniques (31). The thermal glass-transition temperature of poly(vinyl acetal) resins prepared from aliphatic aldehydes can be estimated from empirical relationships such as equation 1 where OH and OAc are the weight percent of vinyl alcohol and vinyl acetate units and C is the number of carbons in the chain derived from the aldehyde. The symbols with subscripts are the corresponding values for a standard (s) resin with known parameters (32). The formula accurately predicts that resin T_g increases as vinyl alcohol content increases, and decreases as vinyl acetate content and aldehyde carbon chain length increases.

$$T_g = T_g s + 1.26(OH - OH_s) - 0.6(OAc - OAc_s) + 46 \ln(C_s/C) \qquad (1)$$

Unformulated poly(vinyl acetal) resins form hard, unpliable materials which are difficult to process without using solvents or plasticizers. The solubility parameter ranges for some commercially available PVF and PVB resins are listed in Table 1. Plasticizers not only aid resin processing but also lower the glass-transition temperature, T_g, and can profoundly change other physical properties of the resins. For example, the mechanical glass-transition temperature of a poly(vinyl acetal) plasticized with dibutyl phthalate is reduced by 1.3°C per part of plasticizer added to 100 parts resin (34). Acetalization with longer chain aldehydes (35) and aldehydes with polyalkylene oxide chains (36) provides a degree of internal plasticization. Long-chain acetal ends tend to reduce T_g without relying exclusively on liquid plasticizers (37).

Table 1. Solubility Parameter Ranges for Some Commercial PVF and PVB Resins[a]

Resin	Low hydrogen bonding solvents	Medium hydrogen bonding solvents	High hydrogen bonding solvents
PVF			
low acetate	9.3–10.0	9.7–10.4	9.9–11.8
PVB			
hydroxyl (9–13%)	9.0–9.8	8.4–12.9	9.7–12.9
hydroxyl (17–21%)	[b]	9.9–12.9	9.7–14.3

[a]Ref. 33.
[b]Insoluble.

Poly(vinyl acetal)s can be formulated with other thermoplastic resins and with a variety of multifunctional cross-linkers. Examples of resins that are at least partially compatible with PVF or PVB resins include some types of polyurethanes (38), polyvinylpyrrolidinone (39), cellulose triacetate (40), nitrocelluloses, poly(vinyl chloride) (41,42), epoxies, isocyanates, phenolics, silicones, and urea–formaldehyde resins. During the curing step with cross-linkers, which is usually carried out at elevated temperatures, covalent bonds are formed between hydroxyl groups on the poly(vinyl acetal) backbone and the reactive centers on the cross-linker. When cross-linking takes place the resin becomes thermoset. Thermosetting generally increases thermal stability, rigidity, and abrasion resistance, and improves resistance to solvents and to acids and bases. It also severely limits processibility by making the resin insoluble and impossible to extrude.

Although they lack commercial importance, many other poly(vinyl acetal)s have been synthesized. These include acetals made from vinyl acetate copolymerized with ethylene (43–46), propylene (47), isobutylene (47), acrylonitrile (48), acrolein (49), acrylates (50,47), allyl ether (51), divinyl ether (52), maleates (53,54), vinyl chloride (55), diallyl phthalate (56), and starch (graft copolymer) (47).

Resins with ionomeric pendent groups have also been prepared by acetalating with an aldehyde containing ionic functionality (57–60). At moderate temperatures the ionic groups cluster to form crystalline-like domains which make the resin stiffer than conventional PVB resins (61). At higher processing temperatures, the crystalline-like domains melt out, resulting in sharper drops in resin viscosity. This reduces stickiness and increases resistance to blocking at ambient temperatures, and lowers melt viscosity and improves flow during higher temperature extrusion and laminating steps compared to conventional PVB resins (62).

Health and Safety Factors

Representative unformulated PVB and PVF resins are practically nontoxic orally (rats) and no more than slightly toxic after skin application (rabbits). No mortalities were produced in rats exposed by single inhalation for four hours and no allergic skin reactions were observed in controlled skin contact studies with human volunteers. More specifically, a representative PVB resin is practically nontoxic by single-dose oral ingestion ($LD_{50} > 10.0$ g/kg) or by single dermal applications ($LD_{50} > 7.9$ g/kg), and it is only slightly irritating to the eyes (2.8 on a scale of 0–110) and nonirritating (0 on a scale of 0–8) to the skin of rabbits tested in standard FHSA tests for irritation (63). Unformulated, the resins appear to have no acute toxicological properties that would require special handling other than good hygienic practices. The highest purity grades of PVB can be formulated to meet the extractibility requirements of the FDA, and some can be used in accordance with CFR regulations as ingredients for can enamels, adhesives, and components for paper and paperboard in contact with aqueous and fatty foods (33).

OSHA and ACGIH have not established specific airborne exposure limits for PVB and PVF resins; however, some products may contain sufficient fines to be considered nuisance dust and present dust explosion potential if sufficient quantities are dispersed in air. Unformulated PVB and PVF resins have flash points above 370°C. The lower explosive limit (lel) for PVB dust in air is about 20 g/m².

Many grades of PVB and PVF resins are made, and most are eventually compounded and used as multicomponent products. Individual product MSDSs need to be consulted prior to handling and each product should be handled appropriately and in accordance with good industrial hygiene and safety practices, which include appropriate skin, respiratory, and eye protection.

Poly(vinyl butyral)

Monsanto, DuPont, Hoechst, Sekisui, and Wacker Chemie are major manufacturers of PVB resins. Several grades are available that differ primarily in residual vinyl alcohol content and molecular weight. Both variables strongly affect solution viscosity, melt flow characteristics, and other physical properties. The physical, mechanical, and thermal properties of various grades of Monsanto's Butvar resins are listed in Tables 2–4, respectively. In general, resin melt and solution viscosity increase with increasing molecular weight and vinyl alcohol content, whereas the tensile strength of materials made from PVB increases with vinyl alcohol content for a given molecular weight.

Table 2. Physical Properties of Butvar Resins

Property	Method	B-72	B-74	B-76	B-90	B-98
molecular mass × 10³ (avg)	[a]	170–250	120–250	90–120	70–100	40–70
viscosity 15 wt %, Pa·s[b]	[c]	7–14	3–7	0.5–1	0.6–1.2	0.2–0.4
viscosity 10 wt %, Pa·s[b]	[d]	1.6–2.5	0.8–1.3	0.2–0.45	0.2–0.4	0.07–0.2
Ostwald soln viscosity, mPa·s[b] (=cP)	[e]	170–260	40–50	18–28	13–17	6–9
specific gravity, 23°C/23°C	ASTM D792-50	1.100	1.100	1.083	1.100	1.100
refractive index	ASTM D542-50	1.490	1.490	1.485	1.490	1.490
vinyl alcohol content, wt %		17–20	17–20	11–13	18–20	18–20
vinyl acetate content, wt %		0–2.5	0–2.5	0–1.5	0–1.5	0–2.5

[a]Determined by size exclusion chromatography in tetrahydrofuran with low angle light scattering.
[b]To convert Pa·s to P, multiply by 10.
[c]Measured in 60:40 toluene:ethanol at 25°C using a Brookfield viscometer.
[d]Measured in 95% ethanol at 25°C using an Ostwald-Cannon-Fenske viscometer.
[e]B-72 in 7.5 wt % anhydrous methanol at 20°C; B-76 and B-79 in 5.0 wt % SD 29 ethanol at 25°C; B-74, B-90, and B-98 in 6.0 wt % anhydrous methanol at 20°C, all using an Ostwald-Cannon-Fenske viscometer.

Table 3. Mechanical Properties of Butvar Resins

Property	ASTM method	B-72	B-74	B-76	B-90	B-98
tensile strength, MPa[a]						
yield	D638-58T	47–54	47–54	40–47	43–50	43–50
break		48–55	48–55	32–39	39–46	39–46
elongation, %						
yield	D638-58T	8	8	8	8	8
break		70	75	110	110	110
modulus of elasticity, GPa[b]		2.28–2.34	2.28–2.34	1.93–2.0	2.07–2.14	2.14–2.21
flexural strength yield, MPa[a]	D790-59T	83–90	83–90	72–79	76–83	76–83
hardness, Rockwell						
M	D785-57	115	115	100	115	110
E		20	20	5	20	20
impact strength, J/m[c]	D256-56[d]	58.7	58.7	42.7	48	37.4

[a]To convert MPa to psi, multiply by 145.
[b]To convert GPa to psi, multiply by 145,000.
[c]To convert J/m to ft·lb/in., divide by 53.38.
[d]Notched Izod (1.27 × 1.27 cm (0.5 × 0.5 in.)).

Table 4. Thermal Properties of Butvar Resins

Property	Method	B-72	B-74	B-76	B-90	B-98
flow temperature, °C, 6.9 MPa[a]	ASTM D569-59	145–155	135–145	110–115	125–130	105–110
T_g, °C	[b]	72–78	72–78	62–72	72–78	72–78
heat distortion temperature, °C	ASTM D648-56	56–60	56–60	50–54	52–56	45–55
heat sealing temperature, °C	[c]	220	220	200	205	200

[a]6.9 MPa = 1000 psi.
[b]By differential scanning calorimetry from 30 to 100°C on dried resin.
[c]Dried film (0.025 mm) on paper cast from 10 wt % resin in 60:40 toluene:ethanol; heat sealer dwell time, 1.5 s at 0.4 MPa (60 psi).

Commercially available PVB resins are generally soluble in lower molecular weight alcohols, glycol ethers, and certain mixtures of polar and nonpolar solvents. A representative list is found in Table 5. Grades with lower vinyl alcohol content are soluble in a wider variety of solvents. A common solvent for all of the Butvar resins is a combination of 60 parts of toluene and 40 parts of ethanol (95%) by weight.

PVB resins are also compatible with a limited number of plasticizers and resins. Plasticizers (qv) improve processability, lower T_g, and increase flexibility and resiliency over a broad temperature range. Useful plasticizers include dibutyl and butyl benzyl phthalates, tricresyl and 2-ethylhexyl diphenyl phosphates, butyl ricinoleate, dibutyl sebacate, dihexyl adipate, triethy-

Table 5. Solubility of Butvar Resins[a]

Solvent	B-72; B-74[b]	B-76; B-79[c]	B-90; B-98[c]
acetic acid (glacial)	S	S	S
acetone	I	S	SW
butyl acetate	I	S	PS
N-butyl alcohol	S	S	S
butyl cellosolve	S	S	S
cyclohexanone	S	S	S
diacetone alcohol	PS	S	S
diisobutyl ketone	I	SW	I
N,N-dimethylacetamide	S	S	S
N,N-dimethylformamide	S	S	S
dimethylsulfoxide	S	S	S
ethyl acetate (99%)	I	S	PS
ethyl alcohol (95%; anhydrous)	S	S	S
ethylene dichloride	SW	S	SW
ethylene glycol	I	I	I
isophorone	PS	S	S
isopropyl alcohol (95%; anhydrous)	S	S	S
isopropyl acetate	I	S	I
methyl acetate	I	S	PS
methyl alcohol	S	SW	S
methyl ethyl ketone	SW	S	PS
methylene chloride	PS	S	S
methyl isobutyl ketone	I	S	I
naphtha (light)	I	SW	I
N-methyl-2-pyrrolidinone	S	S	S
propylene dichloride	SW	S	SW
tetrachloroethylene	SW	SW	SW
tetrahydrofuran	S	S	S
toluene	I	PS	SW
toluene:ethanol, 95% (60:40 by wt)	S	S	S
1,1,1-trichloroethane	SW	S	SW
xylene	I	PS	SW

[a]S, soluble; PS, partially soluble; I, insoluble; and SW, swells.
[b]Employed 5% solids mixture agitated for 24 h at ambient temperature.
[c]Employed 10% solids mixture agitated for 24 h at ambient temperature.

lene glycol di-2-ethylbutyrate, tetraethylene glycol diheptanoate, castor oil, and others (64–73).

PVB combinations with the thermoplastic resins nitrocellulose or shellac have been used as sealers for wood finishing. In these applications the PVB component adds flexibility and adhesion. Tough, optically clear blends have been made with aliphatic polyurethanes (74). Conducting polyacetylene exhibits enhanced stability under ambient conditions when blended with PVB (75). Thermosets are prepared with cross-linkers that form covalent bonds with hydroxyl groups. Cross-linking resins include select isocyanate, phenolic, and epoxy resins. Lower molecular weight organic cross-linkers include dialdehydes, isocyanates, and etherified melamines. Inorganics that can function as cross-linkers include chromic, boric, and phosphoric acid and their derivatives (76).

Manufacture.   PVBs are manufactured by a variety of two-stage heterogeneous processes. In one of these an alcohol solution of poly(vinyl acetate) and an acid catalyst are heated to 60–80°C with strong agitation. As the poly(vinyl alcohol) forms, it precipitates from solution (77). Ethyl acetate, the principle by-product, is stripped off and sold. The precipitated poly(vinyl alcohol) is washed to remove by-products and excess acid. The poly(vinyl alcohol) is then suspended in a mixture of ethyl alcohol, butyraldehyde, and mineral acid at temperatures above 70°C. As the reaction approaches completion the reactants go into solution. When the reaction is complete, the catalyst is neutralized and the PVB is precipitated from solution with water, washed, centrifuged, and dried. Resin from this process has very low residual vinyl acetate and very low levels of gel from intermolecular acetalization.

In the second stage of a representative aqueous process, an aqueous solution of poly(vinyl alcohol) is heated with butyraldehyde [123-72-8] and an acid catalyst (78,79). PVB precipitates from solution as it forms. After 2–3 h at a temperature of about 90°C, when the reaction is complete, the acid is neutralized and the resin is washed, filtered, and dried. While the cost of solvent handling is minimal in this process, it is offset somewhat by the need for separate hydrolysis and acetalization facilities. Because PVB resin precipitates early in the reaction there is a tendency toward high levels of intermolecular acetalization. Cross-linking can be minimized by adding emulsifiers to control particle size (80–84) or substances like ammonium thiocyanate (85), or urea (86) to improve the solubility of PVB in the aqueous phase. To increase the average molecular weight of the resin small quantities of a dialdehyde like glutaraldehyde can be added during the acetalization step (87).

Applications.   During 1994, about 68,000 t of unplasticized PVB was manufactured worldwide. Of this, the overwhelming majority, about 66,000 t, was plasticized and extruded into sheet for use in laminated safety glass. Only about 2,300 t of unplasticized PVB was used for noninterlayer applications.

Laminated Glass.   About 1.5×10^8 m^2 of safety glazing interlayer was manufactured worldwide in 1994. About 75% of this was used for vehicle windshields and most of the rest was used for laminated architectural glazing (88,89) and a variety of security glass applications (90). Major producers of interlayer for laminated glass are Monsanto (Saflex), the largest producer, followed by DuPont (Butacite), Sekisui (S'Lec) and Hüls (Trosofoil).

Plasticized PVB is uniquely suited for safety and security glazing applications. It is easily extruded into sheet. The laminated sheet exhibits high adhesion to glass, optical clarity, stability to sunlight, and high tear strength and impact-absorbing characteristics, all of which are demanded for safety glazing use. In windshields, for example, the interlayer serves several potentially life-saving functions. It adheres to glass shards after a glass-breaking impact, thereby helping to reduce injury from flying glass. After the glass is broken during an impact, the interlayer's high tear strength and resiliency acts like a safety net by absorbing enough energy to resist penetration by a projectile or a vehicle occupant's head (91). In addition to its safety features, laminated glass in architectural applications adds sound attenuation, heat insulation, and break-in security, and also blocks uv radiation.

Most laminated safety glazings are glass–PVB–glass trilayer composites, but bullet- and projectile-resistant laminates. Laminates for other specialty uses may be made with more than three layers. In addition to glass, poly(ethylene terephthalate) (PET), acrylic, and polycarbonate structural components (92) are used in specialty laminated glazing products.

Many grades of interlayer are produced to meet specific length, width, adhesion, stiffness, surface roughness, color (93,94), and other requirements of the laminator and end use. Sheet can be supplied with vinyl alcohol content from 15 to about 23 wt %, depending on the supplier and application. A common interlayer thickness for automobile windshields is 0.76 mm, but interlayer used for architectural or aircraft glazing applications, for example, may be much thinner or thicker. There are also special grades to bond rear-view mirrors to windshields (95,96) and to adhere the components of solar cells (97,98). Multilayer coextruded sheet, each component of which provides a separate property not possible in monolithic sheet, can also be made (99–101).

Adhesion of plasticized PVB to clean glass is very high. It is postulated that both hydrogen bonds and silyl alkyl ether covalent bonds are formed between resin hydroxyl groups and silanol groups at the interlayer–glass interface (30) (Fig. 1). However, the surface chemistry of float glass is complex, and both moisture and low concentrations of ionic substances at the interlayer–glass interface affect adhesion. Thus, interlayer–glass adhesive strength depends not only on the interlayer's formulation and moisture content, but also on the type, age, and condition of the surface of the glass. Salts and moisture are thought to reduce adhesion by competing with resin hydroxyls for bonding sites on the glass surface.

Because there is an inverse relationship between interlayer–glass adhesive strength and laminate penetration resistance, adhesion must be carefully controlled in safety glass applications. If adhesion is too low, glass retention during an impact will be sacrificed even though the interlayer is not penetrated. If adhesion is too high, a projectile can easily penetrate the laminate because cracks made in the glass propagate through the interlayer. In order for the interlayer to

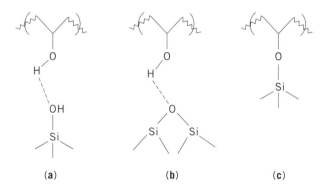

(a)   (b)   (c)

Fig. 1.   Hydrogen and covalent bonds can form between the resin's hydroxyl groups and the surface of glass: (**a**) a hydrogen bond to a silanol group; (**b**) a hydrogen bond to a silyl ether oxygen; and (**c**) an ether-type covalent bond.

cushion an impact, enough debonding must take place in the laminate to allow the interlayer to expand after the glass is cracked.

One measure of impact resistance is the laminate's mean break height (MBH) (102). In the standard test, there is a 50% probability that a five-pound (2.27-kg) ball will not fall through a laminate if the ball is dropped at the MBH. Typical MBHs for 12 in. (~30 cm) square laminates prepared with 30 mil (0.76 mm) thick interlayer are 10 ft (~3 m) at 0°F (−18°C) and 15 ft (~4.6 m) at 70°F (21°C). Figure 2 shows a relationship between adhesion and falling ball penetration resistance measured at 21°C.

To maximize penetration resistance without danger of delamination during impact or aging, adhesion to glass is reduced from inherently high levels (10, in Fig. 2) to intermediate levels (3−7, in Fig. 2) by adding parts per million quantities of Group Ia or IIa alkanoate salts (103−113) or surfactants (114) to the resin. It is not clear how these additives influence adhesion, but they also strongly affect the product's moisture sensitivity and some other performance attributes. Often adhesion fine tuning is carried out on the interlayer after it is extruded by adjusting the interlayer moisture level to between 0.2 and 1%.

Plasticized PVB interlayer is hygroscopic. In addition, T_gs are in the neighborhood of 30°C; thus, interlayer tends to adhere to itself, or block, when rolls or stacks of cut blanks are stored at ambient conditions. For these reasons handling and shipping must be carried out under controlled humidity and at temperatures well below the sheet's T_g. Precut interlayer blanks and rolls are usually stored or shipped refrigerated (3−10°C), and when rolls need to be stored or shipped at ambient conditions, the sheet is interleaved with a thin sheet of nonadhering plastic such as polyethylene.

To help prevent air from being trapped between glass and interlayer surfaces during lamination, the interlayer surface is textured during manufacture (115−120). The textured surface provides pathways for air to escape. Several methods are used to de-air during lamination. In one, the sheet is positioned between two pieces of glass and the prelaminate is tacked together by nipping the sheet between large rubber rolls at about 38°C. The sandwich is then heated to

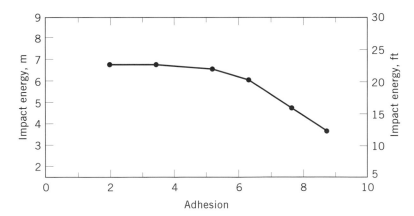

Fig. 2. Typical impact energy vs adhesion (10 is high, 0 is no adhesion) curve obtained from the mean break height test carried out at 21°C with a 2.27 kg ball on 30-cm square laminates. Mean break height dramatically declines when adhesion is at high levels.

about 50–75°C and subjected to an additional nipping step to eliminate most of the remaining air and seal the edges. Excess sheet at the edges is trimmed with a trimming tool. Finally, the laminate is autoclaved at 135–150°C for 30–120 min in pressurized air, 1.3–1.8 MPa (13 to 18 kg/cm^2), or in sealed bags under vacuum (121). During lamination, full adhesion and optical clarity are attained as the softened interlayer flows and wets the glass. Any remaining trapped air dissolves into the sheet.

Interlayer with various widths of gradient color bands are made for automobile windshields (122–124). The band is added by printing the interlayer surface or by coextruding previously mixed dyed or pigmented resin. Before laminating, gradient interlayer is often shaped to accommodate the curvature of the windshield. This is accomplished by differentially stretching sheet heated at about 85–100°C over a tapered shaping drum. The hot sheet is then cooled with cold air, cut into blanks, and stacked prior to laminating.

Noninterlayer Applications. Some categories of nonglazing uses for PVB are as follows:

Phenolic/adhesives
  adhesives/sealants/gaskets
  abrasives and break pad binders
  printed circuit boards
  structural composites/laminates

Metal/glass binders
  metal oxide binders
  retroreflective coatings
  glass optical coatings

Hard copy printing
  inks
  toners
  photoimaging
  reprographics

Coatings/additives
  wash primers
  wood coatings
  paint and varnish additives
  maintenance, metal, and
    industrial coatings
  wire enamels
  can coatings
  foil coatings
  nail polish

Others
  pest strips
  powder coatings
  miscellaneous

For most of these, the resin serves as a film-forming additive, an adhesive, and/or as a binder for pigments, fillers, metal oxides, ceramics, and other materials. The resin provides toughness and flexibility and adheres strongly to additive particles and to glass or other polar surfaces. For example, PVB binds the glass beads used in retroreflective films for licence plates, decals, and road signs. It also helps bind the retroreflective films to the surfaces of which they are applied. PVB is used to bind metal oxide particles during the fabrication of certain glasses, ceramics (125–127), and superconductors (128). PVB resin is also used to make membranes for fuel cells (129), gas–liquid separators (130), and ultrafiltration devices (131).

PVB is used as a film-forming component in corrosion-inhibiting primers for metals called wash primers or metal conditioners. The primers have good adhesion to ferrous metal surfaces and form a good foundation for various types of topcoats. The primers act by stabilizing the metal oxide surface and by continuously supplying corrosion-inhibiting ions such as chromate (132). Formulations containing nontoxic corrosion-inhibiting ions have also been made (133,134). Wash primers are used on storage tanks, bridges, ships, highway guard rails, and on submerged structures (76,135).

Aqueous dispersions of plasticized PVB are used for many coating applications (136,137). The dispersions are manufactured by intensive kneading of the resin in the presence of a plasticizer, surfactant, and water (138–140). Generally, the dispersions are about 50% solids, including about 40 parts of plasticizer per 100 parts of resin. Cast, air-dried films are tough and transparent and adhere to many materials. They are resistant to water and grease, which makes them useful as coatings for grease-proof washable wallpaper, window shades, and packaging materials. Properly formulated, they can be used for strippable or temporary protective coatings.

PVB dispersions are widely used in the textile industry to impart abrasion resistance, durability, and fiber strength. They have been used for finishing nylon webbing for parachute harnesses and seat belts. They can be applied to textiles by spraying, from a dilute bath by impregnation on a padder, or from a thickened dispersion by coating on spreading equipment. The dried dispersion imparts a soft, full-bodied finish to rayon, cotton, or nylon and helps prevent raveling of filament yarns. PVB dispersions have been used to finish curtain and drapery fabrics and upholstery goods, to join fabric to fabric and other materials, and as a component in transparent rug backings.

Poly(vinyl formal)

Estimated worldwide production of poly(vinyl formal) resin was about 2700 t in 1994. PVF resins are currently manufactured by Wacker Chemie (Pioloform F) in Germany and by Chisso (Vinylec) in Japan. Chisso purchased Monsanto's PVF (Formvar) business in 1992. The Vinylec resins are free-flowing white powders with a poly(vinyl formal) content of about 81 wt %. The properties of representative grades are listed in Table 6. Chemical resistance of Vinylec to

Table 6. Properties of Vinylec Poly(vinyl formal) Resins[a]

Property	Value
specific gravity	1.1–1.3
tensile strength, MPa[b]	49–78
flexural strength, MPa[b]	108–127
hardness, Shore	70–80
flow temperature, °C	140–150
heat distortion temperature, °C	84–93
dielectric strength, kV/mm	26–39
dielectric constant, 60 Hz	3.0–3.7
volume resistivity, $\Omega \cdot$cm	10^{14}–10^{16}
dissipation factor, 60 Hz	0.006–0.015

[a]Ref. 141.
[b]To convert MPa to psi, multiply by 145.

acids, bases, and aliphatic hydrocarbons is excellent, and chemical resistance to alcohols, aromatic hydrocarbons, esters, and ketones is good; however, chemical resistance to chlorinated solvents is rated poor (141). Residual vinyl acetate and vinyl alcohol component levels are 9–13 wt % and about 5 wt %, respectively. Grades are available with average molecular weights from about 25,000 to 100,000.

In general, PVF resins are soluble in a limited number of solvents and in certain mixtures of alcohols and aromatic hydrocarbons.

The solubility of representative poly(vinyl formal) resins is as follows in single solvents, where S = soluble and I = insoluble:

acetic acid (glacial)	S	chloroform	S
monochloro acetic acid	S	ethylene dichloride	S
benzyl alcohol	S	acetone	I
dioxane	S	carbon disulfide	I
cresol	S	ethyl alcohol	I
tetrahydrofuran	S	ethyl acetate	I
dimethyl sulfoxide	S	butyl acetate	I
methyl benzoate	S	methyl ethyl ketone	I
N,N-dimethylformamide	S	cyclohexanone	I
N,N-dimethylacetamide	S	nitropropane	I
xylenol	S	hydrocarbons	
furfural	S	aliphatic	I
pyridine	S	aromatic	I

For mixed solvents, PVF is soluble in toluene:ethyl alcohol (95%) (60:40 wt %) and xylene:methyl alcohol (60:40 wt %), but insoluble in toluene:butyl alcohol (60:40 wt %) and xylene:butyl alcohol (60:40 wt %). The solubility of PVF resins in polar solvents increases with increasing proportions of residual vinyl acetate content. Increasing vinyl acetate content also reduces resin stiffness and tensile and impact strength. Solution viscosity increases with increasing vinyl alcohol content and with the average molecular weight of the resin. Although moderate increases in the average molecular weight of the various commercial grades do not substantially affect tensile strength, modulus, and some other physical properties of the unmodified resins, they strongly increase solution viscosity.

PVF resins are generally compatible with phthalate, phosphate, adipate, and dibenzoate plasticizers, and with phenolic, melamine–formaldehyde, urea–formaldehyde, unsaturated polyester, epoxy, polyurethane, and cellulose acetate butylate resins. They are incompatible with polyamide, ethyl cellulose, and poly(vinyl chloride) resins (141).

Commercial PVF is manufactured by a single-stage batch process in acetic acid (142–144). In the single-stage process, hydrolysis and formalization take place concurrently. In one process, poly(vinyl acetate) is dissolved in an aqueous mixture of acetic acid and formaldehyde. Sulfuric acid catalyst is added and the mixture is maintained at 75–85°C for about 6–8 h or until the reactions are completed. The average molecular weight of the product is largely determined by the average molecular weight of the poly(vinyl acetate) charge. The ratio of vinyl acetate and vinyl alcohol components in the acetal product is controlled by the ratio of acetic acid, water, and formaldehyde used. When the reaction is complete, the mineral acid is neutralized. As water is added to the agitated mixture, PVF resin precipitates as fine, off-white granules. Color can be improved by adding antioxidants during formalization (145). The resin is centrifuged and dried after washing with water to remove salt and organic by-products. In this process, hydrolysis of poly(vinyl acetate) is the rate-controlling step.

Applications. PVF resins are used almost exclusively to make electric and magnetic wire insulation. In these applications the PVF resin component helps provide toughness, as well as abrasion and thermal resistance. The resin is combined with phenolic, epoxy, melamine, or other resins capable of cross-linking with hydroxyl groups in suitable vehicles to produce formulations called wire enamels (146–149). A typical wire enamel consists of 100 parts of PVF and 50 parts of a cresol–formaldehyde resin dissolved in a mixture of cresylic acid and naphtha solvent (150). The enamels are coated on copper or aluminum wire and cured in ovens at elevated temperatures to form thermoset coatings with good thermal stability, insulating properties, and abrasion and chemical resistance (151). Magnetic wire insulation made with PVF wire enamels performs well on high speed motor winding equipment and maintains good insulating properties at the elevated temperatures generated during motor overloads.

PVF resins have also been used in a variety of other applications, including conductive films (152), electrophotographic binders (153), as a component for inks (154), and in membranes (155,156), photoimaging (157), solder masks (158), and reprographic toners (159).

BIBLIOGRAPHY

"Poly(vinyl acetals)" in *ECT* 2nd ed., Vol. 21, pp. 304–317, by G. O. Morrison, Technical Consultant; in *ECT* 3rd ed., Vol. 23, pp. 798–816, by E. Lavin and J. A. Snelgrove, Monsanto Co.

1. C. A. Finch, *Polyvinyl Alcohol, Properties and Applications*, John Wiley & Sons, Inc., New York, 1973.
2. Ger. Pat. 480,866 (July 20, 1924); Ger. Pat. 507,962 (Apr. 30, 1927), W. O. Herrmann and W. Haehnel (to Consortium für Elektrochemische Industrie).
3. Ger. Pat. 507,962 (Apr. 30, 1927), W. Haehnel and W. O. Herrmann (to Consortium für Elektrochemische Industrie).
4. U.S. Pat. 2,036,092 (Mar. 31, 1936), reissue 20,430 (June 29, 1937), G. O. Morrison, F. W. Skirrow, and K. G. Blaikie (to Canadian Electro Products).
5. U.S. Pat. 2,114,877 (Apr. 19, 1938), R. W. Hall (to General Electric).
6. U.S. Pat. 2,167,678 (June 13, 1939), H. F. Robertson (to Carbide and Carbon Chemical Corp.).
7. U.S. Pat. 2,168,827 (Aug. 8, 1939), G. O. Morrison and A. F. Price (to Shawinigan Chemicals Ltd.).
8. U.S. Pat. 2,258,410 (Oct. 7, 1941), J. Dahle (to Monsanto Chemical Co.).
9. U.S. Pat. 2,396,209 (Mar. 5, 1946), W. H. Sharkey (to E. I. du Pont de Nemours & Co., Inc.).
10. U.S. Pat. 2,397,548 (Apr. 2, 1946), W. O. Kenyon and W. F. Fowler, Jr. (to Eastman Kodak Co.).

11. U.S. Pat. 2,496,480 (Feb. 7, 1950), E. Lavin, A. T. Marinaro, and W. R. Richard (to Shawinigan Resins Corp.).
12. U.S. Pat. 2,307,063 (Jan. 5, 1943), E. H. Jackson and R. W. Hall (to General Electric Co.).
13. U.S. Pat. 2,162,678 (June 13, 1939); 2,162,680 (June 13, 1939), H. F. Robertson (to Carbide and Carbon Chemical Corp.).
14. R. H. Fariss, *Chemtech*, 38 (Sept. 1993).
15. A. F. Fitzhugh, E. Lavin, and G. O. Morrison, *J. Electrochem. Soc.* **100**, 8 (1953).
16. Y. Ogata, M. Okano, and T. Ganke, *J. Am. Chem. Soc.* **78**, 2962 (1956).
17. G. Smets and B. Petit, *Makromol. Chem.* **33**, 41 (1959).
18. P. H. Flory, *J. Am. Chem. Soc.* **61**, 1518 (1939).
19. *Ibid.* **72**, 5052 (1950).
20. U.S. Pat. 2,179,051 (Nov. 7, 1939), G. O. Morrison and A. F. Price (to Shawinigan Resins Corp.).
21. P. Raghavendrachar and M. Chanda, *Eur. Polym. J.* **19**, 391 (1983).
22. Jpn. Kokai 4317443 (Nov. 9, 1992), K. Ashina, N. Ueda, and H. Omura; 4325503 (Nov. 13, 1992), K. Ashina, N. Ueda, and H. Omura (to Sekisui Chem. Ind.).
23. Eur. Pat. 402,213 (Sept. 21, 1994), D. Dages and D. Klock (to Saint Gobain and DuPont).
24. K. Shibatani, K. Fujii, Y. Oyanagi, J. Ukida, and M. Matsumoto, *J. Polym. Sci. Part C* **23**(Pt. 2), 647 (1968).
25. K. Fujii, J. Ukida, and M. Matsumoto, *Macromol. Chem.* **65**, 86 (1963).
26. K. Fujii and co-workers, *J. Polym. Sci. Polym. Lett. Ed.* **4**, 787 (1966).
27. M. D. Bruch and J. K. Bonesteel, *Macromolecules* **19**, 1622 (1986).
28. B. Lebek, K. Schlothauer, A. Krause, and H. Marschner, *Acta Polym.* **40**, 92 (1989)
29. P. A. Berger, E. E. Remsen, G. C. Leo, and D. J. David, *Macromolecules* **24**, 2189 (1991).
30. D. H. David, *Proceedings*: *Polymer–Solid Interfaces*, First International Conference, Namur, Belgium, 1992, p. 133.
31. A. A. Parker, D. P. Hedrick, and W. M. Ritchey, *J. Appl. Polym. Sci.* **46**, 295 (1992).
32. Technical data, Monsanto Co., Springfield, Mass., 1987.
33. *Butvar*, *Poly(Vinyl Butyral) and Formvar*, *Poly(Vinyl Formal)*, Technical Bulletin No. 6070E, Monsanto Co., St. Louis, Mo., June 1980.
34. A. F. Fitzhugh and R. N. Crozier, *J. Polymer Sci.* **8**, 225 (1952).
35. Jpn. Kokai 62,278,148 (Dec. 3, 1987), K. Morita and T. Ii (to Sekisui Chemical Industries).
36. Eur. Pat. 394,884 (May 11, 1994), M. Gutweiler, R. K. Driscoll, and E. I. Leupold (to Hoechst AG).
37. U.S. Pat. 5,137,954 (Aug. 11, 1992), A. M. DasGupta, D. J. David, and R. J. Tetreault (to Monsanto).
38. U.S. Pat. 5,028,658 (July 2, 1991), D. J. David and T. F. Sincock (to Monsanto).
39. J. R. Isasi, L. Cesteros, and I. Katime, *Polymer* **34**, 2374 (1993).
40. A. V. Varlamov, D. V. Novikov, I. V. Sidorova, and S. S. Mnatsakanov, *Zh. Prikl. Khim.* **64**, 1735 (1991).
41. S. N. Ivanishchuk and co-workers, *Plast. Massy* **9**, 59 (1990).
42. S. N. Ivanishchuk, N. A. Vordyuk, S. Yu Lipatov, and B. S. Koupaev, *Vysokomol. Soedin.* **32**, 1224 (1990).
43. K. Yoezu, N. Tokoh, A. Aoyama, and T. Okya, *Chem. Express* **2**(10), 651 (1987).
44. Jpn. Kokai 62,121,738 (June 3, 1987), T. Sato, J. Yamauchi, and T. Okaya (to Kuraray Co. Ltd.).
45. Jpn. Pat. 94,025,005 (Apr. 6, 1994), H. Maruyama, A. Aoyama, T. Moriya, K. Yonezu, and J. Yamauchi (to Kuraray Co. Ltd.).

46. Jpn. Kokai 63,079,741 (Apr. 9, 1988), H. Maruyama, A. Aoyama, T. Moriya, K. Yonezu, and J. Yamauchi (to Kuraray Co. Ltd.).

47. Jpn. Kokai 61,130,349 (June 18, 1986), T. Sato, J. Yamauchi, and T. Okaya (to Kuraray Co. Ltd.).

48. Ger. Pat. 912,399 (1941), G. F. D'Alelio (to Allgemeine Elektrizitat Gesellschaft).

49. E. Imoto and R. Motoyama, *Kobunshi Kagaku* **11**, 251 (1954).

50. Ger. Pat. 690,332 (1941), W. O. Hermann (to Chemische Forschung GmbH).

51. Jpn. Kokai 60,101,126 (June 5, 1985), K. Maruhashi, T. Oishi, and T. Kawabata (to Nippon Synthetic Chemical Industrial Co., Ltd.).

52. Ital. Pat. 394,607 (1941) (to Compagnia Generale di Electricita).

53. Ger. Pat. 592,233 (1934), A. Voss, E. Dickhauser, and W. Starck (to I. G. Farben Industrie, AG).

54. Jpn. Kokai 57,167,329 (Oct. 15, 1982), Y. Onishi (to Nippon Synthetic Chemical Industries, Inc.).

55. Ger. Pat. 929,643 (1952), H. Bauer, J. Heckmaier, H. Reinecke, and E. Bergmeister (to Wacker Chemie).

56. Ital. Pat. 395,170 (1941) (to Compagnia Generale di Electricita).

57. A. M. DasGupta, D. J. David, and A. Misra, *Front. Polym. Res.* **1**, 571 (1991).

58. A. M. DasGupta, D. J. David, and A. Misra, *J. Appl. Polym. Sci.* **44**, 1213 (1992).

59. A. M. DasGupta, D. J. David, P. A. Berger, and E. E. Remsen, *Polym. Prepr.* **32**, 68 (1991).

60. U.S. Pat. 5,030,688 (July 9, 1991), D. J. David, A. M. DasGupta, and A. Misra (to Monsanto).

61. A. M. DasGupta, D. J. David, and A. Misra, *Polym. Bull.* **25**, 657 (1991).

62. U.S. Pat. 4,968,744 (Nov. 6, 1990), D. J. David, A. M. DasGupta, and A. Misra (to Monsanto).

63. *Butvar, Poly(Vinyl Butyral) and Formvar, Poly(Vinyl Formvar)*, Technical Bulletin No. 6070D, Monsanto Co., St. Louis, Mo., June 1977. See also, individual product MSDSs.

64. *Butvar, Poly(Vinyl Butyral), Formvar, Poly(Vinyl Formal)*, Technical Bulletin No. 6070F, Monsanto Chemical Co., St. Louis, Mo., 1984.

65. U.S. Pat. 3,920,876 (Nov. 18, 1975), R. H. Fariss and J. A. Snelgrove (to Monsanto Co.).

66. U.S. Pat. 4,144,217 (Mar. 13, 1979), J. A. Snelgrove and D. I. Christensen (to Monsanto Co.).

67. U.S. Pat. 3,841,955 (Oct. 15, 1974), A. W. M. Coaker, J. R. Darby, and T. C. Mathis (to Monsanto Co.).

68. U.S. Pat. 4,230,771 (Oct. 28, 1980), T. R. Phillips (to E. I. du Pont de Nemours & Co., Inc.).

69. Jpn. Pat. 71 42,901 (Dec. 18, 1971), K. Takaura, T. Misaka, and S. Ando (to Sekisui Chemical Co. Ltd.).

70. U.S. Pat. 4,128,694 (Dec. 5, 1978), D. A. Fabel, J. A. Snelgrove, and R. H. Fariss (to Monsanto Co.).

71. Eur. Pat. 513,470 (Mar. 17, 1993), D. J. David, R. H. Farriss, D. C. Knowles, and R. T. Tetreault (to Monsanto Co.).

72. A. F. Fitzhugh and R. N. Crozier, *J. Polym. Sci.* **8**, 225 (1952).

73. A. F. Fitzhugh and R. N. Crozier, *J. Polym. Sci.* **9**, 96 (1952).

74. T. F. Sincock and D. J. David, *Polymer* **33**, 4515 (1992).

75. N. S. Sariciftci and co-workers, *Synth. Met.* **53**, 161 (1993).

76. *Butvar, Polyvinyl Butyral Resin*, Technical Bulletin No. 8084A, Monsanto Chemical Co., St. Louis, Mo., 1991.

77. U.S. Pat. 2,496,480 (Feb. 7, 1950), E. Lavin, A. T. Marinaro, and W. R. Richard (to Shawinigan Resins Corp.).

78. U.S. Pat. 2,400,957 (May 28, 1946); 2,422,754 (June 24, 1947), G. S. Stamatoff (to E. I. du Pont de Nemours & Co., Inc.).

79. U.S. Pat. 3,153,009 (Oct. 13, 1964), L. H. Rombach (to E. I. du Pont de Nemours & Co., Inc.).

80. Ger. Pat. 3,526,314 (1986), R. Degeilh (to Saint Gobain Vitrage).

81. Rom. Pat. 64,627 (1978), A. Chifor, V. Dumitrascu, and I. Manu (to Intr Chemica Risnov).

82. Ger. Pat. 2,383,025 (Sept. 10, 1979), P. Dauvergne (to St. Gobain Industrie, SA).

83. Jpn. Kokai 58,067,701 (Apr. 22, 1983); 57,195,706 (Dec. 1, 1982); 57,030,706 (Feb. 19, 1982), S. Nomura, M. Miyagawa, and K. Asahina (to Sekisui Chemical Co. Ltd.).

84. U.S. Pat. 5,349,014 (Sept. 28, 1994), R. Degeilh (to Saint Gobain Vitrage and Du Pont).

85. Pol. Pat. 96,247 (May 31, 1978), H. Pietkiewicz, M. Knypl, and A. Madeja.

86. L. N. Verkhotina, L. S. Gembitskie, E. N. Gubenkova, L. S. Sev'yants, and A. M. Sarkis'yan, *Plast. Massay* (7), 35 (1977).

87. U.S. Pat. 4,902,464 (Feb. 20, 1990); 4,874,814 (Oct. 17, 1989); 4,814,529 (Mar. 21, 1989), G. E. Cartier (to Monsanto Co.); 4,654,179 (Mar. 31, 1987) (to Monsanto Co.).

88. Jpn. Kokai 62,005,849 (Jan. 12, 1987), K. Suzuki and M. Suzuki (to Teijin, Ltd.).

89. U.S. Pat. 3,523,847 (Aug. 11, 1970), J. W. Edwards (to Monsanto Co.).

90. Ger. Pat. 2,903,115 (Mar. 13, 1980), H. Rodeman, H. D. Funk, and G. Breitenberger (to BSF Glassgroup).

91. J. A. Snelgrove, *Nippon Setchaku Kyokaishi* **21**, 489 (1985).

92. Jpn. Kokai 6,155,681 (June 3, 1994), H. Yatani (to Asahi Kasei Kogyo).

93. Jpn. Kokai 6,246,814 (Sept. 6, 1994), T. Yamani and A. Nakajima (to Sekisui Chemical Industries).

94. Jpn. Kokai 6,263,489 (Sept. 20, 1994), J. Miyai (to Sekisui Chemical Industries).

95. U.S. Pat. 5,322,875 (June 21, 1994), D. Dages (to Saint Gobain Vitrage).

96. U.S. Pat. 5,187,217 (Feb. 16, 1993), R. Degeilh and D. Dages (to Saint Gobain Vitrage).

97. K. J. Lewis, in C. G. Gebelein, D. J. Williams, and R. D. Deanin, eds., *Polymers in Solar Energy Utilization*, ACS Symposium Series, American Chemical Society, Washington, D.C., 1983, Vol. 220, Chapt. 23, pp. 367–385.

98. Jpn. Kokai 59,144,178 (Aug. 18, 1984), S. Yamazaki (to Semiconductor Energy Research Instrument Co., Ltd.).

99. Jpn. Kokai 6,139,748 (May 17, 1994), T. Hattori; 6,115,979 (Apr. 26, 1994), Y. Miyake and T. Masaoka; 6,115,980 (Apr. 26, 1994), N. Ueda, K. Asahina, H. Minamino, H. Omura, and J. Miyai; 6,115,981 (Apr. 26, 1994), H. Minamino and M. Suzuki (to Sekiui Chimical Industries).

100. Jpn. Kokai 5,310,449 (Nov. 22, 1993), T. Hattori (to Sekisui Chemical Industries).

101. U.S. Pat. 5,340,654 (Aug. 23, 1994), N. Ueda, K. Asahina, H. Omura, and J. Miyai (to Sekisui Chemical Industries).

102. American National Standards Institute, *Safety Code for Safety Glazing Materials for Motor Vehicles Operating on Land Highways*, No. Z26.1–1983, SDO, New York, 1983.

103. U.S. Pat. 3,262,837 (July 26, 1966), E. Lavin, G. E. Mont, and A. F. Price (to Monsanto Co.).

104. U.S. Pat. 3,249,487 (May 3, 1966), F. T. Buckley and J. S. Nelson (to Monsanto Co.).

105. Jpn. Pat. 75,121,311 (Sept. 23, 1975), I. Karasudani, T. Takashima, and Y. Honda (to Sekisui Chemical Co., Ltd.).

106. Ger. Pat. 2,410,153 (Sept. 4, 1975), R. Beckmann and W. Knackstedt (to Dynamit Nobel, AG).

107. U.S. Pat. 3,718,516 (Feb. 27, 1973), F. T. Buckley, R. F. Riek, and D. I. Christensen (to Monsanto Chemical Co.).

108. Ger. Pat. 2,904,043 (Aug. 9, 1979), H. K. Inskip (to E. I. Du Pont de Nemours & Co., Inc.).
109. Ger. Pat. 2,646,280 (Apr. 20, 1978), H. D. Hermann and J. Ebigt (to Hoechst, AG).
110. Jpn. Kokai 6,228,227 (Aug. 16, 1994), H. Minamino (to Sekisui Chemical Industries).
111. Eur. Pat. Appl. 617,078 (Sept. 28, 1994), H. Fischer (to Hoechst, AG).
112. U.S. Pat. 4,600,655 (July 15, 1986), H. Hermann, K. Fock, K. Kriftel, and J. Ebigt (to Hoechst, AG).
113. U.S. Pat. 4,663,235 (May 5, 1987), K. Fock, H. Hermann, K. Kriftel, and J. Ebigt (to Hoechst, AG).
114. Jpn. Kokai 6,256,043 (Sept. 13, 1994), T. Yoshioka; 6,211,920 (Aug. 2, 1994), T. Kori and K. Nishimura; 6,211,549 (Aug. 2, 1994), T. Yoshioka; 6,127,982 (May 10, 1994), H. Minamino and M. Suzuki (to Sekisui Chemical Industries).
115. U.S. Pat. 3,534,778 (Aug. 18, 1970), W. Jensch, H-G. Groeblinghoff, and R. Beckman (to Dynamit Nobel, AG).
116. Jpn. Kokai 6,127,983 (May 10, 1994), M. Murashima; 6,198,809 (July 19, 1994), M. Murashima and T. Sonaka; 6,210,729 (Aug. 2, 1994), T. Sonaka and M. Murashima (to Sekisui Chemical Industries).
117. U.S. 5,151,234 (Sept. 29, 1992), H. Hori, M. Ishihara, K. Kiminami, S. Takeshita, S. Tatsu, and M. Wakabayashi (to Sekisui Chemical Industries).
118. U.S. Pat. 4,925,725 (May 15, 1990), G. Endo, H. Tateishi, Y. Kawata, I. Karasudani, and H. Omura (to Sekisui Chemical Industries).
119. Eur. Pat. 215,976 (Oct. 19, 1994), G. Endo, I. Karasudani, Y. Kawata, H. Omura, and H. Tateishi (to Sekisui Chemical Industries).
120. U.S. Pat. 4,654,179 (May 31, 1987), G. E. Cartier and P. H. Farmer (to Monsanto).
121. Jpn. Pat. 93,083,590 (Nov. 26, 1993), (to Bridgestone Tire).
122. U.S. Pat. 3,982,984 (Sept. 28, 1976), D. B. Baldridge (to Monsanto Co.).
123. U.S. Pat. 3,973,058 (Aug. 3, 1976), J. L. Grover and W. H. Power (to Monsanto Co.).
124. Ger. Pat. 2,841,287 (Apr. 3, 1980), D. S. Postupack (to PPG Industries, Inc.).
125. J. A. Lewis, A. L. Ogden, D. Schroeder, and K. J. Duchow, *Mater. Res. Soc. Symp. Proc.* **289**, 117 (1993).
126. J. A. Lewis and M. J. Cima, *Mater. Res. Soc. Symp. Proc.* **249**, 363 (1992).
127. Jpn. Kokai 5,295,016 (Nov. 9, 1993), Y. Miyake and T. Masaoka (to Sekisui Chemical Industries).
128. O. M. Prakash and co-workers, *Bull. Mater. Sci.* **14**, 1145 (1991).
129. U.S. Pat. 4,478,776 (Oct. 23, 1984), D. L. Maricle, G. C. Putnam, and R. C. Stewart, Jr.
130. U.S. Pat. 4,444,571 (Apr. 24, 1984), S. L. Matson (to Bend Research, Inc.).
131. M. Gotoh, D. Tamiya, and I. Karube, *J. Appl. Polym. Sci.* **48**, 67 (1993).
132. J. D. Scantlebury and F. H. Karman, *Corros. Sci.* **35**, 1305 (1993).
133. J. L. Nogueira, *Corros. Prot. Mater.* **11**, 11 (1992).
134. T. Foster, G. N. Blenkinsop, P. Blattler, and M. Szandorowski, *J. Coat. Technol.* **63**, 91 (1991).
135. Military specifications: DOD-P-15328D and MIL-C-8514C (ASG), Information Handling Services (IHS), Englewood, Colo.
136. *Butvar, Disperson BR Resin Technical Bulletin*, Publication No. 6019-D, Monsanto Chemical Co., St. Louis, Mo., 1989.
137. Ger. Pat. Appl. 4,235,151 (Apr. 21, 1994), M. Kroggel and H. Schindler (to Hoechst AG).
138. U.S. Pat. 3,234,161 (Feb. 8, 1966), J. A. Snellgrove and W. Whitney (to Monsanto Chemical Co.).
139. U.S. Pat. 2,455,402 (Dec. 7, 1948), U.S. Pat. 2,532,223 (May 28, 1950), W. H. Bromley, Jr. (to Shawinigan Resins Corp.).

140. Brit. Pat. 233,370 (May 7, 1925), W. B. Pratt.

141. *Vinylec, Polyvinyl Formal Resins*, Technical Bulletin, Chisso America Corp., New York, 1994.

142. S. Matsuzawa, T. Imoto, and K. Ogasawara, *Kobunshi Kagaku*, **25**, 173 (1968).

143. A. F. Fitzhugh, E. Lavin, and G. O. Morrison, *J. Electrochem. Soc.* **100**, 351 (1953).

144. S. Matsuzawa, *Kobunshi Kako* **18**(3), 35 (1969) (a review of manufacturing methods, in Japanese).

145. Ger. Pat. 1,071,343 (1957), E. Bergmeister, J. Heckmaier, and H. Zoebelein (to Wacker Chemie, GmbH).

146. U.S. Pat. 2,114,877 (Apr. 19, 1938), R. W. Hall (to General Electric Co.).

147. E. Lavin, A. H. Markhart, and R. W. Ross, *Insulation Libertyville, Ill.* **8**(4), 25 (1967).

148. U.S. Pat. 2,154,057 (Apr. 11, 1939), R. W. Thielking (to Schenectady Varnish Co.).

149. U.S. Pat. 2,307,063 (Jan. 5, 1943), E. H. Jackson and R. W. Hall (to General Electric Co.).

150. U.S. Pat. 2,307,588 (Jan. 5, 1943), E. H. Jackson and R. W. Hall (to General Electric Co.).

151. P. H. Farmer and B. A. Jemmott, in I. Skeist, ed., *Handbook of Adhesives*, 3rd ed., Van Nostrand Reinhold, New York, 1990, p. 433.

152. V. V. R. Rao, R. Narashima, T. S. Rao, and N. N. Das, *J. Phys. Chem. Solids* **47**, 33 (1986).

153. Jpn. Pat. 60 202,448 (Oct. 12, 1985), A. Kojima, J. Hashimoto, and H. Tamura (to Ricoh Co., Ltd.).

154. U.S. Pat. 4,303,718 (Dec. 1, 1981), J. A. Snelgrove (to Monsanto Chemical Co.).

155. T. Uragami, M. Yoshimura, and M. Sugihara, *Technol. Rep. Kansai Univ.* **21**, 119 (1980).

156. E. Ruckenstein and F. Sun, *J. Membr. Sci.* **95**, 207 (1994).

157. U.S. Pat. 4,286,043 (Aug. 25, 1981), H. W. Taylor, Jr. (to Du Pont).

158. Jpn. Pat. 46 33,770 (1971), H. Fujii (to Tokyo Shibaura Electric Co., Ltd.).

159. Jpn. Pat. 55 159,454 (Dec. 11, 1980), T. Kazunori and O. Izumi (to Hitachi Metals).

General References

T. P. Blomstrom in J. I. Kroschwitz, ed., *Encyclopedia of Polymer Science and Engineering*, Vol. 17, Wiley-Interscience, New York, 1989, pp. 136–167.

P. H. Farmer and B. A. Jemmott, in I. Skeist, ed., *Handbook of Adhesives*, 3rd ed., Van Nostrand Reinhold, New York, 1990, pp. 423–436.

JEROME W. KNAPCZYK
Monsanto Company

VINYL ACETATE POLYMERS

Vinyl acetate [*108-05-4*], (VAc), $CH_2=CHOOCCH_3$, the ethenyl ester of acetic acid, is primarily use for the manufacture of poly(vinyl acetate) [*9003-20-7*] (PVAc) and vinyl acetate copolymers. Poly(vinyl acetate) homo- and copolymers are found as components in coatings, paints and sealants, binders (adhesives, nonwovens, construction products, and carpet-backing), and miscellaneous uses such as chewing gum and tablet coatings. Applications have grown over the years in a number of areas (1–4).

Vinyl acetate is a colorless, flammable liquid having an initially pleasant odor which quickly becomes sharp and irritating. Table 1 lists the physical properties of the monomer. Information on properties, safety, and handling of vinyl acetate has been published (5–9). The vapor pressure, heat of vaporization, vapor heat capacity, liquid heat capacity, liquid density, vapor viscosity, liquid viscosity, surface tension, vapor thermal conductivity, and liquid thermal conductivity profile over temperature ranges have also been published (10). Table 2 (11) lists the solubility information for vinyl acetate. Unlike monomers such as styrene, vinyl acetate has a significant level of solubility in water which contributes to unique polymerization behavior. Vinyl acetate forms azeotropic mixtures (Table 3) (12).

The most important chemical reaction of vinyl acetate is free-radical polymerization (13,14). The reaction is summarized as follows:

$$n \; CH_2 = CHOCCH_3 \; \longrightarrow \; -(CH_2 - CH)_n$$

Polymerization can be initiated by organic and inorganic peroxides, azo compounds, redox systems, light, and high energy radiation. Polymerization is inhibited or strongly retarded by aromatic hydroxyl, nitro, or amine compounds and by oxygen, quinone, crotonaldehyde, copper salts, sulfur, conjugated polyolefins, and enynes. Tabulations of quantitative information, eg, polymerization-rate constants, chain-transfer constants, and activation energies for the polymerization reactions of vinyl acetate, are available (13,14). Vinyl acetate has been polymerized by bulk, suspension, solution, and emulsion methods. It copolymerizes readily with some monomers but not with others. Some reactivity ratios are presented in Table 4 for common comonomers. The Q (monomer reactivity factor) and e (electronic factor) values are 0.026 and -0.22, respectively (14).

Hydrolysis of vinyl acetate is catalyzed by acidic and basic catalysts to form acetic acid and vinyl alcohol which rapidly tautomerizes to acetaldehyde. This rate of hydrolysis of vinyl acetate is 1000 times that of its saturated analogue, ethyl acetate, in alkaline media (15). The rate of hydrolysis is minimal at pH 4.44 (16). Other chemical reactions which vinyl acetate may undergo are addition across the double bond, transesterification to other vinyl esters, and oxidation (15–21).

World production of vinyl acetate (VAc) has increased steadily since around 1950 to approximately 3.2 million t (7.1 billion lbs) of VAc in 1994 from plants which had a total capacity of 3.5 million tons (3). Figure 1 summarizes U.S. vinyl acetate capacity and operating rates from 1984 through 1994 along with list prices for the monomer. Other significant VAc-producing regions are Western Europe and Japan, with 19% and 16% of total world capacity, respectively. Table 5 lists principal worldwide producers of vinyl acetate. U.S. demand for VAc was estimated at 2.3 billion lbs in 1994. Figure 2 summarizes historical U.S. demand for VAc in each of the primary polymer applications. Poly(vinyl acetate)

Table 1. Properties and Characteristics of Vinyl Acetate

Property	Value
formula weight	86.09
physical state	liquid
flammable limits in air (101.3 kPa^a), vol %	LEL 2.6, UEL 13.4
flash point, °C	
Tag closed cup (ASTM D56)	−8
Tag open cup (ASTM D1310)	−4
autoignition temperature, °C	426.9
boiling point at 101.3 kPa^a, °C	72.7
relative evaporation rate (n-butyl acetate = 1)	8.9
vapor pressure, kPa^a	
at 60°C	64.9
40°C	29.5
20°C	11.8
Antoine equation coefficients	$\log P = A - (B/(T + C))$
log = base 10, T = °C, P = kPa^a	range = 10–83°C
A	7.51868
B	1452.058
C	240.588
critical temperature, °C	246
critical pressure, kPa^b	3950
color	clear and colorless
specific gravity, 20/20°C	0.934
vapor density (air = 1.00)	2.97
viscosity at 20°C	0.43 cps
freezing point, °C	−92.8
heat of combustion at 25°C	−495.0 kcal/mol
heat of vaporization (1 atm)	87.6 cal/g
heat of formation (liquid at 25°C), kJ/mol^c	−349.4
heat of polarization, kJ/mol^c	89.1
specific heat at 20°C (liquid)	0.46 cal/g °C
odor	not unpleasant, sweetish smell in small quantities
reactivity	reactive with self and variety of other chemicals; stable when property stored and inhibited
light sensitivity	light promotes polymerization
electrical conductivity at 23°C	2.6 × 104 pS/m (1 S = 1 mho)
refractive index, n_{D}^{20}	1.3953
surface tension at 20°C, mN/m(=dyn/cm)	23.6
coefficient of cubical expansion	0.00137/°C at 20°C

^{a}To convert kPa to mm Hg, multiply by 7.5.
^{b}To convert kPa to atm, divide by 101.3.
^{c}To convert kJ/mol to kcal/mol, divide by 4.184.

(PVAc) is the dominant use of vinyl acetate in the United States, accounting for 61% of consumption. VAc consumption patterns are similar in Western Europe. Products that retain their character as poly(vinyl acetate) account for 55–62% of monomer production. Most of this use is in emulsion polymerization processes. Poly(vinyl alcohol) requires 18–20% of the monomer produced, the poly(vinyl

Table 2. Solubilities of Vinyl Acetate[a]

Solubility of	Temperature, °C	Concentration, %
water in vinyl acetate	20	2.0–2.4
	50	2.1–2.5
vinyl acetate in water	20	0.9–1.0
	50	2
2.0 wt % solution of sodium lauryl sulfate	30	4.0
organic liquids		soluble

[a]Ref. 11.

Table 3. Azeotropes of Vinyl Acetate[a]

Second component	Azeotropic boiling point, °C	Vinyl acetate, wt %
water	66.0	92.7
methanol	58.9	63.4
2-propanol	70.8	77.6
cyclohexane	67.4	61.3
heptane	72.0	83.5

[a]Ref. 12.

Table 4. Copolymerization Parameters of Vinyl Acetate (M_1) and Comonomers (M_2)

Comonomer (M_2)	r_1	r_2	Temperature, °C
acrylamide	0.07	7.5	50
acrylic acid	0.01 ± 0.003	10.0 ± 1	70
n-butyl acrylate	0.388	5.529	50
crotonic acid	0.33 ± 0.05	0.01 ± 0.01	68
diethyl maleate	0.17 ± 0.01	0.043 ± 0.005	60
ethylene	1.02 ± 0.02	0.97 ± 0.02	120
2-ethylhexyl acrylate	0.04	7.5	60
maleic anhydride	0.072	0.010	
methyl acrylate	0.029 ± 0.011	6.7 ± 2.2	60
sodium acrylate	0.01	2.0	70
styrene	0.01 ± 0.01	55 ± 10	60
vinyl benzoate	0.6	1.3	60
vinyl chloride	0.6 ± 0.2	1.8 ± 0.6	40
vinylidene chloride	0.0 ± 0.03	3.6 ± 0.5	60
vinylidene cyanide	0.0054	0.11	45
vinyl formate	1.41	0.68	50
vinyl laurate	1.4	0.7	60
vinyl pivalate	0.75 ± 0.04	0.43 ± 0.05	60
N-vinylpyrrolidinone	0.205 ± 0.015	3.30 ± 0.15	50
vinyl stearate	0.90	0.73	70
vinyl Versatate VV9	0.93	0.90	60
vinyl Versatate VV10	0.99	0.92	60

acetal)s ca 8%, and vinyl chloride copolymers ca 8%. Ethylene–vinyl acetate copolymers are a small use but are growing at 5%/yr. Additional quantities are used in polymeric lube-oil additives and in acrylonitrile copolymers for acrylic fibers (see LUBRICATION AND LUBRICANTS).

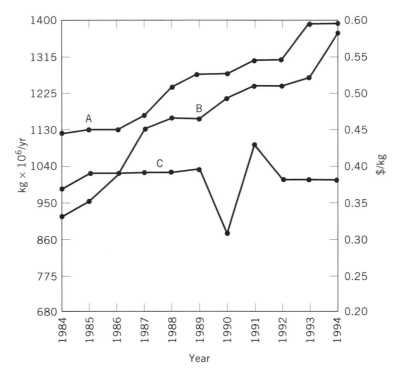

Fig. 1. U.S. production, capacity, and price of vinyl acetate, where A shows total capacity; B, total production; and C, list price.

Table 5. Worldwide Producers of Vinyl Acetate[a]

Country or region	Capacity, 1000 t/yr
Asia	953
Eastern Europe	116
Canada	75
Mexico	100
United States	1500
South America	100
Western Europe	655
World total	*3499*

[a]As of Jan. 1995.

Vinyl acetate monomer is supplied in three grades, which differ in the amount of inhibitor they contain but otherwise have identical specifications. A low p-hydroquinone grade, containing 3–7 ppm p-hydroquinone, is preferred if it is expected to be used within two months of delivery. For monomer stored up to four months before polymerization, the grade containing 12–17 ppm p-hydroquinone is used. When indefinite storage is anticipated, a grade containing 200–300 ppm diphenylamine is supplied. Typical manufacturers' specifications are given in Table 6 (5–9).

In storage vinyl acetate should be kept away from ignition sources. It should be stored in a cool environment away from heat, direct sunlight, oxidizing

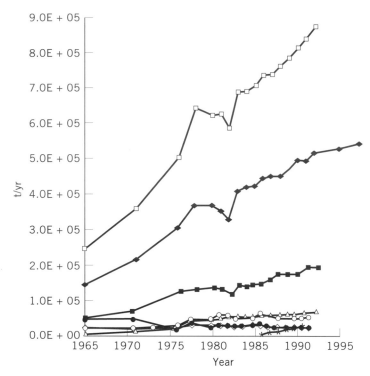

Fig. 2. U.S. consumption of vinyl acetate, where ◆ represents PVAc; ■, poly(vinyl alcohol) (PVA); ▲, ethylene–vinyl acetate (EVA); ○, PVB; *, EVOH; ●, VAM; ◇, other, and □, total (2).

Table 6. Typical Manufacturers' Specifications for Vinyl Acetate

Specification	Value
vinyl acetate, wt %, min	99.8
boiling point, °C	72.0–73.0
acidity as acetic acid, wt %, max	0.007
carbonyls as acetaldehyde, wt %, max	0.013
water, wt %, max	0.04
color, APHA system	0.5
suspended matter	none

materials, and free-radical generating chemicals to avoid rapid uncontrolled polymerization. Bulk storage should be blanketed with dry nitrogen. The monomer is actually more stable at a lower oxygen concentration (22). In the presence of oxygen, there is a tendency to form polyperoxides which could lead to polymerization. Using dry gas helps avoid the potential for hydrolysis to acetic acid and acetaldehyde. Additional information concerning the safe handling and transportation of vinyl acetate is provided in the Material Safety Data Sheet (MSDS) for vinyl acetate (5,23).

Gas chromatography is an excellent method to determine vinyl acetate and its volatile contaminants. If wet chemical techniques are used, vinyl acetate is assayed by adding excess bromide to an aliquot of material, followed by addition of excess potassium iodide and titration with standard sodium thiosulfate. Acidity is determined by direct titration in methanol solution with a standard caustic solution, aldehydes are determined by the addition of excess sodium bisulfite followed by titration with standard iodine solution, and water is determined by the Karl Fischer method. p-Hydroquinone and diphenylamine can be determined by standard titration techniques, or spectrophotometrically in the uv region after evaporation of vinyl acetate. Several companies describe in their product brochures empirical procedures for determining the polymerization activity of the monomer. In these tests, a given amount of monomer is polymerized under standard conditions and the rate of temperature increase or volume shrinkage and induction period are measured. Appropriate ASTM specifications and test procedures are D2190, D2191, D2193, and D2083.

NIOSH recommends a ceiling limit of 4 ppm for 15 minutes of exposure to vinyl acetate vapor. The ACGIH recommends an 8-hour TLV Time Weighted Average of 10 ppm and a 15 minute short-term exposure limit of 15 ppm to its vapor. Vinyl acetate is a severe eye and skin irritant, forming blisters on the skin, and redness, swelling, or corneal burns on the eyes. The vapor is irritating to the nose and throat, and high levels of exposure may result in pulmonary edema. Nausea, headache, or weakness may result with inhalation of mists. Good industrial hygiene practices should be followed and adequate ventilation provided. Repeated or prolonged exposure should be avoided. The ACGIH has designated vinyl acetate as an A3 Animal Carcinogen. This designation refers to an agent which is carcinogenic in experimental animals at a relatively high dose. The International Agency for Research on Cancer (IARC) has evaluated vinyl acetate as a Group 2B material (24). This evaluates VAc as probably carcinogenic to humans based on the sufficient evidence of the carcinogenicity of acetaldehyde in experimental animals. Both VAc and acetaldehyde were found to induce nasal tumors in rats on inhalation. VAc was also found to be genotoxic to human cells *in vitro* and to animal cells *in vivo*.

Vinyl acetate has moderate acute toxicity if ingested. The LD_{50} for oral ingestion in rats is 2.9 g/kg body weight; for absorption through the skin, the LD_{50} in rats is more than 5 mL/kg in 24 h. First-aid procedures to be followed in the event of overexposure to vinyl acetate are as follow:

Type of exposure	Treatment
ingestion	do not induce vomiting; drink large quantities of milk
inhalation	provide fresh air; keep victim warm and quiet; apply artifical respiration if necessary
skin contact	wash with water
eye contact	flush with water for 15 minutes; contact physician immediately

Vinyl Acetate Polymers

Properties.   Poly(vinyl acetate) (PVAc) polymer resins are manufactured in a variety of molecular weights. Some physical properties of the polymer are listed in Table 7 (25–27). With increasing molecular weight, properties vary from viscous liquids to low melting solids to tough, horny materials. They are neutral, water-white to stray-colored, tasteless, odorless, and nontoxic. The resins have no sharply defined melting points but become softer with increasing temperature. Due to their solubility parameter, they are soluble in organic solvents, eg, esters, ketones, aromatics, halogenated hydrocarbons, carboxylic acids, etc, but are insoluble in the lower alcohols (excluding methanol), glycols, water, and nonpolar liquids such as ether, carbon disulfide, aliphatic hydrocarbons, oils, and fats. Alcohols, eg, ethyl, propyl, and butyl, containing 5–10 wt % water dissolve PVAc; butyl alcohol or xylene, both of which only swell the polymer at normal temperatures, dissolve it when heated. The electrical properties are strongly affected by the ability of poly(vinyl acetate) to absorb water. Whereas dried resin has a dielectric constant ϵ of 3.3 and a loss factor ϵ' of 0.08 ± 0.02 at 35°C and 60 Hz, after exposure to 100% relative humidity the numbers become 10 and 0.7, respectively (28,29).

As with many thermoplastic resins (qv), strength properties increase with molecular weight; tensile strengths up to 50.3 MPa (7300 psi) may be obtained. The softening point, as determined by the ring-and-ball method or by the Kraemer and Sarnow method, also increases with molecular weight, as shown in Table 8 (30,31). Poly(vinyl acetate) resin is commercially available in pure, dry form as beads, granules, or lumps and is graded according to viscosity at 20°C of a 1 M solution in benzene (86.09 g or one mole of repeating units of the resin dissolved in benzene to make one liter). In Europe, the Fikentscher K value, also derived from viscosity measurements, is used to characterize commercial resins.

On cooling below room temperature, poly(vinyl acetate)s become brittle. The brittle point may be lowered by plasticization or copolymerization. When heated above room temperature and above the polymer glass-transition temperature (30°C), all viscosity grades become very flexible, and at 50°C become limp. The glass-transition temperature of PVAc is depressed by the incorporation of moisture causing the polymer to be more flexible under humid conditions. Poly(vinyl acetate) can be heated at 125°C for hours without changing, but at 150°C it gradually darkens, and at over 225°C it liberates acetic acid forming a brown insoluble resin which carbonizes at a much higher temperature. The products of thermal decomposition of PVAc are at 150–200°C acetic acid, and at 300–350°C aromatic compounds, ie, benzene, toluene, naphthalene, etc (32). PVAc resists oxidation and degradation by uv and vis radiation, therefore its aging qualities are excellent.

The nmr spectrum of PVAc in carbon tetrachloride solution at 110°C shows absorptions at 4.86 δ (pentad) of the methine proton; 1.78 δ (triad) of the methylene group; and 1.98 δ, 1.96 δ, and 1.94 δ, which are the resonances of the acetate methyls in isotactic, heterotactic, and syndiotactic triads, respectively. Poly(vinyl acetate) produced by normal free-radical polymerization is completely atactic and noncrystalline. The nmr spectra of ethylene vinyl acetate copolymers have also been obtained (33). The ir spectra of the copolymers of vinyl acetate differ from that of the homopolymer depending on the identity of the comonomers and their proportion.

The chemical properties of PVAc are those of an aliphatic ester. Thus, acidic or basic hydrolysis produces poly(vinyl alcohol) and acetic acid or the acetate of

Table 7. Physical Constants for Poly(vinyl acetate)

Property	Value
absorption of water at 20°C for 24–144 h, %	3–6
coefficient of thermal expansion, K^{-1}	
cubic	6.7×10^{-4}
linear, below T_g	7×10^{-5}
above T_g	22×10^{-5}
cohesive energy density, $(MJ/m^3)^{1/2\,a}$	18.6–19.09
compressibility, $cm^3/(g{\cdot}kPa)^b$	17.8×10^{-6}
decomposition temperature, °C	150
density, g/cm^3	
at 20°C	1.191
25°C	1.19
50°C	1.17
120°C	1.11
200°C	1.05
dielectric constant at 2 MHz	
at 50°C	3.5
150°C	8.3
dielectric dissipation factor, at 2 MHz, tan δ	
at 50°C	150
120°C	260
dielectric strength, V/L	
at 30°C	0.394
6°C	0.307
dipole moment, $C{\cdot}m^c$ per monomer unit	
at 20°C	2.30
150°C	1.77
dynamic mechanical loss peak at 100 Hz, °C	70
elongation at break, at 20°C and 0% rh, %	10–20
glass-transition temperature, T_g, °C	28–31
pressure dependence, °C/100 MPa^d	0.22
hardness, at 20°C, Shore units	80–85
heat capacity, at 30°C, J/g^a	1.465
heat distortion point, °C	50
heat of polymerization, kJ/mol^a	87.5
refraction index, n_D	
at 20.7°C	1.4669
30.8°C	1.4657
52.1°C	1.4600
80°C	1.4800
142°C	
interfacial tension, mN/m (=dyn/cm)	
at 20°C with polyethylene	14.5
20°C with polydimethylsiloxane	8.4
20°C with polyisobutylene	9.9
20°C with polystyrene	4.2
internal pressure, $MJ/m^{3\,a}$	
at 0°C	255
28°C	397.8
60°C	418.7
20°C	284.7
40°C	431.3

Table 7. (*Continued*)

Property	Value
modulus of elasticity, GPa^d	1.275–2.256
notched impact strength, J/m^e	102.4
softening temperature, °C	35–50
specific volume, L/kg	
at t = 100–200°C	$0.823 + (6.4 \times 10^{-4})t$
t = 28°C (T_g)	0.84
surface resistance (ohm/cm)	5×10^{11}
surface tension, mN/m(=dyn/cm)	
at 20°C	36.5
140°C	28.6
180°C	25.9
tensile strength, MPa^d	29.4–49.0
thermal conductivity, mW/(m·K)	159
Young's modulus, MPa^d	600

^{a}To convert J to cal, divide by 4.184.
^{b}To convert kPa to atm, divide by 101.3.
^{c}To convert C·m to debye, divide by 3.336×10^{-30}.
^{d}To convert MPa to psi, multiply by 145; GPa to psi, multiply by 145,000.
^{e}To convert J/m to lbf/in., divide by 53.38.

Table 8. Softening Points and Molecular Weights of Commercial Poly(vinyl acetate)^{a,b}

Grade viscosity, mPa·s (=cP)	Fikentsher K value	Softening point, °C		Mol weight
		Kraemer and Sarnow	Ring-and-ball	
1.5	13	65	75	11,000
2.5	19	81	90	18,000
7	32	106	116	45,000
15	42	131	139	90,000
25	38	153	163	140,000
60	58	196		300,000
100	62	230		500,000
800	79			1,500,000

^{a}Refs. 30 and 31.
^{b}A 1 M solution in benzene at 20°C.

the basic cation. Industrially, poly(vinyl alcohol) is produced by a base-catalyzed ester interchange with methanol, where methyl acetate forms in addition to the polymeric product. The chemical properties of PVAc can be modified by copolymerization. When a comonomer having a carboxylic acid group or a sulfuric acid group is used, the copolymer becomes soluble in dilute aqueous alkali or ammonia. These copolymers also adhere better to metals than homopolymers or neutral copolymers because of the interaction between the acid groups and the metal surface. Copolymerization with monomers such as butyl acrylate can improve flexibility and provide specific adhesion to surfaces. Monomers such as *n*-methylol acrylamide copolymerized with vinyl acetate provide sites for cross-linking either during polymerization reaction or in use. Other comonomers such as vinyl Versatate (vinyl neodecanoate) reduce the potential for hydrolytic breakdown of the polymer chain by shielding the VAc group from attack.

In poly(vinyl acetate) copolymer emulsions, the properties are significantly affected by the composition of the aqueous phase and by the stabilizers and buffers used in the preparation of these materials, along with the process conditions (eg, monomer concentrations, pH, agitation, and temperature). The emulsions are milk-white liquids containing ca 55 wt % PVAc, the balance being water and small quantities of wetting agents or protective colloids. The use of poly(vinyl acetate) or copolymer emulsions eliminates the need for expensive, flammable, odorous, or toxic solvents and the need for the recovery of such solvents. They are easy to apply and the equipment is easy to clean with water, if done promptly. Emulsions also offer the advantage of high solids content with fluidity, since the viscosity of emulsions are independent of the molecular weight of the resin in the particles.

The specific gravity at 20°C for all poly(vinyl acetate) emulsions is about 1.1. There generally is a slight odor of residual monomer in standard materials but if this is removed by some means, such as steam stripping or further reaction, the emulsions are virtually odorless. Traces of acetic acid can be neutralized with a base, eg, ammonia, sodium bicarbonate, or triethanolamine, and when the free monomer has been removed, the pH of the resulting emulsion can be adjusted to remain constant at ca 7. The product developer can choose from ingredient and process options to develop properties for the intended end use. These properties include particle size and molecular weight distributions, pH, emulsion viscosity, particle charge, environmental resistance (solvent, moisture, temperature, and uv), film characteristics (modulus, elongation, toughness), and stability to storage, freezing, dilution, mechanical action, and compounding.

Poly(vinyl acetate) polymer and copolymer properties may be modified through the addition of plasticizers and fillers. Cross-linking agents react with copolymers of vinyl acetate containing a cross-linkable group. The material properties can be altered to allow improved flow, rigidity, and toughness. Various resins, plasticizers, thickening agents, solvents, pigments, extenders, and some dyes may be added. Plasticizers are often added to provide increased flexibility to the dried film, but these usually are not incorporated before the final formulation is made because their presence may adversely affect stability, particularly in cold weather.

When the emulsion is applied to a surface, water is lost by evaporation and absorption if the surface is porous. The particles stick together and eventually coalesce to form a tough and somewhat clear continuous coating. Clarity is often improved by the presence of plasticizer, which also enhances the film's resistance to water. This may be due to the improvement in coalescence of the films. Films from most emulsions containing poly(vinyl alcohol) as the protective colloid are likely to re-emulsify on contact with water, unless they contain relatively large quantities of plasticizer or solvent or are costabilized with a surfactant which helps plasticize the polymer. The films are unaffected by light, oxygen, chlorine in moderation, and dilute solutions of acids, alkalies, and salts. The films are also inert to oils, fats, waxes, and greases, unless these are mostly aromatic. Some solvents, namely acetone, alcohol, ethyl acetate, benzene, and toluene, dissolve or at least swell the film. The extent of dissolution depends on the branching in the chain.

Poly(vinyl acetate) emulsion films adhere well to most surfaces that have relatively high surface energy, (eg, wood, paper, glass, and metal), and have good binding capacity for pigments (qv) and fillers (qv). The ability to bond to other surfaces may be enhanced with the use of comonomers. Ethylene copolymers have shown good adhesion to PVC. Elasticized films are strong and flexible at ordinary temperatures.

An important property of a PVAc film is its permeability to water vapor. The permeability to saturated water vapor at 40°C is 2.1 g/(h·m^2) for a film 0.025-mm thick. This allows the film to be laid down on a damp surface with trapped moisture gradually passing through the film without lifting or blistering it.

Poly(vinyl acetate) polymers are environmentally friendly because they easily biodegrade. Poly(vinyl acetate) may be hydrolyzed to poly(vinyl alcohol) which is then assimilated by naturally occurring organisms. In the use of emulsion polymers, the associated components (stabilizers, initiators, etc) should be scrutinized for their effect on the environment. Poly(vinyl acetate) is nontoxic and is approved by the U.S. FDA for food-packaging (qv) applications (CFR 176.170, 175.105). Components in the emulsion polymer system which may migrate from the film into the food may impact the approval of the total package. In food applications the impact on odor and taste of residual low molecular weight components may be important in the selection of a product for use.

Polymerization Processes.   Vinyl acetate has been polymerized industrially by bulk, solution, suspension, and emulsion processes (34). Perhaps 90% of the material identified as poly(vinyl acetate) or copolymers that are predominantly vinyl acetate are made by emulsion techniques. Detailed information is in patent and scientific literature and in procedures available in the brochures from monomer producing companies (15,34).

Emulsion Polymerization.   Poly(vinyl acetate)-based emulsion polymers are produced by the polymerization of an emulsified monomer through free-radicals generated by an initiator system. Descriptions of the technology may be found in several references (35–39).

Recipe.   An emulsion recipe, in general, contains monomer, water, protective colloid or surfactant, initiator, buffer, and perhaps a molecular weight regulator. The recipe may contain 30–70% monomer, but most commercially available emulsions contain ca 55 wt % solids, although copolymer emulsions have been commercially introduced which have greater than 70% solids content. Several monomers are copolymerized commercially with vinyl acetate in emulsion polymerization and numerous others have been copolymerized on a laboratory scale. Among the comonomers most commonly used industrially in emulsion copolymerization with vinyl acetate are ethylene, dibutyl maleate, bis(2-ethylhexyl) maleate, ethyl-, butyl-, and 2-ethylhexyl acrylates, vinyl laurate, and vinyl neodecanoate (13,15). Vinyl hydrogen maleate and vinyl hydrogen fumarate have also been used as comonomers, as have acrylic acid, maleic anhydride, and sodium ethylenesulfonate, in order to incorporate an acidic or ionic group in the polymer. The neutral comonomers are added primarily to decrease the brittle temperature of the polymer below commonly encountered ambient temperatures, since many uses of PVAc require some degree of flexibility in service. They may also improve the compatibility of the polymer with surfaces of different compositions and energies. The monomers that contain acidic groups are added primarily to make the copolymer soluble in basic media, eg, aqueous ammonia. They also help improve adhesion to metallic surfaces. Copolymerization with monomers which lower the glass-transition temperature, gives a polymer which is innately and permanently flexible. In comparison, the lowering of the brittle point accomplished by addition of plasticizers such as dibutyl phthalate, tricresyl phosphate, etc, to the preformed polymer may be lost with the migration of the plasticizer out of the film. The most efficient plasticizing comonomer on a weight or price basis is ethylene. Copolymerization of vinyl acetate with ethylene is the most significant in the adhesive market (40).

Many different combinations of surfactant and protective colloid are used in emulsion polymerizations of vinyl acetate as stabilizers (qv). The properties of the emulsion and the polymeric film depend to a large extent on the identity and quantity of the stabilizers. The choice of stabilizer affects the mean and distribution of particle size which affects the rheology and film formation. The stabilizer system also impacts the stability of the emulsion to mechanical shear, temperature change, and compounding. Characteristics of the coalesced resin affected by the stabilizer include tack, smoothness, opacity, water resistance, and film strength (41,42).

Poly(vinyl acetate) emulsions can be made with a surfactant alone or with a protective colloid alone, but the usual practice is to use a combination of the two. Normally, up to 3 wt % stabilizers may be included in the recipe, but when water sensitivity or tack of the wet film is desired, as in some adhesives, more may be included. The most commonly used surfactants are the anionic sulfates and sulfonates, but cationic emulsifiers and nonionics are also suitable. Indeed, some emulsion compounding formulas require the use of cationic or nonionic surfactants for stable formulations. The most commonly used protective colloids are poly(vinyl alcohol) and hydroxyethyl cellulose, but there are many others, natural and synthetic, which are usable if not preferable for a given application.

Issues to be considered in selecting the best stabilizing system are polymeric chain branching which increases with high temperature and the presence of some stabilizers, polydispersity of the particles produced, and grafting copolymerization, which may occur because of the reaction of vinyl acetate with emulsifiers such as poly(vinyl alcohol) (43,44).

In general, the greater the quantity of stabilizers in a recipe, the smaller the particle size of the emulsion. At the higher emulsifier levels (≥ 1 wt %) used for emulsion polymerization the polymer forms tiny particles that do not settle but remain indefinitely suspended. Particle sizes resulting from high surface-active-agent-to-low-protective-colloid ratios may be 0.005–1 μm; such emulsions contain lower solids and have lower viscosities. Commercial poly(vinyl acetate) emulsions used in adhesives usually are made with higher ratios of protective colloids in the recipes and usually contain ca 55 wt % solids. Average particle sizes are 0.2–10 μm and the viscosity of the emulsion is 400–5000 mPas (=cP). These latter compositions, ie, with higher ratios of protective colloids, are occasionally described as stable dispersions rather than true emulsions (45–49). The term latex is also used to denote these products, particularly those with small, ie, <0.2 μm, particle size.

The initiators used in vinyl acetate polymerizations are the familiar free-radical types, eg, hydrogen peroxide, peroxysulfates, benzoyl peroxide, *t*-butyl hydroperoxide, lauryl peroxide, and redox combinations. In redox combinations reducing agents such as sodium metabisulfite, sodium formaldehyde sulfoxylate, and ascorbic acid are among those commonly used along with transition-metal salts such as ferrous sulfate. Emulsion polymerizations are usually conducted with water-soluble initiators; benzoyl peroxide has been used in emulsion polymerizations with water-soluble initiators, especially where monomer has been added continuously during the reaction. Radiation-induced initiation has been explored because the activation energy is lower (29 kJ/mol), allowing polymerization to take place at lower temperatures (50).

Buffers are frequently added to emulsion recipes and serve two main purposes. The rate of hydrolysis of vinyl acetate and some comonomers is pH-sensitive. Hydrolysis of monomer produces acetic acid, which can affect the initiator, and acetaldehyde which as a chain-transfer agent may lower the molecular weight of the polymer undesirably. The rates of decomposition of some initiators are affected by pH and the buffer is added to stabilize those rates, since decomposition of the initiator frequently changes the pH in an unbuffered system. Vinyl acetate emulsion polymerization recipes are usually buffered to pH 4–5, eg, with phosphate or acetate, but buffering at neutral pH with bicarbonate also gives excellent results. The pH of most commercially available emulsions is 4–6.

Often a chain-transfer agent is added to vinyl acetate polymerizations, whether emulsion, suspension, solution, or bulk, to control the polymer molecular weight. Aldehydes, thiols, carbon tetrachloride, etc, have been added. Some emulsion procedures call for the recipe to include a quantity of preformed PVAc emulsion and sometimes antifoamers must be added (see FOAMS).

Process.   A polymerization process may consist of simply charging all ingredients to the reactor, heating to reflux, and stirring until the reaction is over while controlling the heat removal at the reaction temperature using cooling systems. However, this simple procedure is seldom followed. Typically, only a portion of the monomer and catalyst is initially charged and the remainder is added during the course of the reaction. Better control of the rate of polymerization can be maintained in this fashion, which is particularly important in industrial-scale operations where large quantities of material are involved and heat-transfer capacity may be limited. Continuous monomer addition in emulsion polymerization usually leads to smaller particle size and a more stable dispersion. On the molecular level, monomer fed over the course of the reaction results in more branches in the polymer chain, if the monomer content of the reaction mix is held low. Consequently, the rate of monomer addition has an effect on final film properties. Copolymerizations usually must be conducted with a continuous monomer feed to obtain homogeneous polymer compositions, especially if there is a significant disparity in the reactivities of the two monomers. Emulsifiers also may be added in increments to help stabilize growing particle populations. In a Kuraray comparison of batch to semicontinuous systems (51), films from emulsions made batchwise were found to have better tensile strength than films made from semicontinuous emulsions.

Various finishing techniques are used to reduce the residual monomer in the emulsion polymer. These include raising the temperature, adding a more concentrated initiator, or using mechanical means, such as stripping. Industrially, polymerizations are carried out to over 99% conversion and thus there is no need to reduce the unreacted monomer unless very low levels are required to meet regulatory, product, or workplace requirements. Most poly(vinyl acetate) emulsions contain less than 0.5 wt % unreacted vinyl acetate, which minimizes development of acetic acid and acetaldehyde by monomer hydrolysis on long storage. Rohm and Haas has patented an ultrafiltration (qv) process for recovering and recycling polymer emulsions from dilute wastewater (52).

All of these processes are operated in conventional glass-lined or stainless steel kettles or reactors. The pressure rating of the reactors depends on the volatility of the monomers used. In the homopolymerization of the vinyl acetate and its copolymerization with monomers of low vapor pressure, the use of vessels with a pressure rating close to atmospheric pressure is sufficient. The ethylene–vinyl acetate copolymer (EVA) processes must of necessity be operated under high pressure (40). Agitation design in these systems is important, taking into consideration the need to blend all the ingredients without zones of high concentration and remove the heat from the reaction mass while avoiding subjecting the reaction mixture to conditions of high shear which could destabilize the material. Heat removal in atmospheric pressure reactions, especially those carried out close to or higher than the boiling point of the vinyl acetate uses reflux condensers. In vinyl acetate polymerizations where low temperatures are maintained to avoid chain transfer, which results in lowering the average molecular weight, internal coils, jackets on the reactors, and external heat exchangers are used.

Control of the process is important to ensure reproducibility of the product. Methods have been developed to control the composition of copolymers and their molecular weight (53,54). Calorimetric methods have been described to estimate conversion and provide the basis for control of the reaction. Other properties such as density and particle size can also be monitored to gauge the reaction progress and development of the product properties (55–57). Agitator design is important in controlling particle size and can influence the kinetics, molecular weight, and coagulum formation (58).

Continuous emulsion copolymerization processes for vinyl acetate and vinyl acetate–ethylene copolymer have been reported (59–64). Cyclic variations in the number of particles, conversion, and particle-size distribution have been studied. Control of these variations based on on-line measurements and the use of preformed latex seed particles has been discussed (61,62).

Continuous polymerization systems offer the possibility of several advantages including better heat transfer and cooling capacity, reduction in downtime, more uniform products, and less raw material handling (59,60). In some continuous emulsion homopolymerization processes, materials are added continuously to a first kettle and partially polymerized, then passed into a second reactor where, with additional initiator, the reaction is concluded. Continuous emulsion copolymerizations of vinyl acetate with ethylene have been described (61–64). Recirculating loop reactors which have high heat-transfer rates have found use for the manufacture of latexes for paint applications (59).

Mini-emulsion processes have been developed where the monomer is emulsified under high energy with either a long-chain alcohol or a polymer producing very small droplets. The long-chain alcohol retards the diffusion of the monomer out of the droplets (65). Polymerization takes place primarily in the droplets, allowing the development of fine, narrow particle size distributions (66). The proceedings of a symposium on the emulsion polymerization of vinyl acetate is available (67).

Bulk Polymerizations.   In the bulk polymerization of vinyl acetate the viscosity increases significantly as the polymer forms making it difficult to remove heat from the process. Low molecular weight polymers have been made in this fashion. Continuous processes are known to be used for bulk polymerizations (68).

Suspension Polymerization.   At very low levels of stabilizer, eg, 0.1 wt %, the polymer does not form a creamy dispersion that stays indefinitely suspended in the aqueous phase but forms small beads that settle and may be easily separated by filtration (qv) (69). This suspension or pearl polymerization process has been used to prepare polymers for adhesive and coating applications and for conversion to poly(vinyl alcohol). Products in bead form are available from several commercial suppliers of PVAc resins. Suspension polymerizations are carried out with monomer-soluble initiators predominantly, with low levels of stabilizers. Suspension copolymerization processes for the production of vinyl acetate–ethylene bead products have been described and the properties of the copolymers determined (70). Continuous tubular polymerization of vinyl acetate in suspension (71,72) yields stable dispersions of beads with narrow particle size distributions at high yields.

Solution Polymerization.   Solution polymerization of vinyl acetate is carried out mainly as an intermediate step to the manufacture of poly(vinyl alcohol). A small amount of solution-polymerized vinyl acetate is prepared for the merchant market. When solution polymerization is carried out, the solvent acts as a chain-transfer agent, and depending on its transfer constant, has an effect on the molecular weight of the product. The rate of polymerization is also affected by the solvent but not in the same way as the degree of polymerization. The reactivity of the solvent-derived radical plays an important part. Chain-transfer constants for solvents in vinyl acetate polymerizations have been tabulated (13). Continuous solution polymers of poly(vinyl acetate) in tubular reactors have been prepared at high yield and throughput (73,74).

Mechanisms.   Because of its considerable industrial importance as well as its intrinsic interest, emulsion polymerization of vinyl acetate in the presence of surfactants has been extensively studied (75–77). The Smith-Ewart theory, which describes emulsion polymerization of monomers such as styrene, does not apply to vinyl acetate. Reasons for this are the substantial water solubility of vinyl acetate monomer, and the different reactivities of the vinyl acetate and styrene radicals; the chain transfer to monomer is much higher for vinyl acetate. The kinetics of the polymerization of vinyl acetate has been studied and mechanisms have been proposed (78–82).

Initiation.   Radicals formed from the initiator species, or generated by chain transfer, initiate the dissolved monomer in the water phase as well as in the micelles (aggregates of stabilizer molecules). Homogeneous nucleation, applies to moderately soluble monomers like vinyl acetate (75,77,83). In homogeneous nucleation, oligomers grow from monomers in the water phase. After reaching the solubility limit, they precipitate and partially stabilized primary particles are formed. The particles grow by absorbing and converting monomer to polymer and also by coagulation. Where PVOH is used as a protective colloid, grafting may occur onto the PVOH backbone and the growing polymer may maintain its solubility in water to a higher degree of polymerization (84). The efficiency of the initiator system may be dependent on the stabilizer used. The presence of oxygen may be inhibitory in nature, probably due to the high chain-transfer constant of the oxygen–vinyl acetate adduct (Table 9). pH is also important as the initiator efficiency may be affected, especially in redox systems.

Propagation. The rate of emulsion polymerization has been found to depend on initiator, monomer, and emulsifier concentrations. In a system of vinyl acetate, sodium lauryl sulfate, and potassium persulfate, the following relationship for the rate of polymerization has been suggested (85):

$$R_p = k_p [M]_o^{0.34} [E]_o^{0.13} [I]_o^{0.5}$$

where R_p = g/(cm^3 of H$_2$O) in seconds; $[M]_o$ = monomer concentration; $[E]_o$ = emulsifier concentration; and $[I]_o$ = initiator concentration. The polymerization rate is constant during the period when monomer conversion is between 15 and 80%. The number of particles per unit volume of aqueous phase is given by the following equation:

$$N_p = [E]_o^{0.94} [I]_o^{0.04}$$

Table 9. Chain-Transfer Constants

Compound	$C_s \times 10^4$	°C
n-heptane	17.0	50
toluene	20.75	60
methanol	6	60
ethanol	25	60
tert-butyl alcohol	1.3	60
acetone	1.5	60
acetaldehyde	530	45
propionaldehyde	1000	60
crotonaldehyde	1800	60
methyl acetate	2.5	60
n-butyl mercaptan	$(48 \pm 14) \times 10^4$	60
acetic acid	10	60
lauroyl peroxide	1.0×10^3	60
oxygen–vinyl acetate adduct	2.6×10^3	60
poly(vinyl alcohol)	3.5×10^{-3}	60
poly(vinyl chloride)	2.1×10^{-1}	60
polystyrene	1.5×10^{-3}	60
	1.9×10^{-3}	75
poly(methyl methacrylate)	2.1×10^{-3}	60
	2.6×10^{-3}	75
poly(vinyl acetate)	1.5×10^{-4}	60

The Smith-Ewart theory predicts $R_p = K[I]^{0.4}$. The rate of polymerization of vinyl acetate is virtually independent of emulsifier concentration, depending on the study, whereas the Smith-Ewart theory predicts the rate to be proportional to the 0.6 power of the emulsifier concentration for monomers such as styrene. This may be due to the high chain transfer to the vinyl acetate monomer. The resulting small radical size has higher mobility, allowing diffusion out of the particle into the aqueous phase and causing reinitiation there. This has a significant effect in lowering the molecular weight of the polymer and increasing the overall rate. At high conversions where monomer concentrations are low (due to either a low feed rate of monomer or in later stages of the reaction), the transfer to polymer becomes important, leading to branched chains. Transfer can also occur to the stabilizing species resulting in permanent incorporation into the polymer matrix. An assessment of the best values of the rate constants for propagation and termination is given below (21):

Propagation $k_p[\mathrm{L/(mol \cdot s)}] = 3.2 \times 10^7 \exp[-3150/T]$

Termination $k_t[\mathrm{L/(mol \cdot s)}] = 3.7 \times 10^9 \exp[-1600/T]$

For example, at 60°C, $k_p = 2300$ and $k_t = 2.9 \times 10^7$. An estimate of kinetic chain lifetime, ie, the time from initiation to termination by reaction with another radical, is 1–2 s at 50°C and 4% per hour rate of polymerization. If there are five chain-transfer steps in the course of the kinetic chain, then a PVAc molecule forms in 0.2–0.4 s. Faster rates of conversion give shorter kinetic chain lifetimes in inverse proportion, but an increased percentage of conversion leads to longer chain lifetimes. At 75% conversion and at 60°C, the radical lifetime is ca 10 s.

Vinyl acetate polymerizes chiefly in the usual head-to-tail fashion, but some of the monomers orient head-to-head and tail-to-tail as the chain grows. The fraction of heat-to-head addition increases with temperature. For example, a 1.15 mol % head-to-head structure and a 1.86 mol % structure were determined at 15°C and 110°C, respectively (86).

In vinyl acetate polymerizations, the molecular weights of the products increase with the extent of conversion: the ratio of weight-to-number-average-degree-of-polymerization also changes, becoming larger at higher conversions (87,88). The dilute solution viscosity of poly(vinyl acetate) can be related to the molecular weight by the following equations (28):

$$[\eta] = 0.0102 \, M^{0.72} \text{ dL/g in acetone solvent at } 30°C$$

$$[\eta] = 0.314 \, M^{0.60} \text{ dL/g in methanol solvent at } 30°C$$

These equations apply to linear polymers of $M_w/M_n = 2.0$.

Chain Transfer.   At the molecular scale, vinyl acetate polymerizations generally are understood as free-radical polymerizations, but are characterized in particular by a relatively large amount of chain transfer (13). This high reactivity of the PVAc growing chain radical is attributed to its low degree of resonance stabilization. The high reactivity of the vinyl acetate radical also contributes to the high rate constant for propagation in vinyl acetate polymerization compared to styrene, the acrylates, and the methacrylates. Chain transfer to monomer is an extremely important factor controlling the molecular weight (13). Several determinations of the transfer constant to monomer Cm, ie, the ratio of rate constants of the transfer reaction and the propagation step k_{tr}/k_p show that Cm increases from 1×10^4 to 3×10^4 from 0 to 75°C (13,89,90).

Chain transfer to monomer and to other small molecules leads to lower molecular weight products, but when polymerization occurs in the relative absence of monomer and other transfer agents, such as solvents, chain transfer to polymer becomes more important. As a result, toward the end of batch-suspension or batch-emulsion polymerization reactions, branched polymer chains tend to form. In suspension and emulsion processes where monomer is fed continuously, the products tend to be more branched than when polymerizations are carried out in the presence of a plentiful supply of monomer.

Chain transfer also occurs to the emulsifying agents, leading to their permanent incorporation into the product. Chain transfer to aldehydes, which may be formed as a result of the hydrolysis of the vinyl acetate monomer, tends to lower the molecular weight and slow the polymerization rate because of the lower activity of the radical that is formed. Thus, the presence of acetaldehyde condensates as a poly(vinyl alcohol) impurity strongly retards polymerization (91). Some of the initiators such as lauryl peroxide are also chain-transfer agents and lower the molecular weight of the product.

Investigation has shown that chain transfer to polymer occurs predominantly on the acetate methyl group in preference to the chain backbone; one estimate of the magnitude of the predominance is 40-fold (92,93). The number of branches per molecule of poly(vinyl acetate) polymerized at 60°C is ca 3, at 80% conversion. It rises rapidly thereafter and is ca 15 at 95% conversion and $1-2 \times 10^4$ number-average degrees of polymerization.

Grafting and Stabilizers. The degree of grafting of poly(vinyl acetate) (PVAc) on poly(vinyl alcohol) (PVA) and other stabilizers during emulsion polymerization strongly affects latex properties such as viscosity, rheology, and polymer solubility (94). The composition and structure of poly(vinyl alcohol) significantly affects the emulsion polymerization of vinyl acetate. The chain distribution of residual acetate groups in poly(vinyl alcohol) is important to successful vinyl acetate emulsion polymerization. A block-like distribution is preferred over a random distribution (95–98). The structure of poly(vinyl acetate-co-vinyl alcohol) emulsion particles has been determined by electron microscopy. The latex particle is formed of a great many smaller particles packed together by flocculation processes (99,100).

An extensive series of papers (101–109) have been published on the grafting of vinyl acetate and styrene on poly(vinyl alcohol) studying the effects of type and concentration of the poly(vinyl alcohol), initiator, and reaction temperature. The partially acetylated poly(vinyl alcohol) was the superior grafting base because of the attachment of the poly(vinyl acetate) segments to the polymerizing particle. The resulting latices with the anchored poly(vinyl alcohol) chains had superior stability. The methine hydrogen atom of the poly(vinyl alcohol) was proposed to be a grafting site in all cases as opposed to the acetyl groups. In mixed systems of poly(vinyl alcohol) and ethoxylated cetyl alcohol both benzene-soluble and water-soluble fractions of graft copolymer were identified, with their ratio dependent on the ratio of initially charged-to-delay fed stabilizer and the stirring speed (110).

Emulsion polymerizations of vinyl acetate in the presence of ethylene oxide- or propylene oxide-based surfactants and protective colloids also are characterized by the formation of graft copolymers of vinyl acetate on these materials. This was also observed in mixed systems of hydroxyethyl cellulose and nonylphenol ethoxylates. The oxyethylene chain groups supply the specific site of transfer (111). The concentration of insoluble (grafted) polymer decreases with increase in surfactant ratio, and R_p (max) is observed at an ethoxylation degree of 8 (112).

Porous membranes with selective permeability to organic solvents have been prepared by the extraction of latex films prepared with moderate ratios of PVA–PVAc graft copolymer fractions. The extracted films are made up of a composite of spherical cells of PVA, PVAc microgel, and PVA–PVAc graft copolymers (113).

Poly(vinyl acetate) chains are also stabilized as aqueous-soluble anionic species by complexation with a surfactant. The charge on the water-soluble species prevents their absorption into the particle (114).

The kinetics of vinyl acetate emulsion polymerization in the presence of alkyl phenyl ethoxylate surfactants of various chain lengths indicate that part of the emulsion polymerization occurs in the aqueous phase and part in the particles (115). A study of the emulsion polymerization of vinyl acetate in the presence of sodium lauryl sulfate reveals that a water-soluble poly(vinyl acetate)–sodium dodecyl sulfate polyelectrolyte complex forms, and that latex stability, polymer hydrolysis, and molecular weight are controlled by this phenomenon (116).

Problems with grafting and morphological changes in the particles are reviewed (117). Vinyl acetate has been grafted onto atactic polypropylene (118) and casein (119).

Copolymers.   The effects of emulsion polymerization process type or latex and polymer properties have been studied. Vinyl acetate copolymerizes easily with a few monomers, eg, ethylene, vinyl chloride, and vinyl neodecanoate, which have reactivity ratios close to its own (see Table 4). The kinetics and the effect of process parameters on the copolymerization of vinyl acetate with ethylene have been studied (120,121). With some monomers, eg, maleic anhydride, it tends to form alternating copolymers. Other monomers with high r_2 and low r_1, eg, alkyl acrylates, tend to form copolymers much leaner in vinyl acetate than the initial charge composition with near homopolymer formed by the faster reacting comonomer. A process is required in which the faster reacting monomer is kept at a low concentration relative to the vinyl acetate, preferably by adding it over the course of the reaction. Studies have been carried out to design processes that result in copolymers which have narrow composition distributions (53,54). The morphology of the polymer particle depends on the monomer characteristic and the process. Thus, emulsion copolymerization of vinyl acetate–butyl acrylate comonomer systems by a delayed monomer addition process yields a core-shell structure particle, in which poly(butyl acrylate)-rich copolymer forms the core and vinyl acetate-rich copolymers form the shell (122).

Block copolymers of vinyl acetate with methyl methacrylate, acrylic acid, acrylonitrile, and vinyl pyrrolidinone have been prepared by copolymerization in viscous conditions, with solvents that are poor solvents for the vinyl acetate macroradical (123). Similarly, the copolymerization of vinyl acetate with methyl methacrylate is enhanced by the solvents acetonitrile and acetone and is decreased by propanol (124). Copolymers of vinyl acetate containing cyclic functional groups in the polymer chain have been prepared by copolymerization of vinyl acetate with N,N-diallylcyanamide and N,N-diallylamine (125,126).

Alternating equimolar copolymers of vinyl acetate and ethylene and alternating copolymers of vinyl acetate and acrylonitrile have been reported (127,128). Vinyl acetate and certain copolymers can be produced directly as films on certain metallic substrates by electroinitiation processes in which the substrate functions as one electrode (129).

New terpolymers of vinyl acetate with ethylene and carbon monoxide have been prepared and their uses as additives to improve the curing and flexibility of coating resins, eg, nitrocellulose, asphalt, phenolics, and polystyrene, have been described (130–132). Vinyl acetate and vinylidene cyanide form highly alternating copolymers.

The low vinyl acetate ethylene–vinyl acetate copolymers, ie, those containing 10–40 wt % vinyl acetate, are made by processes similar to those used to make low density polyethylene for which pressures are usually >103 MPa (15,000 psi). A medium, ie, 45 wt % vinyl acetate copolymer with rubber-like properties is made by solution polymerization in t-butyl alcohol at 34.5 MPa (5000 psi). The 70–95 wt % vinyl acetate emulsion copolymers are made in emulsion processes under ethylene pressures of 2.07–10.4 MPa (300–1500 psi).

Blends. Latex film properties are commonly modified through the blending of latexes, eg, a "soft" polymer is made slightly harder by blending with a "hard" latex.

Poly(ethyl methacrylate) (PEMA) yields truly compatible blends with poly(vinyl acetate) up to 20% PEMA concentration (133). Synergistic improvement in material properties was observed. Poly(ethylene oxide) forms compatible homogeneous blends with poly(vinyl acetate) (134). The T_g of the blends and the crystallizability of the PEO depend on the composition. The miscibility window of poly(vinyl acetate) and its copolymers with alkyl acrylates can be broadened through the incorporation of acrylic acid as a third component (135). A description of compatible and incompatible blends of poly(vinyl acetate) and other copolymers has been compiled (136). Blends of poly(vinyl acetate) copolymers with urethanes can provide improved heat resistance to the product providing reduced creep rates in adhesives used for vinyl laminating (137).

Economic Aspects. Prices for PVAc polymers depend on the form of the polymer, ie, whether it is resin or emulsion, homopolymer or copolymer, as well as on the specific product. As of 1994, emulsion prices were $0.57–$0.86/wet kg of resin. Prices of VAE copolymer emulsions tend to be higher than those of the homopolymer priced at $0.97–$1.43/wet kg. Vinyl acrylic copolymers listed for $0.66–$0.88/wet kg of 55% solids emulsion (138). Specialty copolymers generally have a premium price. These price ranges are for large shipments.

Growth in PVAc consumption is illustrated in Figure 3. The emulsions continue to dominate the adhesives and paint markets. It also shows the distribution of PVAc and copolymer usage by market. The companies listed in Table 10 are among the principal suppliers of poly(vinyl acetate)s and vinyl acetate

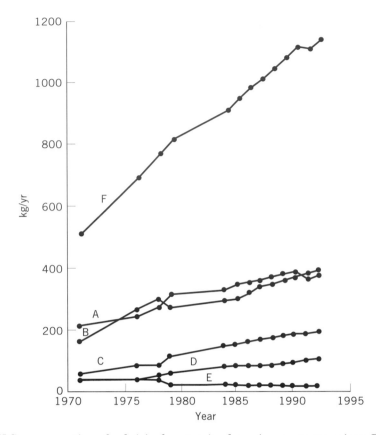

Fig. 3.   U.S. consumption of poly(vinyl acetate), where A represents paints; B, adhesives; C, paper coatings; D, textiles; E, other; and F, total.

copolymers, but there are numerous other suppliers. Many other companies produce these polymers and consume them internally in the formulation of products.

   Specifications and Standards.   Typical specifications of the commercially available emulsions are tabulated in Table 11. However, there are exceptions to the ranges given. For example, most emulsions contain 55–56 wt % solids but some are available at 46–47 wt % with viscosities of 10–15 mPa·s (=cP), and others at 59 wt % solids with viscosities of 200–4500 mPa·s. Specialty copolymer emulsions are available containing 72 wt % solids and viscosities under 2000 mPa·s (139).

   Borax stability is an important property in adhesives, paper (qv), and textile applications where borax is a frequently encountered substance. This is a function of poly(vinyl alcohol) used for stabilization of the product. Other emulsion properties tabulated by manufacturers include tolerance to specific solvents, surface tension, minimum film-forming temperature, dilution stability, freeze–thaw stability, percent soluble polymer, and molecular weight. Properties of films cast from the emulsions are also sometimes listed, including clarity, gloss, light stability, water resistance, flexibility, heat-sealing temperature, specific gravity, and bond strength. Homopolymer resin specifications usually

Table 10. Suppliers of Poly(vinyl acetate)

Country	Company	Product types
United States	Air Products	adhesives, paint, paper, nonwovens
	W. R. Grace	adhesives, textile treatment
	Monsanto	laminating adhesives, paint
	National Starch and Chemical	paper, adhesives, textiles, nonwovens
	Reichhold Chemical	paint, adhesives, paper
	Union Carbide Corp.	paint, adhesives, textiles
	Rohm & Haas	adhesives, paint, paper, textile binders
	Sequa	textile treatments and paper coating
Japan	Kobunshi Chem	
	Nippon Carbide	
	Nippon Gosei	
	Sekisui	
	Showa Denko	
	Sumitomo	
Europe	Imperial Chemical Industries, Ltd.	
	Revertex	
	Unilever	
	Farbwerke Hoechst	
	Wacker Chemie	
	Ebnoether	
	Lonza	
	Rhône-Poulenc	

include viscosity grade (1.0 molar solution in toluene at 20°C), 1.5–800 mPa·s (=cP) ~10%; volatiles, 1–2 wt %; acidity as acetic acid, 0.1–0.3 wt %; and softening point. Data are also given on heat-sealing temperature, tensile strength, elongation, and abrasion resistance. For alkali-soluble resins, the solubility in ammonia and its viscosity are listed (see Table 11).

Poly(vinyl acetate) and its copolymers with ethylene are available as spray-dried emulsion solids with average particle sizes of 2–20 μm; the product can be reconstituted to an emulsion by addition of water or it can be added directly to formulations, eg, concrete. The powders may be used to raise solids of a lower solids latex. Solutions of resin in methyl and ethyl alcohol at 2–50 wt % solids are also available.

Emulsions are shipped in 19 L pails, 209 L drums, 11.0–18.9 m^3 (3000–5000 gal) tank trucks, and 38–76 m^3 (10,000–20,000 gal) railroad tank cars. During storage and handling, care should be taken not to destabilize or otherwise alter the character of the product. Storage tanks must be maintained clean and free from microbial contamination. Latices must not be put into a tank which contains a material with which it is incompatible. Routine cleaning is scheduled to remove skins or other accumulations of dried latex. Some of the large particle size latices may settle if stored for a long time and may need to be agitated to retain consistency. Latices corrode ordinary steel (qv) and must be stored in tanks

Table 11. Poly(vinyl acetate) Emulsion Specifications

Property	Range	Method
solids, wt %	4–72	moisture balance or oven
viscosity, mPa·s(=cP)	200–4500	viscometer
pH	4	pH meter
residual monomer, % max	0.5	titration or gas chromatography
particle size	0.1–3.0	
particle charge	neutral or negative	
density at 25°C, g/cm^3	0.92	pyknometer
borax stability	stable or unstable	stable to borax addition
mechanical stability	good or excellent	maintain stability for set period in high speed high shear blender
freeze–thaw stability	passes 0–5 cycles	remains fluid after multiple cycles of freezing and thawing
accelerated sedimentation, %	<4	sediment after centrifuging diluted sample
coagulum	<50 GMs on 100-mesh screen	filter through a screen or stacked screens
T_g (glass-transition temperature), °C		DS calorimetry

with stainless steel, glass, plastic, or coated surfaces. Suitable coatings are baked phenolic, epoxy phenolic, or PVC. Storage should be at around room temperature, because excessive exposure to high or low temperatures may cause phase separation. Large shipments are insulated against cold. Precautions should be taken to avoid the formation of foam during unloading. If containers are not full during shipment, foam may be generated due to sloshing. Containers should be closed to prevent evaporation of water, which leads to skin formation. Diaphragm pumps and screw pumps are preferred to centrifugal pumps (qv). High shear pumps such as centrifugal pumps may destabilize the latex causing the formation of grit or coagulum. In the case of colloid-protected latices an irreversible reduction of viscosity may occur. Latices are susceptible to biocontamination, largely due to the stabilizer systems. In handling the latices care should be taken to avoid exposure to microbial sources. Producers add microbicidal agents to protect the material against a moderate level of incident contamination. Biocontamination can result in the generation of foul odor, discoloration, and destabilization of the latex. Dry resins should be stored at or below room temperature to prevent caking.

Uses. *Adhesives.* The uses of poly(vinyl acetate) adhesive are packaging (qv) and wood gluing (140,141). Table 12 (142) lists characteristics of poly(vinyl acetate) (PVAc) emulsions important to adhesive applications. PVAc copolymer adhesives are finding application in more diverse areas such as construction and adhesion to more difficult to bond surfaces because of the range of adhesion and the flexibility that may be built into the polymer. The emulsion form of PVAc is especially suitable for adhesives because of several properties that are peculiar to emulsion systems. The stability of PVAc emulsions allows them to accept many types of modifying additives without being damaged. For example,

Table 12. Characterization of Emulsion Polymers[a]

Compound	Properties
emulsion	surface tension
	foaming tendency
	stability (dilution, mechanical, shear, freeze–thaw, heat, bacterial)
	compatibility (plasticizers, thickeners, pigments, extenders, coalescing agents)
	rheological character
polymer	T_g
	molecular weight distribution
	rheological behavior
	solubility
	infrared spectra
film	blocking
	flex
	tensile/elongation
	hardness
	water and chemical resistance
adhesive	heat seal temperature
	bond strength (miscellaneous substrates)
	tack
	peel
	creep

[a]Ref. 142.

solvents, plasticizers, tackifying resins, and fillers can be added directly to homopolymer and copolymer emulsions without requiring pre-emulsification, unlike the elastomeric adhesive latexes. Homopolymer emulsions containing partially hydrolyzed poly(vinyl acetate) as a protective colloid can accept greater amounts of these modifying additives than any other type of emulsion without coagulating.

The stability of the emulsions further permits them to be compounded in simple liquid-blending vessels by means of agitators, eg, marine-type propellers, paddles, or turbines. The adhesives can be adapted to any type of machine application, ie, from spray guns to rollers to extruder-type devices. Different applicators are fairly specific in their viscosity requirements, as are the various substrates receiving the adhesive.

Poly(vinyl acetate) emulsions can be used in high speed gluing equipment. In contrast to aqueous solutions of natural or synthetic products which lose water slowly, the emulsions quickly lose water and invert and set rapidly into a bond.

Poly(vinyl acetate) homopolymers adhere well to porous or cellulosic surfaces, eg, wood, paper, cloth, leather (qv), and ceramics (qv). Homopolymer films tend to creep less than copolymer or terpolymer films. They are especially suitable in adhesives for high speed packaging operations.

Copolymers wet and adhere well to nonporous surfaces, such as plastics and metals. They form soft, flexible films, in contrast to the tough, horny films formed by homopolymers, and are more water-resistant. As the ratio of comonomer to vinyl acetate increases, the variety of plastics to which the copolymer adheres also increases. Comonomers containing functional groups often adhere to specific surfaces; for example, carboxyl containing polymers adhere well to metals.

Plasticization, whether internal (by copolymerization) or external (with additives), is also extremely important for proper performance at the time of application. The ease of coalescence and the wetting characteristics of the polymer emulsion particles are related to their softness and the chemical nature of the plasticizer.

New vinyl acetate–acrylate (VAA) emulsion copolymers stabilized with poly(vinyl alcohol) have been developed. The acrylic component of the VAAs contributes to improved compatibility with tackifiers (143).

Cross-linked polymer emulsions accept water-miscible solvents better than straight-chain emulsions. In the film state, the former resist water and organic solvents better and tend to have higher heat-sealing temperatures, which signify a greater resistance to blocking and cold flow. Straight-chain polymers can have high molecular weights, which contribute to high heat-sealing temperatures. Varying the conditions of polymerization results in either straight-chain or cross-linked polymer emulsions. The incorporation of functional monomers such as n-methylol acrylamide or acrylic acid into the polymer provides the ability to cross-link the product in the application either through heat, catalysis, or by a curing agent. In these cases, the polymer flows onto the substrate surface, then develops its strength on curing. Wood glues which can withstand boiling water immersion have been prepared in this fashion (144).

Tack enables an adhesive to form an immediate bond between contacting surfaces when they are brought together. It permits the alignment of an assembly and prevents the adherends from separating before the adhesive sets. Tack is differentiated into two types: (1) wet tack, also called grab or initial tack, is the tack of an adhesive before the liquid carrier, ie, organic solvent or water, has fully evaporated; (2) dry tack, also called residual tack or pressure sensitivity, is the tack remaining after the liquid carrier has evaporated. Wet tack is often necessary in paper-converting operations. Applicators with little or no pressure on the combining section require emulsion adhesives with enough wet tack to bond strongly at the slightest touch. Emulsions containing poly(vinyl alcohol) as a protective colloid have stronger wet tack than those protected with a cellulose derivative. If more wet tack is required of the poly(vinyl alcohol)-protected adhesive, both a solvent and a plasticizer can be added and the solids content can be increased. Dry tack is needed where two nonporous surfaces are to be bonded. The nature of the adherends does not permit the water in the adhesive to escape by penetration or evaporation, so the adhesive must dry before the adherends are joined, yet still retain enough tack to form a permanent bond. Film-to-film and film-to-foil laminations are good examples of applications requiring dry tack.

The setting speed of an adhesive is the time during which the bond formed by the adhesive becomes permanent. Before the setting of an emulsion adhesive can occur, inversion of the emulsion must take place; that is, it must change from a dispersion of discrete polymer particles in an aqueous continuous phase to a continuous polymer film containing discrete particles of water. The point at which the adhesive has actually set is that point at which an assembly, whether joint or lamination, can no longer be disassembled without damaging one or more of the adherends. Rapid setting speeds are necessary in many types of packaging adhesives intended for use with high speed gluing equipment. The rapid inversion possible with PVAc emulsions and their low viscosities allow them to be compounded into adhesives that not only set rapidly but also machine easily at high speeds. This machinability permits their application by practically any means, including glue guns, rollers, and spray guns.

Increases in setting speed can be achieved by increasing the dispersed phase content; that is, by increasing the amount of water-insoluble substances contained in the emulsions. This crowds the aqueous phase of the emulsion, hastening inversion and setting. Adding tackifying resins is one way to crowd the emulsion; adding plasticizers and solvents is another, but these additives mainly increase the setting speed by softening the polymer particles and hastening their coalescence. Surface-active agents also increase setting speed by helping the water in an adhesive to penetrate porous surfaces more rapidly; they can also retard it by stabilizing the emulsion. The usual way to prepare a high speed adhesive is to add both a solvent and a plasticizer to the base emulsion.

Poly(vinyl acetate) emulsions are excellent bases for water-resistant paper adhesives destined for use in manufacturing bags, tubes, and cartons. Glue-lap adhesives, which require moderate-to-high resistance to water, exemplify this type. When routine water resistance is required, a homopolymer vinyl acetate emulsion containing a cellulosic protective colloid is effective for most purposes. Next effective are emulsions containing fully hydrolyzed poly(vinyl alcohol) as a protective colloid, followed by those containing partially hydrolyzed poly(vinyl acetate).

When more than routine water resistance is required, a copolymer vinyl acetate emulsion can be used. The plasticizing comonomer in the polymer particles increases their intrinsic coalescing ability; thus, they can coalesce more readily than homopolymer particles to a film that has a higher resistance to water. This resistance to water does not extend to the organic solvents, however, which are better resisted by homopolymer films. The soft copolymers have lower solubility parameters than homopolymers and are more readily attacked by solvents of low polarity, eg, hydrocarbons.

Despite their high resistance to water, copolymer emulsions are seldom chosen to bond paper to paper, as the less expensive homopolymer emulsions are effective enough. When copolymers are used in paper adhesives, it is done so chiefly to join coated or uncoated papers to films so as to take advantage of the lower critical surface tension required for wetting the copolymers. Because most packaging adhesives must be able to form strong paper-to-paper bonds, homopolymer vinyl acetate emulsions containing cellulosic protective colloids are always the first choice. The water resistance of any emulsion can be increased by improving its coalescence into a film either by adding solvents and plasticizers to it or by adding specific coalescing agents, eg, diethylene glycol monoethyl ether or diethylene glycol monoethyl ether acetate. Emulsions containing completely hydrolyzed poly(vinyl alcohol) can also be incorporated with cross-linking agents, which further insolubilize them. Cross-linking agents, however, can make a formulation unstable, so that it must often be used within a few hours after it has been compounded.

For any adhesive to be effective, it must first thoroughly wet the surface to be bonded; hence, it must be fluid at the time it is applied. Fluidity is not a problem with emulsion adhesives; wetting, however, can be problematic, especially when slick or coated surfaces are to be joined. An example is the manufacture of high gloss, clay-coated cartons, or the adhering of waxed papers. Copolymer emulsions tend to wet slick surfaces better than homopolymer emulsions because of the extra mobility and softness given to the polymer particles by the plasticizing comonomer. Their dried films also tend to conform better with this type of substrate. Homopolymer emulsions containing poly(vinyl alcohol) as a protective colloid wet paper surfaces well and are the first choice for paper-to-paper packaging adhesives. The setting ability of these emulsions can be increased by incorporating either nonionic wetting agents or partially hydrolyzed poly(vinyl acetate), or both. Small amounts of partially hydrolyzed poly(vinyl acetate) present in an adhesive also help it to wet lightly waxed papers.

The viscosity of an adhesive directly influences its penetration into a substrate; as the viscosity increases, the penetrating power decreases. It also determines the amount of mileage or spread that can be obtained. An optimum viscosity exists for each substrate and each set of machine conditions and must be achieved in order to manufacture an efficient adhesive. Poly(vinyl acetate) emulsions are frequently too low in viscosity to be metered efficiently or to perform well as adhesives by themselves. They must be bodied to working viscosities, eg, by adding thickeners.

Blocking and cold flow are undesirable properties. Blocking refers to dried adhesive surfaces that become sticky, causing unwanted bonding. Cold flow occurs with a polymer having a low softening point; if the temperature of the adhesive assembly warms to the softening range of the polymer, the bond slips. The temperature at which a dried adhesive film forms an instantaneous bond between two surfaces when heat is applied is its heat-sealing temperature. This property is closely related to the blocking temperature and to the temperature at which cold flow or creep can occur.

The heat-sealing temperature of an emulsion is related to the thermoplasticity of the poly(vinyl acetate) particles dispersed in it. Thermoplastic polymers are softened by heat; those of relatively high molecular weight or those that are cross-linked soften at higher temperatures than those of low molecular weight. In addition, poly(vinyl acetate) homopolymers soften at higher temperatures than do copolymers (with monomers that lower the glass-transition temperature) which have similar molecular weight. Another factor which affects the heat-sealing temperature of an emulsion is the amount of poly(vinyl alcohol) in it, if any, since poly(vinyl alcohol) has a high melting point. High heat-sealing temperatures, which are desirable in an emulsion adhesive because they indicate resistance to blocking and cold flow, are usually attainable with emulsions containing large amounts of poly(vinyl alcohol) as a protective colloid.

Fillers are added to emulsion adhesives to build the total solids content, to reduce penetration into a porous substrate, and to lower costs. Plasticizers are added to emulsion adhesives to modify several properties of both the emulsion and the finished adhesive film. By softening the polymer particles dispersed in the emulsion and increasing their mobility, plasticizers cause them to flow together more easily. This usually increases the viscosity of the emulsion and tends to destabilize it for faster breaking and setting speeds at the time it is applied. In addition, the increased softness and mobility help the emulsion to wet smooth, nonporous surfaces, eg, films, foils, and coated papers, thereby increasing its adhesion to them. Also, the softened polymer particles coalesce more rapidly and at a lower temperature than is possible with the unplasticized emulsion. This improved coalescence increases the water resistance of the adhesive film. Plasticizers are usually high boiling esters, eg, phthalates. Phosphate esters are useful as fire-retardant plasticizers.

Solvents are frequently used to perform several functions in emulsion adhesives. Solvents promote adhesion to solvent-sensitive surfaces, they increase the viscosity of the emulsion and intensify the tack of the wet adhesive, and they improve the coalescing properties of the film. Low boiling solvents impart only wet tack to the adhesive film, whereas high boiling solvents confer both dry and wet tack and lower the heat-sealing temperature. The solvent can impart the necessary speed to the wet adhesive but, because it evaporates, it does not cause the dried bond to creep, which often happens with a plasticized film that ages under stress. Solvents promote adhesion to solvent-sensitive adherends by softening and partially dissolving them, thus allowing the adhesive to wet or penetrate the surface or both. Due to the increasing regulatory pressure on solvent emissions, there is a need to develop these properties without volatile materials.

Tackifiers are used to increase the tackiness and the setting speed of adhesives. They increase tackiness by softening the poly(vinyl acetate) polymer in the wet and the dry adhesive film. Tackifiers are usually rosin or its derivatives or phenolic resins. Other additives frequently needed for specific application and service conditions are antifoams, biocides, wetting agents, and humectants.

Specialized copolymer latices, which are inherently and permanently tacky, are available as pressure-sensitive emulsions. They are mechanically stable and have excellent machinability. They are compatible with many other PVAc latices and, therefore, can be easily blended with other resins for modification of surface tack, peel strength, and creep.

Poly(vinyl acetate) dry resins and ethylene–vinyl acetate (EVA) copolymers are used in solvent adhesives, which can be applied by total industrial techniques, eg, brushing, knife-coating, roller-coating, spraying, or dipping. Proper allowances must be made for evaporation of solvent during or before bonding. Poly(vinyl acetate) resins and EVA copolymers containing 21–30 wt % vinyl acetate are widely used in hot-melt adhesive applications. Homopolymers are compounded with 25–35 wt % plasticizer and 25–30 wt % extender resins. These additives increase fluidity and adhesion as well as reduce cost. Ethylene–vinyl acetate resins are mixed with waxes, rubbers, and resin to make the hot-melt adhesive compound. Hot-melt adhesive application processes may be extremely rapid both in application and setting speed, cause no problems resulting from solvent or water evaporation, and provide bonds of high water resistance. The adhesive materials have indefinite shelf lives. Hot-melts are largely used in packaging, laminating, and bookbinding. For the last application, alkali-soluble copolymers are frequently used to facilitate the reclamation of scrap paper.

Film and foil adhesives based on internally plasticized copolymer adhesives have been suggested. For instance, vinyl acetate–ethylene or vinyl acetate–acrylate copolymers may be used for adhesion of films to porous surfaces. For metallic foil adhesion, copolymers containing carboxylate functionality are suggested.

Spray-dried powders find application in adhesives to build solids, increase viscosity, improve tack, and decrease drying time (145). A primary use is in joint compounds with other applications in mastics and grouts, and patching compounds (146).

Paints.   Paints (qv) prepared from poly(vinyl acetate) and its copolymers form flexible, durable films with good adhesion to clean surfaces, including wood, plaster, concrete, stone, brick, cinder blocks, asbestos board, asphalt, tar paper, wallboards, aluminum, and galvanized iron (147). Adherence is also good on painted surfaces if the surfaces are free from dirt, grease, and rust. Developments in emulsion polymerization for paint latices have been reviewed (148).

Poly(vinyl acetate) latex paints are the first choice for interior use (149). Their ability to protect and decorate is reinforced by several advantages belonging exclusively to latex paints: they do not contain solvents so that physiological harm and fire hazards are eliminated; they are odorless; they are easy to apply with spray gun, roller-coater, or brush; and they dry rapidly. The paint can be thinned with water, and brushes or coaters can be cleaned with soap and tepid water. The paint is usually dry in 20 minutes to two hours, and two coats may be applied the same day.

Poly(vinyl acetate) latex paints are also widely used as exterior paints. Their durability, particularly their resistance to chalking, far surpasses that of any conventional oleoresinous paints, which chalk soon after application. The good nonchalking properties of PVAc paints result from their resistance to degradation by uv light. Latex paints that are correctly formulated from quality PVAc latices develop little or no chalk, thus giving maximum tint retention, and they last a long time before repainting becomes necessary. The blister resistance of PVAc paints is another important advantage for their exterior use. Latex paint films are permeable enough to permit water vapor to penetrate them, which prevents blistering and peeling. Their film formation is not impaired if they are painted on damp surfaces or are applied under very humid conditions.

Several types of plasticizing comonomers can be copolymerized with vinyl acetate to produce latices suitable for manufacturing paints. The older, less common comonomers are the alkyl maleates and fumarates, eg, dibutyl maleate and dibutyl fumarate. Acrylic esters, eg, butyl acrylate and 2-ethylhexyl acrylate, as well as ethylene, are the most widely used comonomers in the 1990s. The chief reason for using a comonomer in vinyl acetate polymers is to increase the deformability of the paint film permanently, thus permitting it to expand and contract as the dimensions of the substrate change with changes in temperature. The plasticizing comonomer also softens the polymer particles. For a film to form from a latex paint, the polymer particles must deform and fuse to form a continuous film. This coalescing ability is directly related to the amount of comonomer present. In both interior and exterior paints, the improvement in coalescence that is obtainable by using high comonomer levels results in a general improvement in all the properties necessary for interior and exterior paints. High levels of comonomers, such as ethylene or butyl acrylate, must be added with care. If high comonomer levels are used with low molecular weight polymers, the resulting paint film suffers from excessive dirt pickup at high temperatures because it becomes soft and tacky. It is required, therefore, that high molecular weight polymers be prepared, because only with these is a sufficient amount of comonomer incorporated to give the tint retention without dirt pickup required in exterior paints.

The strength of a polymer increases with its molecular weight; consequently, the toughest paint films are formed from latex polymers having the highest molecular weights. For exterior paints, there is a federal specification of a minimum intrinsic viscosity of 0.45 dL/g. The minimum intrinsic viscosity of commercial paint polymers usually is 0.60–1.0 dL/g and above.

Special vinyl acetate copolymer paints have been developed with greatly improved resistance to blistering or peeling when immersed in water. This property allows better cleaning and use in very humid environments. These lattices exhibit the water resistance of higher priced acrylic resins (150). VAc, vinyl chloride–ethylene terpolymers have been developed which provide the exterior resistance properties of vinyl chloride with the flexibility of the ethylene for exterior paint vehicles (151).

The critical pigment volume concentration (CPVC) of vinyl acetate paints is 48–60%. Generally, the smaller and softer the polymer particles, the higher the CPVC and the greater the pigment-binding capacity of the latex. The most widely used white pigment is titanium dioxide in the form of rutile or of anastase. The color pigments used in latex paints are of two types: the organic pigments (see PIGMENTS, ORGANIC), which are usually hydrophobic, and the inorganic pigments (see PIGMENTS, INORGANIC), which are usually hydrophilic. The hydrophilic pigments are relatively easy to incorporate into water-base paints. The hydrophobic pigments are more difficult to incorporate, but this can be overcome by choosing the correct blend of surfactants for a particular pigment. The organic colorants commonly used are toluidine red, Hansa yellow, phthalocyanine blue and green, pigment green B, and carbon black, among others. Examples of acceptable inorganic colorants are iron oxide red, brown, yellow, black, and chrome oxide green. Colorants that should not be used are those that are reactive, partially water-soluble, or sensitive to pH changes.

In addition to latex and pigment, paint formulations contain dispersants and wetting agents (both surfactants), defoamers, thickeners and protective colloids, freeze–thaw stabilizers, coalescing agents, and biocides. The coalescing agents used in the paint formulations tend to be volatile. In order to reduce the odor levels and reduce solvent emissions, some paint producers are researching latex bases which require less or no solvent to aid in coalescence and yet provide a durable, glossy surface (152). Copolymers are potentially useful in meeting this requirement (153,154).

Paper. Poly(vinyl acetate) emulsions and resins have been used as the binder in coatings for paper and paperboard since 1955 (155–157). The coatings may be clear, colored, or pigmented, and are glossy, odorless, tasteless, greaseproof, nonyellowing, and heat-sealable. Conventional paper-coating equipment is used; formulations normally contain 60–65 wt % solids with a pigment-to-binder ratio of 1:5. Printing quality and ink-pick resistance are excellent. In papermaking, emulsions applied as wet-end additions to the furnish improve the strength and durability of the final product. Plasticized emulsions may be used to give the product toughness and flexibility.

Emulsions used in paper coatings must meet special requirements: the particle size must be small (ca 0.1 μm) and its distribution rather narrow. These properties provide good pigment binding efficiency and high gloss. They must have exceptionally good mechanical stability and freedom from off-sized latex particles or grit, and must exhibit no trace of dilatant flow in a pigment slurry. These properties are required for modern high speed paper-coating applications. They should also be compatible with starch and alginate natural resins frequently used as cobinders (158).

Nonwoven Binders and Textiles. The use of vinyl acetate copolymers as binding agents for nonwoven fabrics has grown rapidly. Vinyl acetate–ethylene copolymer latices have been particularly successful in part because of their low cost but also because of their excellent adhesion to a wide range of substrates, their water and alkali resistance, and their low flammability (159–162). Products made from these binders include paper towels, wipes, and personal hygiene articles. Self-cross-linking polymers are available as are polymers requiring external curing agents. Many of the self-cross-linking polymers release formaldehyde (qv) as part of the reaction. Because formaldehyde has been listed as a potential carcinogen, alternative chemistries have been developed to reduce or eliminate formaldehyde content in the products.

Poly(vinyl acetate) emulsions are widely used as textile finishes because of their low cost and good adhesion to natural and synthetic fibers. In textile piece goods finishing, dispersions diluted with water to 1–3 wt % resin are most often used to obtain a stiff or crisp hand on woven cotton fabrics. Concentrations of 2–20 wt % emulsion are recommended for bodying, stiffening, and bonding. Principal applications include the stiffening of felts. Finishes to improve snag resistance and body or hand of nylon hosiery are based on PVAc emulsions.

Poly(vinyl acetate) emulsions are used to prime-coat fabrics to improve the adhesion of subsequent coatings or to make them adhere better to plastic film. Plasticized emulsions are applied, generally by roller-coating, to the backs of finished rugs and carpets to bind the tufts in place and to impart stiffness and hand. For upholstery fabrics woven from colored yarns, PVAc emulsions may be used to bind the tufts of pile fabrics or to prevent slippage of synthetic yarns.

The emulsion formulations are generally applied to cloth by padding from a bath and squeezing off the excess. Modifying a formulation in the pad box, eg, to increase or decrease firmness, can be easily done by adding an emulsion or softener. The alkali-soluble vinyl acetate copolymers previously mentioned can be used as warp sizes during weaving.

Concrete Additives. Poly(vinyl acetate) was first used in concrete in the 1940s as a thermoplastic polymer to strengthen the concrete matrix. In contrast to other polymers the resistance to water permeation is low due to the hydrolysis of the poly(vinyl acetate) (163,164). Ethylene copolymers have been developed which have improved water resistance and waterproofness. The polymer can be used in the latex form or in a spray-dried form which can be preblended in with the cement (qv) in the proper proportion. The compressive and tensile strength of concrete is improved by addition of PVAc emulsions to the water before mixing.

A polymer-solids-to-total-solids ratio of ca 10:90 is best. The emulsions also aid adhesion between new and old concrete when patching or resurfacing.

Other. Vinyl acetate resins are useful as antishrinking agents for glass fiber-reinforced polyester molding resins (165). Poly(vinyl acetate)s are also used as binders for numerous materials, eg, fibers, leather (qv), asbestos, sawdust, sand, clay, etc, to form compositions that can be shaped with heat and pressure. Joint cements, taping compounds, caulks, and fillers are other uses.

Emulsions containing added poly(vinyl alcohol) and dichromate are used to make light-sensitive stencil screens for textile printing and ceramic decoration. The resins are used in printing inks, nitrocellulose lacquers, and special high gloss coatings. Inks made with PVAc and metallic pigments look like foil because the formulations have a high leafing power, do not induce tarnish, and contribute no unwanted color or aging.

Vinyl acetate polymers have long been used as chewing gum bases. They have been studied as controlled release agents for programmed administration of drugs and as a base for antifouling marine paints (166,167).

BIBLIOGRAPHY

"Vinyl Acetate" under "Vinyl Compounds," in *ECT* 1st ed., Vol. 14, pp. 686–691, by T. P. G. Shaw, pp. 691–698, by K. G. Blaihie and T. P. G. Shaw, pp. 699–709, by K. B. Blaikie and M. S. W. Small, Shawinigan Chemicals, Ltd., "Poly(Vinyl Acetate)" under "Vinyl Polymers," in *ECT* 2nd ed., Vol. 21, pp. 317–353, by D. Rhum, Air Reduction Co.; in *ECT* 3rd ed., Vol. 23, pp. 817–847, by W. E. Daniels, Air Products and Chemical Inc.

1. C. A. Schildknecht, *Vinyl and Related Polymers*, John Wiley & Sons, Inc., New York, 1952, p. 323.
2. *Chemical Economics Handbook Marketing Research Report*, "Vinyl Acetate," (July 1993).
3. C. Sumner and J. R. Zoeller, *Chem. Ind.*, **49**, 225–240 (1993).
4. *Plast. Ind. News*, **41**(4), 2 (Apr. 1995).
5. *Vinyl Acetate, A Guide to Safety and Handling*, compiled by E. I. duPont de Nemours and Co., Inc., Hoechst Celanese Chemical Group, Inc., Quantum Chemical Co., and Union Carbide Corp., 1995.
6. *Vinyl Acetate*, Bulletin No. S-56-3, Celanese Chemical Co., New York, 1969.
7. *Vinyl Acetate Monomer F-41519*, Union Carbide Corp., New York, June 1967.
8. *Vinyl Acetate Monomer BC-6*, Borden Chemical Co., New York, 1969.
9. *Vinyl Acetate Monomer*, Air Reduction Co., New York, 1969.
10. R. W. Gallant, *Hydrocarbon Process.* **47**(10), 115 (1968).
11. S. Okamura and I. Motoyama, *J. Polym. Sci.* **58**, 221 (1962).
12. L. H. Horsley, *Azeotropic Data, II, Advances in Chemistry Series*, No. 35, American Chemical Society, Washington, D.C., 1962.
13. M. K. Lindemann, in G. E. Ham, ed., *Vinyl Polymerization*, Vol. 1, Marcel Dekker, Inc., New York, 1967, Part 1, Chapt. 4.
14. J. Brandrup and E. H. Immergut, eds., *Polymer Handbook*, Interscience Publishers, a division of John Wiley & Sons, Inc., New York, 1966.
15. M. K. Lindemann, in N. M. Bikales, ed., *Encyclopedia of Polymer Science and Technology*, Vol. 15, John Wiley & Sons, Inc., New York, 1971, p. 636.
16. G. O. Morrison and T. P. G. Shaw, *Trans. Electrochem. Soc.* **63**, 425 (1933).

17. D. Swern and E. F. Jordan, Jr., in N. Rabjohn, ed., *Organic Syntheses*, collective Vol. 4, John Wiley & Sons, Inc., New York, 1963, p. 977.
18. H. Hopff and M. A. Osman, *Tetrahedron* **24**, 2205 (1968).
19. C. F. Kohll and R. van Helden, *Rec. Trav. Chim.* **86**, 193 (1967).
20. H. E. Simmons and R. D. Smith, *J. Am. Chem. Soc.* **81**, 4256 (1959).
21. C. Walling, *Free Radicals in Solution*, John Wiley & Sons, Inc., New York, 1957, Chapt. 6.
22. L. B. Levy, *Process Safety Progr.* **12**(1), 47–48 (Jan. 1993).
23. D. W. Butcher and H. M. Sharpe, *MariChem*, 5th ed., 1984, pp. 189–192.
24. *The Evaluation of Carcinogenic Risks to Humans; Dry Cleaning, Some Chlorinated Solvents and Other Industrial Chemicals, IARC Monographs*, **63** (Feb. 1995).
25. R. VanHelden and co-workers, *Rev. Trav. Chim.* **87**, 961 (1968).
26. *Petr. Refiner*, **38**, 304 (1959).
27. H. C. Volger, *Rev. Trav. Chim.* **87**, 501 (1968).
28. D. N. Mead and R. M. Fuoss, *J. Am. Chem. Soc.* **63**, 2832 (1941).
29. S. O. Morgan and Y. A. Yager, *Ind. Eng. Chem.* **32**, 1519 (1940).
30. H. Fikentscher, *Cellul. Chem.* **13**, 71 (1932).
31. E. O. Kraemer, *Ind. Eng. Chem.* **30**, 1200 (1938).
32. A. Ballisteri and co-workers, *J. Polym. Sci. Polym. Chem. Ed.* **18**, 1147 (1980).
33. T. Okada, *Polym. J.* **9**, 121 (1977).
34. H. Bartl, in E. Muller, ed., *Methods of Organic Chemistry (Houben-Weyl), Macromolecular Materials*, Part 1, Georg Thieme Verlag, Stuttgart, Germany, 1961, pp. 905–918.
35. D. R. Bassett and A. E. Hamielec, eds., *Emulsion Polymers and Emulsion Polymerization*, ACS Symposium Series No. 165, American Chemical Society, Washington, D.C., 1981.
36. I. Piirma, ed., *Emulsion Polymerization*, Academic Press, New York, 1982.
37. R. D. Athey, *Emulsion Polymer Technology*, Marcel Dekker, New York, 1991.
38. Q. Wang and co-workers, *Prog. Polym. Sci.* **19**(4), 703–753 (1994).
39. D. C. Blackley, *Emulsion Polymerisation: Theory and Practice*, John Wiley & Sons, Inc., New York, 1975.
40. M. K. Lindemann, *Paint Manuf.* **38**(9), 30 (1968).
41. E. Levine, W. Lindlaw, and J. Vona, *J. Paint Technol.* **41**, 531 (1969).
42. C. Chellappa, *Mod. Paint Coat.* **85**(13), 28–32 (Dec. 1995).
43. A. S. Badran and co-workers, *J. Appl. Polym. Sci.* **49**(2), (1993), 187–96.
44. N. J. Earhart, *The Grafting Reactions of Poly(vinyl alcohol) During the Emulsion Copolymerization of Poly(vinyl acetate–co-butyl acrylate)*, Ph.D. dissertation.
45. *Product List*, Air Reduction Co., New York, 1968.
46. *Elvacet Poly(Vinyl Acetate) Brochure*, E. I. Du Pont de Nemours & Co., Inc., Wilmington, Del., 1968.
47. *Thermoplastics Division Product Directory*, Borden Chemical Co., New York, 1968.
48. *Gelva Poly(Vinyl Acetate) Technical Bulletin*, Publication No. 6103, Monanto Chemical Co., St. Louis, Mo., 1969.
49. E. Tromsdorff and C. E. Schildknecht, in C. E. Schildknecht, ed., *Polymer Processes*, Interscience Publishers, Inc., New York, 1956, pp. 105–109.
50. V. T. Stannett and co-workers, *Prog. Polym. Process. 3, Radiat. Process. Polym.*, 289–317 (1992).
51. T. Okaya and co-workers, *J. Appl. Polym. Sci.* **50**(5), 745–751 (1993).
52. U.S. Pat. 5,171,767 (Dec. 15, 1992), R. G. Buckley and co-workers (to Rohm and Haas).
53. J. Dimitratos and co-workers, *AIChE J.* **40**(12) (Dec. 1994).
54. A. Urretabizkaia and co-workers, *J. Polym. Sci. Part A–Polym. Chem.* **31**(12), 2907–2913 (1993).

55. L. M. Gugliotta and co-workers, *Polymer*, **36**(10), 2019–2023.
56. S. Canegallo and co-workers, *J. Appl. Polym. Sci.* **47**(6), 961–979 (1993).
57. P. D. Gossen and J. F. Macgregor, *J. Coll. Interface Sci.* **160**(1), 24–38 (1993).
58. W. Baade and co-workers, *J. Appl. Sci.* **27**, 2249–2267 (1982).
59. P. Bataille and co-workers, *J. Appl. Polym. Sci.* **38**(12), 2237–2244 (1989).
60. A. Iabbadene and co-workers, *J. Appl. Polym. Sci.* **51**(3), 503–511 (1994).
61. U.S. Pat. 4,164,489 (Aug. 14, 1979), W. E. Lenney and W. E. Daniels (to Air Products and Chemicals, Inc.).
62. C. Kipparissides, J. F. MacGregor, and A. E. Hamiliec, *Can. J. Chem. Eng.* **58**(1), 48 (1980).
63. R. K. Greene, *Continuous Emulsion Polymerization of Vinyl Acetate*, Ph.D. dissertation, 1976.
64. U.S. Pat. 4,035,329 (Nov. 19, 1975), H. Wiest and co-workers (to Wacker Chemie, GmbH).
65. P. L. Tang, *Semicontinuous Polymerization Using Miniemulsions*, M. S. Report, 1989.
66. J. Delgado, *Miniemulsion Copolymerization of Vinyl Acetate and n-Butyl Acrylate*, Ph.D. dissertation, 1986.
67. M. S. El-Aasser and J. W. Vanderhoff, eds., *Emulsion Polymerization of Vinyl Acetate*, Applied Science Publishers, Inc., Englewood, N.J., 1981.
68. R. D. Dunlop and F. E. Reese, *Ind. Eng. Chem.* **40**, 654 (1948).
69. G. Kalfas, H. Yuan, and W. H. Ray, *Ind. Eng. Chem. Res.* **32**(9), 1831–1838 (1993).
70. V. T. Shiriniyan and co-workers, *Zh. Prikl. Khim. (Leningrad)* **44**, 1345 (1977).
71. K. H. Reichert and H. V. Moritz, *J. Appl. Polym. Sci. Polym. Symp.* **36**, 151 (1981).
72. H. V. Moritz and K. H. Reichert, *Chem. Eng. Tech.* **53**(5), 386 (1981).
73. J. W. Harter and W. H. Ray, *Chem. Eng. Sci.* **41**(72), 3083 (1986).
74. *Ibid.*, 3095 (1986).
75. R. G. Gilbert, *Emulsion Polymerization: A Mechanistic Approach*, Academic Press, Inc., New York, 1995.
76. E. Daniels and co-authors, eds., *Polymer Latexes: Preparation, Characterization, and Applications*, ACS Symposium Series Amcrican Chemical Society, Washington, D.C., 1992, p. 492.
77. B. S. Casey, B. R. Morrison, and R. G. Gilbert, *Progr. Polym. Sci.* **18**(6), 1041–1096 (1993).
78. N. Friis and L. Nyhagen, *J. Appl. Polym. Sci.* **17**(8), 2311–2327 (1973).
79. A. Penlidis, J. F. MacGregor, and A. E. Hamielec, *Polym. Process Eng.* **3**(3), 185–218 (1985).
80. M. S. El-Aaser and co-workers, *J. Polym. Sci., Polym. Chem. Ed.* **21**(8), 2363–2382 (1983).
81. J. W. Breitenbach, H. Edelhauser, and R. Hochrainer, *Monatsh. Chem.* **99**, 625 (1968).
82. C. E. Schildknecht and I. Skeist, *Polymerization Processes*, John Wiley & Sons, Inc., New York, 1977.
83. J. Meuldijk and co-workers, *Chem. Eng. Sci.* **47**(9–11) 2603–2608 (1992).
84. C. M. Gilmore and co-workers, *J. Appl. Polym. Sci.* **48**(8), 1449–1460 (1993).
85. M. Nomura and K. Fujita, *Makromol. Chem. Suppl.* **10/11**, 25–42 (1985).
86. P. J. Flory and F. S. Leutner, *J. Polym. Sci.* **3**, 880 (1948); **5**, 267 (1950).
87. D. Stein, *Makromol. Chem.* **76**, 170 (1964).
88. M. Matsumoto and I. Ohyang, *J. Polym. Sci.* **46**, 441 (1960).
89. S. P. Pontis and A. M. Deshpande, *Makromol. Chem.* **125**, 48 (1969).
90. W. W. Graessley, W. C. Uy, and A. Gandhi, *Ind. Eng. Chem. Fundam.* **8**, 697 (1969).
91. V. T. Shiriniyan and co-workers, *Plast. Massy* **8**, 15 (1974).
92. S. Imoto, J. Ukida, and T. Kominami, *Kobunshi Kagaku* **14**, 101 (1957).

93. D. Stein and G. V. Schultz, *Makromol. Chem.* **52**, 249 (1962).
94. I. Gavat, V. Dimonie, and D. Donescu, *J. Polym. Sci. Polym. Symp.* **64**, 125 (1978).
95. K. Noro, *Br. Polym. J.* **2**, 128 (1970).
96. M. Shiraishi, *Br. Polym. J.* **2**, 135 (1970).
97. S. S. Mnatskanov and co-workers, *Vysokmol. Soyed.* **A14**, 4,851 (1972).
98. V. T. Shirininyan and co-workers, *Vysokmol. Soyed.* **A17**, 1,182 (1975).
99. M. Furuta, *J. Polym. Sci. Polym. Lett. Ed.* **11**, 113 (1973).
100. *Ibid.*, **12**, 459 (1974).
101. H. Meissner and G. Heublein, *Acta Polym.* **34**(6), 379 (1983).
102. H. Meissner and co-workers, *Acta Polym.* **35**(3), 250 (1984).
103. G. Heublein and H. Meissner, *Acta Polym.* **35**(12), 744 (1984).
104. *Ibid.*, **36**(5), 245 (1985).
105. *Ibid.*, (6), 343 (1985).
106. H. Meissner, P. Hartschansky, and G. Heublein, *Acta Polym.* **36**(6), 345 (1985).
107. *Ibid.*, **36**(12), 699 (1985).
108. H. Meissner and G. Heublein, *Acta Polym.* **37**(5), 323 (1986).
109. *Ibid.*, **38**(1), 75 (1987).
110. D. Donescu and co-workers, *J. Macromol. Sci. Chem.* **22**(5–7), 931 (1985).
111. D. Donescu, *Rev. Roum. Chem.* **24**(9), 1399 (1979).
112. D. Donescu and co-workers, *Rev. Roum. Chem.* **19**(6), 483 (1984).
113. S. Hayashi and co-workers, *J. Appl. Polym. Sci.* **27**(5), 1607 (1982).
114. V. T. Stannett, *Proc. R. Aust. Chem. Inst.* **42**, 232 (1975).
115. G. F. Lundardon, G. P. Talamini, and V. Grosso, *Eur. Polym. J.* **11**, 437 (1975).
116. P. K. Isaacs and H. A. Edelhauser, *J. Appl. Polym. Sci.* **10**, 171 (1966).
117. H. Warson, *Chem. Ind. London* **6**, 220 (1983).
118. J. Schellenberg and co-workers, *Angew. Makromol. Chem.* **130**, 99 (1985).
119. D. Mohan and co-workers, *J. Polym. Sci. Polym. Chem. Ed.* **21**, 3041 (1983).
120. P. J. Scott, A. Penlidis and G. L. Rempel, *J. Polym. Sci. Part A–Polym. Chemistry*, **31**(9), 1993, 2205–2230.
121. P. J. Scott, A. Penlidis and G. L. Rempel, *Journal of Polymer Science Part A–Polymer Chem.* **31**(2), 403–426 (1993).
122. J. W. Vanderhoff, *J. Polym. Sci. Lett. Ed.* **17**, 567 (1979).
123. R. B. Seymour and G. A. Stahl, *J. Macromol. Sci. Chem.* **11**(1), 53 (1977).
124. W. K. Busfield and R. B. Low, *Eur. Polym. J.* **11**, 309 (1975).
125. A. G. Sayadyan and D. A. Simonyan, *Arm. Khim. Zh.* **21**, 1041 (1968).
126. U.S. Pat. 4,260,533 (Apr. 7, 1981), J. G. Iacoviello and W. E. Daniels (to Air Products and Chemicals, Inc.).
127. T. Yatsu, S. Moriuchi, and H. Fuji, *Makromolecules*, **10**, 243 (1977).
128. C. H. Chen, *J. Polym. Sci.* **14**, 2109 (1976).
129. B. Tidswell and A. W. Train, *Br. Polym. J.* **7**, 409 (1975).
130. U.S. Pat. 4,137,382 (Jan. 30, 1979), C. J. Vetters (to National Distillers and Chemical Corp.).
131. U.S. Pat. 4,172,939 (Oct. 30, 1979), G. J. Hoh (to E. I du Pont de Nemours & Co., Inc.).
132. *Research Disclosure #13816*, Industrial Opportunities, Ltd., Hampshire, U.K., Oct. 1975.
133. J. Y. Olayemi and M. K. Ibiyeve, *J. Appl. Polym. Sci.* **31**(1), 237 (1986).
134. F. Le and co-workers, *J. Polym. Eng.* **7**(2), 113 (1987).
135. R. H. Bott, J. A. Kuphal, L. M. Robeson and D. Sagl, *J. Appl. Polym. Sci.* **58**(9), 1593–1605 (1995).
136. D. R. Paul and co-workers, *Polymer Blends*, Vol. 1, Academic Press, New York, 1978, pp. 71–74.
137. W. R. Furlan, *Adhesives Age* **37**(2), 20–22 (Feb. 1994).

138. R. M. Holmes and co-workers, *CEH Marketing Research Report—Polyvinyl Acetate* (Feb. 1993).

139. M. L. Hausman and co-workers, *Adhesives Age* **38**(11), 49–60 (Oct. 1995).

140. *Working with Vinyl Acetate Based Polymers, An Adhesives Manual*, 3rd ed., Air Products and Chemicals, Inc., Allentown, Pa., 1996.

141. R. A. Weidener, in R. L. Patrick, ed., *Treatise on Adhesion and Adhesives*, Vol. 2, Marcel Dekker, Inc., New York, 1969, Chapt. 10, pp. 432–447, 467–471.

142. I. Skeist, ed., *Handbook of Adhesives*, Van Nostrand Reinhold Co., New York, 1990, p. 468.

143. F. P. Hoenisch and M. C. Bricker, *Adhesives Age*, **36**(8), 20–23 (July 1993).

144. U.S. Pat. 5185308 (1993), J. G. Iacoviello and D. W. Horwat (to Air Products and Chemicals, Inc.).

145. K. W. Rizzi, *Adhesives Age* **38**(8), 24–26 (July 1995).

146. I. Skeist, ed., *Handbook of Adhesives*, Van Nostrand Reinhold Co., New York, 1990, p. 472.

147. *Paint Handbook*, Air Products and Chemicals, Inc., New York, 1969.

148. H. Warson, *Polym. Paint Colour J.* **180**, 473–486 (July 1990).

149. *Am. Paint J.* **53**, 7, 58 (1968).

150. *Am. Paint/Coating J.*, 66 (Nov. 13, 1978).

151. D. B. Farmer and M. H. Edser, *Chem. Ind.*, 228–235 (Mar. 21, 1983).

152. *Chem. Week*, 43 (Aug. 1995).

153. B. Currie, *Mod. Paint Coatings*, **83**(8), 34–40 (Aug. 1993).

154. U.S. Pat. 5,346,948 (1994), F. L. Floyd and G. P. Craun (to Glidden).

155. D. K. Stinebaugh, *Tappi Notes, 1990 Blade Coating Seminar*, Tappi Press, Atlanta, Ga., 1990.

156. J. W. Ramp, *Tappi Notes, 1990 Coating Binders Short Course*, Tappi Press, Atlanta, Ga., 1990.

157. T. F. Walsh and L. A. Gaspar, *TAPPI Monograph Series 37*, 1975, Tappi Press, Atlanta, Ga., Chapt. 5.

158. U.S. Pat. 4,228,047 (July 31, 1981), W. E. Daniels and W. H. Pippen (to Air Products and Chemicals, Inc.).

159. J. R. Halker, *Formed Fabric Industry*, 26 (June 1976).

160. P. L. Rosamilia, *Tappi Paper Synth. Proc.*, 251 (1979).

161. C. C. Chappelow, P. L. Rosamilia, and M. J. Taylor, *INDEX 81 Congress Papers*, Session 2, Washington, D.C.

162. D. A. Brighton, *Non-Wovens Yearbook*, 24–29 (1982).

163. S. Chandra and Y. Ohama, *Polymers in Concrete* CRC Press, Boca Raton, Fla., 1994.

164. L. Holloway, *Polymers and Polymer Composites in Construction*, T. Telford, London, 1990.

165. K. E. Atkins and B. W. Lipinsky, *Auk. Offentliche Jahrestagung Der. 13 Int. Tagung*, *Freudenstadt*, 39/1–4 (Oct. 1976).

166. H. Leeper and H. Benson, *SPE 2nd Ann. Conf.*, Seattle, Wash., Aug. 1976, pp. 141–149.

167. U.S. Pat. 4,143,015 (Jan. 21, 1977), E. Soeterik.

CAJETAN F. CORDEIRO
Air Products and Chemicals, Inc.

VINYL ALCOHOL POLYMERS

Poly(vinyl alcohol) (PVA), a polyhydroxy polymer, is the largest-volume synthetic, water-soluble resin produced in the world. It is commercially manufactured by the hydrolysis of poly(vinyl acetate), because monomeric vinyl alcohol cannot be obtained in quantities and purity that makes polymerization to poly(vinyl alcohol) feasible (1–3).

Poly(vinyl alcohol) [9002-89-5] was discovered through the addition of alkali to a clear alcoholic solution of poly(vinyl acetate), which resulted in the ivory-colored poly(vinyl alcohol) (4). The same discovery has been made by studying the reversible transformation between poly(vinyl alcohol) and poly(vinyl acetate) via esterification and saponification (5). The first scientific reports on poly(vinyl alcohol) were published in 1927 (6,7).

The excellent chemical resistance and physical properties of PVA resins have resulted in broad industrial use. The polymer is an excellent adhesive and possesses solvent-, oil-, and grease-resistant properties matched by few other polymers. Poly(vinyl alcohol) films exhibit high tensile strength, abrasion resistance, and oxygen barrier properties which, under dry conditions, are superior to those of any other known polymer. The polymer's low surface tension provides for excellent emulsification and protective colloid properties.

The main uses of PVA are in textile sizing, adhesives, protective colloids for emulsion polymerization, fibers, production of poly(vinyl butyral), and paper sizing. Significant volumes are also used in the production of concrete additives and joint cements for building construction and water-soluble films for containment bags for hospital laundry, pesticides, herbicides, and fertilizers. Smaller volumes are consumed as emulsifiers for cosmetics, temporary protective film coatings, soil binding to control erosion, and photoprinting plates.

Physical Properties

The physical properties of poly(vinyl alcohol) are highly correlated with the method of preparation. The final properties are affected by the polymerization conditions of the parent poly(vinyl acetate), the hydrolysis conditions, drying, and grinding. Further, the term poly(vinyl alcohol) refers to an array of products that can be considered copolymers of vinyl acetate and vinyl alcohol.

The effect of hydrolysis and molecular weight is illustrated in Figure 1. The variations in properties with molecular weight are for a constant degree of hydrolysis (mol %) (8) and the effect of hydrolysis is at a constant molecular weight. Representative properties are shown in Table 1.

Crystallization and Melting Point. The ability of PVA to crystallize is the single most important physical property of PVA as it controls water solubility, water sensitivity, tensile strength, oxygen barrier properties, and thermoplastic properties. Thus, this feature has been and continues to be a focal point of academic and industrial research (9–50). The degree of crystallinity as measured by x-ray diffraction can be directly correlated to the density of the material or the swelling characteristic of the insoluble part (Fig. 2).

The size of the crystals determines the melting point. Reported values for the melting point of poly(vinyl alcohol) range between 220 and 267°C for fully

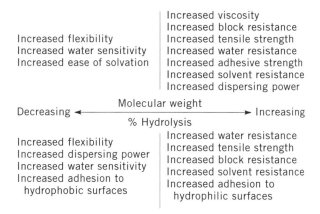

Fig. 1. Effect of molecular weight and hydrolysis on the properties of poly(vinyl alcohol) (8). Courtesy of Air Products and Chemicals, Inc.

hydrolyzed PVA (51–55). Exact determination of the crystalline melting point using normal dta techniques is difficult as decomposition takes place above 140°C. The divergence in reported values may be ascribed to decomposition or prior treatment history. The melting point of poly(vinyl alcohol) containing an appropriate amount of diluent or comonomer is less influenced by decomposition. Thus, the melting point of fully hydrolyzed PVA can be determined by the extrapolation of the measured values to 0% diluent. A more reliable melting point is obtained in this manner. The melting points determined by the diluent method are 255–267°C for commercial superhydrolyzed PVA (greater than 99% hydrolysis). The melting point determined by melting point depression caused by noncrystallizing comonomer units assumes, as a first approximation, that the vinyl acetate units are randomly distributed. This assumption usually does not apply to commercial PVA. The extrapolated values of heat of fusion and melting point obtained with this method are therefore highly dependent on the manufacturing method and the resulting blockiness (Fig. 3). The heat of fusion, determined by either of the above methods, has been calculated as 6.82 ± 2.1 kJ/mol (56–58).

Glass-Transition Temperatures. The glass-transition temperature, T_g, of fully hydrolyzed PVA has been determined to be 85°C for high molecular weight material. The glass transition in case of 87–89% hydrolyzed PVA varies according to the following formula (59):

$$T_g = 58 - (2.0 \times 10^{-3}/\text{DP})°\text{C}$$

Solubility. Poly(vinyl alcohol) is only soluble in highly polar solvents, such as water, dimethyl sulfoxide, acetamide, glycols, and dimethylformamide. The solubility in water is a function of degree of polymerization (DP) and hydrolysis (Fig. 4). Fully hydrolyzed poly(vinyl alcohol) is only completely soluble in hot to boiling water. However, once in solution, it remains soluble even at room temperature. Partially hydrolyzed grades are soluble at room temperature, although grades with a hydrolysis of 70–80% are only soluble at water temperatures of

Table 1. Physical Properties of Poly(Vinyl Alcohol)

Property	Value	Comments
appearance	white to ivory-white granular powder	
specific gravity	1.27–1.31	increases with degree of crystallinity
tensile strength, MPa[a]	67–110[b]	increases with degree of crystallinity (heat treatment) and molecular weight; decreases with increasing humidity
tensile strength, MPa[a]	24–79[c]	increases with molecular weight; decreases with increasing humidity
elongation, %	0–300	increases with increasing humidity
thermal coefficient of expansion $\times 10^{-5}$ per °C	7–12	
specific heat, J/(g·K)[d]	1.67	
thermal conductivity, W/(m·K)	0.2	
glass-transition temperature, K	358	98–99% hydrolyzed
	331	87–89% hydrolyzed
melting point, K	503	98–99% hydrolyzed
	453	87–89% hydrolyzed
electrical resistivity, Ω·cm	$(3.1–3.8) \times 10^7$	
thermal stability	gradual discoloration above 100°C; darkens rapidly above 150°C; rapid decomposition above 200°C	
refractive index, n_D^{20}	1.55	
degree of crystallinity	0–0.54	increases with heat treatment and degree of hydrolysis
storage stability (solid)	indefinite when protected from moisture	
flammability	burns similarly to paper	
stability in sunlight	excellent	

[a]To convert MPa to psi, multiply by 145.
[b]98–99% hydrolyzed.
[c]87–89% hydrolyzed.
[d]To convert J to cal, divide by 4.184.

10–40°C. Above 40°C, the solution first becomes cloudy (cloud point), followed by precipitation of poly(vinyl alcohol).

The hydroxyl groups in poly(vinyl alcohol) contribute to strong hydrogen bonding both intra- and intermolecularly, which reduces solubility in water. The presence of the residual acetate groups in partially hydrolyzed poly(vinyl alcohol) weakens these hydrogen bonds and allow solubility at lower temperatures.

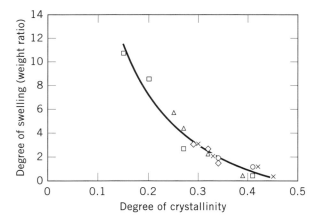

Fig. 2. Relationship between swelling and crystallinity, where □ represents DP 304; △, DP 708; ○, DP 1288; ◇, DP 2317; and ×, DP 4570.

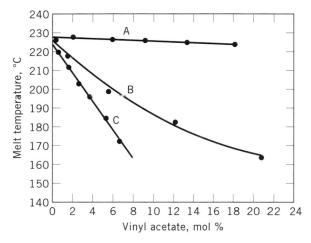

Fig. 3. Influence of vinyl alcohol–vinyl acetate copolymer composition on melting temperature (56), where A represents block copolymers; B, blocky copolymers; and C, random copolymers.

The hydrophobic nature of the acetate groups results in a negative heat of solution (61–64), which increases as the number of acetate groups is increased. This means that the critical temperature or the θ temperature is lower, ie, the solubility decreases as the temperature is increased.

Heat treatment of a few minutes increases the crystallinity and greatly reduces solubility and water sensitivity (Fig. 5). Prolonged heat treatment does not further increase crystallinity. The heat treatment melts the smaller crystals, allowing for diffusion and reformation of crystals with a melting point higher than that of the treatment temperature. The presence of acetate groups reduces the extent of crystallinity; thus heat treatment has little or no influence on low hydrolysis grades. The influence of heat treatment is desirable in applications such as adhesives and paper coatings, where a greater degree of water resistance

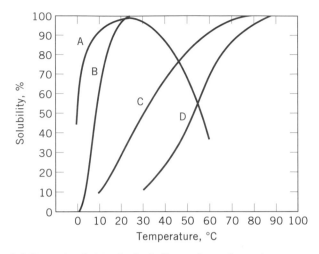

Fig. 4. Water solubility of poly(vinyl alcohol) grades, where A represents 78–81 mol % hydrolyzed, DP = 2000–2100; B, 87–89 mol % hydrolyzed, DP = 500–600; C, 98–99 mol % hydrolyzed, DP = 500–600; and D, 98–99 mol % hydrolyzed, DP = 1700–1800 (60).

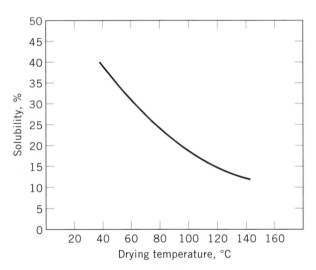

Fig. 5. Influence of heat treatment on solubility at 40°C; DP = 1700, 98–99 mol % hydrolyzed.

is needed, but is highly undesirable in textile warp sizing, where the polymer must be removed after a drying cycle. Poly(vinyl alcohol) solutions also exhibit high tolerance toward many electrolytes (Table 2).

Solution Viscosity. The viscosities of PVA solutions are mainly dependent on molecular weight and solution concentration (Fig. 6). The viscosity increases with increasing degree of hydrolysis and decreases with increasing temperature. Materials with a high degree of hydrolysis tend to show an increase in viscosity on standing and may even gel (60,65–70). The rate of increase depends on dissolution temperature, concentration, and storage temperature. The lower

Table 2. Minimum Salt Concentration for Precipitation of a 5% Poly(Vinyl Alcohol) Solution[a,b]

Salt	g/L
$(NH_4)_2SO_4$	66
Na_2SO_4	50
KSO_4	61
$ZnSO_4$	113
$CaSO_4$	112
$Fe_2(SO_4)_3$	105
$MgSO_4$	60
$Al_2(SO_4)_3$	57
$KAl(SO_4)_2$	58
NH_4NO_3	490
$NaNO_3$	324
KNO_3	264
$Al(NO_3)_3$	255
$NaCl$	210
KCl	194
Na_3PO_4	77
K_2CrO_4	136

[a]Ref. 59.
[b]98% hydrolyzed; DP = 1700–1800.

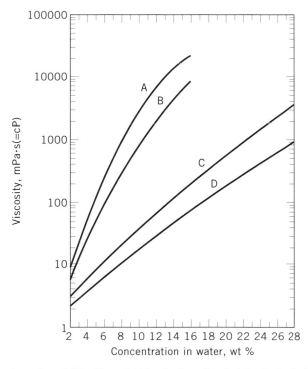

Fig. 6. Solution viscosity of 87–89 mol % hydrolyzed poly(vinyl alcohol) at 20°C. DP for A is 2200; for B, 1500; for C, 550; and for D, 220.

the storage and dissolution temperature and the higher the concentration, the higher the rate of the viscosity increase (Fig. 7). The viscosity can be stabilized to a certain degree by adding small amounts of lower molecular weight aliphatic alcohols (71), urea, or salts such as thiocyanates. The solution viscosity of partially hydrolyzed PVA grades exhibits a greater degree of stability.

Mechanical Properties.   The tensile strength of unplasticized PVA depends on degree of hydrolysis, molecular weight, and relative humidity (Fig. 8). Heat treatment and molecular alignment resulting from drawing increase the tensile strength; plasticizers reduce tensile strength disproportionately, on account of increased water sensitivity.

Tensile elongation of PVA is extremely sensitive to humidity and ranges from <10% when completely dry to 300–400% at 80% rh. Addition of plasticizer can double these values. Elongation is independent of degree of hydrolysis but proportional to the molecular weight. Tear strength increases with increasing relative humidity or with the addition of small amounts of plasticizer.

Solvent Resistance.   Poly(vinyl alcohol) is virtually unaffected by hydrocarbons, chlorinated hydrocarbons, carboxylic acid esters, greases, and animal or vegetable oils. Resistance to organic solvents increases with increasing hydrolysis. This resistance has promoted the use of PVA in the manufacture of gloves for use when handling organic solvents (73).

Gas-Barrier Properties.   The oxygen-barrier properties of PVA at low humidity are the best of any synthetic resin. However, barrier performance dete-

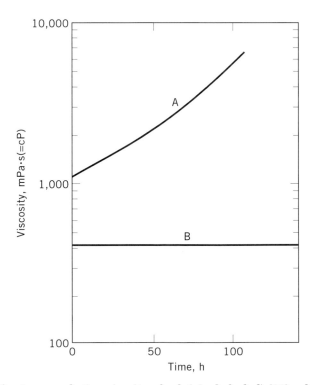

Fig. 7.   Effect of aging on solution vicosity of poly(vinyl alcohol) (60), where A represents DP = 1500, 98–99 mol % hydrolyzed; and B, DP = 1500, 87–89 mol % hydrolyzed.

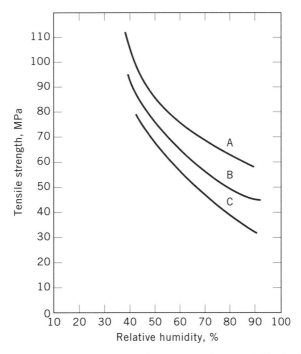

Fig. 8. Tensile strength as a function of relative humidity for fully hydrolyzed poly(vinyl alcohol) films, where A represents DP = 2400; B, DP = 1700; and C, DP = 500 (72). To convert MPa to psi, multiply by 145.

riorates above 60% rh (Fig. 9). No additives or chemical modifiers are known that can effectively reduce moisture sensitivity. The gas-barrier performance is affected by the degree of hydrolysis and rapidly diminishes as the hydrolysis is decreased below 98%.

Surface Tension.   The surface tension of aqueous solutions of PVA varies with concentration (Fig. 10), temperature, degree of hydrolysis, and acetate distribution on the PVA backbone. Random distribution of acetyl groups in the polymer results in solutions having higher surface tension compared to those of polymers in which blocks of acetyl groups are present (74–77). Surface tension decreases slightly as the molecular weight is reduced (Fig. 11).

Intrinsic Viscosity and Molecular Weights.   The relationship between the intrinsic viscosity and molecular weight changes with degree of hydrolysis of the polymer (Table 3). Methods for determining molecular weight and molecular weight distribution using gpc and low angle laser light scattering detection have been developed and utilized (79,80). These methods show a polydispersity of 2.0 or less, which indicates the presence of few long-chain branches. No long-chain branches are found in poly(vinyl alcohol) samples that have a degree of hydrolysis higher than 80 mol %. The number of short-chain branches that can be found is identical to the mol % of short-chain branches generated during the formation of the poly(vinyl acetate), in the range of 0.12–0.17 mol % (81).

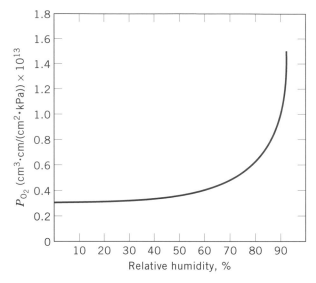

Fig. 9. Oxygen permeability of poly(vinyl alcohol) as a function of humidity, where DP = 1750, 99.9 mol % hydrolyzed (72). See BARRIER POLYMERS for a discussion of permeability units.

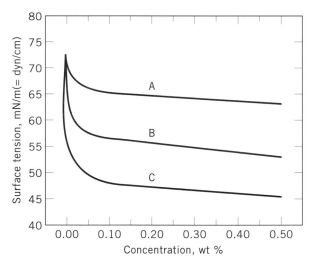

Fig. 10. Surface tension of aqueous poly(vinyl alcohol) solutions at 20°C and DP of 1700, where A represents 98–99 mol % hydrolyzed; B, 87–89 mol % hydrolyzed; and C, 78–81 mol % hydrolyzed.

Chemical Properties

Poly(vinyl alcohol) participates in chemical reactions in a manner similar to other secondary polyhydric alcohols (82–84). Of greatest commercial importance are reactions with aldehydes to form acetals, such as poly(vinyl butyral) and poly(vinyl formal).

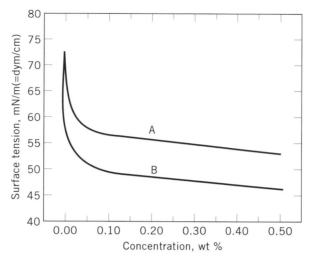

Fig. 11. Surface tension dependence on molecular weight at 20°C and hydrolysis of 87–89 mol %, where A represents DP = 1700 and B, DP = 550.

Table 3. Intrinsic Viscosity–Viscosity Average Molecular Weight Relationship[a] as a Function of Hydrolysis[b]

Hydrolysis, %	$[\eta]^a$
86.8	$8.0 \times 10^{-4} M_v^{0.58}$
93.5	$7.4 \times 10^{-4} M_v^{0.60}$
96.4	$6.9 \times 10^{-4} M_v^{0.61}$
100.0	$5.95 \times 10^{-4} M_v^{0.63}$

[a]Mark-Houwink relationship, $[\eta] = KM^a$.
[b]Ref. 78.

Esterification. *Inorganic Esters.* Boric acid and borax form cyclic esters with poly(vinyl alcohol) (85–100). The reaction is markedly sensitive to pH, boric acid concentration, and the cation-to-boron ratio. An insoluble gel is formed at pH above 4.5–5.0:

Similar complexes are formed between poly(vinyl alcohol) and titanium lactate (96,101), titanyl sulfate (102), or vanadyl compounds (103).

Poly(vinyl nitrate) has been prepared and studied for use in explosives and rocket fuel (104,105). Poly(vinyl alcohol) and sulfur trioxide react to produce poly(vinyl sulfate) (106–111). Poly(vinyl alkane sulfonate)s have been prepared from poly(vinyl alcohol) and alkanesulfonyl chlorides (112–114). In the presence of urea, poly(vinyl alcohol) and phosphorus pentoxide (115) or phosphoric acid (116,117) yield poly(vinyl phosphate)s.

Organic Esters.   An unlimited number of organic esters can be prepared by reactions of poly(vinyl alcohol) employing standard synthesis (82,84). Chloroformate esters react with poly(vinyl alcohol) to yield poly(vinyl carbonates) (118).

$$-CH_2-CH- \ + \ ClCOOR \ \xrightarrow{base} \ -CH_2-CH- \ + \ HCl$$
$$\qquad\qquad\overset{|}{OH} \qquad\qquad\qquad\qquad\quad \overset{|}{\underset{\overset{\|}{O}}{OCOR}}$$

Poly(acrylic acid), poly(methacrylic acid), and maleic anhydride containing polymers reacts with poly(vinyl alcohol) to form insoluble gels (119–122), which are useful as absorbents for water, blood, urine, etc.

Urea and poly(vinyl alcohol) form a polymeric carbamate ester (123–126):

$$-CH_2-CH- \ + \ H_2N-C-NH_2 \ \longrightarrow \ -CH_2-CH- \ + \ NH_3$$

Reaction between poly(vinyl alcohol) and isocyanates yields substituted carbamate esters (127–134):

$$-CH_2-CH- \ + \ OCNR \ \longrightarrow \ -CH_2-CH$$

Etherification.   Ethers of poly(vinyl alcohol) are easily formed. Insoluble internal ethers are formed by the elimination of water, a reaction catalyzed by mineral acids and alkali.

Ethylene oxide reacts with poly(vinyl alcohol) under normal ethoxylation conditions (135–142). The resulting products have properties that make them useful as cold-water-soluble films.

Cationic poly(vinyl alcohol) has been prepared by the reaction of *N*-(3-chloro-2-hydroxypropyl)-*N*,*N*,*N*-trimethylammonium chloride, PVA, and sodium hydroxide (143). Reactions between alkylidene epoxide and PVA in particulate, free-flowing form in an alkaline environment have been reported (144).

Poly(vinyl alcohol) undergoes Michaels addition with compounds containing activated double bonds, including acrylonitrile (145–150), acrylamide (151–153), *N*-methylolacrylamide (154–156), methyl vinyl ketone (157,158), acrolein (157), and sodium 2-acrylamido-2-methylpropanesulfonate (159). The reactions have been carried out under conditions spanning from homogeneous reactions in solvent to heterogeneous reactions occurring in the swollen powder or fiber.

Poly(vinyl alcohol) also reacts with monochloroacetates to yield glycolic acid ethers (160):

$$
\begin{array}{c}
-CH_2-CH- \\
| \\
OH
\end{array}
+ ClCH_2COOR \longrightarrow
\begin{array}{c}
-CH_2-CH- \\
| \\
OCH_2COOR
\end{array}
+ HCl
$$

Acetalization. Poly(vinyl alcohol) and aldehydes form compounds of industrial importance.

Intramolecular acetalization

$$
\begin{array}{c}
-CH_2-CH-CH_2-CH_2- \\
| \qquad\qquad | \\
OH \qquad\quad OH
\end{array}
+ RCHO \longrightarrow
\begin{array}{c}
\sim\!CH_2-CH \quad CH-CH_2\!\sim \\
\diagdown O \quad\; O\diagup \\
CH \\
| \\
R
\end{array}
$$

Intermolecular acetalization

$$
\begin{array}{c}
-CH_2-CH-CH_2-CH_2- \\
| \qquad\qquad\quad | \\
OH \qquad\qquad OH
\end{array}
+ RCHO \longrightarrow
\begin{array}{c}
-CH_2-CH-CH_2-CH- \\
| \qquad\qquad\quad | \\
OH \qquad\qquad O \\
\qquad\qquad\qquad | \\
\qquad\qquad R-CH \\
\qquad\qquad\qquad | \\
OH \qquad\qquad O \\
| \qquad\qquad\quad | \\
-CH_2-CH-CH_2-CH-
\end{array}
$$

Poly(vinyl butyral), prepared by reacting poly(vinyl alcohol) with *n*-butyraldehyde, finds wide application as the interlayer in safety glass and as an adhesive for hydrophilic surfaces (161). Another example is the reaction of poly(vinyl alcohol) with formaldehyde to form poly(vinyl formal), used in the production of synthetic fibers and sponges (162).

Poly(vinyl alcohol) is readily cross-linked with low molecular weight dialdehydes such as glutaraldehyde or glyoxal (163). Alkanol sulfonic acid and poly(vinyl alcohol) yield a sulfonic acid-modified product (164).

Other Reactions. Poly(vinyl alcohol) forms complexes with copper in neutral or slightly basic solutions (165). Sodium hydroxide or potassium hydroxide forms an intermolecular complex with PVA (166,167), causing gelation of the aqueous solution.

Certain organic compounds form reversible gels with poly(vinyl alcohol). Congo red, for example, yields a red gel that melts sharply at about 40°C. Other organic compounds that form temperature-reversible complexes with PVA include azo dyes, resorcinol, catechol, and gallic acid (168–170).

Fully hydrolyzed poly(vinyl alcohol) and iodine form a complex that exhibits a characteristic blue color similar to that formed by iodine and starch (171–173). The color of the complex can be enhanced by the addition of boric acid to the solution consisting of iodine and potassium iodide. This affords a good calorimetric method for the determination of poly(vinyl alcohol). Color intensity of the complex is effected by molecular weight, degree of hydrolysis, extent of branching, stereoregularity, 1,2-glycol content, and reagent composition. The higher the 1,2-glycol content and the lower the molecular weight, the weaker the color (174–176). Partially hydrolyzed PVA and iodine form a red complex, which is sensitive to the sequence distribution of the remaining acetate groups. The color intensity increases with increasing blockiness (177–184).

Cross-Linking.   Poly(vinyl alcohol) can be readily cross-linked using a multifunctional compound that reacts with hydroxyl groups. These types of reactions are of significant industrial importance as they provide ways to obtain improved water resistance of the poly(vinyl alcohol) or to increase the viscosity rapidly. The most commonly used cross-linking agents include glyoxal, glutaraldehyde, urea–formaldehyde, melamine–formaldehyde, trimethylolmelamine sodium borate or boric acid, and isocyanates. Most of the reactions are either acid- or base-catalyzed.

Strongly chelating metal salts of copper and nickel, eg, cupric ammonium complexes, chromium complexes, and organic titanates and dichromates, can be effective insolubilizers for PVA. Heat treatment during drying of the PVA film or coating is generally sufficient to accomplish the cross-linking reaction. The dichromate reaction is catalyzed by ultraviolet light.

Poly(vinyl alcohol), even when insolubilized by cross-linking, swells in water and loses strength on extended exposure. Complete water insensitivity cannot be achieved.

Thermal Decomposition.   The thermal decomposition of poly(vinyl alcohol) in the absence of oxygen occurs in two stages. The first stage begins at about 200°C and is mainly dehydration, accompanied by the formation of volatile products (185–191). The residue is predominantly macromolecules of polyene structure (185,192,193). Further heating to 400–500°C yields carbon and hydrocarbons (185,191). The available data on the nature of the decomposition products show some disagreement (185–190). Differences in manufacturing conditions of the PVA can significantly impact the thermal stability of the PVA, as both molecular structure and the presence of catalyst residues are known sources for decreased thermal stability. The most common decomposition products of vinyl alcohol-vinyl acetate copolymers are shown in Table 4. The formation rate of volatile vapors from PVA pyrolyzed at 185–350°C can be described using a first-order rate equation (192).

The thermal degradation of poly(vinyl alcohol) in the presence of oxygen can be described by the same decomposition scheme as in the absence of oxygen with one modification. Oxidation of the unsaturated polymeric residue from the dehydration introduces ketone groups in the polymer chain. These groups promote the dehydration of the neighboring vinyl units, producing conjugated unsaturated ketone structures (194). Cross-linking under these conditions has also been observed (195). The first-stage degradation is similar to that obtained

Table 4. Thermal Decomposition Products of Vinyl Alcohol–Vinyl Acetate Copolymers[a], %

Product	Hydrolysis, %						
	0	33	50	68	75	84	98
water	14.10	16	20.30	43.57	60.24	59.98	73.88
methanol	1.05	2.0	2.20	7.53	3.80	1.32	0.56
acetone	4.20	2.1	8.90	4.35	2.62	2.62	0.85
ethanol	7.00	6	2.30	1.13	0.43	0.61	1.25
1-propanol	1.10	1.40	1.50	1.0	0.48	1.39	traces
crotonaldehyde		3.05	2.30	traces	traces	0.67	traces
acetic acid	74.00	79.00	59.03	35.13	26.30	26.68	6.98
unidentified compounds		1.98	3.01	1.6	1.83	traces	8.03
benzene	5	traces					

[a]Ref. 188.

during vacuum pyrolysis (193). Above 250°C in the presence of oxygen, induced decomposition with self-ignition may take place (196).

Biodegradation. Poly(vinyl alcohol) is one of the few truly biodegradable synthetic polymers; the degradation products are water and carbon dioxide. At least 55 species or varieties of microorganisms have been shown to degrade or participate in the degradation of PVA, including (197): for bacterium, *Acinetobacter* (two species), *Agrobacterium radiobacter*, *Escherichia coli*, *Enterobacterium* (one species), *Proteus mirabilis*, *Pseudomonas* (19 species), *Bacilius licheniforms*, *Bacterium cadaveris*, *Brevibacterium* (two species), *Aerobacter* (three species), *Alcaligenes viacolatis*, *Alkalegenes* (one species), *Arthrobacter oxydan*, *Coryneform* (one species), *Flavobacterium* (one species), *Fusarium lini*, *Microccus glutamicus*, *Neisseria* (one species), *Sarcina* (two species), and *Xanthomonas* (one species); and for yeast and mold, *Aspergillus niger*, *Endomyces fibuliger*, *Saccharomyces* (three species), *Nadsonia fulvescens*, *Pichia polymorphia*, *Rhodotorula glutinis*, *Lipomyces lipoferus*, *Trichosporon cutaneum*, and *Zygosaccharomyces major*.

Most of these microbes occur in soil, compost, or activated sludge. Poly(vinyl alcohol)-degrading organisms consist not only of 20 different genera of bacteria, but molds, yeast, and fungi as well. The microorganisms that degrade PVA exist in most environments, including activated sludge, facultative ponds, anaerobic digesters, septic systems, compost, aquatic systems, soil, and landfills. The time period for PVA to degrade is dependent on the physical properties of the polymer, the form in which the product exists, and the environment in which it is degraded. PVA is rapidly degraded by activated sludge, especially if the sludge has already been adapted to the PVA molecule (197–201). Many of the microorganisms that degrade PVA can be isolated from soil (202–206).

The mechanism of poly(vinyl alcohol) degradation consists of a random oxidation of a hydroxyl group to a ketone through the influence of a secondary alcohol oxidase (SAO). The random oxidation is continued until a β-diketone is formed. This group is cleaved by an extracellular hydrolase, leading to a reduction in molecular weight and the formation of a carboxylic end group as well as a methyl ketone end group. Continued degradation eventually leads to the formation of acetic acid, which in turn is converted into carbon dioxide and water. The degradation is normally accomplished by means of symbiotic organisms. Symbiotic bacterial paris known to degrade poly(vinyl alcohol) include, for Type I, *Pseudomonas putida* VM 15A, *P. vesicularas* Va., *P. porolyticus* PH, and *P. alkaligenes*; and for Type II, *P. Sp.* VM15C, *P. vesicularis* XL, *P. vesicularis* XL, and *P. vesicularis* PD. More recent studies have shown that isotactic blocks in PVA molecule are more readily biodegraded than those having an atactic structure (207,208). The biological oxygen demand (BOD) for degradation of a 0.1% PVA solution is approximately 5 ppm (203).

Manufacture

Poly(vinyl alcohol) can be derived from the hydrolysis of a variety of poly(vinyl esters), such as poly(vinyl acetate), poly(vinyl formate), and poly(vinyl benzoate), and of poly(vinyl ethers). However, all commercially produced poly(vinyl alcohol) is manufactured by the hydrolysis of poly(vinyl acetate). The manufacturing process can be viewed as one segment that deals with the polymerization of vinyl acetate and another that handles the hydrolysis of poly(vinyl acetate) to poly(vinyl alcohol).

Poly(vinyl acetate) Manufacturing. *Chemistry.* Vinyl acetate is polymerized commercially using free-radical polymerization in either methanol or, in some circumstances, ethanol. Suitable thermal initiators include organic peroxides such as butyl peroxypivalate, di(2-ethylhexyl) peroxydicarbonate, butyl peroxyneodecanoate, benzoyl peroxide, and lauroyl peroxide, and diazo compounds such as 2,2'-azobisisobutyronitrile (205–215). The temperatures of commercial interest range from 55 to 85°C (209,210).

Propagation of the polymerization occurs nearly exclusively by head-to-tail reactions, with only a small fraction of head-to-head reactions. The relative ratio of these two reactions is only a function of temperature and has been found to be independent of molecular weight, polymerization solvent, and method of polymerization. The head-to-head addition yields a 1,2-glycol structure in the resulting poly(vinyl alcohol), which in turn influences the degree of crystallinity, strength, solubility, and thermal stability.

Termination reaction with vinyl acetate is nearly exclusively by disproportionation (216), although there are reports that recombination increases in importance at lower temperatures (217). Termination by disproportionation produces a double bond at the end of the one molecule and a saturated bond at the other. The presence of the unsaturated bond results in an aldehyde end group once the associated acetate group is hydrolyzed (218–220). The resulting aldehyde group decreases the thermal stability of the poly(vinyl alcohol), because the activation energy for splitting off water is reduced on account of the increased ability for formation of conjugated double bonds. The higher color level, observed in lower molecular weight commercial poly(vinyl alcohol) grades, is a result of this reaction.

Chain transfer to monomer is important during vinyl acetate polymerization (217,221–224). The reaction is an abstraction of the hydrogen on the acetyl group of the vinyl acetate (225–227). The growing radical is transferred to the vinyl acetate monomer, which then reinitiates the polymerization. The reaction results in an unsaturated end group, which is still polymerizable. The inclusion of this terminal end group in a growing chain leads to a trifunctional branch point with the incorporation of the entire polymer molecule as a long-chain branch (221,222,224–230).

Chain transfer to polymer is considerable in vinyl acetate polymerization. The intermolecular transfer of a growing radical to a polymer molecule results in a branch point. Several studies have determined the fraction of branch points generated through the acetate group compared to those generated directly on the main chain (229,231–234). The results reported in the literature have been highly variable, ranging from over 0.95 to less than 0.5 at 60°C. The main distinction between branches generated through the acetoxy group and those connected directly to the main chain is that the former is cleavable upon hydrolysis; the latter is not.

It has been shown that intramolecular chain transfer to polymer occurs during the polymerization of vinyl acetate, leading to short-chain branching (81,235–238). The number of short-chain branches has been estimated by nmr to be in the range of 0.12–1.7 mol % (81). The number of short-chain branches increases significantly at low monomer concentration.

Chain transfer to low molecular weight alcohols occurs readily. However, the resulting end groups are not known to affect the physical or chemical properties of the resulting poly(vinyl alcohol). Chain transfer to aldehydes results in the incorporation of a ketone end group. Acetaldehyde is a result of the transesterification between vinyl acetate and the added alcohol solvent. A small amount of an acid (2–50 ppm) may be added to the vinyl acetate in order to limit the transesterification (210,211,239–245). Suitable acids include phosphorous acid, oxalic acid, citric acid, and tartaric acid. The presence of the ketone end group in the resulting poly(vinyl alcohol) facilitates the formation of conjugated double bonds and leads to color.

Commercial Polymerization Processes. *Continuous Polymerization.* A typical continuous flow diagram for the vinyl acetate polymerization is shown in Figure 12. The vinyl acetate is fed to the first reactor vessel, in which the mixture is purged with an inert gas such as nitrogen. Alternatively, the feed may be purged before being introduced to the reactor (209). A methanol solution containing the free-radical initiator is combined with the above stream and passed directly and continuously into the first reactor from which a stream of the polymerization mixture is continuously withdrawn and passed to subsequent reactors. More initiator can be added to these reactors to further increase the conversion.

The polymerization temperatures can range from 45 to 130°C; the preferred range is 55–85°C, which results in operating pressures of 0.1–0.5 MPa (1–5 atm). The heat of reaction is removed by condensing monomer and solvent vapors. Cooling coils are not effective because of the low heat transfer coefficient of the highly viscous material. Viscosity during polymerization can range from 10 to 500 poise. Molecular weight is controlled by the residence time in the reactors, monomer feed rate, solvent concentration, initiator concentration, and the polymerization temperature.

A stripping column is used to remove unpolymerized vinyl acetate from the poly(vinyl acetate) solution leaving the last reactor (Fig. 13). Methanol vapors

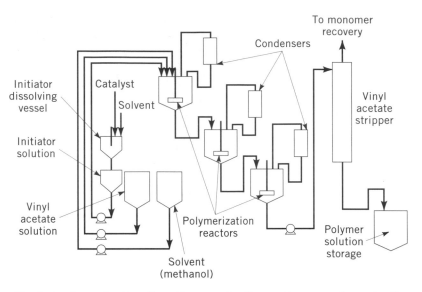

Fig. 12. Typical polymerization flow diagram for the continuous polymerization of vinyl acetate (246).

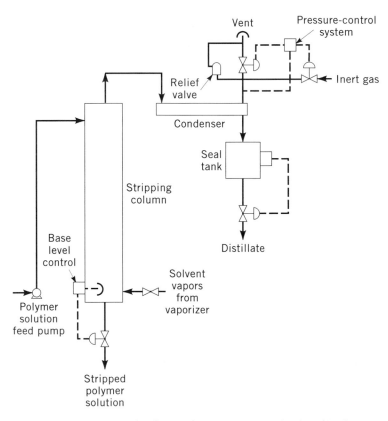

Fig. 13. Typical scheme for monomer stripping (247).

are used to strip vinyl acetate from the solution (247,248). An inhibitor such as hydrazine, hydroquinone, sulfur, or quinone can be added to the solution leaving the last reactor in order to prevent the polymerization of unreacted vinyl acetate in the stripping column (209). The overhead fraction from the stripping column consisting of solvent and vinyl acetate may be passed to a recovery system or directly recycled to the reactor. Vinyl acetate and methanol not recycled to the reactor must be separated and purified. Extractive distillation is normally used because methanol and vinyl acetate form an azeotrope. Suitable solvents include water and ethylene glycol. The vinyl acetate content in the bottom effluent from the stripping column is reduced to less than 0.07 wt %. Higher concentrations of vinyl acetate can lead to yellowing on account of acetaldehyde formation and subsequent aldol condensation during the hydrolysis reaction.

Batch and Semibatch Polymerization. The reactor is normally operated in a semicontinuous mode by delaying vinyl acetate, solvent, and initiator. The same reactor can be used for stripping the poly(vinyl acetate) solution, provided that careful addition of methanol is used in order to prevent the viscosity in the reactor from becoming excessive (249). The disadvantages of batch polymerization are lack of product consistency and unsatisfactory economics in large scale production (250,251). The true batch reaction, where all the reactants are added to the reactor at time zero, yields a product having a very broad molecular weight distribution of limited commercial value.

Poly(vinyl alcohol) Manufacturing. *Chemistry.* Poly(vinyl acetate) can be converted to poly(vinyl alcohol) by transesterification, hydrolysis, or aminolysis. Industrially, the most important reaction is that of transesterification, where a small amount of acid or base is added in catalytic amounts to promote the ester exchange.

The catalysts most often described in the literature (209–211,252) are sodium or potassium hydroxide, methoxide, or ethoxide. The reported ratio of alkali metal hydroxides or metal alcoholates to that of poly(vinyl acetate) needed for conversion ranges from 0.2 to 4.0 wt % (211). Acid catalysts are normally strong mineral acids such as sulfuric or hydrochloric acid (252–254). Acid-catalyzed hydrolysis is much slower than that of the alkaline-catalyzed hydrolysis, a fact that has limited the commercial use of these catalysts.

The solution of poly(vinyl acetate) generated during the stripping operation is normally passed directly through to the alcoholysis system. This has limited the available solvents to methanol and ethanol (209,210). Substituting the alcohol used as solvent with the generated ester limits the hydrolysis and greatly affects how the remaining acetyl groups are distributed on the poly(vinyl alcohol) chain (255–257). This distribution is often referred to as the blockiness of the poly(vinyl alcohol). The higher the methyl acetate concentration, the higher the degree of blockiness.

The presence of catalyst residues, such as alkali hydroxide or alkali acetate, a by-product of the hydrolysis reaction, is known to decrease the thermal stability of poly(vinyl alcohol). Transforming these compounds into more inert compounds and removal through washing are both methods that have been pursued. The use of mineral acids such as sulfuric acid (258), phosphoric acid (259), and *ortho*-phosphoric acid (260) has been reported as means for achieving increased thermal stability of the resulting poly(vinyl alcohol).

Commercial Hydrolysis Process. The process of converting poly(vinyl acetate) to poly(vinyl alcohol) on a commercial scale is complicated on account of the significant physical changes that accompany the conversion. The viscosity of the poly(vinyl acetate) solution increases rapidly as the conversion proceeds, because the resulting poly(vinyl alcohol) is insoluble in the most common solvents used for the polymerization of vinyl acetate. The outcome is the formation of a gel swollen with the resulting acetic acid ester and the alcohol used to effect the transesterification.

Continuous Saponification. There are several types of continuous processes, each with benefits and drawbacks. The basic premise is that continuous mixing is not required after the poly(vinyl acetate) and the caustic is mixed (261). Several designs for the high intensity mixing unit have been suggested in order to obtain efficient mixing and minimize fouling (262,263). Further handling of the reaction mixture on an industrial scale employs such techniques as the belt process, the slurry process, and the screw conveyor process.

In the belt process, the mixture is cast onto the belt or conveyor (high temperature polyethylene (264)) where gelling occurs (261,265). The gel is removed from the surface before syneresis, cut into smaller particles, and passed to a holding or washing tank (Fig. 14). The liquid is removed from the precipitate using common chemical deliquefying methods, and the resulting particles are dried in a continuous dryer.

In the slurry process, the hydrolysis is accomplished using two stirred-tank reactors in series (266). Solutions of poly(vinyl acetate) and catalyst are continuously added to the first reactor, where 90% of the conversion occur, and then transferred to the second reactor to reach full conversion. Alkyl acetate and alcohols are continuously distilled off in order to drive the equilibrium of the reaction. The resulting poly(vinyl alcohol) particles tend to be very fine, resulting in a dusty product. The process has been modified to yield a less dusty product through process changes (267,268) and the use of additives (269). Partially hydrolyzed products having a narrow hydrolysis distribution cannot be prepared by this method.

In the screw conveyor process, solutions of poly(vinyl acetate) and catalyst are mixed in a high intensity mixer and continuously introduced to a screw-type saponification and conveyor system (270). Downstream details are similar to those found in the belt process.

Other methods also exist for handling the reaction mixture. For instance, self-wiping, twin-screw extruder using an intermeshing has been proposed (271). The advantages claimed include higher saponification temperature (less catalyst), increased water solubility (271), and fewer impurities (272).

Batch Saponification. Batch saponification, the oldest PVA manufacturing method (252), is mainly used in the 1990s for the production of specialty products. The process uses a kneader in which the hydrolysis, washing, and drying operations are performed. This is the simplest method of saponification, but the production rates are low, and producing the product quality needed by many end uses is difficult.

Drying and Solids Separation. Separation of the polymer gel from the methanol/methyl acetate liquid is an important step, accomplished by using standard pieces of equipment such as filters, screw presses, or centrifuges.

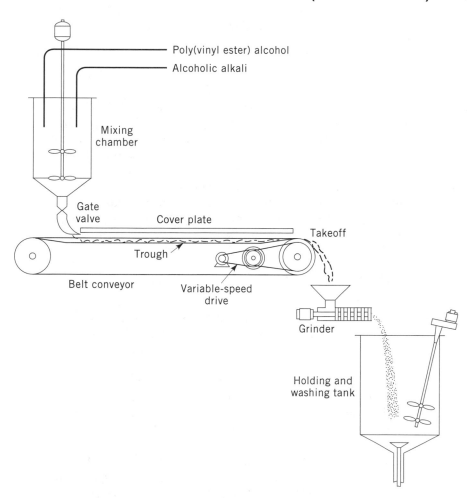

Fig. 14.   The belt saponification process as used in the production of poly(vinyl alcohol) (265).

Drying of the poly(vinyl alcohol) is critical to both the color and solubility of the final product. Excessive drying temperatures result in high product color and an increase in the crystallinity, which in turn reduces the solubility of the product. Drying is initially subjected to a flash regime, where the solvent not contained within the particles is flashed off. This first phase is followed by a period where the rate is controlled by the diffusion rate of solvent from the poly(vinyl alcohol) particles. Because the diffusion rate falls as the material dries, complete drying is not practical. The polymer is therefore generally sold at a specification of 95% solids.

Solvent Recovery.   A mixture of methanol and methyl acetate is obtained after saponification. The methyl acetate can be sold as a solvent or converted back into acetic acid and methanol using a cationic-exchange resin such as a cross-linked styrene–sulfonic acid gel (273–276). The methyl acetate and methanol mixture is separated by extractive distillation using water or ethylene glycol (277–281). Water is preferred if the methyl acetate is to be hydrolyzed to acetic acid. The resulting acetic acid solution is concentrated by extraction or azeotropic distillation.

Copolymers

Numerous vinyl alcohol copolymers have been prepared (282). Copolymers with ethylene and methacrylate are the only copolymers that have found sizable commercial utility. Ethylene–vinyl alcohol (EVOH) copolymers containing 20–30 mol % ethylene are used as an oxygen barrier in food packaging. However, a five-layer coextruded structure with polyolefins is needed in order to protect EVOH from moisture. Vinyl alcohol–methyl methacrylate copolymers are used as sizing agents in the textile industry. The presence of the methacrylate unit disrupts the crystallinity, making the product easier to remove during the desizing operation. The product is especially useful as an alkaline-resistant textile size.

Economic Aspects

The capacity of PVA in 1995 was greater than 550,000 t. Several manufacturers have added capacity since 1987. Air Products & Chemicals added 40,000 t of new capacity; Chang-Chun, Nippon Gohsei, and Hoechst all debottlenecked their plants; and a new small plant was built in Korea by Oriental Chemical Industries (10,000 t).

Approximately two-thirds of capacity are located in Japan, China, and Taiwan, with the remainder in the United States and Europe. Approximately 50,000 t of the Japanese and a large portion of the Chinese production is captively consumed for fiber production. The principal PVA producers in the world (capacity >30,000 t/yr) are shown in Table 5.

The PVA process is highly capital-intensive, as separate facilities are required for the production of poly(vinyl acetate), its saponification to PVA, the recovery of unreacted monomer, and the production of acetic acid from the ester formed during alcoholysis. Capital costs are far in excess of those associated with the traditional production of other vinyl resins.

The PVA price has historically reflected the cost of ethylene, acetic acid, and energy. The price history for a medium molecular weight, fully hydrolyzed grade is $0.77/kg in 1970, $2.20/kg in 1980, $2.75/kg in 1988, and $2.65/kg in 1995.

Table 5. Principal Poly(Vinyl Alcohol) Producers

Producer	Country	Trade name	Capacity, 10^3 t/yr
Air Products and Chemicals, Inc.	United States	Airvol	90
Chang Chun	Taiwan	CCP	55
Chin-Shan Petrochemical	China		33
du Pont	United States	Elvanol	68
Hoechst	Germany	Mowiol	50
Kuraray	Japan	Poval	124
Nippon Goshei	Japan	Gohsenol	65
Shi-Shan Vinylon	China		45

Specifications and Standards

The important commercial grades of PVA are distinguished by the degree of hydrolysis and molecular weight. The resins are most often categorized by degree of hydrolysis, ie, mole percent of alcohol groups in the resin (Table 6). Poly(vinyl alcohol)s having other degrees of hydrolysis are also produced, but maintain a much smaller market share than those shown in Table 6.

Poly(vinyl alcohol) is produced mainly in five molecular weight ranges (Table 7). Several other molecular weight resins are available, but their market shares are relatively low. Industry practice expresses the molecular weight of a particular grade in terms of the viscosity of a 4% aqueous solution. An unlimited number of viscosities can be generated by blending the available molecular weights. Products having different degree of hydrolysis can also be blended to obtain a particular performance characteristic. However, blended products have a broad distribution with respect to molecular weight and, in some cases, hydrolysis, which are undesirable in many applications.

Poly(vinyl alcohol) is an innocuous material having unlimited storage stability. It is most commonly supplied in 20-kg, 22.7-kg (50-lb), and 25-kg bags equipped with a moisture barrier to prevent caking. PVA is also available in bulk or in super sacks. The FDA regulations governing the use of PVA are shown in Table 8. Poly(vinyl alcohol) maintains an exemption for tolerance from the EPA.

Table 6. Hydrolysis of Principal Commercial Poly(Vinyl Alcohol) Grades

Grade	Hydrolysis, mol %
super	99.3+
fully	98.0–98.8
intermediate	95.9–97.0
partially	87.0–89.0
low	79.0–81.0

Table 7. Viscosity and DP of Principal Commercial Poly(Vinyl Alcohol) Grades

Grade	Nominal DP	Viscosity of 4% solution, mPa·s(=cP)
low–low	220	3–4
low	550	5–7
intermediate	900	13–16
medium	1500	28–32
high	2200	55–65

Analytical and Test Methods

The important analytical test methods are those related to the determination of degree of hydrolysis, pH, viscosity, ash, and volatiles. Percent hydrolysis of the PVA is measured by placing the material in a mixture of water and methanol, adding a predetermined quantity of sodium hydroxide, and boiling under reflux

Table 8. FDA Regulations for Poly(Vinyl Alcohol) in Food Applications

Regulation	Description
181.30	manufacture of paper and paperboard products used in packaging of fatty foods
175.105	adhesives; no limitations
176.170	components of paper and paperboard in contact with aqueous and fatty food; extractive limitations
176.180	components of paper and paperboard in contact with dry food; no limitations
177.1200	cellophane coating; no limitations
177.1670	poly(vinyl alcohol) film
177.2260	filters, resin-bonded where fiber is cellulose
177.2600	filters, extractables must be less than 0.08 mg/cm^2
175.200	resinous and polymeric coating
175.320	resinous and polymeric coatings for polyolefin film; net extractable less than 0.08 mg/cm^2
177.2800	textiles and textile fibers; for dry foods only
178.3910	surface lubricants in the manufacture of metallic articles

to hydrolyze residual acetate groups. The moles of sodium hydroxide consumed are equivalent to the number of hydrolyzable acetate groups and determined by back-titration with strong acid. The degree of hydrolysis is calculated by means of the moles of hydroxide consumed (MHC) and the initial sample weight as follows:

$$\text{degree of hydrolysis} = (\text{grams PVA} - 86 \times \text{MHC})/(42 \times \text{MHC} + \text{grams PVA})$$

The pH is measured using a 4% aqueous solution. Viscosity is normally measured using Brookfield viscometer. Alternatively, a capillary-type viscometer or falling ball such as Höppler may be employed. The type of viscometer used must always be noted.

Ash is a measure of residual sodium acetate. A simple method consists of dissolving the PVA in water, diluting to a known concentration of about 0.5 wt %, and measuring the electrical conductivity of the solution at 30°C. The amount of sodium acetate is established by comparing the result to a calibration curve. A more lengthy method involves the extraction of the PVA with methanol using a Soxhlet extractor. The methanol is evaporated and water is added. The solution is titrated using hydrochloric acid in order to determine the amount of sodium acetate.

Volatiles such as residual methanol, methyl acetate, and water are determined as the loss in mass when the polymer is dried at 105 ± 2°C until constant mass is attained. Higher drying temperatures may cause decomposition and related additional weight loss.

Health and Safety Factors

Poly(vinyl alcohol) is a nonhazardous material according to the American Standard for Precautionary Labeling of Hazardous Industrial Chemicals (ANSI

2129.1-1976). Extensive tests indicate a very low order of toxicity when it is administered orally to laboratory animals (203). No toxicity was detected by oral administration of the maximum amount of 1500 mg/kg of two types of PVA (DP 1400, 99.5% hydrolyzed; and DP 1700, 86.8% hydrolyzed) or by sub-cutaneous injection of 3000 mg/kg to mice. PVA injected under the skin or into the lungs is not broken down and remains as a foreign body. The increase in growth and weight of mice given 1000 mg/kg every day during a three-month chronic toxicity test was about the same as that of the control-group animals. No histopathologic tensions were observed.

Poly(vinyl alcohol) has a low oral toxicity rating. The oral LD_{50} is higher than 10,000 mg/kg (rats). Concentrations of up to 10,000 mg/L in water were tested for toxicity to bluegill sunfish. No mortality or response indicative of intoxication was observed (283).

Short-term inhalation of PVA dust has no known health significance, but can cause discomfort and should be avoided in accordance with industry standards for exposure to nuisance dust. The dust is mildly irritating to the eyes. There are no known dermal effects arising from short-term exposure to either solid PVA or its aqueous solutions.

During transport and handling, granular PVA may form an explosive mixture with air, which shows a low severity rating (Bureau of Mines Rating) of 0.1 on a scale in which coal dust has a rating of 1.0 (284). However, the explosive hazard depends on particle size, and extremely fine dust has a higher explosive rating of 1.0–2.0. Residual methanol and methyl acetate can accumulate in the air space of bulk storage tanks; this is especially true at elevated temperatures. Precautions should be taken to ventilate the air space in large vessels and eliminate spark-producing equipment in the area. The issue of residual organic volatiles has been addressed by a few PVA producers, who have proactively implemented manufacturing specifications aimed at obtaining less than 1 wt % organic volatiles in the final product (285).

Processing

Poly(vinyl alcohol) is not considered a thermoplastic polymer because the degradation temperature is below that of the melting point. Thus, industrial applications of poly(vinyl alcohol) are based on and limited by the use of water solutions.

Solution Preparation and Handling. Poly(vinyl alcohol) should be completely dispersed in water at room temperature or lower before heating is commenced. Good agitation is important to prevent lumping during the addition of solid PVA to water. A large diameter, low speed agitator is preferred for providing good mixing at the surface without excessive air entrainment. Agitation requirements increase with solution concentration and with decreasing hydrolysis; the latter is associated with greater solubility in cold water and a tendency to lump.

The temperature of the slurry must be increased to 79–90°C in order to solubilize the poly(vinyl alcohol) fully. Ways to increase the temperature include direct steam injection, jet cooking, or heating with jacket or coil. Aqueous solutions of PVA are stable on storage but must be protected from bacterial growth. An increase in viscosity is commonly observed when storing fully hydrolyzed poly(vinyl alcohol) solutions. Stainless steel or plastic containers are recommended for long-term storage. Many biocides are effective, including FDA-approved compounds (284).

Prolonged heating of the PVA solution has negligible effect on its properties. However, the addition of strong acid or base to solutions of partially hydrolyzed PVA can increase the degree of hydrolysis.

Extrusion. Several attempts to introduce and produce extrudable PVA resins have been made (286–300). However, these types of products have not achieved any significant commercial success. The main obstacles are thermal stability during extrusion and gels in the final product. A wide variety of high boiling water-soluble organic compounds containing hydroxyl groups have been used as plasticizers in order to lower the melt temperature and avoid decomposition. Glycerol and low molecular weight poly(ethylene glycol)s are most widely used. Water is an excellent plasticizer for PVA, although extrusion temperatures below 100°C are required in order to avoid the formation of foam at the extruder outlet.

Uses

The main applications for PVA are in textile sizing, adhesives, polymerization stabilizers, paper coating, poly(vinyl butyral), and PVA fibers. In terms of percentage, and omitting the production of PVA not isolated prior to conversion into poly(vinyl butyral), the principal applications are textile sizes, at 30%; adhesives, including use as a protective colloid, at 25%; fibers, at 15%; paper sizes, at 15%, poly(vinyl butyral), at 10%; and others, at 5%, which include water-soluble films, nonwoven fabric binders, thickeners, slow-release binders for fertilizer, photoprinting plates, sponges for cosmetic, and health care applications.

Textile and Warp Sizing. Warps are sized to obtain increased abrasion resistance and strength in order to avoid breakage when passing through the loom. Yarn breakage during the weaving process reduces weaving efficiency and detracts from the quality of the finished cloth. The sizing material must allow for both easy splitting of the warp yarns after sizing and rapid desizing once the weaving process is complete. The typical sizing process consists of a size application box, squeeze rolls to remove excess size, drying drums, yarn splitting rods, and beam winding equipment.

Poly(vinyl alcohol) is an excellent textile warp size because of superior strength, adhesion, flexibility, and film-forming properties. The adhesion of PVA to natural or synthetic fibers or fiber blends depends on the degree of hydrolysis. Adhesion to such hydrophobic fibers as polyester is enhanced as the degree of hydrolysis is reduced (Fig. 15). Fully hydrolyzed grades adhere well to cotton and other hydrophilic fibers and impart the highest tensile strength for equivalent amounts added. However, there continues to be a trend away from fully hydrolyzed PVA because of its poor adhesion to synthetic fibers and the difficulty encountered when desizing heat-set fabric (209,301,302). The best hydrolysis grade for warp sizing is specific to each manufacturing operation and is controlled by factors such as yarn quality, yarn construction, and speed and type of loom.

The highest molecular weight grades provide the best protection at equivalent concentration for spun yarns. However, low solids sizing solutions are needed

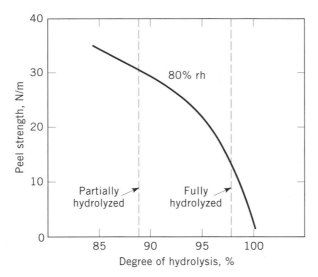

Fig. 15. Adhesion expressed as peel strength of poly(vinyl alcohol) film on polyester film (8). To convert N/m to ppi, divide by 175.

to maintain a manageable viscosity; this increases drying cost and limits production. Desizing of the fabric is difficult because of the lower solubility rate of these grades. Thus, the commonly used warp size grades for spun yarns are the medium to lower molecular weight resins because they balance strength, solution viscosity, and ease of desizing to yield the best possible economics. The best size for filament yarns are those PVA grades that have a low degree of polymerization.

Typical size formulations contain a lubricating wax, starch, and other processing aids. The role of these materials is to provide easy splitting of the yarns after drying, to decrease the sticking tendency of PVA to the drying drums, and to provide lubricity during the weaving process. The additives must to a large degree be incompatible with the dried poly(vinyl alcohol) film to provide the desired function. However, a large degree of incompatibility can result in decreased encapsulation of the yarn and increased shedding during the weaving operation. Poly(vinyl alcohol) is normally used in 3–10 wt % concentrations in the aqueous size solution. Amount of size needed depends on both fabric and loom type, but is usually in the 5–10% range.

The fabric is desized after the weaving operation and then passed through a heated water bath to remove all the size. The rate at which this operation can be accomplished depends to a great degree on solubility rate of the poly(vinyl alcohol). Difficulties encountered in completely removing the lubricating wax, usually tallow wax, has led to the development of several wax-free size compositions (303–311). The main component contained in these blends is PVA in combination with a small amount of a synthetic water-soluble lubricant.

Poly(vinyl alcohol) can be recovered from the desizing liquid by means of commercial ultrafiltration equipment. Recovery rates and effluent losses are inversely proportional to the PVA solution viscosity and independent of the degree of hydrolysis.

Adhesives.   Poly(vinyl alcohol) is used as a component in a wide variety of general-purpose adhesives to bond cellulosic materials, such as paper and paperboard, wood textiles, some metal foils, and porous ceramic surfaces, to each other. It is also an effective binder for pigments and other finely divided powders. Both fully and partially hydrolyzed grades are used. Sensitivity to water increases with decreasing degree of hydrolysis and the addition of plasticizer. Poly(vinyl alcohol) in many applications is employed as an additive to other polymer systems to improve the cohesive strength, film flexibility, moisture resistance, and other properties. It is incorporated into a wide variety of adhesives through its use as a protective colloid in emulsion polymerization.

Adhesives for paper tubes, paperboard, corrugated paperboard, and laminated fiber board are made from dispersions of clays suspended with fully hydrolyzed poly(vinyl alcohol). Addition of boric acid improves wet tack and reduces penetration into porous surfaces (312,313). The tackified grades have higher solution viscosity than unmodified PVA and must be maintained at pH 4.6–4.9 for optimum wet adhesion.

Poly(vinyl alcohol) is employed as a modifier of thermosetting resins used as adhesives in plywood and particle board manufacture (314,315). The polymer is added to urea–formaldehyde or urea–melamine–formaldehyde resins to improve initial grab, to increase viscosity, and, in general, to improve the characteristics of the board.

Poly(vinyl alcohol)s are used as components in vinyl acetate emulsions both as a protective colloid and as a means for improving adhesive properties. Post-addition of PVA to the resulting emulsions is commonly employed to modify viscosity, flow properties, and the rate of formation of the adhesive bond as well as the quality. High addition rates of partially hydrolyzed PVA can be used to prepare remoistenable adhesive formulations. All PVA grades promote the acceptance of starch and clay fillers into a formulation, which in turn prevents excessive adhesive penetration into porous surfaces. The poly(vinyl alcohol) also offers reactive sites that can be used to cross-link the adhesive and improve the water resistance. Cross-linking is accomplished by means of either *N*-methylolacrylamide, copolymerized with the vinyl acetate (316,317), or other commonly used cross-linking agents, such as glyoxal. The performance of PVA-based wood glues is to a great degree dependent on the presence of PVA to provide both the cohesive and the adhesive properties.

Emulsion Polymerization.   Poly(vinyl acetate) and poly(vinyl acetate) copolymer latexes prepared in the presence of PVA find wide applications in adhesives, paints, textile finishes, and coatings. The emulsions show excellent stability to mechanical shear as well as to the addition of electrolytes, and possess excellent machining characteristics.

Partially hydrolyzed poly(vinyl alcohol) grades are preferred because they have a hydrophobic/hydrophilic balance that make them uniquely suited for emulsion polymerization. The compatibility of the residual acetate units with the poly(vinyl acetate) latex particles partly explains the observed stabilization effect. The amount of PVA employed is normally 4–10% on the weight of vinyl acetate monomer. The viscosity of the resulting latex increases with increasing molecular weight and decreasing hydrolysis of the PVA (318).

Paper Coating. The unique binding properties of PVA has made it a primary contributor to the development of higher quality, specialized paper products. Its use spans a great variety of grades, including silicone-coated release liners, greaseproof and glassine packaging papers, food-grade boards, carbonless grades, currency and banknote grades, offset printing papers, high brightness printing and writing grades, offset masters, ink-jet and thermal printing papers, and cigarette filter tip papers.

Poly(vinyl alcohol) is utilized principally as a surface-treating agent, as in clear sizing, pigmented sizing, and pigmented coating. Exceptions include its use as a creping aid in tissue and towel manufacturing, in dye encapsulation for carbonless papers, and in some wet end addition applications. Poly(vinyl alcohol) imparts exceptional strength and outstanding resistance to oils, greases, and organic solvents, and is widely recognized as the strongest paper binder available (Table 9) (319).

The resistance of the polymer to oils and organic solvents can be directly attributed to the hydroxyl functionality and the film-forming properties of the polymer. Treated paper substrates display a significant amount of oil resistance, which make them valuable for packing papers and food-grade paperboard containers. This performance improvement is achieved despite the fact that the 1–3 wt % add-on level, typical of size press applications, is too low to provide a continuous poly(vinyl alcohol) film.

The fully and super hydrolyzed PVA grades are preferred by the paper industry for their superior strength, greater adhesion to cellulose, better water resistance, and better foaming characteristics. However, the intermediate and partially hydrolyzed grades provide better surface filming characteristics on many paper and paperboard substrates.

Poly(vinyl alcohol) is used as a carrier for fluorochemical sizing agents on oil-resistant papers and paperboard. These highly effective, but very expensive, fluoropolymers are used on bags, wrappers, cartons, and trays to prevent oil penetration and wicking. When carried with poly(vinyl alcohol), less amount of fluorochemical can be used to achieve an equivalent performance level. Poly(vinyl alcohol) is widely used on release liners for silicone topcoat holdout, a large and rapidly growing application. The wide diversity in performance requirements of the various release liners has resulted in recommendation for add-on ranging from 1.5–2.0 (320) to 10 g/m². Ultimately, the characteristics of the base paper itself, as well as the level of release required of the finished sheet, determine the best add-on level of the PVA.

Table 9. Relative Strength of Paper Binders

Binder	Parts for equal strength
poly(vinyl alcohol)	1
styrene–butadiene	2–2.5
poly(vinyl acetate)	2–3
soy protein	2.5–3
casein	2.5–3
starch	3–4

The role of poly(vinyl alcohol) in ink-jet printing papers is described in numerous patents. The market is relatively small but the growth rate appears high. Most of the technology has been developed in Japan, which by 1985 had filed 75% of the 400 issued patents (321). Requirements call for a hydrophilic, high porosity surface capable of absorbing ink-jet droplets quickly, with little spreading, wicking, or dye penetration. The coating on the base sheet primarily consists of silica powder, which uses PVA as the binder of choice (322).

Building Products.   Poly(vinyl alcohol) is widely used in connection with spray drying of emulsions, in particular ethylene–vinyl acetate copolymer emulsions. The PVA is added both as a protective colloid during the polymerization and as a redispersion aid prior to the actual spray drying for a total amount of 2–15 wt % (323–326). The poly(vinyl alcohol) contained in these products greatly enhances the adhesion to cementitious materials, improves water retention, and increases strength. The powder, when added to tile grouts, joint compounds, textured compounds, and cementitious repair mortars, improves the bond strength, abrasion resistance, and flexibility of the final construction.

Poly(vinyl alcohol) is used as an additive to dry-wall joint cements and stucco finish compounds. Rapid cold-water solubility, which can be achieved with finely ground PVA, is important in many dry mixed products. Partially hydrolyzed grades are commercially available in fine-particle size under the name S-grades. The main purpose of the poly(vinyl alcohol) is to improve adhesion and act as a water-retention aid.

Fibers.   Poly(vinyl alcohol) fibers possess excellent strength characteristics and provide a pleasant feel in fabrics. The fiber is usually spun by a wet process employing a concentrated aqueous solution of sodium sulfate as the coagulating bath. Water insolubility, even in boiling water, can be obtained by combining stretching, heat treatment, and acetalization with formaldehyde. Super hydrolyzed PVA is the preferred material for fiber production.

PVA fibers have found wide spread industrial use in cement as replacement for asbestos in cement products, reinforcement of rubber material such as conveyor belts and hydraulic rubber hoses used in cars, ropes, fishing nets, etc. Only a small amount of fibers is used in the production of textiles. Several patents (327–329) have been issued that claim processes for production of ultrahigh tensile strength PVA fibers, which have a tensile strength comparable to that of Kevlar.

Other Applications.   Poly(vinyl alcohol) film can be produced by solution casting or extrusion. Film casting is most common, as unplasticized films of all molecular weights and extents of hydrolysis can readily be produced. Water solubility of the film can be controlled by selection of the proper degree of hydrolysis. These films can be used for packaging of detergents, insecticides, and other materials that are to be dissolved in water. Hospital laundry bags produced from PVA eliminate the need for handling contaminated linen when it is placed in the washing machine.

Poly(vinyl alcohol) is useful as a temporary protective coating for metals, plastics, and ceramics. The coating reduces damage from mechanical or chemical agents during manufacturing, transport, and storage. The protective film can be removed by peeling or washing with water.

Ultraviolet cross-linking of PVA with dichromates is the basis for its use in photoengraving, screen printing, printed circuit manufacture, and color-television tube manufacture.

Partially hydrolyzed grades are used in many cosmetic applications for their emulsifying, thickening, and film-forming properties. Poly(vinyl alcohol) is also used as a viscosity builder for aqueous solutions and dispersions.

BIBLIOGRAPHY

"Polyvinyl Alcohol" in *ECT* 1st ed., Vol. 14, pp. 710–715, by T. P. G. Shaw and L. M. Germain, Shawinigan Chemicals, Ltd.; "Poly(Vinyl Alcohol)" in *ECT* 2nd ed., Vol. 21, pp. 353–368, by M. Leeds, Air Reduction Co., Ltd.; in *ECT* 3rd ed., Vol. 23, pp. 848–865, by D. L. Cincera, Air Products and Chemicals, Inc.

1. J. M. Hay and D. Lyon, *Nature*, **216**, 790 (1967).
2. B. Capon, D. S. Watson, and C. Zucco, *J. Am. Chem. Soc.* **103**, 1761 (1987).
3. B. N. Novak and A. K. Cederstav, *J. Am. Chem. Soc.* **116**, 4073 (1994).
4. Ger. Pat. 450,286 (1924), W. Haehnel and W. O. Herrmann (to Consort. f. elecktrochem, GmbH).
5. H. Staudinger, *Arbeitserinnerungen*, Dr. Alfred Hutig Verlag GmbH, Heidelberg, Germany, 1961, p. 196.
6. W. O. Hermann and W. Haehnel, *Ber. Dtsch. Chem. Ges.* **60**, 1658 (1927).
7. H. Staudinger, K. Frey, and W. Stark, *Ber. Dtsch. Chem. Ges.* **60**, 1782 (1927).
8. *Vinol Product Handbook*, Air Product and Chemicals, Inc., Allentown, Pa., 1980.
9. K. Tsuboi and T. Mochizuki, *Kobunshi Kagaku*, **23**, 636 (1966).
10. *Ibid.*, p. 640.
11. *Ibid.*, p. 645.
12. K. Tsuboi and T. Mochizuki, *Kobunshi Kagaku*, **24**, 224 (1967).
13. *Ibid.*, p. 241.
14. *Ibid.*, p. 245.
15. K. Tsuboi, K. Fujii, and T. Mochizuki, *Kobunshi Kagaku*, **24**, 361 (1967).
16. K. Tsuboi and T. Mochizuki, *Kobunshi Kagaku*, **24**, 366 (1967).
17. K. Monobe and Y. Fujiwara, *Kobunshi Kagaku*, **21**, 179 (1964).
18. K. Tsuboi and T. Mochizuki, *J. Poym. Sci. Polym. Lett. Ed.* **1**, 531 (1963).
19. E. N. Gubenkova and co-workers, *Vysokomol. Soedin.* **B9**, 550 (1967).
20. K. Tsuboi and T. Mochizuki, *Kobunshi Kagaku*, **24**, 433 (1967).
21. A. Packter and M. S. Nerurkar, *J. Polym. Sci. Polym. Lett. Ed.* **7**, 761 (1969).
22. C. W. Bunn, *Nature*, **161**, 929 (1948).
23. M. I. Bressonov and A. P. Rudakov, *Fiz. Tverd. Tela*, **6**, 1041 (1964).
24. N. A. Peppas and P. J. Hansen, *J. Appl. Polym. Sci.* **27**, 4787 (1982).
25. N. A. Peppas and E. W. Merrill, *J. Polym. Sci. Polym. Chem. Ed.* **14**, 441 (1976).
26. N. A. Peppas and E. W. Merrill, *J. Appl. Polym. Sci.* **20**, 1457 (1976).
27. N. A. Peppas, *J. Appl. Polym. Sci.* **20**, 1715 (1976).
28. N. A. Peppas, *Eur. Polym. J.* **12**, 495 (1976).
29. K. Fujii, *J. Polym. Sci.* **D1**, 431 (1972).
30. S. Matsuzawa and co-workers, *Makromol. Chem.* **180**, 2009 (1979).
31. J. F. Kenney and G. W. Willcockson, *J. Polym. Sci.* **A4**, 679 (1966).
32. Y. Sone, K. Hirabayashi, and I. Sakurada, *Kabunshi Kagaku*, **10**, 1 (1953).
33. I. Sakurada, Y. Nukushina, and Y. Sone, *Kabunshi Kagaku*, **12**, 506 (1955).
34. I. Sakurada, Y. Nukushina, and Y. Sone, *Ric. Sci.* **25A**, 715 (1955).

35. Y. Sone and I. Sakurada, *Kobunshi Kagaku*, **14**, 92 (1957).
36. Y. Ikada and co-workers, *Radiat. Phys. Chem.* **9**, 633 (1977).
37. A. Packter and M. S. Nerurkar, *Eur. Polym. J.* **4**, 685 (1968).
38. A. Packter and M. S. Nerurkar, *Coll. Polym. Sci.* **253**, 916 (1975).
39. S. Imoto, *Kogyo Kagaku Zasshi*, **64**, 1671 (1961).
40. S. P. Papkov and co-workers, *Vysokomol. Soyed.* **A8**, 1035 (1966).
41. V. N. Lebedeva, G. I. Distler, and Y. I. Kortukova, *Vysokomol. Soyed.* **A9**, 2076 (1967).
42. S. H. Hyon, H. D. Chu, and R. Kitamaru, *Bull. Inst. Chem. Res. Kyoto Univ.* **53**, 367 (1975).
43. Y. Nakanishi, *J. Polym. Sci. Chem. Ed.* **13**, 1223 (1975).
44. A. Booth and J. N. Hay, *Polymer*, **12**, 365 (1971).
45. A. Booth and J. N. Hay, *Br. Polym. J.* **4**, 18 (1972).
46. J. N. Hay, *Br. Polym. J.* **9**, 72 (1977).
47. E. Riande and J. M. Fatou, *Polymer*, **17**, 795 (1976).
48. T. Yamaguchi and M. Amagasa, *Kobunshi Kagaku*, **18**, 653 (1961).
49. N. A. Peppas, *Makromol. Chem.* **178**, 595 (1977).
50. N. A. Peppas and R. E. Benner, *J. Biomaterials*, **1**, 158 (1980).
51. T. Osugi, *Man-Made Fibers*, Vol. 3, Wiley-Interscience, New York, 1968, p. 258.
52. R. K. Tubbs and T. K. Wu, in C. A. Finch, ed., *Polyvinyl Alcohol*, John Wiley & Sons, New York, 1973, p. 169.
53. I. Sakurada, A. Nakajima, and H. Takida, *Kobunshi Kagaku*, **12**, 21 (1955).
54. J. Brandrup and E. H. Immergut, *Polymer Handbook*, 2nd ed., John Wiley & Sons, Inc., New York, 1975.
55. F. Hamada and A. Nakajima, *Kobunshi Kagaku*, **23**, 395 (1966).
56. R. K. Tubbs, *J. Polym. Sci.* **A3**, 4181 (1965).
57. R. K. Tubbs, H. K. Inskip, and P. M. Subramanian, *Properties and Applications of Polyvinyl Alcohol*, Society of Chemical Industry, London, 1968, p. 88.
58. K. Kikukawa, S. Nozakura, and S. Murahashi, *Kobunshi Kagaku*, **25**, 19 (1968).
59. Technical data, Air Products & Chemicals, Allentown, Pa.
60. K. Toyoshima, in Ref. 52, p. 40.
61. I. Sakurada, Y. Sakaguchi, and Y. Ito, *Kobunshi Kagaku*, **14**, 141 (1957).
62. S. N. Timasheff, *J. Amer. Chem. Soc.* **73**, 289 (1951).
63. K. Satake, *Kobunshi Kagaku*, **12**, 122 (1955).
64. I. Sakurada and M. Hosono, *Kobunshi Kagaku*, **2**, 151 (1945).
65. K. Amaya and R. Fujishiro, *Bull. Chem. Soc. Jpn.* **29**, 361 (1956).
66. E. Prokopova, P. Stern, and O. Quadrat, *Coll. Polym. Sci.* **263**, 899 (1985).
67. P. Stern, E. Prokopova, and O. Quadrat, *Coll. Polym. Sci.* **265**, 234 (1987).
68. E. Prokopova, P. Stern, and O Quadrat, *Coll. Polym. Sci.* **265**, 903 (1987).
69. F. F. Vercauteren and co-workers, *Eur. Polym. J.* **23**, 711 (1987).
70. S. Fujishige, *J. Coll. Interface Sci.* **13**, 193 (1958).
71. D. Eagland and co-workers, *Eur. Polym. J.* **22**, 351 (1986).
72. K. Toyoshima, in Ref. 52, p. 339.
73. A. D. Schwope and co-workers, *Am. Ind. Hyg. Assoc. J.* **53**, 352 (1992).
74. K. Noro, *Br. Polym. J.* **2**, 128 (1970).
75. S. Hayashi, C. Nakano, and T. Motoyama, *Kobunshi Kagaku*, **20**, 303 (1963).
76. S. Hayashi, C. Nakano, and T. Motoyama, *Kobunshi Kagaku*, **21**, 300 (1964).
77. M. Matsumoto and Ohyanagi, *J. Polym. Sci.* **31**, 225 (1958).
78. A. Beresniewicz, *J. Polym. Sci.* **39**, 63 (1959).
79. D. J. Nagy, *J. Polym. Sci. Part C*, **24**, 87 (1986).
80. D. C. Bugada and A. Rudin, *J. Appl. Polym. Sci.* **30**, 4137 (1985).
81. Y. Morishima and S. Nozakura, *J. Polym. Sci. Chem. Ed.* **14**, 1277 (1976).
82. S. Noma, *Polyvinyl Alcohol, First Osaka Symposium*, Kobunshi Gakkai, Tokyo, Japan, 1955, pp. 81–103.

83. F. Kainer, *Polyvinylalkohole*, F. Enke, Stuttgart, Germany, 1949, pp. 55–92.
84. C. A. Finch, in Ref. 52, pp. 183–202.
85. Ger. Pat. 606,440, A. Voss and W. Starck (to I. G. Farben).
86. J. P. Lorand and J. O. Edwards, *J. Org. Chem.* **24**, 769 (1959).
87. U.S. Pat. 2,607,765 (Aug. 19, 1952), E. P. Czerwin and L. P. Martin (to E. I. du Pont de Nemours & Co., Inc.).
88. G. L. Roy, A. L. Laferrie, and J. O. Edwards, *J. Inorg. Nuclear Chem.* **4**, 106 (1957).
89. S. Saito and co-workers, *Koll. Z.* **144**, 41 (1955).
90. H. Deuel and H. Neukom, *Makromol. Chem.* **3**, 13 (1949).
91. H. Thiele and H. Lamp, *Kolloid Z.* **173**, 63 (1960).
92. E. P. Irany, *Ind. Eng. Chem.* **35**, 90 (1943).
93. N. Okada and I. Sakurada, *Bull. Inst. Chem. Res. Kyoto Univ.* **26**, 94 (1951).
94. R. K. Schultz and R. R. Myers, *Macromol.* **2**, 281 (1969).
95. U.S. Pat. 2,720,468 (Oct. 11, 1955), C. D. Shacklett (to E. I. du Pont de Nemours & Co., Inc.).
96. U.S. Pat. 3,668,166 (1972), T. G. Kane and W. D. Robinson (to E. I. du Pont de Nemours & Co., Inc.).
97. U.S. Pat. 3,135,648 (June 2, 1964), R. L. Hawkins (to Air Products & Chemicals).
98. Ger. Pat. 529,863 (July 1975) (to Hoechst Aktiengesellschaft).
99. M. Shibayama and co-workers, *Polymer*, **29**, 336 (1988).
100. U.S. Pat. 4,331,781 (May 25, 1982), W. Zimmermann and G. Pospich (to Hoechst Aktiengesellschaft).
101. J. H. Haslam, *Adv. Chem.* **23**, 272 (1950).
102. U.S. Pat. 2,518,193 (Aug. 8, 1950), F. K. Signaigo (to E. I. du Pont de Nemours & Co., Inc.).
103. U.S. Pat. 3,518,242 (June 30, 1970), J. D. Crisp (to E. I. du Pont de Nemours & Co., Inc.).
104. W. Diepold, *Explosivstoffe*, **17**, 2 (1970).
105. K. Noma, S. Oya, and K. Nakamura, *Kobunshi Kagaku*, **4**, 112 (1947).
106. A. Takahashi, M. Nagasawa, and I. Kagawa, *Kogyo Kagaku Zasshi*, **61**, 1614 (1958).
107. R. Asami and W. Tokura, *Kogyo Kagaku Zasshi*, **62**, 1593 (1959).
108. U.S. Pat. 2,623,037 (Dec. 23, 1952), R. V. Jones (to Phillips Petroleum Co.).
109. Ger. Pat. 745,683 (Dec. 16, 1943), W. Heuer and W. Starck (to I. G. Farben).
110. I. M. Finganz and co-workers, *J. Polym. Sci.* **56**, 245 (1962).
111. Jpn. Pat. 12,538 (Aug. 31, 1958), I. Sakurada, Y. Noma, and K. En (to Daiichi Kogyo Seiyaku Co.).
112. D. D. Reynolds and W. O. Kenyon, *J. Am. Chem. Soc.* **72**, 1584 (1950).
113. J. R. Williams and D. G. Borden, *Makromol. Chem.* **73**, 203 (1964).
114. U.S. Pat. 2,395,347 (1942), W. H. Sharkey (to E. I. Du Pont de Nemours & Co., Inc.).
115. R. E. Ferrel, H. S. Olcott, and H. Frenkel-Conrat, *J. Am. Chem. Soc.* **70**, 2101 (1948).
116. G. C. Daul, J. D. Reid, and R. M. Reinhardt, *Ind. Ing. Chem.* **46**, 1042 (1954).
117. Brit. Pat. 995,489 (1965) (to Wacker Chemie).
118. U.S. Pat. 2,592,058 (Apr. 8, 1952), I. E. Muskat and F. Strain (to Columbia-Southern Chemical Co.).
119. U.S. Pat. 2,169,250 (1939), E. F. Izard (to E. I. du Pont de Nemours & Co., Inc.).
120. Jpn. Pat. 059,419 (May 17, 1978) (Sumitomo Chem. Ind. KK).
121. D. Graiver, S. Hyon, and Y. Ikada, *J. Appl. Polym. Sci.* **57**, 1299 (1995).
122. Eur. Pat. Appl. EP-366968 (May 9, 1990), G. Haeubl and W. Scheuchenstuhl (to Chemie Linz GmbH).
123. A. M. Paquin, *Z. Naturforsch.* **1**, 518 (1946).
124. U.S. Pat. 3,193,534 (Feb. 26, 1961), K. Matsubayashi and M. Matsumoto (to Kurashiki Rayon Co.).

125. I. Sakurada, A. Nakajima, and K. Shibatani, *J. Polym. Sci. Part A2*, 3545 (1964).
126. G. O. Yahya and co-workers, *J. Appl. Polym. Sci.* **57**, 343 (1995).
127. U.S. Pat. 4,340,686 (July 20, 1982), R. P. Foss (to E. I. du Pont de Nemours & Co., Inc.).
128. U.S. Pat. 3,776,889 (Dec. 4, 1973), K. C. Pande and S. E. Kallenbach (to Powers Chemco, Inc.).
129. F. Masuo, T. Nakano, and Y. Kimura, *Kogyo Kagaku Zasshi*, **57**, 365 (1954).
130. S. Petersen, *Ann.* **562**, 205 (1949).
131. U.S. Pat. 2,728,745 (Dec. 27, 1955), A. C. Smith and C. C. Unruh (to Eastman Kodak).
132. U.S. Pat. 2,887,469 (May 19, 1959), C. C. Unruh and D. A. Smith (to Eastman Kodak).
133. Ger. Pat. 1,063,802 (1954), W. D. Schellenberg and co-workers (to Bayer).
134. Ger. Pat. 1,067,219 (1955), W. D. Schellenberg and Bartl (to Bayer).
135. U.S. Pat. 3,125,556 (Jan. 2, 1964), J. C. Lukman (to The Borden Co.).
136. S. G. Cohen, H. C. Haas, and H. Slotnick, *J. Polym. Sci.* **11**, 193 (1953).
137. U.S. Pat. 1,971,662 (Aug. 28, 1934), A. Schmidt, G. Balle, and K. Eisfeld (to I. G. Farben).
138. U.S. Pat. 2,844,570 (July 22, 1958), R. E. Broderick (to Union Carbide).
139. U.S. Pat. 2,990,398 (June 27, 1961), H. K. Inskip and W. Klabunde (to E. I. du Pont de Nemours & Co., Inc.).
140. U.S. Pat. 2,434,179 (Jan. 6, 1948), W. H. Sharkey (to E. I. du Pont de Nemours & Co., Inc.).
141. U.S. Pat. 3,052,652 (Sept. 4, 1962), B. D. Halpern and B. O. Krueger.
142. U.S. Pat. 2,844,571 (July 22, 1958), A. E. Broderick (to Union Carbide).
143. U.S. Pat. 4,775,715 (Oct. 4, 1988), A. Beresniewicz and T. Hassall, Jr. (to E. I. du Pont de Nemours & Co., Inc.).
144. U.S. Pat. 4,822,851 (Apr. 18, 1989), R. Stober, E. Kohn, and D. Bischoff (to Degussa Aktiengesellschaft).
145. R. W. Roth, L. J. Patella, and B. L. Williams, *J. Appl. Polym. Sci.* **9**, 1083 (1965).
146. L. Alexandru, M. Opris, and Ciocanel, *J. Polym. Sci.* **59**, 129 (1962).
147. U.S. Pat. 3,194,798 (July 13, 1965), L. W. Frost (to Westinghouse Electric Co.).
148. U.S. Pat. 2,341,553 (Feb. 15, 1944), R. C. Houtz (to E. I. du Pont de Nemours & Co., Inc.).
149. F. Ide, S. Nakano, and K. Nakatsuka, *Kobunshi Kagaku*, **24**, 549 (1967).
150. F. Arranz, M. Sanchez-Chaves, and M. M. Gallego, *Angew. Makromol. Chem.* **218**, 183 (1994).
151. U.S. Pat. 3,505,303 (Apr. 7, 1970), M. K. Lindemann (to Air Reduction Co.).
152. H. Ito, *Kogyo Kagaku Zasshi*, **63**, 338 (1960).
153. U.S. Pat. 5,104,933 (Apr. 14, 1992), P. Shu (to Mobil Oil Corp.).
154. Y. Nakamura and M. Negishi, *Kogyo Kagaku Zasshi*, **68**, 1762 (1965).
155. Y. Nakamura and M. Negishi, *Kogyo Kagaku Zasshi*, **70**, 774 (1967).
156. Y. Nakamura, M. Negishi, and Y. Takekawa, *Kogyo Kagaku Zasshi*, **68**, 1766 (1965).
157. Ger. Pat. 738,869 (July 29, 1943), K. Billig (to I. G. Farben).
158. M. Tsunooka and co-workers, *Kobunshi Kagaku*, **23**, 451 (1966).
159. U.S. Pat. 5,350,801 (Sept. 27, 1994), A. Famili, L. A. Vratsanos, and F. L. Marten (to Air Products & Chemicals, Inc.).
160. M. Hida, *Kogyo Kagaku Zasshi*, **55**, 275 (1952).
161. K. Asahina, in C. A. Finch, ed., *Poly(Vinyl Alcohol) Developments*, John Wiley & Sons, Inc., New York, 1992, pp. 673–688.
162. I. Sakurada, *Poly(Vinyl Alcohol) Fibers*, Marcel Dekker, Inc., New York, 1985.
163. B. Gebben, H. W. A. van den Berg, and D. Bargeman, *Polymer*, **26** (1985).
164. Ger. Pat. DE 3,316,948 (Nov. 15, 1984), W. Zimmerman and W. Eichhorn (to Hoechst AG).

165. Y. Mori, H. Yokoi, and Y. Fujise, *Polym. J.* **3**, 271 (1995).

166. M. Nagano and Y. Yoshioka, *Kobunshi Kagaku*, **9**, 19 (1952).

167. T. Uragami and M. Sugihara, *Ang. Makrom. Chem.* **57**, 123 (1977).

168. C. Dittmar and W. J. Priest, *J. Polym. Sci.* **18**, 275 (1955).

169. R. C. Schulz and J. A. Trisnadi, *Makromol. Chem.* **177**, 1771 (1976).

170. M. Shibayama and co-workers, *Macromolecules*, **27**, 1738 (1994).

171. M. M. Zwick, *J. Appl. Polym. Sci.* **9**, 2393 (1965).

172. M. M. Zwick, *J. Polym. Sci. Part A1*, **4**, 1642 (1966).

173. Y. Choi and K. Miyasaka, *J. Appl. Polym. Sci.* **48**, 313 (1993).

174. K. Shibatini, M. Nakamura, and Y. Oyanagi, *Kobunshi Kagaku*, **26**, 118 (1969).

175. Y. Morishima, K. Fujisawa, and S. Nozakura, *Polymer J.* **10**, 281 (1978).

176. K. Kikukawa, S. Nozakura, and S. Murahashi, *Polymer J.* **2**, 212 (1971).

177. T. Hirai, A. Okazaki, and S. Hayashi, *J. Appl. Polym. Sci.* **32**, 3919 (1986).

178. S. S. Mnatsakanov and co-workers, *Polym. Sci. U.S.S.R.* **14**, 946 (1972).

179. B. J. R. Scholtens and B. H. Bijsterbosch, *J. Polym. Sci. Polym. Phys. Ed.* **17**, 1771 (1979).

180. S. Hayashi, C. Nakano, and T. Moyoyama, *Kobunshi Kagaku*, **20**, 303 (1963).

181. S. Hayashi and co-workers, *J. Polym. Sci. Polym. Lett. Ed.* **20**, 69 (1982).

182. S. Hayashi and co-workers, *J. Polym. Sci. Polym. Chem. Ed.* **20**, 839 (1982).

183. S. Hayashi and co-workers, *Makromol. Chem.* **176**, 3221 (1975).

184. T. Hirai and co-workers, *J. Polym. Sci. Polym. Lett. Ed.* **26**, 299 (1988).

185. Y. Tsuchiya and K. Sumi, *J. Polym. Sci. Part A1*, **7**, 3151 (1969).

186. C. Vasile, C. N. Cascaval, and P. Barbu, *J. Polym. Sci. Polym. Chem. Ed.* **19**, 907 (1981).

187. C. Vasile, S. F. Patachia, and V. Dumitrascu, *J. Polym. Sci. Polym. Chem. Ed.* **21**, 329 (1983).

188. C. Vasile and co-workers, *J. Polym. Sci. Polym. Chem. Ed.* **23**, 2579 (1985).

189. B. Kaesche-Krischer and H. J. Heinrich, *Z. Physik. Chem.* **23**, 292 (1960).

190. K. Ettre and P. F. Varadi, *Anal. Chem.* **35**, 69 (1963).

191. H. Fischer, *Z. Naturforsch.* **19a**, 866 (1964).

192. H. Futama and H. Tanaka, *J. Phys. Soc. Jpn.* **12**, 433 (1957).

193. T. Yamaguchi and M. Amagasa, *Kobunshi Kagaku*, **18**, 645 (1961).

194. A. S. Dunn, R. L. Coley, and B. Duncalf, in *Properties and Applications of Polyvinyl Alcohol*, Society of Chemical Industry, London, U.K., 1968, p. 208.

195. B. Duncalf and A. S. Dunn, Gordon & Breach, New York, 1967, p. 162.

196. B. Kaesche-Krischer and H. J. Heinrich, *Chem. Ing. Tech.* **32**, 740 (1960).

197. R. J. Axelrod and J. H. Phillips, *Proceedings of Plastic Waste Management Workshop*, New Orleans, La., 1991.

198. J. P. Casey and D. G. Manley, *Proceedings of 3rd International Biodegradation Symposium*, Applied Science Publishing, Ltd., London, U.K.

199. W. Schefer and K. Romanin, *Z. Wasser-Abwasser-Forsch.* **22**, 157 (1989).

200. K. Fischer, *Melliand Text. Rep.* **65**, 269–274, 340–345 (1984).

201. U.S. Pat. 3,926,796 (Dec. 16, 1975), Y. Fujita, I. Kawasaki, and H. Nishikawa (to Nippon Gohsei Kagaku Kogyo Kabushiki Kaisha).

202. T. Suzuki and co-workers, *Agr. Biol. Chem.* **37**, 747 (1973).

203. N. Nishikawa and Y. Fujita, *Chem. Econ. Eng. Rev.* **7**, 33 (1975).

204. M. Morita and co-workers, *Agr. Biol. Chem.* **43**, 1225 (1979).

205. K. Sakai and co-workers, *Agr. Biol. Chem.* **45**, 63 (1981).

206. K. Sakai, N. Hamada, and Y. Watanabe, *Agr. Biol. Chem.* **48**, 1093 (1984).

207. R. Fukae and co-workers, *Polym. J.* **26**, 1381 (1994).

208. S. Matsumura, Y. Shimura, and K. Toshima, *Macromol. Chem. Phys.* **196**, 3437 (1995).

209. U.S. Pat. 3,689,469 (Sept. 5, 1972), H. K. Inskip and R. L. Adelman (to E. I. du Pont de Nemours & Co., Inc.).
210. U.S. Pat. 4,618,648 (Oct. 22, 1986), F. L. Marten (to Air Products & Chemicals, Inc.).
211. U.S. Pat. 4,772,663 (Sept. 20, 1988), F. L. Marten, A. Famili, and D. K. Mohanty (to Air Products & Chemicals, Inc.).
212. E. Ger. Pat. DD-134109-A (Feb. 7, 1979), G. Buech, L. Krahnert, and I. Opitz (to VEB Chemische Werke Buna).
213. Jpn. Pat. 63 18,542 (Sept. 17, 1963), K. Noro, G. Morimoto, and E. Uemura (to Nippon Gohsei Co.).
214. W. Baade, H. U. Moritz, and K. H. Reichert, *J. Appl. Polym. Sci.* **27**, 2249 (1982).
215. U.S. Pat. 3,121,705 (Feb. 18, 1964), A. I. Lowell and O. L. Magell (to Wallace and Tiernan Co.).
216. B. L. Funt and W. Pasika, *Can. J. Chem.* **38**, 1865 (1960).
217. C. H. Bamford, R. W. Dyson, and G. C. Eastmond, *Polymer* **10**, 885 (1969).
218. U.S. Pat. 3,679,648 (July 25, 1972), J. E. Bristol (to E. I. du Pont de Nemours & Co., Inc.).
219. U.S. Pat. 3,679,646 (July 25, 1972), J. E. Bristol (to E. I. du Pont de Nemours & Co., Inc.).
220. Brit. Pat. 808,108 (Jan. 28, 1959) (to Celanese Corp.).
221. N. Friis and A. E. Hamielec, *J. Appl. Polym. Sci.* **19**, 97 (1975).
222. N. Friis and co-workers, *J. Appl. Polym. Sci.* **18**, 1247 (1974).
223. I. Sakurada and O. Yoshizaki, *Kobunshi Kagaku*, **14**, 284 (1957).
224. *Ibid.*, 339.
225. H. N. Friedlander, H. E. Harris, and J. G. Pritchard, *J. Polym. Sci.* **4**, 649 (1966).
226. S. Nozakura, Y. Morishima, and S. Murahashi, *J. Polym. Sci. Chem. Ed.* **10**, 2853 (1972).
227. J. T. Clarke, R. O. Howard, and W. H. Stockmeyer, *Makromol. Chem.* **44–46**, 427 (1961).
228. W. W. Graessley and H. M. Mittelhauser, *J. Polym. Sci. Part A2*, **5**, 431 (1967).
229. W. W. Graessley, R. D. Hartung and W. C. Uy, *J. Polym. Sci. Part A2*, **7**, 1919 (1969).
230. K. Nagasubramanian and W. W. Graessley, *Chem. Eng. Sci.* **25**, 1559 (1970).
231. S. Nozakura, Y. Morishima, and S. Murahashi, *J. Polym. Sci. Part A1*, **10**, 2767 (1972).
232. *Ibid.*, p. 2781.
233. *Ibid.*, p. 2853.
234. *Ibid.*, p. 2867.
235. S. Nozakura and co-workers, *J. Polym. Sci. Chem. Ed.* **14**, 759 (1976).
236. Y. Morishima and co-workers, *J. Polym. Sci. Chem. Ed.* **14**, 1269 (1976).
237. Y. Morishima and S. Nozakura, *J. Polym. Sci. Chem. Ed.* **14**, 1277 (1976).
238. Y. Morishima and co-workers, *J. Polym. Sci. Lett.* **13**, 157 (1975).
239. Jpn. Pat. 61 16,446 (Sept. 15, 1961), S. Otsuka, Y. Ono, and S. Masuda (to Denki Kagaku Kogyo Co. Ltd.).
240. Jpn. Pat. 63 14,711 (Aug. 13, 1963), Y. Kotani (to Nippon Gohsei Co. Ltd.).
241. Jpn. Pat. 72 24,533 (July 6, 1972), J. Fujitani (to Denki Kagaku Kogyo Co. Ltd.).
242. Jpn. Pat. 71 14,649 (Apr. 20, 1972), T. Wakasugi and I. Niida (to Kurera Co. Ltd.).
243. Brit. Pat. 917,811 (Feb. 6, 1963) (to Nippon Gohsei Co. Ltd.).
244. J. H. Oak and S. N. Park, *Hwahak Kwa Hwahak Kongop*, **5**, 239 (1971).
245. U.S. Pat. 4,708,999 (Nov. 24, 1987), F. L. Marten (to Air Products & Chemicals, Inc.).
246. Jpn. Pat. 20,745 (1961) (to Kurashiki Rayon Co.).
247. Brit. Pat. 811,535 (1957) (to E. I. du Pont de Nemours & Co., Inc.).
248. U.S. Pat. 2,878,168 (Mar. 17, 1959) (to E. I. du Pont de Nemours & Co., Inc.).
249. E. Ger. Pat. (Jan. 28, 1987) (to VEB Chemische Werke Buna).

250. Ger. Pat. 1,962,986 (June 9, 1970), R. T. Bouchard and co-workers (to Borden Inc.).

251. U.S. Pat. 2,878,168 (Mar. 17, 1959), W. B. Tanner and J. R. Wesel (to E. I. du Pont de Nemours & Co., Inc.).

252. E. Hackel, *Properties and Applications of Polyvinyl Alcohol*, Society of Chemical Industry, London, U.K., 1968, p. 1.

253. U.S. Pat. 2,642,420 (June 16, 1953), W. O. Kenyon, G. P. Waugh, and E. W. Taylor (to Eastman Kodak).

254. U.S. Pat. 2,668,810 (Dec. 9, 1954), E. Bergmeister, W. Gruber, and J. Heckmaier (to Wacker Chemie GmbH).

255. K. Noro, *Brit. Polym. J.* **2**, 128 (1970).

256. F. Gregor and E. Engel, *Chem. Prumysl.* **10**, 53 (1960).

257. U.S. Pat. 3,156,678 (Nov. 10, 1964), H. Dexheimer and co-workers (to Farbwerke Hoechst).

258. Jpn. Pat. 52,026,589 (Aug. 27, 1975) (to Electro Chemical Industry K. K.).

259. U.S. Pat. 3,156,667 (1964) (to Shawinigan Resins Corp.).

260. Rus. Pat. 211,091 (May 7, 1965), L. E. De-Millo, T. V. Kyazeva, and V. A. Kuznetsova.

261. U.S. Pat. 2,643,994 (June 30, 1953), L. M. Germain (to Shawinigan Chemicals Ltd.).

262. Jpn. Pat. 13,141 (1960), (Nippon Gohsei Co.).

263. S. Yamane, in K. Nagano, S. Yamane, and K. Toyoshima, eds., *Poval*, Kobunshi Kankokai, Kyoto, Japan, 1981, p. 114.

264. Brit. Pat. 1,168,757 (Feb. 2, 1969), H. A. Eilers and co-workers (to Farbwerke Hoechst AG).

265. U.S. Pat. 2,642,419 (June 16, 1953), G. P. Waugh and W. O. Kenyon (to Eastman Kodak).

266. U.S. Pat. 2,734,048 (Feb. 7, 1956), J. E. Bristol and W. B. Tanner (to E. I. du Pont de Nemours & Co., Inc.).

267. U.S. Pat. 3,296,236 (1967), W. B. Tanner (to E. I. du Pont de Nemours & Co., Inc.).

268. U.S. Pat. 3,487,060 (Dec. 30, 1969), J. E. Bristol (to E. I. du Pont de Nemours & Co., Inc.).

269. U.S. Pat. 4,389,506 (June 21, 1983), T. Hassall, Jr. (to E. I. du Pont de Nemours & Co., Inc.).

270. U.S. Pat. 3,278,505 (Sept. 10, 1966), T. Kominami (to Kurashiki Rayon Co. Ltd.).

271. U.S. Pat. 4,338,405 (July 6, 1982), R. I. Saxton (to E. I. du Pont de Nemours & Co., Inc.).

272. U.S. Pat. 4,401,790 (Aug. 30, 1983), H. Jung, L. Reihs, and G. Roh (to Hoechst Aktiengesellschaft).

273. U.S. Pat. 4,352,949 (Oct. 10, 1982), R. L. Adelman and R. Segars (to E. I. du Pont de Nemours & Co., Inc.).

274. L. Alexandru, F. Butaciu, and I. Balint, *J. Prakt. Chem.* **16**, 125 (1962).

275. Ger. Pat. 1,211,622 (Mar. 3, 1966), G. Kuenstle and H. Siegl (to Wacker Chemie AG).

276. A. Nakumura, *Kogyo Kagaku Zasshi*, **72**, 626 (1969).

277. U.S. Pat. 2,671,052 (Mar. 2, 1954), R. L. Mitchell and J. W. Walker (to Celanese Corp.).

278. U.S. Pat. 2,636,050 (Apr. 21, 1953), R. L. Hoaglin and W. M. Frankenberger (to Union Carbide Corp.).

279. U.S. Pat. 2,650,249 (Mar. 25, 1953), M. Mention and J. Mercier (to Les Usines de Melle).

280. U.S. Pat. 3,011,954 (Dec. 5, 1961), D. B. Halpern and B. O. Kreuger (to Borden Co.).

281. U.S. Pat. 3,239,572 (Mar. 8, 1966), C. A. Zinsstag (to Lonza Ltd.).

282. T. Okaya, in Ref. 161, pp. 79–103.

283. Elvanol Safe Handling Information, E. I. du Pont de Nemours & Co., Inc., Wilmington, Del.

284. *Airvol Product Handbook*, Air Products and Chemicals, Inc., Allentown, Pa., 1990.

285. *Material Safety Data Sheet for Poly(Vinyl Alcohol) Low Volatiles*, Air Products & Chemicals, Inc., Allentown, Pa., 1995.

286. U.S. Pat. 4,611,019 (Sept. 9, 1986), H. H. Lutzmann and G. W. Miller.

287. U.S. Pat. 4,215,169 (July 29, 1980), R. D. Wysong (to E. I. du Pont de Nemours & Co., Inc.).

288. U.S. Pat. 4,206,101 (June 3, 1980), R. D. Wysong (to E. I. du Pont de Nemours & Co., Inc.).

289. U.S. Pat. 4,267,145 (May 12, 1981), R. D. Wysong (to E. I. du Pont de Nemours & Co., Inc.).

290. U.S. Pat. 4,156,047 (May 22, 1979), R. D. Wysong (to E. I. du Pont de Nemours & Co., Inc.).

291. U.S. Pat. 4,155,971 (May 22, 1979), R. D. Wysong (to E. I. du Pont de Nemours & Co., Inc.).

292. U.S. Pat. 4,119,604 (Oct. 10, 1978), R. D. Wysong (to E. I. du Pont de Nemours & Co., Inc.).

293. Jpn. Pat. J 51,089,562 (Aug. 5, 1976) (to Unitika KK).

294. U.S. Pat. 4,672,087 (June 9, 1987), G. W. Miller and H. Lutzmann.

295. U.S. Pat. 4,369,281 (Jan. 18, 1983), W. Zimmerman and A. Harreus (to Hoechst AG).

296. U.S. Pat. 4,529,666 (July 16, 1985), F. Kleiner, K. Reinking and H. Salzburg (to Bayer AG).

297. U.S. Pat. 3,607,812 (Sept. 9, 1971), T. Hayashi and co-workers (to Denki Kagaku Kogyo).

298. U.S. Pat. 4,469,837 (Sept. 4, 1984), P. Cattaneo.

299. U.S. Pat. 5,051,222 (Sept. 24, 1991), F. L. Marten, A. Famili, and J. F. Nangeroni (to Air Products & Chemicals, Inc.).

300. U.S. Pat. 5,137,969 (Sept. 24, 1991), F. L. Marten, A. Famili, and J. F. Nangeroni (to Air Products & Chemicals, Inc.).

301. J. E. Mooreland, *Text. Chem. Color.* **12**(4), 21 (1980).

302. D. L. Nehrenberg, *Proceedings of AATCC RA73 Research Committee Warp Size Symposium*, Atlanta, Ga., 1981.

303. U.S. Pat. 4,844,709 (July 4, 1989), F. L. Marten (to Air Products and Chemicals, Inc.).

304. U.S. Pat. 4,845,140 (July 4, 1989), F. L. Marten (to Air Products and Chemicals, Inc.).

305. Jpn. Pat. 47 45,634 (Nov. 17, 1972), M. Yamamoto and co-workers (to Unichika Co., Ltd.).

306. U.S. Pat. 4,640,946 (Feb. 3, 1987), D. A. Vasallo and D. W. Zunker (to E. I. du Pont de Nemours & Co., Inc.).

307. U.S. Pat. 4,383,063 (May 10, 1983), R. W. Rees (to E. I. du Pont de Nemours & Co., Inc.).

308. U.S. Pat. 4,399,245 (Aug. 16, 1983), R. Kleber and H. Stuhler (to Hoechst Aktiengesellschaft).

309. U.S. Pat. 4,222,922 (Sept. 16, 1980), R. W. Rees (to E. I. du Pont de Nemours & Co., Inc.).

310. U.S. Pat. 4,309,510 (Jan. 5, 1982), R. Kleber (to Hoechst Aktiengesellschaft).

311. U.S. Pat. 4,251,403 (Feb. 17, 1981), R. W. Rees (to E. I. du Pont de Nemours & Co., Inc.).

312. U.S. Pat. 3,135,648 (June 2, 1964), R. L. Hawkins (to Air Products and Chemicals, Inc.).

313. Eur. Pat. Appl. 306,250 (July 15, 1987), R. M. Marshall and P. L. Krankkala (to H. B. Fuller Co.).

314. Jpn. Pat. 49 12,590 (Mar. 23, 1974), M. Kurata, Y. Hase, and J. Ichimura (to Sugiyama Industrial Chemical Institute).

315. Jpn. Pat. 46 27,256 (Aug. 7, 1971), N. Kobakashi, M. Nomura, and Y. Yamitaki (to Dainippon Ink and Chemicals Inc.).

316. U.S. Pat. 3,941,735 (Mar. 2, 1975), M. K. Lindemann (to Chas. S. Tanner Co.).

317. U.S. Pat. 4,001,160 (Jan. 4, 1977), M. K. Lindemann (to Chas. S. Tanner Co.).

318. M. Shiraishi, *Brit. Polym. J.* **2**, 135 (1970).

319. *Synthetic Binders in Paper Coatings*, Technical Association of the Pulp and Paper Industry, 1975.

320. D. Wolf, *Kunstharz Nachr.* **17**, 34 (1981).

321. M. B. Lyne and J. S. Aspler, *TAPPI J.*, 106 (May, 1985).

322. Jpn. Pat. 88 118,286 (May 23, 1988) (to Asahi Glass Co. Ltd.).

323. U.S. Pat. 3,784,648 (Jan. 8, 1974), E. Bergmeister, P. G. Kirst, and H. Winkler (to Wacker Chemie GmbH).

324. U.S. Pat. 5,118,751 (June 2, 1992), J. Schulze, H. Herold, and J. Hinterwinkler (to Wacker Chemie GmbH).

325. U.S. Pat. 3,883,489 (May 13, 1975), K. Matschke and co-workers (to Hoechst Aktiengesellschaft).

326. U.S. Pat. 5,519,084 (May 21, 1996), H. Pak-Harvey and C. L. Mao (to Air Products and Chemicals, Inc.).

327. U.S. Pat. 4,440,711 (Apr. 3, 1984), Y. D. Kwon, S. Kavesh, and D. C. Prevorsek (to Allied Corp.).

328. U.S. Pat. 4,603,083 (July 29, 1986), H. Tanaka and F. Ueda (to Toray Industries, Inc.).

329. U.S. Pat. 4,551,296 (Nov. 5, 1985), S. Kavesh and D. C. Prevorsek (to Allied Corp.).

General References

C. A. Finch, ed., *Polyvinyl Alcohol Developments*, John Wiley & Sons, Inc., New York, 1992.

I. Sakurada, *Poly(Vinyl Alcohol) Fibers*, Marcel Dekker, Inc., New York, 1985.

M. K. Lindeman, in N. M. Bikales, ed., *Encyclopedia of Polymer Science and Technology*, Vol. 14, John Wiley & Sons, Inc., New York, 1971.

C. A. Finch, ed., *Polyvinyl Alcohol*, John Wiley & Sons, Inc., New York, 1973.

Properties and Applications of Poly(Vinyl Alcohol), Society of the Chemical Industry, London, U.K., 1968.

J. G. Pritchard, *Poly(Vinyl Alcohol) Basic Properties and Uses*, Gordon & Breach, Inc., New York, 1970.

F. Lennart Marten
Air Products and Chemicals, Inc.

VINYL CHLORIDE POLYMERS

Poly(vinyl chloride) [9002-86-2] (PVC), commanding large and broad uses in commerce, is second in volume only to polyethylene, having a volume sales in North America in 1995 of 6.2×10^9 kg (13.7×10^9 lb) (Table 1). This large volume of sales can be attributed to several unique properties. Vinyl compounds usually contain close to 50% chlorine, which not only provides no fuel, but acts to inhibit combustion in the gas phase, thus supplying the vinyl with a high level of combustion resistance, useful in many building as well as electrical housings and electrical insulation applications.

Table 1. 1995 Sales Volume in North America, 10^9 kg[a]

Polymer	United States	Canada	Mexico	North American Total
PVC	5.36	0.45	0.40	6.22
high density polyethylene	5.29	0.73	0.18	6.20
linear low density polyethylene	2.70			2.70
low density polyethylene	3.38	1.31	0.30	4.99
polypropylene	4.84	0.26	0.22	5.32
polystyrene	2.68		0.14	2.81
acrylonitrile–butadiene–styrene	0.66			0.66
thermoplastic polyester	1.78			1.78
polycarbonate	0.34			0.34
acetal	0.15			0.15
acrylic	0.25			0.25
nylon	0.47			0.47

[a]Ref. 1.

PVC has a unique ability to be compounded with a wide variety of additives, making it possible to produce materials that range from flexible elastomers to rigid compounds, that are virtually unbreakable with a notched Izod impact greater than 0.5 J/mm at −40°C, that are weatherable with good property retention for over 30 years, as well as compounds that have stiff melts and little elastic recovery for outstanding dimensional control in profile extrusion, and also low viscosity melts for thin-walled injection molding.

Produced by free radical polymerization, PVC has the structure of $+\mathrm{CH_2CHCl}+_n$, where the degree of polymerization, n, ranges from 500 to 3500. The first discovery of PVC was in 1872, when it was found that exposure of vinyl chloride to sunlight produced a white solid that resisted attack by potassium hydroxide or water and melted with degradation at above 130°C (2,3). From 1912 to 1926, German workers at Chemische Fabrik Griesheim-Electron tried but failed to build machinery that could process PVC and overcome its instability (3). In 1926, a researcher at BFGoodrich, while looking for an adhesive to bond rubber to metal for tank liners, found that boiling PVC in tricresyl phosphate or dibutyl phthalate made it highly elastic (3), thus inventing the first thermoplastic elastomer.

Physical Properties

Morphology as Polymerized. The principal type of polymerization of PVC is the suspension polymerization route. The morphology formed during polymerization strongly influences the processibility and physical properties. Mass-polymerized PVC has a similar morphology to suspension PVC.

In the suspension polymerization of PVC, droplets of monomer 30–150 μm in diameter are dispersed in water by agitation. A thin membrane is formed at the water–monomer interface by dispersants such as poly(vinyl alcohol) or methyl cellulose. This membrane, isolated by dissolving the PVC in tetrahydrofuran and measured at 0.01–0.02-μm thick, has been found to be a graft

copolymer of polyvinyl chloride and poly(vinyl alcohol) (4,5). Early in the polymerization, particles of PVC deposit onto the membrane from both the monomer
and the water sides, forming a skin $0.5-5-\mu m$ thick that can be observed on
grains sectioned after polymerization (4,6). Primary particles, 1 μm in diameter, deposit onto the membrane from the monomer side (Fig. 1), whereas water-
phase polymer, 0.1 μm in diameter, deposits onto the skin from the water side
of the membrane (Fig. 2) (4). These domain-sized water-phase particles may be
one source of the observed domain structure (7).

Mass-polymerized PVC also has a skin of compacted PVC primary particles
very similar in thickness and appearance to the suspension-polymerized PVC
skin, compared in Figure 3. However, mass PVC does not contain the thin-block
copolymer membrane (7).

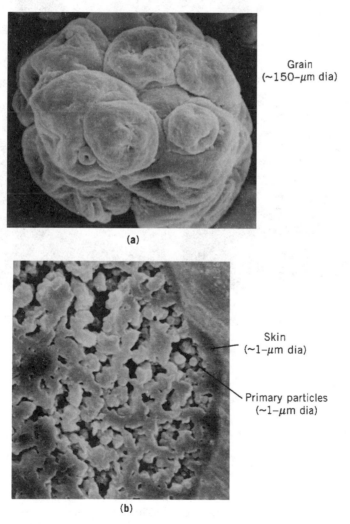

Fig. 1. A grain of (**a**) suspension PVC and (**b**) its cross-section showing the skin and
primary particles.

Fig. 2. Skin of a suspension PVC grain showing 0.1-μm dia particles deposited from the water phase.

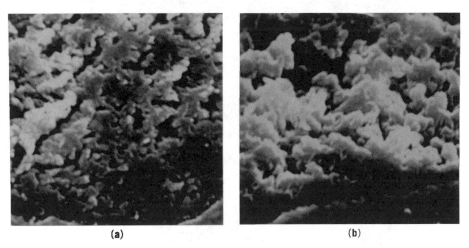

(a)   **(b)**

Fig. 3. Cross-sections of PVC grains from (**a**) suspension and (**b**) mass PVC grains showing similar morphologies.

In suspension PVC polymerization, droplets of polymerizing PVC, 30–150-μm dia agglomerate to form grains at 100–200-μm dia (8). With one droplet per grain, the shape is quite spherical. With several droplets making up the grain, the shape can be quite irregular and knobby (9). The grain shape plays an important role in determining grain packing and bulk density of a powder (9).

For both suspension and mass polymerizations at less than 2% conversion, PVC precipitates from its monomer as stable primary particles, slightly below 1-μm dia (4,10–12). These primary particles are stabilized by a negative chloride charge (4,13). Above 2% conversion, these primary particles agglomerate. Sectioning the PVC grains of either suspension or mass resins readily shows the skins: primary particles at 1-μm dia, and agglomerates of primary particles at 3–10-μm dia (4,7,8,14).

These primary particles also contain smaller internal structures. Electron microscopy reveals a domain structure at about 0.1-μm dia (8,15,16). The origin and consequences of this structure is not well understood. PVC polymerized in the water phase and deposited on the skin may be the source of some of the domain-sized structures. Also, domain-sized flow units may be generated by certain unusual and severe processing conditions, such as high temperature melting at 205°C followed by lower temperature mechanical work at 140–150°C (17), which break down the primary particles further.

On an even smaller scale is the microdomain structure at 0.01-μm spacing. Small-angle x-ray scattering reveals a scattering peak (Fig. 4) corresponding to density fluctuations spaced at about 0.01 μm (18–20). When PVC is swelled by plasticizer or with a poor solvent such as acetone, swelling reaches a limit after which the PVC can no longer absorb plasticizer or acetone (21). This data suggest a structure where the crystallites of about 0.01-μm spacing are tied together by molecules in the amorphous regions. Plasticizer or acetone only swells the amorphous regions without dissolving the crystallites. Also, electron microscopy shows a spacing in plasticized PVC of 0.01 μm (22).

Hierarchical Structure of PVC. PVC has structure that is built upon structure which is, in turn, built upon even more structure. These many layers of structure are all important to performance and are interrelated. A summary of these structures is listed in Table 2; Figure 5 examines a model of these hierarchies on three scales.

Morphology during Processing. The first step in processing is usually powder mixing in a high speed, intensive mixer. PVC resin, stabilizers, plasticizers, lubricants, processing aids, fillers, and pigments are added to the powder blend for distributive mixing. For both suspension and mass PVC resins, intensive mixing above the glass-transition temperature results in a progressive increase in apparent bulk density as mixing temperature rises (23). This increase in apparent bulk density results from the smoothing and rounding of the irregular surface. However, the grains of PVC are largely unchanged; they are neither grossly deformed nor broken down to smaller particles (23).

In plasticized PVC, liquid plasticizers first fill the voids or pores in the PVC grains fairly rapidly during powder mixing. If a large amount of plasticizer is added, the excess plasticizer beyond the capacity of the pores initially remains

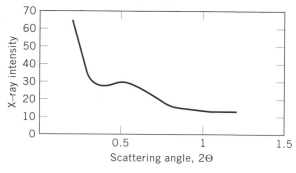

Fig. 4. Small-angle x-ray scattering pattern from PVC plasticized with 20 parts per hundred resin of dioctyl phthalate (18).

Table 2. Summary of Poly(vinyl chloride) Morphology

Feature	Size	Description
droplets	30–150 μm dia	dispersed monomer during suspension polymerization
membranes	0.01–0.02 μm thick	membrane at monomer–water interface in suspension PVC (usually graft copolymer of PVC and dispersant, such as poly(vinyl alcohol))
grains	100–200 μm dia	after polymerization, free-flowing powder usually made up of agglomerated droplets; in mass polymerization, it is free-flowing powder
skins	0.5–5 μm thick	shell on grains made up of PVC deposited onto membrane during suspension polymerization; in mass polymerization, it is PVC compacted on grain surface
primary particles	1 μm dia	formed as single polymerization site in both suspension and mass polymerization by precipitation of polymer from monomer; made up of over a billion molecules, it is often melt flow unit established during melt processing (in emulsion polymerization, it is emulsion particle)
agglomerates of primary particles	3–10 μm dia	formed during polymerization by merging of primary particles
domains	0.1 μm dia	formed under special conditions such as high temperature melting (205°C) followed by lower temperature mechanical work (140–150°C); water-phase polymerization also produces domain-sized structure
microdomains	0.01 μm spacing	crystallite spacing
secondary crystallinity	0.01 μm spacing	crystallinity reformed from amorphous melt and responsible for fusion (gelation)

on the surface of the grains, making the powder somewhat wet and sticky. Continued heating increases the diffusion rate of plasticizer into the PVC mass where the excess liquid is eventually absorbed and the powder dries.

PVC powder compounds are heated, sheared, and deformed during melt processing. During this process, the grains of PVC are broken down. First the skin is torn, exposing the PVC grain's internal structures (24). Subsequently the grains are broken down to agglomerates of primary particles, then to primary particles as the melt flow units. The primary particles seem to be persistent and fairly stable structures in the melt (25–32). This processing window of stable primary particles exists even with continued melt processing. The primary particle is about a billion molecules of PVC held together by a structure of crystallites and tie molecules (21).

The PVC crystallites are small, average 0.7 nm (3 monomer units), in the PVC chain direction, and are packed laterally to a somewhat greater extent (4.1 nm) (21,33). A model of the crystallite is shown in Figure 6. The crystalline

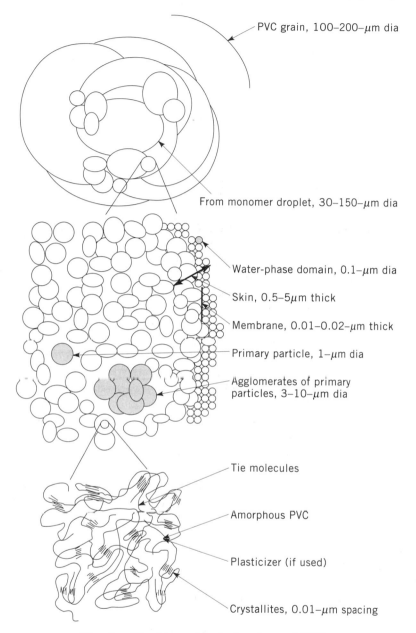

Fig. 5. The hierarchical structure of PVC.

structure of PVC is found to be an orthorhombic system, made of syndiotactic structures, having two monomer units per unit cell and 1.44–1.53 specific gravity (34–37).

PVC Fusion (Gelation). The PVC primary particle flow units (billion molecule bundles) can partially melt, freeing some molecules of PVC that can entangle at the flow unit boundary. These entangled molecules can recrystallize upon cooling, forming secondary crystallites, and tie the flow units together into

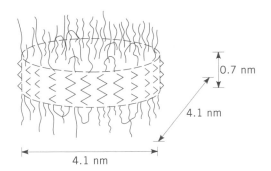

Fig. 6. The crystallite structure of PVC.

a large three-dimensional structure (21,38). This process is known as fusion or gelation.

The strength created by the fusion process is strongly dependent on the previous processing temperature and the molecular weight of the PVC (38–44). The degree of fusion or gelation is measured in several ways. The entrance pressure in capillary rheometry is often used as a measure of the fusion strength (40,42,45–63). Differential scanning calorimetry is also often used as an indication of previous melt temperature and the amount of crystallinity melted and reformed as secondary crystallinity (54,55). X-ray diffraction has been used to measure gelation (55). Acetone or methylene chloride swelling and observation of structural breakdown is widely used as a qualitative measure of fusion (44,56–59). Sometimes the acetone-swollen specimens are sheared between glass slides to further establish the strength of the three-dimensional structure (59). Fusion has also been assessed based on scanning electron microscopy of fractured surfaces (29,30,46), and inverse gas chromatographic measurements have also been useful in accessing the degree of fusion of PVC (60).

The strength of this large three-dimensional fused (gelled) structure has been shown to be critical in determining Izod impact, creep rupture strength, and even flow in rigid injection molding. In these cases, both the melt temperature during processing and the PVC molecular weight play a large role in the Izod and creep rupture (38–41). In plasticized PVC, this large three-dimensional structure, which is also dependent on molecular weight and previous processing temperature, determines tensile strength, creep, and cut resistance. A model for accounting for molecular weight effects and processing temperature effects on PVC fusion is presented in Figure 7.

PVC normally improves in properties with increasing fusion (or increasing melt temperature); however, some observations show a falling off in impact properties with higher melt temperatures (42,56). This has been shown to be caused by melt fracture, when PVC fuses and flows as large melt flow units of multiple primary particles (61). These flow units of fused multiple primary particles can lead to surface roughness during extrusion, in both rigid and plasticized PVC (44,56,62); they can also be responsible for roughness when improperly handling regrind (63).

Plasticized PVC Morphology. Although most of the discussion has been on rigid PVC, plasticized PVC has the same structures as rigid PVC, except

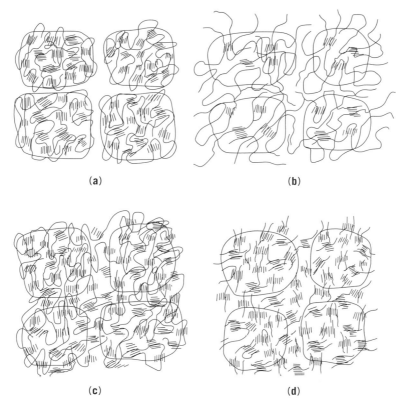

Fig. 7. Model for PVC fusion, accounting for molecular weight effects and processing temperature effects: (**a**) unfused PVC primary particles; (**b**) partially melted PVC primary particles; (**c**) partially melted then recrystallized high molecular weight PVC, showing strong three-dimensional structure; and (**d**) partially melted then recrystallized low molecular weight PVC, showing weak three-dimensional structure.

that plasticizer enters the amorphous phase of PVC and makes the tie molecules elastomeric. The grains break down to 1-μm primary particles which become the melt flow units (44). The crystallites are not destroyed by plasticizer (21). Partial melting allows entanglement at the flow unit boundaries, followed by recrystallization upon cooling to form a strong three-dimensional elastomeric structure (38,45). Table 3 provides a list of the PVC physical parameters.

Chemical Properties

Molecular Structure and Monomer Addition Orientation. The addition of vinyl monomer to a growing PVC chain can be considered to add in a head-to-tail fashion, resulting in a chlorine atom on every other carbon atom, ie,

$$+CH_2CHClCH_2CHClCH_2CHClCH_2CHCl\,\big)_{\overline{n}}$$

or in a head-to-head, tail-to-tail fashion, resulting in chlorine atoms on adjacent carbon atoms, ie,

$$+CH_2CHClCHClCH_2CH_2CHClCHClCH_2\,\big)_{\overline{n}}$$

Table 3. PVC Physical Parameters

PVC property	Value			Reference
crystallographic data	orthorhombic, two monomer units/cell			
	a	b	c	
commercial PVC, nm	1.06	0.54	0.51	34
single crystal, nm	1.024	0.524	0.508	37
crystallinity, %				
as polymerized		19		
from melt		4.9		64
density (uncompounded), g/cc				
whole		1.39		65
crystallites		1.53		37
oxygen permeability, cc/(cm·s)cm^2 cm Hg	$238\,e^{-13.3/RT}$			66
Poisson ratio (rigid PVC)	0.41			
refractive index	1.54			67
glass-transition temperature, °C	83			
coefficient of linear thermal expansion (unplasticized), °C	7×10^{-5}			

specific heat	temp, °C	value, J/g °C[a]	
rigid PVC	23	0.92	68
	50	1.05	
	80	1.45	
	120	1.63	
plasticized PVC (50 phr DOP)	23	1.54	68
	50	1.67	
	80	1.75	
	120	1.88	

	Value	Reference
thermal conductivity (unplasticized), J/(cm·s)°C	17.5×10^{-4}	69
dielectric strength		
kV/mil	0.5	
kV/mm	20	
solubility parameter, (J/cm^3)$^{0.5}$	40.7 (av)	70

[a]To convert J to cal, divide by 4.184.

Dechlorination of head-to-head, tail-to-tail structure can be expected to go to 100% completion. If dechlorination of head-to-tail structure starts at random positions, then 13.5% of the chlorine should remain at the end of reaction. Dilute solutions of PVC treated with zinc removes 87% of the chlorine, proving the head-to-tail structure of PVC (71).

End Groups and Branching. Both saturated and unsaturated end groups can be formed during polymerization by chain transfer to monomer or polymer and by disproportionation. Some of the possible chain end groups are

$$-CH_3 \qquad -CH_2Cl \qquad -CHCl_2$$

$$-CH_2CHClCH=CHCl \qquad -CH_2CHClCCl=CH_2 \qquad -CH_2CHClCH=CH_2$$

PVC polymerization has a high chain-transfer activity to monomer; about 60% of the chains have unsaturated chain ends (72) and the percentage of chain ends containing initiator fragments is low (73). Chain transfer to polymer leads to branching. Branching in PVC has been measured by hydrogenating PVC, removing chlorine with lithium aluminum hydride. The ratio of methyl to methylene groups is measured by infrared spectroscopy using bands at 1378 and 1350, or 1370 and 1386/cm. Conventional PVC resins, made by mass or suspension polymerization at 50–90°C, contain 0.2 to 2 branches per 100 carbon atoms (65,73).

Stereoregularity. The addition of monomer fixes the tacticity of the previous monomer unit. Syndiotactic structure has the adjacent chlorine atoms oriented to opposite sides of the carbon–carbon–carbon plane, whereas isotactic structure has the adjacent chlorine atoms oriented to same side of the carbon–carbon–carbon plane. The potential energy for syndiotactic conformation is 4.2–8.4 kJ/mol (1–2 kcal/mol) lower than for isotactic conformation (74,75). Thus the ratio of propagation rates for syndiotactic to isotactic, k_s/k_i, must increase with decreasing temperature. Consequently, with decreasing polymerization temperature, the degree of syndiotacticity in PVC should increase. Measured amounts of syndiotacticity are illustrated in Figure 8 (76,77).

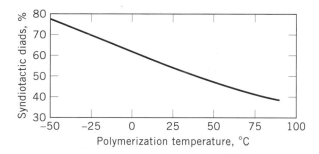

Fig. 8. The syndiotactic structure of PVC (76,77).

Polymerization Kinetics of Mass and Suspension PVC. The polymerization kinetics of mass and suspension PVC are considered together because a droplet of monomer in suspension polymerization can be considered to be a mass polymerization in a very tiny reactor. During polymerization, the polymer precipitates from the monomer when the chain size reaches 10–20 monomer units. The precipitated polymer remains swollen with monomer, but has a reduced radical termination rate. This leads to a higher concentration of radicals in the polymer gel and an increased polymerization rate at higher polymerization conversion.

Reactions in the liquid phase proceed as follows:

Initiation $M \xrightarrow{k_i} M^{\cdot}$

Propagation $R^{\cdot} + M \xrightarrow{k_p} R(M)^{\cdot}$

Chain transfer to monomer $R^{\cdot} + M \longrightarrow P + M^{\cdot}$

Termination $R^{\cdot} + R^{\cdot} \xrightarrow{k_t} P$

and reactions in the polymer gel are

Propagation	$(R^.) + M \longrightarrow (R(M)^.)$
Chain transfer to monomer	$(R^.) + M \longrightarrow (P) + M^.$
Termination	$(R^.) + (R^.) \longrightarrow (P)$

where $R^.$ is a polymer chain radical in liquid monomer; $(R^.)$, a polymer chain radical in the polymer gel phase; M, a monomer molecule; (M), a monomer molecule in the polymer gel phase; P, polymer in monomer; (P), polymer gel; and k_i, k_p, and k_t are reaction rate coefficients for initiation, propagation, and termination, respectively. Values for k_p and k_t at 60°C are 1.23×10^5 and 2.3×10^{10} L/(mol·s), respectively (78).

Polymerization in two phases, the liquid monomer phase and the swollen polymer gel phase, forms the basis for kinetic descriptions of PVC polymerization (79–81). The polymerization rate is slower in the liquid monomer phase than in the swollen polymer gel phase on account of the greater mobility in liquid monomer, which allows for greater termination efficiency. The lack of mobility in the polymer gel phase reduces termination and creates a higher concentration of radicals, thus creating a higher polymerization rate. Thus the polymerization rate increases with conversion to polymer.

Chain transfer to monomer is the main reaction controlling molecular weight and molecular weight distribution. The chain-transfer constant to monomer, C_m, is the ratio of the rate coefficient for transfer to monomer to that of chain propagation. This constant has a value of 6.25×10^{-4} at 30°C and 2.38×10^{-3} at 70°C and a general expression of $5.78\,e^{-2768/T}$. At 30°C, chain transfer to monomer happens once in every 1600 monomer propagation reactions; at 70°C, chain transfer happens once every 420 monomer additions (80,82–84). Thus temperature of polymerization strongly influences PVC molecular weight, ie, the molecular weight increases with lower polymerization temperature (Fig. 9).

PVC molecular weights are usually determined in the United States using inherent viscosity or relative viscosity measured according to ASTM D1243: 0.2 g/100 mL of cyclohexanone at 30°C. In Europe, K values are used, measured at 0.5% in cyclohexanone. The relationship among inherent viscosity, K value,

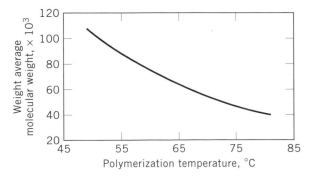

Fig. 9. PVC molecular weight as influenced by polymerization temperature.

number-average molecular weight (M_n), and weight-average molecular weight (M_w) for commercial grades of PVC is shown in Table 4.

Table 4. Commercial PVC Molecular Weights[a]

Inherent viscosity, ASTM D1234	Relative viscosity, ASTM D1234	K value, (DIN 53726)	$M_n \times 10^{-3}$	$M_w \times 10^{-3}$
0.42	1.09	45.0	15.0	30.0
0.47	1.10	47.1	18.0	36.0
0.52	1.11	49.3	20.0	40.0
0.57	1.12	51.3	22.5	45.0
0.62	1.13	53.6	25.0	50.0
0.67	1.14	56.1	27.5	55.0
0.73	1.16	58.2	30.5	61.0
0.78	1.17	60.5	33.0	67.0
0.83	1.18	62.9	36.0	72.0
0.88	1.19	64.9	38.5	78.0
0.92	1.20	67.1	41.0	82.5
0.98	1.22	69.2	44.0	89.5
1.03	1.23	71.5	47.0	95.0
1.08	1.24	73.3	50.0	101.0
1.13	1.25	74.9	52.5	107.5
1.21	1.27	77.5	57.0	117.0
1.30	1.30	80.7	62.5	128.5
1.40	1.32	83.8	68.5	141.0
1.60	1.38	90.8	81.0	168.0
1.80	1.43	96.7	93.5	195.0

[a]Ref. 85.

PVC Resin Manufacturing Processes

Mass Polymerization. Mass or bulk polymerization of PVC is normally difficult. At high conversions the mixture becomes extremely viscous, impeding agitation and heat removal, causing a high polymerization temperature and broad molecular weight distribution (86). A two-stage process that overcomes these problems was originally developed by Saint Gobain (France). The first stage of the process, which forms a skeleton seed grain for polymerization in a second stage, is carried out in a prepolymerizer with flat blade agitator and baffles to about 7–10% conversion. The number of grains remain constant throughout this polymerization (87).

In the second-stage polymerizer, a larger horizontal vessel and more monomer and initiator are added. This vessel is equipped with a slow-moving agitator blade running close to the vessel wall. The reaction proceeds through the liquid stage and at about 25% conversion becomes a powder. Heat removal is achieved, 30% by the jacket, 60% by a condenser, and 10% by the cooled agitator shaft (88). Unreacted monomer is removed by vacuum. Although the mass process saves drying energy, it has remained a minor process when compared to the suspension process.

Suspension Polymerization. Suspension polymerization is carried out in small droplets of monomer suspended in water. The monomer is first finely dispersed in water by vigorous agitation. Suspension stabilizers act to minimize coalescence of droplets by forming a coating at the monomer–water interface. The hydrophobic–hydrophilic properties of the suspension stabilizers are key to resin properties and grain agglomeration (89).

Kinetics of suspension PVC are identical to the kinetics of mass PVC, both increasing in rate with conversion (90). After polymerization to about 80–90% conversion, excess monomer is recovered, the slurry is steam-stripped in a column to a residual monomer level of about 0.0001% (10 ppm), excess water is centrifuged off, and the resin is dried with hot air.

Emulsion Polymerization.   Emulsion polymerization takes place in a soap micelle where a small amount of monomer dissolves in the micelle. The initiator is water-soluble. Polymerization takes place when the radical enters the monomer-swollen micelle (91,92). Additional monomer is supplied by diffusion through the water phase. Termination takes place in the growing micelle by the usual radical–radical interactions. A theory for true emulsion polymerization postulates that the rate is proportional to the number of particles [N]. N depends on the 0.6 power of the soap concentration [S] and the 0.4 power of initiator concentration [I]; the average number of radicals per particle is 0.5 (93).

However, the kinetics of PVC emulsion does not follow the above theory. The rate shows the same increasing behavior with conversion as mass polymerization (94,95). [N] depends on [S], but the relationship varies with the emulsifier type (96,97). However, the rate is nearly independent of [N] (95). The average number of radicals per particle is low, 0.0005 to 0.1 (95). The high solubility of vinyl chloride in water, 0.6 wt %, accounts for a strong deviation from true emulsion behavior. Also, PVC's insolubility in its own monomer accounts for such behavior as a rate dependence on conversion.

Emulsions of up to 0.2-μm dia are sold in liquid form for water-based paints, printing inks, and finishes for paper and fabric. Other versions, 0.3–10-μm dia and dried by spray-drying or coagulation, are used as plastisol resins. Plastisols are dispersions of PVC in plasticizer. Heat allows fast diffusion of plasticizer into the PVC particle, followed by fusion (gelation), to produce a physically cross-linked elastomer, where the physical cross-links are PVC crystallites.

Microsuspension Polymerization.   Whereas emulsion polymerization uses a water-soluble initiator, microsuspension polymerization uses a monomer-soluble initiator. The monomer is homogenized in water along with emulsifiers or suspending agents to control the particle sizes. Microsuspension paste resins at 0.3–1-μm dia are used to make plastisols for flooring, seals, barriers, etc. These plastisols are also dispersions of PVC in liquid plasticizer and are cured by heating. Heating allows plasticizer to diffuse uniformly into the PVC particles; at higher temperatures, the plasticized particles fuse. Microsuspension blending resins at 10–100-μm dia are used as extenders to paste resins in plastisols (98).

Solution Polymerization.   In solution polymerization, a solvent for the monomer is often used to obtain very uniform copolymers. Polymerization rates are normally slower than those for suspension or emulsion PVC. For example, vinyl chloride, vinyl acetate, and sometimes maleic acid are polymerized in a solvent where the resulting polymer is insoluble in the solvent. This makes a uniform copolymer, free of suspending agents, that is used in solution coatings (99).

Copolymerization. Vinyl chloride can be copolymerized with a variety of monomers. Vinyl acetate [*9003-22-9*], the most important commercial comonomer, is used to reduce crystallinity, which aids fusion and allows lower processing temperatures. Copolymers are used in flooring and coatings. This copolymer sometimes contains maleic acid or vinyl alcohol (hydrolyzed from the poly(vinyl acetate)) to improve the coating's adhesion to other materials, including metals. Copolymers with vinylidene chloride are used as barrier films and coatings. Copolymers of vinyl chloride with maleates or fumerates are used to raise heat deflection temperature. Copolymers of vinyl chloride with acrylic esters in latex form are used as film formers in paint, nonwoven fabric binders, adhesives, and coatings. Copolymers with olefins improve thermal stability and melt flow, but at some loss of heat-deflection temperature (100). Copolymerization parameters are listed in Table 5.

Table 5. Copolymerization Parameters of Vinyl Chloride

M_2	r_1	r_2	e	Q	Temperature, °C	Ref.
acrylic acid	0.107	6.8	0.77	1.15	60	101
acrylonitrile	0.04	2.7	1.20	0.60	60	102
butadiene	0.035	8.8	−1.05	2.39	50	103
butene	3.4	0.21				104
n-butyl acrylate	0.07	4.4	1.06	0.50	45	101, 105
diethyl fumarate	0.12	0.47	1.25	0.61	60	106
dimethyl itaconate	0.053	5.0	1.34	1.03	50	107
diethyl maleate	0.77	0.009				104
ethylene	3.21	0.21	−0.20	0.015	50	108
ethylhexyl acrylate	0.16	4.15				104
isobutylene	2.05	0.08	−0.96	0.033	60	109, 110
isoprene			−1.22	3.33		
maleic anhydride	0.296	0.008	2.25	0.23	75	111
methacrylic acid	0.034	23.8	0.65	2.34	60	101
methacrylonitrile			0.68	0.86	60	112
methyl acrylate	0.12	4.4	0.60	0.42	50	113
methyl methacrylate	0.1	10	0.4	0.74	68	114
octyl acrylate	0.12	4.8	1.07	0.35	45	105
propylene	2.27	0.3	−0.78	0.002		115, 110
styrene	0.02	17	−0.80	1.0	60	116
vinyl acetate	1.68	0.23	−0.22	0.026	60	117
N-vinylcarbazole	0.17	4.8	−1.40	0.41	50	118
vinyl chloride			0.20	0.044		
vinyl laurate	7.4	0.2				104
vinylidene chloride	0.3	3.3	0.36	0.22	60	119
vinyl isobutyl ether	2.0	0.02			50	120
N-vinylpyrrolidone	0.53	0.38	−1.14	0.14	50	121

Compounding

The additives found in PVC help make it one of the most versatile, cost-efficient materials in the world. Without additives, literally hundreds of commonly used PVC products would not exist. Many materials are useless until they undergo a similar modification process. Steel, for instance, contains among other things chromium, nickel, and molybdenum. PVCs are tailored to the requirements using sophisticated additives technology.

Stabilizers. Lead stabilizers, particularly tribasic lead sulfate, is commonly used in plasticized wire and cable compounds because of its good nonconducting electrical properties (122).

Organotin stabilizers are commonly used for rigid PVC, including pipe, fittings, windows, siding profiles, packaging, and injection-molded parts. These repair unstable sites on PVC, removing unstable chlorine and replacing it with a ligand from the tin stabilizer molecule (123–125). This produces stability at least an order of magnitude better than without stabilizer. Examples of effective tin compounds are dialkyl tin dilaurate and mono- and dialkyl tin diisooctylthioglycolate. Certain grades of methyl tins and octyl tins are used in food contact applications.

Antimony tris(isooctylthioglycolate) has found use in pipe formulations at low levels. Its disadvantage is that it cross-stains with sulfide-based tin stabilizers (122). Barium–zinc stabilizers have found use in plasticized compounds, replacing barium–cadmium stabilizers. These are used in moldings, profiles, and wire coatings. Cadmium use has decreased because of environmental concerns surrounding certain heavy metals.

Calcium–zinc stabilizers are used in both plasticized PVC and rigid PVC for food contact where it is desired to minimize taste and odor characteristics. Applications include meat wrap, water bottles, and medical uses.

Many stabilizers require costabilizers. Several organic costabilizers are quite useful with barium–zinc and calcium–zinc stabilizers, eg, β-diketones, epoxies, organophosphites, hindered phenols, and polyols (122).

Impact Modifiers. In the early days of plastics, many unplasticized PVC products were brittle. This gave plastics a cheap reputation. It was therefore quite desirable to develop technology to produce tough plastics. In the early 1950s, The Geon company (then a part of BFGoodrich) began adding rubbery polymers to PVC to improve toughness (126). Rubbery particles act as stress concentrators or multiple weak points, leading to crazing or shear-banding under impact load (127). This can result in cavitation and/or cold drawing, thus allowing the PVC to absorb large amounts of energy. Impact modifier choices are listed in Table 6.

Processing Aids. PVC often flows in the form of billions of molecule primary particles. Processing aids glue these particles together before the PVC melts, thus acting as a fusion promoter. Processing aids also modify melt rheology by increasing melt elasticity and die swell; some by reducing melt viscosity and melt fracture. Some processing aids affect dispersion of fillers, impact modifiers, and pigments (129); others lubricate to reduce PVC sticking to metal. The most common processing aids are high molecular weight acrylics based primarily on polymethylmethacrylate copolymers.

Lubricants. Lubricants are often classified as internal or external (130–134). Internal lubricants were considered either to be soluble in PVC, to have little effect on fusion, or to be capable of reducing melt viscosity; external lubricants were considered to retard fusion or to promote metal release (see LUBRICATION AND LUBRICANTS). This system of classifying lubricants has too

Table 6. Impact Modifiers Used for PVC Applications[a,b]

Use	acr	m-acr	ABS	CPE	EVA	MBS	MABS
siding	+	+		+	+		
windows	+	+		+	+		
gutters	+	+		+	+		
pipe and conduit	+	+	+	+	+	+	+
irrigation pipe	+	+		+	+		
fittings	+	+	+	+	+	+	+
interior trim	+	+	+	+	+	+	+
foam	+						
interior profiles	+	+	+	+	+	+	+
clear film			+			+	+
clear bottles			+			+	+
credit cards	+	+	+	+	+	+	+
furniture	+	+		+	+		
appliances	+		+			+	
housings	+		+			+	

[a]Ref. 128.
[b]acr = all acrylic; m-acr = modified acrylic; ABS = acrylonitrile–butadiene–styrene; CPE = chlorinated polyethylene; EVA = ethylene–vinyl acetate; MBS = methyl methacrylate–butadiene–styrene; and MABS = methacrylate–acrylonitrile–butadiene–styrene.

many conflicting measurements to be consistent and useful. Others have shown classifications based on synergy between various lubricants (135,136) but did not explain the nature of that synergy. A model for the lubrication mechanism has been developed that explains synergy between certain lubricants (62). This model treats lubricants as surface-active agents. Thus, some lubricants have polar ends that are attracted to other polar ends and to polar PVC flow units and to polar metal surfaces. These also have nonpolar ends that are repelled by the polar groups. Synergy happens when nonpolar lubricants are added, which are attracted to the nonpolar ends and act as a slip layer. This model is shown in Figure 10.

Plasticizers. It was found in 1926 that solutions of PVC, prepared at elevated temperatures with high boiling solvents, possessed unusual elastic properties when cooled to room temperature (137). Such solutions are flexible, elastic, and exhibit a high degree of chemical inertness and solvent resistance.

This unusual behavior results from unsolvated crystalline regions in the PVC that act as physical cross-links. These allow the PVC to accept large amounts of solvent (plasticizers) in the amorphous regions, lowering its T_g to well below room temperature, thus making it rubbery. PVC was, as a result, the first thermoplastic elastomer (TPE). This rubber-like material has stable properties over a wide temperature range (32,138–140).

A few plasticizers impart specific properties for particular applications. For example, citrate esters are used in food contact applications, benzoates are used for stain resistance, and chlorinated hydrocarbons impart flame resistance and good electrical properties. Aliphatic diesters offer good low temperature flexibility; linear alcohol-based phthalates offer good low temperature flexibility and also have reduced volatility; phosphates improve flame resistance; trimellitates have low volatility, are used for high temperature applications, and also have

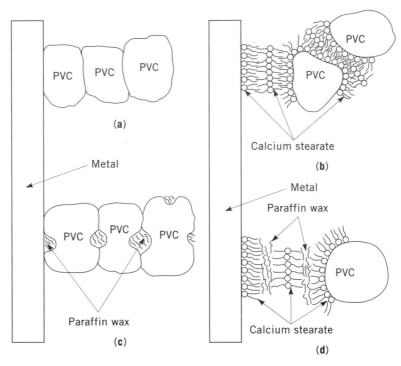

Fig. 10. A model of PVC lubrication mechanism showing (**a**) PVC adhesion to metal without lubricant; (**b**) surface activity of calcium stearate; (**c**) nonmetal releasing character of paraffin only; and (**d**) synergy between calcium stearate and paraffin (62).

good low temperature properties. Polymeric plasticizers do not migrate easily but suffer from poor low temperature flexibility. Epoxy plasticizers are also good plasticizers that have low volatility and act as costabilizers, improving the thermal stability of PVC. Commonly used PVC plasticizers include the following.

Plasticizer	Abbreviation
aliphatic ester	
di(2-ethylhexyl) adipate	DOA
di(2-ethylhexyl) azelate	
di(2-ethylhexyl) sebacate	
phthalate	
di(2-ethylhexyl)	DOP or DEHP
diisooctyl	DIOP
diisodecyl	DIDP
butylbenzyl	BBP
butyloctyl	BOP
diisononyl	DINP
ditridecyl	DTDP
diundecyl	DUP
linear C7-C11	711 phthalate
di(2-ethylhexyl) terephthalate	DOTP

phosphates
    trioctyl   TOP
    cresyl diphenyl   CDP
    tricresyl   TCP
    triphenyl
    tri(2-ethylhexyl)   TEHP
trimellitates
    tris(2-ethylhexyl)   TOTM
    triisooctyl   TIOTM
epoxies
    epoxidized soybean oil   ESO
    epoxidized linseed oil
    epoxy stearate
    2-ethylhexyl epoxytallate

Plasticizers and stabilizers in particular have been researched at length to determine their potential impact on human health and the environment. DEHP (di-2-ethylhexylphthalate) has been used worldwide in applications such as blood bags, saline solutions, meat wraps, and other highly credible uses. However, there has been much debate over that impact because of the differing methods used to evaluate them. Although the U.S. National Toxicology Program and the International Agency for Research on Cancer have classified the plasticizer DEHP as a possible human carcinogen, their methodologies have been criticized for potentially inaccurately ascribing results obtained with rodents to humans (141,142). Mechanistic studies indicate that the carcinogenic response which DEHP produces in rodents is directly related to physiologic and metabolic changes that are specific to that species. Because the evidence indicates the response is an artifact to that species and not a true indication of human hazard, a number of regulatory bodies do not consider DEHP to pose a hazard to humans. The Specialized Experts Working Group of the European Commission, for instance, has concluded that there is no evidence to warrant the classification of DEHP as a carcinogen (143). DEHP is not regulated as a carcinogen by the U.S. Food and Drug Administration, which has long governed the plasticizer's use in medical devices and in food contact applications. DEHP-plasticized PVC is used in medical applications like blood bags where it is known to protect red blood cells from deterioration. Flexible PVC film is considered the most desirable material for wrapping meats, because it is oxygen-permeable, maintains the bright red color needed to make meats salable to consumers, and extends the shelf life of meats.

Fillers. Fillers are used to improve strength and stiffness, to lower cost, and to control gloss. The most common filler is calcium carbonate, which ranges in size from 0.07 to well over 50 μm. Some forms are treated with a stearic acid coating. Clay fillers, such as calcined clay, improve electrical properties. Glass fibers, talc, and mica improve tensile strength and stiffness, but at a loss in ductility.

Pigments. A variety of pigments are added to PVC to give color, including titanium dioxide and carbon black.

Ultraviolet Light Stabilizers.   One form of stabilization is to absorb the ultraviolet light. Both titanium dioxide and carbon black are strong ultraviolet light absorbers and effective in protecting the PVC. Carbon black is a stronger absorber than titanium dioxide and can therefore be used at lower levels in PVC for protection. For ultraviolet light absorption in transparent PVC or for improvement of pigmented systems, various derivatives of benzotriazole are used, such as 2-(2'-hydroxy-3',5'-(di-t-butyl)phenyl)benzotriazole. Where tin carboxylate stabilizers are used instead of tin mercaptide stabilizers, hindered amine light stabilizers, particularly with ultraviolet absorbers, are effective (144).

Biocides.   Although PVC itself and rigid PVC compounds are resistant to attack by microorganisms, plasticized PVC, in specific applications such as flashing and sealing boots on roofs, shower curtains, and swimming pools, may need protection. Many biocides, often containing arsenic compounds, are available for a balance of stability, compatibility, weatherability, and biocidal effectiveness.

Flame Retardants.   Because PVC contains nearly half its weight of chlorine, it is inherently flame-retardant. Not only is chlorine not a fuel, but it acts chemically to inhibit the fast oxidation in the gas phase in a flame. When PVC is diluted with combustible materials, the compound combustibility is also increased. For example, plasticized PVC with >30% plasticizer may require a flame retardant such as antimony oxide, a phosphate-type plasticizer, or chlorinated or brominated hydrocarbons (145,146).

Foaming or Blowing Agents.   Cellular PVC can be made by a variety of techniques, such as whipping air into a plastisol, incorporating a gas under pressure, incorporating a physical blowing agent into the melt, or using a chemical blowing agent which releases a gas when it decomposes with heat.

The most common chemical blowing agent is 1,1'-azobisdicarbonamide, which decomposes with heat to release nitrogen gas. Typically, the closed-cell foams of rigid PVC range down to a density of 0.4 g/cc. Physical blowing agents, such as chlorofluorocarbons, which volatilize without changes in the chemical bonds, are capable of producing foams at down to 0.03 g/cc when used with a copolymer PVC. Because of the damage to the ozone layer in the stratosphere, not all chlorofluorocarbons are acceptable. Newer types of physical blowing agents are becoming available that can minimize ozone depletion.

Economic Aspects

Compound Manufacture.   All PVC must be compounded before used. For the wide variety of requirements where PVC performs, it is usually not feasible to maintain a technical compounding knowledge at the user level. Often processors and original equipment manufacturers (OEMs) purchase compound already balanced to meet the range of properties required and to do the best job economically. Some resin manufacturers are integrated to supply these compounds.

Suppliers of merchant compound sold in North America, as well as their approximate 1995 market share, are as follows (147): Geon, at 24%; Technor Apex, at 10%; Synergistics, at 9%; Georgia Gulf, at 9%; K-Bin, at 8%; Oxychem, at 7%; Vista, at 6%; AlphaGary, at 6%; N. American, at 6%; and eighteen other companies, at 15%. Geon, Georgia Gulf, K-Bin, Oxychem, and Vista are all integrated PVC resin and compound manufacturers. The total volume is not publicly available. These compounds are available in bulk railcars containing 72 metric tons (160,000 lb), in bulk trucks containing 20 t (44,000 lb), in boxes containing 600 kg (1300 lb), or in bags containing 23 kg (50 lb).

For businesses specializing in a large market with a limited product line, it can be economical to streamline the process, from compound manufacture to melt extrusion, to such an extent that compounding can be integrated into the process. Pipe and siding manufacturing are such processes where compounding is integrated in with the extrusion process. Other businesses that integrate compounding to some extent are window manufacturing, barrier and packaging film manufacturing, and wire and cable manufacturing.

PVC Resin Manufacture. PVC resins are manufactured and sold for a wide variety of applications. About 6.5×10^6 tons (14.5×10^9 lb) of resin are manufactured in North America and 24×10^6 tons (53×10^9 lb) worldwide. The 1995 North American PVC resin capacity (1,147,148), at 2.0×10^6 tons (4.5×10^9 lb), is made up of SHINTECH, at 19%; Geon, at 16%: FPC, at 14%; Oxychem, at 10%; Borden, at 9%; Georgia Gulf, at 8%; Vista, at 7%; Westlake, at 4%; Primex, at 3%; Esso, at 2%; Policyd, at 2%; Goodyear, at 1%; and six others, at 5%. The 1995 world capacity (1,147,148), at 7.4×10^6 tons (16.4×10^9 lb), is comprised of Asia, at 30%; North America, at 27%; Western Europe, at 24%; Eastern Europe, at 9%; South America, at 4%; the Middle East, at 3%; Africa, at 2%; and Australia, at 1%. PVC resins are available in bulk railcars containing 72 tons (160,000 lb), in bulk trucks containing 20 tons (44,000 lb), in boxes containing 600 kg (1300 lb), or in bags containing 23 kg (50 lb).

World demand is 22×10^6 tons (49×10^9 lb) in 1995, with the growth accelerating (Fig. 11). Per-capita demand for PVC is high in Canada, United States, Japan, and Western Europe, and shows lots of room for growth in Africa, the Middle East, Asia, Mexico, South America, and Eastern Europe (Fig. 12).

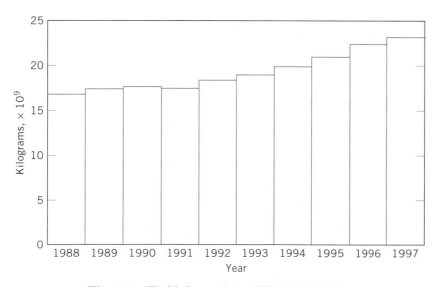

Fig. 11. World demand for PVC (1,147,148).

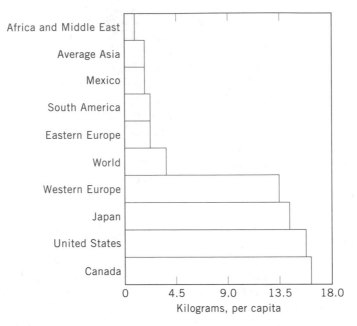

Fig. 12. Per-capita demand for PVC by region (1,147,148).

PVC Resin Price.   PVC resin prices tend to be more stable than other plastics' prices, partly because only about half the molecule is based on hydrocarbon raw material sources. Figure 13 compares PVC prices to other plastics' prices; Figure 14 illustrates PVC prices over several years.

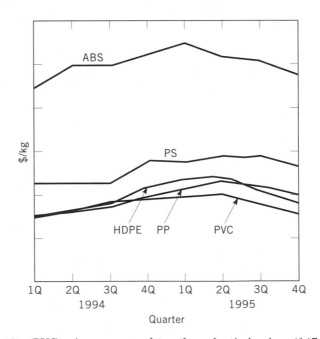

Fig. 13. PVC prices compared to other plastics' prices (147,148).

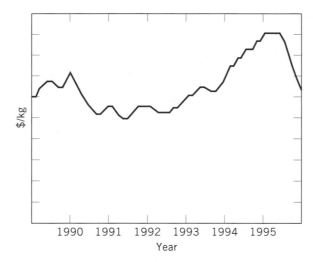

Fig. 14. PVC prices from 1989 to 1996 (147,149).

Uses

Numerous specifications and standards are used to define PVC resins and compounds. Table 7 summarizes a few of these standards.

PVC is so versatile that it can be compounded for a wide range of properties and used in a wide variety of markets. Most of the products are durable goods and have long life spans. Its use in short-term, one-time-use products is limited. A list of all principal uses is found in Table 8.

Pipe and Fittings. Pipe and fittings, a principal market for PVC, are a prime example of PVC as an engineering thermoplastic. These applications are designed with PVC for long-term satisfactory performance and are highly optimized for efficient production at high rates and minimal costs. Pipe manufacturing uses powder compounds, twin-screw extruders, and vacuum sizing/cooling. Because of the volume of products, the process is highly developed. The products, designed to meet appropriate ASTM standards, include pressure pipe of various sizes and drain, waste, and vent applications.

Chlorinated PVC (CPVC) is used in higher temperature applications such as hot-water piping. Because of its superior creep resistance, CPVC is also used in automated fire-safety sprinkler systems.

Weatherable Siding, Windows, and Doors. PVC is accepted commercially as an excellent weathering material. Plastic materials are damaged by the sun, particularly by ultraviolet light. PVC's chemical response to weathering is well understood so that compounds and products can be designed for satisfactory outdoor performance. The mechanism of degradation starts by absorbing the sun's damaging ultraviolet light. Because this absorption is affected by PVC's previous thermal degradation during processing (150,151), it is important to avoid thermal damage to PVC during processing. The absorbed light breaks bonds and forms free radicals, which leads to loss of hydrogen chloride, yellowing, and, at the same time, oxidative bleaching (removal of yellowing). With a proper

Table 7. Specifications and Standards for PVC Resins and Compounds

Standard	Description
ASTM D1755	PVC resin cell classification
ASTM D1784	PVC rigid compound cell classification
ASTM D1785	PVC pipe, schedule 40, 80, 120
ASTM D2287	PVC nonrigid (plasticized) compound cell classification
ASTM D2464	PVC THR fittings
ASTM D2466	PVC socket fittings, schedule 40
ASTM D2467	PVC socket fittings, schedule 80
ASTM D2665	DWV pipe and fittings
ASTM D2729	PVC sewer pipe and fittings
ASTM D2740	PVC tubing
ASTM D2846	CPVC hot water distribution system
ASTM D2949	PVC DWV 7.62 cm (3 in.) thin wall
ASTM D3033	PVC PSP sewer pipe and fittings
ASTM D3034	PVC PSM sewer pipe and fittings
ASTM D4216	PVC rigid building products compounds
NSF 14	PVC potable water, fittings
UL 62	PVC wire insulation
UL 83	PVC T, TW, THW, THWN, THHN wire insulation
UL 444	PVC telecommunications wire insulation jacket
UL 719	PVC NM jacket
UL 758	PVC AWM wire insulation
UL 1272	PVC TC cable jacket

Table 8. U.S. Markets, 1995[a]

Market	Metric tons
construction	
flooring	120,000
pipe and conduit	2,070,000
wire and cable	180,000
siding	640,000
windows and doors	140,000
extrusions	240,000
packaging	
bottles	80,000
extrusions	160,000
calendering	390,000
custom moldings	50,000
paste processes	100,000
textiles and coating	70,000
exports and resale	850,000
other	60,000
Total	*5,360,000*

[a]Ref. 1.

level of titanium dioxide to protect the PVC from the absorption of ultraviolet light (152), the right choice of the type of titanium dioxide to control the rate of the oxidation and bleaching process, and a correct use of other weather-stable ingredients, a well-designed PVC formulation can be very durable to the weather.

As with pipe, siding manufacturing is highly optimized for efficient production at high rates and minimal costs. Siding uses powder compounds, twin-screw extruders, and vacuum sizing/cooling.

Other products, such as windows, may or may not use the powder compound, twin-screw extruder, and vacuum-sizing approach. For windows, this is a much more complex operation than for pipe, requiring large investments to develop and operate the process. Cubed compound, where the PVC grains are already broken down, can be run faster on simpler single-screw extruders because lower melt temperatures are typical. Because elastic swell is reduced at lower melt temperatures, die design is simpler. Either vacuum sizing or air sizing/cooling is possible. The products are designed to meet appropriate ASTM standards. Products include siding, soffits, gutters and down spouts, windows (including all-vinyl windows and vinyl-protected wood windows), and door-glazing applications and garage doors.

Profiles. Complex profiles require specialty manufacturing skills to build, maintain, and operate extrusion dies as well as cooling and sizing equipment to deliver the exact dimensions required. Cubed compound, where the PVC grains are already broken down, can be run faster on simple single-screw extruders on account of the typical low melt temperatures. Because elastic swell is reduced at lower melt temperatures, die design is simpler and is often the same dimensions as the desired profile. Either vacuum sizing or air sizing/cooling is possible.

The low elastic swell, unique to PVC, results from the presence of the billion molecule flow units. Other plastics have higher die swell, which increases at lower melt temperatures (153). These plastics thus pose much more difficulty in designing dies and in maintaining a process to hold dimensions.

Wire and Cable. PVC has been used in wire and cable applications since World War II, when the U.S. Navy demanded lower combustibility materials in construction. These products are manufactured by cross-head extrusion, usually from pellet compounds on single-screw extruders. Some line speeds are 1524 m (5000 ft) per minute (60 mph). The compounds are optimized for the requirements, including low temperature flexibility, high use temperature, especially low combustibility, weatherability, and high resistance to cutthrough.

Injection-Molded Products. Numerous housings, electrical enclosures, and cabinets are injection-molded from rigid PVC. These take advantage of PVC's outstanding UL flammability ratings and easy molding into thin-walled parts. PVC has developed melt flow capabilities to the point where it competes with essentially any other flame-retarded engineering thermoplastic and molds easier than most.

Pipe fittings require quite high tensile and creep resistance; they are thus molded from compounds that have less melt flow than for thin-walled housings. Plasticized compounds are injection-molded into a variety of parts requiring elastomeric properties.

Health and Safety Factors and Toxicology

There are no significant health hazards arising from exposure to poly(vinyl chloride) at ambient temperature (154–158). However, a British study has found a small decrease in breathing capacity for workers who smoked and were exposed to vinyl resin dust (159). This decrease was about one-seventh of that caused by normal aging and about equal to that expected with a one-pack-a-day cigarette smoker.

Because routine inhalation of dust of any kind should be avoided, reduction of exposure to poly(vinyl chloride) dust may be accomplished through the utilization of care when dumping bags, sweeping, mixing, or performing other tasks that can create dust. The use of an approved dust respirator is recommended where adequate ventilation may be unavailable.

At processing temperatures, most polymers emit fumes and vapors that may be irritating to the respiratory tract. This is also true for PVC and its additives. Such irritation may extend to the skin and eyes of sensitive people. Processing emissions exposure can also be greatly reduced or eliminated by the use of properly designed and maintained exhaust ventilation.

Decomposition of plastics, eg, through greatly elevated temperatures above normal operating temperatures, can result in personnel exposure to decomposition or combustion products. In the case of PVC compounds, such decomposition involves hydrogen chloride, which causes irritation of the respiratory tract, eyes, and skin. Depending on the severity of exposure, physiological response can be coughing, pain, or inflammation of the respiratory tract. Fortunately, the pungent odor of hydrogen chloride provides an excellent warning signal, causing exposed personnel to be driven from the area which prohibits long-term exposure. The odor of hydrogen chloride is detectable as low as 1–5 ppm.

Fire and Explosion.   Poly(vinyl chloride) resin has a flash point of approximately 391°C (735°F) and a self-ignition temperature of approximately 454°C (850°F) (ASTM D1929). In general, PVC burns with difficulty because a substantial amount of energy is required to break down the polymer into smaller fragments that can sustain combustion in the gas phase, principally as a consequence of the action of the halogen content of the material. Consequently, PVC is difficult to ignite. Fires tend to extinguish naturally in the absence of a substantial external source of heat or flame. Because hydrogen chloride is generated during combustion, this action serves as a flame-quencher in the vapor phase. Poly(vinyl chloride) releases less heat than many other combustible materials. Precautions should be taken similar to those for other combustible materials, eg, wood or other plastics.

Poly(vinyl chloride) powder has a very low tendency to explode. The minimum ignition energy for explosion is much higher than that of natural materials such as corn starch and flour and also exceeds those of other plastic materials. However, as with any powder materials, care should be taken in addressing ignition sources in working and handling areas if dusting should occur. In addition, walkways and floors should be cleared of PVC dust to prevent slippery footing.

In firefighting where PVC is involved, water, ABC dry chemical, or protein-type air foams should be used as extinguishing media. Carbon dioxide may be ineffective on larger fires because of lack of cooling capacity, which may result in reignition. Firefighters should utilize a self-contained breathing apparatus (SCBA) in positive-pressure mode. In addition, for an enclosed or poorly ventilated area, an SCBA should be worn during cleanup immediately after a fire as well as during the attack phase of firefighting operations.

PVC should not be melt-mixed with acetal polymers. These polymers are chemically incompatible; mixing could cause rapid decomposition and gas evolution.

Toxicology. Toxicology studies have shown poly(vinyl chloride) to be equivocally tumorigenic through oral and implant studies. The International Agency for Research on Cancer (IARC) shows inadequate evidence that poly(vinyl chloride) is carcinogenic in animals or humans and has an overall evaluation of 3 (not classifiable). The United States has no occupational exposure limits for PVC except as particulates not otherwise classified (PNOC). For PNOCs, the American Conference of Governmental Industrial Hygienists (ACGIH) has a threshold limit value (TLV) of 10 mg/m^3 for inhalable and 3 mg/m^3 for respirable, whereas the Occupational Safety and Health Act (OSHA) permissible exposure limits (PELs) are 15 mg/m^3 for total dust and 5 mg/m^3 for the respirable fraction.

Poly(vinyl chloride) is listed on the TSCA inventory and the Canadian Domestic Substances List (DSL) as ethene, chloro-, homopolymer [9002-86-2]. Because polymers do not appear on the European Community Commercial Chemical Substances listing or EINECS, poly(vinyl chloride) is listed through its monomer, vinyl chloride [75-01-4]. In the United States, poly(vinyl chloride) is an EPA hazardous air pollutant under the Clean Air Act Section 112 (40 CFR 61) and is covered under the New Jersey Community Right-to-Know Survey: N.J. Environmental Hazardous Substances (EHS) List as "chloroethylene, polymer" with a reporting threshold of 225 kg (500 lb).

Environmental Considerations and Recycling

Chlorine. Chlorine, the material used to make PVC, is the 20th most common element on earth, found virtually everywhere, in rocks, oceans, plants, animals, and human bodies. It is also essential to human life. Free chlorine is produced geothermally within the earth, and occasionally finds its way to the earth's surface in its elemental state. More usually, however, it reacts with water vapor to form hydrochloric acid. Hydrochloric acid reacts quickly with other elements and compounds, forming stable compounds (usually chloride) such as sodium chloride (common salt), magnesium chloride, and potassium chloride, all found in large quantities in seawater.

The chlorides found in common salt water are an essential element in all body fluids and, in this form, make up about 0.15% of total body weight. A large number of complex organic chlorides and organochlorines are naturally produced chemicals, widely present in nature, and play many essential roles (160–163). Chlorine is also an essential element of naturally occurring antibacterial and antifungal agents such as chlortetracycline, chloramphenicol, and griseofulvin, which have revolutionized the treatment of human bacterial and fungal infections.

Even more complex cyclic organochlorines, including dioxins and furans, are produced from burning wood and other vegetable matter, and are natural by-products of forest fires (164,165). Chlorine-based chemicals are everywhere, and have been so since before the existence of mankind. Both dioxins and furans have been found in lake sediments dating back to 1860 (166), and samples taken from ice cores in Greenland dating back to 1869 showed definite spikes in chlorine content correlating with volcanic activity (167). Given the fact that chlorine compounds are naturally produced in such vast quantities, and so widely distributed in the natural world, banning production of chlorine to keep chlorine compounds out of the environment would be futile, extremely costly, and would deprive the world of hundreds of products critical to society's health and well-being. And the environment would be worse off.

Over 30% of the chlorine produced on a global basis goes to make PVC. Not only is chlorine essential to the chemical composition of PVC, it provides a number of unique properties that give this versatile plastic a distinct advantage in product applications and the marketplace. It makes PVC inherently flame-retardant. PVC is the world's leading electrical material, with over 250×10^3 t (500×10^6 lb) used annually for wire and cable insulation and sheathing, electrical conduit, boxes, and components. PVC is over 50% chlorine and, as a result, one of the most energy-efficient polymers. Chlorine makes PVC far more environmentally acceptable than other materials that are totally dependent on petrochemical feedstocks. In addition, recycling PVC is easier because the chlorine in PVC acts as a marker, enabling automated equipment to sort PVC containers from other plastics in the waste stream (168).

Vinyl Solid Waste and Recycling. Although vinyl is the world's second most widely used plastic, less than one-half percent by weight is found in the municipal solid waste stream. Most of that consists of vinyl packaging, bottles, blister packaging, and flexible film. This is because most vinyl applications are long-term uses, such as pipe and house siding, and are not disposed of quickly. Vinyl wastes are handled by all conventional disposal methods, ie, recycling, landfilling, and incineration (including waste-to-energy).

Vinyl is recycled by at least 170 recyclers in the United States and Canada. In 1994, about 2.9×10^3 t (6.5×10^6 lb) of post-consumer vinyl were recycled in the United States. An estimated additional 135×10^3 t (300×10^6 lb) of vinyl post-industrial scrap was diverted from landfills and recycled into second-generation products. More than 3500 communities accept vinyl products in their recycling programs, which is about 25% of all communities that recycle.

In landfills, vinyl wastes, like all plastics, are extremely resistant to decomposition. In fact, high technology landfills are often lined with thick-gauge vinyl and use PVC pipe to handle liquid leachate and methane gas for environmental protection.

Vinyl compares favorably to other packaging materials. In 1992, a lifecycle assessment comparison of specific packages made from glass, paperboard, paper, and selected plastics concluded that vinyl was the material that has the lowest production energy and carbon dioxide emissions, as well as the lowest fossil fuel and raw material requirements of the plastics studied (169). Vinyl saves more than 34 million Btu per 1000 pounds manufactured compared to the highest energy-consuming plastic (170).

Incinerating PVC Wastes. A study (171) sponsored by the American Society of Mechanical Engineers (ASME), involving the analysis of over 1700 test results from 155 large-scale, commercial incinerator facilities throughout the world, found no relationship between the chlorine content of waste and dioxin emissions from combustion processes. Instead, the study stated, the scientific literature is clear that the operating conditions of combustors are the critical factor in dioxin generation. This work includes and confirms a number of other studies, most notably, the work conducted in 1987 by the New York Energy Research and Development Authority (172). Those tests revealed that the presence or absence of PVC had no effect on the amount of dioxin produced during the incineration process.

Incinerator scrubbing systems can remove about 99% of the hydrogen chloride generated by incinerating vinyl plastics and other chlorine-containing compounds and materials (173). New requirements from the U.S. Environmental Protection Agency make scrubbers mandatory on all incinerators so that they can neutralize a range of acid gases, including sulfur dioxide and nitrogen oxide, which are produced by a variety of materials. Because acid generated in incinerators comes from a variety of sources, including table salt and paper products, scrubbers are necessary whether or not PVC is present in the waste feed (174). A more recent study, conducted by Midwest Research Institute and published by the ASME, concluded that removing vinyl from the waste stream would not eliminate the need for air pollution control devices and monitoring equipment, nor would it influence the choice of incineration equipment (175).

Municipal incinerators are often targeted as a primary cause of acid rain. In fact, power plants burning fossil fuels, which produce sulfur dioxide and nitrogen oxide, are actually the leading cause of acid rain, along with automotive exhaust (176,177). In Europe and Japan, studies show that only about 0.02% of all acid rain can be traced to incineration of PVC (178).

There are many common misconceptions about vinyl. For example, the idea that vinyl is not recycled is untrue. Industrial scrap vinyl has been recycled for years, but in more recent years, post-consumer vinyl recycling is growing, too, with about 3.25×10^3 tons (6.5×10^6 lb) of post-consumer vinyl (primarily bottles) being recycled in the 1990s. When the Council for Solid Waste Solutions (now the American Plastics Council) conducted a nationwide survey in 1991, it found that there were an estimated 1100 municipal recycling programs in place or planned in the United States that include vinyl.

Nor is it true that there is no market for recycled vinyl. In 1989, the University of Toledo identified nearly 100 uses for recycled vinyl. Overall, the potential demand for recycled vinyl is estimated to be over twice the potential supply of all vinyl bottles produced in the United States each year (247×10^3 tons needed vs 103.5×10^3 tons available via recycling of bottles). A more recent directory published by the Vinyl Institute lists nearly 50 companies that make commercial products out of recycled vinyl (179).

It is not true either that vinyl is the problem in municipal recycling because it contaminates other resins. Contamination occurs whether or not vinyl is present. Other resins are just as much a contamination problem as vinyl. Except for commingled plastics applications, different plastic materials cannot be mixed successfully in most recycled products applications. This is why it is crucial to separate efficiently one plastic from another. Because of the chlorine that is present in it, vinyl lends itself very well to automated sorting technology.

Nor is it true that poly(ethylene terephthalate) (PET) and high density polyethylene (HDPE) packaging are listed as 1 and 2 in the Society of the Plastics Industry (SPI) recycling coding system because they are the most recyclable. The numbers assigned to each plastic in the SPI coding system are purely arbitrary and do not reflect the material's recyclability.

The misconception that vinyl gives off dioxin when it is incinerated is misleading. A study conducted by ASME in 1995 (162) found that the presence, or absence, of chlorine-containing wastes in incinerators had no effect on the levels of dioxin produced. Rather, it was found that incinerator operating conditions (primarily temperature) were the key to controlling dioxin formation. More recently, German officials examined the issue of incinerating vinyl waste and decided there was no cause for concern (180).

Misleading also is the idea that vinyl should be banned from incinerators because it contains heavy-metal additives. This is an evolving issue. Most vinyl products do not contain heavy metals and vinyl is a small fraction in feed to incinerators. Reformulation to replace heavy metals is in progress but some use is likely to continue. Banning vinyl from incinerators does not eliminate this problem. Rather, regulations should specify that incinerator residues (ash) be disposed of appropriately.

Furthermore, it is not true that European packagers, grocery stores, and regulators have banned vinyl. There is only one ban on vinyl packaging in Europe, and that is on the use of vinyl-bottled mineral water in Switzerland, a commercial ploy to block the sale of French mineral water in that country. There is a voluntary agreement in Denmark by industry to substitute alternatives to vinyl packaging when feasible. Some municipalities in Germany have restricted the use of certain vinyl products in municipally funded building projects. Industry is working to change those restrictions with several notable reversals, including Berlin and Bielefeld. Government studies on PVC in Belgium and the Netherlands have concluded that there should be no bias against the use of PVC (181–183). Elsewhere in Europe, vinyl packaging continues to be widely used. In Britain, one of the leading retailers, Marks & Spencer, has chosen vinyl over other materials as the chain's most environmentally friendly polymer. In Switzerland, retailer Migros has stated that its whole attitude toward vinyl will change when incinerator scrubber technology is fully employed. The late-1990s trend in Europe, led by the Germans, is to take a comprehensive look at waste reduction and make industry a partner in that process. This involves all industries, not just the vinyl industry. Overall, Europe remains a larger consumer of vinyl packaging than the United States.

It is also not true that vinyl plastics decompose in landfills and give off vinyl chloride monomer, because like all plastics, vinyl is an extremely stable landfill material. It resists chemical attack and degradation, and is so resistant to the conditions present in landfills that it is often used to make landfill liners. On those occasions when vinyl chloride monomer is detected in landfills, it typically can be traced to the presence of other chemicals and solvents.

Furthermore, it is not true that other plastics are more environmentally friendly than vinyl. A more recent study compared vinyl to a number of other packaging materials and found that vinyl consumed the least amount of energy, used the lowest level of fossil fuels, consumed the least amount of raw materials, and produced the lowest levels of carbon dioxide of any of the plastics studied (184). In fact, the Norwegian environmental group Bellona has concluded that a generally reduced use of vinyl plastics can lead to a worsening of the environmental situation (185).

Lastly, it is not true that in a fire, vinyl is unusually hazardous and damaging. The real hazards in a fire are carbon monoxide and heat; these are especially a problem with other materials that readily burn. Because vinyl products contain chlorine, they are inherently flame-retardant and resist ignition. When it does burn, however, vinyl produces carbon monoxide, carbon dioxide, and hydrogen chloride. Of these, the most hazardous is carbon monoxide. Hydrogen chloride is an irritant gas that can be lethal at extremely high levels. However, research indicates that those levels are never reached or even approached in real fires. All organic materials, when burned, release a lengthy list of chemical by-products. For instance, when wood burns, as many as 175 different fire gases may be produced, including benzene and acrolein (186). Burning wool produces hydrogen cyanide. Even the simple act of barbecuing a steak or smoking a cigarette can produce dioxin (187). More importantly, virtually all burning materials produce carbon monoxide, which is by far considered the greatest toxic hazard in fires because of the abundant levels produced and the low levels that can cause death (188).

The fire death rate in the United States is decreasing, dropping from a rate of 76 per million in the 1940s, when most construction and decorative products were made of natural materials, to 29 per million in the 1980s, by which time, PVC had replaced natural materials in numerous applications (189). This downward trend can be attributed in large part to improved building codes and the broader use of sprinkler systems and smoke detectors. However, the increased use of more fire-resistant materials, such as PVC, deserves part of the credit for this improvement.

Hydrogen chloride is produced when PVC burns. A series of tests for the Federal Aviation Administration studied this issue. In those studies, test animals were able to survive exposures to hydrogen chloride reaching 10,000 ppm (190). More recent studies indicate less of a potential for delayed effects on lung function than expected (191). In a typical fire, hydrogen chloride levels rarely exceed 300 ppm, a fact confirmed by the Boston Fire Department and Harvard University (192). In hundreds of autopsies conducted on fire victims in the United States, not one death has been linked to the presence of PVC.

Bell Laboratories studied wire and cable compounds made of PVC or other halogen-based compounds vs halogen-free compounds and found that neither type of material presented a clearcut advantage in a fire, and that the halogenated compounds sometimes outperformed the nonhalogenated products in terms of creating less corrosion (193).

BIBLIOGRAPHY

"Vinyl Chloride and Poly(Vinyl Chloride)" in *ECT* 1st ed., Vol. 14, pp. 723–735, by C. H. Alexander and G. F. Cohen, The B. F. Goodrich Chemical Co.; "Poly(Vinyl Chloride)" in

ECT 2nd ed., Vol. 21, pp. 369–412, by M. J. R. Cantau, Air Reduction Co., Inc.; in *ECT* 3rd ed., pp. 886–936, by J. A. Davidson and K. L. Gardner, BFGoodrich Co.

1. *Mod. Plast.* **7**(1), 70–78 (1996).
2. E. Baumann, *Ann.* **163**, 308 (1872).
3. *Chem. Eng. News*, **62**(25), 38 (1984).
4. J. A. Davidson and D. E. Witenhafer, *J. Polym. Sci.: Polym. Phys. Ed.* **18**, 51 (1980).
5. R. Tregan and A. Bonnemayre, *Rev. Plast. Mod.* **23**, 7 (1970).
6. J. W. Summers and E. B. Rabinovitch, *J. Macromol. Sci.—Phys.* **B29**(2), 219 (1981).
7. J. W. Summers, *J. Vinyl Technol.* **2**(1), 2 (1980).
8. G. R. Johnson, *SPE Technical Papers*, **XXVI**, 379 (1995).
9. P. R. Schwaegerle, *J. Vinyl Technol.* **8**(1), 32 (1986).
10. N. D. Bort, V. G. Marinin, A. Y. Kalinin, and V. A. Kargin, *Vysokomol. Soedin.* **10**, 2574 (1968).
11. G. Palma, G. Talamini, M. Tavan, and M. Carenza, *J. Polym. Sci.—Phys.* **15**, 1537 (1977).
12. M. Carenza, G. Palma, G. Talamini, and M. Tavan, *J. Macromol. Sci.—Chem.* **A11**, 1235 (1977).
13. J. C. Wilson and E. L. Zichy, *Polymer*, **20**, 264 (1979).
14. F. R. Kulas and N. P. Thorshang, *J. Appl. Polym. Sci.* **23**, 1781 (1979).
15. P. H. Geil, *J. Macromol. Sci.—Chem.* **A11**, 1271 (1977).
16. G. Menges and N. Berndtsen, *Kunststoffe*, **66**(1966), 11, 9 (1976).
17. F. N. Cogswell, *Pure Appl. Chem.* **52**, 2031 (1980).
18. C. J. Singleton, T. Stephenson, J. Isner, P. H. Geil, and E. A. Collins, *J. Macromol. Sci.—Phys.* **B14**, 29 (1977).
19. W. Wenig, *J. Polym. Sci.—Phys.* **16**, 1635 (1978).
20. D. J. Blundell, *Polymer*, **20**, 934 (1979).
21. J. W. Summers, *J. Vinyl Technol.* **3**(2), 107 (1981).
22. T. Hattori, K. Tanaka, and M. Matsuo, *Polym. Eng. Sci.* **12**, 199 (1972).
23. E. M. Katchy, *J. Appl. Polym. Sci.* **28**, 1847 (1983).
24. E. B. Rabinovitch, *J. Vinyl Technol.* **4**(2), 62 (1982).
25. F. R. Kulas and N. P. Thorshaug, *J. Appl. Polym. Sci.* **23**, 1781 (1979).
26. H. Munstedt, *J. Macromol. Sci.—Phys.* **B14**, 195 (1977).
27. P. G. Faulkner, *J. Macromol. Sci.—Phys.* **B11**, 251 (1975).
28. A. R. Berens and V. L. Folt, *Polym. Engr. Sci.* **8**, 5 (1968).
29. A. R. Berens and V. L. Folt, *Trans. Soc. Rheol.* **11**(1), 95 (1967).
30. E. B. Rabinovitch and J. W. Summers, *J. Vinyl Technol.* **2**(3), 165 (1980).
31. G. Menges and N. Berndtsen, *Kunststoffe*, **66**(11), 9 (1976).
32. C. Singleton, J. Isner, D. M. Gezovich, P. K. C. Tsou, P. H. Geil, and E. A. Collins, *Polym. Eng. Sci.* **14**, 371 (1974).
33. W. Wenig, *J. Polym. Sci.—Phys.* **16**, 1635 (1978).
34. G. Natta and P. Corradini, *J. Polym. Sci.* **20**, 251 (1956).
35. G. Natta, I. W. Gassi, and P. Corradini, *Rend. Accad. Naz. Lincei.* **31**(1.2), 1 (1961).
36. A. Nakajima and S. Hayashi, *Kolloid Z. U. A. Polymere*, **229**(1), 12 (1969).
37. C. E. Wilkes, V. L. Folt, and S. Krimm, *Macromolecules*, **6**(2), 235 (1973).
38. J. W. Summers and E. B. Rabinovitch, *J. Vinyl Technol.* **13**(1), 54 (1991).
39. L. G. Shaw and A. R. DiLuciano, *J. Vinyl Technol.* **5**, 100 (1983).
40. D. E. Marshall, R. P. Higgs, and O. P. Obande, *Plast. Rubber Proc. Appl.* **3**, 353 (1983).
41. P. J. F. VanderHeuval, *5th International Conference on Plastics and Pipes*, Paper No. 20, 1982.
42. P. Benjamin, *J. Vinyl Technol.* **2**, 254 (1980).
43. K. V. Gotham and M. J. Hitch, *Brit. Polym. J.* **10**, 47 (1978).
44. T. F. Chapman, J. D. Isner, and J. W. Summers, *J. Vinyl Technol.* **1**(3), 131 (1979).

45. J. W. Summers, E. B. Rabinovitch, and P. C. Booth, *J. Vinyl Technol.* **8**(1), 2 (1986).
46. R. J. Krzewki and E. A. Collins, *J. Macromol. Sci.—Phys.* **B20**(4), 443 (1981).
47. *Ibid.*, p. 465.
48. A. Gonze, *Plastica*, **24**(2), 49 (1971).
49. A. Gonze, *Chim. Ind., Genie Chim.* **104**(4–5), 422 (1971).
50. M. Lamberty, *Plast. Mod. Elastomers*, **26**(10), 82, 87 (1974).
51. A. Gray, *PVC Processing, International Conference*, Paper No. 10, Plastics/Rubber Institute, London, 1978.
52. J. Parey and E. Kruger, *Kunststoffe*, **74**, 1 (1984).
53. C. L. Sieglaff, *Pure Appl. Chem.* **53**, 509 (1981).
54. M. Gilbert and J. C. Vyvoda, *Polymer*, **22**, 1135 (1981).
55. M. Gilbert, D. A. Hemsley, and A. Miadonye, *Plast. Rubber Process. Appl.* **3**(4), 343 (1983).
56. J. W. Summers, E. B. Rabinovitch, and J. G. Quisenberry, *J. Vinyl Technol.* **4**(2), 67 (1982).
57. J. W. Summers, J. D. Isner, and E. B. Rabinovitch, *Polym. Engr. Sci.* **20**(2), 155 (1980).
58. M. F. Marx, *J. Vinyl Technol.* **3**(1), 56 (1981).
59. J. W. Summers and E. B. Rabinovitch, *J. Macromol. Sci.—Phys.* **B20**(2), 219 (1981).
60. R. Qin, H. P. Schreiber, and A. Rudin, *J. Appl. Polym. Sci.* **56**, 51 (1995).
61. J. W. Summers, E. B. Rabinovitch, and J. C. Quisenberry, *J. Vinyl Technol.* **7**(1), 32 (1985).
62. E. B. Rabinovitch, E. Lacatus, and J. W. Summers, *J. Vinyl. Technol.* **6**(3), 98 (1984).
63. E. B. Rabinovitch and P. C. Booth, *J. Vinyl Technol.* **12**(1), 43 (1990).
64. S. H. Maron and F. E. Filisko, *J. Macromol. Sci., Part B—Phys.* **B6**(2), 413 (1972).
65. A. Nakajima, H. Hamada, and S. Hayashi, *Makromol. Chem.* **95**, 40 (1966).
66. B. P. Takhomirov, H. B. Hopfenberg, V. Stannett, and J. L. Williams, *Macromol. Chem.* **118**, 117 (1968).
67. R. M. Ogorkiewicz, *Engineering Properties of Thermoplastics*, John Wiley & Sons, Inc., New York, 1970, p. 251.
68. R. Hoffman and W. Knappe, *Kolloid Z. U. Z. Polymere*, **240**, 784 (1970).
69. R. P. Sheldon and K. Lane, *Polymer*, **6**, 77 (1965).
70. E. A. Grulke, in J. Brandrup and E. H. Immergut, eds., *Polymer Handbook*, 3rd ed., John Wiley & Sons, Inc., New York, 1989, pp. VII, 554.
71. C. S. Marvel, J. H. Sample, and M. F. Roy, *J. Am. Chem. Soc.* **61**, 3241 (1939).
72. B. Baum and L. H. Wartman, *J. Polym. Sci.* **28**, 537 (1959).
73. G. Bier and H. Kraemer, *Kunststoffe*, **46**, 498 (1956).
74. J. W. L. Fordham, *J. Polym. Sci.* **39**, 321 (1959).
75. Y. V. Glazkovskii and Y. G. Papulov, *Vysokomol Soedin.* **A10**, 492 (1968).
76. H. Germar, K. H. Hellwege, and K. Johnson, *Makromol. Chem.* **60**, 106 (1963).
77. O. C. Bockman, *Brit. Plast.*, 364 (June 1965).
78. G. M. Burnett and W. W. Wright, *Proc. Roy. Soc. (London)*, **Ser. A 200**, 301 (1950).
79. J. Ugelstad, P. C. Mork, and F. K. Hansen, *Pure Appl. Chem.* **53**, 323 (1981).
80. A. H. Abdel-Alim and A. E. Hamielec, *J. Appl. Polym. Sci.* **16**, 783 (1972).
81. *Ibid.*, **18**, 1603 (1974).
82. J. W. Breitenbach, *Makromol. Chem.* **8**, 147 (1952).
83. F. Danusso, G. Pajaro, and D. Sianesi, *Chim. Ind. (Milan)*, **37**, 278 (1955).
84. *Ibid.*, **41**, 1170 (1959).
85. D. E. Skillicorn, G. A. Perkins, A. Slark, and J. V. Dawkins, *J. Vinyl Technol.* **15**(2), 105 (1993).
86. C. E. Schildknecht, ed., *Polymer Processes*, Interscience Publishers, Inc., New York, 1956, Chapt. 2.
87. D. N. Bort, Y. Y. Rylov, N. A. Okladnov, and V. A. Kargin, *Polym. Sci. USSR*, **9**(2), 334 (1967).

88. N. Fisher and L. Gioran, *Hydrocarbon Process.* **60**(5), 143 (1981).

89. M. H. Lewis and G. R. Johnson, *J. Vinyl Technol.* **3**(2), 102 (1981).

90. A. Crosato-Arnaldi, P. Gasparini, and G. Talamini, *Macromol. Chem.* **117**, 140 (1868).

91. H. Fikentscher and G. Hagen, *Angew. Chem.* **51**, 433 (1938).

92. W. D. Harkins, *J. Am. Chem. Soc.* **69**, 1428 (1947).

93. W. V. Smith and R. H. Ewart, *J. Chem. Phys.* **16**, 592 (1948).

94. G. Talamini and E. Peggion, in G. E. Ham, ed., *Vinyl Polymerization*, Vol. 1, Part 1, Marcel Dekker, Inc., New York, 1967, Chapt. 5.

95. J. Ugelstad, P. C. Mork, P. Dahl, and P. Rangnes, *J. Polym. Sci.* **C27**, 49 (1969).

96. J. T. Lazor, *J. Appl. Polym. Sci.* **1**, 11 (1959).

97. E. Peggion, F. Testa, and G. Talamini, *Makromol. Chem.* **71**, 173 (1964).

98. H. A. Savetnick, *Polyvinyl Chloride*, Van Nostrand Reinhold Co., New York, 1969, pp. 59–63.

99. *Ibid.*, p. 47.

100. *Ibid.*, pp. 24–27.

101. Brit. Pat. 444,257 (Mar. 17, 1936), A. Renfrew, J. W. Walter, and W. E. F. Gates (to Imperial Chemical Industries, Ltd.).

102. U.S. Pat. 2,300,566 (Nov. 3, 1940), H. J. Hahn and E. Brown (to General Aniline & Film Corp.).

103. U.S. Pat. 2,450,000 (Sept. 28, 1948), B. W. Howk and F. L. Johnson (to E. I. du Pont de Nemours & Co., Inc.).

104. P. V. Smallwood, in H. F. Mark and co-workers, eds., *Encyclopedia of Polymer Science and Engineering*, Vol. 17, John Wiley & Sons, Inc., New York, p. 321.

105. U.S. Pat. 2,440,808 (May 4, 1948), H. T. Neher and F. J. Glavis (to Rohm & Haas Co.).

106. U.S. Pat. 2,524,627 (Oct. 3, 1950), W. P. Hohenstein (to Polytechnic Institute of Brooklyn).

107. U.S. Pat. 2,486,855 (Nov. 1, 1949), E. Lavin and C. L. Boyce (to Shawinigan Resins Corp.).

108. U.S. Pat. 2,476,474 (July 19, 1949), M. Baer (to Monsanto Chemical Co.).

109. U.S. Pat. 2,543,094 (Feb. 27, 1951), C. A. Brighton and J. J. P. Staudinger (to The Distillers Co.).

110. A. R. Cain, *Polym. Preprints*, **11**(1), 312 (1970).

111. U.S. Pat. 2,546,207 (Mar. 27, 1951), D. Bandel (to Mathieson Chemical Co.).

112. R. Z. Greenley, in Ref. 70, pp. II, 274.

113. Brit. Pat. 1,275,395 (May 24, 1972) (to Borden Inc.).

114. Brit. Pat. 1,391,597 (Apr. 23, 1975), G. J. Gammon and P. Lewis (to B.P. Chemicals).

115. Brit. Pat. 1,391,598 (Apr. 23, 1975), G. J. Gammon and P. Lewis (to B.P. Chemicals).

116. Ger. Pat. 2,165,369 (July 27, 1972), J. Desilles (to Aquitane-Organico).

117. Brit. Pat. 1,396,703 (June 4, 1975), G. J. Gammon and P. Lewis (to B.P. Chemicals).

118. Brit. Pat. 1,298,636 (Dec. 6, 1972), G. J. Gammon and P. Lewis (to B.P. Chemicals).

119. U.S. Pat. 3,813,373 (May 28, 1974), I. Ito, T. Sekihara, and T. Emura (to Sumitomo Chemical Co.).

120. F. H. Winslow and W. Matreyek, *Ind. Engr. Chem.* **43**, 1108 (1951).

121. E. Farber and M. Koral, *Polym. Engr. Sci.* **8**(1), 11 (1968).

122. R. D. Dworkin, *J. Vinyl Technol.* **11**(1), 15 (1989).

123. A. H. Frye, R. W. Horst, and M. A. Paliobagis, *J. Polym. Sci.* **A2**, 1765 (1964).

124. *Ibid.*, p. 1785.

125. *Ibid.*, p. 1801.

126. J. T. Lutz, Jr. and D. L. Dunkelberger, *Impact Modifiers for PVC; The History and Practice*, John Wiley & Sons, Inc., New York, 1992, p. 34.

127. *Ibid.*, p. 10.

128. *Ibid.*, p. 54.
129. *Encyclopedia of Polymer Science & Engineering*, 2nd ed., Index Vol., John Wiley & Sons, Inc., New York, 1990, p. 307.
130. D. Bower, *Plast. Compound.* **2**(1), 64 (1979).
131. G. Ullman, *SPE J.* **71** (June 1967).
132. B. Pukansky and co-workers, *Muanyagosuestatek harai Kutatasa* (*MKL*), **33**(11), 575.
133. M. C. McMurrer, *Plast. Compound.* **5**(4), 74 (1982).
134. L. F. King and F. Noel, *Polym. Engr. Sci.* **12**(2), 112 (1972).
135. J. E. Hartitz, *SPE Tech. Papers* **XIX**, 362 (1973).
136. D. W. Riley, *SPE Tech. Papers*, **XXIX**, 890 (1983).
137. W. L. Semon and G. A. Stahl, *J. Macromol. Sci., Chem.* **A18**, 2 (1973).
138. T. Alfrey, J. Wiederhorn, R. Stein, and A. V. Tobolsky, *J. Colloid Sci.* **4**, 221 (1949).
139. D. L. Tabb and J. L. Koenig, *Macromolecules*, **8**, 929 (1975).
140. D. M. Gizovich and P. H. Geil, *Int. J. Polym. Notes*, **1**, 223 (1972).
141. *Environmental Profile: Facts About the Safety of PVC Additives*, rev. bull. The Vinyl Institute, a Division of The Society of The Plastics Industry, Inc., Morristown, N.J., July 1993.
142. D. F. Cadogan, *Plasticizers: A Consideration of Their Impact on Health and the Environment*, Plasticizers Sector Group, CEFIC, Brussels, Belgium, 1992.
143. Specialized Experts Working Group of the European Commission, *Official J. Eur. Commun.* (No. C 94/9) (1992).
144. G. Capocci, in E. J. Wickson, ed., *Handbook of Polyvinyl Chloride Formulating*, John Wiley & Sons, Inc., New York, 1993, p. 358.
145. M. L. Dannis and F. L. Ramp, in L. I. Nass, ed., *Encyclopedia of PVC*, Vol. 1, Marcel Dekker, Inc., New York, 1976, Chapt. 6, p. 225.
146. D. P. Miller, *Modern Plastics Encyclopedia*, Vol. 57, McGraw-Hill, Inc., New York, 1981–1982, p. 199.
147. L. Liu and R. Roman, personal communication, The Geon Co., Avon Lake, Ohio, 1996.
148. *1994 World Vinyls Analysis*, CMAI, Chemical Market Associates, Inc., Houston, Tex., 1995.
149. Technical data, Chem Data, Inc., Houston, Tex., 1995.
150. J. D. Isner and J. W. Summers, *Polym. Engr. Sci.* **18**(11), 905 (1978).
151. J. W. Summers and E. B. Rabinovitch, *J. Vinyl Technol.* **5**(3), 91 (1983).
152. J. W. Summers, *J. Vinyl Technol.* **5**(2), 43 (1983).
153. E. B. Rabinovitch, J. W. Summers, and P. C. Booth, *J. Vinyl Technol.* **14**(1), 20 (1992).
154. M. Hross, personal communication, The Geon Co., Avon Lake, Ohio, 1996.
155. *Poly(Vinyl Chloride)*, Material safety data sheets, The Geon Co., Avon Lake, Ohio, 1996.
156. *Registry of Toxic Effects of Chemical Substances*, NIOSH, U.S. Dept. of Health and Public Services, Public Health Service, Centers for Disease Control, Cincinnati, Ohio, (CCINFO disc as format), 1995.
157. *ICRMS North American Database, Version 4*, Ariel Research Corp., Bethesda, Md., 1996.
158. *Guide to Occupational Exposure Values–1995*, American Conference of Governmental Industrial Hygienists, Inc., Cincinnati, Ohio, 1995.
159. M. H. Lloyd, S. Gauld, and C. A. Souter, *Brit. J. Ind. Med.* **41**, 328 (1984).
160. S. L. Niedleman and J. Geigen, *Biohalogenation: Principles, Basic Roles and Applications*, Ellis Horwood Ltd./John Wiley & Sons, Inc., New York, 1986.
161. T. Leisinger, *Experientia*, **39**, 1183 (1983).
162. O. Hutzinger, ed., *Handbook of Environmental Chemistry*, Vol. 1, Part A, Springer–Verlag, Berlin, 1991, pp. 229–254.
163. W. Fenical, *Marine Org. Chem.* **31**, 373 (1981).
164. R. Clement and C. Tashiro, *Abstracts, Dioxin*, **91**, S34 (1991).

165. G. Mariani, *Chemosphere*, **24**(11), 1545 (1992).

166. R. M. Smith and co-workers, *Chemosphere*, **25**, 95 (1992).

167. P. A. Mayewski, *Science* **232**(4753), 975 (May 1986).

168. J. W. Summers, B. K. Mikofalvy, and S. Little, *J. Vinyl Technol.* **12**(3), 161 (1990).

169. *Vinyl Products Lifecycle Assessment*, Chem Systems, Inc., Tarrytown, N.Y., Mar. 1992.

170. *Comparative Energy Evaluation of Plastic Products and Their Alternatives for The Building and Construction and Transportation Industries*, Franklin Associates, Prairie Village, Kans., Mar. 1991.

171. H. G. Rigo, A. J. Chandler, and W. S. Lanier, *The Relationship Between Chlorine In Waste Streams and Dioxin Emissions From Waste Combustor Stacks*, CRTD, Vol. 36, The American Society of Mechanical Engineers, United Engineering Center, New York, 1995.

172. *Results of the Combustion and Emissions Research Project at the Vicon Incinerator Facility in Pittsfield, Massachusetts*, final report, Midwest Research Institute for the New York State Energy Research and Development Authority, New York, June 1987.

173. *Air Emission Tests at Commerce Refuse to Energy Facility, May 26–June 5, 1987*, Vol. I, ESA 20522 449, Energy Systems Associates, Pittsburgh, Pa., July 1987.

174. K. L. Churney, A. E. Ledford, S. S. Bruce, and E. S. Domalski, *The Chlorine Content of Municipal Solid Waste from Baltimore County, Maryland and Brooklyn, New York*, National Bureau of Standards, Gaithersburg, Md., Apr. 1985.

175. D. Randall and B. Suzanne Shoraka-Blair, *An Evaluation of the Cost of Incinerating Wastes Containing PVC*, The American Society of Mechanical Engineers, New York, 1994.

176. "PVC is a Good Bet to Survive Its Global Environmental Travails," *Mod. Plast.* (June 1990).

177. R. S. Magee, *Plastics in Municipal Solid Waste Incineration: A Literature Study*, Hazardous Substance Management Research Center, New Jersey Institute of Technology, Mar. 1989.

178. P. Lightowlers and J. N. Cape, "Does PVC Waste Incineration Contribute to Acid Rain?" *Chem. Ind.* (June 1987).

179. *Directory of Companies Involved in the Recycling of Vinyl (PVC) Plastics*, 3rd ed., The Vinyl Institute, a Division of The Society of the Plastics Industry, Inc., Morristown, N.J., May 1994.

180. W. Lohrer and W. Plehn, *Staub. Reinhalt. Luft.* **47**(7/8), 190 (1987).

181. *Chem. Week (Europe/Mideast News)* (Dec. 14, 1994).

182. *Persbericht*, Nederlandse Federatie voor Kunststoffen, June 25, 1993, p. vv.

183. *Executive Newsline*, 2 (Dec. 26, 1995).

184. *Vinyl Products Life Cycle Assessment*, Chem Systems, Inc., Tarrytown, N.Y., Mar. 1992.

185. B. Bergfald, *PVC*, Bellona Institute Oslo, Norway, May 11, 1990.

186. N. W. Hurst and T. A. Jones, *Fire Mater.* **9**, 1 (1985).

187. *World Environment and PVC*, Vinyl Chloride Industry Association, Tokyo, Japan.

188. G. L. Nelson, D. V. Canfield, and J. B. Larsen, "Carbon Monoxide—Study of Toxicity in Man," *11th International Conference on Fire Safety*, San Francisco, Calif., Jan. 13–17, 1986.

189. Technical data, The National Fire Protection Association, Quincy, Mass., 1990.

190. H. L. Kaplan and co-workers, *J. Fire Sci.* **3**, 228 (1985).

191. H. L. Kaplan, A. Auzeuto, W. G. Switizer, and R. K. Hinderer, *J. Toxicol. Environ. Health*, **23**, 473 (1988).

192. W. A. Burgess, R. D. Treitman, and A. Gold, *Air Contaminants in Structural Fire-fighting*, NFPCA Grant 7X008, Harvard School of Public Health, Cambridge, Mass., 1979.

193. P. R. Dickinson, "Evolving Fire Retardant Materials Issues: A Cable Manufacturer's Perspective," *Fire Technol.* (Nov. 1992).

General References

The following references are useful for information on environmental considerations and recycling.

PVC's Environmental Profile: Fallacy vs. Fact, The Vinyl Institute, a Division of The Society of the Plastics Industry, Inc., Morristown, N.J., July 1995.

Environmental Profile: Facts About Chlorine, The Material Used to Make PVC, The Vinyl Institute, Morristown, N.J., Mar. 1993.

Environmental Profile: Facts About the Economics, Life Cycle Efficiencies, and Performance Benefits, The Vinyl Institute, Morristown, N.J., July 1993.

Environmental Profile: Facts About Safety of Incinerating PVC Wastes, The Vinyl Institute, Morristown, N.J., July 1995.

Environmental Profile: Facts About Recycling PVC, The Vinyl Institute, Morristown, N.J., July 1995.

D. Wisner and F. E. Krause, technical data, The Geon Co., Avon Lake, Ohio, 1996.

JAMES W. SUMMERS
The Geon Company

VINYL ETHER MONOMERS AND POLYMERS

Because of the strong electron-donating oxygen, the polymerization of vinyl ethers (VE) can be readily accomplished using cationic initiators, resulting in polymers and copolymers that have the potential for significant variety. However, only poly(methyl vinyl ether) (PMVE)) achieved commercial success among the homopolymers, and its commercial importance has faded (1). Divinyl ethers are emerging as important ingredients in radiation-cured coatings (2), whereas copolymers of methyl vinyl ether (MVE) and maleic anhydride, easily prepared by free-radical initiation, continue to be valued as ingredients in personal care and pharmaceutical products (3).

The observation in 1949 (4) that isobutyl vinyl ether (IBVE) can be polymerized with stereoregularity ushered in the stereochemical study of polymers, eventually leading to the development of stereoregular polypropylene. In fact, vinyl ethers were key monomers in the early polymer literature. For example, ethyl vinyl ether (EVE) was first polymerized in the presence of iodine in 1878 and the overall polymerization was systematically studied during the 1920s (5). There has been much academic interest in living cationic polymerization of vinyl ethers and in the unusual compatibility of poly(MVE) with polystyrene.

Monomers

The most general commercial process for the manufacture of mono- and divinyl ethers, developed by Reppe in the 1930s at BASF, is by treating alcohols with acetylene under pressure of ≥ 6.8 atm (100 psi) at temperatures of $120-180°C$ in the presence of catalytic amounts of the corresponding metal alcoholate.

Although this is a simple reaction to visualize, the danger of handling acetylene under pressure in concentrated form requires sophisticated equipment and should only be attempted experimentally in an appropriately barricaded high pressure autoclave (6).

$$HC\equiv CH \xrightarrow{ROM} (ROCH = CHM) \xrightarrow{ROH} ROCH = CH_2 + ROM$$

Alternatively, thermal cracking of acetals or metal-catalyzed transvinylation can be employed. Vinyl acetate or MVE can be employed for transvinylation and several references illustrate the preparation especially of higher vinyl ethers by such laboratory techniques. Special catalysts and conditions are required for the synthesis of the phenol vinyl ethers to avoid resinous condensation products (6,7). Direct reaction of ethylene with alcohols has also been investigated (8).

Monomer Properties.   Some physical properties of the lower homologues of vinyl ether are presented in Table 1.

Reactions of Vinyl Ethers.   Vinyl ethers undergo the typical reactions of activated carbon–carbon double bonds. A key reaction of VEs is acid-catalyzed hydrolysis to the corresponding alcohol and acetaldehyde, ie, addition of water followed by decomposition of the hemiacetal. For example, for MVE, the reaction is

$$CH_3OCH{=\!=}CH_2 + H_2O \longrightarrow CH_3O\overset{\displaystyle |}{\underset{\displaystyle OH}{-}CH-}CH_3 \rightleftharpoons CH_3OH + CH_3CHO$$

To avoid this reaction during storage, VEs are often stabilized using small amounts of bases such as triethanol amine.

MVE is a reactive flammable gas and must be handled safely. A detailed review of VE handling issues is available (6).

Table 1. Physical Properties of the Lower Vinyl Ethers[a]

	Methyl	Ethyl	Isopropyl	n-Butyl	Isobutyl
CAS Registry Number	[107-25-5]	[109-92-2]	[926-65-8]	[111-34-2]	
odor	sweet, pleasant	pleasant	pleasant	pleasant	pleasant
boiling point, °C	5.5	35.6	55–56	94.3	83
freezing point, °C	−122	−115.3	−140	−112.7	−132.3
specific gravity at 20/4°C	0.7511	0.753	0.753	0.778	0.767
refractive index, n_D^{25}	1.3947	1.3734	1.3829	1.3997	1.3946
solubility in water at 20°C, wt %	0.97	0.039	0.6	0.1	0.1
flash point, °C	−56[b]	−18[c]		0.55	−9.4
heat of vaporization at 101.3 kPa		367		316	323

[a]Ref. 6.
[b]Cleveland open cup.
[c]Tag open cup (ASTM D1310).

The principal reaction of vinyl ethers to be considered in this article is cationic polymerization.

Homopolymerization

VEs such as MVE polymerize slowly in the presence of free-radical initiators to form low mol wt products of no commercial importance (9). Examples of anionic polymerization are unknown, whereas cationic initiation promotes rapid polymerization to high mol wt polymers in excellent yield and has been extensively studied (10).

A typical cationic polymerization is conducted with highly purified monomer free of moisture and residual alcohol, both of which act as inhibitors, in a suitably dry unreactive solvent such as toluene with a Friedel-Crafts catalyst, eg, boron trifluoride, aluminum trichloride, and stannic chloride. Usually low temperatures (-40 to $-70°C$) are favored in order to prevent chain-transfer or sidereactions.

Complexation of the initiator and/or modification with cocatalysts or activators affords greater polymerization activity (11). Many of the patented processes for commercially available polymers such as poly(MVE) employ BF_3 etherate (12), although vinyl ethers can be polymerized with a variety of acidic compounds, even those unable to initiate other cationic polymerizations of less reactive monomers such as isobutene. Examples are protonic acids (13), Ziegler-Natta catalysts (14), and actinic radiation (15,16).

Initiation is an electrophilic addition of a cation across the double bond, but because of the poor nucleophilicity of the initiator's counterion, propagation is favored over termination. Chain transfer to more nucleophilic species such as fortuitous moisture or impurities such as alcohols either limits propagation or, if such impurities are present in sufficient quantities, totally inhibits polymerization. As the temperature of a particular polymerization is increased, molecular weight drops off quickly because the propagating carbocation can transfer its β-proton to the counterion or to fresh monomer. This results in terminal unsaturation or acetal end groups. The mechanism can be illustrated as follows:

Initiation

$$M^+X^- + CH_2{=}CH \longrightarrow M{-}CH_2{-}CH^+X^-$$
$$\qquad\qquad\quad | \qquad\qquad\qquad\qquad\quad |$$
$$\qquad\qquad\quad OR \qquad\qquad\qquad\qquad\quad OR$$

Propagation

$$M{-}CH_2{-}CH^+X^- + CH_2{=}CH \xrightarrow{\text{kp}} M{-}(CH_2CH)_nCH_2CH^+X^-$$
$$\qquad\quad | \qquad\qquad\qquad\quad | \qquad\qquad\qquad\qquad\quad | \qquad\quad |$$
$$\qquad\quad OR \qquad\qquad\qquad OR \qquad\qquad\qquad\qquad OR \quad OR$$

Chain-transfer termination

$$M{-}(CH_2CH)_nCH_2CHCH_2CH^+B^- \xrightarrow{-H} M{-}CH_2CHCH_2CHCH{=}CH + HB$$
$$\qquad\quad | \qquad\qquad | \qquad | \qquad\qquad\qquad\qquad\qquad | \qquad\quad | \qquad\quad |$$
$$\qquad\quad OR \qquad OR \quad OR \qquad\qquad\qquad\qquad OR \quad OR \quad OR$$

$$\xrightarrow{\text{ROH}} \text{acetal end groups} + HB$$
$$CH_2CH(OR)_2$$

In the presence of cationic initiators, the possibility for loss of pendant ether groups to form free alcohol is another side reaction that usually results in color formation because of the highly conjugated products formed.

$$\text{poly(vinyl ether)} \xrightarrow{\text{several steps} - \text{HOR}} -(CH=CH)_n - CH=CH$$
$$\overset{|}{\underset{OR}{}}$$

This reaction is favored by higher reaction temperatures and polar solvents. Another degradation reaction common to ethers is oxidation, especially when the α-carbon is branched (17). Polymeric ethers of all types must not be exposed to oxygen, especially in the presence of transition metals because formation of peroxides can become significant.

Monomer Reactivity. The nature of the side chain R group exerts considerable influence on the reactivity of vinyl ethers toward cationic polymerization. The rate is fastest when the alkyl substituent is branched and electron-donating. Aromatic vinyl ethers are inherently less reactive and susceptible to side reactions. These observations are shown in Table 2.

The same order of reactivity is also found for reactions such as acid-catalyzed hydrolysis; a closer look at the structure of these vinyl ether homologues suggests the reason.

Table 2. VE Monomer Reactivity

Monomer	CAS Registry Number	Relative reactivity	Formula
Alkyl vinyl ethers[a]			
ethyl vinyl ether	[109-92-2]	1.00	$CH_3CH_2OCH=CH_2$
t-butyl vinyl ether	[926-02-3]	12.5	$(CH_3)_3COCH=CH_2$
isopropyl vinyl ether	[926-65-8]	5.4	$(CH_3)_2CHOCH=CH_2$
2-chloroethyl vinyl ether	[110-75-8]	0.44	$ClCH_2CH_2OCH=CH_2$
Aryl vinyl ethers[b]			
phenyl vinyl ether	[766-94-9]	1.00	$\langle\bigcirc\rangle$—$OCH=CH_2$
p-methoxyphenyl vinyl ether	[4024-19-5]	2.38	CH_3O—$\langle\bigcirc\rangle$—$OCH=CH_2$
p-methylphenyl vinyl ether	[1005-62-5]	1.82	CH_3—$\langle\bigcirc\rangle$—$OCH=CH_2$
p-chlorophenyl vinyl ether	[1074-56-2]	0.278	Cl—$\langle\bigcirc\rangle$—$OCH=CH_2$

[a] Ref. 18.
[b] Ref. 19.

$$\begin{array}{c} \underset{H}{\overset{H}{\diagdown}}C{=}C\underset{O}{\overset{H}{\diagup}} \\ | \\ CH_3 \end{array} \qquad \begin{array}{c} \underset{H}{\overset{H}{\diagdown}}C{=}C\underset{O}{\overset{H}{\diagup}}CH_3 \end{array}$$

syn   anti

Syn- and anti-orientations are possible and there is evidence that the anti-orientation does not favor orbital overlap; such an orientation is favored with larger branched-chain substituents. A ^{13}C-nmr study found that the π-electron density on the vinyl β-carbon is lower as the reactivity of the monomer increases (20). Methyl vinyl ether exists almost entirely in the syn-structure, a favorable orbital overlap situation, and MVE for this reason is less reactive to both polymerization and hydrolysis (21).

This is a puzzling situation because it is expected that a higher electron density on the β-carbon would lead to greater reactivity toward an approaching carbocation. Obviously the reactivity of the carbocation terminus toward an approaching vinyl ether monomer is another factor in determining the reactivity of a particular vinyl ether during cationic polymerization. Most models designed to explain why low electron density on the β-vinyl carbon produces a more reactive vinyl ether suggest that a cyclic π-complex transition state is involved. Such an intermediate allows the α-carbon to benefit from the higher level of electron density as the new carbocation terminus forms (10).

Stereoregular Polymerization.   Chemists at GAF Corporation were first to suggest that stereoregularity or the lack thereof is responsible for both non-tacky and crystalline or tacky and amorphous polymers generated from IBVE with BF_3:$O(C_2H_5)_2$, depending on the reaction conditions (22,23). In addition, it was shown that the crystalline polymer is actually isotactic (24). Subsequently, the reaction conditions necessary to form such polymers have not only been demonstrated, but the stereoregular polymerization has been extended to other monomers, such as methyl vinyl ether (25,26).

In order to generate stereoregular (usually isotactic) polymers, the polymerization is conducted at low temperatures in nonpolar solvents. A variety of soluble initiators can produce isotactic polymers, but there are some initiators, eg, $SnCl_4$, that produce atactic polymers under isotactic conditions (26). The nature of the pendant group can influence tacticity; for example, large, bulky groups are somewhat sensitive to solvent polarity and can promote more crystallinity (14,27).

The low temperature limitation of homogeneous catalysis has been overcome with heterogeneous catalysts such as modified Ziegler-Natta (28) solid-supported protonic acids (29,30) and metal oxides (31). Temperatures as high as 80°C in toluene can be employed to yield, for example, crystalline poly(IBVE) with Cr_2O_3 (31).

It has been suggested that the mechanism of stereoregular vinyl ether polymerization heavily depends on the degree of association of the counterion with the growing terminal carbocation (32,33). Figure 1 illustrates the predicted most-stable configuration of the ultimate and penultimate units of a growing

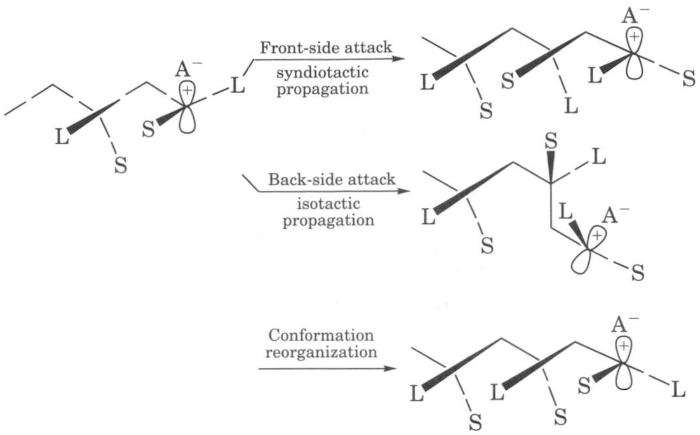

Fig. 1. Effect of ion pairing on stereochemistry of propagation for alkyl vinyl ethers, where L is large substituent; S, small substituent.

chain. It indicates how an incoming monomer can approach the carbocation terminus from either the front- or back-side attack, which attack is prevalent depends on the tightness of the growing ion pair and the steric requirements of the particular vinyl ether monomer. Front-side attack is favored by a loose ion pair associated with polar solvents, whereas a back-side attack is favored by nonpolar solvents where a tight ion pair prevails.

Living Polymerization

Ever since the demonstration that the initiation system of hydrogen iodide and molecular iodine (HI/I$_2$) induces living polymerization of VE monomers (34), numerous studies have been performed to extend the scope of this type of VE polymerization. Living polymerization is characterized by an increasing number-average molecular weight as the monomer is consumed. The rate of M_n increase is inversely proportional to the initial concentration of hydrogen iodide, not iodine, and the molecular weight distribution (MWD) of the polymer is very narrow throughout the course of the polymerization ($M_w/M_n < 1.1$). Thus, this type of polymerization can be stopped and started by consumption or addition of fresh monomer. It is similar to ethylene oxide/propylene oxide (EO/PO) anionic polymerization in this regard, but the initiation system is longer lived (see POLYETHERS).

The mechanism proposed (35) consists of the reaction of HI across the double bond, followed by activation with molecular iodine. This weakens the C–I bond, making the α-carbon cationic enough to insert monomer and propagate polymerization. However, without fresh monomer, the polymerization stops, but without termination. The stability of the terminus of the polymer suppresses side reactions such as chain transfer and termination. The iodine is a weak Lewis acid, but because of the nucleophilicity of the iodide, the carbon–iodide bond remains intact. Other weak Lewis acids can perform the same function; for example, the HI/Zn_2I_2 system (36) and the $C_2H_5AlCl_2$/ester and ether systems (37–39) afford living polymerization at higher temperatures and rates. An overview of the design of initiating systems has been published (40).

Living VE polymerization is usually terminated by addition of alcohols, phenols, amines, etc, that can replace iodide. Without some base present to neutralize generated HI, an aldehyde end group forms if moisture is present because of acid-catalyzed hydrolysis (41).

Details of the mechanism of living polymerization have been refined continually (42), and a dormant species has been shown to be crucial for such polymerization. The dormant species is in equilibrium with the active species and the ratio can determine MWD and hence the quality of the living process (43).

The living polymerization process offers enormous flexibility in the design of polymers (40). It is possible to control terminal functional groups, pendant groups, monomer sequencing along the main chain (including the order of addition and blockiness), steric structure, and spatial shape.

More recent examples include end-functionalized multiarmed poly(vinyl ether) (44), MVE/styrene block copolymers (45), and star-shaped polymers (46–48). With this remarkable control over polymer architecture, the growth of future commercial applications seems entirely likely.

Homopolymer Properties

Physical properties, which depend on molecular weight, the nature of the alkyl group, the nature of the initiator, stereospecificity, and crystallinity, range from viscous liquids, through sticky liquids and rubbery solids, to brittle solids. Polyethers with long alkyl side chains are waxy, however, as the alkyl group in such cases dominates physical properties.

As shown in Table 3, the glass-transition temperatures of the amorphous straight-chain alkyl vinyl ether homopolymers decrease with increasing length

Table 3. Glass-Transition Temperature of Amorphous Poly(Vinyl Ether)s and Melting Points of Crystalline Poly(Vinyl Ether)s[a]

Poly(alkyl vinyl ether)	CAS Registry Number	T_g, °C	Mp, °C
methyl	[34465-52-6]	−34	144
ethyl	[25104-37-4]	−42	
isopropyl	[25585-49-3]	−3	191
n-butyl	[25232-87-5]	−55	
isobutyl	[9003-44-5]	−19	170
2-ethylhexyl	[29160-05-2]	−66	
n-pentyl		−66	
n-hexyl	[25232-88-6]	−77	
n-octyl	[25232-89-7]	−80	
t-butyl	[25655-00-9]		238

[a]Ref. 6.

of the side chain. Also, the melting points of the semicrystalline poly(alkyl vinyl ether)s increase with increasing side-chain branching.

Poly(methyl vinyl ether) [*34465-52-6*], because of its water solubility, continues to generate commercial interest. It is soluble in all proportions and exhibits a well-defined cloud point of 33°C. Like other polybases, ie, polymers capable of accepting acidic protons, such as poly(ethylene oxide) and poly(vinyl pyrrolidone), each monomer unit can accept a proton in the presence of large anions, such as anionic surfactants, HI_3, or polyacids, to form a wide variety of complexes.

Commercial Aspects

Although no longer of significant commercial interest, the characteristics of some of the amorphous homopolymers commercially available at one time or another are illustrated in Table 4. No crystalline polymers are known to have been commercialized. This lack of commercial success results from the economically competitive situation concerning vinyl ether polymers versus other, more readily available polymers such as those based on acrylic and vinyl ester monomers.

The commercial situation in the late-1990s is surprising, considering the inherently inexpensive cost of the monomer raw materials, the apparent simplicity of manufacture, and the wide variety of vinyl ether monomers possible from the vast assortment of commercially available alcohols. Several bright spots in this technology have emerged, however; the field of radiation-curable coating formulations based on vinyl ether monomers is being actively pursued, and living cationic polymerization and polymer–polymer compatibility, especially of PMVE and polystyrene, are intensively studied.

Copolymerization

VEs do not readily enter into copolymerization by simple cationic polymerization techniques; instead, they can be mixed randomly or in blocks with the aid of living polymerization methods. This is on account of the differences in reactivity, resulting in significant rate differentials. Consequently, reactivity ratios must be taken into account if random copolymers, instead of mixtures of homopolymers, are to be obtained by standard cationic polymerization (50,51). Table 5 illustrates this situation for butyl vinyl ether (BVE) copolymerized with other VEs. The rate constants of polymerization (kp) can differ by one or two orders of magnitude, resulting in homopolymerization of each monomer or incorporation of the faster monomer, followed by the slower (assuming no chain transfer).

VEs can also copolymerize by free-radical initiation with a variety of comonomers. According to the Q and e values of 0.023 and -1.77 (isobutyl vinyl ether), VEs are expected to form ideal copolymers with monomers of similar Q and e values or alternating copolymers with monomers such as maleic anhydride (MAN) that have high values of opposite sign ($Q = 0.23$; $e = +2.25$).

Table 4. Commercial Vinyl Ether Homopolymers[a,b]

Vinyl ether	Physical form	Specific viscosity, η_{sp}	Trademark	Manufacturer	Uses
methyl	viscous liquid, balsam-like	0.68	Lutonal M	BASF	plasticizer for coatings; aqueous tackifier
	viscous liquid, balsam-like	0.3–0.5	Gantrez M	ISP	plasticizer for coatings; aqueous tackifier
ethyl	viscous liquid	1.0	Lutonal A	BASF	plasticizer for cellulose nitrate and natural-resin lacquers
	elastomeric solid		PVEE[c]	Union Carbide	pressure-sensitive adhesive base
isobutyl	high polymer viscous liquid	1.0	Lutanol I	BASF	tackifier for adhesives
	viscous liquid	0.1–0.5	Gantrez B	ISP	tackifier for adhesives
	elastomeric solid	2–6	Oppanol C	BASF	pressure-sensitive adhesive base
octadecyl	waxy solid[d]		V-Wax	BASF	polishes and waxes

[a] Refs. 12 and 49.
[b] Some viscous–liquid polymers are also supplied as high solids solutions, eg, 70% in toluene.
[c] Solid and solutions supplied.
[d] Low degree of polymerization, mp 50°C.

Table 5. Reactivity Ratios in the Copolymerization of n-Butyl Vinyl Ether (M_1) with Another Vinyl Ether (M_2)[a,b]

M_2 $CH_2 = CH - OR$	r_1	r_2	$1/r_1$	$\log 1/r_1$
CH_3	5.67 ± 0.02	0.47 ± 0.02	0.18	−0.745
C_4H_6	2.17 ± 0.15	0.24 ± 0.17	0.46	−0.337
C_2H_6	2.05 ± 0.05	0.75 ± 0.02	0.49	−0.310
n-C_3H_7	1.35 ± 0.15	0.99 ± 0.10	0.74	−0.131
n-C_4H_9	1.00	1.00	1.00	0.00
n-C_6H_{13}	0.95 ± 0.02	1.38 ± 0.03	1.05	0.021
i-C_4H_9	0.73 ± 0.15	1.48 ± 0.10	1.37	0.137
$C_6H_5CH_2$	0.72 ± 0.05	1.61 ± 0.08	1.40	0.146
i-C_3H_7	0.38 ± 0.07	2.77 ± 0.10	2.63	0.420
C_6H_{11}	0.29 ± 0.02	3.80 ± 0.02	3.44	0.537
$C_6H_5CHCH_3$	0.38 ± 0.10	1.40 ± 0.10	2.63	0.420
t-C_4H_9	0.19 ± 0.02	9.67 ± 0.05	5.26	0.721

[a]Ref. 51.
[b]Total monomer, 20 mmol; $C_2H_5AlCl_2$, 0.05 mmol; total volume of the reaction mixture, 23 mL in toluene at −78°C.

For bulk copolymerization of methyl, octyl, dodecyl, and octadecyl vinyl ethers using benzoyl peroxide as initiators at 40–100°C with the following comonomers (M_1), where r_2 is 0 in all cases (6), the values of r_1 are

M_1	r_1
acrylonitrile	0.8–1
butyl maleate	0–0.1
maleic anhydride	0.0
methyl acrylate	2.7–3
methyl methacrylate	10
styrene	>50
vinyl acetate	3.4–3.7
vinyl chloride	1.7–2.2
vinyldene chloride	1.3–1.5

Both MAN and VE do not readily homopolymerize by themselves by a free-radical mechanism. If each by itself is initiated with significant quantities of initiator at high temperature and with neat monomer, high yields of oligomer or low mol wt polymer is possible under these extreme conditions. However, when mixed together, the mixture can polymerize explosively even without initiation to high mol wt polymers (52). Early on, workers interested in this alternating polymerization recognized that the resulting polymer, even with varying ratios of monomer, remained 1:1 in monomer ratio, and that the rate of polymerization was at maximum when the monomers were polymerized at a 1:1 mole ratio. In fact, MAN and VE formed a charge-transfer complex (CTC) whose concentration could be measured in the uv. This species is found to be at a maximum when the monomers are at a 1:1 ratio. Hence the idea is that the CTC of MAN + VE is the active *in situ* monomer and this accounts for the complete alternating tendency found in the polymer regardless of monomer feed composition.

Evidence against this idea has fueled a long-enduring controversy. Several detailed review articles have appeared concerning this subject (53–55). However, in trapping experiments in the presence of donor and acceptor monomers that are known to form CTC, no evidence can be found for the concerted addition of the CTC (56,57). This suggests that the mechanism for alternation results from the interaction of an electron-acceptor terminal radical and an electron-donor monomer or the reverse, an electron-donor terminal radical and an electron-acceptor monomer, because of a decrease in the activation energy of cross-propagation (58). Alternatively, a convincing argument has been presented for the formation of diradicals or zwitterionic C_4 intermediates depending on substituents and conditions instead of a CTC (59). The current evidence supports a definite intermediate such as the CTC or C_4 diradical (60). Furthermore, the reality is that such alternating polymerizations can be reliably carried out with monomers such as vinyl ethers and electron-withdrawing monomers such as maleic anhydride, regardless of the actual mechanism. The alternating copolymers of MVE and MAN have achieved commercial success.

MVE/MAN Copolymers. Various mol wt grades of poly(methyl vinyl ether-*co*-maleic anhydride) (PMVEMA) are available from International Specialty Products, Inc. (formerly GAF Corp.), under the trade name of Gantrez. Table 6 illustrates the M_w and MWD found for commercially available polymers. As can be seen, high molecular weights are readily achieved.

PMVEMA, supplied as a white, fluffy powder, is soluble in ketones, esters, pyridine, lactams, and aldehydes, and insoluble in aliphatic, aromatic, or halogenated hydrocarbons, as well as in ethyl ether and nitroparaffins. When the copolymer dissolves in water or alcohols, the anhydride group is cleaved, forming the polymers in free acid form or the half-esters of the corresponding alcohol, respectively. Table 7 illustrates the commercially available alternating copolymers and derivatives.

When hydrolyzed in water, the resulting diacid exhibits two pK_as (~3 and 8); the half ester as expected has one pK_a typical of polycarboxylic acids. A cross-linked version of this polymer, referred to as Stabileze, thickens a variety of formulations by expanding its hydrodynamic volume when neutralized. Because

Table 6. Absolute Molecular Weights and Molecular Weight Distributions of Gantrez AN (PMVEMA)[a,b]

Gantrez sample	M_w	M_n	M_w/M_n
AN-119	2.16×10^5	7.98×10^4	2.71
AN-139	1.08×10^6	3.11×10^5	3.47
AN-149	1.25×10^6	4.85×10^5	2.58
AN-169	1.98×10^6	9.60×10^5	2.06
AN-179	2.40×10^6	1.13×10^6	2.12

[a]Refs. 61 and 62.
[b]Samples from size-exclusion chromatography/low angle laser light scattering (sec/lalls).

Table 7. Commercially Available Alternating Copolymers

Polymer	CAS Registry Number	Trade name	Supplier	Application
poly(methyl vinyl ether-co-maleic anhydride-co-decadiene)	[9011-16-9]	Stabileze	ISP	cosmetic/pharmaceutical thickener
poly(methyl vinyl ether-co-maleic anhydride) (PMVEMA)		Gantrez AN	ISP	adhesives, coatings, pharmaceutical tablet binder
butyl half-ester	[54578-91-5]	Gantrez ES-425	ISP	hairspray fixatives
ethyl half-ester	[50953-57-4]	Gantrez ES-225	ISP	hairspray fixatives
isopropyl half-ester	[31307-95-6]	Gantrez ES-335	ISP	hairspray fixatives
hydrolyzed (free acid)	[25153-40-6]	Gantrez S-95, S-97	ISP	thickener, protective colloid, dispersant
disodium salt	[9019-25-4]	Gantrez DS-1935; Sokalan CP-2	ISP; BASF	phosphate replacement in detergents (sequestrant)
mixed Ca and Na salts		MS-955	ISP	denture adhesive
poly(isobutyl vinyl ether-co-vinyl chloride)	[25154-85-2]	Caroflex MP-45	BASF	marine paints (film former)
poly(isobutyl vinyl ether-co-methyl acrylate-co-acrylonitrile)		Acronal 430D	BASF	polymer additive
poly(octadecyl vinyl ether-co-maleic anhydride	[28214-64-4]	Gantrez AN-8194	ISP	release coatings

of the low pK_a of the vicinal diacids, Stabileze acquires a negative charge at lower pH than competitive cross-linked polyacrylic acids (carbomers). This allows for efficient thickening at lower pHs similar to that found on moist skin, which in turn allows products to perform at optimum pH as desired by skin-care product formulators (63).

The effect of varying the degree of polymer neutralization on the conformations assumed by the poly(alkyl vinyl ether-*co*-maleic acid)s has been studied (64–66). The polyacids of this family with methoxy or ethoxy groups are random coils at all degrees of neutralization. The polyacids with intermediate-size alkoxy, ie, butoxy through octoxy, undergo at a particular degree of neutralization a conformational transition from a compact state, stabilized by hydrophobic forces, to a random coil. With small side groups such as methyl or ethyl, the potential for interaction is minimal. Consequently, as neutralization proceeds even to a small extent, repulsive forces predominate. However, when the pendant groups are hydrophobic, they can interact with each other in the same manner as surfactants, and this energy is sufficient to overcome charge repulsion but only to the point where repulsive forces overcome such hydrophobic interactions. Once this threshold is reached, the polymer uncoils to produce, ultimately, a polyelectrolyte with maximum charge separation.

International Specialty Products (ISP) supplies ethyl, isopropyl, and *n*-butyl half-esters of PMVEMA as 50% solutions in ethanol or 2-propanol. Typical properties are shown in Table 8. These half-esters do not dissolve in water but are soluble in dilute aqueous alkali and in aqueous alcoholic amine solutions. The main application for the half-esters is in hairsprays where they combine excellent hair-holding properties at high humidity without making the hair stiff or harsh. These half-esters are easily removed during shampooing, have a very low order of toxicity, and form tack-free films that exhibit good gloss, luster, and sheen (see HAIR PREPARATIONS).

Table 8. Typical Properties of Alcohol Solutions of the Half-Esters of PMVEMA

Property	Gantrez ES/225; SP/215	Gantrez ES-335-I	Gantrez ES-425 A-425	Gantrez ES-435
alkyl group of monoester	ethyl	isopropyl	butyl	butyl
physical form	clear, viscous liquid	clear, viscous liquid	clear, viscous liquid	clear, viscous liquid
activity, % solids	50 ± 2	50 ± 2	50 ± 2	50 ± 2
solvent	ethanol	2-propanol	ethanol	2-propanol
acid number, 100% solids	275–300	255–285	245–275	245–275
density, g/cm^3	0.983	0.957	0.977	0.962

Health and Safety Factors

Poly(methyl vinyl ether-*co*-maleic anhydride) and their monoalkyl ester derivatives have been shown on rabbits to be neither primary irritants nor primary sensitizers to skin and eyes. The acute oral toxicities on white rats of the two copolymers are, respectively, 29 g/kg and 25 g/kg body weight.

Applications

Radiation-Curable Coatings. The discovery and utilization of onium salt photoinitiators made possible the 1990s enthusiastic commercial interest in cationic radiation-curable vinyl ether coatings (49). The advantage over the older acrylic formulations is insensitivity to air (oxygen), fast cure, and excellent adhesion to metal and wood. In addition, the cationic mechanism allows both vinyl ether and epoxy monomers and oligomers to cure together. Although vinyl ethers are faster than epoxies, they accelerate the rate of overall polymerization even though an interpenetrating network of the two polymers is most likely the end result (67).

A wide variety of monovinyl and divinyl ethers are commercially available for this application, which allows the formulator greater latitude. For example, triethylene glycol divinyl ether [76-12-8] (DVE-3) and 1,4-cyclohexanedimethanol divinyl ether [17351-75-6] (CHVE) can be combined as reactive diluents, with each contributing quite different properties to the subsequently cured coating. CHVE offers hard brittle films, whereas DVE-3 produces films that have greater flexibility. DVE-3 is also a good solvent for the photoinitiator. More recently, the propenyl ether of propylene carbonate [130221-78-2] (PEPC) has been commercialized and specifically developed as a solubilizer for such initiators (68). Combinations of both divinyl ethers offer superior coatings that have controllable levels of flexibility, hardness, and solvent resistance (69). Their trade names are Rapi-Cure DVE-3, CHVE, PEPC, etc.

Vinyl ethers can also be formulated with acrylic and unsaturated polyesters containing maleate or fumarate functionality. Because of their ability to form alternating copolymers by a free-radical polymerization mechanism, such formulations can be cured using free-radical photoinitiators. With acrylic monomers and oligomers, a hybrid approach has been taken using both simultaneous cationic and free-radical initiation. A summary of these approaches can be found in Table 9.

Table 9. Properties of Vinyl Ether-Based Formulations for Radiation-Curing Coating[a]

Property	Cationic	Radical	Hybrid
monomer type	vinyl ether	vinyl ether	vinyl ether
oligomer type	epoxy and vinyl ether	maleate	acrylate
cure speed in			
air	fast	slow	fast
N_2	fast	moderate	fast
adhesion to			
metal	excellent	poor	poor
wood	excellent	excellent	excellent
post-cure	slight	none observed	none observed
formulation latitude	moderate	moderate	wide

[a]Ref. 67.

International Specialty Products has been especially active in promoting and developing this technology. Success with the development of such coatings requires significant help with formulation details and the reader is referred to the appropriate trade literature for greater detail (70).

Polymer–Polymer Compatibility. Frequently when polymers are mixed together they are immiscible because the combinatorial entropy of mixing is too small to overcome the enthalpy changes, which are usually positive. This small entropy of mixing is a result of the high mol wt nature of the component polymers. If the component polymers exhibit a specific interaction such as hydrogen bonding, Van der Waals, or electrostatic, etc, then miscibility can occur (71). In the case of PMVE–polystyrene, the blend presents a lower critical solution temperature (LCST) and the miscibility region depends on the molecular weight of the polymers. The interaction in this case is between the electrons of the ether groups and the aromatic polystyrene ring. In fact, PMVE can function as a diluent for isotactic polystyrene enhancing spherulite formation (72). Depending on the molecular weight and tacticity of the PMVE employed, separated regions of crystallized PMVE can function as reinforcement for polystyrene blends and offer improved plastic properties (73).

Derivatives of styrene such as poly(α-methylstyrene) results in immiscibility with PMVE. Only when hydrogen-bonding sites are incorporated into the poly(α-methylstyrene) can miscibility be re-established (74). Such hydrogen-bonding monomers, eg, vinyl phenol, can be incorporated in a variety of incompatible copolymers, resulting in modified copolymers now miscible with PMVE (75). Small-molecule polymer analogues have been employed with inverse gas chromatography to predict polymer compatibility with PMVE. This predictive technique shows that PMVE is compatible with poly(vinyl propionate) and poly(vinyl butyrate), but not with poly(vinyl acetate) (76). Obviously, small-polymer structural changes can have significant impact on compatibility because small energy differences are involved.

Another mechanism of compatibility is hydrogen bonding. Numerous examples illustrate that homogeneous polymer blends are possible when the polymers can interact by an acid–base mechanism. An interesting example of such interactions can be found in solution. For example, the interaction of strong hydrogen-bonding polymers such as poly(acrylic acid) with polybases such as PMVE or PVP in aqueous solution is strongly influenced by the degree of polyacrylic acid (PAA) neutralization (pH). This is because a certain level of cooperativity is required to form a stable complex. Such complexes can be insoluble in water (and homogeneous upon dry-down) even though each polymer by itself is soluble. When such insoluble complexes are titrated with base, a point is reached where the level of interaction is insufficient to counter electrostatic charge repulsion and the complex dissociates and dissolves. Neutralization therefore disrupts this situation of cooperativity, thus accounting for resolubilization of the precipitated complex. Interestingly, only a few anionic groups are necessary for this effect to take place (77–79).

The strong polybase behavior of PMVE is well documented, and interest in this polymer, especially from academic workers, suggests eventual interest industrially. This feature of PMVE cannot be found in acrylates or other inexpensive polymers but is inherent in PMVE polymers.

BIBLIOGRAPHY

"Reppe Chemistry" in *ECT* 1st ed., Vol. 11, p. 651, by J. M. Wilkinson, Jr., J. Werner, H. B. Haas, and H. Beller, GAF; "Vinyl Ether Monomers and Polymers" in *ECT* 2nd ed., Vol. 21, pp. 412–426, by C. E. Schildknecht, Gettysburg College; "Vinyl Polymers (Vinyl Ether Monomers and Polymers)" in *ECT* 3rd ed., Vol. 23, pp. 937–960, by E. V. Hort and R. C. Gasman, GAF Corp.

1. Bulletins concerning the Gantrez M Line (ISP) and the Lutonal M (BASF) are no longer available; if interested in archival information, however, the reader can contact the prior manufacturers directly.
2. Rapi-Cure Bulletin (VE monomers for Radp-Cure); several bulletins are available, contact ISP Corp., Wayne, N.J.
3. Gantrez ES&S product line brochures; several bulletins are available, contact ISP Corp., Wayne, N.J.
4. C. E. Schildknecht, A. O. Zoss, and F. Grosser, *Ind. Eng. Chem.* **41**, 2891 (1949).
5. J. Wislicenus, *Justus Leibigs Ann. Chem.* **192**, 106 (1878).
6. N. D. Field and D. H. Lorenz, in E. C. Leonard, ed., *Vinyl and Diene Monomer*, Part I, John Wiley & Sons, Inc., 1970, p. 365.
7. Jpn. Kokai Tokkyo Koho 80,02,416 (Jan. 19, 1980), K. Tagaki and C. Motobashi (to Sumitomo).
8. U.S. Pat. 4,057,575 (Nov. 8, 1977), D. L. Klass (to Union Oil of California); U.S. Pat. 4,161,610 (July 17, 1979), D. L. Klass (to Union Oil of California).
9. R. E. Pasquali and F. Rodriguez, *J. Polym. Sci. Part A*, **27**, 2093 (1989).
10. T. Higashimura and M. Sawamoto, in G. Allen and J. Bevington, eds., *Comprehensive Polymer Science* Pergamon, Oxford, U.K., 1989, p. 673.
11. H. Imai, T. Saegusa, and J. Furukawa, *Makromol. Chem.* **61**, 92 (1965).
12. G. Schroder, in *Ullmann's Encyclopedia of Industrial Chemistry*, Vol. A22, VCH Publishers, Weinheim, Germany, 1993.
13. F. Bolza and F. E. Treloar, *Makromol. Chem.* **181**, 83 (1980).
14. M. Delfini and co-workers, *Macromolecules*, **16**, 1212 (1983).
15. K. Hayashi, K. Hayashi, and S. Okamura, *J. Polym. Sci. Part A*, **9**, 2305 (1971).
16. A. Deffieux and co-workers, *Polymer*, **24**, 573 (1983).
17. H. Park, E. M. Pearce, and T. K. Kwei, *Macromolecules*, **23**, 434 (1990).
18. A. Ledwith and H. J. Woods, *J. Chem. Soc. B*, 310 (1970).
19. T. Fueno and co-workers, *J. Polym. Sci. Part A*, **17**, 1447 (1969).
20. H. Yuki and co-workers, *Polym. J.* **1**, 269 (1970).
21. N. L. Owen and H. Shepard, *Trans. Faraday Soc.* **60**, 634 (1964).
22. C. E. Schildknecht, A. O. Zoss, and C. McKinley, *Ind. Eng. Chem.* **39**, 180 (1947).
23. C. E. Schildknecht and co-workers, *Ind. Eng. Chem.* **40**, 2104 (1948).
24. G. Natta, I. Bassi, and P. Corradini, *Makromol. Chem.* **18/19**, 455 (1956).
25. S. Okamura, T. Higashimura, and H. Yamamoto, *J. Polym. Sci.* **33**, 510 (1958).
26. S. Okamura, T. Higashimura, and H. Yamamoto, *J. Polym. Sci.* **39**, 507 (1959).
27. K. Hatada and co-workers, *Poly. J.* **15**, 719 (1983).
28. E. J. Vandenberg, *J. Polym. Sci. Part C*, **1**, 207 (1963).
29. S. Aoki, K. Nakamura, and T. Otsu, *Makromol. Chem.* **115**, 282 (1968).
30. S. Okamura, T. Higashimura, and T. Watanabe, *Makromol. Chem.* **50**, 137 (1961).
31. K. Iwasaki, *J Polym. Sci.* **56**, 27 (1962).
32. T. Kunitake and K. Takarabe, *Makromol. Chem.* **182**, 817 (1981).
33. S. Murahashi and co-workers, *J. Polymer Sci. Part B*, **4**, 59, 65; **3**, 245 (1965).
34. M. Miyamoto, M. Sawamoto, and T. Higashimura, *Macromolecules*, **17**, 265 (1984).
35. T. Higashimura, M. Miyamoto, and M. Sawamoto, *Macromolecules*, **18**, 611 (1985).
36. M. Sawamoto, C. Okamoto, and T. Higashimura, *Macromolecules*, **20**, 2045 (1987).

37. S. Aoshima and T. Higashimura, *Poly. Bull. (Berlin)*, **15**, 417 (1986).
38. Y. Kishimoto, S. Aoshima, and T. Higashimura, *Macromolecules*, **22**, 3877 (1989).
39. T. Higashimura, Y. Kishimoto, and S. Aoshima, *Poly. Bull. (Berlin)*, **18**, 111 (1987).
40. M. Sawamoto and T. Higashimura, *Makromol. Chem. Macromol. Symp.* **32**, 131 (1990).
41. T. Loontjens, F. Derks, and E. Kleuskens, *Polym. Bull (Berlin)*, **28**, 519 (1992).
42. M. Kamigaito and co-workers, *Macromolecules*, **25**, 6400 (1992).
43. M. Kamigaito and co-workers, *Macromolecules*, **26**, 1643 (1993).
44. H. Fukui, M. Sawamoto, and T. Higashimura, *J. Polym. Sci. Part A: Polym. Chem.* **32**, 2699 (1994).
45. T. Ohmura, M. Sawamoto, and T. Higashimura, *Macromolecules*, **27**, 3714 (1994).
46. S. Kanaoka and co-workers, *Macromolecules*, **25**, 6407 (1992).
47. S. Kanaoka, M. Sawamoto, and T. Higashimura, *Macromolecules*, **25**, 6414 (1992).
48. S. Kanaoka and co-workers, *J. Polym. Sci. Part B: Polym. Physics*, **33**, 527 (1995).
49. M. Biswas, A. Mazumdar, and P. Mitra, in H. F. Mark and co-workers, eds., *The Encyclopedia of Polymer Science and Engineering*, 2nd ed., Vol. 17, Wiley-Interscience, New York, 1989, p. 446.
50. J. P. Kennedy and E. Marechal, *Carbocationic Polymerization*, John Wiley & Sons, Inc., N.Y., 1982, Chapt. 5.
51. H. Yuki, K. Hatada, and M. Takeshita, *J. Poly. Sci. Part A*, **7**, 667 (1969).
52. M. L. Hallensleben, *Makromol. Chem.* **144**, 267 (1970).
53. J. M. G. Cowie, *Alternating Copolymers*, Plenum Publishing Corp., New York, 1985.
54. J. M. G. Cowie, in G. Allen and J. Bevington, eds., *Comprehensive Polymer Science*, Pergamon Press, Oxford, U.K., 1989, Chapt. 22.
55. J. Furukawa, in Ref. 49, Vol. 4, p. 233.
56. S. A. Jones and D. A. Tirrell, *Macromolecules*, **19**, 2080 (1986).
57. S. A. Jones and D. A. Tirrell, *J. Polym, Sci. Polym. Chem.* **25**, 3177 (1987).
58. C. C. Price, *J. Polym. Sci.* **3**, 772 (1948).
59. H. K. Hall, Jr., and A. B. Padias, *Macromol. Symp.* **84**, 15 (1994).
60. G. B. Kharas and H. Ajbani, *J. Polym. Sci. Part A*, **31**, 2295 (1993).
61. C. S. Wu, L. Senak, and E. G. Malawer, *J. Liq. Chrom.* **12**(15), 2901 (1989).
62. *Ibid.*, p. 2919.
63. S. Kopolow, Y. T. Kwak, and M. Helioff, *Cosmetics Toiletries* (May, 1993).
64. U. P. Strauss, B. W. Barbieri, and G. Wong, *J. Phys. Chem.* **83**, 2840 (1979).
65. P. J. Martin and U. P. Strauss, *Biophys. Chem.* **11**, 397 (1980).
66. U. P. Strauss and M. S. Schlesinger, *J. Phys. Chem.* **82**, 571 (1978).
67. J. A. Dougherty and co-workers, *Paint and Ink International*, FMJ International Publishing, Ltd, Surrey, U.K., 1994.
68. J. Plotkin and co-workers, *Proceedings of Radtech 1992 Conference*, Boston, Mass., 1992.
69. J. A. Dougherty and F. J. Vara, *Proceedings of Radtech 88-North Americal Conference*, New Orleans, La., 1988.
70. Technical bulletins, ISP, Wayne, N.J., and BASF, Parsippany, N.J., 1997.
71. T. Shiomi and co-workers, *Macromolecules*, **23**, 229 (1990).
72. L. Amelino and co-workers, *Polymer*, **31**, 1051 (1990).
73. G. Beaucage and R. S. Stein, *Polymer*, **35**, 2716 (1994).
74. J. M. G. Cowie and A. Reilly, *Polymer*, **33**, 4814 (1992).
75. Y. Yu and co-workers, *Macromol. Symp.* **84**, 307 (1994).
76. S. Dutta and co-workers, *Polymer*, **34**(16), 3500 (1993).
77. I. Iliopoulis and R. Audebert, *Eur. Polym. J.* **24**(2), 171 (1988).
78. I. Iliopoulis, J. L. Halary, and R. Andebert, *J. Polym. Sci. Part A*, **26**, 275 (1988).
79. *Ibid.*, p. 2093.

General References

ISP Corp. Bulletin 2302-108, ISP Corp., Wayne, N.J.
D. M. Jones and N. F. Woods, *J. Chem. Soc.* 5400 (1964).
A. Ledwith and J. J. Woods, *J. Chem. Soc.* **B**, 753 (1966).
J. V. Crivello, *Adv. Polym. Sci.* **62**, 1 (1984).

ROBERT B. LOGIN
Sybron Chemicals Inc.

N-VINYLAMIDE POLYMERS

N-Vinylamide-based polymers, especially the *N*-vinyllactams, such as poly(*N*-vinyl-2-pyrrolidinone) [9003-39-8] or simply polyvinylpyrrolidinone (PVP), continue to be of major importance to formulators of personal-care, pharmaceutical, agricultural, and industrial products because of desirable performance attributes and very low toxicity profiles. Because of hydrogen bonding of water to the amide group, many of the *N*-vinylamide homopolymers are water-soluble or dispersible. Like proteins, they contain repeating (but pendant) amide (lactam) linkages and share several protein-like characteristics (1). Many studies have actually employed PVP as a substitute for proteins, eg, in simplifying the chemistry of the effects of radiation on polymers (2). Proteins are extremely complicated molecules with not only sequence distribution but tertiary bonding and structural complexity and it is an oversimplification to compare them to PVP, but the effects of radiation on PVP can be more readily studied. PVP can even be considered as a uniform synthetic protein-like analogue. By itself it does not enter into intermolecular hydrogen bonding, thus affording low viscosity concentrates, and also, unlike the proteins, PVP is soluble in polar solvents like alcohol. But even given these differences, the chemistry of PVP, the most commercially successful polymer of the class, is in many respects similar to that of proteins because of amide linkages sharing with them complexation to large anions such as polyphenols, anionic dyes, and surfactants. In addition to the ability to complex, PVP and its analogues along with a large assortment of copolymers are excellent film-formers. They exhibit the ability to interact with a variety of surfaces by hydrogen or electrostatic bonding, resulting in protective coatings and adhesive applications of commercial significance such as hair-spray fixatives, tablet binders, disintegrants, iodophors, antidye redeposition agents in detergents, protective colloids, dispersants, and solubilizers, among many others.

Monomers

N-Vinylamides and *N*-vinylimides can be prepared by reaction of amides and imides with acetylene (3), by dehydration of hydroxyethyl derivatives (4), by pyrolysis of ethylidenebisamides (5), or by vinyl exchange (6), among other methods; the monomers are stable when properly stored.

Only *N*-vinyl-2-pyrrolidinone (VP) [88-12-0] is of significant commercial importance and hence is the principal focus of this article. Vinylcaprolactam is available (BASF) and is growing in importance, and vinyl formamide is available as a developmental monomer (Air Products). Some physical properties are given in Table 1.

N-Vinyl-2-Pyrrolidinone. Commonly called vinylpyrrolidinone or VP, *N*-vinyl-2-pyrrolidinone was developed in Germany at the beginning of World War II. It is a clear, colorless liquid that is miscible in all proportions with water and most organic solvents. It can polymerize slowly by itself but can be easily inhibited by small amounts of ammonia, sodium hydroxide (caustic pellets), or antioxidants such as *N,N'*-di-*sec*-butyl-*p*-phenylenediamine. It is stable in neutral or basic aqueous solution but readily hydrolyzed in the presence of acid to form 2-pyrrolidinone and acetaldehyde. Properties are given in Table 2.

Table 1. Physical Properties of Selected Vinylamides and Vinylimides

Compound	CAS Registry Number	Bp, $°C_{kPa}{}^{a}$	Mp, °C
N-vinylacetamide	[5202-78-8]	107−109	
N,N-methylvinylacetamide	[3195-78-6]	$70_{3.3}$	
N-vinylacetanilide	[4091-14-9]	$102-105_{0.13}$	52
N-vinyl-2-piperidinone	[4370-23-4]	$125-126_{3.3}$	45
N-vinylcaprolactam	[2235-00-9]	$129-130_{2.7}$	34.5
N-vinylphthalimide	[3485-84-5]	$128-130_{3.3}$	86.5
N-vinyl-2-oxazolidinone	[4271-26-5]	$77-78_{0.067}$	
N-vinyl-5-methyl-2-oxazolidinone	[3395-98-0]	$105-108_{0.33}$	

^{a}To convert kPa to mm Hg, multiply by 7.5. Pressure = 101.3 kPa (760 mm Hg) if not shown.

Table 2. Properties of N-Vinyl-2-Pyrrolidinone (Commercial Production)

Property	Value
mol wt	111
assay, %	98.5[a]
moisture content, %	0.2[b]
color (APHA)	100[b]
vapor pressure, Pa[c]	
at 17°C	6.7
24°C	13.3
45°C	67
54°C	133
64°C	266
77°C	667
boiling point at 400 mm Hg	193°C
freezing point	13.5°C
flash point (open cup)	98.4°C
fire point	100.5°C
viscosity at 25°C, mPa·s(=cP)	2.07
specific gravity (25/4°C)	1.04
refractive index, n_D^{25}	1.511
solubility	completely miscible in water and most organic solvents, including methanol, ethyl acetate, methylene chloride, ethyl ether, and hydrocarbons in general
ultraviolet spectrum	no significant absorption at wavelengths longer than 220 nm

[a]Value is minimum.
[b]Value is maximum.
[c]To convert Pa to mm Hg, multiply by 0.0075.

Commercially available VP is usually over 99% pure but does contain several methyl-substituted homologues and 2-pyrrolidinone. Even at this high level of purity, further purification is required if reliable kinetic data concerning rates of polymerization are desired. This can be accomplished only by recrystallization, because distillation will not separate methyl-substituted isomers (7).

Manufacture. The principal manufacturers of N-vinyl-2-pyrrolidinone are ISP and BASF. Both consume most of their production captively as a monomer for the manufacture of PVP and copolymers. The vinylation of 2-pyrrolidinone is carried out under alkaline catalysis analogous to the vinylation of alcohols. 2-Pyrrolidinone is treated with ca 5% potassium hydroxide, then water and some pyrrolidinone are distilled at reduced pressure. A ca 1:1 mixture (by vol) of acetylene and nitrogen is heated at 150–160°C and ca 2 MPa (22 atm). Fresh

2-pyrrolidinone and catalyst are added continuously while product is withdrawn. Conversion is limited to ca 60% to avoid excessive formation of by-products. The N-vinyl-2-pyrrolidinone is distilled at 70–85°C at 670 Pa (5 mm Hg) and the yield is 70–80% (8).

Shipment and Storage; Specifications. N-Vinyl-2-pyrrolidinone is available in tank cars and tank trailers and in drums of various sizes. Shipping containers are normally steel or stainless steel. Tank cars are provided with heating coils to facilitate unloading in cold weather. Rubber, epoxy, and epoxy–phenolic coatings are attacked and must be avoided. Carbon steel has been successfully used for storage tanks, but stainless steel preserves product quality better. Aluminum and certain phenolic coatings are also satisfactory.

Toxicity Data on N-Vinyl-2-Pyrrolidinone. Results of a chronic inhalation study in rats warrant a review of industrial hygiene practices to assure that VP vapor concentrations are maintained at a safe level. One of the manufacturers, ISP, recommends that an appropriate workplace exposure limit be set at 0.1 ppm (vapor) (9). Additionally, normal hygienic practices and precautions are recommended, such as prompt removal from skin and avoidance of ingestion. In case of accidental eye contact, immediately flush with water for at least 15 minutes and seek medical attention. Refer to the manufacturers' Material Safety Data Sheets for more detailed information. Table 3 provides some toxicity data.

Table 3. Summary of Toxicity Data for N-Vinyl-2-Pyrrolidinone

Test	Result
acute oral LD_{50}	1.5 mL/kg (rats)
acute dermal LD_{50}	0.56 g/kg (rabbits)
acute inhalation LC_{50}	700 ± 100 ppm (rats)
eye irritation	severe (rabbits)
primary irritation index (PII)	0.38 (rabbits)
Skin Repeated Insult Patch Test	not a primary irritant or sensitizer (humans)
subacute inhalation	no gross or clinical abnormal effects; subacute/ chronic inflammation of respiratory tract at 16.5 and 66 ppm (rats)
subchronic inhalation	evidence of liver damage at 15, 45, and 120 ppm; no evidence of toxicity at 1 ppm (rats)
chronic inhalation	benign and malignant tumors of the nasal mucosa at the 10 and 20 ppm levels; liver tumors noted at 20 ppm
mutagenicity	negative in a battery of five assays

Homopolymerization of N-Vinyl-2-Pyrrolidinone

VP was originally polymerized in bulk by heating in the presence of small amounts of hydrogen peroxide. This neat polymerization was a difficult process to work up, requiring crushing of the solidified polymer mass and extraction with ether to remove unreacted monomer and by-products. However, it was important to the original application as a blood substitute because it afforded low molecular weight (10,11). Low molecular weight is necessary for excretion from the kidneys (12). Bulk polymerization favors the tendency of VP to undergo chain transfer to monomer. Neat VP polymerized with di-*t*-butyl peroxalate in the presence of a nitroxide scavenger that exclusively traps carbon centered radicals generates considerable nonvinyl radicals by chain transfer (13,14). VP (and presumably PVP) can, under the right circumstances, undergo chain transfer, and this route is more prevalent as the concentration of monomer is increased.

Because VP and PVP are soluble in water, early workers realized that polymerization could be more easily controlled in such a high heat capacity solvent. In the presence of acid, VP readily hydrolyzes, and when initiated with hydrogen peroxide, the pH drops quickly into the acidic range. The problem this presents was solved by buffering with bases. Of all of the bases tried by the early German chemists, ammonia not only prevented hydrolysis by neutralizing acidic by-products, it accelerated the polymerization. The early workers found that with an optimized concentration of monomer and ammonia level, the molecular weight was reproducibly controlled by the hydrogen peroxide level. Even relatively high molecular weight could be achieved by small amounts of hydrogen peroxide, but such levels might easily be compromised by unproductive side reactions. High molecular weight homopolymers are more reliably produced by initiation with organic peroxides and azo initiators.

Ammonia H$_2$O$_2$ Initiation.   The lower molecular weight grades (K-15 and K-30) of PVP are prepared industrially with an ammonia/H$_2$O$_2$ initiation system. Such products are the standards for the pharmaceutical industry and conform to the various national pharmacopeias. Several papers have appeared concerning the mechanism of this polymerization (15).

The proposed rate expression for the ammonia/H$_2$O$_2$ process is as follows:

$$R_p = k[\text{H}_2\text{O}_2]^{1/2}[\text{NH}_3]^{1/4}[\text{VP}]^{3/2} \tag{1}$$

Comparing this to the theoretical expression based on the steady state approximation suggests that the mechanism is not straightforward:

$$R_p = k_p[\text{M}] \frac{(fkd\ [\text{I}])^{1/2}}{kt} \tag{2}$$

Higher than first order for monomer, such as the 3/2 power suggests that VP is involved in initiation (17). If the efficiency of initiation is a function of the monomer concentration, then $f = f^1[\text{M}]$, and substituting in equation 2 gives

$$Rp = k_p[\text{M}]^{3/2} \frac{(f^1kd[\text{I}])^{1/2}}{k_t} \tag{3}$$

The $[\text{I}]^{1/2}$ is reflected in $[\text{H}_2\text{O}_2]^{1/2}$ but $[\text{NH}_3]^{1/4}$ can be explained by the finding (16) that the rate of polymerization is proportional to $[\text{NH}_4^+]^{1/2}$. If the equilibrium expression for $\text{NH}_3/\text{NH}_4^+$ is solved for $[\text{NH}_4^+]$ and this expression substituted, the quarter power for ammonia is apparent.

Several papers (6,18) have appeared that attempt to reconcile the ability of H$_2$O$_2$ to act as a rather strong transfer agent (hydrogen donor), generating the weakly initiating species HOO·, with its ability to act as a source of HO· hydroxy radicals that are known to be active vinyl initiators. Such studies demonstrate that HO· generated by photolysis behaves classically as an initiator for PVP (the rate is, as expected, first-order in VP and half-order in H_2O_2). Polymerization with 2,2'-azobisisobutyronitrile (AIBN) in the presence of H_2O_2 demonstrates that H_2O_2 in this case acts as a proton donor, reducing molecular weight, suggesting it also functions similarly during NH_3/H_2O_2 initiation. Even with other water-soluble initiators, VP behaves classically, with the rate expression being first-order in monomer and half-order in initiator (19,20). This would indicate that VP is a vinyl monomer with normal behavior and hence the H_2O_2/NH_3 initiation system is unusual. The evidence clearly demonstrates that it is a redox initiator system requiring trace amounts of cuprous or ferrous salt (21). Other bases such as NaOH or KOH can be employed to replace NH_3 if precautions are taken to sequester these metal ions, preventing them from being deactivated in the redox complex (22).

In one of the few published studies of H_2O_2-initiation, it is shown that in the case of methacrylamide, the monomer participates in its own initiation by reacting with H_2O_2, forming an intermediate hydroperoxide (23). Subsequently, this hydroperoxide generates hydroxyl radicals capable of initiation. Like that of methacrylamide, VP polymerization is very sensitive to molecular oxygen reacting faster with it than propagation to polymer (18), and a similar reaction might be at work with VP that would be expected to generate an intermediate hydroperoxide capable of entering into a redox initiation system. The proposed mechanism would also explain the formation of 2-pyrrolidinone as a consequence of redox polymerization, dispelling the previous belief that 2-pyrrolidinone was a result of primary radical termination caused by reaction of hydroxyl radicals with the growing chain, followed by hydrolysis of the hemiacetal (24,25) subsequently formed. Hydroxyl radicals afford PVP with one hydroxyl per chain, correlating well with a mechanism that relies on hydroxyl radical initiation and strong H_2O_2 chain transfer (26). In this case, one end of the polymer is not an aldehyde but rather a hydroxyl group, and no evidence for other than proton termination could be found, producing a methylene terminus at the other end of the polymer chain. Figure 1 illustrates the proposed mechanism and explains the formation of acidic by-products responsible for the acidic pH drift during polymerization.

Organic Peroxides and Azo Initiation. The H_2O_2/ammonia initiation system is not employed commercially in the manufacture of higher molecular weight homologues; they are prepared with organic initiators. Such polymerizations follow simple chain theory and are usually performed in water commercially (27). The rate of polymerization is at a maximum in aqueous media at pH 8–10 and at 75 wt % monomer (28,29). Polymerization rates follow the polarity and hydrogen bonding capability of the solvent (30). One possible explanation for this fact is that water is most capable of reducing the apparent negative charge on the beta carbon VP's vinyl group by hydrogen bonding to the pyrrolidinone carbonyl and polar interactions. Such a reduction permits the electron-rich radical terminus to more easily approach another VP and hence allows the acceleration (30). Alternatively, PVP may be somewhat more hydrophobic than VP forming associates ("micelles") capable of enhancing the rate of polymerization by concentrating monomer close to the reacting polymer terminus. This is the reason why

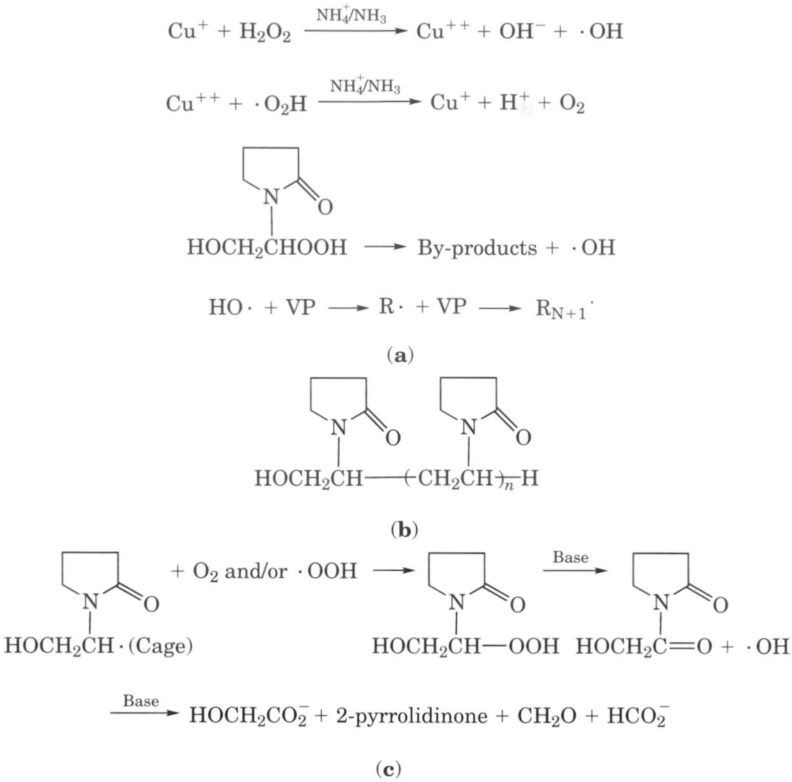

$$\text{Cu}^+ + \text{H}_2\text{O}_2 \xrightarrow{\text{NH}_4^+/\text{NH}_3} \text{Cu}^{++} + \text{OH}^- + \cdot\text{OH}$$

$$\text{Cu}^{++} + \cdot\text{O}_2\text{H} \xrightarrow{\text{NH}_4^+/\text{NH}_3} \text{Cu}^+ + \text{H}^+ + \text{O}_2$$

HOCH₂CHOOH ⟶ By-products + ·OH

HO· + VP ⟶ R· + VP ⟶ R_{N+1}·

(a)

HOCH₂CH——(CH₂CH)ₙ—H

(b)

+ O₂ and/or ·OOH ⟶ ... $\xrightarrow{\text{Base}}$ ...

HOCH₂CH·(Cage)   HOCH₂CH—OOH   HOCH₂C=O + ·OH

$\xrightarrow{\text{Base}}$ HOCH₂CO₂⁻ + 2-pyrrolidinone + CH₂O + HCO₂⁻

(c)

Fig. 1. Mechanism of NH₃/H₂O₂ polymerization of VP: (**a**) initiation; (**b**) end groups; (**c**) 2-pyrrolidinone generation.

even in relatively dilute aqueous solutions the rate can be substantial. The hydrophobic effect accounts for a higher VP concentration at the reactive polymer terminus (31).

Cationic Polymerization. VP polymerizes to low molecular weight (oligomers) with typical cationic initiators, such as boron trifluoride etherate (32). This reaction requires high concentrations, if not neat, of monomer and scrupulously anhydrous conditions for high yields; VP will readily hydrolyze to 2-pyrrolidinone and acetaldehyde even in the presence of trace moisture when catalyzed by strongly acidic reagents. Pyrrolidinone derivatives apparently complex and deactivate cationic polymerization catalysts and generally present an unfavorable environment for polymerization (33). Thus, initiating species are relatively short-lived and readily deactivated, or undergo chain transfer; hence, more than catalytic amounts of initiator are required for high yields.

Interest has been rekindled in cationic polymerization by the discovery that carboxylic acid groups trapped in insoluble matrices like activated carbon or poly(glutamic acid) can generate higher mol wt polymers (34–36). Additionally, oxoaminium salts (37) derived from 2,2,6,6-tetramethylpiperidine-1-oxyl (TEMPO) or anodic polymerization (38) on platinum electrode surfaces will also afford higher molecular weight polymers. Such cationically generated polymers would be expected to afford microstructure and greater tacticity because of the higher activation energy for inversion associated with a growing cationic terminus vs a growing free-radical terminus. Unfortunately, none of the above references present detailed polymer structural characterizations.

Microstructure. Interest in PVP microstructure and the potential for tacticity has been reviewed (39,40). PVP generated by free radicals has been shown to be atactic except when polymerization is conducted in water. In this case some syndiotacticity is observed (40). In the presence of syndiotactic templates of poly(methacrylic acid) (or poly(MAA)), VP will apparently polymerize with syndiotactic microstructure, although proof is lacking (41–45). The reverse, polymerization of MAA in the presence of PVP, affords, as expected, atactic poly(MAA) (46,47).

Advances in VP cationic polymerization hold out the possibility of tacticity, and the study of this route to crystalline homologues continues to be of interest.

Proliferous Polymerization. Early attempts to polymerize VP anionically resulted in proliferous or "popcorn" polymerization (48). This was found to be a special form of free-radical addition polymerization, and not an example of anionic polymerization, as originally thought. VP contains a relatively acidic proton alpha to the pyrrolidinone carbonyl. In the presence of strong base such as sodium hydroxide, VP forms cross-linkers *in situ*, probably by the following mechanism:

EVP (ethylidene vinylpyrrolidinone)

EBVP (ethyldiene-bis-vinylpyrrolidinone)

Both ethylidene vinyl pyrrolidinone (EVP) and ethylidene-bis-vinylpyrrolidinone (EBVP) are generated in about a 10:1 ratio, respectively (24). At the temperature required to generate these cross-linkers and when their concentration reaches some minimum level, usually a few percent, proliferous polymerization begins (49). The same situation can be reached by the addition of a suitable cross-linker (50). Although no initiator is required, the polymerization can be prevented in the presence of typical free-radical inhibitors or initiated by very small amounts of AIBN. Reviews indicate that the rate of polymerization accelerates because the initially formed cross-linked seeds swell to generate active sites by bond homolysis and that growing chains resist termination because of the rigid cross-linked structure. Very high conversions can be achieved and the resulting product, a granular "popcorn" mass, can be freed of residual monomer/soluble polymer by careful washing. Drying results in a free-flowing white powder (51).

Crospovidones are produced commercially by these two processes, ie, *in situ* generation of cross-linker or addition of divinylimidazoline, and they are indistinguishable by ir. Both types exhibit a T_g of 190–195 °C, which is not that much above the 175 °C of high molecular weight, soluble PVP (24). Proliferous polymers prepared with easily hydrolyzed cross-linker containing an imine linkage do not further swell even when the cross-links are hydrolyzed (50). In order to reach the low swell volume typical of these resins with a typical VP/cross-linker polymerized with free-radical initiator, sizeable amounts of cross-linker are actually required, resulting in much higher (240 °C) T_gs. The crospovidones are therefore unusually high molecular weight, highly chain-entangled polymers having covalent cross-links that most likely retard the termination reaction during polymerization and are not entirely responsible for the resulting mechanical properties, such as swell ratio. Even hydrolyzing such cross-links is not sufficient to cause dissolution.

The crospovidones are easily compressed when anhydrous but readily regain their form upon exposure to moisture. This is an ideal situation for use in pharmaceutical tablet disintegration and they have found commercial application in this technology. PVP strongly interacts with polyphenols, the crospovidones can readily remove them from beer, preventing subsequent interaction with beer proteins and the resulting formation of haze. The resin can be recovered and regenerated with dilute caustic.

PVP Hydrogels

Cross-linked versions of water-soluble polymers swollen in aqueous media are broadly referred to as hydrogels (52) and have a growing commercial utility in such applications as oxygen-permeable soft contact lenses (qv) (53) (Table 4) and controlled-release pharmaceutical drug delivery devices (54). Cross-linked PVP and selected copolymers fit this definition and are of interest because of the following structure/performance characteristics:

Structure	Performance	Benefit
nonionic	compatibility with other ingredients	stable formulation

Table 4. Generic Names of Polymeric Compositions Used in Soft Contact Lenses

USAN Generic Name	Polymer composition[a]	Water, %	Trademarks	Manufacturer
Droxifilcon-A	copolymer of HEMA and MA modified with poly(2-vinyl-pyrrolidone)	47	Accugel	Strieter Labs
Lidofilcon-B	copolymer of MMA and 2-vinylpyrro-lidone (VP)	79	Sauflon PW	American Medical Optics
Surfilcon-A	copolymer of MMA, VP, and other methacrylates	74	Permaflex	Cooper Vision, Inc.
Tetrafilcon-A	terpolymer of HEMA, MMA, and VP cross-linked with divinylbenzene	42.5	Aosoft Aquaflex	American Optical Corp. Cooper Vision, Inc.
Vifilcon-A	copolymer of HEMA and MA with PVP cross-linked with EGDM	55	Softcon	American Optical Corp.

[a]HEMA = Hydroxyethylmethacrylate; EGDM = ethyleneglycoldimethacrylate.

pyrrolidinone	low toxicity	nonirritating/nonthrombogenic
	complexation−actives/O_2	controlled release-transport
	high T_g	mechanical stability
	hydrolytic stability	storage-stable
ethylene backbone	nonbiodegradable, hydrolytic stability	resists biocontamination storage-stable
cross-links	swell volume/viscosity	mechanical stability/diffusion control

Cross-linked PVP can be prepared by several routes other than proliferous polymerization PPVP (crospovidones). Although a hydrogel, the swell volume of this type of polymer cannot be controlled over a large increment because the granular particles cannot be formed into larger uniform assemblies. These limitations can be overcome by the polymerization of VP in the presence of a few percent of suitable cross-linker utilizing standard free-radical initiation by, for example, AIBN (55–57) or actinic radiation (gamma rays) (58,59). This results in a lightly cross-linked PVP. If the polymerization is carried out incorrectly, significant amounts of uncross-linked soluble polymer may be present that must be removed before a meaningful physical analysis such as swell ratio can be accomplished. The solution to this problem is to balance the reactivity ratios of the cross-linker and other comonomers with those of VP to obtain uniform copolymerization and cross-linking (60,61). Not only does this reduce the level of soluble uncross-linked polymer but affords crystal-clear hydrogels so important for use in contact lenses. Allyl-substituted sugars used to generate cross-linked

polyacrylic acid gels, carbomers, employed as thickeners in pharmaceutical formulations, have been shown to work well with VP (62).

Cross-linked PVP can also be obtained by cross-linking the preformed polymer chemically (with persulfates, hydrazine, or peroxides) or with actinic radiation (63). This approach requires a source of free radicals capable of hydrogen
abstraction from one or another of the labile hydrogens attached alpha to the
pyrrolidone carbonyl or lactam nitrogen. The subsequently formed PVP radical
can combine with another such radical to form a cross-link or undergo side reactions such as scission or cyclization (64,65), thus:

$n = 2$ or 3

If the starting PVP homopolymer is too low in molecular weight or too
dilute, cyclization or cleavage is preferred (65,66). However, because of the high
T_g of PVP, the backbone is sufficiently rigid to avoid reorientation during bond
homolysis so that the same bond has a good chance of reforming; hence, PVP
yields cross-linked structures in preference to cleavage (67) and PVP hydrogels
formed by E-beam have become commercially important for use as conductive
electrodes for medical applications (68).

Poly(*N*-Vinyl-2-Pyrrolidinone)

Poly(*N*-vinyl-2-pyrrolidinone) (PVP) is undoubtedly the best-characterized and
most widely studied *N*-vinyl polymer. It derives its commercial success from its
biological compatibility, low toxicity, film-forming and adhesive characteristics,
unusual complexing ability, relatively inert behavior toward salts and acids, and
thermal and hydrolytic stability.

First developed in Germany by I. G. Farben (W. Reppe) during the 1930s,
PVP was subsequently widely used in Germany as a blood-plasma substitute
and extender during World War II (69). In the United States, it has been manufactured since 1956 by ISP, and more recently by BASF.

Molecular Weight and K-Value.   Poly(*N*-vinyl-2-pyrrolidinone) is described
in the *United States Pharmacopeia* (70) as consisting of linear *N*-vinyl-2-
pyrrolidinone groups of varying degrees of polymerization. The molecular
weights of PVP samples are determined by size exclusion chromatography

(sec), osmometry, ultracentrifugation, light-scattering, and solution viscosity techniques. The most frequently employed method of determining and reporting the molecular weight of PVP samples utilizes the sec/low angle light scattering (lalls) technique (71,72).

A frequently used and commonly recognized method of distinguishing between different molecular weight grades of PVP is the K value. Its nomenclature is accepted by the USP, FDA, and other authoritative bodies worldwide. The *Pharmacopeia* (USP) specifies that for very low molecular weight, a 5% solution whereas for very high molecular weight, a 0.1% solution be measured. All other molecular weights employ a 1% solution. The relative viscosity is obtained with an Ostwald-Fenske or Cannon-Fenske capillary viscometer, and the K value is derived from Fikentscher's equation (73).

$$\log \frac{\eta_{rel}}{c} = \frac{75 \, K_0^2}{1 + 1.5 K_0 c} + K_0$$

where $K = 1000 \, K_0$, η_{rel} = relative viscosity, and c = concentration of the solution in g/100 mL. Solving directly for K, the Fikentscher equation is converted to:

$$K = [300c \, \log Z + (c + 1.5c \, \log Z)^2 + 1.5c \, \log Z - c]/(0.15c + 0.0003c^2)$$

where $Z = \eta_{rel}$. Table 5 illustrates a correlation chart where the K value is simply read off from a knowledge of η_{rel}.

The intrinsic viscosity $[\eta]$ may be approximated from the Fikentscher equation by:

$$[\eta] = 2.303 \, (0.001 \, K + 0.0000 \, 75 \, K^2)$$

where $[\eta]$ = intrinsic viscosity and K = K value of sample.

Utilizing the Mark-Houwink equation (74)

$$[\eta] = K \overline{M}_v^a$$

it is possible to relate the viscosity-average molecular weight ($\overline{M}_v$) to the K value.

Table 5. K_0 Value vs Relative Viscosity at 1% Concentration (wt/vol)[a]

K Value	Relative viscosity	K Value	Relative viscosity
20	1.120	60	2.031
25	1.175	65	2.258
30	1.243	70	2.527
35	1.325	75	2.846
40	1.423	80	3.225
45	1.539	85	3.678
50	1.677	90	4.219
55	1.839	95	4.870

[a]Ref. 71.

For commercial grades of unfractionated PVP prepared by similar means (presumed to exhibit similar molecular weight distribution (MWD) and degree of branching), the following regression formula can be employed (71):

$$\log \text{mol wt} = 2.82 \log K + 0.594$$

Table 6 indicates mol wt vs K value obtained by this technique. Table 7 lists M_n obtained by osmometry methods. The specifications for Technical and Pharmaceutical grades are given in Tables 8 and 9.

Glass-Transition Temperature.   The T_g of PVP is sensitive to residual moisture (75) and unreacted monomer. It is even sensitive to how the polymer

Table 6. K Value Vs Weight-Average Molecular Weight for PVP[a]

K Value	$\overline{M}_w$, amu	K Value	$\overline{M}_w$, amu
10	2,600	70	626,900
15	8,100	75	761,500
20	18,300	80	913,500
25	34,400	85	1,084,000
30	57,500	90	1,273,000
35	88,800	95	1.483,000
40	129,300	100	1,714,000
45	180,300	105	1,967,000
50	242,700	110	2,242,000
55	317,600	115	2,542,000
60	405,900	120	2,866,000
65	508,600		

[a]Ref. 71.

Table 7. Osmometry Molecular Weights for PVP[a]

Sample	Technique	$\overline{M}_n$
K-90	membrane osmometry	37,4000
K-60	membrane osmometry	67,500
K-30	vapor pressure osmometry	8,430
K-15	vapor pressure osmometry	5,170

[a]Ref. 72.

Table 8. Specifications of Technical PVP Grades

Designation	Form	K Range	Water, % max	Ash, % max	$M_v(\times 10^{-3})^a$
K-15	powder	13–19	5	0.02	10
K-15	aqueous solution	13–19	72		10
K-30	powder	26–34	5	0.02	40
K-60	aqueous solution	50–62	55	0.02	220
K-90	aqueous solution	80–100	80	0.02	630
K-90	powder	88–100	5	0.02	630
K-120	powder	115–125	5	0.02	1,450

[a]Performed at 25°C, in H_2O, using Mark-Houwink constants of $K = 1.4 \times 10^{-4}$ and $a = 0.7$.

Table 9. Specifications of Pharmaceutical PVP Grades (Povidone)

Assay	Value[a]
K value	
10–15	85–115% of stated supplier's value
16–90	90–107% of stated supplier's value
moisture, %	5
pH[b]	3.0–7.0
residue on ignition, %	0.02
aldehydes, %[c]	0.02
N-vinyl-2-pyrrolidinone, %	0.20
lead, ppm	10
arsenic, ppm	1
nitrogen, %	11.5–12.8

[a]All single values are maximum.
[b]Of a 5% solution in distilled water.
[c]Calculated as acetaldehyde.

was prepared, suggesting that MWD, branching, and cross-linking may play a part (76). Polymers presumably with the same molecular weight prepared by bulk polymerization exhibit lower T_gs compared to samples prepared by aqueous solution polymerization, lending credence to an example, in this case, of branching caused by chain-transfer to monomer.

Molecular weight also plays a significant role in T_g, which increases to a limiting value of 180°C for high purity samples above K-90 in molecular weight. The following equation applies:

$$T_g \; (°\mathrm{C}) = 175 - \frac{9685}{K^2}$$

and Table 10 illustrates this relationship with commercially available samples.

Solubility. One of PVP's more outstanding attributes is its solubility in both water and a variety of organic solvents. PVP is soluble in alcohols, acids, ethyl lactate, chlorinated hydrocarbons, amines, glycols, lactams, and nitroparaffins. Solubility means a minimum of 10 wt % PVP dissolves at room temperature (moisture content of PVP can influence solubility). PVP is insoluble in hydrocarbons, ethers, ethyl acetate, *sec*-butyl-4-acetate, 2-butanone, acetone, cyclohexanone, and chlorobenzene. Both solvent polarity and H-bonding strongly influence solubility (77).

Table 10. Glass-Transition Temperatures of PVP

Sample[a]	Measured *K* value	T_g, °C
Plasdone K-15	14.0	126
PVP K-15	14.9	130
Plasdone K-25	22.5	160
PVP K-30	27.5	163
Plasdone K-29/32	28.7	164
PVP K-60	55.5	170
PVP K-90	89.6	174

[a]Courtesy of ISP Corp.

Swelling Behavior.   One way to visualize the interaction of solvents with PVP is to examine the effect the former have on lightly cross-linked PVP, as a model for the linear polymer (78).

Such gels can be prepared from a minimum of diallyl cross-linker and VP to afford products that are mechanically stable and easy to handle as hydrogels. Such samples must be extracted because soluble polymer competes for solvent, affording lower swell volumes than those expected. The extracted and dried samples are swollen to equilibrium in a variety of solvents (Table 11). Three groups of solvents can be distinguished: the first includes those in which the cross-linked polymer swells 15–25 times. VP, H_2O, CH_3OH, C_2H_5OH, benzylamine, and chloroform are examples. Swelling drops off with longer-chain-length alcohols. Aromatic derivatives like benzyl alcohol and aniline afford similar swelling ratios to each other and to aliphatic analogues. This indicates that simple aromatic groups do not interact. When comparing volume instead of weight, water actually causes a 0.36% shrinkage of the gel. Water can therefore cross-link by hydrogen bonding to a limited but measurable extent. The second group consists of acetone, MEK, and dioxane solvents that not only increase the swell ratio (to a much lesser extent) but also increase the swell volume by 60–100%. The third group consists of hydrocarbons, benzene, carbon tetrachloride, isopropyl ether, and triethyl amine. In this case they have little or no effect on swelling. The hallmark of this group is lack of hydrogen-bonding capability.

In water, the swell ratio actually decreases with temperature at a constant rate of −0.12% per degree C. PVP gels therefore swell exothermically in water, and, as expected, heat reverses the process. Cooling back to a lower temperature results in the expected higher swell ratio being reestablished. Alcohols and other hydrogen bonding solvents cause the same effect but to a lesser extent. Nonhydrogen-bonding solvents actually cause an increase in swelling with temperature (Table 12).

Rheology.   PVP solubility in water is limited only by the viscosity of the resulting solution. The heat of solution is −16.61 kJ/mol (−3.97 kcal/mol) (79); aqueous solutions are slightly acidic (pH 4–5). Figure 2 illustrates the kinematic viscosity of PVP in aqueous solution. The kinematic viscosity of PVP K-30 in various organic solvents is given in Table 13.

Table 11. Swelling of Cross-linked Polyvinylpyrrolidone in Various Liquids at 20°C

Liquid	Degree of swelling	Liquid	Degree of swelling
1-propanol	25.6	chloroform	16.7
ethanol	24.8	ethylenediamine	15.6
isoamyl alcohol	24.8	acetone	2.1
methanol	24.0	methyl ethyl ketone	2.0
water	19.7	cyclohexanone	1.9
benzyl alcohol	19.5	dioxane	1.6
n-octanol	19.1	trimethylamine	1.07
benzylamine	17.1	carbon tetrachloride	1.03
prim-phenyl ethyl alcohol	16.9	benzene	1.02
		isopropyl ether	1.0

[a]Ref. 78.

Table 12. Temperature-Dependence of Degree of Swelling in Four Liquids[a]

Liquid	Temperature, °C	Degree of swelling
dioxane	20	1.57
	50	2.34
methyl ethyl ketone	20	1.69
	50	2.86
ethanol	20	18.3
	50	17.3
chloroform	20	24.6
	50	24.0

[a]Ref. 78.

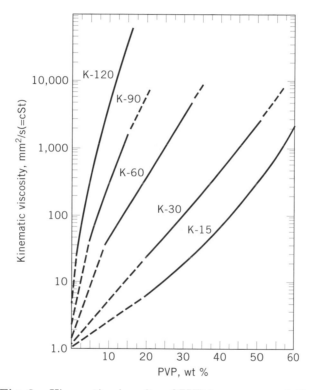

Fig. 2. Kinematic viscosity of PVP in aqueous solution.

Aqueous Solutions of PVP. Although it is soluble in a variety of polar solvents, PVP has generated significant interest because of its aqueous solubility. Water can readily hydrogen-bond to the polar, negatively charged pyrrolidinone carbonyl oxygen because pyrrolidinone, a five-membered planar lactam, affords maximum π, π-orbital overlap. The canonical resonance forms highlight the potential for a partial negative charge to form on oxygen:

Table 13. Kinematic Viscosity of PVP K-30 in Organic Solvents

Solvent	Kinematic viscosity,[a] mm^2/s(=cSt)	
	2% PVP	10% PVP
acetic acid (glacial)	2	12
1,4-butanediol	101	425
butyrolactone	2	8
cyclohexanol	80	376
diacetone alcohol	5	22
diethylene glycol	39	165
ethanol (absolute)	2	6
ethyl lactate	4	18
ethylene glycol	24	95
ethylene glycol monoethyl ether	3	12
glycerol	1,480	2,046
2-propanol	4	12
methyl cyclohexanone	3	10
N-methyl-2-pyrrolidone	2	8
methylene dichloride	1	3
monoethanolamine	27	83
nitroethane	1	3
nonylphenol	3,300	
propylene glycol	66	261
triethanolamine	156	666

[a]Kinematic viscosity = absolute viscosity/density.

The partial charge on nitrogen is sterically shielded by the polymer backbone and the surrounding pyrrolidinone methylenes. Because of high dipole moment and polarity, PVP has a noticeable effect on water structure and various methods have been proposed to measure bound water (80). One study even illustrates the different categories of water generated by freezing aqueous PVP solutions (81). The results are summarized in Figure 3, and, as can be seen, PVP is hydrated with bound water that will not freeze (concentrated solutions >57% will not freeze). PVP is therefore used as a protectorate in cryobiology (82).

The actual amount and structure of this "bound" water has been the subject of debate (83), but the key factor is that in water, PVP and related polymers are water structure organizers, which is a lower entropy situation (84). Therefore, it is not unexpected that water would play a significant role in the homopolymerization of VP, because the polymer and its reactive terminus are more rigidly constrained in this solvent and termination k_t is reduced (85).

Complexation

The combination of electrostatic interaction (induced dipole–dipole interaction) with an increase in entropy resulting from the discharge of bound water is fundamental to PVP's ability to complex with a variety of large anions.

Other factors that can stabilize such a forming complex are hydrophobic bonding by a variety of mechanisms (Van der Waals, Debye, ion-dipole, charge-

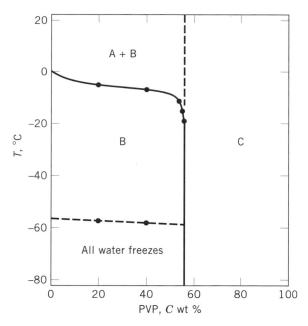

Fig. 3. Phase diagram for the three kinds of water in PVP aqueous solutions (81). A, freezable water; B, bound, nonfreezable water (six per repeat unit); C, nonfreezable water.

transfer, etc). Such forces complement the stronger hydrogen-bonding and electrostatic interactions.

Approximately a minimum $\overline{M}_n$ of 1 to 5,000 is required before complexation is no longer dependent on molecular weight for small anions such as KI_3 and 1-anilinonaphthaline-8-sulfonate (ANS) (86,87). The latter anion is a fluorescent probe that, when bound in hydrophobic environments, will display increased fluorescence and, as expected, shows this effect in the presence of aqueous PVP. PVP, when complexed with HI_3, shrinks in size as it loses hydrodynamic volume, possibly because of interchain complexation. ANS, on the other hand, causes the polymer to swell by charge repulsion because it behaves like a typical polyelectrolyte (88).

Adsorption Isotherms. Equilibrium dialysis studies indicate around 10 repeat VP units (base moles) are required to form favorable complexes (89,90). This figure can rise to several hundred for methyl orange and other anions depending on structure (91,92).

Although hydrophobic bonding is well established as a significant force stabilizing such complexes, some work suggests that such generalizations do not apply to every case (92). However, a study of the complexes of PVP with tetraanionic porphyrins has shown that the reaction of porphyrin with cupric ion is slowed dramatically in the presence of PVP. This is interpreted as demonstrating the existence of hydrophobic pockets preventing a reaction that is clearly favored if both species are in aqueous environments (93). Hydrophobic bonding has been illustrated by comparing competitive binding of butyl orange (BO) with 1-amino-4-methylamino anthraquinone-2-sulfonate (AQ) (94). The thermodynamic data for BO shows that the binding process is athermal and stabilized entirely by the entropy term. On the other hand, AQ exhibits a large enthalpy and small entropy value and its binding is by the stronger and energetic interaction caused by hydrogen bonding (NH groups of AQ) and hydrophobic interaction of the polynuclear aromatic AQ; both structural features are missing from BO (94).

Iodine Complexes.   The small molecule/PVP complex between iodine and PVP is probably the best-known example (95) and can be represented as follows:

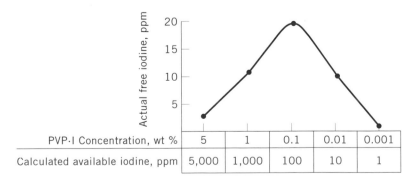

It is widely employed as a disinfectant in medicine (Povidone-iodine) because of its mildness, low toxicity, and water solubility. In actuality, the complex is based on HI_3 since HI is formed *in situ* from iodine during the manufacturing process (96). According to the *U.S. Pharmacopeia*, Povidone-iodine is a free-flowing, brown powder that contains from 9–12% available iodine. It is soluble in water and lower alcohols. When dissolved in water, the uncomplexed free iodine level is very low (Fig. 4) (97); however, the complexed iodine acts as a reservoir and by equilibrium replenishes the free iodine to the equilibrium level. This prevents free iodine from being deactivated because the free form is continually available at effective biocidal levels from this large reservoir (98). The structure of the complex has been studied and in essence is similar to the representation above (98,99). PVP will interact with other small anions and resembles serum albumin and other proteins in this regard (100). It can be "salted in" with anions such as NaSCN or "out" with Na_2SO_4 much like water-soluble proteins (101).

Phenolics.   PVP readily complexes phenolics of all types to some degree, the actual extent depending on structural features such as number and orientation of hydroxyls and electron density of the associated aromatic system. A model has been proposed (102). Complexation with phenolics can result in reduced PVP viscosity and even polymer-complex precipitation (103).

One practical result of this strong interaction is the employment of PVP to remove unwanted phenolics such as bitter tanins from beer and wine. This

PVP·I Concentration, wt %	5	1	0.1	0.01	0.001
Calculated available iodine, ppm	5,000	1,000	100	10	1

Fig. 4.   Free iodine in povidone-iodine aqueous solutions (97).

process is more easily carried out with insoluble crospovidone, which can be regenerated for reuse with dilute base (104). Soluble PVP has been employed to prevent photoyellowing of paper by complexing free phenolic hydroxyl groups in lignin (105).

Dyes.   PVP is currently (ca 1997) employed in a variety of antidye redeposition detergents as a result of its strong interaction with fugitive anionic dyes (106,107). This interaction depends on the structure of the dye. Cationic dyes complex only if they also contain hydrogen-bonding functionality. Anionic dyes complex more easily, depending on the number of anionic groups, size of the aromatic nucleolus, and number and orientation of phenolic hydroxyl groups, etc.

Anionic Surfactants.   PVP also interacts with anionic detergents, another class of large anions (108). This interaction has generated considerable interest because addition of PVP results in the formation of micelles at lower concentration than the critical micelle concentration (CMC) of the free surfactant the mechanism is described as a "necklace" of hemimicelles along the polymer chain, the hemimicelles being surrounded to some extent with PVP (109). The effective lowering of the CMC increases the surfactant's apparent activity at interfaces. PVP will increase foaming of anionic surfactants for this reason.

Because of this interaction, PVP has found application in surfactant formulations, where it functions as a steric stabilizer for example to generate uniform particle-size polystyrene emulsions (110–112). In a variety of formulations, a surfactant's ability to emulsify is augmented by PVP's ability to stabilize colloids sterically and to control rheology.

Polymer/Polymer Complexes.   PVP complexes with other polymers capable of interacting by hydrogen-bonding, ion-dipole, or dispersion forces. For example mixing of PVP with poly(acrylic acid) (PAA) in aqueous solution results in immediate precipitation of an insoluble complex (113). Addition of base results in disruption of hydrogen bonding and dissolution (114–116). Complexes with a variety of poly-acids (117) and polyphenols (118) have been reported. The interest in compatibility on a molecular level, an interesting phenomenon rarely found to exist between dissimilar polymers, is favored by the ability of PVP to form polymer/polymer complexes.

Practical applications have been reported for PVP/cellulosics (108,119,120) and PVP/polysulfones (121,122) in membrane separation technology, eg, in the manufacture of dialysis membranes. Electrically conductive polymers of polyaniline are rendered more soluble and hence easier to process by complexation with PVP (123). Addition of small amounts of PVP to nylon 66 and 610 causes significant morphological changes, resulting in fewer but more regular spherulites (124).

Copolymerization

The Q and e values of VP are 0.088 and -1.62, respectively (125). This indicates resonance interaction of the double bond of the vinyl group with the electrons of the lactam nitrogen, whence the electronegative nature. With high $e+$ monomers such as maleic anhydride, VP forms alternating copolymers, much as expected (126). With other monomers between these Q and e extremes a wide variety of possibilities exist. Table 14 lists reactivity ratios for important comonomers.

Table 14. VP Copolymerization Parameters[a]

Comonomer (M_2)	r_1	r_2	Reference
N-vinyl caprolactam	2.80	1.70	128
maleic anhydride	−0.027	0.074	129
methyl methacrylate	0.01	4.04	130
styrene	0.057	17.2	131
vinyl acetate	0.04	14.6	88
acrylic acid	3.40	0.195	131
methacrylic acid	0.100	0.880	132
dimethylaminoethylmethacrylate	0.07	4.7	133
	0.69	11.16	134

[a]Ref. 127.

Copolymerizations can be conveniently carried out in aqueous solution or in a variety of solvents, depending on monomer/polymer solubilities. Various strategies have been employed to compensate for the divergence in reactivity ratios in order to form uniform (statistical) copolymers such as semibatch or mixed monomer feeds, the goal being to add the more reactive monomer at the rate at which it is being consumed (135). Clearly, if the difference in reactivity is too great, then the amount of more reactive monomer that can be uniformly incorporated is significantly reduced. Of the monomers listed, styrene fits this category (136).

Poly(Vinylpyrrolidinone-*co*-Vinyl Acetate). The first commercially successful class of VP copolymers, poly(vinylpyrrolidinone-co-vinyl acetate) is currently manufactured in sizeable quantities by both ISP and BASF. A wide variety of compositions and molecular weights are available as powders or as solutions in ethanol, isopropanol, or water (if soluble). Properties of some examples of this class of copolymers are listed in Table 15.

Table 15. Properties of PVP/VA Copolymers[a]

	PVP–VA copolymer						
	E-735	E-635	E-535	E-335	I-735	I-535	I-335
physical form at 25°C	← clear liquid →				light yellow liquid		
solvent	SDA-40 anhydrous ethanol				← 2-propanol →		
solids after infrared drying, %	50 ± 2	50 ± 2	50 ± 2	50 ± 2	50 ± 2	50 ± 2	50 ± 2
vinylpyrrolidinone–vinyl acetate ratio	70:30	60:40	50:50	30:70	70:30	50:50	30:70
K value of 1% ethanol solution	30–50	30–50	30–50	25–35	30–40	25–35	20–30
moisture as is, Karl Fischer, % max	0.5	0.5	0.5	0.5	0.5	0.5	0.5
nitrogen, dry basis, Kjeldahl, %	8–9	7–8	5.8–6.8	3.1–4.1	8–9	5.9–6.9	3.9–4.9
specific gravity at 25°C	←			0.955 ± 0.01			→

[a]Ref. 137.

Although reactivity ratios indicate that VP is the more reactive monomer, reaction conditions such as solvent polarity, initiator type, percent conversion, and molecular weight of the growing radical can alter these ratios (138). Therefore, depending on polymerization conditions, copolymers produced by one manufacturer may not be identical to those of another, especially if the end use application of the resin is sensitive to monomer sequence distribution and MWD.

An important reason for the ongoing interest in these copolymers is that vinyl acetate reduces hydrophilicity so that applications that require less moisture-sensitive films such as those employed to set hair are less prone to plasticize and become tacky under high humidity conditions (139) (Fig. 5).

As shown in Figure 6, desirable fixative properties superior to PVP homopolymer can be specified by judicious selection of the amount of vinyl acetate. Hair sprays are limited in the molecular weight of the resin because if they are too high the resulting viscosity of the formulation will result in a poor (coarse) spray pattern. Increasing the VP/VA ratio causes properties to increase in the direction shown by the arrows.

Other applications for VP/VA copolymers are uses as water-soluble or remoistenable hot melt adhesives (140), pharmaceutical tablet coatings, binders, and controlled-release substrates.

Tertiary Amine-Containing Copolymers. Copolymers based on DMAEMA (dimethylaminoethyl methacrylate) in either free amine form or quaternized with diethyl sulfate or methyl chloride have achieved commercial significance as fixatives in hair-styling formulations, especially in the well-publicized "mousses" or as hair-conditioning shampoo additives. This success has occurred because the cationic charge affords substantive resins that strongly adhere to the hair (141).

The most successful of these products contain high ratios of VP to DMAEMA and are partially quaternized with diethyl sulfate (Polyquaternium 11) (142–144). They afford very hard, clear, lustrous, nonflaking films on the hair that are easily removed by shampooing. More recently, copolymers with methylvinylimidazolium chloride (Polyquaternium 16) (145) or MAPTAC

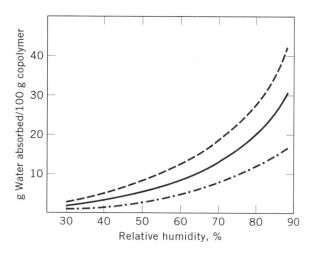

Fig. 5. Hydrophilicity of three wt % monomer ratios of PVP/VA: (– – –), 70/30; (—), 50/50; (– · – ·), 30/70.

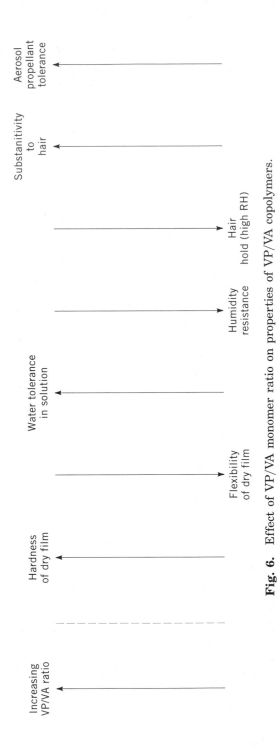

Fig. 6. Effect of VP/VA monomer ratio on properties of VP/VA copolymers.

(methacrylamidopropyltrimethyl ammonium chloride) (Polyquaternium 28) have been introduced. Replacement of the ester group in DMAEMA with an amide analog as in Polyquaternium 28 results in a resin resistant to alkaline hydrolysis and hence greater utility in alkaline permanent-wave and bleach formulations (see QUATERNARY AMMONIUM COMPOUNDS).

Unquaternized DMAEMA copolymers afford resins that are mildly cationic and less hydroscopic. They provide more moisture-resistant fixatives (146). Further refinements have been accomplished by adding a third comonomer such as *N*-vinylcaprolactam (VCl). In this case, replacement of VP with VCl results in a terpolymer (VP/VCl/DMAEMA) with even greater high humidity moisture resistance and curl retention.

Copolymers Containing Carboxylic Groups. A new line of VP/acrylic acid copolymers in powdered form prepared by precipitation polymerization (147) from heptane have been introduced commercially (148). A wide variety of compositions and molecular weights are available, from 75/25 to 25/75 wt % VP/AA and from 20×10^3 to 250×10^3 molecular weights.

The copolymers are insoluble in water unless they are neutralized to some extent with base. They are soluble, however, in various ratios of alcohol and water, suggesting applications where delivery from hydroalcoholic solutions (149) but subsequent insolubility in water is desired, such as in low volatile organic compound (VOC) hair-fixative formulations or tablet coatings. Unneutralized, their T_gs are higher than expected, indicating interchain hydrogen bonding (150).

Miscellaneous Copolymers. VP has been employed as a termonomer with various acrylic monomer–monomer combinations, especially to afford resins useful as hair fixatives. Because of major differences in reactivity, VP can be copolymerized with alpha-olefins, but the products are actually PVP grafted with olefin or olefin oligomers (151,152). Likewise styrene can be polymerized in the presence of PVP and the resulting dispersion is unusually stable, suggesting that this added resistance to separation is caused by some grafting of styrene onto PVP (153). The literature contains innumerable references to other copolymers but at present (ca 1997), those reviewed in this article are the only ones known to have commercial significance.

Applications

An overview of the various product categories is given in Table 16.

The Chemical Abstracts Services Registry Number and IUPAC nomenclature for PVP are [*9003-39-8*] and 1-ethenyl-2-pyrrolidinone homopolymer, respectively; however, it is known by a variety of approved names by foreign and domestic regulatory authorities. For example:

Name	Chemical name
povidone	poly(*N*-vinyl-2-pyrrolidinone)
polyvidone	
polyvidon	poly(*N*-vinylbutyrolactam)
polyvidonum	poly(1-vinyl-2-pyrrolidinone)
	1-vinyl-2-pyrrolidinone polymer
	poly{1-(2-oxo-1-pyrrolinyl)ethylene}

Table 16. Properties and Applications of Commercial PVPs

Polymer	Mfr/Trade name	Grades	Properties/applications
		Homopolymers	
PVP	ISP/PVP, Plasdone	K-15 to K-120	film former, adhesive, binder, complexant, stabilizer, crystallization inhibitor, dye scavenger, detoxicant, viscosity modifier
	BASF/Luviskol, Kollidon	K-12 to K-90	
		Cross-linked	
proliferous polymerization	ISP/Polyclar, Polyplasdone	various (by particle size)	pharmaceutical tablet disintegrant, adsorbent for polyphenols (tanins), beverage clarification
	BASF/Divergan	various (by particle size)	
		Copolymers	
PVP/VA	BASF/ISP/PVP-VA copolymers	various monomer ratios in ethanol, IPA or water in ethanol, IPA or water	film forming adhesives for hairsprays, mousses, gels, shampoos, styling lotions, bio-adhesives, water-remoistenable or removable adhesives
PVP/DMAEMA	ISP/copolymer	845/937/958	mildly cationic, hair styling aids and conditioners, with strong hold; substantive, lustrous film-formers
PVP/DMAEMA DES quaternary	ISP/Gafquat	755N/734	strongly cationic, substantive, mousse and gel hair fixative ingredients
	BASF/Luviquat	PQ11	
PVP/imidazolinum quaternary	BASF/Luviquat FC		
PVP/styrene[a]	ISP/Polectron 430	30% VP	opacifier for personal care products; very stable styrene emulsion
PVP/alpha-olefins[a]	ISP/Ganex	various (olefin chain length and monomer ratios)	surface active film formers; waterproofing of sunscreens
		Terpolymers	
VP/VCI/DMAEMA	ISP/Gaffix VC-713	VC-713	cationic water-soluble hair styling aid
VP/tBMA/MA	BASF/Luviflex	various	hair fixatives

[a]Graft copolymers.

Trade names for nonpharmaceutical grades are PVP, Peregal ST, Albigen A, and Luviskol; for pharmaceutical grades, Plasdone and Kollidon. The insoluble or crospovidones likewise exist as two grades: nonpharmaceutical are Polyclar and Divergan; pharmaceutical, Polyplasdone XL and Kollidon Cl.

With the current prevalence of computer literature searching, it is relatively easy to accomplish patent searches, especially in *Chemical Abstracts* (CAS). In lieu of what would be an extensive list of patent references to the many applications for VP homo- and copolymers, a list of the major categories and the number of patents granted in each of them during the period 1989–1995 is given in Table 17. More detail concerning the patents is provided in Reference 27.

Table 17. Patent References to VP Homo- and Copolymers 1989–June 1995

Percent	Count	Category
30.80	728	pharmaceuticals
13.24	313	radiation, photographic, and photochemistry, and other reprographic processes
11.08	262	personal care and cosmetics
10.96	259	plastics fabrication and uses
10.49	248	coatings, inks, and related products
8.04	190	biochemical methods
4.91	116	chemistry of synthetic high polymers
4.57	108	plastics manufacture and processing
3.26	77	electric phenomena
3.05	72	agrochemical, bioregulators
3.00	71	surface-active agents and detergents
2.41	57	pharmacology
2.41	57	food and feed chemistry
1.95	46	textiles
1.95	46	ceramics
1.52	36	cellulose, lignin, paper, and other wood products
1.44	34	inorganic analytical chemistry
1.44	34	fossil fuels, derivatives, and related products
1.23	29	nonferrous metals and alloys
1.10	26	electrochemical, radiational, and thermal energy technology
1.06	25	immunochemistry

BIBLIOGRAPHY

"Polyvinylpyrrolidone" in *ECT* 1st ed., Vol. 10, pp. 759–764; "Polyvinylpyrrolidone" under "Vinyl Polymers" in *ECT* 2nd ed., Vol. 21, pp. 427–440, by A. S. Wood, GAF Corp.; "*N*-Vinyl Monomers and Polymers" under "Vinyl Polymers" in *ECT* 3rd ed., Vol. 23, pp. 960–979, by E. V. Hort and B. H. Waxman, GAF Corp.

1. P. Molyneux, in G. Starnsby, ed., *The Chemistry and Rheology of Water-Soluble Gums and Colloids*, S.C.I. Monograph No. 24, Society of Chemical Industries, London, 1966, p. 91.
2. A. Charlesby, *Radiat. Phys. Chem.* **V18**(1–2), 59 (1981).
3. W. Reppe and co-workers, *Justus Leibig's Ann. Chem.* **601**, 134 (1956).
4. J. Falbe and H. J. Schulze-Steinem, *Brennst. Chem.* **48**, 136 (1967).

5. D. J. Dawson, R. D. Gless, and R. E. Wingard, Jr., *JACS* **98**, 5996 (1976).

6. T. M. Karaputudze and co-workers, *Vysokomol. Soedin., Ser. B.*, **24**(4), 305 (1982).

7. C. Bramford, E. Schofield, and D. Michael, *Polymer*, **26**, 946 (1985).

8. S. A. Miller, *Acetylene, Its Properties, Manufacture, and Uses*, Vol. 2, Academic Press, New York, 1965, pp. 338–339.

9. *Vinylpyrrolidone (VP) Toxicity*, Update No. 2303-222R' 2M-292, ISP Corp., Wayne, N.J., 1992.

10. G. M. Kline, *Mod. Plastics* **22**, 157 (1945).

11. J. W. Copenhaver and M. H. Bigelow, *Acetylene and Carbon Monoxide Chemistry*, Reinhold Publishing Corp., New York, 1949.

12. B. V. Robinson and co-workers, *PVP, A Critical Review of the Kinetics and Toxicology of Polyvinylpyrrolidone (Povidone)*, Lewis, Chelsea, Mich., 1990.

13. J. W. Breitenbach, *J. Polym. Sci.* **23**, 949 (1957).

14. S. Bottle and co-workers, *Eur. Poly. J.* **23**(7/8), 671 (1989).

15. Y. E. Kirsh, *Polym. Sci.* **35**(2), 98 (1993).

16. R. T. Woodhams, Ph.D. Dissertation, Polytechnic Institute of Brooklyn, Brooklyn, N.Y., 1954.

17. G. Odian, *Principles of Polymerization*, 3rd ed., Wiley-Interscience, p. 217, 1991.

18. V. Shtamm and co-workers, *Zh. Fiz. Khim.* **55**(9), 2289 (1981).

19. M. V. Encinas, E. A. Lissit, and J. Quiroz, *Eur. Polym. J.* **28**(5), 471 (1992).

20. K. C. Gupta, *J. Appl. Polym. Sci.* **53**, 71 (1994).

21. W. Kern and H. Cherdron, in *Houben Weyl, Methoden der Organische Chemie*, 4th ed., Vol. 14, Verlag, Stuttgart, Germany, 1961, p. 1106.

22. Ger. Pat. DE 3,532,747 (1987), A. Nuber and co-workers, (to BASF).

23. S. C. Ng, *Eur. Polm. J.* **18**, 917 (1982).

24. F. Haaf, A. Sanner, F. Straub, *Polym. J.* **17**(1), 143 (1985).

25. V. V. Kopeikin, Y. G. Santuryan, M. Y. Danilova, *Khim. Farm. Zh.* **22**(10), 1253 (1988).

26. I. G. Gulis and co-workers, *Khim. Farm. Zh.* **25**(9), 82 (1991).

27. E. Barabas, in J. I. Kroschwitz, ed., *Encyclopedia of Polymer Science and Engineering,* Vol. 17, Wiley-Interscience, New York, 1989.

28. E. Senogles and R. Thomas, *J. Polym. Sci. Polym. Symp.* **49**, 203 (1975).

29. Ibid, *J. Polym. Sci. Polym. Lett. Ed.* **16**, 555 (1978).

30. T. M. Karaputadze, V. I. Shumskii, and Y. E. Kirsh, *Vysokomol. Soedin. Ser. A*, **20**(8), 1854 (1978).

31. V. R. Gromov and co-workers, *Eur. Poly. J.* **27**(6), 505 (1991).

32. M. Biswas and P. K. Mishra, *Polymer* **16**, 621 (1970).

33. G. D. Jones, in P. H. Plesch, ed., *The Chemistry of Cationic Polymerization*, Macmillan, New York, 1963, Chapt. 14, p. 554.

34. N. Tsubokawa, N. Takeda, and A. Kanamaru, *J. Poly. Sci. Polym. Lett. Ed.* **18**, 625 (1980).

35. N. Tsubokawa, H. Maruyama, and Y. Sone, *J. Macromol. Sci. Chem.* **A25**(2), 171 (1988).

36. N. Tsubokawa and co-workers, *J. Polym. Sci. Polym. Chem. Ed.* **31**, 3193 (1993).

37. E. Yoshida and co-workers, *J. Poly. Sci. Polym. Chem. Ed.* **31**, 1505 (1993).

38. E. Leonard-Stibbe and co-workers, *J. Poly. Sci. Polym. Chem. Ed.* **32**, 1551 (1994).

39. H. N. Cheng, T. E. Smith, and D. M. Vitus, *J. Polym. Sci. Polym. Lett. Ed.* **19**, 29 (1981).

40. J. R. Ebdon, T. N. Hackerby, and E. Senogles, *Polymer* **24**, 339 (1983).

41. T. Bartels, Y. Y. Tan, and G. Challa, *J. Polym. Sci.* **15**, 341 (1977).

42. D. W. Koetsier, G. Challa, and Y. Y. Tan, *Polymer*, 1709 (1980).

43. D. W. Koetsier, Y. Y. Tan, and G. Challa, *J. Polym. Sci.* **18**, 1933 (1980).

44. G. O. R. Alberda van Ekenstein, D. W. Koetsie, and Y. Y. Tan, *Eur. Polym. J.* **17**, 845 (1981).
45. V. S. Rajan and J. Ferguson, *Eur. Polym. J.* **18**, 633 (1982).
46. I. V. Kotlyarskii and co-workers, *Vysokomol. Soedin. Ser. A* **31**(9), 1893 (1989).
47. J. Matuszewska-Czerwik and S. Polowinski, *Eur. Polym. J.* **26**(5), 549 (1990).
48. U.S. Pat. 2,938,017 (1960), F. Grosser (to GAF).
49. U.S. Pat. 5,286,826 (Feb. 15, 1994), J. S. Shih and S. Y. Tseng (to ISP).
50. J. W. Breitenbach and H. Axmann, in N. Platzer, ed., *Polymerization Kinetics and Technology*, Chapt. 7 (Advances in chemistry series 128, ACS, 1973).
51. E. Barabas and C. Adeyeye, "Crospovidone," *Anal. Profiles Drug Subst. Excipients*, **24**, 87–163 (1996).
52. V. Kudela, in J. I. Kroschwitz, ed., *Encyclopedia of Polymer Science and Engineering*, Vol. 7, 1989, pp. 783–807.
53. M. F. ReFogo, in J. I. Kroschwitz, ed., *Encyclopedia of Polymer Science and Engineering*, Vol. 6, Wiley-Interscience, New York, 1989, pp. 720–742.
54. B. I. Da Silveira, *Eur. Polym. J.* **29**(8), 1095 (1993).
55. J. W. Breitenbach, *J. Polym. Sci.* **23**, 949 (1957).
56. J. W. Breitenbach and E. Wolf, *Makromol. Chem.* **18/19**, 217 (1956).
57. J. W. Breitenbach and A. Schmidt, *Monatsh. Chem.* **85**(1), 52 (1954).
58. T. P. Davis and M. B. Huglin, *Makromol. Chem.* **191**, 331 (1990).
59. O. Guven and M. Sen, *Polymer*, **32**(13), 2491 (1991).
60. M. B. Huglin and M. Zakaria, *Polymer*, **25**, 797 (1984).
61. T. Davis, M. Huglin, D. Yip, *Polymer* **29**, 701 (1988).
62. J. Shih, J. C. Chuang, and R. B. Login, *Polym. Mater. Sci. Eng.*, **72**, 374 (1995).
63. C. C. Anderson, F. Rodriguez, D. A. Thurston, *J. Appl. Polym. Sci.* **23**, 2453 (1979).
64. H. Tenhu and F. Sundholm, *Makromol. Chem.* **185**, 2011 (1984).
65. P. Alexander and A. Charlesby, *J. Polym. Sci.* **23**, 355 (1957).
66. A. Chapiro and C. Legris, *Eur. Polym. J.* **25**(3), 305 (1989).
67. J. E. Davis and E. Senogles, *Aust. J. Chem.* **34**, 1413 (1981).
68. U.S. Pat. 4,699,146 (Oct. 13, 1987), D. L. Sieverding (to Valley Lab).
69. J. W. Copenhaver and M. H. Bigelow, *Acetylene and Carbon Monoxide Chemistry*, Reinhold Publishing Corp., New York, 1969, pp. 67–74.
70. *USP 23*, The United States Pharmacopeial Convention, Inc., Rockville, Md., 1995.
71. C. Wu, in C. Wu, ed., *Handbook of Size Exclusion Chromatography*, Marcel Dekker, Inc., New York, 1995, Chapt. 12.
72. L. Senak, C. Wu, and E. Malawer, *J. Liq. Chrom.* **10**(6), 1127 (1987).
73. H. Fikentscher and K. Herrle, *Mod. Plast.* **23**, 157, 212, 214, 216, 218 (1945).
74. J. Brandrup and E. H. Immergut, eds., *Polymer Handbook*, 3rd ed., Interscience Publishers, a division of John Wiley & Sons, Inc., New York, p. VII 18.
75. Y. Y. Tan and G. Challa, *Polymer* **17**, 739 (1976).
76. D. T. Turner and A. Schartz, *Polymer* **26**, 757 (1985).
77. P. Molyneux, in F. Franks, ed., *Water: A Comprehensive Treatise*, Vol. 4, New York, 1975, Chapt. 7, pp. 569–801.
78. J. W. Breitenbach and E. Wolf, *Makromol. Chem.* **18/19**, 217 (1956).
79. R. Meza and L. Gargallo, *Eur. Polymer J.* **13**, 235 (1977).
80. Y. E. Kirsh, *Prog. Polym. Sci.* **18**, 519–542 (1993).
81. N. Shinyashiki and co-workers, *J. Phys. Chem.* **98**, 13612 (1994).
82. S. K. Jain and G. P. Johari, *J. Phys. Chem.* **92**, 5851 (1988).
83. M. J. Blandamer and J. R. Membrey, *J. Chem. Soc. Perkin II*, 1400 (1974).
84. V. V. Kobyakov and co-workers, *Vysokomol. Soedin. Ser. A* **23**(1), 150 (1981).
85. D. A. Topchiev, personal communication, 1995.
86. Y. E. Kirsh and co-workers, *Eur. Polymer J.* **15**, 223 (1979).

87. Y. E. Kirsh and co-workers, *Eur. Polymer J.* **18**(7), 639 (1983).
88. E. Killmann and R. Bittler, *J. Polym. Sci. Part C*, **39**, 247 (1972).
89. B. P. Molyneux and H. P. Frank, *J. Am. Chem. Soc.* **83**, 3169 (1961).
90. H. P. Frank, S. Barkin, and F. R. Eirich, *J. Am. Chem. Soc.* **61**, 1375 (1957).
91. W. Scholtan, *Makromol. Chem.* **11**, 131 (1953).
92. R. L. Reeves, S. A. Harkaway, and A. R. Sochor, *J. Polym. Sci.* **19**, 2427 (1981).
93. F. M. El Torki and co-workers, *J. Phys. Chem.* **91**, 3686 (1987).
94. T. Takagishi and co-workers, *J. Polymer Sci.* **22**, 185 (1984).
95. W. Gottardi, in S. Block, ed., *Disinfection, Sterilization and Preservation*, Lea & Febiger, 1991, Chapt. 8.
96. H. Schenck, P. Simak, and E. Haedicke, *J. Pharm. Sci.* **68**(12), 1505 (1979).
97. J. L. Zamora, *Am. J. Surg.* **151**(3), 400 (1986).
98. D. Horn and W. Ditter, in G. Hierholzer and G. Gortz, eds., *PVP-Iodine in Surgical Medicine*, Springer-Verlag, New York, 1984.
99. H. N. Cheng, T. E. Smith, and D. M. Vitus, *J. Polym. Sci. Polym. Phys. Ed.* **23**, 461 (1985).
100. L. Turker and co-workers, *Colloid Polym. Sci.* **268**, 337 (1990).
101. A. Guner and M. Ataman, *Colloid Polym. Sci.* **272**, 175 (1994).
102. P. Molyneuz and S. Vekavakaynondha, *J. Chem. Soc., Faraday Trans. 1* **82**, 291 (1986).
103. *Ibid.*, **82**, 635 (1986).
104. D. Oechsle and B. Fussnegger, *Brauwelt* **41**, 1780 (1990).
105. M. Ratto and co-workers, *Tappi* **76**(6), 67 (1993).
106. J. C. Hornby, *Tappi*, 88 (Jan. 1995).
107. H. U. Jager and W. Denzinger, *Tenside* **6**, 428 (1991).
108. M. Paillet and co-workers, *Colloid Polym. Sci.* **271**, 311 (1993).
109. Y. J. Nikas and Blankschtein, *Langmuir*, **10**, 3512 (1994).
110. Y. Almog, S. Reich, and M. Levy, *Brit. Polym. J.* **14**, 131 (1982).
111. A. J. Paine, *J. Polym. Sci. Polym. Chem. Ed.* **28**, 2485 (1990).
112. C. M. Tseng and co-workers, *J. Polym. Sci. Polym. Chem. Ed.* **28**, 2569 (1990).
113. H. L. Chen and H. Morawetz, *Eur. Polym. J.* **19**(10/11), 923 (1983).
114. I. Iliopoulos and R. Audebert, *J. Polym. Sci. Polym. Lett. Ed.* **26**, 2093 (1988).
115. I. Iliopoulos, J. L. Halary, and R. Audebert, *J. Polym. Sci. Polym. Chem. Ed.* **26**, 275 (1988).
116. I. Iliopoulos and R. Audebert, *Eur. Polym. J.* **24**(2), 171 (1988).
117. K. R. Shah, *Polymer*, **28**, 1212 (1987).
118. M. J. Fernandez-Berridi and co-workers, *Polymer*, **31**(1), 38 (1993).
119. M. T. Qurashi, H. S. Blair, and S. J. Allen, *J. Appl. Polym. Sci.* **46**, 255, 263 (1992).
120. J. F. Masson and R. St. John Manley, *Macromolecules* **24**, 6670 (1991).
121. C. M. Tam, M. Dal-Cin, and M. D. Guiver, *J. Membrane Sci.* **78**, 123 (1993).
122. C. M. Tam and co-workers, *Desalination* **89**, 275 (1993).
123. W. B. Stockton and M. F. Rubner, *Polym. Preprints* **19**, 319.
124. H. D. Keith, F. J. Padden, Jr., and T. P. Russell, *Macromolecules* **22**, 666 (1989).
125. R. Z. Greenley, in J. Brandrup and E. H. Immergut, eds., *Polymer Handbook*, Interscience Publishers, a division of John Wiley & Sons, Inc., New York, Chapt. II, p. 267.
126. G. Georgiev, C. Konstantinov, and V. Kabaivanov, *Macromolecules* **25**, 6302 (1992).
127. Ref. 125, p. 153.
128. E. E. Skorikova and co-workers, *Vysokomol. Soedin., Ser. B* **27**(11), 869 (1985).
129. W. M. Culbertson, in J. I. Kroschwitz, ed., *Encyclopedia of Polymer Science and Engineering*, Vol. 9, 1987.
130. M. Orbay, R. Laible, and L. Dulog, *Makromol. Chem.* **47**, 183 (1982).

131. J. F. Bork and L. E. Coleman, *J. Polym. Sci.* **43**, 413 (1960).
132. Van Paesschen and G. Smets, *Bull. Soc. Chem. Belg.* **64**, 173 (1955).
133. A. Chapiro and L. D. Trung, *Eur. Polym. J.* **10**, 1103 (1974).
134. K. Deboudt, M. Delporte, and C. Loucheux, *Macromol. Chem. Phys.* **196**, 279 (1995).
135. K. Y. Choi, *J. App. Polym. Sci.* **37**, 1429 (1989).
136. M. B. Huglin and K. S. Khairou, *Eur. Polym. J.* **24**(3), 239 (1988).
137. PVP/VA brochure No. 2302-194, ISP, 1990.
138. Yu. D. Semchikov and co-workers, *Eur. Poly J.* **26**(8), 889 (1990).
139. R. Y. Lochhead, *Cosmet. Toiletries* **103**, 23 (1988).
140. G. Russell, *Eur. Adhes. Sealants* (June 1989).
141. J. Jachowicz, M. Berthiaume, and M. Garcia, *Colloid Polym. Sci.* **263**, 847 (1985).
142. U.S. Pat. 3,910,862 (1975), E. S. Barabas and M. M. Fern (to GAF).
143. U.S. Pat. 3,914,403 (1975), K. Valan (to GAF).
144. Bulletin No. QO394, International Specialty Products, 1994.
145. Ger. Pat. 4,138,763 (1991), H. Meyer and A. Sanner (BASF).
146. U.S. Pat. 4,223,009 (1980), P. M. Chakrabarti (to GAF).
147. U.S. Pat. 5,015,708 (1991), J. S. Shih, T. E. Smith, and R. B. Login (to GAF).
148. J. S. Shih, J. C. Chuang, and R. B. Login, American Chemical Society, Division of Polymeric Materials, **67**, 226 (Fall 1992).
149. U.S. Pat. 5,219,906 (1993), J. S. Shih and T. E. Smith (to ISP).
150. J. C. Hornby, J. D. Pelesko, and D. Jon, *Soap/Cosmet. Specialt.* (June 1993).
151. S. L. Kopolow, R. B. Login, and M. Tazi, in C. G. Gebelein, T. C. Cheng, and U. C. Yang, eds., *Cosmetic and Pharmaceutical Application of Polymers*, Plenum Publishing Co., New York, 1991.
152. *Ganex Bulletin* (2302-191 5M-989), ISP Corp., 1989.
153. *Polectron Emulsion Copolymers Bulletin* (2302-104), GAF Corp., 1984.

ROBERT B. LOGIN
Sybron Chemicals Inc.

VINYL POLYMERS, POLY(VINYL FLUORIDE). See FLUORINE COMPOUNDS, ORGANIC.

VINYLTOLUENE. See STYRENE.

VIRAL INFECTIONS, CHEMOTHERAPY. See ANTIVIRAL AGENTS.